Springer Monographs in Mathematics

This series publishes advanced monographs giving well-written presentations of the "state-of-the-art" in fields of mathematical research that have acquired the maturity needed for such a treatment. They are sufficiently self-contained to be accessible to more than just the intimate specialists of the subject, and sufficiently comprehensive to remain valuable references for many years. Besides the current state of knowledge in its field, an SMM volume should ideally describe its relevance to and interaction with neighbouring fields of mathematics, and give pointers to future directions of research.

More information about this series at http://www.springer.com/series/3733

Katsuro Sakai

Topology of Infinite-Dimensional Manifolds

Springer

Katsuro Sakai
Department of Mathematics,
Faculty of Engineering
Kanagawa University
Yokohama, Kanagawa, Japan

ISSN 1439-7382 ISSN 2196-9922 (electronic)
Springer Monographs in Mathematics
ISBN 978-981-15-7574-7 ISBN 978-981-15-7575-4 (eBook)
https://doi.org/10.1007/978-981-15-7575-4

Mathematics Subject Classification: 57-01, 57-02, 57N17, 57N20, 57N35, 57Q05, 57Q10, 57Q12, 57Q15, 57Q25, 57Q35, 57N99, 54C35, 54C55, 54E55, 54F65, 54H05

This Springer imprint is published by the registered company Springer Nature Singapore Pte Ltd.
The registered company address is: 152 Beach Road, #21-01/04 Gateway East, Singapore 189721, Singapore

There is no limit for our thoughts and imaginations since we are made "in God's image" (Genesis 1:27). Mathematical ability is a gift from God, which enable us to comprehend and to create various abstract concepts. I always thank and praise Jehovah, our Creator. I agree with the following words in the Scriptures:

"The fear of Jehovah is
the beginning of wisdom,
And knowledge of the Most Holy One
is understanding." —Proverbs 9:10

Note: Scripture quotations are from
New World Translation of the Holy Scriptures—Revised 2013
(Watchtower Bible and Tract Society of New York, Inc.)

Preface

An infinite-dimensional manifold is a (topological) manifold modeled on a given infinite-dimensional (homogeneous) space E (called a model space), that is, a paracompact space covered by open sets that are homeomorphic to open sets in E. A manifold modeled on E is simply called an E-manifold. Hilbert space ℓ^2 and the Hilbert cube $\boldsymbol{Q} = [-1, 1]^{\mathbb{N}}$ are typical examples of infinite-dimensional model spaces, where an ℓ^2-manifold and a $\boldsymbol{Q}$-manifold are also called a Hilbert manifold and a Hilbert cube manifold, respectively.[1] We can also take any infinite-dimensional topological linear space as a model space. The direct limit $\mathbb{R}^\infty = \varinjlim \mathbb{R}^n$ of Euclidean spaces is one of them. It is known that the direct limit $\boldsymbol{Q}^\infty = \varinjlim \boldsymbol{Q}^n$ of Hilbert cubes is homeomorphic to some topological linear space. Thus, there are various kinds of infinite-dimensional manifolds, which are research objects of Infinite-Dimensional Topology.[2]

Infinite-dimensional manifolds are more than just generalizations of usual manifolds modeled on Euclidean space $\mathbb{R}^n$. There are many unexpected special phenomena different from finite-dimensional manifolds. From research of infinite-dimensional manifolds, various useful tools, techniques, and ideas have been developed, which are attractive and exciting. In addition, there have been many applications to other fields of Topology.[3] Let us give some examples of outstanding results related to other fields.

In 1928, Fréchet asked whether Hilbert space ℓ^2 is homeomorphic to the countable product of lines $\mathbb{R}^{\mathbb{N}}$. This question was positively answered in 1966 by R.D. Anderson. It was the dawn of Infinite-Dimensional Topology.[4] In 1939,

---

[1] We consider not only ℓ^2 but also non-separable Hilbert spaces as model spaces.

[2] Infinite-Dimensional Topology is a branch of Geometric Topology, which studies infinite-dimensional spaces arising naturally in Topology and Functional Analysis.

[3] For the history of Infinite-Dimensional Topology, refer to the article of T. Koetsier and J. van Mill [96, Sect. 4].

[4] Due to Anderson's essay [9], the starting point is when he answered in [3] affirmatively to a question posed by V. Klee, that is, he proved in 1964 that the product of a triod, T, and the Hilbert cube $\boldsymbol{Q}$ is homeomorphic to $\boldsymbol{Q}$.

M. Wojdysławski conjectured that the hyperspace of a Peano continuum is homeomorphic to $\boldsymbol{Q}$. This conjecture was finally proved in 1978 and it is known as the Curtis–Schori–West Hyperspace Theorem. Simple Homotopy Theory had been established by J.H.C. Whitehead during 1939–1952, where the Whitehead group and Whitehead torsion were introduced. The topological invariance of Whitehead torsion had been a longstanding problem, but T.A. Chapman proved it in 1973 by using $\boldsymbol{Q}$-manifolds. In 1954, K. Borsuk conjectured that a compact absolute neighborhood retract (ANR) has the homotopy type of a finite simplicial complex. This conjecture was proved in 1977 by J.E. West, where $\boldsymbol{Q}$-manifolds were applied. Shape Theory was founded by K. Borsuk in 1968 as the homotopy theory for spaces without a good local behavior. In 1972, Chapman showed that two compacta X and Y in the pseudo-interior $(-1, 1)^{\mathbb{N}}$ of the Hilbert cube $\boldsymbol{Q}$ have the same shape type if and only if their complements $\boldsymbol{Q} \setminus X$ and $\boldsymbol{Q} \setminus Y$ are homeomorphic.

Meanwhile, the theory of infinite-dimensional manifolds was developed and outstanding results had been obtained. In 1969, D.W. Henderson proved the Open Embedding Theorem, that is, every separable ℓ^2-manifold can be embedded into ℓ^2 as an open set. In 1970, he joined with R.M. Schori and J.E. West in establishing the Classification and the Triangulation Theorems for Hilbert manifolds, respectively. Namely, it was shown that two Hilbert manifolds are homeomorphic if they have the same homotopy type and that every Hilbert manifold is homeomorphic to the product of Hilbert space and a locally finite-dimensional simplicial complex with the metric topology. For (compact) $\boldsymbol{Q}$-manifolds, the Triangulation and the Classification Theorems were established in 1973 by Chapman, that is, every compact $\boldsymbol{Q}$-manifold is homeomorphic to the product of $\boldsymbol{Q}$ and a finite simplicial complex, and two compact $\boldsymbol{Q}$-manifolds are homeomorphic if and only if finite simplicial complexes triangulating them have the same simple homotopy type. In 1980 and 1981, H. Toruńczyk succeeded in characterizing Hilbert cube manifolds and Hilbert manifolds topologically.

This book is designed as a textbook for graduate students and researchers in various branches related to Topology to acquire the fundamental results on infinite-dimensional manifolds and various techniques treating infinite-dimensional spaces. This can also be used as a reference book. Up to now, the following six books have been available for the same purpose:

(1) C. Bessaga and A. Pełczyński, *Selected Topics in Infinite-Dimensional Topology*, MM **58** (Polish Sci. Publ., Warsaw, 1975)
(2) T.A. Chapman, *Lectures on Hilbert Cube Manifolds*, CBMS Regional Conf. Ser. in Math. **28** (Amer. Math. Soc., Providence, 1975)
(3) J. van Mill, *Infinite-Dimensional Topology: Prerequisites and Introduction*, North-Holland Math. Library **43** (Elsevier Sci. Publ. B.V., Amsterdam, 1989)
(4) A. Chigogidze, *Inverse Spectra*, North-Holland Math. Library **53** (Elsevier Sci. Publ. B.V., Amsterdam, 1996)
(5) T. Banakh, T. Radul, and M. Zarichnyi, *Absorbing Sets in Infinite-Dimensional Manifolds*, Math. Studies, Monog. Ser. **1** (VNTL Publ., Lviv, 1996)

(6) J. van Mill, *The Infinite-Dimensional Topology of Function Spaces*, North-Holland Math. Library **64** (Elsevier Sci. Publ. B.V., Amsterdam, 2002)

We selected materials from the above books and compiled them into the present book and also included the new results that are not presented in those books. In addition, we have covered the manifolds modeled on the direct limits $\mathbb{R}^\infty = \varinjlim \mathbb{R}^n$ and $\boldsymbol{Q}^\infty = \varinjlim \boldsymbol{Q}^n$. As an infinite-dimensional version of a combinatorial n-manifold, we have defined a combinatorial ∞-manifold, which triangulates an $\mathbb{R}^\infty$-manifold. This is the first book presenting such manifolds.

This book is written to be fairly self-contained if combined with the following book, which is cited as [GAGT]:

- K. Sakai, *Geometric Aspects of General Topology*, Springer Monog. in Math. (Springer, Tokyo, 2013)

To read [GAGT], the readers are required to have the fundamental knowledge on General Topology. For example, it is enough to finish Part I of the following popular textbook:

- J.R. Munkres, *Topology, 2nd ed.* (Prentice Hall, Inc., Upper Saddle River, 2000)

The book [GAGT] also contains some outstanding results in Infinite-Dimensional Topology different from applications mentioned above, that is, the existence of the following spaces or maps: a hereditary infinite-dimensional compact metrizable space (a compactum containing no subspaces of finite-dimension except zero-dimension); an infinite-dimensional compact metrizable space with finite cohomological dimension (Alexandroff's Problem); a cell-like map of a finite-dimensional compactum onto an infinite-dimensional compactum (Cell-Like Mapping Problem); a separable metrizable topological linear space that is not an absolute retract (AR). The first one is contained in Chapter 4 and the other three are in Chapter 7. Those are not necessary for reading the present book.

Almost all required background knowledge is listed in Chap. 1, whose detailed information is founded in [GAGT]. Besides, we need some additional results, for example, some fundamental results in PL Topology, which are also contained with their proofs. Taking a brief look at this first chapter, the reader can recognize what knowledge he or she should take in and where he or she can. Possibly skipping the first chapter, one can start with the second chapter and go back over necessary parts of Chap. 1 when needed. For Chap. 4, we need some results in Simple Homotopy Theory and Wall's Obstruction Theory, covering spaces and Algebraic Topology,[5] which are written in the preliminary part of Chap. 4, not in the first chapter.

Chapter 2 is devoted to the fundamental results on manifolds modeled on an infinite-dimensional normed linear space, the Stability Theorem, the Unknotting Theorem, the Open Embedding Theorem, the Classification Theorem, etc., which are discussed in Bessaga and Pełczynski's book (1). Here, are also proved the fundamental results on Hilbert cube manifolds, which are discussed in van Mill's book (3) and Chapman's lecture notes (2). Furthermore, we prove the Toruńczyk

[5]It is not required for the reader to be familiar with these objects. Only elementary knowledge of Algebraic Topology is necessary.

Factor Theorem, that is, for each completely metrizable ANR X with weight $\leqslant \tau$, the product of X and Hilbert space of weight τ is a Hilbert manifold, which has not been proved in any other book. Combining this theorem with the Classification Theorem, we can easily obtain the Triangulation Theorem for Hilbert manifolds.

Toruńczyk's characterizations of Hilbert manifolds and $\boldsymbol{Q}$-manifolds are proved in Chap. 3. For the characterization of compact $\boldsymbol{Q}$-manifolds, a readable proof is provided in van Mill's book (3), but the non-compact version is not easily derived from the compact case. However, we discuss the non-compact case too. For Hilbert manifolds, we treat not only ℓ^2-manifolds but also non-separable Hilbert manifolds. In this chapter, some applications of the characterization of Hilbert manifolds and $\boldsymbol{Q}$-manifolds are also given. In particular, it is proved that every Fréchet space (= locally convex completely metrizable topological space) is homeomorphic to Hilbert space with the same weight and that every infinite-dimensional compact convex set in a metrizable topological space is homeomorphic to the Hilbert cube $\boldsymbol{Q}$ if it is an AR. It is also proved that the space of all continuous map from a non-discrete compactum to a separable completely metrizable ANR is an ℓ^2-manifold.

As mentioned above, Chap. 2 contains the Triangulation Theorem for Hilbert manifolds but not for $\boldsymbol{Q}$-manifolds. Chapter 4 is devoted to proving the Triangulation Theorem for $\boldsymbol{Q}$-manifolds, which is contained in Chapman's lecture notes (2) but is not in van Mill's book (3). To prove this theorem, we use some algebraic results concerning the Whitehead group and the Wall's finiteness obstruction. They are contained in the first section, but their proofs are not given. From the Triangulation Theorem for compact $\boldsymbol{Q}$-manifolds, we have derived the Borsuk conjecture mentioned above. The topological invariance of Whitehead torsion is also proved in this chapter.

In Chap. 5, we discuss f.d.cap sets and cap sets for ℓ^2-manifolds and $\boldsymbol{Q}$-manifolds, which are manifolds modeled on the following incomplete normed linear spaces:

$$\ell^2_f = \{(x_i)_{i\in\mathbb{N}} \in \ell^2 \mid x_i = 0 \text{ except for finitely many } i \in \mathbb{N}\},$$
$$\ell^2_Q = \{(x_i)_{i\in\mathbb{N}} \in \ell^2 \mid \sup_{i\in\mathbb{N}} 2^i|x_i| < \infty\},$$

which are the linear spans of the orthonormal basis of ℓ^2 and the Hilbert cube $Q = \prod_{n\in\mathbb{N}}[-2^{-n}, 2^{-n}] \subset \ell^2$, respectively. We also consider non-separable versions of f.d.cap sets and cap sets, named absorption bases. As a generalization of (f.d.)cap sets, introducing absorbing sets in Hilbert manifolds, M. Bestvina and J. Mogilski gave characterizations of manifolds modeled on the universal spaces of absolute Borel spaces. Since then, there have been many works on absorbing sets. In this chapter, we also intend to generalize the work of Bestvina and Mogilski to non-separable spaces. Although there are many related results, we treat only a small number of them. For results on absorbing sets, the book of Banakh, Radul, and Zarichinyi (5) can be referred to but non-separable absorbing sets are not treated. Related results also treated in the second book of J. van Mill (6).

Chapter 6 is devoted to manifolds modeled on $\mathbb{R}^\infty$ and $\boldsymbol{Q}^\infty$, which are the direct limits of Euclidean spaces and Hilbert cubes, respectively. Here, we prove Heisey's Theorem, that is, $\boldsymbol{Q}^\infty$ is homeomorphic to a locally convex linear topological space. Characterizations of these manifolds are given and their Classification Theorems are proved. We also discuss simplicial complexes triangulating $\mathbb{R}^\infty$-manifolds. Such a simplicial complex is an infinite-dimensional generalization of combinatorial manifolds, which is called an infinite-dimensional combinatorial manifold (or a combinatorial ∞-manifold). We prove the so-called Hauptvermutung[6] for them. Furthermore, a combinatorial ∞-manifold is characterized as a simplicial complex K, such that every simplex $\sigma \in K$ is a Z-set in $|K|$, equivalently every $\sigma \in K$ has the non-empty contractible link. We also prove that a countable simplicial complex K is a combinatorial ∞-manifold if and only if $|K|$ is an $\mathbb{R}^\infty$-manifold. In the last section, we introduce bi-topological infinite-dimensional manifolds modeled on $(\mathbb{R}^\infty, \mathbb{R}^{\mathbb{N}}_f)$ and $(\boldsymbol{Q}^\infty, \boldsymbol{Q}^{\mathbb{N}}_f)$, which are called $(\mathbb{R}^\infty, \mathbb{R}^{\mathbb{N}}_f)$-manifolds and $(\boldsymbol{Q}^\infty, \boldsymbol{Q}^{\mathbb{N}}_f)$-manifolds, respectively. Every combinatorial ∞-manifold with the weak topology and the metric topology is a manifold modeled on $(\mathbb{R}^\infty, \mathbb{R}^{\mathbb{N}}_f)$.

Acknowledgments The author would like to express his sincere appreciation to his teacher Professor Yukihiro Kodama, who introduced him to Infinite-Dimensional Topology and had been encouraging him to study in this field. He truly thanks his graduate students Yutaka Iwamoto, Yuji Akaike, Shigenori Uehara, Masayuki Kurihara, Masato Yaguchi, Kotaro Mine, Atsushi Yamashita, Atsushi Kogasaka, Katsuhisa Koshino, and Hanbiao Yang for their careful reading of his lecture notes and helpful comments. He appreciates Taras Banakh for his valuable suggestions. He also feels grateful to Tatsukhiko Yagasaki and Yukinobu Yajima for encouraging him to publish this book. The encouragement of Mr. Masayuki Nakamura, the editor at Springer Japan, has been a great help in completing this book. Moreover, he has many people to thank: his wife, Tomoko, his children, his friends, and others. Most of all, he would like to give thanks to our Maker, Jehovah.

Tsukuba, Japan
April 2020

Katsuro Sakai

[6]For meaning of this word, see p. 44.

Note In the text, numbers in brackets [ ] and [( )], respectively, refer to papers in the References and books or texts in the Bibliography at the end of the book. However, the author's first book [(15)] is cited as [GAGT]. Moreover, for a theorem (or proposition, etc.) quoted from [GAGT], its corresponding theorem number in [GAGT] is indicated in a frame box at the end of the statement. For example, [2.9.4] means (Theorem) 2.9.4 in [GAGT].

Contents

Chapter 1
Preliminaries and Background Results

In this chapter, we first introduce the terminology and notation, and then list the background results. The reader may skip this chapter and can read necessary parts later when needed. Several results might be learned in graduate courses, but others are advanced and special. Almost all results are listed without proofs, which can be found in the following book:

- K. Sakai, *Geometric Aspects of General Topology,* Springer Monog. in Math. (Springer, Tokyo, 2013) — [(15)]

That is cited as [GAGT] instead of [(15)], by which the reader can confirm proofs and details. When a theorem (or proposition, etc.) is quoted from [GAGT], each corresponding theorem number in [GAGT] is indicated in a frame box at the end of statement.

Sections 1.1 and 1.2 are almost identical to the same sections of [GAGT]. Contents of Sects. 1.3, 1.4, 1.12, 1.13, and 1.15 come from Chapters 2, 3, 5, 6, and 7 of [GAGT], respectively. Almost all contents of Sections 1.5–1.11 except Sect. 1.9 are contained in Chapter 4 of [GAGT]. Sections 1.9 and 1.14 are supplements for Chapters 4 and 6 of [GAGT], respectively.

Fundamental results on simplicial complexes are described in Chapter 4 of [GAGT]. Besides, additional preliminary results in PL Topology (Combinatorial Topology) are required for $\boldsymbol{Q}$-manifolds and $\mathbb{R}^\infty$-manifolds, which are not covered by [GAGT]. In Sect. 1.8, a version of the PL embedding approximation theorem 1.8.11 is added together with its proof. Section 1.9 is devoted to regular neighborhoods that are effectively used in PL Topology. Although we refer to them too little, PL n-manifolds (or combinatorial n-manifolds) are main subjects in PL Topology. So we provide an appendix for basic results on them at the end of the book. The following is a good textbook:

- C.P. Rourke and B.J. Sanderson, *Introduction to Piecewise-Linear Topology,* Springer Study Edition (Springer-Verlag, Berlin, 1972, 1982) — [(17)]

K. Sakai, *Topology of Infinite-Dimensional Manifolds*, Springer Monographs in Mathematics, https://doi.org/10.1007/978-981-15-7575-4_1

1.1 Terminology and Notation

With respect to terminology and notation, we follow the book [GAGT]. For the standard sets, we use the following notation:

- $\mathbb{N}$ — the set of natural numbers (i.e., positive integers);
- $\omega = \mathbb{N} \cup \{0\}$ — the set of non-negative integers;
- $\mathbb{Z}$ — the set of integers;
- $\mathbb{Q}$ — the set of rationals;
- $\mathbb{R} = (-\infty, \infty)$ — the real line with the usual topology;
- $\mathbb{C}$ — the complex plane;
- $\mathbb{R}_+ = [0, \infty)$ — the half (real) line;
- $\mathbf{I} = [0, 1]$ — the unit closed interval.

A (topological) **space** is assumed to be **Hausdorff** and a **map** is a **continuous** function. A **singleton** is a space consisting of one point. A space is said to be **non-degenerate** if it has at least two points. A compact metrizable space is called a **compactum** and a connected compactum is called a **continuum**.[1] For a space X and $A \subset X$, we use the following notation:

- $\mathrm{cl}_X A$ (or $\mathrm{cl}\, A$) — the closure of A in X;
- $\mathrm{int}_X A$ (or $\mathrm{int}\, A$) — the interior of A in X;
- $\mathrm{bd}_X A$ (or $\mathrm{bd}\, A$) — the boundary of A in X;
- id_X (or id) — the identity map of X.

For a metrizable space X,

- $\mathrm{Metr}(X)$ — the set of all admissible metrics of X.

The cardinality of a set Γ is denoted by $\mathrm{card}\, \Gamma$. The **weight** $w(X)$, the **density** $\mathrm{dens}\, X$, and the **cellularity** $c(X)$ of a space X are defined as follows:

- $w(X) = \min\{\mathrm{card}\, \mathcal{B} \mid \mathcal{B} \text{ is an open basis for } X\}$;
- $\mathrm{dens}\, X = \min\{\mathrm{card}\, D \mid D \text{ is a dense set in } X\}$;
- $c(X) = \sup\{\mathrm{card}\, \mathcal{G} \mid \mathcal{G} \text{ is a pairwise disjoint open collection}\}$.

As is easily observed, $c(X) \leqslant \mathrm{dens}\, X \leqslant w(X)$ in general. In the case where X is metrizable, all these cardinalities coincide (cf. p. 2 of [GAGT]).

For spaces X and Y with subspaces $X_1, \ldots, X_n \subset X$ and $Y_1, \ldots, Y_n \subset Y$,

- $X \approx Y$ means that X and Y are homeomorphic;
- $(X, X_1, \ldots, X_n) \approx (Y, Y_1, \ldots, Y_n)$ means that there exists a homeomorphism $h : X \to Y$ such that $h(X_1) = Y_1, \ldots, h(X_n) = Y_n$;
- $(X, x_0) \approx (Y, y_0)$ means $(X, \{x_0\}) \approx (Y, \{y_0\})$,

where (X, x_0) is called a **pointed space** and x_0 its **base point**.

[1]Their plurals are **compacta** and **continua**, respectively.

For the product space $\prod_{\gamma\in\Gamma} X_\gamma$, the γ-coordinate of each point $x \in \prod_{\gamma\in\Gamma} X_\gamma$ is denoted by $x(\gamma)$, that is, $x = (x(\gamma))_{\gamma\in\Gamma}$. We can regard $x \in \prod_{\gamma\in\Gamma} X_\gamma$ as a function $x : \Gamma \to \bigcup_{\gamma\in\Gamma} X_\gamma$ such that $x(\gamma) \in X_\gamma$ for each $\gamma \in \Gamma$. For each $\gamma \in \Gamma$, the projection $\mathrm{pr}_\gamma : \prod_{\gamma\in\Gamma} X_\gamma \to X_\gamma$ is defined by $\mathrm{pr}_\gamma(x) = x(\gamma)$. For $\Lambda \subset \Gamma$, the projection $\mathrm{pr}_\Lambda : \prod_{\gamma\in\Gamma} X_\gamma \to \prod_{\lambda\in\Lambda} X_\lambda$ is defined by $\mathrm{pr}_\Lambda(x) = x|\Lambda$ $(= (x(\lambda))_{\lambda\in\Lambda})$. In the case $X_\gamma = X$ for every $\gamma \in \Gamma$, we write $\prod_{\gamma\in\Gamma} X_\gamma = X^\Gamma$. In particular, $X^{\mathbb{N}}$ is the product space of countable infinite copies of X. When $\Gamma = \{1, \dots, n\}$, $X^\Gamma = X^n$ is the product space of n copies of X. For the product space $X \times Y$, $\mathrm{pr}_X : X \times Y \to X$ and $\mathrm{pr}_Y : X \times Y \to Y$ denote the projections.

Now, let $X = (X, d)$ be a metric space, $x \in X$, $\varepsilon > 0$, and $A, B \subset X$. We use the following notation:

- $\mathrm{B}_d(x, \varepsilon) = \{y \in X \mid d(x, y) < \varepsilon\}$ — the ε-neighborhood of x in X (or the open ball with center x and radius ε);
- $\overline{\mathrm{B}}_d(x, \varepsilon) = \{y \in X \mid d(x, y) \leqslant \varepsilon\}$ — the closed ε-neighborhood of x in X (or the closed ball with center x and radius ε);
- $\mathrm{diam}_d\, A = \sup\{d(x, y) \mid x, y \in A\}$ — the diameter of A;
- $d(x, A) = \inf\{d(x, y) \mid y \in A\}$ — the distance of x from A;
- $\mathrm{dist}_d(A, B) = \inf\{d(x, y) \mid x \in A,\ y \in B\}$ — the distance of A and B;
- $\mathrm{N}_d(A, \varepsilon) = \{x \in X \mid d(x, A) < \varepsilon\}$ — the ε-neighborhood of A in X;
- $\overline{\mathrm{N}}_d(A, \varepsilon) = \{y \in X \mid d(x, A) \leqslant \varepsilon\}$ — the closed ε-neighborhood of A in X.

It should be noticed that $d(x, A) = \mathrm{dist}_d(\{x\}, A)$, $\mathrm{N}_d(\{x\}, \varepsilon) = \mathrm{B}_d(x, \varepsilon)$, $\overline{\mathrm{N}}_d(\{x\}, \varepsilon) = \overline{\mathrm{B}}_d(x, \varepsilon)$, and $\mathrm{N}_d(A, \varepsilon) = \bigcup_{x\in A} \mathrm{B}_d(x, \varepsilon)$. For a collection $\mathcal{A}$ of subsets of X, the **mesh** of $\mathcal{A}$ is defined as follows:

- $\mathrm{mesh}_d\, \mathcal{A} = \sup\{\mathrm{diam}_d\, A \mid A \in \mathcal{A}\}$.

When there are no possible confusions, we can drop the subscript d and write $\mathrm{B}(x, \varepsilon)$, $\overline{\mathrm{B}}(x, \varepsilon)$, $\mathrm{N}(A, \varepsilon)$, $\mathrm{diam}\, A$, $\mathrm{dist}(A, B)$ and $\mathrm{mesh}\, \mathcal{A}$.

The standard spaces are listed below:

- $\mathbb{R}^n$ — Euclidean n-space with the norm

  $$\|x\| = \sqrt{x(1)^2 + \cdots + x(n)^2},$$

  $\mathbf{0} = (0, \dots, 0) \in \mathbb{R}^n$ — the origin, the zero vector or the zero element, $\mathbf{e}_i \in \mathbb{R}^n$ — the unit vector defined by $\mathbf{e}_i(i) = 1$ and $\mathbf{e}_i(j) = 0$ for $j \neq i$;
- $\mathbb{R}^n_+ = \{x \in \mathbb{R}^n \mid x(n) \geqslant 0\} = \mathbb{R}^{n-1} \times \mathbb{R}_+$ — Euclidean half n-space;
- $\mathbf{S}^{n-1} = \{x \in \mathbb{R}^n \mid \|x\| = 1\}$ — the unit $(n-1)$-sphere;
- $\mathbf{B}^n = \{x \in \mathbb{R}^n \mid \|x\| \leqslant 1\}$ — the unit closed n-ball;
- $\Delta^n = \{x \in (\mathbb{R}_+)^{n+1} \mid \sum_{i=1}^{n+1} x(i) = 1\}$ — the standard n-simplex;
- $\boldsymbol{Q} = [-1, 1]^{\mathbb{N}}$ — the Hilbert cube;
- $\boldsymbol{s} = \mathbb{R}^{\mathbb{N}}$ — the space of sequences.

Note that $\mathbf{S}^{n-1}$, $\mathbf{B}^n$, and Δ^n are not product spaces even though the same notations are used for product spaces, where the indexes $n-1$ and n represent their dimensions.

A separable metrizable space M is called an ***n*-manifold** (or an ***n*-dimensional manifold**)[2] if each $x \in M$ has a neighborhood that is homeomorphic to (an open set in) $\mathbf{I}^n$, where $\mathbf{I}^n$ can be replaced with $\mathbb{R}^n_+$. We call $x \in M$ an **interior point** if it has a neighborhood homeomorphic to (an open set in) $(0,1)^n$ $(\approx \mathbb{R}^n)$. The set $\operatorname{Int} M$ of all interior points of M is called the **interior** of M, which is open in M. We also call $x \in M$ a **boundary point** if it is not an interior point, that is, any neighborhood of x is not homeomorphic to (an open set in) $\mathbb{R}^n$. In other words, $x \in M$ is a boundary point if and only if x has a neighborhood N such that $(N, x) \approx (\mathbf{I}^n, \mathbf{0})$ (or $(N, x) \approx (\mathbb{R}^n_+, \mathbf{0})$). The set ∂M consisting of all boundary points of M is called the **boundary** of M, which is an $(n-1)$-manifold and closed in M. When $\partial M = \emptyset$, M is called an ***n*-manifold without boundary**. A **closed *n*-manifold** is a compact n-manifold without boundary. The closed n-ball $\mathbf{B}^n$ is an n-manifold with $\partial \mathbf{B}^n = \mathbf{S}^{n-1}$ and the n-sphere $\mathbf{S}^n$ is a closed n-manifold.

Let $\mathcal{A}$ and $\mathcal{B}$ be collections of subsets of X and $Y \subset X$. We define

- $\mathcal{A} \wedge \mathcal{B} = \{A \cap B \mid A \in \mathcal{A},\ B \in \mathcal{B}\}$;
- $\mathcal{A}|Y = \{A \cap Y \mid A \in \mathcal{A}\}$;
- $\mathcal{A}[Y] = \{A \in \mathcal{A} \mid A \cap Y \neq \emptyset\}$;
- $\mathcal{A}^{\mathrm{cl}} = \{\operatorname{cl} A \mid A \in \mathcal{A}\}$.

The **star** of Y with respect to $\mathcal{A}$ is defined as follows:

- $\operatorname{st}(Y, \mathcal{A}) = Y \cup \bigcup \mathcal{A}[Y]\ (= Y \cup \bigcup_{A \in \mathcal{A}[Y]} A)$.

When each $A \in \mathcal{A}$ is contained in some $B \in \mathcal{B}$, it is said that $\mathcal{A}$ **refines** $\mathcal{B}$, which is denoted by

$$\mathcal{A} \prec \mathcal{B} \text{ or } \mathcal{B} \succ \mathcal{A}.$$

It is said that $\mathcal{A}$ **covers** Y (or $\mathcal{A}$ is a **cover** of Y in X) if $Y \subset \bigcup \mathcal{A}\ (= \bigcup_{A \in \mathcal{A}} A)$. When $\mathcal{A}$ is a cover of Y in X, $\operatorname{st}(Y, \mathcal{A}) = \bigcup \mathcal{A}[Y]$. A cover of X in X is simply called a **cover** of X.

A cover of Y in X is said to be **open** (resp. **closed**) in X depending on whether its members are open (resp. closed) in X. If $\mathcal{A}$ is an open cover of X, then $\mathcal{A}|Y$ is an open cover of Y and $\mathcal{A}[Y]$ is an open cover of Y in X. When $\mathcal{A}$ and $\mathcal{B}$ are open covers of X, $\mathcal{A} \wedge \mathcal{B}$ is also an open cover of X. For covers $\mathcal{A}$ and $\mathcal{B}$ of X, it is said that $\mathcal{A}$ is a **refinement** of $\mathcal{B}$ if $\mathcal{A} \prec \mathcal{B}$, where $\mathcal{A}$ is an **open** (resp. **closed**) **refinement** if $\mathcal{A}$ is an open (resp. closed) cover.

For a space X, we write

- $\operatorname{cov}(X)$ — the collection of all open covers of X.

[2]The (topological) dimension of M is equal to n.

For $\mathcal{U}, \mathcal{V} \in \operatorname{cov}(X)$, we define

$$\operatorname{st}(\mathcal{U}, \mathcal{V}) = \{\operatorname{st}(U, \mathcal{V}) \mid U \in \mathcal{U}\}.$$

In the case of $\mathcal{V} = \mathcal{U}$, $\operatorname{st}(\mathcal{U}, \mathcal{U})$ is denoted by $\operatorname{st}\mathcal{U}$. When $\operatorname{st}\mathcal{U} \prec \mathcal{V}$, we call $\mathcal{U}$ a **star-refinement** of $\mathcal{V}$. We inductively define $\operatorname{st}^n \mathcal{U}$, $n \in \mathbb{N}$, as follows:

$$\operatorname{st}^n \mathcal{U} = \operatorname{st}(\operatorname{st}^{n-1} \mathcal{U}, \mathcal{U}),$$

where $\operatorname{st}^0 \mathcal{U} = \mathcal{U}$ (so $\operatorname{st}^1 \mathcal{U} = \operatorname{st}\mathcal{U}$).

Let $(X_\gamma)_{\gamma\in\Gamma}$ be a family of (topological) spaces and $X = \bigcup_{\gamma\in\Gamma} X_\gamma$. The **weak topology** on X with respect to $(X_\gamma)_{\gamma\in\Gamma}$ is the topology defined as follows:

$$U \subset X \text{ is open in } X \Leftrightarrow \forall\gamma \in \Gamma, U \cap X_\gamma \text{ is open in } X_\gamma$$

$$\Big(A \subset X \text{ is closed in } X \Leftrightarrow \forall\gamma \in \Gamma, A \cap X_\gamma \text{ is closed in } X_\gamma\Big).$$[3]

Suppose that X has the weak topology with respect to $(X_\gamma)_{\gamma\in\Gamma}$ and the topologies of X_γ and $X_{\gamma'}$ agree on $X_\gamma \cap X_{\gamma'}$ for any $\gamma, \gamma' \in \Gamma$. If $X_\gamma \cap X_{\gamma'}$ is closed (resp. open) in X_γ for any $\gamma, \gamma' \in \Gamma$, then each X_γ is closed (resp. open) in X and the original topology of each X_γ is a subspace topology inherited from X. In the case $X_\gamma \cap X_{\gamma'} = \emptyset$ for $\gamma \neq \gamma'$, X is called the **topological sum** of $(X_\gamma)_{\gamma\in\Gamma}$ and denoted by $X = \bigoplus_{\gamma\in\Gamma} X_\gamma$. In the case $X_\gamma \cap X_{\gamma'} = \{x_0\}$ for $\gamma \neq \gamma'$, X is called the **wedge sum** (or **wedge**) of $(X_\gamma)_{\gamma\in\Gamma}$ at x_0 and denoted by $X = \bigvee_{\gamma\in\Gamma} X_\gamma$.[4] When $\Lambda = \{1, \dots, n\}$, we write as follows:

$$\bigoplus_{i=1}^{n} X_i = X_1 \oplus \cdots \oplus X_n \quad\text{and}\quad \bigvee_{i=1}^{n} X_i = X_1 \vee \cdots \vee X_n.$$

Let $f : A \to Y$ be a map from a closed set A in a space X to another space Y. The **adjunction space** $Y \cup_f X$ is the quotient space $(X \oplus Y)/\sim$, where $X \oplus Y$ is the topological sum and $\sim$ is the equivalence relation corresponding to the decomposition of $X \oplus Y$ into singletons $\{x\}$, $x \in X \setminus A$, and sets $\{y\} \cup f^{-1}(y)$, $y \in Y$ (the latter is a singleton $\{y\}$ if $y \in Y \setminus f(A)$). When Y is a singleton, $Y \cup_f X \approx X/A$. One should note that the adjunction spaces are *not Hausdorff* in general. It is necessary to require some conditions for the adjunction space to be Hausdorff.

[3] I.e., the finest (or largest) topology such that each inclusion $X_\gamma \subset X$ is continuous. (The term "weak topology" is used with a different meaning by functional analysts, etc.)

[4] The wedge (sum) is used for a family of pointed spaces with the common base point.

Let $f : X \to Y$ be a map. For $A \subset X$ and $B \subset Y$, we write

$$f(A) = \{f(x) \mid x \in A\} \text{ and } f^{-1}(B) = \{x \in X \mid f(x) \in B\}.$$

For collections $\mathcal{A}$ and $\mathcal{B}$ of subsets of X and Y, respectively, we write

$$f(\mathcal{A}) = \{f(A) \mid A \in \mathcal{A}\} \text{ and } f^{-1}(\mathcal{B}) = \{f^{-1}(B) \mid B \in \mathcal{B}\}.$$

The restriction of f to $A \subset X$ is denoted by $f|A$. It is said that a map $g : A \to Y$ **extends over** X if there is a map $f : X \to Y$ such that $f|A = g$. Such a map f is called an **extension** of g.

Let $[a, b]$ be a closed interval, where $a < b$ ($\in \mathbb{R}$). A map $f : [a, b] \to X$ is called a **path** (from $f(a)$ to $f(b)$) in X, where it is said that two points $f(a)$ and $f(b)$ are connected by the path f in X. An embedding (i.e., an injective path) $f : [a, b] \to X$ is called an **arc** (from $f(a)$ to $f(b)$) in X, and the image $f([a, b])$ is also called an **arc**. Namely, a space is called an **arc** if it is homeomorphic to $\mathbf{I}$. A space X is **path-connected** (or **arcwise connected**) if each pair of *distinct* points $x, y \in X$ are connected by a path (or an arc). It is said that X is **locally path-connected** (or **locally arcwise connected**) if any neighborhood U of each point $x \in X$ contains a neighborhood V of x such that each pair of distinct points in V are connected by a path (or an arc) in U (i.e., for each $y, z \in V$, there is a path (or an arc) $f : \mathbf{I} \to U$ such that $f(0) = y$ and $f(1) = z$). In this definition, as is easily observed, V may be a path-connected (or an arcwise connected). The (local) arcwise connectedness looks to be stronger than the (local) path-connectedness, but they are the same concepts, that is:

Proposition 1.1.1 *An arbitrary space X is path-connected if and only if X is arcwise connected. Moreover, X is locally path-connected if and only if X is locally arcwise connected.* [5.14.7]

For spaces X and Y, we write

- $\mathrm{C}(X, Y)$ — the set of (continuous) maps from X to Y.

Given subspaces $X_1, \dots, X_n \subset X$ and $Y_1, \dots, Y_n \subset Y$, a map $f : X \to Y$ is said to be a map from $(X, X_1, \dots, X_n)$ to $(Y, Y_1, \dots, Y_n)$ and is written

$$f : (X, X_1, \dots, X_n) \to (Y, Y_1, \dots, Y_n)$$

if $f(X_1) \subset Y_1, \dots, f(X_n) \subset Y_n$. We write

- $\mathrm{C}((X, X_1, \dots, X_n), (Y, Y_1, \dots, Y_n))$ — the set of maps from $(X, X_1, \dots, X_n)$ to $(Y, Y_1, \dots, Y_n)$;
- $\mathrm{C}((X, x_0), (Y, y_0)) = \mathrm{C}((X, \{x_0\}), (Y, \{y_0\}))$.

For maps $f, g : X \to Y$ (i.e., $f, g \in \mathrm{C}(X, Y)$),

- $f \simeq g$ means that f and g are homotopic (or f is homotopic to g), that is, there is a map $h : X \times \mathbf{I} \to Y$ such that $h_0 = f$ and $h_1 = g$, where $h_t : X \to Y$, $t \in \mathbf{I}$, are defined by $h_t(x) = h(x, t)$, and h is called a **homotopy** from f to g (between f and g). When g is a constant map, it is said that f is **null-homotopic**, which is denoted by $f \simeq 0$. For a homotopy $h : X \times \mathbf{I} \to Y$, we call $h(\{x\} \times \mathbf{I})$, $x \in X$, the **tracks** of h, where each $h(\{x\} \times \mathbf{I})$ is the **track** of $x \in X$ by h.

For spaces X and Y,

- $X \simeq Y$ means that X and Y are homotopy equivalent,

that is, there are maps $f : X \to Y$ and $g : Y \to X$ such that $gf \simeq \mathrm{id}_X$ and $fg \simeq \mathrm{id}_Y$, where f is called a **homotopy equivalence** and g is a **homotopy inverse** of f. Then, we also say that X and Y have the **same homotopy type** or X has the **homotopy type** of Y. For each $f, f' \in \mathrm{C}(X, Y)$ and $g, g' \in \mathrm{C}(Y, Z)$, we have the following:

$$f \simeq f', \ g \simeq g' \Rightarrow gf \simeq g'f'.$$

A homotopy h between maps $f, g \in \mathrm{C}((X, X_1, \ldots, X_n), (Y, Y_1, \ldots, Y_n))$ requires the condition that $h_t \in \mathrm{C}((X, X_1, \ldots, X_n), (Y, Y_1, \ldots, Y_n))$ for every $t \in \mathbf{I}$, that is, h is regarded as the map

$$h : (X \times \mathbf{I}, X_1 \times \mathbf{I}, \ldots, X_n \times \mathbf{I}) \to (Y, Y_1, \ldots, Y_n).$$

When there are maps

$$f : (X, X_1, \ldots, X_n) \to (Y, Y_1, \ldots, Y_n),$$
$$g : (Y, Y_1, \ldots, Y_n) \to (X, X_1, \ldots, X_n)$$

such that $gf \simeq \mathrm{id}_X$ and $fg \simeq \mathrm{id}_Y$, we write

- $(X, X_1, \ldots, X_n) \simeq (Y, Y_1, \ldots, Y_n)$;
- $(X, x_0) \simeq (Y, y_0)$ means $(X, \{x_0\}) \simeq (Y, \{y_0\})$.

For $A \subset X$, a homotopy $h : X \times \mathbf{I} \to Y$ is called a **homotopy relative to** A if $h(\{x\} \times \mathbf{I})$ is a singleton for every $x \in A$. When a homotopy from f to g is a homotopy relative to A (where $f|A = g|A$), it is said that f and g are **homotopic relative to** A, which is written as follows:

$$f \simeq g \ \text{ rel. } A.$$

Let $f, g : X \to Y$ be maps and $\mathcal{U}$ a collection of subsets of Y (as usual, $\mathcal{U} \in \mathrm{cov}(Y)$). It is said that f and g are **$\mathcal{U}$-close** (or f is **$\mathcal{U}$-close** to g) if

$$\big\{\{f(x), g(x)\} \bigm| x \in X\big\} \prec \mathcal{U} \cup \big\{\{y\} \bigm| y \in Y\big\},$$

which implies that $\mathcal{U}$ covers the set $\{f(x), g(x) \mid f(x) \neq g(x)\}$. A homotopy h is called a **$\mathcal{U}$-homotopy** if the collection of non-degenerate tracks of h refines $\mathcal{U}$, that is,

$$\{h(\{x\} \times \mathbf{I}) \mid x \in X\} \prec \mathcal{U} \cup \{\{y\} \mid y \in Y\}.$$

In this case, $\mathcal{U}$ covers the set

$$\bigcup \{h(\{x\} \times \mathbf{I}) \mid h(\{x\} \times \mathbf{I}) \text{ is non-degenerate}\}.$$

When a homotopy from f to g is a $\mathcal{U}$-homotopy, it is said that f and g are **$\mathcal{U}$-homotopic** (or f is **$\mathcal{U}$-homotopic** to g), which is written as follows:

$$f \simeq_{\mathcal{U}} g.$$

When $Y = (Y, d)$ is a metric space, we can define a metric d called the **sup-metric** on the set $\mathrm{C}(X, Y)$ as follows:

$$d(f, g) = \sup_{x \in X} \min\{d(f(x), g(x)),\ 1\}.^{5}$$

The metric space $(\mathrm{C}(X, Y), d)$ is denoted by $\mathrm{C}_d(X, Y)$. The topology of $\mathrm{C}_d(X, Y)$ is called the **uniform convergence topology**, where each $f \in \mathrm{C}_d(X, Y)$ has a neighborhood basis consisting of the following:

$$\mathrm{B}_d(f, \varepsilon) = \{g \in \mathrm{C}(X, Y) \mid d(f, g) < \varepsilon\},\ \varepsilon > 0.$$

For $\varepsilon > 0$, it is said that f and g are **ε-close** or f is **ε-close** to g if $d(f, g) < \varepsilon$. A homotopy h is called an **ε-homotopy** if $\operatorname{mesh}\{h(\{x\} \times \mathbf{I}) \mid x \in X\} < \varepsilon$. When a homotopy from f to g is an ε-homotopy, it is said that f and g are **ε-homotopic**, which is written as follows:

$$f \simeq_{\varepsilon} g.$$

The **compact-open topology** on $\mathrm{C}(X, T)$ is generated by the sets

$$\langle K; U\rangle = \{f \in \mathrm{C}(X, Y) \mid f(K) \subset U\},$$

[5] In the case where Y is bounded or X is compact, we can employ the definition $d(f, g) = \sup_{x \in X} d(f(x), g(x))$. But, in general, the case $d(f, g) = \infty$ might occur for this definition.

where K is any compact set in X and U is any open set in Y. With respect to the compact-open topology, we have the following:

Proposition 1.1.2 (Properties of the Compact-Open Topology)

(1) *Every map $f : Z \times X \to Y$ induces the map $\tilde{f} : Z \to \mathrm{C}(X, Y)$ defined by $\tilde{f}(z)(x) = f(x, y)$.* [1.1.1]
(2) *For each $f \in \mathrm{C}(Z, X)$ and $g \in \mathrm{C}(Y, Z)$, the following are continuous:*

$$f^* : \mathrm{C}(X, Y) \to \mathrm{C}(Z, Y),\ f^*(h) = h \circ f;$$

$$g_* : \mathrm{C}(X, Y) \to \mathrm{C}(X, Z),\ g_*(h) = g \circ h.$$ [1.1.3(1)]

(3) *When Y is locally compact, the following composition is continuous:*

$$\mathrm{C}(X, Y) \times \mathrm{C}(Y, Z) \ni (f, g) \mapsto g \circ f \in \mathrm{C}(X, Z).$$ [1.1.3(2)]

(4) *When X is locally compact, the following (evaluation) is continuous:*

$$\mathrm{ev} : \mathrm{C}(X, Y) \times X \to Y,\ \mathrm{ev}(f, x) = f(x).$$

So, every map $f : Z \to \mathrm{C}(X, Y)$ induces the map $\tilde{f} : Z \times X \to Y$ defined by $\tilde{f}(z, x) = f(z)(x)$. [1.1.3(4)]
(5) *When X is locally compact, the following inequalities hold:*

$$w(Y) \leqslant w(\mathrm{C}(X, Y)) \leqslant \aleph_0 w(X) w(Y).$$

In particular, if X is separable locally compact and Y has infinite, then $w(\mathrm{C}(X, Y)) = w(Y)$. [1.1.3(5)]
(6) *When X is compact and $Y = (Y, d)$ is a metric space, the sup-metric on $\mathrm{C}(X, Y)$ is admissible, that is, the compact-open topology is induced by the sup-metric.* [1.1.3(6)]

Regarding $\mathrm{C}(X, Y)$ as a subspace of the product space Y^X, we can introduce another topology on $\mathrm{C}(X, Y)$, which is called the **pointwise convergence topology**. The space $\mathrm{C}(X, Y)$ with the pointwise convergence topology is written as $\mathrm{C}_p(X, Y)$. For each $x \in X$, the **evaluation** $\mathrm{ev}_x : \mathrm{C}(X, Y) \to Y$ is defined by $\mathrm{ev}_x(f) = f(x)$, which is the restriction of the projection $\mathrm{pr}_x : Y^X \to X$. Thus, the pointwise convergence topology is the coarsest topology such that the evaluations ev_x, $x \in X$, are continuous and it is generated by the sets

$$\langle x; U \rangle = \{ f \in \mathrm{C}(X, Y) \mid f(x) \in U \},$$

where x is any point of X and U is any open set in Y. Hence, every open set in $\mathrm{C}_p(X, Y)$ is open in $\mathrm{C}(X, Y)$, that is, the pointwise convergence topology is not finer than the compact-open topology.

Remark 1.1 Proposition 1.1.2(4) does not hold for the space $C_p(X, Y)$ even if X is compact and $Y = \mathbb{R}$. For example, let $X = \{0, 1/n \mid n \in \mathbb{N}\}$ and $c_0 : X \to \mathbb{R}$ be the constant map with $c_0(X) = \{0\}$. For any neighborhood $\mathcal{U}$ of c_0 in $C_p(X, \mathbb{R})$ and any neighborhood V of 0 in X, we can choose a finite set $F \subset X$, $0 < \varepsilon < 1$, and $x_0 \in V \setminus F$ so that $\bigcap_{x \in F} \langle x; (-\varepsilon, \varepsilon)\rangle \subset \mathcal{U}$. Let $f_0 : X \to \mathbb{R}$ be the map defined by $f_0(x_0) = 1$ and $f_0(x) = 0$ for any $x \in X \setminus \{x_0\}$. Then, $(f_0, x_0) \in \mathcal{U} \times V$ but $\mathrm{ev}(f_0, x_0) = f_0(x_0) = 1 \notin (-1, 1)$.

1.2 Banach Spaces in the Product of Real Lines

Throughout this section, let Γ be an infinite set. Here, we review Banach spaces[6] being linear subspaces of the product $\mathbb{R}^\Gamma$. We write

- $\mathrm{Fin}(\Gamma)$ — the set of all non-empty finite subsets of Γ.

Then, note that $\operatorname{card} \mathrm{Fin}(\Gamma) = \operatorname{card} \Gamma$. The product space $\mathbb{R}^\Gamma$ is a linear space with the following scalar multiplication and addition:

$$\mathbb{R}^\Gamma \times \mathbb{R} \ni (x, t) \mapsto tx = (tx(\gamma))_{\gamma \in \Gamma} \in \mathbb{R}^\Gamma;$$

$$\mathbb{R}^\Gamma \times \mathbb{R}^\Gamma \ni (x, y) \mapsto x + y = (x(\gamma) + y(\gamma))_{\gamma \in \Gamma} \in \mathbb{R}^\Gamma.$$

These operations are continuous with respect to the product topology of $\mathbb{R}^\Gamma$. Namely, $\mathbb{R}^\Gamma$ with the product topology is a topological linear space.[7] Note that $w(\mathbb{R}^\Gamma) = \aleph_0 \operatorname{card} \mathrm{Fin}(\Gamma) = \operatorname{card} \Gamma$.

For each $\gamma \in \Gamma$, we define the unit vector $\mathbf{e}_\gamma \in \mathbb{R}^\Gamma$ by $\mathbf{e}_\gamma(\gamma) = 1$ and $\mathbf{e}_\gamma(\gamma') = 0$ for $\gamma' \neq \gamma$. It should be noticed that $\{\mathbf{e}_\gamma \mid \gamma \in \Gamma\}$ is not a Hamel basis for $\mathbb{R}^\Gamma$ and its linear span[8] is the following:

$$\mathbb{R}^\Gamma_f = \left\{x \in \mathbb{R}^\Gamma \;\middle|\; x(\gamma) = 0 \text{ except for finitely many } \gamma \in \Gamma \right\},$$

which is a dense linear subspace of $\mathbb{R}^\Gamma$. The subspace $\mathbb{R}^\mathbb{N}_f$ of $\boldsymbol{s} = \mathbb{R}^\mathbb{N}$ is also denoted by $\boldsymbol{s}_f$, which is consisting of all finite sequences.

As is easily observed, the following are equivalent:

(a) $\mathbb{R}^\Gamma$ is metrizable;
(b) $\mathbb{R}^\Gamma_f$ is metrizable;
(c) $\mathbb{R}^\Gamma_f$ is first countable;
(d) $\operatorname{card} \Gamma \leqslant \aleph_0$.

[6] A **Banach space** is a complete normed linear space.

[7] Refer to p. 23.

[8] The **linear span** of B is the linear subspace generated by a set B.

Thus, when Γ is uncountable, every linear subspace L of $\mathbb{R}^\Gamma$ with $\mathbb{R}^\Gamma_f \subset L$ is non-metrizable. Moreover, $\mathbb{R}^\Gamma$ (or $\mathbb{R}^\Gamma_f$) is metrizable only when Γ is countable. In the case $\operatorname{card}\Gamma = \aleph_0$, $\mathbb{R}^\Gamma$ is linearly homeomorphic to the space of sequences $\boldsymbol{s} = \mathbb{R}^\mathbb{N}$, that is, there exists a linear homeomorphism between $\mathbb{R}^\Gamma$ and $\boldsymbol{s}$. On the other hand, we have the following proposition:

Proposition 1.2.1 *Let Γ be an infinite set. Then, any norm on $\mathbb{R}^\Gamma_f$ does not induce the topology inherited from the product topology of $\mathbb{R}^\Gamma$. Consequently, every linear subspace L of $\mathbb{R}^\Gamma$ with $\mathbb{R}^\Gamma_f \subset L$ is not normable.* 1.2.1

We can consider various norms defined on linear subspaces of $\mathbb{R}^\Gamma$, which are not compatible with the product topology as in Proposition 1.2.1 above. In general, the unit closed ball and the unit sphere of a normed linear space $X = (X, \|\cdot\|)$ are denoted by $\mathbf{B}_X$ and $\mathbf{S}_X$ respectively, that is,

$$\mathbf{B}_X = \big\{x \in X \bigm| \|x\| \leqslant 1\big\} \text{ and } \mathbf{S}_X = \big\{x \in X \bigm| \|x\| = 1\big\}.$$

The zero vector (the zero element) of X is denoted by $\mathbf{0}_X$, or simply by $\mathbf{0}$ if there is no possible confusion.

The Banach space $\ell^\infty(\Gamma)$ and its closed linear subspaces $\boldsymbol{c}(\Gamma) \supset \boldsymbol{c}_0(\Gamma)$ are defined as follows:

- $\ell^\infty(\Gamma) = \big\{x \in \mathbb{R}^\Gamma \bigm| \sup_{\gamma\in\Gamma} |x(\gamma)| < \infty\big\}$ with the sup-norm
$$\|x\|_\infty = \sup_{\gamma\in\Gamma} |x(\gamma)|;$$ [9]
- $\boldsymbol{c}(\Gamma) = \big\{x \in \mathbb{R}^\Gamma \bigm| \exists t \in \mathbb{R} \text{ such that } \forall\varepsilon > 0,\ |x(\gamma) - t| < \varepsilon \text{ except for finitely many } \gamma \in \Gamma\big\}$;
- $\boldsymbol{c}_0(\Gamma) = \big\{x \in \mathbb{R}^\Gamma \bigm| \forall\varepsilon > 0,\ |x(\gamma)| < \varepsilon \text{ except for finitely many } \gamma \in \Gamma\big\}$.

These are linear subspaces of $\mathbb{R}^\Gamma$ but not topological ones as seen above. The space $\boldsymbol{c}(\Gamma)$ is linearly homeomorphic to $\boldsymbol{c}_0(\Gamma) \times \mathbb{R}$ by the following correspondence:

$$\boldsymbol{c}_0(\Gamma) \times \mathbb{R} \ni (x, t) \mapsto (x(\gamma) + t)_{\gamma\in\Gamma} \in \boldsymbol{c}(\Gamma).$$

This correspondence and its inverse are Lipschitz with respect to the norm $\|(x, t)\| = \max\{\|x\|_\infty, |t|\}$.

Furthermore, $\ell^\infty_f(\Gamma)$ denotes $\mathbb{R}^\Gamma_f$ with this norm. Then,

$$\ell^\infty_f(\Gamma) \subset \boldsymbol{c}_0(\Gamma) \subset \boldsymbol{c}(\Gamma) \subset \ell^\infty(\Gamma).$$

[9]In some literature, this space is denoted by $\boldsymbol{m}(\Gamma)$.

For the weight of these spaces, we have the following:

$$w(\ell^\infty(\Gamma)) = 2^{\operatorname{card}\Gamma} \quad \text{but} \quad w(\boldsymbol{c}(\Gamma)) = w(\boldsymbol{c}_0(\Gamma)) = w(\ell^\infty_f(\Gamma)) = \operatorname{card}\Gamma$$

(cf. Proposition 1.2.2 in [GAGT]). The topology of $\ell^\infty_f(\Gamma)$ is different from the topology inherited from the product topology. Indeed, $\{\mathbf{e}_\gamma \mid \gamma \in \Gamma\}$ is discrete in $\ell^\infty_f(\Gamma)$ but $\mathbf{0}$ is a cluster point of this set with respect to the product topology.

In the case $\Gamma = \mathbb{N}$, we write:

- $\ell^\infty(\mathbb{N}) = \ell^\infty$ — the space of bounded sequences;[10]
- $\boldsymbol{c}(\mathbb{N}) = \boldsymbol{c}$ — the space of convergent sequences;
- $\boldsymbol{c}_0(\mathbb{N}) = \boldsymbol{c}_0$ — the space of null-sequences (= sequence tending to 0).

We also write $\ell^\infty_f(\mathbb{N}) = \ell^\infty_f$, where $\ell^\infty_f = \boldsymbol{s}_f$ as sets (linear spaces) but they have different topologies. It should be noted that $\boldsymbol{c}$ and $\boldsymbol{c}_0$ are separable but ℓ^∞ is non-separable. When $\operatorname{card}\Gamma = \aleph_0$, the spaces $\ell^\infty(\Gamma)$, $\boldsymbol{c}(\Gamma)$ and $\boldsymbol{c}_0(\Gamma)$ are linearly isometric to these spaces ℓ^∞, $\boldsymbol{c}$ and $\boldsymbol{c}_0$, respectively.

Here, we regard $\operatorname{Fin}(\Gamma)$ as a directed set by $\subset$. For $x \in \mathbb{R}^\Gamma$, we say that $\sum_{\gamma\in\Gamma} x(\gamma)$ is **convergent** if $\left(\sum_{\gamma\in F} x(\gamma)\right)_{F\in\operatorname{Fin}(\Gamma)}$ is convergent and define

$$\sum_{\gamma\in\Gamma} x(\gamma) = \lim_{F\in\operatorname{Fin}(\Gamma)} \sum_{\gamma\in F} x(\gamma).$$

When $x(\gamma) \geqslant 0$ for all $\gamma \in \Gamma$, in order that $\sum_{\gamma\in\Gamma} x(\gamma)$ is convergent, it is necessary and sufficient that $\left(\sum_{\gamma\in F} x(\gamma)\right)_{F\in\operatorname{Fin}(\Gamma)}$ is upper bounded, and then

$$\sum_{\gamma\in\Gamma} x(\gamma) = \sup_{F\in\operatorname{Fin}(\Gamma)} \sum_{\gamma\in F} x(\gamma).$$

Thus, by $\sum_{\gamma\in\Gamma} x(\gamma) < \infty$, we mean that $\sum_{\gamma\in\Gamma} x(\gamma)$ is convergent.

For $x \in \mathbb{R}^\mathbb{N}$, $\sum_{i\in\mathbb{N}} x(i)$ should be distinguished from $\sum_{i=1}^\infty x(i)$. When the sequence $\left(\sum_{i=1}^n x(i)\right)_{n\in\mathbb{N}}$ is convergent, we say that $\sum_{i=1}^\infty x(i)$ is **convergent** and define

$$\sum_{i=1}^\infty x(i) = \lim_{n\to\infty} \sum_{i=1}^n x(i).$$

Evidently, if $\sum_{i\in\mathbb{N}} x(i)$ is convergent then $\sum_{i=1}^\infty x(i)$ is also convergent and $\sum_{i=1}^\infty x(i) = \sum_{i\in\mathbb{N}} x(i)$. However, $\sum_{i\in\mathbb{N}} x(i)$ is not convergent even if $\sum_{i=1}^\infty x(i)$ is convergent. In fact, due to Proposition 1.2.3 in [GAGT], the following equiva-

[10] In some literature, this space is denoted by $\boldsymbol{m}$.

lence holds:

$$\sum_{i\in\mathbb{N}} x(i) \text{ is convergent } \Leftrightarrow \sum_{i=1}^{\infty} |x(i)| \text{ is convergent.}$$

For each $p \geqslant 1$, the Banach space $\ell^p(\Gamma)$ is defined as follows:

- $\ell^p(\Gamma) = \left\{x \in \mathbb{R}^\Gamma \mid \sum_{\gamma\in\Gamma} |x(\gamma)|^p < \infty\right\}$ with the norm

$$\|x\|_p = \left(\sum_{\gamma\in\Gamma} |x(\gamma)|^p\right)^{1/p}.$$

Like $\ell^\infty_f(\Gamma)$, the space $\mathbb{R}^\Gamma_f$ with this norm is denoted by $\ell^p_f(\Gamma)$.[11]

Similarly to $\boldsymbol{c}_0(\Gamma)$, we have $w(\ell^p(\Gamma)) = \operatorname{card}\Gamma$. When $\operatorname{card}\Gamma = \aleph_0$, the Banach space $\ell^p(\Gamma)$ is linearly isometric to $\ell^p = \ell^p(\mathbb{N})$, which is separable. The space $\ell^2(\Gamma)$ is the **Hilbert space** with the inner product

$$\langle x, y\rangle = \sum_{\gamma\in\Gamma} x(\gamma)y(\gamma),$$

which is well-defined because

$$\sum_{\gamma\in\Gamma} |x(\gamma)y(\gamma)| \leqslant \frac{\|x\|_2^2 + \|y\|_2^2}{2} < \infty.$$

For $1 \leqslant p < q < \infty$, we have

$$\ell^p(\Gamma) \subsetneqq \ell^q(\Gamma) \subsetneqq \boldsymbol{c}_0(\Gamma) \quad \text{as sets (or linear spaces).}$$

These inclusions are continuous because $\|x\|_\infty \leqslant \|x\|_q \leqslant \|x\|_p$ for every $x \in \ell^p(\Gamma)$. When Γ is infinite, the topology of $\ell^p(\Gamma)$ is distinct from the one induced by the norm $\|\cdot\|_q$ or $\|\cdot\|_\infty$ (i.e., the topology inherited from $\ell^q(\Gamma)$ or $\boldsymbol{c}_0(\Gamma)$). In fact, the unit sphere $\mathbf{S}_{\ell^p(\Gamma)}$ is closed in $\ell^p(\Gamma)$ but not closed in $\ell^q(\Gamma)$ for any $q > p$ nor in $\boldsymbol{c}_0(\Gamma)$. — Refer to [GAGT, p. 17].

For $1 \leqslant p \leqslant \infty$, we have

$$\mathbb{R}^\Gamma_f \subset \ell^p(\Gamma) \quad \text{as sets (or linear spaces).}$$

Let $\ell^p_f(\Gamma)$ denote the subspace of $\ell^p(\Gamma)$ with $\ell^p_f(\Gamma) = \mathbb{R}^\Gamma_f$ as sets.

[11]The triangle inequality for $\|x\|_p$ is known as the Minkowski inequality. The proof can be found on pp. 16–17 in [GAGT].

When $\Gamma = \mathbb{N}$, we write $\ell_f^p(\mathbb{N}) = \ell_f^p$. By Proposition 1.2.1, we know $\ell_f^p(\Gamma) \neq \mathbb{R}_f^\Gamma$ as spaces for any infinite set Γ. In the above, the sequence $(x_n)_{n\in\mathbb{N}}$ is contained in the unit sphere $\mathbf{S}_{\ell_f^p(\Gamma)}$ of $\ell_f^p(\Gamma)$, which means that $\mathbf{S}_{\ell_f^p(\Gamma)}$ is not closed in ℓ_f^q, hence $\ell_f^p \neq \ell_f^q$ as spaces for $1 \leqslant p < q \leqslant \infty$. Note that $\mathbf{S}_{\ell_f^p(\Gamma)}$ is a closed subset of ℓ_f^q for $1 \leqslant q < p$.

Concerning the convergence of sequences in $\ell^p(\Gamma)$, we have the following:

Proposition 1.2.2 *For each $p \in \mathbb{N}$ and $x \in \ell^p(\Gamma)$, a sequence $(x_n)_{n\in\mathbb{N}}$ converges to x in $\ell^p(\Gamma)$ if and only if*

$$\|x\|_p = \lim_{n\to\infty} \|x_n\|_p \ \textit{and}\ \forall \gamma \in \Gamma, x(\gamma) = \lim_{n\to\infty} x_n(\gamma). \qquad \boxed{1.2.4}$$

Remark 1.2 It should be noted that Proposition 1.2.2 is valid not only for sequences but also nets, which means that the unit spheres $\mathbf{S}_{\ell^p(\Gamma)}$, $p \in \mathbb{N}$, are subspaces of the product space $\mathbb{R}^\Gamma$, whereas neither $\mathbb{R}^\Gamma$ nor $\mathbb{R}_f^\Gamma$ is metrizable if Γ is uncountable. Therefore, if $1 \leqslant p < q \leqslant \infty$, then $\mathbf{S}_{\ell^p(\Gamma)}$ is also a subspace of $\ell^q(\Gamma)$, while, as mentioned above, $\mathbf{S}_{\ell^p(\Gamma)}$ of $\ell^p(\Gamma)$ is not closed in the space $\ell^q(\Gamma)$. The unit sphere $\mathbf{S}_{\ell_f^p(\Gamma)}$ of $\ell_f^p(\Gamma)$ is a subspace of $\mathbb{R}_f^\Gamma$ ($\subset \mathbb{R}^\Gamma$) and also a subspace of $\ell^q(\Gamma)$ for $1 \leqslant q \leqslant \infty$.

Remark 1.3 The "if" part of Proposition 1.2.2 does not hold for the space $\boldsymbol{c}_0(\Gamma)$ for any infinite set Γ (but the "only if" part obviously does hold). — Refer to Remark 3 on p. 17 of [GAGT].

Concerning the topological classification of $\ell^p(\Gamma)$, we have the following theorem due to S. Mazur [104]:

Theorem 1.2.3 (Mazur) *For each $1 < p < \infty$, $\ell^p(\Gamma)$ is homeomorphic to $\ell^1(\Gamma)$. By the same homeomorphism, $\ell_f^p(\Gamma) \approx \ell_f^1(\Gamma)$, that is, the pair $(\ell^p(\Gamma), \ell_f^p(\Gamma))$ is homeomorphic to the pair $(\ell^1(\Gamma), \ell_f^1(\Gamma))$.* $\boxed{1.2.5}$

For each space X, we simply write $\mathrm{C}_d(X, \mathbb{R}) = \mathrm{C}_u(X)$, where the metric d of $\mathbb{R}$ is the usual metric induced by the absolute value $|\cdot|$, that is, $\mathrm{C}_u(X)$ is the metric space with the sup-metric

$$d(f, g) = \sup_{x\in X} \min\left\{|f(x) - g(x)|,\ 1\right\}.^{12}$$

This metric is not induced by a norm. The topology of $\mathrm{C}_u(X)$ is the uniform convergence topology. It should be noted that $\mathrm{C}_u(X)$ is a linear space but it is not a topological linear space in general. In fact, the scalar multiplication $\mathbb{R} \times \mathrm{C}_u((0, 1]) \to \mathrm{C}_u((0, 1])$ is not continuous with respect to the uniformly convergence topology.

---

[12]See Footnote 5 (p. 8).

Let $f \in C_u((0,1])$ be defined by $f(x) = x^{-1}$ for each $x \in (0,1]$. Then, for any $t > 0$, if $x < t$, then $|tf(x) - 0f(x)| > 1$, that is, $d(tf, 0f) = 1$.

Among subspaces $C_u(X)$, we have the following Banach space:

- $C^B(X) = \{f \in C(X) \mid \sup_{x \in X} |f(x)| < \infty\}$ with the sup-norm

$$\|f\| = \sup_{x \in X} |f(x)|.$$

As is easily observed, $C^B(X)$ is clopen in $C(X)$. Moreover, $C^B(X)$ is a component of the space $C_u(X)$ because $C^B(X)$ is path-connected as a normed linear space. When X is compact, we have $C^B(X) = C_u(X)$. If X is discrete and infinite, then $C^B(X) = \ell^\infty(X)$, so $C^B(\mathbb{N}) = \ell^\infty$ in particular.

The space $C_p(X) = C_p(X, \mathbb{R})$ is a topological linear space as a subspace of the product space $\mathbb{R}^X$. The topology of $C_p(X)$ is the pointwise convergence topology. The space $C_p(\mathbb{N})$ is none other than the space of sequences $\boldsymbol{s} = \mathbb{R}^{\mathbb{N}}$.

1.3 Topological Spaces

The following TIETZE–URYSOHN EXTENSION THEOREM is very useful and applied in various fields:

Theorem 1.3.1 (TIETZE–URYSOHN) *Let A be a closed set in a normal space X. Then, every map $f : A \to \mathbf{I}$ extends over X.* 2.2.2

Let $\mathcal{A}$ be a collection of subsets of a space X. It is said that $\mathcal{A}$ is **locally finite** (resp. **discrete**) in X if each point has a neighborhood U in X which meets only finitely many members (resp. at most one member) of $\mathcal{A}$, that is, $\operatorname{card}\mathcal{A}[U] < \aleph_0$ (resp. $\operatorname{card}\mathcal{A}[U] \leqslant 1$). If $\mathcal{A}$ is locally finite (resp. discrete) in X, then so is $\mathcal{A}^{\mathrm{cl}}$ $(= \{\operatorname{cl} A \mid A \in \mathcal{A}\})$. Moreover, we say that $\mathcal{A}$ is **σ-locally finite** (resp. **σ-discrete**) in X if $\mathcal{A}$ is a countable union of locally finite (resp. discrete) subcollections.

Theorem 1.3.2 (A.H. STONE) *Every open cover of a metrizable space has a locally finite and σ-discrete open refinement.* 2.3.1

Theorem 1.3.3 (BING; NAGATA–SMIRNOV) *For a regular space X, the following conditions are equivalent:*

(a) *X is metrizable;*
(b) *X has a σ-discrete open basis;*
(c) *X has a σ-locally finite open basis.* 2.3.4

The equivalence of (a) and (b) in the above theorem is called the BING METRIZATION THEOREM and the equivalence of (a) and (c) is called the NAGATA–SMIRNOV METRIZATION THEOREM. Separable metrizable spaces are characterized as follows:

Theorem 1.3.4 (Urysohn Metrization) *A space is separable metrizable if and only if it is regular and second countable.* [2.3.5]

For an infinite set Γ, the **hedgehog** $J(\Gamma)$ is defined as the following closed subspace of $\ell^1(\Gamma)$:

$$J(\Gamma) = \bigcup_{\gamma \in \Gamma} \mathbf{I}\mathbf{e}_\gamma = \big\{x \in \ell^1(\Gamma) \bigm| x(\gamma) \in \mathbf{I} \text{ for all } \gamma \in \Gamma,$$
$$x(\gamma) \neq 0 \text{ at most one } \gamma \in \Gamma \big\}.$$

Theorem 1.3.5 *Every metrizable space X can be embedded into $J(\Gamma)^{\mathbb{N}}$, where $w(x) \leqslant \operatorname{card}\Gamma$. If X is separable, then X can be embedded into the Hilbert cube $\boldsymbol{Q}$, and hence into $\mathbf{s} = \mathbb{R}^{\mathbb{N}}$.* [2.3.7] [2.3.8]

A **perfect map** $f : X \to Y$ is a closed map such that $f^{-1}(y)$ is compact for each $y \in Y$. Concerning metrizability, the following is known:

Proposition 1.3.6

(1) *The perfect image of a metrizable space is metrizable, that is, if $f : X \to Y$ is a surjective perfect map of a metrizable space X, then Y is also metrizable.* [2.4.5(1)]
(2) *A space X is metrizable if it is a locally finite union of metrizable closed subspaces.* [2.4.5(2)]

A map $f : X \to Y$ is said to be **proper** if $f^{-1}(K)$ is compact for every compact set $K \subset Y$. In the case that X and Y are metrizable, these concepts coincide, that is:

Proposition 1.3.7 *For a map $f : X \to Y$ between metrizable spaces, the following are equivalent:*

(a) *$f : X \to Y$ is perfect;*
(b) *$f : X \to Y$ is proper;*
(c) *Any sequence $(x_n)_{n\in\mathbb{N}}$ in X has a convergent subsequence if $(f(x_n))_{n\in\mathbb{N}}$ is convergent in Y.* [2.1.5] [2.1.6]

It should be remarked that a map between locally compact spaces is proper if it is close to a proper map, that is:

Proposition 1.3.8 *Let $f : X \to Y$ be a proper map and $\mathcal{U}$ be a locally finite open cover of Y such that every $U \in \mathcal{U}$ has the compact closure in Y (so X and Y should be locally compact). If a map $g : X \to Y$ is $\mathcal{U}$-close to f, then g is also proper.* [2.9.6]

It is said that proper maps $f, g : X \to Y$ are **properly homotopic** (or f is **properly homotopic** to g) if there is a proper homotopy $h : X \times \mathbf{I} \to Y$ from f to g. Then, we write

$$f \simeq^p g.$$

In the above, if h is a proper $\mathcal{U}$-homotopy (resp. a proper homotopy relative to $A \subset X$), we say that f and g are **properly $\mathcal{U}$-homotopic** (resp. **properly homotopic relative to** A) and write

$$f \simeq^p_{\mathcal{U}} g \quad \big(\text{resp. } f \simeq^p g \text{ rel. } A\big).$$

It is said that X and Y are **proper homotopy equivalent** if there are maps $f : X \to Y$ and $g : Y \to X$ such that $gf \simeq^p \mathrm{id}_X$ and $fg \simeq^p \mathrm{id}_Y$, where f is called a **proper homotopy equivalence** and g is a **proper homotopy inverse** of f. Then, it is also said that X and Y have the **same proper homotopy type** or X has the **proper homotopy type** of Y, which is written as follows:

$$X \simeq^p Y.$$

For proper maps $f, f' : X \to Y$ and $g, g' : Y \to Z$,

$$f \simeq^p f', \; g \simeq^p g' \;\Rightarrow\; gf \simeq^p g'f'.$$

Applying Proposition 1.3.8 above, to a homotopy $h : X \times \mathbf{I} \to Y$ and a constant homotopy $h_0 \mathrm{pr}_X : X \times \mathbf{I} \to Y$, we have the following:

Proposition 1.3.9 *Let $\mathcal{U}$ be a locally finite open cover of Y such that every $U \in \mathcal{U}$ has the compact closure in Y. Then, a $\mathcal{U}$-homotopy $h : X \times \mathbf{I} \to Y$ is proper if $h_0 : X \to Y$ is proper.* □

The **(metric) completion** of a metric space $X = (X, d)$ is a complete metric space $\tilde{X} = (\tilde{X}, \tilde{d})$ containing X as a dense set and as a metric subspace (i.e., d is a restriction of $\tilde{d}$).

Theorem 1.3.10 *Every metric space has a completion.* [2.3.10]

A space X is **paracompact** if each open cover of X has a locally finite open refinement.[13] The following is a direct consequence of the Stone Theorem 1.3.2:

- *Every metrizable space is paracompact.*

We have the following characterization:

Theorem 1.3.11 *A space X is paracompact if and only if every open cover of X has an open star-refinement.* [2.6.3]

A space X is **collectionwise normal** if, for each discrete collection $\mathcal{F}$ of closed sets in X, there is a pairwise disjoint collection $\{U_F \mid F \in \mathcal{F}\}$ of open sets in X such that $F \subset U_F$ for each $F \in \mathcal{F}$. Obviously, every collectionwise normal space is normal. In the definition of collectionwise normality, we can take $\{U_F \mid F \in \mathcal{F}\}$ so as to be discrete in X.

[13]Recall that spaces are assumed to be **Hausdorff**.

Indeed, choose an open set V in X so that $\bigcup \mathcal{F} \subset V \subset \mathrm{cl}\, V \subset \bigcup_{F \in \mathcal{F}} U_F$. Then, $F \subset V \cap U_F$ for each $F \in \mathcal{F}$ and $\{V \cap U_F \mid F \in \mathcal{F}\}$ is discrete in X.

It is said that X is **hereditarily paracompact** if every subspace of X is paracompact.

Proposition 1.3.12 (Results on Paracompact Spaces)

(1) *Every paracompact space is collectionwise normal.* 2.6.1
(2) *A space is paracompact if it is a locally finite union of paracompact closed subspaces.* 2.6.7(1)
(3) *Every F_σ subspace of a paracompact space is paracompact.* 2.6.7(2)
(4) *A space X is hereditarily paracompact if every open subspace of X is paracompact.* 2.6.7(3)
(5) *Every locally (completely) metrizable paracompact space is (completely) metrizable.* 2.6.7(4)

From (1) above, it follows that:

- *Every metrizable space is collectionwise normal.*

By virtue of (5), the metrizability in the definition of an n-manifold can be replaced with the paracompactness.

A closed set $A \subset X$ is called a **zero set** in X if $A = f^{-1}(0)$ for some map $f : X \to \mathbb{R}$. The complement of a zero set in X is called a **cozero set**. A normal space X is **perfectly normal** if every closed set in X is G_δ in X (i.e., every open set in X is F_σ in X), which is equivalent to the condition that every closed set in X is a zero set in X, i.e., every open set in X is a cozero set in X (cf. Theorem 2.2.6 in [GAGT]). As is easily observed, we have:

- *Every metrizable space is perfectly normal.*

Moreover, we have:

Theorem 1.3.13 *Every perfectly normal paracompact space is hereditarily paracompact.* 2.6.8

A real-valued function $f : X \to \mathbb{R}$ is said to be **lower semi-continuous** (abbrev. **l.s.c.**) (resp. **upper semi-continuous** (abbrev. **u.s.c.**)) if $f^{-1}((t, \infty))$ (resp. $f^{-1}((-\infty, t))$) is open in X for each $t \in \mathbb{R}$. Evidently, $f : X \to \mathbb{R}$ is continuous if and only if f is l.s.c. and u.s.c.

Theorem 1.3.14 *Let $g, h : X \to \mathbb{R}$ be real-valued functions on a paracompact space X such that g is u.s.c., h is l.s.c., and $g(x) < h(x)$ for each $x \in X$. Then, there exists a map $f : X \to \mathbb{R}$ such that $g(x) < f(x) < h(x)$ for each $x \in X$. Moreover, given a map $f_0 : A \to \mathbb{R}$ of a closed set A in X such that $g(x) < f_0(x) < h(x)$ for each $x \in A$, the map f can be an extension of f_0.* 2.7.6

A space X is called a **Baire space** if X satisfies one of the following equivalent conditions:

(a) The countable intersection of open dense sets in X is dense;
(b) Every countable intersection of dense G_δ sets in X is dense;

(c) The countable union of closed sets in X without interior points has no interior point;
(d) If a countable union of closed sets in X has an interior point, then at least one of them has an interior point.

The following is easily proved:

- *Every open subspace and every dense G_δ subspace of a Baire space are also Baire spaces.*

The following BAIRE CATEGORY THEOREM is very useful and important.

Theorem 1.3.15 (BAIRE) *Every completely metrizable space is a Baire space. Consequently, it cannot be written as a union of countably many closed sets without interior points.* 2.5.1

A metrizable space X is said to be **absolutely G_δ** if X is G_δ in an arbitrary metrizable space which contains X as a subspace. This concept characterizes the complete metrizability, that is:

Theorem 1.3.16 *A metrizable space is completely metrizable if and only if it is absolutely G_δ.* 2.5.2

For the complete metrizability, the following holds:

Theorem 1.3.17 *Let $X = (X, d)$ be a metric space and $A \subset X$.*

(1) *If A is completely metrizable, then A is G_δ in X.*
(2) *If X is complete and A is G_δ in X, then A is completely metrizable.* 2.5.3

The following two extension theorems are due to M. Lavrentieff:

Theorem 1.3.18 (LAVRENTIEFF G_δ-EXTENSION) *Every map $f : A \to Y$ from $A \subset X$ to a completely metrizable space Y extends over a G_δ set G in X such that $A \subset G \subset \operatorname{cl} A$.* 2.5.7

Theorem 1.3.19 (LAVRENTIEFF HOMEOMORPHISM EXTENSION) *Let X and Y be completely metrizable spaces with $A \subset X$ and $B \subset Y$. Then, every homeomorphism $f : A \to B$ extends to a homeomorphism $\tilde{f} : G \to H$ between G_δ sets in X and Y such that $A \subset G \subset \operatorname{cl} A$ and $B \subset H \subset \operatorname{cl} B$.* 2.5.8

Let $\mathcal{P}$ be a property for subsets of a space X. It is said that X has property $\mathcal{P}$ **locally** if each $x \in X$ has a neighborhood U in X which has property $\mathcal{P}$. A property $\mathcal{P}$ for *open* sets in X is said to be ***G*-hereditary** if the following conditions are satisfied:

(G-1) If U has property $\mathcal{P}$, then every open subset of U has $\mathcal{P}$;
(G-2) If U and V have property $\mathcal{P}$, then $U \cup V$ has property $\mathcal{P}$;
(G-3) If $\{U_\lambda \mid \lambda \in \Lambda\}$ is discrete in X and each U_λ has property $\mathcal{P}$, then $\bigcup_{\lambda \in \Lambda} U_\lambda$ has property $\mathcal{P}$.

The following is very useful to show that a space has a certain property:

Theorem 1.3.20 (E. Michael) *Let $\mathcal{P}$ be a G-hereditary property for open sets in a paracompact space X. If X has property $\mathcal{P}$ locally, then X itself has property $\mathcal{P}$.* [2.6.5]

We can apply the above theorem even to a property $\mathcal{P}$ for *closed* sets in X by introducing the property $\mathcal{P}^{\circ}$ as follows:

$$\text{an open set } U \text{ has property } \mathcal{P}^{\circ} \underset{\text{def}}{\Leftrightarrow} \operatorname{cl} U \text{ has property } \mathcal{P}.$$

It is said that $\mathcal{P}$ is ***F*-hereditary** if it satisfies the following conditions:

(F-1) If A has property $\mathcal{P}$, then every closed subset of A has property $\mathcal{P}$;
(F-2) If A and B have property $\mathcal{P}$, then $A \cup B$ has property $\mathcal{P}$;
(F-3) If $\{A_\lambda \mid \lambda \in \Lambda\}$ is discrete in X and each A_λ has property $\mathcal{P}$, then $\bigcup_{\lambda \in \Lambda} A_\lambda$ has property $\mathcal{P}$.

Evidently, if property $\mathcal{P}$ is F-hereditary, then $\mathcal{P}^{\circ}$ is G-hereditary. Therefore, Theorem 1.3.20 yields the following corollary:

Corollary 1.3.21 (E. Michael) *Let $\mathcal{P}$ be an F-hereditary property for closed sets in a paracompact space X. If X has property $\mathcal{P}$ locally, then X itself has property $\mathcal{P}$.* [2.6.6]

Proposition 1.3.22 (Refinements by Open Balls)

(1) *Let X be a metrizable space and $\mathcal{U}$ be an open cover of X. Then, X has an admissible bounded metric ρ such that $\{\mathrm{B}_\rho(x, 1) \mid x \in X\} \prec \mathcal{U}$. Moreover, for given $d \in \mathrm{Metr}(X)$, ρ can be chosen so that $\rho \geqslant d$, which means that if d is complete then ρ is.* [2.7.7(1)]
(2) *Let $X = (X, d)$ be a metric space. For each open cover $\mathcal{U}$ of X, there is a map $\gamma : X \to (0, 1)$ such that $\{\mathrm{B}(x, \gamma(x)) \mid x \in X\} \prec \mathcal{U}$.* [2.7.7(2)]

1.4 Linear Spaces and Convex Sets

Let E be a linear space.[14] It is said that finitely many distinct points $x_1, \ldots, x_n \in E$ are **affinely** (or **geometrically**) **independent** provided that, for $t_1, \ldots, t_n \in \mathbb{R}$,

$$\sum_{i=1}^{n} t_i x_i = \mathbf{0},\ \sum_{i=1}^{n} t_i = 0 \Rightarrow t_1 = \cdots = t_n = 0,$$

[14] Here, we only consider linear spaces over $\mathbb{R}$.

that is, $x_1 - x_n, \dots, x_{n-1} - x_n$ are linearly independent. A set $A \subset E$ is said to be **affinely** (or **geometrically**) **independent** if all finite distinct points of A are affinely independent.

We call $F \subset E$ a **flat**[15] (resp. a **convex set**) if the straight line through (resp. the line segment between) every pair of distinct points of F is contained in F, that is,

$$\forall x, y \in F,\ \forall t \in \mathbb{R} \text{ (resp. } \forall t \in \mathbf{I}),\ (1-t)x + ty \in F.$$

It is easy to see that $F \subset E$ is a flat if and only if $F - x$ is a linear subspace of E for some (or any) $x \in F$, where $F - x = F - y$ for any $x, y \in F$ (cf. [GAGT, Proposition 3.1.1, Corollary 3.1.2]). The **dimension** $\dim F$ of a flat F is defined as the dimension of the linear subspace $F - x$ for some (or any) $x \in F$. In the infinite-dimensional case, $\dim F$ is the cardinal of a maximal affinely independent subset. For a subset $A \subset E$, the smallest flat $\operatorname{fl} A$ containing A is called the **flat hull**[16] of A and the smallest convex set $\langle A \rangle$ containing A is called the **convex hull** of A. When $A = \{x_1, \dots, x_n\}$, we write $\langle A \rangle = \langle x_1, \dots, x_n \rangle$ and

$$\langle x_1, \dots, x_n \rangle = \left\{ \textstyle\sum_{i=1}^n z(i)x_i \;\middle|\; z \in \Delta^{n-1} \right\}.$$

The **dimension** $\dim C$ of a convex set C is defined as the dimension of $\operatorname{fl} C$, where $\operatorname{fl} C$ can be obtained by the following:

Proposition 1.4.1 *For a convex set $C \subset E$,*

$$\operatorname{fl} C = \left\{ (1-t)x + ty \;\middle|\; x, y \in C,\ t \in \mathbb{R} \right\}.$$ [3.2.1]

For a convex set $C \subset E$, the **radial interior** $\operatorname{rint} C$ is defined as follows:

$$\operatorname{rint} C = \left\{ x \in C \;\middle|\; \forall y \in C, \exists \delta > 0 \text{ such that } (1+\delta)x - \delta y \in C \right\}.$$

When $C = \langle x_1, \dots, x_n \rangle$, we have

$$\operatorname{rint}\langle x_1, \dots, x_n \rangle = \left\{ \textstyle\sum_{i=1}^n z(i)x_i \;\middle|\; z \in \Delta^{n-1} \cap (0, \infty)^n \right\}.$$

See [GAGT, p. 79]. In particular, $\operatorname{rint}\{x\} = \{x\}$ and $\operatorname{rint}\langle x, y \rangle = \langle x, y \rangle \setminus \{x, y\}$ if $x \neq y$. The radial interior of C can also be defined as

$$\operatorname{rint} C = \left\{ x \in C \;\middle|\; \forall y \in C, \exists z \in C \text{ such that } x \in \operatorname{rint}\langle y, z \rangle \right\}.$$

[15] A flat is also called an **affine set**, a **linear manifold**, or a **linear variety**.

[16] The flat hull is also called the **affine hull**.

The following set is called the **face** of C at $x \in C$:

$$C_x = \{y \in C \mid \exists \delta > 0 \text{ such that } (1+\delta)x - \delta y \in C\}$$
$$= \{y \in C \mid \exists z \in C \text{ such that } x \in \operatorname{rint}\langle y, z\rangle\}.$$

Then, $\operatorname{rint} C = \{x \in C \mid C_x = C\}$. It is said that a point $x \in E$ is **linearly accessible** from C if $\operatorname{rint}\langle x, y\rangle \subset C$ for some $y \in C$. The **radial closure** $\operatorname{rcl} C$ of C is the set of all linearly accessible points from C. It should be noted that $\operatorname{rcl} C \subset \operatorname{fl} C$ by Proposition 1.4.1, hence $\operatorname{fl}\operatorname{rcl} C = \operatorname{fl} C$. We have the following inclusions:

$$\operatorname{rint} C \subset C \subset \operatorname{rcl} C \subset \operatorname{fl} C.$$

The **radial boundary** ∂C of C is defined as $\partial C = \operatorname{rcl} C \setminus \operatorname{rint} C$.

Proposition 1.4.2 *Let $C \subset E$ be a convex set. If $x \in \operatorname{rint} C$, $y \in \operatorname{rcl} C$, and $0 \leqslant t < 1$, then $(1-t)x + ty \in \operatorname{rint} C$, i.e., $\langle x, y\rangle \setminus \{y\} \subset \operatorname{rint} C$.* [3.2.3]

Proposition 1.4.3 *For every two convex sets $C, D \subset E$,*

$$\operatorname{rint}(C \cap D) = \operatorname{rint} C \cap \operatorname{rint} D \ \textit{if} \ \operatorname{rint} C \cap \operatorname{rint} D \neq \emptyset; \quad [3.2.7(2)]$$

$$(C \cap D)_x = C_x \cap D_x \ \textit{for any}\ x \in C \cap D. \quad [3.2.7(1)]$$

When a convex set C is contained in a flat $F \subset E$, the following set is called the **core** of C in F:[17]

$$\operatorname{core}_F C = \{x \in C \mid \forall y \in F,\ \exists \delta > 0 \text{ such that}$$
$$|t| < \delta \Rightarrow (1-t)x + ty \in C\}$$
$$= \{x \in C \mid \forall y \in F,\ \exists s > 0 \text{ such that } (1-s)x + sy \in C\},$$

where the last equality comes from the convexity of C. Then, we have the following fact (cf. Fact on p. 87 in [GAGT]):

- $\operatorname{core}_F C \neq \emptyset \Leftrightarrow \operatorname{fl} C = F$.

When $F = E$, we can omit the phrase "in E" and simply write $\operatorname{core} C$.

Proposition 1.4.4 *For every convex set $C \subset E$, $\operatorname{core}_{\operatorname{fl} C} C = \operatorname{rint} C$, which is also convex. Then, $\operatorname{core} C \neq \emptyset$ implies $\operatorname{core} C = \operatorname{rint} C$, and $\operatorname{core}\operatorname{core} C = \operatorname{core} C$.* [3.3.2]

For a convex set C in a linear space E with $\mathbf{0} \in \operatorname{core} C$, the **Minkowski functional** $p_C : E \to \mathbb{R}_+$ for C can be defined as follows:

$$p_C(x) = \inf\{s > 0 \mid x \in sC\} = \inf\{s > 0 \mid s^{-1}x \in C\}.$$

[17]The core is defined for any subset of F, but we assume that C is convex.

Then, p_C is sublinear, that is, the following hold:

$$p_C(x+y) \leqslant p_C(x) + p_C(y) \quad \text{for each } x, y \in E,$$
$$p_C(tx) = tp_C(x) \quad \text{for each } x \in E \text{ and } t > 0.$$

Using the Minkowski functional, the radial interior, the radial closure, and the radial boundary can be written as follows:

$$\operatorname{rint} C = \operatorname{core} C = p_C^{-1}([0, 1)), \ \operatorname{rcl} C = p_C^{-1}(\mathbf{I}) \ \text{ and } \ \partial C = p_C^{-1}(1),$$

whose proof can be read in [GAGT, Proposition 3.3.4].

Let X and Y be flats (resp. convex sets) in linear spaces E and E', respectively. A function $f : X \to Y$ is said to be **affine** (or **linear in the affine sense**) provided that

$$f((1-t)x + ty) = (1-t)f(x) + tf(y)$$
$$\text{for every } x, y \in X \text{ and } t \in \mathbb{R} \text{ (resp. } t \in \mathbf{I}).$$

For convex sets $C \subset E$ and $D \subset E'$, every affine function $f : C \to D$ can be uniquely extended to an affine function $\tilde{f} : \operatorname{fl} C \to \operatorname{fl} D$ between their flat hull.

A **topological linear space** E is a linear space equipped with a topology such that the algebraic operations, the addition $(x, y) \mapsto x + y$ and the scalar multiplication $(t, x) \mapsto tx$, are continuous.[18] A neighborhood basis at $\mathbf{0}$ in a topological linear space can be characterized as follows:

Proposition 1.4.5 *Let E be a topological linear space. Then, a neighborhood basis $\mathcal{U}$ at $\mathbf{0}$ in E has the following properties:*

(i) *For each $U, V \in \mathcal{U}$, $U \cap V$ contains some $W \in \mathcal{U}$;*
(ii) *For each $U \in \mathcal{U}$, there is some $V \in \mathcal{U}$ such that $V + V \subset U$;*
(iii) *For each $U \in \mathcal{U}$, there is some $V \in \mathcal{U}$ such that $[-1, 1]V \subset U$;*
(iv) *For each $x \in E$ and $U \in \mathcal{U}$, there is some $a > 0$ such that $x \in aU$;*
(v) $\bigcap \mathcal{U} = \{\mathbf{0}\}$.

Conversely, let $\mathcal{U}$ be a collection of subsets of a linear space E satisfying the above properties (i)–(v). *Then, E has a topology which makes E a topological linear space such that $\mathcal{U}$ is a neighborhood basis at $\mathbf{0}$.* 3.4.1

It is said that $A \subset E$ is **circled** if $[-1, 1]A \subset A$ (i.e., $tA \subset A$ for every $t \in [-1, 1]$). In Proposition 1.4.5 above, $W = [-1, 1]V$ is a circled neighborhood of $\mathbf{0}$ in E.

[18] Recall that topological spaces are assumed to be **Hausdorff**. For topological linear spaces (more generally for topological groups), it suffices to assume the axiom T_0, which implies the **regularity** (cf. Proposition 2.4.2 in [GAGT]).

Concerning convex sets in a topological linear space E, we have the following:

Proposition 1.4.6 *For each convex set C in a topological linear space E with $\mathbf{0} \in \operatorname{core} C$, the Minkowski functional $p_C : E \to \mathbb{R}$ is continuous, $p_C^{-1}([0,1)) = \operatorname{int} C = \operatorname{rint} C$, $p_C^{-1}(\mathbf{I}) = \operatorname{cl} C = \operatorname{rcl} C$, and $p_C^{-1}(1) = \operatorname{bd} C = \partial C$.* 3.4.12

Theorem 1.4.7 *Every open convex set V in a topological linear space E is homeomorphic to E itself.* 3.4.15

Concerning the finite-dimensionality, we have the following:

Proposition 1.4.8 *Every finite-dimensional flat F in any linear space E has the unique topology such that the following operation is continuous:*

$$F \times F \times \mathbb{R} \ni (x, y, t) \mapsto (1-t)x + ty \in F.$$

With respect to this topology, every affine bijection $f : \mathbb{R}^n \to F$ is a homeomorphism, where $n = \dim F$. 3.5.1

As a corollary, we have the following:

Corollary 1.4.9 *Every finite-dimensional linear space E has the unique topology compatible with the algebraic operations (addition and scalar multiplication), and then it is linearly homeomorphic to $\mathbb{R}^n$, where $n = \dim E$.* 3.5.2

Concerning the radial interior of a finite-dimensional convex set, we have the following:

Proposition 1.4.10 *For every finite-dimensional convex set $C \subset E$, $\operatorname{rint} C = \operatorname{int}_{\operatorname{fl} C} C$ with respect to the unique topology for $\operatorname{fl} C$ as in Proposition 1.4.8.* 3.5.5

For the radial closure, we have the following similar result, which was an overlooked result in [GAGT]:

Proposition 1.4.11 *For every finite-dimensional convex set $C \subset E$, $\operatorname{rcl} C = \operatorname{cl}_{\operatorname{fl} C} C$ with respect to the unique topology for $\operatorname{fl} C$ as in Proposition 1.4.8, hence $\partial C = \operatorname{bd}_{\operatorname{fl} C} C$.*

Proof Let $n = \dim C$. Since $\operatorname{fl} C$ is affinely homeomorphic to $\mathbb{R}^n$, we may assume that $C \subset \mathbb{R}^n$. Since $\operatorname{int} C \neq \emptyset$ as in the proof of [GAGT, Proposition 3.5.5], there is $v \in \operatorname{int} C$. Then, choose $\delta > 0$ so that $\|v - u\| < \delta \Rightarrow u \in C$, where $\|\cdot\|$ is the norm for $\mathbb{R}^n$. As is easily observed, $\operatorname{rcl} C \subset \operatorname{cl} C$ [GAGT, Proposition 3.4.7(1)]. To see the converse inclusion, for each $x \in \operatorname{cl} C$ and $0 < t < 1$, let $z = (1-t)x + tv$. Take $y \in C$ so that $\|x - y\| < t\delta/(1-t)$ and let

$$u = \frac{1}{t}z - \frac{1-t}{t}y = \frac{1-t}{t}x + v - \frac{1-t}{t}y.$$

Then, $(1-t)y + tu = z$ and

$$\|u - v\| = \left\| \frac{1-t}{t}x - \frac{1-t}{t}y \right\| = \frac{1-t}{t}\|x - y\| < \delta,$$

which implies $u \in C$, so $z \in C$. Thus, $\operatorname{rint}\langle x, v\rangle \subset C$, hence $x \in \operatorname{rcl} C$. □

The finite-dimensionality of a topological linear space can be topologically characterized as follows:

Theorem 1.4.12 *Let E be a topological linear space. The following are equivalent:*

(a) *E is finite-dimensional;*
(b) *E is locally compact;*
(c) *$\mathbf{0} \in E$ has a totally bounded neighborhood in E.* 3.5.9

In the above, a subset A of a topological linear space E is **totally bounded** provided, for each neighborhood U of $\mathbf{0} \in E$, there exists a finite set $M \subset E$ such that $A \subset U + M = \bigcup_{x \in M}(U + x)$. In this definition, M can be taken as a subset of A. Indeed, there is a neighborhood V of $\mathbf{0}$ in E such that $V - V \subset U$. Then, we can take $x_1, \ldots, x_n \subset E$ so that $A \subset \bigcup_{i=1}^{n}(V + x_i)$ and $(V + x_i) \cap A \neq \emptyset$ for every $i = 1, \ldots, n$. For each $i = 1, \ldots, n$, choose $y_i \in (V + x_i) \cap A$. Since $x_i \in -V_i + y_i$, it follows that

$$V + x_i \subset V - V + y_i = (V - V) + y_i \subset U + y_i.$$

Hence, we have $A \subset \bigcup_{i=1}^{n}(U + y_i)$.

For the metrizability of a topological linear space, we have the following very simple characterization:

Theorem 1.4.13 *A topological linear space E is metrizable if and only if $\mathbf{0} \in E$ has a countable neighborhood basis.* 3.6.1

A metric d on a linear space E is said to be **invariant** if

$$d(x + z, y + z) = d(x, y) \text{ for every } x, y, z \in E.$$

It is equivalent to this condition that $d(x, y) = d(x - y, \mathbf{0})$ for every $x, y \in E$. With respect to an invariant metric on E, addition on E is clearly continuous. Moreover, scalar multiplication on E is continuous if and only if the following three conditions are satisfied:

(i) $d(x_n, \mathbf{0}) \to 0 \Rightarrow \forall t \in \mathbb{R},\ d(tx_n, \mathbf{0}) \to 0$;
(ii) $t_n \to 0 \Rightarrow \forall x \in E,\ d(t_n x, \mathbf{0}) \to 0$;
(iii) $d(x_n, \mathbf{0}) \to 0,\ t_n \to 0 \Rightarrow d(t_n x_n, \mathbf{0}) \to 0$.

An invariant metric d on E satisfying these conditions is called a **linear metric**. A **metric linear space** $E = (E, d)$ is a linear space E with a linear metric d.

Evidently, every linear metric on a linear space E induces a topology of E that makes E a topological linear space. The following is a combination of the Fact on p. 112 of [GAGT] and Theorem 3.6.6 of [GAGT]:

Theorem 1.4.14 *Every admissible invariant metric for a (completely) metrizable topological linear space is a (complete) linear metric.*

A functional $\|\cdot\| : E \to \mathbb{R}$ on a linear space E is called an ***F*-norm**[19] if it satisfies the following conditions:

(F_1) $\|x\| \geqslant 0$;
(F_2) $\|x\| = 0 \Rightarrow x = \mathbf{0}$;
(F_3) $|t| \leqslant 1 \Rightarrow \|tx\| \leqslant \|x\|$;
(F_4) $\|x + y\| \leqslant \|x\| + \|y\|$;
(F_5) $\|x_n\| \to 0 \Rightarrow \|tx_n\| \to 0$;[20]
(F_6) $t_n \to 0 \Rightarrow \|t_n x\| \to 0$.

It should be observed that the converse of (F_2) is true because $\|\mathbf{0}\| = 0$ by the last condition (F_6). Then, $\|x\| = 0$ if and only if $x = \mathbf{0}$. A linear space E given an F-norm $\|\cdot\|$ is called an ***F*-normed linear space**. Every norm is an F-norm, hence every normed linear space is an F-normed space. An F-norm $\|\cdot\|$ induces the linear metric $d(x, y) = \|x - y\|$. Then, every F-normed linear space is a metric linear space. An F-norm on a topological linear space E is said to be **admissible** if it induces the topology for E.

Theorem 1.4.15 *A topological linear space has an admissible F-norm if and only if it is metrizable.* 3.6.3

Every metrizable topological linear space has an admissible F-norm with the following stronger condition than (F_3):

(F_3^*) $x \neq \mathbf{0}$, $|t| < 1 \Rightarrow \|tx\| < \|x\|$,

which implies that $\|sx\| < \|tx\|$ for each $x \neq \mathbf{0}$ and $0 < s < t$.

Due to Theorem 1.4.15, every metrizable topological linear space has an admissible invariant metric. When a topological linear space E is metrizable, the total boundedness of a subset $A \subset E$ is equivalent to the total boundedness with respect to an invariant metric $d \in \mathrm{Metr}(E)$, that is:

Fact *A is totally bounded in E if and only if for each $\varepsilon > 0$, A is covered by finite subsets $A_1, \ldots, A_n$ with* diam $\leqslant \varepsilon$.

In fact, since d is invariant, it follows that

$$\mathrm{B}_d(x, \varepsilon) = \mathrm{B}_d(\mathbf{0}, \varepsilon) + x \quad \text{for each } x \in E \text{ and } \varepsilon > 0.$$

[19] In the Köthe book [[21]], it is called an (F)-norm.

[20] As mentioned on p. 112 of [GAGT], (F_5) is unnecessary because it comes from (F_3) and (F_4). But we follow the definition in Köthe's book [[21]].

If A is totally bounded in E, then for each $\varepsilon > 0$, there is $a_1, \dots, a_n \in A$ such that $A \subset \bigcup_{i=1}^n \mathrm{B}(a_i, \varepsilon/2)$, where $\operatorname{diam} \mathrm{B}(a_i, \varepsilon/2) \leqslant \varepsilon$. Hence, A is totally bounded with respect to d. Conversely, assume A is totally bounded with respect to d. For each open neighborhood U of $\mathbf{0}$ in E, choose $\varepsilon > 0$ so that $\overline{\mathrm{B}}(\mathbf{0}, \varepsilon) \subset U$. Then, A is covered by finite subsets $A_1, \dots, A_n$ with $\operatorname{diam} \leqslant \varepsilon$. Take $a_i \in A_i$, $i = 1, \dots, n$. Since $A_i \subset \overline{\mathrm{B}}(a_i, \varepsilon) \subset U + a_i$, we have $A \subset \bigcup_{i=1}^n (U + a_i)$. Hence A is totally bounded in E.

A topological linear space E is **locally convex** if $\mathbf{0} \in E$ has a neighborhood basis consisting of (open) convex sets; equivalently, open convex sets make up an open basis for E. A completely metrizable locally convex topological linear space is called a **Fréchet space**.

Theorem 1.4.16 (Bartle–Graves–Michael) *Let E be a locally convex metric linear space and F be a linear subspace of E that is complete (so a Fréchet space). Then, $E \approx F \times E/F$. In particular, $E \approx \mathbb{R} \times G$ for some metric linear space G.* [3.8.9]

Theorem 1.4.17 (Banach–Mazur, Klee) *For every Banach space E, there exists a continuous linear surjection $q : \ell^1(\Gamma) \to E$, where* $\operatorname{card} \Gamma = \operatorname{dens} E$. [3.8.11]

As a corollary of the above two theorems, we have the following:

Corollary 1.4.18 *For every Banach space E, there exists a Banach space F such that $E \times F \approx \ell^1(\Gamma)$, where* $\operatorname{card} \Gamma = \operatorname{dens} E$. [3.8.12]

1.5 Cell Complexes and Simplicial Complexes

Let E be a linear space. The convex hull $C = \langle A \rangle$ of a non-empty finite set $A \subset E$ is called a **(convex) linear cell** (or just a **cell**). An n-dimensional cell is called an ***n*-cell**. According to Proposition 1.4.8, the flat hull $\operatorname{fl} C$ ($= \operatorname{fl} A$) has the unique topology such that the following operation is continuous:

$$\operatorname{fl} C \times \operatorname{fl} C \times \mathbb{R} \ni (x, y, t) \mapsto (1-t)x + ty \in \operatorname{fl} C.$$

Then, C is a compact subspace of $\operatorname{fl} C$ with this topology because $C = \langle A \rangle$ is the image of $A \times A \times \mathbf{I}$ by this operation. Hence, $\operatorname{rcl} C = C$ by Proposition 1.4.11. Recall $\operatorname{rint} C = \operatorname{int}_{\operatorname{fl} C} C$ by Propositions 1.4.10. Hence, $\partial C = C \setminus \operatorname{rint} C$.

Note Due to [GAGT, Proposition 3.5.8], a cell $C = \langle A \rangle$ has the unique topology such that the following operation is continuous:

$$C \times C \times \mathbf{I} \ni (x, y, t) \mapsto (1-t)x + ty \in C.$$

Using affine functionals, we can characterize cells as follows:

Proposition 1.5.1 *A non-degenerate $C \subset E$ is a cell if and only if $\operatorname{fl} C$ is finite-dimensional, $x + \mathbb{R}_+(y - x) \not\subset C$ for each distinct points $x, y \in C$, and there* *are finitely many non-constant affine functionals* $f_1, \ldots, f_k : \mathrm{fl}\, C \to \mathbb{R}$ *such that* $C = \mathrm{fl}\, C \cap \bigcap_{i=1}^{k} f_i^{-1}(\mathbb{R}_+)$, *where* $\mathrm{rint}\, C = \mathrm{fl}\, C \cap \bigcap_{i=1}^{k} f_i^{-1}((0, \infty))$.[21] 4.1.8

Every cell C has the smallest finite subset $C^{(0)}$ such that $\langle C^{(0)} \rangle = C$. Then, $C^{(0)} \subset \partial C$. Each point of $C^{(0)}$ is called a **vertex** of C. When $C^{(0)} = \{v_1, \ldots, v_n\}$ is affinely independent, C is called a **simplex** and $\dim C = n - 1$ ($n = \mathrm{card}\, C^{(0)}$). An n-dimensional simplex is called an ***n*-simplex**. A cell D is said to be a **face** of C if $D = C_x$ for some $x \in C$; this is denoted by $D \leqslant C$ or $C \geqslant D$. If $D \leqslant C$, then $D^{(0)} = C^{(0)} \cap D$ (cf. [GAGT, Proposition 4.1.4]). A face $D \leqslant C$ is called a **proper face** of C if $D \neq C$; this is denoted by $D < C$ or $C > D$. Then, $\partial C = \bigcup_{D<C} D$. An n-dimensional face is called an ***n*-face**. For simplexes σ and τ,

$$\sigma \leqslant \tau \;\Leftrightarrow\; \sigma^{(0)} \subset \tau^{(0)} \;\text{ and }\; \sigma < \tau \;\Leftrightarrow\; \sigma^{(0)} \subsetneqq \tau^{(0)}.$$

When a cell C is represented $C = \mathrm{fl}\, C \cap \bigcap_{i=1}^{k} f_i^{-1}(\mathbb{R}_+)$ for some non-constant affine functionals $f_1, \ldots, f_k : \mathrm{fl}\, C \to \mathbb{R}$, if $f_{i_0}^{-1}(0) \cap C \neq \emptyset$, then $D = C \cap f_{i_0}^{-1}(0)$ is a face of C (cf. [GAGT, 3.2.7(4)]), where

$$f_{i_0}^{-1}(0) \cap C = \emptyset \;\Rightarrow\; C = \mathrm{fl}\, C \cap \bigcap_{i \neq i_0} f_i^{-1}(\mathbb{R}_+).$$

If $C \neq \mathrm{fl}\, C \cap \bigcap_{i \neq i_0} f_i^{-1}(\mathbb{R}_+)$, then $D = C \cap f_{i_0}^{-1}(0)$ is a proper face of C and $\mathrm{fl}\, D = \mathrm{fl}\, C \cap f_{i_0}^{-1}(0)$.

Indeed, since f_{i_0} is non-constant, we have $\mathrm{fl}\, C \not\subset f_{i_0}^{-1}(0)$, which implies $C \not\subset f_{i_0}^{-1}(0)$ (cf. [GAGT, Proposition 3.2.2]), that is, $C \cap f_{i_0}^{-1}(0) \neq C$. Due to [GAGT, Proposition 3.2.1],

$$\mathrm{fl}\, D = \big\{(1 - t)x + ty \;\big|\; x, y \in D,\ t \in \mathbb{R}\big\}.$$

Then, it is easy to see that $\mathrm{fl}\, D = \mathrm{fl}\, C \cap f_{i_0}^{-1}(0)$.

Two simplex σ and τ are **joinable** (or σ is **joinable** to τ) if $\sigma \cap \tau = \emptyset$ and $\sigma^{(0)} \cup \tau^{(0)}$ is affinely independent. In this case, $\langle \sigma^{(0)} \cup \tau^{(0)} \rangle$ is a simplex called the **join** of σ and τ, which is written as $\sigma\tau$. Then, it follows that

$$(\sigma\tau)^{(0)} = \sigma^{(0)} \cup \tau^{(0)} \;\text{ and }\; \dim \sigma\tau = \dim \sigma + \dim \tau + 1.$$

For a proper face $\tau < \sigma$, the simplex $\tau' = \langle \sigma^{(0)} \setminus \tau^{(0)} \rangle$ is also a face of σ and σ is the join $\tau\tau'$, where τ' is called the **opposite face** of σ to τ.

[21] Proposition 4.1.8 of [GAGT] does not mention about $\mathrm{rint}\, C$, but this last equality is easily verified.

A collection K of (convex linear) cells in E is called a (**convex linear**) **cell complex** if K satisfies the following two conditions:

(C1) If $C \in K$ and $D \leqslant C$ then $D \in K$.
(C2) For each $C, D \in K$ with $C \cap D \neq \emptyset$, $C \cap D \leqslant C$ (and $C \cap D \leqslant D$).

Under condition (C1), condition (C2) is equivalent to each of the following conditions:

(C2′) For each $C, D \in K$, $C \cap \operatorname{rint} D \neq \emptyset$ implies $D \leqslant C$.
(C2″) For each $C \neq D \in K$, $\operatorname{rint} C \cap \operatorname{rint} D = \emptyset$.

A subcollection L of a cell complex K is called a **subcomplex** if L is a cell complex. For a cell C, the following are cell complexes:

$$F(C) = \{D \mid D \leqslant C\} \quad \text{and} \quad F(\partial C) = \{D \mid D < C\},$$

that is, $F(C)$ is the cell complex consisting of all faces of C and $F(\partial C)$ is the one of all proper faces of C. Then, $F(\partial C)$ is a subcomplex of $F(C)$. A **simplicial complex** is a cell complex consisting of simplexes. For a simplex σ, $F(\sigma)$ and $F(\partial\sigma)$ are simplicial complexes.

For two simplicial complexes K and L in the same linear space, they are **joinable** (or K is **joinable** to L) if every $\sigma \in K$ and $\tau \in L$ are joinable and

(J) $\sigma\tau \cap \sigma'\tau' = (\sigma \cap \sigma')(\tau \cap \tau')$ for each $\sigma, \sigma' \in K$ and $\tau, \tau' \in L$,

where $\emptyset\tau'' = \tau''$ and $\sigma''\emptyset = \sigma''$ for simplexes σ'', τ''. As is easily observed, the above condition is equivalent to the following:

(J′) for $\sigma, \sigma' \in K$ and $\tau, \tau' \in L$, $\sigma \neq \sigma'$ or $\tau \neq \tau'$ implies $\operatorname{rint} \sigma\tau \cap \operatorname{rint} \sigma'\tau' = \emptyset$.

For two joinable simplicial complexes K and L, the following is a simplicial complex:

$$K * L = K \cup L \cup \{\sigma\tau \mid \sigma \in K, \tau \in L\},$$

which is called the **join** of K and L. When $K = \{v_0\}$, we write $v_0 * L$, which is called a **simplicial cone** over L.

When K and L sit in linear spaces E and F respectively, they are regarded as simplical complexes in the linear space $E \times F \times \mathbb{R}$ by the following identification:

$$E = E \times \{\mathbf{0}\} \times \{0\} \quad \text{and} \quad F = \{\mathbf{0}\} \times F \times \{1\}.$$

Then, K and L are joinable and the join $K * L$ can be defined.

It is said that two simplexes σ, τ of a simplicial complex K are **joinable** in K (or σ is **joinable** to τ in K) if they are joinable and $\sigma\tau \in K$. Two subcomplexes K_1, K_2 of K are **joinable** in K (or K_1 is **joinable** to K_2 in K) if they are joinable and $K_1 * K_2 \subset K$.

For a cell (simplicial) complex K, $K^{(0)} = \bigcup_{C \in K} C^{(0)}$ is called the **set of vertices** (or the **0-skeleton**) of K. It is said that K is **finite**, **infinite**, or **countable** corresponding to card K (equivalently card $K^{(0)}$). For each $n \in \mathbb{N}$, the subcomplex $K^{(n)} = \{C \in K \mid \dim C \leqslant n\}$ is called the ***n*-skeleton** of K. The **dimension** of K is defined by $\dim K = \sup_{C \in K} \dim C$ and K is said to be ***n*-dimensional** if $\dim K = n$. Then, $\dim K \leqslant n$ if and only if $K = K^{(n)}$. It is said that K is **finite-dimensional** (denoted by $\dim K < \infty$) if $\dim K \leqslant n$ for some $n \in \mathbb{N}$. Otherwise, K is said to be **infinite-dimensional** (denoted by $\dim K = \infty$).

For a cell (simplicial) complex K, the set $|K| = \bigcup K = \bigcup_{C \in K} C$ is called the **polyhedron**[22] of K. We regard $|K|$ as a space with the weak topology with respect to K,[23] that is,

$$U \subset |K| \text{ is open in } |K| \Leftrightarrow \forall C \in K,\ U \cap C \text{ is open in } C.$$

This topology of $|K|$ is called the **Whitehead** (or **weak**) **topology**. A **polyhedron** (resp. a **topological polyhedron**) is defined as a space P such that $P = |K|$ (resp. $P \approx |K|$) for some cell complex K. A simplicial complex K is called a **triangulation** of a polyhedron (resp. a topological polyhedron) P if $P = |K|$ (resp. $P \approx |K|$), where it is also said that P is **triangulated** by K or K **triangulates** P. A subspace Q of a polyhedron (or a topological polyhedron) P is called a **subpolyhedron**[24] of P if there exists a pair (K, L) of a cell complex and its subcomplex such that $P = |K|$ and $Q = |L|$ (or $(P, Q) \approx (|K|, |L|)$).

Proposition 1.5.2 *Every (topological) polyhedron is perfectly normal and paracompact.* [4.2.2] [4.7.12]

Proposition 1.5.3 *Every subpolyhedron of a polyhedron P is a closed subspace of P.* [4.2.1]

A **full (simplicial) complex** is a simplicial complex K such that $K^{(0)}$ is affinely independent and $\langle v_1, \ldots, v_n \rangle \in K$ for any finite distinct vertices $v_1, \ldots, v_n \in K^{(0)}$. For any simplex σ, $F(\sigma)$ is a full complex. A subcomplex L of a simplicial complex K said to be **full in** K or a **full subcomplex** of K provided that $\sigma \in L$ for any $\sigma \in K$ with $\sigma^{(0)} \subset L^{(0)}$. Let K be an n-dimensional simplicial complex and $\sigma \in K$ an n-simplex. Then, $F(\partial\sigma)$ is not a full complex nor a full subcomplex of K, but it is a full subcomplex of the $(n-1)$-skeleton $K^{(n-1)}$.

Proposition 1.5.4 *For a subcomplex L of a simplicial complex K, the following are equivalent:*

(a) *L is full in K, that is, L is a full subcomplex of K;*
(b) *For $\sigma \in K$, $\sigma \cap |L| \neq \emptyset$ implies $\sigma \cap |L| \leqslant \sigma$;*
(c) *There does not exist $\sigma \in K \setminus L$ such that $\sigma \cap |L| = \partial\sigma$.*

---

[22]The plural is **polyhedra**.

[23]Cf. Footnote 3 (p. 5).

[24]The plural is **subpolyhedra**.

Proof The implication (b) $\Rightarrow$ (c) is obvious. So, we show (c) $\Rightarrow$ (a) $\Rightarrow$ (b).

(c) $\Rightarrow$ (a): If L is not full in K, then there exists a simplex $\tau \in K \setminus L$ with $\tau^{(0)} \subset L^{(0)}$. Let σ be a minimal face of τ such that $\sigma \notin L$. Then, any proper face of σ belongs to L because of the minimality of σ. Hence, $\partial\sigma \subset |L|$, which means that $\sigma \cap |L| = \partial\sigma$. This contradicts condition (c).

(a) $\Rightarrow$ (b): For $\sigma \in K$ with $\sigma \cap |L| \neq \emptyset$, $\sigma' = \langle \sigma^{(0)} \cap |L| \rangle$ is a face of σ. Then, $\sigma' \in L$ because L is full in K. Hence, $\sigma' \subset \sigma \cap |L|$. For $\tau \in L$ with $\tau \cap \sigma \neq \emptyset$, we have $\tau \cap \sigma \leqslant \sigma$. Since $(\tau \cap \sigma)^{(0)} \subset \sigma^{(0)} \cap |L|$, it follows that $\tau \cap \sigma \leqslant \sigma'$. Then,

$$\sigma \cap |L| = \bigcup \{\tau \cap \sigma \mid \tau \in L[\sigma]\} \subset \sigma'.$$

Consequently, $\sigma \cap |L| = \sigma' \leqslant \sigma$. □

The subcomplex $\mathrm{St}(C, K) = \bigcup_{C \leqslant D \in K} F(D)$ of K is called **star** at $C \in K$, where $|\mathrm{St}(C, K)| = \mathrm{st}(\mathrm{rint}\, C, K)$ for each $C \in K$. In particular, for a vertex $v \in K^{(0)}$, $\mathrm{St}(v, K) = \bigcup_{C \in K[v]} F(C)$ and $|\mathrm{St}(v, K)| = \mathrm{st}(v, K)$. For each point $x \in |K|$, let $c_K(x) \in K$ be the smallest cell containing x (so $x \in \mathrm{rint}\, c_K(x)$), which is called the **carrier** of x in K. Then,

$$|\mathrm{St}(c_K(x), K)| = \mathrm{st}(x, K) = \bigcup K[x] = \bigcup_{C \in K[x]} C,$$

which is a (closed) neighborhood of x in $|K|$. On the other hand, the following is an open neighborhood of x in $|K|$:

$$O_K(x) = \bigcup_{C \in K[x]} \mathrm{rint}\, C,$$

which is called the **open star** at x (with respect to K). We have the following special covers of $|K|$:

$$\mathcal{S}_K = \left\{ |\mathrm{St}(v, K)| \;\middle|\; v \in K^{(0)} \right\} \text{ and } \mathcal{O}_K = \left\{ O_K(v) \;\middle|\; v \in K^{(0)} \right\}.$$

When K is a simplicial complex, the **link** at $\sigma \in K$ is defined as follows:

$$\begin{aligned} \mathrm{Lk}(\sigma, K) &= \left\{ \tau \in \mathrm{St}(\sigma, K) \;\middle|\; \tau \cap \sigma = \emptyset \right\} \\ &= \left\{ \tau \in K \;\middle|\; \tau \text{ is joinable to } \sigma \text{ in } K \right\}. \end{aligned}$$

Then, $\mathrm{Lk}(\sigma, K)$ is joinable to $F(\sigma)$ in K and $\mathrm{St}(\sigma, K) = \mathrm{Lk}(\sigma, K) * F(\sigma)$.

Let K be a simplicial complex and $A \subset |K|$. The **simplicial neighborhood** of A in K is defined as the following subcomplex of K:

$$N(A, K) = \left\{ \tau \in K \;\middle|\; \exists \sigma \in K[A] \text{ such that } \tau \leqslant \sigma \right\} = \bigcup_{\sigma \in K[A]} F(\sigma).$$

Then, $|N(A, K)|$ is a neighborhood of $|A|$ in $|K|$. Indeed,

$$A \subset |K| \setminus |C(A, K)| \subset |N(A, K)|,$$

where $C(A, K)$ is a subcomplex of K defined as follows:

$$C(A, K) = K \setminus K[A] = \{\tau \in K \mid \tau \cap |L| = \emptyset\},$$

which is called the **simplicial complement** of A in K. For a subcomplex L of K, we write

$$N(|L|, K) = N(L, K) \text{ and } C(|L|, K) = C(L, K),$$

which are called the **simplicial neighborhood** and the **simplicial complement** of L in K, respectively. Observe that

$$N(L, K) = \bigcup_{v \in L^{(0)}} \mathrm{St}(v, K).$$

A cell complex K is said to be **locally finite** or **locally finite-dimensional** depending on whether the star $\mathrm{St}(v, K)$ at every vertex $v \in K^{(0)}$ is finite or finite-dimensional. The following are elementary:

Proposition 1.5.5 *Let K be a cell complex.*

(1) *$|K|$ is compact metrizable if and only if K is finite;* [4.2.6]
(2) *$|K|$ is separable if and only if K is countable;* [4.2.16(1)]
(3) *$|K|$ is (locally compact) metrizable if and only if K is locally finite.* [4.2.16(2)]

Proposition 1.5.6 *For each pair of cell complexes K_1 and K_2 with $|K_1| = |K_2|$ as sets, $|K_1| = |K_2|$ as spaces if and only if each cell of K_i is covered by finitely many cell of K_{3-i} for $i = 1, 2$.* [4.2.7]

The following theorem is widely used and proved in many textbooks:

Theorem 1.5.7 (Homotopy Extension) *Let L be a subcomplex of a cell complex K and $h : |L| \times \mathbf{I} \to X$ be a homotopy into an arbitrary space X. If h_0 extends to a map $f : |K| \to X$, then h extends to a homotopy $\bar{h} : |K| \times \mathbf{I} \to X$ with $\bar{h}_0 = f$. Moreover, if h is a $\mathcal{U}$-homotopy for an open cover $\mathcal{U}$ of X, then $\bar{h}$ can be taken as a $\mathcal{U}$-homotopy.* [4.3.3]

Each cell complex K itself is a closed cover of the polyhedron $|K|$. It is said that two maps $f, g : X \to |K|$ are **contiguous** if f and g are K-close, that is, for each $x \in X$, there is a cell $C \in K$ such that $f(x), g(x) \in C$. Then, every two contiguous maps $f, g : X \to |K|$ are K-homotopic, i.e., $f \simeq_K g$ because a K-homotopy $h : X \times \mathbf{I} \to |K|$ can be defined as follows:

$$h(x, t) = (1 - t) f(x) + t g(x) \text{ for each } x \in X \text{ and } t \in \mathbf{I}.$$

Such a homotopy is called the **straight line homotopy**.

1.6 Simplicial Subdivisions

A cell complex K' is called a **subdivision** of K (or it is said that K' **subdivides** K) if the following conditions are satisfied:

(S1) Each cell of K is covered by finitely many cells of K';
(S2) Each cell of K' is contained in some cell of K (i.e., K' refines K).

Then, $|K'| = |K|$ as spaces (Proposition 1.5.6). Evidently, if K' is a subdivision of K and K'' is a subdivision of K', then K'' is a subdivision of K. A simplicial complex K' subdividing K is called a **simplicial subdivision**, which can be written as follows:

$$K' \lhd K \quad \text{or} \quad K \rhd K'.$$

The following is Theorem 4.2.10 in [GAGT] but we will give an alternative proof:

Theorem 1.6.1 *Let K be a cell complex with an order on $K^{(0)}$ such that there exists the maximum $v_C \in C^{(0)}$ for each cell $C \in K$ (e.g., a total order on $K^{(0)}$ satisfies this condition). Then, the following is a simplicial subdivision with $K'^{(0)} = K^{(0)}$:*

$$K' = \big\{\langle v_{C_1}, \ldots, v_{C_n}\rangle \;\big|\; C_1 > \cdots > C_n \in K\big\}.$$

Proof First of all, it should be remarked that the representation of simplexes of K' in the above is not unique. For $i < j$, it is possible that $v_{C_i} = v_{C_j}$. In this case, $v_{C_i} = v_C$ for $C_i > C \geqslant C_j$. It should also be remarked that $v_{C_i}, \ldots, v_{C_n} \in C_i^{(0)}$ for each $i = 1, \ldots, n$. Then, in the above definition of K', we can add the following conditions to $C_1 > \cdots > C_n \in K$:

- $v_{C_1} > \cdots > v_{C_n}$ and $v_{C_i} \notin C^{(0)}$ for $C_i > C > C_{i+1}$, $i = 1, \ldots, n$, where the case $i = n$ means $C_n = \{v_{C_n}\}$.

The above additional condition is equivalent to the following:

- Each C_i is the smallest cell of K containing $v_{C_i}, \ldots, v_{C_n}$.

Indeed, $C_n = \{v_{C_n}\}$ is the smallest cell of K containing v_{C_n}. If C_{i+1} is the smallest cell of K containing $v_{C_{i+1}}, \ldots, v_{C_n}$, then C_i is the smallest cell of K containing $v_{C_i}, \ldots, v_{C_n}$ because $v_{C_i} \notin C^{(0)}$ for $C_i > C > C_{i+1}$. By induction, each C_i is the smallest cell of K containing $v_{C_1}, \ldots, v_{C_n}$. The converse implication is trivial.

To show that K' is a simplicial subdivision of K, we have the conditions (C1), (C2″), (S1), and (S2), where (C1) and (S2) are obvious. For (S1), it is enough to show that $|K|$ is covered by K' because each cell of K has only finite vertices. Thus, to see (C2″) and (S1), it suffices to show the following:

(*) For each point $x \in |K|$, there exists a unique sequence $C_1 > \cdots > C_n \in K$ such that $v_{C_1} > \cdots > v_{C_n}$, each C_i is the smallest cell of K containing $v_{C_1}, \ldots, v_{C_n}$, and $x \in \operatorname{rint}\langle v_{C_1}, \ldots, v_{C_n}\rangle$.

(*Uniqueness*) Let $C_1 > \cdots > C_n \in K$ be a sequence as in (*). Then, $\langle v_{C_1}, \ldots, v_{C_n} \rangle \subset (C_1)_x$ by the definition of $(C_1)_x$. From the minimality of C_1, it follows that $C_1 = (C_1)_x$, so $x \in \operatorname{rint} C_1$ [GAGT, Proposition 3.2.5(8)]. Therefore, $C_1 = c_K(x)$ (the carrier of x_1). When $n > 1$, $x_2 \in \partial C_1$ is uniquely determined so that $x \in \operatorname{rint}\langle v_{C_1}, x_2 \rangle$. We also have $x_2' \in \operatorname{rint}\langle v_{C_2}, \ldots, v_{C_n} \rangle$ such that $x \in \operatorname{rint}\langle v_{C_1}, x_2' \rangle$. Since $\langle v_{C_2}, \ldots, v_{C_n} \rangle \subset C_2 \subset \partial C_1$, it follows that $x_2 = x_2' \in \operatorname{rint}\langle v_{C_2}, \ldots, v_{C_n} \rangle$. Then, by the same argument as x, we have $C_2 = c_K(x_2)$. By iterating this operation, we have a unique $x_i \in \partial C_{i-1}$ such that $x_{i-1} \in \operatorname{rint}\langle v_{C_i}, x_i \rangle$, which implies that $x_i \in \operatorname{rint}\langle v_{C_i}, \ldots, v_{C_n} \rangle$ and $C_i = c_K(x_i)$. Thus, a finite sequence $C_1 > \cdots > C_n \in K$ in (*) is uniquely determined.

(*Existence*) For each $x \in |K|$, write $x = x_1$ and let $C_1 = c_K(x_1) \in K$. When $\dim C_1 = 0$, it follows that $C_1 = \{x_1\} = \{v_{C_1}\}$ and $x = x_1 \in \operatorname{rint}\langle v_{C_1} \rangle$. Otherwise, we have $x_2 \in \partial C_1$ such that $x_1 \in \operatorname{rint}\langle v_{C_1}, x_2 \rangle$. Let $C_2 = c_K(x_2) < C_1$. Then, $v_{C_2} < v_{C_1}$ because $v_{C_1} \notin C_2^{(0)}$. If $v_{C_1} \in C^{(0)}$ for some $C_1 > C > C_2$, then $x_1 \in \langle v_{C_1}, x_2 \rangle \subset C$, which contradicts the fact that $C < C_1 = c_K(x_1)$. Therefore, $v_{C_1} \notin C^{(0)}$ for $C_1 > C > C_2$. When $\dim C_2 = 0$, it follows that $C_2 = \{x_2\} = \{v_{C_2}\}$ and $x \in \operatorname{rint}\langle v_{C_1}, v_{C_2} \rangle$. Otherwise, we have $x_3 \in \partial C_2$ such that $x_2 \in \operatorname{rint}\langle v_{C_2}, x_3 \rangle$, which implies

$$x \in \operatorname{rint}\langle v_{C_1}, x_2 \rangle \subset \operatorname{rint}\langle v_{C_1}, v_{C_2}, x_3 \rangle.$$

Let $C_3 = c_K(x_3) < C_2$. Then, $v_{C_2} > v_{C_3}$ because $v_{C_2} \notin C_3^{(0)}$. If $v_{C_2} \in C^{(0)}$ for some $C_2 > C > C_3$, then $x_2 \in \langle v_{C_2}, x_3 \rangle \subset C$, which contradicts the fact that $C < C_2 = c_K(x_2)$. Therefore, $v_{C_2} \notin C^{(0)}$ for $C_2 > C > C_3$. By iterating this operation until $\dim C_n = 0$, we can obtain a finite sequence of cells $C_1 > C_2 > \cdots > C_n = \{v_{C_n}\}$ with $x_i \in C_i$ ($x_1 = x$, $x_n = v_{C_n}$) such that $v_{C_1} > \cdots > v_{C_n}$, $v_{C_i} \notin C^{(0)}$ for $C_i > C > C_{i+1}$, and $x \in \operatorname{rint}\langle v_{C_1}, \ldots, v_{C_n} \rangle$. As we saw above, each C_i is the smallest cell of K containing $v_{C_1}, \ldots, v_{C_n}$ if and only if $v_{C_i} \notin C^{(0)}$ for $C_i > C > C_{i+1}$, $i = 1, \ldots, n$. □

Note The above subdivision K' is the same as the one obtained in [GAGT, Theorem 4.2.10]. Indeed, in its proof, the simplicial subdivision of K is defined as $\bigcup_{n\in\mathbb{N}} L_n$, where $L_1 = K^{(1)}$ and $L_n \lhd K^{(n)}$ is inductively defined as follows:

$$L_n = L_{n-1} \cup \big\{ v_C \sigma \bigm| C \in K^{(n)} \setminus K^{(n-1)},\ \sigma \in L_{n-1}, \\ \text{with } \sigma \subset \partial C \text{ and } v_C \notin c_K(\hat{\sigma})^{(0)} \big\}.$$

Then, L_n satisfies the conditions that $L_n^{(0)} = K^{(0)}$, $L_{n-1} \subset L_n$, and $v_{c_K(\hat{\sigma})} \in \sigma^{(0)}$ for each $\sigma \in L_n$, where $\hat{\sigma}$ is the barycenter of σ (cf. p. 36). To see that this subdivision is equal to the subdivision K' of Theorem 1.6.1, suppose $L_{n-1} \subset K'$. Let $C \in K^{(n)} \setminus K^{(n-1)}$ and $\sigma \in L_{n-1}$ with $\sigma \subset \partial C$ and $v_C \notin c_K(\hat{\sigma})^{(0)}$. By the assumption, σ is written as $\sigma = \langle v_{C_1}, \ldots, v_{C_n} \rangle$, where $C_1 > \cdots > C_n \in K$, $v_{C_1} > \cdots > v_{C_n}$, and each C_i is the smallest cell of K containing $v_{C_1}, \ldots, v_{C_n}$. In the above proof (Uniqueness), let $x = \hat{\sigma}$. Then, $C_1 = c_K(\hat{\sigma})$ and $\operatorname{rint}\sigma \subset \operatorname{rint} C_1$, hence $C > C_1$ and $v_C > v_{C_1}$. Thus, $v_C\sigma = \langle v_C, v_{C_1}, \ldots, v_{C_n} \rangle \in K'$. Therefore, $L_n \subset K'$. Since $\{\operatorname{rint}\sigma \mid \sigma \in \bigcup_{n\in\mathbb{N}} L_n\} = |K|$, it follows that $\bigcup_{n\in\mathbb{N}} L_n = K'$.

A cell complex K with such an order as in Theorem 1.6.1 is called an **ordered cell complex**. In the case where K is simplicial complex, such an order gives a total order on the vertices of each simplex in K. Then, an **ordered simplicial complex** is a simplicial complex K with an order on $K^{(0)}$ such that $\sigma^{(0)}$ is totally ordered for each $\sigma \in K$.

For cell complexes K and L, we define the **product cell complex** $K \times_c L$ as follows:

$$K \times_c L = \{C \times D \mid C \in K,\ D \in L\}.$$

Then, $|K \times_c L| = |K| \times |L|$ as sets but $|K \times_c L| \neq |K| \times |L|$ as spaces in general (cf. [GAGT, Proposition 4.3.2]).

Theorem 1.6.2 *For each pair of cell complexes K and L, $|K \times_c L| = |K| \times |L|$ as spaces if* (1) *both K and L are locally countable or* (2) *one of K and L is locally finite.* [4.3.1]

For simplicial complexes K and L, $K \times_c L$ is not simplicial but does have a simplicial subdivision. In fact, giving an order on $(K \times_c L)^{(0)} = K^{(0)} \times L^{(0)}$ so that $K \times_c L$ is an ordered cell complex, we can obtain a simplicial subdivision of $K \times_c L$ with the same vertices as $K \times_c L$ (Theorem 1.6.1). When K and L are ordered simplicial complexes, such an order on $K^{(0)} \times L^{(0)}$ can be defined as follows:

$$(u, v) \leqslant (u', v') \underset{\text{def}}{\Leftrightarrow} u \leqslant u',\ v \leqslant v'.$$

Let $K \times_s L$ denote the simplicial subdivision of $K \times_c L$ defined by using this order, which can be written as follows:

$$K \times_s L = \big\{\langle (u_1, v_1), \ldots, (u_k, v_k)\rangle \mid \exists \sigma \in K,\ \exists \tau \in L \\ \text{such that } u_1 \leqslant \cdots \leqslant u_k \in \sigma^{(0)},\ v_1 \leqslant \cdots \leqslant v_k \in \tau^{(0)}\big\}.$$

This simplicial complex $K \times_s L$ is called the **product simplicial complex** of K and L.

Overlapping cell complexes, we can obtain a cell complex as in the following proposition:

Proposition 1.6.3 *For each pair of cell complexes K_1 and K_2, the following K is a cell complex with $|K| = |K_1| \cap |K_2|$ as sets:*

$$K = \{C \cap D \mid C \in K_1,\ D \in K_2,\ C \cap D \neq \emptyset\},$$

where if K_0 is a subcomplex of both K_1 and K_2, then K_0 is a subcomplex of K. Moreover, if each cell of K_i is covered by only finitely many cells of K_{3-i} for $i = 1, 2$, then $|K|$ is a closed subspace of both $|K_1|$ and $|K_2|$. [4.2.12]

As a special case, the following can be obtained:

Theorem 1.6.4 *Let K_1 and K_2 be cell complexes such that $|K_1| = |K_2|$ as spaces: Then, K_1 and K_2 have a common subdivision K. In addition, if K_0 is a subcomplex of both K_1 and K_2, then K_0 is also a subcomplex of K.* [4.2.11]

The following is a combination of Theorems 1.6.1 and 1.6.4:

Theorem 1.6.5 *Each pair of cell complexes K_1 and K_2 with $|K_1| = |K_2|$ as spaces have a common simplicial subdivision K. In addition, if they have a common (simplicial) subcomplex K_0, then K can be taken so as to contain K_0 as a subcomplex.* [4.2.13]

The following is obtained from the definition of a subpolyhedron and Theorem 1.6.5 (which was overlooked in [GAGT]):

Proposition 1.6.6 *Let K be a cell complex. For each subpolyhedron P of $|K|$, there exists a simplicial subdivision $K' \lhd K$ such that P is triangulated by a subcomplex L of K', that is, $P = |L|$.* □

The following lemma is useful to construct simplicial subdivision of cell complexes:

Lemma 1.6.7 *Let C be an n-cell and $v_0 \in \operatorname{rint} C$. Given a triangulation L of ∂C (i.e., $L \lhd F(\partial C)$), let $K = v_0 * L$, that is,*

$$K = L \cup \{v_0\} \cup \{v_0\sigma \mid \sigma \in L\}.$$

Then, K is a triangulation of C (i.e., $K \lhd F(C)$) such that L is a full subcomplex of K. [4.6.1]

Using Lemma 1.6.7 in the skeletonwise construction, we can obtain the following:

Proposition 1.6.8 *Let K be a cell complex and L' be a simplicial subdivision of a subcomplex $L \subset K$. Given $v_C \in \operatorname{rint} C$ for each $C \in K \setminus L$, there is a simplicial subdivision K' of K such that L' is a full subcomplex of K' and*

$$K'^{(0)} = L'^{(0)} \cup K^{(0)} \cup \{v_C \mid C \in K \setminus L\}.$$ [4.6.2]

In the case $L = \emptyset$ in Proposition 1.6.8 above, the obtained subdivision is written as follows:

$$K' = \{\langle v_{C_1}, \ldots, v_{C_n}\rangle \mid C_1 < \cdots < C_n \in K\},$$

which is called a **derived subdivision** of K. For a simplicial complex K, as $v_\sigma \in \operatorname{rint}\sigma$, $\sigma \in K$, we can take the **barycenter** $\hat{\sigma} = \sum_{i=1}^{n} n^{-1} v_i$, where $\sigma = \langle v_1, \ldots, v_n\rangle$ ($n = \dim\sigma + 1$). Then, the obtained derived subdivision K' of K is called the **barycentric subdivision** of K and denoted by $\operatorname{Sd} K$.

When $L = L'$ in Proposition 1.6.8, the obtained subdivision K' is called a **derived subdivision** of K **relative to** L, where

$$K' = L \cup \{\langle v_{C_1}, \ldots, v_{C_n}\rangle \mid C_1 < \cdots < C_n \in K \setminus L\}$$
$$\cup \{\langle v_1, \ldots, v_m, v_{C_1}, \ldots, v_{C_n}\rangle \mid \langle v_1, \ldots, v_m\rangle \in L,$$
$$C_1 < \cdots < C_n \in K \setminus L \text{ with } \langle v_1, \ldots, v_m\rangle < C_1\}.$$

When K and L are simplicial and $v_\sigma = \hat{\sigma}$ for each $\sigma \in K \setminus L$, the obtained derived subdivision of K relative to L is called the **barycentric subdivision** of K relative to L and denoted by $\mathrm{Sd}_L\, K$. Note that L is a full subcomplex of $\mathrm{Sd}_L\, K$ (Proposition 1.6.8).

For a simplicial complex K and $x \in |K| \setminus K^{(0)}$, we can define the following simplicial subdivision of K:

$$K_x = (K \setminus K[x]) \cup \{x\sigma \mid \sigma \in \mathrm{St}(c_K(x)) \setminus K[x]\},$$

where $x\sigma$ is the join of x and σ, i.e., $x\sigma = \langle \{x\} \cup \sigma^{(0)}\rangle$. The operation $K \to K_x$ (or K_x itself) is called the **starring** of K at x. Then, as is easily observed,

$$|\,\mathrm{St}(x, K_x)| = |\,\mathrm{St}(c_K(x), K)| \text{ and}$$
$$\mathrm{Lk}(x, K_x) = F(\partial c_K(x)) * \mathrm{Lk}(c_K(x), K).$$

A subdivision obtained by finite starrings is known as a **stellar subdivision**. In general, $(K_x)_y \neq (K_y)_x$ for distinct two points $x, y \in |K| \setminus K^{(0)}$. When K is finite, every derived subdivision of K is a stellar subdivision.

The following theorem is well known and very important (e.g., the proof of the paracompactness of polyhedra is dependent on this theorem). Almost all textbooks mentioning or citing the result do not treat the proof. For the proof, we used to refer to the original article by J.H.C. Whitehead [154]. But now, the complete proof is found in [GAGT, Section 4.7].

Theorem 1.6.9 (J.H.C. WHITEHEAD) *Let K be an arbitrary simplicial complex. For any open cover $\mathcal{U}$ of $|K|$, K has a simplicial subdivision K' such that $\mathcal{S}_{K'} \prec \mathcal{U}$.* [4.7.11]

1.7 The Metric Topology of Polyhedra

Let K be a simplicial complex. There exist maps $\beta_v^K : |K| \to \mathbf{I}$, $v \in K^{(0)}$, such that every β_v^K is affine on each simplex of K,

$$x = \sum_{v \in K^{(0)}} \beta_v^K(x) v, \quad \sum_{v \in K^{(0)}} \beta_v^K(x) = 1, \quad \text{and}$$
$$c_K(x)^{(0)} = \{v \in K^{(0)} \mid \beta_v^K(x) > 0\},$$

where $\beta_v^K(x)$ is called the **barycentric coordinate** of x at v with respect to K. Note that $(\beta_v^K)^{-1}((0,1]) = O_K(v)$ for every $v \in K^{(0)}$. The injection $\beta^K : |K| \to \ell^1(K^{(0)})$ defined by $\beta^K(x) = (\beta_v^K(x))_{v\in K^{(0)}}$ is called the **canonical representation** of K. Observe that $\beta^K(v) = \mathbf{e}_v$ is the unit vector of $\ell^1(K^{(0)})$ for each $v \in K^{(0)}$. We define the metric ρ_K on $|K|$ as follows:

$$\rho_K(x,y) = \|\beta^K(x) - \beta^K(y)\| = \sum_{v\in K^{(0)}} \left|\beta_v^K(x) - \beta_v^K(y)\right|.$$

The topology on $|K|$ induced by the metric ρ_K is called the **metric topology**. The space $|K|$ with this topology is denoted by $|K|_{\mathrm{m}}$. The following fact is easily seen:

Fact *Each map $f : X \to |K|_{\mathrm{m}}$ of an arbitrary space X is continuous if and only if $\beta_v^K f$ is continuous for every $v \in K^{(0)}$.*

The identity $\mathrm{id}_{|K|} : |K| \to |K|_{\mathrm{m}}$ is continuous but not a homeomorphism unless K is locally finite.

Proposition 1.7.1 *For a simplicial complex K, the metric topology on $|K|$ coincides with the Whitehead topology (i.e., $|K|_{\mathrm{m}} = |K|$ as spaces) if and only if K is locally finite.* [4.5.6]

The complete metrizability of $|K|_{\mathrm{m}}$ can be characterized as follows:

Theorem 1.7.2 *For a simplicial complex K, the following are equivalent:*

(a) *$|K|_{\mathrm{m}}$ is completely metrizable;*
(b) *K contains no infinite full complexes as subcomplexes;*
(c) *ρ_K is complete.* [4.5.9]

There exists the following connection between $|K|$ and $|K|_{\mathrm{m}}$:

Theorem 1.7.3 *For every simplicial complex K, the identity $\varphi = \mathrm{id} : |K| \to |K|_{\mathrm{m}}$ is a homotopy equivalence with a homotopy inverse $\psi : |K|_{\mathrm{m}} \to |K|$ such that $\psi\varphi \simeq_K \mathrm{id}$ and $\varphi\psi \simeq_K \mathrm{id}$, where $\psi\varphi \simeq_{O_K} \mathrm{id}$ and $\varphi\psi \simeq_{O_K} \mathrm{id}$ are also valid. These homotopies are realized by the straight-line homotopy.* [4.9.6]

Let K be a simplicial complex and K' be a simplicial subdivision of K. In general, the metric $\rho_{K'}$ is not admissible for $|K|_{\mathrm{m}}$, that is, $|K'|_{\mathrm{m}} \neq |K|_{\mathrm{m}}$. We call K' an **admissible subdivision** of K if the metric $\rho_{K'}$ is admissible for $|K|_{\mathrm{m}}$; equivalently, $|K'|_{\mathrm{m}} = |K|_{\mathrm{m}}$. For example, the barycentric subdivision is admissible. Of course, for a locally finite simplicial complex, every subdivision is admissible. We have the following characterization of admissible subdivisions:

Theorem 1.7.4 *Let K be a simplicial complex and K' be a simplicial subdivision of K. Then, the following are equivalent:*

(a) *K' is an admissible subdivision of K;*
(b) *the open star $O_{K'}(v)$ of every vertex $v \in K'^{(0)}$ is open in $|K|_{\mathrm{m}}$;*
(c) *the set $K'^{(0)}$ of vertices of K' is discrete in $|K|_{\mathrm{m}}$.* [4.8.1] [4.8.4]

The following is the metric topology version of Theorem 1.6.9:

Theorem 1.7.5 (Henderson–Sakai) *Let K be an arbitrary simplicial complex. For any open cover $\mathcal{U}$ of $|K|_{\mathrm{m}}$, there is an admissible subdivision K' of K such that $\mathcal{S}_{K'} \prec \mathcal{U}$.* [4.8.8]

A map $f : X \to Y$ is called a **fine homotopy equivalence** provided that, for each open cover $\mathcal{U} \in \mathrm{cov}(Y)$, there is a map $g : Y \to X$ such that $fg \simeq_{\mathcal{U}} \mathrm{id}$ and $gf \simeq_{f^{-1}(\mathcal{U})} \mathrm{id}$. By virtue of Theorem 1.7.5, Theorem 1.7.3 can be strengthened as follows:

Theorem 1.7.6 *For every simplicial complex K, the identity $\varphi = \mathrm{id} : |K| \to |K|_{\mathrm{m}}$ is a fine homotopy equivalence.*[25]

Proof Let $\mathcal{U} \in \mathrm{cov}(|K|_{\mathrm{m}})$. By Theorem 1.7.5, K has an admissible simplicial subdivision K' such that $\mathcal{S}_{K'} \prec \mathcal{U}$. Then, applying Theorem 1.7.3, we can obtain a map $\psi : |K'|_{\mathrm{m}} \to |K'|$ such that $\psi\varphi \simeq_{K'} \mathrm{id}$ and $\varphi\psi \simeq_{K'} \mathrm{id}$. Since $|K'| = |K|$ and $|K'|_{\mathrm{m}} = |K|_{\mathrm{m}}$ as spaces, and $K \prec \mathcal{U}$, $\psi : |K|_{\mathrm{m}} \to |K|$ is a map such that $\psi\varphi \simeq_{\mathcal{U}} \mathrm{id}$ and $\varphi\psi \simeq_{\varphi^{-1}(\mathcal{U})} \mathrm{id}$. □

1.8 PL Maps and Simplicial Maps

Let K and L be cell complexes. A map $f : |K| \to |L|$ is said to be **piecewise linear** (abbrev., **PL**) if there is a subdivision K' of K such that $f|C$ is affine for each $C \in K'$. In this definition, we can take K' so that $f(K') \prec L$, that is, for each $C \in K'$, $f(C)$ is contained in some cell in L [GAGT, Lemma 4.4.3]. Using this fact, we can easily prove that the composition of PL maps is also PL [GAGT, Proposition 4.4.4]. We have the following characterization of PL maps:

Theorem 1.8.1 *Let P and Q be polyhedra. A map $f : P \to Q$ is PL if and only if the graph $G(f) = \{(x, f(x)) \mid x \in |K|\} \subset P \times Q$ is a polyhedron.* [4.4.2]

Remark 1.4 In the above, when $P = |K|$ and $Q = |L|$ for some cell complexes K and L, $P \times Q \neq |K \times_c L|$ as spaces in general (cf. Theorem 1.6.2). However, we can prove that the topology of the graph $G(f)$ is equal to the one inherited from $|K \times_c L|$ [GAGT, Lemma 4.4.1]. It should also be remarked that the image of a PL map is, in general, not a polyhedron. Such an example can be seen in [GAGT, Remark 4 on p. 158].

For a homeomorphism $f : |K| \to |L|$, if f is PL, then the inverse f^{-1} is also PL by Theorem 1.8.1 above. A homeomorphism $f : |K| \to |L|$ being PL is called a **piecewise linear (PL) homeomorphism**. Every bijective PL map between *compact* polyhedra is a PL homeomorphism but this is not true in general [GAGT, Remark 5

[25] Cf. Notes for Chapter 4 (p. 247) in [GAGT].

on p. 160]. Note that the composition of PL homeomorphisms is also PL. It is said that the polyhedra $|K|$ and $|L|$ are **PL homeomorphic** or $|K|$ is **PL homeomorphic** to $|L|$ if there exists a PL homeomorphism $f : |K| \to |L|$. An embedding $f : |K| \to |L|$ is called a **PL embedding** if $f(|K|)$ is a subpolyhedron of $|L|$ and $f : |K| \to f(|K|)$ is a PL homeomorphism. Note that every PL embedding is a closed embedding.

Let (K, K_0) and (L, L_0) be pairs of cell complexes and subcomplexes. For a map $f : (|K|, |K_0|) \to (|L|, |L_0|)$, if $f : |K| \to |L|$ is PL, then $f||K_0| : |K_0| \to |L_0|$ is also PL. Such a map $f : (|K|, |K_0|) \to (|L|, |L_0|)$ is said to be **piecewise linear** (**PL**). For a homeomorphism $f : (|K|, |K_0|) \to (|L|, |L_0|)$, if $f : |K| \to |L|$ is a PL homeomorphism, then $f||K_0| : |K_0| \to |L_0|$ is also a PL homeomorphism. Such a homeomorphism $f : (|K|, |K_0|) \to (|L|, |L_0|)$ is called a **piecewise linear (PL) homeomorphism**. It is said that the pairs of polyhedra $(|K|, |K_0|)$ and $(|L|, |L_0|)$ are **PL homeomorphic** or $(|K|, |K_0|)$ is **PL homeomorphic** to $(|L|, |L_0|)$ if there exists a PL homeomorphism $f : (|K|, |K_0|) \to (|L|, |L_0|)$. An embedding $f : (|K|, |K_0|) \to (|L|, |L_0|)$ is called a **PL embedding** if f and $f||K_0| : |K_0| \to |L_0|$ are PL embeddings.

Now, let K and L be simplicial complexes. A map $f : |K| \to |L|$ is said to be **simplicial** with respect to K and L (or simply a simplicial map from K to L) if $f|\sigma$ is affine and $f(\sigma) \in L$ for each $\sigma \in K$, where $\dim f(\sigma) \leqslant \dim \sigma$. Evidently, $f(K^{(0)}) \subset L^{(0)}$ and $f'(K) = \{f(\sigma) \mid \sigma \in K\}$ is a subcomplex of L. When $\sigma = \langle v_1, \ldots, v_n \rangle \in K$, we have $f(\sigma) = \langle f(v_1), \ldots, f(v_n) \rangle \in L$ and

$$f\left(\textstyle\sum_{i=1}^n t_i v_i\right) = \sum_{i=1}^n t_i f(v_i) \text{ for } t_1, \ldots, t_n \in \mathbf{I} \text{ with } \sum_{i=1}^n t_i = 1,$$

where it is possible that $f(v_i) = f(v_j)$ for $i \neq j$. Obviously, every simplicial map is PL. A simplicial map $f : |K| \to |L|$ from K to L is also written as $f : K \to L$.[26] Every simplicial map $f : K \to L$ is uniquely determined by the function $f_0 = f||K^{(0)} : K^{(0)} \to L^{(0)}$.

Proposition 1.8.2 *Let K and L be simplicial complexes. For a function $f_0 : K^{(0)} \to L^{(0)}$, the following are equivalent:*

(a) *f_0 extends to a simplicial map;*
(b) *$\langle f_0(\sigma^{(0)}) \rangle \in L$ for each $\sigma \in K$;*
(c) *$\bigcap_{v \in \sigma^{(0)}} O_L(f_0(v)) \neq \emptyset$ for each $\sigma \in K$.*

In this case, the simplicial extension f of f_0 is unique. 4.4.5

[26] Note that K and L are collections of simplexes. This notation $f : K \to L$ means a simplicial map $f : |K| \to |L|$ with respect to K and L, but it does not mean that f itself is a function between collections. Of course, a simplicial map $f : |K| \to |L|$ with respect to K and L induces a function from K to L.

It might be expected that every PL map $f : |K| \to |L|$ is simplicial with respect to some simplicial subdivisions of K and L but this is not true. In fact, there exists a PL map $f : |K| \to |L|$ that is not simplicial with respect to any simplicial subdivisions of K and L [GAGT, Remark 6 on p. 162]. However, we have the following:

Theorem 1.8.3 *Let K and L be cell complexes. A proper map $f : |K| \to |L|$ is PL if and only if f is simplicial with respect to some simplicial subdivisions of K and L.* 4.6.5

For simplicial complexes K and L, if a bijective map $f : |K| \to |L|$ is simplicial with respect to K and L, then the inverse f^{-1} is a simplicial map from L to K. A bijective simplicial map $f : K \to L$ is called a **simplicial isomorphism** or a **simplicial homeomorphism**. It is said that K and L are **simplicially isomorphic** or K is **simplicially isomorphic** to L if there is a simplicial isomorphism $f : K \to L$. An embedding $f : |K| \to |L|$ is called a **simplicial embedding** if $f(K) = \{f(\sigma) \mid \sigma \in K\}$ is a subcomplex of L and $f : K \to f(K)$ is a simplicial isomorphism. Obviously, every simplicial homeomorphism is a PL homeomorphism and every simplicial embedding is a PL embedding. The proof of Theorem 4.4.8 in [GAGT] shows the following:

Theorem 1.8.4 *Let P and Q be polyhedra. In order that $f : P \to Q$ is a PL homeomorphism, it is necessary and sufficient that there exist triangulations K and L of P and Q respectively such that $f : K \to L$ is a simplicial isomorphism.* □

Let (K, K_0) and (L, L_0) be pairs of simplicial complexes and subcomplexes. If $f : |K| \to |L|$ is simplicial with respect to K and L, then $f||K_0| : |K_0| \to |L_0|$ is also simplicial with respect to K_0 and L_0. Such a map $f : (|K|, |K_0|) \to (|L|, |L_0|)$ is said to be **simplicial** with respect to (K, K_0) and (L, L_0) (or from (K, K_0) to (L, L_0)) and is also written as $f : (K, K_0) \to (L, L_0)$. A homeomorphism (a bijective simplicial map) $f : (K, K_0) \to (L, L_0)$ is called a **simplicial homeomorphism** or a **simplicial isomorphism**. Then, it is said that (K, K_0) and (L, L_0) are **simplicially isomorphic** or (K, K_0) is **simplicially isomorphic** to (L, L_0).

Note Let K and L be cell complexes. In a similar way to a simplicial map, we can define a **cellular** map from K to L as a map $f : |K| \to |L|$ such that $f|C$ is affine and $f(C) \in L$ for each $C \in K$.[27] However, such a map is not practical because of the requirement for a cellular map to be affine on each cell. For example, there are no affine homeomorphisms from the unit square $\mathbf{I}^2$ to the rectangle $\langle \mathbf{0}, \mathbf{e}_1, \mathbf{e}_2, v \rangle \subset \mathbb{R}^2$, where $v = (1/2, 1) \in \mathbb{R}^2$. Here, it should be noted that there is a cellular homeomorphism from the boundary $\partial \mathbf{I}^2$ onto $\partial \langle \mathbf{0}, \mathbf{e}_1, \mathbf{e}_2, v \rangle$. As can be seen from this example, even if there exist a bijection $\eta : K \to L$ (between collections of cells) such that for $\sigma, \tau \in K$, $\sigma < \tau$ if and only if $\eta(\sigma) < \eta(\tau)$, there are no cellular homeomorphisms from $|K|$ to $|L|$.

For CW-complexes X and Y, a **cellular** map $f : X \to Y$ is defined as a map such that $f(X^n) \subset Y^n$ for each $n \in \mathbb{N}$. Any cell complex K defines a CW-complex $X = |K|$ with

[27] In fact, this definition is employed in the textbook of C.P. Rourke and B.J. Sanderson [(17)], p. 16.

$X^n = |K^{(n)}|$, $n \in \omega$. Evidently, a cellular map between cell complexes is cellular as a map between CW-complexes but the converse is not true.

For two cell complexes K and L, it is said that K is **congruent** to L, or K and L are **congruent**, written as $K \equiv L$, if there exists a bijection $\eta : K \to L$ (between collections of cells) such that

$$C, D \in K, \ D < C \ \Leftrightarrow \ \eta(D) < \eta(C).^{28}$$

Then, $\eta|K^{(0)} : K^{(0)} \to L^{(0)}$ is a bijection, and for each $C \in K$, $\eta(C)^{(0)} = \eta(C^{(0)})$ and $\dim \eta(C) = \dim C$. For pairs (K, K_0) and (L, L_0) of cell complexes and subcomplexes, when the above bijection $\eta : K \to L$ exists and satisfies $\eta(K_0) = L_0$, it is said that (K, K_0) is **congruent** to (L, L_0), or (K, K_0) and (L, L_0) are **congruent**, written as $(K, K_0) \equiv (L, L_0)$. Obviously, $\equiv$ is an equivalence relation among cell complexes (or pairs of cell complexes).

In [GAGT], the notation $\equiv$ was used only for simplicial complexes, which means "is simplicially isomorphic to." This usage is consistent by the following proposition:

Proposition 1.8.5 *Two simplicial complexes K and L are simplicially isomorphic if and only if they are congruent, that is, $K \equiv L$. For pairs (K, K_0) and (L, L_0) of simplicial complexes and subcomplexes, (K, K_0) and (L, L_0) are simplicially isomorphic if and only if they are congruent, that is, $(K, K_0) \equiv (L, L_0)$.*

Proof Evidently, a simplicial isomorphism $f : K \to L$ induces the bijection η in the definition of congruence. Conversely, assume $K \equiv L$, that is, there exists a bijection $\eta : K \to L$ such that $\sigma, \tau \in K, \sigma < \tau$ if and only if $\eta(\sigma) < \eta(\tau)$. For each $\sigma \in K$, the bijection $\eta|\sigma^{(0)} : \sigma^{(0)} \to \eta(\sigma)^{(0)}$ induces an affine homeomorphism $f_\sigma : \sigma \to \eta(\sigma)$. Then, a simplicial isomorphism $f : K \to L$ is defined by $f|\sigma = f_\sigma$ for each $\sigma \in K$. Thus, we have the first statement of the proposition. It is easy to translate the first statement to the second, namely the pair version. □

Even when K and L are not simplicial complexes, $K \equiv L$ implies that $|K|$ is PL homeomorphic to $|L|$. To see this fact, let $\eta : K \to L$ be a bijection defining $K \equiv L$. Then, we can give orders on $K^{(0)}$ and $L^{(0)}$ so that K and L are ordered cell complexes and $\eta|K^{(0)} : K^{(0)} \to L^{(0)}$ is an order isomorphism. Then, applying Theorem 1.6.1, we can obtain simplicially isomorphic simplicial subdivisions $K' \lhd K$ and $L' \lhd L$ such that $K'^{(0)} = K^{(0)}$ and $L'^{(0)} = L^{(0)}$. In fact,

$$K' = \big\{\langle v_{C_1}, \ldots, v_{C_n}\rangle \bigm| C_1 > \cdots > C_n \in K\big\},$$

where $v_C = \max C^{(0)}$ for each $C \in K$. Since $\eta|K^{(0)}$ is an order isomorphism,

$$\eta(v_C) = \max \eta(C^{(0)}) = \max \eta(C)^{(0)} = v_{\eta(C)} \ \text{ for each } C \in K.$$

[28] Here, η is a function between collections K and L but not a map from $|K|$ to $|L|$.

Therefore, we have

$$L' = \{\langle v_{\eta(C_1)}, \ldots, v_{\eta(C_n)} \rangle \mid \eta(C_1) > \cdots > \eta(C_n) \in L\}$$
$$= \{\langle \eta(v_{C_1}), \ldots, \eta(v_{C_n}) \rangle \mid C_1 > \cdots > C_n \in K\}.$$

Then, it follows that η induces a bijection $\eta' : K' \to L'$ defined as follows:

$$\eta'(\langle v_{C_1}, \ldots, v_{C_n} \rangle) = \langle \eta(v_{C_1}), \ldots, \eta(v_{C_n}) \rangle \text{ for } C_1 > \cdots > C_n \in K.$$

Evidently, for $\sigma, \tau \in K'$, $\sigma \leqslant \tau$ if and only if $\eta'(\sigma) \leqslant \eta'(\tau)$. Hence, we have $K' \equiv L'$, so K' is simplicially isomorphic to L' by Proposition 1.8.5 above. Namely, $|K|$ is PL homeomorphic to $|L|$.

The above argument is valid for pairs of cell complexes and subcomplexes. Thus, we have the following:

Proposition 1.8.6 *If cell complexes K and L are congruent, then $|K|$ and $|L|$ are PL homeomorphic. If pairs (K, K_0) and (L, L_0) of cell complexes and subcomplexes are congruent, then $(|K|, |K_0|)$ and $(|L|, |L_0|)$ are PL homeomorphic.* □

For cell complexes K and L, it is said that K is **combinatorially equivalent** to L or K and L are **combinatorially equivalent** and denoted by $K \cong L$ if they have subdivisions which are congruent. As we just saw above, subdivisions can be simplicial, that is,

$$K \cong L \Leftrightarrow \exists K' \lhd K,\ \exists L' \lhd L \text{ such that } K' \equiv L'.$$

For pairs (K, K_0) and (L, L_0) of cell complexes and subcomplexes, it is said that (K, K_0) is **combinatorially equivalent** to (L, L_0) or (K, K_0) and (L, L_0) are **combinatorially equivalent** and denoted by $(K, K_0) \cong (L, L_0)$ if they have (simplicial) subdivisions which are congruent. Namely, we have the following equivalence:

$$(K, K_0) \cong (L, L_0) \Leftrightarrow \exists (K', K_0') \lhd (K, K_0),\ \exists (L', L_0') \lhd (L, L_0)$$
$$\text{such that } (K', K_0') \equiv (L', L_0').$$

As is easily observed, Theorem 4.4.8 in [GAGT] is valid for cell complexes, that is, we have the following theorem, from which it follows that $\cong$ is an equivalence relation.

Theorem 1.8.7 *Two cell complexes K and L are combinatorially equivalent if and only if their polyhedra $|K|$ and $|L|$ are PL homeomorphic.* [4.4.8]

Remark 1.5 The above can be easily translated to pairs (K, K_0) and (L, L_0) of cell complexes and subcomplexes, that is,

$$(K, K_0) \cong (L, L_0) \Leftrightarrow (|K|, |K_0|) \text{ and } (|L|, |L_0|) \text{ are PL homeomorphic.}$$

For cell (or simplicial) complexes K and L, the following implications are trivial:

$$K \equiv L \Rightarrow K \cong L \Rightarrow |K| \approx |L|.$$

It goes without saying that the converse of the first implication does not hold. But, the converse of the second implication is non-trivial. It was well known as a long-standing problem and is called the **Hauptvermutung** (*the main conjecture*).[29] It had taken a long time to show that this conjecture is false. First, the Hauptvermutung for polyhedra was disproved in 1961 by Milnor [107], and then Kirby and Siebenmann [94] demonstrated in 1969 that this conjecture is false for triangulations of n-manifolds for $n \geqslant 5$. However, it should be noted that $|K| = |L|$ implies $K \cong L$ by Theorems 1.6.1 and 1.6.4.

Let K and L be simplicial complexes. A simplicial map $g : K \to L$ is called a **simplicial approximation** of a map $f : |K| \to |L|$ if each $g(x)$ is contained in the carrier $c_L(f(x))$ of $f(x)$ in L, where $f \simeq_L g$ by the straight line homotopy. When K and L are locally finite, if f is proper, then g is also proper. Indeed, since $|\operatorname{St}(v, L)|$ is compact for each $v \in L^{(0)}$ and $\mathcal{O}_L^{\mathrm{cl}} = \mathcal{S}_L$, we can apply Proposition 1.3.8. Thus, we can assert as follows:

- *Every simplicial approximation of a proper map is proper.*

As easy applications of Theorem 1.6.9, the following approximation theorems can be obtained:

Theorem 1.8.8 (Simplicial Approximation) *Let K and L be simplicial complexes. Then, each map $f : |K| \to |L|$ has a simplicial approximation $g : K' \to L$ for some simplicial subdivision $K' \lhd K$.* [4.7.14]

Theorem 1.8.9 (PL Approximation) *Let P and Q be polyhedra. For each map $f : P \to Q$ and any open cover $\mathcal{U}$ of Q, there exists a PL map $g : P \to Q$ that is $\mathcal{U}$-homotopic to f.* [4.7.15]

The following General Position Lemma is an important tool in both PL Topology and Dimension Theory, and we will also use it in Chap. 6:

Lemma 1.8.10 (General Position) *Let $\{U_i \mid i \in \mathbb{N}\}$ be a countable open collection in $\mathbb{R}^n$ and $A \subset \mathbb{R}^n$ with* $\operatorname{card} A \leqslant \aleph_0$ *such that all $n+1$ many points of A are affinely independent. Then, there exists $B = \{v_i \mid i \in \mathbb{N}\}$ such that $v_i \in U_i \setminus A$ for each $i \in \mathbb{N}$ and all $n+1$ many points of $A \cup B$ are affinely independent.* [5.8.4]

In the same way as the General Position Lemma 1.8.10 above, we can prove the following PL Embedding Approximation Theorem (that is not treated in [GAGT]):

[29] This German terminology was introduced in 1921 by Kneser, but Poincaré claimed this toward the end of nineteenth century and it was formulated as a conjecture in 1908 by Steiniz and Tieze. Refer to Introduction of [113].

Theorem 1.8.11 (PL Embedding Approximation for Products) *Let P be a compact polyhedron with $P_0 \subset P$ a subpolyhedron, and let Q be another polyhedron. Suppose that $f : P \to Q \times \mathbf{I}^n$ is a map such that $f|P_0$ is a PL embedding and $n \geqslant 2\dim P + 1$. For each $\mathcal{U} \in \operatorname{cov}(Q \times \mathbf{I})$, there exists a PL embedding $h : P \to Q \times \mathbf{I}^n$ such that $h|P_0 = f|P_0$ and $h \simeq_{\mathcal{U}} f$ rel. P_0.*

Proof Because $\mathrm{pr}_Q f(P)$ is compact, it is contained in some compact subpolyhedron of Q, which is PL homeomorphic to a subpolyhedron of $\mathbb{R}^m$ for some $m \in \mathbb{N}$. Then, Q itself can be regarded as a subpolyhedron of $\mathbb{R}^m$. As an admissible metric for $Q \times \mathbf{I}^n$, we employ the metric induced from the Euclidean metric for $\mathbb{R}^m \times \mathbf{I} \subset \mathbb{R}^{m+n}$. Let $\varepsilon > 0$ be a Lebesgue number for $\mathcal{U}$.

By Proposition 1.6.8 and Theorem 1.6.9, we have a simplicial complex K and its subcomplex K_0 such that $|K| = P$, $|K_0| = P_0$, $f|\sigma$ is affine on each $\sigma \in K_0$, and $\operatorname{diam} f(\sigma) < \varepsilon/4$ for every $\sigma \in K$. Let $K^{(0)} \setminus K_0^{(0)} = \{v_i \mid i = 1, \ldots, k\}$. By induction, using the Baire Category Theorem 1.3.15 at each step, we can obtain a map $g_0 : K^{(0)} \to \mathbf{I}^n$ such that $g_0|K_0^{(0)} = \mathrm{pr}_{\mathbf{I}^n} f|K_0^{(0)}$, $\|g_0(v_i) - \mathrm{pr}_{\mathbf{I}^n} f(v_i)\| < \varepsilon/8$, $i = 1, \ldots, k$, and

$$g_0(v_i) \notin \bigcup \big\{ \mathrm{fl}\{x_1, \ldots, x_n\} \bigm| x_i \in \mathrm{pr}_{\mathbf{I}^n} f(K_0^{(0)}) \cup \{g_0(v_1), \ldots, g_0(v_{i-1})\}\big\},$$

where the above set is nowhere dense in $\mathbf{I}^n$. Then, it follows that, for each $\sigma \in K \setminus K_0$ and $\tau \in K$, $g_0(\sigma^{(0)}) \cup g_0(\tau^{(0)})$ is affinely independent. Note that $g_0|K^{(0)} \setminus K_0^{(0)}$ is an injection and $g_0(K^{(0)} \setminus K_0^{(0)}) \cap g_0(K_0^{(0)}) = \emptyset$, but $g_0|K_0^{(0)}$ is not injective in general.

We define a map $h_0 : K^{(0)} \to Q \times \mathbf{I}^n$ by

$$h_0(v) = (\mathrm{pr}_Q f(v), g_0(v)) \quad \text{for each } v \in K^{(0)}.$$

Then, h_0 is an injection, $h_0|K_0^{(0)} = f|K_0^{(0)}$, and $\|h_0(v) - f(v)\| < \varepsilon/8$ for each $v \in K^{(0)} \setminus K_0^{(0)}$. For each $\sigma \in K \setminus K_0$ and $\tau \in K$, $h_0(\sigma^{(0)}) \cup h_0(\tau^{(0)})$ is affinely independent, where $h_0(\sigma^{(0)}) \cup h_0(\tau^{(0)})$ might be affinely dependent for some distinct $\sigma, \tau \in K_0$. This injection h_0 extends to a PL embedding $h : P = |K| \to Q \times \mathbf{I}^n$ such that $h|\sigma$ is affine on each $\sigma \in K$. For each $\sigma \in K$,

$$\begin{aligned} \operatorname{diam} h(\sigma) &= \operatorname{diam} h_0(\sigma^{(0)}) < \operatorname{diam} f(\sigma^{(0)}) + \varepsilon/4 \\ &\leqslant \operatorname{diam} f(\sigma) + \varepsilon/4 < \varepsilon/2. \end{aligned}$$

If $\sigma \in K_0$, then $h|\sigma = f|\sigma$ because they are affine on $\sigma = \langle \sigma^{(0)} \rangle$ and $h|\sigma^{(0)} = h_0|\sigma^{(0)} = f|\sigma^{(0)}$. Since $P_0 = |K_0|$, it follows that $g|P_0 = f|P_0$. For each $x \in P = |K|$, let $\sigma = c_K(x) \in K$ be the carrier of x and $v \in \sigma^{(0)}$. Then, it follows that

$$\begin{aligned} \|h(x) - f(x)\| &\leqslant \operatorname{diam} h(\sigma) + \|h(v) - f(v)\| + \operatorname{diam} f(\sigma) \\ &< \varepsilon/2 + \varepsilon/8 + \varepsilon/4 < \varepsilon. \end{aligned}$$

Consequently, $h \simeq_{\mathcal{U}} f$ rel. P_0 by the straight line homotopy. □

In Theorem 1.8.11 above, the product $Q \times \mathbf{I}^n$ can be replaced with a PL n-manifold. which is proved in the Appendix (Theorem A.4.1).

1.9 Derived and Regular Neighborhoods

The natural triangulation of the unit interval $\mathbf{I} = [0, 1]$ is the simplicial complex $I = \{0, 1, \mathbf{I}\}$. Let K be a simplicial complex. For a subcomplex L of K, we have the simplicial map $\varphi_L : K \to I$ defined by

$$\varphi_L(v) = \begin{cases} 0 & \text{if } v \in L^{(0)}, \\ 1 & \text{if } v \in K^{(0)} \setminus L^{(0)}, \end{cases}$$

that is, $\varphi_L : |K| \to \mathbf{I}$ is defined by

$$\varphi_L(x) = \sum_{v \in K^{(0)} \setminus L^{(0)}} \beta_v^K(x) \quad \text{for each } x \in |K|.$$

Then, $L^{(0)} = K^{(0)} \cap \varphi_L^{-1}(0) = (\varphi_L|K^{(0)})^{-1}(0)$ and $\varphi_L(|L|) = 0$, so $|L| \subset \varphi_L^{-1}(0)$. However, $|L| \neq \varphi_L^{-1}(0)$ if L is not full in K. Using this map, we can characterize a full subcomplex of K as follows:

Proposition 1.9.1 *Let K be a simplicial complex. A subcomplex L of K is full in K (i.e., L is a full subcomplex of K) if and only if $|L| = \varphi_L^{-1}(0)$.*

Proof To see the "only if" part, let $x \in \varphi_L^{-1}(0)$. Since φ_L is simplicial, we have $\varphi_L(x) = \sum_{v \in c_K(x)^{(0)}} \beta_v^K(x)\varphi_L(v)$, where $c_K(x) \in K$ is the carrier of x and $c_K(x)^{(0)} = \{v \in K^{(0)} \mid \beta_v^K(x) > 0\}$. Hence, $c_K(x)^{(0)} \subset \varphi_L^{-1}(0) = L^{(0)}$. Since L is full in K, it follows that $c_K(x) \in L$, so $x \in |L|$. Thus, $\varphi_L^{-1}(0) \subset |L|$, which implies that $|L| = \varphi_L^{-1}(0)$.

To see the "if" part, let $\sigma \in K$ with $\sigma^{(0)} \subset L^{(0)}$. Since $\varphi_L(\sigma^{(0)}) = 0$ and φ_L is simplicial, we have $\varphi_L(\sigma) = 0$. Hence, $\sigma \subset \varphi_L^{-1}(0) = |L|$, which means that $\sigma \in L$. Thus, L is full in K. □

Remark 1.6 Observe that the complement $C(L, K)$ of L in K is a full subcomplex of K. Since $K^{(0)} \setminus C(L, K)^{(0)} = L^{(0)}$, it follows that

$$\varphi_{C(L,K)}(x) = \sum_{v \in L^{(0)}} \beta_v^K(x) = 1 - \sum_{v \in K^{(0)} \setminus L^{(0)}} \beta_v^K(x) = 1 - \varphi_L(x).$$

Hence, $|C(L, K)| = \varphi_{C(L,K)}^{-1}(0) = \varphi_L^{-1}(1)$.

For a simplicial complex K and its full subcomplex L, let K' be a derived subdivision relative to $L \cup C(L, K)$. Then, $N(L, K')$ is called a **derived neighborhood** of L in K. The following uniqueness of derived neighborhoods easily follows from the definition:

Lemma 1.9.2 *Let K be a simplicial complex and L a full subcomplex of K. For any two derived subdivisions K' and K'' of K relative to $L \cup C(L, K)$, there exists a simplicial isomorphism $h : K' \to K''$ such that*

$$h|L \cup C(L, K) = \mathrm{id} \ \text{and} \ h(N(L, K')) = N(L, K'').$$

In particular, $N(L, K') \equiv N(L, K'')$, that is, these derived neighborhoods are simplicial isomorphic. □

For $0 < \varepsilon < 1$, the following is called the **ε-neighborhood** of L in K:

$$N_\varepsilon(L, K) = \left\{\sigma \cap \varphi_L^{-1}(\tau) \;\middle|\; \sigma \in K[|L|],\ \tau \leqslant [0, \varepsilon]\right\},$$

which is a cell complex with $L \subset N_\varepsilon(L, K)$ and $|N_\varepsilon(L, K)| = \varphi_L^{-1}([0, \varepsilon])$ (Fig. 1.1). We have also the following cell complex:

$$C_\varepsilon(L, K) = \left\{\sigma \cap \varphi_L^{-1}(\tau) \;\middle|\; \sigma \in K \setminus L,\ \tau \leqslant [\varepsilon, 1]\right\},$$

where $C(L, K) \subset C_\varepsilon(L, K)$ and $|C_\varepsilon(L, K)| = \varphi_L^{-1}([\varepsilon, 1])$. Then, $N_\varepsilon(L, K)$ and $C_\varepsilon(L, K)$ are subcomplexes of the following cell complex subdividing K:

$$K_L^\varepsilon = N_\varepsilon(L, K) \cup C_\varepsilon(L, K).$$

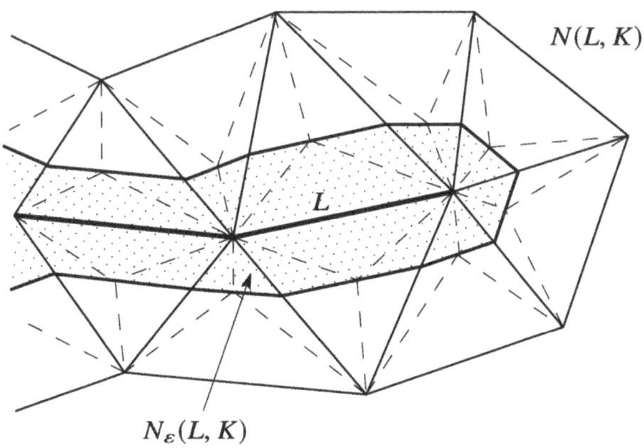

Fig. 1.1 $N(L, K') \lhd N_\varepsilon(L, K)$

As is easily observed, $|N_\varepsilon(L, K)| \setminus |L| = \varphi_L^{-1}((0, \varepsilon])$ and

$$\mathrm{bd}\,|N_\varepsilon(L, K)| = \varphi_L^{-1}(\varepsilon) = |N_\varepsilon(L, K)| \cap |C_\varepsilon(L, K)|.$$

It should be noted that $K \setminus (L \cup C(L, K)) = K[|L|] \setminus L$. Due to Proposition 1.6.8, choosing $v_\sigma \in \operatorname{rint}\sigma \cap \varphi_L^{-1}(\varepsilon)$ for each $\sigma \in K[|L|] \setminus L$, we can define a derived subdivision K' of K relative to $L \cup C(L, K)$. Then, as is easily observed, $K' \lhd K_L^\varepsilon$ and $N(L, K') \lhd N_\varepsilon(L, K)$, where $N(L, K')$ is a derived neighborhood of L in K.

Lemma 1.9.3 *Let K be a locally finite simplicial complex and let L be a finite full subcomplex of K. Suppose that K_1 is a simplicial subdivision of K and L_1 is the simplicial subdivision of L induced from K_1, where it follows that L_1 is full in K_1. Then, there exist derived subdivisions K' and K_1' of K and K_1 relative to $L \cup C(L, K)$ and $L_1 \cup C(L_1, K_1)$, respectively, such that $|N(L, K')| = |N(L_1, K_1')|$.*

Proof Since K_1 is locally finite and L_1 is finite, we can find $\varepsilon > 0$ such that

$$\varphi_L^{-1}([0, \varepsilon]) \cap |C(L_1, K_1)| = \emptyset, \quad \text{i.e., } \varphi_L^{-1}([0, \varepsilon]) \subset \operatorname{int}|N(L_1, K_1)|.$$

As mentioned above, by choosing all the new vertices on $\varphi_L^{-1}(\varepsilon)$, we can define derived subdivisions $K' \lhd K$ and $K_1' \lhd K_1$ relative to $L \cup C(L, K)$ and $L_1 \cup C(L_1, K_1)$, respectively. Both $N(L, K')$ and $N(L_1, K_1')$ are simplicial subdivisions of $N_\varepsilon(L, K)$. Hence, $|N(L, K')| = |N(L_1, K_1')| = |N_\varepsilon(L, K)|$. □

Let P be a polyhedron with a subpolyhedron $Q \subset P$. A subpolyhedron $N \subset P$ is called a **regular neighborhood** of Q in P if there exist a triangulation K of P, a full subcomplex $L \subset K$ with $|L| = Q$, and a derived subdivision K' of K relative to $L \cup C(L, K)$ such that $P = |K|$, $Q = |L|$, and $N = |N(L, K')|$, where Propositions 1.6.6 and 1.6.8 guarantee the existence of such a pair (K, L). Roughly speaking, a regular neighborhood is a polyhedral neighborhood triangulated by a derived neighborhood. As mentioned before Lemma 1.9.3, for any simplicial complex K and any full subcomplex L of K, and $0 < \varepsilon < 1$, we can take a derived subdivision K' of K relative to $L \cup C(L, K)$ such that $N(L, K') \lhd N_\varepsilon(L, K)$. Thus, the polyhedron of an ε-neighborhood is a regular neighborhood, that is:

Lemma 1.9.4 *For any simplicial complex K, any full subcomplex L of K, and $0 < \varepsilon < 1$, $|N_\varepsilon(L, K)|$ is a regular neighborhood of $|L|$ in $|K|$.* □

Concerning regular neighborhoods, we have the following uniqueness:

Theorem 1.9.5 (Uniqueness of Regular Neighborhoods) *Let P be a locally compact polyhedron with a compact subpolyhedron $Q \subset P$. If N_1 and N_2 are regular neighborhoods of a subpolyhedron Q in P, then there is a PL homeomorphism $h : P \to P$ such that $h(N_1) = N_2$, $h|Q = \mathrm{id}$, and $h|P \setminus U = \mathrm{id}$ for some open neighborhood U of Q in P.*

Proof Let (K_i, L_i), $i = 1, 2$, be pairs of simplicial complexes and their full subcomplexes and let K_i' be a derived subdivision of K_i relative to $L_i \cup C(L_i, K_i)$ such that $P = |K_i|$, $Q = |L_i|$, and $N_i = |N(L_i, K_i')|$. By Theorem 1.6.5, P has a triangulation K that is a common subdivision of K_1 and K_2, where K is locally finite because P is locally compact. Let L be a subcomplex of K with $|L| = Q$. Then, L is finite because Q is compact. As is easily observed, L is full in K.

By Lemma 1.9.3, we can obtain derived subdivisions K' and K_1'' of K and K_1 relative to $L \cup C(L, K)$ and $L_1 \cup C(L_1, K_1)$, respectively, such that $|N(L, K')| = |N(L_1, K_1'')|$. Similarly, we have derived subdivisions K'' and K_2'' of K and K_2 relative to $L \cup C(L, K)$ and $L_2 \cup C(L_2, K_2)$, respectively, such that $|N(L, K'')| = |N(L_2, K_2'')|$.

By Lemma 1.9.2, there exist simplicial isomorphisms $f_i : K_i' \to K_i''$, $i = 1, 2$, and $g : K' \to K''$ such that

$$f_i|L_i \cup C(L_i, K_i) = \mathrm{id}, \ \ f_i(N(L_i, K_i')) = N(L_i, K_i''),$$
$$g|L \cup C(L, K) = \mathrm{id}, \ \text{and} \ \ g(N(L, K')) = N(L, K'').$$

Then, $h = f_2^{-1} g f_1 : P \to P$ is the desired PL homeomorphism. Indeed,

$$\begin{aligned} h(N_1) &= f_2^{-1} g f_1(|N(L_1, K_1')|) = f_2^{-1} g(|N(L_1, K_1'')|) \\ &= f_2^{-1} g(|N(L, K')|) = f_2^{-1}(|N(L, K'')|) \\ &= f_2^{-1}(|N(L_2, K_2'')|) = |N(L_2, K_2')| = N_2. \end{aligned}$$

Because $Q = |L_1| = |L_2| = |L|$, we have

$$h|Q = f_2^{-1} g f_1|Q = f_2^{-1} g|Q = f_2^{-1}|Q = \mathrm{id}.$$

Moreover, the following is an open neighborhood of Q in P:

$$U = P \setminus (|C(L_1, K_1)| \cap |C(L_2, K_2)|).$$

Then, $P \setminus U \subset |C(L_i, K)| \subset |C(L, K)|$, $i = 1, 2$. Hence,

$$h|P \setminus U = f_2^{-1} g f_1|P \setminus U = f_2^{-1} g|P \setminus U = f_2^{-1}|P \setminus U = \mathrm{id}.$$

This completes the proof. □

A subpolyhedron A of P is said to be **PL bi-collared** (resp. **PL collared**) in P if there is a PL embedding $k : A \times [-1, 1] \to P$ (resp. $k : A \times \mathbf{I} \to P$) such that $k(x, 0) = x$ for every $x \in A$ and $k(A \times (-1, 1))$ (resp. $k(A \times [0, 1))$) is open in P (i.e., $k|A \times (-1, 1)$ (resp. $k|A \times [0, 1)$) is an open embedding in P), where k is called a **PL bi-collar** (resp. a **PL collar**) in P.[30] A subset $A \subset X$ is called a **strong deformation retract** of X if there is a homotopy $h : X \times \mathbf{I} \to X$ such that $h_0 = \mathrm{id}$, $h_t|A = \mathrm{id}$ for each $t \in \mathbf{I}$, and $h_1(X) = A$.[31]

Theorem 1.9.6 *Let P be a locally compact polyhedron and let N be a regular neighborhood of a compact subpolyhedron Q in P. The boundary* $\mathrm{bd}\, N$ *is PL bi-collared in P (and PL collared in N). Furthermore, we have a homeomorphism $h : \mathrm{bd}\, N \times \mathbb{R}_+ \to N \setminus Q$ such that $h(x, 0) = x$ for every $x \in \mathrm{bd}\, N$, so* $\mathrm{bd}\, N$ *is a strong deformation retract of $N \setminus Q$.*

Proof We have a triangulation K of P and a full subcomplex L with $|L| = Q$. Let $0 < \varepsilon < 1$. Take a derived subdivision K' of K relative to $L \cup C(L, K)$ so that $K'^{(0)} \setminus K^{(0)} \subset \varphi_L^{-1}(\varepsilon)$. As observed before, $|N(L, K')| = |N_\varepsilon(L, K)|$. By the Uniqueness of Regular Neighborhoods (Theorem 1.9.5), we may assume that $N = |N_\varepsilon(L, K)| = \varphi_L^{-1}([0, \varepsilon])$. Then, $\mathrm{bd}\, N = \varphi_L^{-1}(\varepsilon)$.

We have the following subcomplex of $N_\varepsilon(L, K)$:

$$\partial N_\varepsilon(L, K) = \left\{\sigma \cap \varphi_L^{-1}(\varepsilon) \;\middle|\; \sigma \in K[|L|] \setminus L\right\}.$$

Let $J = \{\pm 1, [-1, 1]\}$ be the natural triangulation of $[-1, 1]$ and let

$$M = \partial N_\varepsilon(L, K) \times_c J.$$

Then, $|\partial N_\varepsilon(L, K)| = \mathrm{bd}\, N$ and $|M| = \mathrm{bd}\, N \times [-1, 1]$. For each $C \in M$, we define $v_C \in \mathrm{rint}\, C$ as follows:

$$v_C = \begin{cases} (v_\sigma, 0) & \text{if } C = (\sigma \cap \varphi_L^{-1}(\varepsilon)) \times [-1, 1], \\ (v_\sigma, \pm 1) & \text{if } C = (\sigma \cap \varphi_L^{-1}(\varepsilon)) \times \{\pm 1\}, \end{cases}$$

where $\sigma \in K[|L|] \setminus L$. Let M' be the derived subdivision of M defined by using these points. Then, $M'_0 = \{\tau \in M' \mid \tau \subset \mathrm{bd}\, N \times \{0\}\}$ is a subcomplex of M' with $|M'_0| = \mathrm{bd}\, N \times \{0\}$.

Now, take $0 < a < \varepsilon < b < 1$ and define the cell complex as follows:

$$B = \left\{\sigma \cap \varphi_L^{-1}(\tau) \;\middle|\; \sigma \in K[|L|] \setminus L,\ \tau \leqslant [a, b]\right\}.$$

Then, $|B| = \varphi_L^{-1}([a, b])$, which is a subpolyhedron of P. Let B' be a derived subdivision of B, where we choose $v_D = v_\sigma \in \mathrm{rint}\, \sigma \cap \varphi_L^{-1}(\varepsilon)$ for $D =$

[30]The definitions of a **collar** and a **collared set** are given on p. 129. A PL collar is a closed collar, where a **closed collar** is defined on p. 135. It is known that every collared closed set in a paracompact space has a closed collar (Proposition 2.5.5). For a bi-collar and a bi-collared set, refer to Remark 2.7.

[31]A deformation retract will be defined on p. 68. These concepts will be discussed in Sect. 1.13.

$\sigma \cap \varphi_L^{-1}([a,b])$, $\sigma \in K[|L|] \setminus L$. Then, $B'_0 = \{\tau \in B' \mid \tau \subset \varphi_L^{-1}(\varepsilon)\}$ is a subcomplex of B' with $|B'_0| = \operatorname{bd} N = \varphi_L^{-1}(\varepsilon)$.

We have a bijection $\eta : M \to B$ defined by

$$\eta((\sigma \cap \varphi_L^{-1}(\varepsilon)) \times [-1,1]) = \sigma \cap \varphi_L^{-1}([a,b]),$$
$$\eta((\sigma \cap \varphi_L^{-1}(\varepsilon)) \times \{1\}) = \sigma \cap \varphi_L^{-1}(a), \text{ and}$$
$$\eta((\sigma \cap \varphi_L^{-1}(\varepsilon)) \times \{-1\}) = \sigma \cap \varphi_L^{-1}(b),$$

where $\sigma \in K[|L|] \setminus L$. Observe that for $C, D \in M$,

$$C \leqslant D \Leftrightarrow \eta(C) \leqslant \eta(D).$$

Hence, we can define a simplicial isomorphism $k : M' \to B'$ as follows:

$$k(v_C) = v_{\eta(C)} \text{ for each } C \in M.$$

When $C = (\sigma \cap \varphi_L^{-1}(\varepsilon)) \times [-1,1]$, $\sigma \in K[|L|] \setminus L$, since $v_C = (v_\sigma, 0)$ and $\eta(C) = \sigma \cap \varphi_L^{-1}([a,b])$, we have $v_{\eta(C)} = v_\sigma$. Because k is simplicial, it follows that $k(x,0) = x$ for every $x \in \operatorname{bd} N$. Consequently, $k : \operatorname{bd} N \times [-1,1] \to P$ is a PL bi-collar, so $\operatorname{bd} N$ is PL bi-collared in P. Moreover, since $k(\operatorname{bd} N \times [0,1]) = \varphi_L^{-1}([a,\varepsilon]) \subset \varphi_L^{-1}([0,\varepsilon]) = N$, the restriction $k|\operatorname{bd} N \times [0,1] : \operatorname{bd} N \times [0,1] \to N$ is a PL collar in N.

To see the additional statement, take $a = a_1 > a_2 > \cdots > 0$ so that $\inf_{n\in\mathbb{N}} a_n = 0$, and let $a_0 = \varepsilon$. Then,

$$N \setminus Q = \varphi_L^{-1}((0,\varepsilon]) = \bigcup_{n\in\mathbb{N}} \varphi_L^{-1}([a_n, a_{n-1}]).$$

As shown above, we have homeomorphisms

$$k_n : \varphi_L^{-1}(a_{n-1}) \times \mathbf{I} \to \varphi_L^{-1}([a_n, a_{n-1}]),\ n \in \mathbb{N},$$

such that $k_n(x,0) = x$ for every $x \in \varphi_L^{-1}(a_{n-1})$ and $k_n(\varphi_L^{-1}(a_{n-1}) \times \{1\}) = \varphi_L^{-1}(a_n)$. By induction, homeomorphisms

$$h_n : \big(\operatorname{bd} N \times [n-1,n], \operatorname{bd} N \times \{n\}\big) \to \big(\varphi_L^{-1}([a_n,a_{n-1}]), \varphi_L^{-1}(a_n)\big),\ n \in \mathbb{N},$$

can be defined as follows: for each $(x,t) \in \operatorname{bd} N \times [n-1,n]$,

$$h_n(x,t) = k_n(h_{n-1}(x,n-1), t-n+1).$$

where $h_0(x, 0) = x$, so $h_1(x, t) = k_1(x, t)$. Then, for every $x \in \operatorname{bd} N$ and $n \in \mathbb{N}$,

$$h_n(x, n-1) = k_n(h_{n-1}(x, n-1), 0) = h_{n-1}(x, n-1).$$

So, we can define a homeomorphism $h : \operatorname{bd} N \times \mathbb{R}_+ \to N \setminus Q$ by

$$h|\operatorname{bd} N \times [n-1, n] = h_n, \ n \in \mathbb{N}.$$

Then, $h(x, 0) = k_1(x, 0) = x$ for every $x \in \operatorname{bd} N$. □

Remark 1.7 In the above, observe that $|N(L, K)| \setminus |L \cup C(L, K)| = \varphi_L^{-1}((0, 1))$. Since $\varphi_{C(L,K)}(x) = 1 - \varphi_L(x)$ (cf. Remark 1.6), it follows that

$$|N_{1-\varepsilon}(C(L, K), K)| = \varphi_{C(L,K)}^{-1}([0, 1-\varepsilon]) = \varphi_L^{-1}([\varepsilon, 1]),$$

$$\operatorname{bd} |N_{1-\varepsilon}(C(L, K), K)| = \varphi_{C(L,K)}^{-1}(1-\varepsilon) = \varphi_L^{-1}(\varepsilon) = \operatorname{bd} N, \text{ and}$$

$$|N_{1-\varepsilon}(C(L, K), K)| \setminus |C(L, K)| = \varphi_{C(L,K)}^{-1}((0, 1-\varepsilon]) = \varphi_L^{-1}([\varepsilon, 1)).$$

By the above result, we have a homeomorphism $h' : \operatorname{bd} N \times \mathbb{R}_+ \to \varphi_L^{-1}([\varepsilon, 1))$ such that $h'(x, 0) = x$ for every $x \in \operatorname{bd} N$. Combining h above with this h', we can easily define a homeomorphism

$$\tilde{h} : \operatorname{bd} N \times \mathbb{R} \to \varphi_L^{-1}((0, 1)) = |N(L, K)| \setminus |L \cup C(L, K)|$$

such that $\tilde{h}(x, 0) = x$ for every $x \in \operatorname{bd} N = \varphi_L^{-1}(\varepsilon)$. Thus, we have the following:

$$|N(L, K)| \setminus |L \cup C(L, K)| \approx \operatorname{bd} N \times \mathbb{R}.$$

So, it follows that $\operatorname{bd} N$ is a strong deformation retract of $|N(L, K)| \setminus |L \cup C(L, K)|$.

For regular neighborhoods, we can also prove the following:

Theorem 1.9.7 *Let P be a locally compact polyhedron. Then, each compact subpolyhedron Q of P is a strong deformation retract of any regular neighborhood N of Q in P.*

Proof As in the above proof, we may assume that $N = |N_\varepsilon(L, K)| = \varphi_L^{-1}([0, \varepsilon])$, where $0 < \varepsilon < 1$, K is a triangulation of P, and L is its full subcomplex with $|L| = Q$. Recall that $\varphi_L : |K| \to \mathbf{I}$ is defined by $\varphi(x) = \sum_{v \in K^{(0)} \setminus L^{(0)}} \beta_v^K(x)$. Then, $1 - \varphi_L(x) = \sum_{v \in L^{(0)}} \beta_v^K(x)$. Since $Q = \varphi_L^{-1}(0)$ and $0 < \varphi_L(x) \leqslant \varepsilon$ for $x \in N \setminus Q$, we can define a homotopy $h : N \times \mathbf{I} \to N$ as follows: for each $t \in \mathbf{I}$, $h_t|Q = \mathrm{id}$ and

$$\begin{aligned} h(x,t) &= (1-t)x + \frac{t}{1-\varphi_L(x)} \sum_{v \in L^{(0)}} \beta_v^K(x)v \\ &= (1-t) \sum_{v \in K^{(0)} \setminus L^{(0)}} \beta_v^K(x)v + \left(1 + \frac{t\varphi_L(x)}{1-\varphi_L(x)}\right) \sum_{v \in L^{(0)}} \beta_v^K(x)v \\ &\qquad\qquad \text{if } x \in N \setminus Q. \end{aligned}$$

Then, $h_0 = \mathrm{id}$ and $h_1 : N \to Q$ is a retraction. Therefore, Q is a strong deformation retract of N. □

1.10 The Homotopy Type of Simplicial Complexes

Given a map $f : X \to Y$, the **mapping cylinder** M_f of f is defined as the following adjunction space:

$$M_f = Y \cup_{f \mathrm{pr}_X | X \times \{0\}} X \times \mathbf{I}.$$

The **collapsing map** $c_f : M_f \to Y$ is defined by $c_f|Y = \mathrm{id}_Y$ and $c_f|X \times (0,1] = f\mathrm{pr}_X$. Let $i_f : X \to M_f$ be the natural embedding defined by $i_f(x) = (x,1)$. Then, $c_f i_f = f$ and $c_f \simeq \mathrm{id}$ rel. Y in M_f by the homotopy $h^f : M_f \times \mathbf{I} \to M_f$ defined by $h_0^f = c_f$, $h_t^f|Y = \mathrm{id}$ for every $t \in \mathbf{I}$, and $h_t^f(x,s) = (x,st)$ for $(x,s) \times (0,1]$ and $t \in (0,1]$. When Y is a singleton, the mapping cylinder M_f is naturally homeomorphic to the **cone** $C_X = (X \times \mathbf{I})/(X \times \{0\})$ over X. As is easily observed, if $f : X \to Y$ is a proper map, then the collapsing map $c_f : M_f \to Y$ is proper, and the homotopy $h^f : M_f \times \mathbf{I} \to M_f$ is also proper, hence $c_f \simeq^p \mathrm{id}$ rel. Y in M_f. For a proper (perfect) map $f : X \to Y$ between metrizable spaces, the mapping cylinder M_f is metrizable because it is the perfect image of a metrizable space $Y \oplus X \times \mathbf{I}$ (Theorem 1.3.6(1)).

For spaces X and Y with $A \subset X \cap Y$, we write

$$X \simeq Y \quad \text{rel. } A$$

if there exist homotopy equivalences $f : X \to Y$ and $g : Y \to X$ such that $f|A = g|A = \mathrm{id}$, $gf \simeq \mathrm{id}_X$ rel. A, and $fg \simeq \mathrm{id}_Y$ rel. A. In the above, if f and g are proper, $gf \simeq^p \mathrm{id}_X$ rel. A, and $fg \simeq^p \mathrm{id}_Y$ rel. A, we write

$$X \simeq^p Y \quad \text{rel. } A.$$

Proposition 1.10.1 *Concerning the mapping cylinders, the following hold:*

(1) *For (proper) maps* $f, g : X \to Y$,

$$f \simeq g \;\Leftrightarrow\; M_f \simeq M_g \;\; \textit{rel. } Y \cup (X \times \{1\})$$
$$\big(f \simeq^p g \;\Leftrightarrow\; M_f \simeq^p M_g \;\; \textit{rel. } Y \cup (X \times \{1\})\big). \qquad \boxed{4.11.1}$$

(2) *For (proper) maps* $f : X \to Y$ *and* $g : Y \to Z$,

$$M_{gf} \simeq M_g \cup_{i_g} M_f \;\; \textit{rel. } Z \cup (X \times \{1\})$$
$$\big(M_{gf} \simeq^p M_g \cup_{i_g} M_f \;\; \textit{rel. } Z \cup (X \times \{1\})\big). \qquad \boxed{4.11.2}$$

In Theorems 4.11.1 and 4.11.2 of [GAGT], the proper cases are not mentioned but the same proofs are valid, where all homotopies are proper. Let A be a closed set in a space X with a closed collar $k : A \times \mathbf{I} \to X$ (i.e., k is a closed embedding such that $k(A \times [0, 1))$ is open in X and $k(x, 0) = x$ for every $x \in A$).[32] Then, it follows that

$$\big(A \times \mathbf{I}) \cup (X \times \{1\}), A \times \{0\}\big) \approx \big(X, A\big).$$

For a map $f : A \to Y$ from A to another space Y, the adjunction space $Y \cup_f X$ can be regarded as the space $M_f \cup (X \times \{1\})$. As a corollary of Proposition 1.10.1(1), we have the following:

Corollary 1.10.2 *Let A be a closed set in a space X with a closed collar $k : A \times \mathbf{I} \to X$ and let $f, g : A \to Y$ be maps to another space Y. If $f \simeq g$, then $Y \cup_f X \simeq Y \cup_g X$ rel. $Y \cup (X \setminus k(A \times [0, 1))$.* □

Let K and L be simplicial complexes, where $K^{(0)}$ is given an order so that K is an *ordered simplicial complex*. For a simplicial map $f : K \to L$, the **simplicial mapping cylinder** Z_f of f is defined as the following simplicial complex:

$$Z_f = K \cup L \cup \big\{\langle f(v_1), \ldots, f(v_{m'}), v_m, \ldots, v_n\rangle \;\big|\; \langle v_1, \ldots, v_n\rangle,\; v_1 < \cdots < v_n,\; 1 \leqslant m' \leqslant m \leqslant n\big\},$$ [33]

The **simplicial collapsing map** $\bar{c}_f : Z_f \to L$ is defined by $\bar{c}_f(v) = f(v)$ for $v \in K^{(0)}$ and $\bar{c}_f(v) = v$ for $L^{(0)}$. When $L = \{v_0\}$ (hence f is a constant map), the simplicial mapping cylinder Z_f is equal to the simplicial cone $v_0 * K$ over K. If K and L are locally finite and f is a proper simplicial map, then the simplicial mapping cylinder Z_f is locally finite.

---

[32] See Footnote 30 (p. 50).

[33] The simplicial mapping cylinder is dependent on such an order of $K^{(0)}$. See Fig. 4.15 (p. 217) of [GAGT].

For an ordered simplicial complex K and a subdivision $K' \lhd K$, we define the simplicial subdivision $I(K', K)$ of the product simplicial complex $K \times_s I$ as follows:

$$I(K', K) = (K' \times \{0\}) \cup (K \times \{1\}) \\ \cup \big\{\langle \sigma \times \{0\} \cup \langle v_m, \dots, v_n \rangle \times \{1\} \rangle \bigm| \sigma \in K' \\ \langle v_1, \dots, v_n \rangle \in K,\ v_1 < \dots < v_n,\ \sigma \subset \langle v_1, \dots, v_m \rangle \big\},$$

where $K' \times \{0\} = \{\sigma \times \{0\} \mid \sigma \in K'\}$ and $K \times \{1\} = \{\tau \times \{1\} \mid \tau \in K\}$. For a simplicial map $f : K' \to L$, give an order on $K'^{(0)}$ so that K' is an ordered simplicial complex and define the simplicial mapping cylinder Z_f. Identifying $K' \subset Z_f$ with $K' \times \{0\} \subset I(K', K)$, we can obtain the simplicial complex:

$$Z_f^K = Z_f \cup_{K' = K' \times \{0\}} I(K', K).$$

Then, $|Z_f^K| = |Z_f| \cup_{\mathrm{pr}_{|K|} | |K| \times \{0\}} |K| \times \mathbf{I}$. If K and L are locally finite and f is a proper simplicial map, then Z_f^K is also locally finite.

Theorem 1.10.3 *Let K and K' be ordered simplicial complexes such that $K' \lhd K$ and let L be a simplicial complex. For any simplicial approximation $f : K' \to L$ of a map $g : |K| \to |L|$,*

$$|Z_f^K| \simeq M_g \quad \text{rel. } |L| \cup (|K| \times \{1\}).$$

When K and L are locally finite and g is proper,

$$|Z_f^K| \simeq^p M_g \quad \text{rel. } |L| \cup (|K| \times \{1\}).^{34}$$ 4.11.3

It is said that X is **homotopy dominated** by Y (or Y **homotopy dominates** X) if there are maps $f : X \to Y$ and $g : Y \to X$ such that $gf \simeq \mathrm{id}$, where $g : Y \to X$ is called a **homotopy domination**. Moreover, if $gf \simeq_{\mathcal{U}} \mathrm{id}$ for a given open cover $\mathcal{U} \in \mathrm{cov}(X)$, it is said that X is **$\mathcal{U}$-homotopy dominated** by Y (or Y **$\mathcal{U}$-homotopy dominates** X) and $g : Y \to X$ is called a **$\mathcal{U}$-homotopy domination**. The **homotopy type of a simplicial complex** K means the homotopy type of the polyhedron $|K|$ (or $|K|_{\mathrm{m}}$). We say that X is **homotopy dominated by a simplicial complex** K if X is homotopy dominated by the polyhedron $|K|$ (or $|K|_{\mathrm{m}}$). The following theorem on the homotopy type of simplicial complexes is called the WHITEHEAD–MILNOR THEOREM:

---

[34] The proper case of Theorem 1.10.3 is not mentioned in Theorem 4.11.3 of [GAGT] but the same proof is valid.

Theorem 1.10.4 (J.H.C. Whitehead; Milnor) *A space X homotopy dominated by a simplicial complex has the homotopy type of a simplicial complex K with* $\operatorname{card} K^{(0)} = \operatorname{dens}|K| \leqslant \operatorname{dens} X$. *If X is separable, X has the homotopy type of a countable complex.* [4.12.1]

Moreover, we have the following:

Theorem 1.10.5 *Every simplicial complex has the homotopy type of a locally finite-dimensional simplicial complex with the same density. In addition, every countable simplicial complex has the homotopy type of a countable locally finite simplicial complex with the same density.* [4.12.3]

The following is a combination of the above two theorems:

Corollary 1.10.6 *A space X homotopy dominated by a simplicial complex has the homotopy type of a locally finite-dimensional simplicial complex K with* $\operatorname{card} K^{(0)} = \operatorname{dens}|K| \leqslant \operatorname{dens} X$. *If X is separable, X has the homotopy type of a countable locally finite simplicial complex.* [4.12.4]

It is said that X is **proper homotopy dominated** by Y (or Y **proper homotopy dominates** X) if there are proper maps $f : X \to Y$ and $g : Y \to X$ such that $gf \simeq^p \mathrm{id}$, where $g : Y \to X$ is called a **proper homotopy domination**. For a locally compact space, we can show the following proper homotopy version of Theorem 1.10.4:

Theorem 1.10.7 *Let X be a locally compact space. If X is proper homotopy dominated by a locally finite simplicial complex K, then $X \times \mathbb{R}_+$ has the proper homotopy type of a locally finite simplicial complex L.*

Proof To prove that $X \times \mathbb{R}$ has the proper homotopy type of a locally finite simplicial complex L, we can perform the same proof as Theorem 4.12.1 of [GAGT]. For the reader's convenience, we give its outline below:

Let $f : X \to |K|$ and $g : |K| \to X$ be proper maps such that $gf \simeq^p \mathrm{id}$. By the Simplicial Approximation Theorem 1.8.8, fg has a simplicial approximation $\varphi : K' \to K$ for some $K' \lhd K$ Since fg is proper, φ is also proper. Giving orders on $K'^{(0)}$ and $K^{(0)}$ so that K' and K are ordered simplicial complexes, we can obtain the simplicial complex Z_φ^K. Since K is locally finite and φ is proper, Z_φ^K is locally finite. For each $n \in \mathbb{Z}$, let L_n be a copy of Z_φ^K. Identifying $K \subset L_{n+1}$ with $K \times \{1\} \subset L_n$ for each $n \in \mathbb{Z}$, we have a locally finite simplicial complex $L = \bigcup_{n \in \mathbb{Z}} L_n$. For each $n \in \mathbb{Z}$, let M_{2n-1} and M_{2n} be copies of M_f and M_g, respectively. Identifying $X \times \{1\} \subset M_{2n-1}$ and $|K| \times \{1\} \subset M_{2n}]$ with $X \subset M_{2n}$ and $|K| \subset M_{2n+1}$ respectively, we define $M = \bigcup_{n \in \mathbb{Z}} M_n$. Then, by using Proposition 1.10.1 and Theorem 1.10.3 we can prove that $|L| \simeq^p |M| \simeq^p X \times \mathbb{R}$. Refer to Fig. 4.19 (p. 223) of [GAGT].

As is easily observed (see Fig. 1.2), we have

$$\mathbf{I} \times \mathbb{R}_+ \approx \mathbf{I} \times [0, 1) \approx (0, 1) \times [0, 1) \approx \mathbb{R} \times \mathbb{R}_+.$$

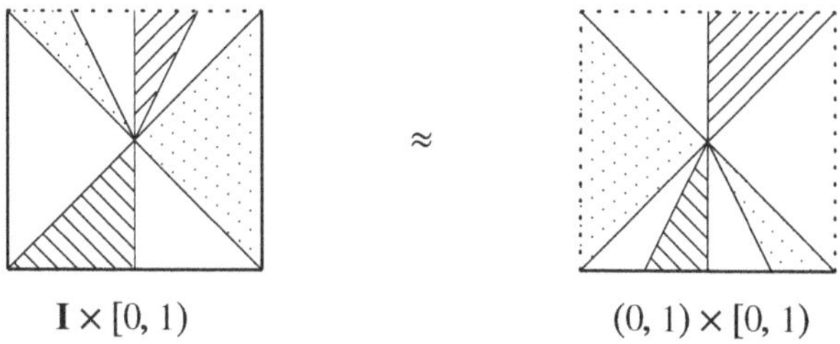

Fig. 1.2 $\mathbf{I} \times [0,1) \approx (0,1) \times [0,1)$

Since $X \times \mathbb{R}_+ \simeq^p X \times \mathbf{I} \times \mathbb{R}_+$, it follows that

$$X \times \mathbb{R}_+ \simeq^p X \times \mathbb{R} \times \mathbb{R}_+ \simeq^p |L| \times \mathbb{R}_+ = |L \times_s R_+|,$$

where R_+ is the natural triangulation of $\mathbb{R}_+$, i.e., $R_+ = \{[n, n+1], n \mid n \in \omega\}$. Thus, $X \times \mathbb{R}_+$ has the proper homotopy type of $L \times_s R_+$, which is a locally finite simplicial complex. □

Remark 1.8 In Theorem 1.10.7, if X is connected, then L is countable because L is also connected.

Remark 1.9 Relating to Theorem 1.10.4 and Corollary 1.10.6, the question arises whether a compact space homotopy dominated by a simplicial complex has the homotopy type of a finite simplicial complex. This is equivalent to asking whether a compact space having the homotopy type of a simplicial complex has the homotopy type of a finite simplicial complex. As will be seen in Sect. 1.13, every ANR has the homotopy type of a simplicial complex and every simplicial complex has the homotopy type of an ANR. Thus, the question is equivalent to asking whether a compact ANR has the homotopy type of a finite simplicial complex. The positive answer has been expected and it is called the BORSUK CONJECTURE (cf. [33]). This conjecture was proved by J. West [152, 153]. We will discuss this conjecture in Chap. 4.

1.11 The Nerves of Open Covers

Let V be an arbitrary set. Recall that $\mathrm{Fin}(V)$ is the collection of all non-empty finite subsets of V. An **abstract complex** $\mathcal{K}$ over V is a subcollection $\mathcal{K} \subset \mathrm{Fin}(V)$ satisfying the following condition:

(AC) if $A \in \mathcal{K}$ and $\emptyset \neq B \subset A$ then $B \in \mathcal{K}$.

A subcollection $\mathcal{L} \subset \mathcal{K}$ satisfying (AC) is called a **subcomplex** of $\mathcal{K}$. In particular, $\mathrm{Fin}(V)$ is an abstract complex and every abstract complex $\mathcal{K}$ over V is a subcomplex of Fin(V). For each $n \in \omega$, the ***n*-skeleton** $\mathcal{K}^{(n)}$ of $\mathcal{K}$ is defined by

$$\mathcal{K}^{(n)} = \{A \in \mathcal{K} \mid \operatorname{card} A \leqslant n + 1\},$$

where we regard $\mathcal{K}^{(0)} \subset V$, and so $\mathrm{Fin}(V)^{(0)} = V$. Each $\mathcal{K}^{(n)}$ is a subcomplex of $\mathcal{K}$. If $\mathcal{K} = \mathcal{K}^{(n)}$, we say that $\mathcal{K}$ is **at most *n*-dimensional** and write $\dim \mathcal{K} \leqslant n$. It is said that $\mathcal{K}$ is ***n*-dimensional** (written as $\dim \mathcal{K} = n$) if $\dim \mathcal{K} \leqslant n$ and $\dim \mathcal{K} \not\leqslant n - 1$. Note that every abstract complex $\mathcal{K}$ over V with $\dim \mathcal{K} \leqslant n$ is a subcomplex of $\mathrm{Fin}(V)^{(n)}$.

For any simplicial complex K, $\mathcal{K} = \{\sigma^{(0)} \mid \sigma \in K\} \subset \mathrm{Fin}(K^{(0)})$ is an abstract complex, which is called the **abstract complex defined by** K. Each $\mathcal{K}^{(n)}$ is defined by $K^{(n)}$. In particular, $\mathcal{K}^{(0)} = K^{(0)}$.

Conversely, each abstract complex $\mathcal{K}$ over V is defined by some simplicial complex. In fact, consider the linear space $\mathbb{R}^V_f$. By identifying each $v \in V$ with $\mathbf{e}_v \in \mathbb{R}^V_f$ defined by $\mathbf{e}_v(v) = 1$ and $\mathbf{e}_v(v') = 0$ if $v \neq v'$, we can regard V as a Hamel basis for $\mathbb{R}^V_f$. Then, $K = \{\langle A \rangle \mid A \in \mathcal{K}\}$ is a simplicial complex that defines $\mathcal{K}$. This K is called the **simplicial complex defined by** $\mathcal{K}$. Then each $K^{(n)}$ is defined by $\mathcal{K}^{(n)}$ and $K^{(0)} = \mathcal{K}^{(0)}$. Note that the full simplicial complex $F(V)$ is the simplicial complex defined by Fin(V).

Remark 1.10 When V is a subset of a linear space E, for an abstract complex $\mathcal{K}$ over V, $K = \{\langle A \rangle \mid A \in \mathcal{K}\}$ is a simplicial complex that defines $\mathcal{K}$ if and only if each $A \in \mathcal{K}$ is affinely independent and

$$\langle A \rangle \cap \langle A' \rangle = \langle A \cap A' \rangle \quad \text{for each } A, A' \in \mathcal{K}.$$

In particular, K is a simplicial complex if $\mathcal{K}$ satisfies the condition:

($\sharp$) $A \cup A'$ is affinely independent for each $A, A' \in \mathcal{K}$.

Due to the General Position Lemma 1.8.10, there exists a countable (discrete) set V in $\mathbb{R}^{2n+1}$ such that all $2n + 2$ many points of V are affinely independent. Then, it follows that every countable complex K with $\dim K \leqslant n$ is simplicially isomorphic to a simplicial complex in $\mathbb{R}^{2n+1}$.

Now, consider two abstract complexes $\mathcal{K}$ and $\mathcal{L}$ over V and W, respectively. Let K and L be the simplicial complexes defined by $\mathcal{K}$ and $\mathcal{L}$, respectively. Recall $\mathcal{K}^{(0)} = K^{(0)}$ and $\mathcal{L}^{(0)} = L^{(0)}$. Suppose that a function $\varphi : K^{(0)} \to L^{(0)}$ satisfies the following condition:

($*$) $A \in \mathcal{K}$ implies $\varphi(A) \in \mathcal{L}$.

Then, $\varphi : K^{(0)} \to L^{(0)}$ induces the simplicial map $f : K \to L$ with $f|K^{(0)} = \varphi$. Conversely, for any simplicial map $f : K \to L$, the restriction $\varphi = f|K^{(0)} : K^{(0)} \to L^{(0)}$ satisfies condition ($*$) and f itself is the simplicial map induced by this φ. Such a function $\varphi : K^{(0)} \to L^{(0)}$ is also called a **simplicial map** from $\mathcal{K}$ to $\mathcal{L}$, and is written as $\varphi : \mathcal{K} \to \mathcal{L}$. If a bijection $\varphi : K^{(0)} \to L^{(0)}$ satisfies the condition that $A \in \mathcal{K}$ if and only if $\varphi(A) \in \mathcal{L}$, then φ induces the simplicial isomorphism $f : K \to L$ with $f|K^{(0)} = \varphi$, and any simplicial isomorphism $f : K \to L$ is induced by such a bijection $\varphi : K^{(0)} \to L^{(0)}$. Such a bijection $\varphi : K^{(0)} \to L^{(0)}$ is also called a **simplicial isomorphism** from $\mathcal{K}$ to $\mathcal{L}$. It is said that $\mathcal{K}$ is **simplicially isomorphic** to $\mathcal{L}$ (denoted by $\mathcal{K} \equiv \mathcal{L}$) if there is a simplicial isomorphism from $\mathcal{K}$ to $\mathcal{L}$.

For any open cover $\mathcal{U}$ of a space X, we define the abstract complex $\mathfrak{N}(\mathcal{U})$ over $\mathcal{U} \setminus \{\emptyset\}$ as follows:

$$\mathfrak{N}(\mathcal{U}) = \big\{\{U_1, \cdots, U_n\} \in \mathrm{Fin}(\mathcal{U}) \bigm| U_1 \cap \cdots \cap U_n \neq \emptyset\big\}.$$

The simplicial complex $N(\mathcal{U})$ defined by $\mathfrak{N}(\mathcal{U})$ is called the **nerve** of $\mathcal{U}$. A map $f : X \to |N(\mathcal{U})|$ (or $f : X \to |N(\mathcal{U})|_{\mathrm{m}}$) is called a **canonical map** for $\mathcal{U}$ if

$$f^{-1}(O_{N(\mathcal{U})}(U)) \subset U \quad \text{for each } U \in N(\mathcal{U})^{(0)} = \mathcal{U}.$$

Observe that every compact set in $|N(\mathcal{U})|$ meets $O_{N(\mathcal{U})}(U)$ for only finitely many $U \in \mathcal{U}$. Hence, if every $U \in \mathcal{U}$ has the compact closure in X, then each canonical map $f : X \to |N(\mathcal{U})|$ is proper. We can characterize canonical maps as follows:

Proposition 1.11.1 *For an open cover $\mathcal{U}$ of X, a map $f : X \to |N(\mathcal{U})|$ (or $f : X \to |N(\mathcal{U})|_{\mathrm{m}}$) is a canonical map if and only if*

$$c_{N(\mathcal{U})}(f(x))^{(0)} \subset \mathcal{U}[x] \ \textit{for each } x \in X,$$

where $c_{N(\mathcal{U})}(f(x)) \in N(\mathcal{U})$ is the carrier of $f(x)$. If $\mathcal{U}[x]$ is finite, this condition is equivalent to $f(x) \in \langle \mathcal{U}[x] \rangle \in N(\mathcal{U})$. [4.9.1]

Proposition 1.11.1 yields the following:

Corollary 1.11.2 *Let $\mathcal{U}$ be an open cover of X. Then, any two canonical maps $f, g : X \to |N(\mathcal{U})|$ (or $f, g : X \to |N(\mathcal{U})|_{\mathrm{m}}$) are contiguous.* [4.9.2]

For each open refinement $\mathcal{V}$ of $\mathcal{U} \in \mathrm{cov}(X)$, we have a simplicial map $\varphi : N(\mathcal{V}) \to N(\mathcal{U})$ such that $V \subset \varphi(V)$ for each $V \in \mathcal{V} = N(\mathcal{V})^{(0)}$. Such a simplicial map is called a **refining simplicial map**.

Corollary 1.11.3 *Let $\mathcal{U}$ and $\mathcal{V}$ be open covers of X with $\mathcal{V} \prec \mathcal{U}$ and $\varphi : N(\mathcal{V}) \to N(\mathcal{U})$ be a refining simplicial map. If $f : X \to |N(\mathcal{V})|$ (or $f : X \to |N(\mathcal{V})|_{\mathrm{m}}$) is a canonical map for $\mathcal{V}$, then $\varphi f : X \to |N(\mathcal{U})|$ (or $\varphi f : X \to |N(\mathcal{U})|_{\mathrm{m}}$) is also a canonical map for $\mathcal{U}$.* [4.9.3]

Concerning the existence of canonical maps, we have the following:

Theorem 1.11.4 *For every open cover $\mathcal{U}$ of a paracompact space X, there exists a canonical map $f : X \to |N(\mathcal{U})|$ such that $f : X \to |N(\mathcal{U})|_{\mathrm{m}}$ is also a canonical map.* [4.9.5]

The following is trivial:

Proposition 1.11.5 *Let K be a simplicial complex and X an arbitrary space. If two maps $f, g : X \to |K|_{\mathrm{m}}$ are contiguous then $f \simeq_K g$ by the straight line homotopy.*

In Proposition 1.11.5, $|K|_{\mathrm{m}}$ cannot be replaced by $|K|$ in general. Such an example can be seen in Remark 2 for Proposition 4.3.4 in [GAGT]. When X is locally compact (more generally it is a k-space), $|K|_{\mathrm{m}}$ can be replaced by $|K|$. If we abandon the straight line homotopy, then we can obtain the following by combining Proposition 1.11.5 with Theorem 1.11.4:

Proposition 1.11.6 *Let K be a simplicial complex and X an arbitrary space. If two maps $f, g : X \to |K|$ are contiguous, then $f \simeq_K g$.* 4.9.7

Combining Corollary 1.11.2 with Proposition 1.11.6 (or 1.11.5), we have the following corollary:

Corollary 1.11.7 *Let $\mathcal{U}$ be an open cover of a space X. Then, $f \simeq_{N(\mathcal{U})} g$ for any two canonical maps $f, g : X \to |N(\mathcal{U})|$ (or $f, g : X \to |N(\mathcal{U})|_{\mathrm{m}}$).* 4.9.8

An open cover $\mathcal{U}$ of a space X is said to be **star-finite** if $\mathcal{U}[U]$ is finite for each $U \in \mathcal{U}$, which is equivalent to the condition that the nerve $N(\mathcal{U})$ is locally finite. In fact, $\mathrm{St}(U, N(\mathcal{U}))^{(0)} = \mathcal{U}[U]$ for each $U \in \mathcal{U} = N(\mathcal{U})^{(0)}$. Thus, the star-finiteness of an open cover characterizes the local finiteness of its nerve. On the other hand, the nerve $N(\mathcal{U})$ is locally finite-dimensional (l.f.d.) if and only if $\sup_{x \in U} \operatorname{card} \mathcal{U}[x] < \infty$ for each $U \in \mathcal{U}$. In this case, we have

$$\sup_{x \in U} \operatorname{card} \mathcal{U}[x] = \dim \mathrm{St}(U, N(\mathcal{U})) + 1.$$

Note that every star-finite open cover is locally finite and its nerve is locally finite-dimensional, and that if an open cover is locally finite or its nerve is locally finite-dimensional then it is point-finite, that is, we have the following implications:

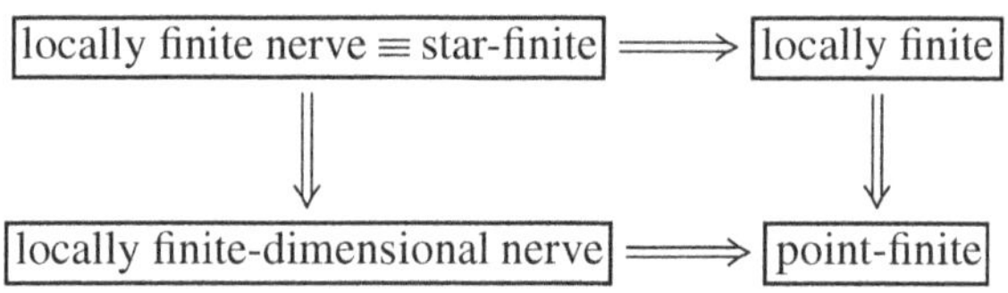

In the above, the converse implications do not hold and there are no connections between the local finiteness of an open cover and the local finite-dimensionality of its nerve. For details, refer to [GAGT, p. 201].

Theorem 1.11.8 *Every open cover of a paracompact space has a locally finite σ-discrete open refinement with the locally finite-dimensional nerve.* 4.9.9

When X is a locally compact paracompact space, each $\mathcal{U} \in \operatorname{cov}(X)$ has a locally finite open refinement $\mathcal{V}$ such that $\operatorname{cl} V$ is compact for each $V \in \mathcal{V}$. Then, as is easily observed, $\mathcal{V}$ is star-finite, so the nerve $N(\mathcal{V})$ is locally finite. Thus, we have the following theorem:

Theorem 1.11.9 *Every open cover of a locally compact paracompact space has a star-finite open refinement whose nerve is locally finite.* [4.9.10]

In addition, we can also show that:

Theorem 1.11.10 *Every open cover of a regular Lindelöf space has a countable star-finite open refinement whose nerve is countable and locally finite.* [4.9.11]

1.12 Dimensions

For an open cover $\mathcal{U}$ of a space X, $\operatorname{ord}\mathcal{U} = \sup\{\operatorname{card}\mathcal{U}[x] \mid x \in X\}$ is called the **order** of $\mathcal{U}$. Note that $\operatorname{ord}\mathcal{U} = \dim N(\mathcal{U}) + 1$, where $N(\mathcal{U})$ is the nerve of $\mathcal{U}$. The **(covering) dimension** of X (denoted by $\dim X$) is defined as follows: $\dim X \leqslant n$ if each *finite* open cover $\mathcal{U}$ of X has a *finite* open refinement $\mathcal{V}$ with $\operatorname{ord}\mathcal{V} \leqslant n+1$. It can be proved that the finiteness of $\mathcal{V}$ is not necessary in the definition (cf. [GAGT, Lemma 5.2.1]). When X is *paracompact*, not only $\mathcal{V}$ but also $\mathcal{U}$ is not required to be finite, that is,

$$\dim X \leqslant n \ \Leftrightarrow\ \forall \mathcal{U} \in \operatorname{cov}(X),\ \exists \mathcal{V} \in \operatorname{cov}(X) \quad \text{such that } \mathcal{V} \prec \mathcal{U},\ \operatorname{ord}\mathcal{V} \leqslant n+1,$$

where $\mathcal{V}$ can be locally finite σ-discrete in X (cf. [GAGT, Theorem 5.2.4]). For a *metric* space X, the following characterization is valid (cf. [GAGT, Theorem 5.3.1]):

$$\dim X \leqslant n \ \Leftrightarrow\ \exists \mathcal{U}_1 \succ \mathcal{U}_2 \succ \cdots \in \operatorname{cov}(X) \ \text{ such that } \lim_{i\to\infty} \operatorname{mesh}\mathcal{U}_i = 0,\ \operatorname{ord}\mathcal{U}_i \leqslant n+1,$$

where each $\mathcal{U}_i$ can be locally finite σ-discrete in X like the paracompact case.

We define $\dim X = n$ if $\dim X \leqslant n$ and $\dim X \not\leqslant n$. It is said that X is **finite-dimensional (f.d.)** ($\dim X < \infty$) if $\dim X \leqslant n$ for some $n \in \omega$. Otherwise, X is said to be **infinite-dimensional (i.d.)** ($\dim X = \infty$).

The dimension of a normal space can be characterized by extendability of maps from closed sets to the sphere, that is:

Theorem 1.12.1 *For a normal space X and $n \in \omega$,* $\dim X \leqslant n$ *if and only if each map $f : A \to \mathbf{S}^n$ of any closed set A in X extends over X.* [5.2.3]

As a matter of course, the following hold:

Fact (1) *For each* $n \in \mathbb{N}$, $\dim \mathbf{B}^n = \dim \mathbb{R}^n = n$. [5.2.7] [5.2.11]
(2) *For any simplicial complex* K, $\dim K = \dim |K| = \dim |K|_{\mathrm{m}}$. [5.2.10]

In particular, $\dim \mathbf{B}^n \not\leq n - 1$ is obtained by combining Theorem 1.12.1 with the following NO RETRACTION THEOREM:

Theorem 1.12.2 (NO RETRACTION) *For any* $n \in \mathbb{N}$, *there does not exist any map* $r : \mathbf{B}^n \to \mathbf{S}^{n-1}$ *such that* $r|\mathbf{S}^{n-1} = \mathrm{id}$.[35] [5.1.5]

The No Retraction Theorem 1.12.2 is also applied to prove the following characterization of boundary points of a closed subset $\mathbb{R}^n$:

Theorem 1.12.3 *Let* X *be a closed set in* $\mathbb{R}^n$ *and* $x \in X$. *In order that* $x \in \mathrm{bd}\, X$, *it is necessary and sufficient that each neighborhood* U *of* x *in* X *contains a neighborhood* V *of* x *in* X *such that every map* $f : X \setminus V \to \mathbf{S}^{n-1}$ *extends to a map* $\tilde{f} : X \to \mathbf{S}^{n-1}$. [5.1.7]

We can use this theorem to prove the following:

Theorem 1.12.4 *For any open set* $U \subset \mathbb{R}^n$, *every continuous injection* $f : U \to \mathbb{R}^n$ *is an open embedding.*

Proof[36] Let $x \in U$ and take any compact neighborhood C in $\mathbb{R}^n$ with $C \subset U$. Then, $f|C : C \to h(C)$ is a homeomorphism. We show that $f(x) \in \mathrm{int}\, f(C)$. On the contrary, assume that $f(x) \in \mathrm{bd}\, f(C)$. For each neighborhood V of x in C, since $f(V)$ is a neighborhood of $f(x)$ in $f(C)$, we can apply Theorem 1.12.3 to find a neighborhood W of $f(x)$ in $f(C)$ such that $W \subset h(V)$ and every map $g : f(C) \setminus W \to \mathbf{S}^{n-1}$ extends to a map $\tilde{g} : f(C) \to \mathbf{S}^{n-1}$. Then, $f^{-1}(W)$ is a neighborhood of x in C with $f^{-1}(W) \subset V$. For every map $h : C \setminus f^{-1}(W) \to \mathbf{S}^{n-1}$, $hf^{-1} : f(C) \setminus W \to \mathbf{S}^{n-1}$ extends to a map $\tilde{g} : f(C) \to \mathbf{S}^{n-1}$. Thus, we have a map $\tilde{g}f : C \to \mathbf{S}^{n-1}$ such that

$$
\begin{aligned}
\tilde{g}f|C \setminus f^{-1}(W) &= (\tilde{g}|f(C) \setminus W)(f|C \setminus f^{-1}(W)) \\
&= hf^{-1}(f|C \setminus f^{-1}(W)) = h.
\end{aligned}
$$

Due to Theorem 1.12.3, this means that $x \in \mathrm{bd}\, C$, which is a contradiction. Therefore, $f(C)$ is a neighborhood of x in $\mathbb{R}^n$ and $f(C) \subset f(U)$. Consequently, $f(U)$ is open in $\mathbb{R}^n$ and $f : U \to f(U)$ is a homeomorphism, hence f is an open embedding. □

The INVARIANCE OF DOMAIN can be obtained as a corollary of the above theorem:

Corollary 1.12.5 (INVARIANCE OF DOMAIN) *For each pair of subsets* $X, Y \subset \mathbb{R}^n$, $X \approx Y$ *implies* $\mathrm{int}\, X \approx \mathrm{int}\, Y$. [5.1.8]

---

[35] Such a map r is called a **retraction**, which is discussed in Sect. 1.13.

[36] The proof is the same as the Invariance of Domain in [GAGT, p. E2].

Indeed, let $h : X \to Y$ be a homeomorphism. By Theorem 1.12.4, $h|\operatorname{int} X : \operatorname{int} X \to \mathbb{R}^n$ and $h^{-1}|\operatorname{int} Y : \operatorname{int} Y \to \mathbb{R}^n$ are open embeddings. Then, it follows that $h(\operatorname{int} X) \subset \operatorname{int} Y$ and $h^{-1}(\operatorname{int} Y) \subset \operatorname{int} X$. Therefore, $h(\operatorname{int} X) = \operatorname{int} Y$, which means that $\operatorname{int} X \approx \operatorname{int} Y$.

Let A and B be disjoint closed sets in a space X. A closed set C in X is called a **partition** between A and B in X if there exist disjoint open sets U and V in X such that $A \subset U$, $B \subset V$ and $X \setminus C = U \cup V$. A family $(A_\gamma, B_\gamma)_{\gamma\in\Gamma}$ of pairs of disjoint closed sets in X is **inessential** in X if there are partitions L_γ between A_γ and B_γ with $\bigcap_{\gamma\in\Gamma} L_\gamma = \emptyset$. Note that if one of A_γ or B_γ is empty then (A_γ, B_γ) is inessential. If $(A_\gamma, B_\gamma)_{\gamma\in\Gamma}$ is not inessential in X (i.e., $\bigcap_{\gamma\in\Gamma} L_\gamma \neq \emptyset$ for any partitions L_γ between A_γ and B_γ), it is said to be **essential** in X. For example, the n-cube $\mathbf{I}^n$ has an essential family $(\mathrm{pr}_i^{-1}(0), \mathrm{pr}_i^{-1}(1))_{i=1}^n$, where $\mathrm{pr}_i : \mathbf{I}^n \to \mathbf{I}$ is the projection onto the i-th factor (refer to [GAGT, Corollary 5.2.16]). Using essential families, we can also characterize dimension as follows:

Theorem 1.12.6 (Eilenberg–Otto Characterization) *For a normal space X and $n \in \mathbb{N}$,* $\dim X \geqslant n$ *if and only if X has an essential family of n pairs of disjoint closed sets.* [5.2.17]

By definition, it is obvious that $\dim Y \leqslant \dim X$ if Y is a *closed* subspace of X. However, there is a zero-dimensional compact space X containing a subspace Y with $\dim Y > 0$ (cf. [GAGT, Theorem 5.5.3]). For metrizable spaces, the following Subset Theorem holds:

Theorem 1.12.7 (Subset) *For every subset Y of a metrizable space X,* $\dim Y \leqslant \dim X$. [5.3.3]

Additional fundamental theorems on Dimension Theory are listed below:

Theorem 1.12.8 (Countable Sum) *Let $X = \bigcup_{i\in\mathbb{N}} F_i$ be normal and $n \in \omega$, where each F_i is closed in X. If* $\dim F_i \leqslant n$ *for every $i \in \mathbb{N}$, then* $\dim X \leqslant n$. [5.4.1]

Theorem 1.12.9 (Locally Finite Sum) *Let X be normal and $n \in \omega$. If X has a locally finite closed cover $\{F_\lambda \mid \lambda \in \Lambda\}$ such that* $\dim F_\lambda \leqslant n$ *for every $\lambda \in \Lambda$, then* $\dim X \leqslant n$. [5.4.2]

Theorem 1.12.10 (Decomposition) *Let X be metrizable and $n \in \omega$. Then,* $\dim X \leqslant n$ *if and only if X is covered by $n+1$ many subsets $X_1, \ldots, X_{n+1}$ with* $\dim X_i \leqslant 0$. [5.4.5]

In the Decomposition Theorem 1.12.10 above, each X_i is not closed in X in general. In fact, a countable union of zero-dimensional closed subspaces is zero-dimensional by Theorem 1.12.8.

Theorem 1.12.11 (Addition) *For any two subspaces X and Y of a metrizable space,*

$$\dim X \cup Y \leqslant \dim X + \dim Y + 1. \qquad [5.4.8]$$

Theorem 1.12.12 (Product) *For any two metrizable spaces X and Y,*

$$\dim X \times Y \leqslant \dim X + \dim Y. \qquad \boxed{5.4.9}$$

In the above, the equality $\dim X \times Y = \dim X + \dim Y$ does not hold. However, if one of X or Y is a locally compact polyhedron (or a metric polyhedron), then the equality does hold. This fact can be proved by using the cohomological dimension (cf. Theorem 7.9.7 of [GAGT]).

According to the Eilenberg–Otto Characterization Theorem 1.12.6, the infinite-dimensionality can be characterized as follows:

$$\dim X = \infty \Leftrightarrow \forall n \in \mathbb{N}, X \text{ has an essential family of } n \text{ pairs of disjoint closed sets.}$$

A space X is said to be **strongly infinite-dimensional (s.i.d.)** if X has an infinite essential family of pairs of disjoint closed sets. For example, the Hilbert cube $\boldsymbol{Q} = [-1, 1]^{\mathbb{N}}$ is strongly infinite-dimensional because it has an essential family $(\mathrm{pr}_i^{-1}(-1), \mathrm{pr}_i^{-1}(1))_{i \in \mathbb{N}}$, where $\mathrm{pr}_i : \boldsymbol{Q} \to [-1, 1]$ is the projection onto the i-th factor (refer to [GAGT, Theorem 5.6.1]). It easily follows from the definition that a space is strongly infinite-dimensional if it contains an strongly infinite-dimensional closed subspace. Thus, $\mathbb{R}^{\mathbb{N}}$ and Hilbert space ℓ^2 is also strongly infinite-dimensional because $\mathbb{R}^{\mathbb{N}} \supset \boldsymbol{Q}$ and $\ell^2 \supset \prod_{n \in \mathbb{N}}[-2^{-n}, 2^{-n}] \approx \boldsymbol{Q}^{\mathbb{N}}$. Obviously, if X is s.i.d. then $\dim X = \infty$. It is said that X is **weakly infinite-dimensional (w.i.d.)** if $\dim X = \infty$ and X is not s.i.d., that is, for every family $(A_i, B_i)_{i \in \mathbb{N}}$ of pairs of disjoint closed sets in X, there are partitions L_i between A_i and B_i such that $\bigcap_{i \in \mathbb{N}} L_i = \emptyset$.[37]

It is said that X is **(strongly) countable-dimensional ((s.)c.d.)** if X is a countable union of f.d. normal (closed) subspaces. Obviously, every strongly countable-dimensional space is countable-dimensional. By virtue of Theorem 1.12.10, we have the following characterization of the countable-dimensionality:

- A metrizable space is countable-dimensional if and only if it is a countable union of zero-dimensional subspaces.

It should be remarked that a countable-dimensional space is never strong infinite-dimensional, that is:

Theorem 1.12.13 *A countable-dimensional metrizable space X with* $\dim X = \infty$ *is weakly infinite-dimensional. Equivalently, any strongly infinite-dimensional metrizable space is not countable-dimensional.* $\boxed{5.6.2}$

The following Embedding Theorem will be used in Chaps. 5 and 6:

[37] In many articles, the infinite-dimensionality is not assumed, i.e., w.i.d. = not s.i.d., so f.d. implies w.i.d. However, here we assume the infinite-dimensionality because we discuss the difference among infinite-dimensional spaces.

Theorem 1.12.14 (EMBEDDING) *Every separable metrizable space with* dim $\leqslant n$ *can be embedded in* $\mathbf{I}^{2n+1}$, *hence it can be embedded in the Euclidean space* $\mathbb{R}^{2n+1}$. [5.8.1]

This Embedding Theorem can be obtained as a corollary of the following strengthened result:

Theorem 1.12.15 (EMBEDDING APPROXIMATION) *For a separable metrizable space X with* dim $X \leqslant n$, *the following hold:*

(1) *If X is compact, then every map $f : X \to \mathbf{I}^{2n+1}$ can be approximated by embeddings, that is, for each $\varepsilon > 0$, there is an embedding $h : X \to \mathbf{I}^{2n+1}$ that is ε-close to f.* [5.8.5]
(2) *If X is locally compact, then every proper map $f : X \to \mathbb{R}^{2n+1}$ can be approximated by closed embeddings, that is, for each $\mathcal{U} \in$ cov$(\mathbb{R}^{2n+1})$, there is a closed embedding $h : X \to \mathbb{R}^{2n+1}$ that is $\mathcal{U}$-close to f.* [5.8.10]

1.13 Absolute Neighborhood Retracts

A subset A of a space X is called a **retract** of X if there is a map $r : X \to A$ such that $r|A = \mathrm{id}$, which is called a **retraction**. As is easily observed, every retract of a space X is *closed* in X. A **neighborhood retract** of X is a *closed* set in X that is a retract of some neighborhood in X. A *metrizable* space X is called an **absolute neighborhood retract** (**ANR**) (resp. an **absolute retract** (**AR**)) if X is a neighborhood retract (resp. a retract) of an arbitrary metrizable space containing X as a closed subspace. A space Y is called an **absolute neighborhood extensor for metrizable spaces** (**ANE**) if each map $f : A \to Y$ from any closed set A in an arbitrary metrizable space X extends over some neighborhood U of A in X. When f can always be extended over X (i.e., $U = X$ in the above), we call Y an **absolute extensor for metrizable spaces** (**AE**). As is easily observed, every *metrizable* ANE (resp. a *metrizable* AE) is an ANR (resp. an AR). Moreover, every neighborhood retract of an ANE is an ANE and every retract of an AE is an AE.

The following theorems are fundamental to ANR theory.

Theorem 1.13.1 (DUGUNDJI EXTENSION) *Let E be a locally convex topological linear space and X be a metrizable space with a closed set $A \subset X$. Then, each map $f : A \to E$ can be extended to a map $\tilde{f} : X \to E$ such that the image $\tilde{f}(X)$ is contained in the convex hull $\langle f(A) \rangle$ of $f(A)$, hence every convex set is an AE.* [6.1.1]

Remark 1.11 In the above, when $E = (E, \|\cdot\|)$ is a normed linear space and f is bounded, the extension $\tilde{f}$ can be taken so as to be bounded and have the same sup-norm as f, (i.e., $\|\tilde{f}\| = \|f\|$). Cf. Remark 2 in [GAGT, Chapter 6].

Theorem 1.13.2 (ARENS–EELLS EMBEDDING) *Every (complete) metric space X can be isometrically embedded in a (complete) normed linear space E with* dens $E = \aleph_0$ dens X *as a linearly independent closed set.* [6.2.1]

By combining these two theorems, we can obtain the following:

- a *metrizable* space X is an ANR (resp. an AR) if and only if X is an ANE (resp. an AE).

Let X be a space and $A \subset X$. It is said that A is **contractible** in X if the inclusion $A \subset X$ is null-homotopic, that is, there is a homotopy $h : X \times \mathbf{I} \to X$ such that $h_0 = \mathrm{id}$ and h_1 is constant. Such a homotopy h is called a **contraction** of A in X. When $A = X$, we say that X is **contractible** and h is called a **contraction** of X. As is easily observed, X is contractible if and only if $X \simeq 0$ (X is homotopic to a singleton). It is said that X is **locally contractible** if any neighborhood U of each point $x \in X$ contains a neighborhood of x that is contractible in U.

Proposition 1.13.3 (1) *A contractible ANE is an AE.* [6.1.5]
(2) *A metrizable space is an AR if and only if it is a contractible ANR.* [6.2.9]
(3) *Every ANR is locally contractible.* [6.2.8]

The converse of the above (3) does not hold in general, that is, locally contractible metrizable space is not necessary to be an ANR. For such an example, refer to [GAGT, Theorem 6.3.8].

For simplicial complexes, we have the following:

Theorem 1.13.4 *For any simplicial complex K, the polyhedron $|K|$ is an ANE and $|K|_{\mathrm{m}}$ is an ANR.* [6.1.4] [6.2.6]

Basic properties of ANEs are listed below:

Proposition 1.13.5 (Basic Properties of ANEs)

(1) *An arbitrary product of AEs is an AE and a finite product of ANEs is an ANE.* [6.1.9(1)]
(2) *A retract of an AE is an AE and a neighborhood retract of an ANE is an ANE.* [6.1.9(2)]
(3) *Any open set in an ANE is also an ANE.* [6.1.9(3)]
(4) (Hanner Theorem) *A paracompact space is an ANE if it is locally an ANE, that is, each point has an ANE neighborhood.* [6.1.9(6)]
(5) *Let $X = X_1 \cup X_2$, where X_1 and X_2 are closed in X. If X_1, X_2, and $X_1 \cap X_2$ are ANEs (AEs), then so is X. If X and $X_1 \cap X_2$ are ANEs (AEs), then so are X_1 and X_2.* [6.1.9(7)]

Since a metrizable space is an ANR (resp. an AR) if and only if it is an ANE (resp. an AE), the above list can be translated to ANRs as follows:

Proposition 1.13.6 (Basic Properties of ANRs)

(1) *An countable product of ARs is an AR and a finite product of ANRs is an ANR.* [6.2.10(1)]
(2) *A retract of an AR is an AR and a neighborhood retract of an ANR is an ANR.* [6.2.10(2)]

(3) *Any open set in an ANR is also an ANR.* 6.2.10(3)

(4) (HANNER THEOREM) *A paracompact space is an ANR if it is locally an ANR, that is, each point has an ANR neighborhood.* 6.2.10(4)

(5) *Let $X = X_1 \cup X_2$, where X_1 and X_2 are closed in X. If X_1, X_2, and $X_1 \cap X_2$ are ANRs (ARs), then so is X. If X and $X_1 \cap X_2$ are ANRs (ARs), then so are X_1 and X_2.* 6.2.10(5)

The following is a very useful procedure to extend homeomorphisms:

Theorem 1.13.7 (KLEE'S TRICK) *Let E and F be metrizable topological linear spaces which are AEs and let A and B be homeomorphic closed sets in E and F, respectively. Then, each homeomorphism $f : A \times \{\mathbf{0}\} \to \{\mathbf{0}\} \times B$ can be extended to a homeomorphism $\tilde{f} : E \times F \to E \times F$.* 6.2.2

Klee's Trick can be applied to prove Hausdorff's Metric Extension Theorem:

Theorem 1.13.8 (HAUSDORFF'S METRIC EXTENSION) *Let A be a closed set in a (completely) metrizable space X. Every admissible (complete) metric on A extends to an admissible (complete) metric on X.* 6.2.3

In Theorem 1.13.2, if an embedding is not required to be an isometry, we have the following alteration:

Theorem 1.13.9 *Every completely metrizable space can be embedded in a Hilbert space with the same density as a closed set. And every metrizable space can be embedded in a pre-Hilbert space (that is, a linear subspace of a Hilbert space) with the same density as a closed set.* 6.2.4

Let $\mathcal{V}$ be an open refinement of an open cover $\mathcal{U}$ of a space X. We call $\mathcal{V}$ an **h-refinement** (resp. **$\hbar$-refinement**) of $\mathcal{U}$ if any two $\mathcal{V}$-close maps $f, g : Y \to X$ defined on an arbitrary space Y are $\mathcal{U}$-homotopic (resp. $\mathcal{U}$-homotopic rel. $\{y \in Y \mid f(y) = g(y)\}$), where we write

$$\mathcal{V} \underset{h}{\prec} \mathcal{U} \quad \text{or} \quad \mathcal{U} \underset{h}{\succ} \mathcal{V} \quad \left(\text{resp. } \mathcal{V} \underset{\hbar}{\prec} \mathcal{U} \quad \text{or} \quad \mathcal{U} \underset{\hbar}{\succ} \mathcal{V}\right).$$

Theorem 1.13.10 *Every open cover of an ANR has an $\hbar$-refinement (hence it has an h-refinement).* 6.3.5

The following is frequently used in this book:

Theorem 1.13.11 (HOMOTOPY EXTENSION) *Let Y be an ANE, $\mathcal{U}$ be an open cover of Y, and $h : A \times \mathbf{I} \to Y$ be a $\mathcal{U}$-homotopy of a closed set A in a metrizable space X. If h_0 is extended to a map $f : X \to Y$, then h can be extended to a $\mathcal{U}$-homotopy $\tilde{h} : X \times \mathbf{I} \to Y$ with $\tilde{h}_0 = f$.* 6.4.1

Let X be a subspace of Y. A homotopy $h : X \times \mathbf{I} \to Y$ with $h_0 = \mathrm{id}$ is called **deformation** of X in Y. When $X = Y$, h is called a **deformation** of X. A subset A of a space X is said to be a **deformation retract** of X if there is a deformation $h : X \times \mathbf{I} \to X$ such that h_1 is a retraction of X onto A, where h_1 is called a

deformation retraction of X onto A. When $h_t|A = \mathrm{id}$ for all $t \in \mathbf{I}$ in the above, we call A a **strong deformation retract** of X and h_1 a **strong deformation retraction** of X onto A.[38] A retraction $r : X \to A$ is a deformation retraction (resp. strong deformation retraction) if and only if $r \simeq \mathrm{id}$ (resp. $r \simeq \mathrm{id}$ rel. A). A *closed* set $A \subset X$ is a **neighborhood deformation retract** of X if A has a neighborhood U in X with a homotopy $h : U \times \mathbf{I} \to X$ such that $h_0 = \mathrm{id}$ and h_1 is a retraction of U onto A, where h_1 is called a **deformation retraction** of U onto A in X. When $h_t|A = \mathrm{id}$ for all $t \in \mathbf{I}$ in the above, we call A a **strong neighborhood deformation retract** of X and h_1 a **strong deformation retraction** of U onto A in X.

Proposition 1.13.12 *A retract of an AR is a strong deformation retract and a neighborhood retract of an ANR is a strong neighborhood deformation retract.* 6.2.7

Using the Homotopy Extension Theorem 1.13.11, we can also show the following:

Proposition 1.13.13 *A deformation retract of an ANR is a strong deformation retract.* 6.4.4

It is said that X is **deformable** into A ($\subset X$) if there is a deformation $h : X \times \mathbf{I} \to X$ with $h_1(X) \subset A$. Obviously, if A is a deformation retract of X, then X is deformable into A and A is a retract of X. The converse also holds, that is:

Proposition 1.13.14 *A subset A of a space X is a deformation retract of X if and only if X is deformable into A and A is a retract of X.*[39] p. E3

The following can be easily proved:

Proposition 1.13.15 *Let A be an ANR that is a closed subset of an ANR X. Then, the inclusion $A \subset X$ is a homotopy equivalence if and only if A is a strong deformation retract of X.*

Proof The "if" part is trivial. In fact, a strong deformation retraction of X onto A is a homotopy inverse of the inclusion $A \subset X$.

To see the "only if", let $f : X \to A$ be a homotopy inverse of the inclusion $A \subset X$. Since $f|A \simeq \mathrm{id}$ in A, we can apply the Homotopy Extension Theorem 1.13.11 to obtain a retraction $r : X \to A$ such that $r \simeq f$ in A. Moreover, because $f \simeq \mathrm{id}$ in X, X is deformable into A. Hence, A is a deformation retract of X by Proposition 1.13.14. Then, it follows from Proposition 1.13.13 that A is a strong deformation retract of X. □

The following is very useful:

Theorem 1.13.16 (Kruse–Liebnitz) *Let X be metrizable and A be a strong neighborhood deformation retract of X. If A and $X \setminus A$ are ANRs, then so is X.* 6.5.2

[38]A strong deformation retract was defined on p. 50.

[39]Refer to Erratum of [GAGT], Comment for p. 348.

For a map $f : X \to Y$, we define the **mapping cylinder** $M(f)$ as the space $Y \cup (X \times (0, 1])$ (the disjoint union) with the topology generated by open sets in $X \times (0, 1]$ and sets $V \cup f^{-1}(V) \times (0, \varepsilon)$, where V is open in Y and $0 < \varepsilon < 1$. If X and Y are metrizable, then $M(f)$ is also metrizable, but M_f is not in general. Then, we call $M(f)$ the **metrizable mapping cylinder** to distinguish it from M_f. For the following facts, refer to Section 6.5 (pp. 362–363) in [GAGT].

Fact (1) *For each map $f : X \to Y$ between completely metrizable spaces, $M(f)$ is also completely metrizable.*
(2) *For a perfect map $f : X \to Y$ between metrizable spaces, M_f is metrizable and $M_f \approx M(f)$.*

As an easy application of the Kruse–Liebnitz Theorem 1.13.16 above, we have the following:

Corollary 1.13.17 *For any map $f : X \to Y$ between ANRs, the mapping cylinder $M(f)$ is an ANR.* [6.5.4]

The (**metrizable**) **cone** $C(X)$ (resp. the (**metrizable**) **open cone** $C^o(X)$) over a (metrizable) space X is defined as the space $\{0\} \cup X \times (0, 1]$ (resp. $C^o(X) = \{0\} \cup X \times (0, 1)$) with the topology generated by open sets in $X \times (0, 1]$ (resp. $X \times (0, 1)$) and sets $\{0\} \cup X \times (0, \varepsilon)$, $0 < \varepsilon < 1$. In other words, the cone $C(X)$ is the mapping cylinder $M(c_0)$ of the constant map $c_0 : X \to \{0\}$. Recall that the usual cone C_X is defined as the quotient space $C_X = (X \times \mathbf{I})/(X \times \{0\})$, which is another mapping cylinder M_{c_0} of the constant map $c_0 : X \to \{0\}$. If X is metrizable, then $C(X)$ is also metrizable but C_X is not in general. Moreover, if X is completely metrizable, then so are $C(X)$ and $C^o(X)$. When X is compact, $C_X \approx C(X)$. The following is the special case of Corollary 1.13.17:

Corollary 1.13.18 *The cone $C(X)$ over any ANR X is an AR, hence so is the open cone $C^o(X)$.* [6.5.7]

Let X be a space and $\mathcal{U} \in \operatorname{cov}(X)$. Let K be a simplicial complex and L be a subcomplex of K with $K^{(0)} \subset L$. Then, a map $f : |L| \to X$ is called a **partial $\mathcal{U}$-realization** of K in X if $\{f(\sigma \cap |L|) \mid \sigma \in K\} \prec \mathcal{U}$. A **full $\mathcal{U}$-realization** of K in X is a map $f : |K| \to X$ such that $\{f(\sigma) \mid \sigma \in K\} \prec \mathcal{U}$. We call $\mathcal{V} \in \operatorname{cov}(X)$ a **Lefschetz refinement** of $\mathcal{U}$ and denote

$$\mathcal{V} \underset{L}{\prec} \mathcal{U} \quad \text{or} \quad \mathcal{U} \underset{L}{\succ} \mathcal{V}$$

if $\mathcal{V} \prec \mathcal{U}$ and any partial $\mathcal{V}$-realization of an arbitrary simplicial complex in X extends to a full $\mathcal{U}$-realization in X. The following is called LEFSCHETZ'S CHARACTERIZATION of ANRs:

Theorem 1.13.19 (LEFSCHETZ) *A metrizable space X is an ANR if and only if any open cover of X admits a Lefschetz refinement.* [6.6.1]

Recall that given $\mathcal{U} \in \mathrm{cov}(X)$ (resp. $\varepsilon > 0$), X is **$\mathcal{U}$-homotopy dominated** (resp. **ε-homotopy dominated**) by Y if there are maps $f : X \to Y$ and $g : Y \to X$ such that $gf \simeq_{\mathcal{U}} \mathrm{id}_X$ (resp. $gf \simeq_\varepsilon \mathrm{id}_X$). The following is called HANNER'S CHARACTERIZATION of ANRs:

Theorem 1.13.20 (HANNER) *For a metric space $X = (X, d)$, the following are equivalent:*

(a) *X is an ANR;*
(b) *For each open cover $\mathcal{U}$ of X, there is a simplicial complex K such that X is $\mathcal{U}$-homotopy dominated by $|K|$;*
(c) *For any $\varepsilon > 0$, X is ε-homotopy dominated by an ANE.* [6.6.2]

Remark 1.12 In condition (b) above, we can require K to be locally finite-dimensional and $\mathrm{card}\, K^{(0)} \leqslant w(X)$. If X is compact, then K can be replaced by a finite simplicial complex. Indeed, let $f : X \to |K|$ and $g : |K| \to X$ be maps such that $gf \simeq_{\mathcal{U}} \mathrm{id}_X$. Then, because of compactness, $f(X)$ is contained in $|L|$ for some finite subcomplex L of K. Hence, X is $\mathcal{U}$-homotopy dominated by L. Moreover, when, X is separable, we can take a locally finite, countable simplicial complex as K. Indeed, in the proof of (a) $\Rightarrow$ (b) of Theorem 6.6.2 in [GAGT], we can take a countable star-finite open cover $\mathcal{W}'$ by Theorem 1.11.10. — Refer to Remark 9 for Theorem 6.6.2 in [GAGT].

Theorem 1.13.21 *Let $f : X \to Y$ be a map from a paracompact space X to an ANR Y and $\mathcal{U}$ be an open cover of Y. Then, each open cover $\mathcal{V}$ of X has an open refinement $\mathcal{W}$ with a map $\psi : |N(\mathcal{W})| \to Y$ such that $\psi\varphi \simeq_{\mathcal{U}} f$ for any canonical map $\varphi : X \to |N(\mathcal{W})|$.* [6.6.3]

Remark 1.13 In the above theorem, we can take a locally finite σ-discrete open refinement $\mathcal{W}$ of $\mathcal{V}$ such that the nerve $N(\mathcal{W})$ is locally finite-dimensional. When X is separable, we can take a star-finite countable open refinement $\mathcal{W}$ of $\mathcal{V}$ such that the nerve $N(\mathcal{W})$ is locally finite. If X is compact, then a finite open refinement $\mathcal{W}$ can be taken. — Refer to Remark 10 for Corollary 6.6.3 in [GAGT].

By Corollary 1.10.6 and Theorem 1.13.20, we obtain the following:

Theorem 1.13.22 *Every ANR X has the homotopy type of a locally finite-dimensional simplicial complex K with* $\mathrm{card}\, K^{(0)} \leqslant w(X)$. *Moreover, every separable ANR has the homotopy type of a countable locally finite simplicial complex.* [6.6.4]

For locally compact ANRs, the following can be proved in the same way as the implication (a) $\Rightarrow$ (b) of Theorem 1.13.20:

Theorem 1.13.23 *Let X be a locally compact ANR. For each open cover $\mathcal{U}$ of X, there is a locally finite simplicial complex K such that X is proper $\mathcal{U}$-homotopy dominated by $|K|$.*

Sketch of Proof Take open refinements as follows:

$$\mathcal{U} \succ \mathcal{U}' \underset{h}{\succ} \mathcal{V} \overset{*}{\succ} \mathcal{V}' \underset{L}{\succ} \mathcal{W} \overset{*}{\succ} \mathcal{W}',$$

where $\mathcal{U}'$ is locally finite, every $U \in \mathcal{U}'$ has the compact closure, and $\mathcal{W}'$ is star-finite (cf. Theorem 1.11.9). Then, the nerve K of $\mathcal{W}'$ is locally finite. Let $f : X \to |K|$ be a canonical map. Choosing $g'(W) \in W \in \mathcal{W}'$, we have a partial $\mathcal{W}$-realization $g' : K^{(0)} \to X$, which can be extended to a full $\mathcal{V}'$-realization $g : |K| \to X$ Then, observe that f and g are proper, and $gf \simeq^{p}_{\mathcal{W}'} \mathrm{id}$ (cf. Proposition 1.3.9).

Combining Theorems 1.10.7 and 1.13.23, we have the following:

Theorem 1.13.24 *For every locally compact ANR X, the product space $X \times \mathbb{R}_+$ has the proper homotopy type of a locally finite simplicial complex.* □

A subset A of a space X said to be **homotopy dense** in X if there exists a homotopy $h : X \times \mathbf{I} \to X$ such that $h_0 = \mathrm{id}$ and $h(X \times (0, 1]) \subset A$.

Proposition 1.13.25 *A subset X of a metrizable space Y is homotopy dense in Y if and only if the inclusion map $X \subset Y$ is a fine homotopy equivalence.* 6.7.1

Theorem 1.13.26 *Let X be a metrizable space and A be a homotopy dense subset of X. Then, X is an ANR if and only if A is an ANR.* 6.6.7

Proposition 1.13.27 *Let $x_0 \in A \subset X$. If A is homotopy dense in X and X is contractible, then the following set $A^{\mathbb{N}}_f$ is homotopy dense in $X^{\mathbb{N}}$:*

$$A^{\mathbb{N}}_f = \left\{ x \in A^{\mathbb{N}} \mid x(n) = x_0 \ \textit{except for finitely many } n \in \mathbb{N} \right\}.$$ 6.6.8

A space X is said to be **homotopically trivial** provided, for each $n \in \mathbb{N}$, every map $f : \mathbf{S}^{n-1} \to X$ is null-homotopic, that is, f extends over $\mathbf{B}^n$. Proposition 1.13.3(2) can be improved as follows:

Theorem 1.13.28 *A metrizable space is an AR if and only if it is a homotopically trivial ANR.* 6.6.10

A map $f : X \to Y$ is called a **weak homotopy equivalence**[40] provided that, for each $n \in \mathbb{N}$ and each map $\alpha : \mathbf{S}^{n-1} \to X$, if $f\alpha$ extends to a map $\beta : \mathbf{B}^n \to Y$, then α extends to a map $\tilde{\alpha} : \mathbf{B}^n \to X$ such that $f\tilde{\alpha} \simeq \beta$ in Y rel. $\mathbf{S}^{n-1}$:

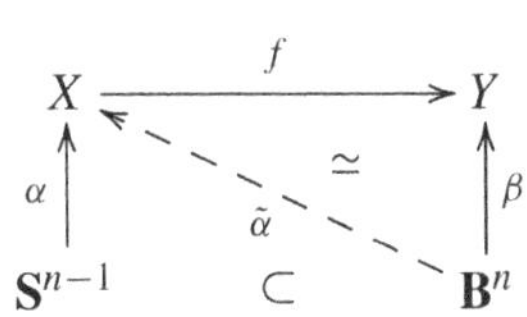

In some literature, the following is adopted as the definition:

[40] It is also called an ∞-equivalence because it is an n-equivalence for every $n \in \mathbb{N}$.

Theorem 1.13.29 *A map $f : X \to Y$ is a weak homotopy equivalence if and only if f induces the bijection between the path-components and the isomorphisms between the homotopy groups.* [4.14.12]

Theorem 1.13.30 *Let X and Y be ANRs. Then, every weak homotopy equivalence $f : X \to Y$ is a homotopy equivalence.* [6.6.6]

Let $X = (X, d_X)$ and $Y = (Y, d_Y)$ be metric spaces and A be a closed set in X. A map $f : X \to Y$ is said to be **uniformly continuous at** A if, for each $\varepsilon > 0$, there is some $\delta > 0$ such that for each $a \in A$ and $x \in X$, $d_X(a, x) < \delta$ implies $d_Y(f(a), f(x)) < \varepsilon$. We call A a **uniform retract** of X if there is a retraction $r : X \to A$ that is uniformly continuous at A. A subset $U \subset X$ such that $\mathrm{dist}(A, X \setminus U) > 0$ is called a **uniform neighborhood** of A in X. When A has a uniform neighborhood U in X and A is a uniform retract of U, A is called a **uniform neighborhood retract** of X. As is easily observed, every compact (neighborhood) retract of X is a uniform (neighborhood) retract.

A metric space X is called a **uniform AR** (resp. **uniform ANR**) if X is a uniform retract (resp. a uniform neighborhood retract) of an arbitrary metric space containing X as a closed set. It should be noted that every compact AR (ANR) is a uniform AR (ANR).

Theorem 1.13.31 *Every convex set in a locally convex metric linear space is a uniform AR.* [6.8.5]

By virtue of the following modification of Theorem 1.13.10, uniform ANRs are very useful:

Theorem 1.13.32 *Let X be a uniform ANR. For each $\varepsilon > 0$, there exists $\delta > 0$ such that any two δ-close maps of an arbitrary space to X are ε-homotopic.* [6.8.6]

We need the following results concerning uniform ANRs.

Theorem 1.13.33 *A uniform ANR is a uniform AR if and only if it is homotopically trivial.* [6.8.7]

Proposition 1.13.34 *A uniform ANR X is homotopy dense in every metric space Z that contains X isometrically as a dense subset.* [6.8.8]

Theorem 1.13.35 *A metric space X is a uniform ANR (resp. a uniform AR) if and only if the metric completion $\widetilde{X}$ of X is a uniform ANR (resp. a uniform AR) and X is homotopy dense in $\widetilde{X}$.* [6.8.10]

Proposition 1.13.36 *Let X be a subset of a metric space $M = (M, d)$. If X is an ANR, then X is homotopy dense in some G_δ-set Y in M, hence Y is also an ANR.* [6.8.2]

Theorem 1.13.37 *For any admissible metric d on an AR (resp. ANR) X, X has an admissible metric $\rho \geqslant d$ such that (X, ρ) is a uniform AR (resp. uniform ANR). If d is bounded, then so is ρ.* [6.8.11]

Every ANR is locally contractible but the converse does not hold in general. However, we have the following:

Theorem 1.13.38 (Haver) *Every countable-dimensional locally contractible metrizable space is an ANR.* [6.10.1]

1.14 Locally Equi-Connected Spaces

This section contains complementary results of Section 6.3 of the book [GAGT], which will be referred to in Sect. 6.13 of the present book.

For a space X, let Δ_X denote the **diagonal** of X^2, that is,

$$\Delta_X = \{(x, x) \mid x \in X\} \subset X^2.$$

For a each $A \subset X^2$ and $x \in X$, we write $A(x) = \{y \in X \mid (x, y) \in A\}$. Each neighborhood U of Δ_X in X^2 gives every $x \in X$ its neighborhood $U(x)$ simultaneously. Given $\mathcal{U} \in \operatorname{cov}(X)$, $W = \bigcup_{U \in \mathcal{U}} U^2$ is an open neighborhood of Δ_X in X^2, where $W(x) = \operatorname{st}(x, \mathcal{U})$ for each $x \in X$.

A space X is said to be **locally equi-connected** (**LEC**) if the diagonal Δ_X of X^2 has a neighborhood U with a map $\lambda : U \times \mathbf{I} \to X$ such that

$$\lambda(x, y, 0) = x \text{ and } \lambda(x, y, 0) = y \text{ for each } (x, y) \in U \text{ and}$$
$$\lambda(x, x, t) = x \text{ for each } x \in X \text{ and } t \in \mathbf{I},$$

that is, $\mathbf{I} \ni t \mapsto \lambda(x, y, t) \in X$ is a path from x to y in X, where a map λ is called an **equi-connecting map** for X. When $U = X^2$, X is said to be **equi-connected** (**EC**). Every topological linear space E is EC with the equi-connecting map defined by

$$E^2 \times \mathbf{I} \ni (x, y, t) \mapsto (1 - t)x + ty \in E.$$ [41]

Basic results are listed below:

Proposition 1.14.1 (Basic Property of LEC Spaces)

(1) *An AR is EC and an ANR is LEC.* [6.3.1]
(2) *A space X is EC (LEC) if and only if Δ_X is a strong deformation (neighborhood) retract of X^2.* [6.3.2]

[41] More generally, it is known that every semi-locally contractible topological group is LEC (cf. [GAGT, p. 349]).

(3) *A metrizable space X is EC if and only if X is contractible and LEC.*[42] 6.3.3
(4) *A space X is LEC if and only if each open cover of X has an $\hbar$-refinement.* 6.3.4
(5) *An arbitrary product of EC spaces is EC and a finite product of LEC spaces is LEC.* 6.3.9(1)
(6) *A retract of an EC space is EC and a neighborhood retract of an LEC space is LEC.* 6.3.9(2)
(7) *Any open set in an LEC space is also LEC.* 6.3.9(3)
(8) *A metrizable space is LEC if each point of X has an LEC neighborhood in X.* 6.3.9(5)

For a metrizable LEC space X, Lefschetz's Characterization of ANRs (Theorem 1.13.20) can be weakened as follows:

Theorem 1.14.2 (Dugundji) *A metrizable LEC space X is an ANR if and only if each open cover $\mathcal{V}$ has an open refinement $\mathcal{W}$ such that for an arbitrarily simplicial complex K, every partial $\mathcal{W}$-realization $f : |K^{(0)}| \to X$ of K extends to a full $\mathcal{V}$-realization $\tilde{f} : |K| \to X$ of K.*

Proof It is enough to show the "if" part. By Hanner's Characterization of ANRs (Theorem 1.13.19), it suffices to show that for each $\mathcal{V} \in \operatorname{cov}(X)$, there is a simplicial complex K such that X is $\mathcal{V}$-homotopy dominated by $|K|$. Take $\mathcal{V}', \mathcal{V}'' \in \operatorname{cov}(X)$ so that $\operatorname{st}\mathcal{V}' \prec \mathcal{V}''$ and $\mathcal{V}''$ is an $\hbar$-refinement of $\mathcal{V}$. By the condition, we have $\mathcal{W}' \in \operatorname{cov}(X)$ such that for an arbitrary simplicial complex K, every partial $\mathcal{W}'$-realization $g : |K^{(0)}| \to X$ extends to a full $\mathcal{V}'$-realization $\tilde{g} : |K| \to X$. Let $\mathcal{W} \in \operatorname{cov}(X)$ be a star-refinement of $\mathcal{W}'$.

Now, let K be the nerve of $\mathcal{W}$ with $f : X \to |K|$ a canonical map. We will show that X is $\mathcal{V}$-homotopy dominated by $|K|$. Choosing a point $g_0(W) \in W$ for each $W \in \mathcal{W} = K^{(0)}$, we define a map $g_0 : |K^{(0)}| \to X$. For each simplex $\sigma = \langle W_0, \dots, W_n \rangle \in K$, we can take $x_0 \in \bigcap_{i=0}^n W_i$ $(\neq \emptyset)$. Then, $g_0(\sigma^{(0)}) \subset \bigcup_{i=0}^n W_i \subset \operatorname{st}(x_0, \mathcal{W})$. Since $\mathcal{W}$ is a star-refinement of $\mathcal{W}'$, $g_0(\sigma^{(0)})$ is contained in some member of $\mathcal{W}'$. Thus, g_0 is a partial $\mathcal{W}'$-realization of K. Hence, g_0 extends to a full $\mathcal{V}'$-realization $g : |K| \to X$ of K. To see that $gf \simeq_{\mathcal{V}} \mathrm{id}$, it suffices to show that gf is $\mathcal{V}''$-close to id.

Due to Proposition 1.11.1, for each $x \in X$, $c_K(f(x))^{(0)} = \{W_0, \dots, W_n\} \subset \mathcal{W}[x]$. As we saw above, $g(c_K(f(x))^{(0)}) \subset \operatorname{st}(x, \mathcal{W})$. Since g is a full $\mathcal{V}'$-realization of K, $g(c_K(f(x)))$ is contained in some $V \in \mathcal{V}'$. Hence, $x, gf(x) \in \operatorname{st}(V, \mathcal{W}) \subset \operatorname{st}(V, \mathcal{V}')$. Because $\operatorname{st}\mathcal{V}' \prec \mathcal{V}''$, it follows that gf is $\mathcal{V}''$-close to id. The proof is completed. □

Remark 1.14 In the above proof, it suffices to take $\mathcal{V}''$ as an h-refinement of $\mathcal{V}$ instead of $\hbar$-refinement. Therefore, Theorem 1.14.2 is valid for a metrizable unified locally contractile (ULC) space X. Refer to Theorem 6.3.6 in [GAGT].[43]

[42] More generally, this is valid for a space X such that X^2 is normal.

[43] It is said that X is **unified locally contractile** (**ULC**) if each neighborhood U of Δ_X in X^2 contains a neighborhood V of Δ_X with a homotopy $h : V \times \mathbf{I} \to U$ such that $h_0 = \mathrm{id}$,

Let X be an LEC space with an equi-connecting map $\lambda : U \times \mathbf{I} \to X$. A subset $A \subset X$ is said to be **λ-convex** if $A^2 \subset U$ and $\lambda(A^2 \times \mathbf{I}) \subset A$, that is, $\lambda(x, y, t) \in A$ for every $x, y \in A$ and $t \in \mathbf{I}$. For $A \subset X$ with $A^2 \subset U$, the **λ-extension** of $A \subset X$ is the smallest subset $A^\lambda \subset X$ such that $A \times A^\lambda \subset U$, and $\lambda(A \times A^\lambda \times \mathbf{I}) \subset A^\lambda$, where $A \subset \lambda(A \times A^\lambda \times \mathbf{I}) = A^\lambda$ because $A^\lambda \subset \lambda(A \times A^\lambda \times \mathbf{I}) \subset A^\lambda$.

We can construct such an A^λ as follows: First, let $A_0 = A$ and inductively define $A_n = \lambda(A \times A_{n-1} \times \mathbf{I})$. Then, we have $A^\lambda = \bigcup_{n\in\omega} A_n$. Indeed,

$$\lambda(A \times A^\lambda \times \mathbf{I}) = \bigcup_{n\in\omega} \lambda(A \times A_n \times \mathbf{I}) \subset \bigcup_{n\in\mathbb{N}} A_n \subset A^\lambda.$$

If $\lambda(A \times B \times \mathbf{I}) \subset B$, then $A_0 = A \subset B$ and it inductively follows that

$$A_n = \lambda(A \times A_{n-1} \times \mathbf{I}) \subset \lambda(A \times B \times \mathbf{I}) \subset B.$$

Hence, $A^\lambda = \bigcup_{n\in\omega} A_n \subset B$.

For $A \subset B \subset X$ such that $A^2 \subset U$, it is said that A is **λ-stable** in B if $A^\lambda \subset B$. Evidently, each λ-convex set A is λ-stable in itself, hence in any B containing A. We define a **λ-stable refinement** of an open cover $\mathcal{V}$ of X as an open refinement $\mathcal{W}$ such that $W^2 \subset U$ for every $W \in \mathcal{W}$ and each $W \in \mathcal{W}$ is λ-stable in some $V \in \mathcal{V}$.

Theorem 1.14.3 (Dugundji) *Let X be a metrizable LEC space with an equi-connecting map $\lambda : U \times \mathbf{I} \to X$. If every open cover $\mathcal{V} \in \operatorname{cov}(X)$ has a λ-stable refinement $\mathcal{W} \in \operatorname{cov}(X)$, then X is an ANR.*

Proof By virtue of Theorem 1.14.2, it suffices to show that if $\mathcal{W}$ is a λ-stable refinement of $\mathcal{V} \in \operatorname{cov}(X)$ and K is a simplicial complex, then every partial $\mathcal{W}$-realization $f : |K^{(0)}| \to X$ extends to a full $\mathcal{V}$-realization $\tilde{f} : |K| \to X$.

Let $f : |K^{(0)}| \to X$ be a partial $\mathcal{W}$-realization of K, where we may assume that K is an ordered simplicial complex. Each $\sigma \in K$ with $\dim\sigma > 0$ can be written $\sigma = v_\sigma\sigma_*$, where $v_\sigma = \min\sigma^{(0)}$ and $\sigma_* < \sigma$ is the opposite face of σ to v_σ. If τ is an $(n-1)$-face of σ and $\tau \neq \sigma_*$, then $v_\tau = v_\sigma$ and $\tau_* < \sigma_*$. We can inductively define maps $f_n : \sigma \to X$, $n \in \omega$, as follows:

$$f_n((1-t)v_\sigma + tx) = \lambda(f(v_\sigma), x, t)$$
$$\text{for each } n\text{-simplex } \sigma \in K, x \in \sigma_*, \text{ and } t \in \mathbf{I},$$

where $f_0 = f$. It can be inductively shown that f_n is well-defined and $f_n||K^{(n-1)}| = f_{n-1}$. Because $f_n(\sigma) = \lambda(\{v_\sigma\} \times \sigma_* \times \mathbf{I})$ for every n-simplex $\sigma \in K$, we can see by induction that $f(\sigma^{(0)}) \subset W$ implies $f_n(\sigma) \subset W^\lambda$ for every n-simplex $\sigma \in K$. Since f is a partial $\mathcal{W}$-realization of K and $\mathcal{W}$ is a λ-stable

$h_1(V) = \Delta_X$, and $\mathrm{pr}_1 h_t = \mathrm{pr}_1|V$ for every $t \in \mathbf{I}$. In Theorem 6.3.6 in [GAGT], it is proved that a paracompact space X is ULC if and only if each open cover of X has an h-refinement.

refinement of $\mathcal{V}$, the desired full $\mathcal{V}$-realization $\tilde{f} : |K| \to X$ can be defined by $\tilde{f}||K^{(n)}| = f_n$, $n \in \omega$. □

Let X be a metrizable LEC space with an equi-connecting map $\lambda : U \times \mathbf{I} \to X$. Suppose that each point $x \in X$ has arbitrarily small λ-convex neighborhoods, that is, for each open neighborhood V of x, there is a λ-convex neighborhood W of x such that $W \subset V$. Then, $(\operatorname{int} W)^\lambda \subset W^\lambda = W$. Hence, every open cover $\mathcal{V}$ has a λ-stable open refinement $\mathcal{W}$. Thus, we have the following:

Corollary 1.14.4 *A metrizable LEC (resp. EC) space X is an ANR (resp. AR) if X has an equi-connecting map λ such that each point of X has arbitrarily small λ-convex neighborhoods.* □

1.15 Cell-Like Maps and Fine Homotopy Equivalences

A compact set A in a space X is said to be **cell-like** in X if A is contractible in every neighborhood of A in X. A compactum X is **cell-like** if X is cell-like in some *metrizable* space that contains X as a subspace. A **cell-like (CE) map** is a *perfect* map $f : X \to Y$ such that every fiber $f^{-1}(y)$ is cell-like. There is a close relationship between cell-like maps and fine homotopy equivalences. A fine homotopy equivalence is defined on p. 39, for which the perfectness is not required.

As a related concept, a **soft map** $f : X \to Y$ is defined as a map such that for any metrizable space Z (with $\dim Z \leqslant n$), and any map $g : C \to X$ of a closed set $C \subset Z$, if fg extends to a map $h : Z \to Y$, then g extends to a map $\tilde{g} : Z \to X$ with $f\tilde{g} = h$:

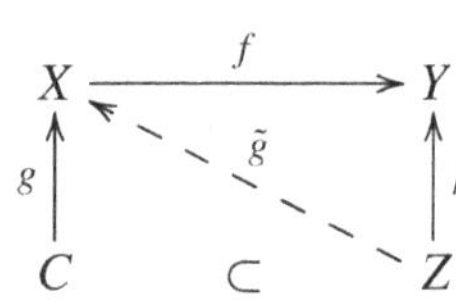

In the above, evidently, if Y is an AE or an ANE, then so is X.

We have several varieties of soft maps. In the definition of a soft map, replacing (Z, C) with the pair of a polyhedron and its subpolyhderon C, a **polyhedrally soft map** $f : X \to Y$ can be defined. A **homotopically soft map** (resp. an **approximately soft map**) $f : X \to Y$ is defined as a map such that for any metrizable space Z, any map $g : C \to X$ of a closed set $C \subset Z$, and each open cover $\mathcal{U}$ of Y, if fg extends to a map $h : Z \to Y$, then g extends to a map $\tilde{g} : Z \to X$ with $f\tilde{g} \simeq_{\mathcal{U}} h$ (resp. $f\tilde{g}$ is $\mathcal{U}$-close to h). In this definition, replacing (Z, C) with the pair of a polyhedron and its subpolyhderon C, a **polyhedrally homotopically soft map** (resp. a **polyhedrally approximately soft map**) $f : X \to Y$ can be defined:

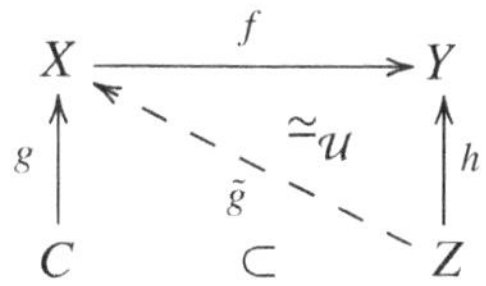

We call a map $f : X \to Y$ a **hereditary weak homotopy equivalence** if $f|f^{-1}(U) : f^{-1}(U) \to U$ is a weak homotopy equivalence for every open set $U \subset Y$. A map $f : X \to Y$ is called a **local *n*-connection** if $f(X)$ is dense in Y and every neighborhood of each $y \in Y$ contains a neighborhood V of y such that, for each $0 \leqslant i < n$, every map $g : \mathbf{S}^i \to f^{-1}(V)$ is null-homotopic in $f^{-1}(U)$, and, for each map $g : \mathbf{S}^n \to f^{-1}(V)$, fg is null-homotopic in U. A **local ∞-connection** is a map $f : X \to Y$ that is a local n-connection for every $n \in \omega$. A map $f : X \to Y$ is called a **local $*$-connection** if $f(X)$ is dense in Y and every neighborhood of each $y \in Y$ contains a neighborhood V of y such that $f^{-1}(V)$ is contractible in $f^{-1}(U)$.

By using concepts introduced above, fine homotopy equivalences between ANRs can be characterized as follows:

Theorem 1.15.1 *For a map $f : X \to Y$ between ANRs, the following are equivalent:*

(a) *f is a fine homotopy equivalence;*
(b) *f is approximately (= homotopically) soft;*
(c) *f is polyhedrally approximately (= homotopically) soft;*
(d) *f is a hereditary weak homotopy equivalence;*
(e) *f is a local ∞-connection;*
(f) *f is a local $*$-connection.* [7.4.3]

Remark 1.15 In the above theorem, instead of assuming that X is an ANR, it suffices to assume that X is an ANE with the following property:

($*$) For each $\mathcal{U} \in \operatorname{cov}(X)$, X is $\mathcal{U}$-homotopy dominated by a polyhedron.

Indeed, Proposition 7.4.2 in [GAGT] is valid even if "ANR" in (1) is replaced by "ANE" and "both X and Y are ANRs" in (2) is replaced by "X is an ANE and Y is an ANR" (by the same proof). Hence, the implications of the diagram in the proof of Theorem 7.4.3 in [GAGT] (p. 442) hold except (c) $\Rightarrow$ (a). By Hanner's Characterization 1.13.20, every ANR has property ($*$), which is the property of X required in the proof of (c) $\Rightarrow$ (a). In particular, Theorem 1.15.1 is valid for a (non-metrizable) polyhedron X and an ANR Y.[44]

The above theorem has the following corollaries:

Corollary 1.15.2 *Let $f : X \to Y$ be a fine homotopy equivalence between ANRs. Then, for every open set U in Y, $f|f^{-1}(U) : f^{-1}(U) \to U$ is also a fine homotopy equivalence.* [7.4.4]

---

[44]More generally, Theorem 1.15.1 is valid if X and Y are ANRs for stratifiable spaces [23, Proposition 2].

Corollary 1.15.3 *Let $f : X \to Y$ be a fine homotopy equivalence between ANRs. If $A \subset Y$ is contractible in an open neighborhood U in Y, then $f^{-1}(A)$ is contractible in $f^{-1}(U)$.* [7.4.5]

If $f_i : X_i \to Y_i$, $i = 1, 2$ are local $*$-connections, then $f_1 \times f_2 : X_1 \times X_2 \to Y_1 \times Y_2$ is clearly a local $*$-connection. Then, we have the following:

Corollary 1.15.4 *Let $f_i : X_i \to Y_i$, $i = 1, 2$ be fine homotopy equivalences between ANRs. Then, $f_1 \times f_2 : X_1 \times X_2 \to Y_1 \times Y_2$ is a fine homotopy equivalence.*[45]

Remark 1.16 The above corollary is valid even if X_1 and X_2 are ANEs with the property (*) in Remark 1.15.

Due to Proposition 1.13.25, a subset X of a metrizable space Y is homotopy dense in Y if and only if the inclusion map $X \subset Y$ is a fine homotopy equivalent. As another corollary of Theorem 1.15.1, we have the following characterization of homotopy denseness:

Corollary 1.15.5 *Let X and Y be ANRs such that X is a dense subset of Y. Then, the following are equivalent:*

(a) *X is homotopy dense in Y;*
(b) *For each open set U in Y, the inclusion $U \cap X \subset U$ is a weak homotopy equivalence, i.e., for each $n \in \mathbb{N}$, if a map $\alpha : \mathbf{S}^{n-1} \to U \cap X$ extends to a map $\beta : \mathbf{B}^n \to U$ then α extends to a map $\tilde{\alpha} : \mathbf{B}^n \to U \cap X$ such that $\tilde{\alpha} \simeq \beta$ in U rel. $\mathbf{S}^{n-1}$;*
(c) *Every neighborhood U of each y in Y contains a neighborhood V of y in Y such that every map $\alpha : \mathbf{S}^{n-1} \to V \cap X$ is null-homotopic in $U \cap X$ for each $n \in \mathbb{N}$.* [7.4.6]

From the equivalence of (a) and (b) in the above corollary, the following is easily derived:

Corollary 1.15.6 *Let X be a homotopy dense set in an ANR Y. Then, for every open set U in Y, $X \cap U$ is homotopy dense in U.* □

Applying Corollary 1.15.5, we can show that every locally homotopy dense subset of an ANR is homotopy dense, that is:

Proposition 1.15.7 *Let X be a subset of an ANR Y. If Y has an open cover $\mathcal{U}$ such that for each $U \in \mathcal{U}$, $X \cap U$ is homotopy dense in U, then X is homotopy dense in Y.*

Proof First of all, note that each $U \in \mathcal{U}$ is an ANR as an open set in an ANR. It follows from Theorem 1.13.26 that $X \cap U$ is an ANR. By Proposition 1.13.6(4) (Hannar's Theorem), X is an ANR.

[45] This corollary does not appear in [GAGT].

Let U be an open neighborhood of $y \in Y$. We have $U' \in \mathcal{U}$ such that $y \in U'$. As we saw above, U' and $X \cap U'$ are ANRs. By Corollary 1.15.5, $U \cap U'$ contains a neighborhood V of y in U' such that every map $\alpha : \mathbf{S}^{n-1} \to V \cap X \cap U' \ (= V \cap X)$ is null-homotopic in $U \cap U'$ for each $n \in \mathbb{N}$. Then, every map $\alpha : \mathbf{S}^{n-1} \to V \cap X$ is null-homotopic in U for each $n \in \mathbb{N}$. Due to Corollary 1.15.5, this means that X is homotopy dense in Y. □

To Corollary 1.15.5, we can add one more condition, that is:

Proposition 1.15.8 *Let X and Y be ANRs with $X \subset Y$ and $d \in \mathrm{Metr}(Y)$. Then, X is homotopy dense in Y if and only if, for each compact set A in Y and $\varepsilon > 0$, there is a map $f : A \to X$ that is ε-close to* id.

Proof The "only if" part is trivial. To see the "if" part, take U, α, and β as in condition (b) of Corollary 1.15.5. Since $U \cap X$ is an ANR and $\mathbf{S}^{n-1}$ is compact, we have $\delta > 0$ such that a map $\alpha' : \mathbf{S}^{n-1} \to U \cap X$, $\alpha' \simeq \alpha$ in $U \cap X$ if α' is δ-close to α. Choose $\varepsilon > 0$ so that

$$\varepsilon < \min\{\delta,\ \mathrm{dist}(\beta(\mathbf{B}^n), Y \setminus U)\}.$$

Applying the condition to $\beta(\mathbf{B}^n)$ and $\varepsilon > 0$, we can find a map $f : \beta(\mathbf{B}^n) \to X$ that is ε-close to id. Thus, we can obtain a map $f\beta : \mathbf{B}^n \to X \cap U$ such that $f\beta|\mathbf{S}^{n-1}$ is δ-close to $\beta|\mathbf{S}^{n-1} = \alpha$, which implies that $f\beta|\mathbf{S}^{n-1} \simeq \alpha$ in $U \cap X$. We can apply the Homotopy Extension Theorem 1.13.11 to extend α to a map $\tilde{\alpha} : \mathbf{B}^n \to U \cap X$. Therefore, X is homotopy dense in Y by Corollary 1.15.5. □

It is said that a closed set $A \subset X$ has **Property UV^n** in X (or simply A is **UV^n** in X) if each neighborhood U of A in X contains a neighborhood V of A such that, for each $0 \leqslant i \leqslant n$, every map $\alpha : \mathbf{S}^i \to V$ is null-homotopic in U. We say that A has **Property UV^∞** in X (or simply A is **UV^∞** in X) if it is UV^n in X for every $n \in \omega$. It is said that A has **Property UV^*** in X (or simply A is **UV^*** in X) if each neighborhood U of A in X contains a neighborhood V of A that is contractible in U. A surjective map $f : X \to Y$ is called a **UV^* map**, a **UV^∞ map**, or a **UV^n map** depending on if each fiber $f^{-1}(y)$ is UV^*, UV^∞, or UV^n.

For a map $f : X \to Y$ and a space Z, let $f^* : [Y, Z] \to [X, Z]$ be the function defined by $f^*([g]) = [gf]$ for each map $g : Y \to Z$:

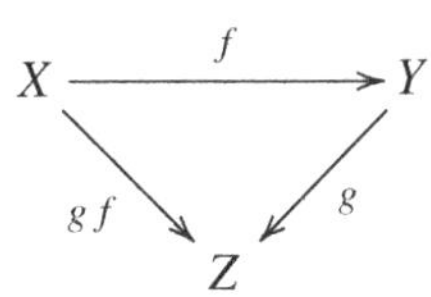

A map $f : X \to Y$ is called a **shape equivalence** if the function $f^* : [Y, P] \to [X, P]$ is a bijection for every ANR P (equiv. for every polyhedron P). We call a map $f : X \to Y$ a **hereditary shape equivalence** if $f|f^{-1}(A) : f^{-1}(A) \to A$ is a shape equivalence for any closed set $A \subset Y$.

By using the above concepts, fine homotopy equivalence (or cell-like maps) between ANRs are characterized as follows:

Theorem 1.15.9 *For a closed surjective map $f : X \to Y$ between ANRs X and Y, the following are equivalent:*

(a) *f is a fine homotopy equivalence;*
(b) *f is a UV^∞ map (= a local ∞-connection);*
(c) *f is a UV^* map (= a local $*$-connection);*
(d) *f is a hereditary shape equivalence.*

In the case where f is a perfect map, the following is also equivalent to the above:

(e) *f is a cell-like map.* [7.5.4]

As a corollary of the above characterization, the following can be obtained:

Corollary 1.15.10 *Let X, Y and Z be ANRs. For each two cell-like maps $f : X \to Y$ and $g : Y \to Z$, the composition $gf : X \to Y$ is also a cell-like map.* [7.5.5]

Concerning the uniform limit of fine homotopy equivalences, we have the following:

Theorem 1.15.11 *Let X and Y be ANRs and $f = \lim_{i\to\infty} f_i : X \to Y$ the uniform limit of fine homotopy equivalences with respect to some admissible metric $d \in \mathrm{Metr}(Y)$. Then, f is also a fine homotopy equivalence.* [7.4.7]

Notes for Chapter 1

As mentioned at the beginning of this chapter, almost all results are treated in the book [GAGT] but we need more results in PL Topology (Combinatorial Topology) which are not covered by [GAGT]. For notes concerning results of this chapter, refer to Notes for corresponding chapters of [GAGT].

In Sect. 1.9, we proved basic results on regular neighborhoods, which are very useful and basic tools in PL Topology but missing in [GAGT]. The concept of regular neighborhoods is introduced by J.H.C. Whitehead [154]. For more information for them, refer to the books [(17)], [(18)], and [(19)]. For PL manifolds and combinatorial manifolds, refer to Appendix at the end of the book.

Section 1.14 is a supplement to Section 6.3 of [GAGT]. The local equiconnectedness (LEC-ness) was introduced by R.H. Fox [67]. Theorems 1.14.2 and 1.14.3 were proved by J. Dugundji [57]. For related results, refer to papers [50, 58, 86, 106, 118], etc.

Chapter 2
Fundamental Results on Infinite-Dimensional Manifolds

Given a paracompact (homogenuous) space E (called the **model space**), an **E-manifold** is a (topological) manifold modeled on E, that is, a paracompact space covered by open sets that are homeomorphic to open sets in E. Additionally, we postulate the following condition:

- *An E-manifold has the same density as E.*

When E is separable, every E-manifold is separable. An $\mathbb{R}^n$-manifold (an $\mathbf{I}^n$-manifold) is a usual n-manifold (an n-manifold with boundary). An $\ell^2(\Gamma)$-manifold and a $\boldsymbol{Q}$-manifold are respectively called a **Hilbert manifold** and a **Hilbert cube manifold**, where Γ is an infinite set. Throughout this chapter, as a model space E, we consider the following space:

- *Let $E = (E, \|\cdot\|)$ be a normed linear space*[1] *such that*

$$E \approx E^{\mathbb{N}} \quad \text{or} \quad E \approx E^{\mathbb{N}}_f.$$

We are going to see that the Hilbert space $\ell^2(\Gamma)$ satisfies this condition (the first homeomorphy), that is, $\ell^2(\Gamma) \approx \ell^2(\Gamma)^{\mathbb{N}}$ (Corollary 2.2.13). In Theorem 5.5.10 (p. 407), it will be shown that normed linear spaces $\ell^2_f(\Gamma)$, $\ell^2_f(\Gamma) \times \ell^2_Q$, $\ell^2_f(\Gamma) \times \ell^2$, and $\ell^2(\Gamma) \times \ell^2_f$ satisfy this condition (the second homeomorphy).[2] It should be noted that every E-manifold is metrizable because every locally metrizable paracompact space is metrizable (Proposition 1.3.12(5)).

[1]In this chapter, many results are valid for locally convex metrizable topological linear spaces. In the latter half, we will use the fact that E is homeomorphic to the open cone. To prove this fact, we assume that E is a normed linear space.

[2]For $\ell^2_f(\Gamma)$ and ℓ^2_Q, see the definition on p. 353.

K. Sakai, *Topology of Infinite-Dimensional Manifolds*, Springer Monographs in Mathematics, https://doi.org/10.1007/978-981-15-7575-4_2

In this chapter, we give fundamental results on E-manifolds (e.g., $\ell^2(\Gamma)$-manifolds) and $\boldsymbol{Q}$-manifolds. In particular, it will be seen that every E-manifold is homeomorphic to some open set in E, and that two E-manifolds are homeomorphic if they have the same homotopy type. For $\boldsymbol{Q}$-manifolds M such that $M \times \mathbb{R}_+ \approx M$, similar results will be proved. We prove the Toruńczyk Factor Theorem 2.2.12 (resp. 2.2.14), that is, every completely metrizable AR (resp. ANR) X with $\operatorname{dens}(X) \leqslant \operatorname{card} \Gamma$ is a topological factor of $\ell^2(\Gamma)$ (resp. an $\ell^2(\Gamma)$-manifold), i.e., $X \times \ell^2(\Gamma) \approx \ell^2(\Gamma)$ (resp. $X \times \ell^2(\Gamma)$ is an $\ell^2(\Gamma)$-manifold).

2.1 Remarks on the Model Spaces and Isotopies

First, we remark that the model spaces E and $\boldsymbol{Q}$ are ARs by the Dugundji Extension Theorem 1.13.1. By Hanner's Theorem 1.13.6(4), E-manifolds and $\boldsymbol{Q}$-manifolds are ANRs. Recall that the weight, density and cellularity are equal in metrizable spaces (see Sect. 1.1).

The density postulate on E-manifolds is equivalent to the following:

- *The cardinality of components of an E-manifold is not greater than* $\operatorname{dens} E$.

This equivalence follows from the following lemma:

Lemma 2.1.1 *Let X be a paracompact space covered by open sets that are homeomorphic to open sets in E. If X is connected, then X has the same density as E. In general, every component of X has the same density as E.*

Proof Let $\mathcal{U}$ be a locally finite open cover of X such that each $U \in \mathcal{U}$ is homeomorphic to a non-empty open set in E. Note that each non-empty open set in E has the same density as E. Each $U \in \mathcal{U}$ has a dense subset D_U with $\operatorname{card} D_U = \operatorname{dens} E$. Then, $D = \bigcup_{U \in \mathcal{U}} D_U$ is dense in X. Let $U_0 \in \mathcal{U}$ and $D_0 = D_{U_0}$. For each $n \in \mathbb{N}$, we define $U_n = \operatorname{st}^n(U_0, \mathcal{U})$ $(= \bigcup \mathcal{U}[U_{n-1}])$ and $D_n = \bigcup_{U \in \mathcal{U}[U_{n-1}]} D_U$. Since X is connected, it follows that $X = \bigcup_{n\in\mathbb{N}} U_n = \bigcup_{n\in\mathbb{N}} \operatorname{st}^n(U_0, \mathcal{U})$, hence $D = \bigcup_{n\in\mathbb{N}} D_n$. Because D_n is dense in U_n, we have $\mathcal{U}[U_n] = \mathcal{U}[D_n]$. Since $\mathcal{U}$ is locally finite in X, it follows that

$$\begin{aligned}\operatorname{card} D_{n-1} \leqslant \operatorname{card} D_n &\leqslant \operatorname{card} \mathcal{U}[D_{n-1}] \operatorname{dens} E \\ &\leqslant \aleph_0 \operatorname{card} D_{n-1} \operatorname{dens} E = \operatorname{card} D_{n-1} \operatorname{card} D_0 \\ &\leqslant \operatorname{card} D_{n-1} \operatorname{card} D_{n-1} = \operatorname{card} D_{n-1}.\end{aligned}$$

By induction, we have $\operatorname{card} D_n = \operatorname{card} D_0 = \operatorname{dens} E$ for every $n \in \mathbb{N}$. Therefore,

$$\operatorname{dens} E = \operatorname{dens} U_0 \leqslant \operatorname{dens} X \leqslant \operatorname{card} D = \sup_{n\in\mathbb{N}} \operatorname{card} D_n = \operatorname{dens} E.$$

Thus, we have $\operatorname{dens} X = \operatorname{dens} E$. □

When E is a Banach space (= a complete normed linear space), every E-manifold is completely metrizable (Proposition 1.3.12(5)). On the other hand, every $\boldsymbol{Q}$-manifold is locally compact. Let $\mathfrak{M}_0$ and $\mathfrak{M}_1$ denote the classes of all compacta (= compact metrizable spaces) and of all complete metrizable spaces, respectively. For infinite cardinal τ, let $\mathfrak{M}_1(\tau)$ be the subclass of $\mathfrak{M}_1$ consisting of spaces with density $\leqslant \tau$. In particular, $\mathfrak{M}_1(\aleph_0)$ is the class of separable completely metrizable spaces. For a space X, let $\mathcal{F}_X$ denote the class of spaces that can be embedded in X as closed sets. Then, $\mathcal{F}_{\boldsymbol{Q}} = \mathfrak{M}_0$ and $\mathcal{F}_{\ell^2(\Gamma)} = \mathfrak{M}_1(\tau)$ ($\mathcal{F}_{\ell^2} = \mathfrak{M}_1(\aleph_0)$), where $\tau = \operatorname{card} \Gamma$ (Theorem 1.13.9).

A space X is said to be (topologically) **homogeneous** if for every pair of points $x_0, x_1 \in X$, there is a homeomorphism $h : X \to X$ such that $h(x_0) = x_1$. Every topological group is homogeneous. In particular, E is homogeneous. Although finite-dimensional cubes $\mathbf{I}^n$ are not homogeneous, the Hilbert cube $\boldsymbol{Q}$ is homogeneous. To prove this, we use the metric $d \in \operatorname{Metr}(\boldsymbol{Q})$ defined by

$$d(x, y) = \sum_{n \in \mathbb{N}} 2^{-n} |x(n) - y(n)|.$$

Let $\operatorname{Homeo}(\boldsymbol{Q})$ be the space of all homeomorphisms of $\boldsymbol{Q}$ onto itself with the sup-metric

$$d(f, g) = \sup_{x \in \boldsymbol{Q}} d(f(x), g(x)).$$

Then, $\operatorname{Homeo}(\boldsymbol{Q})$ is completely metrizable because it has the following admissible complete metric:

$$d^*(f, g) = d(f, g) + d(f^{-1}, g^{-1}).$$

The admissibility of d^* can be shown as follows:

Given $f \in \operatorname{Homeo}(\boldsymbol{Q})$ and $\varepsilon > 0$, we have $0 < \delta < \varepsilon/2$ such that $d(f^{-1}(x), f^{-1}(y)) < \varepsilon/2$ for every $x, y \in \boldsymbol{Q}$ with $d(x, y) < \delta$. For $g \in \operatorname{Homeo}(\boldsymbol{Q})$ with $d(f, g) < \delta$,

$$d(f^{-1}, g^{-1}) = d(f^{-1}g, \mathrm{id}) = d(f^{-1}g, f^{-1}f) \leqslant \varepsilon/2,$$

hence $d^*(f, g) < \varepsilon$.

The completeness of d^* can be seen as follows: Let $(f_n)_{n \in \mathbb{N}}$ be a d^*-Cauchy sequence of $\operatorname{Homeo}(\boldsymbol{Q})$. Then, $(f_n)_{n \in \mathbb{N}}$ and $(f_n^{-1})_{n \in \mathbb{N}}$ are d-Cauchy, hence they converge to f and g in $\mathrm{C}(\boldsymbol{Q}, \boldsymbol{Q})$ because $\mathrm{C}(\boldsymbol{Q}, \boldsymbol{Q})$ is d-complete. It easy to see that $f \in \operatorname{Homeo}(\boldsymbol{Q})$, $g = f^{-1}$, and $d^*(f_n, f) \to 0$ as $n \to \infty$.

The subset $(-1,1)^{\mathbb{N}} \subset \boldsymbol{Q}$ is called the **pseudo-interior** of $\boldsymbol{Q}$ and denoted by $I(\boldsymbol{Q})$. Evidently, $I(\boldsymbol{Q}) \approx \mathbb{R}^{\mathbb{N}} = \boldsymbol{s}$.[3] The complement $\boldsymbol{Q} \setminus I(\boldsymbol{Q})$ is called the **pseudo-boundary** of $\boldsymbol{Q}$ and denoted by $B(\boldsymbol{Q})$.

Theorem 2.1.2 *The Hilbert cube $\boldsymbol{Q}$ is homogeneous.*

Proof For every pair of points $x_0, x_1 \in \boldsymbol{Q}$, we have to construct a homeomorphism $h : \boldsymbol{Q} \to \boldsymbol{Q}$ such that $h(x_0) = x_1$. In the case that $x_0, x_1 \in I(\boldsymbol{Q})$, for each $n \in \mathbb{N}$, there is a PL-homeomorphism $h_n : [-1,1] \to [-1,1]$ such that $h_n(\pm 1) = \pm 1$ and $h_n(x_0(n)) = x_1(n)$. The desired homeomorphism $h : \boldsymbol{Q} \to \boldsymbol{Q}$ can be defined by $h(x) = (h_n(x(n)))_{n\in\mathbb{N}}$ for each $x \in \boldsymbol{Q}$.

In order to reduce the general case into the case above, it suffices to show that, for each $x_0 \in B(\boldsymbol{Q})$, there is a homeomorphism $h : \boldsymbol{Q} \to \boldsymbol{Q}$ such that $h(x_0) \in I(\boldsymbol{Q})$. For each $n \in \mathbb{N}$, let

$$G_n = \big\{h \in \operatorname{Homeo}(\boldsymbol{Q}) \bigm| \operatorname{pr}_n h(x_0) \in (-1,1)\big\},$$

where $\operatorname{pr}_n : \boldsymbol{Q} \to [-1,1]$ is the projection onto the n-th factor (i.e., $\operatorname{pr}_n(x) = x(n)$). Evidently, each G_n is open in $\operatorname{Homeo}(\boldsymbol{Q})$. We shall show that each G_n is dense in $\operatorname{Homeo}(\boldsymbol{Q})$. Then, $\bigcap_{n\in\mathbb{N}} G_n$ is dense in $\operatorname{Homeo}(\boldsymbol{Q})$ by the Baire Category Theorem 1.3.15, hence $h \in \bigcap_{n\in\mathbb{N}} G_n$ is the desired homeomorphism.

To show that each G_n is dense in $\operatorname{Homeo}(\boldsymbol{Q})$, let $f \in \operatorname{Homeo}(\boldsymbol{Q})$ and $0 < \varepsilon < 1$. It suffices to find $h \in \operatorname{Homeo}(\boldsymbol{Q})$ such that $d(h, \mathrm{id}) < \varepsilon$ and $\operatorname{pr}_n hf(x_0) \in (-1,1)$, where it may be assumed that $|\operatorname{pr}_n f(x_0)| = 1$. Take a homeomorphism $\varphi : [-1,1]^2 \to [-1,1]^2$ (as illustrated in Fig. 2.1) such that

$$\varphi|[-1+\varepsilon/2, 1-\varepsilon/2] \times [-1,1] = \mathrm{id},\ \varphi(-x) = -\varphi(x) \ \text{ and}$$
$$\varphi(\{1\} \times [-1,1]) = [1 - 3\varepsilon/8, 1 - \varepsilon/8] \times \{1\}.$$

Choosing $m > n$ so that $2^{-m+1} < \varepsilon/2$, we define $h \in \operatorname{Homeo}(\boldsymbol{Q})$ as follows:

$$\operatorname{pr}_i h = \operatorname{pr}_i \ \text{ for every } i \in \mathbb{N} \setminus \{n, m\} \text{ and}$$
$$(\operatorname{pr}_n h(x), \operatorname{pr}_m h(x)) = \varphi(x(n), x(m)) \ \text{ for each } x \in \boldsymbol{Q}.$$

Then, $d(h, \mathrm{id}) < \varepsilon$ and $\operatorname{pr}_n hf(x_0) \in (-1,1)$. This completes the proof. □

By the homogeneity of $\boldsymbol{Q}$ (Theorem 2.1.2), we can show the following:

Proposition 2.1.3 $\mathcal{F}_{\boldsymbol{Q}\setminus\{0\}} = \mathcal{F}_{\boldsymbol{Q}\times[0,1)}$ *is the class of locally compact separable metrizable spaces.*[4]

---

[3]In the book [(1)], it is denoted by P. In much of the literature, the pseudo-interior is denoted by $\boldsymbol{s}$. In fact, we can naturally identify $I(\boldsymbol{Q}) = \boldsymbol{s}$. However, to avoid confusion, we distinguish them (i.e., $I(\boldsymbol{Q}) = (-1,1)^{\mathbb{N}} \neq \mathbb{R}^{\mathbb{N}} = \boldsymbol{s}$).

[4]It will be shown in Sect. 2.7 that $\boldsymbol{Q} \times [0,1) \approx \boldsymbol{Q} \setminus \{0\}$.

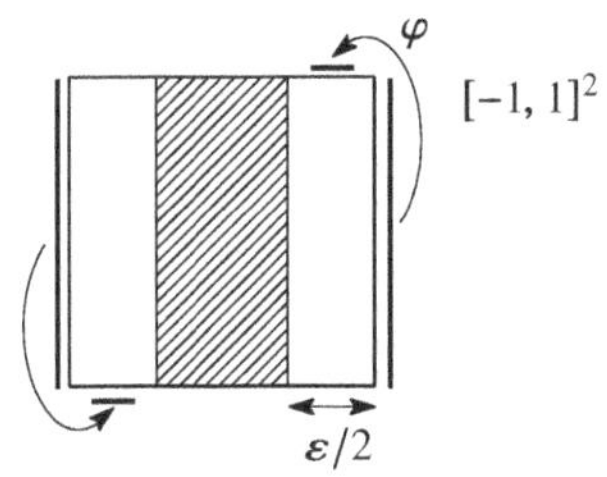

Fig. 2.1 The homeomorphism φ

Proof A space homeomorphic to a closed subspace of $\boldsymbol{Q}\setminus\{0\}$ or $\boldsymbol{Q}\times[0, 1)$ is locally compact separable metrizable. Conversely, let X be a locally compact separable metrizable space. Then, we have an embedding $f : \alpha X \to \boldsymbol{Q}$ of the one-point compactification $\alpha X = X \cup \{\infty\}$ of X. By the homogeneity of $\boldsymbol{Q}$, we may assume that $f(\infty) = 0$. Then, $f|X : X \to \boldsymbol{Q} \setminus \{0\}$ is a closed embedding. Moreover, using a map $k : \alpha M \to \mathbf{I}$ with $k^{-1}(0) = \{\infty\}$, we can define a closed embedding $h : M \to \boldsymbol{Q} \times [0, 1)$ by $h(x) = (f(x), 1 - k(x))$ for each $x \in M$. □

From Fig. 2.2, we can grasp the homeomorphisms in the following lemma:

Lemma 2.1.4 $[0, 1) \times (0, 1) \approx [0, 1) \times [0, 1) \approx [0, 1) \times \mathbf{I}$. □

By using the above lemma, we can easily show that the Hilbert cube $\boldsymbol{Q}$ has an open basis consisting of open sets homeomorphic to $\boldsymbol{Q} \times [0, 1)$. Hence, we have the following:

Proposition 2.1.5 *Every $\boldsymbol{Q}$-manifold can be covered by arbitrarily small open sets homeomorphic to $\boldsymbol{Q} \times [0, 1)$.* □

For each $x \in E$ and $\varepsilon > 0$, the ε-neighborhood $\mathrm{B}(x, \varepsilon)$ ($\approx \mathbf{B}_E \setminus \mathbf{S}_E$) is homeomorphic to E. Thus, we have the following:

Proposition 2.1.6 *Every E-manifold is covered by arbitrarily small open sets homeomorphic to E.* □

We show Riesz's Lemma as follows:

Lemma 2.1.7 (Riesz) *Let $L = (L, \|\cdot\|)$ be a normed linear space, F be a closed linear subspace of L with $F \subsetneq L$, and $0 < \delta < 1$. Then, there exists $v \in L$ such that $\|v\| = 1$ and* $\mathrm{dist}(v, F) \geqslant \delta$.

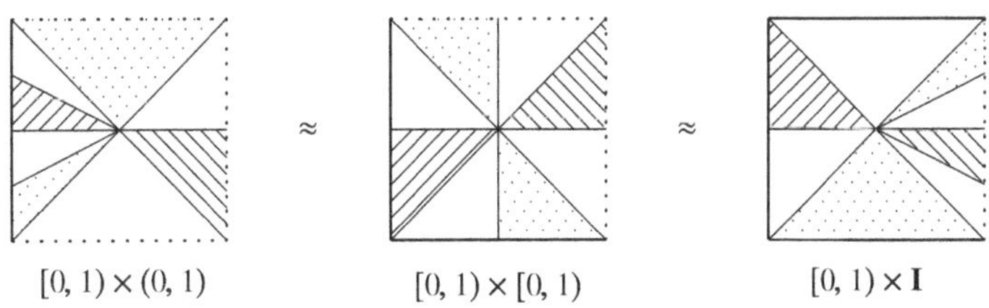

Fig. 2.2 The homeomorphisms in Lemma 2.1.4

Proof Take $x \in L \setminus F$ and let $r = \operatorname{dist}(x, F) > 0$. Since $0 < \delta < 1$, we have $y \in F$ such that $\|x - y\| < \delta^{-1} r$. Then, $v = \|x - y\|^{-1}(x - y) \in L$ is the desired one. Indeed, for each $z \in F$,

$$\begin{aligned} \|v - z\| &= \big\| \|x - y\|^{-1}(x - y) - z \big\| \\ &= \|x - y\|^{-1} \big\| x - y - \|x - y\| z \big\| \\ &> r^{-1}\delta \cdot \operatorname{dist}(x, F) = \delta. \end{aligned}$$

□

Let $X = (X, d)$ be a metric space, $A \subset X$, and $\delta > 0$. It is said that A is **δ-discrete** if $d(x, y) \geqslant \delta$ for each pair of distinct points $x, y \in A$. We say that A is **uniformly discrete** if it is δ-discrete for some $\delta > 0$. As an application of Riesz's Lemma, we can see that the unit sphere $\mathbf{S}_L$ of a normed linear space L contains a uniformly discrete infinite set.

Proposition 2.1.8 *Let $L = (L, \|\cdot\|)$ be an infinite-dimensional normed linear space. For each $0 < \delta < 1$, the unit sphere $\mathbf{S}_L$ of L contains a δ-discrete linearly independent set D with* $\operatorname{card} D = \operatorname{dens} L \; (= w(L))$.

Proof Using Zorn's Lemma, we can find a maximal δ-discrete linearly independent subset $D \subset \mathbf{S}_L$. Then, $\operatorname{card} D = \operatorname{dens} L$. Indeed, if $\operatorname{card} D < \operatorname{dens} L$, then the linear span F of D is not dense in L, because D is finite or $\operatorname{dens} F = \operatorname{card} D$. Applying Riesz's Lemma 2.1.7 to the closed linear subspace $\operatorname{cl} F \subsetneq L$, we have $x \in \mathbf{S}_L$ such that $\operatorname{dist}(x, \operatorname{cl} F) = \operatorname{dist}(x, F) \geqslant \delta$. Then, $D \cup \{x\}$ is a δ-discrete linearly independent subset of $\mathbf{S}_L$, which contradicts the maximality of D. □

According to Theorem 1.4.12, no infinite-dimensional normed linear spaces are locally compact. For normed linear spaces, this fact also follows from Proposition 2.1.8 above.

The following lemma is simple but important.

Lemma 2.1.9 *In the case $E \approx E^{\mathbb{N}}$, $E \times \mathbb{R}^{\mathbb{N}} \approx E$; in the case $E \approx E^{\mathbb{N}}_f$, $E \times \mathbb{R}^{\mathbb{N}}_f \approx E$. Under our assumption ($E \approx E^{\mathbb{N}}$ or $E \approx E^{\mathbb{N}}_f$), it follows that $E \times \mathbb{R} \approx E$.*

Proof Because of their similarity, we show only the case $E \approx E^{\mathbb{N}}$. By the Bartle–Graves–Michael Theorem 1.4.16, $E \approx F \times \mathbb{R}$ for some space F. Then,

$$\begin{aligned} E \times \mathbb{R}^{\mathbb{N}} &\approx E^{\mathbb{N}} \times \mathbb{R}^{\mathbb{N}} \approx (F \times \mathbb{R})^{\mathbb{N}} \times \mathbb{R}^{\mathbb{N}} \\ &\approx F^{\mathbb{N}} \times \mathbb{R}^{\mathbb{N}} \approx (F \times \mathbb{R})^{\mathbb{N}} \approx E^{\mathbb{N}} \approx E. \end{aligned}$$

□

Lemma 2.1.10 *There is a discrete open collection $\mathcal{U}$ in E such that* $\operatorname{card} \mathcal{U} = \operatorname{dens} E$ *and every $U \in \mathcal{U}$ is homeomorphic to E.*

Proof For each $n \in \mathbb{N}$, we can use Zorn's Lemma to find a maximal 2^{-n}-discrete subset D_n of E. Then, $\mathrm{B}(x, 2^{-n}/3) \approx E$ for each $x \in D_n$ and $\{\mathrm{B}(x, 2^{-n}/3) \mid x \in D_n\}$ is discrete in X. Since $\bigcup_{n\in\mathbb{N}} D_n$ is dense in E, it follows that

$$\operatorname{dens} E \leqslant \operatorname{card} \bigcup_{n\in\mathbb{N}} D_n = \sup_{n\in\mathbb{N}} \operatorname{card} D_n \leqslant c(E) = \operatorname{dens} E,$$

hence $\sup_{n\in\mathbb{N}} \operatorname{card} D_n = w(E)$. Let $h : E \times \mathbb{R} \to E$ be a homeomorphism. Then, we have the following discrete open collection in E:

$$\mathcal{U} = \big\{h(\mathrm{B}(x, 2^{-n}/3) \times (2n-1, 2n)) \bigm| n \in \mathbb{N},\ x \in D_n\big\},$$

where $\operatorname{card} \mathcal{U} = \sup_{n\in\mathbb{N}} \operatorname{card} D_n = w(E)$ and

$$h(\mathrm{B}(x, 2^{-n}/3) \times (2n-1, 2n)) \approx E \times \mathbb{R} \approx E. \qquad \square$$

Next, we show the following:

Theorem 2.1.11 *Every E-manifold M can be embedded in E as a closed set, namely $M \in \mathcal{F}_E$.*[5]

Proof We apply Michael's Theorem on local properties (Corollary 1.3.21). Let $\mathcal{P}$ be the property for closed sets A in M such that $A \in \mathcal{F}_E$. Then, M has $\mathcal{P}$ locally, that is, each point $x \in M$ has a closed neighborhood having $\mathcal{P}$. To see that M has $\mathcal{P}$ (i.e., $M \in \mathcal{F}_E$), it suffices to show that $\mathcal{P}$ is F-hereditary, where (F-1) is obvious.

(F-2): Let A_1, A_2 be closed sets in M with $A_1, A_2 \in \mathcal{F}_E$, that is, there are closed embeddings $h_i : A_i \to E$, $i = 1, 2$. Since $E^2 \times \mathbb{R} \approx E$, it suffices to construct a closed embedding $h : A_1 \cup A_2 \to E^2 \times \mathbb{R}$. Using Klee's Trick (1.13.7), we can obtain a homeomorphism $f : E^2 \to E^2$ such that $f(h_1(x), \mathbf{0}) = (\mathbf{0}, h_2(x))$ for every $x \in A_1 \cap A_2$. Let $g : A_1 \to \mathbf{I}$ be a map with $g^{-1}(0) = A_1 \cap A_2$. Then, a closed embedding $h : A_1 \cup A_2 \to E^2 \times \mathbb{R}$ can be defined as follows:

$$h(x) = \begin{cases} (f(h_1(x), \mathbf{0}), g(x)) & \text{if } x \in A_1, \\ ((\mathbf{0}, h_2(x)), 0) & \text{if } x \in A_2. \end{cases}$$

(F-3): Let $\{A_\lambda \mid \lambda \in \Lambda\}$ be a discrete collection of closed sets in M such that $A_\lambda \in \mathcal{F}_E$ for every $\lambda \in \Lambda$. Then, $\operatorname{card} \Lambda \leqslant w(E)$ because $c(M) = w(M) = w(E)$ by Lemma 2.1.1. On the other hand, by Lemma 2.1.10, E has a discrete open collection $\mathcal{U}$ such that $\operatorname{card} \mathcal{U} = w(E)$ and every $U \in \mathcal{U}$ is homeomorphic to E

[5]Recall our postulate that every E-manifold has the same density as E.

itself. Since $E \times \mathbb{R} \approx E$, it suffices to show that, for each $U \in \mathcal{U}$, every $A \in \mathcal{F}_E$ can be embedded in $U \times \mathbb{R}$ as a closed set in $E \times \mathbb{R}$.

Since $U \approx E$, we have a closed embedding $f : A \to U$. However, it should be noted that $f(A)$ is not closed in E. We define an embedding $h : A \to U \times \mathbb{R}$ as follows:

$$h(x) = \big(f(x), \operatorname{dist}(f(x), E \setminus U)^{-1}\big).$$

Then, it is easy to see that h is a closed embedding into $E \times \mathbb{R}$. □

It should be remarked that a $\boldsymbol{Q}$-manifold cannot be embedded in $\boldsymbol{Q}$ unless it is compact. However, every $\boldsymbol{Q}$-manifold M can be embedded in $\boldsymbol{Q} \times [0, 1)$ as a closed set, because it is locally compact, separable, and metrizable (Proposition 2.1.3).

A homotopy $h : X \times \mathbf{I} \to Y$ is called an **isotopy** if each $h_t : X \to Y$ is a homeomorphism. For homeomorphisms $f, g : X \to Y$, it is said that f is **isotopic** to g if there is an isotopy $h : X \times \mathbf{I} \to Y$ with $h_0 = f$ and $h_1 = g$, where h is also called an **isotopy** from f to g. Like "homotopic," if f is isotopic to g, then g is isotopic to f. For an open collection $\mathcal{U}$ in Y, an isotopy h is called a **$\mathcal{U}$-isotopy** if the collection of its tracks refine $\mathcal{U}$. Then, it is said that h_0 is **$\mathcal{U}$-isotopic** to h_1.

An isotopy $h : X \times \mathbf{I} \to Y$ is said to be **invertible** if $h^* : Y \times \mathbf{I} \to X$ defined by $h^*(y, t) = h_t^{-1}(y)$ is continuous (hence h^* is also an isotopy). Equivalently, $\bar{h} : X \times \mathbf{I} \to Y \times \mathbf{I}$ defined by $\bar{h}(x, t) = (h_t(x), t)$ is a homeomorphism. A map $\varphi : X \times \mathbf{I} \to Y \times \mathbf{I}$ is **I-preserving** if $\mathrm{pr}_{\mathbf{I}}\varphi = \mathrm{pr}_{\mathbf{I}}$. For such a map φ, we define maps $\varphi_t : X \to Y$, $t \in \mathbf{I}$, by $\varphi_t(x) = \mathrm{pr}_Y\varphi(x, t)$ for each $x \in X$, so $\varphi(x, t) = (\varphi_t(x), t)$ for each $(x, t) \in X \times \mathbf{I}$. Here, the notation φ_t is used differently than before. When $\varphi : X \times \mathbf{I} \to Y \times \mathbf{I}$ is considered as a homotopy or an isotopy, φ_t is defined by $\varphi_t(x) = \varphi(x, t)$, hence that is a map from X to $Y \times \mathbf{I}$. However, there would be no confusion by the context. The homeomorphism $\bar{h}$ defined above is **I**-preserving and $\bar{h}_t = h_t$ for $t \in \mathbf{I}$. For an **I**-preserving homeomorphism $\varphi : X \times \mathbf{I} \to Y \times \mathbf{I}$, $\mathrm{pr}_Y\varphi : X \times \mathbf{I} \to Y$ is an invertible isotopy. When an isotopy h from f to g is invertible, f is said to be **invertibly isotopic** to g, where f^{-1} is isotopic to g^{-1} by the isotopy h^* defined above. It should be remarked that, even if h is not invertible, an isotopy from g^{-1} to f^{-1} can also be defined by $g^{-1}h_t f^{-1}$ (or $f^{-1}h_t g^{-1}$). Then, observe that

$$\big\{g^{-1}h_t f^{-1}(x) \bigm| t \in \mathbf{I}\big\} = g^{-1}(h(\{f^{-1}(x)\} \times \mathbf{I}))$$
$$\Big(\text{or } \big\{f^{-1}h_t g^{-1}(x) \bigm| t \in \mathbf{I}\big\} = f^{-1}(h(\{g^{-1}(x)\} \times \mathbf{I}))\Big).$$

Thus, if f is $\mathcal{U}$-isotopic to g, then f is $g^{-1}(\mathcal{U})$-isotopic (or $f^{-1}(\mathcal{U})$-isotopic) to g.

Proposition 2.1.12 *There exists a non-invertible isotopy of ℓ^1.*

Proof For each $i \in \mathbb{N}$, let $f_i : \mathbf{I} \to [1/i, 1]$ be a surjective map such that $f_i([0, 1/(i+1)]) = f_i([1/i, 1]) = \{1\}$. We define a non-invertible isotopy of $h : \ell^1 \times \mathbf{I} \to \ell^1$ by $h(x, t) = (x(i)f_i(t))_{i\in\mathbb{N}}$. Then, each $h_t : \ell^1 \to \ell^1$ is a homeomorphism with the inverse $h_t^{-1}(y) = (y(i)/f_i(t))_{i\in\mathbb{N}}$. However, the map $\bar{h}$ defined above is not a homeomorphism. Indeed, let $x_i \in \ell^1$ such that $x_i(i) = 1/i$ and $x_i(j) = 0$ for $j \neq i$, and let $t_i \in f_i^{-1}(1/i)$. Then, $\|x_i\| = 1/i \to 0$ and $0 < t_i < 1/i \to 0$ $(i \to \infty)$, but $\bar{h}^{-1}(x_i, t_i) = (\mathbf{e}_i, t_i)$, where $\mathbf{e}_i(i) = 1$ and $\mathbf{e}_i(j) = 0$ for $j \neq i$. Hence, $\bar{h}^{-1}$ is not continuous at $(\mathbf{0}, 0) \in \ell^1 \times \mathbf{I}$. □

The following is another example:

Proposition 2.1.13 *There exists an AR X in Euclidean plane $\mathbb{R}^2$ which has a non-invertible isotopy.*

Proof Let X be the subspace of $\mathbb{R}^2$ defined as follows (see the left-hand side of Fig. 2.3):

$$X = \mathbf{I} \times \{0\} \cup \bigcup_{n\in\mathbb{N}} \{1/n\} \times \mathbf{I}.$$

For each $n \in \mathbb{N}$, let $p_n = (1/n, 1/3n)$, $q_n = (1/n, 2/3) \in X$. We can define a non-invertible isotopy $h : X \times \mathbf{I} \to X$ as follows: Let $h_0 = h_{2^{-2n+2}} = \mathrm{id}$ for each $n \in \mathbb{N}$. If $2^{-2n} \leqslant t \leqslant 2^{-2n+2}$, then $h_t|X \setminus (\{1/n\} \times \mathbf{I}) = \mathrm{id}$ and $h_t|\{1/n\} \times \mathbf{I}$ is a homeomorphism of $\{1/n\} \times \mathbf{I}$ onto itself such that h_t is linear on each $\{1/n\} \times [0, 2/3]$ and $\{1/n\} \times [2/3, 1]$, and the point $h_t(q_n)$ is sliding on $\{1/n\} \times \mathbf{I}$ from q_n down to p_n and back to q_n as t moves from 2^{-2n+2} to 2^{-2n} (see the right-hand side of Fig. 2.3). The precise definition is as follows:

$$h_t(q_n) = \begin{cases} \left(\dfrac{1}{n}, \dfrac{2 - 2^{2n-1}t}{3n} + \dfrac{2(2^{2n-1}t - 1)}{3}\right) & \text{if } 2^{-2n+1} \leqslant t \leqslant 2^{-2n+2}, \\ \left(\dfrac{1}{n}, \dfrac{2(2 - 2^{2n}t)}{3} + \dfrac{2^{2n}t - 1}{3n}\right) & \text{if } 2^{-2n} \leqslant t \leqslant 2^{-2n+1}. \end{cases}$$

The continuity of h at each point except $(\mathbf{0}, 0) = ((0, 0), 0)$ is trivial. Note that $h(\{1/n\} \times [0, s] \times \mathbf{I}) \subset \{1/n\} \times [0, s]$ and $h([0, s] \times \{0\} \times \mathbf{I}) = [0, s] \times \{0\}$. Clearly, $h(\mathrm{B}(\mathbf{0}, \varepsilon) \times \mathbf{I}) \subset \mathrm{B}(\mathbf{0}, \varepsilon)$ for any $\varepsilon > 0$, hence h is also continuous at $(\mathbf{0}, 0)$. Thus, h is an isotopy of X. On the other hand, $(p_n, 2^{-2n+1}) \to (\mathbf{0}, 0)$ in $X \times \mathbf{I}$ as $n \to \infty$,

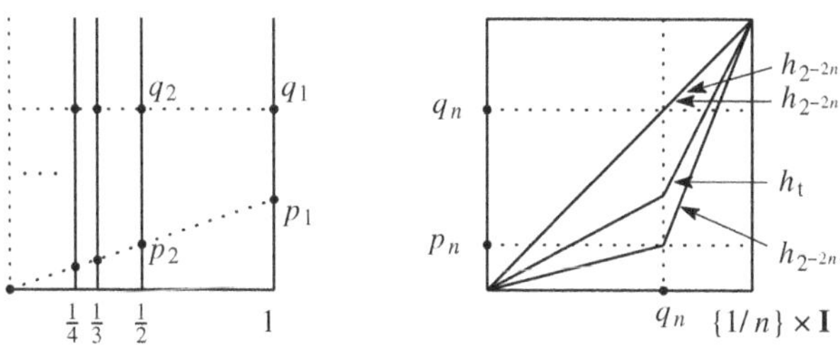

Fig. 2.3 The space X and the graph of h_t $(2^{-2n} \leqslant t \leqslant 2^{-2n+2})$

but $h'(p_n, 2^{-2n+1}) = h^{-1}_{2^{-2n+1}}(p_n) = q_n$ is not convergent in X, that is, h is not invertible. □

When X and Y are compact, every isotopy $h : X \times \mathbf{I} \to Y$ is clearly invertible. This fact holds for locally compact spaces.

Proposition 2.1.14 *If X and Y are locally compact, then every isotopy $h : X \times \mathbf{I} \to Y$ is invertible.*

Proof We define $\bar{h} : X \times \mathbf{I} \to Y \times \mathbf{I}$ by $\bar{h}(x, t) = (h_t(x), t)$. It suffices to show that $\bar{h}$ is open. Let G be an open set in $X \times \mathbf{I}$. For each $(x, t) \in G$, choose open sets $U \subset X$ and $S \subset \mathbf{I}$ so that $\operatorname{cl} U$ is compact and $(x, t) \in U \times S \subset G$. Since h_t is a homeomorphism, we have an open neighborhood V of $h_t(x)$ in Y such that $\operatorname{cl} V \subset h_t(U)$. Then, $h(\operatorname{bd} U \times \{t\}) \subset Y \setminus \operatorname{cl} V$. Since $\operatorname{bd} U$ is compact, we have a neighborhood W of x in X and a connected neighborhood J of t in $\mathbf{I}$ such that

$$J \subset S,\ h(W \times J) \subset V \ \text{ and } \ h(\operatorname{bd} U \times J) \subset Y \setminus \operatorname{cl} V.$$

Then, $h_t(W) \times J$ is a neighborhood of $\bar{h}(x, t) = (h_t(x), t)$ in $Y \times \mathbf{I}$. For each $z \in W$ and $s \in J$, $h_s^{-1}(h(\{z\} \times J))$ is connected,

$$h_s^{-1}(h(\{z\} \times J)) \subset h_s^{-1}(\operatorname{cl} V) \subset X \setminus \operatorname{bd} U = U \cup (X \setminus \operatorname{cl} U) \ \text{ and}$$
$$z \in h_s^{-1}(h(\{z\} \times J)) \cap h_t^{-1}(V) \subset h_s^{-1}(h(\{z\} \times J)) \cap U,$$

hence $h_s^{-1}(h(\{z\} \times J)) \subset U$. In particular, $h_s^{-1}h_t(z) \in U$, that is, $h_t(z) \in h_s(U)$. Thus, $h_t(W) \subset h_s(U)$ for each $s \in J$. Then, we have

$$h_t(W) \times J \subset \Big(\bigcap_{s \in J} h_s(U)\Big) \times J \subset \bigcup_{s \in S} \big(h_s(U) \times \{s\}\big) = \bar{h}(U \times S) \subset \bar{h}(G),$$

hence $\bar{h}(G)$ is open in $Y \times \mathbf{I}$. □

For an invertible isotopy $h : X \times \mathbf{I} \to X$ with $h_0 = \mathrm{id}_X$, let $h^* : X \times \mathbf{I} \to X$ be the isotopy defined by h_t^{-1}, $t \in \mathbf{I}$. For their tracks, it should be remarked that:

- h^* has tracks different from h in general.

Moreover, for an open collection $\mathcal{U}$ in X,

- even if h is a $\mathcal{U}$-isotopy, h^* need not be a $\mathcal{U}$-isotopy.

For example, let $h : \mathbb{R}^2 \times \mathbf{I} \to \mathbb{R}^2$ be the isotopy defined as follows:

$$h_t(x, y) = \begin{cases} (x, y + 2t) & \text{if } 0 \leqslant t \leqslant 1/2, \\ (x + 2t - 1, y + 1) & \text{if } 1/2 \leqslant t \leqslant 1. \end{cases}$$

Then, the set $(\{0\} \times \mathbf{I}) \cup (\mathbf{I} \times \{1\})$ is the track of the origin $\mathbf{0} = (0, 0) \in \mathbb{R}^2$ by h. Observe that

$$h_t^*(x, y) = h_t^{-1}(x, y) = \begin{cases} (x, y - 2t) & \text{if } 0 \leqslant t \leqslant 1/2, \\ (x - 2t + 1, y - 1) & \text{if } 1/2 \leqslant t \leqslant 1. \end{cases}$$

The set $(\{1\} \times \mathbf{I}) \cup (\mathbf{I} \times \{0\})$ is the track of $(1, 1) = h_1(\mathbf{0})$ by the isotopy defined by h^*. Moreover, let $\mathcal{U} = \{U(x, y) \mid (x, y) \in \mathbb{R}^2\} \in \operatorname{cov}(\mathbb{R}^2)$, where

$$U(x, y) = ((x - 1/2, x + 1/2) \times (y - 1/2, y + 3/2)) \\ \cup ((x - 1/2, x + 3/2) \times (y + 1/2, y + 3/2)).$$

The above h is a $\mathcal{U}$-isotopy but h^* is not a $\mathcal{U}$-isotopy.

Nevertheless, the following holds:

Proposition 2.1.15 *Let $\mathcal{U}$ be an open collection in X. For an invertible $\mathcal{U}$-isotopy $h : X \times \mathbf{I} \to X$ with $h_0 = \mathrm{id}_X$, the isotopy $h^* : X \times \mathbf{I} \to X$ defined by h_t^{-1}, $t \in \mathbf{I}$, is a* st *$\mathcal{U}$-isotopy.*

Proof Let $x \in X$ such that $h^*(\{x\} \times \mathbf{I})$ is non-degenerate. It suffices to show that $h^*(\{x\} \times \mathbf{I}) \subset \operatorname{st}(x, \mathcal{U})$. For each $t \in \mathbf{I}$, there is $U_t \in \mathcal{U}$ such that $h(\{h_t^{-1}(x)\} \times \mathbf{I}) \subset U_t$. Since $x = h_t(h_t^{-1}(x)) \in h(\{h_t^{-1}(x)\} \times \mathbf{I})$, it follows that $U_t \in \mathcal{U}[x]$, that is, $U_t \subset \operatorname{st}(x, \mathcal{U})$. Observe that

$$h_t^{-1}(x) = h_0(h_t^{-1}(x)) \in h(\{h_t^{-1}(x)\} \times \mathbf{I}) \subset U_t \subset \operatorname{st}(x, \mathcal{U}).$$

Hence, we have $h^*(\{x\} \times \mathbf{I}) = \{h_t^{-1}(x) \mid t \in \mathbf{I}\} \subset \operatorname{st}(x, \mathcal{U})$. □

For embeddings $f, g : X \to Y$ (resp. subsets $A, B \subset Y$), it is said that f is **ambiently isotopic** to g in Y (resp. A is **ambiently isotopic** to B in Y) if there is an isotopy $\varphi : Y \times \mathbf{I} \to Y$ with $\varphi_0 = \mathrm{id}$ and $\varphi_1 f = g$ (resp. $\varphi_1(A) = B$). This is called an **ambient isotopy** from f to g (resp. from A to B). For an ambient isotopy φ from f to g (resp. from A to B), an ambient isotopy from g to f (resp. from B to A) is defined by $\varphi_{1-t}\varphi_1^{-1}$,[6] so g is ambiently isotopic to f (resp. B is ambiently isotopic to A in Y). When φ is an invertible isotopy, it is said that f is **ambiently invertibly isotopic** to g in Y (resp. A is **ambiently invertibly isotopic** to B in Y), where an ambient isotopy from g to f (resp. from B to A) can also be defined by φ_t^{-1} because φ is invertible. Indeed, $\varphi_0^{-1} = \mathrm{id}$ and $\varphi_1^{-1} g = f$

When $f, g : X \to Y$ are homeomorphisms, f is isotopic to g if and only if f is ambiently isotopic to g. Indeed, for an ambient isotopy φ from f to g, an isotopy from f to g is defined by $\varphi_t f$. Conversely, for an isotopy h from f to g, an ambient isotopy from f to g is defined by $h_t f^{-1}$.

[6] Such an isotopy can also be defined by $\varphi_1^{-1}\varphi_{1-t}$. However, when φ is a $\mathcal{U}$-isotopy, this definition does not give a $\mathcal{U}$-isotopy but the above does.

We shall show the existence of very useful open covers of an open set in a metrizable space:

Lemma 2.1.16 *Every open set W in a metrizable space Y has an open cover $\mathcal{U}$ with the following properties:*

(i) *Given a map $f : X \to Y$, if a map $g : f^{-1}(W) \to W$ is $\mathcal{U}$-close to $f|f^{-1}(W)$ then g can be extended to the map $\tilde{g} : X \to Y$ by $\tilde{g}|X \setminus f^{-1}(W) = f|X \setminus f^{-1}(W)$;*
(ii) *If a homeomorphism $h : W \to W$ is $\mathcal{U}$-close to* id, *then h can be extended to the homeomorphism $\tilde{h} : Y \to Y$ by $\tilde{h}|Y \setminus W =$ id;*
(iii) *Given a subset $X \subset Y$, if an embedding $h : X \cap W \to W$ is $\mathcal{U}$-close to* id, *then h can be extended to the embedding $\tilde{h} : X \cup (Y \setminus W) \to Y$ by $\tilde{h}|Y \setminus W =$ id. Moreover, if h is a closed embedding, i.e., $h(X \cap W)$ is closed in W, then $\tilde{h}$ is also a closed embedding.*

Proof Take $d \in \mathrm{Metr}(Y)$ and define

$$\mathcal{U} = \left\{\mathrm{B}(y, 2^{-1}d(y, Y \setminus W)) \;\middle|\; y \in W\right\} \in \mathrm{cov}(W).$$

To see that $\mathcal{U}$ satisfies (i), it suffices to show the continuity of $\tilde{g}$ in (i) at each $x \in \mathrm{bd}\, f^{-1}(W)$. For each $\varepsilon > 0$, x has a neighborhood V in X such that $f(V) \subset \mathrm{B}(f(x), \varepsilon/3)$. For each $x' \in V \cap f^{-1}(W)$, there is some $y \in W$ such that $g(x'), f(x') \in \mathrm{B}(y, 2^{-1}d(y, Y \setminus W))$. Since $d(f(x'), f(x)) < \varepsilon/3$ and $f(x) \in Y \setminus W$, it follows that

$$\begin{aligned} d(y, Y \setminus W) &\leqslant d(y, f(x)) \leqslant d(y, f(x')) + d(f(x'), f(x)) \\ &< 2^{-1}d(y, Y \setminus W) + \varepsilon/3, \end{aligned}$$

which implies $2^{-1}d(y, Y \setminus W) < \varepsilon/3$. Then, we have

$$\begin{aligned} d(\tilde{g}(x'), \tilde{g}(x)) &\leqslant d(g(x'), f(x')) + d(f(x'), f(x)) \\ &< d(y, Y \setminus W) + \varepsilon/3 < \varepsilon. \end{aligned}$$

Thus, $\tilde{g}$ is continuous at x.

Now, we show that (ii) and (iii) follow from (i). First, applying (i) to $f = \mathrm{id}_Y$, we can extend h in (ii) to the map $\tilde{h} : Y \to Y$. The inverse h^{-1} in (ii) is also $\mathcal{U}$-close to id, so $\tilde{h}^{-1}$ is continuous by (i). Thus, $\tilde{h}$ is a homeomorphism.

For (iii), h can be extended to the map $\tilde{h} : X \cup (Y \setminus W) \to Y$ by $\tilde{h}|Y \setminus W = \mathrm{id}$. Then, $\tilde{h}$ is injective,

$$\tilde{h}(X \cup (Y \setminus W)) = h(X \cap W) \cup (Y \setminus W) \;\text{ and}$$

$$\tilde{h}(X \cup (Y \setminus W)) \cap W = h(X \cap W).$$

Since $h^{-1} : h(X \cap W) \to W$ is $\mathcal{U}$-close to id, it follows from (1) that $\tilde{h}^{-1} : \tilde{h}(X \cup (Y \setminus W)) \to Y$ is continuous, hence $\tilde{h}$ is an embedding. If $h(X \cap W)$ is closed in W, then $W \cap \mathrm{cl}_Y h(X \cap W) = h(X \cap W)$, hence

$$\tilde{h}(X \cup (Y \setminus W)) = \mathrm{cl}_Y h(X \cap W) \cup (Y \setminus W).$$

Therefore, $\tilde{h}(X \cup (Y \setminus W))$ is closed in Y. □

For an open set W in a space Y, an open cover $\mathcal{U} \in \mathrm{cov}(W)$ is said to be **fitting** in Y if it satisfies the conditions of Lemma 2.1.16.[7] As we saw in the above proof, if $\mathcal{U} \in \mathrm{cov}(W)$ satisfies condition (i), then it satisfies conditions (ii) and (iii), too. Obviously, every refinement $\mathcal{V} \in \mathrm{cov}(W)$ of $\mathcal{U}$ is also fitting in Y. Then, we have the following:

Proposition 2.1.17 *Let Y be a metrizable space with W an open set. Every open cover $\mathcal{U}$ of W has an open refinement $\mathcal{V}$ that is fitting in Y.* □

In the theory of $\boldsymbol{Q}$-manifolds, it is important whether maps and homotopies are proper or not. When only compact $\boldsymbol{Q}$-manifolds are considered, there is no problem because every map and homotopy are proper. However, for non-compact $\boldsymbol{Q}$-manifolds, we must check to see if maps and homotopies are proper. It should be recalled that a proper map between metrizable spaces is identical to a perfect map (Proposition 1.3.7). Moreover, it really helps to keep in mind that, for an open over $\mathcal{U}$ of a $\boldsymbol{Q}$-manifold M such that every $U \in \mathcal{U}$ has the compact closure in M, any map to M is proper if it $\mathcal{U}$-close to a proper map (Proposition 1.3.8).

2.2 The Toruńczyk Factor Theorem

In this section, let $L = (L, \|\cdot\|)$ be a normed linear space with $\dim L > 0$. Recall that $L^{\mathbb{N}}$ denotes the countable infinite product of L with the product topology, which is a topological linear space and contains the following dense linear subspace:

$$L_f^{\mathbb{N}} = \{x \in L^{\mathbb{N}} \mid x(i) = \mathbf{0} \text{ except for finitely many } i \in \mathbb{N}\}.$$

Then, $\mathbb{R}_f^{\mathbb{N}}$ can be embedded into $L_f^{\mathbb{N}}$ as a topological linear subspace because L contains a one-dimensional linear subspace (that is linearly isomorphic to $\mathbb{R}$). Since $\mathbb{R}_f^{\mathbb{N}}$ is not normable (Proposition 1.2.1), it follows that any linear subspace F of $L^{\mathbb{N}}$ with $L_f^{\mathbb{N}} \subset F$ has no admissible norm.

[7] In [17], such an open cover is said to be **normal** to $Y \setminus W$.

We define the **ℓ^1-product** $\Pi_{\ell^1} L$ of L as the following normed linear space:

$$\Pi_{\ell^1} L = \{x \in L^{\mathbb{N}} \mid (\|x(i)\|)_{i\in\mathbb{N}} \in \ell^1\} = \{x \in L^{\mathbb{N}} \mid \textstyle\sum_{i\in\mathbb{N}} \|x(i)\| < \infty\}$$

with the norm $\|x\|_1 = \sum_{i\in\mathbb{N}} \|x(i)\|$. Then, $L_f^{\mathbb{N}} \subset \Pi_{\ell^1} L \subset L^{\mathbb{N}}$, but the topology of $\Pi_{\ell^1} L$ is different from the product topology, as mentioned above. The normed linear subspace $L_f^{\mathbb{N}}$ of $\Pi_{\ell^1} L$ by $\Pi_{\ell_f^1} L$, that is,

$$\Pi_{\ell_f^1} L = \{x \in L^{\mathbb{N}} \mid (\|x(i)\|)_{i\in\mathbb{N}} \in \ell_f^1\}$$

with the norm $\|x\|_1 = \sum_{i\in\mathbb{N}} \|x(i)\|$. We call $\Pi_{\ell_f^1} L$ the **ℓ_f^1-product**.[8]

The following can be proved in the same way as the completeness of ℓ^1.

Proposition 2.2.1 *If L is a Banach space, then the ℓ^1-product $\Pi_{\ell^1} L$ is also a Banach space.*

Sketch of Proof For each Cauchy sequence $(x_n)_{n\in\mathbb{N}}$ in $\Pi_{\ell^1} L$, we have $x \in L^{\mathbb{N}}$ such that $x(i) = \lim_{n\to\infty} x_n(i)$ for each $i \in \mathbb{N}$ because $(x_n(i))_{n\in\mathbb{N}}$ is Cauchy in L. Then, show that $x \in \Pi_{\ell^1} L$ and $\lim_{n\to\infty} \|x_n - x\|_1 = 0$.

Evidently, the ℓ^1-product and the ℓ_f^1-product of $\mathbb{R} = (\mathbb{R}, |\cdot|)$ are ℓ^1 and ℓ_f^1, respectively. It is easy to construct a linear isometry from the ℓ^1-product $\Pi_{\ell^1} \ell^1(\Gamma)$ onto $\ell^1(\mathbb{N} \times \Gamma)$, which sends the ℓ_f^1-product $\Pi_{\ell_f^1} \ell_f^1(\Gamma)$ onto $\ell_f^1(\mathbb{N} \times \Gamma)$. Thus, we have the following:

Proposition 2.2.2 *The pair $\big(\Pi_{\ell^1} \ell^1(\Gamma), \Pi_{\ell_f^1} \ell_f^1(\Gamma)\big)$ is linearly isometric to the pair $(\ell^1(\Gamma), \ell_f^1(\Gamma))$ if Γ is infinite.* □

Remark 2.1 More generally, for a sequence $(L_i)_{i\in\mathbb{N}}$ of normed linear spaces $L_i = (L_i, \|\cdot\|_i)$, $i \in \mathbb{N}$, we can similarly define the ℓ^1-product and the ℓ_f^1-product as the following normed linear spaces:

$$\Pi_{\ell^1} L_i = \{x \in L^{\mathbb{N}} \mid (\|x(i)\|_i)_{i\in\mathbb{N}} \in \ell^1\},$$
$$\Pi_{\ell_f^1} L_i = \{x \in L^{\mathbb{N}} \mid (\|x(i)\|_i)_{i\in\mathbb{N}} \in \ell_f^1\},$$

with the norm $|||x|||_1 = \sum_{i\in\mathbb{N}} \|x(i)\|_i$. This definition is available even if L_i, $i \in \mathbb{N}$, are F-normed spaces (cf. Sect. 1.4), where $|||\cdot|||_1$ is an F-norm. Using ℓ^p instead of ℓ^1, the ℓ^p-product and the ℓ_f^p-product can also be defined.

[8]The ℓ_f^1-product $\Pi_{\ell_f^1} L$ is written as $\sum_{\ell^1} L$ in Toruńczyk's papers [139, 140]. In the book [(2)] of Bessaga and Pełczyński (p. 208, p. 255), the notations $\Sigma_{\ell^1} L$ and $\Sigma_{\ell_f^1} L$ are used instead of $\Pi_{\ell^1} L$ and $\Pi_{\ell_f^1} L$, respectively.

Recall that X is called a **uniform retract** of L if there is a retraction $r : L \to X$ that is uniformly continuous at X,[9] that is, for each $\varepsilon > 0$, there is $\delta > 0$ such that for each $x \in X$ and $y \in L$, $\|x - y\| < \delta$ implies $\|x - r(y)\| < \varepsilon$ (cf. Sect. 6.8 in [GAGT]). The following is the main theorem of this section:

Theorem 2.2.3 (Toruńczyk) *Let X be a retract of a normed linear space $L = (L, \|\cdot\|)$ with a retraction $r : L \to X$ that is uniformly continuous at X. Then, $X \times \Pi_{\ell^1_f} L \approx \Pi_{\ell^1_f} L$. If X is $\|\cdot\|$-complete, then*

$$\left(X \times \Pi_{\ell^1} L, X \times \Pi_{\ell^1_f} L\right) \approx \left(\Pi_{\ell^1} L, \Pi_{\ell^1_f} L\right).$$

To prove Theorem 2.2.3, we may assume that $\mathbf{0} \in X$. Since the case $X = \{\mathbf{0}\}$ is obvious, we may also assume that $X \neq \{\mathbf{0}\}$. We will introduce a modification to the ℓ^1-product. A function $v : \mathbb{R}_+ \to \mathbb{R}_+$ is said to be **sub-additive** if

$$v(s + t) \leqslant v(s) + v(t) \quad \text{for each } s, t \in \mathbb{R}_+.$$

If $v : \mathbb{R}_+ \to \mathbb{R}_+$ is a homeomorphism, then $v(0) = 0$ and v is increasing, i.e., $s < t \Rightarrow v(s) < v(t)$. The continuity at 0 means that $t_n \to 0 \Rightarrow v(t_n) \to 0$. Therefore, given a sub-additive homeomorphism $v : \mathbb{R}_+ \to \mathbb{R}_+$, for a normed linear space $L = (L, \|\cdot\|)$, an F-norm can be defined by $v(\|x\|)$, $x \in L$.

For a sub-additive homeomorphism $v : \mathbb{R}_+ \to \mathbb{R}_+$, we define

$$\Pi_{(\ell^1,v)} L = \left\{x \in L^{\mathbb{N}} \mid (2^i v(\|x(i)\|))_{i\in\mathbb{N}} \in \ell^1\right\},$$
$$\Pi_{(\ell^1_f,v)} L = \left\{x \in L^{\mathbb{N}} \mid (2^i v(\|x(i)\|))_{i\in\mathbb{N}} \in \ell^1_f\right\},$$

and $\|x\|_v = \sum_{i\in\mathbb{N}} 2^i v(\|x(i)\|)$ for each $x \in \Pi_{(\ell^1,v)} L$ (refer to Remark 2.1). For $x, y \in \Pi_{(\ell^1,v)} L$, we have $-x, x + y \in \Pi_{(\ell^1,v)} L$ and

$$\|-x\|_v = \|x\|_v \quad \text{and} \quad \|x + y\|_v \leqslant \|x\|_v + \|y\|_v.$$

By induction, $nx \in \Pi_{(\ell^1,v)} L$ for every $n \in \mathbb{N}$. Moreover, if $0 < t < 1$, then $tx \in \Pi_{(\ell^1,v)} L$ and $\|tx\|_v < \|x\|_v$. For every $a > 0$, choose $n \in \mathbb{N}$ so that $a < n$. Then, $ax = (a/n)(nx) \in \Pi_{(\ell^1,v)} L$ and $(-a)x = -(ax) \in \Pi_{(\ell^1,v)} L$. Thus, $\Pi_{(\ell^1,v)} L$ is a linear space. It should be remarked that $\|\cdot\|_v$ is not a norm, but an F-norm (cf. Sect. 1.4), which induces the linear metric $d(x, y) = \|x - y\|_v$. We regard $\Pi_{(\ell^1,v)} L$ as a metric linear space with this metric. Then, each projection $\mathrm{pr}_n : \Pi_{(\ell^1,v)} L \to L$ is continuous.

[9] In [138] and [139], r is called a **regular retraction**.

Lemma 2.2.4 *For any sub-additive homeomorphisms* $v : \mathbb{R}_+ \to \mathbb{R}_+$, $i \in \mathbb{N}$,

$$\big(\Pi_{(\ell^1,v)} L, \Pi_{(\ell^1_f,v)} L\big) \approx \big(\Pi_{\ell^1} L, \Pi_{\ell^1_f} L\big).^{10}$$

Proof We define

$$\varphi : \big(\Pi_{(\ell^1,v)} L,\ \Pi_{(\ell^1_f,v)} L\big) \to \big(\Pi_{\ell^1} L,\ \Pi_{\ell^1_f} L\big) \quad \text{and}$$
$$\psi : \big(\Pi_{\ell^1} L,\ \Pi_{\ell^1_f} L\big) \to \big(\Pi_{(\ell^1,v)} L,\ \Pi_{(\ell^1_f,v)} L\big)$$

as follows:

$$\varphi(x)(i) = \begin{cases} \dfrac{2^i v(\|x(i)\|)}{\|x(i)\|} x(i) & \text{if } x(i) \neq \mathbf{0}, \\ \mathbf{0} & \text{if } x(i) = \mathbf{0}; \end{cases}$$

$$\psi(x)(i) = \begin{cases} \dfrac{v^{-1}(2^{-i}\|x(i)\|)}{\|x(i)\|} x(i) & \text{if } x(i) \neq \mathbf{0}, \\ \mathbf{0} & \text{if } x(i) = \mathbf{0}. \end{cases}$$

Since $\|\varphi(x)(i)\| = 2^i v(\|x(i)\|)$ and $2^i v(\|\psi(x)(i)\|) = \|x(i)\|$, it follows that φ and ψ are well-defined, $\varphi\psi = \mathrm{id}$, and $\psi\varphi = \mathrm{id}$. It remains to verify the continuity of φ and ψ.

(*Continuity of* φ) For each $x \in \Pi_{(\ell^1,v)} L$ and $\varepsilon > 0$, choose $m \in \mathbb{N}$ so that $\sum_{i \geqslant m} 2^i v(\|x(i)\|) < \varepsilon/4$. Since each $\mathrm{pr}_i\varphi$ is continuous, we have $\delta > 0$ such that

$$y \in \Pi_{(\ell^1,v)} L,\ \|x - y\|_v < \delta \Rightarrow \sum_{i<m} \|\varphi(x)(i) - \varphi(y)(i)\| < \varepsilon/4.$$

For $y \in \Pi_{(\ell^1,v)} L$, if $\|x - y\|_v < \varepsilon/4$ then

$$\begin{aligned}\sum_{i \geqslant m} 2^i v(\|y(i)\|) &\leqslant \sum_{i \geqslant m} 2^i v(\|x(i)\|) + \sum_{i \geqslant m} 2^i v(\|x(i) - y(i)\|) \\ &< \varepsilon/4 + \|x - y\|_v < \varepsilon/2.\end{aligned}$$

---

[10] More generally, replacing $2^i v$, $i \in \mathbb{N}$, with sub-additive homeomorphisms $v_i : \mathbb{R}_+ \to \mathbb{R}_+$, we have F-normed spaces $L_i = (L, v_i(\|\cdot\|))$ and their ℓ^1-product $\Pi_{\ell^1} L_i$ and ℓ^1_f-product $\Pi_{\ell^1_f} L_i$ (see Remark 2.1). Then, we can also prove the following:

$$\big(\Pi_{\ell^1} L_i,\ \Pi_{\ell^1_f} L_i\big) \approx \big(\Pi_{\ell^1} L,\ \Pi_{\ell^1_f} L\big).$$

Hence, $\|x - y\|_v < \min\{\delta, \varepsilon/4\}$ implies

$$\begin{aligned}\|\varphi(x) - \varphi(y)\|_1 &\leqslant \sum_{i<m} \|\varphi(x)(i) - \varphi(y)(i)\| \\ &\qquad + \sum_{i\geqslant m} \|\varphi(x)(i)\| + \sum_{i\geqslant m} \|\varphi(y)(i)\| \\ &< \varepsilon/4 + \sum_{i\geqslant m} 2^i v(\|x(i)\|) + \sum_{i\geqslant m} 2^i v(\|y(i)\|) \\ &< \varepsilon/4 + \varepsilon/4 + \varepsilon/2 = \varepsilon.\end{aligned}$$

(*Continuity of* ψ) For each $x \in \Pi_{\ell^1} L$ and $\varepsilon > 0$, choose $m \in \mathbb{N}$ so that $\sum_{i\geqslant m} \|x(i)\| < \varepsilon/4$. From each $\mathrm{pr}_i \psi$ and $2^i v$ are continuous, we have $\delta > 0$ such that

$$y \in \Pi_{\ell^1} L,\ \|x - y\|_1 < \delta \Rightarrow \sum_{i<m} 2^i v(\|\psi(x)(i) - \psi(y)(i)\|) < \varepsilon/4.$$

For $y \in \Pi_{\ell^1} L$, if $\|x - y\|_1 < \varepsilon/4$ then

$$\sum_{i\geqslant m} \|y(i)\| \leqslant \sum_{i\geqslant m} \|x(i)\| + \sum_{i\geqslant m} \|x(i) - y(i)\| < \varepsilon/4 + \|x - y\|_1 < \varepsilon/2.$$

Hence, $\|x - y\|_1 < \min\{\delta, \varepsilon/4\}$ implies

$$\begin{aligned}\|\psi(x) - \psi(y)\|_v &\leqslant \sum_{i<m} 2^i v(\|\psi(x)(i) - \psi(y)(i)\|) \\ &\qquad + \sum_{i\geqslant m} 2^i v(\|\psi(x)(i)\|) + \sum_{i\geqslant m} 2^i v(\|\psi(y)(i)\|) \\ &< \varepsilon/4 + \sum_{i\geqslant m} \|x(i)\| + \sum_{i\geqslant m} \|y(i)\| \\ &< \varepsilon/4 + \varepsilon/4 + \varepsilon/2 = \varepsilon.\end{aligned}$$

□

Recall that $r : L \to X$ is a retraction that is uniformly continuous at X.

Lemma 2.2.5 *There exists a sub-additive homeomorphism* $v : \mathbb{R}_+ \to \mathbb{R}_+$ *and* $\gamma > 0$ *such that the following condition* $(*)$ *is satisfied:*

$(*)$ $x \in X,\ y \in L,\ v(\|y\|) < \gamma \Rightarrow \|y\| \leqslant v(\|y\|),\ v(\|x - r(x + y)\|) \leqslant 2v(\|y\|)$.

Proof First, we define $w : \mathbb{R}_+ \to \mathbb{R}_+$ as follows:

$$w(s) = \sup\left\{\|x - r(x + y)\| \;\middle|\; x \in X,\ \|y\| \leqslant s\right\}.$$

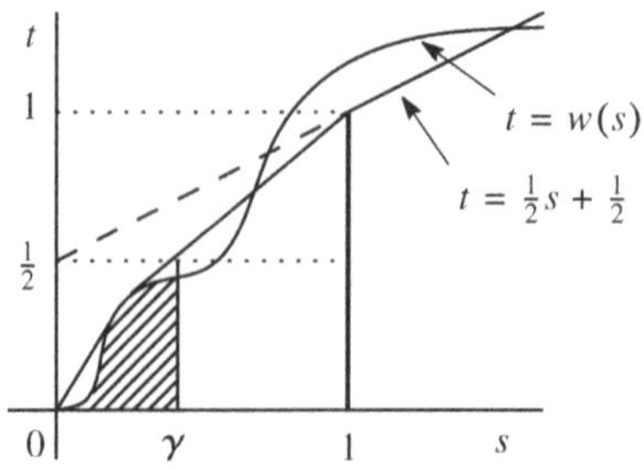

Fig. 2.4 Definition of u

Then, $w^{-1}(0) = \{0\}$. Indeed, $w(0) = 0$ is obvious. Recall that $\mathbf{0} \in X$ and $X \setminus \{\mathbf{0}\} \neq \emptyset$. By the path-connectedness of X, for each $s > 0$, we can find $y \in X \setminus \{0\}$ with $\|y\| \leqslant s$, hence $w(s) \geqslant \|r(y)\| = \|y\| > 0$. Moreover, by the uniform continuity of r at X, it can be easily seen that w is continuous at 0. In particular, we have $\gamma \in (0, 1)$ such that $w(s) < 1/2$ for $0 \leqslant s < \gamma$. Let K be the convex hull of the following set in $\mathbb{R}^2$ (see Fig. 2.4):[11]

$$\bigcup_{s\in[0,\gamma)} \big(\{s\} \times [0, w(s)]\big) \cup \big(\{1\} \times [0, 1]\big).$$

We define $u : \mathbb{R}_+ \to \mathbb{R}_+$ as follows:

$$u(s) = \begin{cases} \sup\big\{t \bigm| (s, t) \in K\big\} & \text{for } 0 \leqslant s \leqslant 1, \\ \dfrac{1}{2}s + \dfrac{1}{2} & \text{for } s \geqslant 1. \end{cases}$$

Then, u is a sub-additive homeomorphism. Indeed, by definition, we have $u^{-1}(0) = \{0\}$. It is easy to see that $u|\mathbf{I}$ is strictly increasing and continuous, so it is a homeomorphism of $\mathbf{I}$ onto itself, hence u is a homeomorphism. It follows from the convexity of $\bigcup_{s\in\mathbb{R}_+}\{s\} \times [0, u(s)]$ that

$$\frac{u(s+t)}{s+t}s \leqslant u(s) \quad \text{and} \quad \frac{u(s+t)}{s+t}t \leqslant u(t),$$

[11]Recall that the convex hull of a set A in a linear space E is denoted by $\langle A\rangle$. Using this notation, we can write as follows:

$$K = \Big\langle \bigcup_{s\in[0,\gamma)} \big(\{s\} \times [0, w(s)]\big) \cup \big(\{1\} \times [0, 1]\big)\Big\rangle.$$

hence $u(s+t) \leqslant u(s) + u(t)$ for each $s, t \in \mathbb{R}_+$. Moreover, observe that

$$w(s) \leqslant u(s) \quad \text{for } 0 \leqslant s < \gamma.$$

Since $s \leqslant u(s) \leqslant 1$ for $0 \leqslant s \leqslant 1$, it follows that

$$s \leqslant u(s) \leqslant u^2(s) \leqslant \cdots \leqslant 1 \quad \text{for } 0 \leqslant s \leqslant 1.$$

Since $s \geqslant u(s) \geqslant 1$ for $s \geqslant 1$, it follows that

$$s \geqslant u(s) \geqslant u^2(s) \geqslant \cdots \geqslant 1 \quad \text{for } s \geqslant 1.$$

Now, we can define a homeomorphism $v : \mathbb{R}_+ \to \mathbb{R}_+$ by $v = \sum_{i\in\mathbb{N}} 2^{-i}u^i$, where $u^1 = u$ and $u^{i+1} = u \circ u^i$. To see the condition $(*)$, let $x \in X$ and $y \in L$ with $v(\|y\|) < \gamma$ (< 1). Then $\|y\| \leqslant u(\|y\|) \leqslant v(\|y\|) < \gamma$, so we have

$$\|x - r(x+y)\| \leqslant w(\|y\|) \leqslant u(\|y\|),$$

which implies

$$\begin{aligned} v(\|x - r(x+y)\|) \leqslant v(u(\|y\|)) &= \sum_{i\in\mathbb{N}} 2^{-i}u^{i+1}(\|y\|) \\ &= 2\sum_{i=2}^{\infty} 2^{-i}u^i(\|y\|) \leqslant 2v(\|y\|). \end{aligned}$$

This completes the proof. □

Using the sub-additive homeomorphism v with the condition $(*)$, we define $f : X \times \Pi_{(\ell^1,v)}L \to \Pi_{(\ell^1,v)}L$ as follows:

$$\begin{aligned} f(x,y) = (y(1) + x,\ y(2) + x - r(y(1) + x), \ldots, \\ y(i) + x - r(y(i-1) + x), \ldots). \end{aligned}$$

Then, the following is easily observed:

$$f\big(X \times \Pi_{(\ell^1_f,v)}L\big) \subset \Pi_{(\ell^1_f,v)}L.$$

Lemma 2.2.6 *The function f is well-defined and continuous.*

Proof (*Well-Definedness of f*) Let $(x, y) \in X \times \Pi_{(\ell^1,v)}L$. We can take $m \in \mathbb{N}$ so that $v(\|y(i)\|) < \gamma$ for $i \geqslant m$ because $\sum_{i=1}^{\infty} 2^i v(\|y(i)\|) < \infty$. By condition $(*)$, we have

$$v(\|x - r(y(i-1) + x)\|) \leqslant 2v(\|y(i-1)\|) \quad \text{for each } i > m.$$

Then, it follows that

$$
\begin{aligned}
&2v(\|y(1)+x\|)+\sum_{i=2}^{\infty}2^i v(\|y(i)+x-r(y(i-1)+x)\|)\\
&\quad\leqslant 2v(\|y(1)+x\|)+\sum_{i=2}^{m}2^i v(\|y(i)+x-r(y(i-1)+x)\|)\\
&\qquad+\sum_{i=m+1}^{\infty}2^i v(\|y(i)\|)+4\sum_{i=m+1}^{\infty}2^{i-1}v(\|y(i-1)\|)<\infty,
\end{aligned}
$$

that is, $f(x,y)\in\Pi_{(\ell^1,v)}L$.

(*Continuity of f*) Let $(x,y)\in X\times\Pi_{(\ell^1,v)}L$ and $\varepsilon>0$. Since $y\in\Pi_{(\ell^1,v)}L$, we can choose $m\in\mathbb{N}$ so that

$$
v(\|y(i)\|)<\gamma/2 \text{ for each } i\geqslant m \quad\text{and}\quad \sum_{i=m}^{\infty}2^i v(\|y(i)\|)<\varepsilon/30.
$$

Since each $\mathrm{pr}_i f$ is continuous, we can find $\delta>0$ such that

$$
\begin{aligned}
&\|x-x'\|<\delta \text{ and } \|y-y'\|_v<\delta\\
&\qquad\Rightarrow\ 2v(\|(y(1)+x)-(y'(1)+x')\|)\\
&\qquad\qquad+\sum_{i=2}^{m}2^i v(\|(y(i)+x-r(y(i-1)+x))\\
&\qquad\qquad\qquad-(y'(i)+x'-r(y'(i-1)+x'))\|)<\varepsilon/3.
\end{aligned}
$$

Note that if $\|y-y'\|_v<\gamma/2$ and $i\geqslant m$, then

$$
v(\|y'(i)\|)\leqslant v(\|y(i)\|)+v(\|y(i)-y'(i)\|)<\gamma.
$$

Now, assume that $\|x-x'\|<\delta$ and $\|y-y'\|_v<\min\{\delta,\gamma/2,\varepsilon/15\}$. Then, using condition $(*)$, we can show that $\|f(x,y)-f(x',y')\|_v<\varepsilon$ as follows:

$$
\begin{aligned}
\|f(x,y)-f(x',y')\|_v<\varepsilon/3+&\sum_{i=m+1}^{\infty}2^i v(\|y(i)+x-r(y(i-1)+x)\|)\\
&+\sum_{i=m+1}^{\infty}2^i v(\|y'(i)+x'-r(y'(i-1)+x')\|)
\end{aligned}
$$

$$
\begin{aligned}
&\leqslant \varepsilon/3 + \sum_{i=m+1}^{\infty} 2^i v(\|y(i)\|) + 4 \sum_{i=m+1}^{\infty} 2^{i-1} v(\|y(i-1)\|) \\
&\quad + \sum_{i=m+1}^{\infty} 2^i v(\|y'(i)\|) + 4 \sum_{i=m+1}^{\infty} 2^{i-1} v(\|y'(i-1)\|) \\
&\leqslant \varepsilon/3 + 5 \sum_{i=m}^{\infty} 2^i v(\|y(i)\|) + 5 \sum_{i=m}^{\infty} 2^i v(\|y'(i)\|) \\
&\leqslant \varepsilon/3 + 10 \sum_{i=m}^{\infty} 2^i v(\|y(i)\|) + 5 \sum_{i=m}^{\infty} 2^i v(\|y(i) - y'(i)\|) \\
&< \varepsilon/3 + \varepsilon/3 + \varepsilon/3 = \varepsilon.
\end{aligned}
$$

Therefore, f is continuous. □

By induction, we define $g_n : \Pi_{(\ell^1,v)} L \to L$, $n \in \mathbb{N}$, as follows:

$$g_1(z) = z(1),\ g_2(z) = z(2) + r(z(1)), \ldots,\ g_n(z) = z(n) + r g_{n-1}(z), \ldots,$$

that is, $g_1 = \mathrm{pr}_1$ and $g_n = \mathrm{pr}_n + r \circ g_{n-1}$ for $n > 1$.

Since each projection $\mathrm{pr}_n : \Pi_{(\ell^1,v)} L \to L$ is continuous, the lemma below can be proved by induction:

Lemma 2.2.7 *Each $g_n : \Pi_{(\ell^1,v)} L \to L$ is continuous.* □

For each $z \in \Pi_{(\ell^1_f,v)} L$, there is some $m \in \mathbb{N}$ such that $z(i) = 0$ for $i \geqslant m$, hence $g_m(z) = r g_{m-1}(z) \in X$ and $g_{n+1}(z) = r g_n(z) = g_n(z)$ for $n \geqslant m$. Thus, $\big(g_n | \Pi_{(\ell^1_f,v)} L\big)_{n \in \mathbb{N}}$ converges to $g'_\infty : \Pi_{(\ell^1_f,v)} L \to X$. When X is $\| \cdot \|$-complete, g'_∞ can be extended to $g_\infty : \Pi_{(\ell^1,v)} L \to X$. In fact, by the next lemma, g_∞ can be defined as follows:

$$g_\infty(z) = \lim_{i \to \infty} r g_i(z) \quad \text{for each } z \in \Pi_{(\ell^1,v)} L.$$

Lemma 2.2.8 *For each $z \in \Pi_{(\ell^1,v)} L$, $(r g_n(z))_{n \in \mathbb{N}}$ is $\| \cdot \|$-Cauchy.*

Proof For each $\varepsilon > 0$, choose $m \in \mathbb{N}$ so that

$$\sum_{i=m+1}^{\infty} 2^i v(\|z(i)\|) < \min\{\varepsilon, \gamma\}.$$

Then, $v(\|z(i)\|) < \gamma$ for $i > m$. For each $n' > n \geqslant m$,

$$\begin{aligned}\|rg_n(z) - rg_{n'}(z)\| &\leqslant v(\|rg_n(z) - rg_{n'}(z)\|)\\ &\leqslant v(\|rg_n(z) - rg_{n+1}(z)\|) + \cdots\\ &\qquad + v(\|rg_{n'-1}(z) - rg_{n'}(z)\|)\\ &= v(\|rg_n(z) - r(z(n+1) + rg_n(z))\|) + \cdots\\ &\qquad + v(\|rg_{n'-1}(z) - r(z(n') + rg_{n'-1}(z))\|)\\ &\leqslant 2v(\|z(n+1)\|) + \cdots + 2v(\|z(n')\|)\\ &\leqslant \sum_{i=m+1}^{\infty} 2^i v(\|z(i)\|) < \min\{\varepsilon, \gamma\} \leqslant \varepsilon.\end{aligned}$$

Therefore, $(rg_n(z))_{n\in\mathbb{N}}$ is $\|\cdot\|$-Cauchy. □

From the proof of Lemma 2.2.8, the following can also be obtained:

Lemma 2.2.9 *When X is $\|\cdot\|$-complete, if $v(\|z(i)\|) < \gamma$ for $i > m$, then*

$$\begin{aligned}\|rg_m(z) - g_\infty(z)\| &\leqslant v(\|rg_m(z) - g_\infty(z)\|)\\ &\leqslant 2\sum_{i=m+1}^{\infty} v(\|z(i)\|) \left(\leqslant \sum_{i=m+1}^{\infty} 2^i v(\|z(i)\|)\right).\end{aligned}$$ □

Now, we can prove the continuity of g'_∞ and g_∞.

Lemma 2.2.10 *The function g'_∞ is continuous. When X is $\|\cdot\|$-complete, g_∞ is also continuous.*

Proof Because of their similarity, we only prove the continuity of g_∞. Let $z \in \Pi_{(\ell^1,v)}L$ and $\varepsilon > 0$. Choose $m \in \mathbb{N}$ so that

$$v(\|g_\infty(z) - rg_m(z)\|),\ \sum_{i=m}^{\infty} 2^i v(\|z(i)\|) < \min\left\{\varepsilon/4,\ \gamma\right\}.$$

Then, $\|g_\infty(z) - rg_m(z)\| < \varepsilon/4$ and $v(\|z(i)\|) < \gamma/2$ for $i \geqslant m$. Since the continuity of rg_m follows from Lemma 2.2.7, we can find $\delta > 0$ such that

$$\|z - z'\|_v < \delta \Rightarrow \|rg_m(z) - rg_m(z')\| < \varepsilon/4.$$

If $\|z - z'\|_v < \gamma/2$ and $i \geqslant m$, then

$$v(\|z'(i)\|) \leqslant v(\|z(i)\|) + v(\|z(i) - z'(i)\|) < \gamma.$$

On the other hand, if $\|z - z'\|_v < \varepsilon/4$, then

$$\sum_{i=m+1}^{\infty} 2^i v(\|z'(i)\|) \leqslant \sum_{i=m+1}^{\infty} 2^i v(\|z(i)\|) + \|z - z'\|_v < \varepsilon/4 + \varepsilon/4 = \varepsilon/2.$$

Hence, by Lemma 2.2.9, if $\|z - z'\|_v < \min\{\delta,\ \gamma/2,\ \varepsilon/4\}$, then

$$\begin{aligned}\|g_\infty(z) - g_\infty(z')\| &\leqslant \|g_\infty(z) - rg_m(z)\| + \|rg_m(z) - rg_m(z')\| \\ &\qquad + \|g_\infty(z') - rg_m(z')\| \\ &< \varepsilon/4 + \varepsilon/4 + \sum_{i=m+1}^{\infty} 2^i v(\|z'(i)\|) < \varepsilon/2 + \varepsilon/2 = \varepsilon.\end{aligned}$$

Therefore, g_∞ is continuous. □

We define $g' : \Pi_{(\ell^1_f, v)} L \to X \times \Pi_{(\ell^1_f, v)} L$ as follows:

$$g'(z) = (g'_\infty(z),\ g_1(z) - g'_\infty(z), \ldots,\ g_i(z) - g'_\infty(z), \ldots).$$

When X is $\|\cdot\|$-complete, $g : \Pi_{(\ell^1, v)} L \to X \times \Pi_{(\ell^1, v)} L$ is defined by

$$g(z) = (g_\infty(z),\ g_1(z) - g_\infty(z), \ldots,\ g_i(z) - g_\infty(z), \ldots).$$

Then, g is an extension of g'.

Lemma 2.2.11 *The function g' is continuous. When X is $\|\cdot\|$-complete, g is well-defined and continuous.*

Proof Since the continuity of g' can be proved in the same way as for g, we only prove that g is well-defined and continuous. Evidently, it suffices to show the well-definedness and continuity of $pg : \Pi_{(\ell^1, v)} L \to \Pi_{(\ell^1, v)} L$, where $p : X \times \Pi_{(\ell^1, v)} L \to \Pi_{(\ell^1, v)} L$ is the projection.

(*Well-Definedness of pg*) For each $z \in \Pi_{(\ell^1, v)} L$, choose $m \in \mathbb{N}$ so that $\sum_{i=m}^{\infty} 2^i v(\|z(i)\|) < \gamma$. Since $v(\|z(i)\|) < \gamma$ for $i \geqslant m$, it follows from Lemma 2.2.9 that

$$\begin{aligned}\sum_{i=m}^{\infty} 2^i v(\|g_i(z) - g_\infty(z)\|) &\leqslant \sum_{i=m}^{\infty} 2^i v(\|g_i(z) - rg_{i-1}(z)\|) \\ &\qquad + \sum_{i=m}^{\infty} 2^i v(\|g_\infty(z) - rg_{i-1}(z)\|)\end{aligned}$$

$$\leqslant \sum_{i=m}^{\infty} 2^i v(\|z(i)\|) + 2\sum_{i=m}^{\infty} 2^i \Big(\sum_{j=i}^{\infty} v(\|z(j)\|)\Big)$$

$$< \gamma + 2\sum_{j=m}^{\infty} \Big(\sum_{i=m}^{j} 2^i\Big) v(\|z(j)\|)$$

$$\leqslant \gamma + 4\sum_{j=m}^{\infty} 2^j v(\|z(j)\|) < \gamma + 4\gamma = 5\gamma < \infty.$$

Therefore, pg is well-defined.

(*Continuity of pg*) For each $z \in \Pi_{(\ell^1,v)}L$ and $\varepsilon > 0$, choose $m \in \mathbb{N}$ so that

$$\sum_{i=m}^{\infty} 2^i v(\|z(i)\|) < \min\{\varepsilon/20,\ \gamma\}.$$

Then, $v(\|z(i)\|) < \gamma/2$ for $i \geqslant m$. Replacing γ by $\min\{\varepsilon/20,\ \gamma\}$ in the proof of the well-definedness of pg above, we have

$$\sum_{i=m}^{\infty} 2^i v(\|g_i(z) - g_\infty(z)\|) < \varepsilon/4.$$

If $\|z - z'\|_v < \varepsilon/20$, then

$$\sum_{i=m}^{\infty} 2^i v(\|z'(i)\|) \leqslant \sum_{i=m}^{\infty} 2^i v(\|z(i)\|) + \sum_{i=m}^{\infty} 2^i v(\|z(i) - z'(i)\|)$$
$$< \varepsilon/20 + \|z - z'\|_v < \varepsilon/10.$$

If $\|z - z'\|_v < \gamma/2$ and $i \geqslant m$, then

$$v(\|z'(i)\|) \leqslant v(\|z(i)\|) + v(\|z(i) - z'(i)\|) < \gamma/2 + \|z - z'\|_v < \gamma.$$

Thus, $\|z - z'\|_v < \min\{\varepsilon/20,\ \gamma/2\}$ implies

$$\sum_{i=m}^{\infty} 2^i v(\|z'(i)\|) < \varepsilon/10 \quad \text{and} \quad v(\|z'(i)\|) < \gamma \ \text{ for } i \geqslant m.$$

In the same way as above, we can obtain

$$\sum_{i=m}^{\infty} 2^i v(\|g_i(z') - g_\infty(z')\|) < \varepsilon/2.$$

By the continuity of v, g_∞ and $g_i - g_\infty$, $i = 1, \ldots, m-1$, we can find $\delta > 0$ such that

$$\|z - z'\|_v < \delta \Rightarrow \sum_{i=1}^{m-1} 2^i v(\|(g_i(z) - g_\infty(z)) - (g_i(z') - g_\infty(z'))\|) < \varepsilon/4.$$

Hence, if $\|z - z'\|_v < \min\{\delta,\ \varepsilon/20,\ \gamma/2\}$, then

$$\begin{aligned}\|pg(z) - pg(z')\|_v &\leqslant \sum_{i=1}^{m-1} 2^i v\big(\|(g_i(z) - g_\infty(z)) - (g_i(z') - g_\infty(z'))\|\big)\\ &\qquad + \sum_{i=m}^{\infty} 2^i v(\|g_i(z) - g_\infty(z)\|) + \sum_{i=m}^{\infty} 2^i v(\|g_i(z') - g_\infty(z')\|)\\ &< \varepsilon/4 + \varepsilon/4 + \varepsilon/2 = \varepsilon.\end{aligned}$$

Therefore, pg is continuous. □

Now, we can prove Theorem 2.2.3.

Proof (Theorem 2.2.3) For each $x \in \Pi_{(\ell_f^1, v)} L$,

$$\begin{aligned}fg'(x) &= f(g'_\infty(x),\ g_1(x) - g'_\infty(x),\ g_2(x) - g'_\infty(x), \ldots)\\ &= (g_1(x),\ g_2(x) - rg_1(x),\ g_3(x) - rg_2(x), \ldots)\\ &= (x(1), x(2), x(3), \ldots) = x,\end{aligned}$$

that is, $fg' = \mathrm{id}$. For each $(x, y) \in X \times \Pi_{(\ell_f^1, v)} L$,

$$g_1(f(x, y)) = y(1) + x,\ g_2(f(x, y)) = y(2) + x, \ldots,$$

so $g'_\infty(f(x, y)) = g_n(f(x, y)) = y(n) + x$ for sufficiently large $n \in \mathbb{N}$. Since $y(n) = 0$ for sufficiently large $n \in \mathbb{N}$, we have $g'_\infty(f(x, y)) = x$. Then, it follows that

$$\begin{aligned}g' f(x, y) &= (g'_\infty(f(x, y)),\ g_1(f(x, y)) - g'_\infty(f(x, y)),\\ &\qquad\qquad g_2(f(x, y)) - g'_\infty(f(x, y)), \ldots)\\ &= (x, y(1), y(2), \ldots) = (x, y).\end{aligned}$$

Hence, $g' f|\Pi_{(\ell_f^1, v)} L = \mathrm{id}$. Thus, we have $X \times \Pi_{(\ell_f^1, v)} L \approx \Pi_{(\ell_f^1, v)} L$.

If X is $\|\cdot\|$-complete, then $fg = \mathrm{id}$ and $gf = \mathrm{id}$ because $\Pi_{(\ell^1_f,v)}L$ is dense in $\Pi_{(\ell^1,v)}L$. In this case, we have

$$\left(X \times \Pi_{(\ell^1,v)}L, X \times \Pi_{(\ell^1_f,v)}L\right) \approx \left(\Pi_{(\ell^1,v)}L, \Pi_{(\ell^1_f,v)}L\right).$$

Then, Theorem 2.2.3 follows from Lemma 2.2.4. □

As a corollary of Theorem 2.2.3, we have the following result on completely metrizable ARs, which is called the TORUŃCZYK FACTOR THEOREM:

Theorem 2.2.12 (TORUŃCZYK) *For every non-degenerate completely metrizable AR X, $X \times \ell^1(\Gamma) \approx \ell^1(\Gamma)$, where* $\operatorname{dens} X = \operatorname{card}\Gamma$.[12] *In general, for an arbitrary non-degenerate AR X, there exists a normed linear space L such that* $\operatorname{dens} X = \operatorname{dens} L$ *and $X \times L \approx L$.*

Proof In the completely metrizable case, by virtue of Theorem 1.13.37, we have an admissible complete metric d of X such that (X, d) is a uniform AR. By the Arens–Eells Embedding Theorem 1.13.2, $X = (X, d)$ can be embedded isometrically into some Banach space L with $\operatorname{dens} X = \operatorname{dens} L$ as a closed set. Then, there is a retraction $r : L \to X$ that is uniformly continuous at X. By Theorem 2.2.3, $X \times \Pi_{\ell^1}L \approx \Pi_{\ell^1}L$. Since $\Pi_{\ell^1}L$ is a Banach space, we can apply Corollary 1.4.18 to obtain a Banach space F such that $\left(\Pi_{\ell^1}L\right) \times F \approx \ell^1(\Gamma)$, where

$$\operatorname{card}\Gamma = \operatorname{dens}\Pi_{\ell^1}L = \operatorname{dens} L = \operatorname{dens} X.$$

Hence, we have the first statement.

Without assuming the completeness of $X = (X, d)$, X can be embedded isometrically into some normed linear space L with $\operatorname{dens} X = \operatorname{dens} L$ as a closed set. In this case, we have $X \times \Pi_{\ell^1_f}L \approx \Pi_{\ell^1_f}L$. Thus, the general case also holds. □

Because $\ell^1(\Gamma) \approx \ell^2(\Gamma)$ (Theorem 1.2.3), $\ell^1(\Gamma)$ in Theorem 2.2.12 can be replaced with $\ell^2(\Gamma)$. Since $\ell^2(\Gamma)^{\mathbb{N}}$ is a completely metrizable AR, we have

$$\ell^2(\Gamma)^{\mathbb{N}} \approx \ell^2(\Gamma)^{\mathbb{N}} \times \ell^2(\Gamma) \approx \ell^2(\Gamma).$$

Thus, we have the following:

Corollary 2.2.13 *For an arbitrary infinite set Γ, $\ell^2(\Gamma)^{\mathbb{N}} \approx \ell^2(\Gamma)$.* □

Recall that the **metrizable cone** $C(X)$ over a metrizable space X is the space $\{0\} \cup (X \times (0, 1])$ with the topology generated by open sets in the product space $X \times (0, 1]$ and sets $\{0\} \cup (X \times (0, \varepsilon))$, $\varepsilon \in (0, 1)$. Then, $C(X)$ has the same density as X if X is not finite. If X is compact, then $C(X)$ is homeomorphic to the usual cone C_X that is defined as the quotient space $(X \times \mathbf{I})/(X \times \{0\})$. The **metrizable**

[12] Note that $\operatorname{dens} X$ is infinite because X is non-degenerate.

open cone over X is the subspace $\{0\} \cup (X \times (0, 1))$ of $C(X)$ and denoted by $C^o(X)$. It should be noted that if X is completely metrizable, then so are $C(X)$ and $C^o(X)$. Moreover, the cone $C(X)$ and the open cone $C^o(X)$ over an ANR X are ARs by Corollary 1.13.18.

By using the metrizable open cone, we can easily prove the following theorem, which is also called the TORUŃCZYK FACTOR THEOREM:

Theorem 2.2.14 (TORUŃCZYK) *For every completely metrizable ANR X with* $\operatorname{dens} X \leqslant \operatorname{card}\Gamma$, *$X \times \ell^2(\Gamma)$ is an $\ell^2(\Gamma)$-manifold. In fact, $X \times \ell^2(\Gamma)$ is homeomorphic to an open set in $\ell^2(\Gamma)$.*

Proof As mentioned above, the open cone $C^o(X)$ is a completely metrizable AR and clearly $\operatorname{dens} C^o(X) \leqslant \operatorname{card}\Gamma$. Then, $C^o(X) \times \ell^2(\Gamma) \approx \ell^2(\Gamma)$ by Theorem 2.2.12. Since $\ell^2(\Gamma) \times (0, 1) \approx \ell^2(\Gamma)$, $X \times \ell^2(\Gamma)$ is homeomorphic to the open set $(0, 1) \times X \times \ell^2(\Gamma)$ in $C^o(X) \times \ell^2(\Gamma)$. Thus, $X \times \ell^2(\Gamma)$ can be embedded in $\ell^2(\Gamma)$ as an open set, hence it is an $\ell^2(\Gamma)$-manifold. □

For a locally finite-dimensional simplicial complex K, $|K|_{\mathrm{m}}$ is a completely metrizable ANR by Theorems 1.13.4 and 1.7.2. Recall that $\ell^1(\Gamma) \approx \ell^2(\Gamma)$ (Theorem 1.2.3). Hence, we have the following corollary:

Corollary 2.2.15 *For every locally finite-dimensional simplicial complex K with* $\operatorname{card} K^{(0)} \leqslant \operatorname{card}\Gamma$, *$|K|_{\mathrm{m}} \times \ell^2(\Gamma)$ is an $\ell^2(\Gamma)$-manifold.* □

For the subspace $\ell^2_f(\Gamma)$ of $\ell^2(\Gamma)$, we have the following result:

Proposition 2.2.16 *For every full simplicial complex F with* $\operatorname{card} F^{(0)} \leqslant \operatorname{card}\Gamma$, $|F|_{\mathrm{m}} \times \ell^2_f(\Gamma) \approx \ell^2_f(\Gamma)$.

Proof Identifying $F^{(0)}$ is a subset of the unit vectors of $\ell^1_f(\Gamma)$, we can regard $|F|_{\mathrm{m}} \subset \ell^1_f(\Gamma)$. Then, $|F|$ is a convex set $\ell^1_f(\Gamma)$ and the metric ρ_F is induced by the norm of $\ell^1_f(\Gamma)$. Because $(|F|, \rho_F)$ is a uniform AR by Theorem 1.13.31, there is a retraction $r : \ell^1_f(\Gamma) \to |F|$ that is uniformly continuous at $|F|$. Applying Theorem 2.2.3, we have $|F|_{\mathrm{m}} \times \Pi_{\ell^1_f} \ell^1_f(\Gamma) \approx \Pi_{\ell^1_f} \ell^1_f(\Gamma)$. Since $\Pi_{\ell^1_f} \ell^1_f(\Gamma) \approx \ell^1_f(\Gamma) \approx \ell^2_f(\Gamma)$ by Proposition 2.2.2 and Theorem 1.2.3, it follows that $|F|_{\mathrm{m}} \times \ell^2_f(\Gamma) \approx \ell^2_f(\Gamma)$. □

Moreover, we have the following:

Proposition 2.2.17 *For every locally finite simplicial complex K with* $\operatorname{card} K^{(0)} \leqslant \operatorname{card}\Gamma$, *$|K| \times \ell^2_f(\Gamma)$ is an $\ell^2_f(\Gamma)$-manifold. If K is a contractible finite simplicial complex, then $|K| \times \ell^2_f(\Gamma) \approx \ell^2_f(\Gamma)$.*[13]

Proof First, we show the case where K is a contractible finite simplicial complex. As in the proof of Proposition 2.2.16, we can regard $|K| \subset \ell^1_f(\Gamma)$ and the metric

[13] $|K|_{\mathrm{m}} = |K|$ for a locally finite simplicial complex K (Proposition 1.7.1).

ρ_K is induced by the norm of $\ell_f^1(\Gamma)$. Since $(|K|, \rho_K)$ is a compact AR (hence a uniform AR), there is a retraction $r : \ell_f^1(\Gamma) \to |K|$ that is uniformly continuous at $|K|$. By the same argument as the proof of Proposition 2.2.16, we can obtain $|K| \times \ell_f^2(\Gamma) \approx \ell_f^2(\Gamma)$.

Now, we show the general case. For every $v \in K^{(0)}$, since $\mathrm{St}(v, K)$ is a contractible finite simplicial complex, $|\mathrm{St}(v, K)| \times \ell_f^2(\Gamma) \approx \ell_f^2(\Gamma)$ by the above case. Hence, $O_K(v) \times \ell_f^2(\Gamma)$ is homeomorphic to an open set in $\ell_f^2(\Gamma)$. Thus, $|K| \times \ell_f^2(\Gamma)$ is covered by open sets $O_K(v) \times \ell_f^2(\Gamma)$, $v \in K^{(0)}$. Namely, $|K| \times \ell_f^2(\Gamma)$ is an $\ell_f^2(\Gamma)$-manifold. □

Remark 2.2 Proposition 2.2.17 holds without the local finite condition, which will be proved in Chapter 5 (combination of Theorems 5.10.18, 5.5.1, Proposition 5.10.3(2)).

2.3 Stability and Deficiency

A map $f : X \to Y$ is called a **near-homeomorphism** provided, for each $\mathcal{U} \in \mathrm{cov}(Y)$, f is $\mathcal{U}$-close to a homeomorphism $h : X \to Y$. A space X is said to be ***E*-stable** if $X \times E \approx X$. A subset A of X said to be ***E*-deficient** if there is a homeomorphism $h : X \to X \times E$ such that $h(A) \subset X \times \{0\}$. In this section, it is proved that if X is E-stable, then the projection $\mathrm{pr}_X : X \times E \to X$ is a near-homeomorphism and every open set in X is also E-stable. Moreover, we prove that every E-manifold is E-stable, and observe some properties of E-deficient sets.

First, we introduce the reduced products, which will also be used in the following sections. Let X and Y be spaces and A be a closed set in X. The product of X and Y **reduced over** A is the space

$$(X \times Y)_A = \big((X \setminus A) \times Y\big) \cup A,$$

with the topology generated by open sets in the product space $(X \setminus A) \times Y$ and sets

$$\big((U \setminus A) \times Y\big) \cup (U \cap A),$$

where U is open in X. Then, the projection $\mathrm{pr}_X : X \times Y \to X$ is factored through $(X \times Y)_A$ into two natural maps $q_A : X \times Y \to (X \times Y)_A$ and $p_A : (X \times Y)_A \to X$ defined by

$$\begin{cases} q_A|(X \setminus A) \times Y = \mathrm{id}, \\ q_A|A \times Y = \mathrm{pr}_A \end{cases} \quad \text{and} \quad \begin{cases} p_A|(X \setminus A) \times Y = \mathrm{pr}_{X \setminus A}, \\ p_A|A = \mathrm{id}. \end{cases}$$

Thus, we have the following commutative diagram:

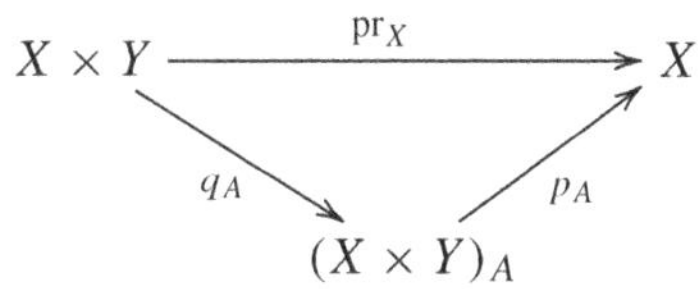

The topology of $(X \times Y)_A$ is the coarsest topology such that the map p_A is continuous and the product space $(X \setminus A) \times Y$ is its subspace. By the Bing Metrization Theorem 1.3.3, if X and Y are metrizable, then $(X \times Y)_A$ is also metrizable. In the case where Y is compact, $(X \times Y)_A$ is homeomorphic to the adjunction space $A \cup_{\mathrm{pr}_A} (X \times Y)$, where $\mathrm{pr}_A : A \times Y \to A$ is the projection. It should also be remarked that $(\mathbf{I} \times X)_{\{0\}}$ is naturally homeomorphic to the (metrizable) cone $C(X)$, so we sometimes regard $C(X) = (\mathbf{I} \times X)_{\{0\}}$.

To prove the stability of E-manifolds, our main tool is the coordinate-switching pseudo-isotopy. For simplicity, we use the following notational convention:

- Given a space X, let X^* denote $X \times E^{\mathbb{N}}$ or $X \times E_f^{\mathbb{N}}$ depending on $E \approx E^{\mathbb{N}}$ or $E \approx E_f^{\mathbb{N}}$, and let $p_0 : X^* \to X$ and $p_n : X^* \to X \times E^n$, $n \in \mathbb{N}$, be the projections. Additionally, let $X_0 = X \times \{\mathbf{0}\} \subset X^*$.

We define a homotopy $\theta^E : X^* \times E \times \mathbf{I} \to X^*$ as follows: For $(x, y_1, y_2, \dots) \in X^*$ ($x \in X$, $y_i \in E$, $i \in \mathbb{N}$) and $z \in E$,

$$
\begin{aligned}
\theta_1^E(x, y_1, y_2, \dots, z) &= (x, z, -y_1, -y_2, \dots), \\
\theta_{2^{-1}}^E(x, y_1, y_2, \dots, z) &= (x, y_1, z, -y_2, -y_3, \dots), \\
&\vdots \\
\theta_{2^{-n}}^E(x, y_1, y_2, \dots, z) &= (x, y_1, \dots, y_n, z, -y_{n+1}, -y_{n+2}, \dots), \\
\theta_{2^{-(n+1)}}^E(x, y_1, y_2, \dots, z) &= (x, y_1, \dots, y_n, y_{n+1}, z, , -y_{n+2}, \dots), \\
&\vdots \\
\theta_0^E(x, y_1, y_2, \dots, z) &= (x, y_1, y_2, \dots),
\end{aligned}
$$

and for $2^{-(n+1)} \leqslant t \leqslant 2^{-n}$, $n \in \omega$,

$$
\begin{aligned}
\theta_t^E(x, y_1, y_2, \dots, z) = (x, y_1, \dots, y_n,\ & z \cos \pi(1 - 2^n t) + y_{n+1} \sin \pi(1 - 2^n t), \\
& z \sin \pi(1 - 2^n t) - y_{n+1} \cos \pi(1 - 2^n t), \\
& - y_{n+2}, -y_{n+3}, \dots).
\end{aligned}
$$

Then, it should be remarked that:

($\natural$) $\theta_t^E(x, \mathbf{0}) = x$ for each $x \in X_0$ and $t \in \mathbf{I}$.
($\sharp$) $t \leqslant 2^{-n} \Rightarrow p_n\theta_t^E = p_n \mathrm{pr}_{X^*}$.

In addition, θ^E induces a homeomorphism $\tilde{\theta}^E : X^* \times E \times (0, 1] \to X^* \times (0, 1]$ defined by

$$\tilde{\theta}^E(x, y, z, t) = (\theta^E(x, y, z, t), t) \ \text{ for each } (x, y) \in X^*, z \in E, \text{and } t \in (0, 1].$$

This means that $(\theta^E)_t^{-1} : X^* \to X^* \times E$ is also continuous with respect to $t \in (0, 1]$. We call θ^E the **coordinate-switching pseudo-isotopy** for X^*.

Given a map $\tau : X^* \to \mathbf{I}$, we define a map $\theta_\tau^E : (X^* \times E)_{\tau^{-1}(0)} \to X^*$ by

$$\theta_\tau^E|\tau^{-1}(0) = \mathrm{id} \ \text{ and } \ \theta_\tau^E(w, z) = \theta_{\tau(w)}^E(w, z) \ \text{ if } \ \tau(w) \neq 0.$$

Then, by ($\natural$), the following is satisfied:

$$\theta_\tau^E(x, \mathbf{0}) = x \ \text{ for every } x \in X_0 \setminus \tau^{-1}(0).$$

The continuity of θ_τ^E at $w \in \tau^{-1}(0)$ follows from ($\sharp$). Indeed, each neighborhood U of $\theta_\tau^E(w) = w$ in X^* contains $p_n^{-1}(V)$, where V is an open neighborhood of $p_n(w)$ in $X \times E^n$. Then, the following is an open neighborhood of w in $(X^* \times E)_{\tau^{-1}(0)}$:

$$W = \big((p_n^{-1}(V) \cap \tau^{-1}((0, 2^{-n}))) \times E\big) \cup (p_n^{-1}(V) \cap \tau^{-1}(0)).$$

Since $p_n\theta_\tau^E|\tau^{-1}((0, 2^{-n})) = p_n \mathrm{pr}_{X^*}|\tau^{-1}((0, 2^{-n}))$ by ($\sharp$), we have

$$\theta_\tau^E((p_n^{-1}(V) \cap \tau^{-1}((0, 2^{-n}))) \times E) \subset p_n^{-1}(V).$$

Hence, it follows that $\theta_\tau^E(W) \subset p_n^{-1}(V) \subset U$.

Lemma 2.3.1 *Let $\tau : X^* \to \mathbf{I}$ be a map. Suppose $X^* \setminus \tau^{-1}(0) = \bigcup_{n\in\mathbb{N}} D_n$ for some $\emptyset = D_0 \subset D_1 \subset \cdots$ such that:*

(i) $\tau(w) \leqslant 2^{-n-1}$ *for each* $w \in X^* \setminus D_n$ *and*
(ii) $w \in D_n,\ p_n(w) = p_n(w') \Rightarrow \tau(w) = \tau(w')$.

Then, θ_τ^E is a homeomorphism.

Proof For simplicity, let $A = \tau^{-1}(0)$. First, observe the following two facts:
(1) $\tau(\theta_\tau^E(w, z)) = \tau(w)$ for each $(w, z) \in (X^* \setminus A) \times E$.
(2) $\tau(\mathrm{pr}_{X^*}(\theta_{\tau(w)}^E)^{-1}(w)) = \tau(w)$ for each $w \in X^* \setminus A$.

(1): Let $(w, z) \in (X^* \setminus A) \times E$. Then, $w \in D_n \setminus D_{n-1}$ for some $n \in \mathbb{N}$, whence $\tau(w) \leqslant 2^{-n}$ by (i). Thus, it follows from (♯) that

$$p_n(\theta_\tau^E(w, z)) = p_n\theta_{\tau(w)}^E(w, z) = p_n\mathrm{pr}_{X^*}(w, z) = p_n(w),$$

which implies $\tau(\theta_\tau^E(w, z)) = \tau(w)$ by (ii).

(2): Each $w \in X^* \setminus A$ is contained in some $D_n \setminus D_{n-1}$. In the same way as the above, we have $p_n\theta_{\tau(w)}^E = p_n\mathrm{pr}_{X^*}$. Then, $p_n(w) = p_n(\mathrm{pr}_{X^*}(\theta_{\tau(w)}^E)^{-1}(w))$, which implies $\tau(\mathrm{pr}_{X^*}(\theta_{\tau(w)}^E)^{-1}(w)) = \tau(w)$ by (ii).

Now, we define $\eta_\tau^E : X^* \to (X^* \times E)_A$ by $\eta_\tau^E|A = \mathrm{id}$ and $\eta_\tau^E(w) = (\theta_{\tau(w)}^E)^{-1}(w)$ for $w \in X^* \setminus A$ $(= X^* \setminus \tau^{-1}(0))$. Then, by (2),

$$w \in X^* \setminus A \Rightarrow \eta_\tau^E(w) = (\theta_{\tau(w)}^E)^{-1}(w) \in (X^* \setminus A) \times E = (X^* \times E)_A \setminus A.$$

Since $\tau(\mathrm{pr}_{X^*}\eta_\tau^E(w)) = \tau(w)$ for $w \in X^* \setminus A$, it is easy to see that $\eta_\tau^E\theta_\tau^E = \mathrm{id}$ and $\theta_\tau^E\eta_\tau^E = \mathrm{id}$, namely that η_τ^E is the inverse of θ_τ^E.

The continuity of η_τ^E remains to be proved. It follows from the continuity of $(\tilde{\theta}^E)^{-1}$ that η_τ^E is continuous at each point of $X^* \setminus A$. We shall show the continuity of η_τ^E at $w \in A$. Each neighborhood of $w = \eta_\tau^E(w)$ in $(X^* \times E)_A$ contains $(W \cap A) \cup ((W \setminus A) \times E)$ for some open neighborhood W of w in X^*. We can find $n \in \mathbb{N}$ and an open neighborhood W_0 of $p_n(w)$ in $X \times E^n$ such that $p_n^{-1}(W_0) \subset W$. Then, $V = p_n^{-1}(W_0) \cap \tau^{-1}([0, 2^{-n}))$ is an open neighborhood of w in X^*. Moreover,

$$\eta_\tau^E(V) \subset (W \cap A) \cup ((W \setminus A) \times E).$$

Indeed, $\eta_\tau^E(V \cap A) = V \cap A \subset p_n^{-1}(W_0) \cap A \subset W \cap A$. For each $v \in V \setminus A$, we write

$$\eta_\tau^E(v) = (\theta_{\tau(v)}^E)^{-1}(v) = (v', z) \in (X^* \setminus A) \times E.$$

Since $\tau(v') = \tau(v) < 2^{-n}$ by (2), it follows from (♯) that

$$p_n(v') = p_n\theta_{\tau(v')}^E(v', z) = p_n\theta_{\tau(v)}^E(v', z) = p_n(v) \in W_0,$$

hence $v' \in W \setminus A$. Thus, we have $\eta_\tau^E(v) = (v', z) \in (W \setminus A) \times E$. □

Lemma 2.3.2 *Let X be an E-stable perfectly normal space, A be an E-deficient set, $A_0 \subset A_1$ be closed sets, and W be an open set in X such that $\mathrm{cl}(A_1 \setminus A_0) \subset W$. Then, there is a homeomorphism $h : (X \times E)_{A_0} \to (X \times E)_{A_1}$ such that*

(i) $h(x, \mathbf{0}) = (x, \mathbf{0})$ *for each* $x \in A \setminus A_1$,
(ii) $h(x, \mathbf{0}) = x$ *for each* $x \in A \cap (A_1 \setminus A_0)$, *and*
(iii) $h|(X \setminus (A_0 \cup W)) \times E \cup A_0 = \mathrm{id}$.

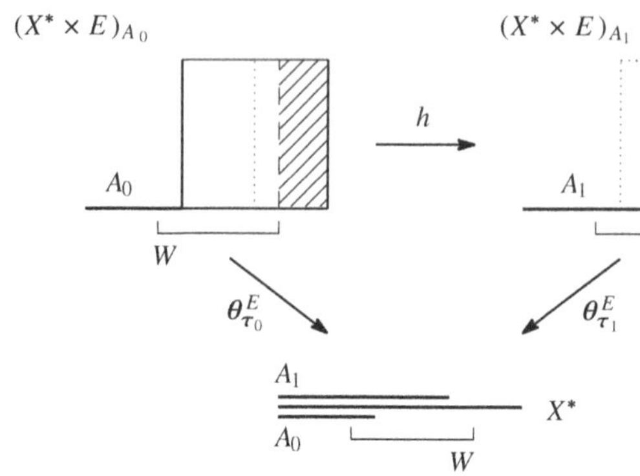

Fig. 2.5 Homeomorphism h

Proof Since $X \times E \approx X$ and $E \approx E^{\mathbb{N}}$ or $E \approx E^{\mathbb{N}}_f$, we may replace X by X^*, where A can be regarded as a subset of $X_0 = X \times \{\mathbf{0}\} \subset X^*$ because A is E-deficient. For $j = 0, 1$, we shall construct maps $\tau_j : X^* \to \mathbf{I}$ and $\emptyset = D^j_0 \subset D^j_1 \subset \cdots$ so that $X^* \setminus \tau_j^{-1}(0) = \bigcup_{n \in \mathbb{N}} D^j_n$ and those satisfy the conditions of Lemma 2.3.1 and the following additional conditions:

(0) $\tau_0|X^* \setminus W = \tau_1|X^* \setminus W$ and $\tau_0^{-1}(0) = A_0 \subset A_1 = \tau_1^{-1}(0)$.

Then, we would have homeomorphisms $\theta^E_{\tau_j} : (X^* \times E)_{A_i} \to X^*$, $j = 0, 1$, that satisfy the following:

(1) $\theta^E_{\tau_j}(x, 0) = x$ for every $x \in X_0 \setminus A_j$,
(2) $\theta^E_{\tau_j}|A_j = \mathrm{id}_{A_j}$, and
(3) $\theta^E_{\tau_0}|(X^* \setminus (A_0 \cup W)) \times E = \theta^E_{\tau_1}|(X^* \setminus (A_0 \cup W)) \times E$.

Then, $h = (\theta^E_{\tau_1})^{-1}\theta^E_{\tau_0}$ is the desired homeomorphism (Fig. 2.5).

We write $A_2 = A_0 \cup (X^* \setminus W)$. Then, $X^* \setminus A_2 = W \setminus A_0$. Now, we take $\mathcal{B}_i \in \mathrm{cov}(X^* \setminus A_i)$, $i = 1, 2$, which consist of basic open sets in X^*.[14] Here, $B \subset X^*$ is said to be ***n*-basic** if $p_n^{-1}(p_n(B)) = B$. Note that each n-basic subset of X^* is m-basic for every $m \geqslant n$. For each $n \in \mathbb{N}$, let

$$B^{[n]}_i = \bigcup \{B \in \mathcal{B}_i \mid B \text{ is } n\text{-basic}\},\ i = 1, 2.$$

Then, $B^{[n]}_i$ is n-basic and $p_n(B^{[n]}_i)$ is open in $X \times E^n$. Since $X \times E^n \approx X$ is perfectly normal, we have a map $k_{i,n} : X \times E^n \to \mathbf{I}$ such that

$$k^{-1}_{i,n}(0) = (X \times E^n) \setminus p_n(B^{[n]}_i) = p_n(X^* \setminus B^{[n]}_i).$$

[14] I.e., open sets $p_n^{-1}(U)$, $n \in \mathbb{N}$, where U is open in $X \times E^n$.

For each $m \in \mathbb{N}$, let $D^n_{i,m} = k^{-1}_{i,n}([1/m, 1])$. Then, $p_n(B^{[n]}_i) = \bigcup_{m\in\mathbb{N}} D^n_{i,m}$, each $D^n_{i,m}$ is closed in $X \times E^n$, and $D^n_{i,m} \subset \operatorname{int} D^n_{i,m+1}$. For each $n \in \mathbb{N}$, we define

$$D^{[n]}_i = \bigcup \{p^{-1}_m(D^m_{i,n}) \mid m \leqslant n\},\ i = 1, 2.$$

Observe that each $D^{[n]}_i$ is n-basic, $D^{[n]}_i \subset \operatorname{int} D^{[n+1]}_i$, and $\bigcup_{n\in\mathbb{N}} D^{[n]}_i = X^* \setminus A_i$.

For each $n \in \mathbb{N}$, let

$$H^{[n]}_i = D^{[n]}_i \setminus \operatorname{int} D^{[n-1]}_i,\ i = 1, 2,$$

where $D^{[0]}_i = \emptyset$. Then, it follows that

$$\bigcup_{n\in\mathbb{N}} H^{[n]}_i = \bigcup_{n\in\mathbb{N}} D^{[n]}_i = X^* \setminus A_i,$$

$$X^* \setminus D^{[n]}_i \subset A_i \cup \bigcup_{m>n} H^{[m]}_i \text{ and}$$

$$H^{[n]}_i \cap H^{[n+1]}_i = \operatorname{bd} D^{[n]}_i \text{ for each } n \in \mathbb{N}.$$

Moreover, $H^{[m]}_i \cap H^{[n]}_i = \emptyset$ if $|m-n| > 1$. Since $p_n(\operatorname{bd} D^{[n]}_i)$ and $p_n(\operatorname{bd} D^{[n-1]}_i)$ are disjoint closed sets in $p_n(H^{[n]}_i)$, we can take maps $\tau^{[n]}_i : p_n(H^{[n]}_i) \to [2^{-n-1}, 2^{-n}]$, $i = 1, 2$, such that

$$\tau^{[n]}_i(p_n(\operatorname{bd} D^{[n]}_i)) = 2^{-n-1} \text{ and } \tau^{[n]}_i(p_n(\operatorname{bd} D^{[n-1]}_i)) = 2^{-n}.$$

We define $\tau_i : X^* \to \mathbf{I}$, $i = 1, 2$, as follows:

$$\tau_i(A_i) = 0 \text{ and } \tau_i|H^{[n]}_i = \tau^{[n]}_i p_n|H^{[n]}_i.$$

It is easy to see that τ_i is continuous at each point of $X^* \setminus A_i$. To verify the continuity of τ_i at $x \in A_i$, for each $\varepsilon > 0$, choose $n \in \mathbb{N}$ so that $2^{-n} < \varepsilon$. Then, $U = X^* \setminus D^{[n]}_i$ is a neighborhood of x in X^*. Note that $U \setminus A_i \subset \bigcup_{m>n} H^{[m]}_i$. If $y \in U \setminus A_i$, then $y \in H^{[m]}_i$ for some $m > n$, hence

$$\tau_i(y) = \tau^{[m]}_i p_m(y) \leqslant 2^{-m} < 2^{-n} < \varepsilon.$$

Therefore, $\tau_i(U) \subset [0, \varepsilon)$. It follows from the definitions that τ_i and $(D^{[n]}_i)_{n\in\mathbb{N}}$ satisfy the conditions of Lemma 2.3.1. Moreover, $\tau^{-1}_i(0) = A_i$.

For each $n \in \mathbb{N}$, let $D^{[n]}_0 = D^{[n]}_1 \cup D^{[n]}_2$. Define $\tau_0 : X^* \to \mathbf{I}$ by $\tau_0(x) = \max\{\tau_1(x), \tau_2(x)\}$. Then, τ_j and $(D^{[n]}_j)_{n\in\mathbb{N}}$, $j = 0, 1$, satisfy the required conditions. Indeed, $\tau^{-1}_0(0) = \tau^{-1}_1(0) \cap \tau^{-1}_2(0) = A_1 \cap A_2 = A_0$.

Since $X^* \setminus W \subset A_2 = \tau_2^{-1}(0)$, $\tau_0(x) = \tau_1(x)$ for each $x \in X^* \setminus W$, that is, $\tau_0|X^* \setminus W = \tau_1|X^* \setminus W$. Thus, those satisfy condition (0). Finally, using the fact that τ_i and $(D_i^{[n]})_{n\in\mathbb{N}}$, $i = 1, 2$, satisfy the conditions of Lemma 2.3.1, we can easily show that τ_0 and $(D_0^{[n]})_{n\in\mathbb{N}}$ also satisfy the same conditions. □

Remark 2.3 In the above proof, each $x \in X^* \setminus A_1$ is contained in some $D_1^{[n]}$. Then, there is some $m \leqslant n$ such that $p_m(x) \in D_{1,n}^m \subset p_m(B_1^{[m]})$, hence $x \in B_1^{[m]}$. Thus, x is contained in some m-basic $B \in \mathcal{B}_1$. Since $\tau_1(x) \leqslant 2^{-m}$, it follows from (♯) that $p_m\theta^E_{t\tau_1(x)}(x, z) = p_m(x)$, so $\theta^E_{t\tau_1(x)}(x, z) \in B$ for every $z \in E$ and $t \in \mathbf{I}$, where $\theta^E_0 = \mathrm{pr}_{X^*}$. By choosing $\mathcal{B}_1$ to refine a given $\mathcal{U} \in \mathrm{cov}(X^* \setminus A_1)$, the homeomorphism $\theta^E_{\tau_1} : (X^* \times E)_{A_1} \to X^*$ is $\mathcal{U}$-homotopic to pr_{X^*} by the homotopy $\theta^E_{t\tau_1}$, $t \in \mathbf{I}$. Recall that we have the natural map $q_{A_1} : X^* \times E \to (X^* \times E)_{A_1}$ defined by $q_{A_1}|(X^* \setminus A_1) \times E = \mathrm{id}$ and $q_{A_1}|A_1 \times E = \mathrm{pr}_{A_1}$. Then, $f = \theta^E_{\tau_1} q_{A_1} : X^* \times E \to X^*$ is a map such that $f|A_1 \times E = \mathrm{pr}_{A_1}$ and $f|(X^* \setminus A_1) \times E$ is a homeomorphism onto $X^* \setminus A_1$ that is $\mathcal{U}$-homotopic to $\mathrm{pr}_{X^* \setminus A_1}$.

In this remark, for an open set $U \subset X$, let $A_1 = X \setminus U$. Then, we have the following:

Proposition 2.3.3 *Let X be an E-stable perfectly normal space, A be an E-deficient set in X, and U be an open set in X. Then, for each open cover $\mathcal{U}$ of U, there exists a map $f : X \times E \to X$ such that $f(x, \mathbf{0}) = x$ for each $x \in A \cap U$, $f|(X \setminus U) \times E = \mathrm{pr}_{X \setminus U}$ and $f|U \times E$ is a homeomorphism onto U that is $\mathcal{U}$-homotopic to pr_U.* □

Corollary 2.3.4 *Let X be an E-stable perfectly normal space and A be an E-deficient set in X. Then, the following hold:*

(1) *Every open set in X is E-stable.*
(2) *For each open set U in X, $A \cap U$ is E-deficient in U.*
(3) *The projection $\mathrm{pr}_X : X \times E \to X$ is a near-homeomorphism.*
(4) *For any open cover $\mathcal{U}$ of X, there is a homeomorphism $f : X \times E \to X$ such that $f(x, \mathbf{0}) = x$ for each $x \in A$ and $f \simeq_{\mathcal{U}} \mathrm{pr}_X$.* □

Replacing $E = (E, \mathbf{0})$ with a pointed space $L = (L, *)$, the L-stability and the L-deficiency are similarly defined, that is, a space X is ***L*-stable** if $X \times L \approx X$, and a subset A of X is ***L*-deficient** if there is a homeomorphism $h : X \to X \times L$ such that $h(A) \subset X \times \{*\}$.

Corollary 2.3.5 *Let X be an E-stable perfectly normal paracompact space, A be an E-deficient set in X, and $L = (L, *)$ be a pointed space such that $E \times L \approx E$. Then, for each open cover $\mathcal{U}$ of X, there exists a homeomorphism $h : X \times L \to X$ such that $h(x, *) = x$ for each $x \in A$ and $h \simeq_{\mathcal{U}} \mathrm{pr}_X$. Consequently, the projection $\mathrm{pr}_X : X \times L \to X$ is a near-homeomorphism and every E-deficient set in X is L-deficient.*

Proof Let $\varphi : E \times L \to E$ be a homeomorphism with $\varphi(\mathbf{0}, *) = \mathbf{0}$. Since X is paracompact, there exists $\mathcal{V} \in \operatorname{cov}(X)$ such that $\operatorname{st}\mathcal{V} \prec \mathcal{U}$. By Corollary 2.3.4(4) above, we have a homeomorphism $f : X \times E \to X$ such that $f(x, \mathbf{0}) = x$ for each $x \in A$ and $f \simeq_{\mathcal{V}} \mathrm{pr}_X$. Then, the desired homeomorphism $h : X \times L \to X$ can be defined as the composition of the following homeomorphisms:

$$X \times L \xrightarrow{f^{-1}\times \mathrm{id}_L} X \times E \times L \xrightarrow{\mathrm{id}_X \times \varphi} X \times E \xrightarrow{f} X$$

Indeed, it is easy to see that $h(x, *) = x$ for each $x \in A$ and

$$h \simeq_{\mathcal{V}} \mathrm{pr}_X(\mathrm{id}_X \times \varphi)(f^{-1} \times \mathrm{id}_L) = \mathrm{pr}_X(f^{-1} \times \mathrm{id}_L) = \mathrm{pr}_X f^{-1}\mathrm{pr}_X,$$

where the same notation pr_X stands for three different projections. Since $\mathrm{pr}_X f^{-1} \simeq_{\mathcal{V}} ff^{-1} = \mathrm{id}_X$, it follows that $h \simeq_{\operatorname{st}\mathcal{V}} \mathrm{pr}_X$, hence $h \simeq_{\mathcal{U}} \mathrm{pr}_X$. □

It is said that X is **locally *E*-stable** if each $x \in X$ has an E-stable neighborhood in X. Applying Michael's Theorem 1.3.20 on local property, we prove the following theorem:

Theorem 2.3.6 *For every perfectly normal paracompact space X, if X is locally E-stable, then X is E-stable and the projection $\mathrm{pr}_X : X \times E \to X$ is a near-homeomorphism.*

Proof Let $\mathcal{P}$ be the property for open sets in X to be E-stable. The result follows from Michael's Theorem 1.3.20 on local properties if $\mathcal{P}$ is G-hereditary. The condition (G-1) is Corollary 2.3.4(1), and (G-3) is obvious. It remains to show (G-2).

(G-2): For any E-stable open sets U_1 and U_2 in X, let $U = U_1 \cup U_2$. Since U is normal, we can choose open sets V_1 and V_2 in U so that $\mathrm{cl}_U V_1 \cap \mathrm{cl}_U V_2 = \emptyset$, $U \setminus U_2 \subset V_1$, and $U \setminus U_1 \subset V_2$. Applying Lemma 2.3.2 to $A_0 = \emptyset$, $A_1 = \mathrm{cl}_U V_1$, $W = U_1 \setminus \mathrm{cl}_U V_2$, and $X = U_1$, we have a homeomorphism $h_1 : U_1 \times E \to (U_1 \times E)_{\mathrm{cl}_U V_1}$ such that $h_1|(U_1 \cap \mathrm{cl}_U V_2) \times E = \mathrm{id}$. Then, h_1 can be extended to a homeomorphism $\bar{h}_1 : U \times E \to (U \times E)_{\mathrm{cl}_U V_1}$ by $\bar{h}_1|(U \setminus U_1) \times E = \mathrm{id}$ (Fig. 2.6).

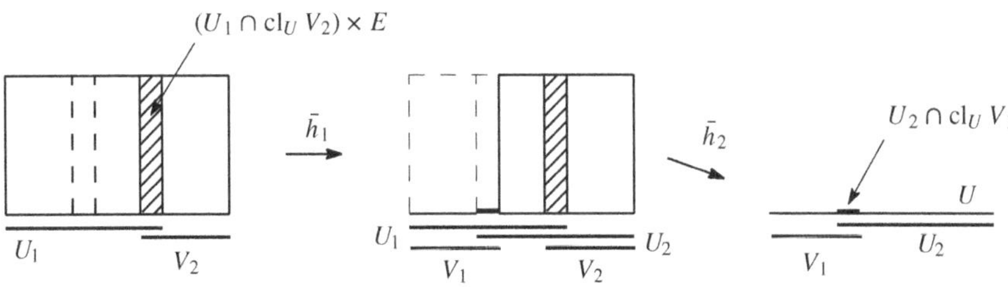

Fig. 2.6 Homeomorphisms $\bar{h}_1$ and $\bar{h}_2$

Next, applying Lemma 2.3.2 to $A_0 = U_2 \cap \mathrm{cl}_U V_1$, $A_1 = U_2$, and $W = X = U_2$, we have a homeomorphism $h_2 : (U_2 \times E)_{U_2 \cap \mathrm{cl}_U V_1} \to U_2$ such that $h_2|U_2 \cap \mathrm{cl}_U V_1 = \mathrm{id}$. Then, h_2 can be extended to a homeomorphism $\bar{h}_2 : (U \times E)_{\mathrm{cl}_U V_1} \to U$ by $\bar{h}_2|U \setminus U_2 = \mathrm{id}$. Thus, we have a homeomorphism $\bar{h}_2\bar{h}_1 : U \times E \to U$, that is, U is E-stable. □

Note that every E-manifold is perfectly normal and paracompact because it is metrizable. Since each open set in E is E-stable by Corollary 2.3.4(1), every E-manifold is locally E-stable. Thus, as a corollary of Theorem 2.3.6 above, we have the following STABILITY THEOREM:

Theorem 2.3.7 (STABILITY) *Every E-manifold M is E-stable and the projection $\mathrm{pr}_M : M \times E \to M$ is a near-homeomorphism.* □

It is said that a subset A of an E-stable space X is **locally E-deficient** in X if each $a \in A$ has a neighborhood U in X such that $A \cap U$ is E-deficient in U. By making small changes in the proof of Theorem 2.3.6, we can prove the following:

Proposition 2.3.8 *Let X be an E-stable perfectly normal paracompact space. Then, every locally E-deficient closed set A in X is E-deficient in X. As the result, if a closed set $A \subset X$ is E-deficient in an open set U in X, then A is E-deficient in X.*

Proof Let $\mathcal{P}$ be the property for open sets U in M such that $A \cap U$ is E-deficient in U. This property implies that U is E-stable. Actually, an open set U with $A \cap U = \emptyset$ has property $\mathcal{P}$ if U is E-stable. Then, X has $\mathcal{P}$ locally.[15] In the proof of Theorem 2.3.6, applying Corollary 2.3.4(2) instead of (1) gives a homeomorphism $h : X \times E \to X$ such that $h(x, 0) = x$ for each $x \in A$. □

Now, we consider $\boldsymbol{Q}$-stability. Because $[-1, 1]^2 \approx \mathbf{B}^2$, there exists an isotopy $\varphi : [-1, 1]^2 \times \mathbf{I} \to [-1, 1]^2$ such that

$$\varphi_0 = \mathrm{id}, \;\; \varphi_t(0, 0) = (0, 0) \;\text{ for each } t \in \mathbf{I}, \text{ and}$$

$$\varphi_1(x, y) = (-y, x) \;\text{ for each } (x, y) \in [-1, 1]^2,$$

that is, φ is induced by the anti-clockwise rotation of $\mathbf{B}^2$. We define the isotopy $\varphi^{\boldsymbol{Q}} : \boldsymbol{Q}^2 \times \mathbf{I} \to \boldsymbol{Q}^2$ as follows:

$$\varphi_t^{\boldsymbol{Q}}(x, y) = \psi\big((\varphi_t(x(i), y(i)))_{i \in \mathbb{N}}\big),$$

where $\psi : \big([-1, 1]^2\big)^{\mathbb{N}} \to \boldsymbol{Q}^2$ is the homeomorphism defined by

$$\psi\big((x(i), y(i))_{i \in \mathbb{N}}\big) = \big((x(i))_{i \in \mathbb{N}}, (y(i))_{i \in \mathbb{N}}\big).$$

---

[15]When A is not closed in M, it is unknown whether a point $a \in \mathrm{cl}\, A \setminus A$ has an open neighborhood U such that $A \cap U$ is E-deficient in U.

Then, $\varphi_0^{\boldsymbol{Q}} = \mathrm{id}$, $\varphi_t^{\boldsymbol{Q}}(\mathbf{0}, \mathbf{0}) = (\mathbf{0}, \mathbf{0})$ for each $t \in \mathbf{I}$, and $\varphi_1^{\boldsymbol{Q}}(x, y) = (-y, x)$ for each $(x, y) \in \boldsymbol{Q}^2$. Using this isotopy, we can define the **coordinate-switching pseudo-isotopy**

$$\theta^{\boldsymbol{Q}} : X \times \boldsymbol{Q}^{\mathbb{N}} \times \boldsymbol{Q} \times \mathbf{I} \to X \times \boldsymbol{Q}^{\mathbb{N}}$$

in the same way as $\theta^E : X^* \times E \times \mathbf{I} \to X^*$ ($X^* = X \times E^{\mathbb{N}}$). Then, $\theta^{\boldsymbol{Q}}$ satisfies the following conditions:

($\natural$) $\theta_t^{\boldsymbol{Q}}(x, \mathbf{0}) = x$ for each $x \in X \times \{\mathbf{0}\} \subset X \times \boldsymbol{Q}^{\mathbb{N}}$ and $t \in \mathbf{I}$.

($\sharp$) $t \leqslant 2^{-n} \Rightarrow p_n \theta_t^{\boldsymbol{Q}} = p_n \mathrm{pr}_{X \times \boldsymbol{Q}^{\mathbb{N}}}$,

where $p_n : X \times \boldsymbol{Q}^{\mathbb{N}} \to X \times \boldsymbol{Q}^n$ is the projection. Note that $\boldsymbol{Q}^{\mathbb{N}} \approx \boldsymbol{Q}$. By the same arguments as above, we have the following $\boldsymbol{Q}$-stable versions of Proposition 2.3.3 and Corollary 2.3.4:

Proposition 2.3.9 *Let X be a $\boldsymbol{Q}$-stable perfectly normal space, A be a $\boldsymbol{Q}$-deficient set in X, and U be an open set in X. Then, for each open cover $\mathcal{U}$ of U, there exists a map $f : X \times \boldsymbol{Q} \to X$ such that $f(x, \mathbf{0}) = x$ for each $x \in A \cap U$, $f|(X \setminus U) \times \boldsymbol{Q} = \mathrm{pr}_{X \setminus U}$ and $f|U \times \boldsymbol{Q}$ is a homeomorphism onto U that is $\mathcal{U}$-homotopic to pr_U.* □

Corollary 2.3.10 *Let X be a $\boldsymbol{Q}$-stable perfectly normal space and A be a $\boldsymbol{Q}$-deficient set in X. Then, the following hold:*

(1) *Every open set in X is $\boldsymbol{Q}$-stable.*
(2) *For each open set U in X, $A \cap U$ is $\boldsymbol{Q}$-deficient in U.*
(3) *The projection $\mathrm{pr}_X : X \times \boldsymbol{Q} \to X$ is a near-homeomorphism.*
(4) *For any open cover $\mathcal{U}$ of X, there is a homeomorphism $f : X \times \boldsymbol{Q} \to X$ such that $f(x, \mathbf{0}) = x$ for each $x \in A$ and $f \simeq_{\mathcal{U}} \mathrm{pr}_X$.* □

Since $\boldsymbol{Q} \times \mathbf{I}^n \approx \boldsymbol{Q}$ for every $n \in \mathbb{N}$, the following can be obtained in the same way as Corollary 2.3.5:

Corollary 2.3.11 *For every $\boldsymbol{Q}$-stable perfectly normal paracompact space X and $n \in \mathbb{N}$, the projection $\mathrm{pr}_X : X \times \mathbf{I}^n \to X$ is a near-homeomorphism.* □

It is said that X is **locally $\boldsymbol{Q}$-stable** if each $x \in X$ has a $\boldsymbol{Q}$-stable neighborhood in X. The following is the $\boldsymbol{Q}$-manifold version of Theorem 2.3.6:

Theorem 2.3.12 *For every perfectly normal paracompact space X, if X is locally $\boldsymbol{Q}$-stable, then X is $\boldsymbol{Q}$-stable and the projection $\mathrm{pr}_X : X \times \boldsymbol{Q} \to X$ is a near-homeomorphism.* □

Thus, we have the following $\boldsymbol{Q}$-manifold version of Theorem 2.3.7:

Theorem 2.3.13 (Stability for $\boldsymbol{Q}$-manifolds) *Every $\boldsymbol{Q}$-manifold M is $\boldsymbol{Q}$-stable and the projection $\mathrm{pr}_M : M \times \boldsymbol{Q} \to M$ is a near-homeomorphism.*

Additionally, for each $n \in \mathbb{N}$, the projection $\mathrm{pr}_M : M \times \mathbf{I}^n \to X$ *is also a near-homeomorphism.* □

In the same way, locally $\boldsymbol{Q}$-deficient sets and $\boldsymbol{Q}$-deficient embeddings are defined. Then, the following can also be obtained:

Proposition 2.3.14 *Let X be a perfectly normal paracompact space. Then, every locally $\boldsymbol{Q}$-deficient closed set in X is $\boldsymbol{Q}$-deficient in X.* □

Note In the above, to prove Proposition 2.3.9, we translated the arguments for E to $\boldsymbol{Q}$, where $\boldsymbol{Q}^{\mathbb{N}}$ was used instead of $E^{\mathbb{N}}$. However, because $\boldsymbol{Q}$ itself has infinitely many coordinates, it is possible to define the **coordinate-switching pseudo-isotopy**

$$\theta^{\boldsymbol{Q}} : X \times \boldsymbol{Q} \times \boldsymbol{Q} \times \mathbf{I} \to X \times \boldsymbol{Q}$$

so that $\theta_0^{\boldsymbol{Q}}$ is the projection, $\theta_t^{\boldsymbol{Q}}$ is a homeomorphism for $t > 0$, and

(♮) $\theta_t^{\boldsymbol{Q}}(x, \mathbf{0}) = x$ for each $x \in X \times \{\mathbf{0}\} \subset X \times \boldsymbol{Q}$ and $t \in \mathbf{I}$.

Here, we use an isotopy $\varphi : [-1, 1]^2 \times \mathbf{I} \to [-1, 1]^2$ such that

$$\varphi_0 = \mathrm{id}, \ \varphi_t(0, 0) = (0, 0) \ \text{ for each } t \in \mathbf{I}, \text{ and}$$

$$\varphi_1(x, y) = (y, -x) \ \text{ for each } (x, y) \in [-1, 1]^2,$$

that is, φ is induced by the clockwise rotation of $\mathbf{B}^2 \approx [-1, 1]^2$. Then, we define $\theta^{\boldsymbol{Q}}$ as follows: For $(x, y, z) \in X \times \boldsymbol{Q} \times \boldsymbol{Q}$,

$$\theta_1^{\boldsymbol{Q}}(x, y, z) = (x, z(1), y(1), z(2), y(2), z(3), y(3), \dots)$$

$$\theta_{2^{-1}}^{\boldsymbol{Q}}(x, y, z) = (x, y(1), -z(1), y(2), -z(2), y(3), -z(3), \dots),$$

$$\vdots$$

$$\theta_{2^{-n}}^{\boldsymbol{Q}}(x, y, z) = (x, y(1), \dots, y(n), (-1)^n z(1), y(n+1), (-1)^n z(2), y(n+2), \dots),$$

$$\theta_{2^{-(n+1)}}^{\boldsymbol{Q}}(x, y, z) = (x, y(1), \dots, y(n), y(n+1), (-1)^{n+1} z(1), y(n+2), (-1)^{n+1} z(2), \dots),$$

$$\vdots$$

$$\theta_0^{\boldsymbol{Q}}(x, y, z) = (x, y) = (x, y(1), y(2), y(3), \dots)$$

and for $2^{-(n+1)} \leqslant t \leqslant 2^{-n}$,

$$\theta_t^{\boldsymbol{Q}}(x, y, z) = (x, y(1), \dots, y(n), \varphi_{2-2^{n+1}t}((-1)^n z(1), y(n+1)), \varphi_{2-2^{n+1}t}((-1)^n z(2), y(n+2)), \dots).$$

Then, (♮) and the following are satisfied:

(♯) $\quad t \leqslant 2^{-n} \Rightarrow p_n\theta_t^{Q} = p_n \mathrm{pr}_{X\times Q}$,

where $p_n : X \times Q \to X \times [-1, 1]^n$ is the projection. Let $d \in \mathrm{Metr}(Q)$ be the metric defined by

$$d(x, y) = \sup_{n\in\mathbb{N}} 2^{-n}|x(n) - y(n)|.$$

Taking any $d \in \mathrm{Metr}(X)$, we define $\rho \in \mathrm{Metr}(X \times Q)$ as follows:

$$\rho((x, y), (x', y')) = d_X(x, x') + d(y, y').$$

Then, $\rho(\theta_t^{Q}, \mathrm{pr}_{X\times Q}) \leqslant 2^{-n}$ if $t \leqslant 2^{-n}$.

For the space $s \approx \mathbb{R}^{\mathbb{N}}$, we can analogously define the **coordinate-switching pseudo-isotopy**

$$\theta^s : X \times s \times s \times \mathbf{I} \to X \times s.$$

An embedding $f : Y \to X$ is said to be ***E*-deficient** if $f(Y)$ is E-deficient in X. Recall that $\mathcal{F}_E$ is the class of spaces which can be embedded in E as closed sets. Applying the Stability Theorem 2.3.7, we can prove the following theorem, which is also known as the Closed Embedding Approximation Theorem:

Theorem 2.3.15 (Strong Universality of E-manifolds) *Let M be an E-manifold and $A \in \mathcal{F}_E$, with a closed set $B \subset A$ and a map $f : A \to M$, such that $f|B$ is an E-deficient closed embedding. Then, for each $\mathcal{U} \in \mathrm{cov}(M)$, there exists an E-deficient closed embedding $h : A \to M$ such that $h|B = f|B$ and $h \simeq_{\mathcal{U}} f$.*

Proof First, note that $E^3 \approx E$ and M is an ANR. By Corollary 2.3.4(4), we have a homeomorphism $g : M \times E^3 \to M$ such that

$$g(f(x), \mathbf{0}, \mathbf{0}, \mathbf{0}) = f(x) \text{ for each } x \in B \text{ and } g \simeq_{\mathcal{U}} \mathrm{pr}_M.$$

From the assumption, there is a closed embedding $j : A \to E$. Take $d \in \mathrm{Metr}(A)$ and $d_M \in \mathrm{Metr}(M)$ such that $d, d_M \leqslant 1/2$, and define $d_A \in \mathrm{Metr}(A)$ as follows:

$$d_A(x, y) = d(x, y) + d_M(f(x), f(y)) \text{ for each } x, y \in A.$$

Then, $d_M(f(x), f(y)) \leqslant d_A(x, y) \leqslant 1$ for each $x, y \in A$. Let $k : A \to \mathbf{I}$ be the map defined by $k(x) = d_A(x, B)$. Let $e \neq 0 \in E$ be fixed. The desired embedding $h : A \to M$ can be defined by

$$h(x) = g(f(x), k(x)e, k(x)j(x), \mathbf{0}).$$

Indeed, $h|B = f|B$ by definition, and $h(A)$ is E-deficient in M because $M \times E^2 \approx M$. Since $g \simeq_{\mathcal{U}} \mathrm{pr}_M$, it follows that $h \simeq_{\mathcal{U}} f$. If $x \neq y \in A$, then $k(x) \neq k(y)$,

$k(x)j(x) \neq k(y)j(y)$, or $k(x) = k(y) = 0$. In the third case, $x, y \in B$, which implies $f(x) \neq f(y)$. Thus, h is injective.

It remains to see that h is closed. Let $x_n \in A$, $n \in \mathbb{N}$, such that $(h(x_n))_{n\in\mathbb{N}}$ is convergent in M. Then, it is enough to show that $(x_n)_{n\in\mathbb{N}}$ is convergent. The sequences $(f(x_n))_{n\in\mathbb{N}}$, $(k(x_n))_{n\in\mathbb{N}}$, and $(k(x_n)j(x_n))_{n\in\mathbb{N}}$ are convergent.

When $k(x_n) \to t \neq 0$, $k(x_n) > 0$ for sufficiently large n and $k(x_n)^{-1} \to t^{-1}$, so it follows that $(j(x_n))_{n\in\mathbb{N}}$ is convergent. Since j is a closed embedding, $(x_n)_{n\in\mathbb{N}}$ is convergent.

When $k(x_n) \to 0$, we can choose $y_n \in B$ so that $d_A(x_n, y_n) \to 0$, hence $d_M(f(x_n), f(y_n)) \to 0$ by the definition of d_A. Then, $(f(y_n))_{n\in\mathbb{N}}$ is convergent. Since $f|B$ is a closed embedding, $(y_n)_{n\in\mathbb{N}}$ is convergent, and hence so is $(x_n)_{n\in\mathbb{N}}$.

□

Note that the class $\mathfrak{M}_0$ of compacta (= compact metrizable spaces) is equal to the class $\mathcal{F}_{\boldsymbol{Q}}$ of spaces which can be embedded into $\boldsymbol{Q}$. By the same proof as Theorem 2.3.15 above, we can prove the following:

Theorem 2.3.16 (STRONG UNIVERSALITY OF $\boldsymbol{Q}$-MANIFOLDS) *Let M be a $\boldsymbol{Q}$-manifold and A be a compactum, with a closed set $B \subset A$ and a map $f : A \to M$, such that $f|B$ is a $\boldsymbol{Q}$-deficient closed embedding of a closed set $B \subset A$. Then, for each $\mathcal{U} \in \operatorname{cov}(M)$, there exists a $\boldsymbol{Q}$-deficient closed embedding $h : A \to M$ such that $h|B = f|B$ and $h \simeq_{\mathcal{U}} f$.* □

The following is a non-compact version of the above:

Theorem 2.3.17 (CLOSED EMBEDDING APPROXIMATION) *Let M be a $\boldsymbol{Q}$-manifold, and A be a locally compact separable metrizable space, with a closed set $B \subset A$ and a proper map $f : A \to M$, such that $f|B$ is a $\boldsymbol{Q}$-deficient closed embedding. Then, for each $\mathcal{U} \in \operatorname{cov}(M)$, there exists a $\boldsymbol{Q}$-deficient closed embedding $h : A \to M$ such that $h|B = f|B$ and $h \simeq^p_{\mathcal{U}} f$.*[16]

Proof Replacing $\mathcal{U}$ by a refinement, we can assume that $\mathcal{U}$ is locally finite and every $U \in \mathcal{U}$ has the compact closure in M. By Corollary 2.3.10(4), we have a homeomorphism $g : M \times \boldsymbol{Q}^2 \to M$ such that $g(f(x), \mathbf{0}, \mathbf{0}) = f(x)$ for each $x \in B$ and $g \simeq_{\mathcal{U}} \mathrm{pr}_M$. We have an embedding $j : A \to \boldsymbol{Q}$ with $j(A) \subset \mathrm{pr}_1^{-1}(1)$ ($\approx \boldsymbol{Q}$),[17] where $\mathrm{pr}_1 : \boldsymbol{Q} \to [-1, 1]$ is the projection of the first factor. Let $k : A \to \mathbf{I}$ be a map with $k^{-1}(0) = B$. The desired embedding $h : A \to M$ can be defined as follows:

$$h(x) = g(f(x), k(x)j(x), \mathbf{0}).$$

[16] Recall that $h \simeq^p_{\mathcal{U}} f$ means that h is properly $\mathcal{U}$-homotopic to f.

[17] This embedding is not closed unless A is compact. An embedding of A can be obtained by restricting an embedding of the one-point compactification $\alpha A = A \cup \{\infty\}$.

Indeed, by definition, $h|B = f|B$ and $h(A)$ is $\boldsymbol{Q}$-deficient in M. Let $x \neq x' \in A$. If $k(x) \neq k(x')$, then $k(x)j(x) \neq k(x')j(x')$ because

$$\mathrm{pr}_1(k(x)j(x)) = k(x) \neq k(x') = \mathrm{pr}_1(k(x')j(x')).$$

If $k(x) = k(x') > 0$, then $k(x)j(x) \neq k(x')j(x')$ because $j(x) \neq j(x')$. If $k(x) = k(x') = 0$, then $x, x' \in B$, so $h(x) = f(x) \neq f(x') = h(x')$ because $f|B$ is an embedding. In any case, $h(x) \neq h(x')$. Thus, h is injective.

Since $g \simeq_{\mathcal{U}} \mathrm{pr}_M$, it easily follows that $h \simeq_{\mathcal{U}} f$. Then, because f is proper, it follows from Proposition 1.3.8 that $h \simeq^p_{\mathcal{U}} f$. In particular, h is proper, hence it is closed (cf. Proposition 1.3.7(c)). Therefore, h is a closed embedding. □

Remark 2.4 In Theorem 2.3.15, it suffices to assume that M is E-stable like Proposition 2.3.8. In Theorems 2.3.16 and 2.3.17, it suffices to assume that M is $\boldsymbol{Q}$-stable like Proposition 2.3.14.

2.4 Negligibility and Deficiency

In this section, we will prove that if A is an E-deficient closed[18] set in an E-stable space X, then $X \setminus A \approx X$. We will also prove that E is homeomorphic to its open cone when E is a normed linear space. Additionally, it is shown that every finite union of E-deficient sets is also E-deficient. First, we prove the following:

Proposition 2.4.1 *There exists an* $\mathbf{I}$*-preserving homeomorphism*

$$\zeta : \left(\mathbb{R}^{\mathbb{N}} \times \mathbf{I} \setminus \{(0,0)\}, \mathbb{R}^{\mathbb{N}}_f \times \mathbf{I} \setminus \{(0,0)\}\right) \to \left(\mathbb{R}^{\mathbb{N}} \times \mathbf{I}, \mathbb{R}^{\mathbb{N}}_f \times \mathbf{I}\right).$$

Proof For each $n \in \mathbb{N}$, let $\varphi^{(n)} : \mathbb{R}^2 \times \mathbf{I} \to \mathbb{R}^2$ be an ambient invertible isotopy (as illustrated in Fig. 2.7) such that

$$\varphi^{(n)}_0 = \mathrm{id},\ \varphi^{(n)}_1(n-1, y) = (n+y, 0)\ \text{ if }\ y \geqslant -2^{-n}\ \text{ and}$$
$$\varphi^{(n)}_t|(-\infty, (n-1) - 2^{-(n-1)}] \times \mathbb{R} = \mathrm{id}\ \text{ for each }\ t \in \mathbf{I}.$$

The second condition above means that $\varphi^{(n)}_1$ maps the vertical ray $\{n-1\} \times [-2^{-n}, \infty)$ onto the horizontal ray $[n - 2^{-n}, \infty) \times \{0\}$ isometrically.

[18]The condition being closed is weakened to being *locally* closed (see Theorem 2.4.3).

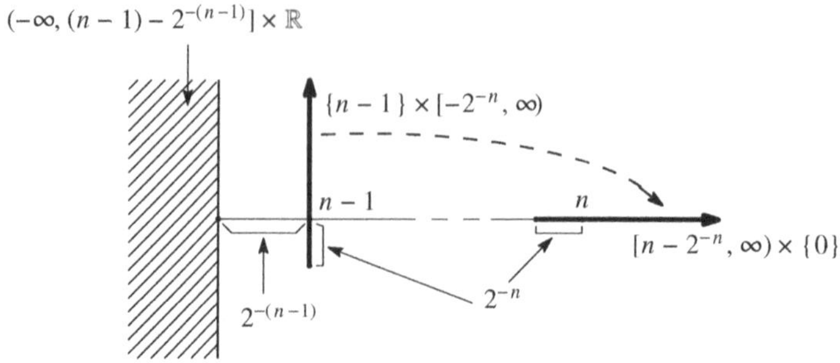

Fig. 2.7 The ambient invertible isotopy $\varphi^{(n)}$

We define an ambient invertible isotopy $\psi^{(n)} : \mathbb{R}^{\mathbb{N}} \times \mathbf{I} \to \mathbb{R}^{\mathbb{N}}$, $(x, t) \mapsto y = \psi_t^{(n)}(x)$, as follows:

$$y(i) = x(i) \ \text{ if } i \neq 1, n+1, \quad \text{and}$$
$$(y(1), y(n+1)) = \varphi^{(n)}_{t\gamma_n(x)}(x(1), x(n+1)),$$

where $\gamma_n : \mathbb{R}^{\mathbb{N}} \to \mathbf{I}$ is a map defined by $\gamma_1(x) = 1$ and, for $n \geqslant 2$,

$$\gamma_n(x) = \max\left\{0, 1 - 2^{n-1}\max\{|x(2)|, \ldots, |x(n)|\}\right\}.$$

It should be remarked that $2^{-n+1}\gamma_n(x)$ is the distance of $(x(2), \ldots, x(n)) \in \mathbb{R}^{n-1}$ from the complement of $[-2^{-n+1}, 2^{-n+1}]^{n-1}$ (the 2^{-n+1}-neighborhood of $\mathbf{0} \in \mathbb{R}^{n-1}$) with respect to the norm $\|y\|_\infty = \max\{|y(1)|, \ldots, |y(n-1)|\}$.[19]

For each $n \in \mathbb{N}$, let $h_n = \psi_1^{(n)} \cdots \psi_1^{(1)} : \mathbb{R}^{\mathbb{N}} \to \mathbb{R}^{\mathbb{N}}$. For the sake of convenience, write $h_0 = \mathrm{id}_{\mathbb{R}^{\mathbb{N}}}$. If there exists a homeomorphism $h_\infty : \mathbb{R}^{\mathbb{N}} \setminus \{0\} \to \mathbb{R}^{\mathbb{N}}$ such that each $x \in \mathbb{R}^{\mathbb{N}} \setminus \{0\}$ has a neighborhood U in $\mathbb{R}^{\mathbb{N}} \setminus \{0\}$ such that $h_\infty|U = \psi_t^{(n+1)}h_n|U = h_n|U$ for a sufficiently large $n \in \mathbb{N}$ and any $t \in \mathbf{I}$, and each $y \in \mathbb{R}^{\mathbb{N}}$ has a neighborhood V such that $h_\infty^{-1}|V = h_n^{-1}(\psi_t^{(n+1)})^{-1}|V = h_n^{-1}|V$ for a sufficiently large $n \in \mathbb{N}$ and any $t \in \mathbf{I}$, then the desired homeomorphism ζ can be defined as follows:

$$\zeta(x, t) = \begin{cases} (\psi^{(n)}_{2-2^n t}h_{n-1}(x), t) & \text{if } 2^{-n} \leqslant t \leqslant 2^{-(n-1)} \\ (h_\infty(x), 0) & \text{if } t = 0. \end{cases}$$

[19]This norm of $\mathbb{R}^{n-1}$ can be replaced with $\|y\|_2 = \left(\sum_{i=1}^{n-1} y(i)^2\right)^{1/2}$ or $\|y\|_1 = \sum_{i=1}^{n-1} |y(i)|$. In this case, γ_n is defined by $\gamma_n(x) = \max\left\{0, 1 - 2^{n-1}\left(\sum_2^n x(i)^2\right)^{1/2}\right\}$ or $\gamma_n(x) = \max\left\{0, 1 - 2^{n-1}\sum_2^n |x(i)|\right\}$.

To see the existence of h_∞, it suffices to prove the following two facts:

(1) *Each* $x \in \mathbb{R}^{\mathbb{N}} \setminus \{0\}$ *has a neighborhood* U *in* $\mathbb{R}^{\mathbb{N}} \setminus \{0\}$ *and* $n \in \mathbb{N}$ *such that* $\psi_t^{(j)}|h_n(U) = \mathrm{id}$ for $j > n$;

(2) *Each* $y \in \mathbb{R}^{\mathbb{N}}$ *has a neighborhood* V *in* $\mathbb{R}^{\mathbb{N}}$ *and* $n \in \mathbb{N}$ *such that* $h_n^{-1}(V) \subset \mathbb{R}^{\mathbb{N}} \setminus \{0\}$ *and* $(\psi_t^{(j)})^{-1}|V = \mathrm{id}$ *for* $j > n$.

(1): For each $x \in \mathbb{R}^{\mathbb{N}} \setminus \{0\}$, let $m = \min\{i \in \mathbb{N} \mid x(i) \neq 0\}$. In other words, $x(m)$ is the first nonzero coordinate of x. We write $y = h_{m-1}(x)$ and $z = h_m(x) = \psi_1^{(m)}(y)$.

When $m = 1$, $y = h_0(x) = x$ and $z = \psi_1^{(1)}(x)$. Since $x(1) \neq 0$, it follows that $(z(1), z(2)) \notin [2^{-1}, \infty) \times \{0\}$, that is, $z(1) < 2^{-1}$ or $z(2) \neq 0$.

When $m > 1$, observe that

$$\begin{aligned} h_{m-1}(x) = y = (y(1), 0, \ldots, 0, y(m), x(m+1), x(m+2), \ldots) \text{ and} \\ h_{m-1}(x) \neq h_{m-1}(0, 0, \ldots, 0, x(m+1), x(m+2), \ldots) \\ = (m-1, 0, \ldots, 0, x(m+1), x(m+2), \ldots). \end{aligned}$$

Thus, we have $y(1) \neq m-1$ or $y(m) \neq 0$.

In the case $y(1) \neq m-1$, since $(z(1), z(m+1)) \notin [m - 2^{-m}, \infty) \times \{0\}$, it follows that $z(1) < m - 2^{-m}$ or $z(m+1) \neq 0$. If $y(m) \neq 0$, then $z(m) = y(m) \neq 0$. Anyway, we have

$$z(1) < m - 2^{-m} \quad \text{or} \quad \max\{|z(2)|, \ldots, |z(m+1)|\} \neq 0.$$

When $z(1) < m - 2^{-m}$, $V = \{z' \in \mathbb{R}^{\mathbb{N}} \mid z'(1) < m - 2^{-m}\}$ is a neighborhood of z in $\mathbb{R}^{\mathbb{N}}$ and $\psi_t^{(j)}|V = \mathrm{id}$ for every $j > m$. Then, $U = h_m^{-1}(V)$ is the desired neighborhood of x in $\mathbb{R}^{\mathbb{N}}$.

When $\max\{|z(2)|, \ldots, |z(m+1)|\} \neq 0$, choose $n \geqslant m+1$ so that $2^{-n} < \max\{|z(2)|, \ldots, |z(m+1)|\}$. Let $z^* = h_n(x)$. Since $z^*(2) = z(2), \ldots, z^*(m+1) = z(m+1)$, we have

$$\max\{|z^*(2)|, \ldots, |z^*(n)|\} \geqslant \max\{|z(2)|, \ldots, |z(m+1)|\} > 2^{-n}.$$

So, x has a neighborhood U in $\mathbb{R}^{\mathbb{N}}$ such that

$$z' \in h_n(U) \Rightarrow \max\{|z'(2)|, \ldots, |z'(n)|\} > 2^{-n}.$$

Then, $\gamma_n(z') = 0$ for each $z' \in h_n(U)$. Hence, $\psi_t^{(j)}|h_n(U) = \mathrm{id}$ for every $j > n$.

(2): For each $y \in \mathbb{R}^{\mathbb{N}}$, choose $n \in \mathbb{N}$ so that $y(1) < n - 2^{-n}$. Then, $V = \{z \in \mathbb{R}^{\mathbb{N}} \mid z(1) < n - 2^{-n}\}$ is a neighborhood of y in $\mathbb{R}^{\mathbb{N}}$ and $h_n(0) = (n, 0, 0, \ldots) \notin V$, that is, $h_n^{-1}(V) \subset \mathbb{R}^{\mathbb{N}} \setminus \{0\}$. For each $j > n$, we have $\psi_t^{(j)}|V = \mathrm{id}$, that is, $(\psi_t^{(j)})^{-1}|V = \mathrm{id}$. □

Using Proposition 2.4.1 above, we prove the following:

Proposition 2.4.2 *Let X be an E-stable perfectly normal paracompact space and A be an E-deficient closed set in X. Then, for each open cover $\mathcal{U}$ of X, there exists an $\mathbf{I}$-preserving homeomorphism*

$$h : X \times \mathbf{I} \to X \times \mathbf{I} \setminus A \times \{0\}$$

such that each h_t is $\mathcal{U}$-close to id_X. *In particular, $h_0 : X \to X \setminus A$ is a homeomorphism that is $\mathcal{U}$-close to* id_X.

Proof For simplicity, we write $F = \mathbb{R}^{\mathbb{N}}$ or $\mathbb{R}^{\mathbb{N}}_f$ depending on $E \approx E^{\mathbb{N}}$ or $E \approx E^{\mathbb{N}}_f$. Since $E \approx E \times F$ by Lemma 2.1.9, X is F-stable and A is F-deficient in X (Corollary 2.3.5). Let $\zeta : F \times \mathbf{I} \setminus \{(\mathbf{0}, 0)\} \to F \times \mathbf{I}$ be the $\mathbf{I}$-preserving homeomorphism obtained in Proposition 2.4.1. Then, we have an $\mathbf{I}$-preserving homeomorphism

$$\varphi : (X \times F \times \mathbf{I}) \setminus (A \times \{\mathbf{0}\} \times \{0\}) \to X \times F \times \mathbf{I}$$

defined by $\varphi(x, y, t) = (x, \zeta_{\max\{t,\alpha(x)\}}(y), t)$, where $\alpha : X \to \mathbf{I}$ is a map such that $\alpha^{-1}(0) = A$. Let $\mathcal{V}$ be an open star-refinement of $\mathcal{U}$. By Corollary 2.3.4(4), we have a homeomorphism $f : X \times F \to X$ such that $f(x, 0) = x$ for each $x \in A$ and f is $\mathcal{V}$-close to pr_X, so $\mathrm{pr}_X f^{-1}$ is $\mathcal{V}$-close to id_X. The desired $\mathbf{I}$-preserving homeomorphism h can be defined by $h(x, t) = (f\varphi_t^{-1} f^{-1}(x), t)$. Indeed, each $h_t = f\varphi_t^{-1} f^{-1}$ is $\mathcal{V}$-close to $\mathrm{pr}_X \varphi_t^{-1} f^{-1} = \mathrm{pr}_X f^{-1}$ because φ preserves X-coordinates. Then, it follows that h_t is $\mathcal{U}$-close to id_X. □

A set A in a space X is said to be **negligible** in X if $X \setminus A \approx X$. It is said that A is **locally closed** in X if each $x \in A$ has a neighborhood U in X such that $A \cap U$ is closed in U. It is easy to see that A is locally closed in X if and only if $A = W \cap \mathrm{cl}_X A$ for some open set W in X. Now, we have the following Negligibility Theorem:

Theorem 2.4.3 (Negligibility) *Let X be an E-stable metrizable space. If A is an E-deficient locally closed set in X, then the inclusion $X \setminus A \subset X$ is a near-homeomorphism, hence A is negligible in X.*

Proof Let $\mathcal{U} \in \mathrm{cov}(X)$ and $d \in \mathrm{Metr}(X)$. Choose an open set W in X so that $A \subset W$ and A is closed in W. Due to Corollary 2.3.4(1), (2), W is E-stable and A is E-deficient. By Proposition 2.1.17, W has an open cover $\mathcal{V}$ that is fitting in X and refines $\mathcal{U}$. By Proposition 2.4.2, we have a homeomorphism $h : W \to W \setminus A$ that is $\mathcal{V}$-close to id. Then, h can be extended to a homeomorphism $\tilde{h} : X \to X \setminus A$ by $\tilde{h}|X \setminus W = \mathrm{id}$. The inverse homeomorphism $\tilde{h}^{-1} : X \setminus A \to X$ is $\mathcal{U}$-close to the inclusion $X \setminus A \subset X$. □

A locally closed set in a metrizable space X is F_σ in X, because it is the intersection of a closed set and an open set. Hence, an E-deficient locally closed set in X is a countable union of E-deficient closed sets in X, where it should be remarked that a countable union of E-deficient locally closed sets in X can be written as a countable union of E-deficient closed sets in X. In the case where X is completely metrizable, Theorem 2.4.3 can be strengthened slightly as follows:

Theorem 2.4.4 (NEGLIGIBILITY) *Let X be an E-stable completely metrizable space. If A is a countable union of E-deficient closed sets in X, then the inclusion $X \setminus A \subset X$ is a near-homeomorphism, hence A is negligible in X.*

Proof Let $A = \bigcup_{n\in\mathbb{N}} A_n$, where each A_n is an E-deficient closed set in X. For each $n \in \mathbb{N}$, let $X_n = X \setminus \bigcup_{i=1}^{n} A_i$. Note that X_n is open in X, and is hence completely metrizable. For each $\mathcal{U} \in \operatorname{cov}(X)$, X has an admissible complete metric d_0 such that $\{\overline{\mathrm{B}}_{d_0}(x, 1) \mid x \in X\} \prec \mathcal{U}$ (cf. 1.3.22(1)).

Since the inclusion $X_1 \subset X$ is a near-homeomorphism by Theorem 2.4.3, we have a homeomorphism $f_1 : X \to X_1$ with $d_0(f_1, \mathrm{id}) < 2^{-1}$ (hence $d_0(f_1^{-1}, \mathrm{id}) < 2^{-1}$). Choose an admissible complete metric d_1 for X_1 so that

$$d_1(x, x') \geqslant \max\{d_0(x, x'),\ d_0(f_1^{-1}(x), f_1^{-1}(x'))\} \quad \text{for each } x, x' \in X_1.$$

Since $A_2 \cap X_1$ is an E-deficient closed set in X_1 and $X_2 = X_1 \setminus A_2$, the inclusion $X_2 \subset X_1$ is a near-homeomorphism by Theorem 2.4.3. Then, we have a homeomorphism $f_2 : X_1 \to X_2$ with $d_1(f_2, \mathrm{id}) < 2^{-2}$ (hence $d_1(f_2^{-1}, \mathrm{id}) < 2^{-2}$). By induction, we can obtain homeomorphisms $f_n : X_{n-1} \to X_n$, and admissible complete metrics d_n for X_n, such that $d_{n-1}(f_n, \mathrm{id}) < 2^{-n}$ (hence $d_{n-1}(f_n^{-1}, \mathrm{id}) < 2^{-n}$) and

$$d_n(x, x') \geqslant \max\{d_{n-1}(x, x'),\ d_{n-1}(f_n^{-1}(x), f_n^{-1}(x'))\} \quad \text{for each } x, x' \in X_n,$$

where $X_0 = X$. Then, $(f_n f_{n-1} \cdots f_1)_{n\in\mathbb{N}}$ is a Cauchy sequence in $\mathrm{C}(X, X)$ with respect to the sup-metric induced by d_0, so it converges to a map

$$f = \lim_{n\to\infty} f_n f_{n-1} \cdots f_1 : X \to X.$$

Since $d_0(f, \mathrm{id}) \leqslant \sum_{n\in\mathbb{N}} 2^{-n} = 1$, it follows that f is $\mathcal{U}$-close to id. For each $x \in X$ and $i \in \mathbb{N}$, let $y = f_i \cdots f_1(x) \in X_i$. Then, $(f_n f_{n-1} \cdots f_{i+1}(y))_{n>i}$ is a Cauchy sequence in X_i with respect to d_i, which implies

$$f(x) = \lim_{n>i} f_n f_{n-1} \cdots f_{i+1}(y) \in X_i.$$

Therefore, $f(X) \subset \bigcap_{i\in\mathbb{N}} X_i = X \setminus A$. It remains to show that $f : X \to X \setminus A$ is a homeomorphism.

To construct the inverse of $f : X \to X \setminus A$, observe that for each $n \in \mathbb{N}$ and $x \in X \setminus A$,

$$\begin{aligned} &d_0(f_1^{-1} \cdots f_{n-1}^{-1} f_n^{-1}(x), f_1^{-1} \cdots f_{n-1}^{-1}(x)) \\ &\quad \leqslant d_1(f_2^{-1} \cdots f_{n-1}^{-1} f_n^{-1}(x), f_2^{-1} \cdots f_{n-1}^{-1}(x)) \leqslant \cdots \\ &\qquad \leqslant d_{n-2}(f_{n-1}^{-1} f_n^{-1}(x), f_{n-1}^{-1}(x)) \leqslant d_{n-1}(f_n^{-1}(x), x) < 2^{-n}. \end{aligned}$$

Hence, $(f_1^{-1} \cdots f_{n-1}^{-1} f_n^{-1} | X \setminus A)_{n \in \mathbb{N}}$ is a Cauchy sequence in $\mathrm{C}(X \setminus A, X)$ with respect to the sup-metric induced by d_0, and this converges to a map

$$g = \lim_{n \to \infty} f_1^{-1} \cdots f_{n-1}^{-1} f_n^{-1} : X \setminus A \to X.$$

For each $x \in X \setminus A$ and $\varepsilon > 0$, choose $n \in \mathbb{N}$ so that

$$d_0(fg(x), ff_1^{-1} \cdots f_{n-1}^{-1} f_n^{-1}(x)) < \varepsilon/2 \ \text{ and } \ d_0(f, f_n f_{n-1} \cdots f_1) < \varepsilon/2,$$

which implies that $d_0(fg(x), x) < \varepsilon$. Hence, $fg(x) = x$ for each $x \in X \setminus A$.

Now, we shall show that $gf(x) = x$ for each $x \in X$.[20] For each $\varepsilon > 0$, choose $n \in \mathbb{N}$ so that

$$\begin{aligned} d_0(g, f_1^{-1} \cdots f_{n-1}^{-1} f_n^{-1} | X \setminus A) < \varepsilon/3 \ \text{ and} \\ d_0(f_1^{-1} \cdots f_{n-1}^{-1} f_n^{-1} | X_m, f_1^{-1} \cdots f_{m-1}^{-1} f_m^{-1}) < \varepsilon/3 \ \text{ for every } m \geqslant n. \end{aligned}$$

By the continuity of $f_1^{-1} \cdots f_{n-1}^{-1} f_n^{-1}$ at $f(x)$, we have $\delta > 0$ such that

$$\begin{aligned} &y \in X_n, \ d_n(f(x), y) < \delta \\ &\qquad \Rightarrow d_0(f_1^{-1} \cdots f_{n-1}^{-1} f_n^{-1} f(x), f_1^{-1} \cdots f_{n-1}^{-1} f_n^{-1}(y)) < \varepsilon/3. \end{aligned}$$

Choose $m \geqslant n$ so that $2^{-m} < \delta$. Then,

$$d_n(f(x), f_m f_{m-1} \cdots f_1(x)) \leqslant \sum_{i=1}^{\infty} d_{m+i}(f_{m+i}, \mathrm{id}) < 2^{-m} < \delta,$$

which implies that

$$d_0(f_1^{-1} \cdots f_{n-1}^{-1} f_n^{-1} f(x), f_1^{-1} \cdots f_{n-1}^{-1} f_n^{-1} f_m f_{m-1} \cdots f_1(x)) < \varepsilon/3.$$

---

[20] Note that g is defined on $\bigcap_{n \in \mathbb{N}} X_n = X \setminus A$ but the image $f_n f_{n-1} \cdots f_1(X)$ is not contained in $X \setminus A$.

It should be remarked that

$$\begin{aligned} &d_0(f_1^{-1}\cdots f_{n-1}^{-1}f_n^{-1}f_m f_{m-1}\cdots f_1, \mathrm{id}) \\ &\qquad = d_0(f_1^{-1}\cdots f_{n-1}^{-1}f_n^{-1}|X_m, f_1^{-1}\cdots f_{m-1}^{-1}f_m^{-1}) < \varepsilon/3. \end{aligned}$$

Then, it follows that

$$\begin{aligned} d_0(gf(x), x) &\leqslant d_0(gf(x), f_1^{-1}\cdots f_{n-1}^{-1}f_n^{-1}f(x)) \\ &\quad + d_0(f_1^{-1}\cdots f_{n-1}^{-1}f_n^{-1}f(x), f_1^{-1}\cdots f_{n-1}^{-1}f_n^{-1}f_m f_{m-1}\cdots f_1(x)) \\ &\quad + d_0(f_1^{-1}\cdots f_{n-1}^{-1}f_n^{-1}f_m f_{m-1}\cdots f_1(x), x) \\ &< \varepsilon/3 + \varepsilon/3 + \varepsilon/3 = \varepsilon. \end{aligned}$$

Therefore, $gf(x) = x$. Thus, f is a homeomorphism with $f^{-1} = g$. □

We consider $C(X) = (\mathbf{I} \times X)_{\{0\}}$ and

$$C^o(X) = ([0, 1) \times X)_{\{0\}} \approx ([0, \infty) \times X)_{\{0\}},$$

where $C^o(X)$ is called the **open cone** over X.

Proposition 2.4.5 $E \approx C^o(E)$.

Proof Let $\mathbf{S}_E$ be the unit sphere of E. Then, we have homeomorphisms:

$$E \approx E \setminus \{\mathbf{0}\} \overset{f}{\approx} \mathbf{S}_E \times (0, \infty) \approx \mathbf{S}_E \times \mathbb{R},$$

where the first homeomorphism is obtained by the Negligibility Theorem 2.4.3 above (or Proposition 2.4.2) and the second homeomorphism f is defined by $f(x) = (\|x\|^{-1}x, \|x\|)$. Note that $E \times \mathbb{R}$ is E-stable and $\{0\} \times \mathbb{R}$ is E-deficient in $E \times \mathbb{R}$. By the Negligibility Theorem 2.4.3, we have

$$E \times \mathbb{R} \approx E \times \mathbb{R} \setminus \big(\{\mathbf{0}\} \times (\mathbb{R} \setminus \{0\})\big) = (E \setminus \{\mathbf{0}\}) \times \mathbb{R} \cup \{(\mathbf{0}, 0)\}.$$

The desired homeomorphism is obtained by the following diagram:

$$\begin{array}{ccc} E \approx E \times \mathbb{R} & \approx & (E \setminus \{\mathbf{0}\}) \times \mathbb{R} \cup \{(\mathbf{0}, 0)\} \\ \approx\big\downarrow & & \big\downarrow g \\ C^o(E) \approx C^o(\mathbf{S}_E \times \mathbb{R}) & \approx & \big([0, \infty) \times (\mathbf{S}_E \times \mathbb{R})\big)_{\{0\}}, \end{array}$$

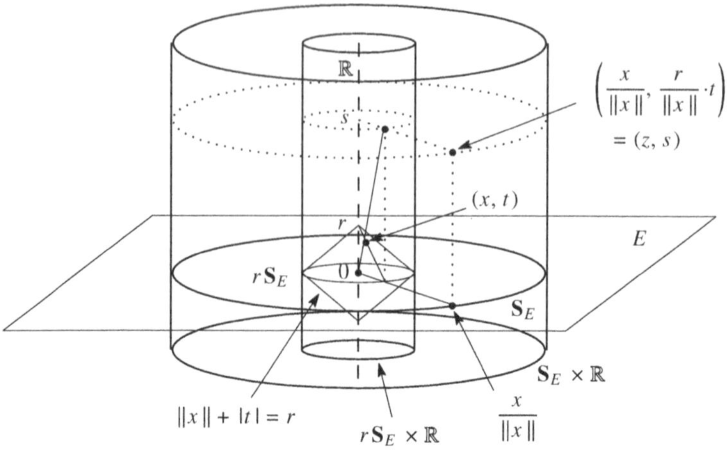

Fig. 2.8 $(E \setminus \{\mathbf{0}\}) \times \mathbb{R} \cup \{(\mathbf{0}, 0)\} \approx \big([0, \infty) \times (\mathbf{S}_E \times \mathbb{R})\big)_{\{0\}}$

where g is defined as follows:

$$g(\mathbf{0}, 0) = \mathbf{0} \text{ and } g(x, t) = \left(\|x\| + |t|,\ \frac{x}{\|x\|},\ \frac{\|x\| + |t|}{\|x\|} \cdot t\right).$$

It is easy to see that g is continuous. The inverse of g can be defined by

$$g^{-1}(\mathbf{0}) = (\mathbf{0}, 0) \text{ and } g^{-1}(r, z, s) = \left(\frac{r^2}{r + |s|} \cdot z,\ \frac{rs}{r + |s|}\right).$$

The continuity of g^{-1} at $\mathbf{0}$ comes from the following:

$$\left\|\frac{r^2}{r + |s|} \cdot z\right\| \leqslant r \text{ and } \left|\frac{rs}{r + |s|}\right| \leqslant r \text{ for } (r, z, s) \in (0, \infty) \times \mathbf{S}_E \times \mathbb{R}.$$

The proof is complete (Fig. 2.8). □

For a closed set X in E, the open cone $C^o(X)$ is closed in $C^o(E) \approx E$ and the cone $C(X) = (\mathbf{I} \times X)_{\{0\}}$ is closed in $([0, \infty) \times E)_{\{0\}} \approx C^o(E) \approx E$. Thus, the following holds:

Corollary 2.4.6 *For every $X \in \mathcal{F}_E$, $C^o(X) \in \mathcal{F}_E$ and $C(X) \in \mathcal{F}_E$.* □

Moreover, we have the following:

Corollary 2.4.7 *For every locally finite-dimensional simplicial complex K with* card $K^{(0)} \leqslant w(E)$, *the polyhedron* $|K|_{\mathrm{m}}$ *with the metric topology can be embedded in E as a closed set, i.e.,* $|K|_{\mathrm{m}} \in \mathcal{F}_E$.

Proof First, suppose that $\dim K < \infty$. Then, by induction on $\dim K$, we shall show that $|K|_{\mathrm{m}} \in \mathcal{F}_E$. Since E contains a discrete set D with card $D = w(E)$ by Lemma 2.1.10, we have the case $\dim K = 0$. Assuming the $(n-1)$-dimensional case, we prove the case $\dim K = n$. For each $v \in K^{(0)}$, $|\operatorname{Lk}(v, K)|_{\mathrm{m}} \in \mathcal{F}_E$ and $|\operatorname{St}(v, K)|_{\mathrm{m}}$ is homeomorphic to the metrizable cone $C(|\operatorname{Lk}(v, K)|_{\mathrm{m}})$, hence $|\operatorname{St}(v, K)|_{\mathrm{m}} \in \mathcal{F}_E$ by Corollary 2.4.6 above. In the same way as Theorem 2.1.11, we can apply Michael's Theorem on local properties (Corollary 1.3.21) to obtain $|K|_{\mathrm{m}} \in \mathcal{F}_E$.

When K is locally finite-dimensional, it follows from the above finite-dimensional case that $|\operatorname{St}(v, K)|_{\mathrm{m}} \in \mathcal{F}_E$ for each $v \in K^{(0)}$. Again, we can apply Michael's Theorem (Corollary 1.3.21) to obtain $|K|_{\mathrm{m}} \in \mathcal{F}_E$. □

Corollary 2.4.8 $E \times [0, 1) \approx E$.

Proof First, note that $E \times [0, 1)$ is E-stable. Then, by Lemma 2.3.2 and Proposition 2.4.5, we have

$$\begin{aligned} E \times [0,1) &\approx ((E \times [0,1)) \times E)_{E \times \{0\}} = E \times ([0,1) \times E)_{\{0\}} \\ &= E \times C^o(E) \approx E \times E \approx E. \end{aligned}$$ □

A subset A in a space X is said to be **collared** in X if there is an open embedding $k : A \times [0, 1) \to X$ such that $k(x, 0) = x$ for all $x \in A$. Such an embedding k is called a **collar** of A in X. The following can be easily observed:

Fact *Every collared set in X is locally closed in X.*

Using collars, we can characterize the E-deficiency as follows:

Theorem 2.4.9 *Let X be an E-stable perfectly normal paracompact space with $A \subset X$. Suppose that E is a normed linear space. Then, the following are equivalent:*

(a) *A is E-deficient in X;*
(b) *The closure* cl A *is contained in some collared set in X;*
(c) *There is a homeomorphism $h : X \to X \times [0, 1)$ such that $h(A) \subset X \times \{0\}$.*

Proof The implication (c) ⇒ (b) is trivial, because $h(\operatorname{cl} A) = \operatorname{cl} h(A) \subset X \times \{0\}$ in the condition (c).

(a) ⇒ (c): Since $E \times [0, 1) \approx E$ (Corollary 2.4.8), we have a homeomorphism $h : X \to X \times [0, 1)$ such that $h(A) \subset X \times \{0\}$ by Corollary 2.3.5.

(b) ⇒ (a): Let C be a collared set in X with $\operatorname{cl} A \subset C$. Then, $C \times [0, 1)$ is homeomorphic to an open set in X, hence it is E-stable by Corollary 2.3.4(1). It suffices to show that $C \times \{0\}$ is E-deficient in $C \times [0, 1)$. Indeed, this implies that $\operatorname{cl} A \times \{0\}$ is E-deficient in $C \times [0, 1)$. Since every locally E-deficient closed set in X is E-deficient (Proposition 2.3.8), $\operatorname{cl} A$ is E-deficient in X, which implies (a). By Lemma 2.3.2, we have

$$\begin{aligned}\big(C \times [0, 1), C \times \{0\}\big) &\approx \big(((C \times [0, 1)) \times E)_{C \times \{0\}}, C \times \{0\}\big)\\ &= \big(C \times ([0, 1) \times E)_{\{0\}}, C \times \{0\}\big)\end{aligned}$$

On the other hand, $([0, 1) \times E)_{\{0\}} = C^o(E) \approx E \approx [0, 1) \times E$ by Proposition 2.4.5 and Corollary 2.4.8. Then, it follows from the homogeneity of E that

$$\big(C \times [0, 1), C \times \{0\}\big) \approx \big(C \times [0, 1) \times E, C \times \{0\} \times \{0\}\big),$$

which implies that $C \times \{0\}$ is E-deficient in $C \times [0, 1)$. □

Corollary 2.4.10 *$C(E) \approx E$ and $E \times \mathbf{I} \approx E$.*

Proof By Proposition 2.4.5 and Corollary 2.4.8, $C(E) = C^o(E) \cup (0, 1] \times E$ is an E-manifold, hence it is E-stable (Theorem 2.3.7). Since a collared set in an E-stable space is negligible (Theorems 2.4.9 and 2.4.3), we have

$$C(E) \approx C(E) \setminus E \times \{1\} = C^o(E) \approx E$$

by Proposition 2.4.5. On the other hand, $E \times \mathbf{I}$ is also an E-manifold by Corollary 2.4.8. Similarly, we have $E \times \mathbf{I} \approx E \times [0, 1) \approx E$. □

We define $\boldsymbol{Q}_f = [-1, 1]^{\mathbb{N}}_f \subset [-1, 1]^{\mathbb{N}} = \boldsymbol{Q}$.

Corollary 2.4.11 *Depending on $E \approx E^{\mathbb{N}}$ or $E \approx E^{\mathbb{N}}_f$,*

$$E \times \boldsymbol{Q} \approx E \quad or \quad E \times \boldsymbol{Q}_f \approx E. \qquad \square$$

Remark 2.5 Since $E \times \mathbf{I} \approx E$ by Corollary 2.4.10, Theorem 2.4.9 is valid even if $[0, 1)$ in (c) can be replaced with $\mathbf{I}$. Moreover, under the assumption of Theorem 2.4.9, A is E-deficient in X if and only if A is $\boldsymbol{Q}$-deficient (resp. $\boldsymbol{Q}_f$-deficient) in X in the case that $E^{\mathbb{N}} \approx E$ (resp. $E^{\mathbb{N}}_f \approx E$) by Corollaries 2.3.5 and 2.4.11.

As with the model space E (Corollary 2.4.10), the Hilbert cube $\boldsymbol{Q}$ is homeomorphic to the cone $C(\boldsymbol{Q})$. This will be proved in a different way in Sect. 2.7. Moreover, the $\boldsymbol{Q}$-deficient closed set in a $\boldsymbol{Q}$-manifold can be characterized in the same way as in Theorem 2.4.9, where E and $[0, 1)$ are replaced with $\boldsymbol{Q}$ and $\mathbf{I}$. This will be shown in Theorem 2.10.9.

Combining the Stability Theorem 2.3.7 with Corollaries 2.4.8, 2.4.10, and 2.4.11, we have the following:

Corollary 2.4.12 *For every E-manifold M, $M \times [0, 1) \approx M \times \mathbf{I} \approx M$. And, depending on $E \approx E^{\mathbb{N}}$ or $E \approx E_f^{\mathbb{N}}$, $M \times \boldsymbol{Q} \approx M$ or $M \times \boldsymbol{Q}_f \approx M$.* □

In the rest of this section, we shall prove that every finite union of E-deficient sets is also E-deficient. The following is a key lemma:

Lemma 2.4.13 *There exists a homeomorphism $h : \mathbf{I} \times E \to (\mathbf{I} \times E)_{\{0\}}$ such that $h|(0, 1] \times \{\mathbf{0}\} = \mathrm{id}$ and $h(0, \mathbf{0}) = 0$.*

Proof Since $E \approx C(E) = (\mathbf{I} \times E)_{\{0\}}$ by Corollary 2.4.10 and E is homogeneous, there is a homeomorphism $f : E \to (\mathbf{I} \times E)_{\{0\}}$ with $f(\mathbf{0}) = 0$. Then, we have a homeomorphism

$$f' : (\mathbf{I} \times E)_{\{0\}} \to (\mathbf{I} \times (\mathbf{I} \times E)_{\{0\}})_{\{0\}}$$

such that $f'(0) = 0$ and $f'|(0, 1] \times E = \mathrm{id} \times f$. As is easily observed,

$$\mathbf{I}^2 \approx \left\{x \in \mathbf{I}^2 \;\middle|\; x(1) \leqslant x(2)\right\} \approx (\mathbf{I} \times \mathbf{I})_{\{0\}}.$$

We can easily construct a homeomorphism $g : \mathbf{I} \times \mathbf{I} \to (\mathbf{I} \times \mathbf{I})_{\{0\}}$ such that $g(0, 0) = 0$ and $g|(0, 1] \times \{0\} = \mathrm{id}$, which induces the homeomorphism

$$g' : ((\mathbf{I} \times \mathbf{I}) \times E)_{\mathbf{I} \times \{0\}} \to ((\mathbf{I} \times \mathbf{I})_{\{0\}} \times E)_{\{0\} \cup (0,1] \times \{0\}}.$$

Then, $g'|\mathbf{I} \times \{0\} = g|\mathbf{I} \times \{0\}$ (i.e., $g'(0, 0) = 0$ and $g'|(0, 1] \times \{0\} = \mathrm{id}$), and $g'|(\mathbf{I} \times (0, 1]) \times E = (g|\mathbf{I} \times (0, 1]) \times \mathrm{id}_E$. We define the desired homeomorphism h as the composition of three homeomorphisms in the following diagram:

$$\begin{array}{ccc}
\mathbf{I} \times E & \xrightarrow{\quad h \quad} & (\mathbf{I} \times E)_{\{0\}} \\
{\scriptstyle \mathrm{id}_{\mathbf{I}} \times f}\big\downarrow & & \big\uparrow{\scriptstyle f'^{-1}} \\
\mathbf{I} \times (\mathbf{I} \times E)_{\{0\}} & & (\mathbf{I} \times (\mathbf{I} \times E)_{\{0\}})_{\{0\}} \\
\Big\| & & \Big\| \\
((\mathbf{I} \times \mathbf{I}) \times E)_{\mathbf{I} \times \{0\}} & \xrightarrow[\quad g' \quad]{} & ((\mathbf{I} \times \mathbf{I})_{\{0\}} \times E)_{\{0\} \cup (0,1] \times \{0\}}
\end{array}$$

For each $t \in (0, 1]$, since $f'h(t, \mathbf{0}) = g'(t, f(\mathbf{0})) = g'(t, 0) = (t, 0)$ and $f'(t, \mathbf{0}) = (t, f(\mathbf{0})) = (t, 0)$, it follows that $h(t, \mathbf{0}) = (t, 0)$. Moreover, $f'h(0, \mathbf{0}) = g'(0, f(\mathbf{0})) = g'(0, 0) = 0$. □

Proposition 2.4.14 *Let X be an E-stable perfectly normal space. Every finite union of E-deficient sets in X is also E-deficient.*

Proof It suffices to show that the union of two E-deficient sets A and B in X is also E-deficient, where we may assume that A and B are closed in X because their closures in X are E-deficient.

Since $\mathbf{I} \times E \approx E$ by Corollary 2.4.10, we have a homeomorphism $f : X \to X \times \mathbf{I}$ such that $f(A) \subset X \times \{0\}$ (Corollary 2.3.5). Let $C = f^{-1}(X \times \{0\})$. Then, $A \subset C$ and f induces the homeomorphism

$$f' : (X \times E)_C \to ((X \times \mathbf{I}) \times E)_{X \times \{0\}} = X \times (\mathbf{I} \times E)_{\{0\}}.$$

Then, $f'|C = f|C$ and $f'|(X \setminus C) \times E = (f|X \setminus C) \times \mathrm{id}_E$. By Lemma 2.4.13 above, we have a homeomorphism $h : \mathbf{I} \times E \to (\mathbf{I} \times E)_{\{0\}}$ such that $h(t, \mathbf{0}) = (t, \mathbf{0})$ for every $t \in (0, 1]$ and $h(0, \mathbf{0}) = 0$. Let $h' : X \times E \to (X \times E)_C$ be the homeomorphism defined by the following diagram:

$$\begin{array}{ccc} X \times E & \xrightarrow{\quad h' \quad} & (X \times E)_C \\ {\scriptstyle f \times \mathrm{id}_E} \downarrow & & \uparrow {\scriptstyle f'^{-1}} \\ X \times \mathbf{I} \times E & \xrightarrow[\mathrm{id}_X \times h]{} & X \times (\mathbf{I} \times E)_{\{0\}} \end{array}$$

For each $x \in C$, $h'(x, \mathbf{0}) = x$ because $f(x) = f'(x) \in X \times \{\mathbf{0}\}$. For each $x \in X \setminus C$, let $f(x) = (x', t)$. Then, $t > 0$. Since $f'h'(x, \mathbf{0}) = (x', h(t, \mathbf{0})) = (x', t, \mathbf{0})$ and $f'(x, \mathbf{0}) = (f(x), \mathbf{0}) = (x', t, \mathbf{0})$, we have $h'(x, \mathbf{0}) = (x, \mathbf{0})$.

Applying Lemma 2.3.2 ($A = B$, $A_0 = C$, $A_1 = X$), we have a homeomorphism $g : (X \times E)_C \to X$ such that $g(x, \mathbf{0}) = x$ for each $x \in B \setminus C$ and $g|C = \mathrm{id}$. Then, $gh' : X \times E \to X$ is a homeomorphism. For each $x \in C$, $gh'(x, \mathbf{0}) = g(x) = x$. For each $x \in B \setminus C$, $gh'(x, \mathbf{0}) = g(x, \mathbf{0}) = x$. Thus, $gh'(x, \mathbf{0}) = x$ for each $x \in A \cup B$. □

2.5 The Collaring and Unknotting Theorems

In this section, we show that an E-manifold closed subset of an E-manifold is collared if it is E-deficient, that every E-manifold can be embedded in E as an open set, and that a homeomorphism between E-deficient closed sets in an E-manifold M can be extended to a homeomorphism of M onto itself if it is homotopic to id in M. We frequently use the following lemma in this section:

Lemma 2.5.1 *Let A be a closed set in a perfectly normal paracompact space X, $a \leqslant 0 \leqslant b$, and $\mathcal{U}$ be an open cover of $A \times [a, b]$ in $X \times \mathbb{R}$ such that*

$$\big\{\{x\} \times [a, b] \bigm| x \in A\big\} \prec \mathcal{U}.$$

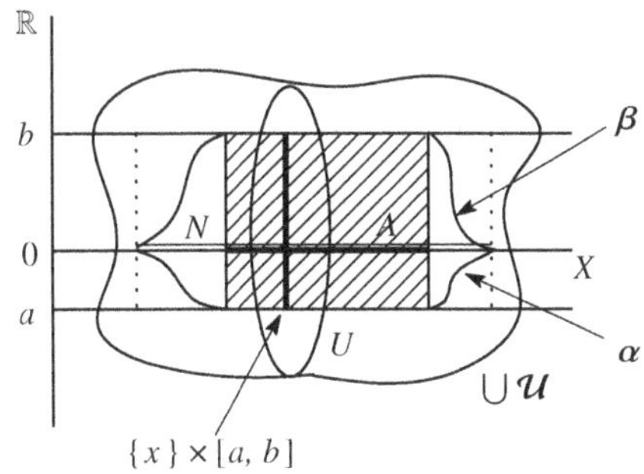

Fig. 2.9 The maps α and β

Then, there exist maps $\alpha, \beta, \gamma : X \to \mathbb{R}$ such that:

(i) $a \leqslant \alpha(x) \leqslant 0 \leqslant \beta(x) \leqslant b$ *for* $x \in X$,
(ii) $\alpha(x) = a$ *and* $\beta(x) = b$ *if* $x \in A$,
(iii) $\alpha(x) = \beta(x) = \gamma(x) = 0$ *if* $(x, 0) \notin \bigcup \mathcal{U}$,
(iv) $\gamma(x) > 0$ *if* $(x, 0) \in \bigcup \mathcal{U}$,
(v) $\big\{\{x\} \times [\alpha(x) - \gamma(x), \beta(x) + \gamma(x)] \bigm| (x, 0) \in \bigcup \mathcal{U}\big\} \prec \mathcal{U}$.

Proof First, observe that A is contained in the following open set:

$$N = \big\{x \in X \bigm| \{x\} \times [a, b] \subset U \text{ for some } U \in \mathcal{U}\big\}.$$

Let $k : X \to \mathbf{I}$ be an Urysohn map with $k(X \setminus N) = 0$ and $k(A) = 1$. We define maps $\alpha, \beta : X \to \mathbb{R}$ by $\alpha(x) = ak(x)$ and $\beta(x) = bk(x)$. Then, α and β satisfy (i), (ii), and (iii). For each $x \in N$, $\{x\} \times [\alpha(x), \beta(x)]$ is contained in some $U \in \mathcal{U}$ (Fig. 2.9).

Let $U_0 = \{x \in X \mid (x, 0) \in \bigcup \mathcal{U}\}$. Then, U_0 is an open set in X and $\bigcup \mathcal{U} \cap (X \times \{0\}) = U_0 \times \{0\}$. We define $\gamma' : U_0 \to (0, 1]$ as follows:

$$\gamma'(x) = \sup \big\{t \in (0, 1) \bigm| \exists U \in \mathcal{U} \text{ such that } \{x\} \times [\alpha(x) - t, \beta(x) + t] \subset U\big\}.$$

Then, γ' is l.s.c. Indeed, if $\gamma'(x) > t > s$ then $\{x\} \times [\alpha(x) - t, \beta(x) + t]$ is contained in some $U \in \mathcal{U}$, so x has a neighborhood V in X such that

$$V \times [\alpha(x) - t, \beta(x) + t] \subset U.$$

Since α and β are continuous, x has a neighborhood W in X such that $W \subset V$ and $|\alpha(x) - \alpha(y)|, |\beta(x) - \beta(y)| < t - s$ for $y \in W$. Hence,

$$\{y\} \times [\alpha(y) - s, \beta(y) + s] \subset U \text{ for every } y \in W,$$

that is, $\gamma'(y) > s$ for $y \in W$.

Note that U_0 is paracompact because X is hereditarily paracompact by Theorem 1.3.13. Applying Theorem 1.3.14, we have a map $\gamma_0 : U_0 \to (0, 1)$ such that $0 < \gamma_0(x) < \gamma'(x)$ for each $x \in U_0$. Since X is perfectly normal, $X \setminus U_0 = k_0^{-1}(0)$ for some map $k_0 : X \to \mathbf{I}$. We can define a map $\gamma : X \to \mathbf{I}$ by $\gamma(X \setminus U_0) = \{0\}$ and $\gamma(x) = k_0(x)\gamma_0(x)$ for each $x \in U_0$. It is easy to see that γ satisfies conditions (iii), (iv), and (v). □

Remark 2.6 In Lemma 2.5.1 above, if X is not assumed to be perfectly normal, we define a map γ using k instead of k_0. This does not satisfy (iv), but satisfies the following weaker condition:

(iv)′ $\gamma(x) > 0$ if $\alpha(x) < \beta(x)$ (equivalently $\alpha(x) < 0$ or $\beta(x) > 0$).

When $a = b = 0$ in Lemma 2.5.1, we have the following:

Lemma 2.5.2 *Let X be a perfectly normal paracompact space.*[21] *For each open collection $\mathcal{U}$ in $X \times \mathbf{I}$, there exists a map $\gamma : X \to \mathbf{I}$ such that $\gamma(x) > 0$ if and only if $(x, 0) \in \bigcup \mathcal{U}$, and*

$$\big\{\{x\} \times [0, \gamma(x)] \bigm| x \in X,\ (x, 0) \in \bigcup \mathcal{U}\big\} \prec \mathcal{U}.$$

In particular, for an open set U in $X \times \mathbf{I}$, there is a map $\gamma : X \to \mathbf{I}$ such that $\gamma(x) > 0$ if and only if $(x, 0) \in U$, and

$$\big\{(x, t) \in X \times \mathbf{I} \bigm| \gamma(x) > 0,\ t \leqslant \gamma(x)\big\} \subset U.$$

□

In the above, if $\mathcal{U}$ is an open cover (or U is an open neighborhood) of $X \times \{0\}$ in $X \times \mathbf{I}$, without assuming the perfect normality of X, we can prove the same result, that is:

Lemma 2.5.3 *Let X be a paracompact space. For each open cover $\mathcal{U}$ of $X \times \{0\}$ in $X \times \mathbf{I}$, there exists a map $\gamma : X \to \mathbf{I}$ such that $\gamma(x) > 0$ if and only if*

$$\big\{\{x\} \times [0, \gamma(x)] \bigm| x \in X,\ \gamma(x) > 0\big\} \prec \mathcal{U}.$$

In particular, for an open neighborhood U of $X \times \{0\}$ in $X \times \mathbf{I}$, there is a map $\gamma : X \to \mathbf{I}$ such that $\gamma(x) > 0$ if and only if $(x, 0) \in U$, and

$$\big\{(x, t) \in X \times \mathbf{I} \bigm| \gamma(x) > 0,\ t \leqslant \gamma(x)\big\} \subset U.$$

□

Lemma 2.5.4 *Let X be a hereditarily paracompact space, that is, every open set in X is paracompact (cf. Theorem 1.3.13). If A is a collared set in X and U is an open set in X with $A \cap U \neq \emptyset$, then $A \cap U$ is collared in U.*

[21] If $\mathcal{U}$ is an open cover (or U is an open neighborhood) of $X \times \{0\}$ in $X \times \mathbf{I}$, the perfect normality of X need not be assumed.

Proof Let $k : A \times [0, 1) \to X$ be a collar of A in X. Applying Lemma 2.5.3, we have a map $\gamma : A \cap U \to (0, 1)$ such that

$$\{(x, t) \in (A \cap U) \times [0, 1) \mid t < \gamma(x)\} \subset k^{-1}(U).$$

We define $h : (A \cap U) \times [0, 1) \to U$ by $h(x, t) = k(x, \gamma(x)t)$. Then, h is a collar of $A \cap U$ in U. □

For a collared closed set A in a space X, a collar $k : A \times [0, 1) \to X$ may have a defect such that $k(A \times [0, t])$ need not be closed in X for $0 < t < 1$. For example, $(0, \infty) \times \{0\}$ is a closed set in $[0, \infty)^2 \setminus \{(0, 0)\}$ that has the natural collar with the image $(0, \infty) \times [0, 1)$. On this account, we introduce a collar without such a defect. A **closed collar** of A in X is a closed embedding $k : A \times \mathbf{I} \to X$ such that $k|A \times [0, 1)$ is a collar of A in X.

Proposition 2.5.5 *Every collared closed set A in a paracompact space X has a closed collar in X. Moreover, for each open cover $\mathcal{U}$ of X, there is a closed collar $k : A \times \mathbf{I} \to X$ such that $\{k(\{x\} \times \mathbf{I}) \mid x \in A\} \prec \mathcal{U}$.*

Proof Let $h : A \times [0, 1) \to X$ be a collar of A in X. Choose an open set U in X so that

$$A \subset U \subset \operatorname{cl} U \subset h(A \times [0, 1)).$$

By Lemma 2.5.3, we have a map $\gamma : A \to (0, 1)$ such that

$$A_\gamma = \{(x, t) \in A \times [0, 1) \mid t \leqslant \gamma(x)\} \subset h^{-1}(U).$$

Since A_γ is closed in $A \times [0, 1)$ and $h(A_\gamma) \subset \operatorname{cl} U \subset h(A \times [0, 1))$, it follows that $h(A_\gamma)$ is closed in $\operatorname{cl} U$, hence it is closed in X. Then, a closed collar $k : A \times \mathbf{I} \to X$ of A can be defined by $k(x, t) = h(x, \gamma(x)t)$.

When an open cover $\mathcal{U}$ of X is given above, γ can be taken by Lemma 2.5.3 so as to satisfy

$$\{\{x\} \times [0, \gamma(x)] \mid x \in A\} \prec h^{-1}(\mathcal{U}).$$

Then, we have $\{k(\{x\} \times \mathbf{I}) \mid x \in A\} \prec \mathcal{U}$. Thus, the additional statement is also true. □

Remark 2.7 A subset B of a space X is said to be **bi-collared** in X if there is an open embedding $k : B \times (-1, 1) \to X$ such that $k(x, 0) = 0$ for every $x \in B$, where k is called a **bi-collar** of B in X. Then, B is collared in $X \setminus k(B \times (-1, 0))$ and in $X \setminus k(B \times (0, 1))$. Every collared closed set A in a paracompact space X has a closed collar $k : A \times \mathbf{I} \to X$ such that $\operatorname{bd} k(A \times \mathbf{I}) = k(A \times \{1\})$ is bi-collared in X. Indeed, using a closed collar $k' : A \times \mathbf{I} \to X$ obtained by Proposition 2.5.5, such a closed collar $k : A \times \mathbf{I} \to X$ can be defined by $k(x, t) = k'(x, t/2)$.

When X is metrizable, we can show the following:

Lemma 2.5.6 *Let A be a collared set in a metrizable space X and U be an open set in X such that $A \subset U$. Then, there is a closed map $h : \mathrm{cl}\, A \times \mathbf{I} \to X$ such that $h|A \times \mathbf{I}$ is an embedding into U, $h|A \times [0, 1)$ is a collar of A in X, and $h(\{x\} \times \mathbf{I}) = \{x\}$ for each $x \in \mathrm{cl}\, A \setminus A$.*

Proof Since A is locally closed in X, we can assume that $A = U \cap \mathrm{cl}\, A$ by replacing U with a small open set. We have a collar $k : A \times [0, 1) \to U$ by Lemma 2.5.4. Then, note that

$$k(A \times [0, 1)) \cap (\mathrm{cl}\, A \setminus A) = \emptyset.$$

Take $d \in \mathrm{Metr}(X)$ and let

$$V = \{k(x, t) \mid (x, t) \in A \times [0, 1),\ d(k(x, t), x) < 2^{-1} d(x, X \setminus k(A \times [0, 1)))\}.$$

Then, $A \subset V \subset U$ and V is open in X because $k(A \times [0, 1))$ is open in X. By Lemma 2.5.2, there is a map $\gamma : A \to (0, 1)$ (as illustrated in the left-hand side of Fig. 2.10) such that

$$\{(x, t) \in A \times [0, 1) \mid t \leqslant \gamma(x)\} \subset k^{-1}(V).$$

For each $(x, t) \in A \times \mathbf{I}$ and $y \in X \setminus k(A \times [0, 1))$, since $k(x, t\gamma(x)) \in V$, it follows that $d(k(x, t\gamma(x)), x) < 2^{-1} d(x, y)$, hence

$$\begin{aligned} d(x, y) &\leqslant d(k(x, t\gamma(x)), x) + d(k(x, t\gamma(x)), y) \\ &< 2^{-1} d(x, y) + d(k(x, t\gamma(x)), y) \quad \text{and} \end{aligned}$$
$$d(k(x, t\gamma(x)), y) \leqslant d(k(x, t\gamma(x)), x) + d(x, y) < 2^{-1} d(x, y) + d(x, y).$$

Thus, we have the following:

$$2^{-1} d(x, y) < d(k(x, t\gamma(x)), y) < 2d(x, y).$$

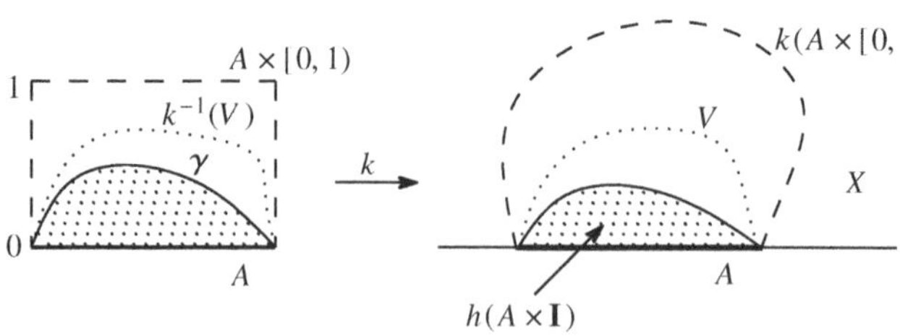

Fig. 2.10 Refining a collar $k : A \times \mathbf{I} \to X$

By this inequality, the desired map $h : \mathrm{cl}\, A \times \mathbf{I} \to X$ can be defined as follows (see Fig. 2.10):

$$h(x,t) = \begin{cases} k(x, t\gamma(x)) & \text{if } x \in A, \\ x & \text{if } x \in \mathrm{cl}\, A \setminus A. \end{cases}$$

Indeed, the continuity of h at $(x,t) \in (\mathrm{cl}\, A \setminus A) \times \mathbf{I}$ can be seen as follows: Take $(x_n, t_n) \in A \times \mathbf{I}$, $n \in \mathbb{N}$, so that $(x_n, t_n) \to (x, t)$ as $n \to \infty$. Since $x \in \mathrm{cl}\, A \setminus A$, it follows that

$$d(h(x_n, t_n), h(x,t)) = d(k(x_n, t_n\gamma(x_n)), x) < 2d(x_n, x) \to 0.$$

To show that h is closed, let $(x_n, t_n) \in \mathrm{cl}\, A \times \mathbf{I}$, $n \in \mathbb{N}$, and assume $h(x_n, t_n) \to y$ for some $y \in X$. Then, it suffices to show that (x_n, t_n) has a convergent subsequence. Since $\mathbf{I}$ is compact, we may assume that $t_n \to t$ for some $t \in \mathbf{I}$.

When $y \in k(A \times [0,1))$, since k is an open embedding, it follows that (x_n, t_n) is convergent in $A \times [0,1)$.

When $y \notin k(A \times [0,1))$, if $(x_n, t_n) \notin A \times [0,1)$ for infinitely many $n \in \mathbb{N}$, then $(x_n)_{n\in\mathbb{N}}$ has a subsequence that converges to y. Otherwise, we may assume that $(x_n, t_n) \in A \times [0,1)$ for every $n \in \mathbb{N}$. Then, it follows that

$$2^{-1}d(x_n, y) < d(k(x_n, t_n\gamma(x_n)), y) = d(h(x_n, t_n), y) \to 0,$$

hence $x_n \to y$.

It is easy to see that h satisfies the other conditions. □

A subset A in a space X is said to be **locally collared** in X if each $x \in A$ has a neighborhood in A that is collared in X. Obviously, a collared set is locally collared, but the converse is also true, that is, the following theorem holds:

Theorem 2.5.7 (M. Brown) *Every locally collared set A in a metrizable space X is collared in X.*

Proof First, observe that A is locally closed in X, that is, closed in some open set in X. Since a collared set in an open set in X is collared in X, we may assume that A is closed in X. To apply Michael's Theorem 1.3.20, it suffices to show the following:

(G-1) If U is open in A and collared in X, then each open set V in U is also collared in X;
(G-2) If U and V are open in A and collared in X, then $U \cup V$ is collared in X;
(G-3) If $\{U_\lambda \mid \lambda \in \Lambda\}$ is a discrete collection of open sets in A such that every U_λ is collared in X, then $\bigcup_{\lambda\in\Lambda} U_\lambda$ is also collared in X.

Note that (G-1) easily follows from Lemma 2.5.4. For (G-3), we have a discrete collection $\{W_\lambda \mid \lambda \in \Lambda\}$ of open sets in X such that $U_\lambda \subset W_\lambda$ (cf. the definition of collectionwise normality and its remark before Proposition 1.3.12(1)). Then, (G-3) can be obtained by Lemma 2.5.4.

To show (G-2), let U and V be open sets in A that are collared in X. Take an open set $\widetilde{U}$ in X so that $U = A \cap \widetilde{U}$ and apply Lemma 2.5.6 to obtain a closed map $h : \operatorname{cl} U \times \mathbf{I} \to X$ such that $h|U \times \mathbf{I}$ is an embedding into $\widetilde{U}$, $h|U \times [0, 1)$ is a collar of U in X, and $h(\{x\} \times \mathbf{I}) = \{x\}$ for $x \in \operatorname{cl} U \setminus U$. Observe

$$A \cap h(\operatorname{cl} U \times \mathbf{I}) = (A \cap h(U \times \mathbf{I})) \cup (A \cap \operatorname{cl} U) \subset (A \cap \widetilde{U}) \cup \operatorname{cl} U = \operatorname{cl} U,$$

hence $A \cap h(\operatorname{cl} U \times \mathbf{I}) = \operatorname{cl} U$. Then, we have the homeomorphism

$$\varphi : X \to X \setminus h(U \times [0, 1/2)) \subset X$$

defined by $\varphi|X \setminus h(U \times [0, 1)) = \mathrm{id}$ and $\varphi(h(x, t)) = h(x, t/2 + 1/2)$ for each $(x, t) \in \operatorname{cl} U \times \mathbf{I}$, where $\varphi(x) = h(x, 1/2)$ for each $x \in \operatorname{cl} U$.

We have a collar $k : V \times [0, 1) \to X$ of V in X, where it should be noted that $\varphi k \varphi^{-1}|\varphi(V) \times \mathrm{id}_{[0,1)}$ is a collar of $\varphi(V)$ in $\varphi(X)$. Let $\alpha : U \cup V \to \mathbf{I}$ be a map with $\alpha^{-1}(0) = V \setminus U$ and $\alpha^{-1}(1) = U \setminus V$. Then, we can define a map $k^* : (U \cup V) \times [0, 1) \to X$ as follows (see Fig. 2.11):

$$k^*(x, t) = \begin{cases} h\left(x, \dfrac{t}{2\alpha(x)}\right) & \text{if } x \in U \text{ and } t \leqslant \alpha(x), \\ \varphi k(x, t - \alpha(x)) & \text{if } x \in V \text{ and } t \geqslant \alpha(x). \end{cases}$$

Then, k^* is injective and $k^*(x, 0) = x$ for every $x \in U \cup V$. To see that k^* is a collar of $U \cup V$ in X, it remains to be verified that k^* is an open map.

Evidently, k^* is continuous at every point of the following open sets:

$$(U \cup V) \times (0, 1),\ U \times [0, 1),\ (V \setminus \operatorname{cl} U) \times [0, 1).$$

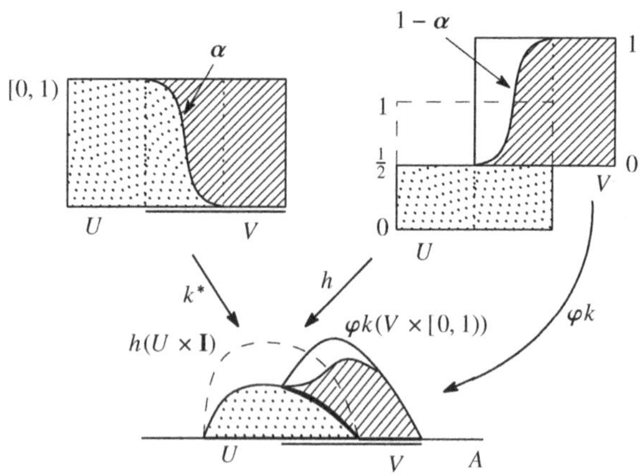

Fig. 2.11 A collar of $U \cup V$

We shall show the continuity of k^* at $(x, 0)$, where $x \in V \cap (\operatorname{cl} U \setminus U)$. Let W be a neighborhood of $x = k^*(x, 0)$ in X. Since $h(\{x\} \times \mathbf{I}) = \{x\}$, we have some neighborhood N of x in V such that $h((N \cap \operatorname{cl} U) \times \mathbf{I}) \subset W$. Observe that $\varphi k(x, 0) = \varphi(x) = h(x, 1/2) = x$. Then, we have a neighborhood N' of x in V with $0 < \delta < 1$ such that $\varphi k(N' \times [0, \delta)) \subset W$. It follows that

$$k^*((N \cap N') \times [0, \delta)) \subset h((N \cap \operatorname{cl} U) \times [0, 1/2]) \cup \varphi k(N' \times [0, \delta)) \subset W.$$

Finally, we shall prove that k^* is open. Let W be a neighborhood of $(x, t) \in (U \cup V) \times [0, 1)$.

When $t < \alpha(x)$, it follows that $x \in U$. Then,

$$W^- = \{(x', t'/2\alpha(x')) \mid (x', t') \in W,\ t' < \alpha(x')\}$$

is a neighborhood of $(x, t/2\alpha(x))$ in $U \times [0, 1/2)$ and $h(W^-) \subset k^*(W)$. Hence, $k^*(W)$ is a neighborhood of $h(x, t/2\alpha(x)) = k^*(x, t)$ in X.

When $t > \alpha(x)$, we have $x \in V$. Then,

$$W^+ = \{(x', t' - \alpha(x')) \mid (x', t') \in W,\ t' > \alpha(x')\}$$

is a neighborhood of $(x, t - \alpha(x))$ in $V \times (0, 1)$ and $\varphi k(W^+) \subset k^*(W)$. Hence, $k^*(W)$ is a neighborhood of $\varphi k(x, t - \alpha(x)) = k^*(x, t)$ in X.

If $t = \alpha(x) > 0$, then $x \in U \cap V$. Let

$$W_- = \{(x', t'/2\alpha(x')) \mid (x', t') \in W,\ t' \leqslant \alpha(x')\} \text{ and}$$
$$W_+ = \{(x', t' - \alpha(x')) \mid (x', t') \in W,\ t' \geqslant \alpha(x')\}.$$

Then, $h(W_-)$ is a neighborhood of $k^*(x, t)$ in $h(\operatorname{cl} U \times [0, 1/2])$ and $\varphi k(W_+)$ is a neighborhood of $k^*(x, t)$ in $\varphi(X) = X \setminus h(U \times [0, 1/2))$. Hence, $k^*(W) = h(W_-) \cup \varphi k(W_+)$ is a neighborhood of $k^*(x, t)$ in X.

When $t = \alpha(x) = 0$ and $x \in V \setminus \operatorname{cl} U$, we have $(x, 0) \notin (\varphi k)^{-1} h(\operatorname{cl} U \times \mathbf{I})$ because $h(\operatorname{cl} U \times \mathbf{I}) \cap A = \operatorname{cl} U$. Choose a neighborhood N of x in $V \setminus \operatorname{cl} U$ and $0 < \delta < 1$ so that

$$W_0 = N \times [0, \delta) \subset \big((V \times [0, 1)) \setminus (\varphi k)^{-1} h(\operatorname{cl} U \times \mathbf{I})\big) \cap W.$$

Then, $k(W_0) = \varphi k(W_0)$ is a neighborhood of $k(x, 0) = k^*(x, t)$ in the open set $k((V \setminus \operatorname{cl} U) \times [0, 1))$ in X, hence it is a neighborhood of $k^*(x, t)$ in X. Since $\varphi k(W_0) \subset k^*(W)$, $k^*(W)$ is also a neighborhood of $k^*(x, t)$ in X.

When $t = \alpha(x) = 0$ and $x \in V \cap \operatorname{bd} U$, choose an open neighborhood N of x in V and $0 < \delta < 1$ so that $N \times [0, \delta) \subset W$ and $\alpha(N) \subset [0, \delta)$. Then, $h((N \cap \operatorname{cl} U) \times [0, 1/2])$ is a neighborhood of x in $h(\operatorname{cl} U \times [0, 1/2])$. Observe that

$$M = \{(x', t' - \alpha(x')) \mid x' \in N,\ \alpha(x') \leqslant t' < \delta\}$$

is a neighborhood of $(x, 0)$ in $V \times [0, 1)$, hence $\varphi k(M)$ is a neighborhood of x in $\varphi(X) = X \setminus h(U \times [0, 1/2))$. Since

$$k^*(W) \supset h((N \cap \operatorname{cl} U) \times [0, 1/2]) \cup \varphi k(M),$$

it follows that $k^*(W)$ is a neighborhood of x. □

Since $E^2 \approx E$, the following version of Klee's Trick 1.13.7 can be easily obtained:

Theorem 2.5.8 (KLEE'S TRICK) *Every homeomorphism $h : A \to B$ between E-deficient closed sets in E can be extended to a homeomorphism $\tilde{h} : E \to E$.* □

Using this form of Klee's Trick, we show the following:

Theorem 2.5.9 (COLLARING) *Let M and N be E-manifolds such that N is E-deficient closed in M. Then, N is collared in M. Moreover, for each open cover $\mathcal{U}$ of M, there is a closed collar $k : N \times \mathbf{I} \to M$ such that*

$$\{k(\{x\} \times \mathbf{I}) \mid x \in N\} \prec \mathcal{U}.$$

Proof Since a locally collared set is collared (Theorem 2.5.7), it suffices to show that N is locally collared in M. As is easily observed, each $x \in N$ has an open neighborhood U in M such that $U \cap N$ and U are homeomorphic to open sets in E. Let $f : U \cap N \to E \times \{0\}$ ($\subset E \times [0, 1)$) and $g : U \to E \times [0, 1)$ be open embeddings. Choose an open neighborhood V of x in U so that $\operatorname{cl} V \subset U$ and $f(N \cap \operatorname{cl} V)$, $g(\operatorname{cl} V)$ are closed in $E \times \{0\}$ and $E \times [0, 1)$, respectively. Since $N \cap \operatorname{cl} V$ is E-deficient in U by Corollary 2.3.4(2), $g(N \cap \operatorname{cl} V)$ is an E-deficient closed set in $E \times [0, 1)$ by Proposition 2.3.8. On the other hand, $f(N \cap \operatorname{cl} V)$ is also an E-deficient closed set in $E \times [0, 1)$ by Theorem 2.4.9. Using Klee's Trick above, $f(g|N \cap \operatorname{cl} V)^{-1}$ can be extended to a homeomorphism $h : E \times [0, 1) \to E \times [0, 1)$. Then, $hg|V : V \to E \times [0, 1)$ is an open embedding such that $hg(N \cap V) = f(N \cap V)$ is open in $E \times \{0\}$. Since $E \times \{0\}$ is collared in $E \times [0, 1)$, $hg(N \cap V)$ is also collared in $hg(V)$ by Lemma 2.5.4. Hence, $N \cap V$ is collared in V (Fig. 2.12).

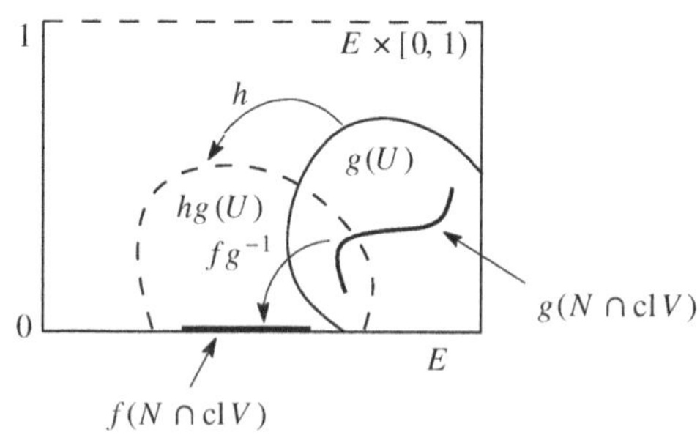

Fig. 2.12 A collar of $U \cap V$

Due to Proposition 2.5.5, the additional statement follows from the fact that N is a collared closed set in M. □

Now, we have the following:

Theorem 2.5.10 (Open Embedding) *Every E-manifold M can be embedded in E as an open set.*[22] *Moreover, if A is an E-deficient closed set in M, then there is an open embedding $g : M \to E$ such that $g(A)$ is E-deficient and closed in E.*

Proof Note that $M \in \mathcal{F}_E$ (Theorem 2.1.11). Then, by the Strong Universality Theorem 2.3.15, we have an E-deficient closed embedding $h : M \to E$, hence $h(M)$ has a collar $k : h(M) \times [0, 1) \to E$ by the Collaring Theorem 2.5.9. On the other hand, by Theorem 2.4.9, we have a homeomorphism $f : M \to M \times [0, 1)$ such that $f(A) \subset M \times \{0\}$. Then, $g = k(h \times \mathrm{id})f : M \to E$ is the desired open embedding because $g(A)$ is closed in the closed subspace $k(h(M) \times \{0\}) = h(M)$ of E. □

In the above proof, $h(M)$ has a closed collar $k : h(M) \times \mathbf{I} \to E$ by Proposition 2.7. Furthermore, k can be taken so that $\mathrm{bd}\, k(h(M)\times\mathbf{I}) = k(h(M)\times\{1\})$ is bi-collared in E by Remark 2.7. Since $M \times \mathbf{I} \approx M$ (Corollary 2.4.12), we have a homeomorphism $f : M \to M \times \mathbf{I}$ Then, $g = k(h \times \mathrm{id})f : M \to E$ is a closed embedding and $\mathrm{bd}\, g(M) = \mathrm{bd}\, k(h(M) \times \mathbf{I}) = k(h(M) \times \{1\})$ is bi-collared in E. Thus, we have the following:

Theorem 2.5.11 *Every E-manifold M can be embedded in E as a closed set having an E-manifold bi-collared boundary.*[23] □

Remark 2.8 Theorems 2.5.10 and 2.5.11 are valid for any E-manifold M such that $w(M) = w(E)$; equivalently, M has at most $w(E)$-many components. Indeed, by Lemma 2.1.10, E has a discrete open collection $\mathcal{U}$ such that $\mathrm{card}\, \mathcal{U} = w(E)$ and every $U \in \mathcal{U}$ is homeomorphic to E itself.

Theorem 2.5.8 can be extended to E-manifolds as follows:

Theorem 2.5.12 (Unknotting) *Let $h : A \to B$ be a homeomorphism between E-deficient closed sets in an E-manifold M and $\mathcal{U}$ be an open cover of M. If there is a homotopy $f : A \times \mathbf{I} \to M$ with $f_0 = \mathrm{id}$ and $f_1 = h$, then h can be extended to a homeomorphism $\tilde{h} : M \to M$ that is ambiently invertibly $\mathcal{U}^*$-isotopic to* id, *where $\mathcal{U}^* = \{\mathrm{st}(f(\{x\} \times \mathbf{I}), \mathcal{U}) \mid x \in A\}$.*

Proof Take $\mathcal{V} \in \mathrm{cov}(M)$ so that $\mathrm{st}^3\, \mathcal{V} \prec \mathcal{U}$ and let $\mathcal{V}[B] = \{V \in \mathcal{V} \mid V \cap B \neq \emptyset\}$. First, we will construct an ambient invertible $\mathcal{V}[B]$-isotopy $\psi : M \times \mathbf{I} \to M$ such that $\psi_0 = \mathrm{id}$ and $A \cap \psi_1(B) = \emptyset$. Let $\mathcal{V}'$ be an open star-refinement of $\mathcal{V}$. Since $A \cup B$ is E-deficient in M by Proposition 2.4.14, we apply Corollary 2.3.4(4) to obtain a homeomorphism $k : M \times E \to M$ such that $k(x, 0) = x$ for each $x \in A \cup B$

[22] Recall we assume that every E-manifold has the same density as E.

[23] Recall our assumption that every E-manifold has the same density as E.

and k is $\mathcal{V}'$-close to pr_M. Taking $v \in E \setminus \{0\}$ and an Urysohn map $\alpha : M \to \mathbf{I}$ with

$$\alpha(k^{-1}(B)) = \{1\} \text{ and } \alpha(M \setminus k^{-1}(\mathrm{st}(B, \mathcal{V}'))) = \{0\},$$

we define an ambient invertible isotopy $\psi' : M \times E \times \mathbf{I} \to M \times E$ by

$$\psi'(x, y, t) = (x, y + t\alpha(x)v) \text{ for each } (x, y, t) \in M \times E \times \mathbf{I}.$$

The isotopy ψ can be defined by $\psi_t = k\psi'_t k^{-1}$ for each $t \in \mathbf{I}$.

To verify that ψ is a $\mathcal{V}[B]$-isotopy, for each $x \in M$, let $k^{-1}(x) = (x', y)$. If $x' \notin \mathrm{st}(B, \mathcal{V}')$, then $\alpha(x') = 0$, hence

$$\psi_t(x) = k\psi'(x', y, t) = k(x', y) = x \text{ for every } t \in \mathbf{I},$$

that is, $\psi(\{x\} \times \mathbf{I}) = \{x\}$. When $x' \in \mathrm{st}(B, \mathcal{V}')$, we have $V' \in \mathcal{V}'$ such that $x' \in V'$ and $V' \cap B \neq \emptyset$. Since $\mathrm{st}\,\mathcal{V}' \prec \mathcal{V}$, there is some $V \in \mathcal{V}$ such that $\mathrm{st}(V', \mathcal{V}') \subset V$, where $V \cap B \neq \emptyset$, that is, $V \in \mathcal{V}[B]$. Because k is $\mathcal{V}'$-close to pr_M, it follows that

$$\psi(\{x\} \times \mathbf{I}) \subset k(\{x'\} \times E) \subset \mathrm{st}(x', \mathcal{V}') \subset \mathrm{st}(V', \mathcal{V}') \subset V.$$

It should be remarked that $\psi_t^{-1} = k\psi_t'^{-1}k^{-1}$ for each $t \in \mathbf{I}$, where

$$\psi_t'^{-1}(x, y) = (x, y - t\alpha(x)v) \text{ for each } (x, y, t) \in M \times E \times \mathbf{I}.$$

Then, by the same argument as above, the isotopy defined by ψ_t^{-1}, $t \in \mathbf{I}$, is a $\mathcal{V}[B]$-isotopy.[24]

Let $f' : A \times \mathbf{I} \to M$ be the homotopy defined by

$$f'_t = \begin{cases} f_{2t} & \text{if } 0 \leqslant t \leqslant 1/2, \\ \psi_{2t-1}h & \text{if } 1/2 \leqslant t \leqslant 1. \end{cases}$$

Since $f'_0 = \mathrm{id}$ and $f'_1 = \psi_1 h$, $f'|A \times \{0, 1\}$ is an E-deficient closed embedding into M. Moreover,

$$f'(\{x\} \times \mathbf{I}) \subset \mathrm{st}(f(\{x\} \times \mathbf{I}), \mathcal{V}) \text{ for each } x \in A.$$

By the Strong Universality Theorem 2.3.15, there is an E-deficient closed embedding $f'' : A \times \mathbf{I} \to M$ such that $f''|A \times \{0\} = f'|A \times \{0\} = \mathrm{id}$, $f''|A \times \{1\} = f'|A \times \{1\} = \psi_1 h$ and f'' is $\mathcal{V}$-close to f'. Then,

$$f''(\{x\} \times \mathbf{I}) \subset \mathrm{st}(f'(\{x\} \times \mathbf{I}), \mathcal{V}) \text{ for each } x \in A.$$

[24]Cf. Proposition 2.1.15.

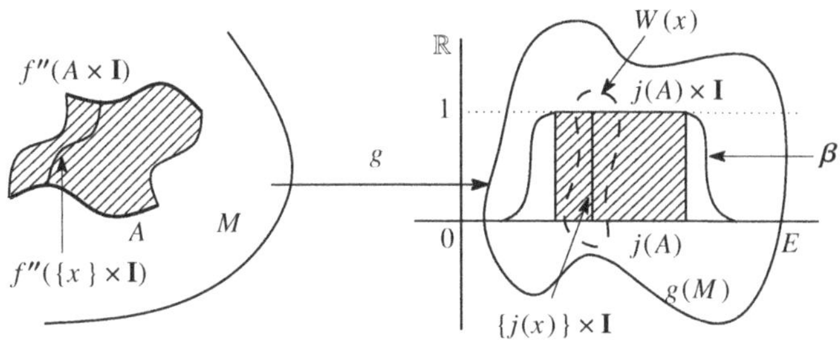

Fig. 2.13 The open embedding g and the open collection $\mathcal{W}$

On the other hand, we have an E-deficient closed embedding $j : A \to E$. Then, $j(A) \times \mathbf{I}$ is an E-deficient closed set in $E \times \mathbb{R}$. Using the Open Embedding Theorem 2.5.10 and Klee's Trick 2.5.8, we can construct an open embedding $g : M \to E \times \mathbb{R}$ such that $g(f''(x,t)) = (j(x), t)$ for $(x,t) \in A \times \mathbf{I}$ (cf. Fig. 2.13).

For each $x \in A$, define

$$W(x) = g(\mathrm{st}(f'(\{x\} \times \mathbf{I}), \mathcal{V})) \subset g(M)$$

and let $\mathcal{W} = \{W(x) \mid x \in A\}$. Note that

$$\{j(x)\} \times \mathbf{I} = gf''(\{x\} \times \mathbf{I}) \subset W(x) \quad \text{for each } x \in A.$$

By Lemma 2.5.1, we have maps $\beta, \gamma : E \to \mathbb{R}$ such that:

(i) $\beta(x) \geqslant 0$ for $x \in E$,
(ii) $\beta(x) = 1$ and $\gamma(x) > 0$ if $x \in j(A)$,
(iii) $\beta(x) = \gamma(x) = 0$ if $(x,0) \notin \bigcup_{x \in A} W(x)$,
(iv) $\gamma(x) > 0$ if $\beta(x) > 0$,
(v) $\{\{x\} \times [-\gamma(x), \beta(x) + \gamma(x)] \mid (x,0) \in \bigcup_{x \in A} W(x)\} \prec \mathcal{W}$.

On the right-hand side of Fig. 2.13, we have an ambient invertible $\mathcal{W}$-isotopy $\varphi' : E \times \mathbb{R} \times \mathbf{I} \to E \times \mathbb{R}$ sliding points vertically so as to move $\{0\} \times E$ to the graph of β. The following is the precise definition of φ':

$$\varphi'_t(x,s) = \begin{cases} \left(x, t\beta(x) + \dfrac{((1-t)\beta(x) + \gamma(x))s}{\beta(x) + \gamma(x)}\right) & \text{if } 0 \leqslant s \leqslant \beta(x) + \gamma(x), \\ \left(x, s + \dfrac{t\beta(x)(s + \gamma(x))}{\gamma(x)}\right) & \text{if } -\gamma(x) \leqslant s \leqslant 0, \\ (x,s) & \text{otherwise.} \end{cases}$$

Then, $\varphi'_0 = \mathrm{id}$ and $\varphi'_1 g|A = g\psi_1 h$. Indeed, for every $x \in A$,

$$\begin{aligned} \varphi'_1 g(x) &= \varphi'_1 gf'(x,0) = \varphi'_1 gf''(x,0) = \varphi'_1(j(x),0) \\ &= (j(x),1) = gf''(x,1) = gf'(x,1) = g\psi_1 h(x). \end{aligned}$$

Since $\varphi'_t(g(M)) = g(M)$ for each $t \in \mathbf{I}$, we have an ambient invertible $g^{-1}(\mathcal{W})$-isotopy $\varphi : M \times \mathbf{I} \to M$ defined by $\varphi_t = g^{-1}\varphi'_t g$. Then, $\varphi_0 = \mathrm{id}$ and $\varphi_1|A = g^{-1}\varphi'_1 g|A = \psi_1 h$. Thus, h is extended to a homeomorphism $\tilde{h} = \psi_1^{-1}\varphi_1 : M \to M$. Moreover, we can define an ambient invertible isotopy $\eta : M \times \mathbf{I} \to M$ by

$$\eta_t = \begin{cases} \varphi_{2t} & \text{if } 0 \leqslant t \leqslant 1/2, \\ \psi_{2t-1}^{-1}\varphi_1 & \text{if } 1/2 \leqslant t \leqslant 1. \end{cases}$$

Then, $\eta_0 = \mathrm{id}$ and $\eta_1 = \psi_1^{-1}\varphi_1 = \tilde{h}$.

To verify that η is a $\mathcal{U}^*$-isotopy, let $x \in M$. When $\varphi(\{x\} \times \mathbf{I})$ is not contained in $g^{-1}(W(y))$ for any $y \in A$, since φ is a $g^{-1}(\mathcal{W})$-isotopy, it follows that $\varphi(\{x\} \times \mathbf{I}) = \{x\}$. Then,

$$x \notin \bigcup_{y \in A} g^{-1}(W(y)) = \mathrm{st}(f'(A \times \mathbf{I}), \mathcal{V}),$$

hence $x \notin \mathrm{st}(f'_{1/2}(A), \mathcal{V}) = \mathrm{st}(B, \mathcal{V})$. In this case, for each $t \in \mathbf{I}$, $\psi_t(x) = x$, that is, $\psi_t^{-1}(x) = x$. Thus, we have $\eta(\{x\} \times \mathbf{I}) = \{x\}$.

When $\varphi(\{x\} \times \mathbf{I})$ is contained in $g^{-1}(W(y))$ for some $y \in A$, since the isotopy defined by ψ_t^{-1}, $t \in \mathbf{I}$, is a $\mathcal{V}[B]$-isotopy, it follows that

$$\begin{aligned} \eta(\{x\} \times \mathbf{I}) &\subset \mathrm{st}(g^{-1}(W(y)), \mathcal{V}) = \mathrm{st}(\mathrm{st}(f'(\{y\} \times \mathbf{I}), \mathcal{V}), \mathcal{V}) \\ &\subset \mathrm{st}(f(\{y\} \times \mathbf{I}), \mathrm{st}\,\mathcal{V}) \subset \mathrm{st}(f(\{y\} \times \mathbf{I}), \mathcal{U}) \in \mathcal{U}^*. \end{aligned}$$

Therefore, η is a $\mathcal{U}^*$-isotopy. □

The Open Embedding Theorem 2.5.10 can be generalized as follows:

Theorem 2.5.13 (Open Embedding Approximation) *Let M and N be E-manifolds, A be an E-deficient closed set in M and $f : M \to N$ be a map such that $f|A$ is an E-deficient closed embedding into N. Then, for each open cover $\mathcal{U}$ of N, there exists an open embedding $g : M \to N$ such that $g|A = f|A$ and g is $\mathcal{U}$-homotopic to f.*

Proof Let $\mathcal{V} \in \mathrm{cov}(N)$ such that $\mathrm{st}\,\mathcal{V} \prec \mathcal{U}$. Recall $M \in \mathcal{F}_E$ (Theorem 2.1.11). By the Strong Universality Theorem 2.3.15, we have an E-deficient closed embedding $h : M \to N$ such that $h|A = f|A$ and h is $\mathcal{V}$-homotopic to f. By Corollary 2.3.5, we have a homeomorphism $\varphi : M \times [0, 1) \to M$ such that $\varphi(x, 0) = x$ for each $x \in A$ and φ is $h^{-1}(\mathcal{V})$-homotopic to pr_M. Since $h(M)$ is collared in N by the Collaring Theorem 2.5.9, we have an open embedding $k : M \times [0, 1) \to N$ such that $k(x, 0) = h(x)$ for each $x \in M$. By Lemma 2.5.2, we can require k to satisfy the condition that

$$\big\{k(\{x\} \times [0, 1)) \bigm| x \in M\big\} \prec \mathcal{V},$$

which implies that k is $\mathcal{V}$-homotopic to $h\circ\mathrm{pr}_M$. Since $h\circ\mathrm{pr}_M$ is $\mathcal{V}$-homotopic to $h\varphi$ and h is $\mathcal{V}$-homotopic to f, k is $\mathrm{st}\,\mathcal{V}$-homotopic, so $\mathcal{U}$-homotopic to $f\varphi$. Thus, we have an open embedding $g = k\circ\varphi^{-1} : M \to N$ which is $\mathcal{U}$-homotopic to f. Moreover,

$$g(x) = k\varphi^{-1}(x) = k(x, 0) = h(x) = f(x) \text{ for each } x \in A,$$

that is, $g|A = f|A$. □

2.6 Classification of E-Manifolds

In this section, we establish the following:

Theorem 2.6.1 (Classification) *Two E-manifolds are homeomorphic if they have the same homotopy type.*

As a corollary, we have the following:

Corollary 2.6.2 *Every contractible E-manifold is homeomorphic to E.* □

Recall that a map $f : X \to Y$ is a **$\mathcal{U}$-homotopy equivalence** for $\mathcal{U} \in \mathrm{cov}(Y)$ if there is a map $g : Y \to X$ (called a **$\mathcal{U}$-homotopy inverse** of f) such that fg is $\mathcal{U}$-homotopic to id_Y and gf is $f^{-1}(\mathcal{U})$-homotopic to id_X. A **fine homotopy equivalence** $f : X \to Y$ is a $\mathcal{U}$-homotopy equivalence for every $\mathcal{U} \in \mathrm{cov}(Y)$. The Classification Theorem 2.6.1 is a direct consequence of the following Homeomorphism Approximation Theorem:

Theorem 2.6.3 (Homeomorphism Approximation) *Let M and N be E-manifolds and $\mathcal{U}$ be an open cover of N. Then, every $\mathcal{U}$-homotopy equivalence $f : M \to N$ is $\mathrm{st}^3\,\mathcal{U}$-close ($\mathrm{st}^6\,\mathcal{U}$-homotopic) to a homeomorphism. In particular, every fine homotopy equivalence of M to N is a near-homeomorphism.*

We adjust the mapping cylinder for use in the proof. For each map $f : X \to Y$ and $a < b < c$, the **fat mapping cylinder** $M_a^{b,c}(f)$ of f is defined as follows:

$$M_a^{b,c}(f) = X \times [a, b) \cup Y \times [b, c]$$

with the topology generated by open sets in the product spaces $X \times [a, b)$ and $Y \times [b, c]$, and the following sets:

$$(f^{-1}(U) \times (b - \varepsilon, b)) \cup (U \times [b, b + \varepsilon)),$$

where U is open in Y and $0 < \varepsilon < \min\{b - a, c - b\}$. That is the copy of the following space:

$$M(f) \cup_{Y = Y\times\{0\}} Y \times \mathbf{I},$$

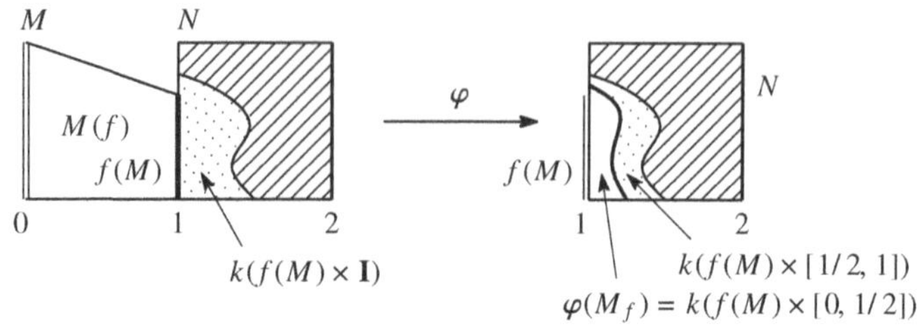

Fig. 2.14 The homeomorphism φ

where $Y \subset M(f)$ is identified with $Y \times \{0\} \subset Y \times \mathbf{I}$.

Lemma 2.6.4 *Let M and N be E-manifolds and $f : M \to N$ a closed embedding. Then, for each $\mathcal{V} \in \operatorname{cov}(N \times \mathbf{I})$, there exists a homeomorphism $\varphi : M_0^{1,2}(f) \to N \times [1, 2]$ such that $\varphi(x, 0) = (f(x), 1)$ for each $x \in M$, $\varphi|N \times \{2\} = \mathrm{id}$ and*

$$\varphi \simeq_{\mathcal{V}} r \quad \textit{rel. } M \times \{0\} \cup N \times \{2\},$$

where $r : M_0^{1,2}(f) \to N \times [1, 2]$ is the retraction defined by $r(x, t) = (f(x), 1)$ for each $(x, t) \in M \times [0, 1)$ and $r|N \times [1, 2] = \mathrm{id}$.

Proof It follows from Theorem 2.4.9 that $f(M) \times \{1\}$ is an E-manifold that is E-deficient and closed in $N \times [1, 2]$. By the Collaring Theorem 2.5.9, $f(M)$ has a closed collar $k : f(M) \times \mathbf{I} \to N \times [1, 2]$ such that

$$\{k(\{f(x)\} \times \mathbf{I}) \mid x \in M\} \prec \mathcal{V}.$$

We can define a $\mathcal{V}$-homotopy $h : M_0^{1,2}(f) \times \mathbf{I} \to N \times [1, 2]$ as follows:

$$h_t(x, s) = \begin{cases} k(f(x), st/2) & \text{if } (x, s) \in M \times [0, 1), \\ k(f(x'), (s' + t)/2) & \text{if } (x, s) = k(f(x'), s') \in k(f(M) \times [0, t]), \\ (x, s) & \text{if } (x, s) \in (N \times [1, 2]) \setminus k(f(M) \times [0, t]). \end{cases}$$

Refer to Fig. 2.14. Then, $h_0 = r$, $h_t|M \times \{0\} = f \times \mathrm{id}_{\{0\}}$ and $h_t|N \times \{2\} = \mathrm{id}$ for each $t \in \mathbf{I}$, and h_t is a homeomorphism for $t > 0$. Therefore, $\varphi = h_1$ is the desired homeomorphism. □

Lemma 2.6.5 *Let M and N be E-manifolds, and let $f : M \to N$ and $g : N \to M$ be closed embeddings. Suppose that there exists a homotopy $h : M \times \mathbf{I} \to M$ with $h_0 = \mathrm{id}_M$ and $h_1 = gf$. Then, for each $\mathcal{U} \in \operatorname{cov}(M)$, there exists a homeomorphism*

$$\psi : M_0^{1,2}(f) \cup M_2^{3,4}(g) \to M \times [0, 4]$$

such that $\psi|M \times \{0,4\} = \mathrm{id}$ *and* $\mathrm{pr}_M\psi \simeq_{\mathcal{U}^*} \theta$, *where* $M_0^{1,2}(f) \cup M_2^{3,4}(g)$ *is the space obtained by pasting* $M_0^{1,2}(f)$ *and* $M_2^{3,4}(g)$ *along* $N \times \{2\}$, *and* $\theta : M_0^{1,2}(f) \cup M_2^{3,4}(g) \to M$ *is the map defined as follows:*

$$\theta|M \times [0,1) = gf\mathrm{pr}_M, \ \theta|N \times [1,3) = g\mathrm{pr}_N, \ \textit{and} \ \theta|M \times [3,4] = \mathrm{pr}_M,$$

and $\mathcal{U}^* = \{\mathrm{st}(h(\{x\} \times \mathbf{I}), \mathcal{U}) \mid x \in M\}$.

Proof Let $\eta : M \times [3,4] \to M \times [0,4]$ be the homeomorphism defined by $\eta(x,t) = (x, 4t-12)$. Take $\mathcal{V} \in \mathrm{cov}(M \times [0,4])$ and $\mathcal{W} \in \mathrm{cov}(M \times [3,4])$ so that $\mathrm{st}\,\mathcal{V} \prec \mathrm{pr}_M^{-1}(\mathcal{U})$ and $\mathrm{st}\,\mathcal{W} \prec \eta^{-1}(\mathcal{V})$. Observe that

$$(N \times [1,2]) \cup M_2^{3,4}(g) = M_1^{3,4}(g).$$

Applying Lemma 2.6.4 to g, we can obtain a homeomorphism $\varphi' : M_1^{3,4}(g) \to M \times [3,4]$ such that $\varphi'(x,1) = (g(x),3)$ for each $x \in N$, $\varphi'|M \times \{4\} = \mathrm{id}$, and

$$\varphi' \simeq_{\mathcal{W}} r' \ \text{ rel. } N \times \{1\} \cup M \times \{4\},$$

where $r' : M_1^{3,4}(g) \to M \times [3,4]$ is the retraction defined by $r'(x,t) = (g(x),3)$ for each $(x,t) \in N \times [1,3)$ and $r'|M \times [3,4] = \mathrm{id}$.

Extending the homeomorphism obtained in Lemma 2.6.4, we can obtain a homeomorphism

$$\varphi'' : M_0^{1,2}(f) \cup M_2^{3,4}(g) \to (N \times [1,2]) \cup M_2^{3,4}(g) = M_1^{3,4}(g)$$

such that $\varphi''(x,0) = (f(x),1)$ for each $x \in M$, $\varphi''|M_2^{3,4}(g) = \mathrm{id}$, and

$$\varphi'' \simeq_{\varphi'^{-1}(\mathcal{W})} r'' \ \text{ rel. } M \times \{0\} \cup M_2^{3,4}(g),$$

where $r'' : M_0^{1,2}(f) \cup M_2^{3,4}(g) \to M_1^{3,4}(g)$ is the retraction defined by $r''(x,t) = (f(x),1)$ for $(x,t) \in M \times [0,1)$ and $r''|M_1^{3,4}(g) = \mathrm{id}$.

Now, we have the following homeomorphism:

$$\varphi = \varphi'\varphi'' : M_0^{1,2}(f) \cup M_2^{3,4}(g) \to M \times [3,4].$$

Refer to Fig. 2.15. Then, $\varphi(x,0) = (gf(x),3)$ for each $x \in M$, $\varphi|M \times \{4\} = \mathrm{id}$, and

$$\varphi \simeq_{\mathrm{st}\,\mathcal{W}} r = r'r'' \ \text{ rel. } M \times \{0\} \cup M \times \{4\},$$

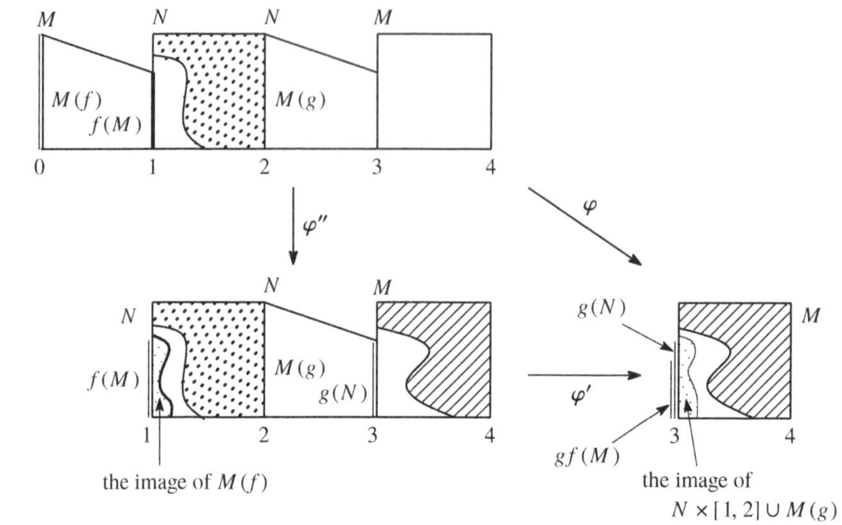

Fig. 2.15 The homeomorphisms φ', φ'', and φ

where $r(x,t) = (gf(x), 3)$ for $(x,t) \in M \times [0,1)$, $r(x,t) = (g(x), 3)$ for each $(x,t) \in N \times [1,3)$, and $r|M \times [3,4] = \mathrm{id}$. It should be remarked that

$$\eta\varphi|M \times \{0\} = gf \times \mathrm{id}_{\{0\}}, \ \eta\varphi|M \times \{4\} = \mathrm{id},$$
$$\eta(\mathrm{st}\,\mathcal{W}) \prec \mathcal{V} \text{ and } \mathrm{pr}_M \eta r = \mathrm{pr}_M r = \theta.$$

Recall that $h : M \times \mathbf{I} \to M$ is a homotopy with $h_0 = \mathrm{id}_M$ and $h_1 = gf$. By Theorem 2.4.9, $gf(M) \times \{0\}$, $M \times \{0\}$, and $M \times \{4\}$ are E-deficient in $M \times [0,4]$. So, we can apply the Unknotting Theorem 2.5.12 to obtain a homeomorphism $\zeta : M \times [0,4] \to M \times [0,4]$ such that $\zeta|M \times \{0\} = gf \times \mathrm{id}_{\{0\}}$, $\zeta|M \times \{4\} = \mathrm{id}$, and ζ is $\mathcal{V}^*$-isotopic to id, where

$$\mathcal{V}^* = \left\{ \mathrm{st}(h(\{x\} \times \mathbf{I}) \times \{0\}, \mathcal{V}) \;\middle|\; x \in M \right\}.$$

Then, the following is the desired homeomorphism:

$$\psi = \zeta^{-1}\eta\varphi : M_0^{1,2}(f) \cup M_2^{3,4}(g) \to M \times [0,4].$$

Indeed, $\psi|M \times \{4\} = \zeta^{-1}\eta\varphi|M \times \{4\} = \mathrm{id}$ and

$$\psi|M \times \{0\} = \zeta^{-1}\eta\varphi|M \times \{0\} = \zeta^{-1}(gf \times \mathrm{id}_{\{0\}}) = \zeta^{-1}\zeta|M \times \{0\} = \mathrm{id}.$$

Since $\psi = \zeta^{-1}\eta\varphi \simeq_{\mathcal{V}^*} \eta\varphi \simeq_{\mathcal{V}} \eta r$ and

$$\begin{aligned}\mathrm{pr}_M(\mathrm{st}(\mathcal{V}^*, \mathcal{V})) &= \left\{\mathrm{pr}_M(\mathrm{st}^2(h(\{x\} \times \mathbf{I}) \times \{0\}, \mathcal{V})) \;\middle|\; x \in M\right\} \\ &\prec \left\{\mathrm{pr}_M(\mathrm{st}(h(\{x\} \times \mathbf{I}) \times \{0\}, \mathrm{st}\,\mathcal{V})) \;\middle|\; x \in M\right\} \\ &\prec \left\{\mathrm{st}(h(\{x\} \times \mathbf{I}), \mathcal{U})) \;\middle|\; x \in M\right\} = \mathcal{U}^*,\end{aligned}$$

it follows that $\mathrm{pr}_M\psi \simeq_{\mathcal{U}^*} \mathrm{pr}_M\eta r = \theta$. □

We now prove Theorem 2.6.3:

Proof (Theorem 2.6.3) Since f is a $\mathcal{U}$-homotopy equivalence, f has a $\mathcal{U}$-homotopy inverse $g : N \to M$, that is, $fg \simeq_{\mathcal{U}} \mathrm{id}_N$ and $gf \simeq_{f^{-1}(\mathcal{U})} \mathrm{id}_M$. Take $\mathcal{V}, \mathcal{W}, \mathcal{W}' \in \mathrm{cov}(N)$ so that

$$\mathrm{st}\,\mathcal{V} \prec \mathcal{U},\ \mathrm{st}^2\,\mathcal{W} \prec \mathcal{V},\ \mathcal{W}' \prec \mathcal{W},\ \text{and}\ g(\mathcal{W}') \prec f^{-1}(\mathcal{W}).$$

By the Strong Universality Theorem 2.3.15, there exist closed embeddings $f_0 : M \to N$ and $g_0 : N \to M$ such that $f_0 \simeq_{\mathcal{W}'} f$ and $g_0 \simeq_{f^{-1}(\mathcal{W})} g$. Then, it follows that

$$\begin{gathered} f_0 g_0 \simeq_{\mathcal{W}'} f g_0 \simeq_{\mathcal{W}} fg \simeq_{\mathcal{U}} \mathrm{id}_N \ \text{ and} \\ g_0 f_0 \simeq_{f^{-1}(\mathcal{W})} g f_0 \simeq_{g(\mathcal{W}')} gf \simeq_{f^{-1}(\mathcal{U})} \mathrm{id}_M. \end{gathered}$$

Thus, we have $f_0 g_0 \simeq_{\mathrm{st}(\mathcal{U}, \mathrm{st}\,\mathcal{W})} \mathrm{id}_N$ and $g_0 f_0 \simeq_{f^{-1}(\mathrm{st}(\mathcal{U}, \mathrm{st}\,\mathcal{W}))} \mathrm{id}_M$.

For simplicity, we write as follows:

$$X_{2n-1} = M_{2n-2}^{2n-3/2, 2n-1}(f_0),\ X_{2n} = M_{2n-1}^{2n-1/2, 2n}(g_0),\ n \in \mathbb{N}.$$

Then, each X_{2n-1} and X_{2n} are copies of $M(f_0) \cup_{N = N \times \{0\}} (N \times \mathbf{I})$ and $M(g_0) \cup_{M = M \times \{0\}} (M \times \mathbf{I})$, respectively. We define $X = \bigcup_{n \in \mathbb{N}} X_n$, that is illustrated as the middle space in Fig. 2.16.

Observe that

$$\begin{gathered} \mathrm{st}(\mathrm{st}(\mathcal{U}, \mathrm{st}\,\mathcal{W}), \mathcal{W}) \prec \mathrm{st}(\mathcal{U}, \mathrm{st}^2\,\mathcal{W}) \prec \mathrm{st}(\mathcal{U}, \mathcal{V}) \ \text{ and} \\ \mathrm{st}(f^{-1}(\mathrm{st}(\mathcal{U}, \mathrm{st}\,\mathcal{W})), f^{-1}(\mathcal{W})) \prec f^{-1}(\mathrm{st}(\mathcal{U}, \mathrm{st}\,\mathcal{V})). \end{gathered}$$

For each $n \in \mathbb{N}$, applying Lemma 2.6.5, we obtain homeomorphisms

$$\begin{gathered} \varphi_n : X_{2n-1} \cup X_{2n} \to M \times [2n-2, 2n] \ \text{ and} \\ \psi_n : X_{2n} \cup X_{2n+1} \to N \times [2n-1, 2n+1] \end{gathered}$$

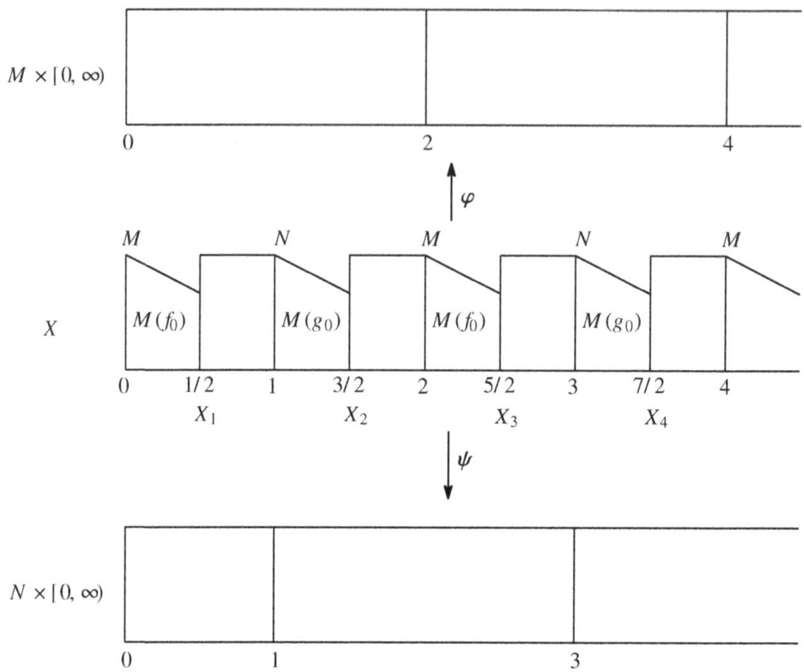

Fig. 2.16 The homeomorphisms φ and ψ

satisfying the following properties:

$$\varphi_n | M \times \{2n-2, 2n\} = \mathrm{id}, \;\; \psi_n | N \times \{2n-1, 2n+1\} = \mathrm{id},$$
$$\mathrm{pr}_M \varphi_n \simeq_{f^{-1}(\mathrm{st}(\mathcal{U},\mathcal{V}))} \theta_n \;\text{ and }\; \mathrm{pr}_N \psi_n \simeq_{\mathrm{st}(\mathcal{U},\mathcal{V})} \vartheta_n,$$

where θ_n and ϑ_n are defined as follows:

$$\begin{cases} \theta_n | M \times [2n-2, 2n-3/2) = g_0 f_0 \mathrm{pr}_M, \\ \theta_n | N \times [2n-3/2, 2n-1/2) = g_0 \mathrm{pr}_N, \\ \theta_n | M \times [2n-1/2, 2n] = \mathrm{pr}_M, \end{cases}$$

$$\begin{cases} \vartheta_n | N \times [2n-1, 2n-1/2) = f_0 g_0 \mathrm{pr}_N, \\ \vartheta_n | M \times [2n-1/2, 2n+1/2) = f_0 \mathrm{pr}_M, \\ \vartheta_n | N \times [2n+1/2, 2n+1] = \mathrm{pr}_N. \end{cases}$$

Additionally, applying Lemma 2.6.4, we have a homeomorphism $\psi_0 : X_1 \to N \times [0, 1]$ such that $\psi_0|M \times \{0\} = f_0 \times \mathrm{id}_{\{0\}}$, $\psi_0|N \times \{1\} = \mathrm{id}$, and $\mathrm{pr}_N \psi_0 \simeq_{\mathcal{V}} \vartheta_0$, where ϑ_0 is defined by

$$\vartheta_0|M \times [0, 1/2) = f_0 \mathrm{pr}_M \text{ and } \vartheta_0|N \times [1/2, 1] = \mathrm{pr}_N.$$

Then, we can define homeomorphisms

$$\varphi : X \to M \times [0, \infty) \text{ and } \psi : X \to N \times [0, \infty)$$

by $\varphi|X_{2n-1} \cup X_{2n} = \varphi_n$, $\psi|X_{2n} \cup X_{2n+1} = \psi_n$ for each $n \in \mathbb{N}$, and $\psi|X_1 = \psi_0$ (see Fig. 2.16).

Dividing the domain into the following three parts:

(1) $X_{2n} = M_{2n-1}^{2n-1/2,2n}(g_0)$, $n \in \mathbb{N}$,
(2) $X_{2n+1} = M_{2n}^{2n+1/2,2n+2}(f_0)$, $n \in \mathbb{N}$, and
(3) $X_1 = M_0^{1/2,1}(f_0)$, we will prove that $\mathrm{pr}_N \psi$ is $\mathrm{st}(\mathrm{st}^2 \mathcal{U}, \mathcal{V})$-close to $f \mathrm{pr}_M \varphi$, which means that $\mathrm{pr}_N \psi \varphi^{-1} k^{-1}$ is $\mathrm{st}(\mathrm{st}^2 \mathcal{U}, \mathcal{V})$-close to $f \mathrm{pr}_M k^{-1}$ $(= f \mathrm{pr}_M \varphi \varphi^{-1} k^{-1})$.

(1): The restriction $f \mathrm{pr}_M \varphi|X_{2n} = f \mathrm{pr}_M \varphi_n|X_{2n}$ is $\mathrm{st}(\mathcal{U}, \mathcal{V})$-close to $f\theta_n|X_{2n}$ and $\mathrm{pr}_N \psi|X_{2n} = \mathrm{pr}_N \psi_n|X_{2n}$ is $\mathrm{st}(\mathcal{U}, \mathcal{V})$-close to $\vartheta_n|X_{2n}$. Since f is $\mathcal{V}$-close to f_0, it follows that $f\theta_n|N \times [2n-1, 2n-1/2) = f g_0 \mathrm{pr}_N$ and $f\theta_n|M \times [2n-1/2, 2n] = f \mathrm{pr}_M$ are $\mathcal{V}$-close to $\vartheta_n|N \times [2n-1, 2n-1/2) = f_0 g_0 \mathrm{pr}_N$ and $\vartheta_n|M \times [2n-1/2, 2n] = f_0 \mathrm{pr}_M$, respectively. Thus, $f\theta_n|X_{2n}$ is $\mathcal{V}$-close to $\vartheta_n|X_{2n}$. Because $\mathrm{st}\,\mathcal{V} \prec \mathcal{U}$, we have

$$\mathrm{st}(\mathcal{V}, \mathrm{st}(\mathcal{U}, \mathcal{V})) \prec \mathrm{st}(\mathrm{st}\,\mathcal{U}, \mathcal{V}).$$

Therefore, $f \mathrm{pr}_M \varphi|X_{2n}$ is $\mathrm{st}(\mathrm{st}\,\mathcal{U}, \mathcal{V})$-close to $\mathrm{pr}_N \psi|X_{2n}$.

(2): The restriction $f \mathrm{pr}_M \varphi|X_{2n+1} = f \mathrm{pr}_M \varphi_{n+1}|X_{2n+1}$ is $\mathrm{st}(\mathcal{U}, \mathcal{V})$-close to $f\theta_{n+1}|X_{2n+1}$ and $\mathrm{pr}_N \psi|X_{2n+1} = \mathrm{pr}_N \psi_n|X_{2n+1}$ is $\mathrm{st}(\mathcal{U}, \mathcal{V})$-close to $\vartheta_n|X_{2n+1}$. Since $f g_0$ is $\mathrm{st}(\mathcal{U}, \mathcal{V})$-close to id_N, it follows that $f\theta_{n+1}|M \times [2n, 2n+1/2) = f g_0 f_0 \mathrm{pr}_M$ and $f\theta_{n+1}|N \times [2n+1/2, 2n+1] = f g_0 \mathrm{pr}_N$ are $\mathrm{st}(\mathcal{U}, \mathcal{V})$-close to $\vartheta_n|M \times [2n, 2n+1/2) = f_0 \mathrm{pr}_M$ and $\vartheta_n|N \times [2n+1/2, 2n+1] = \mathrm{pr}_N$, respectively. Thus, $f\theta_n|X_{2n+1}$ is $\mathrm{st}(\mathcal{U}, \mathcal{V})$-close to $\vartheta_n|X_{2n+1}$. Because $\mathrm{st}\,\mathcal{V} \prec \mathcal{U}$, we have

$$\mathrm{st}(\mathrm{st}(\mathcal{U}, \mathcal{V}), \mathrm{st}(\mathcal{U}, \mathcal{V})) \prec \mathrm{st}(\mathrm{st}(\mathrm{st}\,\mathcal{U}, \mathcal{U}), \mathcal{V}) = \mathrm{st}(\mathrm{st}^2 \mathcal{U}, \mathcal{V}).$$

Therefore, $f \mathrm{pr}_M \varphi|X_{2n+1}$ is $\mathrm{st}(\mathrm{st}^2 \mathcal{U}, \mathcal{V})$-close to $\mathrm{pr}_N \psi|X_{2n+1}$.

(3): The restriction $f \mathrm{pr}_M \varphi|X_1 = f \mathrm{pr}_M \varphi_1|X_1$ is $\mathrm{st}(\mathcal{U}, \mathcal{V})$-close to $f\theta_1|X_1$ and $\mathrm{pr}_N \psi|X_1 = \mathrm{pr}_N \psi_0|X_1$ is $\mathcal{V}$-close to $\vartheta_0|X_1$. Since $f g_0$ is $\mathrm{st}(\mathcal{U}, \mathcal{V})$-close to id_N, it follows that $f\theta_1|M \times [0, 1/2) = f g_0 f_0 \mathrm{pr}_M$ and $f\theta_1|N \times [1/2, 1] = f g_0 \mathrm{pr}_N$

are $\mathrm{st}(\mathcal{U}, \mathcal{V})$-close to $\vartheta_0|M \times [0, 1/2) = f_0\mathrm{pr}_M$ and $\vartheta_0|N \times [1/2, 1] = \mathrm{pr}_N$, respectively. Because $\mathrm{st}\,\mathcal{V} \prec \mathcal{U}$, we have

$$\mathrm{st}(\mathrm{st}(\mathrm{st}(\mathrm{st}\,\mathcal{V}, \mathcal{U}), \mathcal{V}), \mathcal{V}) \prec \mathrm{st}(\mathrm{st}(\mathcal{U}, \mathcal{U}), \mathrm{st}\,\mathcal{V}) \prec \mathrm{st}(\mathrm{st}\,\mathcal{U}, \mathcal{U}) = \mathrm{st}^2\,\mathcal{U}.$$

Hence, $f\mathrm{pr}_M\varphi|X_1$ is $\mathrm{st}^2\,\mathcal{U}$-close to $\mathrm{pr}_N\psi|X_{2n+1}$.

Since $[0, \infty) \times E \approx E$, we can apply the Stability Theorem 2.3.7 to obtain homeomorphisms $k : M \times [0, \infty) \to M$ and $h : N \times [0, \infty) \to N$ such that $k \simeq_{f^{-1}(\mathcal{V})} \mathrm{pr}_M$ and $h \simeq_{\mathcal{V}} \mathrm{pr}_N$. Then, $h\psi\varphi^{-1}k^{-1} : M \to N$ is the desired homeomorphism. Indeed, $h\psi\varphi^{-1}k^{-1}$ is $\mathcal{V}$-close to $\mathrm{pr}_N\psi\varphi^{-1}k^{-1}$, $\mathrm{pr}_N\psi\varphi^{-1}k^{-1}$ is $\mathrm{st}(\mathrm{st}^2\,\mathcal{U}, \mathcal{V})$-close to $f\mathrm{pr}_M k^{-1}$ as we saw above, and $f\mathrm{pr}_M k^{-1}$ is $\mathcal{V}$-close to $fkk^{-1} = f$. Since $\mathrm{st}\,\mathcal{V} \prec \mathcal{U}$, we have

$$\mathrm{st}(\mathrm{st}(\mathrm{st}^2\,\mathcal{U}, \mathcal{V}), \mathcal{V}) \prec \mathrm{st}(\mathrm{st}^2\,\mathcal{U}, \mathrm{st}\,\mathcal{V}) \prec \mathrm{st}^3\,\mathcal{U}.$$

Hence, $h\psi\varphi^{-1}k^{-1}$ is $\mathrm{st}^3\,\mathcal{U}$-close to f.

Furthermore, let $r : M \times [0, \infty) \to M \times \{0\}$ be the retraction defined by $r(x, t) = (x, 0)$ for each $(x, t) \in M \times [0, \infty)$. Then, we have

$$\begin{aligned}\mathrm{pr}_N\psi\varphi^{-1}r &= (\mathrm{pr}_N\psi\varphi^{-1}|M \times \{0\})r = (\mathrm{pr}_N\psi|M \times \{0\})r \\ &= (f_0\mathrm{pr}_M|M \times \{0\})r = f_0\mathrm{pr}_M r = f_0\mathrm{pr}_M;\end{aligned}$$

$$\begin{aligned}\left\{\mathrm{pr}_N\psi\varphi^{-1}(\{x\} \times [0, \infty)) \;\middle|\; x \in M\right\} &\prec \left\{\mathrm{st}(f(x), \mathrm{st}(\mathrm{st}^2\,\mathcal{U}, \mathcal{V})) \;\middle|\; x \in M\right\} \\ &\prec \mathrm{st}(\mathrm{st}^5\,\mathcal{U}, \mathcal{V}).\end{aligned}$$

Now, it follows that

$$\begin{aligned}h\psi\varphi^{-1}k^{-1} &\simeq_{\mathcal{V}} \mathrm{pr}_N\psi\varphi^{-1}k^{-1} \simeq_{\mathrm{st}(\mathrm{st}^5\,\mathcal{U}, \mathcal{V})} \mathrm{pr}_N\psi\varphi^{-1}rk^{-1} = f_0\mathrm{pr}_M k^{-1} \\ &\simeq_{\mathcal{V}} f\mathrm{pr}_M k^{-1} \simeq_{\mathcal{V}} fkk^{-1} = f.\end{aligned}$$

Observe that

$$\mathrm{st}^2(\mathrm{st}(\mathrm{st}^5\,\mathcal{U}, \mathcal{V}), \mathcal{V}) = \mathrm{st}(\mathrm{st}^5\,\mathcal{U}, \mathrm{st}\,\mathcal{V}) \prec \mathrm{st}^6\,\mathcal{U}.$$

Therefore, $h\psi\varphi^{-1}k^{-1} \simeq_{\mathrm{st}^6\,\mathcal{U}} f$. □

Remark 2.9 Later in Sect. 2.11, the $\boldsymbol{Q}$-manifold version of Theorem 2.6.3 will be proved. It will be helpful to review the above proof. We have first applied the Strong Universality Theorem 2.3.15 to approximate a $\mathcal{U}$-homotopy equivalence and its $\mathcal{U}$-homotopy inverse with closed embeddings, and then used Lemmas 2.6.4 and 2.6.5 to construct the homeomorphism $\psi\varphi^{-1} : M \times \mathbb{R}_+ \to N \times \mathbb{R}_+$ so that $\mathrm{pr}_N\psi\varphi^{-1}$ is $\mathrm{st}(\mathrm{st}^2\,\mathcal{U}, \mathcal{V})$-close to $f\mathrm{pr}_M$, where $\mathcal{V}$ is an open star-refinement of $\mathcal{U}$. Finally, using the fact that the projections $\mathrm{pr}_M : M \times \mathbb{R}_+ \to M$ and $\mathrm{pr}_N : N \times \mathbb{R}_+ \to N$ are near-homeomorphisms, we have obtained Theorem 2.6.3.

To prove Lemma 2.6.4, we only use the fact that a closed submanifold is collared if it is contained in some collared set. This fact follows from Theorem 2.4.9 and the Collaring Theorem 2.5.9. Those theorems will be proved for $\boldsymbol{Q}$-manifolds in Sects. 2.10 and 2.11. For Lemma 2.6.5, we use also the Unknotting Theorem 2.5.12, which will be proved for $\boldsymbol{Q}$-manifolds in Sect. 2.11. On the other hand, the $\boldsymbol{Q}$-manifold version of the Strong Universality Theorem 2.3.15 has been obtained as the Closed Embedding Approximation Theorem 2.3.17.

Because $\ell_1(\Gamma) \approx \ell_2(\Gamma)$ (Theorem 1.2.3), Corollary 2.2.13 is restated that $\ell^2(\Gamma)^{\mathbb{N}} \approx \ell^2(\Gamma)$. Every $\ell^2(\Gamma)$-manifold M is an ANR with $w(M) = \operatorname{card}\Gamma = \tau$, which has the homotopy type of some locally finite-dimensional simplicial complex K with $\operatorname{card} K^{(0)} \leqslant \tau$ by Theorem 1.13.22. Due to Corollary 2.2.15, $|K|_{\mathrm{m}} \times \ell^2(\Gamma)$ is an $\ell^2(\Gamma)$-manifold. By virtue of the Classification Theorem 2.6.1, we have $M \approx |K|_{\mathrm{m}} \times \ell^2(\Gamma)$. When M is an ℓ^2-manifold, the above K is countable locally finite (Theorem 1.13.22), where $|K|_{\mathrm{m}} = |K|$. Then, $M \approx |K| \times \ell^2$. Thus, we have the following Triangulation Theorem for $\ell^2(\Gamma)$-manifolds:

Theorem 2.6.6 (TRIANGULATION) *Each $\ell^2(\Gamma)$-manifold is homeomorphic to $|K|_{\mathrm{m}} \times \ell^2(\Gamma)$ for some locally finite-dimensional simplicial complex K with $\operatorname{card} K^{(0)} \leqslant \operatorname{card}\Gamma = \tau$. In particular, each ℓ^2-manifold is homeomorphic to $|K| \times \ell^2$ for some countable locally finite simplicial complex K.* □

Similarly, the Triangulation Theorem for ℓ^2_f-manifolds can be obtained as a combination of the Classification Theorem 2.6.1 and Corollary 2.2.17:

Theorem 2.6.7 (TRIANGULATION) *Each ℓ^2_f-manifold is homeomorphic to $|K| \times \ell^2_f$ for some countable locally finite simplicial complex K with $\operatorname{card} K^{(0)} \leqslant \operatorname{card}\Gamma = \tau$.* □

Remark 2.10 It is difficult to prove the Triangulation Theorem for $\boldsymbol{Q}$-manifolds, which will be done in Chap. 4.

2.7 The Bing Shrinking Criterion

In this section, we prove the following very useful criterion to examine whether maps are near-homeomorphisms:

Theorem 2.7.1 (BING SHRINKING CRITERION) *A map $\pi : X \to Y$ between completely metrizable spaces is a near-homeomorphism if $\pi(X)$ is dense in Y and it satisfies the following condition:*

(Bi) *For each $\mathcal{U} \in \operatorname{cov}(X)$ and $\mathcal{V} \in \operatorname{cov}(Y)$, there exist $\mathcal{W} \in \operatorname{cov}(Y)$ and a homeomorphism $f : X \to X$ such that πf is $\mathcal{V}$-close to π and $f(\pi^{-1}(\mathcal{W})) \prec \mathcal{U}$:*

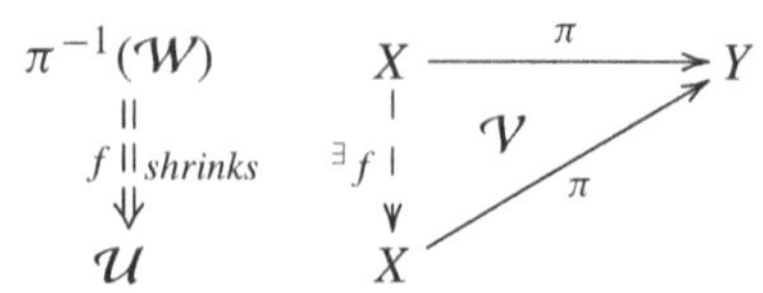

Proof For each $\mathcal{V} \in \operatorname{cov}(Y)$, we construct a homeomorphism $g : X \to Y$ that is $\mathcal{V}$-close to π. Take complete metrics $d \in \operatorname{Metr}(X)$ and $\rho \in \operatorname{Metr}(Y)$ so that $\operatorname{diam}_d X \leqslant 1/2$ and $\{\overline{\mathrm{B}}_\rho(y, 1) \mid y \in Y\} \prec \mathcal{V}$ (cf. 1.3.22(1)). By induction, we construct $\mathcal{W}_n \in \operatorname{cov}(Y)$ and homeomorphisms $f_n : X \to X$, $n \in \mathbb{N}$, so that

(1) $\operatorname{mesh} \mathcal{W}_n \leqslant 2^{-n-1}$;
(2) $\operatorname{mesh} f_n(\pi^{-1}(\mathcal{W}_n)) \leqslant 2^{-n-1}$;
(3) $d(f_n, f_{n-1}) \leqslant 2^{-n}$;
(4) $\rho(\pi f_n^{-1}, \pi f_{n-1}^{-1}) \leqslant 2^{-n}$,

where $f_0 = \mathrm{id}_X$ and $\mathcal{W}_0 = \{\mathrm{B}_\rho(y, \frac{1}{4}) \mid y \in Y\} \in \operatorname{cov}(Y)$. Assume that f_{n-1} and $\mathcal{W}_{n-1}$ are defined. Take $\mathcal{W}'_n \in \operatorname{cov}(X)$ with $\operatorname{mesh} \mathcal{W}'_n \leqslant 2^{-n-1}$. We apply (Bi) to obtain $\mathcal{W}_n \in \operatorname{cov}(Y)$ and a homeomorphism $h : X \to X$ such that $\operatorname{mesh} \mathcal{W}_n \leqslant 2^{-n-1}$, $h(\pi^{-1}(\mathcal{W}_n)) \prec f_{n-1}^{-1}(\mathcal{W}'_n)$, and h is $\pi^{-1}(\mathcal{W}_{n-1})$-close to id. Let $f_n = f_{n-1}h$ and see the following diagram:

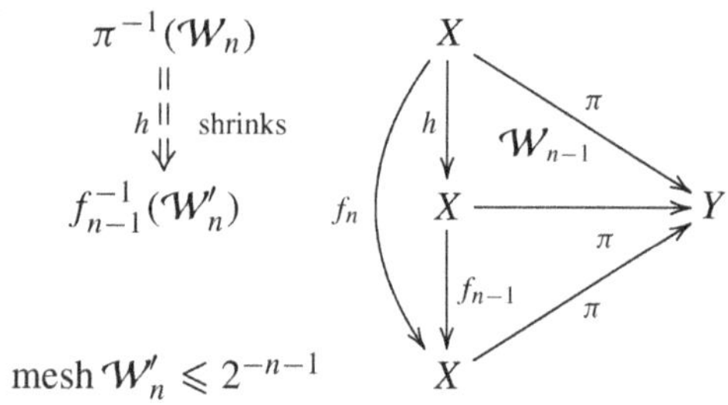

Then, $\mathcal{W}_n$ and f_n satisfy (1) and (2). Since f_{n-1} satisfies (2), we have

$$d(f_n, f_{n-1}) = d(f_{n-1}h, f_{n-1}) \leqslant \operatorname{mesh} f_{n-1}(\pi^{-1}(\mathcal{W}_{n-1})) \leqslant 2^{-n}.$$

Since $\mathcal{W}_{n-1}$ satisfies (1), we also have

$$\rho(\pi f_n^{-1}, \pi f_{n-1}^{-1}) = \rho(\pi h^{-1}, \pi) \leqslant \operatorname{mesh} \mathcal{W}_{n-1} \leqslant 2^{-n}.$$

Thus, (3) and (4) are also satisfied.

Since $X = (X, d)$ and $Y = (Y, \rho)$ are complete, $(f_n)_{n\in\mathbb{N}}$ uniformly converges to a map $f : X \to X$ by (3) and $(\pi f_n^{-1})_{n\in\mathbb{N}}$ uniformly converges to a map $g : X \to Y$ by (4). For each $n \in \mathbb{N}$, $d(f, f_n) \leqslant 2^{-n+1}$ by (3), hence $\operatorname{mesh} f(\pi^{-1}(\mathcal{W}_n)) < 2^{-n+3}$ by (2). See the following diagram:

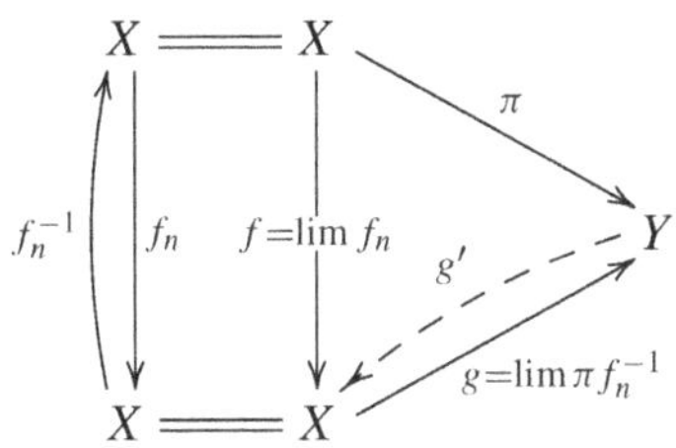

Since $\pi(X)$ is dense in Y and $X = (X, d)$ is complete, we have a map $g' : Y \to X$ such that

$$\{g'(y)\} = \bigcap_{n \in \mathbb{N}} \operatorname{cl} f(\pi^{-1}(\operatorname{st}(y, \mathcal{W}_n))) \quad \text{for each } y \in Y.$$

The continuity of g' is easy to see. Observe that $g'\pi = f$. By (4), we have

$$\rho(\pi f_n^{-1} f_m, \pi) = \rho(\pi f_n^{-1}, \pi f_m^{-1}) \leqslant 2^{-n+1} \quad \text{for each } n \leqslant m,$$

which implies $\rho(\pi f_n^{-1} f, \pi) \leqslant 2^{-n+1}$ for each $n \in \mathbb{N}$, hence $\rho(gf, \pi) = 0$, i.e., $gf = \pi$. Since $gg'\pi = gf = \pi$ and $\pi(X)$ is dense in Y, it follows that $gg' = \mathrm{id}$. Since $g'gf = g'\pi = f$ and $f(X)$ is dense in X, we have $g'g = \mathrm{id}$. Therefore, g is a homeomorphism. Moreover,

$$\rho(g, \pi) \leqslant \sum_{n \in \mathbb{N}} \rho(\pi f_n^{-1}, \pi f_{n-1}^{-1}) \leqslant \sum_{n \in \mathbb{N}} 2^{-n} = 1,$$

hence g is $\mathcal{V}$-close to π. □

Remark 2.11 In condition (Bi), a homeomorphism f can be replaced by a near-homeomorphism. In fact, for each $\mathcal{U} \in \operatorname{cov}(X)$ and $\mathcal{V} \in \operatorname{cov}(Y)$, take $\mathcal{U}' \in \operatorname{cov}(X)$ and $\mathcal{V}' \in \operatorname{cov}(Y)$ such that $\operatorname{st}\mathcal{U}' \prec \mathcal{U}$, $\mathcal{U}' \prec \pi^{-1}(\mathcal{V}')$, and $\operatorname{st}\mathcal{V}' \prec \mathcal{V}$. Assume that there exists a near-homeomorphism $f : X \to X$ with $\mathcal{W} \in \operatorname{cov}(Y)$ such that πf is $\mathcal{V}'$-close to π and $f(\pi^{-1}(\mathcal{W})) \prec \mathcal{U}'$ (and $f|A = \mathrm{id}$). Then, f is $\mathcal{U}'$-close to a homeomorphism $h : X \to X$, which satisfies $h(\pi^{-1}(\mathcal{W})) \prec \operatorname{st}(f(\pi^{-1}(\mathcal{W})), \mathcal{U}') \prec \operatorname{st}\mathcal{U}' \prec \mathcal{U}$ and πh is $\mathcal{V}'$-close to πf, so $\mathcal{V}$-close to π.

In the above theorem, let $A \subset X$. If f in (Bi) can be taken so as to satisfy the additional condition that $f|A = \mathrm{id}$, then π can be approximated by a homeomorphism $g : X \to Y$ such that $g|A = \pi|A$. In fact, we can take f_n in the proof so as to satisfy $f_n|A = \mathrm{id}$, hence $\pi f_n^{-1}|A = \pi|A$. Taking the limit, $g|A = \pi|A$. Thus, we have the following:

Theorem 2.7.2 *Let $\pi : X \to Y$ be a map between completely metrizable spaces and $A \subset X$. Assume that $\pi(X)$ is dense in Y and the following condition:*

(Bi$_A$) *For each* $\mathcal{U} \in \operatorname{cov}(X)$ *and* $\mathcal{V} \in \operatorname{cov}(Y)$*, there exist* $\mathcal{W} \in \operatorname{cov}(Y)$ *and a homeomorphism* $f : X \to X$ *such that* $f|A = \mathrm{id}$*,* πf *is* $\mathcal{V}$*-close to* π*, and* $f(\pi^{-1}(\mathcal{W})) \prec \mathcal{U}$*:*

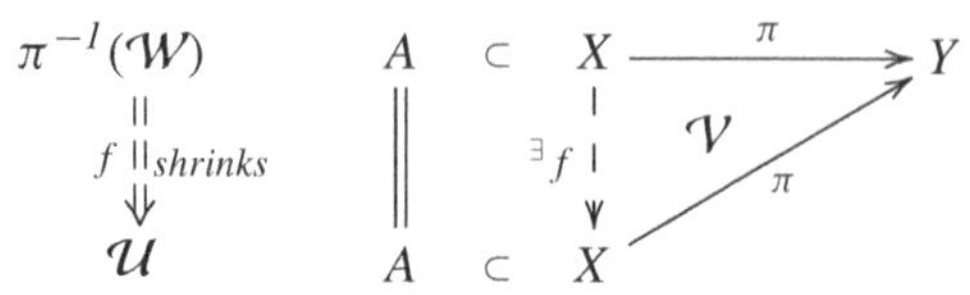

Then, for each $\mathcal{V} \in \operatorname{cov}(Y)$*,* π *is* $\mathcal{V}$*-close to a homeomorphism* $g : X \to Y$ *such that* $g|A = \pi|A$. □

Remark 2.12 When π is a closed map, for each open neighborhood U of $\pi^{-1}(y)$ in X, y has an open neighborhood V in Y such that $\pi^{-1}(V) \subset U$. In Theorems 2.7.1 and 2.7.2 above, the condition $f(\pi^{-1}(\mathcal{W})) \prec \mathcal{U}$ in (Bi) and (Bi$_A$) can be weakened to the condition $\{f\pi^{-1}(y) \mid y \in Y\} \prec \mathcal{U}$, that is, each $f\pi^{-1}(y)$ is contained in some $U \in \mathcal{U}$.

For incomplete spaces, we have the following:

Theorem 2.7.3 *Let* $\pi : X \to Y$ *be a map between metrizable spaces and* A *a closed set in* X *such that* $B = \pi(A)$ *is closed in* Y*. Assume that* $\pi(X) = Y$ *and the following condition:*

(Bi$'_A$) *For each* $\mathcal{U} \in \operatorname{cov}(X)$ *and* $\mathcal{V} \in \operatorname{cov}(Y)$*, there exist* $\mathcal{W} \in \operatorname{cov}(Y)$ *and a homeomorphism* $f : X \to X$ *such that* $f|A \cup (X \setminus \pi^{-1}(\operatorname{st}(B, \mathcal{V}))) = \mathrm{id}$*,* πf *is* $\mathcal{V}$*-close to* π*, and* $f(\pi^{-1}(\mathcal{W})) \prec \mathcal{U}$*:*

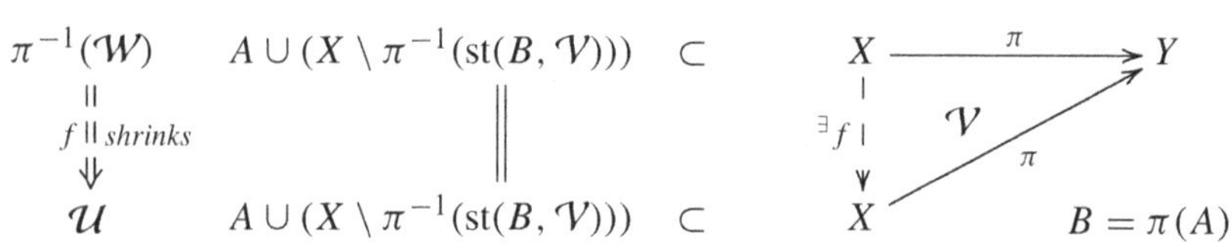

Then, for each $\mathcal{V} \in \operatorname{cov}(Y)$*,* π *is* $\mathcal{V}$*-close to a homeomorphism* $g : X \to Y$ *such that*

$$g|A \cup (X \setminus \pi^{-1}(\operatorname{st}(B, \mathcal{V}))) = \pi|A \cup (X \setminus \pi^{-1}(\operatorname{st}(B, \mathcal{V}))).$$

Proof In the proof of Theorem 2.7.1, the homeomorphisms f_n, $n \in \mathbb{N}$, can be taken so as to satisfy the following condition in addition to (1)–(4):

(5) $f_n|A \cup (X \setminus \pi^{-1}(\operatorname{st}(B, \mathcal{W}_{n-1}))) = f_{n-1}|A \cup (X \setminus \pi^{-1}(\operatorname{st}(B, \mathcal{W}_{n-1})))$,

from which we have $f_n|A \cup (X \setminus \pi^{-1}(\operatorname{st}(B, \mathcal{V}))) = \mathrm{id}$. Using this additional condition (5) instead of the completeness, we show that $(f_n)_{n\in\mathbb{N}}$ and $(\pi f_n^{-1})_{n\in\mathbb{N}}$ converge to maps f and g, respectively. Since $(f_n)_{n\in\mathbb{N}}$ and $(\pi f_n^{-1})_{n\in\mathbb{N}}$ are uniformly Cauchy by (3) and (4), it suffices to show that $(f_n(x))_{n\in\mathbb{N}}$ and $(\pi f_n^{-1}(x))_{n\in\mathbb{N}}$ are convergent for each $x \in X$.

First, we show that $(f_n(x))_{n\in\mathbb{N}}$ is convergent in X. Due to (1), each $x \in X \setminus \pi^{-1}(B)$ is contained in $X \setminus \pi^{-1}(\mathrm{st}(B, \mathcal{W}_{n-1}))$ for sufficiently large n, which implies that the sequence $(f_n(x))_{n\in\mathbb{N}}$ is eventually constant. For each $x \in \pi^{-1}(B)$, there is $a \in A$ such that $\pi(a) = \pi(x)$, hence $(f_n(x))_{n\in\mathbb{N}}$ converges to a by (2) and (5). Thus, we have $f(x) = \lim_{n\to\infty} f_n(x) \in X$ for every $x \in X$.

Next, we show that $(\pi f_n^{-1}(x))_{n\in\mathbb{N}}$ is convergent in Y. By virtue of (5), for each $x \in A$ and $n \in \mathbb{N}$, $f_n(x) = x$, so $\pi f_n^{-1}(x) = \pi(x)$. For each $x \in X \setminus A$, choose $m \in \mathbb{N}$ so that $2^{-m-1} < d(x, A)$. Then,

$$x' = f_m^{-1}(x) \in X \setminus \pi^{-1}(\mathrm{st}(B, \mathcal{W}_m)).$$

Otherwise, we have $W \in \mathcal{W}_m$ and $a \in A$ such that $\pi(x'), \pi(a) \in W$, hence

$$d(x, a) \leqslant d(f_m(x'), f_m(a)) \leqslant \operatorname{diam} f_m(\pi^{-1}(W)) \leqslant 2^{-m-1} < d(x, A),$$

which is a contradiction. By (5), $f_{m+1}(x') = f_m(x') = x$. By induction, we have $f_n(x') = x$ for any $n \geqslant m$, hence $(\pi f_n^{-1}(x))_{n\in\mathbb{N}}$ is eventually equal to $\pi(x')$. Thus, we can define $g : Y \to X$ by $g(x) = \lim_{n\to\infty} \pi f_n^{-1}(x) \in X$ for every $x \in X$.

In the proof of Theorem 2.7.1, we used a map g' to show that g is a homeomorphism. In order to define g' without completeness, we need to assume that $\pi(X) = Y$. Indeed, $f(\pi^{-1}(y)) \neq \emptyset$ for every $y \in Y$. As observed in the proof of Theorem 2.7.1, $f(\pi^{-1}(\mathcal{W}_n)) < 2^{-n+3}$. Then, g' can be defined by $\{g'(y)\} = f(\pi^{-1}(y))$. □

For compact metric spaces, the Bing Shrinking Criterion 2.7.1 can be easily rewritten using Remark 2.12 as follows:

Theorem 2.7.4 (Bing Shrinking Criterion) *A map $\pi : X \to Y$ between compact metric spaces is a near-homeomorphism if $\pi(X) = Y$ and π satisfies the following condition:*

(Bi$_0$) *For each $\varepsilon > 0$, there is a homeomorphism $f : X \to X$ such that $\operatorname{diam} f(\pi^{-1}(y)) < \varepsilon$ for each $y \in Y$ and πf is ε-close to π.* □

Remark 2.13 By the same argument as in Remark 2.11, a homeomorphism f in (Bi$_0$) can be replaced by a near-homeomorphism. Furthermore, as Theorem 2.7.2, given $A \subset X$, if f in (Bi$_0$) can be taken so as to satisfy the condition $f|A = \mathrm{id}$, then π can be approximated by a homeomorphism $g : X \to Y$ such that $g|A = \pi|A$.

Applying the above Bing Shrinking Criterion 2.7.4, we show the following:

Theorem 2.7.5 *The Hilbert cube $\boldsymbol{Q}$ is homeomorphic to its cone $C(\boldsymbol{Q})$, which is homeomorphic to the usual cone $C_{\boldsymbol{Q}} = (\mathbf{I} \times \boldsymbol{Q})/(\{0\} \times \boldsymbol{Q})$ because $\boldsymbol{Q}$ is compact.*

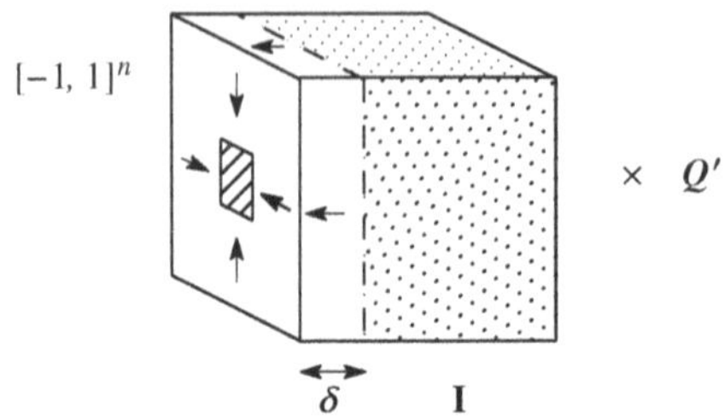

Fig. 2.17 The homeomorphism φ

Proof Let $\pi : \mathbf{I} \times \boldsymbol{Q} \to C(\boldsymbol{Q}) = (\mathbf{I} \times \boldsymbol{Q})_{\{0\}}$ be the natural map. We use the metric $d \in \mathrm{Metr}(\mathbf{I} \times \boldsymbol{Q})$ defined by

$$d((t, x), (s, y)) = |t - s| + \sum_{n \in \mathbb{N}} 2^{-n} |x(n) - y(n)|.$$

For each $\varepsilon > 0$ choose $n \in \mathbb{N}$ and $0 < \delta < 1$ so that

$$2^{-n} < \varepsilon/4 \text{ and } (0, \delta] \times \boldsymbol{Q} \subset \mathrm{B}(0, \varepsilon/2) \subset C(\boldsymbol{Q}),$$

where $C(\boldsymbol{Q})$ is any given admissible metric. Represent $\boldsymbol{Q} = [-1, 1]^n \times \boldsymbol{Q}'$ and let $q : \mathbf{I} \times \boldsymbol{Q} \to \boldsymbol{Q}'$ be the projection. It is easy to construct $\varphi \in \mathrm{Homeo}(\mathbf{I} \times \boldsymbol{Q})$ (as illustrated in Fig. 2.17) such that

$$\varphi|[\delta, 1] \times \boldsymbol{Q} = \mathrm{id}, \ q\varphi = q \text{ and}$$
$$\varphi(\{0\} \times \boldsymbol{Q}) = \{0\} \times [-\varepsilon/4, \varepsilon/4]^n \times \boldsymbol{Q}'.$$

Then, $\operatorname{diam} \varphi(\{0\} \times \boldsymbol{Q}) < \varepsilon$ and $\pi\varphi$ is ε-close to π. By the Bing Shrinking Criterion 2.7.4, π is a near-homeomorphism. Thus, we have $\boldsymbol{Q} \approx \mathbf{I} \times \boldsymbol{Q} \approx C(\boldsymbol{Q})$. □

Combining Theorem 2.7.5 with the homogeneity of $\boldsymbol{Q}$ (Theorem 2.1.2), we have the following:

Corollary 2.7.6 $\boldsymbol{Q} \setminus \{\mathbf{0}\} \approx \boldsymbol{Q} \times [0, 1)$. □

As another application of the Bing Shrinking Criterion 2.7.4 (with Remark 2.13), we can prove the following GENERALIZED SCHOENFLIES THEOREM, which will be used in the proof of Theorem 4.5.1 concerning the existence of an immersion of punctured n-torus into $\mathbb{R}^n$:

Theorem 2.7.7 (GENERALIZED SCHOENFLIES) *Let $h : \mathbf{S}^{n-1} \to \mathbf{S}^n$ be an embedding such that $h(\mathbf{S}^{n-1})$ is bi-collared in $\mathbf{S}^n$. Then, h extends to a homeomorphism $\tilde{h} : \mathbf{S}^n \to \mathbf{S}^n$, where $\mathbf{S}^{n-1}$ is identified with $\mathbf{S}^n \cap (\mathbb{R}^n \times \{0\}) \subset \mathbf{S}^n$. As* *a consequence, the complement* $\mathbf{S}^n \setminus h(\mathbf{S}^{n-1})$ *has precisely two components whose closures are* n*-cells.*

As preparation of the proof of Theorem 2.7.7, we introduce cellular sets. A homeomorphic image of $\mathbf{B}^n$ is called a **(topological)** $\boldsymbol{n}$**-cell** (or just a **cell**).[25] When $(C, D) \approx (\mathbf{B}^n, \mathbf{S}^{n-1})$, D is called the **boundary** of a cell C and denoted by ∂C. Let X be an n-dimensional compactum. It is said that $A \subset X$ is **cellular** in X if there exist n-cells $C_1, C_2, \ldots$ in X such that $C_{i+1} \subset C_i \setminus \partial C_i$, $i \in \mathbb{N}$, and $A = \bigcap_{i \in \mathbb{N}} C_i$. This is equivalent to the condition that for each neighborhood U of A in X, there is an n-cell C such that $A \subset C \setminus \partial C \subset C \subset U$.

Lemma 2.7.8 *Let* $f : \mathbf{B}^n \to \mathbf{S}^n$ *be a map with* $A \subset \mathbf{B}^n \setminus \mathbf{S}^{n-1}$ *such that* $f(A)$ *is a finite set,* $A = f^{-1}(f(A))$, *and* $f|\mathbf{B}^n \setminus A$ *is injective. Then, there exists a component* D *of* $\mathbf{S}^n \setminus f(\mathbf{S}^{n-1})$ *such that* $f(\mathbf{B}^n) = f(\mathbf{S}^{n-1}) \cup D$.

Proof First, note that $f(\mathbf{B}^n) \not\subset f(\mathbf{S}^{n-1})$. Indeed, since $h = f|\mathbf{S}^{n-1} : \mathbf{S}^{n-1} \to f(\mathbf{S}^{n-1})$ is a homeomorphism, we have a retraction $r = h^{-1}f : \mathbf{B}^n \to \mathbf{S}^{n-1}$, which contradicts the No Retraction Theorem 1.12.2. Hence, $f(\mathbf{B}^n)$ meets a component D of $\mathbf{S}^n \setminus f(\mathbf{S}^{n-1})$. Since $f(\mathbf{S}^{n-1})$ does not separate $f(\mathbf{B}^n)$, it follows that $f(\mathbf{B}^n) \subset \operatorname{cl} D$, so we have $f(\mathbf{B}^n) \subset f(\mathbf{S}^{n-1}) \cup D$.

Note that $(\mathbf{B}^n \setminus \mathbf{S}^{n-1}) \setminus A$ is an open set in $\mathbb{R}^n$ and $f|(\mathbf{B}^n \setminus \mathbf{S}^{n-1}) \setminus A$ is a continuous injection. Since D is regarded as an open set in $\mathbb{R}^n$, it follows from Theorem 1.12.4 that $f|(\mathbf{B}^n \setminus \mathbf{S}^{n-1}) \setminus A$ is an open embedding. Hence, $f((\mathbf{B}^n \setminus \mathbf{S}^{n-1}) \setminus A)$ is open in $D \setminus f(A)$, which is also closed in $D \setminus f(A)$. When $n > 1$, any finite set does not separate D, so $D \setminus f(A)$ is connected. Then, it follows that $f((\mathbf{B}^n \setminus \mathbf{S}^{n-1}) \setminus A) = D \setminus f(A)$, hence $f(\mathbf{B}^n) \supset D$. Thus, we have $f(\mathbf{B}^n) = f(\mathbf{S}^{n-1}) \cup D$. When $n = 1$, since $f(\mathbf{B}^1)$ $(= f([-1, 1]))$ is an arc connecting $f(-1)$ and $f(1)$, it follows that $f(\mathbf{B}^1) = f(\mathbf{S}^0) \cup D$. □

Lemma 2.7.9 *Let* $f : \mathbf{B}^n \to \mathbf{S}^n$ *be a map with* $A \subset \mathbf{B}^n \setminus \mathbf{S}^{n-1}$ *such that* $f(A)$ *is a singleton,* $A = f^{-1}(f(A))$, *and* $f|\mathbf{B}^n \setminus A$ *is injective. Then,* A *is cellular in* $\mathbf{B}^n$.

Proof By Lemma 2.7.8, there is a component D of $\mathbf{S}^n \setminus f(\mathbf{S}^{n-1})$ such that $f(\mathbf{B}^n) = f(\mathbf{S}^{n-1}) \cup D$. Let U be any open set in $\mathbf{B}^n$ such that $A \subset U \subset \mathbf{B}^n \setminus \mathbf{S}^{n-1}$. Then, $f(U)$ is open in D and $f(A) \subset f(U)$. Since $f(A)$ is a singleton, we can take a homeomorphism $h : \mathbf{S}^n \to \mathbf{S}^n$ and a neighborhood V of $f(A)$ in $\mathbf{S}^n$ such that $V \subset f(U)$, $h|V = \mathrm{id}$, and $h(D) \subset f(U)$. Applying Theorem 1.12.4 as in the proof of Lemma 2.7.8, we can see that $g = f|\mathbf{B}^n \setminus A : \mathbf{B}^n \setminus A \to f(\mathbf{B}^n \setminus A)$ is a homeomorphism. Then, we can define a map $\varphi : \mathbf{B}^n \to \mathbf{B}^n$ as follows:

$$\varphi(x) = \begin{cases} x & \text{if } x \in A, \\ g^{-1}hf(x) & \text{if } x \in \mathbf{B}^n \setminus A. \end{cases}$$

[25] In Chapter 1, a (convex) linear cell is called a cell (see Sect. 1.5, p. 27).

Since $f^{-1}(V)$ is a neighborhood of A in $\mathbf{B}^n$ and $g^{-1}hf|f^{-1}(V)\setminus A = \mathrm{id}$, it follows that φ is a well-defined embedding and $\varphi|f^{-1}(V) = \mathrm{id}$. Because $hf(\mathbf{B}^n) \subset f(U)$, we have $f^{-1}(V) \subset \varphi(\mathbf{B}^n) \subset U$. Consequently, A is cellular in $\mathbf{B}^n$. □

Lemma 2.7.10 *Let $f : \mathbf{S}^n \to \mathbf{S}^n$ be a map with $A, B \subset \mathbf{S}^n$ such that $f(A)$, $f(B)$ are singletons, $A = f^{-1}(f(A))$, $B = f^{-1}(f(B))$, and $f|\mathbf{S}^n \setminus (A \cup B)$ is injective. Then, both A and B are cellular in $\mathbf{S}^n$.*

Proof Decomposing $\mathbf{S}^n$ into two n-cells C_0, C_1 with the common boundary $\partial C_0 = \partial C_1 = C_0 \cap C_1$ so that $A \cup B \subset \mathbf{S}^n \setminus C_0 = C_1 \setminus \partial C_1$, we can write $\mathbf{S}^n = C_0 \cup C_1$. Applying Lemma 2.7.8 to $f|C_1$, we have a component D of $\mathbf{S}^n \setminus f(\partial C_0)$ such that $f(C_1) = f(\partial C_1) \cup D$. Since $f(A)$ and $f(B)$ are singletons contained in D, there is an open set U in D such that $f(A) \subset U$ and $\mathrm{cl}\, U \cap f(B) = \emptyset$. Then, we can take a homeomorphism $h : \mathbf{S}^n \to \mathbf{S}^n$ and a neighborhood V of $f(A)$ in $\mathbf{S}^n$ such that $V \subset U$, $h|V = \mathrm{id}$, and $h(f(C_1)) \subset U$. Applying Theorem 1.12.4 as before, we can see that $g = f|f^{-1}(U) \setminus A : f^{-1}(U) \setminus A \to U \setminus f(A)$ is a homeomorphism. We define a map $\varphi : C_1 \to \mathbf{S}^n$ as follows:

$$\varphi(x) = \begin{cases} x & \text{if } x \in A, \\ g^{-1}hf(x) & \text{if } x \in C_1 \setminus A. \end{cases}$$

Then, $\varphi(B) = g^{-1}hf(B)$ is a singleton, $\varphi^{-1}\varphi(B) = f^{-1}f(B) = B$, and $\varphi|C_1 \setminus B$ is injective. So, it follows from Lemma 2.7.9 that B is cellular in C_1. By the definition of a cellular set, B is also cellular in $\mathbf{S}^n$. In the same way, it can be shown that A is cellular in $\mathbf{S}^n$. □

Proof (Theorem 2.7.7) By shrinking a bi-collar, we can obtain a closed bi-collar $k : h(\mathbf{S}^{n-1}) \times [-1, 1] \to \mathbf{S}^n$ of $h(\mathbf{S}^{n-1})$ in $\mathbf{S}^n$. We define an embedding $h_* : \mathbf{S}^{n-1} \times [-1, 1] \to \mathbf{S}^n$ by

$$h_*(x, t) = k(h(x), t) \quad \text{for each } (x, t) \in \mathbf{S}^{n-1} \times [-1, 1].$$

Then, $\mathbf{S}^n \setminus h_*(\mathbf{S}^{n-1} \times (-1, 1))$ has two components A and B with

$$h_*(\mathbf{S}^{n-1} \times \{1\}) \subset A \quad \text{and} \quad h_*(\mathbf{S}^{n-1} \times \{-1\}) \subset B.$$

Consider the suspension $\Sigma\mathbf{S}^{n-1}$ of $\mathbf{S}^{n-1}$ defined as the following quotient space:[26]

$$\Sigma\mathbf{S}^{n-1} = \mathbf{S}^{n-1} \times [-1, 1]/\{\mathbf{S}^{n-1} \times \{-1\}, \mathbf{S}^{n-1} \times \{1\}\},$$

where $\mathbf{S}^{n-1} \times \{-1\}$, $\mathbf{S}^{n-1} \times \{1\}$ are regarded as two distinct points a, b, respectively. Let $\pi : \mathbf{S}^{n-1} \times [-1, 1] \to \Sigma\mathbf{S}^{n-1}$ be the quotient map. We can define a map

[26] We change the coordinates in the definition of the suspension in [GAGT, p. 453].

$\varphi : \mathbf{S}^n \to \Sigma\mathbf{S}^{n-1}$ as follows:

$$\varphi(x) = \begin{cases} \pi h_*^{-1}(x) & \text{if } x \in h_*(\mathbf{S}^{n-1} \times [-1, 1]), \\ a & \text{if } x \in A, \\ b & \text{if } x \in B. \end{cases}$$

Identifying $\Sigma\mathbf{S}^{n-1} = \mathbf{S}^n$, we have a map $\varphi : \mathbf{S}^n \to \mathbf{S}^n$ such that $\varphi(A) = \{a\}$, $\varphi(B) = \{b\}$, $\varphi^{-1}(a) = A$, $\varphi^{-1}(b) = B$ and $\varphi|\mathbf{S}^n \setminus (A \cup B)$ is injective. Then, both A and B are cellular by Lemma 2.7.10.

For each $\varepsilon > 0$, a and b have open neighborhoods $U$and V in $\mathbf{S}^n$ respectively such that $U \cap V = \emptyset$ and $\operatorname{diam} U$, $\operatorname{diam} V < \varepsilon$. Since A and B are cellular, there exist n-cells C_i, D_i, $i = 0, 1$ in $\mathbf{S}^n$ such that

$$A \subset C_0 \setminus \partial C_0 \subset C_0 \subset C_1 \setminus \partial C_1 \subset C_1 \subset \varphi^{-1}(U) \text{ and}$$

$$B \subset D_0 \setminus \partial D_0 \subset D_0 \subset D_1 \setminus \partial D_1 \subset D_1 \subset \varphi^{-1}(V).$$

Then, we can take a homeomorphism $\psi : \mathbf{S}^n \to \mathbf{S}^n$ such that

$$\psi|\mathbf{S}^n \setminus (C_1 \cup D_1) = \mathrm{id}; \text{ and } \operatorname{diam} \psi(C_0), \operatorname{diam} \psi(D_0) < \varepsilon.$$

Observe that $\varphi\psi$ is ε-close to φ,

$$\operatorname{diam} \psi(\varphi^{-1}(a)), \operatorname{diam} \psi(\varphi^{-1}(b)) < \varepsilon, \text{ and } \psi|h(\mathbf{S}^{n-1}) = \mathrm{id}.$$

Now, applying the Bing Shrinking Criterion 2.7.4 with Remark 2.13, we can obtain a homeomorphism $g : \mathbf{S}^n \to \mathbf{S}^n$ such that $g|h(\mathbf{S}^{n-1}) = \varphi|h(\mathbf{S}^{n-1})$, hence

$$g(h(x)) = \pi h_*^{-1}(h_*(x, 0)) = \pi(x, 0) = x \text{ for each } x \in \mathbf{S}^{n-1},$$

where we identify $(\Sigma\mathbf{S}^n, \mathbf{S}^{n-1} \times \{0\}) = (\mathbf{S}^n, \mathbf{S}^{n-1})$. Then, $\tilde{h} = g^{-1} : \mathbf{S}^n \to \mathbf{S}^n$ is a homeomorphism and $\tilde{h}(x) = g^{-1}(x) = h(x)$ for each $x \in \mathbf{S}^{n-1}$, that is, $\tilde{h}|\mathbf{S}^{n-1} = h$. □

In Theorem 2.7.7, decomposing $\mathbf{S}^n$ into two n-cells C and D, we can write $\mathbf{S}^n = C \cup D$, where $h(\mathbf{S}^{n-1}) = C \cap D = \partial C = \partial D$. Given $a \in C$ and $b \in D$, h extends to a homeomorphism $\tilde{h} : \mathbf{S}^n \to \mathbf{S}^n$ with $\tilde{h}(a) = a$ and $\tilde{h}(b) = b$. Indeed, for each $x_0, x_1 \in \mathbf{B}^n \setminus \mathbf{S}^{n-1}$, there exists a homeomorphism $h : \mathbf{B}^n \to \mathbf{B}^n$ such that $h|\mathbf{S}^{n-1} = \mathrm{id}$ and $h(x_0) = x_1$. Using this fact, we can arrange the homeomorphism $\tilde{h}$ obtained in Theorem 2.7.7 so as to satisfy the condition $\tilde{h}(a) = a$ and $\tilde{h}(b) = b$. Since $\mathbf{S}^n$ is the one-point compactification of $\mathbb{R}^n$, the following can be obtained:

Corollary 2.7.11 *Let $h : \mathbf{S}^{n-1} \to \mathbb{R}^n$ be an embedding such that $h(\mathbf{S}^{n-1})$ is bi-collared in $\mathbb{R}^n$. Then, h extends to a homeomorphism $\tilde{h} : \mathbb{R}^n \to \mathbb{R}^n$. As a* *consequence, the complement $\mathbb{R}^n \setminus h(\mathbf{S}^{n-1})$ has precisely two components and the closure of the bounded component is an n-cell.* □

2.8 Z-Sets and Strong Z-Sets in ANRs

A closed set A in a space X is called a **Z-set** in X if, for each $\mathcal{U} \in \mathrm{cov}(X)$, there is a map $f : X \to X \setminus A$ that is $\mathcal{U}$-close to id. If $A \cap \mathrm{cl}\, f(X) = \emptyset$ in the above, A is called a **strong Z-set** in X. A countable union of Z-sets (resp. strong Z-sets) is called a $\mathbf{Z}_\sigma$ **set** (resp. **strong** $\mathbf{Z}_\sigma$**-set**). Every closed subset of a (strong) Z-set is also a (strong) Z-set and every F_σ-set contained in a (strong) Z_σ-set is a (strong) Z_σ-set. Obviously a strong Z-set is a Z-set. The converse is also true in a locally compact paracompact space, that is:

Proposition 2.8.1 *Every Z-set in a locally compact paracompact space X is a strong Z-set. In particular, every Z-set in a $\mathbf{Q}$-manifold is a strong Z-set.*

Proof Each $\mathcal{U} \in \mathrm{cov}(X)$ has a locally finite open refinement $\mathcal{V} \in \mathrm{cov}(X)$ such that $\mathrm{cl}\, V$ is compact for each $V \in \mathcal{V}$. Then, id_X is $\mathcal{V}$-close to a map $f : X \to X \setminus A$. Since id_X is a proper map, f is also proper by Proposition 1.3.8, hence it is perfect by Proposition 1.3.7. Thus, $f(X)$ is closed in X, which implies that $A \cap \mathrm{cl}\, f(X) = A \cap f(X) = \emptyset$. □

In general, a (compact) Z-set is not a strong Z-set even in an AR. For example, let

$$X = \mathbf{I} \times \{0\} \cup \{n^{-1} \mid n \in \mathbb{N}\} \times \mathbf{I} \subset \mathbb{R}^2 \text{ and } A = \{(0,0)\} \subset X.$$

Then, X is an AR, in which A is a Z-set but not a strong Z-set (Fig. 2.18).

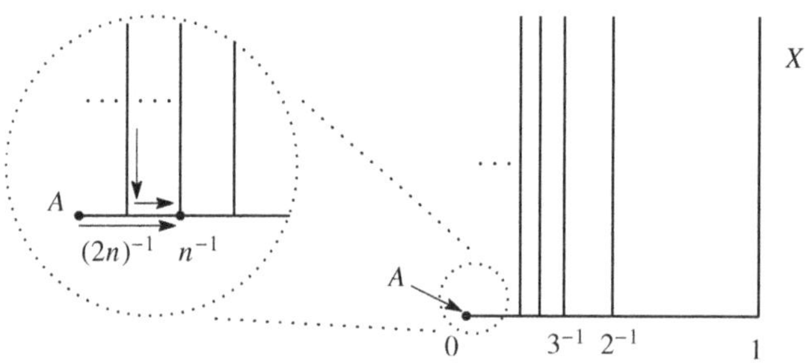

Fig. 2.18 A Z-set being not a strong Z-set

Indeed, for each $n \in \mathbb{N}$, we have a map $f_n : X \to X \setminus A$ defined as follows:

$$f_n(s,t) = \begin{cases} (s,t) & \text{if } s \geqslant n^{-1} \text{ or } t \geqslant n^{-1}, \\ (s, 2t - n^{-1}) & \text{if } s < n^{-1} \text{ and } (2n)^{-1} \leqslant t \leqslant n^{-1}, \\ (2nts + (1-2nt)n^{-1}, 0) & \text{if } s < n^{-1} \text{ and } 0 < t \leqslant (2n)^{-1}, \\ (n^{-1}, 0) & \text{if } s \leqslant n^{-1} \text{ and } t = 0, \end{cases}$$

where $t > 0$ implies $s = m^{-1}$ for some $m \in \mathbb{N}$. Then, f_n is $\sqrt{2}n^{-1}$-close to id_X with respect to the Euclidean metric. Hence, A is a Z-set in X. On the other hand, assume that a map $f : X \to X$ is $\mathcal{U}$-close to id_X, where

$$\mathcal{U} = \left\{\{n^{-1}\} \times (2^{-1}, 1] \;\middle|\; n \in \mathbb{N}\right\} \cup \left\{X \cap [0,1)^2,\ (2^{-1}, 1] \times [0,1)\right\} \in \mathrm{cov}(X).$$

For each $n \in \mathbb{N}$, let $v_n = (n^{-1}, 1) \in X$. Since $f(v_n) \in \{n^{-1}\} \times (2^{-1}, 1]$ and $f(v_1) \in (2^{-1}, 1] \times [0,1)$, the image $f(X)$ contains every point $(n^{-1}, 0) \in X$, which implies $(0,0) \in \mathrm{cl}_X f(X)$. Thus, A is not a strong Z-set in X.

Lemma 2.8.2 *Let A be a closed set in a metrizable space X. If A is contained in some collared set C in X, then A is a strong Z-set.*

Proof Let $k : C \times [0,1) \to X$ be a collar of C in X. Then, A has an open neighborhood W in X with $\mathrm{cl}\, W \subset k(C \times [0,1))$. For each $\mathcal{U} \in \mathrm{cov}(X)$, choose a map $\gamma : C \to \mathbf{I}$ so that $\gamma(x) > 0$ for $x \in A$, $\gamma(x) = 0$ for $x \in C \setminus W$, and $\{\{x\} \times [0, \gamma(x)] \mid x \in C \cap W\} \prec k^{-1}(\mathcal{U}|W)$ (Lemma 2.5.2). Let $V = \{k(x,t) \mid t < \gamma(x)\} \subset W$. Observe that V is open in X, $A \subset V$, and $\mathrm{cl}\, V = \{k(x,t) \mid t \leqslant \gamma(x)\}$. We can define a retraction $f : X \to X \setminus V$ as follows: $f|X \setminus V = \mathrm{id}$ and $f(k(x,t)) = k(x, \gamma(x))$ if $t \leqslant \gamma(x)$. Then, f is $\mathcal{U}$-close to id and $f(X) = X \setminus V$ is a closed set in X that misses A. Hence, A is a strong Z-set in X. □

In a metrizable space, every collared set is an F_σ-set because it is locally closed, that is, closed in an open set. Thus, by Lemma 2.8.2 above, we have the following:

Proposition 2.8.3 *Every collared set in a metrizable space is a strong Z_σ-set.* □

The following is a combination of Theorem 2.4.9 and Lemma 2.8.2:

Proposition 2.8.4 *Every E-deficient closed (or F_σ) set in an E-stable space is a strong Z-set (or a strong Z_σ-set).* □

Let $n \in \mathbb{N}$. A closed set A in a space X is called a $\mathbf{Z_n}$**-set** if, for each $\mathcal{U} \in \mathrm{cov}(X)$, every map $f : \mathbf{I}^n \to X$ is $\mathcal{U}$-close to a map $f' : \mathbf{I}^n \to X \setminus A$. For every $k \leqslant n \in \mathbb{N}$, each Z_n-set in X is a Z_k-set in X. Indeed, in the case $k < n$, let $p = \mathrm{pr}_{\mathbf{I}^k} : \mathbf{I}^n = \mathbf{I}^k \times \mathbf{I}^{n-k} \to \mathbf{I}^k$ be the projection and $i : \mathbf{I}^k \to \mathbf{I}^n$ be the embedding defined by $i(x) = (x, 0) \in \mathbf{I}^k \times \mathbf{I}^{n-k}$ for each $x \in \mathbf{I}^k$. For each map $f : \mathbf{I}^k \to X$ and $\mathcal{U} \in \mathrm{cov}(X)$, we have a map $f' : \mathbf{I}^n \to X \setminus A$ that is $\mathcal{U}$-close to fp. Then,

$f'i : \mathbf{I}^k \to X \setminus A$ is $\mathcal{U}$-close to $f = fpi$:

$$\begin{array}{ccc} \mathbf{I}^k & \xrightarrow{\ f\ } & X \\ p \uparrow \downarrow i & & \cup \\ \mathbf{I}^n & \overset{f'}{\dashrightarrow} & X \setminus A \end{array}$$

When A is a Z_n-set in X for every $n \in \mathbb{N}$, we call A a $\mathbf{Z_\infty}$**-set** in X.

Proposition 2.8.5 *Every locally finite union of Z_n-sets (resp. Z_∞-sets) in a paracompact space X is also a Z_n-set (resp. a Z_∞-set).*

Proof First of all, observe that if the Z_n-set case holds for every $n \in \mathbb{N}$, then the Z_∞-set case follows from the definition.

By induction, the finite case reduces to showing that if A and B are Z_n-sets in X then $A \cup B$ is also a Z_n-set in X. For each $\mathcal{U} \in \operatorname{cov}(X)$, let $\mathcal{V} \in \operatorname{cov}(X)$ be a star-refinement of $\mathcal{U}$. Each map $f : \mathbf{I}^n \to X$ is $\mathcal{V}$-close to a map $f' : \mathbf{I}^n \to X \setminus A$. Let $\mathcal{W} \in \operatorname{cov}(X)$ be a common refinement of $\mathcal{V}$ and $\{X \setminus A,\ X \setminus f'(\mathbf{I}^n)\}$. Then, f' is $\mathcal{W}$-close to a map $f'' : \mathbf{I}^n \to X \setminus B$. Since $f'(\mathbf{I}^n) \subset X \setminus A$, it follows that $f''(\mathbf{I}^n) \subset X \setminus A$. Note that f'' is $\mathcal{U}$-close to f. Thus, $A \cup B$ is a Z_n-set in X.

We now show the locally finite case. Let $A = \bigcup_{\lambda \in \Lambda} A_\lambda$ be a locally finite union of Z_n-sets A_λ in X and $\mathcal{U} \in \operatorname{cov}(X)$. For each map $f : \mathbf{I}^n \to X$, since $f(\mathbf{I}^n)$ is compact, $f(I^n)$ has an open neighborhood U that meets only finitely many A_λ, that is, there are $\lambda_1, \ldots, \lambda_n \in \Lambda$ such that

$$U \cap A = U \cap (A_{\lambda_1} \cup \cdots \cup A_{\lambda_n}).$$

Let $\mathcal{V} \in \operatorname{cov}(X)$ be a common refinement of $\mathcal{U}$ and $\{U, X \setminus f(\mathbf{I}^n)\}$. Since $A_{\lambda_1} \cup \cdots \cup A_{\lambda_n}$ is a Z_n-set in X, f is $\mathcal{V}$-close to a map

$$f' : \mathbf{I}^n \to X \setminus (A_{\lambda_1} \cup \cdots \cup A_{\lambda_n}).$$

Then, $f'(\mathbf{I}^n) \subset X \setminus A$ because $f'(\mathbf{I}^n) \subset U$. Hence, A is a Z_n-set in X. □

In an ANR, Z-sets are characterized as follows:

Theorem 2.8.6 *For a closed set A in an ANR X, the following are equivalent:*

(a) *A is a Z-set in X;*
(b) *For each $\varepsilon > 0$, each map $f : \boldsymbol{Q} \to X$ is ε-close to a map $g : \boldsymbol{Q} \to X \setminus A$ with respect to any given metric $d \in \operatorname{Metr}(X)$;*
(c) *A is a Z_∞-set in X;*
(d) *For an arbitrary simplicial complex K and any $\mathcal{U} \in \operatorname{cov}(X)$, each map $f : |K| \to X$ is $\mathcal{U}$-close to a map $g : |K| \to X \setminus A$;*
(e) *For a finite simplicial complex K and any $\mathcal{U} \in \operatorname{cov}(X)$, each map $f : |K| \to X$ is $\mathcal{U}$-close to a map $g : |K| \to X \setminus A$;*

(f) *The inclusion $X \setminus A \subset X$ is a hereditary weak homotopy equivalence, i.e., for every open set U in X, the inclusion $U \setminus A \subset U$ is a weak homotopy equivalence;*
(g) *For every open set U in X, the inclusion $U \setminus A \subset U$ is a homotopy equivalence;*
(h) *$X \setminus A$ is homotopy dense in X;*
(i) *$X \setminus A$ is dense in X and, for each $a \in A$, each neighborhood U of a in X contains a neighborhood V in X such that every map $\alpha : \mathbf{S}^{n-1} \to V \setminus A$ is null-homotopic in $U \setminus A$ for every $n \in \mathbb{N}$.*

Proof The implications (a) $\Rightarrow$ (b) $\Rightarrow$ (c) and (d) $\Rightarrow$ (e) $\Rightarrow$ (c) are trivial. The equivalence (f) $\Leftrightarrow$ (g) follows from the fact that every weak homotopy equivalence between ANRs is a homotopy equivalence (Corollary 1.13.30). The equivalences (f) $\Leftrightarrow$ (h) $\Leftrightarrow$ (i) are direct consequences of Corollary 1.15.5. By Hanner's Characterization of ANRs (Theorem 1.13.20), for each $\mathcal{U} \in \operatorname{cov}(X)$, we have a simplicial complex K with maps $f : X \to |K|$ and $g : |K| \to X$ such that gf is $\mathcal{U}$-close to id_X. Then, it is easy to see the implication (d) $\Rightarrow$ (a). We have to show the implication (c) $\Rightarrow$ (d) and the equivalence (e) $\Leftrightarrow$ (f). Refer to the following diagram of implications:

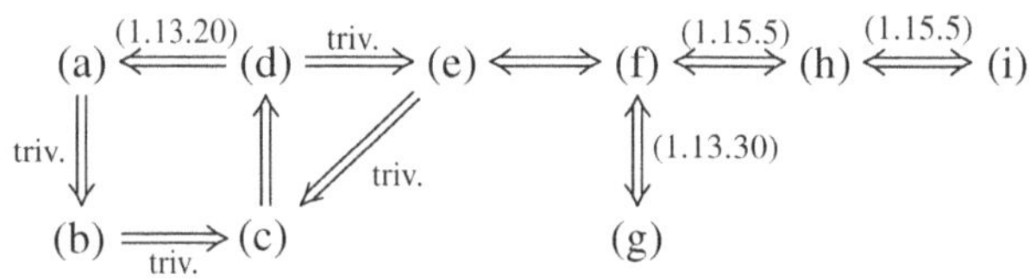

(c) $\Rightarrow$ (d): For each $\mathcal{U} \in \operatorname{cov}(X)$, we adopt a metric $d \in \operatorname{Metr}(X)$ such that $\{\overline{\mathrm{B}}_d(x, 1) \mid x \in X\} \prec \mathcal{U}$ (1.3.22). Then, it suffices to show that each map $f : |K| \to X$ is 1-close to a map $g : |K| \to X \setminus A$. We shall construct maps $g_n : |K| \to X$, $n \in \omega$, such that

$$g_n(|K^{(n)}|) \subset X \setminus A, \ g_n||K^{(n-1)}| = g_{n-1}||K^{(n-1)}|$$

and g_n is 2^{-n-1}-close to g_{n-1}, where $g_{-1} = f$. Then, $(g_n)_{n\in\mathbb{N}}$ is uniformly convergent to the map $g : |K| \to X$ defined by $g||K^{(n)}| = g_n||K^{(n)}|$ for each $n \in \omega$, whence g is 1-close to f and

$$g(|K|) = \bigcup_{n\in\omega} g_n(|K^{(n)}|) \subset X \setminus A.$$

Assume that $g_{-1}, g_0, \ldots, g_{n-1}$ have been obtained. By Corollary 1.13.10, we have $\mathcal{V} \in \operatorname{cov}(X)$ that is an h-refinement of an open cover of X with mesh $< 2^{-n-1}$. Then, any two $\mathcal{V}$-close maps are 2^{-n-1}-homotopic. For each n-simplex $\sigma \in K$, choose $\varepsilon > 0$ so that $\{\mathrm{B}(g_{n-1}(x), \varepsilon) \mid x \in \sigma\} \prec \mathcal{V}$ and let

$$\delta = \min\left\{\varepsilon/2, \ \operatorname{dist}(g_{n-1}(\partial\sigma), A)\right\} > 0.$$

Since X is an ANR, we apply (c) to obtain a map $g_\sigma : \sigma \to X \setminus A$ with $g_\sigma \simeq_\delta g_{n-1}|\sigma$ in X. It follows from the definition of δ that $g_\sigma|\partial\sigma \simeq_\delta g_{n-1}|\partial\sigma$ in $X \setminus A$. Since $X \setminus A$ is also an ANR, we can apply the Homotopy Extension Theorem 1.13.11 to extend $g_{n-1}|\partial\sigma$ to a map $g'_\sigma : \sigma \to X \setminus A$ with $g'_\sigma \simeq_\delta g_\sigma$. Note that g'_σ and $g_{n-1}|\sigma$ are $\mathcal{V}$-close because $d(g'_\sigma(x), g_{n-1}(x)) < 2\delta \leqslant \varepsilon$ for each $x \in \sigma$. Then, we have a map $g'_n : |K^{(n)}| \to X \setminus A$ defined by $g'_n||K^{(n-1)}| = g_{n-1}||K^{(n-1)}|$ and $g'_n|\sigma = g'_\sigma$ for each n-simplex $\sigma \in K$, whence g'_n is $\mathcal{V}$-close to $g_{n-1}||K^{(n)}|$, which implies $g'_n \simeq_{2^{-n-1}} g_{n-1}||K^{(n)}|$ in X. Applying the Homotopy Extension Theorem 1.13.11 again, we can extend g'_n to a map $g_n : |K| \to X$ that is 2^{-n-1}-close to g_{n-1}. This completes the induction.

(e) $\Rightarrow$ (f): For each open set U in X and $i \in \mathbb{N}$, let $\beta : \mathbf{B}^i \to U$ be a map with $\alpha = \beta|\mathbf{S}^{i-1} : \mathbf{S}^{i-1} \to U \setminus A$. Since U is an ANR and $\beta(\mathbf{B}^i)$ is compact, we can choose $\varepsilon > 0$ so that a map of $\mathbf{B}^i$ to U extending α is homotopic to β rel. $\mathbf{S}^{i-1}$ whenever it is ε-close to β (Theorem 1.13.10). Let

$$\delta = \min\left\{\varepsilon/2,\ \mathrm{dist}(\beta(\mathbf{B}^i), X \setminus U),\ \mathrm{dist}(\alpha(\mathbf{S}^{i-1}), A)\right\} > 0.$$

We apply (e) with Theorem 1.13.10 to obtain a map $\gamma : \mathbf{B}^i \to X \setminus A$ that is δ-homotopic to β in X, hence in $U \setminus A$. Observe that $\gamma|\mathbf{S}^{i-1}$ is δ-homotopic to α in $U \setminus A$. Since $U \setminus A$ is an ANR, we can apply the Homotopy Extension Theorem 1.13.11 to extend α to a map $\bar{\alpha} : \mathbf{B}^i \to U \setminus A$ that is δ-homotopic to γ. Since $\bar{\alpha}$ is ε-close to β, $\bar{\alpha} \simeq \beta$ rel. $\mathbf{S}^{i-1}$. This means that the inclusion $U \setminus A \subset U$ is a weak homotopy equivalence.

(f) $\Rightarrow$ (e): Let $\dim K = n$. For each $\mathcal{U} \in \mathrm{cov}(X)$, we take open covers of X as follows:

$$\mathcal{U} = \mathcal{U}_n \overset{*}{\succ} \mathcal{U}_{n-1} \overset{*}{\succ} \cdots \overset{*}{\succ} \mathcal{U}_1 \overset{*}{\succ} \mathcal{U}_0 \overset{*}{\succ} \mathcal{U}_{-1}.$$

For each map $f : |K| \to X$, by subdividing K, we can assume that $K \prec f^{-1}(\mathcal{U}_{-1})$, that is, $f(K) = \{f(\sigma) \mid \sigma \in K\} \prec \mathcal{U}_{-1}$ (Theorem 1.6.9).

Let $f_{-1} = f$, and assume that we have a map $f_{i-1} : |K| \to X$ such that

$$f_{i-1}(|K^{(i-1)}|) \subset X \setminus A,\ f_{i-1}(K) \prec \mathcal{U}_{i-1} \text{ and } f_{i-1} \simeq_{\mathcal{U}_{i-1}} f.$$

For each i-simplex $\sigma \in K$, we can choose $U_\sigma \in \mathcal{U}_{i-1}$ so that $f_{i-1}(\sigma) \subset U_\sigma$, whence $f_{i-1}(\partial\sigma) \subset U_\sigma \setminus A$. Since the inclusion $U_\sigma \setminus A \subset U_\sigma$ is an i-homotopy equivalence, there is a map $f_\sigma : \sigma \to U_\sigma \setminus A$ such that $f_\sigma|\partial\sigma = f_{i-1}|\partial\sigma$ and $f_\sigma \simeq f_{i-1}|\sigma$ rel. $\partial\sigma$ in U_σ. We define a map $f'_i : |K^{(i)}| \to X \setminus A$ by $f'_i||K^{(i-1)}| = f_{i-1}||K^{(i-1)}|$ and $f'_i|\sigma = f_\sigma$ for each i-simplex $\sigma \in K$. Observe that $f'_i \simeq_{\mathcal{U}_{i-1}} f_{i-1}||K^{(i)}|$ rel. $|K^{(i-1)}|$ in X. Then, by the Homotopy Extension Theorem 1.13.11, we can extend f'_i to a map $f_i : |K| \to X$ that is $\mathcal{U}_{i-1}$-homotopic to f_{i-1}, so $\mathcal{U}_i$-homotopic to f. Moreover,

$$f_i(K) \prec \mathrm{st}(f_{i-1}(K), \mathcal{U}_{i-1}) \prec \mathrm{st}\,\mathcal{U}_{i-1} \prec \mathcal{U}_i.$$

By induction, we have a map $f_n : |K| = |K^{(n)}| \to X \setminus A$ that is $\mathcal{U}$-close to f. Thus, (e) holds. □

Remark 2.14 When X is an E-manifold, each neighborhood U of any point of X contains a homotopically trivial open neighborhood V, where V is said to be **homotopically trivial** if for each $n \in \omega$, every map $\alpha : \mathbf{S}^n \to V$ is null-homotopic. Then, (i) is derived from the following condition, which is weaker than (f) (or (g)):

(z) If U is a homotopically trivial open set in X, then $U \setminus A$ is also homotopically trivial.

Thus, the above (z) characterizes Z-sets in an E-manifold, and this was the first definition due to R.D. Anderson.

Using the above characterization of Z-sets (2.8.6), the following is easily shown:

Corollary 2.8.7 *Let X be an ANR.*

(1) *A locally finite union of Z-sets in X is a Z-set.*
(2) *For each Z-set A in X and each open set U in X, $A \cap U$ is a Z-set in U.*
(3) *A closed set $A \subset X$ is a Z-set in X if each $x \in A$ has a neighborhood U in X such that $A \cap U$ is a Z-set in U.* □

Proposition 2.8.8 *A closed Z_σ-set A in an ANR X is a Z-set if A is completely metrizable (i.e., absolutely G_δ). In particular, every compact Z_σ-set in an ANR is a Z-set.*

Proof Since a finite union of Z-sets is also a Z-set (Corollary 2.8.7(1)), we have a tower $A_1 \subset A_2 \subset \cdots$ of Z-sets in X such that $A = \bigcup_{n\in\mathbb{N}} A_n$. Since A is closed in X and A has an admissible complete metric, we apply Hausdorff's Metric Extension Theorem 1.13.8 to obtain $d \in \mathrm{Metr}(X)$ such that d is complete on A. To see that A is a Z-set in X, we verify that A satisfies condition (b) in Theorem 2.8.6.

Let $f : \boldsymbol{Q} \to X$ be a map and $0 < \varepsilon < 1$. Using the fact that each A_n is a Z-set in X, we can inductively construct maps $f_n : \boldsymbol{Q} \to X$, $n \in \mathbb{N}$, so as to satisfy the following conditions:

(1) $f_n(\boldsymbol{Q}) \cap A_n = \emptyset$,
(2) $\varepsilon_n = \min\{\mathrm{dist}(f_n(\boldsymbol{Q}), A_n), \varepsilon_{n-1}/2\}$, where $\varepsilon_0 = \varepsilon < 1$,
(3) $d(f_n, f_{n-1}) < \varepsilon_{n-1}/2$, where $f_0 = f$,
(4) $f_n|D_{n-1} = f_{n-1}|D_{n-1}$, where $D_{n-1} = \{x \in \boldsymbol{Q} \mid d(f_{n-1}(x), A) \geqslant 2^{-(n-1)}$.

Suppose that $f_0, \ldots, f_{n-1}$ have been constructed, where $\varepsilon_0, \ldots, \varepsilon_{n-1}$ are defined in (2) and $D_1, \ldots, D_{n-1}$ are defined in (4). Take $\delta > 0$ such that a map $g : \boldsymbol{Q} \to X$ with $d(g, f_{n-1}) < \delta$ is $\varepsilon_{n-1}/2$-homotopic to f_{n-1}. Since A_n is a Z-set in X, we have a map $g : \boldsymbol{Q} \to X \setminus A_n$ with $d(g, f_{n-1}) < \delta$. Then, we have an $\varepsilon_{n-1}/2$-homotopy $h : \boldsymbol{Q} \times \mathbf{I} \to X$ such that $h_0 = f_{n-1}$ and $h_1 = g$. Let

$$U = \left\{x \in \boldsymbol{Q} \;\middle|\; d(h_t(x), A) > 2^{-n} \text{ for every } t \in \mathbf{I}\right\}.$$

Note that $\varepsilon_{n-1}/2 < 2^{-n}$ because $\varepsilon_0 = \varepsilon < 1$. For each $x \in D_{n-1}$ and $t \in \mathbf{I}$,

$$\begin{aligned} d(h_t(x), A) &\geqslant d(f_{n-1}(x), A) - d(h_t(x), f_{n-1}(x) \\ &> 2^{-(n-1)} - \varepsilon_{n-1}/2 > 2^{-n}, \end{aligned}$$

which implies that U is an open neighborhood of D_{n-1} in $\boldsymbol{Q}$. Let $\alpha : \boldsymbol{Q} \to \mathbf{I}$ be an Urysohn map with $\alpha(D_{n-1}) = \{0\}$ and $\alpha(\boldsymbol{Q} \setminus U) = \{1\}$. We define $f_n : \boldsymbol{Q} \to X$ by $f_n(x) = h(x, \alpha(x))$. Then, $f_n(\boldsymbol{Q}) \cap A_n = \emptyset$, $d(f_n, f_{n-1}) < \varepsilon_{n-1}/2$, and $f_n|D_{n-1} = f_{n-1}|D_{n-1}$.

Note that $D_{n-1} \subset D_n$ for each $n \in \mathbb{N}$. Indeed, for each $x \in D_{n-1}$,

$$\begin{aligned} d(f_n(x), A) &\geqslant d(f_{n-1}(x), A) - d(f_n(x), f_{n-1}(x)) \\ &> 2^{-(n-1)} - \varepsilon_{n-1}/2 > 2^{-n}. \end{aligned}$$

If $\boldsymbol{Q} = \bigcup_{n\in\mathbb{N}} D_n$, we have a map $\tilde{f} = \lim_{n\to\infty} f_n : \boldsymbol{Q} \to X \setminus A$ (i.e., $\tilde{f}|D_n = f_n|D_n$ for each $n \in \mathbb{N}$) and $d(\tilde{f}, f) < \varepsilon$, where it should be noted that $(f_n)_{n\in\mathbb{N}}$ is uniformly Cauchy but d might be non-complete on X. To show that $\boldsymbol{Q} = \bigcup_{n\in\mathbb{N}} D_n$, suppose that there exists $x_0 \in \boldsymbol{Q} \setminus \bigcup_{n\in\mathbb{N}} D_n$. Then, we can take $y_n \in A$, $n \in \mathbb{N}$, such that $f_n(x_0), y_n) < 2^{-n}$. Since $(f_n(x_0))_{n\in\mathbb{N}}$ is Cauchy, so is $(y_n)_{n\in\mathbb{N}}$. By the completeness of A, $(y_n)_{n\in\mathbb{N}}$ converges to $y_0 \in A$. Then, $\lim_{n\to\infty} f_n(x_0) = y_0$ and $y_0 \in A_m$ for some $m \in \mathbb{N}$. For each $n > m$,

$$\begin{aligned} d(f_m(x_0), y_0) &\leqslant d(f_m(x_0), f_n(x_0)) + d(f_n(x_0), y_0) \\ &< \sum_{i=0}^{n-m-1} \varepsilon_{m+i}/2 + d(f_n(x_0), y_0). \end{aligned}$$

Since $\lim_{n\to\infty} d(f_n(x_0), y_0) = 0$ and $\sum_{i=0}^{\infty} \varepsilon_{m+i}/2 < \sum_{i=0}^{\infty} 2^{-i-1}\varepsilon_m = \varepsilon_m$, it follows that $d(f_m(x_0), y_0) < \varepsilon_m$, which contradicts

$$d(f_m(x_0), y_0) \geqslant \operatorname{dist}(f_m(\boldsymbol{Q}), A_m) \geqslant \varepsilon_m.$$

Therefore, $\boldsymbol{Q} = \bigcup_{n\in\mathbb{N}} D_n$. The proof is completed. □

Remark 2.15 In the above, a closed Z_σ-set in an ANR is not always a Z-set. Indeed, each $\mathbf{I}^n$ is a strong Z-set in $\mathbf{I}_f^{\mathbb{N}}$ by Lemma 2.8.2, where we identify $\mathbf{I}^n = \mathbf{I}^n \times \{\mathbf{0}\} \subset \mathbf{I}^{\mathbb{N}}$. Then, $\mathbf{I}_f^{\mathbb{N}} = \bigcup_{n\in\mathbb{N}} \mathbf{I}^n$ is a strong Z_σ-set in itself but $\mathbf{I}_f^{\mathbb{N}}$ is not a Z-set in itself.

In the rest of this section, we shall show some properties of strong Z-sets. Unlike the Z-set case, the following can be directly derived from the definition:

Proposition 2.8.9 (1) *A finite union of strong Z-sets in a paracompact space X is also a strong Z-set.*

(2) *A closed set A in a metrizable space X is a strong Z-set in X if A is a strong Z-set in an open neighborhood U of A in X.*

Proof (1): It suffices to show that $A \cup B$ is a strong Z-set in X for strong Z-sets A and B in X. For each open cover $\mathcal{U} \in \operatorname{cov}(X)$, let $\mathcal{V} \in \operatorname{cov}(X)$ be an open star-refinement of $\mathcal{U}$. We have a map $f : X \to X$ and an open neighborhood V of A in X such that f is $\mathcal{V}$-close to id and $\operatorname{cl} V \cap \operatorname{cl} f(X) = \emptyset$. Choose $\mathcal{W} \in \operatorname{cov}(X)$ so that $\mathcal{W} \prec \mathcal{V}$ and $\mathcal{W} \prec \{X \setminus \operatorname{cl} f(X), X \setminus clV\}$. Now, we have a map $g : X \to X$ such that g is $\mathcal{W}$-close to id and $B \cap \operatorname{cl} g(X) = \emptyset$. Then, gf is $\mathcal{U}$-close to id and $B \cap \operatorname{cl} gf(X) = \emptyset$. Since g is $\{X \setminus \operatorname{cl} f(X), X \setminus \operatorname{cl} V\}$-close to f, we have $gf(X) \subset X \setminus \operatorname{cl} V$, hence $A \cap \operatorname{cl} gf(X) = \emptyset$. Thus, it follows that $(A \cup B) \cap \operatorname{cl} gf(X) = \emptyset$.

(2): For each open cover $\mathcal{U} \in \operatorname{cov}(X)$, we can take $\mathcal{V} \in \operatorname{cov}(U)$ such that $\mathcal{V} \prec \mathcal{U}|U$ and $\mathcal{V}$ is fitting in X by Proposition 2.1.17. Since A is a strong Z-set in U, we have a map $f : U \to U$ and an open neighborhood W of A in U such that f is $\mathcal{V}$-close to id and $W \cap f(U) = \emptyset$. Then, f can be extended to a map $\tilde{f} : X \to X$ by $\tilde{f}|X \setminus U = \mathrm{id}$, where $\tilde{f}$ is $\mathcal{U}$-close to id and $W \cap \tilde{f}(X) = W \cap U \cap \tilde{f}(X) = W \cap f(U) = \emptyset$. Therefore, A is a strong Z-set in X. □

To see that strong Z-sets have the same properties as in Corollary 2.8.7 above, it is necessary to introduce a generalization of closed maps. A map $f : X \to Y$ is said to be **closed over** $C \subset Y$ if $C \cap f(A) = C \cap \operatorname{cl} f(A)$ for every closed set A in X. Note that $C \cap f(A) = C \cap \operatorname{cl} f(A)$ means that $C \cap f(A) = \emptyset$ implies $C \cap \operatorname{cl} f(A) = \emptyset$.

Lemma 2.8.10 *A map $f : X \to Y$ is closed over $C \subset Y$ if and only if, for each $y \in C$ and each neighborhood U of $f^{-1}(y)$ in X (which might be empty), there exists a neighborhood V of y in Y such that $f^{-1}(V) \subset U$.*

Proof To see the "if" part, let A be a closed set in X. For each $y \in C \setminus f(A)$, since $X \setminus A$ is a neighborhood of $f^{-1}(y)$, y has a neighborhood V in Y such that $f^{-1}(V) \subset X \setminus A$. This implies that $V \cap f(A) = \emptyset$, so $y \in C \setminus \operatorname{cl} f(A)$. Thus, $C \setminus f(A) \subset C \setminus \operatorname{cl} f(A)$, that is, $f(A) \cap C \supset \operatorname{cl} f(A) \cap C$. Therefore, $f(A) \cap C = \operatorname{cl} f(A) \cap C$.

To see the "only if" part, let $y \in C$ and U be an open neighborhood of $f^{-1}(y)$ in X. Since $A = X \setminus U$ is a closed set in X, we have $f(A) \cap C = \operatorname{cl} f(A) \cap C$. Since $y \in C \setminus f(A) \subset Y \setminus \operatorname{cl} f(A)$, $V = Y \setminus \operatorname{cl} f(A)$ is a neighborhood of y in Y. Then, $f^{-1}(V) \cap A = \emptyset$, that is, $f^{-1}(V) \subset X \setminus A = U$. □

The following can easily be observed:

Lemma 2.8.11 *Let $f : X \to Y$ be a map that is closed over $C \subset Y$.*

(1) *For each closed set A in X, $f|A : A \to Y$ is closed over C.*
(2) *For each $W \subset Y$, $f|f^{-1}(W) : f^{-1}(W) \to W$ is closed over $C \cap W$.* □

Proposition 2.8.12 *Let $f : X \to Y$ be a map of a metrizable space X into an ANR Y and A be a closed set in X such that $f|A$ is a closed embedding. For each* *strong Z-set C in Y and $\mathcal{U} \in \operatorname{cov}(Y)$, f is $\mathcal{U}$-close to a map $g : X \to Y$ such that $f|A = g|A$, $g(X \setminus A) \subset Y \setminus C$, and g is closed over C.*

Proof We can take $d \in \operatorname{Metr}(X)$ and $\rho \in \operatorname{Metr}(Y)$ such that $\{\overline{\mathrm{B}}_\rho(y, 1) \mid y \in Y\} \prec \mathcal{U}$ (1.3.22). We inductively construct maps $f_i : X \to Y$ and closed sets X_i in X, $i \in \mathbb{N}$, so as to satisfy the following:

(1) For each $x \in X \setminus X_i$, there exists $a \in A$ such that $d(x, a) < 2^{-i}$ and $\rho(f_{i-1}(x), f_{i-1}(a)) < 2^{-i}$,
(2) $X_{i-1} \subset \operatorname{int} X_i \subset X_i \subset X \setminus A$,
(3) $\rho(f_i, f_{i-1}) < 2^{-i}$, $f_i|A \cup X_{i-1} = f_{i-1}|A \cup X_{i-1}$ and $\operatorname{cl} f_i(X_i) \cap C = \emptyset$,

where $f_0 = f$ and $X_0 = \emptyset$.

Suppose that f_{i-1} and X_{i-1} have been obtained. Let

$$U = \{x \in X \mid d(x, a),\ \rho(f_{i-1}(x), f_{i-1}(a)) < 2^{-i} \text{ for some } a \in A\}.$$

Then, U is an open neighborhood of A in X. Since $(X \setminus U) \cup X_{i-1} \subset X \setminus A$, we can find a closed set X_i in X such that

$$(X \setminus U) \cup X_{i-1} \subset \operatorname{int} X_i \subset X_i \subset X \setminus A.$$

Then, X_i satisfies conditions (1) and (2). Since $\operatorname{cl} f_{i-1}(X_{i-1}) \cap C = \emptyset$, it follows that $f_{i-1}(X_{i-1}) \cap N_1 = \emptyset$ for some closed neighborhood N_1 of C in Y. Since C is a strong Z-set in Y, we have a 2^{-i}-homotopy $h : X \times \mathbf{I} \to Y$ such that

$$h_0 = f_{i-1},\ h((X_{i-1}) \times \mathbf{I}) \cap N_1 = \emptyset \text{ and } \operatorname{cl} h_1(X) \cap C = \emptyset.$$

Then, C has a closed neighborhood N in Y such that $h_1(X) \cap N = \emptyset$ and $N \subset N_1$, which implies

$$\big((X_{i-1} \times \mathbf{I}) \cup (X \times \{1\})\big) \cap h^{-1}(N) = \emptyset.$$

Moreover, X_{i-1} has an open neighborhood V in X such that $(V \times \mathbf{I}) \cap h^{-1}(N) = \emptyset$. Take an Urysohn map $\alpha : X \to \mathbf{I}$ such that $\alpha(A \cup X_{i-1}) = 0$ and $\alpha(X_i \setminus V) = 1$. Then, $(x, \alpha(x)) \notin h^{-1}(N)$ for every $x \in X_i$. We define a map $f_i : X \to Y$ by $f_i(x) = h(x, \alpha(x))$. Then, f_i satisfies condition (3) (Fig. 2.19).

Now, we have $X \setminus A = \bigcup_{i \in \mathbb{N}} X_i$ by (1) and (2). By virtue of (3), we can define $g : X \to Y$ by $g|A = f|A$ and $g|X_i = f_i|X_i$ for each $i \in \mathbb{N}$. Since $(f_i)_{i \in \mathbb{N}}$ is uniformly Cauchy by (3), g is continuous as the uniform limit of $(f_i)_{i \in \mathbb{N}}$. It also follows that $\rho(g, f_i) < 2^{-i}$. In particular, $\rho(g, f) = \rho(g, f_0) < 1$, hence g is $\mathcal{U}$-close to f. Moreover,

$$g(X \setminus A) = g\big(\textstyle\bigcup_{i \in \mathbb{N}} X_i\big) = \displaystyle\bigcup_{i \in \mathbb{N}} f_i(X_i) \subset Y \setminus C.$$

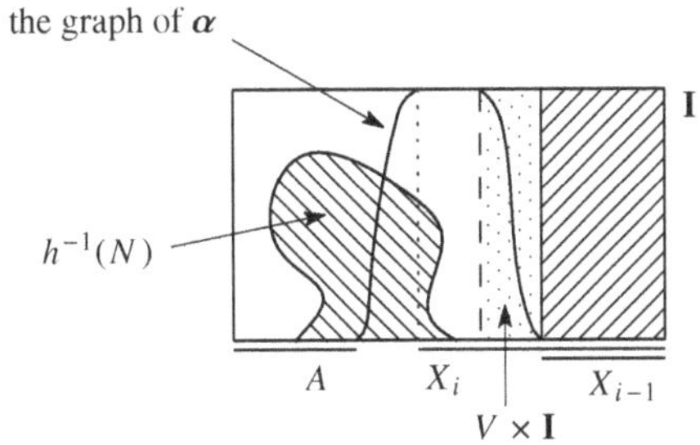

Fig. 2.19 A map α

To see that g is closed over C, assume, on the contrary, that there exists $z \in C$ with an open neighborhood V of $g^{-1}(z)$ in X such that $g^{-1}(W) \not\subset V$ for any neighborhood W of z in Y (Lemma 2.8.10). Then, we have a sequence $(x_n)_{n\in\mathbb{N}}$ in $X \setminus V$ such that $(g(x_n))_{n\in\mathbb{N}}$ converges to z. Due to (3), for each $i \in \mathbb{N}$, there is a neighborhood W_i of z in Y that is disjoint from $f_i(X_i) = g(X_i)$. This implies that $(x_n)_{n\in\mathbb{N}}$ has a subsequence $(x_{n_i})_{i\in\mathbb{N}}$ such that $x_{n_i} \notin X_i$ for each $i \in \mathbb{N}$. By (1), we can take $a_i \in A$, $i \in \mathbb{N}$, so that $d(x_{n_i}, a_i) < 2^{-i}$ and $\rho(f_{i-1}(x_{n_i}), f_{i-1}(a_i)) < 2^{-i}$. Since $\rho(g, f_{i-1}) < 2^{-i+1}$, it follows that $\rho(g(x_{n_i}), g(a_i)) < 2^{-i+3}$, which means that $(g(a_i))_{\in\mathbb{N}}$ converges to z. Since $g|A = f|A$ is a closed embedding, it follows that $(a_i)_{\in\mathbb{N}}$ converges to some $a \in A$. Recall that $d(x_{n_i}, a_i) < 2^{-i}$, hence $(x_{n_i})_{i\in\mathbb{N}}$ also converges to some $a \in A$, which means that $a \in X \setminus V$ and $z = g(a)$. However, $g^{-1}(z) \subset V$, which is a contradiction. Consequently, g is closed over C. □

In Theorem 2.8.6, it was shown that a closed set A in an ANR X is a Z-set if and only if $X \setminus A$ is homotopy dense in X. For strong Z-sets, we have the following:

Proposition 2.8.13 *A closed set A in an ANR X is a strong Z-set if and only if there is a homotopy $h : X \times \mathbf{I} \to X$ such that*

$$h_0 = \mathrm{id},\ h(X \times (0, 1]) \subset X \setminus A,\ \textit{and } h \textit{ is closed over } A.$$

Proof If A is a strong Z-set, then we can obtain the desired homotopy h by applying Proposition 2.8.12 to $\mathrm{pr}_X : X \times \mathbf{I} \to X$ and $X \times \{0\} \subset X \times \mathbf{I}$. Thus, we have the "only if" part.

To see the "if" part, suppose the homotopy h, as defined above, is given. For each $\mathcal{U} \in \mathrm{cov}(X)$, we apply Lemma 2.5.2 to obtain a map $\alpha : X \to (0, 1]$ so that

$$\big\{\{x\} \times [0, \alpha(x)] \bigm| x \in X\big\} \prec h^{-1}(\mathcal{U}).$$

We can define a map $f : X \to X$ by $f(x) = h(x, \alpha(x))$. Then, f is $\mathcal{U}$-close to id. Since h is closed over A, it follows from Lemma 2.8.11(1) that f is also closed over A, hence we have $A \cap \mathrm{cl}\, f(X) = A \cap f(X) = \emptyset$. □

Now, we can show the following:

Proposition 2.8.14 *Let A be a strong Z-set in an ANR X and U be an open set in X. Then, $A \cap U$ is a strong Z-set in U.*

Proof By Proposition 2.8.13, we have a homotopy $h : X \times \mathbf{I} \to X$ such that $h_0 = \mathrm{id}$, $h(X \times (0, 1]) \subset X \setminus A$, and h is closed over A. Then, note that $h^{-1}(U)$ is an open set in $X \times \mathbf{I}$ with $h^{-1}(U) \cap (X \times \{0\}) = U \times \{0\}$. Applying Lemma 2.5.2, we can obtain a map $\alpha : X \to \mathbf{I}$ such that $\alpha(U) \subset (0, 1]$ and

$$\{x\} \times [0, \alpha(x)] \subset h^{-1}(U) \quad \text{for each } x \in U.$$

We can define a homotopy $g : U \times \mathbf{I} \to U$ by $g(x, t) = h(x, t\alpha(x))$. Then, $g_0 = \mathrm{id}$ and $g(U \times (0, 1]) \subset U \setminus A$. It follows from Lemma 2.8.11 that g is closed over $A \cap U$. By Proposition 2.8.13, $A \cap U$ is a strong Z-set in U. □

Remark 2.16 In the above, given $\mathcal{U} \in \mathrm{cov}(U)$, we have a map $\tilde{\alpha} : X \to \mathbf{I}$ such that $\tilde{\alpha}^{-1}(0) = X \setminus U$, $\tilde{\alpha}|U \leqslant \alpha$, and $\{\{x\} \times [0, \tilde{\alpha}(x)] \mid x \in U\} \prec \mathcal{U}$. We define $\tilde{g} : X \times \mathbf{I} \to X$ by $\tilde{g}(x, t) = h(x, t\tilde{\alpha}(x))$. Then, $\tilde{g}$ is a $\mathcal{U}$-homotopy such that $\tilde{g}_0 = \mathrm{id}$, $\tilde{g}(U \times (0, 1]) \subset U \setminus A$, and $\tilde{g}_t|X \setminus U = \mathrm{id}$ for every $t \in \mathbf{I}$. Moreover, $\tilde{g}|U \times \mathbf{I}$ is closed over $A \cap U$, hence $\mathrm{cl}\, \tilde{g}_1(U) \cap A \cap U = \tilde{g}_1(U) \cap A \cap U = \emptyset$.[27]

To prove the following proposition, we can apply Michael's Theorem 1.3.20 on local properties.

Proposition 2.8.15 *A closed set A in an ANR X is a strong Z-set in X if each $a \in A$ has an open neighborhood U in X such that $A \cap U$ is a strong Z-set in U.*

Proof To apply Michael's Theorem 1.3.20, it is enough to prove the following:

(G-1) For open sets $U \supset V$ in X, if $A \cap U$ is a strong Z-set in U, then $A \cap V$ is a strong Z-set in V;

(G-2) For two open sets U_1 and U_2 in X, if $A \cap U_i$ is a strong Z-set in U_i, $i = 1, 2$, then $A \cap (U_1 \cup U_2)$ is a strong Z-set in $U_1 \cup U_2$;

(G-3) For a discrete collection $\{U_\lambda\}_{\lambda \in \Lambda}$ of open sets in X, if each $A \cap U_\lambda$ is a strong Z-set in U_λ, then $A \cap \bigcup_{\lambda \in \Lambda} U_\lambda$ is a strong Z-set in $\bigcup_{\lambda \in \Lambda} U_\lambda$.

Since (G-1) is just Proposition 2.8.14 and (G-3) is trivial, we only have to show (G-2).

Under the assumption given by (G-2), let $U = U_1 \cap U_2$. We can write $A \cap U = A_1 \cup A_2$ such that $A_i \subset U_i$ and A_i is closed in U. Indeed, since $A \cap U \setminus U_2$ and $A \cap U \setminus U_1$ are disjoint closed sets in U, we have open sets U_1' and U_2' in U such that $A \cap U \setminus U_2 \subset U_1'$, $A \cap U \setminus U_1 \subset U_2'$, and $U_1' \cap U_2' = \emptyset$. Let $A_1 = A \cap U \setminus U_2'$ and $A_2 = A \cap U \setminus U_1'$. Then, A_1 and A_2 are closed in U, $A_1 \subset U_1$, $A_2 \subset U_2$, and $A \cap U = A_1 \cup A_2$. By Proposition 2.8.9(2) and the assumption in (G-2), we can see that A_1 and A_2 are strong Z-sets in U. Then, $A \cap U = A_1 \cup A_2$ is a strong Z-set in U by Proposition 2.8.9(1). Thus, we have (G-2). □

[27] We can also apply Proposition 2.1.17 to obtain such a homotopy.

Combining Propositions 2.8.9(1) and 2.8.15, we have the following:

Corollary 2.8.16 *A locally finite union of strong Z-sets in an ANR is also a strong Z-set.* □

For (strong) Z-sets in a separable ANR X, we have the following:

Proposition 2.8.17 *Let X be a separable ANR and $A \subset X$. If A is a Z-set and a strong Z_σ-set in X, then A is a strong Z-set.*

Proof Since A is a strong Z_σ-set in X, we can use Proposition 2.8.9(1) to obtain strong Z-sets $A_1 \subset A_2 \subset \cdots$ in X such that $A = \bigcup_{n\in\mathbb{N}} A_n$. For each $\mathcal{U} \in \operatorname{cov}(X)$, we can take $d \in \operatorname{Metr}(X)$ so that $\{\overline{\mathrm{B}}(x, 1) \mid x \in X\} \prec \mathcal{U}$ (Proposition 1.3.22(1)). By Hanner's Characterization of ANRs Theorem 1.13.20 with Remark 1.12, there exists a locally finite countable simplicial complex K with maps $f : X \to |K|$ and $g : |K| \to X$ such that $gf \simeq_{2^{-1}} \mathrm{id}$. Since K is countable, we have finite subcomplexes $K_1 \subset K_2 \subset \cdots$ with $K = \bigcup_{n\in\mathbb{N}} K_n$. Then, each $P_n = |K_n|$ is compact and $|K| = \bigcup_{n\in\mathbb{N}} P_n$.

We can inductively define sequences of open covers $\mathcal{U}_n \in \operatorname{cov}(X)$, maps $g_n : |K| \to X$, and open sets $W_n \subset X$, $n \in \mathbb{N}$, with the following conditions:

(1) $\operatorname{st}\mathcal{U}_n \prec \mathcal{U}_{n-1} \wedge \{X \setminus \operatorname{cl} g_{n-1}(|K|), X \setminus \operatorname{cl} W_{n-1}\}$, $\operatorname{mesh}\mathcal{U}_n < 2^{-n-1}$,
(2) g_n is $\mathcal{U}_n$-close to g_{n-1}, $g_n|P_{n-1} = g_{n-1}|P_{n-1}$,
(3) $A_n \cap \operatorname{cl} g_n(|K|) = \emptyset$, $A \cap g_n(|K|) = \emptyset$, and
(4) $A_n \subset W_n \subset \operatorname{cl} W_n \subset X \setminus \operatorname{cl} g_n(|K|)$,

where $g_0 = g$, $W_0 = \emptyset$, and $\mathcal{U}_0$ is any open cover of X. Indeed, suppose that $\mathcal{U}_{n-1}$, g_{n-1}, and W_{n-1} have been obtained. Note that $g_{n-1}(P_{n-1})$ is closed in X because it is compact. Choose $\mathcal{U}_n \in \operatorname{cov}(X)$ so as to satisfy (1) and let $\mathcal{V}$ be an open cover of $X \setminus g_{n-1}(P_{n-1})$ such that $\operatorname{st}\mathcal{V}$ is fitting in X and $\operatorname{st}\mathcal{V} \prec \mathcal{U}_n$ (Proposition 2.1.17). Since A_n is a strong Z-set in $X \setminus g_{n-1}(P_{n-1})$ by Proposition 2.8.14, we have a map $h' : X \setminus g_{n-1}(P_{n-1}) \to X \setminus g_{n-1}(P_{n-1})$ such that h' is $\mathcal{V}$-close to id and

$$A_n \cap \operatorname{cl} h'(X \setminus g_{n-1}(P_{n-1})) = \emptyset.$$

Choose open sets V, W in $X \setminus g_{n-1}(P_{n-1})$ so that $\operatorname{cl} V \cap \operatorname{cl} W = \emptyset$, $A_n \subset V$, and $\operatorname{cl} h'(X \setminus g_{n-1}(P_{n-1})) \subset W$. Let

$$\mathcal{W} = \mathcal{V} \wedge \big\{X \setminus \operatorname{cl} V,\ X \setminus \operatorname{cl} W\big\} \in \operatorname{cov}(X \setminus g_{n-1}(P_{n-1})).$$

Since A is a Z-set in $X \setminus g_{n-1}(P_{n-1})$ by Corollary 2.8.7(2), we have a map $h'' : X \setminus g_{n-1}(P_{n-1}) \to X \setminus g_{n-1}(P_{n-1})$ such that h' is $\mathcal{W}$-close to id and

$$A \cap h''(X \setminus g_{n-1}(P_{n-1})) = \emptyset.$$

Then, $h''(\operatorname{cl} W') \subset X \setminus \operatorname{cl} W$, which implies that $\operatorname{cl} h''(W') \subset X \setminus W \subset X \setminus A_n$, so $\operatorname{cl} h''(W') \cap A_n = \emptyset$. Since $h''h'$ is $\operatorname{st}\mathcal{V}$-close to id and $\operatorname{st}\mathcal{V}$ is fitting in $X \setminus$

$g_{n-1}(P_{n-1})$, we can be extend $h''h'$ to a map $h : X \to X$ by $h|g_{n-1}(P_{n-1}) = \mathrm{id}$. Then, it follows that

$$\begin{aligned} A \cap h(X) &= \big(A \cap g_{n-1}(P_{n-1})\big) \cup \big(A \cap h''h'(X \setminus g_{n-1}(P_{n-1}))\big) \\ &\subset \big(A \cap g_{n-1}(|K|)\big) \cup \big(A \cap h''(X \setminus g_{n-1}(P_{n-1}))\big) = \emptyset; \\ A_n \cap \mathrm{cl}\, h(X) &= \big(A_n \cap g_{n-1}(P_{n-1})\big) \cup \big(A_n \cap \mathrm{cl}\, h''h'(X \setminus g_{n-1}(P_{n-1}))\big) \\ &\subset \big(A \cap g_{n-1}(|K|)\big) \cup \big(A_n \cap \mathrm{cl}\, h''(W')\big) = \emptyset. \end{aligned}$$

Thus, we have the map $g_n = hg_{n-1} : |K| \to X$ satisfying (2) and (3). Then, an open set $W_n \subset X$ can be taken so as to satisfy (4).

Since $h_m|P_n = h_n|P_n$ for each $m > n$ by (2), we can define $g_\infty : |K| \to X$ by $g_\infty(x) = g_n(x)$ for $x \in P_n$, $n \in \mathbb{N}$. Since $d(g_n, g_{n+1}) < 2^{-n-2}$ by (1) and (2), we have $d(g_\infty, g_n) < 2^{-n-1}$, which implies that $(g_n)_{n\in\omega}$ is uniformly convergent to g_∞. Hence, g_∞ is continuous. Moreover, $d(g_\infty, g) < 2^{-1}$. Hence, we have

$$d(g_\infty f, \mathrm{id}) \leqslant d(g_\infty f, gf) + d(gf, \mathrm{id}) < 2^{-1} + 2^{-1} = 1,$$

which implies that $g_\infty f$ is $\mathcal{U}$-close to id. From (1) and (2), it follows that g_∞ is $\mathrm{st}\,\mathcal{U}_{n+1}$-close to g_n. Since $\mathrm{st}\,\mathcal{U}_{n+1} \prec \{X \setminus \mathrm{cl}\, g_n(|K|), X \setminus \mathrm{cl}\, W_n\}$, we have $g_\infty(|K|) \subset X \setminus \mathrm{cl}\, W_n$, which implies that $g_\infty(|K|) \cap W_n = \emptyset$. Therefore, $g_\infty(|K|) \cap \bigcup_{n\in\omega} W_n = \emptyset$. Consequently, $A \cap \mathrm{cl}\, g_\infty f(X) = \emptyset$. Thus, A is a strong Z-set in X. □

Remark 2.17 It is not known whether Proposition 2.8.17 is valid for a non-separable ANR X, or not. In the above proof, note that $g_{n-1}(P_{n-1})$ is closed in X, where $P_{n-1} = |K_{n-1}|$ is compact because K_{n-1} is finite.

The following is a combination of Propositions 2.8.8 and 2.8.17:

Corollary 2.8.18 *Let X be a separable ANR. A closed strong Z_σ-set in X is a strong Z-set if it is completely metrizable (i.e., absolutely G_δ).* □

2.9 Z-Sets and Strong Z-Sets in E-Manifolds

In this section, we show that the three conditions for a closed set in an E-manifold to be a Z-set, a strong Z-set, E-deficient are equivalent. First, we prove the following:

Proposition 2.9.1 *Suppose that $X \times E$ is an E-manifold and A is a strong Z-set in X. For each open cover $\mathcal{U}$ of $(X \times E)_A$, there is a homeomorphism $h : X \times E \to (X \times E)_A$ such that h is $\mathcal{U}$-close to the natural map and $h(x, \mathbf{0}) = x$ for each $x \in A$. In particular, $(X \times E)_A$ is also an E-manifold.*

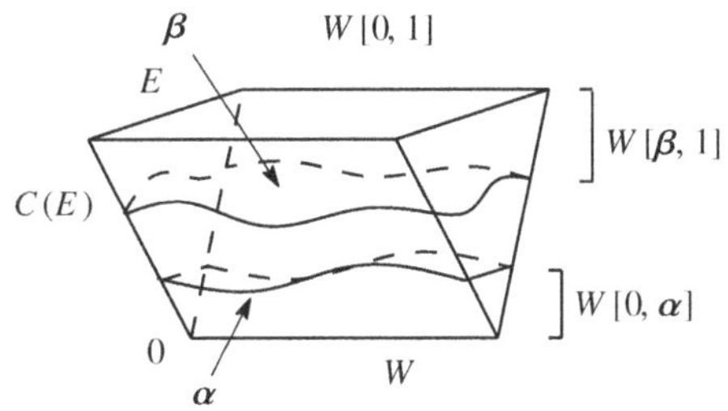

Fig. 2.20 $W[0, \alpha]$ and $W[\beta, 1]$

Proof First, note that $\operatorname{dens} X \leqslant \operatorname{dens} X \times E = \operatorname{dens} E$ by Lemma 2.1.1. Since $E \approx C(E) = (\mathbf{I} \times E)_{\{0\}}$ by Corollary 2.4.10, we can replace E by $C(E)$, where

$$X \times C(E) = (X \times (0, 1] \times E) \cup (X \times \{0\}).$$

We simplify the notation as follows: For each $W \subset X$, we write $W[0, 1] = W \times C(E)$, $W[0] = W \times \{0\}$ and $W[1] = W \times \{1\} \times E$. For each $\alpha \in \mathrm{C}(W, (0, 1])$ and $\beta \in \mathrm{C}(W \times E, (0, 1))$, we write

$$W[0, \alpha] = W[0] \cup \big\{(x, t) \in W \times (0, 1] \bigm| t \leqslant \alpha(x)\big\} \times E \quad \text{and}$$
$$W[\beta, 1] = \big\{(x, t, y) \in W \times (0, 1] \times E \bigm| \beta(x, y) \leqslant t\big\}.$$

Refer to Fig. 2.20.

Due to Theorem 2.7.3, it suffices to show that the following natural map satisfies (Bi'_A):

$$q_A : X[0, 1] = X \times C(E) \to q_A(X[0, 1]) = (X \times C(E))_A,$$

that is, for each $\mathcal{U} \in \operatorname{cov}(X[0, 1])$ and $\mathcal{V} \in \operatorname{cov}(q_A(X[0, 1]))$, there exist a homeomorphism $\varphi : X[0, 1] \to X[0, 1]$ and $\mathcal{W} \in \operatorname{cov}(q_A(X[0, 1]))$ such that $q_A \circ \varphi$ is $\mathcal{V}$-close to q_A, $\varphi(q_A^{-1}(\mathcal{W})) \prec \mathcal{U}$, and

$$\varphi|A[0] \cup (X[0, 1] \setminus q_A^{-1}(\operatorname{st}(A, \mathcal{V}))) = \mathrm{id}.$$

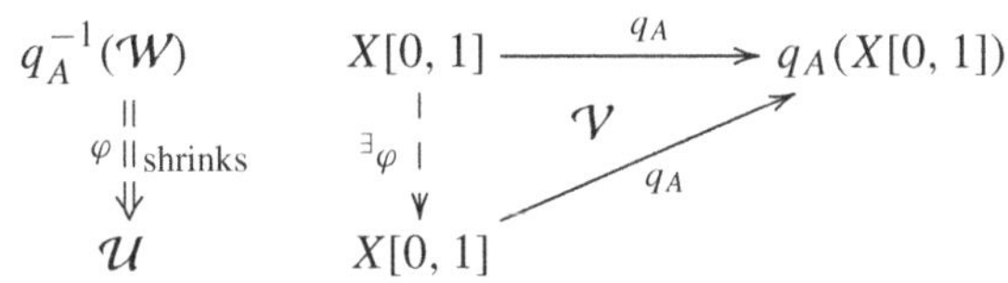

For each $x \in X$, let

$$\alpha'(x) = \sup\big\{t \in \mathbf{I} \bigm| \exists N : \text{ a neighborhood of } x \text{ in } X,\\ \exists U \in \mathcal{U} \text{ such that } N[0,t] \subset U\big\}.$$

Since $\alpha' : X \to (0, 1]$ is l.s.c., we have $\alpha \in \mathrm{C}(X, (0, 1))$ such that $\alpha < \alpha'$. Then, there exists $\mathcal{N} \in \operatorname{cov}(X)$ such that

$$\big\{N[0, \alpha] \bigm| N \in \mathcal{N}\big\} \prec \mathcal{U} \text{ and } \big\{N[0, 1] \bigm| N \in \mathcal{N}[A]\big\} \prec q_A^{-1}(\mathcal{V}).$$

Let $N_0 = \operatorname{st}(A, \mathcal{N})$ and choose an open neighborhood N_1 of A in X so that $\operatorname{cl} N_1 \subset N_0$. Since $N_0 \times E$ is an E-manifold and $w(N_0) \leqslant w(X) \leqslant w(E)$, $N_0 \times E \in \mathcal{F}_E$ by Theorem 2.1.11, and hence $N_0 \in \mathcal{F}_E$. Take $\mathcal{W} \in \operatorname{cov}(N_0)$ so that $\operatorname{st}^5 \mathcal{W} \prec \mathcal{N}$ and let

$$\widetilde{\mathcal{W}} = \big\{W[0, 1] \bigm| W \in \mathcal{W}\big\} \in \operatorname{cov}(N_0[0, 1]).$$

By the remark for Proposition 2.8.14, we have a map $f : X \to X$ such that $f|X \setminus N_1 = \mathrm{id}$, $f \simeq_{\mathcal{W}} \mathrm{id}$ rel. $X \setminus N_1$, $\operatorname{cl} f(N_1) \subset N_0$, and $A \cap \operatorname{cl} f(X) = \emptyset$. Then, A has an open neighborhood N_2 in X such that $\operatorname{cl} N_2 \cap \operatorname{cl} f(X) = \emptyset$. By Theorem 2.4.9, $(N_0 \setminus N_1)[1]$ is an E-deficient closed set in $N_0[0, 1]$. By the Strong Universality Theorem 2.3.15, there is an E-deficient closed embedding $g_0 : N_0[1] \to N_0[0, 1]$ such that

$$g_0|(N_0 \setminus N_1)[1] = \mathrm{id},\ g_0 \simeq_{\widetilde{\mathcal{W}}} (f \times \mathrm{id})|N_0[1] \text{ rel. } (N_0 \setminus N_1)[1],\\ g_0(\operatorname{cl} N_1[1]) \subset N_0[0, 1] \setminus N_0[0] \text{ and } g_0(N_0[1]) \cap \operatorname{cl} N_2[0, 1] = \emptyset.$$

We can extend g_0 to the map $g_1 : X[1] \to X[0, 1]$ by $g_1|(X \setminus N_0)[1] = \mathrm{id}$. Then, g_1 is an E-deficient closed embedding. Indeed, note that $g_1|\operatorname{cl} N_1[1]$ is a closed embedding into $X[0, 1]$. Then, it is easy to see that g_1 is a closed embedding into $X[0, 1]$. Since

$$g_1(X[1]) = (X \setminus N_1)[1] \cup g_0(\operatorname{cl} N_1[1]),$$

$g_1(X[1])$ is E-deficient in $X[0, 1]$ by Theorem 2.4.9 an Propositions 2.3.8 and 2.4.14. Moreover, observe

$$g_1|(X \setminus N_1)[1] = \mathrm{id},\ g_1 \simeq_{\widetilde{\mathcal{W}}} f \times \mathrm{id} \text{ rel. } (X \setminus N_1)[1],\\ g_1(X[1]) \cap X[0] = \emptyset \text{ and } g_1(X[1]) \cap \operatorname{cl} N_2[0, 1] = \emptyset.$$

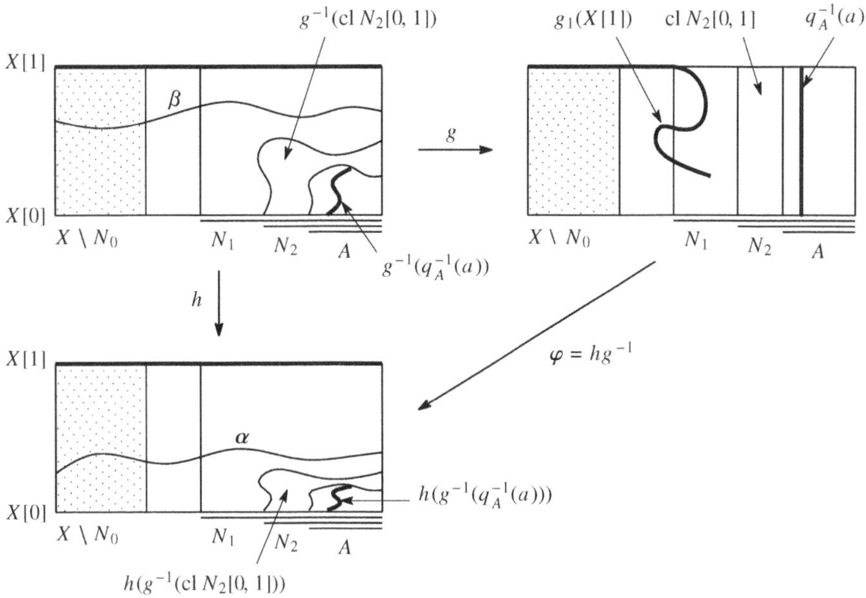

Fig. 2.21 $q_A(X[0,1]) = (X \times C(E))_A$

On the other hand, $X[1]$ is an E-deficient closed set in $X[0,1]$ by Theorem 2.4.9 and

$$g_1 \simeq_{\mathrm{st}\,\widetilde{\mathcal{W}}} \mathrm{id} \text{ rel. } (X \setminus N_1)[1].$$

By the Unknotting Theorem 2.5.12, g_1 can be extended to a homeomorphism $g : X[0,1] \to X[0,1]$ such that

$$g \simeq_{\mathrm{st}^2\,\widetilde{\mathcal{W}}} \mathrm{id}, \;\; g|X[0] = \mathrm{id} \;\text{ and }\; g|(X \setminus N_0)[0,1] = \mathrm{id}.$$

Refer to Fig. 2.21. Note that $X[1] \cap g^{-1}(\mathrm{cl}\, N_2[0,1]) = \emptyset$ and

$$g^{-1}(\widetilde{\mathcal{W}}) \prec \mathrm{st}(\widetilde{\mathcal{W}}, \mathrm{st}^2\,\widetilde{\mathcal{W}}) = \mathrm{st}^5\,\widetilde{\mathcal{W}} \prec \mathcal{N}.$$

For each $(x, y) \in X \times E$, let

$$\beta'(x, y) = \inf\bigl\{t \in [1/2, 1] \bigm| s \geqslant t \Rightarrow (x, s, y) \notin \mathrm{cl}\, g^{-1}(N_2[0,1])\bigr\}.$$

Since $\beta' : X \times E \to (0,1)$ is u.s.c., we have a map $\beta : X \times E \to [0,1)$ such that $\beta > \beta'$. Then, it follows that

$$g^{-1}(N_2[0,1]) \cap X[\beta, 1] = \emptyset.$$

Sliding the **I**-coordinates and using an Urysohn map (see Fig. 2.21), we can easily define a homeomorphism $h : X[0, 1] \to X[0, 1]$ such that

$$\mathrm{pr}_X \circ h = \mathrm{pr}_X, \; h|X[0] \cup X[1] \cup (X \setminus N_0)[0, 1] = \mathrm{id} \text{ and}$$
$$h(x, \beta(x, y), y) = (x, \alpha(x), y) \text{ for each } x \in N_1 \text{ and } y \in E.$$

Let $\varphi = hg^{-1}$. For each $W \in \mathcal{W}$, $W^* = q_A((W \cap N_2)[0, 1])$ is open in $q_A(X[0, 1])$ and there are some $N \in \mathcal{N}$ and $U \in \mathcal{U}$ such that

$$\varphi(q_A^{-1}(W^*)) = \varphi((W \cap N_2)[0, 1]) \subset N[0, \alpha] \subset U.$$

Thus, φ is the desired homeomorphism with the open cover

$$\left\{W^* \mid W \in \mathcal{W}\right\} \cup \varphi^{-1}(\mathcal{U}|(X \setminus A)[0, 1]) \in \mathrm{cov}(q_A(X[0, 1])).$$

This completes the proof (Fig. 2.21). □

When $X \setminus A$ is an E-manifold, the projection $\mathrm{pr}_{X\setminus A} : (X \setminus A) \times E \to X \setminus A$ is a near-homeomorphism by the Stability Theorem 2.3.7. Take $d \in \mathrm{Metr}(X)$ and let

$$\mathcal{U} = \left\{\mathrm{B}_d(x, 2^{-1}d(x, A)) \mid x \in X \setminus A\right\} \in \mathrm{cov}(X \setminus A).$$

Then, we have a homeomorphism $h : (X \setminus A) \times E \to X \setminus A$ that is $\mathcal{U}$-close to $\mathrm{pr}_{X\setminus A}$. Let $p_A : (X \times E)_A \to X$ be the natural map, that is, $p_A|(X \setminus A) \times E = \mathrm{pr}_{X\setminus A}$ and $p_A|A = \mathrm{id}$. As is seen in the proof of Lemma 2.1.16(i), $\mathcal{U}$ is fitting in X, hence h can be extended to a map $\tilde{h} : (X \times E)_A \to X$ by $\tilde{h}|A = \mathrm{id}$. Then, $\tilde{h}$ is a homeomorphism. To see this, it suffices to show that, for each open set U in X, the following is open in X:

$$\tilde{h}\big(((U \setminus A) \times E) \cup (U \cap A)\big) = h((U \setminus A) \times E) \cup (U \cap A).$$

Assume that this is not open in X, that is, this is not a neighborhood of $x_0 \in U \cap A$ in X because $h((U \setminus A) \times E)$ is open in X. Then, we have

$$x_n \in X \setminus (h((U \setminus A) \times E) \cup (U \cap A)), \; n \in \mathbb{N},$$

such that $x_n \to x_0$ $(n \to \infty)$. For every $n \in \mathbb{N}$, we may assume that $x_n \in X \setminus A$, so $h^{-1}(x_n) \in (X \setminus A) \times E$. Since h is $\mathcal{U}$-close to $\mathrm{pr}_{X\setminus A}$, we have $y_n \in X \setminus A$ such that

$$x_n, \mathrm{pr}_{X\setminus A} h^{-1}(x_n) \in \mathrm{B}_d(y_n, 2^{-1}d(y_n, A)).$$

Then, observe that

$$d(y_n, A) \leqslant d(y_n, x_0) \leqslant d(x_n, y_n) + d(x_n, x_0) < 2^{-1}d(y_n, A) + d(x_n, x_0),$$

which implies that $2^{-1}d(y_n, A) < d(x_n, x_0)$. Since $d(\mathrm{pr}_{X\setminus A}h^{-1}(x_n), x_n) < d(y_n, A)$ and $x_n \to x_0$ $(n \to \infty)$, it follows that $\mathrm{pr}_{X\setminus A}h^{-1}(x_n) \to x_0$ $(n \to \infty)$. Note that $\mathrm{pr}_{X\setminus A}h^{-1}(x_n) \in X \setminus A$. For sufficiently large $n \in \mathbb{N}$, $\mathrm{pr}_{X\setminus A}h^{-1}(x_n) \in U \setminus A$, hence $x_n \in h((U \setminus A) \times E)$. This is a contradiction. Consequently, $(X \times E)_A \approx X$. Thus, we have the following:

Corollary 2.9.2 *Let A be a strong Z-set in X. If $X \times E$ and $X \setminus A$ are E-manifolds, then X is also an E-manifold.* □

Remark 2.18 In Corollary 2.9.2 and Proposition 2.9.1, "strong Z-set" cannot be replaced by "Z-set." Indeed, consider the example defined on p. 162 (see Fig. 2.18), that is,

$$X = \mathbf{I} \times \{0\} \cup \{n^{-1} \mid n \in \mathbb{N}\} \times \mathbf{I} \subset \mathbb{R}^2 \text{ and } A = \{(0,0)\} \subset X.$$

Then, X is a separable completely metrizable AR, in which A is a Z-set but not a strong Z-set. Similarly, in the reduced product $(X \times \ell^2)_A$, A is a Z-set but not a strong Z-set. Due to the following theorem, every Z-set in an ℓ^2-manifold is a strong Z-set. Therefore, $(X \times \ell^2)_A$ is not an ℓ^2-manifold. However, by Toruńczyk's Factor Theorem 2.2.12,

$$(X \times \ell^2)_A \times \ell^2 \approx \ell^2 \text{ and } (X \times \ell^2)_A \setminus A = (X \setminus A) \times \ell^2 \approx \ell^2.$$

Moreover, $X \times \ell^2 \approx \ell^2 \not\approx (X \times \ell^2)_A$.

For Z-sets in an E-manifold, we have the following characterization:

Theorem 2.9.3 *For a closed set A in an E-manifold M, the following are equivalent:*

(a) *A is a Z-set in M;*
(b) *A is a strong Z-set in M;*
(c) *A is E-deficient in M;*
(d) *There is a homeomorphism $h : M \to M \times [0,1)$ such that $h(A) \subset M \times \{0\}$;*
(e) *A is contained in some collared set in M.*

Proof The equivalence among (c), (d), and (e) is due to Theorem 2.4.9 and the implication (e) (or (d)) ⇒ (b) is due to Lemma 2.8.2. The implication (b) ⇒ (a) is obvious. Note that an E-manifold is E-stable. By Lemma 2.3.2, there is a homeomorphism $f : (M \times E)_A \to M$ such that $f|A = \mathrm{id}$. Then, (b) ⇒ (c) can be obtained by Proposition 2.9.1. It remains to prove (a) ⇒ (b).

(a) ⇒ (b): For each $\mathcal{U} \in \mathrm{cov}(M)$, choose $\mathcal{V} \in \mathrm{cov}(M)$ so that $\mathrm{st}^3 \mathcal{V} \prec \mathcal{U}$. From Remark 1.12 on Hanner's characterization of ANRs (Theorem 1.13.20), we have a locally finite-dimensional simplicial complex K with $\mathrm{card}\, K^{(0)} \leqslant w(M)$ and maps $f : M \to |K|$ and $g : |K| \to M$ such that $gf \simeq_{\mathcal{V}} \mathrm{id}_M$, where K is the nerve of an open refinement of $\mathcal{V}$ and f is a canonical map. By Theorem 1.7.3, $\varphi = \mathrm{id} : |K| \to |K|_{\mathrm{m}}$ is an $\mathcal{O}_K$-homotopy equivalence. Let $\psi : |K|_{\mathrm{m}} \to |K|$ be an $\mathcal{O}_K$-homotopy inverse of φ. Since $f^{-1}(\mathcal{O}_K) \prec \mathcal{V}$, it follows that $g(\mathcal{O}_K) \prec gf(\mathcal{V}) \prec \operatorname{st}\mathcal{V}$, hence $g\psi\varphi f \simeq_{\operatorname{st}\mathcal{V}} gf \simeq_{\mathcal{V}} \mathrm{id}$. Since $|K|_{\mathrm{m}} \in \mathcal{F}_E$ by Corollary 2.4.7, $g\psi$ is $\mathcal{V}$-close to a closed embedding $h : |K|_{\mathrm{m}} \to M$ by the Strong Universality Theorem 2.3.15. Then, $h\varphi f$ is $\operatorname{st}^2\mathcal{V}$-close to id. Thus, it suffices to construct a map $k : |K|_{\mathrm{m}} \to M$ such that k is $\mathcal{V}$-close to h and $A \cap \operatorname{cl} k(|K|) = \emptyset$.

Since h is a closed embedding, there exists $\mathcal{W} \in \operatorname{cov}(M)$ such that $\mathcal{W} \prec \mathcal{V}$ and $h^{-1}(\operatorname{st}\mathcal{W}) \prec \mathcal{O}_K \prec \mathcal{S}_K$. Take open covers of M as follows:

$$\mathcal{W} \overset{*}{\succ} \mathcal{W}_0 \overset{*}{\succ} \mathcal{W}_1 \overset{*}{\succ} \cdots .$$

By induction, we shall construct E-deficient closed embeddings $h_n : |K|_{\mathrm{m}} \to M$, $n \in \omega$, such that h_n is $\mathcal{W}_n$-close to h_{n-1}, $h_n||K^{(n-1)}| = h_{n-1}||K^{(n-1)}|$, and $h_n(|K^{(n)}|) \cap A = \emptyset$, where $h_{-1} = h$.

Since $K^{(0)}$ is discrete in $|K|_{\mathrm{m}}$ and h is a closed embedding, M has a discrete open collection $\{U_v \mid v \in K^{(0)}\}$ such that $h(v) \in U_v$. Choose $\mathcal{W}'_0 \in \operatorname{cov}(M)$ so that

$$\operatorname{st}\mathcal{W}'_0 \prec \mathcal{W}_0 \text{ and } \mathcal{W}'_0 \prec \{U_v \mid v \in K^{(0)}\} \cup \{M \setminus h(K^{(0)})\}.$$

Then, $\{\operatorname{st}(h(v), \mathcal{W}'_0) \mid v \in K^{(0)}\}$ is discrete in M. Since A is a Z-set in M, h is $\mathcal{W}'_0$-close to a map $k_0 : |K|_{\mathrm{m}} \to M\setminus A$. Then, $k_0(K^{(0)})$ is discrete in M. Since $k_0(|K^{(0)}|)$ is E-deficient in M by Proposition 2.3.8, it follows that $k_0||K^{(0)}|$ is an E-deficient closed embedding into M. By the Strong Universality Theorem 2.3.15, we have an E-deficient closed embedding $h_0 : |K|_{\mathrm{m}} \to M$ such that $h_0||K^{(0)}| = k_0||K^{(0)}|$ and h_0 is $\mathcal{W}'_0$-close to k_0, hence h_0 is $\mathcal{W}_0$-close to h.

Now, assume that h_{n-1} has been constructed. Since $h_{n-1}(|K^{(n-1)}|)$ and A are disjoint closed sets in M, we have an open set U in M such that

$$h_{n-1}(|K^{(n-1)}|) \subset U \subset \operatorname{cl} U \subset M \setminus A.$$

For each n-simplex $\sigma \in K$, let

$$C_\sigma = \sigma \setminus h_{n-1}^{-1}(U) \subset \operatorname{rint}\sigma \subset O_{\operatorname{Sd} K}(\hat{\sigma}).$$

Then, $\{C_\sigma \mid \sigma \in K^{(n)} \setminus K^{(n-1)}\}$ is discrete in $|K|_{\mathrm{m}}$. Since h_{n-1} is a closed embedding, we can use a method similar to the above to choose an open star-refinement $\mathcal{W}'_n$ of $\mathcal{W}_n$ such that

$$\left\{ \operatorname{st}(h_{n-1}(C_\sigma), \mathcal{W}'_n) \;\middle|\; \sigma \in K^{(n)} \setminus K^{(n-1)} \right\} \text{ is discrete in } M.$$

Since A is a Z-set in M, we apply the Homotopy Extension Theorem 1.13.11 to obtain a map $k_n : |K|_{\mathrm{m}} \to M \setminus A$ such that $k_n|h_{n-1}^{-1}(\operatorname{cl} U) = h_{n-1}|h_{n-1}^{-1}(\operatorname{cl} U)$ and

$k_n \simeq_{\mathcal{W}'_n} h_{n-1}$. Then, $k_n(|K^{(n)}|)$ is closed in M. Indeed,

$$k_n(|K^{(n)}|) = \bigcup_{\sigma \in K^{(n)} \setminus K^{(n-1)}} k_n(C_\sigma) \cup h_{n-1}\big(|K^{(n)}| \cap h_{n-1}^{-1}(\operatorname{cl} U)\big)$$

and $\{k_n(C_\sigma) \mid \sigma \in K^{(n)} \setminus K^{(n-1)}\}$ is discrete in M. By the Strong Universality Theorem 2.3.15, we have an E-deficient closed embedding $h_n : |K|_{\mathrm{m}} \to M$ such that $h_n||K^{(n-1)}| = h_{n-1}||K^{(n-1)}|$, $h_n(|K^{(n)}|) \cap A = \emptyset$, and h_n is $\mathcal{W}'_n$-close to k_n, hence h_n is $\mathcal{W}_n$-close to h_{n-1}.

Since K is locally finite-dimensional, we have a map $k : |K|_{\mathrm{m}} \to M$ defined by $k||K^{(n)}| = h_n||K^{(n)}|$ that is $\mathcal{W}$-close (so $\mathcal{V}$-close) to h. Each $a \in A$ is contained in some $W \in \mathcal{W}$. Since $h^{-1}(\operatorname{st} \mathcal{W}) \prec \mathcal{S}_K$, we have $v \in K^{(0)}$ such that $h^{-1}(\operatorname{st}(W, \mathcal{W})) \subset |\operatorname{St}(v, K)|$. Because of the local finite-dimensionality of K, $\operatorname{St}(v, K)$ is contained in some $K^{(n)}$. For each $x \in k^{-1}(W)$, we have $h(x) \in \operatorname{st}(W, \mathcal{W})$, so $x \in h^{-1}(\operatorname{st}(W, \mathcal{W})) \subset |K^{(n)}|$. Then, we have $k(|K|) \cap W \subset k(|K^{(n)}|)$. Since $k(|K^{(n)}|) = h_n(|K^{(n)}|)$ is closed in M and $h_n(|K^{(n)}|) \cap A = \emptyset$, it follows that $V = W \setminus h_n(|K^{(n)}|)$ is an open neighborhood of a in M and $V \cap k(|K|) = \emptyset$. Thus, $A \cap \operatorname{cl} k(|K|) = \emptyset$. This completes the proof. □

For a closed set in a homotopy dense subset of an E-manifold, conditions (a) and (b) in Theorem 2.9.3 above are equivalent to each other, that is,

Proposition 2.9.4 *Let X be a homotopy dense subset of an ANR M. Suppose that every Z-set in M is a strong Z-set in M. Then, every Z-set in X is also a strong Z-set in X.*

Proof Let $A \subset X$ be a Z-set in X. For each open cover $\mathcal{U}$ of X, we have a collection $\widetilde{\mathcal{U}}$ of open sets in M such that $\widetilde{\mathcal{U}}|X = \mathcal{U}$. Then $U = \bigcup \widetilde{\mathcal{U}}$ is open in M and X is homotopy dense in U. Let $\mathcal{V} \in \operatorname{cov}(X)$ be a star-refinement of $\widetilde{\mathcal{U}}$. Since X is homotopy dense in M, it is easy to see that $\operatorname{cl}_M A$ is a Z_∞-set in M, which means that it is a Z-set in M by Theorem 2.8.6, and hence it is a strong Z-set in M by the assumption. Then, $\operatorname{cl}_U A = U \cap \operatorname{cl}_M A$ is a strong Z-set in U by Proposition 2.8.14. Thus, there is a map $f : U \to U$ such that f is $\mathcal{V}$-close to id_U and $\operatorname{cl}_U f(U) \cap \operatorname{cl}_U A = \emptyset$. Choose an open refinement $\mathcal{W}$ of $\mathcal{V}$ such that if a map $f' : U \to U$ is $\mathcal{W}$-close to f then $\operatorname{cl}_U f'(U) \cap \operatorname{cl}_U A = \emptyset$. Since X is homotopy dense in U, there exists a map $g : U \to X$ that is $\mathcal{W}$-close to id_U. Then, the map $h = gf|_X$ is $\mathcal{U}$-close to id_X. Since gf is $\mathcal{W}$-close to f, we have $\operatorname{cl}_U gf(U) \cap \operatorname{cl}_U A = \emptyset$, hence $\operatorname{cl}_X h(X) \cap A = \emptyset$. Therefore, A is a strong Z-set in X. □

An embedding $h : A \to X$ is called a **Z-embedding** (resp. a **strong Z-embedding**) if $h(A)$ is a Z-set (resp. a strong Z-set) in X. By Theorem 2.9.3, an E-deficient closed embedding into an E-manifold is simply a Z-embedding. A **Z-submanifold** of an E-manifold M is a Z-set in M that is an E-manifold.

By Theorem 2.9.3, the Collaring Theorem 2.5.9 can be reformulated (in a slightly stronger manner) as follows:

Theorem 2.9.5 (COLLARING) *Let M and N be E-manifolds such that $N \subset M$. Then, N is a Z-submanifold of M if and only if N is a collared closed set in M.* □

Remark 2.19 Due to Theorem 2.9.5 above, a Z-submanifold of M may be considered as a boundary of M, but it is not unique.

By Theorem 2.9.3, we can restate the Strong Universality Theorem 2.3.15 and the Unknotting Theorem 2.5.12 as follows:

Theorem 2.9.6 (STRONG UNIVERSALITY) *Let M be an E-manifold and $A \in \mathcal{F}_E$ with a closed set $B \subset A$ and a map $f : A \to M$ such that $f|B$ is a Z-embedding. Then, for each $\mathcal{U} \in \mathrm{cov}(M)$, there exists a Z-embedding $h : A \to M$ such that $h|B = f|B$ and h is $\mathcal{U}$-close to f.* □

Theorem 2.9.7 (Z-SET UNKNOTTING) *Let A be a Z-set in an E-manifold M and $h : A \to M$ be a Z-embedding with $h \simeq \mathrm{id}$ by a homotopy $f : A \times \mathbf{I} \to M$. Then, for each open cover $\mathcal{U}$ of M, h can be extended to a homeomorphism $\tilde{h} : M \to M$ that is ambiently invertibly $\mathcal{U}^*$-isotopic to* id, *where $\mathcal{U}^* = \{\mathrm{st}(f(\{x\} \times \mathbf{I}), \mathcal{U}) \mid x \in A\}$.* □

As a corollary of the Negligibility Theorem 2.4.3, we have the following:

Theorem 2.9.8 (NEGLIGIBILITY OF Z-SETS) *For every Z-set A in an E-manifold M, the inclusion $M \setminus A \subset M$ is a near-homeomorphism.* □

This can also be obtained from Theorems 2.6.3 and 2.8.6. When E is a complete normed linear space, every E-manifold is completely metrizable. Hence, the following is obtained from Theorem 2.4.4:

Theorem 2.9.9 (NEGLIGIBILITY OF Z_σ-SETS) *When E is complete, if A is a Z_σ-set in an E-manifold M, the inclusion $M \setminus A \subset M$ is a near-homeomorphism.* □

Remark 2.20 In $\boldsymbol{Q}$-manifolds, every Z-set is a strong Z-set because of the local compactness. The $\boldsymbol{Q}$-manifold version of Theorem 2.9.3 will be proved in the next section. The results for $\boldsymbol{Q}$-manifolds corresponding to Theorems 2.9.5, 2.9.6, and 2.9.7 will also be established in Sect. 2.11. On the other hand, Theorem 2.9.8 has no $\boldsymbol{Q}$-manifold versions. In fact, $\boldsymbol{Q} \times \mathbf{I} \not\approx \boldsymbol{Q} \times [0, 1) \not\approx \boldsymbol{Q} \times (0, 1)$.

2.10 Z-Sets in the Hilbert Cube and Q-Manifolds

In this section, we show that a closed set in $\boldsymbol{Q}$ is a Z-set if and only if it is $\boldsymbol{Q}$-deficient, and that every homeomorphism between Z-sets in $\boldsymbol{Q}$ can be extended to a homeomorphism of $\boldsymbol{Q}$ onto itself. Although we have denoted by pr_i the projection of a product space onto the i-th factor, when several product spaces are discussed,

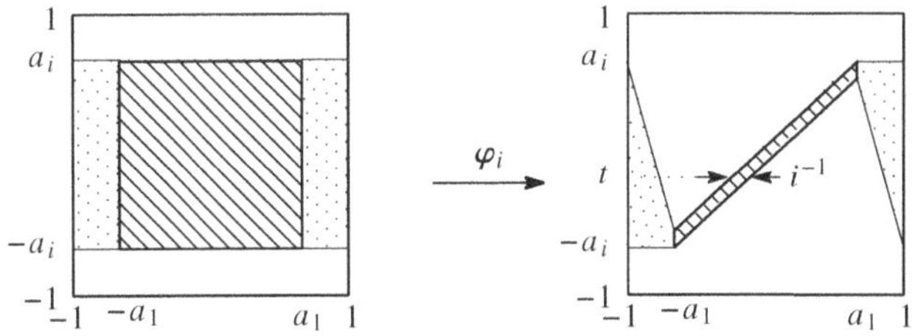

Fig. 2.22 The homeomorphism $\varphi_i : [-1, 1]^2 \to [-1, 1]^2$

the same pr_i is used for deferent projections. So, in this section, to avoid confusion, the projection of $\boldsymbol{Q} = [-1, 1]^{\mathbb{N}}$ onto the i-th factor is denoted by π_i instead of pr_i.

For the Hilbert cube $\boldsymbol{Q} = [-1, 1]^{\mathbb{N}}$, we use $d \in \mathrm{Metr}(\boldsymbol{Q})$ defined as follows:

$$d(x, y) = \sum_{i \in \mathbb{N}} 2^{-i} |x(i) - y(i)|.$$

Then, $\boldsymbol{Q}$ is isometric to the subspace $\prod_{n \in \mathbb{N}} [-2^{-n}, 2^{-n}]$ of the Banach space ℓ^1 by the natural bijection $x \mapsto (2^{-i} x(i))_{i \in \mathbb{N}}$. Recall that $I(\boldsymbol{Q}) = (-1, 1)^{\mathbb{N}}$ is the pseudo-interior of $\boldsymbol{Q}$.

Lemma 2.10.1 *For every compact set $A \subset I(\boldsymbol{Q})$, there exists a homeomorphism $h : \boldsymbol{Q} \to \boldsymbol{Q}$ such that* $\mathrm{pr}_1 h(A) = \{0\}$ *and* $h(I(\boldsymbol{Q})) = I(\boldsymbol{Q})$.

Proof Due to its compactness, $A \subset \prod_{i \in \mathbb{N}} [-a_i, a_i]$ for some $0 < a_i < 1$, $i \in \mathbb{N}$. Then, we may assume that $A = \prod_{i \in \mathbb{N}} [-a_i, a_i]$. For each $i > 1$, as illustrated in Fig. 2.22, we have a homeomorphism $\varphi_i : [-1, 1]^2 \to [-1, 1]^2$ with the following properties:

$$\mathrm{pr}_1 \varphi_i = \mathrm{pr}_1, \ \ \varphi_i | \partial [-1, 1]^2 = \mathrm{id} \ \ \text{and}$$

$$\operatorname{diam} \varphi_i([-a_1, a_1] \times [-a_i, a_i]) \cap \mathrm{pr}_2^{-1}(t) \leqslant i^{-1} \ \ \text{for each } t \in [-1, 1].$$

We define a map $\varphi : \boldsymbol{Q} \to \boldsymbol{Q}$ as follows:

$$(\mathrm{pr}_1 \varphi(x), \mathrm{pr}_i \varphi(x)) = \varphi_i(x(1), x(i)) \ \ \text{for every } i > 1.$$

It is easy to see that φ is a homeomorphism and $\varphi(I(\boldsymbol{Q})) = I(\boldsymbol{Q})$.

We write $\boldsymbol{Q} = [-1, 1] \times Q_1$, where $Q_1 = [-1, 1]^{\mathbb{N} \setminus \{1\}}$. Let $p : \boldsymbol{Q} \to Q_1$ be the projection, that is,

$$p(x) = (x(2), x(3), \dots).$$

For each $x, y \in \varphi(A)$, $p(x) = p(y)$ implies $x(1) = y(1)$. Indeed, for every $i \in \mathbb{N}$, we have $|x(1) - y(1)| \leqslant i^{-1}$ because $x(i) = y(i)$. Then, $p|\varphi(A) : \varphi(A) \to Q_1$ is

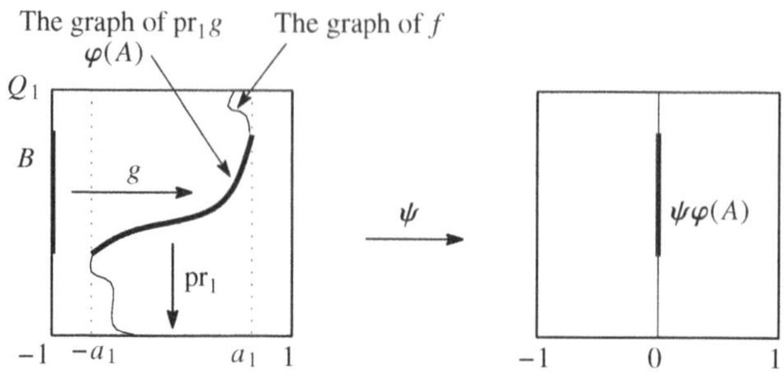

Fig. 2.23 The homeomorphism ψ

injective, hence it is an embedding. Let $B = p\varphi(A)$ and $g = (p|\varphi(A))^{-1} : B \to \varphi(A)$. Note that $\varphi(A)$ is the graph of the map $\mathrm{pr}_1 g : B \to [-a_1, a_1]$. By the Tietze Extension Theorem 1.3.1, $\mathrm{pr}_1 g$ can be extended to a map $f : Q_1 \to [-a_1, a_1]$. Then, there exists a homeomorphism $\psi : \boldsymbol{Q} \to \boldsymbol{Q}$ such that $p\psi = p$, $\psi|\{\pm 1\} \times Q_1 = \mathrm{id}$, and $\psi(f(x), x) = (0, x)$ for every $x \in Q_1$. The homeomorphism $\psi\varphi$ is the desired one. Refer to Fig. 2.23. □

Lemma 2.10.2 *For every compactum $A \subset I(\boldsymbol{Q})$ and $\varepsilon > 0$, there exists a homeomorphism $h : \boldsymbol{Q} \to \boldsymbol{Q}^2 = \boldsymbol{Q} \times \boldsymbol{Q}$ such that*

$$h(A) \subset I(\boldsymbol{Q}) \times \{\mathbf{0}\},\ h(I(\boldsymbol{Q})) = I(\boldsymbol{Q})^2 = I(\boldsymbol{Q}) \times I(\boldsymbol{Q}),$$

and $d(\mathrm{pr}_1 h, \mathrm{id}) < \varepsilon$ (equivalently, $d(h^{-1}, \mathrm{pr}_1) < \varepsilon$), where $\mathrm{pr}_1 : \boldsymbol{Q}^2 \to \boldsymbol{Q}$ is the projection onto the first factor of $\boldsymbol{Q}^2$. In particular, A is $\boldsymbol{Q}$-deficient in $\boldsymbol{Q}$ and it is also $I(\boldsymbol{Q})$-deficient in $I(\boldsymbol{Q})$.

Proof Choose $n \in \mathbb{N}$ so that $2^{-n+1} < \varepsilon$. Let $p_n : \boldsymbol{Q} \to [-1, 1]^n$ denote the projection onto the first n-coordinates. Let $\varphi : \boldsymbol{Q} \to [-1, 1]^n \times \boldsymbol{Q}^{\mathbb{N}}$ be a homeomorphism such that $\varphi(I(\boldsymbol{Q})) = (-1, 1)^n \times I(\boldsymbol{Q})^{\mathbb{N}}$ and $q_0\varphi = p_n$, where $q_0 : [-1, 1]^n \times \boldsymbol{Q}^{\mathbb{N}} \to [-1, 1]^n$ is the projection. For each $i \in \mathbb{N}$, let $q_i : [-1, 1]^n \times \boldsymbol{Q}^{\mathbb{N}} \to \boldsymbol{Q}$ be the projection onto the i-th factor of $\boldsymbol{Q}^{\mathbb{N}}$. Then,

$$\varphi(A) \subset \prod_{i \in \omega} q_i(\varphi(A)) \subset (-1, 1)^n \times I(\boldsymbol{Q})^{\mathbb{N}}.$$

By Lemma 2.10.1, we have homeomorphisms $h_i : \boldsymbol{Q} \to \boldsymbol{Q}$, $i \in \mathbb{N}$, such that

$$p_1 h_i(q_i(\varphi(A))) = \{0\} \text{ and } h_i(I(\boldsymbol{Q})) = I(\boldsymbol{Q}),$$

where $p_1 = \pi_1$ ($= \mathrm{pr}_1$) is the projection onto the first factor of $\boldsymbol{Q} = [-1, 1]^{\mathbb{N}}$. We define a homeomorphism $h : \boldsymbol{Q}^{\mathbb{N}} \to \boldsymbol{Q}^{\mathbb{N}}$ by $h(x) = (h_i(x(i)))_{i \in \mathbb{N}}$. Let ψ :

$[-1, 1]^n \times \boldsymbol{Q}^{\mathbb{N}} \to \boldsymbol{Q}^2$ be a homeomorphism such that

$$p_n \mathrm{pr}_1 \psi = q_0, \ \psi((-1,1)^n \times I(\boldsymbol{Q})) = I(\boldsymbol{Q})^2 \ \text{ and}$$
$$\mathrm{pr}_2 \psi(x) = (p_1(q_i(x)))_{i\in\mathbb{N}} \ \text{ for every } x \in [-1,1]^n \times \boldsymbol{Q}^{\mathbb{N}},$$

where $\mathrm{pr}_1, \mathrm{pr}_2 : \boldsymbol{Q}^2 \to \boldsymbol{Q}$ are the projections onto the first and the second factors of $\boldsymbol{Q}^2$ respectively. Then, $\psi(\mathrm{id} \times h)\varphi : \boldsymbol{Q} \to \boldsymbol{Q}^2$ is the desired homeomorphism. Indeed,

$$\psi(\mathrm{id} \times h)\varphi(A) \subset \psi\big(q_0\varphi(A) \times \textstyle\prod_{i\in\mathbb{N}} h_i q_i \varphi(A)\big) \subset \boldsymbol{Q} \times \{\mathbf{0}\}.$$

Since $p_n \mathrm{pr}_1 \psi(\mathrm{id} \times h)\varphi = q_0(\mathrm{id} \times h)\varphi = q_0\varphi = p_n$, it follows that

$$d(\mathrm{pr}_1 \psi(\mathrm{id} \times h)\varphi, \mathrm{id}) \leqslant 2^{-n+1} < \varepsilon. \qquad \square$$

We will prove that every homeomorphism $h : A \to B$ between compacta in $I(\boldsymbol{Q})$ can be extended to a homeomorphism $\tilde{h} : \boldsymbol{Q} \to \boldsymbol{Q}$. In the proof, we have to modify Klee's Trick (Theorem 1.13.7). To this end, we need the following:

Lemma 2.10.3 *For each map $f : \boldsymbol{Q} \to I(\boldsymbol{Q})$, there exists a homeomorphism $\varphi : \boldsymbol{Q}^2 \to \boldsymbol{Q}^2$ such that $\mathrm{pr}_1\varphi = \mathrm{pr}_1$, $\varphi(x, \mathbf{0}) = (x, f(x))$ for each $x \in \boldsymbol{Q}$, and $\varphi(I(\boldsymbol{Q})^2) = I(\boldsymbol{Q})^2$.*

Proof We can define $\varphi, \psi : \boldsymbol{Q}^2 \to \boldsymbol{Q}^2$ as follows: for each $(x, y) \in \boldsymbol{Q}^2$, $\mathrm{pr}_1\varphi(x, y) = \mathrm{pr}_1\psi(x, y) = x$,

$$\mathrm{pr}_2\varphi(x,y)(n) = \begin{cases} (1 + f(x)(n))y(n) + f(x)(n) & \text{if } -1 \leqslant y(n) \leqslant 0, \\ (1 - f(x)(n))y(n) + f(x)(n) & \text{if } 0 \leqslant y(n) \leqslant 1, \end{cases}$$

$$\mathrm{pr}_2\psi(x,y)(n) = \begin{cases} \dfrac{y(n) - f(x)(n)}{1 + f(x)(n)} & \text{if } -1 \leqslant y(n) \leqslant f(x)(n), \\[2ex] \dfrac{y(n) - f(x)(n)}{1 - f(x)(n)} & \text{if } f(x)(n) \leqslant y(n) \leqslant 1. \end{cases}$$

Then, φ and ψ are continuous, $\psi\varphi = \mathrm{id}$ and $\varphi\psi = \mathrm{id}$. Therefore, φ is a homeomorphism with $\varphi^{-1} = \psi$. The other required conditions come from the definition. $\square$

Theorem 2.10.4 *Let $h : A \to B$ be a homeomorphism between compacta in $I(\boldsymbol{Q})$ with $d(h, \mathrm{id}) = \sup_{x\in A} d(h(x), x) < \varepsilon$. Then, h can be extended to a homeomorphism $\tilde{h} : \boldsymbol{Q} \to \boldsymbol{Q}$ such that $d(\tilde{h}, \mathrm{id}) = \sup_{x\in \boldsymbol{Q}} d(\tilde{h}(x), x) < \varepsilon$ and $\tilde{h}(I(\boldsymbol{Q})) = I(\boldsymbol{Q})$.*

Proof We may regard $\boldsymbol{Q} = \prod_{n\in\mathbb{N}}[-2^{-n}, 2^{-n}] \subset \ell^1$, and the pseudo-interior $I(\boldsymbol{Q})$ of $\boldsymbol{Q}$ is identified with $\prod_{n\in\mathbb{N}}(-2^{-n}, 2^{-n})$. Let $\delta = \varepsilon - d(h, \mathrm{id}) > 0$. By

Lemma 2.10.2, we have a homeomorphism $\varphi : \boldsymbol{Q} \to \boldsymbol{Q}^2$ such that $d(\mathrm{pr}_1\varphi, \mathrm{id}) < \delta/3$ and

$$\varphi(A \cup B) \subset \prod_{n \in \mathbb{N}} (-2^{-n}, 2^{-n}) \times \{\mathbf{0}\}.$$

Let $A' = \mathrm{pr}_1(\varphi(A))$ and $B' = \mathrm{pr}_1(\varphi(B))$. Then, $\varphi(A) = A' \times \{\mathbf{0}\}$ and $\varphi(B) = B' \times \{\mathbf{0}\}$. We have the homeomorphism

$$f = \mathrm{pr}_1\varphi h \varphi^{-1} i | A' : A' \to B',$$

where $i : \boldsymbol{Q} \to \boldsymbol{Q} \times \{\mathbf{0}\} \subset \boldsymbol{Q}^2$ is the natural injection (i.e., $i(x) = (x, \mathbf{0})$ for each $x \in \boldsymbol{Q}$). Choose $0 < a_n < 2^{-n}$, $n \in \mathbb{N}$, so that $A' \cup B' \subset \prod_{n \in \mathbb{N}} [-a_n, a_n]$ and $\sum_{n \in \mathbb{N}} (2^{-n} - a_n) < \delta/3$. We have the retraction $r : \ell^1 \to \prod_{n \in \mathbb{N}} [-a_n, a_n]$ defined as follows:

$$r(x)(n) = \begin{cases} \min\{x(n), a_n\} & \text{if } x(n) \geqslant 0, \\ \max\{x(n), -a_n\} & \text{if } x(n) < 0. \end{cases}$$

Then, for each $x, y \in \ell^1$, $\|r(x) - r(y)\| \leqslant \|x - y\|$ because $|r(x)(n) - r(y)(n)| \leqslant |x(n) - y(n)|$ for every $n \in \mathbb{N}$. Since $\sum_{n \in \mathbb{N}} (2^{-n} - a_n) < \delta/3$, it follows that $d(r|\boldsymbol{Q}, \mathrm{id}) < \delta/3$.

We can apply the Dugundji Extension Theorem 1.13.1 with Remark 1.11 to obtain a map $g : \boldsymbol{Q} \to \ell^1$ such that $g|A' = f - \mathrm{id}$ and $\|g\| = \|f - \mathrm{id}\| = d(f, \mathrm{id})$, where $\|\cdot\|$ is the sup-norm. Then, $\tilde{f} = r(g + \mathrm{id}) : \boldsymbol{Q} \to \prod_{n \in \mathbb{N}} [-a_n, a_n] \subset \boldsymbol{Q}$ is an extension of f with $\|\tilde{f} - \mathrm{id}\| = d(f, \mathrm{id})$. Indeed, for each $x \in \boldsymbol{Q}$,

$$\|\tilde{f}(x) - x\| = \|r(g(x) + x) - r(x)\| \leqslant \|(g(x) + x) - x\| = \|g(x)\|,$$

hence $\|f - \mathrm{id}\| \leqslant \|\tilde{f} - \mathrm{id}\| \leqslant \|f - \mathrm{id}\|$, that is, $\|\tilde{f} - \mathrm{id}\| = \|f - \mathrm{id}\| = d(f, \mathrm{id})$, and if $x \in A'$, then $\tilde{f}(x) = r(g(x) + x) = rf(x) = f(x)$, that is, $\tilde{f}|A' = f$. Moreover, note that

$$d(\tilde{f}, r|\boldsymbol{Q}) \leqslant d(\tilde{f}, \mathrm{id}) + d(r|\boldsymbol{Q}, \mathrm{id}) < d(f, \mathrm{id}) + \delta/3.$$

On the other hand, by the Dugundji Extension Theorem 1.13.1, f^{-1} can be extended to a map $f^* : \boldsymbol{Q} \to \prod_{n \in \mathbb{N}} [-a_n, a_n]$. By Lemma 2.10.3, we have homeomorphisms $\psi_i : \boldsymbol{Q}^2 \to \boldsymbol{Q}^2$, $i = 1, 2, 3$ (Fig. 2.24), such that:

(1) $\mathrm{pr}_1\psi_1 = \mathrm{pr}_1$, $\psi_1(x, 0) = (x, r(x))$ for each $x \in \boldsymbol{Q}$,
(2) $\mathrm{pr}_2\psi_2 = \mathrm{pr}_2$, $\psi_2(r(x), x) = (\tilde{f}(x), x)$ for each $x \in \boldsymbol{Q}$, $\|\mathrm{pr}_1\psi_2 - \mathrm{pr}_1\| = \|\tilde{f} - r\| < d(f, \mathrm{id}) + \delta/3$,
(3) $\mathrm{pr}_1\psi_3 = \mathrm{pr}_1$, $\psi_3(x, f^*(x)) = (x, 0)$ for each $x \in \boldsymbol{Q}$,
(4) $\psi_i\big(\big(\prod_{n \in \mathbb{N}} (-2^{-n}, 2^{-n})\big)^2\big) = \big(\prod_{n \in \mathbb{N}} (-2^{-n}, 2^{-n})\big)^2$.

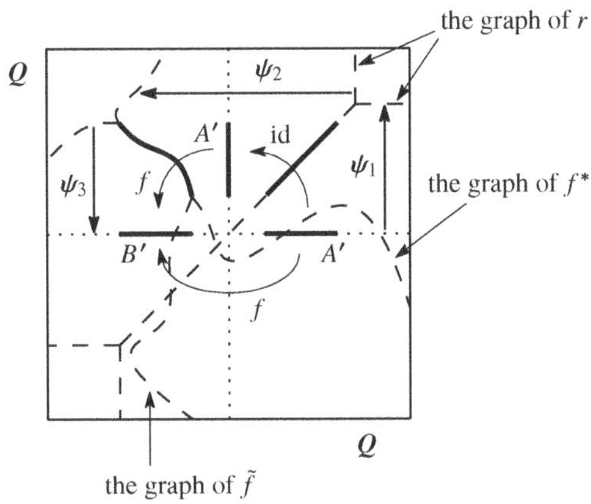

Fig. 2.24 The homeomorphisms ψ_1, ψ_2, and ψ_3

Thus, we have a homeomorphism $\tilde{h} = \varphi^{-1}\psi_3\psi_2\psi_1\varphi : \boldsymbol{Q} \to \boldsymbol{Q}$. Observe that

$$\tilde{h}\big(\textstyle\prod_{n\in\mathbb{N}}(-2^{-n}, 2^{-n})\big) = \displaystyle\prod_{n\in\mathbb{N}}(-2^{-n}, 2^{-n}).$$

For each $x \in A'$,

$$\begin{aligned}\psi_3\psi_2\psi_1(x,0) &= \psi_3\psi_2(x, r(x)) = \psi_3\psi_2(x,x) \\ &= \psi_3\psi_2(r(x), x) = \psi_3(f(x), x) \\ &= \psi_3(f(x), f^{-1}(f(x))) = (f(x), 0) = \varphi h \varphi^{-1}(x, 0).\end{aligned}$$

Hence, $\tilde{h}|A = \varphi^{-1}\psi_3\psi_2\psi_1\varphi|A = h$. Since $d(\varphi^{-1}, \mathrm{pr}_1) = d(\mathrm{pr}_1\varphi, \mathrm{id}) < \delta/3$, it follows that $\tilde{h}$ is $\delta/3$-close to $\mathrm{pr}_1\psi_3\psi_2\psi_1\varphi = \mathrm{pr}_1\psi_2\psi_1\varphi$. Therefore,

$$\begin{aligned}d(\tilde{h}, \mathrm{id}) &\leqslant d(\tilde{h}, \mathrm{pr}_1\psi_2\psi_1\varphi) + d(\mathrm{pr}_1\psi_2\psi_1\varphi, \mathrm{pr}_1\psi_1\varphi) + d(\mathrm{pr}_1\psi_1\varphi, \mathrm{id}) \\ &< \delta/3 + d(\mathrm{pr}_1\psi_2, \mathrm{pr}_1) + d(\mathrm{pr}_1\varphi, \mathrm{id}) \\ &< d(f, \mathrm{id}) + \delta < \varepsilon.\end{aligned}$$

Thus, $\tilde{h}$ is the desired one. □

Let Homeo($\boldsymbol{Q}$) be the space of all homeomorphisms of $\boldsymbol{Q}$ onto itself with the sup-metric $d(f, g) = \sup_{x\in\boldsymbol{Q}} d(f(x), g(x))$, where the sup-metric d is not complete but Homeo($\boldsymbol{Q}$) is completely metrizable (as is shown in Sect. 2.1).

Lemma 2.10.5 *Let $W = \{1\} \times [-1, 1]^{\mathbb{N}\setminus\{1\}} \subset \boldsymbol{Q}$. Then, there is a homeomorphism $h : \boldsymbol{Q} \to \boldsymbol{Q}$ such that $h(W) \subset I(\boldsymbol{Q})$.*

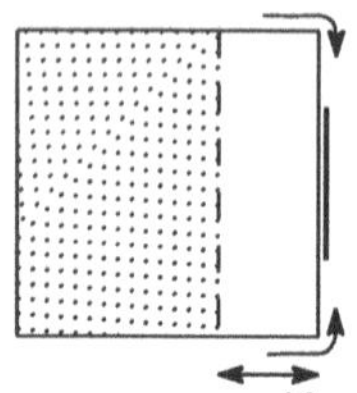

Fig. 2.25 The homeomorphism φ

Proof First, we shall find a homeomorphism $g_1 : \boldsymbol{Q} \to \boldsymbol{Q}$ such that $g_1(W) \subset \{1\} \times (-1, 1)^{\mathbb{N}\setminus\{1\}}$. We consider the following closed subspace of Homeo($\boldsymbol{Q}$):

$$H = \big\{ f \in \mathrm{Homeo}(\boldsymbol{Q}) \;\big|\; f(W) \subset W \big\}.$$

For each $n > 1$, the following set is open in H:

$$H_n = \big\{ f \in H \;\big|\; \mathrm{pr}_n f(W) \subset (-1, 1) \big\}.$$

We shall show that each H_n is dense in H. Then, $\bigcap_{n\in\mathbb{N}} H_n$ would be dense in H by the Baire Category Theorem 1.3.15. Thus, the desired g_1 could be found in $\bigcap_{n\in\mathbb{N}} H_n$.

To see that H_n is dense in H, let $f \in H$ and $\varepsilon > 0$. It suffices to find $h \in H_n$ with $d(h, \mathrm{id}) < \varepsilon$, because $hf \in H_n$ and $d(hf, f) < \varepsilon$. Take a homeomorphism $\varphi : [-1, 1]^2 \to [-1, 1]^2$ as illustrated in Fig. 2.25 such that

$$\varphi|[-1, 1 - \tfrac{1}{2}\varepsilon] \times [-1, 1] = \mathrm{id},$$
$$|\mathrm{pr}_2\varphi(x) - \mathrm{pr}_2(x)| \leqslant \varepsilon \text{ for each } x \in [-1, 1]^2 \text{ and}$$
$$\varphi(\{1\} \times [-1, 1]) = \{1\} \times [-1 + \tfrac{1}{2}\varepsilon, 1 - \tfrac{1}{2}\varepsilon].$$

Then, the desired $h \in H_n$ can be defined as follows:

$$h(x)(i) = x(i),\; i \in \mathbb{N} \setminus \{1, n\}, \text{ and}$$
$$(h(x)(x), h(x)(n)) = \varphi(x(1), x(n)) \text{ for each } x \in \boldsymbol{Q}.$$

We have now obtained a homeomorphism $g_1 : \boldsymbol{Q} \to \boldsymbol{Q}$ such that $g_1(W) \subset \{1\} \times (-1, 1)^{\mathbb{N}\setminus\{1\}}$. Then, applying Lemma 2.10.2, we can obtain a homeomorphism $g_2 : \boldsymbol{Q} \to \boldsymbol{Q}^2$ such that

$$g_2 g_1(W) \subset \{(1, 0, 0, \dots)\} \times I(\boldsymbol{Q}).$$

By the homogeneity of $\boldsymbol{Q}$ (Theorem 2.1.2), there is a homeomorphism $g_3 : \boldsymbol{Q} \to \boldsymbol{Q}$ such that $g_3(1, 0, 0, \dots) = \mathbf{0}$. Take a homeomorphism $\psi : \boldsymbol{Q}^2 \to \boldsymbol{Q}$ with

$\psi(I(\boldsymbol{Q})^2) = I(\boldsymbol{Q})$ and define a homeomorphism

$$h = \psi(g_3 \times \mathrm{id})g_2g_1 : \boldsymbol{Q} \to \boldsymbol{Q}.$$

Then, $h(W) \subset I(\boldsymbol{Q})$. □

Theorem 2.10.6 *For a closed set A in $\boldsymbol{Q}$, the following are equivalent:*

(a) *A is a Z-set in $\boldsymbol{Q}$;*
(b) *A is $\boldsymbol{Q}$-deficient in $\boldsymbol{Q}$;*
(c) *There exists a homeomorphism $h : \boldsymbol{Q} \to \boldsymbol{Q}$ such that $h(A) \subset I(\boldsymbol{Q})$;*
(d) *For each $\varepsilon > 0$, there exists a homeomorphism $h : \boldsymbol{Q} \to \boldsymbol{Q}$ such that $h(A) \subset I(\boldsymbol{Q})$ and $d(h, \mathrm{id}) < \varepsilon$.*

Proof It is easy to see that $\{0\}$ is a Z-set in $\boldsymbol{Q}$, hence $\boldsymbol{Q} \times \{0\}$ is a Z-set in $\boldsymbol{Q} \times \boldsymbol{Q}$. Then, we have the implication (b) ⇒ (a). The implication (c) ⇒ (b) follows from Lemma 2.10.2 and the implication (d) ⇒ (c) is trivial.

(a) ⇒ (d): For each $n \in \mathbb{N}$, let

$$W_{2n-1} = \mathrm{pr}_n^{-1}(1) \ \text{ and } \ W_{2n} = \mathrm{pr}_n^{-1}(-1).$$

We consider the following subsets of Homeo($\boldsymbol{Q}$):

$$H_n = \left\{h \in \mathrm{Homeo}(\boldsymbol{Q}) \;\middle|\; h(A) \cap W_n = \emptyset\right\}.$$

Then, each H_n is open in Homeo($\boldsymbol{Q}$). We shall show that each H_n is dense in Homeo($\boldsymbol{Q}$). Then, $\bigcap_{n\in\mathbb{N}} H_n$ is dense in Homeo($\boldsymbol{Q}$) by the Baire Category Theorem 1.3.15, which gives condition (d).

To see that H_n is dense in Homeo($\boldsymbol{Q}$), let $f \in$ Homeo($\boldsymbol{Q}$) and $\varepsilon > 0$. By Lemma 2.10.5, we have a homeomorphism $\varphi \in$ Homeo($\boldsymbol{Q}$) such that $\varphi(W_n) \subset I(\boldsymbol{Q})$. By the uniform continuity of φ^{-1}, there is some $\delta > 0$ such that $d(x, y) < \delta$ implies $d(\varphi^{-1}(x), \varphi^{-1}(y)) < \varepsilon$. Since $\varphi f(A)$ is a Z-set in $\boldsymbol{Q}$ and $I(\boldsymbol{Q})$ is homotopy dense in $\boldsymbol{Q}$, we have a map $g' : \varphi(W_n) \to I(\boldsymbol{Q}) \setminus \varphi f(A)$ such that $d(g', \mathrm{id}) < \delta/2$. Choose $m \in \mathbb{N}$ so that

$$2^{-m+1} < \min\left\{\delta/2,\ \mathrm{dist}(g'\varphi(W_n), \varphi f(A))\right\}$$

and take an embedding

$$k : \varphi(W_n) \to (-1, 1)^{\mathbb{N}\setminus\{1,\ldots,m\}}.$$

Then, we can define an embedding $g : \varphi(W_n) \to I(\boldsymbol{Q}) \setminus \varphi f(A)$ by

$$g(x) = (\mathrm{pr}_1 g'(x), \ldots, \mathrm{pr}_m g'(x), k(x)).$$

Since $d(g, g') \leqslant 2^{-m+1} < \delta/2$, it follows that $d(g, \mathrm{id}) < \delta$. By Theorem 2.10.4, g can be extended to a homeomorphism $\tilde{g} : \boldsymbol{Q} \to \boldsymbol{Q}$ such that $d(\tilde{g}, \mathrm{id}) < \delta$, which implies

$$d(\varphi^{-1}\tilde{g}^{-1}\varphi f, f) = d(\varphi^{-1}, \varphi^{-1}\tilde{g}) < \varepsilon.$$

Since $\tilde{g}\varphi(W_n) \cap \varphi f(A) = \emptyset$, we have $W_n \cap \varphi^{-1}\tilde{g}^{-1}\varphi f(A) = \emptyset$, which means $\varphi^{-1}\tilde{g}^{-1}\varphi f \in H_n$. Hence, H_n is dense in $\mathrm{Homeo}(\boldsymbol{Q})$. This completes the proof. □

The following is a combination of Theorems 2.10.6 and 2.10.4:

Corollary 2.10.7 *Every homeomorphism $h : A \to B$ between Z-sets in $\boldsymbol{Q}$ can be extended to a homeomorphism $\tilde{h} : \boldsymbol{Q} \to \boldsymbol{Q}$.* □

Corollary 2.10.8 *Every homeomorphism $h : A \to B$ between Z-sets in $\boldsymbol{Q} \setminus \{\mathbf{0}\}$ can be extended to a homeomorphism $\tilde{h} : \boldsymbol{Q} \setminus \{\mathbf{0}\} \to \boldsymbol{Q} \setminus \{\mathbf{0}\}$.*

Proof Since $A \cup \{\mathbf{0}\}$ and $B \cup \{\mathbf{0}\}$ are closed in $\boldsymbol{Q}$ and $\{\mathbf{0}\}$ is a Z-set in $\boldsymbol{Q}$, it is easy to see that $A \cup \{\mathbf{0}\}$ and $B \cup \{\mathbf{0}\}$ are Z-sets in $\boldsymbol{Q}$. As is easily observed, h can be extended to the homeomorphism $h' : A \cup \{\mathbf{0}\} \to B \cup \{\mathbf{0}\}$ by $h'(\mathbf{0}) = \mathbf{0}$. By Corollary 2.10.7, we can extend h' to a homeomorphism $\bar{h} : \boldsymbol{Q} \to \boldsymbol{Q}$. The desired $\tilde{h}$ can be obtained as the restriction of $\bar{h}$. □

Recall that every Z-set in a $\boldsymbol{Q}$-manifold is a strong Z-set by Proposition 2.8.1. The following is the $\boldsymbol{Q}$-manifold version of Theorem 2.9.3:

Theorem 2.10.9 *For a closed set A in a $\boldsymbol{Q}$-manifold M, the following are equivalent:*

(a) *A is a Z-set in M;*
(b) *A is $\boldsymbol{Q}$-deficient in M;*
(c) *There is a homeomorphism $h : M \to M \times \mathbf{I}$ such that $h(x) = (x, 0)$ for each $x \in A$;*
(d) *A is contained in some collared set in M.*

Proof The implications (c) ⇒ (d) ⇒ (a) are easy. We shall show the implications (a) ⇒ (b) ⇒ (c).

(a) ⇒ (b): Each $a \in A$ has an open neighborhood U in M with an open embedding $\varphi : U \to \boldsymbol{Q}$. Take another open neighborhood V of a in M so that $\mathrm{cl}\, V$ is compact and $\mathrm{cl}\, V \subset U$. By Theorem 2.8.7(2), $\varphi(A \cap U)$ is a Z-set in $\varphi(U)$, hence $\varphi(A \cap \mathrm{cl}\, V)$ is also a Z-set in $\varphi(U)$. Since $\varphi(A \cap \mathrm{cl}\, V)$ is compact, it is a Z-set in $\boldsymbol{Q}$ by Theorem 2.8.7(3). It follows from Theorem 2.10.6 that $\varphi(A \cap \mathrm{cl}\, V)$ is $\boldsymbol{Q}$-deficient in $\boldsymbol{Q}$. By Corollary 2.3.10(2), $\varphi(A \cap V) = \varphi(A \cap \mathrm{cl}\, V) \cap \varphi(V)$ is $\boldsymbol{Q}$-deficient in $\varphi(V)$. Hence, $A \cap V$ is $\boldsymbol{Q}$-deficient in V. This means that A is locally $\boldsymbol{Q}$-deficient in M. It follows from Proposition 2.3.14 that A is $\boldsymbol{Q}$-deficient in M.

(b) ⇒ (c): Let $\varphi : M \to M \times \boldsymbol{Q}$ be a homeomorphism such that $\varphi(x) = (x, \mathbf{0})$ for each $x \in A$ (Corollary 2.3.10(4)). On the other hand, $\boldsymbol{Q} \times \mathbf{I} \approx \boldsymbol{Q}$ is homogeneous (Theorem 2.1.2), hence we have a homeomorphism $\psi : \boldsymbol{Q} \to \boldsymbol{Q} \times \mathbf{I}$ such that

$\psi(\mathbf{0}) = (\mathbf{0}, 0)$. The following is the desired homeomorphism:

$$h = (\varphi^{-1} \times \mathrm{id}_{\mathbf{I}})(\mathrm{id}_M \times \psi)\varphi : M \to M \times \mathbf{I}. \qquad \square$$

Using Theorem 2.10.9, the Strong Universality Theorem 2.3.16 and the Closed Embedding Approximation Theorem 2.3.17 can be rewritten as follows:

Theorem 2.10.10 (STRONG UNIVERSALITY) *Let M be a $\boldsymbol{Q}$-manifold and $A \in \mathfrak{M}_0$ with a closed set $B \subset A$ and a map $f : A \to M$ such that $f|B$ is a Z-embedding. Then, for each $\mathcal{U} \in \mathrm{cov}(M)$, there exists a Z-embedding $h : A \to M$ such that $h|B = f|B$ and h is $\mathcal{U}$-homotopic to f.* □

Theorem 2.10.11 (Z-EMBEDDING APPROXIMATION) *Let M be a $\boldsymbol{Q}$-manifold and A be a locally compact separable metrizable space with a closed set $B \subset A$ and a proper map $f : A \to M$ such that $f|B$ is a Z-embedding. Then, for each $\mathcal{U} \in \mathrm{cov}(M)$, there exists a Z-embedding $h : A \to M$ such that $h|B = f|B$ and $h \simeq^p_{\mathcal{U}} f$ (i.e., h is properly $\mathcal{U}$-homotopic to f).* □

2.11 Complementary Basic Results on $\boldsymbol{Q}$-Manifolds

In this section, we establish the $\boldsymbol{Q}$-manifold versions of the Collaring Theorem 2.9.5 and the Unknotting Theorem 2.9.7. We also show the Classification and Open Embedding Theorems for $\mathbb{R}_+$-stable $\boldsymbol{Q}$-manifolds, where a $\boldsymbol{Q}$-manifold M is said to be **$\mathbb{R}_+$-stable** if $M \times \mathbb{R}_+ \approx M$. Moreover, we give a characterization of $\mathbb{R}_+$-stability for $\boldsymbol{Q}$-manifolds. The Classification Theorem for all $\boldsymbol{Q}$-manifolds will be shown in the next chapter.

First of all, we shall prove the following Collaring Theorem:

Theorem 2.11.1 (COLLARING) *Let M and N be $\boldsymbol{Q}$-manifolds such that N is closed in M. Then, N is a Z-submanifold of M if and only if N is collared in M.*

Proof The "if" part is trivial. We show the "only if" part. Since a locally collared set is collared (Theorem 2.5.7), it suffices to show that N is locally collared in M. Each $x \in N$ has an open neighborhood U in M with an open embedding $\varphi : U \to \boldsymbol{Q}$. Let $W = \mathrm{pr}_1^{-1}(1) \subset \boldsymbol{Q}$, where $\mathrm{pr}_1 : \boldsymbol{Q} \to [-1, 1]$ is the projection onto the first factor. We can easily find an open neighborhood V of x in N and an embedding $\psi : \mathrm{cl}\, V \to W$ such that $\mathrm{cl}\, V$ is compact, $\mathrm{cl}\, V \subset U$, and $\psi(V)$ is open in W. By Theorem 2.8.7(2), $\mathrm{cl}\, V$ is a Z-set in U, hence $\varphi(\mathrm{cl}\, V)$ is also a Z-set in $\varphi(U)$. Since $\varphi(\mathrm{cl}\, V)$ is compact, it is a Z-set in $\boldsymbol{Q}$ by Theorem 2.8.7(3). On the other hand, since W is a Z-set in $\boldsymbol{Q}$, $\psi(\mathrm{cl}\, V)$ is also a Z-set in $\boldsymbol{Q}$. Applying Corollary 2.10.7, we can obtain a homeomorphism $h : \boldsymbol{Q} \to \boldsymbol{Q}$ such that $h\varphi|\,\mathrm{cl}\, V = \psi$. Then, $h\varphi(U)$ is an open set in $\boldsymbol{Q}$ and $\psi(V) = h\varphi(V)$ is open in $h\varphi(U) \cap W$. Hence, $h\varphi(V)$ is collared in $h\varphi(U)$ (cf. (G-1) in the proof of Theorem 2.5.7), which implies that V is collared in U. Thus, N is locally collared in M. □

Due to the above theorem, any Z-submanifold of each $\boldsymbol{Q}$-manifold M can be regarded as a boundary of M, but it is not unique (cf. Remark 2.19, p. 182).

Note that the Open Embedding Theorem does not hold for $\boldsymbol{Q}$-manifolds, because any compact $\boldsymbol{Q}$-manifold cannot be embedded into $\boldsymbol{Q}$ as an open set unless it is homeomorphic to $\boldsymbol{Q}$. However, combining the Collaring Theorem 2.11.1 with Theorem 2.10.11, we can see that every compact $\boldsymbol{Q}$-manifold M can be embedded into $\boldsymbol{Q}$ as a collared closed set. Since $M \times \mathbf{I} \approx M$ by Theorem 2.3.13, we can prove the following $\boldsymbol{Q}$-manifold version of Theorem 2.5.11 in the same way:

Theorem 2.11.2 *Every compact $\boldsymbol{Q}$-manifold M can be embedded in $\boldsymbol{Q}$ as a closed set having a $\boldsymbol{Q}$-manifold bi-collared boundary.* □

For a non-compact $\boldsymbol{Q}$-manifold M, let $\alpha M = M \cup \{\infty\}$ be the one-point compactification of M. Since αM is compact metrizable, we have a Z-embedding $h : \alpha M \to \boldsymbol{Q}$. Then, $h(M)$ is a Z-submanifold of $\boldsymbol{Q} \setminus \{h(\infty)\}$, hence it is collared in $\boldsymbol{Q} \setminus \{h(\infty)\}$ by the Collaring Theorem 2.11.1. By the homogeneity of $\boldsymbol{Q}$ (Theorem 2.1.2), $\boldsymbol{Q} \setminus \{h(\infty)\} \approx \boldsymbol{Q} \setminus \{\mathbf{0}\}$. The following can be obtained similarly:

Theorem 2.11.3 *Every $\boldsymbol{Q}$-manifold M can be embedded in $\boldsymbol{Q} \setminus \{\mathbf{0}\}$ as a closed set whose boundary is bi-collared and also a $\boldsymbol{Q}$-manifold.* □

By the above argument, the following $\boldsymbol{Q}$-manifold version of the Open Embedding Theorem 2.5.10 can also be obtained:

Theorem 2.11.4 (Open Embedding) *For every $\boldsymbol{Q}$-manifold M, $M \times [0, 1)$ can be embedded in $\boldsymbol{Q}$ as an open set.* □

In the Open Embedding Theorem 2.11.4, an open embedding of $M \times [0, 1)$ into $\boldsymbol{Q}$ is defined as the composition of the homeomorphism $(h \times \mathrm{id})|M \times [0, 1)$ and a collar $k : h(M) \times [0, 1) \to \boldsymbol{Q} \setminus \{\mathbf{0}\}$ of $h(M)$, where $h(\infty) = \mathbf{0}$. Consider the following:

$$M \approx M \times \mathbf{I} \quad \supset \quad M \times [0, 1) \underset{\approx}{\xrightarrow{h \times \mathrm{id}}} h(M) \times [0, 1) \xrightarrow{k} \boldsymbol{Q} \setminus \{\mathbf{0}\}.$$

For each Z-set A in M, $h(A)$ is a Z-set in $\boldsymbol{Q} \setminus \{\mathbf{0}\}$. On the other hand, by Theorem 2.10.9, we have a homeomorphism $f : M \to M \times \mathbf{I}$ such that $f(x) = (x, 0)$ for each $x \in A$. Then, $U = f^{-1}(M \times [0, 1))$ is a homotopy dense open set in M with $A \subset U$ and $k(h \times \mathrm{id})f|U : U \to \boldsymbol{Q} \setminus \{\mathbf{0}\}$ is an open embedding. Thus, we have the following:

Corollary 2.11.5 *Let M be a $\boldsymbol{Q}$-manifold and A a Z-set in M. Then, there is an open dense set U in M with an open embedding $g : U \to \boldsymbol{Q} \setminus \{\mathbf{0}\}$ such that $A \subset U$ and $g(A)$ is a Z-set in $\boldsymbol{Q} \setminus \{\mathbf{0}\}$.* □

Modifying the proof of the Unknotting Theorem 2.5.12, we can prove the following Z-Set Unknotting Theorem for $\boldsymbol{Q}$-manifolds:

Theorem 2.11.6 (Z-SET UNKNOTTING) *Let M be a $\boldsymbol{Q}$-manifold and $h : A \to B$ a homeomorphism between Z-sets in M. If h is properly homotopic to* id *in M, then h can be extended to a homeomorphism $\tilde{h} : M \to M$. Moreover, if $f : A \times \mathbf{I} \to M$ is a proper homotopy with $f_0 =$* id *and $f_1 = h$ and $\mathcal{U}$ is an open cover of $f(A \times \mathbf{I})$ in M, then $\tilde{h}$ can be chosen so as to be ambiently invertibly $\mathcal{U}^*$-isotopic to* id, *where $\mathcal{U}^* = \{\mathrm{st}(f(\{x\} \times \mathbf{I}), \mathcal{U}) \mid x \in A\}$.*

Proof Take $\mathcal{V} \in \mathrm{cov}(M)$ such that $\mathrm{st}^3 \mathcal{V} \prec \mathcal{U}$, $\mathcal{V}$ is locally finite, and $\mathrm{cl}\, V$ is compact for each $V \in \mathcal{V}$. Observe that $\mathcal{V}[\mathrm{cl}\, V]$ is finite for each $V \in \mathcal{V}$. Then, it follows that $\mathrm{st}^i \mathcal{V}$, $i = 1, 2, 3$, are also locally finite and the closures of these members are compact.

First, we shall construct an ambient invertible $\mathrm{st}\,\mathcal{V}$-isotopy $\psi : M \times \mathbf{I} \to M$ such that $A \cap \psi_1(B) = \emptyset$. Since $A \cup B$ is a Z-set in M, it is $\boldsymbol{Q}$-deficient in M. By Corollary 2.3.10(4), we have a homeomorphism $k : M \times \boldsymbol{Q} \to M$ such that $k(x, \mathbf{0}) = x$ for each $x \in A \cup B$ and k is $\mathcal{V}$-close to pr_M. We define an ambient invertible isotopy $\zeta : \boldsymbol{Q} \times \mathbf{I} \to \boldsymbol{Q}$ as follows:

$$\zeta_t(x) = \begin{cases} ((1 - t/2)x(1) + t/2, x(2), x(3), \dots) & \text{if } x(1) \geqslant 0, \\ ((1 + t/2)x(1) + t/2, x(2), x(3), \dots) & \text{if } x(1) \leqslant 0. \end{cases}$$

Then, $\zeta_0 = \mathrm{id}$ and $\zeta_1(0) = (1/2, 0, 0, \dots)$. The desired isotopy $\psi : M \times \mathbf{I} \to M$ can be defined by $\psi_t = k(\mathrm{id} \times \zeta_t)k^{-1}$. In the same way as in the proof of Theorem 2.5.12, we have a proper homotopy $f' : A \times \mathbf{I} \to M$ from id to $\psi_1 h$ (cf. Proposition 1.3.9). Then, by Theorem 2.10.11, we have a Z-embedding $f'' : A \times \mathbf{I} \to M$ such that $f''|A \times \{0, 1\} = f'|A \times \{0, 1\}$ and f'' is $\mathcal{V}$-close to f'.

On the other hand, we have a Z-embedding $j : A \to \boldsymbol{Q} \setminus \{\mathbf{0}\}$ by Corollary 2.11.5. Then, $j(A) \times \mathbf{I}$ is a Z-set in $(\boldsymbol{Q} \setminus \{\mathbf{0}\}) \times [-1, 2]$ and $(\boldsymbol{Q} \setminus \{\mathbf{0}\}) \times [-1, 2] \approx \boldsymbol{Q} \setminus \{\mathbf{0}\}$ by the Stability Theorem 2.3.13. Using Corollary 2.11.5, we have an open neighborhood U of $f''(A \times \mathbf{I})$ in M with an open embedding $g' : U \to (\boldsymbol{Q} \setminus \{\mathbf{0}\}) \times [-1, 2]$ such that $g' f''(A \times \mathbf{I})$ is a Z-set in $(\boldsymbol{Q} \setminus \{\mathbf{0}\}) \times [-1, 2]$. Using Corollary 2.10.8, we can obtain a homeomorphism

$$g'' : (\boldsymbol{Q} \setminus \{\mathbf{0}\}) \times [-1, 2] \to (\boldsymbol{Q} \setminus \{\mathbf{0}\}) \times [-1, 2]$$

such that $g'' g' f''(x, t) = (j(x), t)$ for each $(x, t) \in A \times \mathbf{I}$. Then, note that

$$\{j(x)\} \times \mathbf{I} = g'' g' f''(\{x\} \times \mathbf{I}) \subset g'' g'(\mathrm{st}(f'(\{x\} \times \mathbf{I}), \mathcal{V})).$$

Let W_0 be an open neighborhood of $g'' g' f''(A \times \mathbf{I})$ in $(\boldsymbol{Q} \setminus \{\mathbf{0}\}) \times [-1, 2]$ such that $\mathrm{cl}\, W_0 \subset g'' g'(U)$. For each $x \in A$, define

$$W(x) = W_0 \cap g'' g'(\mathrm{st}(f'(\{x\} \times \mathbf{I}), \mathcal{V}))$$

and let $\mathcal{W} = \{W(x) \mid x \in A\}$. In the same way as the proof of Theorem 2.5.12, we can construct an ambient invertible $\mathcal{W}$-isotopy

$$\varphi' : (\boldsymbol{Q} \setminus \{\boldsymbol{0}\}) \times [-1, 2] \times \mathbf{I} \to (\boldsymbol{Q} \setminus \{\boldsymbol{0}\}) \times [-1, 2]$$

such that $\varphi'_0 = \mathrm{id}$, $\varphi'(\{x\} \times \mathbf{I} \times \mathbf{I}) \subset W(x)$, and $\varphi'_1(j(x), 0) = (j(x), 1)$ for every $x \in A$, where the last condition means that $\varphi'_1 g'' g'|A = g'' g' \psi_1 h$. Since

$$\varphi'_t|(\boldsymbol{Q} \setminus \{\boldsymbol{0}\}) \times [-1, 2] \setminus W_0 = \mathrm{id},$$

we can define an ambient invertible $(g''g')^{-1}(\mathcal{W})$-isotopy $\varphi : M \times \mathbf{I} \to M$ by

$$\varphi_t|U = (g''g')^{-1}\varphi'_t g'' g' \ \text{ and } \ \varphi_t|M \setminus U = \mathrm{id}.$$

Then, $\varphi_0 = \mathrm{id}$ and $\varphi_1|A = (g''g')^{-1}\varphi'_1 g'' g'|A = \psi_1 h$. Hence, h can be extended to a homeomorphism $\tilde{h} = \psi_1^{-1}\varphi_1 : M \to M$. In the same way as for Theorem 2.5.12, we can define an ambient invertible $\mathcal{U}^*$-isotopy η from id to $\tilde{h} = \psi_1^{-1}\varphi_1$. Thus, we have the result. □

Remark 2.21 In the above, the properness of the homotopy is necessary. In fact, $A = \mathbb{R} \times \{0, 1\} \times \boldsymbol{Q}$ is a Z-set in the $\boldsymbol{Q}$-manifold $M = \mathbb{R} \times \mathbf{I} \times \boldsymbol{Q}$. We define a homeomorphism $h : A \to A$ by $h(t, i, x) = ((-1)^i t, i, x)$ for each $(t, i, x) \in A = \mathbb{R} \times \{0, 1\} \times \boldsymbol{Q}$. Then, $h \simeq \mathrm{id}$ in $M = \mathbb{R} \times \mathbf{I} \times \boldsymbol{Q}$, but h cannot be extended to a homeomorphism of M onto itself. Indeed, assume that h is extended to a homeomorphism $\tilde{h} : M \to M$. Let $U_+ = (0, \infty) \times \mathbf{I} \times \boldsymbol{Q}$, $U_- = (-\infty, 0) \times \mathbf{I} \times \boldsymbol{Q}$ and $L = \{0\} \times \mathbf{I} \times \boldsymbol{Q}$. Choose $\beta > 0$ so that $\beta \notin \mathrm{pr}_{\mathbb{R}}\tilde{h}(L)$. Then, $\{\beta\} \times \mathbf{I} \times \boldsymbol{Q} \subset \tilde{h}(U_+) \cup \tilde{h}(U_-)$, $(\beta, 0, \boldsymbol{0}) = h(\beta, 0, \boldsymbol{0}) \in \tilde{h}(U_+)$ and $(\beta, 1, \boldsymbol{0}) = h(-\beta, 1, \boldsymbol{0}) \in \tilde{h}(U_-)$, which contradicts the connectedness of $\{\beta\} \times \mathbf{I} \times \boldsymbol{Q}$.

Suppose that $f : M \to N$ is a proper $\mathcal{U}$-homotopy equivalence between $\boldsymbol{Q}$-manifolds, where $\mathcal{U} \in \mathrm{cov}(N)$. As observed in Sect. 2.6, Remark 2.9, we can apply the same construction as in the proof of Theorem 2.6.3 to $\boldsymbol{Q}$-manifolds, where Theorem 2.4.9, the Strong Universality Theorem 2.3.15, the Collaring Theorem 2.5.9, and the Unknotting Theorem 2.5.12 are replaced by Theorem 2.10.9, the Closed Embedding Approximation 2.3.17, the Collaring Theorem 2.11.1, and the Z-set-Unknotting Theorem 2.11.6, respectively. Thus, we can obtain the homeomorphism

$$h = \psi^{-1}\varphi : M \times \mathbb{R}_+ \to N \times \mathbb{R}_+$$

such that $\mathrm{pr}_N h$ is $\mathrm{st}(\mathrm{st}^2\,\mathcal{U}, \mathcal{V})$-close to $f\mathrm{pr}_M$, where $\mathcal{V}$ is an open star-refinement of $\mathcal{U}$. Thus, we have the following:

Proposition 2.11.7 *Let M and N be $\boldsymbol{Q}$-manifolds and $\mathcal{U}$ be an open cover of N. For each proper $\mathcal{U}$-homotopy equivalence $f : M \to N$, there exists a* *homeomorphism* $h : M \times \mathbb{R}_+ \to N \times \mathbb{R}_+$ *such that* $\mathrm{pr}_N h$ *is* $\mathrm{st}^5\,\mathcal{U}$*-homotopic to* $f\mathrm{pr}_M$ *and* $h(M \times \{0\}) \subset N \times \{0\}$. □

By the way, we have the following:

Proposition 2.11.8 *Let X and Y be connected locally compact spaces. If* $f : X \to Y$ *is a homotopy equivalence, then there is a proper homotopy equivalence* $f_* : X \times \mathbb{R}_+ \to Y \times \mathbb{R}_+$ *such that* $\mathrm{pr}_Y f_* = f\mathrm{pr}_X$. *In particular,*

$$X \simeq Y \ \Rightarrow X \times \mathbb{R}_+ \simeq^p Y \times \mathbb{R}_+.$$

Proof Let $g : Y \to X$ be a homotopy inverse of f, that is, $gf \simeq \mathrm{id}$ and $fg \simeq \mathrm{id}$. Let $\varphi : X \times \mathbf{I} \to X$ and $\psi : Y \times \mathbf{I} \to Y$ be homotopies such that $\varphi_0 = \mathrm{id}$, $\varphi_1 = gf$, $\psi_0 = \mathrm{id}$, and $\psi_1 = fg$. Since X and Y be connected locally compact, we can write $X = \bigcup_{n\in\mathbb{N}} X_n$ and $Y = \bigcup_{n\in\mathbb{N}} Y_n$, where each X_n and Y_n are compact and contained in $\mathrm{int}\, X_{n+1}$ and $\mathrm{int}\, Y_{n+1}$, respectively. Using Urysohn maps, we can construct maps $\alpha : X \to \mathbb{R}_+$ and $\beta : Y \to \mathbb{R}_+$ so that

$$\alpha(\mathrm{bd}\, X_n) = n,\ \ \alpha(X_n \setminus \mathrm{int}\, X_{n-1}) = [n-1, n],$$
$$\beta(\mathrm{bd}\, Y_n) = n,\ \ \beta(Y_n \setminus \mathrm{int}\, Y_{n-1}) = [n-1, n].$$

We define maps $f_* : X \times \mathbb{R}_+ \to Y \times \mathbb{R}_+$ and $g_* : Y \times \mathbb{R}_+ \to X \times \mathbb{R}_+$ as follows:

$$f_*(x, t) = (f(x), \max\{t, \alpha(x)\}),\ \ g_*(y, t) = (g(x), \max\{t, \beta(x)\}).$$

Then, $\mathrm{pr}_Y f_* = f\mathrm{pr}_X$. Each compact set B in $Y \times \mathbb{R}_+$ is contained in some $Y_n \times [0, n]$. If $x \notin X_{n+1}$ or $t \notin [0, n+1]$, then $\max\{t, \alpha(x)\} \geqslant n+1$, which implies that $f_*(x, t) \notin Y_n \times [0, n]$. Hence,

$$f_*^{-1}(B) \subset f_*^{-1}(Y_n \times [0, n]) \subset X_{n+1} \times [0, n+1],$$

which implies that $f_*^{-1}(B)$ is compact. Thus, f_* is proper. Similarly, it can be shown that g_* is proper.

To show that $g_* f_* \simeq^p \mathrm{id}$ and $f_* g_* \simeq^p \mathrm{id}$, we define a homotopies $\tilde{\varphi} : X \times \mathbb{R}_+ \times \mathbf{I} \to X \times \mathbb{R}_+$ and $\tilde{\psi} : Y \times \mathbb{R}_+ \times \mathbf{I} \to Y \times \mathbb{R}_+$ as follows:

$$\tilde{\varphi}(x, t, s) = \begin{cases} (\varphi(x, 2s-1), \max\{t, \alpha(x), \beta(f(x))\}) & \text{if } 1/2 \leqslant s \leqslant 1, \\ (x, (1-2s)t + 2s\max\{t, \alpha(x), \beta(f(x))\}) & \text{if } 0 \leqslant s \leqslant 1/2, \end{cases}$$

$$\tilde{\psi}(y, t, s) = \begin{cases} (\psi(y, 2s-1), \max\{t, \beta(y), \alpha(g(y))\}) & \text{if } 1/2 \leqslant s \leqslant 1, \\ (y, (1-2s)t + 2s\max\{t, \beta(y), \alpha(g(y))\}) & \text{if } 0 \leqslant s \leqslant 1/2. \end{cases}$$

Then, $\tilde{\varphi}_0 = \mathrm{id}$ and $\tilde{\psi}_0 = \mathrm{id}$. For each $(x, t) \in X \times \mathbb{R}_+$,

$$\tilde{\varphi}_1(x, t) = (\varphi_1(x), \max\{t, \alpha(x), \beta(f(x))\}) = g_* f_*(x, t).$$

Hence, $\tilde{\varphi}_1 = g_* f_*$. Similarly, $\tilde{\psi}_1 = f_* g_*$. To see that $\tilde{\varphi}$ is proper, let A be a compact set in $X \times \mathbb{R}_+$. Then, $A \subset X_n \times [0, n]$ for some $n \in \mathbb{N}$. If $x \notin X_{n+1}$ or $t \notin [0, n+1]$, then

$$\max\{t, \alpha(x), \beta(f(x))\} \geqslant \max\{t, \alpha(x)\} \geqslant n + 1,$$

which implies that $\tilde{\varphi}(x, t, s) \notin X_n \times [0, n]$ for each $1/2 \leqslant s \leqslant 1$. Hence,

$$\tilde{\varphi}_s^{-1}(X_n \times [0, n]) \subset X_{n+1} \times [0, n+1] \text{ for each } 1/2 \leqslant s \leqslant 1.$$

If $x \notin X_n$ or $t \notin [0, n]$, then for each $0 \leqslant s \leqslant 1/2$,

$$\tilde{\varphi}(x, t, s) = (x, (1 - 2s)t + 2s \max\{t, \alpha(x), \beta(f(x))\}) \notin X_n \times [0, n].$$

Hence, it follows that

$$\tilde{\varphi}_s^{-1}(X_n \times [0, n]) \subset X_n \times [0, n] \text{ for each } 0 \leqslant s \leqslant 1/2.$$

Thus, we have $\tilde{\varphi}^{-1}(A) \subset X_{n+1} \times [0, n + 1] \times \mathbf{I}$, which implies that $\tilde{\varphi}^{-1}(A)$ is compact. Similarly, we can show that $\tilde{\psi}$ is proper. Consequently, $g_* f_* \simeq^p \mathrm{id}$ and $f_* g_* \simeq^p \mathrm{id}$. □

For every $\boldsymbol{Q}$-manifold M,

$$M \times \mathbb{R}_+ \times \mathbb{R}_+ \approx M \times \mathbf{I} \times \mathbb{R}_+ \approx M \times \mathbb{R}_+,$$

where $\mathbb{R}_+ \approx [0, 1)$ and $[0, 1)^2 \approx [0, 1) \times \mathbf{I}$ by Lemma 2.1.4. Combining Propositions 2.11.7 and 2.11.8, we have:

Theorem 2.11.9 *Let M and N be $\boldsymbol{Q}$-manifolds. For each homotopy equivalence $f : M \to N$, there exists a homeomorphism $h : M \times \mathbb{R}_+ \to N \times \mathbb{R}_+$ such that $\mathrm{pr}_N h \simeq f \mathrm{pr}_M$.* □

Recall that a space X is $\mathbb{R}_+$-stable if $X \times \mathbb{R}_+ \approx X$. As we saw above, for every $\boldsymbol{Q}$-manifold M, $M \times \mathbb{R}_+$ is an $\mathbb{R}_+$-stable $\boldsymbol{Q}$-manifold.

Lemma 2.11.10 *For each $\mathbb{R}_+$-stable $\boldsymbol{Q}$-manifold M, there is a homeomorphism $h : M \times \mathbb{R}_+ \to M$ that is homotopic to the projection.*

Proof In the following, we replace $\mathbb{R}_+$ with $[0, 1)$. Let $\varphi : M \to M \times [0, 1)$ be a homeomorphism. Since $M \times [0, 1)$ is a $\boldsymbol{Q}$-manifold and $\boldsymbol{Q} \times \mathbf{I} \approx \boldsymbol{Q}$, we apply the Stability Theorem 2.3.13 to obtain a homeomorphism

$$f : M \times [0, 1) \times \mathbf{I} \to M \times [0, 1)$$

that is homotopic to the projection. On the other hand, as is observed above, we have a homeomorphism $\psi : [0, 1)^2 \to [0, 1) \times \mathbf{I}$ by Lemma 2.1.4. Then, we can define a homeomorphism $h : M \times [0, 1) \to M$ so as to make the following diagram commutative:

$$\begin{array}{ccc} M \times [0,1) & \xrightarrow{\varphi \times \mathrm{id}_{[0,1)}} & M \times [0,1) \times [0,1) \\ & & \downarrow \mathrm{id}_M \times \psi \\ \downarrow h & & M \times [0,1) \times \mathbf{I} \\ & & \downarrow f \\ M & \xrightarrow[\varphi]{} & M \times [0,1) \end{array}$$

Since $[0, 1)$ is contractible, it follows that $\mathrm{pr}_{[0,1)}\psi \simeq \mathrm{pr}_{[0,1)}$, hence the following diagram is commutative up to homotopy:

$$\begin{array}{ccc} M \times [0,1) \times [0,1) & & \\ \downarrow \mathrm{id}_M \times \psi & \searrow^{\mathrm{pr}_{M \times [0,1)}} & \\ M \times [0,1) \times \mathbf{I} & \xrightarrow[\mathrm{pr}_{M \times [0,1)}]{} & M \times [0,1) \end{array}$$

Note that the bottom $\mathrm{pr}_{M \times [0,1)}$ is homotopic to f. Thus, we have

$$\begin{aligned} h &= \varphi^{-1} \circ f \circ \mathrm{id}_M \times \psi \circ \varphi \times \mathrm{id}_{[0,1)} \\ &\simeq \varphi^{-1} \circ \mathrm{pr}_{M \times [0,1)} \circ \mathrm{id}_M \times \psi \circ \varphi \times \mathrm{id}_{[0,1)} \\ &\simeq \varphi^{-1} \circ \mathrm{pr}_{M \times [0,1)} \circ \varphi \times \mathrm{id}_{[0,1)} = \varphi^{-1} \circ \varphi \circ \mathrm{pr}_M = \mathrm{pr}_M. \end{aligned}$$ □

Combining Theorem 2.11.9 and Lemma 2.11.10, we can easily obtain the following Classification Theorem for $\mathbb{R}_+$-stable $\boldsymbol{Q}$-manifolds:

Theorem 2.11.11 (Classification) *Let M and N be $\mathbb{R}_+$-stable $\boldsymbol{Q}$-manifolds. Then, every homotopy equivalence $f : M \to N$ is homotopic to a homeomorphism. Thus, two $\mathbb{R}_+$-stable $\boldsymbol{Q}$-manifolds are homeomorphic if they have the same homotopy type.* □

By the Open Embedding Theorem 2.11.4, we have the following:

Theorem 2.11.12 (Open Embedding) *Every $\mathbb{R}_+$-stable $\boldsymbol{Q}$-manifold can be embedded in $\boldsymbol{Q}$ as an open set.* □

Due to the Classification Theorem 2.11.11 and the Open Embedding Theorem 2.11.12 above, $\mathbb{R}_+$-stable $\boldsymbol{Q}$-manifolds are tractable among $\boldsymbol{Q}$-manifolds. In order to characterize such $\boldsymbol{Q}$-manifolds, we introduce the notion of proper contractibility to infinity. A space X is said to be **properly contractible to infinity** provided, for each compact set C in X, there is a proper homotopy $h : X \times \mathbf{I} \to X$ with $h_0 = \mathrm{id}_X$ and $h_1(X) \subset X \setminus C$.

Theorem 2.11.13 *A connected $\boldsymbol{Q}$-manifold M is $\mathbb{R}_+$-stable if and only if M is properly contractible to infinity.*

Proof The "only if" part follows from the fact that $M \times \mathbb{R}_+$ is properly contractible to infinity. This fact can be shown as follows: For each compact set $C \subset M \times \mathbb{R}_+$, choose $a > 0$ so that $C \subset M \times [0, a)$. The desired proper homotopy $h : M \times \mathbb{R}_+ \times \mathbf{I} \to M \times \mathbb{R}_+$ can be defined by $h(x, s, t) = (x, s + at)$ for every $(x, s, t) \in M \times \mathbb{R}_+ \times \mathbf{I}$.

To see the "if" part, it suffices to prove that $M \times \mathbf{I} \approx M \times [0, 1)$, because $M \approx M \times \mathbf{I}$ by the Stability Theorem 2.3.13. Since M is locally compact and σ-compact, there are compact sets X_n, $n \in \mathbb{N}$, in M such that $X_n \subset \operatorname{int} X_{n+1}$ and $M = \bigcup_{n\in\mathbb{N}} X_n$. Take $0 < t_1 < t_2 < \cdots < 1$ so that $\sup_{n\in\mathbb{N}} t_n = 1$. We will inductively construct homeomorphisms $h_n : M \times \mathbf{I} \to M \times \mathbf{I}$, $n \in \mathbb{N}$, so as to satisfy the following:

(1) $h_n h_{n-1} \cdots h_1(M \times \{1\}) \subset (M \setminus X_n) \times \{1\}$,
(2) $h_n h_{n-1} \cdots h_1|M \times [0, t_n] = h_{n-1} \cdots h_1|M \times [0, t_n]$,
(3) $h_n|X_{n-1} \times \mathbf{I} = \mathrm{id}$.

In the following, keep in mind that every proper map of M into itself can be approximated by closed embeddings (the Z-Embedding Approximation Theorem 2.10.11). Since M is properly contractible to infinity, we have a closed embedding $f_1 : M \times \{1\} \to (M \setminus X_1) \times \{1\}$ that is properly homotopic to id. By Theorem 2.10.9, $M \times \{1\}$ is a Z-set in $M \times [t_1, 1]$. Applying the Z-Set Unknotting Theorem 2.11.6, we can extend f_1 to a homeomorphism $h_1 : M \times \mathbf{I} \to M \times \mathbf{I}$ such that $h_1|M \times [0, t_1] = \mathrm{id}$. Then, $h_1(M \times \{1\}) \subset (M \setminus X_1) \times \{1\}$.

Suppose that homeomorphisms $h_1, \ldots, h_{n-1}$ have been defined so as to satisfy (1), (2), and (3). As is easily observed,

$$(h_{n-1} \cdots h_1)^{-1}(X_{n-1} \times \mathbf{I}) \cap (M \times \{1\}) = \emptyset.$$

Choose $s \in (t_n, 1)$ so that

$$(h_{n-1} \cdots h_1)^{-1}(X_{n-1} \times \mathbf{I}) \subset M \times [0, s].$$

Since M is properly contractible to infinity, we have a closed embedding

$$f_n : M \times \{1\} \to (M \times \{1\}) \setminus (h_{n-1} \cdots h_1)^{-1}(X_n \times \mathbf{I})$$

that is properly homotopic to id. Applying the Z-Set Unknotting Theorem 2.11.6, we can extend f_n to a homeomorphism $g_n : M \times \mathbf{I} \to M \times \mathbf{I}$ such that $g_n|M \times [0, s] = \mathrm{id}$ (cf. Fig. 2.26). Then,

$$g_n|(h_{n-1} \cdots h_1)^{-1}(X_{n-1} \times \mathbf{I}) = \mathrm{id}. \qquad (*)$$

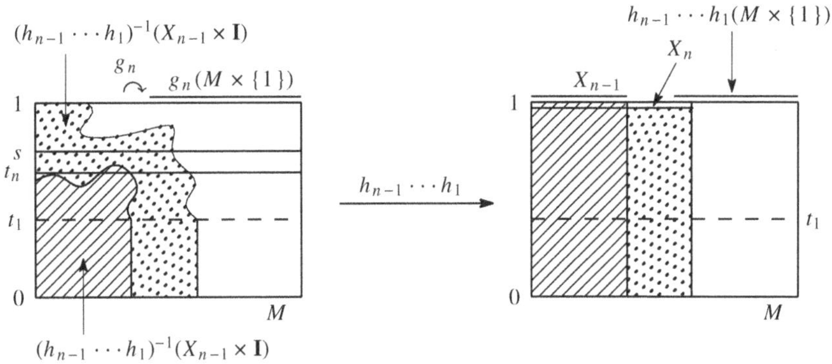

Fig. 2.26 Homeomorphisms $h_{n-1} \cdots h_1$ and g_n

Since g_n is an extension of f_n, we also have

$$(h_{n-1} \cdots h_1)g_n(M \times \{1\}) \subset (M \setminus X_n) \times \{1\}.$$

The following homeomorphism satisfies (1), (2), and (3):

$$h_n = (h_{n-1} \cdots h_1)g_n(h_{n-1} \cdots h_1)^{-1} : M \times \mathbf{I} \to M \times \mathbf{I}.$$

Indeed, this can be verified as follows:

$$h_n h_{n-1} \cdots h_1(M \times \{1\}) = (h_{n-1} \cdots h_1)g_n(M \times \{1\}) \subset (M \setminus X_n) \times \{1\}.$$

Since $g_n|M \times [0, s] = \mathrm{id}$ and $t_n < s$, we have

$$\begin{aligned} h_n h_{n-1} \cdots h_1|M \times [0, t_n] &= (h_{n-1} \cdots h_1)g_n|M \times [0, t_n] \\ &= h_{n-1} \cdots h_1|M \times [0, t_n]. \end{aligned}$$

Moreover, it follows from $(*)$ that

$$h_n|X_{n-1} \times \mathbf{I} = (h_{n-1} \cdots h_1)g_n(h_{n-1} \cdots h_1)^{-1}|X_{n-1} \times \mathbf{I} = \mathrm{id}.$$

Now, by induction, we have homeomorphisms h_n, $n \in \mathbb{N}$, satisfying (1), (2), and (3). Then, by virtue of (2), we can define $h : M \times [0, 1) \to M \times \mathbf{I}$ by

$$h(x) = \lim_{n \in \mathbb{N}} h_n h_{n-1} \cdots h_1(x) \quad \text{for each } (x, t) \in M \times [0, 1).$$

Since $M \times [0, 1) = \bigcup_{n \in \mathbb{N}} M \times [0, t_n)$ and

$$h|M \times [0, t_n) = h_{n-1} \cdots h_1|M \times [0, t_n)$$

is an open embedding into $M \times \mathbf{I}$ for each $n \in \mathbb{N}$, it is easy to see that h is continuous, open, and injective. For each $y \in M \times \mathbf{I} = \bigcup_{n \in \mathbb{N}} X_n \times \mathbf{I}$, choose $n \in \mathbb{N}$ so that $y \in X_n \times \mathbf{I}$. Then, we have $x = (h_n \cdots h_1)^{-1}(y) \in M \times [0, 1)$. Indeed, $h_n \cdots h_1(x) = y \notin (M \setminus X_n) \times \{1\}$, which implies $x \notin M \times \{1\}$ by (1). Since $h_n \cdots h_1(x) \in X_n \times \mathbf{I}$ and $h_i | X_n \times \mathbf{I} = \mathrm{id}$ for every $i > n$, it follows that $h(x) = h_n \cdots h_1(x) = y$. Hence, h is surjective. Consequently, h is a homeomorphism. □

Notes for Chapter 2

For topics related to this chapter, refer to the book [(1)] of Bessaga and Pełczyński. Concerning $\boldsymbol{Q}$-manifolds, see the textbook [(3)] of J. van Mill and Chapman's lecture notes [(2)]. In this chapter, we have treated almost all the contents of the book of van Mill except for the characterization of compact $\boldsymbol{Q}$-manifolds. Without compactness, the characterization of $\boldsymbol{Q}$-manifolds will be treated in the next chapter together with the one of $\ell^2(\Gamma)$-manifolds. As a well-known application to Shape Theory, Chapman established in [42] the Complement Theorem, that is, two Z-sets X and Y in the Hilbert cube $\boldsymbol{Q}$ have the same shape type if and only if their complements $\boldsymbol{Q} \setminus X$ and $\boldsymbol{Q} \setminus Y$ are homeomorphic. As another remarkable application to Simple Homotopy Theory, Chapman proved in [45] the topological invariance of Whitehead torsion, which will be proved in Chapter 4. For Simple Homotopy Theory, refer to Cohen's book [(20)].

The weight version of Lemma 2.1.1 is due to Cutler (cf. Lemma 1.1 in [80]). Theorem 2.1.2 was established by Keller [92]. Theorem 2.1.11 is due to Henderson [80] (cf. [85] and [81]). In this section, we need not assume that E is normable.

The non-invertible isotopy in the proof of Proposition 2.1.12 is due to Anderson and Bing [12]; that of Proposition 2.1.13 is a modification of the one constructed by Crowell in [49]. Note that Crowell's example is not locally connected. Lemma 2.1.16 is due to Anderon, Henderson, and West [17].

The results of Sect. 2.2 are due to Toruńczyk [139] (the compact case was proved in [138]). In [140], the results have been generalized to metric linear spaces, where proofs are not elementary. The maps f and g in Sect. 2.2 were first defined on $X \times \varinjlim \mathbb{R}^n$ and $\varinjlim \mathbb{R}^n$ by D.W. Henderson in [82] so as to prove that $X \times \varinjlim \mathbb{R}^n \approx \varinjlim \mathbb{R}^n$.

The Stability Theorem 2.3.7 was first established for ℓ_2-manifolds by Anderson and Schori [15] and generalized as Theorem 2.3.7 by Schori [131] (cf. [85]). For the Stability Theorem, we can assert more (cf. [69] and [114]). The results can be extended to local trivial bundles with E-manifold fibers (cf. [115]). Reformulating Lemma 2.3.1 by Yamashita, the proofs of Lemmas 2.3.1 and 2.3.2 were made slim. The proof of the Strong Universality Theorem 2.3.15 is due to [114]. In this section, only the assumption that $E \approx E^{\mathbb{N}}$ or $\approx E_f^{\mathbb{N}}$ is used.

Proposition 2.4.1 is due to Anderson and Bing [12], which is generalized to an infinite-dimensional normed linear space E, that is, there exists an $\mathbf{I}$-preserving homeomorphism ζ : $E \times \mathbf{I} \setminus \{(0,0)\} \to E \times \mathbf{I}$. This can be proved by using a non-complete norm. Refer to Bessaga and Pełczyński's book [(1)], Ch. III, §5 (cf. [32]). The Negligibility Theorem was first established by Anderson, Henderson and West [17] and extended to Theorem 2.4.3 by Cutler [53]. In [4], Anderson proved that σ-compact sets are negligible in ℓ^2 and $\mathbb{R}^{\mathbb{N}}$, that is a key in his proof of $\ell^2 \approx \mathbb{R}^{\mathbb{N}}$. The proof of Proposition 2.4.14 is due to [115].

Theorem 2.5.7 was established by M. Brown [35]. Our proof of Theorem 2.5.7 is bases on Connelly's proof [48]. The Open Embedding Theorem 2.5.10 and the Unknotting Theorem 2.5.12 were respectively established by Henderson [79] (cf. [81]) and Anderson and McCharen [14] for ℓ_2-manifolds. They were extended in [84] (cf. [81]) and [40], respectively. The proof is due to [114]. These results can be extended to local trivial bundles with E-manifold fibers (cf. [116]).

The Classification Theorem 2.6.1 was established by Henderson and Schori [84] (cf. [85] and [81]). Theorem 2.6.3 is due to Ferry [64]. The Triangulation Theorem for ℓ^2-manifolds was first obtained by D.W. Henderson [79] (cf. [77, 78]), as a corollary of the Open Embedding Theorem by virtue of a triangulation result on open subsets of Hilbert space proved in [77]. Here, it is obtained as a combination of the Classification Theorem 2.6.1 and the Toruńczyk Factor Theorem 2.2.14 (Corollary 2.2.15).

The Bing Shrinking Criterion arose in [31] and has been generalized by several authors (cf. [103]). Theorem 2.7.1 here is due to Toruńczyk [146]. The Generalized Schoenflies Theorem 2.7.7 was proved by B. Mazur [105] and M. Brown [34]. Although Lemmas 2.7.8, 2.7.9, 2.7.10 are due to Brown [34], our proof of Theorem 2.7.7 is different from Brown's proof in [34]. The case $n = 2$ is established by A. Schoenflies [130] without the bi-collared condition. The bi-collared condition is necessary for $n \geqslant 3$. In fact, J.W. Alexander [2] constructed an embedding of $\mathbf{S}^2$ in $\mathbb{R}^3$ such that the unbounded component of the complement is not simply connected. This example is called the **Alexander horned sphere**.

The condition (z) in 2.14 on Theorem 2.8.6 is called Property Z in [6]. In [14], a closed set with Property Z is called a Z-set. T.A. Chapman [40] showed the equivalence of Property Z and E-deficiency (infinite deficiency) in E-manifolds (Theorem 2.9.3). The necessity of strong Z-sets was emphasized in the paper [30], where the example (p. 162) of a Z-set being not a strong Z-set was given (cf. [146]). Proposition 2.8.8 is due to Curtis, Dobrowolski, and Mogilski [51, Lemma 2.4]. The notation of "closed over a set" was introduced by Bestvina and Mogilski [29], who adopt Lemma 2.8.10 as the definition. Propositions 2.8.12, 2.8.13, and 2.8.14 were proved in [29]. Yamashita simplified the proof of Proposition 2.8.12, which is presented here. Proposition 2.8.17 was proved by Bestvina and Mogilski [29].

Proposition 2.9.1 is due to Toruńczyk [142] (cf. [146]). The example in Remark 2.18 was given in [30]. In the case where E is locally convex, Z-sets in an E-manifold are strong Z-sets, for which refer to the paper [83]. The non-locally convex case is not known. In [83], by using proper subdivisions of simplicial complexes, Henderson proved the implication (a) $\Rightarrow$ (b) in Theorem 2.9.3. In our proof, we avoid proper subdivisions. Theorem 2.9.9 was established by Anderson [7].

In [5], R.D. Anderson showed that every compact set in the pseudo-interior $I(\boldsymbol{Q})$ is $\boldsymbol{Q}$-deficient (infinite deficient) in $\boldsymbol{Q}$ (Theorem 2.10.4), where Lemmas 2.10.1, 2.10.2, and 2.10.3 were also proved. In [6], he showed the equivalence of Property Z and $\boldsymbol{Q}$-deficiency (infinite deficiency) in $\boldsymbol{Q}$ (Theorem 2.10.6), where Lemma 2.10.5 was also proved. In [41], using Theorem 2.9.3, Chapman proved the equivalence of Property Z and $\boldsymbol{Q}$-deficiency (infinite deficiency) in $\boldsymbol{Q}$-manifolds (Theorem 2.10.9).

The Z-set Unknotting Theorem (Theorem 2.11.6) was established by Anderson and Chapman [13]. Theorem 2.11.9, the Classification Theorem for $\mathbb{R}_+$-stable $\boldsymbol{Q}$-manifolds (Theorem 2.11.11), and the Open Embedding Theorems (Theorems 2.11.4 and 2.11.12) are due to Chapman [43]. R.Y.T. Wong characterized in [160] the $\mathbb{R}_+$-stability of $\boldsymbol{Q}$-manifolds as in Theorem 2.11.13.

Chapter 3
Characterizations of Hilbert Manifolds and Hilbert Cube Manifolds

Throughout this chapter,

- *let* Γ *be an infinite set (or an infinite discrete space) with* $\operatorname{card}\Gamma = \tau$.

In this chapter, we prove the Toruńczyk characterizations of Hilbert manifolds (i.e., $\ell^2(\Gamma)$-manifolds) and Hilbert cube manifolds (i.e., $\boldsymbol{Q}$-manifolds). By using the characterization of Hilbert space, we show that every Fréchet space is homeomorphic to Hilbert space with the same weight. In particular, $s = \mathbb{R}^{\mathbb{N}} \approx \ell^2$. Additionally, it is shown that if X is a non-discrete compactum and Y is a completely metrizable separable ANR Y without isolated points, then the space $\mathrm{C}(X, Y)$ of maps from X to Y is an ℓ^2-manifold. To prove the characterization of $\ell^2(\Gamma)$-manifolds, we require the Toruńczyk Factor Theorem 2.2.14, that is, for each completely metrizable ANR X with $w(X) \leqslant \tau$, the product $X \times \ell^2(\Gamma)$ is an $\ell^2(\Gamma)$-manifold. For the characterization of $\boldsymbol{Q}$-manifolds, we need to prove its $\boldsymbol{Q}$-version, called the Edwards Factor Theorem 3.8.1, that is, for each locally compact ANR X, the product $X \times \boldsymbol{Q}$ is a $\boldsymbol{Q}$-manifold. By using the characterization of $\boldsymbol{Q}$, we prove Keller's Theorem, that is, every infinite-dimensional compact metrizable convex set in a locally convex topological linear space is homeomorphic to $\boldsymbol{Q}$.

3.1 (Strong) Universality and $\mathcal{U}$-Maps

Given a class $\mathfrak{C}$ of spaces, a space X is said to be **$\mathfrak{C}$-universal** or **universal** for $\mathfrak{C}$ if any map $f : C \to X$ from $C \in \mathfrak{C}$ can be approximated by closed embeddings, that is, for each open cover $\mathcal{U}$ of X, f is $\mathcal{U}$-close to a closed embedding.[1] It is said that

[1]This is a little stronger than a universal space in Dimension Theory, where a space X is called a **universal space** for $\mathfrak{C}$ if $X \in \mathfrak{C}$ and every space in $\mathfrak{C}$ can be embedded into X, where it is not required to be embedded into X as a closed set (cf. Sect. 5.9 of [GAGT]).

K. Sakai, *Topology of Infinite-Dimensional Manifolds*, Springer Monographs in Mathematics, https://doi.org/10.1007/978-981-15-7575-4_3

X is **strongly $\mathfrak{C}$-universal** or **strongly universal** for $\mathfrak{C}$ if the following condition is satisfied:

($\mathrm{SU}_{\mathfrak{C}}$) For each $C \in \mathfrak{C}$ and each closed set $D \subset C$, if $f : C \to X$ is a map such that $f|D$ is a strong Z-embedding, then, for each open cover $\mathcal{U}$ of X, there is a strong Z-embedding $h : C \to X$ such that $h|D = f|D$ and h is $\mathcal{U}$-close to f.

Obviously, the strong $\mathfrak{C}$-universality implies the $\mathfrak{C}$-universality.

In this section and the next, we discuss and characterize the (strong) universality for the class $\mathfrak{M}_1(\tau)$ of completely metrizable spaces with density $\leqslant \tau$ (equivalently, weight $\leqslant \tau$),[2] and the class $\mathfrak{M}_0$ of compacta (= compact metrizable spaces). Note that $\mathfrak{M}_0 = \mathcal{F}_Q$ and $\mathfrak{M}_1(\tau) = \mathcal{F}_{\ell^2(\Gamma)}$ (Theorem 1.13.9). For a completely metrizable ANR X, it will be proved that the $\mathfrak{M}_1(\tau)$-universality implies the strong $\mathfrak{M}_1(\tau)$-universality, so they are equivalent (Theorem 3.1.19).

For each $f \in \mathrm{C}(X, Y)$ and $\mathcal{U} \in \mathrm{cov}(Y)$, we define

$$\mathcal{U}(f) = \{g \in \mathrm{C}(X, Y) \mid g \text{ is } \mathcal{U}\text{-close to } f\}.$$

Observe that, if $\mathcal{V} \in \mathrm{cov}(Y)$ is a star-refinement of $\mathcal{U}$, then $\mathcal{V}(g) \subset \mathcal{U}(f)$ for each $g \in \mathcal{V}(f)$. When Y is *paracompact*, $\mathrm{C}(X, Y)$ has a topology such that $\{\mathcal{U}(f) \mid \mathcal{U} \in \mathrm{cov}(Y)\}$ is a neighborhood basis of f. Such a topology is called the **limitation topology**.

Let $\mathrm{Emb}(X, Y)$ denote the subspace of $\mathrm{C}(X, Y)$ consisting of all **closed** embeddings of X into Y. Then, $f \in \mathrm{cl}\,\mathrm{Emb}(X, Y)$ means that f can be approximated by closed embeddings. Thus, the definition of the $\mathfrak{C}$-universality can be rewritten as follows:

- X is $\mathfrak{C}$-universal if $\mathrm{Emb}(Y, X)$ is dense in the space $\mathrm{C}(Y, X)$ with the limitation topology for every $Y \in \mathfrak{C}$.

For the limitation topology, refer to Sect. 2.9 in [GAGT].

For each metrizable space Y, let $\mathrm{Metr}^B(Y)$ be the set of all admissible bounded metrics for Y. When Y is completely metrizable, let $\mathrm{Metr}^c(Y)$ be the set of all admissible bounded complete metric for Y. For each $f \in \mathrm{C}(X, Y)$ and $d \in \mathrm{Metr}^B(Y)$, let

$$U_d(f) = \{g \in \mathrm{C}(X, Y) \mid \sup_{x \in X} d(f(x), g(x)) < 1\}.$$

Proposition 3.1.1 *For each metrizable space Y, $\{U_d(f) \mid d \in \mathrm{Metr}^B(Y)\}$ is a neighborhood basis of $f \in \mathrm{C}(X, Y)$ in the space $\mathrm{C}(X, Y)$ with the limitation topology. When Y is completely metrizable, $\{U_d(f) \mid d \in \mathrm{Metr}^c(Y)\}$ is a neighborhood basis of $f \in \mathrm{C}(X, Y)$.* [2.9.1]

[2]Recall that $\mathfrak{M}_1$ denotes the class of all completely metrizable spaces. The class of all separable completely metrizable spaces is denoted by $\mathfrak{M}_1(\aleph_0)$.

When $Y = (Y, d)$ is a metric space, for each $f \in \mathrm{C}(X, Y)$ and $\alpha \in \mathrm{C}(Y, (0, \infty))$, let

$$N_\alpha(f) = \{g \in \mathrm{C}(X, Y) \mid \forall x \in X,\ d(f(x), g(x)) < \alpha f(x)\}.$$

Proposition 3.1.2 *When $Y = (Y, d)$ is a metric space, $\{N_\alpha(f) \mid \alpha \in \mathrm{C}(Y, (0, \infty))\}$ is a neighborhood basis of $f \in \mathrm{C}(X, Y)$ in the space $\mathrm{C}(X, Y)$ with the limitation topology.* [2.9.3]

In the above proposition, we can consider only Lipschitz maps with the Lipschitz constant $\leqslant 1$ as $\alpha \in \mathrm{C}(Y, (0, \infty))$. In fact, this is derived from the following lemma:

Lemma 3.1.3 *Let $\mathcal{U}$ be a locally finite open cover of a metric space $Y = (Y, d)$ such that $\operatorname{diam} \operatorname{st}(y, \mathcal{U}) < \infty$ for each $y \in Y$. Then, the map $\alpha : Y \to (0, \infty)$ defined by $\alpha(y) = \sup_{U \in \mathcal{U}} \operatorname{dist}(y, Y \setminus U)$ is a Lipschitz map with the Lipschitz constant $\leqslant 1$, that is,*

$$|\alpha(y) - \alpha(y')| \leqslant d(y, y') \text{ for every } y, y' \in Y.$$

Proof Let $y, y' \in Y$. Without loss of generality, it can be assumed that $\alpha(y) > \alpha(y')$. We can take $U \in \mathcal{U}$ so that $\operatorname{dist}(y, Y \setminus U) > \alpha(y')$. Since $\alpha(y') \geqslant \operatorname{dist}(y', Y \setminus U)$, we have

$$\operatorname{dist}(y, Y \setminus U) - \alpha(y') \leqslant \operatorname{dist}(y, Y \setminus U) - \operatorname{dist}(y', Y \setminus U) \leqslant d(y, y').$$

Thus, it follows that

$$\alpha(y) - \alpha(y') = \sup_{U \in \mathcal{U}} \operatorname{dist}(y, Y \setminus U) - \alpha(y') \leqslant d(y, y'). \qquad \square$$

In the above lemma, as is easily observed,

$$\alpha(y) = \sup\{r > 0 \mid \exists U \in \mathcal{U} \text{ such that } \mathrm{B}(y, r) \subset U\}.$$

From the definition of α, for each $y, y' \in Y$,

$$d(y, y') < \alpha(y) \Rightarrow \exists U \in \mathcal{U} \text{ such that } y, y' \in U.$$

The limitation topology is favorable because of the following properties:

Proposition 3.1.4 (Basic Properties of the Limitation Topology) *Let X be an arbitrary space but Y and Z be paracompact.*

(1) *The evaluation map*

$$\mathrm{ev} : X \times \mathrm{C}(X, Y) \ni (x, f) \mapsto f(x) \in Y$$

is continuous with respect to the limitation topology. [2.9.9(1)]

(2) *The composition*

$$\mathrm{C}(X, Y) \times \mathrm{C}(Y, Z) \ni (f, g) \mapsto gf \in \mathrm{C}(X, Z)$$

is continuous with respect to the limitation topology. [2.9.9(2)]

(3) *The inverse operation*

$$\mathrm{Homeo}(Y) \ni h \mapsto h^{-1} \in \mathrm{Homeo}(Y)$$

is continuous with respect to the limitation topology. Combining this with (2) above, the group Homeo(Y) *with the limitation topology is a topological group.* [2.9.9(3)]

Hereafter, we assume that the space $\mathrm{C}(X, Y)$ and its subspace have the *limitation topology*.

Theorem 3.1.5 *For a completely metrizable space Y, the space* $\mathrm{C}(X, Y)$ *is a Baire space, that is, the intersection of countably many dense open sets (or dense G_δ-sets) in* $\mathrm{C}(X, Y)$ *is also dense.* [2.9.4]

The following strong version can be proved in the same way as above:

Theorem 3.1.6 *Let Y be a completely metrizable space, $f \in \mathrm{C}(X, Y)$, and $A \subset X$ be a closed set. Then, the following space is also a Baire space:*

$$\mathrm{C}_{f|A}(X, Y) = \{g \in \mathrm{C}(X, Y) \mid g|A = f|A\}.$$

Sketch of Proof. Let $G_n, n \in \mathbb{N}$, be open dense sets in $\mathrm{C}_{f|A}(X, Y)$. For each $g \in \mathrm{C}_{f|A}(X, Y)$ and a complete bounded metric $d \in \mathrm{Metr}^c(Y)$, we can inductively choose $g_n \in \mathrm{C}_{f|A}(X, Y)$ and $d_n \in \mathrm{Metr}(Y)$, $n \in \mathbb{N}$, so that

$$g_n \in U_{2d_{n-1}}(g_{n-1}) \cap G_n, \ U_{d_n}(g_n) \subset G_n, \text{ and } d_n \geqslant d_{n-1},$$

where $g_0 = g$ and $d_0 = d$. Then, it is easy to see that $(g_n)_{n \in \mathbb{N}}$ is Cauchy with respect to the sup-metric induced by d. By the completeness, $(g_n)_{n \in \mathbb{N}}$ is uniformly convergent to $g_\infty \in \mathrm{C}_{f|A}(X, Y)$. It is easy to calculate $d_0(g, g_\infty) < 1$ and $d_n(g_n, g_\infty) < 1$, which means $g_\infty \in U_d(g)$ and $g_\infty \in U_{d_n}(g_n) \subset G_n$. Hence, $g_\infty \in U_d(g) \cap \bigcap_{n \in \mathbb{N}} G_n$.

Let $\mathrm{C}^P(X, Y)$ be the subspace of $\mathrm{C}(X, Y)$ consisting of all proper maps. Due to Proposition 1.3.7, *when X and Y are metrizable, a map $f : X \to Y$ is proper if and only if it is perfect.*[3] The following is the direct consequence of Proposition 1.3.8:

Proposition 3.1.7 *If Y is locally compact and paracompact, then* $\mathrm{C}^P(X, Y)$ *is clopen in* $\mathrm{C}(X, Y)$*, where X is locally compact if* $\mathrm{C}^P(X, Y) \neq \emptyset$. [2.9.7]

[3]Even when Y is non-metrizable, if Y is locally compact, then this equivalence also holds [GAGT, Proposition 2.1.5].

We also consider the class $\mathfrak{L}$ of all locally compact metrizable spaces. Let $\mathfrak{L}(\tau)$ be the subclass of $\mathfrak{L}$ consisting of spaces with density $\leqslant \tau$ (equivalently, weight $\leqslant \tau$). Then, $\mathfrak{L}(\aleph_0) = \mathcal{F}_{Q\setminus\{\mathbf{0}\}}$ (Proposition 2.1.3). According to our definition, X is $\mathfrak{L}(\tau)$-universal if $\mathrm{Emb}(Y, X)$ is dense in $\mathrm{C}(Y, X)$ for every $Y \in \mathfrak{L}(\tau)$. However, the following fact should be remarked:

Fact *Any locally compact metrizable space X is not $\mathfrak{L}(\aleph_0)$-universal.*

Indeed, let $f : \mathbb{R} \to X$ be a constant map and let $\mathcal{U} = \{U, X \setminus f(\mathbb{R})\} \in \mathrm{cov}(X)$, where $\mathrm{cl}\, U$ is compact. If f is $\mathcal{U}$-close to g, then $g(\mathbb{R}) \subset U$, which implies that $\mathrm{cl}\, g(\mathbb{R})$ is compact. Hence, g is not a closed embedding.

On this account, we define that X is **properly $\mathfrak{L}(\tau)$-universal** (or **properly universal** for $\mathfrak{L}(\tau)$) if $\mathrm{Emb}(Y, X)$ is dense in $\mathrm{C}^P(Y, X) \neq \emptyset$ for every $Y \in \mathfrak{L}(\tau)$, where $\mathrm{C}^P(Y, X) \neq \emptyset$ if and only if $\mathrm{Emb}(Y, X) \neq \emptyset$. If X is properly $\mathfrak{L}(\tau)$-universal, then the discrete space Γ with $\mathrm{card}\, \Gamma = \tau$ can be embedded in X as a closed set. Hence, we have the following:

Lemma 3.1.8 *If X is properly $\mathfrak{L}(\tau)$-universal, then $w(X) \geqslant \tau$.* □

The following proposition can easily proved by the same method as the well-known fact that every component of a locally connected space is clopen.

Proposition 3.1.9 *Every locally compact metrizable space X is a discrete union of locally compact separable subspaces, that is, a topological sum of locally compact separable metrizable spaces.*

Sketch of Proof. Take a locally finite open cover $\mathcal{U} \in \mathit{cov}(X)$ so that $\mathrm{cl}\, U$ is compact for every $U \in \mathcal{U}$. Then, $\mathcal{U}$ is star-finite (i.e., $\mathcal{U}[U]$ is finite for every $U \in \mathcal{U}$). For each $U \in \mathcal{U}$, we define $U^* = \bigcup_{n\in\mathbb{N}} \mathrm{st}^n(U, \mathcal{U})$, where each $\mathrm{st}^n(U, \mathcal{U}) = \mathrm{st}(\mathrm{st}^{n-1}(U, \mathcal{U}), \mathcal{U})$, $\mathrm{st}^0(U, \mathcal{U}) = U$ (so $\mathrm{st}^1(U, \mathcal{U}) = \mathrm{st}(U, \mathcal{U})$). Then, U^* is an equivalence class with respect to the equivalence relation $\sim$ on X defined as follows:

$$x \sim y \underset{\text{def}}{\Leftrightarrow} \exists U_1, \dots, U_n \in \mathcal{U} \text{ such that}$$
$$x \in U_1,\ y \in U_n,\ U_i \cap U_{i+1} \neq \emptyset, i = 1, \dots, n-1.$$

Thus, for each $U, V \in \mathcal{U}$, $U^* \neq V^* \Leftrightarrow U^* \cap V^* = \emptyset$. Hence, U^* is clopen in X for each $U \in \mathcal{U}$. As is easily observed, for every $U \in \mathcal{U}$, $U^* = \bigcup_{n\in\mathbb{N}} \mathrm{cl}\, \mathrm{st}^n(U, \mathcal{U})$ and $\mathrm{cl}\, \mathrm{st}^n(U, \mathcal{U})$, $n \in \mathbb{N}$, are compact. Hence, U^* is σ-compact and separable.[4] Since U^* is open in X, U^* is locally compact.

By virtue of Proposition 3.1.9 above, the following holds:

Proposition 3.1.10 *A locally compact metrizable space X is properly $\mathfrak{L}(\tau)$-universal if and only if X is properly $\mathfrak{L}(\aleph_0)$-universal and $w(X) \geqslant \tau$.*

Proof The "only if" part follows from Lemma 3.1.8 and the fact $\mathfrak{L}(\aleph_0) \subset \mathfrak{L}(\tau)$. To see the "if" part, let $f : Y \to X$ be a proper map from $Y \in \mathfrak{L}(\tau)$. By

[4] A locally compact metrizable space is separable if and only if it is σ-compact.

Proposition 3.1.9, we can write $X = \bigoplus_{\lambda\in\Lambda} X_\lambda$, where each X_λ is a locally compact separable metrizable space. Then, $Y = \bigoplus_{\lambda\in\Lambda} f^{-1}(X_\lambda)$. Each X_λ has a countable compact cover $\{X_{\lambda,n} \mid n \in \mathbb{N}\}$. Then, $f^{-1}(X_\lambda) = \bigcup_{n\in\mathbb{N}} f^{-1}(X_{\lambda,n})$ and each $f^{-1}(X_{\lambda,n})$ is compact because f is proper. Hence, $f^{-1}(X_\lambda)$ is a locally compact separable metrizable space and $f|f^{-1}(X_\lambda)$ is a proper map. For each $\mathcal{U} \in \mathrm{cov}(X)$, we can choose an open refinement $\mathcal{V} \in \mathrm{cov}(X)$ of $\mathcal{U}$ so that each $V \in \mathcal{V}$ meets only one of X_λ, $\lambda \in \Lambda$, that is, $\mathrm{st}(X_\lambda, \mathcal{V}) = X_\lambda$ for each $\lambda \in \Lambda$. For each $\lambda \in \Lambda$, we have a closed embedding $h_\lambda : f^{-1}(X_\lambda) \to X$ that is $\mathcal{V}$-close to $f|f^{-1}(X_\lambda)$. Since $h_\lambda(f^{-1}(X_\lambda)) \subset \mathrm{st}(X_\lambda, \mathcal{V}) = X_\lambda$ for each $\lambda \in \Lambda$, we can define a closed embedding $h : Y \to X$ by $h|f^{-1}(X_\lambda) = h_\lambda$ for each $\lambda \in \Lambda$. Since $\mathcal{V} \prec \mathcal{U}$, h is $\mathcal{U}$-close to f. □

In order to discuss the proper $\mathfrak{C}(\tau)$-universality of a locally compact metrizable space, we need the following theorem that can be obtained by combining Proposition 3.1.7 with Theorem 3.1.5:

Theorem 3.1.11 *For every locally compact metrizable spaces X and Y, the space $\mathrm{C}^P(X, Y)$ is a Baire space.* [2.9.8]

The following is a strong version corresponding to Theorem 3.1.6:

Theorem 3.1.12 *Let X and Y be locally compact metrizable spaces. For each $f \in \mathrm{C}^P(X, Y)$ and each closed set $A \subset X$, the following space is also a Baire space:*

$$\mathrm{C}^P_{f|A}(X, Y) = \{g \in \mathrm{C}^P(X, Y) \mid g|A = f|A\}.$$ □

Given $\mathcal{U} \in \mathrm{cov}(X)$, a map $f : X \to Y$ is called a **$\mathcal{U}$-map** if $f^{-1}(\mathcal{V}) \prec \mathcal{U}$ for some $\mathcal{V} \in \mathrm{cov}(Y)$. Let $\mathrm{C}_\mathcal{U}(X, Y)$ denote the subspace of $\mathrm{C}(X, Y)$ consisting of all $\mathcal{U}$-maps. Then, $\mathrm{Emb}(X, Y) \subset \mathrm{C}_\mathcal{U}(X, Y)$.[5]

Lemma 3.1.13 *Let Y be paracompact and $\mathcal{U} \in \mathrm{cov}(X)$. Then, $\mathrm{C}_\mathcal{U}(X, Y)$ is open in the space $\mathrm{C}(X, Y)$.* [5.8.6]

Sketch of Proof. Let $f \in \mathrm{C}_\mathcal{U}(X, Y)$. Then, $f^{-1}(\mathcal{V}) \prec \mathcal{U}$ for some $\mathcal{V} \in \mathrm{cov}(Y)$. For a star-refinement $\mathcal{W} \in \mathrm{cov}(Y)$ of $\mathcal{V}$, $\mathcal{W}(f) \subset \mathrm{C}_\mathcal{U}(X, Y)$.

Lemma 3.1.14 *Let $X = (X, d)$ be a complete metric space and $\mathcal{U}_n \in \mathrm{cov}(X)$, $n \in \mathbb{N}$, with $\mathrm{mesh}\,\mathcal{U}_n < 2^{-n}$. Then, $\mathrm{Emb}(X, Y) = \bigcap_{n\in\mathbb{N}} \mathrm{C}_{\mathcal{U}_n}(X, Y)$. Thus, when X is a completely metrizable space, $\mathrm{Emb}(X, Y)$ is a G_δ-set in the space $\mathrm{C}(X, Y)$.* [5.8.7]

Sketch of Proof. Each $f \in \bigcap_{n\in\mathbb{N}} \mathrm{C}_{\mathcal{U}_n}(X, Y)$ is injective. For $x_n \in X$, $n \in \mathbb{N}$, if $(f(x_n))_{n\in\mathbb{N}}$ is convergent, then $(x_n)_{n\in\mathbb{N}}$ is Cauchy, so convergent. Hence, $f \in \mathrm{Emb}(X, Y)$.

By combining the above two lemmas with Theorem 3.1.5, the following can be obtained:

---

[5]For $\mathcal{U}$-maps, refer to Sect. 5.8 of [GAGT].

Proposition 3.1.15 *Let X and Y be completely metrizable spaces. Suppose that, for each $\mathcal{U} \in \mathrm{cov}(X)$, $\mathrm{C}_{\mathcal{U}}(X, Y)$ is dense in the space $\mathrm{C}(X, Y)$. Then, $\mathrm{Emb}(X, Y)$ is also dense in $\mathrm{C}(X, Y)$.* 5.8.8

The following is easy to see:

Lemma 3.1.16 *Let $\mathcal{U}$ be an open cover of X consisting of open sets with compact closure. Then, $\mathrm{C}_{\mathcal{U}}(X, Y) \subset \mathrm{C}^P(X, Y)$.* □

We have the following locally compact version of Proposition 3.1.15:

Proposition 3.1.17 *Let X and Y be locally compact metrizable. Suppose that, for each open cover $\mathcal{U}$ of X consisting of open sets with the compact closure, $\mathrm{C}_{\mathcal{U}}(X, Y)$ is dense in the space $\mathrm{C}^P(X, Y)$. Then, $\mathrm{Emb}(X, Y)$ is also dense in $\mathrm{C}^P(X, Y)$.* 5.8.9

When a map is a $\mathcal{U}$-map on a closed set, it is a $\mathcal{U}$-map on a neighborhood, that is:

Proposition 3.1.18 *Let X and Y be paracompact spaces, $f : X \to Y$ be a map, $A \subset X$ be a closed set and $\mathcal{U} \in \mathrm{cov}(X)$. If $f|A$ is a $(\mathcal{U}|A)$-map, then A has a neighborhood N in X such that $f|N$ is a $(\mathcal{U}|N)$-map.*

Proof Choose a locally finite open cover $\mathcal{V}$ of Y so that

$$(f|A)^{-1}(\mathcal{V}^{\mathrm{cl}}) = f^{-1}(\mathcal{V}^{\mathrm{cl}})|A \prec \mathcal{U}|A \prec \mathcal{U}.$$

Let $\mathcal{V}_0$ be an open star-refinement of $\mathcal{V}$. By the continuity of f and the local finiteness of $(f|A)^{-1}(\mathcal{V}^{\mathrm{cl}})$, we can obtain $\mathcal{W} \in \mathrm{cov}(X)$ such that

$$f(\mathcal{W}) \prec \mathcal{V}_0 \text{ and } \left\{ \mathrm{st}(f^{-1}(\mathrm{cl}\, V) \cap A, \mathcal{W}) \;\middle|\; V \in \mathcal{V} \right\} \prec \mathcal{U}.$$

Then, $N = \mathrm{st}(A, \mathcal{W})$ is a neighborhood of A in X. We shall show that $(f|N)^{-1}(\mathcal{V}_0) \prec \mathcal{U}$, which means that $f|N$ is a $(\mathcal{U}|N)$-map. For each $V_0 \in \mathcal{V}_0$, choose $V \in \mathcal{V}$ so that $\mathrm{st}(V_0, \mathcal{V}_0) \subset V$. Then, it suffices to see

$$(f|N)^{-1}(V_0) \subset \mathrm{st}(f^{-1}(V) \cap A, \mathcal{W}).$$

For each $x \in (f|N)^{-1}(V_0) = f^{-1}(V_0) \cap N$, we have $a \in A$ such that $x, a \in W$ for some $W \in \mathcal{W}$, which implies that $f(x), f(a) \in V_1$ for some $V_1 \in \mathcal{V}_0$. Then, it follows that $f(a) \in \mathrm{st}(V_0, \mathcal{V}_0) \subset V$, so $a \in f^{-1}(V) \cap A$. Consequently, we have $x \in \mathrm{st}(f^{-1}(V) \cap A, \mathcal{W})$. □

For an ANR $X \in \mathfrak{M}_1(\tau)$, we can prove the $\mathfrak{M}_1(\tau)$-universality implies the strong $\mathfrak{M}_1(\tau)$-universality, so they are equivalent, that is:

Theorem 3.1.19 *A completely metrizable ANR X with weight τ is strongly $\mathfrak{M}_1(\tau)$-universal if and only if X is $\mathfrak{M}_1(\tau)$-universal.*

To prove Theorem 3.1.19, we need the following lemma:

Lemma 3.1.20 *Let $X \in \mathfrak{M}_1(\tau)$ be an ANR and A a strong Z-set in X. If X is $\mathfrak{M}_1(\tau)$-universal, then the identity id_X can be approximated by strong Z-embeddings $h : X \to X$ such that $h|A = \mathrm{id}$.*

Proof First, we will construct a map $\varphi : X \times (0,1] \to X \times (0,1]$ so that:

(1) $\mathrm{pr}_X\varphi|X \times [a,1]$ is a closed embedding for each $a \in (0,1)$,
(2) $\mathrm{pr}_X\varphi(X \times (0,1]) \cap A = \emptyset$, and
(3) $d(\mathrm{pr}_X\varphi(x,t), x) < t$ for each $(x,t) \in X \times (0,1]$.

For simplicity, we write

$$M = \mathrm{C}(X \times (0,1], X \times (0,1]),$$

where we consider M as the space with the limitation topology. Then, M is a Baire space by Theorem 3.1.5.

For each $n \in \mathbb{N}$, let

$$G_n = \big\{ f \in M \bigm| \mathrm{pr}_X f|X \times [2^{-n},1] \in \mathrm{Emb}(X \times [2^{-n},1], X), \\ \mathrm{pr}_X f(X \times [2^{-n},1]) \cap A = \emptyset \big\},$$

where it should be noted that $\mathrm{pr}_X f(X \times [2^{-n},1])$ is closed in X. Recall that $\mathrm{Emb}(X \times [2^{-n},1], X)$ is a G_δ set in $\mathrm{C}(X \times [2^{-n},1], X)$ by Lemma 3.1.14. Moreover, the following is open in $\mathrm{C}(X \times [2^{-n},1], X)$:

$$O_n = \big\{ g \in \mathrm{C}(X \times [2^{-n},1], X) \bigm| \mathrm{cl}\, g(X \times [2^{-n},1]) \cap A = \emptyset \big\}.$$

Since the correspondence $f \mapsto \mathrm{pr}_X f|X \times [2^{-n},1]$ is continuous (3.1.4(2)) and G_n is the inverse image of a G_δ-set $\mathrm{Emb}(X \times [2^{-n},1], X) \cap O_n$ by this correspondence, it follows that G_n is a G_δ-set in M.

To see that G_n is dense in M, let $f \in M$ and $\mathcal{U} \in \mathrm{cov}(X \times (0,1])$. By the compactness of $[2^{-n-1},1]$, we have $\mathcal{W} \in \mathrm{cov}(X)$ such that

$$\big\{ W \times \{t\} \bigm| W \in \mathcal{W},\ t \in [2^{-n-1},1] \big\} \prec \mathcal{U}.$$

Since A is a strong Z-set in X and X is an $\mathfrak{M}_1(\tau)$-universal ANR, we can easily obtain a closed embedding $g : X \times [2^{-n},1] \to X$ such that

$$g(X \times [2^{-n},1]) \cap A = \emptyset \ \text{ and } \ g \simeq_{\mathcal{W}} \mathrm{pr}_X f|X \times [2^{-n},1].$$

By the Homotopy Extension Theorem 1.13.11, g can be extended to a map $\tilde{g} : X \times (0,1] \to X$ such that

$$\tilde{g}|X \times (0,2^{-n-1}] = \mathrm{pr}_X f|X \times (0,2^{-n-1}] \ \text{ and}$$

$$\tilde{g} \simeq_{\mathcal{W}} \mathrm{pr}_X f \ \text{ rel. } X \times (0,2^{-n-1}].$$

Then, f is $\mathcal{U}$-close to the map $f' \in G_n$ defined by $\mathrm{pr}_X f' = \tilde{g}$ and $\mathrm{pr}_{(0,1]} f' = \mathrm{pr}_{(0,1]} f$, that is,

$$f'(x,t) = (\tilde{g}(x,t), \mathrm{pr}_{(0,1]} f(x,t)) \quad \text{for each } (x,t) \in X \times (0,1].$$

Thus, G_n is dense in M. Consequently, G_n is a dense G_δ set in M.

Consider the following neighborhood of $\mathrm{id} \in M$:

$$V = \{f \in M \mid \forall (x,t) \in X \times (0,1],\ d(\mathrm{pr}_X f(x,t), x) < t\}.$$

Since M is a Baire space, we can obtain the desired $\varphi \in V \cap (\bigcap_{n\in\mathbb{N}} G_n)$, which satisfies (1), (2), and (3). Here, note that $\mathrm{pr}_X \varphi$ is injective by (1). For any open cover $\mathcal{U} \in \mathrm{cov}(X)$, using this φ, we will construct a strong Z-embedding $h : X \to X$ such that $h|A = \mathrm{id}$ and h is $\mathcal{U}$-close to id.

Due to Lemma 2.1.16, there exists $\mathcal{V} \in \mathrm{cov}(X \setminus A)$ which is fitting in X and refines $\mathcal{U}$. Take $\alpha \in \mathrm{C}(X \setminus A, (0,1])$ so that $N_\alpha(\mathrm{id}_{X\setminus A}) \subset \mathcal{V}(\mathrm{id}_{X\setminus A})$ (Proposition 3.1.2), and define $h_\alpha : X \to X$ as follows:

$$h_\alpha(x) = \begin{cases} \mathrm{pr}_X \varphi(x, \alpha(x)) & \text{if } x \in X \setminus A, \\ x & \text{if } x \in A. \end{cases}$$

Then, $h_\alpha(X \setminus A) \subset X \setminus A$ and $d(h_\alpha(x), x) < \alpha(x)$ for each $x \in X \setminus A$, which means $h_\alpha|X \setminus A \in N_\alpha(\mathrm{id}_{X\setminus A})$. Thus, $h_\alpha|X \setminus A$ is $\mathcal{V}$-close to $\mathrm{id}_{X\setminus A}$. Since $h_\alpha|A = \mathrm{id}$ and $\mathcal{V}$ is fitting in X, it follows that h_α is continuous, where h_α is a closed embedding if $h_\alpha|X \setminus A$ is a closed embedding into $X \setminus A$ (2.1.16(iii)). Because $\mathcal{V} \prec \mathcal{U}$ and $h_\alpha|A = \mathrm{id}$, h_α is $\mathcal{U}$-close to id_X. We have to show that h_α is a strong Z-embedding, that is, h_α is a closed embedding and $h_\alpha(X)$ is a strong Z-set in X.

We show that $h_\alpha|X \setminus A$ is a closed embedding into $X \setminus A$. Then, as mentioned above, it follows that h_α is a closed embedding. Since $\mathrm{pr}_X \varphi$ is injective, $h_\alpha|X \setminus A$ is an injection into $X \setminus A$. To see that $h_\alpha|X \setminus A$ is closed, let $x_i \in X \setminus A$, $i \in \mathbb{N}$, such that $(h_\alpha(x_i))_{i\in\mathbb{N}}$ converges to $y \in X \setminus A$. If $\inf_{i\in\mathbb{N}} \alpha(x_i) = 0$, then $(x_i)_{i\in\mathbb{N}}$ has a subsequence $(x_{i_j})_{j\in\mathbb{N}}$ such that $\lim_{j\to\infty} \alpha(x_{i_j}) = 0$. In this case, $\lim_{j\to\infty} x_{i_j} = y$ because $d(h_\alpha(x_{i_j}), x_{i_j}) < \alpha(x_{i_j}) \to 0\ (j \to \infty)$. Then, $\alpha(y) = \lim_{j\to\infty} \alpha(x_{i_j}) = 0$, which is a contradiction. Therefore, $\inf_{n\in\mathbb{N}} \alpha(x_n) > 0$. Choose $n \in \mathbb{N}$ so that $2^{-n} < \inf_{i\in\mathbb{N}} \alpha(x_i)$. Because $\varphi \in G_n$, $\mathrm{pr}_X \varphi|X \times [2^{-n}, 1]$ is a closed embedding into $X \setminus A$. Since $\mathrm{pr}_X \varphi(x_i, \alpha(x_i)) = h_\alpha(x_i) \to y\ (i \to \infty)$, it follows that $(x_i)_{i\in\mathbb{N}}$ is convergent in $X \setminus A$. Thus, $h_\alpha|X \setminus A$ is a closed embedding.

Finally, we show that $h_\alpha(X)$ is a strong Z-set in X. For each $\mathcal{W} \in \mathrm{cov}(X)$, let $\mathcal{W}' \in \mathrm{cov}(X)$ be a star-refinement of $\mathcal{W}$. Since A is a strong Z-set in X, there is a map $g : X \to X$ such that g is $\mathcal{W}'$-close to id_X and $A \cap \mathrm{cl}\, g(X) = \emptyset$. Choose an open set U in X so that $A \cap \mathrm{cl}\, U = \emptyset$ and $\mathrm{cl}\, g(X) \subset U$. Let

$$\mathcal{W}'' = \mathcal{W}' \wedge \{U,\ X \setminus \mathrm{cl}\, U,\ X \setminus (A \cup \mathrm{cl}\, g(X))\} \in \mathrm{cov}(X).$$

Take $\beta \in \mathrm{C}(X \setminus A, (0, 1])$ so that $\beta < \alpha$ and $N_\beta(\mathrm{id}_{X\setminus A}) \subset \mathcal{W}''(\mathrm{id}_{X\setminus A})$ (Proposition 3.1.2). Replacing α with this β, we can obtain a closed embedding h_β such that $d(h_\beta(x), x) < \beta(x)$ for all $x \in X$. Thus, we have a map $h_\beta g : X \to X$, which is $\mathcal{W}''$-close to g because $d(h_\beta g(x), g(x)) < \beta(g(x))$ for each $x \in X$. Since $\mathrm{st}(\mathcal{W}', \mathcal{W}'') \prec \mathcal{W}$, it follows that $h_\beta g$ is $\mathcal{W}$-close to id_X. Moreover,

$$\mathrm{cl}\, h_\beta g(X) \subset \mathrm{cl}\, \mathrm{st}(g(X), \mathcal{W}'') \subset \mathrm{cl}\, U \subset X \setminus A.$$

Since $h_\beta(X \setminus A)$ is closed in $X \setminus A$, $\mathrm{pr}_X\varphi$ is injective, and $\beta(x) < \alpha(x)$ for each $x \in X$, it follows that

$$\begin{aligned} h_\alpha(X) \cap \mathrm{cl}\, h_\beta g(X) &= (A \cup h_\alpha(X \setminus A)) \cap \mathrm{cl}\, h_\beta g(X) \\ &\subset h_\alpha(X \setminus A) \cap \mathrm{cl}\, h_\beta(X \setminus A) \\ &= h_\alpha(X \setminus A) \cap h_\beta(X \setminus A) \\ &= \left\{\mathrm{pr}_X\varphi(x, t) \;\middle|\; \alpha(x) = \beta(x) = t\right\} \\ &= \emptyset. \end{aligned}$$

Therefore, $h_\alpha(X)$ is a strong Z-set in X. □

Remark 3.1 In the above, by (3), $\mathrm{pr}_X\varphi$ extends to the homotopy $\tilde{\varphi} : X \times \mathbf{I} \to X$ by $\tilde{\varphi}_0 = \mathrm{id}$, that is a deformation. For any $\alpha \in \mathrm{C}(X, (0, 1])$, we can define a map $h_\alpha : X \to X$ $h_\alpha(x) = \tilde{\varphi}(x, \alpha_A(x))$, where $\alpha_A(x) = \min\{\alpha(x), d(x, A)\}$. Then, $h_{\alpha_A} \in N_\alpha(\mathrm{id}_X)$ and $h_{\alpha_A}|A = \mathrm{id}$. It can be proved that h_{α_A} is a strong Z-embedding, that is, h_{α_A} is a closed embedding and $h_{\alpha_A}(X)$ is a strong Z-set in X. The proof is a good exercise.

With easy modification of the above proof, the following version of Lemma 3.1.20 can be shown:

Lemma 3.1.21 *Let $X \in \mathfrak{L}(\tau)$ be an ANR and A a (strong) Z-set in X. If X is properly $\mathfrak{L}(\tau)$-universal, then the identity* id_X *can be approximated by strong Z-embeddings $h : X \to X$ such that $h|A = \mathrm{id}$.*

Modification in Proof of Lemma 3.1.20. Replace $\mathrm{C}(X \times (0, 1], X \times (0, 1])$, $\mathrm{C}(X \times [2^{-n}, 1], X)$, $\mathrm{C}_{f|D}(C, X)$, and "$\mathfrak{M}_1(\tau)$-universal" with $\mathrm{C}^P(X \times (0, 1], X \times (0, 1])$, $\mathrm{C}^P(X \times [2^{-n}, 1], X)$, $\mathrm{C}^P_{f|D}(C, X)$, and "proper $\mathfrak{L}(\tau)$-universal," respectively. Then, Theorem 3.1.11 substitutes for Theorem 3.1.5 and Lemma 3.1.14 can be applied together with Lemma 3.1.16. To see that G_n is dense in C, note that $\mathrm{pr}_X f|X \times [2^{-n}, 1]$ is proper because f is proper. Choose $\mathcal{U} \in \mathrm{cov}(X \times (0, 1])$ so that $\mathrm{cl}\, U$ is compact for each $U \in \mathcal{U}$. Then, f' is also proper because it is $\mathcal{U}$-close to f.

Now, we prove Theorem 3.1.19:

Proof (Theorem 3.1.19) It suffices to show the "if" part. Let $f : C \to X$ be a map of $C \in \mathfrak{M}_1(\tau)$ such that $f|D$ is a strong X-embedding. By virtue of Lemma 3.1.20, it suffices to show that f can be approximated by closed embeddings $g : Y \to X$

such that $f|D = g|D$, that is, the following set is dense in $\mathrm{C}_{f|D}(C, X)$:

$$\mathrm{Emb}_{f|D}(C, X) = \mathrm{C}_{f|D}(C, X) \cap \mathrm{Emb}(C, X).$$

For each $n \in \mathbb{N}$, take $\mathcal{U}_n \in \mathrm{cov}(C)$ so that $\mathrm{mesh}\,\mathcal{U}_n < 2^{-n}$ for some $d \in \mathrm{Metr}(C)$, and let

$$F_n = \mathrm{C}_{f|D}(C, X) \cap \mathrm{C}_{\mathcal{U}_n}(C, X).$$

Then, each F_n is open in $\mathrm{C}_{f|D}(C, X)$ by Lemma 3.1.13 and $\mathrm{Emb}_{f|D}(C, X) = \bigcap_{n\in\mathbb{N}} F_n$. Since $\mathrm{C}_{f|D}(C, X)$ is a Baire space by Theorem 3.1.6, it is enough to show that each F_n is dense in $\mathrm{C}_{f|D}(C, X)$.

Let $g \in \mathrm{C}_{f|D}(C, X)$ and $\mathcal{U} \in \mathrm{cov}(X)$. Take a star-refinement $\mathcal{U}' \in \mathrm{cov}(X)$ of $\mathcal{U}$. We apply Proposition 2.8.12 to obtain a map $g' : C \to X$ such that $g'|D = g|D = f|D$, $g'(C \setminus D) \subset X \setminus f(D)$, g' is closed over $f(D)$, and g' is $\mathcal{U}'$-close to g. Because $g'|D = f|D$ is a closed embedding, it is a $(\mathcal{U}_n|D)$-map. Due to Proposition 3.1.18, D has an open neighborhood N in C such that $g'|N$ is a $(\mathcal{U}_n|N)$-map. Since g' is closed over $f(D)$ and $C \setminus N$ is closed in C, it follows that

$$f(D) \cap \mathrm{cl}\, g'(C \setminus N) = f(D) \cap g'(C \setminus N) \subset f(D) \cap g'(C \setminus D) = \emptyset.$$

Then, $O = X \setminus \mathrm{cl}\, g'(C \setminus N)$ is open set in X and $D \subset g'^{-1}(O) \subset N$.

Since $\mathrm{C}_{\mathcal{U}_n|N}(N, X)$ is open in $\mathrm{C}(N, X)$ by Lemma 3.1.13, there is a locally finite open cover $\mathcal{W}$ of X such that $\mathcal{W} \prec \mathcal{U}'$, $\mathcal{W}(g'|N) \subset \mathrm{C}_{\mathcal{U}_n|N}(N, X)$, and

$$\mathrm{cl}\,\mathrm{st}(\mathrm{cl}\, g'(C \setminus N), \mathrm{st}\,\mathcal{W}) \cap f(D) = \emptyset.$$

Observe that $\mathrm{st}(W, \mathcal{W}) \cap f(D) = \emptyset$ for each $W \in \mathcal{W}[X \setminus O] = \mathcal{W}[\mathrm{cl}\, g'(C \setminus N)]$. Let

$$D' = g'^{-1}(\mathrm{cl}\,\mathrm{st}(\mathrm{cl}\, g'(C \setminus N), \mathrm{st}\,\mathcal{W})).$$

Then, D' is closed in C and $D \cap D' = \emptyset$. Because X is an $\mathfrak{M}_1(\tau)$-universal ANR, there exists a closed embedding $h' : D' \to X$ such that $h' \simeq_{\mathcal{W}} g'|D'$. By using the Homotopy Extension Theorem 1.13.11, h' can be extended to a map $h : C \to X$ such that $h|D = g'|D = f|D$ and $h \simeq_{\mathcal{W}} g'$. Then, h is $\mathcal{U}'$-close to g', so $\mathcal{U}$-close to g. It remains to be shown that $h \in F_n$.

To see $h \in F_n$, we may prove that h is a $\mathcal{U}_n$-map. Since the closed embedding $h|D' = h'$ is a $(\mathcal{U}_n|D')$-map, we have $\mathcal{V}' \in \mathrm{cov}(X)$ such that $\mathcal{V}' \prec \mathcal{W}$ and $(h|D')^{-1}(\mathcal{V}') \prec \mathcal{U}_n|D'$. Let $V \in \mathcal{V}'$ such that

$$V \cap \mathrm{st}(\mathrm{cl}\, g'(C \setminus N), \mathcal{W}) \neq \emptyset, \text{ i.e., } \mathrm{st}(V, \mathcal{W}) \cap \mathrm{cl}\, g'(C \setminus N) \neq \emptyset.$$

Since h is $\mathcal{W}$-close to g' and $\mathcal{V}' \prec \mathcal{W}$, it follows that

$$h^{-1}(V) \subset g'^{-1}(\mathrm{st}(V, \mathcal{W})) \subset g'^{-1}(\mathrm{st}(\mathrm{cl}\, g'(C \setminus N), \mathrm{st}\, \mathcal{W})) = D',$$

which implies that $h^{-1}(V) = h^{-1}(V) \cap D' = (h|D')^{-1}(V)$. Therefore,

$$h^{-1}(\mathcal{V}'[\mathrm{st}(\mathrm{cl}\, g'(Y \setminus N), \mathcal{W})]) \subset (h|D')^{-1}(\mathcal{V}') \prec \mathcal{U}_n.$$

Because $h|N \in \mathcal{W}(g'|N)$, $h|N$ is a $(\mathcal{U}_n|N)$-map. So, there is $\mathcal{V}'' \in \mathrm{cov}(X)$ such that $\mathcal{V}'' \prec \mathcal{W}$ and $(h|N)^{-1}(\mathcal{V}'') \prec \mathcal{U}_n|N$. Let $V \in \mathcal{V}''$ such that

$$V \cap \mathrm{st}(\mathrm{cl}\, g'(C \setminus N), \mathcal{W}) = \emptyset, \text{ i.e., } \mathrm{st}(V, \mathcal{W}) \cap g'(C \setminus N) = \emptyset.$$

Since h is $\mathcal{W}$-close to g', it follows that $h^{-1}(V) \subset g'^{-1}(\mathrm{st}(V, \mathcal{W})) \subset N$. Then, $h^{-1}(V) = h^{-1}(V) \cap N = (h|N)^{-1}(V)$. Hence,

$$h^{-1}(\mathcal{V}'' \setminus \mathcal{V}''[\mathrm{st}(\mathrm{cl}\, g'(C \setminus N), \mathcal{W})]) \prec \mathcal{U}_n.$$

For a common refinement $\mathcal{V} \in \mathrm{cov}(X)$ of $\mathcal{V}'$ and $\mathcal{V}''$, we have $h^{-1}(\mathcal{V}) \prec \mathcal{U}_n$. Consequently, h is a $\mathcal{U}_n$-map. □

It is said that X is **strongly properly $\mathfrak{L}(\tau)$-universal** or **strongly properly universal for** $\mathfrak{L}(\tau)$ if X satisfies the following condition:

($\mathrm{SU}^p_{\mathfrak{L}(\tau)}$) For each $C \in \mathfrak{L}(\tau)$ and each closed set $D \subset C$, if $f : C \to X$ is a proper map and $f|D$ is a strong Z-embedding, then $f : C \to X$ can be approximated by strong Z-embeddings $g : C \to X$ such that $g|D = f|D$.

In the above condition ($\mathrm{SU}^p_{\mathfrak{L}(\tau)}$), if X is locally compact, then "strong Z-embedding" can be replaced with "Z-embedding" because every Z-set in X is a strong Z-set (Proposition 2.8.1). The following version of Theorem 3.1.19 can be proved by the same method with suitable modification:

Theorem 3.1.22 *A locally compact ANR X with weight τ is strongly properly $\mathfrak{L}(\tau)$-universal if and only if X is properly $\mathfrak{L}(\tau)$-universal.*

Modification in Proof of Theorem 3.1.19. Let $f : C \to X$ be a proper map of $C \in \mathfrak{L}(\tau)$ such that $f|D$ is a (strong) Z-embedding. Lemma 3.1.21 substitutes for Lemma 3.1.20. Replace $\mathrm{C}_{f|D}(C, X)$ with $\mathrm{C}^P_{f|D}(C, X)$ and take $\mathcal{U}_n \in \mathrm{cov}(C)$ so as to satisfy the additional condition that $\mathrm{cl}\, U$ is compact for every $U \in \mathcal{U}_n$. Lemma 3.1.13 can used together with Lemma 3.1.16. To see that F_n is dense in $\mathrm{C}^P_{f|D}(C, X)$, take $\mathcal{U} \in \mathrm{cov}(X)$ so that $\mathrm{cl}\, U$ is compact for each $U \in \mathcal{U}$. Then, h is also proper because it is $\mathcal{U}$-close to g.

Remark 3.2 In Theorem 3.1.22, "properly $\mathfrak{L}(\tau)$-universal" can be replaced with "properly $\mathfrak{L}(\aleph_0)$-universal and $w(X) = \tau$" (Proposition 3.1.10).

For completely metrizable ANRs, we characterize the $\mathfrak{M}_1(\tau)$-universality as follows:

Theorem 3.1.23 *For a completely metrizable ANR X, in order to be $\mathfrak{M}_1(\tau)$-universal (resp. $\mathfrak{M}_1(\aleph_0)$-universal or $\mathfrak{M}_0$-universal), the following condition is necessary and sufficient:*

- *For any locally finite-dimensional simplicial complex K with* $\operatorname{card} K^{(0)} \leqslant \tau$ *(resp. any countable locally finite simplicial complex K or any finite simplicial complex K), each map $f : |K| \to X$ can be approximated by O_K-maps.*[6]

Proof (*Necessity*) When X is $\mathfrak{M}_1(\aleph_0)$-universal or $\mathfrak{M}_0$-universal, let K be a countable locally finite simplicial complex or a finite simplicial complex, respectively. Then, $|K| \in \mathfrak{M}_1(\aleph_0)$ or $|K| \in \mathfrak{M}_0$, respectively. Since every closed embedding from $|K|$ to X is an O_K-map, each map $f : |K| \to X$ can be approximated by O_K-maps.

When X is $\mathfrak{M}_1(\tau)$-universal, let K be a locally finite-dimensional simplicial complex with $\operatorname{card} K^{(0)} \leqslant \tau$ and let $f : |K| \to X$ be a map. Each $\mathcal{U} \in \operatorname{cov}(X)$ has an open star-refinement $\mathcal{V}$. We have a subdivision $K' \lhd K$ such that $K' \prec f^{-1}(\mathcal{V})$ (Theorem 1.6.9). By Theorem 1.7.3, $\varphi = \mathrm{id} : |K| \to |K'|_{\mathrm{m}}$ has a homotopy inverse $\psi : |K'|_{\mathrm{m}} \to |K|$ such that $\psi\varphi \simeq_{K'} \mathrm{id}$ and $\varphi\psi \simeq_{K'} \mathrm{id}$. Then, $f\psi\varphi$ is $\mathcal{V}$-close to f and $|K'|_{\mathrm{m}} \in \mathfrak{M}_1(\tau)$. By the $\mathfrak{M}_1(\tau)$-universality, $f\psi$ is $\mathcal{V}$-close to a closed embedding $h : |K'|_{\mathrm{m}} \to X$. Then, $h\varphi$ is $\mathcal{V}$-close to $f\psi\varphi$, so $\mathcal{U}$-close to f. Note that $h(O_K) = \mathcal{W}|h(|K|)$ for some $\mathcal{W} \in \operatorname{cov}(X)$. Then, $(h\varphi)^{-1}(\mathcal{W}) = O_K$, that is, $h\varphi : |K| \to X$ is an O_K-map.

(*Sufficiency*) Let $Y \in \mathfrak{M}_1(\tau)$ (resp. $Y \in \mathfrak{M}_1(\aleph_0)$ or $Y \in \mathfrak{M}_0$). By Proposition 3.1.15, it suffices to show that $\mathrm{C}_{\mathcal{V}}(Y, X)$ is dense in $\mathrm{C}(Y, X)$ for any $\mathcal{V} \in \operatorname{cov}(Y)$, that is, each $f \in \mathrm{C}(Y, X)$ can be approximated by $\mathcal{V}$-maps. For each $\mathcal{U} \in \operatorname{cov}(X)$, take an open star-refinement $\mathcal{U}'$ of $\mathcal{U}$. Applying Theorem 1.13.21 with Remark 1.13, we can obtain a σ-discrete (resp. countable star-finite or finite) refinement $\mathcal{W} \prec \mathcal{V}$ ($\in \operatorname{cov}(Y)$) with the locally finite-dimensional nerve $N(\mathcal{W})$ and a map $\psi : |N(\mathcal{W})| \to X$ such that $\psi\varphi \simeq_{\mathcal{U}'} f$, where $\varphi : Y \to |N(\mathcal{W})|$ is a canonical map for $\mathcal{W}$, that is, $\varphi^{-1}(O_{N(\mathcal{W})}(W)) \subset W$ for every $W \in \mathcal{W}$. Since $\mathcal{W}$ is σ-discrete (resp. countable star-finite or finite), it follows that $\operatorname{card} N(\mathcal{W})^{(0)} = \operatorname{card} \mathcal{W} \leqslant \aleph_0 w(Y) \leqslant \tau$ (resp. $N(\mathcal{W})$ is countable locally finite or finite). Thus, we can apply the condition to obtain an $O_{N(\mathcal{W})}$-map $\psi' : |N(\mathcal{W})| \to X$ that is $\mathcal{U}'$-close to ψ. Then, $\psi'\varphi$ is a $\mathcal{W}$-map, so it is a $\mathcal{V}$-map. Since $\psi'\varphi$ is $\mathcal{U}'$-close to $\psi\varphi$, it is $\mathcal{U}$-close to f. □

The following is the locally compact version of Theorem 3.1.23 above:

Theorem 3.1.24 *For a locally compact ANR X with density τ, in order to be properly $\mathfrak{L}(\aleph_0)$-universal (hence properly $\mathfrak{L}(\tau)$-universal), the following condition is necessary and sufficient:*

[6] Recall that O_K is the open star cover of K defined in Sect. 1.5.

- *For any countable locally finite simplicial complex K, each proper map $f : |K| \to X$ can be approximated by O_K-maps.*

Proof This can be proved by modifying the proof of Theorem 3.1.23. Just like the $\mathfrak{M}_1(\aleph_0)$-case, the necessity can be easily seen.

To prove the sufficiency, we apply Proposition 3.1.17 instead of Proposition 3.1.15. Let Y be a locally compact separable metrizable space, $f : Y \to X$ be a proper map, and $\mathcal{U} \in \operatorname{cov}(X)$ such that $\operatorname{cl} U$ is compact for every $U \in \mathcal{U}$. Moreover, let $\mathcal{U}' \in \operatorname{cov}(X)$ be an open star-refinement of $\mathcal{U}$. By Theorem 1.13.21 with Remark 1.13, every $\mathcal{V} \in \operatorname{cov}(Y)$ has a countable star-finite refinement $\mathcal{W} \in \operatorname{cov}(Y)$ with a map $\psi : |N(\mathcal{W})| \to X$ such that $\operatorname{cl} W$ is compact for every $W \in \mathcal{W}$ and $\psi\varphi \simeq_{\mathcal{U}'} f$, where $\varphi : Y \to |N(\mathcal{W})|$ is a canonical map for $\mathcal{W}$. Then, φ is proper because $\varphi^{-1}(O_{N(\mathcal{W})}(W)) \subset W$ for every $W \in \mathcal{W}$. The star-finiteness of $\mathcal{W}$ means the local compactness of $|N(\mathcal{W})|$, hence φ is perfect (Proposition 1.3.7), so it is closed. Since $\psi\varphi$ is $\mathcal{U}'$-close to the proper map f, it is proper. Then, it follows that $\psi|\varphi(Y)$ is proper. For each $U \in \mathcal{U}$, $\psi^{-1}(\operatorname{cl} U) \cap \varphi(Y)$ is compact. It is easy to find an open set U^* in $|N(\mathcal{W})|$ such that $U^* \cap \varphi(Y) = \psi^{-1}(U) \cap \varphi(Y)$ and $\operatorname{cl} U^*$ is compact. Since $\varphi(Y) \subset \bigcup_{U \in \mathcal{U}} U^*$, we have a subcomplex K of some small subdivision of $N(\mathcal{W})$ such that $\varphi(Y) \subset |K| \subset \bigcup_{U \in \mathcal{U}} U^*$. Then, $\psi||K|$ is proper. Indeed, each compact set $A \subset X$ is covered by finitely many $U_1, \ldots, U_n \in \mathcal{U}$. Since

$$\psi^{-1}(A) \cap |K| \subset \bigcup_{i=1}^{n} \psi^{-1}(U_i) \cap |K| \subset \bigcup_{i=1}^{n} \operatorname{cl} U_i^*,$$

it follows that $\psi^{-1}(A) \cap |K|$ is compact. Thus, we can apply the condition to obtain an O_K-map $\psi' : |K| \to X$ that is $\mathcal{U}'$-close to ψ. Then, $\psi'\varphi$ is a $\mathcal{W}$-map, so a $\mathcal{V}$-map. Since $\psi'\varphi$ is $\mathcal{U}'$-close to $\psi\varphi$, it is $\mathcal{U}$-close to f. □

3.2 The Discrete (or Disjoint) Cells Property

In this section, improving Theorems 3.1.23 and 3.1.24, we give characterizations of the universalities for $\mathfrak{M}_0$ and $\mathfrak{M}_1(\tau)$.

An indexed family $(A_\lambda)_{\lambda \in \Lambda}$ of subsets of a space X is said to be **discrete** (resp. **locally finite**) in X if each $x \in X$ has a neighborhood U such that

$$\operatorname{card}\left\{\lambda \in \Lambda \;\middle|\; U \cap A_\lambda \neq \emptyset\right\} \leqslant 1 \;\;(\text{resp. } < \aleph_0).^7$$

[7] This is not the same as saying that the collection $\{A_\lambda \mid \lambda \in \Lambda\}$ is discrete (resp. locally finite) in X. Indeed, when $A_\lambda = A \neq \emptyset$ for every $\lambda \in \Lambda$, $\{A_\lambda \mid \lambda \in \Lambda\}$ is discrete in X, but $(A_\lambda)_{\lambda \in \Lambda}$ is not discrete (resp. locally finite) in X if $\operatorname{card} \Lambda > 1$ (resp. $\geqslant \aleph_0$). Evidently, if $(A_\lambda)_{\lambda \in \Lambda}$ is discrete (resp. locally finite) in X, then $\{A_\lambda \mid \lambda \in \Lambda\}$ is discrete (resp. locally finite) in X.

For each $n \in \omega$, it is said that X has the **τ-discrete** (resp. **τ-locally finite**) **n-cells property** if every map $f : \mathbf{I}^n \times \Gamma \to X$ can be approximated by maps $g : \mathbf{I}^n \times \Gamma \to X$ such that $(g_\gamma(\mathbf{I}^n))_{\gamma \in \Gamma}$ is discrete (resp. locally finite) in X, where Γ is a discrete space with $\operatorname{card} \Gamma = \tau$, $g_\gamma : \mathbf{I}^n \to X$ is defined by $g_\gamma(x) = g(x, \gamma)$, and $\mathbf{I}^0 = \{0\}$ when $n = 0$. The $\aleph_0$-discrete (resp. $\aleph_0$-locally finite) n-cells property is simply called the **countable discrete** (resp. **countable locally finite**) **n-cells property**. When X has the τ-discrete (resp. τ-locally finite) n-cells property for every $n \in \mathbb{N}$, it is said that X has the **τ-discrete** (resp. **τ-locally finite**) **cells property**. The $\aleph_0$-discrete (resp. $\aleph_0$-locally finite) cells property is simply called the **countable discrete** (resp. **countable locally finite**) **cells property**. Evidently, if X has the τ-locally finite n-cells property, then $w(X) \geqslant \tau$. The following are easily observed:

Fact (1) *The τ-discrete n-cells property implies the disjoint n-cells property and the τ'-discrete n'-cells property for every $\aleph_0 \leqslant \tau' \leqslant \tau$ and $0 \leqslant n' \leqslant n$.*
(2) *A paracompact space X has the τ-discrete 0-cells property if and only if every neighborhood of each point of X contains a discrete set D with* $\operatorname{card} D = \tau$.

The "if" part of (2) can be shown as follows: When $\tau = \aleph_0$, we regard $\Gamma = (\mathbb{N}, \leqslant)$, where $\leqslant$ is the natural order. When $\tau > \aleph_0$, we can assume that $\Gamma = (\Gamma, \leqslant)$ is a well-ordered set such that $\operatorname{card} \Gamma(\gamma) < \operatorname{card} \Gamma = \tau$ for every $\gamma \in \Gamma$, where $\Gamma(\gamma) = \{\gamma' \in \Gamma \mid \gamma' < \gamma\}$. Indeed, let $\Gamma = (\Gamma, \leqslant)$ be a well-ordered set with $\operatorname{card} \Gamma = \tau$. If $\operatorname{card} \Gamma(\gamma) = \tau$ for some $\gamma \in \Gamma$, take the minimum element $\gamma_0 \in \Gamma$ such that $\operatorname{card} \Gamma(\gamma_0) = \tau$ and replace Γ with $\Gamma(\gamma_0)$.

Each $\mathcal{U} \in \operatorname{cov}(X)$ has a locally finite refinement $\mathcal{V} \in \operatorname{cov}(X)$. Given a net $(x_\gamma)_{\gamma \in \Gamma} \in X^\Gamma$, choose $V_\gamma \in \mathcal{V}$, $\gamma \in \Gamma$, so that $x_\gamma \in V_\gamma$. Each V_γ contains a discrete set D_γ with $\operatorname{card} D_\gamma = \tau$, where it might occur that $D_\gamma = D_{\gamma'}$ even if $V_\gamma \neq V_{\gamma'}$. Since $\{D_\gamma \mid \gamma \in \Gamma\}$ is locally finite, $D = \bigcup_{\gamma \in \Gamma} D_\gamma$ is discrete in X. By transfinite induction, we can choose $y_\gamma \in D_\gamma \setminus \{y_{\gamma'} \mid \gamma' < \gamma\} \subset D$, $\gamma \in \Gamma$, because $\operatorname{card} \Gamma(\gamma) < \tau = \operatorname{card} D_\gamma$. Then, $(y_\gamma)_{\gamma \in \Gamma}$ is discrete in X and $\{\{x_\gamma, y_\gamma\} \mid \gamma \in \Gamma\} \prec \mathcal{U}$.

Obviously the τ-discrete n-cells property implies the τ-locally finite n-cells property. However, the converse is also true as in the following proposition. Therefore, these concepts are equivalent.

Proposition 3.2.1 *For each $n \in \omega$, the τ-locally finite n-cells property implies the τ-discrete n-cells property.*

Proof As in the above proof of the "if" part of Fact (2), we can assume that $\Gamma = (\mathbb{N}, \leqslant)$ ($\leqslant$ is the natural order) or $\Gamma = (\Gamma, \leqslant)$ is a well-ordered set such that $\operatorname{card} \Gamma(\gamma) < \operatorname{card} \Gamma = \tau$ for every $\gamma \in \Gamma$, where $\Gamma(\gamma) = \{\gamma' \in \Gamma \mid \gamma' < \gamma\}$.

Let $f : \mathbf{I}^n \times \Gamma \to X$ be a map and $\mathcal{U} \in \operatorname{cov}(X)$. Since $\operatorname{card} \Gamma \geqslant \aleph_0$, we have $\operatorname{card} \Gamma^2 = \operatorname{card} \Gamma$. Applying the τ-locally finite n-cells property to the map $\tilde{f} : \mathbf{I}^n \times \Gamma^2 \to X$ defined by $\tilde{f}(z, \gamma, \gamma') = f(z, \gamma)$, we can obtain a map $\tilde{g}$ such that $\tilde{g}$ is $\mathcal{U}$-close to $\tilde{f}$ and $(\tilde{g}(\mathbf{I}^n \times \{(\gamma, \gamma')\}))_{(\gamma, \gamma') \in \Gamma^2}$ is locally finite in X. By the compactness, each $\tilde{g}(\mathbf{I}^n \times \{(\gamma, \gamma')\})$ meets only finitely many other $\tilde{g}(\mathbf{I}^n \times \{(\delta, \delta')\})$.

By transfinite induction (or induction), we can choose $\delta(\gamma) \in \Gamma$ for each $\gamma \in \Gamma$ so that

$$\tilde{g}\big(\mathbf{I}^n \times \{(\gamma, \delta(\gamma))\}\big) \cap \tilde{g}\big(\mathbf{I}^n \times \{(\gamma', \delta(\gamma'))\}\big) = \emptyset \ \text{ for } \gamma' < \gamma.$$

Then, $(\tilde{g}(\mathbf{I}^n \times \{(\gamma, \delta(\gamma))\}))_{\gamma\in\Gamma}$ is discrete in X because it is locally finite and mutually disjoint. Hence, the desired map $g : \mathbf{I}^n \times \Gamma \to X$ can be defined by $g(z, \gamma) = \tilde{g}(z, \gamma, \delta(\gamma))$. □

A map $\varphi : X \to Y$ is said to be **polyhedrally approximately soft** provided that for any polyhedron Z, any map $f : C \to X$ of a subpolyhedron C of Z, and any open cover $\mathcal{U} \in \operatorname{cov}(Y)$, if φf extends to a map $g : Z \to Y$, then f extends to $\tilde{f} : Z \to X$ such that $\varphi\tilde{f}$ is $\mathcal{U}$-close to g:

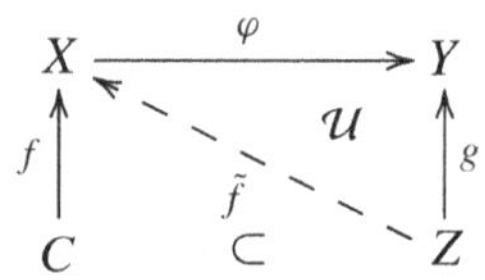

In the above, when the condition $\dim Z \leqslant n$ is added, φ is said to be **polyhedrally approximately *n*-soft**. For soft maps, refer to Section 7.2 in [GAGT].

Proposition 3.2.2 *Let $\varphi : X \to Y$ be a polyhedrally approximately n-soft proper map and $n \in \omega$. If X has the τ-discrete n-cells property, then Y has also the τ-discrete n-cells property.*

Proof By virtue of Proposition 3.2.1 above, the τ-discrete n-cells property can be replaced with the τ-locally finite n-cells property. Let $f : \mathbf{I}^n \times \Gamma \to Y$ be a map. For any $\mathcal{U} \in \operatorname{cov}(Y)$, let $\mathcal{V} \in \operatorname{cov}(Y)$ be a star-refinement of $\mathcal{U}$. Since φ is polyhedrally approximately n-soft, we have a map $g : \mathbf{I}^n \times \Gamma \to X$ such that φg is $\mathcal{V}$-close to f. By the τ-locally finite n-cells property of X, g is $\mathcal{V}$-close to a map $g' : \mathbf{I}^n \times \Gamma \to X$ such that $(g'_\gamma(\mathbf{I}^n))_{\gamma\in\Gamma}$ is locally finite in X. Then, it follows that $\varphi g'_\gamma$ is $\mathcal{U}$-close to f. Moreover, $(\varphi g'_\gamma(\mathbf{I}^n))_{\gamma\in\Gamma}$ is locally finite in Y. Indeed, if it is not locally finite at $y \in Y$, then we can find $x_i \in \mathbf{I}^n$ and $\gamma_i \in \Gamma$ for each $i \in \mathbb{N}$ so that $\gamma_i \neq \gamma_j$ if $i \neq j$, and $(\varphi g'_{\gamma_i}(x_i))_{i\in\mathbb{N}}$ converges to y. Because φ is proper, it follows from Proposition 1.3.7 that $(g'_{\gamma_i}(x_i))_{i\in\mathbb{N}}$ has a convergent subsequence in X, which contradicts the local finiteness of $(g'_\gamma(\mathbf{I}^n))_{\gamma\in\Gamma}$. □

For an ANR, the τ-discrete n-cells property is open hereditary, that is:

Proposition 3.2.3 *Let X be an ANR and $n \in \omega$. If X has the τ-discrete (or τ-locally finite) n-cells property, then every open set W in X has the τ-discrete (or τ-locally finite) n-cells property.*

Proof By virtue of Proposition 3.2.1, it suffices to show that W has the τ-locally finite n-cells property.

Let $f : \mathbf{I}^n \times \Gamma \to W$ be a map and $\mathcal{U} \in \operatorname{cov}(W)$. We can write $W = \bigcup_{i\in\mathbb{N}} W_i$, where each W_i is open in X and $\operatorname{cl} W_i \subset W_{i+1}$. For each $i \in \mathbb{N}$, let $R_i = f^{-1}(\operatorname{cl} W_i \setminus W_{i-1})$, where $W_0 = \emptyset$. Then, R_i is closed in $\mathbf{I}^n \times \Gamma$ and $f(R_i) \subset \operatorname{cl} W_i \setminus W_{i-1}$. Take a star-refinement $\mathcal{U}_0 \in \operatorname{cov}(W)$ of $\mathcal{U}$ and define $\mathcal{U}' \in \operatorname{cov}(W)$ and $\mathcal{V}_i \in \operatorname{cov}(X)$, $i \in \mathbb{N}$, as follows:

$$\mathcal{U}' = \bigcup_{i\in\mathbb{N}} \mathcal{U}_0|W_{i+1} \setminus \operatorname{cl} W_{i-1} \quad \text{and} \quad \mathcal{V}_i = \mathcal{U}'|W_{2i} \cup \{X \setminus \operatorname{cl} W_{2i-1}\},$$

where $W_{-1} = \emptyset$. By the τ-discrete n-cells property of X, we have maps $g_i : \mathbf{I}^n \times \Gamma \to X$, $i \in \mathbb{N}$, such that $g_i \simeq_{\mathcal{V}_i} f$ and $(g_i(\mathbf{I}^n \times \{\gamma\})_{\gamma\in\Gamma}$ is discrete in X. Then, $g_i|R_{2i-1} \simeq_{\mathcal{U}'} f|R_{2i-1}$ (in W) for each $i \in \mathbb{N}$. Using the Homotopy Extension Theorem 1.13.11, we can obtain a map $g : \mathbf{I}^n \times \Gamma \to W$ such that $g|R_{2i-1} = g_i|R_{2i-1}$ for each $i \in \mathbb{N}$ and $g \simeq_{\mathcal{U}'} f$. Then, it follows that $g(R_i) \subset W_{i+1} \setminus \operatorname{cl} W_{i-2}$ for each $i \in \mathbb{N}$. Let

$$A_{\gamma,i} = (\mathbf{I}^n \times \{\gamma\}) \cap R_{2i-1}, \; \gamma \in \Gamma, \; i \in \mathbb{N}.$$

As is easily observed, $(g(A_{\gamma,i}))_{\gamma\in\Gamma}$ is discrete in $W_{2i+1} \setminus \operatorname{cl} W_{2i-2}$, which implies that $(g(A_{\gamma,i}))_{\gamma\in\Gamma,i\in\mathbb{N}}$ is locally finite in W.

For each $x \in W$, let V_x be an open neighborhood of x in W such that V_x meets only finitely many $g(A_{\gamma,i})$, $\gamma \in \Gamma$, $i \times \mathbb{N}$. Take $\mathcal{V}' \in \operatorname{cov}(W)$ so that $\operatorname{st}\mathcal{V}' \prec \{V_x \mid x \in W\}$ and $\mathcal{V}' \prec \mathcal{U}'$. If h is $\mathcal{V}'$-close to g, then $(h(A_{\gamma,i})_{\gamma\in\Gamma,i\in\mathbb{N}}$ is locally finite in W. Indeed, for each $V \in \mathcal{V}'$, $\operatorname{st}(V, \mathcal{V}')$ is contained in some V_x. If $V \cap h(A_{\gamma,i}) \neq \emptyset$, then $\operatorname{st}(V, \mathcal{V}') \cap g(A_{\gamma,i}) \neq \emptyset$, which implies that $V_x \cap g(A_{\gamma,i}) \neq \emptyset$. Such V_x meets only finitely many $g(A_{\gamma,i})$, V meets only finitely many $h(A_{\gamma,i})$. Recall that $g(R_i) \subset W_{i+1} \setminus \operatorname{cl} W_{i-2}$ for each $i \in \mathbb{N}$. Let

$$B_{\gamma,i} = (\mathbf{I}^n \times \{\gamma\}) \cap R_{2i}, \; \gamma \in \Gamma, \; i \in \mathbb{N}.$$

Then, $g(B_{\gamma,i}) \subset W_{2i+1} \setminus \operatorname{cl} W_{2i-2}$. In the same way as the above g, we can construct a map $h : \mathbf{I}^n \times \Gamma \to W$ so that $h \simeq_{\mathcal{V}'} g$ and $(h(B_{\gamma,i}))_{\gamma\in\Gamma}$ is discrete in $W_{2i+2} \setminus \operatorname{cl} W_{2i-3}$, which implies that $(h(B_{\gamma,i}))_{\gamma\in\Gamma,i\in\mathbb{N}}$ is locally finite in W. Since $h \simeq_{\mathcal{V}'} g$, $(h(A_{\gamma,i}))_{\gamma\in\Gamma,i\in\mathbb{N}}$ is also locally finite in W. Observe

$$\mathbf{I}^n \times \{\gamma\} = \bigcup_{i\in\mathbb{N}} \big((\mathbf{I}^n \times \{\gamma\}) \cap R_i\big) = \bigcup_{i\in\mathbb{N}} (A_i \cup B_i).$$

Therefore, $(h(\mathbf{I}^n \times \{\gamma\}))_{\gamma\in\Gamma,i\in\mathbb{N}}$ is locally finite in W. □

Remark 3.3 Let $A = \bigoplus_{\lambda\in\Lambda} A_\lambda$ be a topological sum of metrizable spaces. Suppose that an ANR X has the following property:

(*) every map $f : A \to X$ can be approximated by maps $g : A \to X$ such that $(g(A_\lambda))_{\lambda\in\Lambda}$ is locally finite in X.

Then, by the same proof as Proposition 3.2.3, it can be proved that every open set in X has property (*).

It is said that X has the **disjoint n-cells property** if every two maps $f, g : \mathbf{I}^n \to X$ can be approximated by maps $f', g' : \mathbf{I}^n \to X$ such that $f'(\mathbf{I}^n) \cap g'(\mathbf{I}^n) = \emptyset$. Namely, the disjoint n-cells property can be defined by replacing Γ with $\mathbf{2} = \{0, 1\}$. If X has the disjoint n-cells property for every $n \in \mathbb{N}$, it is said that X has the **disjoint cells property**.

The open heritability of the disjoint n-cells property can be directly proved, which is simpler than Proposition 3.2.3.

Proposition 3.2.4 *Let X be a paracompact space and $n, n_1, \ldots, n_k \in \mathbb{N}$ $(k \geqslant 2)$ with $n_1, \ldots, n_k \leqslant n$. If X has the disjoint n-cells property, then all maps $f_i : \mathbf{I}^{n_i} \to X$, $i = 1, \ldots, k$, can be approximated by maps $f'_i : \mathbf{I}^{n_i} \to X$ such that $f'_i(\mathbf{I}^{n_i}) \cap f'_j(\mathbf{I}^{n_j}) = \emptyset$ for $i \neq j$.*

Proof We shall prove the proposition by induction on k. To see the case $k = 2$, for $m < n$, let $p_m : \mathbf{I}^n \to \mathbf{I}^m$ and $i_m : \mathbf{I}^m \to \mathbf{I}^n$ be the maps defined by

$$p_m(x) = (x(1), \ldots, x(m)) \text{ and } i_m(x) = (x(1), \ldots, x(m), 0, \ldots, 0).$$

We write $p_n = i_n = \mathrm{id}_{\mathbf{I}^n}$ for convenience. Let $f_j : \mathbf{I}^{n_j} \to X$, $j = 1, 2$, be maps and $\mathcal{U} \in \operatorname{cov}(X)$. By the disjoint n-cells property, the maps $f_j p_{n_j} : \mathbf{I}^n \to X$ are $\mathcal{U}$-close to maps $g_j : \mathbf{I}^n \to X$ such that $g_1(\mathbf{I}^n) \cap g_2(\mathbf{I}^n) = \emptyset$. Then, the maps $f'_j = g_j i_{n_j} : \mathbf{I}^{n_j} \to X$ are $\mathcal{U}$-close to $f_j = f_j p_{n_j} i_{n_j}$ and $f'_1(\mathbf{I}^n) \cap f'_2(\mathbf{I}^n) = \emptyset$.

Now, assuming the case $k - 1$, we prove the case k. Let $f_i : \mathbf{I}^{n_i} \to X$, $i = 1, \ldots, k$, be maps. For each $\mathcal{U} \in \operatorname{cov}(X)$, take its star-refinement $\mathcal{V} \in \operatorname{cov}(X)$. By the inductive assumption, the maps $f_1, \ldots, f_{k-1}$ are $\mathcal{V}$-close to maps $f'_i : \mathbf{I}^{n_i} \to X$, $i = 1, \ldots, k - 1$, such that $f'_i(\mathbf{I}^{n_i}) \cap f'_j(\mathbf{I}^{n_j}) = \emptyset$ for $1 \leqslant i < j < k$. Take $\mathcal{V}' \in \operatorname{cov}(X)$ so that

$$\mathcal{V}' \prec \mathcal{V} \wedge \left\{X \setminus f'_1(\mathbf{I}^{n_1}),\ X \setminus \textstyle\bigcup_{i=2}^{k-1} f'_i(\mathbf{I}^{n_i})\right\}.$$

Again by the inductive assumption, the maps $f'_2, \ldots, f'_{k-1}, f_k$ are $\mathcal{V}'$-close to maps $f''_i : \mathbf{I}^{n_i} \to X$, $i = 2, \ldots, k$, such that $f''_i(\mathbf{I}^{n_i}) \cap f''_j(\mathbf{I}^{n_j}) = \emptyset$ for $1 < i < j \leqslant k$. Then, observe that $f'_1(\mathbf{I}^{n_1}) \cap f''_i(\mathbf{I}^n) = \emptyset$ for every $i = 2, \ldots, k - 1$. Take $\mathcal{V}'' \in \operatorname{cov}(X)$ so that

$$\mathcal{V}'' \prec \mathcal{V} \wedge \left\{X \setminus (f'_1(\mathbf{I}^{n_1}) \cup f''_k(\mathbf{I}^{n_k})),\ X \setminus \textstyle\bigcup_{i=2}^{k-1} f''_i(\mathbf{I}^{n_i})\right\}.$$

By the case $k = 2$, the maps f'_1 and f''_k are $\mathcal{V}''$-close to maps $f'''_1 : \mathbf{I}^{n_1} \to X$ and $f'''_k : \mathbf{I}^{n_k} \to X$ such that $f'''_1(\mathbf{I}^{n_1}) \cap f'''_k(\mathbf{I}^{n_k}) = \emptyset$. Then, observe that $f'''_1(\mathbf{I}^{n_1}) \cap f''_i(\mathbf{I}^{n_i}) = \emptyset$ and $f'''_k(\mathbf{I}^{n_1}) \cap f''_i(\mathbf{I}^{n_i}) = \emptyset$ for every $i = 2, \ldots, k - 1$. For each $1 < i < k$, let $f'''_i = f''_i$. Then, it follows that $f'''_i(\mathbf{I}^n) \cap f'''_j(\mathbf{I}^n) = \emptyset$ for $1 \leqslant i < j \leqslant k$. Moreover, every f'''_i is $\mathcal{U}$-close to f_i because $\operatorname{st} \mathcal{V} \prec \mathcal{U}$. □

Since $\mathbf{I}^n \times \Gamma \in \mathfrak{M}_1(\tau)$ and $\mathbf{I}^n \times \{0, 1\} \in \mathfrak{M}_0$, we have the following:

Fact (1) *The $\mathfrak{M}_1(\tau)$-universality implies the τ-discrete cells property.*
(2) *The $\mathfrak{M}_0$-universality implies the disjoint cells property.*

Remark 3.4 Let E be an infinite-dimensional normed linear space with dens $E \geqslant \tau$. By Proposition 2.1.8, we can take Γ as a δ-discrete set in the sphere $\mathbf{S}_E$ of E, where $0 < \delta < 1$. Taking an embedding $h : \mathbf{I}^n \to E$ with $h(\mathbf{I}^n) \subset (\delta/3)\mathbf{B}_E$, we can define a closed embedding $f : \mathbf{I}^n \times \Gamma \to E$ by $f(x, \gamma) = h(x) + \gamma$. Thus, each $\mathbf{I}^n \times \Gamma$ can be embedded in E as a closed set, that is, $\mathbf{I}^n \times \Gamma \in \mathcal{F}_E$. Consequently, the $\mathcal{F}_E$-universality implies the τ-discrete cells property.

For a simplex σ with $\dim \sigma > 0$ and $t \in \mathbf{I}$, the following notation is useful:

$$\sigma[t] = \{(1-s)\hat{\sigma} + sy \mid y \in \partial\sigma,\ 0 \leqslant s \leqslant t\}.$$

Then, $\sigma[0] = \{\hat{\sigma}\}$, $\sigma[1] = \sigma$, and $\sigma[t] \subset \operatorname{rint} \sigma$ for every $t < 1$.

Lemma 3.2.5 *Let X be an ANR with the τ-discrete cells property and K be a finite-dimensional simplicial complex with* card $K^{(0)} \leqslant \tau$. *Suppose that all maps from $\mathbf{I}^n$, $n \in \mathbb{N}$, to X can be approximated by maps with strong Z-set images. Then, each map $f : |K| \to X$ can be approximated by O_K-maps.*

Proof By induction on $\dim K$, we shall show the proposition. It is easy to see that the τ-discrete 0-cells property implies the case $\dim K = 0$. Assume that the proposition is valid for any $(n-1)$-dimensional simplicial complex. Let $\dim K = n$ and $f : |K| \to X$ be a map. For each $\mathcal{U} \in \operatorname{cov}(X)$, let $\mathcal{V} \in \operatorname{cov}(X)$ be a star-refinement of $\mathcal{U}$.

First, by the inductive assumption and the Homotopy Extension Theorem 1.13.11, we have a map $f' : |K| \to X$ such that $f'||K^{(n-1)}|$ is an $O_{K^{(n-1)}}$-map and $f' \simeq_{\mathcal{V}} f$. Since $O_{K^{(n-1)}} = O_K|K^{(n-1)}$, we apply Proposition 3.1.18 to obtain an open neighborhood R of $|K^{(n-1)}|$ in $|K|$ such that $f'|R$ is an $(O_K|R)$-map, that is, f' belongs to the following subset of $\mathrm{C}(|K|, X)$:

$$\{g \in \mathrm{C}(|K|, X) \mid g|R \text{ is an } (O_K|R)\text{-map}\}.$$

This set is open in $\mathrm{C}(|K|, X)$ because $\mathrm{C}_{O_K|R}(R, X)$ is open in $\mathrm{C}(R, X)$ by Lemma 3.1.13 and the restriction operator of $\mathrm{C}(|K|, X)$ to $\mathrm{C}(R, X)$ is continuous by Proposition 3.1.4(2). Hence, we can find $\mathcal{V}' \in \operatorname{cov}(X)$ such that $\operatorname{st} \mathcal{V}' \prec \mathcal{V}$ and if a map $g' : |K| \to X$ is $\operatorname{st} \mathcal{V}'$-close to f', then $g'|R$ is also an $(O_K|R)$-map.

Next, using the τ-discrete n-cells property, we can obtain maps $f_\sigma : \sigma \to X$, $\sigma \in K \setminus K^{(n-1)}$, such that $\{f_\sigma(\sigma) \mid \sigma \in K \setminus K^{(n-1)}\}$ is discrete in X, $f_\sigma \simeq_{\mathcal{V}'} f'|\sigma$, and $f_\sigma(\sigma)$ is a strong Z-set in X. For each $\sigma \in K \setminus K^{(n-1)}$, let $h^\sigma : \sigma \times \mathbf{I} \to X$ be a $\mathcal{V}'$-homotopy with $h^\sigma_0 = f'|\sigma$ and $h^\sigma_1 = f_\sigma$. For each $\sigma \in K \setminus K^{(n-1)}$, choose $t_\sigma \in (0,1)$ so that $\mathrm{cl}(\sigma \setminus \sigma[t_\sigma]) \subset R$ and define a map $g_\sigma : \sigma \to X$ by $g_\sigma|\sigma[t_\sigma] = f_\sigma|\sigma[t_\sigma]$ and

$$g_\sigma((1-s)\hat{\sigma} + sy) = h^\sigma\left((1-s)\hat{\sigma} + sy, \frac{1-s}{1-t_\sigma}\right) \text{ for } (y,s) \in \partial\sigma \times [t_\sigma, 1].$$

Then, $\{g_\sigma(\sigma[t_\sigma]) \mid \sigma \in K \setminus K^{(n-1)}\}$ is discrete in X, and each $g_\sigma(\sigma[t_\sigma])$ is a strong Z-set in X. Observe that

$$g_\sigma|\partial\sigma = f'|\partial\sigma \text{ and } g_\sigma \simeq_{\mathcal{V}'} f'|\sigma \text{ rel. } \partial\sigma.$$

Let $g' = \mathrm{pr}_X\tilde{g} : |K| \to X$, that is, $g'|\sigma = g_\sigma$ for each $\sigma \in K \setminus K^{(n-1)}$. Then, $g' \simeq_{\mathcal{V}'} f'$ and $\{g'(\sigma[t_\sigma]) \mid \sigma \in K \setminus K^{(n-1)}\}$ is discrete in X. Let

$$Z = \bigcup_{\sigma \in K \setminus K^{(n-1)}} g'(\sigma[t_\sigma]).$$

Since each $g'(\sigma[t_\sigma]) = f_\sigma(\sigma[t_\sigma])$ is a strong Z-set in X, their discrete union Z is also a strong Z-set in X by Corollary 2.8.16. Due to Proposition 2.8.13, there is a homotopy $\varphi : X \times \mathbf{I} \to X$ such that $\varphi_0 = \mathrm{id}$, $\varphi(X \times (0,1]) \cap Z = \emptyset$, and φ is closed over Z, that is, $Z \cap \mathrm{cl}\,\varphi(A) = Z \cap \varphi(A)$ for any closed set A in $X \times \mathbf{I}$. By Lemma 2.5.2, we have a map $\alpha : X \to (0,1]$ such that

$$\big\{\{x\} \times [0, \alpha(x)] \bigm| x \in X\big\} \prec \varphi^{-1}(\mathcal{V}').$$

Since $\{g'(\sigma[t_\sigma]) \mid \sigma \in K \setminus K^{(n-1)}\}$ is discrete in X, X has a discrete open collection $\{W_\sigma \mid \sigma \in K \setminus K^{(n-1)}\}$ such that $g'(\sigma[t_\sigma]) \subset W_\sigma$ for every $\sigma \in K \setminus K^{(n-1)}$. Then, $\sigma[t_\sigma] \subset g'^{-1}(W_\sigma)$. Since $|K|$ is perfectly normal by Proposition 1.5.2, there is a map $\beta : |K| \to \mathbf{I}$ such that

$$\beta^{-1}(0) = \bigcup_{\sigma \in K \setminus K^{(n-1)}} \sigma[t_\sigma] \text{ and}$$

$$\beta^{-1}(1) = |K| \setminus \bigcup_{\sigma \in K \setminus K^{(n-1)}} \big(\mathrm{rint}\,\sigma \cap g'^{-1}(W_\sigma)\big),$$

which implies

$$\beta^{-1}((0,1]) = |K| \setminus \bigcup_{\sigma \in K \setminus K^{(n-1)}} \sigma[t_\sigma] = \bigcup_{\sigma \in K \setminus K^{(n-1)}} (\sigma \setminus \sigma[t_\sigma])$$

$$= \bigcup_{\sigma \in K \setminus K^{(n-1)}} k_\sigma(\partial\sigma \times [0,1)) \subset R \text{ and}$$

$$\beta^{-1}([0,1)) = \bigcup_{\sigma \in K \setminus K^{(n-1)}} \left(\operatorname{rint}\sigma \cap g'^{-1}(W_\sigma)\right).$$

We now define a map $g : |K| \to X$ by

$$g(x) = \varphi(g'(x), \alpha g'(x) \cdot \beta(x)) \quad \text{for each } x \in |K|.$$

Then, $g|\sigma[t_\sigma] = g'|\sigma[t_\sigma]$ for each $\sigma \in K \setminus K^{(n-1)}$, so we have

$$Z = \bigcup_{\sigma \in K \setminus K^{(n-1)}} g(\sigma[t_\sigma]).$$

Moreover, g is $\mathcal{V}$-close to f because g is $\mathcal{V}'$-close to g', g' is $\mathcal{V}'$-close to f', f' is $\mathcal{V}$-close to f, $\mathcal{V}' \prec \mathcal{V}$, and $\operatorname{st}\mathcal{V} \prec \mathcal{U}$. Since g is $\operatorname{st}\mathcal{V}'$-close to f', it follows that $g|R$ is also an O_K-map.

Finally, we will show that g is an O_K-map. Each $x \in X$ has an open neighborhood W in X such that $(g|R)^{-1}(W)$ is contained in $O_K(v)$ for some $v \in K^{(0)}$ (the case $(g|R)^{-1}(W) = \emptyset$ is possible). When $x \in X \setminus Z$, $W \setminus Z = W \cap (X \setminus Z)$ is an open neighborhood of x in X and

$$\begin{aligned} g^{-1}(W \setminus Z) &= g^{-1}(W) \setminus \bigcup_{\sigma \in K \setminus K^{(n-1)}} g^{-1}(g(\sigma[t_\sigma])) \\ &\subset g^{-1}(W) \setminus \bigcup_{\sigma \in K \setminus K^{(n-1)}} \sigma[t_\sigma] \\ &= g^{-1}(W) \cap \left(|K| \setminus \bigcup_{\sigma \in K \setminus K^{(n-1)}} \sigma[t_\sigma]\right) \\ &= g^{-1}(W) \cap \bigcup_{\sigma \in K \setminus K^{(n-1)}} (\sigma \setminus \sigma[t_\sigma]) \\ &\subset g^{-1}(W) \cap R = (g|R)^{-1}(W) \subset O_K(v). \end{aligned}$$

When $x \in Z$, x is contained in only one $g(\sigma[t_\sigma])$. Then, W_σ is an open neighborhood of x in X. Note that $\varphi(x, 0) = x$ and $x \notin \varphi(X \times (0, 1])$ because $x \in Z$. Hence, $\varphi^{-1}(x) = \{(x, 0)\}$. Consider the following closed set in $X \times \mathbf{I}$:

$$A = \{(x', t) \in W_\sigma \times \mathbf{I} \mid t \geqslant \alpha(x')\}.$$

Then, $Z \cap \varphi(A) = \emptyset$. Since $\varphi : X \times \mathbf{I} \to X$ is closed over Z, it follows that $Z \cap \operatorname{cl}\varphi(A) = \emptyset$. Hence, x has an open neighborhood W_0 in X such that $W_0 \subset W_\sigma$ and $W_0 \cap \varphi(A) = \emptyset$, so

$$\varphi^{-1}(W_0) \subset \left\{(x', t) \in W_\sigma \times \mathbf{I} \;\middle|\; t < \alpha(x')\right\}.$$

For each $y \in g^{-1}(W_0)$, $(g'(y), \alpha g'(y)\beta(y)) \in \varphi^{-1}(W_0)$, which means $\beta(y) < 1$. Since $g'(y) \in W_0 \subset W_\sigma$, it follows that $y \in \operatorname{rint}\sigma \cap g'^{-1}(W_\sigma)$. Hence, $g^{-1}(W_0) \subset \operatorname{rint}\sigma \subset O_K(v)$ for $v \in \sigma^{(0)}$. Thus, g is an O_K-map. □

Remark 3.5 In Lemma 3.2.5, if the projection $\mathrm{pr}_X : X \times \mathbf{I} \to X$ is a near-homeomorphism, e.g., X is $\boldsymbol{Q}$-stable (Corollary 2.3.11), then every map $f : C \to X$ from a compactum C can be approximated by maps with strong Z-set images. Indeed, for each $\mathcal{U} \in \operatorname{cov}(X)$, take a homeomorphism $\varphi : X \times \mathbf{I} \to X$ that is $\mathcal{U}$-close to pr_X. Then, f is $\mathcal{U}$-close to the map $f' : C \to X$ defined as follows:

$$f'(x) = \varphi(f(x), 0) \quad \text{for each } x \in C.$$

Then, $f'(C)$ is a strong Z-set in X because it is collared (Lemma 2.8.2).

Note that the disjoint cells property means the m-discrete cells property for every $m \in \mathbb{N}$ (cf. Proposition 3.2.4). Hence, the proof above is available, even if the τ-discrete cells property and the condition $\operatorname{card} K^{(0)} \leqslant \tau$ are replaced by the disjoint cells property and the condition $\operatorname{card} K^{(0)} < \aleph_0$ (i.e., K is a finite simplicial complex). Thus, we have the following:

Lemma 3.2.6 *Let X be a separable completely metrizable ANR and K a finite simplicial complex. Suppose that all maps from $\mathbf{I}^n$, $n \in \mathbb{N}$, to X can be approximated by maps with strong Z-set images. Then, for any $\mathcal{U} \in \operatorname{cov}(X)$, each map $f : |K| \to X$ is $\mathcal{U}$-close to an O_K-map.* □

In the above lemma, as is easily observed, the assumption implies that X has the disjoint cells property. In fact, we have this following characterization of the disjoint cells property for a separable completely metrizable ANR X:

Proposition 3.2.7 *For each separable completely metrizable ANR X, the following are equivalent:*

(a) *X has the disjoint cells property;*
(b) *Each pair of maps $f, g : \boldsymbol{Q} \to X$ are arbitrarily close to maps $f', g' : \boldsymbol{Q} \to X$ with $f'(\boldsymbol{Q}) \cap g'(\boldsymbol{Q}) = \emptyset$;*
(c) *$\mathrm{C}(\boldsymbol{Q}, X)$ has no isolated points and $\mathrm{C}(\boldsymbol{Q}, X)$ contains a countable dense set $\{f_i \mid i \in \mathbb{N}\}$ such that $f_i(\boldsymbol{Q}) \cap f_j(\boldsymbol{Q}) = \emptyset$ if $i \neq j$;*
(d) *Every map $f : \boldsymbol{Q} \to X$ can be approximated by maps $g : \boldsymbol{Q} \to X$ such that $g(\boldsymbol{Q})$ is a Z-set in X;*
(e) *For each $n \in \mathbb{N}$, every map $f : \mathbf{I}^n \to X$ can be approximated by maps $g : \mathbf{I}^n \to X$ such that $g(\mathbf{I}^n)$ is a Z-set in X;*
(f) *For each $n \in \mathbb{N}$, every map $f : \mathbf{I}^n \to X$ can be approximated by maps $g : \mathbf{I}^n \to X$ such that $g(\mathbf{I}^n)$ is a Z_n-set in X.*

Proof The implications (e) ⇒ (f) ⇒ (a) are trivial. For each $n \in \mathbb{N}$, let $p_n : \boldsymbol{Q} \to \mathbf{I}^n$ and $i_n : \mathbf{I}^n \to \boldsymbol{Q}$ be the maps defined by

$$p_n(x) = (x(1), \ldots, x(n)) \quad \text{and} \quad i_n(x) = (x(1), \ldots, x(n), 0, 0, \ldots).$$

Then, $p_n i_n = \mathrm{id}$, from which the implication (d) ⇒ (e) easily follows. On the other hand, $i_n p_n$ converges to $\mathrm{id}_{\boldsymbol{Q}}$ as $n \to \infty$. We can use this fact to see the implication (a) ⇒ (b). In condition (c), $\{f_i \mid i \in \mathbb{N}\} \setminus \{f_j\}$ is dense in $\mathrm{C}(\boldsymbol{Q}, X)$ for each $j \in \mathbb{N}$ because f_j is not isolated in $\mathrm{C}(\boldsymbol{Q}, X)$. Then, each $f_j(\boldsymbol{Q})$ satisfies condition (b) of Theorem 2.8.6, which means that $f_j(\boldsymbol{Q})$ is a Z-set in X. So, we have the implication (c) ⇒ (d).

(b) ⇒ (c): We may assume that $X = (X, d)$ is a separable complete metric space. The space $\mathrm{C}(\boldsymbol{Q}, X)$ with the sup-metric is complete, where the same letter d stands for the sup-metric. It easily follows from (b) that $\mathrm{C}(\boldsymbol{Q}, X)$ has no isolated points. Since $\mathrm{C}(\boldsymbol{Q}, X)$ is separable metrizable,[8] it has a countable dense set $\{g_i \mid i \in \mathbb{N}\}$. Then, $\{g_i \mid i > k\}$ is also dense in $\mathrm{C}(\boldsymbol{Q}, X)$ for each $k \in \mathbb{N}$. We define

$$M = \left\{(f_i)_{i\in\mathbb{N}} \in \mathrm{C}(\boldsymbol{Q}, X)^{\mathbb{N}} \;\middle|\; d(f_i, g_i) < 2^{-i} \text{ for each } i \in \mathbb{N}\right\}.$$

Since M is a G_δ-set in the product space $\mathrm{C}(\boldsymbol{Q}, X)^{\mathbb{N}}$, M is also completely metrizable. For each $j < k$, we have the following open set:

$$U_{j,k} = \{(f_i)_{i\in\mathbb{N}} \in M \mid f_j(\boldsymbol{Q}) \cap f_k(\boldsymbol{Q}) = \emptyset\},$$

which is dense in M by (b). By the Baire Category Theorem 1.3.15, $\bigcap_{j<k} U_{j,k}$ is dense in M. In particular, we have $(f_i)_{i\in\mathbb{N}} \in \bigcap_{j<k} U_{j,k}$. Then, the set $\{f_i \mid i \in \mathbb{N}\}$ satisfies condition (c). □

Now, we can prove that the disjoint cells property characterizes the $\mathfrak{M}_0$-universality for separable completely metrizable ANRs, that is:

Proposition 3.2.8 *Let X be a separable completely metrizable ANR such that every Z-set in X is a strong Z-set. Then, X is $\mathfrak{M}_0$-universal if and only if X has the disjoint cells property.*

Proof Since $\mathbf{I}^n \times \{0, 1\} \in \mathfrak{M}_0$, the "only if" part is trivial. If X has the disjoint cells property, then all maps from $\mathbf{I}^n$, $n \in \mathbb{N}$, to X can be approximated by maps with Z-set images by Proposition 3.2.7. Because we assume that every Z-set in X is a strong Z-set, the "if" part follows from Lemma 3.2.6 and Theorem 3.1.23. □

When X is a locally compact ANR, every Z-set in X is a strong Z-set (Proposition 2.8.1) and each component of X is clopen and separable (Proposition 3.1.9). Thus, it follows from Proposition 3.2.8 that a locally compact ANR X is $\mathfrak{M}_0$-universal if and only if X has the disjoint cells property. Furthermore, we have the following:

Theorem 3.2.9 *For a locally compact ANR X with weight τ, the following are equivalent:*

(a) *X has the disjoint cells property;*

[8] In general, $w(\mathrm{C}(\boldsymbol{Q}, X)) = w(X)$. — Refer to [GAGT, 1.1.3(5)].

(b) *X is $\mathfrak{M}_0$-universal;*
(c) *X is strongly $\mathfrak{M}_0$-universal;*
(d) *X is properly $\mathfrak{L}(\aleph_0)$-universal;*
(e) *X is strongly properly $\mathfrak{L}(\tau)$-universal.*

Proof The equivalence (a) $\Leftrightarrow$ (b) has been mentioned above, the equivalence (d) $\Leftrightarrow$ (e) is Theorem 3.1.22, and the implications (e) $\Rightarrow$ (c) $\Rightarrow$ (b) are trivial. We will show the implication (a) $\Rightarrow$ (d).

(a) $\Rightarrow$ (d): Due to Theorem 3.1.24, it suffices to show that $\mathrm{C}_{\mathcal{O}_K}(|K|, X)$ is dense in $\mathrm{C}^P(|K|, X)$ for any countable locally finite simplicial complex K, where $\mathrm{C}_{\mathcal{O}_K}(|K|, X) \subset \mathrm{C}^P(|K|, X)$ because K is locally finite. Note that the case where K is finite has already been obtained by the equivalence (a) $\Leftrightarrow$ (b). Hence, we may consider the case where K is infinite, i.e., $\operatorname{card} K = \aleph_0$. Then, we can write $K = \{\sigma_i \mid i \in \mathbb{N}\}$. For each $n \in \mathbb{N}$, we define

$$\mathcal{A}_n = \big\{\sigma_i[1 - 2^{-n}] \bigm| i \leqslant n\big\},$$

where the notation $\sigma[t]$ was defined just before Lemma 3.2.5 and we regard $\sigma_i[1 - 2^{-n}] = \sigma_i$ if $\dim \sigma_i = 0$. Then, $\mathcal{A}_n$ is a pairwise disjoint finite collection of cells in $|K|$. Note that the space $\mathrm{C}^P(|K|, X)$ of all proper maps is open in $\mathrm{C}(|K|, X)$ (Proposition 3.1.7). Since X has the disjoint cells property and is an ANR, we can apply Proposition 3.2.4 and the Homotopy Extension Theorem 1.13.11 to see that the following set is dense in $\mathrm{C}^P(|K|, X)$:

$$F_n = \big\{f \in \mathrm{C}^P(|K|, X) \bigm| f(\mathcal{A}_n) \text{ is a pairwise disjoint collection}\big\}.$$

It should be observed that F_n is open in $\mathrm{C}^P(|K|, X)$. Since $\mathrm{C}^P(|K|, X)$ is a Baire by Theorem 3.1.11, $\bigcap_{n\in\mathbb{N}} F_n$ is dense in $\mathrm{C}^P(|K|, X)$.

For each $g \in \bigcap_{n\in\mathbb{N}} F_n$ and $x \in g(|K|)$, there is a unique σ_i such that $g^{-1}(x) \subset \operatorname{rint} \sigma_i$. Indeed, $x \in g(\operatorname{rint} \sigma_i) \cap g(\operatorname{rint} \sigma_j)$ for some $i \neq j$. Then,

$$x \in g(\sigma_i[1 - 2^{-n}]) \cap g(\sigma_j[1 - 2^{-n}]) \text{ for some } n \geqslant i,\ j.$$

This contradicts the fact that $g \in F_n$ (i.e., $g(\mathcal{A}_n)$ is pairwise disjoint). Thus, we have $\{g^{-1}(x) \mid x \in g(|K|)\} \prec \mathcal{O}_K$, hence g is an $\mathcal{O}_K$-map because g is a closed map (Proposition 1.3.7). Therefore, $\bigcap_{n\in\mathbb{N}} F_n \subset \mathrm{C}_{\mathcal{O}_K}(|K|, X)$, hence $\mathrm{C}_{\mathcal{O}_K}(|K|, X)$ is dense in $\mathrm{C}^P(|K|, X)$. □

For an ANR X, the discrete cells property implies that all maps from $\mathbf{I}^n$, $n \in \mathbb{N}$, to X can be approximated by maps with Z-set images. In fact, every compact set in X is a Z-set by the following proposition:

Proposition 3.2.10 *Let X be an ANR with the τ-discrete cells property. For a closed set A in X, if $w(A) < \tau$ or A is compact, then A is a Z-set in X. In particular, if X is an ANR with the countable discrete cells property, then every compact set in X is a Z-set.*

Proof Recall that Γ is a discrete space with $\operatorname{card}\Gamma = \tau \geqslant \aleph_0$ in our convention. For each $n \in \mathbb{N}$ and each map $f : \mathbf{I}^n \to X$, let $\tilde{f} : \mathbf{I}^n \times \Gamma \to X$ be the map defined by $\tilde{f}(x, \gamma) = f(x)$. For each open cover $\mathcal{U}$ of X, $\tilde{f}$ is $\mathcal{U}$-close to a map $g : \mathbf{I}^n \times \Gamma \to X$ such that $(g_\gamma(\mathbf{I}^n))_{\gamma\in\Gamma}$ is discrete in X. Since $w(A) < \tau$ or A is compact, it is easy to see that $A \cap g_\gamma(\mathbf{I}^n) = \emptyset$ for some $\gamma \in \Gamma$, where g_γ is $\mathcal{U}$-close to f. Then, A is a Z-set in X by Theorem 2.8.6. □

In the rest of this section, we shall show the following proposition, which will be used in Chap. 5:

Proposition 3.2.11 *Let X be an ANR with $w(X) = \tau$. Suppose that X has the τ-discrete cells property. If A is a Z-set and also a strong Z_σ-set in X, then A is a strong Z-set in X.*

Proof We can write $A = \bigcup_{i\in\omega} A_i$, where $A_0 \subset A_1 \subset A_2 \subset \cdots$ are strong Z-sets in X. For each open cover $\mathcal{U}$ of X, let $\mathcal{U}_{-1}$ be an open star-refinement of $\mathcal{U}$. Since X is an ANR, we have a locally finite-dimensional simplicial complex K with $\operatorname{card} K^{(0)} \leqslant w(X)$, $f : X \to |K|$ and $g : |K| \to X$ such that gf is $\mathcal{U}_{-1}$-close to id_X, where $|K|$ has the weak (Whitehead) topology.

We inductively construct $\mathcal{U}_i \in \operatorname{cov}(X)$, maps $h_i : |K| \to X$, open sets V_i, V_i' in X and discrete collections $\mathcal{W}_i = \{W_\sigma \mid \sigma \in K^{(i)} \setminus K^{(i-1)}\}$, $\mathcal{W}_i' = \{W_\sigma' \mid \sigma \in K^{(i)} \setminus K^{(i-1)}\}$ of open sets in X, $i \in \omega$, such that:

(1) $\operatorname{mesh}\mathcal{U}_i < 2^{-i}$, $\operatorname{st}\mathcal{U}_i \prec \mathcal{U}_{i-1} \wedge \{V_{i-1}, X \setminus \operatorname{cl} V_{i-1}'\}$, $\operatorname{st}(W_\sigma', \mathcal{U}_i) \subset W_\sigma$ for each $\sigma \in K^{(i-1)}$,
(2) h_i is $\mathcal{U}_i$-close to h_{i-1}, $h_i||K^{(i-1)}| = h_{i-1}||K^{(i-1)}|$,
(3) $A_i \subset V_i' \subset \operatorname{cl} V_i' \subset V_i \subset X \setminus h_i(|K|)$, and
(4) $\operatorname{cl} W_\sigma' \subset W_\sigma \subset X \setminus A$, and $h_i(\sigma) \subset \bigcup_{\sigma'\leqslant\sigma} W_{\sigma'}'$ for each $\sigma \in K^{(i)}$,

where $h_{-1} = g$. Since $\{W_\sigma' \mid \sigma \in K^{(i)}\}$ is locally finite in X, condition (4) implies the following condition:

(5) $\operatorname{cl} h_i(|K^{(i)}|) \subset \operatorname{cl} \bigcup_{\sigma\in K^{(i)}} W_\sigma' = \bigcup_{\sigma\in K^{(i)}} \operatorname{cl} W_\sigma' \subset X \setminus A.$

Assume that $\mathcal{U}_j$, h_j, V_j, V_j', $\mathcal{W}_j$, and $\mathcal{W}_j'$ have been obtained for $j < i$. Since $\operatorname{cl} V_{i-1}' \subset V_{i-1}$, $\operatorname{cl} W_\sigma' \subset W_\sigma$ for each $\sigma \in K^{(i-1)}$, and $\mathcal{W}_{i-1}$ is discrete in X, we can choose an open cover $\mathcal{U}_i$ of X so as to satisfy condition (1). Let $\mathcal{U}_i'$ be an open star-refinement of $\mathcal{U}_i$. Since $\operatorname{cl} h_{i-1}(|K^{(i-1)}|) \cap A_i = \emptyset$ and A_i is a strong Z-set in X, we have a map $h_i' : |K| \to X$ and open neighborhoods V_i, V_i' of A_i in X such that:

(6) h_i' is $\mathcal{U}_i'$-close to h_{i-1},
(7) $h_i'||K^{(i-1)}| = h_{i-1}||K^{(i-1)}|$, and
(8) $\operatorname{cl} V_i' \subset V_i \subset \operatorname{cl} V_i \subset X \setminus \operatorname{cl} h_i'(|K|)$.

Let $\mathcal{U}_i^*$ be an open refinement of $\mathcal{U}_i'$ such that:

(9) $\mathcal{U}_i^* \prec \big\{V_i,\ X \setminus (\operatorname{cl} h_i'(|K|) \cup \operatorname{cl} V_i'),\ X \setminus \operatorname{cl} V_i\big\}$.

Since X is an ANR, $\mathcal{U}_i^*$ has an $\hbar$-refinement $\mathcal{U}_i''$ by Theorem 1.13.10.

For each i-simplex $\sigma \in K$, $U_\sigma = \bigcup_{\sigma' < \sigma} h_{i-1}^{-1}(W'_{\sigma'})$ is an open neighborhood of $\partial\sigma$ in $|K|$. Choose an i-cell c_σ so that $\sigma \setminus U_\sigma \subset c_\sigma \subset \operatorname{rint}\sigma$ and $\{c_\sigma \mid \sigma \in K^{(i)} \setminus K^{(i-1)}\}$ is discrete in $|K|$. Since X has the $w(X)$-discrete i-cells property and A is a Z-set in X, we can obtain a map $h_i'' : \bigcup_{\sigma \in K^{(i)} \setminus K^{(i-1)}} c_\sigma \to X$ such that:

(10) $h_i''\big(\bigcup_{\sigma \in K^{(i)} \setminus K^{(i-1)}} c_\sigma\big) \cap A = \emptyset$,
(11) h_i'' is $\mathcal{U}_i''$-close to $h_i'|\bigcup_{\sigma \in K^{(i)} \setminus K^{(i-1)}} c_\sigma$, and
(12) $\{h_i''(c_\sigma) \mid \sigma \in K^{(i)} \setminus K^{(i-1)}\}$ is discrete in X.

By using the Homotopy Extension Theorem 1.13.11, we can obtain a map $h_i : |K| \to X$ such that:

(13) $h_i|\bigcup_{\sigma \in K^{(i)} \setminus K^{(i-1)}} c_\sigma = h_i''|\bigcup_{\sigma \in K^{(i)} \setminus K^{(i-1)}} c_\sigma$,
(14) $h_i||K^{(i-1)}| = h_i'||K^{(i-1)}|$, and
(15) h_i is $\mathcal{U}_i^*$-homotopic to h_i'.

Then, it follows that $h_i||K^{(i-1)}| = h_{i-1}||K^{(i-1)}|$ and h_i is $\mathcal{U}_i$-close to h_{i-1}, that is, h_i satisfies condition (2). Since h_i is $\mathcal{U}_i^*$-close to h_i', it follows from (9) that $h_i(|K|) \subset X \setminus \operatorname{cl} V_i$, that is, $\operatorname{cl} V_i \subset X \setminus h_i(|K|)$. Thus, condition (3) is satisfied.

By (10), (12), and (13), for each i-simplex $\sigma \in K$, $h_i(c_\sigma)$ has open neighborhoods W_σ, W'_σ in X such that $\operatorname{cl} W'_\sigma \subset W_\sigma \subset X \setminus A$ and $\mathcal{W}_i = \{W_\sigma \mid \sigma \in K^{(i)} \setminus K^{(i-1)}\}$ is discrete in X, hence $\mathcal{W}_i' = \{W'_\sigma \mid \sigma \in K^{(i)} \setminus K^{(i-1)}\}$ is also discrete in X. For each i-simplex $\sigma \in K^{(i)}$, $h_i(c_\sigma) \subset W'_\sigma$. For each $x \in \sigma \setminus C_\sigma \subset \sigma \cap U_\sigma$, choose $\sigma' < \sigma$ so that $h_{i-1}(x) \in W'_{\sigma'}$. Since h_i is $\mathcal{U}_i$-close to h_{i-1}, it follows that $h_i(x) \in \operatorname{st}(W'_{\sigma'}, \mathcal{U}_i) \subset W_{\sigma'}$. Therefore, $h_i(\sigma) \subset \bigcup_{\sigma' \leqslant \sigma} W_{\sigma'}$. Then, condition (4) is also satisfied.

By induction, we can obtain $\mathcal{U}_i$, h_i, V_i, and $\mathcal{W}_i$ for all $i \in \omega$. Because $h_j||K^{(i)}| = h_i||K^{(i)}|$ for every $j > i$ by (2) and $|K|$ has the weak topology, we can define a map $h : |K| \to X$ by $h||K^{(i)}| = h_i||K^{(i)}|$. It follows from (1) and (2) that h is $\operatorname{st}\mathcal{U}_i$-close to h_{i-1}. In particular, h is $\operatorname{st}\mathcal{U}_0$-close to $h_{-1} = g$, hence hf is $\mathcal{U}_{-1}$-close to gf. Thus, hf is $\mathcal{U}$-close to id_X. Moreover,

$$hf(X) \subset h(|K|) \subset \operatorname{st}(h_i(|K|), \operatorname{st}\mathcal{U}_{i+1}).$$

Since $\operatorname{st}\mathcal{U}_{i+1} \prec \{V_i, X \setminus \operatorname{cl} V_i'\}$ by (1), it follows from (3) that

$$\operatorname{st}(h_i(|K|), \operatorname{st}\mathcal{U}_{i+1}) \subset X \setminus \operatorname{cl} V_i'.$$

Therefore, $hf(X) \cap \bigcup_{i \in \omega} V_i' = \emptyset$. Since $\bigcup_{i \in \omega} V_i'$ is an open neighborhood of A in X, we have $\operatorname{cl} hf(X) \cap A = \emptyset$. □

In Proposition 3.2.11 above, when X is separable (i.e., $w(X) = \aleph_0$), it is not necessary to assume that X has the countable discrete cells property, that is, the following holds:

Proposition 3.2.12 *Let X be a separable ANR. If A is a Z-set and also a strong Z_σ-set in X, then A is a strong Z-set in X.*

Indeed, in the proof of Proposition 3.2.11, K can be taken so as to be countable (and locally finite). Then, write $K = \{\sigma_i \mid i \in \omega\}$ so that $\dim \sigma_i \leqslant \dim \sigma_j$ if $i < j$. Instead of skeleton-wise induction, we can apply simplex-wise induction, where the discrete cells property is no longer necessary. Details are left to the reader.

3.3 The Discrete F.D. Polyhedra Property

To characterize $\mathfrak{M}_1(\tau)$-universality, in addition to the τ-discrete cells property (or the τ-locally finite cells property), we introduce the following property that is called the **discrete** (resp. **locally finite**) **f.d. τ-polyhedra property**:

- Let $(K_i)_{i\in\mathbb{N}}$ be a sequence of finite-dimensional simplicial complexes with $\operatorname{card} K_i \leqslant \tau$ (equiv. $\operatorname{card} K_i^{(0)} \leqslant \tau$). Then, each map $f : \bigoplus_{i\in\mathbb{N}} |K_i| \to X$ can be approximated by maps $g : \bigoplus_{i\in\mathbb{N}} |K_i| \to X$ such that $(g(|K_i|))_{i\in\mathbb{N}}$ is discrete (resp. locally finite) in X.

Obviously, the discrete f.d. τ-polyhedra property implies the locally finite f.d. τ-polyhedra property.

The discrete (resp. locally finite) f.d. $\aleph_0$-polyhedra property is also called the **discrete** (resp. **locally finite**) **f.d. separable polyhedra property**. In the above definitions, restricting only sequences of finite simplicial complexes (that is, replacing "$\operatorname{card} K_i \leqslant \tau$" with "$\operatorname{card} K_i < \aleph_0$"), we define the **discrete** (resp. **locally finite**) **compact polyhedra property**.

Remark 3.6 In the above definition, the case $\sup_{i\in\mathbb{N}} \dim K_i = \infty$ is essential (even for a sequence of finite simplicial complexes) because it follows from Lemma 3.2.5 that the τ-discrete cells property implies the case $\sup_{i\in\mathbb{N}} \dim K_i < \infty$ under the assumption that every compact set is a strong Z-set. Indeed, let $K = \bigoplus_{i\in\mathbb{N}} K_i$. Then, $|K| = \bigoplus_{i\in\mathbb{N}} |K_i|$ and $\dim K = \sup_{i\in\mathbb{N}} \dim K_i < \infty$. For any $\mathcal{U} \in \operatorname{cov}(X)$, each map $f : |K| \to X$ is $\mathcal{U}$-close to an O_K-map $g : |K| \to X$ by Lemma 3.2.5. Each $x \in X$ has an open neighborhood V such that $g^{-1}(V) \subset O(v, K)$ for some $v \in K^{(0)}$. Then, v is a vertex of some K_i. Since $g^{-1}(V) \subset O(v, K) = O(v, K_i) \subset |K_i|$, it follows that $V \cap g(|K_j|) = \emptyset$ for any $j \in \mathbb{N} \setminus \{i\}$. This means that $(g(|K_i|))_{i\in\mathbb{N}}$ is discrete in X.

In the above definition, whenever K_i is non-locally finite, $|K_i| \neq |K_i|_{\mathrm{m}}$, that is, the weak topology of $|K_i|$ is different from the metric topology. However, the weak topology of $|K_i|$ can be replaced with the metric topology, that is:

Proposition 3.3.1 *The discrete (resp. locally finite) f.d. τ-polyhedra property is equivalent to the following metric topology version:*

- *Let $(K_i)_{i\in\mathbb{N}}$ be a sequence of finite-dimensional simplicial complexes with* $\operatorname{card} K_i \leqslant \tau$. *Then, each map $f : \bigoplus_{i\in\mathbb{N}} |K_i|_{\mathrm{m}} \to X$ can be approximated by maps $g : \bigoplus_{i\in\mathbb{N}} |K_i|_{\mathrm{m}} \to X$ such that $(g(|K_i|))_{i\in\mathbb{N}}$ is discrete (resp. locally finite) in X.*

Proof First, assume that X has the discrete (resp. locally finite) f.d. τ-polyhedra property. For a sequence $(K_i)_{i\in\mathbb{N}}$ of finite-dimensional simplicial complexes with $\operatorname{card} K_i \leqslant \tau$, let $K = \bigoplus_{i\in\mathbb{N}} K_i$. Then, $|K| = \bigoplus_{i\in\mathbb{N}} |K_i|$ and $|K|_{\mathrm{m}} = \bigoplus_{i\in\mathbb{N}} |K_i|_{\mathrm{m}}$. Let $f : |K|_{\mathrm{m}} \to X$ be a map. For each $\mathcal{U} \in \operatorname{cov}(X)$, let $\mathcal{V} \in \operatorname{cov}(X)$ be a star-refinement of $\mathcal{U}$. Due to Theorem 1.7.6, the identity map $\varphi = \mathrm{id} : |K| \to |K|_{\mathrm{m}}$ is a fine homotopy equivalence. Hence, we have a map $\psi : |K|_{\mathrm{m}} \to |K|$ such that $\psi\varphi \simeq_{f^{-1}(\mathcal{V})} \mathrm{id}$ and $\varphi\psi \simeq_{f^{-1}(\mathcal{V})} \mathrm{id}$. Since X has the discrete (resp. locally finite) f.d. τ-polyhedra property, $f\varphi : |K| \to X$ is $\mathcal{V}$-close to a map $g : |K| \to X$ such that $(g(|K_i|))_{i\in\mathbb{N}}$ is discrete (resp. locally finite) in X. Then, $g\psi : |K|_{\mathrm{m}} \to X$ is a map that is $\mathcal{V}$-close to $f\varphi\psi$, hence $\mathcal{U}$-close to f, because $\operatorname{st}\mathcal{V} \prec \mathcal{U}$. Since $\psi(|K_i|) \subset |K_i|$ for every $i \in \mathbb{N}$, $(g\psi(|K_i|))_{i\in\mathbb{N}}$ is discrete (resp. locally finite) in X.

Conversely, assume that X has the metric topology version of the discrete (resp. locally finite) f.d. τ-polyhedra property. For a sequence $(K_i)_{i\in\mathbb{N}}$ of finite-dimensional simplicial complexes with $\operatorname{card} K_i \leqslant \tau$, let $K = \bigoplus_{i\in\mathbb{N}} K_i$. Then, $|K| = \bigoplus_{i\in\mathbb{N}} |K_i|$ and $|K|_{\mathrm{m}} = \bigoplus_{i\in\mathbb{N}} |K_i|_{\mathrm{m}}$. Let $f : |K| \to X$ be a map. For each $\mathcal{U} \in \operatorname{cov}(X)$, let $\mathcal{V} \in \operatorname{cov}(X)$ be a star-refinement of $\mathcal{U}$. By Theorem 1.6.9, K has a subdivision K' such that $\mathcal{S}_{K'} \prec f^{-1}(\mathcal{V})$, where $|K| = |K'|$ but $|K|_{\mathrm{m}} \neq |K'|_{\mathrm{m}}$. Due to Theorem 1.7.3, the identity map $\varphi = \mathrm{id} : |K'| \to |K'|_{\mathrm{m}}$ has a homotopy inverse $\psi : |K'|_{\mathrm{m}} \to |K|$ such that $\psi\varphi \simeq_{K'} \mathrm{id}$ and $\varphi\psi \simeq_{K'} \mathrm{id}$. Since X has the metric version of the discrete (resp. locally finite) f.d. τ-polyhedra property, $f\psi : |K'|_{\mathrm{m}} \to X$ is $\mathcal{V}$-close to a map $g : |K'|_{\mathrm{m}} \to X$ such that $(g(|K_i|))_{i\in\mathbb{N}}$ is discrete (resp. locally finite) in X. Then, $g\varphi : |K| = |K'| \to X$ is a map that is $\mathcal{V}$-close to $f\psi\varphi$. Since $\psi\varphi \simeq_{K'} \mathrm{id}$ and $K' \prec \mathcal{S}_{K'} \prec f^{-1}(\mathcal{V})$, it follows that $f\psi\varphi$ is $\mathcal{V}$-close to f, Therefore, $g\varphi$ is $\mathcal{U}$-close to f, because $\operatorname{st}\mathcal{V} \prec \mathcal{U}$. Evidently, $\varphi(|K_i|) = |K_i|$ for every $i \in \mathbb{N}$. Hence, $(g\varphi(|K_i|))_{i\in\mathbb{N}}$ is discrete (resp. locally finite) in X. □

In the above proposition, $\bigoplus_{i\in\mathbb{N}} |K_i|_{\mathrm{m}} \in \mathfrak{M}_1(\tau)$ because each K_i is finite-dimensional (Theorem 1.7.2). It follows that the $\mathfrak{M}_1(\tau)$-universality implies the metric version of the discrete f.d. τ-polyhedra property, which means the following:

- *The $\mathfrak{M}_1(\tau)$-universality implies the discrete f.d. τ-polyhedra property.*

Moreover, by the above proposition, the proof of Proposition 3.2.3 can be applied to prove the following proposition (cf. Remark 3.3):

Proposition 3.3.2 *Let X be an ANR with the locally finite f.d. τ-polyhedra property. Then, every open set in X has the locally finite f.d. τ-polyhedra property.* □

To characterize $\mathfrak{M}_1(\tau)$-universality, we need the following lemma:

Lemma 3.3.3 *Suppose that X has either the τ-discrete n-cells property or the discrete f.d. τ-polyhedra property. Then, the product space $X \times \boldsymbol{Q}$ has the same property.*

This is easily derived from the following lemma:

Lemma 3.3.4 *Let Z be a compact space. For each open cover $\mathcal{U}$ of the product space $Y \times Z$, there exists an open cover $\mathcal{V}$ of Y such that, for a map $f : X \to Y \times Z$, if $\mathrm{pr}_Y f$ is $\mathcal{V}$-close to a map $g : X \to Y$, then f is $\mathcal{U}$-close to the map $\tilde{g} : X \to Y \times Z$ defined by $\tilde{g}(x) = (g(x), \mathrm{pr}_Z f(x))$ for each $x \in X$.*

Proof For each $(y, z) \in Y \times Z$, choose $U(y, z) \in \mathcal{U}$ so that $(y, z) \in U(y, z)$, and take open sets $V(y, z) \subset Y$ and $W(y, z) \subset Z$ so that $(y, z) \in V(y, z) \times W(y, z) \subset U(y, z)$. Since Z is compact, there is a finite set $F(y) \subset Z$ such that $Z = \bigcup_{z \in F(y)} W(y, z)$. Then, we have an open neighborhood $V_y = \bigcap_{z \in F(y)} V(y, z)$ of y in Y. The open cover $\mathcal{V} = \{V_y \mid y \in Y\} \in \mathrm{cov}(Y)$ satisfies the required condition. Indeed, let $f : X \to Y \times Z$ and $g : X \to Y$ be maps such that $\mathrm{pr}_Y f$ is $\mathcal{V}$-close to g. For each $x \in X$, we have $y \in Y$ such that $\mathrm{pr}_Y f(x), g(x) \in V_y$. Then, we can find $z \in F(y)$ such that $\mathrm{pr}_Z f(x) \in W(y, z)$, which implies that

$$f(x),\ (g(x), \mathrm{pr}_Z f(x)) \in V(y, z) \times W(y, z) \subset U(y, z).$$

This completes the proof. □

Using the above lemma, we can also show the following lemma:

Lemma 3.3.5 *Let X be a completely metrizable ANR with the τ-discrete cells property and K be a simplicial complex with $\mathrm{card}\, K^{(0)} \leqslant \tau$. Then, the space $\mathrm{C}(|K|_{\mathrm{m}}, X \times \boldsymbol{Q})$ has a dense G_δ set F such that*

$$\bigl\{f^{-1}(\{x\} \times \boldsymbol{Q}) \bigm| x \in X\bigr\} \prec \mathcal{O}_K \ \textit{for every}\ f \in F.$$

Proof For each $n \in \mathbb{N}$, we define

$$\mathcal{A}_n = \bigl\{\sigma[1 - 2^{-n}] \bigm| \sigma \in K^{(n)}\bigr\},$$

where the notation $\sigma[t]$ was defined before Proposition 3.2.5. Then, $\mathcal{A}_n$ is discrete in $|K|_{\mathrm{m}}$ and $\mathrm{card}\, \mathcal{A}_n \leqslant \mathrm{card}\, K \leqslant \tau$.

We will show that the following set is dense and open in $\mathrm{C}(|K|_{\mathrm{m}}, X \times \boldsymbol{Q})$:

$$F_n = \bigl\{f \in \mathrm{C}(|K|_{\mathrm{m}}, X \times \boldsymbol{Q}) \bigm| \mathrm{pr}_X f(\mathcal{A}_n) \ \text{is discrete in}\ X\bigr\}.$$

For each $\mathcal{U} \in \mathrm{cov}(X \times \boldsymbol{Q})$, take $\mathcal{V} \in \mathrm{cov}(X)$ as in Lemma 3.3.4. For each $f \in \mathrm{C}(|K|_{\mathrm{m}}, X \times \boldsymbol{Q})$, we use the τ-discrete cells property X and the Homotopy Extension Theorem 1.13.11 to construct a map $g : |K|_{\mathrm{m}} \to X$ such that $g \simeq_{\mathcal{V}} \mathrm{pr}_X f$ and $g(\mathcal{A}_n)$ is discrete in X. Then, f is $\mathcal{U}$-close to the map $\tilde{g} : |K|_{\mathrm{m}} \to X \times \boldsymbol{Q}$ defined by $\tilde{g}(x) = (g(x), \mathrm{pr}_{\boldsymbol{Q}} f(x))$ for each $x \in |K|_{\mathrm{m}}$, where $\mathrm{pr}_X \tilde{g}(\mathcal{A}_n) = g(\mathcal{A}_n)$

is discrete in X, that is, $\tilde{g} \in F_n$. Hence, F_n is dense in $\mathrm{C}(|K|_{\mathrm{m}}, X \times \boldsymbol{Q})$. Moreover, for each $h \in F_n$, there exists $\mathcal{W} \in \mathrm{cov}(X)$ such that each $W \in \mathcal{W}$ meets at most one member of $h(\mathcal{A}_n)$. Let $\widetilde{\mathcal{W}} = \mathrm{pr}_X^{-1}(\mathcal{W}) \in \mathrm{cov}(X \times \boldsymbol{Q})$. Then, it follows that $\widetilde{\mathcal{W}}(h) \subset F_n$. Therefore, F_n is open in $\mathrm{C}(|K|_{\mathrm{m}}, X \times \boldsymbol{Q})$.

Since $\mathrm{C}(|K|_{\mathrm{m}}, X \times \boldsymbol{Q})$ is a Baire space (Proposition 3.1.5), $F = \bigcap_{n\in\mathbb{N}} F_n$ is a dense G_δ set in $\mathrm{C}(|K|_{\mathrm{m}}, X \times \boldsymbol{Q})$. To see that $\{f^{-1}(\{x\} \times \boldsymbol{Q}) \mid x \in X\} \prec O_K$, it suffices to show that, for each $f \in F$ and $x \in \mathrm{pr}_X f(|K|)$, there is a unique simplex $\sigma \in K$ such that $f^{-1}(\{x\} \times \boldsymbol{Q}) \subset \mathrm{rint}\,\sigma$. Since $x \in \mathrm{pr}_X f(|K|)$, it follows that $x \in \mathrm{pr}_X f(\mathrm{rint}\,\sigma)$ for some $\sigma \in K$. Then, $x \notin \mathrm{pr}_X f(\mathrm{rint}\,\tau)$ for any other $\tau \in |K|$. Indeed, if $x \in \mathrm{pr}_X f(\mathrm{rint}\,\tau)$ for any other $\tau \in |K|$, then we can take $n \in \mathbb{N}$ such that $\sigma, \tau \in K^{(n)}$ and

$$x \in \mathrm{pr}_X f(\sigma[1-2^{-n}]) \cap \mathrm{pr}_X f(\tau[1-2^{-n}]),$$

which contradicts the fact that $f \in F_n$. Thus, we have only one $\sigma \in K$ such that $x \in \mathrm{pr}_X f(\mathrm{rint}\,\sigma)$, which implies that

$$f^{-1}(\{x\} \times \boldsymbol{Q}) = (\mathrm{pr}_X f)^{-1}(x) \subset \mathrm{rint}\,\sigma.$$

The proof is complete. □

We can now characterize the $\mathfrak{M}_1(\tau)$-universality as follows:

Theorem 3.3.6 *Let X be a completely metrizable ANR with $w(X) = \tau$. Then, X is $\mathfrak{M}_1(\tau)$-universal if and only if X has the τ-discrete cells property and the discrete f.d. τ-polyhedra property.*

Proof As already observed, the $\mathfrak{M}_1(\tau)$-universality implies both the τ-discrete cells property and the discrete f.d. τ-polyhedra property. Hence, it suffices to prove the "if" part.

First, we shall show the case where X is $\boldsymbol{Q}$-stable. Let K be a locally finite-dimensional simplicial complex with $\mathrm{card}\, K^{(0)} \leqslant \tau$ and $\mathcal{U} \in \mathrm{cov}(X)$. We shall show that each map $f : |K| \to X$ is $\mathcal{U}$-close to an O_K-map. Then, the $\mathfrak{M}_1(\tau)$-universality follows from Theorem 3.1.23.

For each $n \in \omega$, let

$$W_n = \left\{x \in |K| \;\middle|\; \exists v \in K^{(0)} \text{ such that } \dim \mathrm{St}(v, K) \leqslant n,\ \beta_v^K(x) > 2^{-n}\right\},$$

where $\beta_v^K(x)$ is the barycentric coordinate of x at v. Then, $W_0 = \emptyset$, $\dim W_n \leqslant n$, and $\mathrm{cl}\, W_{n-1} \subset W_n$ for each $n \in \mathbb{N}$. Since K is locally finite-dimensional, it follows that $\bigcup_{n\in\mathbb{N}} W_n = |K|$. Note that $\{\mathrm{cl}\, W_{2i+1} \setminus W_{2i} \mid i \in \omega\}$ is discrete in $|K|$ and each $\mathrm{cl}\, W_{2i+1} \setminus W_{2i}$ is a finite-dimensional polyhedron. Let

$$S_1 = \bigcup_{i\in\omega} (\mathrm{cl}\, W_{2i+1} \setminus W_{2i}) = \bigoplus_{i\in\omega} (\mathrm{cl}\, W_{2i+1} \setminus W_{2i}).$$

Take $\mathcal{V}_1, \mathcal{V}'_1 \in \operatorname{cov}(X)$ so that $\operatorname{st}\mathcal{V}'_1 \prec \mathcal{V}_1$ and $\operatorname{st}\mathcal{V}_1 \prec \mathcal{U}$. Applying the discrete f.d. τ-polyhedra property and the Homotopy Extension Theorem 1.5.7, we can obtain a map $g : |K| \to X$ such that $g \simeq_{\mathcal{V}'_1} f$ and $\{g(\operatorname{cl} W_{2i+1} \setminus W_{2i}) \mid i \in \omega\}$ is discrete in X. Choose $\mathcal{V} \in \operatorname{cov}(X)$ so that $\mathcal{V} \prec \mathcal{V}'_1$ and, if a map $h : |K| \to X$ is $\mathcal{V}$-close to g, then $\{h(\operatorname{cl} W_{2i+1} \setminus W_{2i}) \mid i \in \omega\}$ is discrete in X. By Lemma 3.2.5 with Remark 3.5, we have $\mathcal{O}_{K^{(2i+1)}}$-maps $h_i : |K^{(2i+1)}| \to X$, $i \in \mathbb{N}$, such that $h_i \simeq_{\mathcal{V}} g||K^{(2i+1)}|$. Observe that $\operatorname{cl} W_{2i+1} \setminus W_{2i} \subset |K^{(2i+1)}|$ for each $i \in \mathbb{N}$. Using the Homotopy Extension Theorem 1.5.7, we can obtain a map $g_1 : |K| \to X$ such that

$$g_1|\operatorname{cl} W_{2i+1} \setminus W_{2i} = h_i|\operatorname{cl} W_{2i+1} \setminus W_{2i} \ \text{ and } \ g_1 \simeq_{\mathcal{V}} g.$$

Then, $\{g_1(\operatorname{cl} W_{2i+1} \setminus W_{2i}) \mid i \in \omega\}$ is discrete in X. For each $x \in X \setminus \operatorname{cl} g_1(S_1)$, $V_x = X \setminus \operatorname{cl} g_1(S_1)$ is an open neighborhood of x in X and $(g_1|S_1)^{-1}(V_x) = g_1^{-1}(V_x) \cap S_1 = \emptyset$. For each $x \in \operatorname{cl} g_1(S_1)$, we have an open neighborhood V'_x of x in X and $i_x \in \omega$ such that

$$V'_x \cap g_1(\operatorname{cl} W_{2i_x+1} \setminus W_{2i_x}) \neq \emptyset \ \text{ and } \ V'_x \cap g_1(\operatorname{cl} W_{2i+1} \setminus W_{2i}) = \emptyset \ \text{ if } i \neq i_x.$$

Since $\operatorname{cl} W_{2i_x+1} \setminus W_{2i_x} \subset |K^{(2i_x+1)}|$, it follows that

$$g_1|\operatorname{cl} W_{2i_x+1} \setminus W_{2i_x} = h_i|\operatorname{cl} W_{2i_x+1} \setminus W_{2i_x},$$

which is an $(\mathcal{O}_K|\operatorname{cl} W_{2i_x+1} \setminus W_{2i_x})$-map. Hence, we have $v \in K^{(0)}$ and an open neighborhood V_x of x in X such that $V_x \subset V'_x$ and

$$\begin{aligned} g_1^{-1}(V_x) \cap (\operatorname{cl} W_{2i_x+1} \setminus W_{2i_x}) &= (g_1|\operatorname{cl} W_{2i_x+1} \setminus W_{2i_x})^{-1}(V_x) \\ &\subset O_K(v) \cap (\operatorname{cl} W_{2i_x+1} \setminus W_{2i_x}) \subset O_K(v) \cap S_1. \end{aligned}$$

Since $V_x \subset V'_x$, $g_1^{-1}(V_x) \cap (\operatorname{cl} W_{2i+1} \setminus W_{2i}) = \emptyset$ for $i \neq i_x$, which implies

$$(g_1|S_1)^{-1}(V_x) = g_1^{-1}(V_x) \cap S_1 = g_1^{-1}(V_x) \cap (\operatorname{cl} W_{2i_x+1} \setminus W_{2i_x}).$$

Therefore, $g_1|S_1$ is an $(\mathcal{O}_K|S_1)$-map. Moreover, $g_1 \simeq_{\mathcal{V}_1} f$ because $\operatorname{st}\mathcal{V}'_1 \prec \mathcal{V}_1$.

Choose $\mathcal{V}_2 \in \operatorname{cov}(X)$ so that $\operatorname{st}\mathcal{V}_2 \prec \mathcal{V}$ and, if a map $h : |K| \to X$ is $\mathcal{V}_2$-close to g_1, then $h|S_1$ is an $(\mathcal{O}_K|S_1)$-map. In the same construction as g_1, we can construct a map $g_2 : |K| \to X$ such that $g_2 \simeq_{\mathcal{V}_2} g_1$ and $g_2|S_2$ is an $(\mathcal{O}_K|S_2)$-map, where

$$S_2 = \bigcup_{i \in \mathbb{N}} (\operatorname{cl} W_{2i} \setminus W_{2i-1}) = \bigoplus_{i \in \mathbb{N}} (\operatorname{cl} W_{2i} \setminus W_{2i-1}).$$

Since g_2 is $\mathcal{V}_2$-close to g_1, it follows that $g_2|S_1$ is an $(O_K|S_1)$-map. By Proposition 3.1.18, we have an open neighborhood N of S_1 in $|K|$ such that $g_2|N$ is an $(O_K|N)$-map.

Recall that we assume the $\boldsymbol{Q}$-stability, i.e., $X \times \boldsymbol{Q} \approx X$. There is a homeomorphism $\varphi : X \times \mathbf{I} \to X$ that is $\mathcal{V}_2$-close to pr_X by Corollary 2.3.11. Let $\lambda : |K| \to \mathbf{I}$ be an Urysohn map with $\lambda(|K| \setminus N) = 0$ and $\lambda(S_1) = 1$. We define a map $h : |K| \to X$ by $h(x) = \varphi(g_2(x), \lambda(x))$ for each $x \in |K|$. Then, h is $\mathcal{V}_2$-close to g_2. Since $\mathrm{st}\,\mathcal{V}_2 \prec \mathcal{V} \prec \mathcal{V}_1$ and $\mathrm{st}\,\mathcal{V}_1 \prec \mathcal{U}$, it follows that h is $\mathcal{U}$-close to f.

To see that h is an O_K-map, it suffices to show that $\tilde{h} = \varphi^{-1}h : |K| \to X \times \mathbf{I}$ is an O_K-map because φ is a homeomorphism. Let $(x, t) \in X \times \mathbf{I}$. When $t = 0$, note that $\tilde{h}^{-1}(X \times [0, 1)) \cap S_1 = \emptyset$, that is, $h^{-1}(X \times [0, 1)) \subset S_2$. Since $g_2|S_2$ is an $(O_K|S_2)$-map, x has an open neighborhood V_x in X such that $g_2^{-1}(V_x) \cap S_2 = (g_2|S_2)^{-1}(V_x)$ is contained in $O_K(v)$ for some $v \in K^{(0)}$. Then, $V_x \times [0, 1)$ is a neighborhood of $(x, t) = (x, 0)$ in $X \times \mathbf{I}$ and

$$h^{-1}(V_x \times [0, 1)) \subset g_2^{-1}(V_x) \cap S_2 \subset O_K(v).$$

When $t > 0$, note that $\tilde{h}^{-1}(X \times (0, 1]) \subset N$. Since $g_2|N$ is an $(O_K|N)$-map, x has an open neighborhood V_x in X such that $g_2^{-1}(V_x) \cap N = (g_2|N)^{-1}(V_x)$ is contained in $O_K(v)$ for some $v \in K^{(0)}$. Then, $V_x \times (0, 1]$ is a neighborhood of (x, t) in $X \times \mathbf{I}$ and

$$h^{-1}(V_x \times (0, 1]) \subset g_2^{-1}(V_x) \cap N \subset O_K(v).$$

Consequently, $\tilde{h} : |K| \to X \times \mathbf{I}$ is an O_K-map. Thus, it follows that h is an O_K-map.

Finally, we shall show the "if" part without assuming the $\boldsymbol{Q}$-stability of X. As in the $\boldsymbol{Q}$-stable case, let K be a locally finite-dimensional simplicial complex with $\mathrm{card}\, K^{(0)} \leqslant \tau$. It suffices to show that each map $f : |K| \to X$ can be approximated by O_K-maps.

For each $\mathcal{U} \in \mathrm{cov}(X)$, take $\mathcal{V} \in \mathrm{cov}(X)$ such that $\mathrm{st}\,\mathcal{V} \prec \mathcal{U}$. Then, K has a subdivision K' such that $O_{K'} \prec f^{-1}(\mathcal{V})$ (Theorem 1.6.9). Let $\varphi = \mathrm{id} : |K| = |K'| \to |K'|_{\mathrm{m}}$. Then, there is a map $\psi : |K'|_{\mathrm{m}} \to |K'|$ such that $\varphi\psi \simeq_{O_{K'}} \mathrm{id}$ and $\psi\varphi \simeq_{O_{K'}} \mathrm{id}$ (cf. Theorem 1.7.3). If $f\psi : |K'|_{\mathrm{m}} \to X$ is $\mathcal{V}$-close to an $O_{K'}$-map $g : |K'|_{\mathrm{m}} \to X$, then $g\varphi : |K'| \to X$ is also an $O_{K'}$-map that is $\mathcal{V}$-close to $f\psi\varphi$. Since $\psi\varphi$ is $f^{-1}(\mathcal{V})$-close to id, $g\varphi$ is $\mathcal{U}$-close to f. Thus, replacing K by K', it suffices to show that $\mathrm{C}_{O_K}(|K|_{\mathrm{m}}, X)$ is dense in the space $\mathrm{C}(|K|_{\mathrm{m}}, X)$.

Since K is locally finite-dimensional, ρ_K is complete (Theorem 1.7.2), hence $|K|_{\mathrm{m}}$ is completely metrizable. By Lemma 3.3.3, $X \times \boldsymbol{Q}$ satisfies the same condition as X. Since $X \times \boldsymbol{Q}$ is $\boldsymbol{Q}$-stable, $X \times \boldsymbol{Q}$ is $\mathfrak{M}_1(\tau)$-universal by the $\boldsymbol{Q}$-stable case, that is, $\mathrm{Emb}(|K|_{\mathrm{m}}, X \times \boldsymbol{Q})$ is dense in $\mathrm{C}(|K|_{\mathrm{m}}, X \times \boldsymbol{Q})$. Note that $\mathrm{Emb}(|K|_{\mathrm{m}}, X \times \boldsymbol{Q})$ is also G_δ in $\mathrm{C}(|K|_{\mathrm{m}}, X \times \boldsymbol{Q})$ by Lemma 3.1.14. On the other hand, due to Lemma 3.3.5, $\mathrm{C}(|K|_{\mathrm{m}}, X \times \boldsymbol{Q})$ has a dense G_δ set F such that $\{f^{-1}(\{x\} \times \boldsymbol{Q}) \mid$

$x \in X\} \prec \mathcal{O}_K$ for every $f \in F$. Then, $F_0 = F \cap \mathrm{Emb}(|K|_\mathrm{m}, X \times \boldsymbol{Q})$ is a dense G_δ set in $\mathrm{C}(|K|_\mathrm{m}, X \times \boldsymbol{Q})$ and

$$\{f^{-1}(\{x\} \times \boldsymbol{Q}) \mid x \in X\} \prec \mathcal{O}_K \quad \text{for every } f \in F_0.$$

For each $f \in F_0$, $\mathrm{pr}_X f$ is a closed map because $f \in F_0$ is a closed embedding. Then, it is easy to see that $\mathrm{pr}_X f$ is an $\mathcal{O}_K$-map. Therefore, $\mathrm{C}_{\mathcal{O}_K}(|K|_\mathrm{m}, X)$ is dense in $\mathrm{C}(|K|_\mathrm{m}, X)$. □

In the above proof of the "if" part (both the $\boldsymbol{Q}$-stable case and the general case), when $\tau = \aleph_0$, that is, X is separable, it is enough to assume that K is a *countable locally finite* simplicial complex. In the $\boldsymbol{Q}$-stable case of the proof, we can write $K = \bigcup_{n \in \mathbb{N}} K_n$, where K_n is finite and $|K_n| \subset \mathrm{int}\,|K_{n+1}|$. Then, the proof is valid even if each W_n is defined as $W_n = \mathrm{int}\,|K_n|$. Since each $\mathrm{cl}\, W_n \setminus W_{n-1} = |K_n| \setminus \mathrm{int}\,|K_{n-1}|$ is a *compact* polyhedron, the discrete $\aleph_0$-f.d.polyhedra property can be replaced by the discrete compact polyhedra property. Obviously, this property implies the countable discrete cells property. Thus, we have the following:

Theorem 3.3.7 *For a separable completely metrizable ANR, the $\mathfrak{M}_1(\aleph_0)$-universality is equivalent to the discrete compact polyhedra property.* □

In the above, the discrete compact polyhedra property can be replaced by the locally finite compact polyhedra property. In the same way as Proposition 3.2.1, it can be shown that the locally finite compact polyhedra property implies the discrete compact polyhedra property.

Indeed, let $(K_i)_{i \in \mathbb{N}}$ be a sequence of finite-simplicial complexes. For each map $f : \bigoplus_{i \in \mathbb{N}} |K_i| \to X$, let $\tilde{f} = f \mathrm{pr}_1 : \bigoplus_{i \in \mathbb{N}} |K_i| \times \mathbb{N} \to X$. Applying the locally finite compact polyhedra property, we have a map $\tilde{g} : \bigoplus_{i \in \mathbb{N}} |K_i| \times \mathbb{N} \to X$ arbitrarily close to $\tilde{f}$ such that $(\tilde{g}(|K_i| \times \{n\}))_{(i,n) \in \mathbb{N}^2}$ is locally finite in X. By compactness, each $\tilde{g}(|K_i| \times \{n\})$ meets only finitely many other $\tilde{g}(|K_{i'}| \times \{n'\})$. By induction, we can choose $n(i) \in \mathbb{N}$, $i \in \mathbb{N}$, so that

$$\tilde{g}(|K_i| \times \{n(i)\}) \cap \tilde{g}(|K_{i'}| \times \{n(i')\}) = \emptyset \ \text{ for } i' < i.$$

Then, $(\tilde{g}(|K_i| \times \{n(i)\}))_{i \in \mathbb{N}}$ is discrete in X. The desired map $g : \bigoplus_{i \in \mathbb{N}} |K_i| \to X$ can be defined by $g(z) = \tilde{g}(z, n(i))$ if $z \in |K_i|$, $i \in \mathbb{N}$.

Thus, Theorem 3.3.7 can be rewritten as follows:

Theorem 3.3.8 *For a separable completely metrizable ANR, the $\mathfrak{M}_1(\aleph_0)$-universality is equivalent to the locally finite compact polyhedra property.* □

To the locally finite f.d. τ-polyhedra property, the above argument cannot be applied because non-compact polyhedra are discussed. Nevertheless, the following version of Theorem 3.3.6 can be proved:

Theorem 3.3.9 *Let X be a completely metrizable ANR with $w(X) = \tau$. Then, X is $\mathfrak{M}_1(\tau)$-universal if and only if X has the τ-locally finite cells property and the locally finite f.d. τ-polyhedra property.*

In the above theorem, the τ-locally finite cells property implies the τ-discrete cells property (Proposition 3.2.1). Hence, the above theorem is obtained from the following proposition:

Proposition 3.3.10 *For a completely metrizable ANR with the τ-discrete cells property, the locally finite f.d. τ-polyhedra property implies the discrete f.d. τ-polyhedra property.*

Proof Let X be a completely metrizable ANR with the τ-discrete cells property and the locally finite f.d. τ-polyhedra property. Then, we prove that X has the discrete f.d. τ-polyhedra property. For a sequence $(K_i)_{i\in\mathbb{N}}$ of finite-dimensional simplicial complexes with card $K_i^{(0)} \leqslant \tau$, let $K = \bigoplus_{i\in\mathbb{N}} K_i$, where we are regarding $K_i \cap K_j = \emptyset$ for $i \neq j$. Since X has the locally finite f.d. τ-polyhedra property, it suffices to show that if $f : |K| \to X$ is a map such that $(f(|K_i|))_{i\in\mathbb{N}}$ is locally finite in X, then f can be approximated by maps $g : |K| \to X$ such that $(g(|K_i|))_{i\in\mathbb{N}}$ is discrete in X.

Let $f : |K| \to X$ be a map such that $(f(|K_i|))_{i\in\mathbb{N}}$ is locally finite in X. Then, there is $\mathcal{V} \in \operatorname{cov}(X)$ such that if a map $g : |K| \to X$ is $\mathcal{V}$-close to f then $(g(|K_i|))_{i\in\mathbb{N}}$ is also locally finite in X. For each $\mathcal{U} \in \operatorname{cov}(X)$, take $\mathcal{W} \in \operatorname{cov}(X)$ such that $\operatorname{st}\mathcal{W} \prec \mathcal{U} \wedge \mathcal{V}$. Then, K has a subdivision K' such that $\mathcal{O}_{K'} \prec f^{-1}(\mathcal{W})$ (Theorem 1.6.9). For each $i \in \mathbb{N}$, let K_i' be the subdivision of K_i induced by K'. Let $\varphi = \mathrm{id} : |K| = |K'| \to |K'|_{\mathrm{m}}$. Then, there is a map $\psi : |K'|_{\mathrm{m}} \to |K'|$ such that $\varphi\psi \simeq_{\mathcal{O}_{K'}} \mathrm{id}$ and $\psi\varphi \simeq_{\mathcal{O}_{K'}} \mathrm{id}$ (cf. Theorem 1.7.3). Since $\mathcal{O}_{K'} \prec f^{-1}(\mathcal{W})$, it follows that $f\psi\varphi \simeq_{\mathcal{W}} f$. Let $i_0 : X \to X \times \boldsymbol{Q}$ be the map defined by $i_0(x) = (x, \mathbf{0})$ for every $x \in X$. Due to Lemma 3.3.5, there is a map $g' : |K'|_{\mathrm{m}} \to X \times \boldsymbol{Q}$ such that g' is $\mathrm{pr}_X^{-1}(\mathcal{W})$-close to $i_0 f\psi$ and

$$\left\{g'^{-1}(\{x\} \times \boldsymbol{Q}) \mid x \in X\right\} \prec \mathcal{O}_{K'} \prec f^{-1}(\mathcal{W}).$$

Then, $g = \mathrm{pr}_X g'\varphi : |K| \to X$ is a map such that g is $\mathcal{W}$-close to $f\psi\varphi$ and

$$\left\{g^{-1}(x) \mid x \in X\right\} \prec \mathcal{O}_{K'}.$$

Note that $\mathcal{O}_{K'}(v) = \mathcal{O}_{K_i'}(v) \subset |K_i'|$ for some $i \in \mathbb{N}$ because $K' = \bigoplus_{i\in\mathbb{N}} K_i'$. Hence, for each $x \in g(|K|)$, $g^{-1}(x)$ is contained in only one of $|K_i|$. This means that $(g(|K_i|))_{i\in\mathbb{N}}$ is mutually disjoint. Since g is $\operatorname{st}\mathcal{W}$-close to f, g is $\mathcal{V}$-close to f, so $(g(|K_i|))_{i\in\mathbb{N}}$ is locally finite in X. Then, it follows that $(g(|K_i|))_{i\in\mathbb{N}}$ is discrete in X. Moreover, g is $\mathcal{U}$-close to f. Thus, the proof is completed. □

Combining Propositions 3.3.2 and 3.3.10, we have the following:

Corollary 3.3.11 *Let X be a completely metrizable ANR with the τ-discrete cells property. If X has the discrete (or locally finite) f.d. τ-polyhedra property, then every open set in X has the discrete (or locally finite) f.d. τ-polyhedra property.* □

Moreover, the following can be obtained by the above corollary, Proposition 3.2.3, and Theorem 3.3.6:

Corollary 3.3.12 *Let X be a completely metrizable ANR with $w(X) = \tau$. If X is $\mathfrak{M}_1(\tau)$-universal, then every open set in X is $\mathfrak{M}_1(\tau)$-universal.* □

Remark 3.7 The discrete (or locally finite) f.d. τ-polyhedra property is equivalent to the following property:

- For each simplicial complex K with $\operatorname{card} K^{(0)} \leqslant \tau$, each map $f : |K| \to X$ and $\mathcal{U} \in \operatorname{cov}(X)$, there exist maps $g_i : |K^{(i)}| \to X$, $i \in \mathbb{N}$, such that each g_i is $\mathcal{U}$-close to $f||K^{(i)}|$ and $(g_i(|K^{(i)}|)_{i\in\mathbb{N}}$ is discrete (or locally finite) in X.

The proof of this equivalence is left to the reader, which is a good exercise.

The following property is called the **discrete** (resp. **locally finite**) **self-copies property**:

- For each $\mathcal{U} \in \operatorname{cov}(X)$, there exist closed embeddings $h_i : X \to X$, $i \in \mathbb{N}$, such that every h_i is $\mathcal{U}$-close to id_X and $(h_i(X))_{i\in\mathbb{N}}$ is discrete (resp. locally finite) in X.

The discrete self-copies property can also be defined as follows:

- The projection $\mathrm{pr}_X : X \times \mathbb{N} \to X$ can be approximated by closed embeddings.

Due to Proposition 3.2.10, for an ANR X, the τ-discrete cells property implies that a closed set A in X is a Z-set if $w(A) < \tau$ or A is compact. If X has the discrete self-copies property, then the converse also holds, that is:

Theorem 3.3.13 *For an ANR X with the locally finite self-copies property, the τ-discrete cells property is equivalent to the following condition:*

- *Every compact set and every closed set A in X with $w(A) < \tau$ are Z-sets.*

Proof As mentioned above, the τ-discrete cells property implies the above Z-set condition. Due to Proposition 3.2.1 the τ-locally finite cells property implies the τ-discrete cells property. Thus, it suffices to prove that X has the τ-locally finite cells property if X satisfies the condition. We may assume that $\Gamma = (\mathbb{N}, \leqslant)$ ($\leqslant$ is the natural order) or $\Gamma = (\Gamma, \leqslant)$ is a well-ordered set as in the proof of Proposition 3.2.1, that is, $\operatorname{card}\Gamma(\gamma) < \operatorname{card}\Gamma = \tau$ for every $\gamma \in \Gamma$, where $\Gamma(\gamma) = \{\gamma' \in \Gamma \mid \gamma' < \gamma\}$.

For each map $f : \mathbf{I}^n \times \Gamma \to X$, let $f_\gamma : \mathbf{I}^n \to X$ be the map defined by $f_\gamma(z) = f(z, \gamma)$. For each $\mathcal{U} \in \operatorname{cov}(X)$, let $\mathcal{V} \in \operatorname{cov}(X)$ be a star-refinement of $\mathcal{U}$. By transfinite induction, we can construct maps $g_\gamma : \mathbf{I}^n \to X$, $\gamma \in \Gamma$, such that g_γ is $\mathcal{V}$-close to f_γ and $g_\gamma(\mathbf{I}^n) \subset X \setminus A_\gamma$, where

$$A_\gamma = \operatorname{cl} \bigcup_{\gamma' < \gamma} g_{\gamma'}(\mathbf{I}^n).$$

Since A_γ is compact or $w(A_\gamma) \leqslant \aleph_0 \operatorname{card} \Gamma(\gamma) < \operatorname{card} \Gamma = \tau$, it follows from the assumption that every A_γ is a Z-set in X, which guarantees to perform the above transfinite induction. Take $d \in \operatorname{Metr}(X)$ and let

$$\Gamma_i = \left\{\gamma \in \Gamma \mid 2^{-i} < \operatorname{dist}_d(g_\gamma(\mathbf{I}^n), A_\gamma) \leqslant 2^{-i+1}\right\} \text{ for each } i \in \mathbb{N}.$$

Then, $\Gamma = \bigoplus_{i \in \mathbb{N}} \Gamma_i$.

By the locally finite self-copies property, we have closed embeddings $h_i : X \to X$, $i \in \mathbb{N}$, such that every h_i is $\mathcal{V}$-close to id_X and $(h_i(X))_{i \in \mathbb{N}}$ is locally finite in X. Let $g : \mathbf{I}^n \times \Gamma = \bigoplus_{i \in \mathbb{N}}(\mathbf{I}^n \times \Gamma_i) \to X$ be the map defined as follows:

$$g(z, \gamma) = h_i g_\gamma(z) \text{ if } (x, \gamma) \in \mathbf{I}^n \times \Gamma_i.$$

For each $\gamma \in \Gamma$, $g|\mathbf{I}^n \times \{\gamma\}$ is $\mathcal{U}$-close to $f|\mathbf{I}^n \times \{\gamma\}$ because it is $\mathcal{V}$-close to $g_\gamma \mathrm{pr}_{\mathbf{I}^n}|\mathbf{I}^n \times \{\gamma\}$. Hence, g is $\mathcal{U}$-close to f. Moreover, $(g(\mathbf{I}^n \times \{\gamma\}))_{\gamma \in \Gamma}$ is locally finite in X. Indeed, $g(\mathbf{I}^n \times \{\gamma\}) = h_i g_\gamma(\mathbf{I}^n)$ if $\gamma \in \Gamma_i$. Each $x \in X$ has a neighborhood V in X such that $\{i \in \mathbb{N} \mid V \cap h_i(X) \neq \emptyset\}$ is finite. Let $\{i \in \mathbb{N} \mid V \cap h_i(X) \neq \emptyset\} = \{i_1, \ldots, i_k\}$. Since $(h_{i_j} g_\gamma(\mathbf{I}^n)_{\gamma \in \Gamma_i}$ is discrete in X for each $j = 1, \ldots, k$, we can take a neighborhood $W \subset V$ of x in X which meets $h_{i_j} g_\gamma(\mathbf{I}^n)$ for at most one $\gamma \in \Gamma_{i_j}$. It follows that $W \cap g(\mathbf{I}^n \times \{\gamma\})$ for at most k many $\gamma \in \Gamma$. □

For any space $X \in \mathfrak{M}_1(\tau)$, the $\mathfrak{M}_1(\tau)$-universality implies the discrete self-copies property because $X \times \mathbb{N} \in \mathfrak{M}_1(\tau)$. Obviously, the discrete (resp. locally finite) self-copies property implies the discrete (resp. locally finite) f.d. τ-polyhedra property. Thus, combining the above theorem with Theorem 3.3.9, we have:

Corollary 3.3.14 *Let X be a completely metrizable ANR with $w(X) = \tau$. Then, X is $\mathfrak{M}_1(\tau)$-universal if and only if X has the locally finite self-copies property and every compact set and every closed set A in X with $w(A) < \tau$ are Z-sets.* □

Note $\ell^2 \times \mathbb{R} \approx \ell^2$. Due to Corollary 2.3.5, if X is ℓ^2-stable (i.e., $X \times \ell^2 \approx X$), then $\mathrm{pr}_X : X \times \mathbb{R} \to X$ is a near-homeomorphism. Hence, $\mathrm{pr}_X : X \times \mathbb{N} \to X$ can be approximated by closed embeddings, that is, X has the discrete self-copies property. Thus,

- *If X is ℓ^2-stable, then X has the discrete self-copies property.*

Here, we introduce a property similar to the locally finite self-copies property. The following is called the **locally finite displacing property**:[9]

- For each $\mathcal{U} \in \operatorname{cov}(X)$, the projection $\mathrm{pr}_X : X \times \mathbb{N} \to X$ is $\mathcal{U}$-close to a map $g : X \times \mathbb{N} \to X$ such that $(g(X \times \{n\}))_{n \in \mathbb{N}}$ is locally finite in X.

[9] This property is called the **countable locally finite approximation property** in [21] and [24].

Evidently, the locally finite self-copies property implies the locally finite displacing property, and the locally finite displacing property implies the locally finite f.d. τ-polyhedra property for every cardinal τ. In Theorem 3.3.9, the locally finite f.d. τ-polyhedra property can be replaced with the locally finite displacing property, which will be shown in the next section.

Remark 3.8 In Theorem 3.3.13, when $\tau < \mathfrak{d}$,[10] the locally finite self-copies property can be replaced by the locally finite displacing property, which was proved in [24].

3.4 Characterization of $\ell^2(\Gamma)$-Manifolds

In this section, we prove the Toruńczyk characterization of Hilbert manifolds (i.e., $\ell^2(\Gamma)$-manifolds). We start by proving the following characterization:

Theorem 3.4.1 (Toruńczyk) *A completely metrizable ANR X with $w(X) = \tau$ is an $\ell^2(\Gamma)$-manifold if and only if X is strongly $\mathfrak{M}_1(\tau)$-universal.*

Proof Because $\mathcal{F}_{\ell^2(\Gamma)} = \mathfrak{M}_1(\tau)$ (Theorem 1.13.9), the "only if" part follows from the Strong Universality Theorem 2.3.15 together with Theorem 2.9.3. To prove the "if" part, we apply the Toruńczyk Factor Theorem 2.2.14 and the Bing Shrinking Criterion 2.7.1.

Since $C(\ell^2(\Gamma)) \approx \ell^2(\Gamma)$ by Corollary 2.4.10, $X \times C(\ell^2(\Gamma))$ is an $\ell^2(\Gamma)$-manifold by the Toruńczyk Factor Theorem 2.2.14. Hence, it is enough to prove that the projection $\mathrm{pr}_X : X \times C(\ell^2(\Gamma)) \to X$ is a near-homeomorphism. For simplicity, we use the same notation as in the proof of Proposition 2.9.1. Moreover, for $\alpha < \beta \in \mathrm{C}(X \times \ell^2(\Gamma), (0, 1))$, let

$$\begin{aligned} X[\alpha, \beta] &= \big\{(x, t, y) \in X[0, 1] \bigm| \alpha(x, y) \leqslant t \leqslant \beta(x, y)\big\} \\ &\subset X \times (0, 1] \times \ell^2(\Gamma) \subset X \times C(\ell^2(\Gamma)). \end{aligned}$$

For $M \subset X \times \ell^2(\Gamma)$, we write $M[\alpha, \beta] = X[\alpha, \beta] \cap p^{-1}(M)$, where $p : X \times (0, 1] \times \ell^2(\Gamma) \to X \times \ell^2(\Gamma)$ is the projection. According to this notation, $X[\alpha, \beta]$ should be denoted by $(X \times \ell^2(\Gamma))[\alpha, \beta]$, but we here abuse the notation.

We apply the Bing Shrinking Criterion 2.7.1 to see that $\mathrm{pr}_X : X[0, 1] \to X$ is a near-homeomorphism, where $X[0, 1] = X \times C(\ell^2(\Gamma))$ is an $\ell^2(\Gamma)$-manifold. For each $\mathcal{U} \in \mathrm{cov}(X[0, 1])$ and $\mathcal{V} \in \mathrm{cov}(X)$, we shall construct a near-homeomorphism $h : X[0, 1] \to X[0, 1]$ and $\mathcal{W} \in \mathrm{cov}(X)$ such that $h(\mathrm{pr}_X^{-1}(\mathcal{W})) \prec \mathcal{U}$ and $\mathrm{pr}_X h$ is $\mathcal{V}$-close to pr_X (cf. Remark 2.11):

[10]The cardinal $\mathfrak{d}$ is well-known in Set Theory as the **dominating number**, which can be equivalently defined as the compact covering number of ℓ^2 (cf. Introduction of [24]).

$$\begin{array}{ccc} \mathrm{pr}_X^{-1}(\mathcal{W}) & & X[0,1] \xrightarrow{\ \mathrm{pr}_X\ } X \\ \| & & \\ h\ \|\ \text{shrinks} & & h \downarrow \quad \mathcal{V} \nearrow \mathrm{pr}_X \\ \Downarrow & & \\ \mathcal{U} & & X[0,1] \end{array}$$

As in the proof of Proposition 2.9.1, let $\alpha \in \mathrm{C}(X, (0,1))$ and $\mathcal{V}_0 \in \mathrm{cov}(X)$ such that $\mathcal{V}_0 \prec \mathcal{V}$ and $\{V[0,\alpha] \mid V \in \mathcal{V}_0\} \prec \mathcal{U}$. Moreover, we have

$$\alpha_0 < \alpha_1 < \alpha_2 < \cdots \in \mathrm{C}(X \times \ell^2(\Gamma), (0,1)) \text{ and } \mathcal{U}_0 \in \mathrm{cov}(X \times \ell^2(\Gamma))$$

such that $\alpha_1 < \alpha \mathrm{pr}_X$, $\lim_{n\to\infty} \alpha_n(x,y) = 1$ and

$$\{U[\alpha_{n-1}, \alpha_{n+1}] \mid U \in \mathcal{U}_0,\ n \in \mathbb{N}\} \prec \mathcal{U}.$$

Indeed, for each $(x,y) \in X \times \ell^2(\Gamma)$, let

$$\gamma'(x,y) = \sup\{s \in (0,1) \mid \forall t \in [\alpha(x)/2, 1],\ \exists U \in \mathcal{U} \\ \text{such that } \{x\} \times ([t, t+s] \cap \mathbf{I}) \times \{y\} \subset U\}.$$

To see the lower semi-continuity of $\gamma' : X \times \ell^2(\Gamma) \to (0,1]$, let $(x,y) \in \gamma'^{-1}((s,1])$, $s \in (0,1)$ (i.e., $0 < s < \gamma'(x,y)$). Then, we take $s < s' < \gamma'(x,y)$. Since $[\alpha(x)/2, 1]$ is compact, we can choose open sets $V_i \subset X$, $W_i \subset \ell^2(\Gamma)$, $J_i \subset \mathbf{I}$, $i = 1, \ldots, n$, so that $x \in V_i$, $y \in W_i$, $[\alpha(x)/2, 1] \subset \bigcup_{i=1}^n J_i$, and

$$\exists U \in \mathcal{U} \text{ such that } \forall x' \in V_i,\ \forall y' \in W_i,\ \forall t' \in J_i, \\ \{x'\} \times ([t', t'+s'] \cap \mathbf{I}) \times \{y'\} \subset U.$$

Moreover, choose a neighborhood V_0 of x in X so that $[\alpha(x')/2, 1] \subset \bigcup_{i=1}^n J_i$ for every $x' \in V_0$. Then, for each $x' \in \bigcap_{i=0}^n V_i$ and $y' \in \bigcap_{i=1}^n W_i$,

$$\forall t' \in [\alpha(x')/2, 1],\ \exists U \in \mathcal{U} \text{ such that } \{x'\} \times ([t', t'+s'] \cap \mathbf{I}) \times \{y'\} \subset U,$$

which means that $\gamma'(x', y') \geqslant s' > s$. Therefore, γ' is lower semi-continuous.

By Theorem 1.3.14, we have a map $\gamma : X \times \ell^2(\Gamma) \to (0,1)$ with $2\gamma(x,y) < \gamma'(x,y)$ and $\gamma(x,y) < \alpha(x)$. The maps $\alpha_0 < \alpha_1 < \alpha_2 < \cdots$ can be defined as follows:

$$\alpha_n(x,y) = \frac{\alpha(x) + (n-1)\gamma(x,y)}{1 + \alpha(x) + (n-1)\gamma(x,y)}.$$

Then, $\alpha_1(x, y) < \alpha(x)$ and $\lim_{n\to\infty} \alpha_n(x, y) = 1$. For each $n \in \mathbb{N}$,

$$\begin{aligned} 0 &< \alpha_n(x, y) - \alpha_{n-1}(x, y) \\ &= \frac{\alpha(x) + (n-1)\gamma(x, y)}{1 + \alpha(x) + (n-1)\gamma(x, y)} - \frac{\alpha(x) + (n-2)\gamma(x, y)}{1 + \alpha(x) + (n-2)\gamma(x, y)} \\ &< \frac{\gamma(x, y)}{1 + \alpha(x) + (n-1)\gamma(x, y)} < \gamma(x, y) < \gamma'(x, y)/2. \end{aligned}$$

For each $(x, y) \in X \times \ell^2(\Gamma)$, we have $n \in \mathbb{N}$ such that $1 - \alpha_n(x, y) < \gamma(x, y)$. Then, we can easily obtain $\mathcal{U}_0 \in \operatorname{cov}(X \times \ell^2(\Gamma))$ such that

$$\left\{U[\alpha_{n-1}, \alpha_{n+1}] \;\middle|\; U \in \mathcal{U}_0,\ n \in \mathbb{N}\right\} \prec \mathcal{U}.$$

Let $\mathcal{V}_1, \mathcal{V}_2 \in \operatorname{cov}(X)$ such that

$$\operatorname{st}^3 \mathcal{V}_2 \prec \mathcal{V}_1 \overset{*}{\prec} \mathcal{V}_0.$$

By the strong $\mathfrak{M}_1(\tau)$-universality of X, there exists a strong Z-embedding $v : X[1] \to X$ such that $v \simeq_{\mathcal{V}_2} \operatorname{pr}_X|X[1]$. Let $A = v(X[1])$ and let $q_A : X[0, 1] \to X[0, 1]_A$ and $p_A : X[0, 1]_A \to X$ be the natural maps. Then, $p_A q_A = \operatorname{pr}_X$. Since q_A is a near-homeomorphism by Proposition 2.9.1, we have a homeomorphism $f : X[0, 1] \to X[0, 1]_A$ such that

$$f|A[0] = q_A|A[0] \text{ and } f \simeq_{p_A^{-1}(\mathcal{V}_2)} q_A.$$

Observe that $f^{-1}(v(X[1])) = f^{-1}(A) = A[0]$ is a (strong) Z-set in $X[0, 1]$, hence $f^{-1}v : X[1] \to X[0, 1]$ is a (strong) Z-embedding. Moreover, we have

$$\operatorname{pr}_X f^{-1} v = p_A q_A f^{-1} v \simeq_{\mathcal{V}_2} p_A v = v \simeq_{\mathcal{V}_2} \operatorname{pr}_X|X[1].$$

Since $C(\ell^2(\Gamma))$ is contractible, we have $i_0 \operatorname{pr}_X \simeq \operatorname{id}_{X[0,1]}$ by an X-preserving homotopy, where $i_0 : X \to X[0]$ is the natural injection. Therefore,

$$f^{-1} v \simeq_{\operatorname{pr}_X^{-1}(\operatorname{st}\mathcal{V}_2)} \operatorname{id}_{X[1]} \text{ in } X[0, 1].$$

Note that $X[1]$ is a (strong) Z-set in $X[0, 1]$ (Lemma 2.8.2). By the Z-set Unknotting Theorem 2.9.7 (cf. 2.5.12), we have a homeomorphism $g : X[0, 1] \to X[0, 1]$ such that

$$g|X[1] = f^{-1} v \text{ and } g \simeq_{\operatorname{pr}_X^{-1}(\operatorname{st}^2 \mathcal{V}_2)} \operatorname{id}.$$

Then, $fg(X[1]) = A$. Note that $\widetilde{\mathcal{U}}_0 = p^{-1}(\mathcal{U}_0)$ is an open cover of $X[1]$ in $X[0, 1]$, hence $fg(\widetilde{\mathcal{U}}_0)$ is an open cover of A in $X[0, 1]_A$. Let $\mathcal{W}_0$ be an open cover of A in

X such that

$$q_A(\mathrm{pr}_X^{-1}(\mathcal{W}_0)) = p_A^{-1}(\mathcal{W}_0) = \{W[0,1]_A \mid W \in \mathcal{W}_0\} \prec fg(\widetilde{\mathcal{U}}_0).$$

Thus, we have a near-homeomorphism $\varphi = g^{-1}f^{-1}q_A : X[0,1] \to X[0,1]$ such that $\varphi(\mathrm{pr}_X^{-1}(\mathcal{W}_0)) \prec \widetilde{\mathcal{U}}_0$. Observe

$$\begin{aligned}\mathrm{pr}_X\varphi = \mathrm{pr}_X g^{-1}f^{-1}q_A &\simeq_{\mathrm{st}^2\mathcal{V}_2} \mathrm{pr}_X f^{-1}q_A = p_A q_A f^{-1} q_A \\ &\simeq_{\mathcal{V}_2} p_A q_A = \mathrm{pr}_X.\end{aligned}$$

Then, it follows that $\mathrm{pr}_X\varphi$ is $\mathcal{V}_1$-close to pr_X. — Refer to Fig. 3.1.

Let $M_0 = \bigcup \mathcal{W}_0$ and choose an open neighborhood N_0 of A in X so that $\mathrm{cl}\, N_0 \subset M_0$. Then, $\varphi(N_0[0,1]) = g^{-1}f^{-1}p_A^{-1}(N_0)$ is an open neighborhood of $X[1] = \varphi(A[0,1])$ in $X[0,1]$. See the lower left of Fig. 3.1. As in the proof of Proposition 2.9.1, we have $\beta_0 \in \mathrm{C}(X \times \ell^2(\Gamma), (0,1))$ such that $X[\beta_0, 1] \subset \varphi(N_0[0,1])$. Then, it follows that

$$\varphi((X \setminus N_0)[0,1]) = X[0,1] \setminus \varphi(N_0[0,1]) \subset X[0,\beta_0].$$

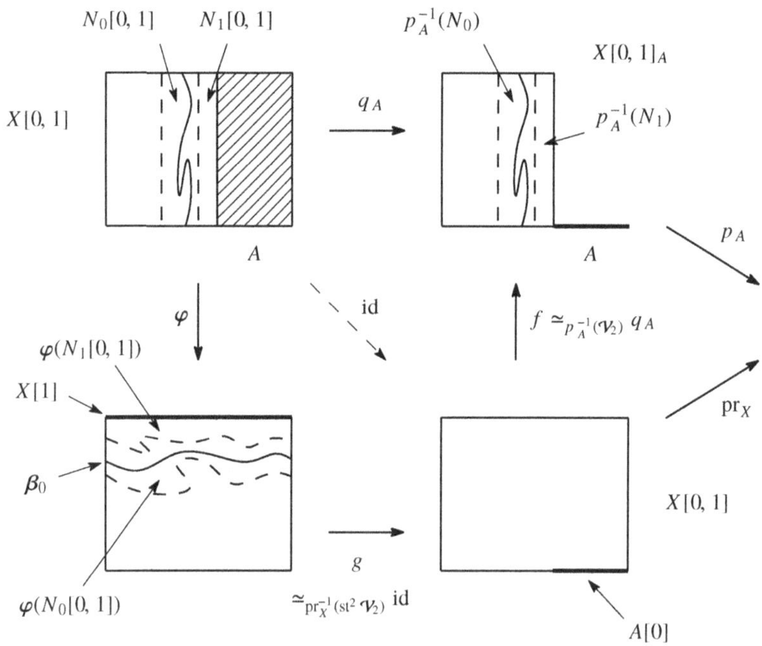

Fig. 3.1 Homeomorphisms f and g

Since $fg(X[\beta_0, 1])$ is a neighborhood of A in $X[0,1]_A$, A has open neighborhoods N_1 and M_1 in X such that $\operatorname{cl} N_1 \subset M_1 \subset \operatorname{cl} M_1 \subset N_0$ and $p_A^{-1}(\operatorname{cl} M_1) \subset fg(X[\beta_0, 1])$. Then, $\varphi(N_1[0,1]) = g^{-1}f^{-1}p_A^{-1}(N_1)$ is an open neighborhood of $X[1] = \varphi(A[0,1])$ in $X[0,1]$ and $\varphi((\operatorname{cl} M_1)[0,1]) \subset X[\beta_0, 1]$. We have $\beta_1 \in \mathrm{C}(X \times \ell^2(\Gamma), (0,1))$ such that $X[\beta_1, 1] \subset \varphi(N_1[0,1])$ and $\beta_1 > (1+\beta_0)/2$ $(> \beta_0)$. It should be noted that

$$\varphi((X \setminus N_1)[0,1]) = X[0,1] \setminus \varphi(N_1[0,1]) \subset X[0, \beta_1].$$

Thus, we can inductively choose open neighborhoods N_i, M_i of A in X and $\beta_i \in \mathrm{C}(X \times \ell^2(\Gamma), (0,1))$, $i \in \omega$, such that

$$\operatorname{cl} N_i \subset M_i \subset \operatorname{cl} M_i \subset N_{i-1}, \ \bigcap_{i \in \omega} N_i = A, \ \beta_i > \frac{1+\beta_{i-1}}{2} \ \text{ and}$$

$$X[\beta_i, 1] \subset \varphi(N_i[0,1]) \subset \varphi((\operatorname{cl} M_i)[0,1]) \subset X[\beta_{i-1}, 1].$$

Then, it follows that

$$\varphi((X \setminus \operatorname{cl} N_1)[0,1]) \subset X[0, \beta_1] \ \text{ and}$$
$$\varphi((M_i \setminus \operatorname{cl} N_{i+1})[0,1]) \subset X[\beta_{i-1}, \beta_{i+1}] \ \text{ for each } i \in \mathbb{N}.$$

Sliding $\mathbf{I}$-coordinates, we can obtain a homeomorphism $\psi : X[0,1] \to X[0,1]$ such that

$$\mathrm{pr}_X \psi = \mathrm{pr}_X, \ \psi | X[0] \cup X[1] = \mathrm{id} \ \text{ and}$$
$$\psi(x, \beta_i(x,y), y) = (x, \alpha_i(x,y), y) \ \text{ for } (x,y) \in X \times \ell^2(\Gamma) \text{ and } i \in \omega.$$

See Fig. 3.2. Thus, we have a near-homeomorphism $h = \psi\varphi : X[0,1] \to X[0,1]$ such that $\mathrm{pr}_X h = \mathrm{pr}_X \psi\varphi$ is $\mathcal{V}_1$-close (hence $\mathcal{V}$-close) to pr_X. To see that h is the desired near-homeomorphism, it remains to be verified that each $x \in X$ has a neighborhood W_x in X such that $h(\mathrm{pr}_X^{-1}(W_x)) = h(W_x[0,1])$ is contained in some member of $\mathcal{U}$.

When $x \in X \setminus \operatorname{cl} N_1$, choose $V_x \in \mathcal{V}_1$ so that $x \in V_x$ and let $W_x = V_x \setminus \operatorname{cl} N_1$. Since $\mathrm{pr}_X h = \mathrm{pr}_X \psi\varphi = \mathrm{pr}_X \varphi$ is $\mathcal{V}_1$-close to pr_X, we have

$$\mathrm{pr}_X h(\mathrm{pr}_X^{-1}(W_x)) = \mathrm{pr}_X \psi\varphi(W_x[0,1]) \subset \operatorname{st}(V_x, \mathcal{V}_1),$$

which is contained in some $V \in \mathcal{V}_0$. Then, it follows that

$$h(\mathrm{pr}_X^{-1}(W_x)) = \psi\varphi(W_x[0,1]) \subset \psi(V[0,\beta_1]) = V[0,\alpha_1] \subset V[0,\alpha],$$

which is contained in some member of $\mathcal{U}$.

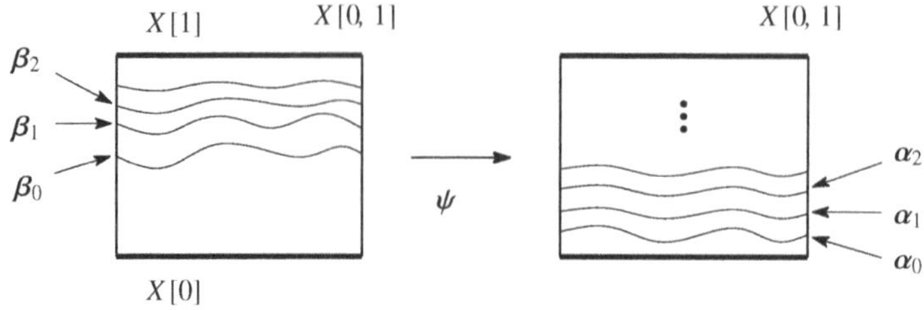

Fig. 3.2 The homeomorphism ψ

When $x \in M_1 \setminus A$, $x \in M_i \setminus \operatorname{cl} N_{i+1}$ for some $i \in \mathbb{N}$. Choose $W'_x \in \mathcal{W}_0$ so that $x \in W'_x$ and let $W_x = W'_x \cap (M_i \setminus \operatorname{cl} N_{i+1})$. Then, $\varphi(W'_x[0, 1]) \subset U[0, 1]$ for some $U \in \mathcal{U}_0$. It follows that

$$h(\mathrm{pr}_X^{-1}(W_x)) = \psi\varphi(W_x[0, 1]) \subset \psi(U[\beta_{i-1}, \beta_{i+1}]) = U[\alpha_{i-1}, \alpha_{i+1}],$$

which is contained in some member of $\mathcal{U}$.

When $x \in A$, $\psi(g^{-1} f^{-1}(x)) = g^{-1} f^{-1}(x)$ is contained in some $U \in \mathcal{U}$. By the continuity of $g^{-1} f^{-1} : X[0, 1]_A \to X[0, 1]$ and the definition of a neighborhood basis of $x \in A$ in $X[0, 1]_A$, x has an open neighborhood W_x in X such that

$$h(\mathrm{pr}_X^{-1}(W_x)) = \psi\varphi(W_x[0, 1]) = \psi(g^{-1} f^{-1}(p_A^{-1}(W_x))) \subset U.$$

Thus, h is the desired near-homeomorphism. □

In the previous section, we gave some characterizations of (strong) $\mathfrak{M}_1(\tau)$-universality. Now, summing them up together with Theorem 3.4.1 above, we have the following characterization:

Theorem 3.4.2 (Toruńczyk Characterization) *For a completely metrizable ANR X with $w(X) = \tau$, the following are equivalent:*

(a) *X is an $\ell^2(\Gamma)$-manifold;*
(b) *X is strongly $\mathfrak{M}_1(\tau)$-universal;*
(c) *X is $\mathfrak{M}_1(\tau)$-universal;*
(d) *X has the τ-discrete cells property and the discrete f.d. τ-polyhedra property;*
(e) *X has the τ-locally finite cells property and the locally finite f.d. τ-polyhedra property;*
(f) *X has the τ-locally finite cells property and the locally finite displacing property;*
(g) *X has the locally finite self-copies property and every compact set and every closed set A in X with $w(A) < \tau$ are Z-sets.*

Proof The equivalences (a) ⇔ (b) ⇔ (c) ⇔ (d) are just Theorems 3.4.1, 3.1.19, and 3.3.6. The equivalences (c) ⇔ (e) and (c) ⇔ (g) are Theorem 3.3.9 and

Corollary 3.3.14, respectively. Obviously, the combination (e) and (g) implies (f), so we have (c) ⇒ (f). The implication (f) ⇒ (e) is trivial. □

We shall weaken the locally finite f.d. τ-polyhedra property of condition (e) in the above theorem. Let $F(\Gamma)$ be the simplicial complex defined by the abstract simplicial complex $\mathrm{Fin}(\Gamma)$ consisting of all non-empty finite subsets of Γ. Namely, $F(\Gamma)$ is the full complex with $\operatorname{card} F(\Gamma) = \tau$ ($F(\Gamma)^{(0)} = \Gamma$). For each $n \in \mathbb{N}$, let $P_\Gamma^n = |F(\Gamma)^{(n)}|$ and define $P_\Gamma = \bigoplus_{n\in\mathbb{N}} P_\Gamma^n$. The following is called the **locally finite τ-skeletal tower property**:

- Each map $f : P_\Gamma \to X$ can be approximated by maps $g : P_\Gamma \to X$ such that $(g(P_\Gamma^n))_{n\in\mathbb{N}}$ is locally finite in X.

The combination of this property with the τ-locally finite cells property characterizes $\ell^2(\Gamma)$-manifolds, that is:

Theorem 3.4.3 (Characterization) *A completely metrizable ANR X with $w(X) = \tau$ is an $\ell^2(\Gamma)$-manifold if and only if X has both the τ-locally finite cells property and the locally finite τ-skeletal tower property.*

To prove this theorem, we need the following lemma:

Lemma 3.4.4 *Let W be an open set in an ANR X that is contractible in X. If X has the locally finite τ-skeletal tower property, then W has the locally finite f.d. τ-polyhedra property.*

Proof By Proposition 3.3.1, we may show that W has the metric version of the locally finite f.d. τ-polyhedra property. Let $(K_i)_{i\in\mathbb{N}}$ be a sequence of finite-dimensional simplicial complexes with $\operatorname{card} K_i^{(0)} \leqslant \tau = \operatorname{card}\Gamma$ and let $K = \bigoplus_{i\in\mathbb{N}} K_i$. Note that the identity map $\varphi = \mathrm{id} : |K| \to |K|_{\mathrm{m}}$ is a fine homotopy equivalence (Theorem 1.7.6). Let $f : |K|_{\mathrm{m}} \to W$ be a map and $\mathcal{U} \in \operatorname{cov}(W)$. Take a metric $d \in \operatorname{Metr}(X)$ and define open sets

$$W_i = \{x \in X \mid d(x, X \setminus W) > 1/i\},\ i \in \mathbb{N}.$$

Then, $W = \bigcup_{i\in\mathbb{N}} W_i = \bigcup_{i\in\mathbb{N}} \operatorname{cl} W_i$ and $\operatorname{cl} W_i \subset W_{i+1}$ for each $i \in \mathbb{N}$. We define

$$R_i = f^{-1}(\operatorname{cl} W_i \setminus W_{i-1}) \subset |K|_{\mathrm{m}},\ i \in \mathbb{N}, \text{ and}$$
$$\mathcal{U}' = \bigcup_{i\in\mathbb{N}} \mathcal{U}|(W_i \setminus \operatorname{cl} W_{i-2}) \in \operatorname{cov}(W),$$

where $W_0 = W_{-1} = \emptyset$. Then, $|K|_{\mathrm{m}} = \bigcup_{i\in\mathbb{N}} R_i$ and $\mathcal{U}' \prec \mathcal{U}$.

We can regard each K_i as a subcomplex of $F(\Gamma)^{(n_i)}$, where $n_1 < n_2 < \cdots$ ($\in \mathbb{N}$). Then, $|K_i| \subset P_\Gamma^{n_i}$, $i \in \mathbb{N}$, which means $|K| = \bigoplus_{i\in\mathbb{N}} |K_i| \subset \bigoplus_{i\in\mathbb{N}} P_\Gamma^{n_i} \subset P_\Gamma$. Since W is contractible in X, we can apply the Homotopy Extension Theorem 1.5.7

to extend $f\varphi$ to a map $\tilde{f} : P_\Gamma \to X$. For each $k \in \mathbb{N}$, let $\mathcal{U}_k \in \operatorname{cov}(X)$ be a star-refinement of

$$\{X \setminus \operatorname{cl} W_{2k}\} \cup \bigcup_{i=1}^{2k+1} \mathcal{U}|(W_i \setminus \operatorname{cl} W_{i-2}) \in \operatorname{cov}(X).$$

Then, observe that $\operatorname{st} \mathcal{U}_k | W_{2k} \prec \mathcal{U}'$ and

$$\operatorname{st}(f(R_{2k-1}), \operatorname{st} \mathcal{U}_k) \subset \operatorname{st}(\operatorname{cl} W_{2k-1} \setminus W_{2k-2}, \operatorname{st} U_k) \subset W_{2k} \setminus \operatorname{cl} W_{2k-3}.$$

By the locally finite τ-skeletal tower property of X, there exists a map $g_k : P_\Gamma \to X$ such that $g_k \simeq_{\mathcal{U}_k} \tilde{f}$ and $(g_k(P_\Gamma^n))_{n\in\mathbb{N}}$ is locally finite in X. Then, $g_k(R_{2k-1}) \subset W_{2k} \setminus \operatorname{cl} W_{2k-3}$.

For each $k \in \mathbb{N}$, let $\psi_k : |K|_{\mathrm{m}} \to |K|$ be an $f^{-1}(\mathcal{U}_k)$-homotopy inverse of $\varphi = \mathrm{id} : |K| \to |K|_{\mathrm{m}}$. Then,

$$g_k\psi_k|R_{2k-1} \simeq_{\mathcal{U}_k} \tilde{f}\psi_k|R_{2k-1} = f\varphi\psi_k|R_{2k-1} \simeq_{\mathcal{U}_k} f|R_{2k-1},$$

which implies that $g_k\psi_k|R_{2k-1} \simeq_{\mathcal{U}'} f|R_{2k-1}$ because $\operatorname{st}(f(R_{2k-1}), \operatorname{st} \mathcal{U}_k) \subset W_{2k}$ and $\operatorname{st} \mathcal{U}_k|W_{2k} \prec \mathcal{U}'$. By the Homotopy Extension Theorem 1.13.11, we have a map $g : |K|_{\mathrm{m}} \to W$ such that $g \simeq_{\mathcal{U}'} f$ and $g|R_{2k-1} = g_k\psi_k|R_{2k-1}$ for each $k \in \mathbb{N}$:

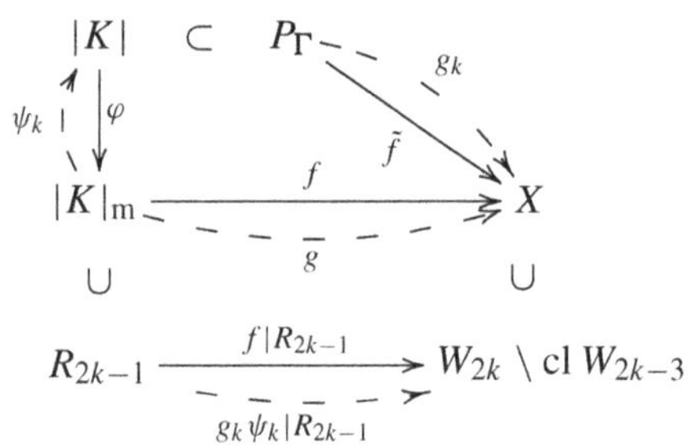

Then, for each $n, k \in \mathbb{N}$,

$$g(|K_n| \cap R_{2k-1}) = g_k\psi_k(|K_n| \cap R_{2k-1}) \subset g_k(P_\Gamma^n).$$

Thus, each $(g(|K_n| \cap R_{2k-1}))_{n\in\mathbb{N}}$ is locally finite in $W_{2k} \setminus \operatorname{cl} W_{2k-3}$. Hence, it follows that $(g(|K_n| \cap R_{2k-1}))_{(k,n)\in\mathbb{N}^2}$ is locally finite in W.

Now, choose $\mathcal{V} \in \operatorname{cov}(W)$ so that $\mathcal{V} \prec \mathcal{U}$ and $(h(|K_n| \cap R_{2i-1}))_{(i,n)\in\mathbb{N}^2}$ is locally finite in W for every $h \in \mathcal{V}(g)$, and define

$$\mathcal{V}' = \bigcup_{i\in\mathbb{N}} \mathcal{V}|(W_i \setminus \operatorname{cl} W_{i-2}) \in \operatorname{cov}(W).$$

As in the case of f, we can extend $g\varphi$ to a map $\tilde{g} : P_\Gamma \to X$. For each $k \in \mathbb{N}$, let $\mathcal{V}_k \in \operatorname{cov}(X)$ be a star-refinement of

$$\{X \setminus \operatorname{cl} W_{2k+1}\} \cup \bigcup_{i=1}^{2k+2} \mathcal{V}|(W_i \setminus \operatorname{cl} W_{i-2}) \in \operatorname{cov}(X).$$

By the locally finite τ-skeletal tower property of X, there exists a map $h_k : P_\Gamma \to X$ such that $h_k \simeq_{\mathcal{V}_k} g$ and $(h_k(P_\Gamma^n))_{n\in\mathbb{N}}$ is locally finite in X. Let $\psi_k' : |K|_{\mathrm{m}} \to |K|$ be a $g^{-1}(\mathcal{U}_k)$-homotopy inverse of φ. By the same construction as g, we use the Homotopy Extension Theorem 1.13.11 to construct a map $h : |K|_{\mathrm{m}} \to W$ such that $h \simeq_{\mathcal{V}'} g$ and $h|R_{2k} = h_k\psi_k'|R_{2k}$ for each $k \in \mathbb{N}$. By the same argument as g, it follows that $(h(|K_n| \cap R_{2i}))_{(i,n)\in\mathbb{N}^2}$ is locally finite in W. On the other hand, since h is $\mathcal{V}$-close to g, $(h(|K_n| \cap R_{2i-1}))_{(i,n)\in\mathbb{N}^2}$ is also locally finite in W. Therefore, $(h(|K_n| \cap R_i))_{(i,n)\in\mathbb{N}^2}$ is locally finite in W, which implies that $(h(|K_n|))_{n\in\mathbb{N}}$ is locally finite in W. Note that $h||K|$ is $\operatorname{st}\mathcal{U}$-close to f. Thus, W has the metric version of the locally finite f.d. τ-polyhedra property. As mentioned at the beginning of the proof, we have the result by Proposition 3.3.1. □

Now, we shall prove Theorem 3.4.3.

Proof (Theorem 3.4.3) Since the "only if" part is trivial, we shall show the "if" part. Because the local contractibility, X has an open cover $\mathcal{W}$ such that each $W \in \mathcal{W}$ is contractible in X. By Lemma 3.4.4 above, each $W \in \mathcal{W}$ has the locally finite f.d. τ-polyhedra property. Moreover, due to Proposition 3.2.3, W has the τ-locally finite cells property. Hence, W is an $\ell^2(\Gamma)$-manifold by Theorem 3.4.2. Thus, X is an $\ell^2(\Gamma)$-manifold. □

Since every contractible $\ell^2(\Gamma)$-manifold is homeomorphic to $\ell^2(\Gamma)$ (Corollary 2.6.2), we have the following:

Corollary 3.4.5 (Characterization of $\ell_2(\Gamma)$) *For a completely metrizable AR X with $w(X) = \tau$, the following are equivalent:*

(a) $X \approx \ell^2(\Gamma)$;
(b) *X is $\mathfrak{M}_1(\tau)$-universal;*
(c) *X has both the τ-locally finite cells property and the locally finite τ-skeletal tower property;*
(d) *X has both the τ-locally finite cells property and the locally finite displacing property;*
(e) *X has the locally finite self-copies property and every compact set and every closed set A in X with $w(A) < \tau$ are Z-sets.* □

The following property is called the **discrete** (resp. **locally finite**) **cell-tower property**:[11]

---

[11]The discrete cell-tower is known as the **discrete approximation property**.

- Each map $f : \bigoplus_{n\in\mathbb{N}} \mathbf{I}^n \to X$ can be approximated by maps $g : \bigoplus_{n\in\mathbb{N}} \mathbf{I}^n \to X$ such that $(g(\mathbf{I}^n))_{n\in\mathbb{N}}$ is discrete (resp. locally finite) in X.

It should be remarked that the countable discrete (resp. locally finite) *cells* property is a little weaker than the discrete (resp. locally finite) *cell-tower* property. Moreover, it is easy to see that the discrete (resp. locally finite) cell-tower property is equivalent to the following **countable discrete** (resp. **locally finite**) **Hilbert cubes property**:

- Each map $f : Q \times \mathbb{N} \to X$ can be approximated by maps $g : Q \times \mathbb{N} \to X$ such that $(g(Q \times \{n\}))_{n\in\mathbb{N}}$ is discrete (resp. locally finite) in X.

Since every compact polyhedron can be embedded into some $\mathbf{I}^n$ as a subpolyhedron, the following can be proved in the same way as Lemma 3.4.4:

Lemma 3.4.6 *Let W be an open set in an ANR X that is contractible in X. If X has the locally finite cell-tower property, then W has the locally finite compact polyhedra property.* □

In the separable case, we have the following characterization:

Theorem 3.4.7 (Toruńczyk Characterization) *For a separable completely metrizable ANR (resp. AR) X, the following are equivalent:*

(a) *X is an ℓ^2-manifold (resp. $X \approx \ell^2$);*
(b) *X is $\mathfrak{M}_1(\aleph_0)$-universal;*
(c) *X has the discrete cell-tower property;*
(d) *X has the locally finite cell-tower property.*

Proof The equivalence (a) ⇔ (b) is Theorem 3.4.1. Since $\bigoplus_{n\in\mathbb{N}} \mathbf{I}^n \in \mathfrak{M}_1(\aleph_0)$, we have (b) ⇒ (c). The implication (c) ⇒ (d) is obvious. The implication (d) ⇒ (b) can be proved in the same way as Theorem 3.4.3. Take $\mathcal{W} \in \operatorname{cov}(X)$ such that each $W \in \mathcal{W}$ is contractible in X. Due to Lemma 3.4.6, each $W \in \mathcal{W}$ has the locally finite compact polyhedra property. Then, it follows from Theorem 3.3.8 that W is $\mathfrak{M}_1(\aleph_0)$-universal. □

Although the following theorem is a direct consequence of Theorems 2.4.4 and 2.9.3, it can also be proved by using the characterization of $\ell^2(\Gamma)$-manifolds.

Theorem 3.4.8 (Negligibility of Z_σ-Sets) *For every Z_σ-set A in an $\ell^2(\Gamma)$-manifold M, the inclusion $M \setminus A \subset M$ is a near-homeomorphism.*

Due to Theorem 2.8.6, the complement of a Z-set in an $\ell^2(\Gamma)$-manifold is homotopy dense. Using the completeness, it can be easily shown that the complement of a Z_σ-set in an $\ell^2(\Gamma)$-manifold is also homotopy dense. Hence, the above Negligibility Theorem is equivalent to the following theorem, which can be proved by applying the characterization Theorem 3.4.7:

Theorem 3.4.9 *For every homotopy dense G_δ set X in an $\ell^2(\Gamma)$-manifold M, the inclusion $X \hookrightarrow M$ is a near-homeomorphism.*

Proof If X is also an $\ell^2(\Gamma)$-manifold, then the result follows from Theorem 2.6.3. It suffices to show that X is an $\ell^2(\Gamma)$-manifold. Note that X is a completely metrizable ANR (cf. Theorems 1.3.17 and 1.13.26). For each $\mathcal{U} \in \mathrm{cov}(X)$, let $\mathcal{V}$ be an open collection in M such that $\mathcal{V}|X \in \mathrm{cov}(X)$ is a star-refinement of $\mathcal{U}$. Note that $W = \bigcup \mathcal{V}$ is an $\ell^2(\Gamma)$-manifold because it is open in M. Every map $f : \mathbf{I}^n \times \Gamma \to X$ $(\subset W)$ is $\mathcal{V}$-close to a map $f' : \mathbf{I}^n \times \Gamma \to W$ such that $(f'(\mathbf{I}^n \times \{\gamma\}))_{\gamma \in \Gamma}$ is discrete in W. Choose $\mathcal{V}' \in \mathrm{cov}(W)$ so that $\mathcal{V}' \prec \mathcal{V}$ and, if a map $f'' : \mathbf{I}^n \times \Gamma \to W$ is $\mathcal{V}'$-close to f', then $(f''(\mathbf{I}^n \times \{\gamma\}))_{\gamma \in \Gamma}$ is discrete in W. Since X is homotopy dense in M, the inclusion $X \hookrightarrow M$ is a fine homotopy equivalence. Hence, we have a map $h : M \to X$ such that $h|X \simeq_{\mathcal{V}'|X} \mathrm{id}_X$ and $h \simeq_{\mathcal{V}'} \mathrm{id}_M$. Then, $(hf'(\mathbf{I}^n \times \{\gamma\}))_{\gamma \in \Gamma}$ is discrete in W, hence in X. Since $\mathrm{st}(\mathcal{V}, \mathcal{V}') \prec \mathcal{U}$, the map $hf' : \mathbf{I}^n \times \Gamma \to X$ is $\mathcal{U}$-close to f. Therefore, X has the τ-discrete cells property. It can be similarly proved that X has the discrete f.d. τ-polyhedra property. By the above characterization Theorem 3.4.7, X is an $\ell^2(\Gamma)$-manifold. □

Remark 3.9 In Theorem 3.4.2 (resp. Corollary 3.4.5), when $w(X) = \tau < \mathfrak{d}$, "the locally finite self-copies property" in condition (f) (resp. (d)) can be weaken to "the locally finite displacing property" (cf. Remark 3.8, p. 239), that is, a complete metrizable connected ANR (resp. AR) X with $w(X) = \tau < \mathfrak{d}$ is an $\ell^2(\Gamma)$-manifold (resp. homeomorphic to $\ell^2(\Gamma)$) if and only if X has the locally finite displacing property and every closed set $A \subset X$ with $w(A) < \tau$ is a Z-set. This was proved by T. Banakh and I. Zarichnyy [24].

Remark 3.10 In [16], it was shown that there exists a completely metrizable separable metric AR $X = (X, d)$ which has the following properties:

(i) for each map $f : \boldsymbol{Q} \times \mathbb{N} \to X$ and $\varepsilon > 0$, there is a map $g : \boldsymbol{Q} \times \mathbb{N} \to X$ such that g is ε-close to f and $(g(\boldsymbol{Q} \times \{n\}))_{n \in \mathbb{N}}$ is discrete in X;
(ii) X is embedded as a convex set in ℓ^2;
(iii) every compact set in X is a Z-set;
(iv) $X \times X \approx \ell^2$;
(v) X is homogeneous;
(vi) $X \approx X \setminus G$ for every countable set $G \subset X$;

but which is *not* homeomorphic to ℓ^2 because of the following property:

(vii) $X \not\approx X \setminus C$ for any Cantor set $C \subset X$.

It should be noted that the above (i) is weaker than the countable discrete Hilbert cubes property. In fact, if the phrases "$\varepsilon > 0$" and "ε-close" in (i) were changed to "$\mathcal{U} \in \mathrm{cov}(X)$" and "$\mathcal{U}$-close" respectively, then (i) would be just the countable discrete Hilbert cubes property (equiv. the discrete cell-tower property).

3.5 Fréchet Spaces and the Countable Product of ARs

In this section, it is shown that every Fréchet space is homeomorphic to a Hilbert space with the same weight, and that the countable product of completely metrizable ARs is homeomorphic to a Hilbert space under some condition. We also modify the conditions characterizing $\ell^2(\Gamma)$-manifolds (or $\ell^2(\Gamma)$). The following is called the **uniformly separating cells property**:

(USC) For each $\varepsilon > 0$, there exists $\delta > 0$ such that any map $f : \mathbf{I}^n \to X$ ($n \in \mathbb{N}$) is ε-homotopic to a map $g : \mathbf{I}^n \to X$ with $\operatorname{dist}(f(\mathbf{I}^n), g(\mathbf{I}^n)) > \delta$.

First, we show the following:

Lemma 3.5.1 *Let $X = (X, d)$ be a locally path-connected metric space. If X has the uniformly separating cells property, then every open set in X has the locally finite cell-tower property (i.e., the discrete cell-tower property).*

Proof We may assume that d is bounded. Let W be an open set in X. It suffices to show that each component has the discrete cell-tower property. Hence, it can be assumed that W and X are connected, which means path-connected because of local path-connectedness. Denote $D = \bigoplus_{n\in\mathbb{N}} \mathbf{I}^n$ and take maps $f : D \to W$ and $\alpha : W \to (0, 1)$, where we may require α to satisfy the following condition:

$$\alpha(x) < d(x, X \setminus W) \ \text{ for each } x \in W.$$

For each $k \in \mathbb{N}$, let

$$D_k = \left\{x \in D \mid \alpha f(x) \geqslant 3^{-k}\right\} = (\alpha f)^{-1}([3^{-k}, 1]).$$

Then, $D_1 \subset D_2 \subset \cdots$ and $D \setminus \bigcup_{k\in\mathbb{N}} D_k = (\alpha f)^{-1}(0)$. We construct sequences $g_k : D \to X$ and $\varepsilon_k > 0$, $k \in \mathbb{N}$, so as to satisfy the following conditions:

(1) $g_k|D \setminus D_{k+1} = f|D \setminus D_{k+1}$ and $g_k|D_{k-1} = g_{k-1}|D_{k-1}$;
(2) $\operatorname{dist}(g_k(\mathbf{I}^n \cap D_k), g_k(\mathbf{I}^m)) > \varepsilon_k$ if $m < n$;
(3) $d(g_k, g_{k-1}) < \frac{1}{3}\varepsilon_{k-1}$;
(4) $\varepsilon_k \leqslant \frac{1}{3}\varepsilon_{k-1}$,

where $\varepsilon_0 = 3^{-2}$, $g_0 = f$ and $D_0 = \emptyset$. Note that $\varepsilon_k \leqslant 3^{-k-2}$ for each $k \in \mathbb{N}$.

Assume g_{k-1} and ε_{k-1} have been constructed. Choose a neighborhood U of D_{k-1} in D so that

(5) $\operatorname{dist}(g_{k-1}(\mathbf{I}^n \cap U), g_{k-1}(\mathbf{I}^m)) > \varepsilon_{k-1}$ if $m < n$.

For $\varepsilon = \frac{1}{3}\varepsilon_{k-1} > 0$, take $\delta > 0$ as in (USC) and define $\varepsilon_k = \min\{\varepsilon, \delta\} \leqslant \frac{1}{3}\varepsilon_{k-1}$. Then, (4) is obvious. By induction on n, we define a map $g_k|\mathbf{I}^n$ as follows: Let $g_k|\mathbf{I}^1 = g_{k-1}|\mathbf{I}^1$. Assume that $g_k|\mathbf{I}^1 \oplus \cdots \oplus \mathbf{I}^{n-1}$ is defined so as to satisfy (1), (2),

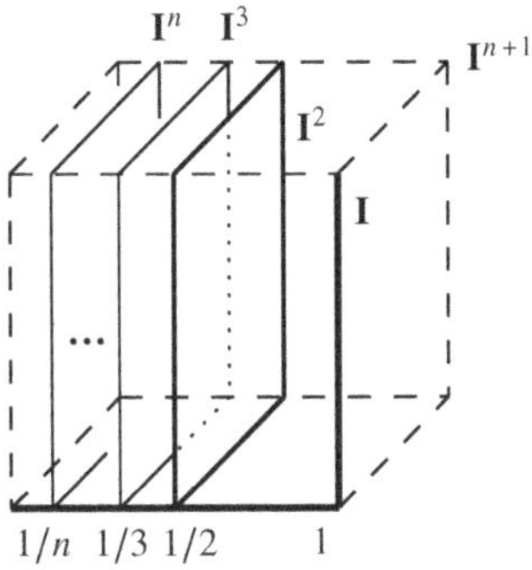

Fig. 3.3 $F = \mathbf{I} \times \{0\} \cup \bigcup_{i=1}^{n} \{1/i\} \times \mathbf{I}^i$

and (3), where D is replaced by $\mathbf{I}^1 \oplus \cdots \oplus \mathbf{I}^{n-1}$. Regarding $\mathbf{I} \subset \mathbf{I}^2 \subset \cdots \subset \mathbf{I}^n$, we define

$$F = \mathbf{I} \times \{0\} \cup \bigcup_{i=1}^{n} \{1/i\} \times \mathbf{I}^i \subset \mathbf{I}^{n+1},$$

where $\bigcup_{i=1}^{n} \{1/i\} \times \mathbf{I}^i$ can be identified with $\mathbf{I}^1 \oplus \cdots \oplus \mathbf{I}^n$ (see Fig. 3.3). Since X is assumed to be path-connected, we have a map $v : F \to W$ such that $v(1/i, x) = g_k(x)$ for $x \in \mathbf{I}^i$, $i < n$, and $v(1/n, x) = g_{k-1}(x)$ for $x \in \mathbf{I}^n$. Combining with a retraction of $\mathbf{I}^{n+1}$ onto F, we can extend v to a map $w : \mathbf{I}^{n+1} \to W$. As the restriction of an ε-homotopy obtained by applying (USC) to w, we have an ε-homotopy $s : \mathbf{I}^n \times \mathbf{I} \to X$ such that $s_0 = g_{k-1}|\mathbf{I}^n$ and

(6) $\operatorname{dist}(s_1(\mathbf{I}^n), g_k(\mathbf{I}^1 \oplus \cdots \oplus \mathbf{I}^{n-1})) > \delta$.

Recall that $D_k \cap \operatorname{cl}(D \setminus D_{k+1}) = \emptyset$ and $D_{k-1} \subset U$. Let $\lambda : D \to \mathbf{I}$ be an Urysohn map with

$$\lambda(D_k \setminus U) = 1 \ \text{ and } \ \lambda(\operatorname{cl}(D \setminus D_{k+1}) \cup D_{k-1}) = 0.$$

We define $g_k|\mathbf{I}^n : \mathbf{I}^n \to X$ by

$$g_k(x) = s(x, \lambda(x)) \ \text{ for each } x \in \mathbf{I}^n.$$

Then, $g_k|\mathbf{I}^1 \oplus \cdots \oplus \mathbf{I}^n$ satisfies (1) and (3). To see (2), let $m < n$, $x \in \mathbf{I}^n \cap D_k$, and $y \in \mathbf{I}^m$. If $x \notin U$, then $g_k(x) = s_1(x)$, hence it follows from (6) that $d(g_k(x), g_k(y)) > \delta \geqslant \varepsilon_k$. If $x \in U$, then $d(g_{k-1}(x), g_{k-1}(y)) > \varepsilon_{k-1}$ by (5), $d(g_k(x), g_{k-1}(x)) < \frac{1}{3}\varepsilon_{k-1}$, and $d(g_k(y), g_{k-1}(y)) < \frac{1}{3}\varepsilon_{k-1}$ by (3), hence it follows that $d(g_k(x), g_k(y)) > \frac{1}{3}\varepsilon_{k-1} \geqslant \varepsilon_k$.

By (1) and (3), we have a map $g : D \to X$ as the limit of g_k, that is, $g|D_k = g_k|D_k$ for each $k \in \mathbb{N}$. For each $x \in D$, choose $k \in \mathbb{N}$ so that $x \in D_{k+1} \setminus D_k$. Then, by the definition of g, (1), and (3),

$$d(g(x), f(x)) = d(g_{k+1}(x), g_{k-1}(x)) < \tfrac{1}{3}\varepsilon_k + \tfrac{1}{3}\varepsilon_{k-1} < \tfrac{1}{2}\varepsilon_{k-1} < \tfrac{1}{2}3^{-k-1} < 3^{-k-1} \leqslant \alpha f(x),$$

which implies $g(x) \in W$. Hence, $g(D) \subset W$, so $g : D \to W$, and $g \in N_\alpha(f)$.

We shall show that $(g(\mathbf{I}^n))_{n\in\mathbb{N}}$ is discrete in W. It follows from (2) that $(g(\mathbf{I}^n))_{n\in\mathbb{N}}$ is disjoint. So, it suffices to show that $(g(\mathbf{I}^n))_{n\in\mathbb{N}}$ is locally finite in W. Assume that $(g(\mathbf{I}^n))_{n\in\mathbb{N}}$ is not locally finite in W. We can take $y \in W$ and $x_i \in \mathbf{I}^{n_i}$, $i \in \mathbb{N}$ so that $\lim_{i\to\infty} g(x_i) = y \in W$ and $n_i < n_{i+1}$. Then, $\inf_{i\in\mathbb{N}} \alpha f(x_i) > 0$. Otherwise $(f(x_i))_{i\in\mathbb{N}}$ contains a subsequence $(y_i)_{i\in\mathbb{N}}$ such that $\alpha(y_i) \to 0$ and $y_i \to y$, which contradicts $\alpha(y) > 0$. Therefore, we can choose $k \in \mathbb{N}$ so that $x_i \in D_k$ for all $i \in \mathbb{N}$. Then, by (2),

$$d(g(x_i), g(x_j)) = d(g_k(x_i), g_k(x_j)) > \varepsilon_k \text{ if } i \neq j.$$

This is contrary to the assumption. Thus, it follows that $(g(\mathbf{I}^n))_{n\in\mathbb{N}}$ is locally finite in W. □

With minor modification of the above proof, we can obtain the following:

Lemma 3.5.2 *Let $X = (X, d)$ be a locally path-connected metric space with a tower $M_1 \subset M_2 \subset \cdots \subset X$ satisfying the following properties:*

(i) *Given a compactum $A \subset \mathbf{I}^n$, a map $f : \mathbf{I}^n \to X$ with $f(A) \subset M_i$, and $\varepsilon > 0$, there is a map $g : \mathbf{I}^n \to M_j$ for some $j \geqslant i$ such that $f|A = g|A$ and $d(f, g) < \varepsilon$;*
(ii) *Given $\varepsilon > 0$, there is some $\delta > 0$ such that any map $f : \mathbf{I}^n \to M_i$ is ε-homotopic to a map $g : \mathbf{I}^n \to M_j$ for some $j \geqslant i$ such that $\operatorname{dist}(f(\mathbf{I}^n), g(\mathbf{I}^n)) > \delta$.*

Then, X has the discrete cell-tower property.

Modification in Proof of Lemma 3.5.1. Because of the condition (i), replacing f with an arbitrarily close map, we can assume that each $f(\mathbf{I}^n)$ is contained in some $M_{i(n)}$, where $i(1) < i(2) < \cdots$ Then, replacing the condition (USC) with (ii), we can perform the proof, that is, we can construct sequences $g_k : D \to X$ and $\varepsilon_k > 0$, $k \in \mathbb{N}$, so as to satisfy the conditions (1)–(4) and the following additional condition:

(0) $g_k(\mathbf{I}^n) \subset M_{j(n)}$, where $j(n) \geqslant j(n-1), i(n)$.

In the construction of g_k and ε_k, let $\delta > 0$ be as the above (ii) instead of (USC). Moreover, we assume that $g_k(\mathbf{I}^1 \oplus \cdots \oplus \mathbf{I}^{n-1})$ satisfies (0), hence $g_k(\mathbf{I}^1 \oplus \cdots \oplus \mathbf{I}^{n-1}) \subset M_{j(n-1)}$. Then, applying (ii) with (i) instead of (USC),[12] we have an ε-homotopy $s : \mathbf{I}^n \times \mathbf{I} \to M_{j(n)}$ such that $s_0 = g_{k-1}|\mathbf{I}^n$, $j(n) \geqslant \max\{i(n), j(n-1)\}$ and the condition (6) is satisfied. The map $g_k|\mathbf{I}^n : \mathbf{I} \to X$ obtained by the same definition satisfies $g_k(\mathbf{I}^n) \subset M_{j(n)}$, that is, the condition (0) is also satisfied.

[12] We use (i) to adjust the image of a homotopy so as to be contained in some M_j.

We will apply Lemma 3.5.1 to topological groups. Let G be a topological group, where $e \in G$ is the unit of G. A subset A of G is **totally bounded** provided, for each neighborhood U of e, there exists some finite set $F \subset G$ such that $A \subset UF = \bigcup_{x \in F} Ux = \{yx \mid x \in F, y \in U\}$, where F can be taken as a subset of A because there is a neighborhood V of e such that $VV^{-1} \subset U$ (cf. p. 25). When G is metrizable, G has an admissible *right invariant metric* d, i.e., $d(x, y) = d(xz, yz)$ for any $x, y, z \in G$ (cf. Theorem 3.6.2 in [GAGT]). Then, the total boundedness defined above coincides with the one in a metric space (G, d), that is:

- A is totally bounded in the sense of the above definition if and only if for each $\varepsilon > 0$, G is covered by finitely many subsets with diam $\leqslant \varepsilon$.

In fact, since d is right invariant, we have

$$\mathrm{B}_d(e, \varepsilon)x = \mathrm{B}_d(x, \varepsilon) \text{ for each } x \in G \text{ and } \varepsilon > 0,$$

from which the above fact easily follows. — Refer to p. 26.

It is said that a subset X of G is **multiplicative** if X contains the unit $e \in G$ and $xy \in X$ for any $x, y \in X$.

Proposition 3.5.3 *Let X be a path-connected multiplicative subset of a topological group G with an admissible right invariant metric d. If X is locally path-connected at the unit e and any neighborhood of e in X is not totally bounded, then (X, d) has the uniformly separating cells property, hence every open set in X has the discrete cell-tower property.*

Proof For each $\varepsilon > 0$, choose $\eta > 0$ so that $x \in X$ with $d(x, e) < \eta$ is connected to e by a path in X with diam $< \varepsilon$. Also choose $\delta > 0$ so that $\mathrm{B}(e, \eta) \cap X$ has no finite cover consisting of open balls with radius 2δ. For any map $f : \mathbf{I}^n \to X$, the following A is compact:

$$A = \left\{ f(x) \cdot f(y)^{-1} \;\middle|\; x, y \in \mathbf{I}^n \right\} \subset G.$$

Then, there exists some $a \in \mathrm{B}(e, \eta) \cap X$ so that $d(a, A) > \delta$. Otherwise, $\mathrm{B}(e, \eta) \cap X \subset \bigcup_{y \in A} \overline{\mathrm{B}}(y, \delta)$. Since A is compact, we have $y_1, \dots, y_k \in A$ such that $A \subset \bigcup_{i=1}^k \mathrm{B}(y_i, \delta)$. Each $y \in A$ is contained in some $\mathrm{B}(y_i, \delta)$, which implies that $\overline{\mathrm{B}}(y, \delta) \subset \mathrm{B}(y_i, 2\delta)$. Hence, $\mathrm{B}(e, \eta) \cap X \subset \bigcup_{i=1}^k \mathrm{B}(y_i, 2\delta)$, which contradicts the choice of δ.

Let $\rho : \mathbf{I} \to X$ be a path such that $\rho(0) = e$, $\rho(1) = a$, and diam $\rho(\mathbf{I}) < \varepsilon$. Since X is multiplicative, we can define an ε-homotopy $h : \mathbf{I}^n \times \mathbf{I} \to X$ by

$$h(x, t) = \rho(t) \cdot f(x) \text{ for each } x \in \mathbf{I}^n \text{ and } t \in \mathbf{I}.$$

Then, $h_0 = f$ and, for each $x, y \in \mathbf{I}^n$,

$$d(h_1(x), f(y)) = d(a, f(y) \cdot f(x)^{-1}) \geqslant d(a, A),$$

that is, $\operatorname{dist}(f(\mathbf{I}^n), h_1(\mathbf{I}^n)) \geqslant d(a, A) > \delta$. □

When a topological group G is locally path-connected (equiv., locally connected at the unit $e \in G$ because of homogeneity), each path-component of G is clopen in G and homeomorphic to the identity component G_0 (= the component of the unit $e \in G$) that is a clopen subgroup of G. Applying Proposition 3.5.3 to the case $X = G_0$, we have the following:

Corollary 3.5.4 *Let G is a locally path-connected metrizable topological group. If any neighborhood of the unit $e \in G$ is not totally bounded, then every open set in G has the discrete cell-tower property.* □

Due to Theorem 1.4.12, a topological linear space is finite-dimensional if and only if $\mathbf{0}$ has a totally bounded neighborhood. Combining the above corollary and the Toruńczyk Characterization Theorem 3.4.7, we have the following:

Theorem 3.5.5 (Dobrowolski–Toruńczyk) *A topological linear space is homeomorphic to ℓ^2 if and only if it is an infinite-dimensional separable completely metrizable AR.* □

Since a Fréchet space (= a locally convex completely metrizable topological linear space) is an AR, the following Kadec–Anderson Theorem can be obtained as a special case.

Theorem 3.5.6 (Kadec–Anderson) *Every infinite-dimensional separable Fréchet space is homeomorphic to ℓ^2. In particular, $\mathbb{R}^{\mathbb{N}} \approx \ell^2$.* □

Now, applying Proposition 3.5.3 to convex sets, we will generalize Theorem 3.5.5 as follows:

Theorem 3.5.7 (Dobrowolski–Toruńczyk) *A separable completely metrizable convex set C in a metrizable topological linear space E is homeomorphic to ℓ^2 if C is an AR and some point of C has no totally bounded neighborhoods in C.*

By the Toruńczyk Characterization Theorem 3.4.7, this theorem can be reduced to the following lemma:

Lemma 3.5.8 *A separable convex set C in a metrizable topological linear space E has the discrete cell-tower property if some point of C has no totally bounded neighborhoods in C.*

Proof We may assume that $\mathbf{0} \in C$ has no totally bounded neighborhoods in C. First, we show that $C \times \mathbb{R}$ has the discrete cell-tower property. To this end, consider the following subset of the product topological linear space $E \times \mathbb{R}$:

$$C^* = \left\{(tx, t) \;\middle|\; x \in C,\ t \in \mathbb{R}_+\right\} = \bigcup_{r \geqslant 0} rC \times \{r\}.$$

In order to see that $C \times \mathbb{R} \approx C^* \setminus \{(\mathbf{0}, 0)\}$ has the discrete cell-tower property, we can apply Proposition 3.5.3. So, we have to show that any neighborhood V

of $(\mathbf{0}, 0)$ in C^* is not totally bounded. On the contrary, assume that V is totally bounded. Then, choose a circled neighborhood W of $\mathbf{0}$ in E and $r \in (0, 1)$ so that $(W \times [0, r]) \cap C^* \subset V$. Since $rW \subset W$, it follows that

$$r(W \cap C) \times \{r\} = (rW \times \{r\}) \cap (rC \times \{r\}) \subset (W \times [0, r]) \cap C^* \subset V.$$

Hence, $r(W \cap C)$ is totally bounded. For any neighborhood U of $\mathbf{0}$ in E, there is a finite set A in $r(W \cap C)$ such that $r(W \cap C) \subset A + rU$. Then, $W \cap C \subset r^{-1}A + U$, where $r^{-1}A$ is a finite subset in $W \cap C$. Thus, $W \cap C$ is totally bounded, which contradicts the assumption that $\mathbf{0} \in C$ has no totally bounded neighborhoods in C.

Now, using the discrete cell-tower property of $C \times \mathbb{R}$, we show the locally finite cell-tower property of C, which implies the discrete cell-tower property. Let $f : \bigoplus_{n\in\mathbb{N}} \mathbf{I}^n \to C$ be a map. For each $\mathcal{U} \in \operatorname{cov}(C)$, let

$$\mathcal{U}^* = \{U \times (-1, 1) \mid U \in \mathcal{U}\} \cup \{C \times (\mathbb{R} \setminus \{0\})\}.$$

The map $f' : \bigoplus_{n\in\mathbb{N}} \mathbf{I}^n \to C \times \mathbb{R}$ defined by $f'(x) = (f(x), 0)$ is $\mathcal{U}^*$-close to a map $g' : \bigoplus_{n\in\mathbb{N}} \mathbf{I}^n \to C \times \mathbb{R}$ such that $(g'(\mathbf{I}^n))_{n\in\mathbb{N}}$ is discrete in $C \times \mathbb{R}$. Then, $g = \mathrm{pr}_C g' : \bigoplus_{n\in\mathbb{N}} \mathbf{I}^n \to C$ is $\mathcal{U}$-close to f. Since each $g'(\mathbf{I}^n)$ is contained in $C \times [-1, 1]$ and $\mathrm{pr}_C|C \times [-1, 1]$ is a proper map onto C, it easily follows that $(g(\mathbf{I}^n))_{n\in\mathbb{N}} = (\mathrm{pr}_C g'(\mathbf{I}^n))_{n\in\mathbb{N}}$ is locally finite in C. □

The converse of Theorem 3.5.7 above does not hold. In fact, the convex set $\prod_{i\in\mathbb{N}}(-2^{-i}, 2^{-i}) \subset \ell^2$ is homeomorphic to $\mathbb{R}^{\mathbb{N}} \approx \ell^2$. However, when E is completely metrizable and C is a closed convex set in E, the converse of Theorem 3.5.7 clearly holds, that is, we have the following characterization:[13]

Theorem 3.5.9 *A closed convex set C in a completely metrizable topological linear space is homeomorphic to ℓ^2 if and only if C is a separable AR and not locally compact.* □

Next, we apply Lemma 3.5.1 to show that infinite products have the locally finite cell-tower property.

Proposition 3.5.10 *Let $X = \prod_{i\in\mathbb{N}} X_i$ be the countable product of path-connected completely metrizable spaces. If infinitely many X_i's are non-compact then X has the uniform cells separation for some admissible metric, which implies that X has the discrete cell-tower property.*

Proof For each $i \in \mathbb{N}$, let $d_i \in \operatorname{Metr}(X_i)$. We assume that d_i is not totally bounded if X_i is non-compact. Namely, each non-compact X_i has no finite cover with mesh $<$ ε_i for some $\varepsilon_i > 0$. Then, we can assume that $\varepsilon_i \geqslant 4$. Because of this assumption,

[13] A metric space is compact if and only if it is totally bounded and complete.

if X_i is non-compact and $A \subset X_i$ is compact, then $d_i(x, A) \geqslant 1$ for some $x \in X_i$. We define a metric $d \in \mathrm{Metr}(X)$ as follows:

$$d(x, y) = \sum_{i \in \mathbb{N}} \min\left\{2^{-i}, d_i(x(i), y(i))\right\}.$$

We show that (X, d) has the uniform cells separation. For each $\varepsilon > 0$, choose $k \in \mathbb{N}$ so that $2^{-k} < \varepsilon/2$ and X_k is non-compact. For each map $f : \mathbf{I}^n \to X = \prod_{i \in \mathbb{N}} X_i$ and $i \in \mathbb{N}$, let $f_i = \mathrm{pr}_i f : \mathbf{I}^n \to X_i$, where $\mathrm{pr}_i : X \to X_i$ is the projection. Since X_k is non-compact, there is some $a \in X_k$ with $d_k(a, f_k(\mathbf{I}^n)) \geqslant 1$. Since $\mathbf{I}^n$ is contractible and X_k is path-connected, we have a homotopy $h : \mathbf{I}^n \times \mathbf{I} \to X_k$ such that $h_0 = f_k$ and $h_1(\mathbf{I}^n) = a$. We define a homotopy $h^* : \mathbf{I}^n \times \mathbf{I} \to X = \prod_{i \in \mathbb{N}} X_i$ as follows:

$$h^*(x, t) = (f_1(x), \ldots, f_{k-1}(x), h(x, t), f_{k+1}(x), \ldots).$$

Then, $h^*_0 = f$ and $g = h^*_1$ is the desired map. Indeed, for each $x \in \mathbf{I}^n$ and $t \in \mathbf{I}$,

$$d(f(x), h^*(x, t)) = \min\left\{2^{-k}, d_k(f_k(x), h(x, t))\right\} < \varepsilon/2,$$

hence h^* is an ε-homotopy. For each $x, y \in \mathbf{I}^n$,

$$\begin{aligned} d(f(x), g(y)) = d(f(x), h^*_1(y)) &\geqslant \min\left\{2^{-k}, d_k(f_k(x), h_1(y))\right\} \\ &\geqslant \min\left\{2^{-k}, d_k(a, f_k(\mathbf{I}^n))\right\} = 2^{-k}, \end{aligned}$$

that is, $\mathrm{dist}(f(\mathbf{I}^n), g(\mathbf{I}^n)) \geqslant 2^{-k}$. □

As an application, we can easily prove the following theorem (the details are left to the reader):

Theorem 3.5.11 *The countable product space $X = \prod_{i \in \mathbb{N}} X_i$ is homeomorphic to the separable Hilbert space ℓ^2 if and only if all X_i's are separable completely metrizable ARs and infinitely many X_i's are non-compact.*

From now, let us consider the non-separable case. We start with the following:

Lemma 3.5.12 *Assume that X has the discrete (or locally finite) self-copies property. Let $f : \bigoplus_{\gamma \in \Gamma} Y_\gamma \to X$ be a map such that $(f(Y_\gamma))_{\gamma \in \Gamma}$ is σ-discrete (or σ-locally finite) in X. Then, for each $\mathcal{U} \in \mathrm{cov}(X)$, f is $\mathcal{U}$-close to a map g such that $(g(Y_\gamma))_{\gamma \in \Gamma}$ is discrete (or locally finite) in X.*

Proof By the discrete (or locally finite) self-copies property, we have closed embeddings $h_i : X \to X$, $i \in \mathbb{N}$, such that every h_i is $\mathcal{U}$-close to id_X and $(h_i(X))_{i \in \mathbb{N}}$ is discrete (or locally finite) in X. We can write $\Gamma = \bigoplus_{i \in \mathbb{N}} \Gamma_i$ such that each $(f(Y_\gamma))_{\gamma \in \Gamma_i}$ is discrete (or locally finite) in X. The desired map

$g : \bigoplus_{\gamma\in\Gamma} Y_\gamma \to X$ can be defined by $g(y) = h_i(f(y), i)$ for $y \in Y_\gamma$, $\gamma \in \Gamma_i$, $i \in \mathbb{N}$. □

Lemma 3.5.13 *Let X be a completely metrizable ANR with $w(X) = \tau$. Then, X is an $\ell^2(\Gamma)$-manifold if X has the locally finite self-copies property and satisfies the following condition with respect to some $d \in \mathrm{Metr}(X)$:*

(∗) *For any $n \in \mathbb{N}$ and $\varepsilon > 0$, each $f : \mathbf{I}^n \times \Gamma \to X$ is ε-close to a map g such that $(g(\mathbf{I}^n \times \{\gamma\}))_{\gamma\in\Gamma}$ is σ-locally finite in X.*

Proof It is easy to see that the locally finite self-copies property implies the locally finite f.d. τ-polyhedra property (or the locally finite τ-skeletal tower property). We will show that X has the τ-locally finite cells property. Then, we can apply Theorem 3.4.2 (or 3.4.3) to have the result.

To see the τ-locally finite cells property, let $f : \mathbf{I}^n \times \Gamma \to X$ and $\alpha : X \to (0, 1)$ be maps. By Lemma 3.5.12, it suffices to find a map $g : \mathbf{I}^n \times \Gamma \to X$ such that $g \in N_\alpha(f)$ and $(g(\mathbf{I}^n \times \{\gamma\}))_{\gamma\in\Gamma}$ is σ-locally finite in X. For each $i \in \mathbb{N}$, let

$$\Gamma_i = \left\{\gamma \in \Gamma \;\middle|\; 2^{-i} \leqslant \inf\{\alpha f(z, \gamma) \mid z \in \mathbf{I}^n\} < 2^{-i+1}\right\}.$$

Then, $\Gamma = \bigoplus_{i\in\mathbb{N}} \Gamma_i$. By condition (∗), we have maps $g_i : \mathbf{I}^n \times \Gamma_i \to X$, $i \in \mathbb{N}$, such that $d(g_i, f|\mathbf{I}^n \times \Gamma_i) < 2^{-i}$ and $(g_i(\mathbf{I}^n \times \{\gamma\}))_{\gamma\in\Gamma_i}$ is σ-locally finite in X. Let $g : \mathbf{I}^n \times \Gamma \to X$ be the map defined by $g|\mathbf{I}^n \times \Gamma_i = g_i$. Then, g is clearly the desired one. □

Remark 3.11 In Remark 3.10, when $\tau = \aleph_0$ (i.e., $\Gamma = \mathbb{N}$), condition (i) implies (∗). Then, the space X of Remark 3.10 has no locally finite self-copies property because $X \not\approx \ell^2$. Hence, X is not ℓ^2-stable, i.e., $X \times \ell^2 \not\approx X$.

The following lemma will be used in the proof of the non-separable version of Theorem 3.5.11.

Lemma 3.5.14 *Let $X = \prod_{i\in\mathbb{N}} X_i$ be the countable product of completely metrizable connected ANRs (or ARs) such that $w(X) = \tau$. If each X_i is an AR containing a closed copy of $\ell^2(\Gamma)$ except for finitely many $i \in \mathbb{N}$, then X is an $\ell^2(\Gamma)$-manifold (or $X \approx \ell^2(\Gamma)$).*

Proof By virtue of Theorem 3.4.2, it suffices to verify $\mathfrak{M}_1(\tau)$-universality. We may assume that X_i is an AR containing a closed copy of $\ell^2(\Gamma)$ for every $i > 1$. Let $f : Y \to X$ be a map of $Y \in \mathfrak{M}_1(\tau)$ and $\alpha : X \to (0, 1)$ be a map. We use the same metric d for X as in the proof of Proposition 3.5.10, that is,

$$d(x, y) = \sum_{i\in\mathbb{N}} \min\left\{2^{-i}, d_i(x(i), y(i))\right\},$$

where $d_i \in \mathrm{Metr}(X_i)$ for each $i \in \mathbb{N}$. Since X_i contains a closed copy of $\ell^2(\Gamma)$, we have a closed embedding $h_i : Y \to X_i$ for each $i > 1$. For each $i > 1$, let $\lambda_i : X_i^2 \times \mathbf{I} \to X_i$ be an equi-connecting map for X_i, that is, $\lambda_i(x, y, 0) = x$,

$\lambda_i(x, y, 1) = y$, and $\lambda_i(x, x, t) = x$ for each $x, y \in X$ and $t \in \mathbf{I}$. We define a map $g : Y \to X$ as follows:

$$g(y) = (\mathrm{pr}_1 f(y), \ldots, \mathrm{pr}_i f(y),\\ \lambda_{i+1}(\mathrm{pr}_{i+1} f(y), h_{i+1}(y), 2^i\alpha(f(y)) - 1),\\ h_{i+2}(y), h_{i+3}(y), \ldots) \text{ if } 2^{-i} < \alpha(f(y)) \leqslant 2^{-i+1}.$$

Then, $d(f(y), g(y)) < \alpha(f(y))$ for each $y \in Y$.

We show that g is a closed embedding. Assume that $g(y_n) \to x$ $(n \to \infty)$. Let $s = \inf_{n\in\mathbb{N}} \alpha(f(y_n))$. If $s = 0$ then $\alpha(f(y_{n_i})) \to 0$ for some $n_1 < n_2 < \cdots$, whence $f(y_{n_i}) \to x$, which means $\alpha(x) = 0$. This is a contradiction. Then, we can choose $k \in \mathbb{N}$ so that $2^{-k} < s$. For each $n \in \mathbb{N}$, $\alpha(f(y_n)) > 2^{-k}$, hence $\mathrm{pr}_{k+2} g(y_n) = h_{k+2}(y_n)$. Since h_{k+2} is a closed embedding, $(y_n)_{n\in\mathbb{N}}$ is convergent in Y. Thus, g is a closed embedding. □

Now, we prove the non-separable version of Theorem 3.5.11:

Theorem 3.5.15 *Let $X = \prod_{i\in\mathbb{N}} X_i$ be the countable product of metrizable spaces such that $w(X) = \tau = \sup_{i\geqslant n} w(X_i)$ for each $n \in \mathbb{N}$. Then, $X \approx \ell^2(\Gamma)$ if and only if all X_i's are completely metrizable ARs and infinitely many X_i's are non-compact.*

Proof The "only if" part is trivial. To see the "if" part, it suffices to show that X contains a closed copy of $\ell^2(\Gamma)$. In fact, considering products of infinitely many X_i instead of X_n, we can assume that each X_n contains a closed copy of $\ell^2(\Gamma)$, hence the result follows from the lemma above.

As is well known, every path-connected space is arcwise connected (cf. [GAGT, Corollary 5.14.7]). Then, it is easy to see that the product $Y_1 \times Y_2$ of non-compact path-connected completely metrizable spaces contains a closed copy of the half real line $\mathbb{R}_+ = [0, \infty)$. By Theorem 3.5.11, $\mathbb{R}_+^{\mathbb{N}} \approx \ell^2$. Without loss of generality, we may assume that $w(X) = \lim_{i\in\mathbb{N}} w(X_{2i})$, where every X_{2i} is an AR. Then, $Y = \prod_{i\in\mathbb{N}} X_{2i}$ is a completely metrizable AR with $w(Y) = w(X)$ and X contains a closed copy of $Y \times \ell^2$.

We show that $Y \times \ell^2 \approx \ell^2(\Gamma)$. To this end, it suffices to verify condition $(*)$ in Lemma 3.5.13. We use the following metric for $Y \times \ell^2$:

$$d((y, z), (y', z')) = \sum_{i\in\mathbb{N}} \min\left\{2^{-i}, d_i(y(i), y'(i))\right\} + \|z - z'\|.$$

Let $n \in \mathbb{N}$ and $f : \mathbf{I}^n \times \Gamma \to Y \times \ell^2$ be a map. For each $\varepsilon > 0$, choose $k \in \mathbb{N}$ so that $2^{-k} < \varepsilon$. Let $Z = \prod_{i>k} X_{2i}$. Since $w(Z) = w(X) = \operatorname{card}\Gamma$, Z has a σ-discrete open basis $\mathcal{B} = \{B_\gamma \mid \gamma \in \Gamma\}$. Then, we have a map $\varphi : \Gamma \to Z$ such that $\varphi(\gamma) \in B_\gamma$ for each $\gamma \in \Gamma$. We define a map $g : \mathbf{I}^n \times \Gamma \to Y \times \ell^2$ as follows:

$$g(x, \gamma) = (p_k(f(x, \gamma)), \varphi(\gamma), q(f(x, \gamma))) \in \prod_{i=1}^{k} X_{2i} \times Z \times \ell^2 = Y \times \ell^2,$$

where $p_k : Y \times \ell^2 \to \prod_{i=1}^{k} X_{2i}$ and $q : Y \times \ell^2 \to \ell^2$ are the projections. Since $\mathcal{B}$ is σ-discrete in Z, it follows that $\varphi(\Gamma)$ is σ-discrete in Z, which implies that $(g(\mathbf{I}^n \times \{\gamma\}))_{\gamma\in\Gamma}$ is σ-discrete in $Y \times \ell^2$. Thus, $Y \times \ell^2$ satisfies condition $(*)$ in Lemma 3.5.13, hence $Y \times \ell^2 \approx \ell^2(\Gamma)$. □

As a special case of Theorems 3.5.11 and 3.5.15, we have the following:

Corollary 3.5.16 *For every infinite set* Γ, $J(\Gamma)^{\mathbb{N}} \approx \ell^2(\Gamma)$. □

The Kadec–Anderson Theorem 3.5.6 can be extended to the non-separable case as follows:

Theorem 3.5.17 (Kadec–Anderson–Toruńczyk) *Every infinite-dimensional Fréchet space is homeomorphic to a Hilbert space with the same weight.*

Proof Since the separable case has been proved, we prove the non-separable case. Let X be an infinite-dimensional non-separable Fréchet space with $w(X) = \tau > \aleph_0$. We may assume that X has an admissible F-norm $\|\cdot\|$ (Theorem 1.4.15). First, observe that X contains an infinite-dimensional separable closed linear subspace X_0. Indeed, such an X_0 can be obtained as the closure of the linear span of a linearly independent countable subset $A \subset X$. Note that X_0 is a separable Fréchet space, which is homeomorphic to ℓ^2 by the Kadec–Anderson Theorem 3.5.6. By the Bartle–Graves–Michael Theorem 1.4.16, $X \approx X_1 \times \ell^2$ for some X_1, hence

$$X \times \ell^2 \approx X_1 \times \ell^2 \times \ell^2 \approx X_1 \times \ell^2 \approx X.$$

Then, as mentioned after Corollary 3.3.14, it follows that X has the discrete self-copies property.

By virtue of Corollary 3.4.5, it suffices to see that any closed set A in X with $w(A) < \tau$ is a Z-set. This will be proved in more general setting as the following proposition. □

Proposition 3.5.18 *Let C be a non-separable convex set in a metrizable topological linear space E. Then, every closed set A in C with* $\operatorname{dens} A < \operatorname{dens} C$ *is a Z-set in C.*

Proof Due to Theorem 1.4.15, E has an admissible F-norm $\|\cdot\|$. Without loss of generality, we may assume that $\mathbf{0} \in C$. Let $n \in \mathbb{N}$ and $\varepsilon > 0$. It suffices to show that each map $f : \mathbf{I}^n \to C$ is ε-close to a map $g : \mathbf{I}^n \to C \setminus A$. Observe that

$$\operatorname{dens} \operatorname{fl}(A \cup f(\mathbf{I}^n)) = \operatorname{dens}(A \cup f(\mathbf{I}^n)) = \operatorname{dens} A < \operatorname{dens} C.$$

Hence, there exists $v \in C \setminus \operatorname{cl} \operatorname{fl}(A \cup f(\mathbf{I}^n))$. Because $f(\mathbf{I}^n) - v \approx f(\mathbf{I}^n)$ is compact, we can find $\delta > 0$ such that $\delta(f(\mathbf{I}^n) - v) \subset \mathrm{B}(\mathbf{0}, \varepsilon)$. We define a map $g : \mathbf{I}^n \to C \setminus \operatorname{fl} A$ as follows:

$$g(x) = (1 - \delta) f(x) + \delta v \quad \text{for every } x \in \mathbf{I}^n.$$

Then, g is ε-close to f. Indeed, for each $x \in \mathbf{I}^n$, $f(x) - g(x) = \delta(f(x) - v) \in \mathrm{B}(\mathbf{0}, \varepsilon)$, which means that $\|f(x) - g(x)\| < \varepsilon$. □

Remark 3.12 As Theorem 3.5.6 was generalized to Theorems 3.5.7 and 3.5.17 can be generalized to convex sets in Fréchet spaces. By virtue of Proposition 3.5.18, it suffices to show that every non-separable convex G_δ set in a Fréchet space has the locally finite self-copies property. T. Banakh and I. Zarichnyy [24] showed that the locally finite self-copies property can be weakened to the locally finite displacing property (cf. Remark 3.9, p. 249). Using the result of [20], T. Banakh and R. Cauty [21] showed that every non-separable convex G_δ set in a Fréchet space has the locally finite displacing property (cf. Remark 3.8, p. 239). Thus, every non-separable convex G_δ set in a Fréchet space is homeomorphic to a Hilbert space. For details, refer to the papers [20, 21], and [24].

3.6 The Function Space C(X, Y)

In this section, it is proved that the space $\mathrm{C}(X, Y)$ of maps from a non-discrete compactum X to a completely metrizable ANR Y without isolated points is an ℓ^2-manifold. Recall that the **compact-open topology** on $\mathrm{C}(X, Y)$ is generated by the sets

$$\langle K; U\rangle = \{f \in \mathrm{C}(X, Y) \mid f(K) \subset U\},$$

where K is any compact set in X and U is any open set in Y. On the other hand, the **pointwise convergence topology** on $\mathrm{C}(X, Y)$ is generated by the sets

$$\langle x; U\rangle = \{f \in \mathrm{C}(X, Y) \mid f(x) \in U\},$$

where x is any point of X and U is any open set in Y. For a subset $\mathcal{F} \subset \mathrm{C}(X, Y)$, let $\mathrm{cl}_p \mathcal{F}$ denote the closure of $\mathcal{F}$ with respect to the pointwise convergence topology. Then, it follows that $\mathrm{cl}\,\mathcal{F} \subset \mathrm{cl}_p \mathcal{F}$.

When $Y = (Y, d)$ is a metric space, a subset $\mathcal{F} \subset \mathrm{C}(X, Y)$ is said to be **equi-continuous** at $x_0 \in X$ provided that for each $\varepsilon > 0$, there is a neighborhood U of x_0 in X such that $f(U) \subset \mathrm{B}_d(f(x_0), \varepsilon)$ for all $f \in \mathcal{F}$. It is said that $\mathcal{F}$ is **equi-continuous** on X if it is equi-continuous at each point of X.

Proposition 3.6.1 *Let X be a locally compact space and $Y = (Y, d)$ a metric space. Every compact subset $\mathcal{F} \subset \mathrm{C}(X, Y)$ is equi-continuous on X.*

Proof To show that $\mathcal{F}$ is equi-continuous at $x_0 \in X$, let $\varepsilon > 0$. For each $f \in \mathcal{F}$, since X is locally compact and f is continuous, x_0 has a compact neighborhood N_f in X such that $f(N_f) \subset \mathrm{B}(f(x_0), \varepsilon/2)$. Note that $\langle N_f; \mathrm{B}(f(x_0), \varepsilon/2)\rangle$ is an open neighborhood of f in $\mathrm{C}(X, Y)$. Since $\mathcal{F}$ is compact, we have $f_1, \ldots, f_n \in \mathcal{F}$ such that

$$\mathcal{F} \subset \bigcup_{i=1}^{n} \langle N_{f_i}; \mathrm{B}(f_i(x_0), \varepsilon/2)\rangle.$$

Then, $N = \bigcap_{i=1}^{n} N_{f_i}$ is a neighborhood of x_0 in X. Each $f \in \mathcal{F}$ is contained in some $\langle N_{f_i}; \mathrm{B}(f_i(x_0), \varepsilon/2)\rangle$, that is, $f(N_{f_i}) \subset \mathrm{B}(f_i(x_0), \varepsilon/2)$. For each $x \in N$, since $f(x), f(x_0) \in \mathrm{B}(f_i(x_0), \varepsilon/2)$, it follows that

$$\begin{aligned} d(f(x), f(x_0)) &\leqslant d(f(x), f_i(x_0)) + d(f_i(x_0), f(x_0)) \\ &< \varepsilon/2 + \varepsilon/2 = \varepsilon. \end{aligned}$$

Hence, $f(N) \subset \mathrm{B}(f(x_0), \varepsilon)$. This means that $\mathcal{F}$ is equi-continuous at x_0. □

Proposition 3.6.2 *Let X be an arbitrary space and $Y = (Y, d)$ a metric space. If $\mathcal{F} \subset \mathrm{C}(X, Y)$ is equi-continuous on X, then $\mathrm{cl}_p \mathcal{F}$ is also equi-continuous on X.*

Proof We show the equi-continuity of $\mathrm{cl}_p \mathcal{F}$ at each $x_0 \in X$. For each $\varepsilon > 0$, since $\mathcal{F}$ is equi-continuous at x_0, there is a neighborhood U of x_0 in X such that $f(U) \subset \mathrm{B}_d(f(x_0), \varepsilon/3)$ for all $f \in \mathcal{F}$. Then, it suffices to show that $g(U) \subset \mathrm{B}_d(g(x_0), \varepsilon)$ for all $g \in \mathrm{cl}_p \mathcal{F}$. On the contrary, assume that $g(U) \not\subset \mathrm{B}_d(g(x_0), \varepsilon)$ for some $g \in \mathrm{cl}_p \mathcal{F}$. Then, we have $x_1 \in U$ such that $d(g(x_0), g(x_1)) \geqslant \varepsilon$. Because $g \in \mathrm{cl}_p \mathcal{F}$,

$$\langle x_0, \mathrm{B}_d(g(x_0), \varepsilon/3)\rangle \cap \langle x_1, \mathrm{B}_d(g(x_1), \varepsilon/3)\rangle \cap \mathcal{F} \neq \emptyset,$$

that is, there is some $f \in \mathcal{F}$ such that $d(f(x_0), g(x_0)), d(f(x_1), g(x_1)) < \varepsilon/3$. Since $x_1 \in U$, it follows that $d(f(x_0), f(x_1)) < \varepsilon/3$. Then,

$$\begin{aligned} d(g(x_0), g(x_1)) &\leqslant d(g(x_0), f(x_0)) + d(f(x_0), f(x_1)) + d(f(x_1, g(x_1)) \\ &< \varepsilon/3 + \varepsilon/3 + \varepsilon/3 = \varepsilon, \end{aligned}$$

which is a contradiction because $d(g(x_0), g(x_1)) \geqslant \varepsilon$. Therefore, $g(U) \subset \mathrm{B}_d(g(x_0), \varepsilon)$ for all $g \in \mathrm{cl}_p \mathcal{F}$. □

Using the above proposition, we can show the following well-known theorem:

Theorem 3.6.3 (ARZELÀ–ASCOLI) *Let X be an arbitrary space and $Y = (Y, d)$ a metric space. Suppose that $\mathcal{F} \subset \mathrm{C}(X, Y)$ satisfies the following conditions:*

(i) *$\mathcal{F}$ is equi-continuous on X,*
(ii) *$\mathrm{cl}_Y \mathcal{F}(x) = \mathrm{cl}_Y\{f(x) \mid f \in \mathcal{F}\}$ is compact for each $x \in X$.*

Then, $\mathrm{cl}\, \mathcal{F}$ is compact, that is, $\mathcal{F}$ has the compact closure with respect to the compact-open topology.

Proof The product space $\prod_{x \in X} \mathrm{cl}_Y \mathcal{F}(x)$ is compact by the Tychonoff Theorem [GAGT, Theorem 2.1.1]. Since $\mathrm{ev}_x : \mathrm{C}_p(X, Y) \to Y$ is continuous, we have

$$(\mathrm{cl}_p \mathcal{F})(x) = \mathrm{ev}_x(\mathrm{cl}_p \mathcal{F}) \subset \mathrm{cl}_Y \mathrm{ev}_x(\mathcal{F}) = \mathrm{cl}_Y \mathcal{F}(x),$$

which implies $\mathrm{cl}_p \mathcal{F} \subset \prod_{x \in X} \mathrm{cl}_Y \mathcal{F}(x)$. Hence, the subspace $\mathrm{cl}_p \mathcal{F}$ of $\mathrm{C}_p(X, Y)$ is compact. Thus, it suffices to show that $\mathrm{cl}\,\mathcal{F} = \mathrm{cl}_p \mathcal{F}$ as spaces. Since $\mathrm{id} : \mathrm{C}(X, Y) \to \mathrm{C}_p(X, Y)$ is continuous, we have $\mathrm{cl}\,\mathcal{F} \subset \mathrm{cl}_p \mathcal{F}$. Then, it suffices to show the continuity of the inclusion of $\mathrm{C}_p(X, Y)$ with the pointwise convergent topology to $\mathrm{C}(X, Y)$ with the compact-open topology.

By Proposition 3.6.2, (i) implies that $\mathrm{cl}_p \mathcal{F}$ is equi-continuous on X. The evaluation $\mathrm{ev} : \mathrm{C}_p(X, Y) \times X \to Y$ is not continuous in general (cf. Remark 1.1). However, the restriction $\mathrm{ev}|\,\mathrm{cl}_p \mathcal{F} \times X$ is continuous, where $\mathrm{cl}_p \mathcal{F}$ has the pointwise convergent topology. Indeed, let $(f, x) \in \mathrm{cl}_p \mathcal{F} \times X$. For each $\varepsilon > 0$, $\langle x; \mathrm{B}_d(f(x), \varepsilon/2)\rangle$ is a neighborhood of f in $\mathrm{C}_p(X, Y)$. By the equi-continuity of $\mathrm{cl}_p \mathcal{F}$, there is a neighborhood U of x in X such that $f'(U) \subset \mathrm{B}_d(f'(x), \varepsilon/2)$ for all $f' \in \mathrm{cl}_p \mathcal{F}$. Then, for each $(f', x') \in \langle x; \mathrm{B}_d(f(x), \varepsilon/2)\rangle \times U$,

$$d(f'(x'), f(x)) \leqslant d(f'(x'), f'(x)) + d(f'(x), f(x)) < \varepsilon/2 + \varepsilon/2 = \varepsilon,$$

hence $\mathrm{ev}(f', x') = f'(x') \in \mathrm{B}_d(f(x), \varepsilon) = \mathrm{B}_d(\mathrm{ev}(f, x), \varepsilon)$. By Proposition 1.1.2(1), $\mathrm{ev}|\,\mathrm{cl}_p \mathcal{F} \times X$ induces the map of $\mathrm{cl}_p \mathcal{F}$ to $\mathrm{C}(X, Y)$, which is the inclusion. Hence, $\mathrm{cl}_p \mathcal{F} \subset \mathrm{cl}\,\mathcal{F}$ and the compact-open topology is not finer than the pointwise convergence topology on $\mathrm{cl}_p \mathcal{F}$. Consequently, $\mathrm{cl}_p \mathcal{F} = \mathrm{cl}\,\mathcal{F}$ as spaces. □

Theorem 3.6.4 *Let X be a compactum and Y a metrizable space.*

(1) *If Y is completely metrizable, then so is* $\mathrm{C}(X, Y)$.
(2) *If Y is separable, then so is* $\mathrm{C}(X, Y)$.
(3) *If Y is an ANR, then so is* $\mathrm{C}(X, Y)$.

Proof (1): Let d be an admissible complete metric on Y. By Proposition 1.1.2(5), the sup-metric induced by d is admissible for $\mathrm{C}(X, Y)$. Let $(f_n)_{n \in \mathbb{N}}$ be a Cauchy sequence in $\mathrm{C}(X, Y)$ with respect to the sup-metric. For each $x \in X$, $(f_n(x))_{n \in \mathbb{N}}$ is a Cauchy sequence of $Y = (Y, d)$. Then, we can define $f(x) = \lim_{n \to \infty} f_n(x)$ for each $x \in X$. We will show that $f \in \mathrm{C}(X, Y)$ and $\lim_{n \to \infty} f_n = f$.

To see that $f \in \mathrm{C}(X, Y)$, let $x \in X$ and $\varepsilon > 0$. First, choose $n_0 \in \mathbb{N}$ so that $d(f_i, f_j) < \varepsilon/3$ for each $i, j \geqslant n_0$. By the continuity of f_{n_0}, x has a neighborhood U in X such that $f_{n_0}(U) \subset \mathrm{B}_d(f_{n_0}(x), \varepsilon/9)$. Since $f(x) = \lim_{n \to \infty} f_n(x)$, we have $n_x \in \mathbb{N}$ such that $d(f_i(x), f(x)) < \varepsilon/9$ for each $i \geqslant n_x$. Similarly, for each

$x' \in U$, we have $n_{x'} \in \mathbb{N}$ such that $d(f_i(x'), f(x')) < \varepsilon/9$ for each $i \geqslant n_{x'}$. Let $i = \max\{n_0, n_x, n_{x'}\}$. Then, it follows that

$$\begin{aligned} d(f(x'), f(x)) &\leqslant d(f'(x'), f_i(x')) + d(f_i(x'), f_{n_0}(x')) \\ &\quad + d(f_{n_0}(x'), f_{n_0}(x)) + d(f_{n_0}(x), f_i(x)) + d(f_i(x), f(x)) \\ &< \varepsilon/9 + \varepsilon/3 + \varepsilon/9 + \varepsilon/3 + \varepsilon/9 = \varepsilon. \end{aligned}$$

Hence, f is continuous.

Finally, we show that $\lim_{n\to\infty} f_n = f$. For each $\varepsilon > 0$, take the same $n_0 \in \mathbb{N}$ as above. Then, we show that $d(f_n, f) < \varepsilon$ for each $n \geqslant n_0$. For each $x \in X$, take the same $n_x \in \mathbb{N}$ as above. Let $i = \max\{n_x, n\}$, Then, we have

$$\begin{aligned} d(f_n(x), f(x)) &\leqslant d(f_n(x), f_i(x)) + d(f_i(x), f(x)) \\ &< \varepsilon/3 + \varepsilon/9 = 4\varepsilon/9. \end{aligned}$$

Hence, $d(f_n, f) \leqslant 4\varepsilon/9 < \varepsilon$.

(2): From Proposition 1.1.2(5), it follows that C(X, Y) is second countable, so it is separable.

(3): Let Z be a metrizable space, A a closed set in Z, and let $f : A \to$ C(X, Y) be a map. Then, f induces a map $\tilde{f} : A \times X \to Y$ defined by $\tilde{f}(z, x) = f(z)(x)$ for each $(z, x) \in A \times X$. Since Y is an ANR, $A \times X$ has a neighborhood W in $Z \times X$ and $\tilde{f}$ extends to a map $f' : W \to Y$. Because X is compact, there is a neighborhood U of A in Z such that $U \times X \subset W$. Then, $f'|U \times X$ induces a map $\bar{f} : U \to$ C(X, Y) defined by $\bar{f}(z)(x) = f'(z, x)$ for each $(z, x) \in U \times X$. Observe that $\bar{f}$ is an extension of f. □

As another application of the Toruńczyk Characterization of ℓ^2-manifolds, we prove the following:

Theorem 3.6.5 *Let X be a non-discrete compactum and Y be a separable completely metrizable ANR without isolated points. Then, the space* C(X, Y) *with the compact-open topology is an ℓ^2-manifold.*

Proof By Theorem 3.6.4, C(X, Y) is a completely metrizable separable ANR. Hence, it suffices that C(X, Y) has the discrete cell-tower property. To this end, we will apply Lemma 3.5.2.

Since X is non-discrete, X has a cluster points x_∞. Take a sequence $(x_i)_{i\in\mathbb{N}}$ of distinct points of X converging to x_∞, and define a tower $M_1 \subset M_2 \subset \cdots \subset$

$\mathrm{C}(X, Y)$ as follows:

$$M_i = \{ f \in \mathrm{C}(X, Y) \mid f(x_j) = f(x_\infty) \ \text{ for all } j \geqslant i \}.$$

Since Y has no isolated points, Y has an admissible complete metric d such that every component of Y has the diameter > 1 (cf. Proposition 1.3.22(1)). By Theorem 1.13.37, we can assume that $Y = (Y, d)$ is a uniform ANR. Adopting the sup-metric induced by d, we verify the conditions (i) and (ii) of Lemma 3.5.2.

(i): Let $A \subset \mathbf{I}^n$ be a compactum, $f : \mathbf{I}^n \to \mathrm{C}(X, Y)$ a map with $f(A) \subset M_{i_0}$, and $\varepsilon > 0$. Then, f induces a map $\tilde{f} : \mathbf{I}^n \times X \to Y$ defined by $\tilde{f}(z, x) = f(z)(x)$. Since Y is a uniform ANR, we have $\delta > 0$ such that any two δ-close maps to Y are ε-homotopic (Theorem 1.13.32). Since $f(\mathbf{I}^n)$ is a compact subset of $\mathrm{C}(X, Y)$, it is equi-continuous by Proposition 3.6.1, that is, there exists a neighborhood U of x_∞ in X such that $f(z)(U) \subset \mathrm{B}(f(z)(x_\infty), \delta)$, that is, $\tilde{f}(\{z\} \times U) \subset \mathrm{B}(\tilde{f}(z, x_\infty), \delta)$. Then, we have $i_1 \geqslant i_0$ such that $x_i \in U$ for all $i \geqslant i_1$. Note that $\tilde{f}(z, x_i) = \tilde{f}(z, x_\infty)$ for every $z \in A$ and $i \geqslant i_0$, and

$$d(\tilde{f}(z, x_i), \tilde{f}(z, x_\infty)) < \delta \ \text{ for every } z \in \mathbf{I}^n \text{ and } i \geqslant i_1.$$

Using the Homotopy Extension Theorem 1.13.11, we can easily obtain a map $g' : \mathbf{I}^n \times X \to Y$ such that $g' \simeq_\varepsilon \tilde{f}$, $g'|A \times X = \tilde{f}|A \times X$, and $g'(z, x_i) = g'(z, x_\infty) = \tilde{f}(z, x_\infty)$ for every $z \in \mathbf{I}^n$ and $i \geqslant i_1$. Then, g' induces a map $g : \mathbf{I}^n \to \mathrm{C}(X, Y)$ defined by $g(z)(x) = g'(z, x)$, where g is ε-close to f and $g|A = f|A$. Moreover, $g(z)(x_i) = g'(z, x_i) = g'(z, x_\infty) = g(z)(x_\infty)$ for every $z \in \mathbf{I}^n$ and $i \geqslant i_1$, which means that $g(\mathbf{I}^n) \subset M_{i_1}$.

(ii): For each $\varepsilon > 0$, we take $\delta > 0$ as in (i), where we may assume that $\delta < 1$. Each map $f : \mathbf{I}^n \to M_{i_0}$ induces a map $\tilde{f} : \mathbf{I}^n \times X \to Y$ defined by $\tilde{f}(z, x) = f(z)(x)$, where $\tilde{f}(z, x_i) = \tilde{f}(z, x_\infty)$ for every $z \in \mathbf{I}^n$ and $i \geqslant i_0$. By the compactness of $\mathbf{I}^n \times X$, there exist $y_1, \ldots, y_m \in Y$ such that $\tilde{f}(\mathbf{I}^n \times X) \subset \bigcup_{j=1}^m \mathrm{B}(y_j, \delta/3)$. Since every component of Y has the diameter $> \delta$, we can choose $v_j \in \mathrm{B}(y_j, \delta) \setminus \mathrm{B}(y_j, 2\delta/3)$ for each $j = 1, \ldots, m$. Using the Homotopy Extension Theorem 1.13.11, we can obtain a map $g_j : Y \to Y$ such that $g_j \simeq_\varepsilon \mathrm{id}$, $g_j|Y \setminus \mathrm{B}(y_j, \varepsilon) = \mathrm{id}$, and $g_j(\overline{\mathrm{B}}(y_j, \delta/3)) = \{v_j\}$. Then, the following holds:

(*) $\max\{d(y, g_j(y)) \mid j = 1, \ldots, m\} > \delta/3$ for every $y \in \tilde{f}(\mathbf{I}^n \times X)$.

Again using the Homotopy Extension Theorem 1.13.11, we can easily obtain a map $g' : \mathbf{I}^n \times X \to Y$ such that $g' \simeq_\varepsilon \tilde{f}$,

$$g'(z, x_\infty) = \tilde{f}(z, x_\infty), \ g'(z, x_i) = \tilde{f}(z, x_i) \ \text{ for } i > i_0 + m, \text{ and}$$

$$g'(z, x_{i_0+j}) = g_j \tilde{f}(z, x_{i_0+j}) \ \text{ for } j = 1, \ldots, m.$$

Then, g' induces a map $g : \mathbf{I}^n \to \mathrm{C}(X, Y)$ defined by $g(z)(x) = g'(z, x)$, where the homotopy $g' \simeq_\varepsilon \tilde{f}$ induces a homotopy $g \simeq_\varepsilon f$. Moreover, $g(\mathbf{I}^n) \subset M_{i_0+m}$ because

$$g(z)(x_i) = g'(z, x_i) = \tilde{f}(z, x_i) = f(z)(x_\infty) = g(z)(x_\infty) \text{ for } i > i_0 + m.$$

It remains to show that $\mathrm{dist}(f(\mathbf{I}^n), g(\mathbf{I}^n)) \geqslant \delta/6$. Conversely, assume that $\mathrm{dist}(f(\mathbf{I}^n), g(\mathbf{I}^n)) < \delta/6$. Then, we have $z, z' \in \mathbf{I}^n$ such that $d(f(z), g(z')) < \delta/6$. Observe that

$$\begin{aligned} d(\tilde{f}(z, x_\infty), \tilde{f}(z', x_\infty)) &= d(\tilde{f}(z, x_\infty), g'(z', x_\infty)) \\ &= d(f(z)(x_\infty), g(z')(x_\infty)) \\ &\leqslant d(f(z), g(z')) < \delta/6. \end{aligned}$$

Moreover, for each $j = 1, \ldots, m$,

$$\begin{aligned} d(\tilde{f}(z, x_\infty), g_j \tilde{f}(z', x_\infty)) &= d(\tilde{f}(z, x_{i_0+j}), g_j \tilde{f}(z', x_{i_0+j})) \\ &= d(\tilde{f}(z, x_{i_0+j}), g'(z', x_{i_0+j})) \\ &\leqslant d(f(z), g(z')) < \delta/6. \end{aligned}$$

Consequently,

$$\begin{aligned} d(\tilde{f}(z', x_\infty), g_j \tilde{f}(z', x_\infty)) &\leqslant d(\tilde{f}(z', x_\infty), \tilde{f}(z, x_\infty)) \\ &\quad + d(\tilde{f}(z, x_\infty), g_j \tilde{f}(z', x_\infty)) \\ &< \delta/6 + \delta/6 = \delta/3, \end{aligned}$$

which contradicts (*). Thus, we have $\mathrm{dist}(f(\mathbf{I}^n), g(\mathbf{I}^n)) \geqslant \delta/6$. □

Remark 3.13 In Theorem 3.6.5, if X is discrete, then X has only finitely many points. In this case, $\mathrm{C}(X, Y) = Y^X \approx Y^n$, where $n = \mathrm{card}\, X < \infty$. If Y has an isolated point y_0, then the constant map $c_{y_0} : X \to Y$ with $c_{y_0}(X) = \{y_0\}$ is an isolated point of $\mathrm{C}(X, Y)$.

Remark 3.14 When X is non-compact and $Y = (Y, d)$ is a separable complete metric space, it was proved by A. Yamashita [162] that (1) if Y is an ANRU (= ANR in the category of metric spaces and uniform continuous mappings) and the diameters of components of Y are bounded away from zero, then the space $\mathrm{C}(X, Y)$ with the sup-metric is an $\ell^2(2^{\aleph_0})$-manifold; (2) if Y is a connected complete Riemannian manifold, then the subspace $\mathrm{C}^B(X, Y)$ of bounded maps is an $\ell^2(2^{\aleph_0})$-manifold.

At the end of this section, we will introduce one of the famous longstanding problems, called the Homeomorphism Group Problem. For a compactum X, let $\mathrm{Homeo}(X)$ be the homeomorphism group of X, that is, the subspace of $\mathrm{C}(X, X)$ consisting of all homeomorphisms of X onto itself, which is a topological group with function composition as the group operation (cf. Proposition 1.1.2(3)). For a closed set $A \subset X$, let $\mathrm{Homeo}_A(X) = \{h \in \mathrm{Homeo}(X) \mid h|A = \mathrm{id}\}$, where $\mathrm{Homeo}_\emptyset(X) = \mathrm{Homeo}(X)$. For a metric $d \in \mathrm{Metr}(X)$, we can define a metric d^* on $\mathrm{Homeo}(X)$ as follows:

$$\begin{aligned} d^*(f, g) &= d(f, g) + d(f^{-1}, g^{-1}) \\ &= \sup_{x \in X} d(f(x), g(x)) + \sup_{x \in X} d(f^{-1}(x), g^{-1}(x)). \end{aligned}$$

Then, d^* is an admissible complete metric. For an n-manifold M with the boundary ∂M, we write $\mathrm{Homeo}_{\partial M}(M) = \mathrm{Homeo}_\partial(M)$. The HOMEOMORPHISM GROUP PROBLEM is the following fascinating problem:

? HOMEOMORPHISM GROUP PROBLEM

For a compact n-manifold M ($n \in \mathbb{N}$), is $\mathrm{Homeo}_\partial(M)$ an ℓ^2-manifold?

When $n = 1, 2$, this problem is affirmative. Indeed, R.D. Anderson proved in [11] that $\mathrm{Homeo}_\partial(X)$ is an ℓ^2-manifold for every non-discrete finite graph X (i.e., $X = |K|$ for some one-dimensional finite simplicial complex K). The proof of $\mathrm{Homeo}_\partial(\mathbf{I}) \approx \ell^2$ can be found in the paper [91]. The two-dimensional case was proved in [100] by R. Luke and W.K. Mason, that is, $\mathrm{Homeo}_\partial(M)$ is an ℓ^2-manifold for every compact surface (two-dimensional manifold) M. Generalizing this, W. Jakobsche [88] showed that $\mathrm{Homeo}(X)$ is an ℓ^2-manifold for a compact two-dimensional polyhedron X.

The Homeomorphism Group Problem is still open for $n > 2$. The ℓ^2-stability of $\mathrm{Homeo}_\partial(X)$ was independently proved by R. Geoghegan [68] and J. Keesling [91] under some assumption, that is, X has an open set (called a cone path) admitting some kind of structure in [68] or a nontrivial flow in [91], where the former implies the latter. Thus, by virtue of the Toruńczyk Factor Theorem 2.2.14, the Homeomorphism Group Problem is reduced to the problem of whether $\mathrm{Homeo}_\partial(M)$ is an ANR or not. R.D. Edwards and R.C. Kirby [60] and A.V. Černavskiĭ [39] independently showed that $\mathrm{Homeo}(M)$ is locally contractible for every compact n-manifold. Furthermore, S. Ferry [64] and H. Toruńczyk [141] independently proved the following:

Theorem 3.6.6 *For every compact Q-dimensional manifold M,* $\mathrm{Homeo}(M)$ *is an ANR (hence an ℓ^2-manifold).*

By using the result of Edwards and Kirby in [60], Haver [70] reduced the Homeomorphism Group Problem to the following problem:

? Reduced Homeomorphism Group Problem

For $n > 2$, is $\mathrm{Homeo}_\partial(\mathbf{B}^n)$ an ANR?

The following result (or the method for proving) is called the ALEXANDER TRICK, which was demonstrated by J.W. Alexander in [1]:

Theorem 3.6.7 (ALEXANDER TRICK) *For every $n \in \mathbb{N}$, $\mathrm{Homeo}_\partial(\mathbf{B}^n)$ is contractible. Namely, a homeomorphism of the n-ball $\mathbf{B}^n$ fixing the boundary sphere $\mathbf{S}^{n-1}$ is canonically isotopic to the identity.*

Proof A contraction $\varphi : \mathrm{Homeo}_\partial(\mathbf{B}^n) \times \mathbf{I} \to \mathrm{Homeo}_\partial(\mathbf{B}^n)$ can be defined as follows:

$$\varphi(h,t)(x) = \begin{cases} x & \text{if } \|x\| \geqslant 1-t, \\ (1-t)h((1-t)^{-1}x) & \text{if } \|x\| \leqslant t < 1. \end{cases}$$

Then, $\varphi_1(h) = \mathrm{id}$ and $\varphi_1(h) = h$ for every $h \in \mathrm{Homeo}_\partial(\mathbf{B}^n)$. □

In the above proof, $\varphi_t(\mathrm{B}_d(\mathrm{id}, \varepsilon)) \subset \mathrm{B}_d(\mathrm{id}, \varepsilon)$ for each $t \in \mathbf{I}$, where d is the sup-metric induced by the usual norm. Indeed, $d(\varphi_t(h), \mathrm{id}) \leqslant (1-t)d(h, \mathrm{id})$ for each $h \in \mathrm{Homeo}_\partial(\mathbf{B}^n)$. Since $\mathrm{Homeo}_\partial(\mathbf{B}^n)$ is homogeneous as a topological group, it follows that $\mathrm{Homeo}_\partial(\mathbf{B}^n)$ is locally contractible.

3.7 Cell-Like Images of $\boldsymbol{Q}$-Manifolds

We want to characterize $\boldsymbol{Q}$-manifolds in the same way as the Characterization of ℓ^2-Manifolds 3.4.7. Since $\boldsymbol{Q} \approx C(\boldsymbol{Q})$ by Theorem 2.7.5 and we have obtained $\boldsymbol{Q}$-manifold versions of all results used in the proof of Proposition 2.9.1, the $\boldsymbol{Q}$-manifold version of Proposition 2.9.1 is also valid. However, it cannot be applied because we have not proved that $X \times \boldsymbol{Q}$ is a $\boldsymbol{Q}$-manifold for a locally compact ANR X. This will be proved in the next section. To prove this and the characterization of $\boldsymbol{Q}$-manifolds, we need a more general result.

Lemma 3.7.1 *Let X be a separable locally compact ANR with the disjoint cells property. Then, the space $\mathrm{C}(\boldsymbol{Q}, X)$ has a countable dense set $\{\alpha_i \mid i \in \mathbb{N}\}$ consisting of Z-embeddings such that $\alpha_i(\boldsymbol{Q}) \cap \alpha_j(\boldsymbol{Q}) = \emptyset$ if $i \neq j$.*

Proof Due to Proposition 3.2.8, X is $\mathfrak{M}_0$-universal. Then, condition (b) of Proposition 3.2.7 can be strengthened as follows:

(b') All maps $f, g : \boldsymbol{Q} \to X$ are arbitrarily close to Z-embeddings $f', g' : \boldsymbol{Q} \to X$ with $f'(\boldsymbol{Q}) \cap g'(\boldsymbol{Q}) = \emptyset$.

Because of the separability, $\mathrm{C}(\boldsymbol{Q}, X)$ has a countable dense set $\{f_i \mid i \in \mathbb{N}\}$. As for the implication (b) $\Rightarrow$ (c) of Proposition 3.2.7, we can apply the Baire Category Theorem 1.3.15 (using the above (b') instead of (b)) to find $(\alpha_i)_{i\in\mathbb{N}} \in \mathrm{C}(\boldsymbol{Q}, X)^{\mathbb{N}}$ such that each α_i is an embedding with $d(\alpha_i, f_i) < 2^{-i}$ and $\alpha_i(\boldsymbol{Q}) \cap \alpha_j(\boldsymbol{Q}) = \emptyset$ for $i \neq j$, where d is the sup-metric with respect to an admissible complete metric for X. Then, each α_i is a Z-embedding and $\{\alpha_i \mid i \in \mathbb{N}\}$ is dense in $\mathrm{C}(\boldsymbol{Q}, X)$. □

For each map $f : X \to Y$, we define

$$N(f) = \{x \in X \mid f^{-1}(x) \text{ is non-degenerate}\}.$$

Proposition 3.7.2 *Let $f : M \to X$ be a cell-like map from a $\boldsymbol{Q}$-manifold M onto an ANR X. If $f^{-1}(N(f))$ is contained in a Z-set A in M, then f is a near-homeomorphism and $f(A)$ is a Z-set in X (hence so is* $\mathrm{cl}\, N(f)$*).*

Proof First, note that f is a fine homotopy equivalence by Theorem 1.15.9 and $f(M \setminus A) = X \setminus f(A)$ because $f^{-1}(N(f)) \subset A$. Since A is a Z-set in M and $f(A)$ is closed in X, it is easy to see that $f(A)$ is a Z-set in X. As the proper image of a locally compact space, X is locally compact, hence it is completely metrizable. To show that f is a near-homeomorphism, we apply the Bing Shrinking Criterion 2.7.1. Since f is a closed map, for each $\mathcal{U} \in \mathrm{cov}(M)$ and $\mathcal{V} \in \mathrm{cov}(X)$, it suffices to construct a homeomorphism $\varphi : M \to M$ such that $\{\varphi f^{-1}(x) \mid x \in X\} \prec \mathcal{U}$ and $f\varphi$ is $\mathcal{V}$-close to f (cf. Remark 2.12):

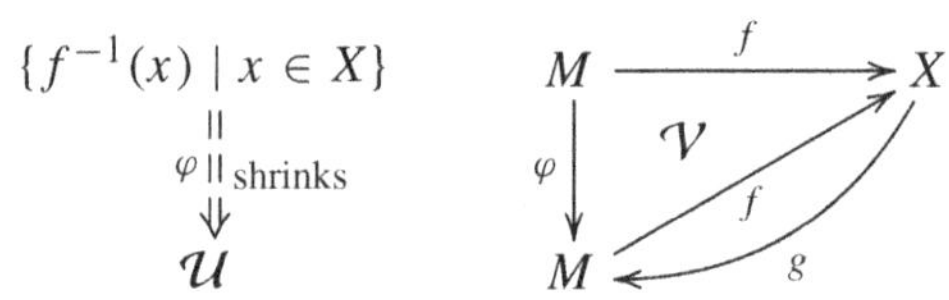

Choose locally finite open covers $\mathcal{U}' \in \mathrm{cov}(M)$ and $\mathcal{V}' \in \mathrm{cov}(X)$ so that the closures of each $U \in \mathcal{U}'$ and $V \in \mathcal{V}'$ are compact,

$$\mathrm{st}^2\, \mathcal{V}' \prec \mathcal{V} \text{ and } \mathrm{st}\, \mathcal{U}' \prec \mathcal{U} \wedge f^{-1}(\mathcal{V}').$$

Let $g : X \to M$ be a $\mathcal{V}'$-homotopy inverse of f and $h : M \times \mathbf{I} \to M$ an $f^{-1}(\mathcal{V}')$-homotopy such that $h_0 = \mathrm{id}_M$ and $h_1 = gf$, where h is proper by Proposition 1.3.8. By Theorem 2.10.11 and Proposition 1.3.8, $h_1|A$ is properly $\mathcal{U}'$-homotopic to a Z-embedding, which extends to a homeomorphism $\varphi : M \to M$ such that φ is $\mathrm{st}^2\, f^{-1}(\mathcal{V}')$-isotopic to id by Theorem 2.11.6. Then, $f\varphi$ is $\mathcal{V}$-close to f. If $f^{-1}(x)$ is non-degenerate, then $h_1(f^{-1}(x)) = gf(f^{-1}(x)) = g(x)$ and $f^{-1}(x) \subset A$, hence $\varphi f^{-1}(x) \subset \mathrm{st}(g(x), \mathcal{U}')$. Then, $\{\varphi f^{-1}(x) \mid x \in X\} \prec \mathcal{U}$. This completes the proof. □

For each perfect map $f : X \to Y$ between ANRs X and Y, the mapping cylinder M_f $(= M(f))$ is an ANR by Corollary 1.13.17. If f is cell-like, then so is the quotient map $q : X \times \mathbf{I} \to M_f$ and every non-degenerate $q^{-1}(y)$ is contained in the Z-set $X \times \{0\}$ in $X \times \mathbf{I}$. By the above proposition, we have the following MAPPING CYLINDER THEOREM that is a generalization of Theorem 2.7.5:

Theorem 3.7.3 (MAPPING CYLINDER) *Let $f : M \to X$ be a cell-like map from a $\boldsymbol{Q}$-manifold M onto an ANR X. Then, the mapping cylinder $M(f)$ $(= M_f)$ is a $\boldsymbol{Q}$-manifold with X a Z-set in $M(f)$ and the natural (quotient) map $q_f : M \times \mathbf{I} \to M(f)$ is a near-homeomorphism.* □

Strengthening Proposition 3.7.2 a little, we have the following generalization of the $\boldsymbol{Q}$-manifold version of Proposition 2.9.1:

Proposition 3.7.4 *Let $f : M \to X$ be a cell-like map from a $\boldsymbol{Q}$-manifold M onto an ANR X. If $\operatorname{cl} N(f)$ is a Z-set in X, then f is a near-homeomorphism.*

Proof For each $\mathcal{U} \in \operatorname{cov}(X)$, X has an admissible complete metric d such that $\{\overline{\mathrm{B}}_d(x, 1) \mid x \in X\} \prec \mathcal{U}$ and each $\overline{\mathrm{B}}_d(x, 1)$ is compact (cf. 1.3.22(1)). We shall construct a cell-like map $g : M \to X$ such that $d(f, g) \leqslant 1/2$, $N(g) \subset \operatorname{cl} N(f)$, and $\operatorname{cl} g^{-1}(N(g))$ is a Z-set in M. Then, we can apply Proposition 3.7.2 to obtain a homeomorphism that is $1/2$-close to g, hence it is $\mathcal{U}$-close to f. Thus, f is a near-homeomorphism.

Let $A = \operatorname{cl} N(f)$. Note that $f|M \setminus f^{-1}(A) : M \setminus f^{-1}(A) \to X \setminus A$ is a homeomorphism, hence $X \setminus A$ is a $\boldsymbol{Q}$-manifold. By Lemma 3.7.1, $\mathrm{C}(\boldsymbol{Q}, M)$ has a countable dense set $\{\alpha_i \mid i \in \mathbb{N}\}$ consisting of Z-embeddings. Note that a closed set in M is a Z-set if it misses all $\alpha_i(\boldsymbol{Q})$. Then, it suffices to construct a cell-like map $g : M \to X$ so that $N(g) \subset A$ and $g^{-1}(A) \cap \alpha_i(\boldsymbol{Q}) = \emptyset$, i.e., $g\alpha_i(\boldsymbol{Q}) \cap A = \emptyset$ for all $i \in \mathbb{N}$.

Let $\varepsilon_0 = 2^{-1}$, $g_0 = f$, and take an locally finite open cover $\mathcal{V}$ of X with $\operatorname{mesh} \mathcal{V} < \varepsilon_0/10$. Since A is a Z-set in X and $X \setminus A$ is a $\boldsymbol{Q}$-manifold, we apply the Strong Universality Theorem 2.10.10 (or 2.10.11) to obtain a $\mathcal{V}$-homotopy $h : \boldsymbol{Q} \times \mathbf{I} \to X$ such that $h_0 = g_0\alpha_1$, $h_1 : \boldsymbol{Q} \to X \setminus A$ is a Z-embedding and

$$h_t|\boldsymbol{Q} \setminus \alpha_1^{-1} g_0^{-1}(\mathrm{N}_d(A, \varepsilon_0)) = g_0\alpha_1|\boldsymbol{Q} \setminus \alpha_1^{-1} g_0^{-1}(\mathrm{N}_d(A, \varepsilon_0)) \text{ for every } t \in \mathbf{I}.$$

Then, $\beta_1 = g_0^{-1} h_1 : \boldsymbol{Q} \to M \setminus g_0^{-1}(A) \subset M$ is a Z-embedding and

$$\beta_1|\boldsymbol{Q} \setminus \alpha_1^{-1} g_0^{-1}(\mathrm{N}_d(A, \varepsilon_0)) = \alpha_1|\boldsymbol{Q} \setminus \alpha_1^{-1} g_0^{-1}(\mathrm{N}_d(A, \varepsilon_0)).$$

By the approximate softness of the cell-like map $g_0 = f$ (Theorems 1.15.9 and 1.15.1), we have a homotopy $\tilde{h} : \boldsymbol{Q} \times \mathbf{I} \to M$ such that $\tilde{h}_0 = \alpha_1$, $\tilde{h}_1 = \beta_1$,

$$\tilde{h}_t|\boldsymbol{Q} \setminus \alpha_1^{-1} g_0^{-1}(\mathrm{N}_d(A, \varepsilon_0)) = \alpha_1|\boldsymbol{Q} \setminus \alpha_1^{-1} g_0^{-1}(\mathrm{N}_d(A, \varepsilon_0)) \text{ for every } t \in \mathbf{I},$$

and $g_0\tilde{h}$ is $\mathcal{V}$-close to h. The last condition implies that $\tilde{h}$ is a $g_0^{-1}(\operatorname{st} \mathcal{V})$-homotopy. We can apply the Z-Set Unknotting Theorem 2.11.6 to obtain a homeomorphism $\varphi_1 : M \to M$ such that $\varphi_1\alpha_1 = \beta_1$, φ_1 is $f^{-1}(\operatorname{st}^2 \mathcal{V})$-close to id, and $\varphi_1|M \setminus$

$g_0^{-1}(\mathrm{N}_d(A, \varepsilon_0)) = \mathrm{id}$. Thus, we have a cell-like map $g_1 = g_0\varphi_1 : M \to X$ such that $d(g_1, g_0) < \varepsilon_0/2$ and

$$g_1|M \setminus g_0^{-1}(\mathrm{N}_d(A, \varepsilon_0)) = g_0|M \setminus g_0^{-1}(\mathrm{N}_d(A, \varepsilon_0)).$$

Since $g_1^{-1}(x) = \varphi_1^{-1}g_0^{-1}(x)$ for every $x \in X$ and φ is a homeomorphism, we have $N(g_1) = N(g_0)$, so $\mathrm{cl}\, N(g_1) = A$. Since $g_1\alpha_1 = h_1$, it follows that $g_1\alpha_1(\boldsymbol{Q}) \cap A = \emptyset$. We define

$$\varepsilon_1 = \tfrac{1}{2}\min\left\{\varepsilon_0,\ \mathrm{dist}(g_1\alpha_1(\boldsymbol{Q}), A)\right\} > 0.$$

By the same construction as above, we can inductively obtain cell-like maps $g_i : M \to X$ and $\varepsilon_i > 0$, $i \in \mathbb{N}$, such that $\mathrm{cl}\, N(g_i) = A$, $d(g_i, g_{i-1}) < \varepsilon_{i-1}/2$,

$$g_i|M \setminus g_{i-1}^{-1}(\mathrm{N}_d(A, 2^{-i})) = g_{i-1}|M \setminus g_{i-1}^{-1}(\mathrm{N}_d(A, 2^{-i})),$$

$\varepsilon_i \leqslant \varepsilon_{i-1}/2$, and $\varepsilon_i < \mathrm{dist}(g_i\alpha_j(\boldsymbol{Q}), A)$ for each $j \leqslant i$. Then, $(g_i)_{i\in\mathbb{N}}$ is uniformly convergent to a map $g : M \to X$, which is a fine homotopy equivalence by Theorem 1.15.11. Moreover,

$$d(g, g_i) \leqslant \sum_{j=1}^{\infty} \varepsilon_{i+j}/2 \leqslant \sum_{j=1}^{\infty} 2^{-j}\varepsilon_i = \varepsilon_i,$$

which implies that $g\alpha_j(\boldsymbol{Q}) \cap A = \emptyset$ for each $j \leqslant i$. Then, $d(g, f) = d(g, g_0) \leqslant \varepsilon_0 = 1/2$ and $g^{-1}(A)$ misses all $\alpha_i(\boldsymbol{Q})$. Since g is proper by Proposition 1.3.8, g is cell-like by Theorem 1.15.9.

To see $N(g) \subset A$, let $x \in X \setminus A$ and choose $i \in \mathbb{N}$ so that $2^{-i} < d(x, A)$, that is, $x \in X \setminus \mathrm{N}_d(A, 2^{-i})$. Let $y \in g^{-1}(x)$. If $y \in g_i^{-1}(\mathrm{N}_d(A, 2^{-(i+1)}))$, then $d(g(y), g_i(y)) < \varepsilon_i \leqslant 2^{-(i+1)}$, hence $x = g(y) \in \mathrm{N}_d(A, 2^{-i})$. This is a contradiction. Therefore, $y \in M \setminus g_i^{-1}(\mathrm{N}_d(A, 2^{-(i+1)}))$. Since $g_j(y) = g_i(y)$ for every $j \geqslant i$, it follows that $x = g(y) = g_i(y)$, that is, $y \in g_i^{-1}(x)$. Thus, we have $g^{-1}(x) \subset g_i^{-1}(x)$. Since $N(g_i) \subset A$, $g_i^{-1}(x)$ is a singleton, hence so is $g^{-1}(x)$. Consequently, $N(g) \subset A$. □

As mentioned at the beginning of this section, we need a substitution of the $\boldsymbol{Q}$-manifold version of Proposition 2.9.1. For a map $f : M \to X$ and a closed set $A \subset X$, we define the **mapping cylinder reduced over** A to be the space

$$M(f)_A = X \cup \left(f^{-1}(X \setminus A) \times (0, 1]\right) = X \cup \left((M \setminus f^{-1}(A)) \times (0, 1]\right)$$

that admits the topology generated by open sets in

$$(X \setminus A) \cup \left((M \setminus f^{-1}(A)) \times (0, 1]\right) = M(f|M \setminus f^{-1}(A))\ (\subset M(f))$$

and sets $U \cup (f^{-1}(U \setminus A) \times (0, 1])$, where U is open in X. Then, the natural map $\pi_A : M(f) \to M(f)_A$ can be defined as follows:

$$\pi_A | X \cup (f^{-1}(X \setminus A) \times (0, 1]) = \mathrm{id} \text{ and}$$
$$\pi_A | f^{-1}(A) \times (0, 1] = f \operatorname{pr}_{f^{-1}(A)} = c_f.$$

Let $\tau_A : M(f)_A \to X$ be the map induced by the collapsing, that is, $\tau_A \pi_A = c_f$. Then, $\tau_A \simeq \mathrm{id}$ rel. X in $M(f)_A$ by the homotopy $h : M(f)_A \times \mathbf{I} \to M(f)_A$ defined by $h_0 = \tau_A$, $h_t|X = \mathrm{id}$ for each $t \in \mathbf{I}$ and

$$h_t(x, s) = (x, st) \text{ for each } t \in (0, 1] \text{ and } (x, s) \in (M \setminus f^{-1}(A)) \times (0, 1].$$

Thus, X is a strong deformation retract of $M(f)_A$.

In the case where $f = \mathrm{id}_X$, $M(\mathrm{id}_X)_A$ coincides with the reduced product $(X \times \mathbf{I})_A$. The product space $X \times C(E)$ can be regarded as the mapping cylinder $M(\operatorname{pr}_X)$ of the projection $\operatorname{pr}_X : X \times E \to X$, and the reduced product $(X \times C(E))_A$ as corresponding to the reduced mapping cylinder $M(\operatorname{pr}_X)_A$. The following Reduced Mapping Cylinder Theorem is a substitution of the $\boldsymbol{Q}$-manifold version of Proposition 2.9.1:

Theorem 3.7.5 (Reduced Mapping Cylinder) *Let $f : M \to X$ be a cell-like map from a $\boldsymbol{Q}$-manifold M onto an ANR X and A be a Z-set in X. Then, the reduced mapping cylinder $M(f)_A$ is a $\boldsymbol{Q}$-manifold with A a Z-set in $M(f)_A$ and the natural map $\pi_A : M(f) \to M(f)_A$ is a near-homeomorphism.*

Proof First of all, note that the mapping cylinder $M(f)$ is a $\boldsymbol{Q}$-manifold by the Mapping Cylinder Theorem 3.7.3 above, and that the map π_A is cell-like and $A = N(\pi_A) = \operatorname{cl} N(\pi_A)$. As observed above, X is a strong deformation retract of $M(f)_A$. Since $M(f)_A \setminus X = (M \setminus f^{-1}(A)) \times (0, 1]$ and X are ANRs, $M(f)_A$ is an ANR by Theorem 1.13.16. Then, Proposition 3.7.4 can be applied, that is, it suffices to show that A is a Z-set in $M(f)_A$.

For each open over $\mathcal{U} \in \operatorname{cov}(M(f)_A)$, there is an open cover $\mathcal{V}$ of A in X such that

$$\{V \cup (f^{-1}(V \setminus A) \times (0, 1]) \mid V \in \mathcal{V}\} \prec \mathcal{U}.$$

Take open sets W_0, W_1 in X so that

$$A \subset W_1 \subset \operatorname{cl} W_1 \subset W_0 \subset \operatorname{cl} W_0 \subset \bigcup \mathcal{V}$$

and let $\varphi, \psi : X \to \mathbf{I}$ be Urysohn maps with $\varphi(\operatorname{cl} W_0) = 0$, $\varphi(X \setminus \bigcup \mathcal{V}) = 1$, $\psi(X \setminus W_0) = 0$, and $\psi(\operatorname{cl} W_1) = 1$. Since A is a Z-set in X, $X \setminus A$ is homotopy dense in X by Theorem 2.8.6, hence there is a homotopy $h : X \times \mathbf{I} \to X$ such that $h_0 = \mathrm{id}$ and $h_t(X) \subset X \setminus A$ for each $t \in (0, 1]$. By Lemma 2.5.2, it can be assumed that h is a $\mathcal{V}$-homotopy, where $h_t|X \setminus \bigcup \mathcal{V} = \mathrm{id}$. We can define a map $g : M(f)_A \to M(f)_A \setminus A$ as follows:

$$g(x) = h(x, \psi(x)) \quad \text{for each } x \in X \text{ and}$$

$$g(x,t) = \begin{cases} (x,t) & \text{if } 0 < t \leqslant \varphi f(x), \\ (x, \varphi f(x)) & \text{if } t > \varphi f(x) > 0, \\ h(f(x), \psi f(x)) & \text{if } t \geqslant \varphi f(x) = 0. \end{cases}$$

Then, g is $\mathcal{U}$-close to id. □

Proposition 3.7.6 *Let $f : M \to X$ be a cell-like map from a $\boldsymbol{Q}$-manifold M onto an ANR X with the disjoint cells property. Then, the collapsing $c_f : M(f) \to X$ of the mapping cylinder $M(f)$ $(= M_f)$ of f is a near-homeomorphism, and hence X is also a $\boldsymbol{Q}$-manifold.*

Proof First of all, notice that the disjoint cells property of X implies that every proper map from a locally compact separable metrizable space can be approximated by Z-embeddings (Proposition 3.2.9). Moreover, the mapping cylinder $M(f)$ is a $\boldsymbol{Q}$-manifold by the Mapping Cylinder Theorem 3.7.3. Since f is a proper (= perfect) map, the collapsing $c_f : M(f) \to X$ is also a cell-like map.

To prove that c_f is a near-homeomorphism, we can apply the Bing Shrinking Criterion 2.7.1 with Remark 2.11. For each $\mathcal{U} \in \mathrm{cov}(M(f))$ and $\mathcal{V} \in \mathrm{cov}(X)$, it suffices to find a near-homeomorphism $\varphi : M(f) \to M(f)$ such that $c_f\varphi$ is $\mathcal{V}$-close to c_f and $\{\varphi c_f^{-1}(x) \mid x \in X\} \prec \mathcal{U}$. Our strategy is the same as the proof of (b) ⇒ (a) in Theorem 3.4.2.

We introduce a similar notation as in the proof of Proposition 2.9.1 and Theorem 3.4.2. For each $W \subset X$ and $\alpha \in \mathrm{C}(X, (0, 1])$, we write

$$W[0,\alpha] = W \cup \big\{(x,t) \in c_f^{-1}(W) \times (0,1] \bigm| t \leqslant \alpha f(x)\big\}.$$

For each $N \subset M$ and $\beta \leqslant \gamma \in \mathrm{C}(M, (0, 1])$,

$$N[\beta,\gamma] = \big\{(x,t) \in N \times (0,1] \bigm| \beta(x) \leqslant t \leqslant \gamma(x)\big\}.$$

For simplicity, let $M[1] = M \times \{1\}$. We can easily find $\alpha \in \mathrm{C}(X, (0, 1))$ and $\mathcal{V}_0 \in \mathrm{cov}(X)$ such that $\mathcal{V}_0 \prec \mathcal{V}$ and $\{V[0, \alpha] \mid V \in \mathcal{V}_0\} \prec \mathcal{U}$. Moreover, we have

$$\alpha_0 < \alpha_1 < \alpha_2 < \cdots \in \mathrm{C}(M, (0,1)) \quad \text{and} \quad \mathcal{U}_0 \in \mathrm{cov}(M)$$

such that $\alpha_1 < \alpha f$, $\lim_{n\to\infty} \alpha_n(x) = 1$, and

$$\big\{U[\alpha_{n-1}, \alpha_{n+1}] \bigm| U \in \mathcal{U}_0,\ n \in \mathbb{N}\big\} \prec \mathcal{U}.$$

Indeed, for each $x \in M$, let

$$\gamma'(x) = \sup\big\{s \in (0,1) \bigm| \forall t \in [\alpha(x)/2, 1],\ \exists U \in \mathcal{U} \\ \text{such that } \{x\} \times ([t, t+s] \cap \mathbf{I}) \subset U\big\}.$$

Then, as in the proof of Theorem 3.4.2, it can be seen that γ' is lower semi-continuous. By Theorem 1.3.14, we have a map $\gamma : M \to (0,1)$ with $2\gamma(x) < \gamma'(x)$ and $\gamma(x) < \alpha f(x)$. The maps $\alpha_0 < \alpha_1 < \alpha_2 < \cdots$ can be defined as in the proof of Theorem 3.4.2, that is,

$$\alpha_n(x) = \frac{\alpha f(x) + (n-1)\gamma(x)}{1 + \alpha f(x) + (n-1)\gamma(x)}.$$

Then, $\alpha_1(x) < \alpha f(x)$, $\lim_{n\to\infty} \alpha_n(x) = 1$, and

$$0 < \alpha_n(x) - \alpha_{n-1}(x) < \gamma(x) < \gamma'(x)/2 \quad \text{for each } n \in \mathbb{N}.$$

For each $x \in M$, we have $n \in \mathbb{N}$ such that $1 - \alpha_n(x) < \gamma(x)$. Then, we can easily obtain $\mathcal{U}_0 \in \operatorname{cov}(M)$ with the desired property.

Let $\mathcal{V}_1, \mathcal{V}_2 \in \operatorname{cov}(X)$ be locally finite open refinements of $\mathcal{V}_0$ such that each member of both of $\mathcal{V}_1$ has the compact closure and

$$\operatorname{st}^3 \mathcal{V}_2 \prec \mathcal{V}_1 \overset{*}{\prec} \mathcal{V}_0.$$

By Proposition 3.2.9, we have a Z-embedding $v : M[1] \to X$ such that $v \simeq_{\mathcal{V}_2} c_f$. Let $A = v(M[1])$, and let $\pi : M(f) \to M(f)_A$ be the natural map and $\tau : M(f)_A \to X$ be the map induced by the collapsing, i.e., $\tau_A \pi = c_f$. Due to the Reduced Mapping Cylinder Theorem 3.7.5, A is a Z-set in $M(f)_A$ and π is a near-homeomorphism. Using the Z-set Unknotting Theorem 2.11.6, we have a homeomorphism $h : M(f) \to M(f)_A$ such that

$$h \simeq_{\tau^{-1}(\mathcal{V}_2)} \pi \quad \text{and} \quad h|A = \mathrm{id}.$$

Due to the Mapping Cylinder Theorem 3.7.3, X is a Z-set in $M(f)$, hence so is $h^{-1}(A) = A$. Then, $h^{-1}v : M[1] \to M(f)$ is a Z-embedding and

$$c_f h^{-1} v = \tau\pi h^{-1} v \simeq_{\mathcal{V}_2} \tau v = v \simeq_{\mathcal{V}_2} c_f|M[1],$$

where the above two $\mathcal{V}_2$-homotopies are proper by Proposition 1.3.8. Since $c_f \simeq \mathrm{id}$ by the homotopy h_f with $c_f h_f = c_f \mathrm{pr}_{M(f)}$, it follows that $h^{-1}v$ is properly $c_f^{-1}(\operatorname{st}\mathcal{V}_2)$-homotopic to id. By the Z-set Unknotting Theorem 2.11.6, we have a homeomorphism $g : M(f) \to M(f)$ such that

$$g|M[1] = h^{-1}v \text{ and } g \simeq_{c_f^{-1}(\mathrm{st}^2 \mathcal{V}_2)} \mathrm{id}.$$

Then, $hg(M[1]) = A = v(M[1])$. Note that $\tilde{\mathcal{U}}_0 = \{U \times (0,1] \mid U \in \mathcal{U}_0\}$ is an open cover of $M[1]$ in $M(f)$, hence $hg(\tilde{\mathcal{U}}_0)$ is an open cover of A in $M(f)_A$. Let $\mathcal{W}_0$ be an open cover of A in X such that

$$\pi(c_f^{-1}(\mathcal{W}_0)) = \tau^{-1}(\mathcal{W}_0) = \{W[0,1]_A \mid W \in \mathcal{W}_0\} \prec hg(\tilde{\mathcal{U}}_0).$$

Thus, we have a near-homeomorphism $\eta = g^{-1}h^{-1}\pi : M(f) \to M(f)$ such that $\eta(c_f^{-1}(\mathcal{W}_0)) \prec \tilde{\mathcal{U}}_0$ (Fig. 3.4). Since

$$c_f\eta = c_f g^{-1}h^{-1}\pi \simeq_{\mathrm{st}^2\mathcal{V}_2} c_f h^{-1}\pi = \tau\pi h^{-1}\pi \simeq_{\mathcal{V}_2} \tau\pi = c_f,$$

it follows that $c_f\eta$ is $\mathcal{V}_1$-close to c_f.

Let $M_0 = \bigcup \mathcal{W}_0$ and choose an open neighborhood N_0 of A in X so that $\mathrm{cl}\, N_0 \subset M_0$. Then, $\eta(c_f^{-1}(N_0)) = g^{-1}h^{-1}\tau^{-1}(N_0)$ is an open neighborhood of $i_f(M) = \eta(c_f^{-1}(A))$ in $M(f)$. As in the proof of the Reduced Mapping Cylinder Theorem 3.7.5, we have $\beta_0 \in \mathrm{C}(M,(0,1))$ such that $M[\beta_0, 1] \subset \eta(c_f^{-1}(N_0))$.

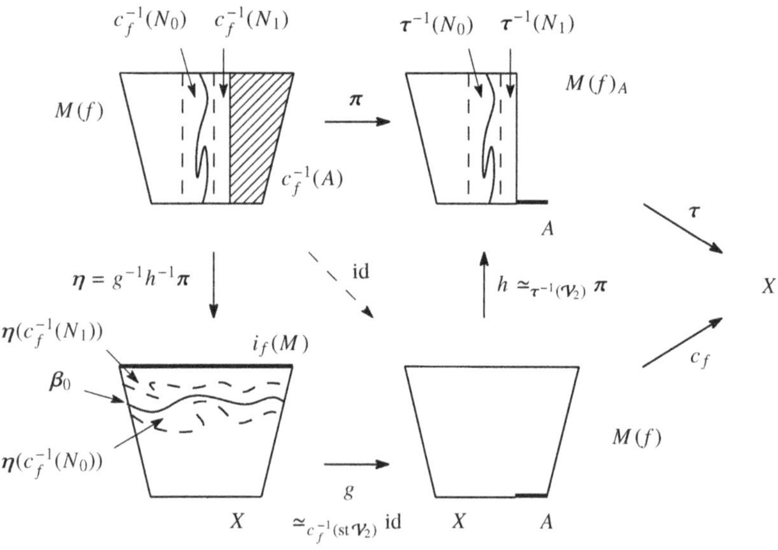

Fig. 3.4 The homeomorphism η

Then,

$$\eta(M(f) \setminus c_f^{-1}(N_0)) = M(f) \setminus \eta(c_f^{-1}(N_0)) \subset M[0, \beta_0].$$

Since $hg(M[\beta_0, 1])$ is a neighborhood of A in $M(f)_A$, A has open neighborhoods N_1 and M_1 in X such that $\operatorname{cl} N_1 \subset M_1 \subset \operatorname{cl} M_1 \subset N_0$ and $\tau^{-1}(\operatorname{cl} M_1) \subset hg(M[\beta_0, 1])$. Then, $\eta(c_f^{-1}(N_1)) = g^{-1}h^{-1}\tau^{-1}(N_1)$ is an open neighborhood of $i_f(M) = \eta(c_f^{-1}(A))$ in $M(f)$ and $\eta(c_f^{-1}(\operatorname{cl} M_1)) \subset M[\beta_0, 1]$. Hence, we have $\beta_1 \in \mathrm{C}(M, (0, 1))$ such that $M[\beta_1, 1] \subset \eta(c_f^{-1}(N_1))$ and $\beta_1 > (1 + \beta_0)/2\ (> \beta_0)$, where it should be noted that

$$\eta(c_f^{-1}(X \setminus N_1)) = M(f) \setminus \eta(c_f^{-1}(N_1)) \subset M[0, \beta_1].$$

Thus, we can inductively choose open neighborhoods N_i, M_i of A in X and $\beta_i \in \mathrm{C}(M, (0, 1))$, $i \in \omega$, such that

$$\operatorname{cl} N_i \subset M_i \subset \operatorname{cl} M_i \subset N_{i-1}, \ \bigcap_{i\in\omega} N_i = A, \ \beta_i > \frac{1 + \beta_{i-1}}{2} \ \text{and}$$

$$M[\beta_i, 1] \subset \eta(c_f^{-1}(N_i)) \subset \eta(c_f^{-1}(\operatorname{cl} M_i)) \subset M[\beta_{i-1}, 1].$$

Then, it follows that

$$\eta(c_f^{-1}(X \setminus \operatorname{cl} N_1)) \subset M[0, \beta_1] \ \text{and}$$

$$\eta(c_f^{-1}(M_i \setminus \operatorname{cl} N_{i+1})) \subset M[\beta_{i-1}, \beta_{i+1}] \ \text{for each } i \in \mathbb{N}.$$

Sliding along the $(0, 1]$-coordinates as illustrating in Fig. 3.5, we can obtain a homeomorphism $\xi : M(f) \to M(f)$ such that $c_f\xi = c_f$, $\xi | X \cup i_f(M) = \mathrm{id}$, and

$$\psi(x, \beta_i(x)) = (x, \alpha_i(x)) \ \text{for } x \in M \text{ and } i \in \omega.$$

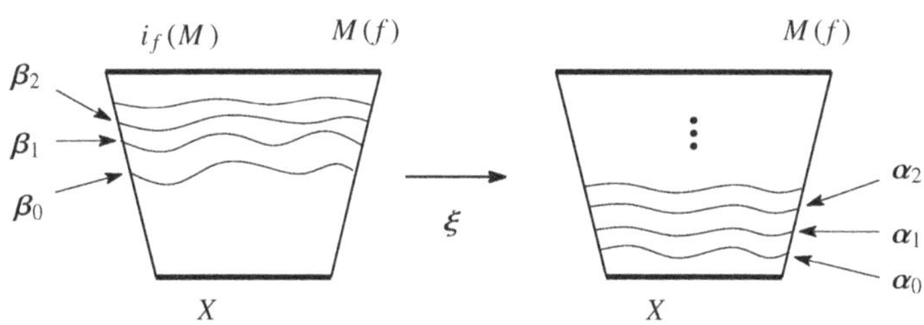

Fig. 3.5 The homeomorphism ξ

Then, $\xi\eta : M(f) \to M(f)$ is a near-homeomorphism such that $c_f\xi\eta$ is $\mathcal{V}_1$-close (hence $\mathcal{V}$-close) to c_f. We verify that each $x \in X$ has a neighborhood W_x in X such that $\xi\eta(c_f^{-1}(W_x)) = \xi\eta(c_f(W_x))$ is contained in some member of $\mathcal{U}$.

If $x \in X \setminus \operatorname{cl} N_1$, choose $V_x \in \mathcal{V}_1$ so that $x \in V_x$ and let $W_x = V_x \setminus \operatorname{cl} N_1$. Since $c_f\xi\eta = c_f\eta$ is $\mathcal{V}_1$-close to c_f, we have

$$c_f\xi\eta(W_x[0,1]) \subset \operatorname{st}(V_x, \mathcal{V}_1),$$

which is contained in some $V \in \mathcal{V}_0$. Then, it follows that

$$\xi\eta(c_f^{-1}(W_x)) \subset \xi(V[0,\beta_1]) = V[0,\alpha_1] \subset V[0,\alpha],$$

which is contained in some member of $\mathcal{U}$.

When $x \in M_1 \setminus A$, $x \in M_i \setminus \operatorname{cl} N_{i+1}$ for some $i \in \mathbb{N}$. Choose $W'_x \in \mathcal{W}_0$ so that $x \in W'_x$ and let $W_x = W'_x \cap (M_i \setminus \operatorname{cl} N_{i+1})$. Then, $\eta(c_f^{-1}(W'_x)) \subset c_f^{-1}(U)$ for some $U \in \mathcal{U}_0$. It follows that

$$\xi\eta(c_f^{-1}(W_x)) \subset \xi(U[\beta_{i-1},\beta_{i+1}]) = U[\alpha_{i-1},\alpha_{i+1}],$$

which is contained in some member of $\mathcal{U}$.

When $x \in A$, $\xi(g^{-1}h^{-1}(\tau^{-1}(x))) = g^{-1}h^{-1}(\tau^{-1}(x))$ is a singleton, hence it is contained in some $U \in \mathcal{U}$. Then, x has an open neighborhood W_x in X such that

$$\xi\eta(c_f^{-1}(W_x)) = \xi(g^{-1}h^{-1}(\tau^{-1}(W_x))) \subset U.$$

Thus, $\xi\eta$ is the desired near-homeomorphism. □

By the Z-Embedding Approximation Theorem 2.10.11, the following is a direct consequence of Proposition 3.7.6:

Corollary 3.7.7 *Let $f : M \to N$ be a cell-like map between $\boldsymbol{Q}$-manifolds. Then, the collapsing $c_f : M(f) \to N$ is a near-homeomorphism.* □

Now, we can prove the following:

Theorem 3.7.8 *Every cell-like map $f : M \to N$ between $\boldsymbol{Q}$-manifolds is a near-homeomorphism.*

Proof Consider the following commutative diagram:

$$\begin{array}{ccc} M \times \mathbf{I} & \xrightarrow{q_f} & M(f) \\ {\scriptstyle \mathrm{pr}_M}\downarrow & & \downarrow{\scriptstyle c_f} \\ M & \xrightarrow[f]{} & N, \end{array}$$

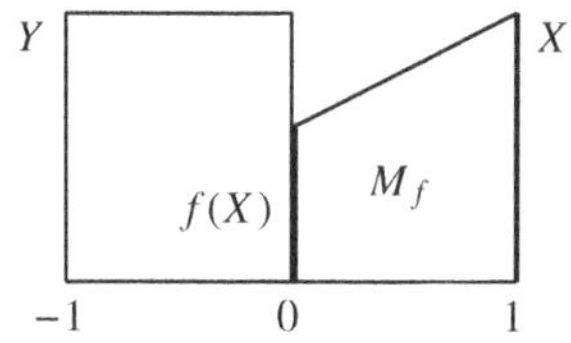

Fig. 3.6 The mapping cylinder M_{f^*}

where the projection pr_M, the natural (quotient) map q_f, and the collapsing c_f are near-homeomorphisms by the Stability Theorem 2.3.13 with Corollary 2.3.11, the Mapping Cylinder Theorem 3.7.3, and Corollary 3.7.7, respectively. Thus, it follows that f is a near-homeomorphism. □

By the above theorem, the following can be easily obtained:

Theorem 3.7.9 *Every compact contractible $\boldsymbol{Q}$-manifold M is homeomorphic to the Hilbert cube $\boldsymbol{Q}$.*

Proof The projection $\mathrm{pr}_2 : M \times \boldsymbol{Q} \to \boldsymbol{Q}$ is cell-like, hence it is a near-homeomorphism by Theorem 1.15.9. On the other hand, $M \times \boldsymbol{Q} \approx M$ by the Stability Theorem 2.3.13. Consequently, we have $M \approx \boldsymbol{Q}$. □

3.8 Characterization and Classification of $\boldsymbol{Q}$-Manifolds

In this section, we establish the Toruńczyki characterization of $\boldsymbol{Q}$-manifolds. To this end, we first prove the EDWARDS FACTOR THEOREM, which is the $\boldsymbol{Q}$-version of the Toruńczyk Factor Theorem 2.2.14, that is:

Theorem 3.8.1 (EDWARDS) *For every locally compact ANR X, $X \times \boldsymbol{Q}$ is a $\boldsymbol{Q}$-manifold.*

Since a compact contractible $\boldsymbol{Q}$-manifold M is homeomorphic to $\boldsymbol{Q}$ (Theorem 3.7.9), we have the following:

Corollary 3.8.2 *For every compact AR X, $X \times \boldsymbol{Q}$ is homeomorphic to the Hilbert cube $\boldsymbol{Q}$.* □

To prove the Edwards Factor Theorem 3.8.1 above, we use the **fat mapping cylinder**. For a map $f : X \to Y$, let $f^* : X \to Y \times [-1, 0]$ be the map defined by $f^*(x) = (f(x), 0)$ for each $x \in X$. The mapping cylinder M_{f^*} can be regarded as the space $(Y \times [-1, 0]) \cup M_f$, where $Y \times \{0\}$ is identified with $Y \subset M_f$.[14] The collapsing c_{f^*} is the unique extension of c_f (Fig. 3.6).

[14]When f is a proper map, this is the same space $M(f) \cup_{Y = Y \times \{0\}} Y \times \mathbf{I}$ defined in Sect. 2.6 (see p. 145).

Lemma 3.8.3 *Let $r : \boldsymbol{Q} \to X$ be a retraction of $\boldsymbol{Q}$ onto X and d be an admissible metric for $\boldsymbol{Q}$. For each $\varepsilon > 0$, there is a cell-like map $\beta : \boldsymbol{Q} \times [-1, 1] \to M_{r^*}$ such that $\beta^{-1}(x, 1) = \{(x, 1)\}$ for each $x \in X$, $\beta|\boldsymbol{Q} \times \{-1\} = \mathrm{id}$, and $d(\mathrm{pr}_X \bar{r}\beta, \mathrm{pr}_X(r \times \mathrm{id})) < \varepsilon$, where $\bar{r} : M_{r^*} \to X \times [-1, 1]$ is the map defined by $\bar{r}(x, t) = (r(x), t)$ for each $(x, t) \in (\boldsymbol{Q} \times [-1, 0]) \cup (\boldsymbol{Q} \times (0, 1])$.*

Proof By Corollary 2.3.10(4) and Stability Theorem 2.3.13, for any small $\delta_1 > 0$, we have a homeomorphism

$$h : \boldsymbol{Q} \times [-1, 1] \times \boldsymbol{Q} \to \boldsymbol{Q} \times [-1, 1]$$

such that h is δ_1-close to the projection and $h(x, \pm 1, \mathbf{0}) = (x, \pm 1)$ for every $x \in \boldsymbol{Q}$.

Let $k : \boldsymbol{Q} \to \boldsymbol{Q}$ be a map such that $k^{-1}(\mathbf{0}) = X$ and $k|\boldsymbol{Q} \setminus X$ is injective. We define $\zeta : \boldsymbol{Q} \to \boldsymbol{Q} \times [-1, 0] \times \boldsymbol{Q}$ by $\zeta(x) = (r(x), 0, k(x))$ for each $x \in \boldsymbol{Q}$. Since ζ is injective and $\boldsymbol{Q} \times \{1\} \times \boldsymbol{Q}$ is a Z-set in $\boldsymbol{Q} \times \mathbf{I} \times \boldsymbol{Q}$, it follows that ζ is a Z-embedding. Note that $\zeta(x) = (x, 0, \mathbf{0})$ for every $x \in X$ and $\mathrm{pr}_1 \zeta = r$, where $\mathrm{pr}_1 : \boldsymbol{Q} \times [-1, 0] \times \boldsymbol{Q} \to \boldsymbol{Q}$ is the projection onto the first factor.

We define a homeomorphism

$$\xi : \boldsymbol{Q} \times [-1, 1] \times \boldsymbol{Q} \to \boldsymbol{Q} \times [-1, 0] \times \boldsymbol{Q} \subset M_\zeta$$

by $\xi(x, t, y) = (x, (t - 1)/2, y)$ (Fig. 3.7).

Using the Z-Set Unknotting Theorem 2.11.6, we can easily construct a homeomorphism $\varphi : \boldsymbol{Q} \times [-1, 1] \to M_\zeta$ such that $\varphi(\boldsymbol{Q} \times \{-1\}) = \boldsymbol{Q} \times \{-1\} \times \boldsymbol{Q}$, $\varphi(x, 0) = \zeta(x)$ for each $x \in \boldsymbol{Q}$, and $\varphi|\boldsymbol{Q} \times (0, 1] = \mathrm{id}$. For any small $\delta_2 > 0$, we can obtain a homeomorphism $\eta : \boldsymbol{Q} \times [-1, 1] \to \boldsymbol{Q} \times [-1, 0]$ such that $\eta|\boldsymbol{Q} \times [-1, -\delta_2] = \mathrm{id}$ and $\eta(\{x\} \times [-\delta_2, 1]) = \{x\} \times [-\delta_2, 0]$ for each $x \in \boldsymbol{Q}$.

We have a cell-like map $p : M_\zeta \to M_r^*$ such that $p|\boldsymbol{Q} \times (0, 1] = \mathrm{id}$ and $p|\boldsymbol{Q} \times [-1, 0] \times \boldsymbol{Q}$ is the projection onto $\boldsymbol{Q} \times [-1, 0]$. Thus, we have a cell-like map

$$\beta = p\varphi\eta^{-1}\varphi^{-1}\xi h^{-1} : \boldsymbol{Q} \times [-1, 1] \to M_r^*.$$

Observe that $\beta|\boldsymbol{Q} \times \{-1\} = \mathrm{id}$ and $\beta^{-1}(x, 1) = \{(x, 1)\}$ for each $x \in X$. We can choose $\delta_2 > 0$ so small that $\mathrm{pr}_X \bar{r} p\varphi\eta\varphi^{-1}$ is $\varepsilon/2$-close to $\mathrm{pr}_X \bar{r} p c_\zeta = \mathrm{pr}_X \bar{r} p$. Then, it follows that

$$\begin{aligned} d(\mathrm{pr}_X \bar{r}\beta,\ r\,\mathrm{pr}_Q p\xi h^{-1}) &= d(\mathrm{pr}_X \bar{r}\beta,\ \mathrm{pr}_X \bar{r} p\xi h^{-1}) \\ &= d(\mathrm{pr}_X \bar{r} p\varphi\eta^{-1}\varphi^{-1}\xi h^{-1},\ \mathrm{pr}_X \bar{r} p\varphi\eta\varphi^{-1}\varphi\eta^{-1}\varphi^{-1}\xi h^{-1}) < \varepsilon/2. \end{aligned}$$

Observe that $\mathrm{pr}_Q p\xi$ is the projection of $\boldsymbol{Q} \times [-1, 1] \times \boldsymbol{Q}$ onto the first factor. We can choose $\delta_1 > 0$ so small that $\mathrm{pr}_X \bar{r} h = r\,\mathrm{pr}_Q h$ is $\varepsilon/2$-close to $\mathrm{pr}_X \bar{r} p\xi = r\,\mathrm{pr}_Q p\xi$, hence $d(\mathrm{pr}_X \bar{r} p\xi h^{-1}, \mathrm{pr}_X(r \times \mathrm{id})) < \varepsilon/2$. Consequently, $d(\mathrm{pr}_X \bar{r}\beta, \mathrm{pr}_X(r \times \mathrm{id})) < \varepsilon$. This completes the proof. □

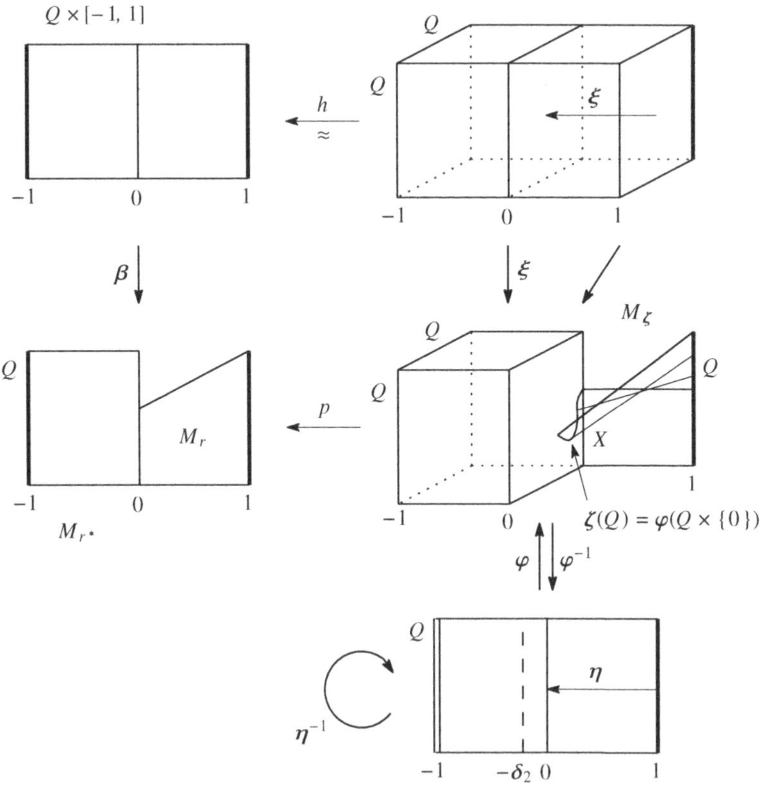

Fig. 3.7 M_ζ and constructing β

Lemma 3.8.4 *For a retraction* $r : \boldsymbol{Q} \to X$, *there is a cell-like map* $\varphi : M_r \to X \times \mathbf{I}$ *such that* $\varphi(x) = (x, 0)$ *for each* $x \in X$.

Proof Take any metric $d \in \mathrm{Metr}(X)$ and define the admissible metric ρ for $X \times \mathbf{I}$ as follows:

$$\rho((x, t), (y, s)) = \max \left\{ d(x, y),\ |t - s| \right\}.$$

For each $n \in \omega$, we define the space $M^n = \bigcup_{i=1}^{2^n} M_i^n$, where each M_i^n is a copy of the mapping cylinder M defined by

$$M_i^n = \left(X \times \left\{ \frac{i-1}{2^n} \right\} \right) \cup_{r \times \mathrm{id}_{\{(i-1)/2^n\}}} \left(\boldsymbol{Q} \times \left[\frac{i-1}{2^n}, \frac{i}{2^n} \right] \right).$$

Then, $X \times \mathbf{I}$ is contained in the space M^n as a retract with the retraction $r_n : M^n \to X \times \mathbf{I}$ defined by $r_n(x, t) = (r(x), t)$ for each $(x, t) \in M^n$ (Fig. 3.8). The mapping cylinder M_r can be regarded as $M^0 = M_1^0$.

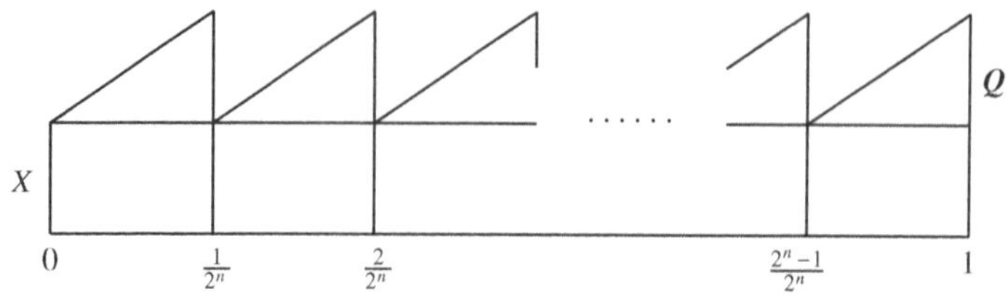

Fig. 3.8 The space M^n

We shall construct cell-like maps $f_n : M^n \to M^{n+1}$, $n \in \omega$. Observe that

$$M_i^n = M_{4i-3}^{n+2} \cup \left(\boldsymbol{Q} \times \left[\frac{4i-3}{2^{n+2}}, \frac{i}{2^n} \right] \right),$$

and the following set is a copy of $M_{r^*} = (\boldsymbol{Q} \times [-1, 0]) \cup M_r$ in Lemma 3.8.3:

$$\left(\boldsymbol{Q} \times \left[\frac{4i-3}{2^{n+2}}, \frac{2i-1}{2^{n+1}} \right] \right) \cup M_{2i}^{n+1}, \ n \in \mathbb{N}, \ i = 1, \ldots, 2^n.$$

We can apply Lemma 3.8.3 to obtain a cell-like map

$$f_{n,i} : \boldsymbol{Q} \times \left[\frac{4i-3}{2^{n+2}}, \frac{i}{2^n} \right] \to \left(\boldsymbol{Q} \times \left[\frac{4i-3}{2^{n+2}}, \frac{2i-1}{2^{n+1}} \right] \right) \cup M_{2i}^{n+1}$$

such that $f_{n,i} | \boldsymbol{Q} \times \{(4i-3)/2^{n+2}\} = \mathrm{id}$, $f_{n,i}^{-1}(x, i/2^n) = \{(x, i/2^n)\}$ for each $x \in X$, and $d(\mathrm{pr}_X r_{n+1} f_{n,i}, \mathrm{pr}_X (r \times \mathrm{id})) < 2^{-n}$, which extends over M_i^n by $f_{n,i} | M_{4i-3}^{n+2} = \mathrm{id}$. Now, we have a cell-like map $f_n : M^n \to M^{n+1}$ defined by $f_n | M_i^n = f_{n,i}$ for each $i = 1, \ldots, 2^n$ (Fig. 3.9). Since $d(\mathrm{pr}_X r_{n+1} f_n, \mathrm{pr}_X r_n) < 2^{-n}$, it follows that $\rho(r_{n+1} f_n, r_n) < 2^{-n}$.

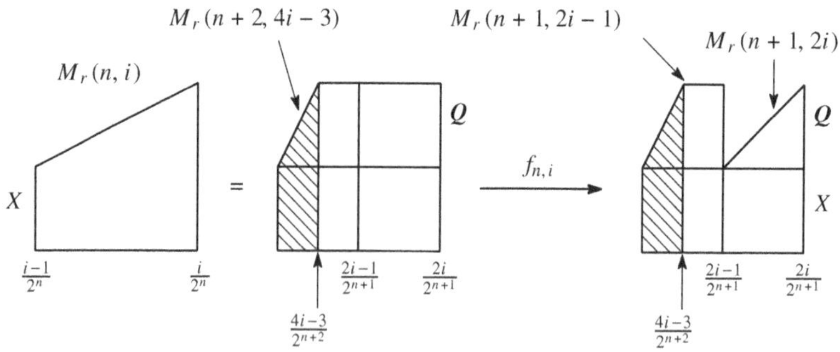

Fig. 3.9 Construction of the cell-like map f_n

For each $n \in \omega$, since $\mathrm{C}(M^n, X \times \mathbf{I})$ is complete with respect to the sup-metric, we can obtain a map $\varphi_n : M^n \to X \times \mathbf{I}$ as the limit of the Cauchy sequence

$$\left(r_{n+j+1} f_{n+i} \cdots f_{n+1} f_n\right)_{j \in \mathbb{N}}.$$

Then, $\rho(r_n, \varphi_n) < \sum_{j=0}^{\infty} 2^{-(n+j)} < 2^{-n+1}$. It should be noted that

$$\varphi_n | X \times \{i/2^n\} = \mathrm{id} \ \text{ for each } i = 0, \ldots, 2^n$$

because each $f_{n+j} | X \times \{i/2^n\} = \mathrm{id}$.

To see that $\varphi = \varphi_0 : M^0 = M_r \to X \times \mathbf{I}$ is the desired one, we have to show that φ_0 is cell-like. Let $(x, t) \in X \times \mathbf{I}$. For each neighborhood U of (x, t) in $X \times \mathbf{I}$, we have $\varepsilon > \delta > 0$ such that $\mathrm{B}_\rho((x, t), \varepsilon) \subset U$ and $\mathrm{B}_d(x, \delta)$ is contractible in $\mathrm{B}_d(x, \varepsilon)$. Choose $n \in \mathbb{N}$ so that $2^{-n+3} < \delta$ and $n \geqslant 3$.

Claim *$A = \varphi_n^{-1}(x, t)$ is contractible in $\varphi_n^{-1}(U)$.*

This Claim implies that $\varphi^{-1}(x, t)$ is contractible in $\varphi^{-1}(U)$. In fact, $\varphi = \varphi_n f_{n-1} \cdots f_0$, where $f_{n-1} \cdots f_0$ is a fine homotopy equivalence by Corollary 1.15.10 and Theorem 1.15.9. Then, by Corollary 1.15.3, $\varphi^{-1}(x, t) = (f_{n-1} \cdots f_0)^{-1}(A)$ is contractible in $\varphi^{-1}(U) = (f_{n-1} \cdots f_0)^{-1}(\varphi_n^{-1}(U))$. Thus, it follows that φ is cell-like.

Now, we shall prove the Claim. For each $a, a' \in A$, since $\varphi_n(a) = \varphi_n(a') = x$, it follows that

$$\begin{aligned}\rho(r_n(a), r_n(a')) &\leqslant \rho(r_n(a), \varphi_n(a)) + \rho(\varphi_n(x'), r_n(x')) \\ &< 2^{-n+1} + 2^{-n+1} = 4 \cdot 2^{-n}.\end{aligned}$$

This means that A is contained in the union of some five successive M_i^n, that is, $A \subset \bigcup_{i=j}^{j+4} M_i^n$ for some $j = 1, \ldots, 2^n - 4$. Since each M_i^n is a copy of the mapping cylinder M_r, we can obtain a strong deformation retraction of $\bigcup_{i=j}^{j+4} M_i^n$ onto $X \times \{(j-1)/2^n\}$ as the composition of collapsings. Thus, there is a homotopy $h : \bigcup_{i=j}^{j+4} M_i^n \times \mathbf{I} \to \bigcup_{i=j}^{j+4} M_i^n$ such that $h_0 = \mathrm{id}$, h_1 is a retraction onto $X \times \{(j - 1)/2^n\}$, and, $(z, s) \in M_i^n$ $(j \leqslant i \leqslant j+4)$ implies

$$h(\{(z, s)\} \times \mathbf{I}) \subset \left(\{r(z)\} \times \left[\frac{j-1}{2^n}, \frac{i-1}{2^n}\right]\right) \cup \left(\{z\} \times \left(\frac{i-1}{2^n}, \frac{i}{2^n}\right]\right).$$

When $(z, s) \in A \cap M_i^n$, since $\rho(\varphi_n(z, s), (r(z), s)) \leqslant \rho(\varphi_n, r_n) < 2^{-n+1}$, it follows that

$$\operatorname{diam}_\rho \varphi_n h(\{(z, s)\} \times \mathbf{I}) \leqslant \frac{5}{2^n} + \frac{1}{2^n} + \frac{2}{2^n} = \frac{8}{2^n} = 2^{-n+3} < \varepsilon.$$

Since $\varphi_n(z, s) = (x, t)$, we have

$$\varphi_n h(\{(z, s)\} \times \mathbf{I}) \subset \mathrm{B}_\rho((x, t), \varepsilon) \subset U,$$

hence $h(\{(z, s)\} \times \mathbf{I}) \subset \varphi_n^{-1}(U)$. Therefore, $h(A \times \mathbf{I}) \subset \varphi_n^{-1}(U)$. Moreover, since $d(x, r(z)) < 2^{-n+1} < \delta$ in the above, it follows that

$$h_1(A) \subset \mathrm{B}_d(x, \delta) \times \{(j-1)/2^n\} \subset \mathrm{B}_d(x, \varepsilon) \times \{(j-1)/2^n\},$$

where the second set is contractible in the third one. Here, recall that $\varphi_n|X \times \{(j-1)/2^n\} = \mathrm{id}$. Then,

$$\mathrm{B}_d(x, \varepsilon) \times \{(j-1)/2^n\} \subset \varphi_n^{-1}(\mathrm{B}_d(x, \varepsilon) \times \{(j-1)/2^n\}) \subset \varphi_n^{-1}(U).$$

Consequently, $h_1(A)$ is contractible in $\varphi_n^{-1}(U)$, hence so is A. This completes the proof. □

Lemma 3.8.5 *Let X be a non-compact separable locally compact ANR. Then, the one-point compactification $\alpha(X \times [0, 1)) = (X \times [0, 1)) \cup \{\infty\}$ of $X \times [0, 1)$ is a compact AR. When X is an AR, $\{\infty\}$ is a Z-set in $\alpha(X \times [0, 1))$.*

Proof Since X is separable and locally compact, we can apply the Urysohn Metrization Theorem 1.3.4 to see that $\alpha(X \times [0, 1))$ is metrizable. We can regard $\alpha(X \times [0, 1))$ as the following quotient space:

$$(\alpha X \times \mathbf{I})/((\{*\} \times \mathbf{I}) \cup (\alpha X \times \mathbf{I})),$$

where $\alpha X = X \cup \{*\}$ is the one-point compactification of X. Let $q : \alpha X \times \mathbf{I} \to \alpha(X \times [0, 1))$ be the quotient map. Then, there exists the deformation $h : (\alpha X \times \mathbf{I}) \times \mathbf{I} \to \alpha X \times \mathbf{I}$ defined by $h_t(x, s) = (x, 1 - (1-t)(1-s))$, which induces the contraction of $\alpha(X \times [0, 1))$ onto $\{\infty\}$ fixing $\{\infty\}$. Thus, $\{\infty\}$ is a strong deformation retract of $\alpha(X \times [0, 1))$. By the Kruse–Liebnitz Theorem 1.13.16, $\alpha(X \times [0, 1))$ is an ANR, hence it is an AR because it is contractible.

When X is an AR, we shall show that $\{\infty\}$ is a Z-set in $\alpha(X \times [0, 1))$. For each open cover $\mathcal{U}$ of $\alpha(X \times [0, 1))$, we take $U \in \mathcal{U}$ such that $\infty \in U$. Choose open neighborhoods V and W of $*$ in αX so that $V \times \mathbf{I} \subset q^{-1}(U)$ and $\mathrm{cl}\, W \subset V$. Let $k : \alpha X \to \mathbf{I}$ be an Urysohn map with $k((\alpha X) \setminus V) = 0$ and $k(\mathrm{cl}\, W) = 1$. Define a map $r : \alpha X \times \mathbf{I} \to \alpha X \times \mathbf{I}$ as follows:

$$r(x, t) = \begin{cases} (x, t) & \text{if } t \geqslant k(x), \\ (x, k(x)) & \text{if } t < k(x). \end{cases}$$

Choose $0 < t_0 < t_1 < 1$ so that $\alpha X \times [t_0, 1] \subset q^{-1}(U)$. Let $\varphi : X \times \mathbf{I} \to X$ be a contraction with $\varphi_1(X) = \{x_0\}$. Define a map $f : r(\alpha X \times \mathbf{I}) \to X \times [0, 1)$

as follows:

$$f(x,t) = \begin{cases} (x_0, t_1) & \text{if } t_1 \leqslant t \leqslant 1, \\ \left(\varphi\left(x, \dfrac{t - t_0}{t_1 - t_0}\right), t\right) & \text{if } t_0 \leqslant t \leqslant t_1, \\ (x, t) & \text{if } 0 \leqslant t \leqslant t_0. \end{cases}$$

Then, the map $fr : \alpha X \times \mathbf{I} \to X \times [0, 1)$ induces the map $g : \alpha(X \times [0, 1)) \to X \times [0, 1)$, where $g|(X \setminus V) \times [0, t_0] = \mathrm{id}$ and

$$g(\alpha(X \times [0, 1)) \setminus ((X \setminus V) \times [0, t_0])) \subset U,$$

which implies that g is $\mathcal{U}$-close to id. Thus, it follows that $\{\infty\}$ is a Z-set in $\alpha(X \times [0, 1))$. □

We can now prove the Edwards Factor Theorem 3.8.1. Here, we also use Proposition 3.7.6.

Proof (Edwards Factor Theorem 3.8.1) First, note that if $X \times [0, 1) \times \boldsymbol{Q}$ ($\approx X \times (0, 1] \times \boldsymbol{Q}$) is a $\boldsymbol{Q}$-manifold, then so is $X \times \boldsymbol{Q}$ because

$$X \times \mathbf{I} \times \boldsymbol{Q} = (X \times [0, 1) \times \boldsymbol{Q}) \cup (X \times (0, 1] \times \boldsymbol{Q}).$$

When X is a compact AR, we may assume that X is a closed set in $\boldsymbol{Q}$. Then, there is a retraction $r : \boldsymbol{Q} \to X$ whose mapping cylinder M_r is an AR (Corollaries 1.13.17 and 1.13.3). We define

$$S = M_{r^*} = (\boldsymbol{Q} \times [-1, 0]) \cup M_r,$$

where $X \times \{0\} \subset \boldsymbol{Q} \times [-1, 0]$ is identified with $X = X \times \{1\} \subset M_r$. Then, S is an AR (1.13.6(5)). On the other hand, we have a cell-like map $f : \boldsymbol{Q} \times [-1, 1] \to S$ by Lemma 3.8.3. Let

$$T = (\boldsymbol{Q} \times [-1, 0]) \cup (X \times [0, 1]),$$

where T can be regarded as a subspace of S. By Lemma 3.8.4, we have a cell-like map $g : M_r \to X \times [0, 1]$ such that $g(x) = (x, 0)$ for each $x \in X$, which can be extended to the cell-like map $\tilde{g} : S \to T$ by $\tilde{g}|\boldsymbol{Q} \times [-1, 0] = \mathrm{id}$. Then, we have the cell-like map

$$h = \tilde{g}f \times \mathrm{id}_{\boldsymbol{Q}} : \boldsymbol{Q} \times [-1, 1] \times \boldsymbol{Q} \to T \times \boldsymbol{Q}.$$

Since $T \times \boldsymbol{Q}$ is a compact AR with the disjoint-cells property, we can apply Proposition 3.7.6 to obtain

$$T \times \boldsymbol{Q} \approx \boldsymbol{Q} \times [-1, 1] \times \boldsymbol{Q} \approx \boldsymbol{Q}.$$

Hence, $X \times (0, 1] \times \boldsymbol{Q}$ is a $\boldsymbol{Q}$-manifold as an open set in $T \times \boldsymbol{Q}$.

In the general case, it suffices to show the theorem for each component of X. Then, we may assume that X is connected, and hence separable. Due to Lemma 3.8.5, the one-point compactification $\alpha(X \times [0, 1))$ is a compact AR. Since $\alpha(X \times [0, 1)) \times \boldsymbol{Q}$ is a $\boldsymbol{Q}$-manifold, its open set $X \times [0, 1) \times \boldsymbol{Q}$ is also a $\boldsymbol{Q}$-manifold. □

By combining the Edwards Factor Theorem 3.8.1 and Proposition 3.7.6, we can establish the following characterization of $\boldsymbol{Q}$-manifolds:

Theorem 3.8.6 (CHARACTERIZATION OF $\boldsymbol{Q}$-MANIFOLDS) *For a separable locally compact ANR X, the following are equivalent:*

(a) *X is a $\boldsymbol{Q}$-manifold;*
(b) *X has the disjoint cells property;*
(c) *X is $\mathfrak{M}_0$-universal.*

Proof Since the equivalence (b) $\Leftrightarrow$ (c) was proved in Theorem 3.2.9 and the implication (a) $\Rightarrow$ (c) follows from the Strong Universality of $\boldsymbol{Q}$-manifolds (Theorem 2.10.10 or 2.3.16), it suffices to prove the implication (b) $\Rightarrow$ (a). By the Edwards Factor Theorem 3.8.1 above, $X \times \boldsymbol{Q}$ is a $\boldsymbol{Q}$-manifold. Since the projection $\mathrm{pr}_X : X \times \boldsymbol{Q} \to X$ is a cell-like map, the result follows from Proposition 3.7.6. □

Since every compact contractible $\boldsymbol{Q}$-manifold M is homeomorphic to $\boldsymbol{Q}$ (Theorem 3.7.9), we have the following characterization of the Hilbert cube:

Theorem 3.8.7 (CHARACTERIZATION OF $\boldsymbol{Q}$) *A space X is homeomorphic to the Hilbert cube if and only if X is a compact AR with the disjoint cells property.* □

To establish the Classification Theorem for $\boldsymbol{Q}$-manifolds, we introduce the notion of CE equivalence, which is a generalization of simple homotopy equivalence. It is said that ANRs X and Y are **CE equivalent** if there exist cell-like maps $f : Z \to X$ and $g : Z \to Y$ of an ANR Z.

Theorem 3.8.8 (CLASSIFICATION) *Two $\boldsymbol{Q}$-manifolds M and N are homeomorphic if and only if they are CE equivalent.*

Proof The "only if" part is trivial. To see the "if" part, let $f : Z \to M$ and $g : Z \to N$ be cell-like maps of an ANR Z. Since f (or g) is proper, Z is locally compact. By the Edwards Factor Theorem 3.8.1, $Z \times \boldsymbol{Q}$ is a $\boldsymbol{Q}$-manifold. Then, $f\mathrm{pr}_Z : Z \times \boldsymbol{Q} \to M$ is a cell-like map, hence it is a near-homeomorphism by Theorem 3.7.8. Similarly, $g\mathrm{pr}_Z : Z \times \boldsymbol{Q} \to N$ is also a near-homeomorphism. Consequently, $M \approx Z \times \boldsymbol{Q} \approx N$. □

The Triangulation Theorem for $\boldsymbol{Q}$-manifolds will be proved in Chap. 4, where not only some more results on $\boldsymbol{Q}$-manifolds but also some results on Algebraic Topology are required. However, the Triangulation Theorem for $\mathbb{R}_+$-stable $\boldsymbol{Q}$-manifolds can be easily obtained:

Theorem 3.8.9 (Triangulation) *For each $\mathbb{R}_+$-stable $\boldsymbol{Q}$-manifold M, there exists a locally finite simplicial complex K such that $M \approx |K| \times \boldsymbol{Q}$.*

Proof Due to Theorem 1.13.24, M has the proper homotopy type of a locally finite simplicial complex K. Then, $|K| \times \boldsymbol{Q}$ is a $\boldsymbol{Q}$-manifold by the Edwards Factor Theorem 3.8.1 and it is also $\mathbb{R}_+$-stable by Theorem 2.11.13. Hence, $M \approx |K| \times \boldsymbol{Q}$ by the Classification Theorem 2.11.11 for $\mathbb{R}_+$-stable $\boldsymbol{Q}$-manifolds: □

As a corollary, we have the following:

Corollary 3.8.10 *For each $\boldsymbol{Q}$-manifold M, there exists a locally finite simplicial complex K such that $M \times \mathbb{R}_+ \approx |K| \times \boldsymbol{Q}$.* □

3.9 Keller's Theorem and the Countable Product of ARs

In this section, we apply characterization of $\boldsymbol{Q}$ to prove Keller's Theorem on infinite-dimensional compact convex sets and the compact version of Theorem 3.5.11. We start to show the latter:

Theorem 3.9.1 *The countable product $X = \prod_{i \in \mathbb{N}} X_i$ of non-degenerate spaces is homeomorphic to the Hilbert cube if and only if all X_i, $i \in \mathbb{N}$, are compact ARs.*

Proof The "only if" part is trivial. To prove the "if" part, since X is a compact AR, it suffices to show that X has the disjoint cells property.

For each $i \in \mathbb{N}$, let $d_i \in \operatorname{Metr}(X_i)$. We define a metric d for X as follows:

$$d(x, y) = \sum_{i \in \mathbb{N}} \min\left\{2^{-i}, d_i(x(i), y(i))\right\}.$$

Let $f, g : \mathbf{I}^n \to X$ be maps. For each $\varepsilon > 0$, choose $i_0 \in \mathbb{N}$ so that $2^{-i_0} < \varepsilon$. Because X_{i_0} is non-degenerate, there are two distinct points $a, b \in X_{i_0}$. We can define maps $f', g' : \mathbf{I}^n \to X$ as follows:

$$\operatorname{pr}_i f'(x) = \begin{cases} a & \text{if } i = i_0, \\ \operatorname{pr}_i f(x) & \text{otherwise,} \end{cases}$$

$$\operatorname{pr}_i g'(x) = \begin{cases} b & \text{if } i = i_0, \\ \operatorname{pr}_i g(x) & \text{otherwise.} \end{cases}$$

Then, $f'(\mathbf{I}^n) \cap g'(\mathbf{I}^n) = \emptyset$, and f' and g' are ε-close to f and g, respectively. Hence, X has the disjoint cells property. □

Next, we prove the following Generalized Keller's Theorem:

Theorem 3.9.2 (KELLER–DOBROWOLSKI–TORUŃCZYK) *A convex set C in a metrizable topological linear space E is homeomorphic to $\boldsymbol{Q}$ if and only if it is an infinite-dimensional compact AR.*

When $E = \ell^2$, the above theorem is called KELLER'S THEOREM. To prove the above theorem, we use the following lemma:

Lemma 3.9.3 *Let $E = (E, \|\cdot\|)$ be an F-normed linear space, C a convex set in E and P a compact polyhedron. For any $\varepsilon > 0$, each map $f : P \to C$ is ε-close to a map $g : P \to C$ such that $g(P)$ is contained in a (convex) linear cell in C.*

Proof Let $n = \dim P$ and take a triangulation K of P so that $\operatorname{diam} f(\sigma) < \varepsilon/(n+1)$ for every $\sigma \in K$. Since C is convex, we can define $g : P = |K| \to C$ on each $\sigma = \langle v_1, \ldots, v_k \rangle \in K$ as follows:

$$g(x) = \sum_{v \in K^{(0)}} t_i f(v_i) \ \text{ if } x = \sum_{i=1}^{k} t_i v_i,\ t_1, \ldots, t_k \in \mathbf{I} \text{ with } \sum_{i=1}^{k} t_i = 1.$$

Then, it follows that $g(P) \subset \langle f(K^{(0)}) \rangle$ and

$$\begin{aligned}\|f(x) - g(x)\| &\leqslant \sum_{i=1}^{k} \|t_i(f(x) - f(v_i))\| \leqslant \sum_{i=1}^{k} \|f(x) - f(v_i)\| \\ &\leqslant (n+1) \operatorname{diam} f(\sigma) < \varepsilon,\end{aligned}$$

that is, f is ε-close to g. □

Proof (Theorem 3.9.2) Due to the Characterization of $\boldsymbol{Q}$ (Theorem 3.8.7), it suffices to show that C has the disjoint cells property. By Theorem 1.4.15, E has an admissible F-norm $\|\cdot\|$. Let $f, g : \mathbf{I}^n \to C$ be maps and $\mathcal{U} \in \operatorname{cov}(C)$ and $\varepsilon > 0$. By Lemma 3.9.3, we may assume that $f(\mathbf{I}^n) \cup g(\mathbf{I}^n)$ is contained in a cell D in C. Since C is infinite-dimensional and D is finite-dimensional, we have $v \in C \setminus \operatorname{fl} D$. Because $C - C$ is compact, we can find $\delta > 0$ so that $\delta(C - C) \subset \mathrm{B}(\mathbf{0}, \varepsilon)$. We define a map $f' : \mathbf{I}^n \to C$ as follows:

$$f'(x) = (1 - \delta) f(x) + \delta v \quad \text{for each } x \in \mathbf{I}^n.$$

Then, $f(x) - f'(x) = \delta(f(x) - v) \in \mathrm{B}(\mathbf{0}, \varepsilon)$, which means that $\|f(x) - f'(x)\| < \varepsilon$. Namely, f' is ε-close to f. Let $g' = g : \mathbf{I}^n \to C$. Since $g'(\mathbf{I}^n) \subset D$, it follows that $f(\mathbf{I}^n) \cap g'(\mathbf{I}^n) = \emptyset$. □

In the above proof, the proof of the disjoint cells property of C can be performed without the compactness of C. Then, we have also the following:

Theorem 3.9.4 *A convex set in a metrizable topological linear space is a $\boldsymbol{Q}$-manifold if and only if it is an infinite-dimensional locally compact ANR (which implies an AR).* □

The following classification of infinite-dimensional locally compact closed convex sets was proved by V. Klee [95]:

Theorem 3.9.5 (KLEE) *An infinite-dimensional locally compact closed convex set in a normed linear space is homeomorphic to one of* $\boldsymbol{Q}$, $\boldsymbol{Q} \times \mathbb{R}_+$ *or* $\boldsymbol{Q} \times \mathbb{R}^n$, $n \in \mathbb{N}$.

Here, we do not give a proof of this theorem. For details, one can refer to the book of Bessaga and Pełczyński [(1)], Chap. III, §7.

The Toruńczyk characterization of $\boldsymbol{Q}$ can also be applied to Hyperspace Theory. For a metric space $X = (X, d)$, the hyperspace of all non-empty compacta in X is denoted by $\mathrm{Comp}(X)$, which has the **Hausdorff metric** d_H defined as follows:[15]

$$d_H(A, B) = \max\left\{ \sup_{x\in A} d(x, B),\ \sup_{y\in B} d(y, A)\right\}$$
$$= \inf\left\{r > 0 \,\middle|\, A \subset \mathrm{N}_d(B, r),\ B \subset \mathrm{N}_d(A, r)\right\}.^{16}$$

Let $\mathrm{Cont}(X)$ be the subspace of $\mathrm{Comp}(X)$ consisting of all continua in X. It is well known that $\mathrm{Cont}(\mathbf{I}) \approx \mathbf{B}^2$.

In fact, $\mathrm{Cont}(\mathbf{I})$ consists of all closed subintervals of $\mathbf{I}$, that is,

$$\mathrm{Cont}(\mathbf{I}) = \left\{[a, b] \,\middle|\, 0 \leqslant a \leqslant b \leqslant 1\right\}.$$

We can define $\varphi : \mathrm{Cont}(\mathbf{I}) \to \mathbb{R}^2$ as follows:

$$\varphi([a, b]) = ((a + b)/2, (b - a)/2) \in \mathbb{R}^2 \text{ each } [a, b] \in \mathrm{Cont}(\mathbf{I}).$$

As is easily observed, φ is a continuous injection. Because of the compactness of $\mathrm{Cont}(\mathbf{I})$, φ is an embedding. Since $\varphi(\mathrm{Cont}(\mathbf{I}))$ is a triangle with vertices $\mathbf{0} = (0, 0)$, $(1/2, 1/2)$, $\mathbf{e}_1 = (1, 0)$ in the plane $\mathbb{R}^2$, it follows that $\mathrm{Cont}(\mathbf{I}) \approx \mathbf{B}^2$.

An arc A in X (i.e., $A \approx \mathbf{I}$) is called a **free arc** in X if $\operatorname{int} A = A \setminus \partial A$, where $(A, \partial A) \approx (\mathbf{I}, \partial\mathbf{I})$ $(= (\mathbf{I}, \{0, 1\}))$. If X has a free arc, then $\mathrm{Cont}(X)$ contains a two-dimensional open set. The following CURTIS–SCHORI–WEST HYPERSPACE THEOREM is a celebrated application, which was conjectured by M. Wojdysławski [158] in 1938. In [159], Wojdysławski proved that $\mathrm{Comp}(X)$ and $\mathrm{Cont}(X)$ are ARs for every Peano continuum X, whose shorter proof was given by J.L. Kelley [93].

Theorem 3.9.6 (CURTIS–SCHORI–WEST) *For every non-degenerate metric Peano continuum* X, $\mathrm{Comp}(X)$ *is homeomorphic to* $\boldsymbol{Q}$. *Additionally, if* X *has no free arcs, then* $\mathrm{Cont}(X)$ *is also homeomorphic to* $\boldsymbol{Q}$.

Here, we do not give the proof of the above theorem (Appendix of [143]). For the hyperspace $\mathrm{Comp}(X)$, one can refer to the book of Jan van Mill [(3)].

---

[15]When d is bounded (i.e., $\mathrm{diam}_d\, X < \infty$), this definition can be acceptable for the space $\mathrm{Cld}(X)$ of all non-empty closed sets in X.

[16]Recall that $d(x, B) = \inf_{y\in B} d(x, y)$ and $\mathrm{N}_d(B, r) = \{x \in X \mid d(x, B) < r\}$.

Notes for Chapter 3

In this chapter, we concentrated our attention on Toruńczyk's characterizations of Q-manifolds and $\ell_2(\Gamma)$-manifolds. The main results were established in [143], [145], and [146]. For survey, refer to [144], [102, p. 250].

The disjoint n-cells property was introduced by Toruńczyk to characterize Q-manifolds in [143]. In [37], J.W. Cannon independently introduced the disjoint disk property to distinguish topological manifolds from homological manifolds, which was generalized by R.D. Edwards. The discrete n-cells property was introduced by Toruńczyk to characterize $\ell^2(\Gamma)$-manifolds in [145].

Proposition 3.2.3 was pointed out by Koshino during the writing of the next chapter. Proposition 3.2.12 was proved by Bestvina and Mogilski [29] and its non-separable version, Proposition 3.2.11, is due to Sakai and Yaguchi [129].

The characterizing conditions of Hilbert manifolds given in [145] (cf. [146]) are little modified in Theorem 3.4.2. Theorem 3.4.7 was given in [145] but, as its non-separable version, Theorem 3.4.3 is formulated and proved in the present book. In [30], an alternative proof for the characterization separable Hilbert manifolds was given.

Lemma 3.5.1 is due to Bessaga, Dobrowolski, and Toruńczyk [55], which is modified to Lemma 3.5.2 by Sakai [126]. M.I. Kadec [89, 90] proved that all separable infinite-dimensional Banach spaces are homeomorphic. R.D. Anderson [4, 5] proved that $\mathbb{R}^{\mathbb{N}}$ is homeomorphic to Hilbert space ℓ^2. As a combination of these two results and earlier results obtained by Bessaga and Pełczyński in [26] and [27], the Kadec–Anderson Theorem 3.5.6 was established. T. Dobrowolski and H. Toruńczyk generalized this result to Theorem 3.5.5 in [54] and later to Theorem 3.5.7 in [55]. The non-separable version of the Kadec–Anderson Theorems 3.5.6 and 3.5.17, was obtained by Toruńczyk in [145].

Theorem 3.6.5 was proved in a more general formulation in [126]. In [142], H. Toruńczyk proved the same result when X has an open set U such that bd U is compact and collared in cl U (cf. Geoghegan's result in [68]).

The characterization of compact Q-manifolds is also treated in the book [(3)] of Jan van Mill, where its alternative proof is given. However, we do not assume the compactness in this chapter. It is not trivial to derive the non-compact version from the characterization of compact Q-manifolds. In [148], J.J. Walsh also gave another alternative proof, where Miller's theorem was used. Our approach is a combination of these two. In [59], one more proof due to R.D. Edwards can be found.

The Mapping Cylinder Theorem 3.7.3 was proved by West [151]. Fathi, Marin, and Visetti [62] established Lemma 3.7.2, which produces the Mapping Cylinder Theorem 3.7.3.

Theorem 3.8.1 is established by R.D. Edwards but his own proof was not published; the first published proof is in Chapman's Lecture Notes [(2)]. The proof of its compact case is in the book of van Mill [(3)], where Lemmas 3.8.3 and 3.8.4 are key lemmas. The first part of Lemma 3.8.5 was first proved by Fathi and Visetti [61] in the case where X is an AR, but the proof presented here is due to Sakai and Wong [128]. In the proof of Theorem 3.8.1, the reduction of the locally compact case to the compact case is due to Fathi and Visetti [61].

Theorem 3.9.1 was established by J.E. West [149] when each X_i is a non-degenerate Q-factor, i.e., $X_i \times Q \approx Q$ (cf. Theorem 3.8.1). Theorem 3.9.2 was originally proved by O.H. Keller [92] when a topological linear space is completely metrizable and locally convex (i.e., a Fréchet space) and generalized to the present form by T. Dobrowolski and H. Toruńczyk [54]. The proof presented here is due to [54] but there is a gap in the proof of Lemma 3 in [54]. Theorem 3.9.6 was finally proved by D.W. Curtis and R.M. Schori [52] but in advance, R.M. Schori and J.E. West [132] had first proved that Comp($\mathbf{I}$) $\approx Q$.

Chapter 4
Triangulation of Hilbert Cube Manifolds and Related Topics

In this chapter, we focus on proving the Triangulation Theorem for $\boldsymbol{Q}$-manifolds. In particular, for each compact $\boldsymbol{Q}$-manifold M, we have to find a finite simplicial complex K such that $|K| \times \boldsymbol{Q} \approx M$. This is bound up with the Borsuk conjecture asserting that every compact ANR has the homotopy type of a finite simplicial complex. It is related to the homotopy type problem of whether a space has the homotopy type of a finite simplicial complex if it is homotopy dominated by a finite simplicial complex, where the first simplicial complex might be different from the second (cf. the Whitehead–Milnor Theorem 1.10.4 and Remark 1.9 on p. 57). To prove the Triangulation Theorem for $\boldsymbol{Q}$-manifolds, we will use Wall's work on the homotopy type of a finite simplicial complex (Theorem 4.1.6) and a result on simple homotopy equivalences (Theorem 4.1.5). Combining the Triangulation Theorem for compact $\boldsymbol{Q}$-manifolds with the Edwards Factor Theorem 3.8.1, we know that the Borsuk conjecture is true. We will also prove the topological invariance of Whitehead torsion, which had been a longstanding problem.

As is mentioned in Preface, this chapter requires further results other than results contained in [GAGT] and added in Chap. 1. In Sect. 4.1, we give a brief introduction of the simple homotopy type and write down some results from Simple Homotopy Theory without proofs, which will be used in Sects. 4.4 and 4.6. Concerning this theory, Cohen's excellent textbook [(20)] is our source of information. At the end of Sect. 4.1, we write down a result from Wall's Obstruction Theory without proof, which will be used in Sect. 4.6. Since the covering space arguments will be applied to $\boldsymbol{Q}$-manifolds in Sect. 4.6, we add Sect. 4.2 concerning covering spaces, which is not contained in [GAGT]. Additionally, we give in Sect. 4.2 some elementary results in Algebraic Topology (the homotopy and homology groups), which will be used in Sect. 4.3. The contents in Sect. 4.2 are contained in any standard textbook on Algebraic Topology.

K. Sakai, *Topology of Infinite-Dimensional Manifolds*, Springer Monographs in Mathematics, https://doi.org/10.1007/978-981-15-7575-4_4

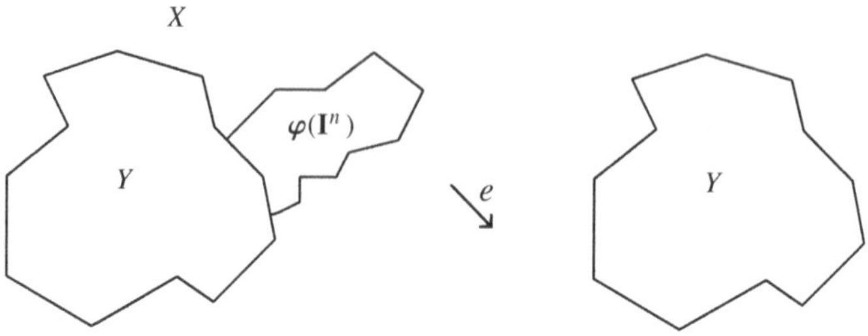

Fig. 4.1 An elementary collapse

4.1 Simple Homotopy Equivalences

Let X and Y be compact polyhedra.[1] It is said that X **collapses** to Y by an **elementary collapse**, written $X \searrow^e Y$, if Y is a subpolyhedron of X and there is a PL embedding $\varphi : \mathbf{I}^n \to X$ for some $n \in \mathbb{N}$ such that $X = Y \cup \varphi(\mathbf{I}^n)$ and $\varphi(\mathbf{I}^n) \cap Y = \varphi(\mathbf{I}^{n-1} \times \{0\})$ (see Fig. 4.1), where it should be noted that $\mathrm{bd}_X\, \varphi(\mathbf{I}^n) = \varphi(\mathbf{I}^{n-1} \times \{0\})$. Then, a strong deformation retraction $r_\varphi : X \to Y$ can be defined as follows: $r_\varphi|Y = \mathrm{id}$ and

$$r_\varphi(\varphi(x)) = \varphi(x(1), \ldots, x(n-1), 0) \quad \text{for each } x \in \mathbf{I}^n.$$

Note that $r_\varphi^{-1}(y) = \{y\}$ or $\approx \mathbf{I}$ for each $y \in Y$. This PL map r_φ is called an **elementary collapse**. When $X \searrow^e Y$, we also say that Y **expands** to X by an **elementary expansion** and write $Y \,{}^e\!\nearrow X$, where the inclusion $Y \subset X$ is a homotopy equivalence and called an **elementary expansion**.

We say that X **collapses** to Y and write $X \searrow Y$ if there is a finite sequence of elementary collapses

$$X = X_0 \searrow^e X_1 \searrow^e \cdots \searrow^e X_n = Y,$$

which also means that Y **expands** to X, written $Y \nearrow X$. When Y is a singleton, X is said to be **collapsible**, written $X \searrow 0$. Note that if X is collapsible, then X is contractible.

It is said that X and Y have **the same simple homotopy type** or X has **the simple homotopy type** of Y (X is **simple homotopy equivalent** to Y) if there is a finite sequence of compact polyhedra X_i, $i = 0, 1, \ldots, n$, such that $X_0 = K$, $X_n = Y$, and $X_{n-1} \searrow X_n$ or $X_{n-1} \nearrow X_n$. A map $f : X \to Y$ is called a **simple homotopy**

[1] More generally, X and Y might be CW complexes. In fact, the Simple Homotopy Theory has been developed for CW complexes. Refer to Cohen's book [(20)].

equivalence if f is homotopic to a finite composition

$$X = X_0 \xrightarrow{f_1} X_1 \xrightarrow{f_2} \cdots \longrightarrow X_{n-1} \xrightarrow{f_n} X_n = Y,$$

where each X_i is a compact polyhedron and each f_i is an elementary collapse or an elementary expansion. Evidently, if X and Y have the same simple homotopy type, then they must have the same homotopy type, and every simple homotopy equivalence is a homotopy equivalence. But, the converse is a problem.

Obviously, a compact polyhedron has the same simple homotopy type of a singleton if it is collapsible. However, the converse is wrong. To give such an example and to show that simple homotopy type is a useful concept, consider the two-dimensional polyhedron

$$\begin{aligned} H = \big(&\partial\mathbf{I}^3 \cup \big([0, 1/2] \times \partial([0, 1/3] \times [0, 1/2])\big) \\ &\cup \big(\{1/2\} \times \mathbf{I}^2\big) \cup \big([1/2, 1] \times \partial([2/3, 1] \times [0, 1/2])\big)\big) \\ &\setminus \big((\{0, 1/2\} \times (0, 1/3) \times (0, 1/2)) \cup (\{1/2, 1\} \times (2/3, 1) \times (0, 1/2))\big), \end{aligned}$$

that is a house with two rooms and two entrances illustrated in Fig. 4.2, where the entrance to the right-hand room is on the left and the entrance to the left-hand room is on the right. Then, H is not collapsible because $\mathrm{bd}_H\, \varphi(\mathbf{I}^2) \neq \varphi(\mathbf{I} \times \{0\})$ for any PL embedding $\varphi : \mathbf{I}^2 \to H$. However, $\mathbf{I}^3$ is collapsible and $\mathbf{I}^3 \searrow H$ (a rigorous proof is messy), equivalently $H \nearrow \mathbf{I}^3$, so H has the simple homotopy type of a singleton $\{\mathbf{0}\}$. In particular, $H \simeq \{\mathbf{0}\}$, that is, H is contractible. This example also cancels the apparent restrictiveness of the definition of simple homotopy type.

The following is evident from the definition:

Proposition 4.1.1 *Let $f : X \to Y$, $g : Y \to Z$, and $h : X \to Z$ be maps of compact polyhedra such that $h \simeq gf$. If any two of the maps f, g, and h are simple homotopy equivalences, then so is the third one.* □

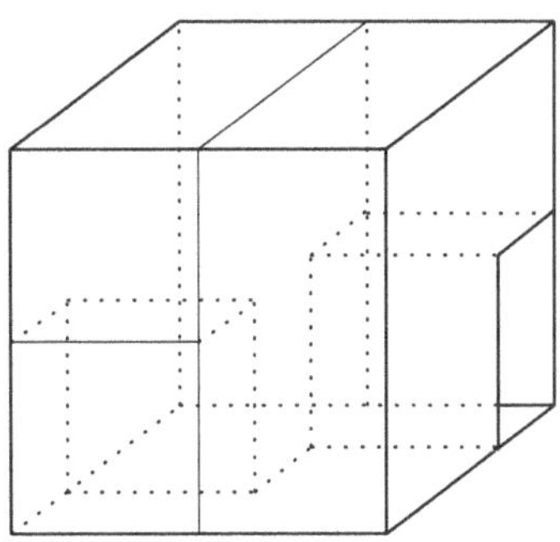

Fig. 4.2 A house with two rooms

Combining the Edwards Factor Theorem 3.8.1 with 3.7.8, we have the following:

Theorem 4.1.2 *Let $f : X \to Y$ be a simple homotopy equivalence between compact polyhedra. Then, $f \times \mathrm{id} : X \times \boldsymbol{Q} \to Y \times \boldsymbol{Q}$ is homotopic to a homeomorphism.*

Proof By the Edwards Factor Theorem 3.8.1, $X \times \boldsymbol{Q}$ and $Y \times \boldsymbol{Q}$ are $\boldsymbol{Q}$-manifolds. When f is an elementary collapse, since $f \times \mathrm{id} : X \times \boldsymbol{Q} \to Y \times \boldsymbol{Q}$ is cell-like, $f \times \mathrm{id}$ is a near-homeomorphism by Theorem 3.7.8. If f is an elementary expansion, then f is the inclusion and there is an elementary collapse $g : Y \to X$, where f is a homotopy inverse of g because g is a strong deformation retraction. Since $g \times \mathrm{id}$ is homotopic to a homeomorphism $h : Y \times \boldsymbol{Q} \to X \times \boldsymbol{Q}$, it follows that $f \times \mathrm{id}$ is homotopic to h^{-1}. Indeed,

$$f \times \mathrm{id} = h^{-1}h(f \times \mathrm{id}) \simeq h^{-1}(g \times \mathrm{id})(f \times \mathrm{id}) = h^{-1}.$$

The result easily follows from the definition of a simple homotopy. □

The converse of the above theorem is also valid, which will be proved in Sect. 4.6 (Theorem 4.6.5).

For each compact polyhedron X, an Abelian group $\mathrm{Wh}(X)$ called the **Whitehead group** of X can be defined, and for each homotopy equivalence $f : Y \to X$ from another compact polyhedron Y to X, an element $\tau(f) \in \mathrm{Wh}(X)$ named the **Whitehead torsion** of f can be defined so that

- f is a simple homotopy equivalence if and only if $\tau(f) = 0$.

Then, $\tau(f) = \tau(f')$ for any homotopy equivalence $f' : Y \to X$ homotopic to f. It is known that the Whitehead group $\mathrm{Wh}(X)$ is determined by the fundamental group $\pi_1(X)$. — Refer to §6 and §22 of Cohen's book [(20)].

The following TOPOLOGICAL INVARIANCE OF WHITEHEAD TORSION had been a longstanding problem raised by J.H.C. Whitehead [156] and was proved by T.A. Chapman [45] by using $\boldsymbol{Q}$-manifolds:

Theorem 4.1.3 (TOPOLOGICAL INVARIANCE OF WHITEHEAD TORSION) *Every homeomorphism $f : X \to Y$ between compact polyhedra is a simple homotopy equivalence, that is, $\tau(f) = 0$.*

In Theorem 4.6.5, using $\boldsymbol{Q}$-manifolds, we will characterize simple homotopy equivalences. Theorem 4.1.3 above can be obtained as a particular case. To prove Theorem 4.6.5, we will establish the Handle Straightening Theorem 4.6.2 and apply the following result on Whitehead torsion:

Theorem 4.1.4 *Let $f : X \to Y$ be a map between compact polyhedra. Suppose that $X = X_1 \cup X_2$, $Y = Y_1 \cup Y_2$, each X_i and Y_i are subpolyhedra of X and Y, respectively, such that $f(X_i) \subset Y_i$, and let $X_0 = X_1 \cap X_2$ and $Y_0 = Y_1 \cap Y_2$. If the restrictions $f|X_i : X_i \to Y_i$, $i = 0, 1, 2$, are simple homotopy equivalences (i.e.,*

$\tau(f|X_i) = 0$ *for each* $i = 0, 1, 2$*), then* $f : X \to Y$ *is also a simple homotopy equivalence (i.e.,* $\tau(f) = 0$*).*[2]

The Handle Straightening Theorem 4.6.2 also will play a very important role in our proof of the Triangulation Theorem 4.7.3 for compact $\boldsymbol{Q}$-manifolds. To prove the Handle Straightening Theorem 4.6.2, we will establish the Splitting Theorem 4.4.1. In order to prove these theorems, we need the following two theorems, whose proofs cannot be given in this book because they are heavily reliant on algebra. The first is the following theorem concerning the Whitehead group:

Theorem 4.1.5 *For a compact connected polyhedron* X*, if* $\pi_1(X)$ *is free or free Abelian, then* $\mathrm{Wh}(X) = 0$*, hence every homotopy equivalence* $f : Y \to X$ *of another compact connected polyhedron* Y *to* X *is a simple homotopy equivalence.*[3]

The second is related to Wall's finiteness obstruction:

Theorem 4.1.6 *Let* X *be a compact connected space that is homotopy dominated by a finite simplicial complex. If* $\pi_1(X)$ *is free or free Abelian, then* X *has the homotopy type of a finite simplicial complex.*[4]

The following theorem will play an important role in our proof of the Triangulation Theorem for $\boldsymbol{Q}$-manifolds, which is a combination of Theorems 4.1.2 and 4.1.5:

Theorem 4.1.7 *Let* X *and* Y *be compact connected polyhedra with the same homotopy type such that the fundamental group of* X *(and* Y*) is free or free Abelian. Then, every homotopy equivalence* $f : X \times \boldsymbol{Q} \to Y \times \boldsymbol{Q}$ *is homotopic to a homeomorphism.*

Proof We can easily obtain a homotopy equivalence $f_0 : X \to Y$ such that $f \simeq f_0 \times \mathrm{id}_{\boldsymbol{Q}}$. Due to Theorem 4.1.5, f_0 is a simple homotopy equivalence. Then, we have the result by Theorem 4.1.2. □

In Remark 1.9 of Sect. 1.10, p. 57, we mentioned the question of whether a compact space homotopy dominated by a simplicial complex has the homotopy type of a finite simplicial complex. Note that if a compact space is homotopy dominated by a simplicial complex K, then it is homotopy dominated by a finite subcomplex of K (cf. Remark 1.12, p. 70). In [147], C.T.C. Wall answered a more general question of whether a space homotopy dominated by a finite simplicial complex has

---

[2]This is a direct consequence of the Sum Theorem (23.1) in Cohen's book [(20)]. As mentioned in Chapman's lecture notes [(2)], p. 56, there is a geometric proof which is implicit in Section 6 in Cohen's book [(20)].

[3]In Cohen's book [(20)], it is shown in (21.1) that $\mathrm{Wh}(X)$ is equal to the Whitehead group $\mathrm{Wh}(\pi_1(X))$ for the group $\pi_1(X)$, which reduces the free Abelian case to (11.3) and the free case to a combination of (11.2) and (11.6), but the proofs of (11.3) and (11.6) are not given because of difficulty, where two papers [25] and [134] are cited for (11.3) and (11.6), respectively. The proofs of $\mathrm{Wh}(\{0\}) = 0$ (11.1) and $\mathrm{Wh}(\mathbb{Z}) = 0$ (11.2) are given in the book [(20)].

[4]This theorem will be used in the proof of the Splitting Theorem 4.4.1, which is the key to proving the Handle Straightening Theorem 4.6.2.

the homotopy type of a finite simplicial complex.[5] In fact, let X be a connected space homotopy dominated by a finite simplicial complex. Then, he defined an invariant $[X]$ in an Abelian group $\tilde{K}_0(\mathbb{Z}\pi_1(X))$ determined by the fundamental group $\pi_1(X)$ such that $[X] = 0$ if and only if X has the homotopy type of a finite simplicial complex. Thus, Theorem 4.1.6 follows from the algebraic result that $\tilde{K}_0(\mathbb{Z}\pi_1(X)) = 0$ if $\pi_1(X)$ is free or free Abelian.

As is the case with the topological invariance of Whitehead torsion, the proof of the Borsuk conjecture also depends on the Triangulation Theorem for compact $\boldsymbol{Q}$-manifolds (Theorem 4.7.3). Thus, to prove both the Borsuk conjecture and the topological invariance of Whitehead torsion, we need the above algebraic results. So, the author is awaiting geometric proofs of Theorems 4.1.5 and 4.1.6. By the way, S. Ferry gave in [65] a geometrical approach to Wall's finiteness obstruction as an analogy of Cohen's approach to the Whitehead group, where an invariant $[X]$ can be realized as an element of $\mathrm{Wh}(X \times \mathbf{S}^1)$.

Note For finite simplicial complexes K and L, it is said that K **collapses** to L by an **elementary simplicial collapse**, written $K \searrow^e L$, if L is a subcomplex of K and $K \setminus L = \{\sigma, v\sigma\}$, where $v\sigma$ is the join of a vertex $v \in L^{(0)}$ and a simplex $\sigma \in K \setminus L$. When $K \searrow^e L$, it is also said that L **expands** to L by an **elementary simplicial expansion**, written $L \,{}^e\!\nearrow K$. Obviously, $K \searrow^e L$ implies $|K| \searrow^e |L|$ ($L \,{}^e\!\nearrow K$ implies $|L| \,{}^e\!\nearrow |K|$), where the converse does not hold. Thus, $K \searrow^e L$ and $|K| \searrow^e |L|$ ($L \,{}^e\!\nearrow K$ and $|L| \,{}^e\!\nearrow |K|$) have different meanings even if the same notation $\searrow^e$ (${}^e\!\nearrow$) is used. It is said that K **simplicially collapses** to L (or L **simplicially expands** to K), written $K \searrow L$ (or $L \nearrow K$), if there is a finite sequence of elementary simplicial collapses

$$K = K_0 \searrow^e K_1 \searrow^e \cdots \searrow^e K_n = L.$$

It is said that K and L have the **same simple-homotopy type** or K has **the simple-homotopy type** of L (K is **simple-homotopy equivalent** to L) if there is a finite sequence of simplicial complexes K_i, $i = 0, 1, \ldots, n$, such that $K_0 = K$, $K_n = L$, and $K_{n-1} \searrow K_n$ or $K_{n-1} \nearrow K_n$.

It can be easily seen (but a proof is bothersome) that, if $X \searrow^e Y$, then there is a simplicial complex K with a subcomplex $L \subset K$ such that $X = |K|$, $Y = |L|$ and $K \searrow L$. Thus, it follows that $|K| \searrow |L|$ if and only if there are simplicial subdivisions $K' \lhd K$ and $L' \lhd L$ such that $K' \searrow L'$. Consequently, we have the following:

- *For compact polyhedra X and Y, $X \searrow Y$ if and only if $K \searrow L$ for some triangulations K and L of X and Y, respectively.*

[5] Refer to Sect. 1 of [111]: "Historical background."

4.2 Covering Spaces and Algebraic Preliminaries

As observed in the previous section, the fundamental group is important in Simple Homotopy Theory and Wall Obstruction Theory. In calculation of this group, the covering space is indispensable.

Let $p : E \to B$ be a surjective map. An open set U in B is said to be **evenly covered** by p if $p^{-1}(U)$ is a disjoint union of open sets, each of which is mapped by p homeomorphically onto U. When every point $x \in B$ has an open neighborhood that is evenly covered by p, E is called a **covering space** of B and p is called a **covering projection (map)**.[6] The most typical example is the **exponential map** $e : \mathbb{R} \to \mathbf{S}^1$ defined as follows:

$$e(x) = (\cos \pi x, \sin \pi x).$$

When every $(x, y) \in \mathbb{R}^2$ is identified with $x + iy \in \mathbb{C}$ (i.e., we regard $\mathbb{R}^2 = \mathbb{C}$), we can write $\mathbf{S}^1 = \{z \in \mathbb{C} \mid |z| = 1\}$ and

$$e(t) = \exp(\pi i t) = \cos \pi t + i \sin \pi t \quad \text{for every } t \in \mathbb{R}.$$

One can picture e as a function wrapping the real line $\mathbb{R}$ around the circle $\mathbf{S}^1$, where each interval $[2n - 1, 2n + 1]$ makes a circuit of $\mathbf{S}^1$.[7] The ***n*-torus (*n*-dimensional torus)**[8] $\mathbf{T}^n$ is defined as the n-hold product of the unit sphere $\mathbf{S}^1$, that is,

$$\mathbf{T}^n = (\mathbf{S}^1)^n = \underbrace{\mathbf{S}^1 \times \cdots \times \mathbf{S}^1}_{n \text{ many}}.$$

Then, $\mathbb{R}^n$ is a covering space of $\mathbf{T}^n$ with the covering projection $e^n : \mathbb{R}^n \to \mathbf{T}^n$ defined by

$$e^n(x) = (e(x(1)), \ldots, e(x(n))) \quad \text{for each } x \in \mathbb{R}^n.$$

This can be shown by using the following fact inductively:

Fact *For covering projections $p : E \to B$ and $p' : E' \to B'$, $p \times p' : E \times E' \to B \times B'$ is also a covering projection.*

---

[6]For a covering projection, the book [GAGT] gives its definition (p. 498) but not any other information. One can refer to Sect. 53 of Munkres' book [[7]].

[7]It is easy to check details (cf. Theorem 53.1 of Munkres' book [[7]]).

[8]The plural is **tori**.

Let $p : E \to B$ be a map. For a map $f : X \to B$, a **lifting** of f is a map $\tilde{f} : X \to E$ such that $p\tilde{f} = f$, that is, the following diagram commutes:

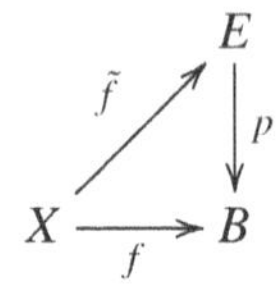

Then, it is said that f is **lifted** to $\tilde{f}$. Every covering projection has the **unique path lifting property**, that is, the following lemma holds:

Lemma 4.2.1 (Unique Path Lifting) *Let $p : E \to B$ be a covering projection with $a_0 \in p^{-1}(b_0)$. Every path $f : \mathbf{I} \to B$ with $f(0) = b_0$ is uniquely lifted to a path $\tilde{f} : \mathbf{I} \to E$ with $\tilde{f}(x_0) = a_0$.*

Proof Let $\mathcal{U}$ be an open cover of B such that each open set in $\mathcal{U}$ is evenly covered by p. Because of the compactness of $\mathbf{I}$, there is a sequence $s_0 = 0 < s_1 < \cdots < s_n = 1$ such that each $[s_{i-1}, s_i]$ is contained in $f^{-1}(U)$ for some $U \in \mathcal{U}$.

When $\tilde{f}(s_{i-1})$ has been defined, we can take $U \in \mathcal{U}$ and an open set V in E such that $f([s_{i-1}, s_i]) \subset U$, $\tilde{f}(s_{i-1}) \in V$, and $p|V : V \to U$ is a homeomorphism. If f' is a lifting of $f|[s_{i-1}, s_i]$ with $f'(s_{i-1}) = \tilde{f}(s_{i-1})$, then $f'([s_{i-1}, s_i]) \subset V$ because of connectedness. Hence, we have $f' = (p|V)^{-1} f|[s_{i-1}, s_i]$, which means that $\tilde{f}|[s_{i-1}, s_i]$ is uniquely defined by $\tilde{f}|[s_{i-1}, s_i] = (p|V)^{-1} f|[s_{i-1}, s_i]$. So, first define $\tilde{f}(0) = a_0$ and iterate this process n times. Then, we can define $\tilde{f}$ on all of $\mathbf{I}$ and it is unique. □

Lemma 4.2.2 *Let $p : E \to B$ be a covering projection with $a_0 \in p^{-1}(b_0)$. Every map $f : \mathbf{I}^2 \to B$ with $f(0, 0) = b_0$ is uniquely lifted to a map $\tilde{f} : \mathbf{I}^2 \to E$ with $\tilde{f}(0, 0) = a_0$.*

Proof As in the proof of the previous lemma, take $\mathcal{U} \in \operatorname{cov}(B)$ so that each open set in $\mathcal{U}$ is evenly covered by p. Because of the compactness of $\mathbf{I}^2$, there are sequences $s_0 = 0 < s_1 < \cdots < s_m = 1$ and $t_0 = 0 < t_1 < \cdots < t_n = 1$ such that each $D_{i,j} = [s_{i-1}, s_i] \times [t_{j-1}, t_j]$ is contained in $f^{-1}(U)$ for some $U \in \mathcal{U}$.

When $\tilde{f}(s_{i-1}, t_{j-1})$ has been defined, we can take $U \in \mathcal{U}$ and an open set V in E such that $f(D_{i,j}) \subset U$, $\tilde{f}(s_{i-1}, t_{j-1}) \in V$, and $p|V : V \to U$ is a homeomorphism. If f' is a lifting of $f|D_{i,j}$ with $f'(s_{i-1}, t_{j-1}) = \tilde{f}(s_{i-1}, t_{j-1})$, then $f'(D_{i,j}) \subset V$ because of connectedness. Hence, we have $f' = (p|V)^{-1} f|D_{i,j}$. This means that $\tilde{f}|D_{i,j}$ is uniquely defined by $\tilde{f}|D_{i,j} = (p|V)^{-1} f|D_{i,j}$, where $\tilde{f}|\{s_{i-1}\} \times [t_{j-1}, t_i]$ and $\tilde{f}|[s_{i-1}, s_i] \times \{t_{j-1}\}$ are also uniquely determined. So, first define $\tilde{f}(0, 0) = a_0$ and iterate this process mn times in the following order:

$$(i', j') < (i, j) \underset{\text{def.}}{\Longleftrightarrow} \begin{cases} j' < j, \\ j' = j,\ i' < i. \end{cases}$$

Then, we can define $\tilde{f}$ on all of $\mathbf{I}$ and it is unique. □

Proposition 4.2.3 *Let $p : E \to B$ be a covering projection and $\alpha, \beta : \mathbf{I} \to E$ be paths from a_0 to a_1 in E. Then, $p\alpha \simeq p\beta$ rel. $\partial\mathbf{I}$ implies $\alpha \simeq \beta$ rel. $\partial\mathbf{I}$.*

Proof Let $h : \mathbf{I} \times \mathbf{I} \to B$ be a homotopy with $h_t(0) = p(a_0)$ and $h_t(1) = p(a_1)$ for each $t \in \mathbf{I}$, $h_0 = p\alpha$, and $h_1 = p\beta$. By Lemma 4.2.2, h is lifted to a homotopy $\tilde{h} : \mathbf{I} \times \mathbf{I} \to E$ such that $\tilde{h}(0, 0) = a_0$. From the uniqueness of liftings of $h|\{0\} \times \mathbf{I}$ and $h|\{1\} \times \mathbf{I}$, h_0 and h_1, it follows that $\tilde{h}(\{0\} \times \mathbf{I}) = \{a_0\}$ and $\tilde{h}(\{1\} \times \mathbf{I}) = \{a_1\}$. $\tilde{h}_0 = \alpha$, and $\tilde{h}_1 = \beta$. Hence, $\alpha \simeq \beta$ rel. $\partial\mathbf{I}$. □

Recall that a **loop** in X (at x_0) is a path $\alpha : \mathbf{I} \to X$ such that $\alpha(0) = \alpha(1)$ $(= x_0)$. For the fundamental groups, we use the same notation as Sect. 4.14 of [GAGT].

Proposition 4.2.4 *For a covering projection $p : E \to B$ with $a_0 \in p^{-1}(b_0)$, the following hold:*

(1) *The induced homomorphism $p_\sharp : \pi_1(E, a_0) \to \pi_1(B, b_0)$ is injective.*
(2) *A loop α in B at b_0 is lifted to a loop in E at a_0 if and only if $[\alpha] \in p_\sharp(\pi_1(E, a_0))$.*

Proof In Proposition 4.2.3, replacing paths by loops, we have (1). The "only if" part of (2) is trivial.

To see the "if" part of (2), assume that $[\alpha] \in p_\sharp(\pi_1(E, a_0))$, that is, there is a loop $\beta : \mathbf{I} \to E$ at a_0 such that $p\beta \simeq \alpha$ rel. $\partial\mathbf{I}$. Let $h : \mathbf{I} \times \mathbf{I} \to B$ be a homotopy with $h(\partial\mathbf{I} \times \mathbf{I}) = \{b_0\}$, $h_0 = \alpha$, and $h_1 = p\beta$. By Lemma 4.2.2, h is lifted to a homotopy $\tilde{h} : \mathbf{I} \times \mathbf{I} \to E$ with $\tilde{h}(0, 0) = a_0$. From the uniqueness of liftings of $h|\{0\} \times \mathbf{I}$ and $h|\{1\} \times \mathbf{I}$, and h_1, it follows that $\tilde{h}(\partial\mathbf{I} \times \mathbf{I}) = \{a_0\}$ and $\tilde{h}_1 = \beta$. In particular, $\tilde{h}_0$ is a loop in E at a_0, which is a lift of $h_0 = \alpha$. □

The following theorem is a generalization of Lemma 4.2.1:

Theorem 4.2.5 (Lifting Criterion) *Let $p : E \to B$ be a covering projection with $a_0 \in p^{-1}(b_0)$ and let X be a connected, locally path-connected space. A map $f : X \to B$ with $x_0 \in f^{-1}(b_0)$ can be lifted to a map $\tilde{f}(x_0) = a_0$ if and only if $f_\sharp(\pi_1(X, x_0)) \subset p_\sharp(\pi_1(E, a_0))$. Furthermore, a lifting $\tilde{f}$ is unique if it exists.*

Proof If the lifting $\tilde{f}$ exists, then

$$f_\sharp(\pi_1(X, x_0)) = p_\sharp(\tilde{f}_\sharp(\pi_1(X, x_0))) \subset p_\sharp(\pi_1(E, a_0)).$$

This gives the "only if" part.

The uniqueness of $\tilde{f}$ can also be proved as follows: For each $x \in X$, we can take a path α from x_0 to x in X.[9] By Lemma 4.2.1 above, the path $f\alpha$ can be lifted to a unique path γ in E with $\gamma(0) = a_0$. If a lifting $\tilde{f}$ of f exists, then $\tilde{f}(x) = \gamma(1)$.

[9]A locally path-connected space is path-connected if it is connected.

Indeed, since $\tilde{f}\alpha$ is a lifting of $f\alpha$, it follows from Lemma 4.2.1 that $\tilde{f}\alpha = \gamma$, which implies $\tilde{f}(x) = \tilde{f}\alpha(1) = \gamma(1)$. Thus, we can see the uniqueness of $\tilde{f}$.

Now, we will show the "if" part. By the argument above, if the lifting $\tilde{f}$ exists, then for each $x \in X$, $\tilde{f}(x)$ must be defined as follows: Take a path α from x_0 to x in X, and lift the path $f\alpha$ to a path γ with $\gamma(0) = a_0$, and then define $\tilde{f}(x) = \gamma(1)$. We have to show that this is well-defined, that is, the point $\gamma(1)$ is uniquely determined independent from the choice of α, where γ is uniquely determined as a lifting of $f\alpha$.

Given $x \in X$, let α and β be paths from x_0 to x. Then, $f\alpha$ is lifted to γ with $\gamma(0) = a_0$ and $f\beta^{\leftarrow}$ is lifted to ζ with $\zeta(0) = \gamma(1)$, where $\beta^{\leftarrow}$ is the inverse path of β that is defined by $\beta^{\leftarrow}(t) = \beta(1-t)$. The join $\gamma * \zeta$ is a lifting of the loop $f\circ(\alpha * \beta^{\leftarrow})$ at b_0 and $\gamma * \zeta = \gamma(0) = a_0$. Due to Proposition 4.2.4, the condition $f_\sharp(\pi_1(X, x_0)) \subset p_\sharp(\pi_1(E, a_0))$ implies that the loop $f\circ(\alpha * \beta^{\leftarrow})$ is lifted to a loop in E at a_0. By uniqueness of liftings, $\gamma * \zeta$ must be a loop at a_0. Hence, $\zeta^{\leftarrow}$ is a lifting of $f\beta$ with $\zeta^{\leftarrow}(0) = \zeta(1) = a_0$ and $\zeta^{\leftarrow}(1) = \zeta(0) = \gamma(1)$. This means that the point $\gamma(1)$ is uniquely determined independent from the choice of α.

We have to show the continuity of $\tilde{f}$ at $x \in X$. As above, let α be a path from x_0 to x in X, and γ be a lifting of the path $f\alpha$ with $\gamma(0) = a_0$. Then, $\tilde{f}(x) = \gamma(1)$. Since p is a covering projection and $p\tilde{f}(x) = f(x)$, $f(x)$ and $\tilde{f}(x)$ have open neighborhoods U_0 and V_0 respectively such that $p|V_0 : V_0 \to U_0$ is a homeomorphism. For each open neighborhood V of $\tilde{f}(x)$ in E, $p(V \cap V_0)$ is an open neighborhood of $f(x)$ in B. Since f is continuous and X is locally path-connected, x has a path-connected open neighborhood W in X such that $f(W) \subset p(V \cap V_0)$. For each $y \in W$, we have a path β from x to y in W. Then, $f\beta$ is a path from $f(x)$ to $f(y)$ in $p(V \cap V_0)$. Since $p|V_0$ is a homeomorphism, $\zeta = (p|V_0)^{-1} f\beta$ is a path in $V \cap V_0$, which is a lifting of $f\beta$ with $\zeta(0) = \tilde{f}(x) = \gamma(1)$. Then, $\gamma * \zeta$ is a lifting of $f\alpha * f\beta = f\circ(\alpha * \beta)$ with $\gamma * \zeta(0) = \gamma(0) = a_0$, where $\alpha * \beta$ is a path from x_0 to y in X. Hence $\tilde{f}(y) = \gamma * \zeta(1) = \zeta(1) \in V \cap V_0 \subset V$. Thus, we have $\tilde{f}(W) \subset V$. Therefore, $\tilde{f}$ is continuous. □

In the rest of this chapter, we will also use elementary results on the homotopy groups and the homology groups. The reader can refer to any standard textbook on Algebraic Topology, e.g., [(22)], [(23)], etc.[10]

Theorem 4.2.6 (Seifert–van Kampen) *Let $X = \operatorname{int} X_1 \cup \operatorname{int} X_2$ and $x_0 \in X_0 = X_1 \cap X_2$, where X_0, X_1 and X_2 are path-connected (hence X is also path-connected). For $i = 1, 2$, let $j_i : \pi_1(X_i, x_0) \to \pi_1(X, x_0)$ and $j_{0,i} : \pi_1(X_0, x_0) \to \pi_1(X_i, x_0)$ denote the homomorphisms induced by the inclusions. Then, for each group H with homomorphisms $f_i : \pi_1(X_i, x_0) \to H$, $i = 1, 2$, such that*

[10]The Seifert–van Kampen Theorem can be found in not only such textbooks but also Chap. 70 of Munkres' textbook [(7)].

$f_1 j_{0,1} = f_2 j_{0,2}$, *there exists a unique homomorphism* $f : \pi_1(X, x_0) \to H$ *such that* $f j_1 = f_1$ *and* $f j_2 = f_2$.

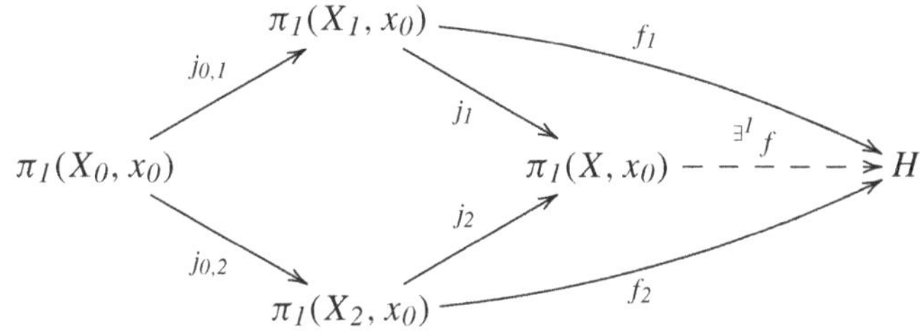

Concerning the homology groups, it is enough to know the simplicial homology groups.[11] Indeed, we treat only ANRs and every ANR has the homotopy type of some simplicial complex (Theorem 1.13.22).

A sequence of groups and homomorphisms

$$G_1 \xrightarrow{f_1} G_2 \xrightarrow{f_2} \cdots \xrightarrow{f_r} G_{r+1}$$

is said to be **exact** if $\operatorname{im} f_i = \ker f_{i+1}$ for each $i = 1, \ldots, r$. A homomorphism $f : G \to H$ is injective (resp. surjective) if and only if

$$0 \longrightarrow G \xrightarrow{f} H \quad (\text{resp. } G \xrightarrow{f} H \longrightarrow 0)$$

is exact. An injective homomorphism is called a **monomorphism** and a surjective homomorphism is called an **epimorphism**. If H is a subgroup of G then

$$0 \longrightarrow H \xrightarrow{\subset} G \xrightarrow{p} G/H \longrightarrow 0$$

is exact, where p is the quotient homomorphism.

The following theorem is very useful to calculate the homology of spaces:

Theorem 4.2.7 (Mayer–Vietoris) *Let* $X = \operatorname{int} X_1 \cup \operatorname{int} X_2$ *and* $X_0 = X_1 \cap X_2$. *Moreover, let*

$$\varphi_n : H_n(X_0) \to H_n(X_1) \oplus H_n(X_2),$$
$$\psi_n : H_n(X_1) \oplus H_n(X_2) \to H_n(X), \ n \in \omega$$

be the homomorphisms defined by

$$\varphi_n(x) = (j_{0,1}(x), j_{0,2}(x)) \text{ and } \psi_n(x_1, x_2) = j_1(x_1) - j_2(x_2),$$

[11] Nevertheless, for an arbitrary space, the singular homology or the Čech homology can be used.

where $j_{0,i} : H_n(X_0) \to H_n(X_i)$, $j_i : H_n(X_i) \to H_n(X)$, $i = 1, 2$, *denote the homomorphism induced by the inclusions. Then, there are homomorphisms* $\delta_n : H_n(X) \to H_{n-1}(X_0)$, $n \in \omega$, *such that the following sequence is exact:*

$$\cdots \xrightarrow{\delta_{n+1}} H_n(X_0) \xrightarrow{\varphi_n} H_n(X_1) \oplus H_n(X_2) \xrightarrow{\psi_n} H_n(X)$$
$$\xrightarrow{\delta_n} H_{n-1}(X_0) \xrightarrow{\varphi_{n-1}} \cdots \xrightarrow{\psi_1} H_1(X) \xrightarrow{\delta_1} H_0(X_0)$$
$$\xrightarrow{\varphi_0} H_0(X_1) \oplus H_0(X_2) \xrightarrow{\psi_0} H_0(X) \xrightarrow{\delta_0} 0.$$

The exact sequence in Theorem 4.2.7 above is called the **Mayer–Vietoris exact sequence** for $(X; X_1, X_2)$.

For any space X and $n \in \mathbb{N}$, there exists a group homomorphism $h_* : \pi_n(X) \to H_n(X)$ called the **Hurewicz homomorphism** from the nth homotopy group to the nth homology group (with integer coefficients), which is equivalent to the canonical Abelianization map $h_* : \pi_1(X) \to \pi_1(X)/[\pi_1(X), \pi_1(X)]$ for $n = 1$ and a path-connected space X. The Hurewicz theorem states that if X is $(n-1)$-connected, then the Hurewicz map is an isomorphism for all $k \leqslant n$ when $n \geqslant 2$ and Abelianization for $n = 1$. In particular, this theorem says that the Abelianization of the first homotopy group (the fundamental group) is isomorphic to the first homology group, that is, $H_1(X) \cong \pi_1(X)/[\pi_1(X), \pi_1(X)]$.

Theorem 4.2.8 (Hurewicz) *Let X be a path-connected space such that* $\pi_i(X) = 0$ *for* $i = 1, \ldots, n-1$, *where* $n \geqslant 2$. *Then,* $H_i(X) = 0$ *for* $i = 1, \ldots, n-1$ *and* $\pi_n(X) \cong H_n(X)$.

4.3 The Procedure for Killing Homotopy Groups

A path-connected space X is said to be ***n*-connected** if $\pi_i(X) = \{0\}$ for every $i = 1, \ldots, n$. This is equivalent to the condition that any map $\alpha : \mathbf{S}^i \to X$ is null-homotopic for every $i = 1, \ldots, n$. When X is 1-connected, it is also said that X is **simply connected**. A space X is homotopically trivial if and only if X is n-connected for every $n \in \mathbb{N}$. Recall that an ANR X is contractible ($X \simeq 0$) if and only if X is homotopically trivial (Theorem 1.13.28). In this section, we will prove the following two theorems. The first theorem is well known but its proof is given for completeness and for the readers' convenience. And then, using this, we will prove the second theorem. In the following, $\bigoplus_{i=1}^m \mathbf{S}_i^n$ and $\bigoplus_{i=1}^m \mathbf{B}_i^{n+1}$ are the topological sums of m copies of the unit n-sphere $\mathbf{S}^n$ and the unit $(n+1)$-ball $\mathbf{B}^{n+1}$.

Theorem 4.3.1 *Let X be a connected ANR that is homotopy dominated by a finite simplicial complex. Then, there exists a tower* $X = X_0 \subset X_1 \subset \cdots \subset X_{k_0}$ *and maps* $\varphi_n : \bigoplus_{i=1}^{m_n} \mathbf{S}_i^n \to X_{n-1}$, $n = 1, \ldots, k_0$, *such that* $X_{k_0} \simeq 0$, *each* X_n *is* n-

connected, and $X_n = X_{n-1} \cup_{\varphi_n} \bigoplus_{i=1}^{m_n} \mathbf{B}_i^{n+1}$*, where if* X_{n-1} *is n-connected then* $m_n = 0$ *and* $X_n = X_{n-1}$.

Theorem 4.3.2 *For each compact connected* $\boldsymbol{Q}$*-manifold M, there exists a tower* $M = M_0 \subset M_1 \subset \cdots \subset M_{k_0}$ *of compact* $\boldsymbol{Q}$*-manifolds and Z-embeddings* $\varphi_n : \bigoplus_{j=1}^{m_n}(\mathbf{S}_i^n \times \boldsymbol{Q}) \to X_{n-1}$, $n = 1, \dots, k_0$, *such that* $M_{k_0} \approx \boldsymbol{Q}$*, each* M_n *is n-connected, and*

$$M_n = M_{n-1} \cup_{\varphi_n} \bigoplus_{i=1}^{m_n}(\mathbf{B}_i^{n+1} \times \boldsymbol{Q}),$$

where if M_{n-1} *is n-connected then* $m_n = 0$ *and* $M_n = M_{n-1}$.

As mentioned in Remark 1.12, the following is obtained from Hanner's Characterization of ANRs (Theorem 1.13.20):

Proposition 4.3.3 *Every compact ANR X is homotopy dominated by a finite simplicial complex.* □

The following is an alternative familiar proof of the above proposition:

Proof (Proposition 4.3.3) By the universality of $\boldsymbol{Q}$, we may assume that X is a closed set in $\boldsymbol{Q}$. Then, we have an open neighborhood U of X in $\boldsymbol{Q}$ with a retraction $r : U \to X$. Because of the compactness of X, we can easily find $n \in \mathbb{N}$ and an open set V in $[-1, 1]^n$ such that $X \subset p_n^{-1}(V) \subset U$, where $p_n : \boldsymbol{Q} \to [-1, 1]^n$ is the projection defined by $p_n(x) = (x(1), \dots, x(n))$ for each $x \in \boldsymbol{Q}$ (cf. Wallace's Theorem 2.1.2 in [GAGT]). Note that $p_n(X) \subset V$. Take a small triangulation L of $[-1, 1]^n$ so that each $\sigma \in L[p_n(X)]$ is contained in V. Let $K = \bigcup_{\sigma \in L[p_n(X)]} F(\sigma)$. Then, K is a subcomplex of L and $p_n(X) \subset |K| \subset V$. Let $f = p_n|X : X \to |K|$ and define a map $g : |K| \to X$ by $g(x) = r(x(1), \dots, x(n), 0, 0, \dots)$ for each $x \in |K|$ ($\subset [-1, 1]^n$). Since $gf(x) = r(x(1), \dots, x(n), 0, 0, \dots)$ for each $x \in X$, it follows that $gf \simeq \mathrm{id}_X$ by the homotopy $h : X \times \mathbf{I} \to X$ defined as follows:

$$h(x, t) = r(x(1), \dots, x(n), tx(n+1), tx(x+2), \dots) \text{ for each } (x, t) \in X \times \mathbf{I}.$$

Thus, X is homotopy dominated by K. □

The following is easily proved:

Proposition 4.3.4 *If a connected (resp. path-connected) X is homotopy dominated by Y, then X is homotopy dominated by a component (resp. path-component) of Y.*

Proof We have maps $f : Y \to X$ and $g : X \to Y$ such that $fg \simeq \mathrm{id}_X$. Since $g(X)$ is connected (resp. path-connected), it is contained in some component (resp. path-component) Y_0 of Y. Then, $f|Y_0 : Y_0 \to X$ is a homotopy domination because $(f|Y_0)g = fg \simeq \mathrm{id}_X$. □

The following is a combination of Propositions 4.3.3 and 4.3.4:

Corollary 4.3.5 *Every compact connected ANR X is homotopy dominated by a connected finite simplicial complex.* □

A group G is said to be **finitely generated** if G has finite elements $g_1, \dots, g_n$ such that every element $x \in G$ can be written as a product of powers of g_i, $i = 1, \dots, n$. The following is elementary and well-known:

Fact *For every connected finite simplicial complex K, the fundamental group $\pi_1(|K|)$ is finitely generated.*[12]

For the readers' convenience, we give a proof of the above fact:

Proof The 1-skeleton $K^{(1)}$ is a finite connected graph, which has a maximal tree T (i.e., T is a tree with $T^{(0)} = K^{(0)}$). Let $K^{(1)} \setminus T = \{e_i \mid i = 1, \dots, n\}$ and take $v_0 \in K^{(0)}$ as a base point. For each $i = 1, \dots, n$, we have a PL loop $\alpha_i : \mathbf{I} \to |K^{(1)}|$ such that $\alpha_i(0) = \alpha_i(1) = v_0$, $e_1 = \langle \alpha_i(1/3), \alpha_i(2/3) \rangle$, and $\alpha_i([0, i/3])$ is the unique arc from v_0 to $\alpha(i/3)$ in $|T|$ for $i = 1, 2$. By the Simplicial Approximation Theorem 1.8.8, for each loop $\alpha : \mathbf{I} \to |K|$ at v_0, there exist $I' \lhd I = \{0, 1, \mathbf{I}\}$ and a simplicial loop $\alpha' : I' \to |K^{(1)}|$ at v_0 such that $\alpha \simeq \alpha'$ rel. $\partial\mathbf{I}$. Then, α' is homotopic to a join of some members of $\{\alpha_i, \alpha_i^{\leftarrow} \mid i = 1, \dots, n\}$, where $\alpha_i^{\leftarrow}$ is the inverse loop of α_i. This means that $\pi_1(|K|)$ is finitely generated. □

The following is a direct consequence of the definition of the simplicial homologies:

Fact *For every finite simplicial complex K, $H_k(K)$ is finitely generated for $0 \leqslant k \leqslant \dim K$ and $H_k(K) = 0$ if $k > \dim K$. Additionally, $H_0(X)$ and $H_{\dim K}(K)$ is free, where the rank of $H_0(X)$ is equal to the number of components of K.*

Using the above two facts, we have the following:

Proposition 4.3.6 *Let X be a connected ANR which is homotopy dominated by a finite simplicial complex K. Then, $\pi_1(X)$ and $H_k(X)$, $0 < k \leqslant \dim K$, are finitely generated and $H_k(X) = 0$ for every $k > \dim K$. Additionally, $H_{\dim K}(X)$ is free.*

Proof There are maps $f : |K| \to X$ and $g : X \to |K|$ such that $fg \simeq \mathrm{id}$. Then, f induces the epimorphisms $f_\sharp : \pi_1(|K|) \to \pi_1(X)$ and $f_* : H_k(K) \to H_k(X)$, $k \in \mathbb{N}$, because $f_\sharp g_\sharp = \mathrm{id}$ and $f_* g_* = \mathrm{id}$. For $0 < k \leqslant \dim k$, since $H_k(K)$ is finitely generated, it follows that $H_*(X)$ is also finitely generated. For every $k > \dim K$, since $H_k(K) = 0$, we have $H_k(X) = 0$. Furthermore, since $g_* : H_{\dim K}(X) \to H_{\dim K}(K)$ is a monomorphism and $H_{\dim K}(K)$ is free, it follows that $H_{\dim K}(X)$ is free. □

Let $\mathbf{e}_1 = (1, 0, \dots, 0) \in \mathbf{S}^n \subset \mathbb{R}^{n+1}$. Observe

$$(\mathbf{I}^n/\partial\mathbf{I}^n, \partial\mathbf{I}^n/\partial\mathbf{I}^n) \approx (\mathbf{S}^n, \mathbf{e}_1).$$

Then, we have a map $\theta : (\mathbf{I}^n, \partial\mathbf{I}^n) \to (\mathbf{S}^n, \mathbf{e}_1)$ such that $\theta(\partial\mathbf{I}^n) = \mathbf{e}_1$ and $\theta|\operatorname{rint}\mathbf{I}^n \to \mathbf{S}^n \setminus \{\mathbf{e}_1\}$ is a homeomorphism. Each map $\alpha : (\mathbf{I}^n, \partial\mathbf{I}^n) \to (X, *)$ induces the map $\tilde{\alpha} : (\mathbf{S}^n, \mathbf{e}_1) \to (X, *)$ such that $\alpha = \tilde{\alpha}\theta$. In the proof of the following lemma, we use the Seifert–van Kampen Theorem 4.2.6, the Mayer–Vietoris Theorem 4.2.7, and the Hurewicz Isomorphism Theorem 4.2.8.

[12]Since $|K|$ is path-connected, we can omit a base point.

Lemma 4.3.7 *Let X be an ANR that is $(n-1)$-connected but not n-connected, where $n > 0$. Suppose that $\pi_n(X)$ is generated by m many generators and $H_{n+1}(X)$ is also finitely generated. Then, there exists a map $\varphi : \bigoplus_{i=1}^m \mathbf{S}_i^n \to X$ such that the adjunction space $X_\varphi = X \cup_\varphi \bigoplus_{i=1}^m \mathbf{B}_i^n$ is n-connected, $\pi_{n+1}(X_\varphi) = H_{n+1}(X_\varphi)$ is finitely generated, and $H_k(X_\varphi) \cong H_k(X)$ for $k > n+1$, where if $H_{n+1}(X)$ is free then $\pi_{n+1}(X_\varphi) = H_{n+1}(X_\varphi)$ is also free, and if $\pi_n(X)$ is free and $H_{n+1}(X) = 0$ then $\pi_{n+1}(X_\varphi) = H_{n+1}(X_\varphi) = 0$.*

Proof Take maps $\alpha_1, \ldots, \alpha_m : (\mathbf{I}^n, \partial\mathbf{I}^n) \to (X, *)$ so that $\pi_n(X)$ is generated by $[\alpha_1], \ldots, [\alpha_m]$. Let $(\mathbf{B}_i^{n+1}, \mathbf{S}_i^n, *_i)$, $i = 1, \ldots, m$, be disjoint copies of $(\mathbf{B}^{n+1}, \mathbf{S}^n, \mathbf{e}_1)$. For each $i = 1, \ldots, m$, there is a map $\theta_i : (\mathbf{I}^n, \partial\mathbf{I}^n) \to (\mathbf{S}_i^n, *_i)$ such that $\theta_i(\partial\mathbf{I}^n) = *_i$ and $\theta_i | \operatorname{rint} \mathbf{I}^n \to \mathbf{S}_i^n \setminus \{*_i\}$ is a homeomorphism. Then, each α_i induces a map $\tilde{\alpha}_i : (\mathbf{S}_i^n, *_i) \to (X, *)$ such that $\alpha_i = \tilde{\alpha}_i\theta_i$. The desired map $\varphi : \bigoplus_{i=1}^m \mathbf{S}_j^n \to X$ is defined by $\varphi|\mathbf{S}_i^n = \tilde{\alpha}_i$, $i = 1, \ldots, m$. We will show that $X_\varphi = X \cup_\varphi \bigoplus_{i=1}^m \mathbf{B}_i^n$ satisfies the required conditions.

Let $q : X \oplus \bigoplus_{i=1}^m \mathbf{B}_i^{n+1} \to X_\varphi$ be the quotient map. For each $i = 1, \ldots, m$, we define

$$B_i = \left\{(1-t)x + t*_i \;\middle|\; x \in \mathbf{S}_i^n,\ 0 \leqslant t \leqslant 1/2\right\},$$

$$S_i = \partial B_i = \left\{\tfrac{1}{2}x + \tfrac{1}{2}*_i \;\middle|\; x \in \mathbf{S}_i^n\right\}, \quad \text{and}$$

$$C_i = \mathbf{B}_i^{n+1} \setminus \operatorname{rint} B_i = \left\{(1-t)x + t*_i \;\middle|\; x \in \mathbf{S}_i^n,\ 1/2 \leqslant t \leqslant 1\right\}.$$

Then, we have retraction $r_i : C_i \to \mathbf{S}_i^n$ defined by

$$r_i((1-t)x + t*_i) = x \quad \text{for } x \in \mathbf{S}_i^n, 1/2 \leqslant t \leqslant 1,$$

where $r_i \simeq \text{id rel. } \mathbf{S}_i^n$. Since $r_i|S_i : S_i \to \mathbf{S}_i^n$ is a homeomorphism, there is a map $\theta_i' : (\mathbf{I}^n, \partial\mathbf{I}^n) \to (S_i, *_i)$ such that $\theta_i = r_i\theta_i'$, so $\alpha_i = \tilde{\alpha}_i r_i \theta_i'$. Let $Y = q(\bigoplus_{i=1}^m S_i)$, $Z = q(\bigoplus_{i=1}^m B_i)$, and

$$M = X_\varphi \setminus (Z \setminus Y) = X \cup q(\textstyle\bigoplus_{i=1}^m C_i).$$

These are subspaces of X_φ such that $X_\varphi = M \cup Z$ and $Y = M \cap Z$. Note that $Y = \bigvee_{i=1}^m S_i \approx \bigvee_{i=1}^m \mathbf{S}_i^n$ and $Z = \bigvee_{i=1}^m B_i \approx \bigvee_{i=1}^m \mathbf{B}_i^{n+1}$.[13] Moreover, we have a retraction $r : M \to X$ defined by $r|X = \text{id}$ and $rq|C_i = \varphi r_i|C_i = \tilde{\alpha}_i r_i|C_i$ for each $i = 1, \ldots, m$. Then, as is easily observed, $r \simeq \text{id rel. } X$ (hence X is a strong deformation retract of M). It follows that r induces an isomorphism $r_\sharp : \pi_n(M) \to \pi_n(X)$, whose inverse is induced by the inclusion $X \subset M$. The map $r|Y : Y \to X$ induces an epimorphism $(r|Y)_\sharp : \pi_n(Y) \to \pi_n(X)$ because $rq\theta_i' = \tilde{\alpha}_i r_i \theta_i' = \alpha_i$ for $i = 1, \ldots, m$, that is, the generators $[\alpha_1], \ldots, [\alpha_m] \in \pi_n(X)$ are the images

[13] For the wedge (sum) $\bigvee_{i=1}^m X_i$, refer to p. 5.

of $[q\theta'_1], \ldots, [q\theta'_m] \in \pi_n(Y)$. Since $r|Y$ is the composition of r and the inclusion $Y \subset M$, it follows that the inclusion $Y \subset M$ induces an epimorphism from $\pi_n(Y)$ to $\pi_n(M)$.

In the following commutative diagram, the four arrows of the right rectangle are the homomorphisms induced by the inclusions:

$$\begin{array}{ccccc} & & \pi_1(Y) & \longrightarrow & \pi_1(Z) = 0 \\ & \swarrow^{(r|Y)_\sharp} & \downarrow & & \downarrow \\ \pi_1(X) & \underset{r_\sharp}{\overset{\cong}{\longleftarrow}} & \pi_1(M) & \longrightarrow & \pi_1(X_\varphi) \end{array}$$

By the Seifert–van Kampen Theorem 4.2.6, $\pi_1(X_\varphi)$ is generated by the union of the images of $\pi_1(M)$ and $\pi_1(Z)$. Since $\pi_1(Z) = 0$, the bottom arrow of the rectangle is an epimorphism. When $n = 1$, as observed above, the left vertical arrow is also an epimorphism. Hence, the inclusion $Y \subset X_\varphi$ induces an epimorphism from $\pi_1(Y)$ to $\pi_1(X_\varphi)$. Since $Y \subset Z \subset X_\varphi$ and $\pi_1(Z) = 0$, it follows that $\pi_1(X_\varphi) = 0$, that is, X_φ is 1-connected. When $n > 1$, since $\pi_1(M) \cong \pi_1(X) = 0$, we have $\pi_1(X_\varphi) = 0$. Note that $H_1(X_\varphi) = 0$ by the Hurewicz Theorem 4.2.8.

By the Mayer–Vietoris Theorem 4.2.7, we have the following exact sequence of homology groups:

$$\cdots \longrightarrow H_k(Y) \overset{(i,j)}{\longrightarrow} H_k(M) \oplus H_k(Z) \overset{i'-j'}{\longrightarrow} H_k(X_\varphi)$$
$$\overset{\delta}{\longrightarrow} H_{k-1}(Y) \longrightarrow \cdots \longrightarrow H_1(M) \oplus H_1(Z) \overset{i'-j'}{\longrightarrow} H_1(X_\varphi) = 0,$$

where i, i' and j, j' are the homomorphisms induced by the inclusions $Y \subset M \subset X_\varphi$ and $Y \subset Z \subset X_\varphi$. Since X is $(n-1)$-connected and $M \simeq X$, M is also $(n-1)$-connected. By the Hurewicz Theorem 4.2.8, we have $H_k(M) \cong \pi_k(M) = 0$ for every $0 < k < n$ and $H_n(M) \cong \pi_n(M)$. Since $H_k(Z) = 0$ for every $k > 0$ and

$$H_k(Y) \cong H_k(\textstyle\bigvee_{i=1}^m \mathbf{S}_i^n) \cong \begin{cases} \underbrace{\mathbb{Z} \oplus \cdots \oplus \mathbb{Z}}_{m \text{ many}} & \text{if } k = n, \\ 0 & \text{if } k \neq 0, n, \end{cases}$$

it follows that

$$H_k(X_\varphi) \cong H_k(M) = 0 \text{ for } 0 < k < n \text{ and}$$
$$H_k(X_\varphi) \cong H_k(M) \cong H_k(X) \text{ for } k > n+1.$$

Inductively, using the Hurewicz Theorem 4.2.8, we can see that $\pi_k(X_\varphi) \cong H_k(X_\varphi) = 0$ for $0 < k < n$, that is, X_φ is also $(n-1)$-connected. In the following diagram, the bottom sequence is exact and two vertical arrows are Hurewicz isomorphisms:

$$\begin{array}{ccccccc} & & \pi_n(Y) & \xrightarrow{\text{epi}} & \pi_n(M) & & \\ & & \cong\downarrow & & \downarrow\cong & & \\ \cdots & \longrightarrow & H_n(Y) & \longrightarrow & H_n(M) & \xrightarrow{i} & H_n(X_\varphi) \longrightarrow 0 \end{array}$$

As already observed, the top arrow of the rectangle is an epimorphism. Hence, the bottom arrow is also an epimorphism, which implies that $H_n(X_\varphi) = 0$. Again, using the Hurewicz Theorem 4.2.8, we have $\pi_n(X_\varphi) \cong H_n(X_\varphi) = 0$. Consequently, X_φ is n-connected.

Moreover, we have the following short exact sequence:

$$0 \longrightarrow H_{n+1}(M) \xrightarrow{i} H_{n+1}(X_\varphi) \longrightarrow \operatorname{im}\delta \longrightarrow 0,$$

where $\operatorname{im}\delta$ is free as a subgroup of the free Abelian group $H_n(Y)$. So, the above short exact sequence is split. Therefore, we have

$$H_{n+1}(X_\varphi) \cong H_{n+1}(M) \oplus \operatorname{im}\delta,$$

which is finitely generated.

In the above, when $H_{n+1}(X)$ is free, since $H_{n+1}(M) \cong H_{n+1}(X)$, it follows that $H_{n+1}(X_\varphi)$ is also free. When $\pi_n(X)$ is free, $\varphi_\sharp : \pi_n(Y) \to \pi_n(X)$ is an isomorphism because the generators $[\alpha_1], \ldots, [\alpha_m] \in \pi_n(X)$ are the images of $[\theta_1], \ldots, [\theta_m] \in \pi_n(Y)$. Therefore, the inclusion $Y \subset M$ induces an isomorphism from $\pi_n(Y)$ to $\pi_n(M)$, hence it induces an isomorphism from $H_n(Y)$ to $H_n(M)$. When $H_{n+1}(X) = 0$, we have $\operatorname{im}\delta = 0$. Since $H_{n+1}(M) \cong H_{n+1}(X) = 0$, it follows that $H_{n+1}(X_\varphi) = 0$. This completes the proof. □

Now, we can prove Theorem 4.3.1.

Proof (Theorem 4.3.1) By virtue of Proposition 4.3.6, we can iteratively apply Lemma 4.3.7 to obtain a tower $X = X_0 \subset X_1 \subset \cdots \subset X_{k_0}$ and maps $\varphi_n : \bigoplus_{i=1}^{m_n} \mathbf{S}_i^n \to X_{n-1}$, $n = 1, \ldots, k_0$, such that $X_{k_0} \simeq 0$, each X_n is n-connected, and $X_n = X_{n-1} \cup_{\varphi_n} \bigoplus_{i=1}^{m_n} \mathbf{B}_i^{n+1}$, where if X_{n-1} is n-connected then $m_n = 0$ and $X_n = X_{n-1}$. □

In Theorem 4.3.2, we use an adjunction space $X \cup_\varphi \bigoplus_{i=1}^m (\mathbf{B}_i^{n+1} \times \boldsymbol{Q})$ by a Z-embedding $\varphi : \bigoplus_{i=1}^m (\mathbf{S}_i^n \times \boldsymbol{Q}) \to X$ instead of $X \cup_\varphi \bigoplus_{i=1}^m \mathbf{B}_i^{n+1}$ by a map $\varphi : \bigoplus_{i=1}^m \mathbf{S}_i^n \to X$.

Lemma 4.3.8 *Let X be an ANR and let $\mathrm{pr}_1 : \bigoplus_{i=1}^m (\mathbf{S}_i^n \times \boldsymbol{Q}) \to \bigoplus_{i=1}^m \mathbf{S}_i^n$ be the projection.*

(1) *For each map $f : \bigoplus_{i=1}^m \mathbf{S}_i^n \to X$,*

$$X \cup_{f\mathrm{pr}_1} \bigoplus_{i=1}^m (\mathbf{B}_i^{n+1} \times \boldsymbol{Q}) \simeq X \cup_f \bigoplus_{i=1}^m \mathbf{B}_i^{n+1}.$$

(2) *For each map $f : \bigoplus_{i=1}^m (\mathbf{S}_i^n \times \boldsymbol{Q}) \to X$,*

$$X \cup_f \bigoplus_{i=1}^m (\mathbf{B}_i^{n+1} \times \boldsymbol{Q}) \approx M_f \cup \bigoplus_{i=1}^m (\mathbf{B}_i^{n+1} \times \boldsymbol{Q}) \times \{1\}.$$

(3) *For two maps $f, g : \bigoplus_{i=1}^m (\mathbf{S}_i^n \times \boldsymbol{Q}) \to X$,*

$$f \simeq g \;\Rightarrow\; X \cup_f \bigoplus_{i=1}^m (\mathbf{B}_i^{n+1} \times \boldsymbol{Q}) \simeq X \cup_g \bigoplus_{i=1}^m (\mathbf{B}_i^{n+1} \times \boldsymbol{Q}).$$

Proof (1): We have a cell-like map

$$\psi : X \cup_{f\mathrm{pr}_1} \bigoplus_{i=1}^m (\mathbf{B}_i^{n+1} \times \boldsymbol{Q}) \to X \cup_f \bigoplus_{i=1}^m \mathbf{B}_i^{n+1}$$

defined by $\psi|X = \mathrm{id}$ and $\psi|\bigoplus_{i=1}^m (\mathrm{rint}\,\mathbf{B}_i^{n+1} \times \boldsymbol{Q}) = \mathrm{pr}_1$. Then, ψ is a fine homotopy equivalence by Theorem 1.15.9.

(2): It suffices to see that

$$\Bigl(\bigoplus_{i=1}^m (\mathbf{S}_i^n \times \boldsymbol{Q} \times \mathbf{I}) \cup \bigoplus_{i=1}^m (\mathbf{B}_i^{n+1} \times \boldsymbol{Q} \times \{1\}),\ \bigoplus_{i=1}^m (\mathbf{S}_i^n \times \boldsymbol{Q} \times \{0\})\Bigr) \approx \Bigl(\bigoplus_{i=1}^m \mathbf{B}_i^{n+1} \times \boldsymbol{Q},\ \bigoplus_{i=1}^m \mathbf{S}_i^n \times \boldsymbol{Q}\Bigr).$$

This can be obtained by the following:

$$\bigl((\mathbf{S}^n \times \mathbf{I}) \cup (\mathbf{B}^{n+1} \times \{1\}), \mathbf{S}^n \times \{0\}\bigr) \approx \bigl(\mathbf{B}^{n+1}, \mathbf{S}^n\bigr).$$

(3): By Proposition 1.10.1(1),

$$M_f \simeq M_g \ \text{ rel. } X \cup \Bigl(\bigoplus_{i=1}^m (\mathbf{S}_i^{n+1} \times \boldsymbol{Q}) \times \{1\}\Bigr).$$

Hence, it follows from the above (2) that

$$\begin{aligned} X \cup_f \bigoplus_{i=1}^{m} (\mathbf{B}_i^{n+1} \times \boldsymbol{Q}) &\simeq M_f \cup \bigoplus_{i=1}^{m} (\mathbf{B}_i^{n+1} \times \boldsymbol{Q}) \times \{1\} \\ &\simeq M_g \cup \bigoplus_{i=1}^{m} (\mathbf{B}_i^{n+1} \times \boldsymbol{Q}) \times \{1\} \\ &\simeq X \cup_g \bigoplus_{i=1}^{m} (\mathbf{B}_i^{n+1} \times \boldsymbol{Q}). \end{aligned}$$ □

In Lemma 4.3.7, assume that X is a $\boldsymbol{Q}$-manifold. Then, applying the Z-Embedding Approximation Theorem 2.10.11, we have a Z-embedding $\psi : \bigoplus_{i=1}^{m} (\mathbf{S}^n \times \boldsymbol{Q}) \to X$, which is homotopic to $\varphi \mathrm{pr}_1 : \bigoplus_{i=1}^{m} (\mathbf{S}^n \times \boldsymbol{Q}) \to X$. It follows from Lemma 4.3.8 (1) and (3) that

$$X \cup_\varphi \bigoplus_{i=1}^{m} \mathbf{B}^{n+1} \simeq X \cup_\psi \bigoplus_{i=1}^{m} (\mathbf{B}^{n+1} \times \boldsymbol{Q}),$$

where the right-hand side of the above equation is a $\boldsymbol{Q}$-manifold because $\psi(\bigoplus_{i=1}^{m} (\mathbf{S}^n \times \boldsymbol{Q}))$ is collared in X. Using this modification, we can prove Theorem 4.3.2 in the same way as Theorem 4.3.1.

Proof (Theorem 4.3.2) By virtue of Proposition 4.3.6, we can iteratively apply the above modification of Lemma 4.3.7 to obtain a tower $M = M_0 \subset M_1 \subset \cdots \subset M_{k_0}$ of compact $\boldsymbol{Q}$-manifolds and Z-embeddings $\varphi_n : \bigoplus_{i=1}^{m_n} (\mathbf{S}_i^n \times \boldsymbol{Q}) \to M_{n-1}$, $n = 1, \ldots, k_0$, such that $M_{k_0} \simeq 0$, each M_n is n-connected, and

$$M_n = M_{n-1} \cup_{\varphi_n} \bigoplus_{i=1}^{m_n} (\mathbf{B}_i^{n+1} \times \boldsymbol{Q}),$$

where if M_{n-1} is n-connected then $m_n = 0$ and $M_n = M_{n-1}$. Since a compact contractible $\boldsymbol{Q}$-manifold is homeomorphic to $\boldsymbol{Q}$ (Theorem 3.7.9), we have $M_{k_0} \approx \boldsymbol{Q}$. □

In the above proof, since φ_n is a Z-embedding, we have

$$\bigcup_{i=1}^{m_n} \varphi_n(\mathbf{S}_i^n \times \boldsymbol{Q}) = \bigoplus_{i=1}^{m_n} \varphi_n(\mathbf{S}_i^n \times \boldsymbol{Q}),$$

which is collared in M_{n-1} by the Collaring Theorem 2.11.1. Then, it follows that the above set is the boundary $\mathrm{bd}_{M_n} M_{n-1}$. Thus, in Theorem 4.3.2, attaching one piece $\mathbf{B}_i^{n+1} \times \boldsymbol{Q}$ of $\bigoplus_{i=1}^{m_n} (\mathbf{B}_i^{n+1} \times \boldsymbol{Q})$ at a time, we have the following:

Corollary 4.3.9 *For a compact connected $\boldsymbol{Q}$-manifold M, there exists a tower $M = M_0 \subset M_1 \subset \cdots \subset M_k$ of compact $\boldsymbol{Q}$-manifolds and $0 < n_1 \leqslant n_2 \leqslant \cdots \leqslant n_k$ such that $M_k \approx \boldsymbol{Q}$, each $\mathrm{bd}_{M_i} M_{i-1}$ is a Z-submanifold of M_{i-1} and*

$$\left(M_i \setminus \mathrm{int}_{M_i} M_{i-1}, \mathrm{bd}_{M_i} M_{i-1}\right) \approx \left(\mathbf{B}^{n_i+1} \times \boldsymbol{Q}, \mathbf{S}^{n_i} \times \boldsymbol{Q}\right).$$

Thus, $\mathrm{bd}_{M_i} M_{i-1}$ is collared in both M_{i-1} and $M_i \setminus \mathrm{int}_{M_i} M_{i-1}$, so it is bi-collared in M_i. □

Concerning the tower of Corollary 4.3.9 above, assume that $M_i = M_{i-1} \cup_{\varphi_i} \mathbf{B}^{n_i+1} \times \boldsymbol{Q} \approx P_i \times \boldsymbol{Q}$ for some compact polyhedron P_i. If we could find a compact polyhedron P_{i-1} such that $M_{i-1} \approx P_{i-1} \times \boldsymbol{Q}$, then the downward induction would provide us with a compact polyhedron P such that $M \approx P \times \boldsymbol{Q}$. This is our strategy to prove the Triangulation Theorem for compact $\boldsymbol{Q}$-manifolds. In this process, we need the following:

Theorem 4.3.10 *Let X be an ANR that is homotopy dominated by a finite simplicial complex. Then, there exists a finite simplicial complex K and a homotopy domination $f : |K| \to X$, which induces a bijection between the components of $|K|$ and X, and which induces isomorphisms between the fundamental groups of their corresponding components.*

In the rest of this section, we will prove this theorem.

Proposition 4.3.11 *Let X be homotopy dominated by Y. Then, every path-component of X is homotopy dominated by a path-component of Y. Moreover, let $f : Y \to X$ and $g : Y \to X$ be maps such that $fg \simeq \mathrm{id}_X$ (i.e., f is a homotopy domination) and let Y' be the union of path-components meeting $g(X)$. Then, the restriction $f' = f|Y' : Y' \to X$ is a homotopy domination that induces the bijection $f'_\sharp : \pi_0(Y') \to \pi_0(X)$ between the sets of path-components.*

Proof Let $X_0 \in \pi_0(X)$ (i.e., X_0 is a path-component of X). Since $g(X_0)$ is path-connected, it is contained in some path-component Y_0 of Y, where $Y_0 \in \pi_0(Y')$. Note that $f(Y_0)$ is path-connected and $fg(X_0) \subset f(Y_0)$, where $fg(X_0) \subset X_0$ because $fg \simeq \mathrm{id}_X$. Then, it follows that $f(Y_0) \subset X_0$ and $(f|Y_0)(g|X_0) = fg|X_0 \simeq \mathrm{id}_{X_0}$ in X_0. Therefore, X_0 is homotopy dominated by Y_0.

For each $Y_0 \in \pi_0(Y')$, $f(Y_0)$ is contained in a unique $X_0 \in \pi_0(X)$, which is $f'_\sharp(Y_0)$. In the above, we have shown that for each $X_0 \in \pi_0(X)$, $Y_0 \in \pi_0(Y')$ with $f(Y_0) \subset X_0$ is uniquely determined,[14] that is, $f'_\sharp$ is bijective. □

Applying Propositions 4.3.3 and 4.3.11, we can obtain the following:

Corollary 4.3.12 *For each compact ANR X, there exists a finite simplicial complex K with a homotopy domination $f : |K| \to X$ such that f induces $f_\sharp : \pi_0(|K|) \to \pi_0(X)$.* □

[14] In general, $Y_0 \in \pi_0(Y)$ with $f(Y_0) \subset X_0$ is not unique for $X_0 \in \pi_0(X)$.

By Corollary 4.3.12 and Proposition 4.3.3, Theorem 4.3.10 is reduced to the connected case, that is:

Theorem 4.3.13 *If a connected ANR X is homotopy dominated by a finite simplicial complex, then there exists a connected finite simplicial complex K and a homotopy domination $f : |K| \to X$, which induces an isomorphism between their fundamental groups.*

Proof Let $f : |L| \to X$ be a homotopy domination, where L is a finite simplicial complex. By Proposition 4.3.12, L can be taken so as to be connected. Then, there is a map $g : X \to |L|$ such that $fg \simeq \mathrm{id}_X$. It follows that f induces the epimorphism[15] $f_\sharp : \pi_1(|L|) \to \pi_1(X)$ because $f_\sharp g_\sharp = \mathrm{id}$. Since $\pi_1(|L|)$ is finitely generated, the kernel of $f_\sharp$ is also finitely generated. Take $v_0 \in L^{(0)}$ as a base point. Then, there exist loops $\alpha_1, \dots, \alpha_n : \mathbf{I} \to |L|$ at v_0 such that for a loop $\alpha : \mathbf{I} \to |L|$, if $f\alpha \simeq 0$ rel. $\partial\mathbf{I}$, then α is homotopic to a join of some members of $\{\alpha_i, \alpha_i^{\leftarrow} \mid i = 1, \dots, n\}$. Let $q : \mathbf{I} \to \mathbf{I}/\partial\mathbf{I}$ be the quotient map. For each $i = 1, \dots, n$, applying the Simplicial Approximation Theorem 1.8.8, we can obtain a simplicial complex L_i with a homeomorphism $\varphi_i : \mathbf{I}/\partial\mathbf{I} \to |L_i|$ and a simplicial map $\tilde{\alpha}_i : L_i \to L$ such that $\tilde{\alpha}_i\varphi_i q \simeq \alpha_i$ rel. $\partial\mathbf{I}$. Since $f\tilde{\alpha}_i\varphi_i q \simeq f\alpha_i \simeq 0$ rel. $\partial\mathbf{I}$, it follows that $f\tilde{\alpha}_i \simeq 0$, hence $f\tilde{\alpha}_i$ extends to the map $f_i : |v_i * L_i| \to X$, where $v_i * L_i$ is the simplicial cone over L_i defined by $v_i * L_i = \bigcup_{\sigma \in L_i} F(v_i\sigma)$. Then, we have the following diagram, where the right-hand square is commutative but the left-hand square is commutative up to $\simeq$ rel. $\partial\mathbf{I}$:

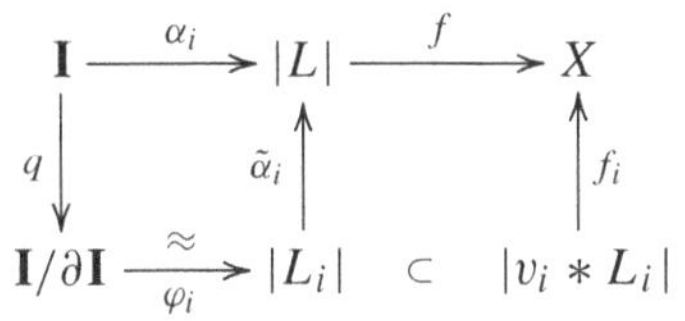

We define a finite simplicial complex $K = \bigcup_{i=1}^n (Z_{\tilde{\alpha}_i} \cup |v_i * L_i|)$, where

$$\left(Z_{\tilde{\alpha}_i} \cup |v_i * L_i|\right) \cap \left(Z_{\tilde{\alpha}_j} \cup |v_j * L_j|\right) = |L| \quad \text{if } i \neq j.$$

Since $f\bar{c}_{\tilde{\alpha}_i}||L_i| = f\tilde{\alpha}_i = f_i||L_i|$, we can extend f to a map $\tilde{f} : |K| \to X$ as follows:

$$\tilde{f}||Z_{\tilde{\alpha}_i}| = f\bar{c}_{\tilde{\alpha}_i} \text{ and } \tilde{f}||v_i * L_i| = f_i \text{ for each } i = 1, \dots, n.$$

Then, $\tilde{f}$ is a homotopy domination because $\tilde{f}g = fg \simeq \mathrm{id}_X$. So $\tilde{f}$ induces the epimorphism $\tilde{f}_\sharp : \pi_1(|K|) \to \pi_1(X)$.

[15] A surjective homomorphism is called an epimorphism.

To see that $\tilde{f}_\sharp$ is a monomorphism,[16] let $\alpha : \mathbf{I} \to |K|$ be a loop at v_0 such that $\tilde{f}\alpha \simeq 0$ rel. $\partial\mathbf{I}$. By the Simplicial Approximation Theorem 1.8.8, we may assume α is simplicial with respect to some $I' \lhd I = F(\mathbf{I})$ and K. Moreover, it can be assumed that $\alpha(\mathbf{I})$ does not contain any of $v_1, \ldots, v_n$, which means that $\alpha(\mathbf{I}) \subset \bigcup_{i=1}^n |Z_{\tilde{\alpha}_i}|$. Since $|L|$ is a strong deformation retract of $\bigcup_{i=1}^n |Z_{\tilde{\alpha}_i}|$, there is $\alpha' : \mathbf{I} \to |L|$ such that $\alpha' \simeq \alpha$ rel. $\partial\mathbf{I}$. Then, $f\alpha' = \tilde{f}\alpha' \simeq \tilde{f}\alpha \simeq 0$ rel. $\partial\mathbf{I}$, which implies that α' is homotopic to a join of some members of $\{\alpha_i, \alpha_i^{\leftarrow} \mid i = 1, \ldots, n\}$. For each $i = 1, \ldots, n$, $\alpha_i \simeq 0$ rel. $\partial\mathbf{I}$ in $|K|$ because $\alpha_i \simeq \tilde{\alpha}_i\varphi_i q$ rel. $\partial\mathbf{I}$ and $\tilde{\alpha}_i \simeq 0$ in $|K|$. Consequently, $\alpha \simeq \alpha' \simeq 0$ rel. $\partial\mathbf{I}$ in $|K|$. Therefore, $\tilde{f}_\sharp$ is a monomorphism. Thus, $\tilde{f} : |K| \to X$ is a homotopy domination such that $\tilde{f}_\sharp : \pi_1(|\tilde{K}|) \to \pi_1(X)$ is an isomorphism. □

4.4 The Splitting Theorem for $\boldsymbol{Q}$-Manifolds

In this section, we proceed to a discussion in the following setting:

- *Let X and Y be locally compact polyhedra, where X is compact and connected, and let M be an open set in $Y \times \boldsymbol{Q}$ that is homeomorphic to $X \times \boldsymbol{Q} \times \mathbb{R}$ by a homeomorphism $h : X \times \boldsymbol{Q} \times \mathbb{R} \to M$.*

One should keep in mind that $X \times \boldsymbol{Q}$ is embedded into $Y \times \boldsymbol{Q}$ as a bi-collared set, and M is the image of its bi-collar. For each $n \in \mathbb{N}$, we write $\boldsymbol{Q} = [-1, 1]^n \times \boldsymbol{Q}_n$. Let

$$p_n : Y \times \boldsymbol{Q} = Y \times [-1, 1]^n \times \boldsymbol{Q}_n \to Y \times [-1, 1]^n \quad \text{and}$$
$$i_n : Y \times [-1, 1]^n \to Y \times [-1, 1]^n \times \{\mathbf{0}\} \subset Y \times \boldsymbol{Q}$$

be the projection and the natural injection, respectively.

In the above setting, we define a **splitting** of M as a decomposition, $M = M_1 \cup M_2$ (cf. Fig. 4.3), which satisfies the following:

(i) M_1 and M_2 are non-compact $\boldsymbol{Q}$-manifolds which are closed in M,
(ii) $M_0 = M_1 \cap M_2 = p_n^{-1}(p_n(M_0)) = p_n(M_0) \times \boldsymbol{Q}_n$ for some $n \in \mathbb{N}$,
(iii) $A = p_n(M_0)$ is a PL bi-collared compact subpolyhedron of $Y \times [-1, 1]^n$,

where A is **PL bi-collared** in $Y \times [-1, 1]^n$ if there is a PL embedding $k : A \times [-1, 1] \to Y \times [-1, 1]^n$ that is a closed bi-collar in $Y \times [-1, 1]^n$, which is called a

[16] An injective homomorphism is called an monomorphism.

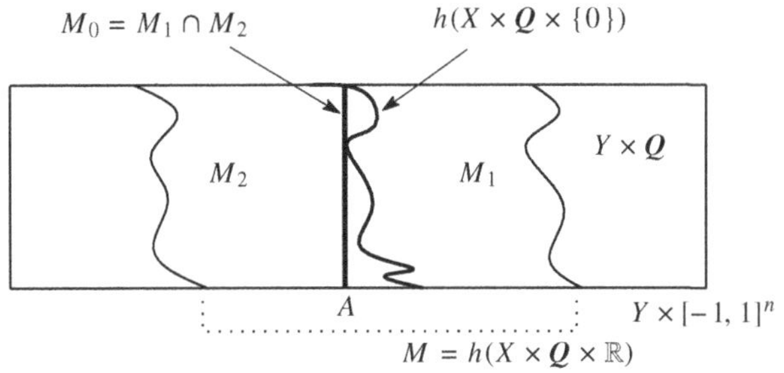

Fig. 4.3 A splitting of M

PL bi-collar of A. Then, we can assume that

$$p_n^{-1}(k(A \times [0, 1])) = k(A \times [0, 1]) \times \boldsymbol{Q}_n \subset M_1 \text{ and}$$
$$p_n^{-1}(k(A \times [-1, 0])) = k(A \times [-1, 0]) \times \boldsymbol{Q}_n \subset M_2.$$

Indeed, since A is compact and

$$p_n^{-1}(k(A \times \{0\})) = k(A \times \{0\}) \times \boldsymbol{Q}_n = A \times \boldsymbol{Q}_n = M_0 \subset M,$$

we can find $0 < \delta \leqslant 1$ such that $p_n^{-1}(k(A \times [-\delta, \delta])) \subset M$. Changing the scale of $[-1, 1]$, we can assume that $p_n^{-1}(k(A \times [-1, 1])) \subset M$. Let C be a component of A. Then, each of $p_n^{-1}(k(C \times (0, 1]))$ and $p_n^{-1}(k(C \times [-1, 0)))$ is contained in one of $M_1 \setminus M_2$ or $M_2 \setminus M_1$ and both of them are not contained in the same side. With respect to each component of A, changing the positive side and the negative side if necessary, we make k so as to satisfy the above.

In this section, applying Theorem 4.1.6, we will prove the following Splitting Theorem:

Theorem 4.4.1 (SPLITTING) *If $\pi_1(X)$ is free or free Abelian, then there exists a splitting $M = M_1 \cup M_2$ such that $M_0 = M_1 \cap M_2$ is a strong deformation retract of M, that is, the inclusion $M_0 \subset M$ is a homotopy equivalence.*

We start by showing the existence of a splitting of M.

Proposition 4.4.2 *M has a splitting.*

Proof Since $h(X \times \boldsymbol{Q} \times \{0\})$ is compact, we can find $n \in \mathbb{N}$ and an open set U in $Y \times [-1, 1]^n$ such that

$$h(X \times \boldsymbol{Q} \times \{0\}) \subset p_n^{-1}(U) = U \times \boldsymbol{Q}_n \subset M.$$

Take a triangulation K of $Y \times [-1, 1]^n$ so that

$$\mathrm{st}(p_n h(X \times \boldsymbol{Q} \times \{0\}), K) \subset U \ (\subset p_n(M)),$$

and let L be the smallest subcomplex of K containing $p_n h(X \times \boldsymbol{Q} \times \{0\})$, that is,

$$L = \{\sigma \in K \mid \exists \tau \in K[p_n h(X \times \boldsymbol{Q} \times \{0\})] \text{ such that } \sigma \leqslant \tau\}.$$

Then, $|L| \subset U$ and L is finite because $p_n h(X \times \boldsymbol{Q} \times \{0\})$ is compact. By Proposition 1.6.8, we can assume that L is a full subcomplex of K_1. Choose $0 < \varepsilon < 1$ so that $|N_\varepsilon(L, K)| \subset U$, where $N_\varepsilon(L, K)$ is an ε-neighborhood of L in K (cf. Sect. 1.9, p. 47). According to Lemma 1.9.3, $N = |N_\varepsilon(L, K)|$ is a regular neighborhood of $|L|$ in $|K|$, which is contained in U. Due to Theorem 1.9.6, bd N is bi-collared in $Y \times [-1, 1]^n$. Note that

$$\operatorname{bd} N \subset p_n k(X \times \boldsymbol{Q} \times (-\infty, 0)) \cup p_n k(X \times \boldsymbol{Q} \times (0, \infty)).$$

We have the following compact PL bi-collared subpolyhedron of $Y \times [-1, 1]^n$ (see Fig. 4.4):

$$\begin{aligned} A &= \operatorname{bd} N \cap p_n k(X \times \boldsymbol{Q} \times (0, \infty)) \\ &= h(h^{-1}(p_n^{-1}(\operatorname{bd} N)) \cap X \times \boldsymbol{Q} \times (0, \infty)). \end{aligned}$$

Then, we have a splitting $M = M_1 \cup M_2$, where

$$\begin{aligned} M_1 &= h(X \times \boldsymbol{Q} \times (0, \infty)) \setminus p_n^{-1}(\operatorname{int} N) \text{ and} \\ M_2 &= h(X \times \boldsymbol{Q} \times (-\infty, 0]) \cup p_n^{-1}(N). \end{aligned}$$

Indeed, M_1 and M_2 are closed in M and non-compact $\boldsymbol{Q}$-manifolds. Observe that $p_n(M_1 \cap M_2) = A$ and $p_n^{-1}(A) = M_1 \cap M_2$. □

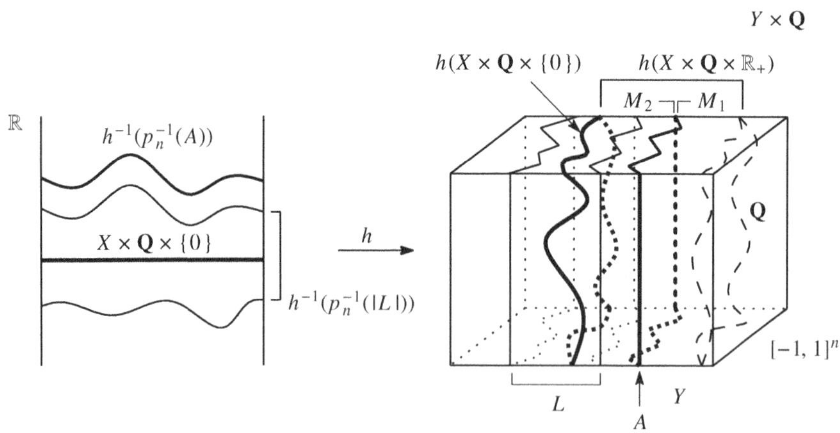

Fig. 4.4 Construction of a splitting

Lemma 4.4.3 *For any splitting $M = M_1 \cup M_2$, each of M_1 and M_2 is homotopy dominated by a finite simplicial complex.*

Proof Since $M_0 = M_1 \cap M_2$ is compact, we can find $a > 0$ such that $M_0 \subset h(X \times \boldsymbol{Q} \times (-a, a))$. Because of connectedness, each of $h(X \times \boldsymbol{Q} \times [a, \infty))$ and $h(X \times \boldsymbol{Q} \times (-\infty, -a])$ is contained in one of $M_1 \setminus M_0$ or $M_2 \setminus M_0$. Moreover, both of them are not contained in the same side. Otherwise, either one of M_1 and M_2 would be contained in $h(X \times \boldsymbol{Q} \times [-a, a])$, which contradicts the non-compactness of M_1 and M_2. We have a strong deformation retraction

$$r : M = h(X \times \boldsymbol{Q} \times \mathbb{R}) \to h(X \times \boldsymbol{Q} \times [-a, a])$$

such that, for every $(x, y) \in X \times \boldsymbol{Q}$,

$$r(h(\{(x, y)\} \times [a, \infty)) = \{h(x, y, a)\} \text{ and}$$
$$r(h(\{(x, y)\} \times (-\infty, -a]) = \{h(x, y, -a)\}.$$

Then, the restrictions $r|M_i : M_i \to M_i \cap h(X \times \boldsymbol{Q} \times [-a, a])$, $i = 1, 2$, are strong deformation retractions. Thus, M_i is homotopy equivalent to

$$M_i \cap h(X \times \boldsymbol{Q} \times [-a, a]),$$

which is a compact ANR. Since each compact ANR is homotopy dominated by a finite simplicial complex (Proposition 4.3.3), we have the result. □

Let K and L be locally finite simplicial complexes, where $K^{(0)}$ is given an order so that K is an ordered simplicial complex. For a proper (equiv. perfect) simplicial map $f : K \to L$ (i.e., $f^{-1}(v)$ is compact for every $v \in L^{(0)}$), the simplicial mapping cylinder Z_f is also locally finite (cf. Sect. 1.10). We use the simplicial mapping cylinder in the proof of the following proposition:

Lemma 4.4.4 *Let $M = M_1 \cup M_2$ be a splitting. Suppose that M_1 is homotopy dominated by a finite simplicial complex K with a homotopy domination $f : |K| \to M_1$. Then, there exists a splitting $M = N_1 \cup N_2$ with the following properties:*

(i) *$N_1 \subset \operatorname{int} M_1$ (equiv. $\operatorname{int} N_2 \supset M_2$),*
(ii) *$N_0 = N_1 \cap N_2$ is strong deformation retract of $M_1 \cap N_2 = N_2 \setminus \operatorname{int} M_2$,*
(iii) *there exists a homotopy equivalence $\alpha : |K| \to N_0$ such that $\alpha \simeq f$ in M_1.*

Proof Since $M = M_1 \cup M_2$ is a splitting, $p_n^{-1}(p_n(M_1 \cap M_2)) = M_1 \cap M_2$ for some $n \in \mathbb{N}$ and $A = p_n(M_1 \cap M_2)$ is a PL bi-collared compact subpolyhedron of $Y \times [-1, 1]^n$. Since $f : |K| \to M_1$ is a homotopy domination, there is a map $g : M_1 \to |K|$ such that $fg \simeq \mathrm{id}_{M_1}$. Note that $i_n(A) \subset M_1$. Applying the Simplicial Approximation Theorem 1.8.8, we have a triangulation L of A and a simplicial map $\psi : L \to K$ that is homotopic to gi_n. By giving an order on $L^{(0)}$ such that L is considered an ordered simplicial complex, we can define the simplicial mapping cylinder Z_ψ of ψ with the simplicial collapsing $\bar{c}_\psi : Z_\psi \to K$ (cf. Sect. 1.10).

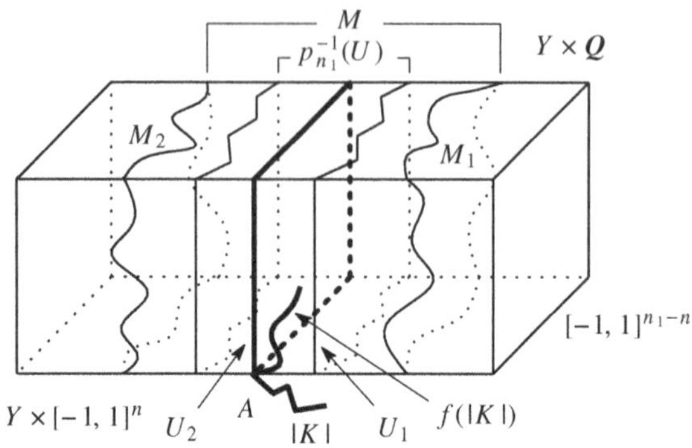

Fig. 4.5 N and bd N

Then, $A = |L| \subset |Z_\psi|$ and $f\bar{c}_\psi|A = f\psi \simeq fgi_n|A \simeq i_n|A$. We can apply the Homotopy Extension Theorem 1.5.7 or 1.13.11 to obtain a map $f' : |Z_\psi| \to M_1$ such that $f'|A = i_n$ and $f' \simeq f\bar{c}_\psi$. It follows that $f'g \simeq f\bar{c}_\psi g = f(\bar{c}_\psi||K|)g = fg \simeq \mathrm{id}_{M_1}$, which means that f' is a homotopy domination. Replacing K and f by Z_ψ and f', we may assume that A is a subpolyhedron of $|K|$ and $f|A = i_n|A$.

Since $f(|K|) \cup p_n^{-1}(A)$ is a compact subset of the open set M in $Y \times \boldsymbol{Q}$, we can find $n_1 \geqslant n$ and an open set U in $Y \times [-1, 1]^{n_1}$ such that

$$f(|K|) \cup p_n^{-1}(A) \subset p_{n_1}^{-1}(U) \subset M.$$

Note that $U \subset p_{n_1}(M)$. Let $U_i = U \cap p_{n_1}(M_i)$, $i = 1, 2$. Refer to Fig. 4.5. Then, $U = U_1 \cup U_2$, $p_{n_1}^{-1}(U_i) \subset M_i$, $i = 1, 2$, and

$$U_1 \cap U_2 = A \times [-1, 1]^{n_1-n} \subset Y \times [-1, 1]^{n_1}.$$

Moreover, $A \times [-1, 1]^{n_1-n}$ is PL bi-collared in $Y \times [-1, 1]^{n_1}$, so in U, which implies that $A \times [-1, 1]^{n_1-n}$ is PL collared in both U_1 and U_2.

Let $m = 2\dim K + 1$ and $n_2 = n_1 + m$. Note that $p_{n_2}f|A = p_{n_2}i_n|A$ is a PL embedding of A into $U_1 \times [-1, 1]^m$ ($\subset Y \times [-1, 1]^{n_2}$). Applying the PL Embedding Approximation Theorem 1.8.11, we can obtain a PL embedding $\varphi : |K| \to U_1 \times [-1, 1]^m$ ($\subset Y \times [-1, 1]^{n_2}$) such that $\varphi|A = p_{n_2}f|A$ and $\varphi \simeq p_{n_2}f$ rel. A in $U_1 \times [-1, 1]^m$. Since $A \times [-1, 1]^{n_2-n}$ is PL collared in $U_1 \times [-1, 1]^m$, we can push $\varphi(|K| \setminus A)$ off $A \times [-1, 1]^{n_2-n}$ so as to satisfy

$$\varphi(|K|) \cap (A \times [-1, 1]^{n_2-n}) = \varphi(A) = A \times \{\mathbf{0}\},$$

where $\mathbf{0} \in [-1, 1]^{n_2-n}$. Consider the following subpolyhedron of $Y \times [-1, 1]^{n_2}$:

$$P = (A \times [-1, 1]^{n_2-n+1}) \cup (\varphi(|K|) \times \{1\}).$$

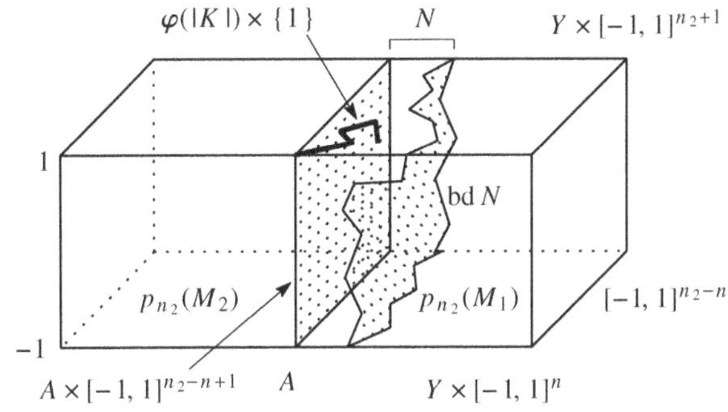

Fig. 4.6 P and bd N

Then, $\varphi(|K|)\times\{1\}$ is a strong deformation retract of P, hence the inclusion $\varphi(|K|)\times\{1\} \subset P$ is a homotopy equivalence. Let N be a regular neighborhood of P in $U_1 \times [-1, 1]^{m+1}$. Then, P is a strong deformation retract of N, hence the inclusion $P \subset N$ is a homotopy equivalence. Moreover, $p_{n_2+1}^{-1}(N) \cap M_2 = M_1 \cap M_2 = p_n^{-1}(A)$. Refer to Fig. 4.6.

Now, the desired splitting $M = N_1 \cup N_2$ can be defined as follows:

$$N_2 = M_2 \cup p_{n_2+1}^{-1}(N) \quad \text{and} \quad N_1 = M \setminus \operatorname{int} N_2.$$

Then, $N_1 \subset M \setminus M_2 = \operatorname{int} M_1$, that is, (i) holds. Observe that $N_1 \cap N_2 = p_{n_2+1}^{-1}(\operatorname{bd} N)$ and $N_2 \setminus \operatorname{int} M_2 = p_{n_2+1}^{-1}(N)$. Hence, to see (ii), we have to show that bd N is a strong deformation retract of N. Since N is a regular neighborhood of P, bd N is a strong deformation retract of $N \setminus P$ (Theorem 1.9.6). Since P is a Z-set in N, $N \setminus P$ is homotopy dense in N. Then, it is easy to see that bd N is a strong deformation retract of N. Thus, we have (ii). For (iii), recall that $N_1 \cap N_2 = p_{n_2+1}^{-1}(\operatorname{bd} N)$. Let $\alpha : |K| \to p_{n_2+1}^{-1}(\operatorname{bd} N)$ be the following composition of homotopy equivalences:

$$|K| \xrightarrow{\varphi} \varphi(|K|) = \varphi(|K|) \times \{1\} \subset P \subset N \xrightarrow{r} \operatorname{bd} N \xrightarrow{i_{n_2+1}} p_{n_2+1}^{-1}(\operatorname{bd} N),$$

where $r : N \to \operatorname{bd} N$ is a strong deformation retraction. □

Proposition 4.4.5 *There exists a splitting $M = M_1 \cup M_2$ such that $M_0 = M_1 \cap M_2$, M_1, and M_2 are connected and the inclusions $M_0 \subset M_i$, $i = 1, 2$, induce isomorphisms between their fundamental groups.*

Proof Let $M = M_1' \cup M_2'$ be any splitting (the existence is guaranteed by Proposition 4.4.2). Then, M_1' is homotopy dominated by a finite simplicial complex by Proposition 4.4.3. By Theorem 4.3.10, we have a finite simplicial complex K and a homotopy domination $f : |K| \to M_1'$ that induces a bijection between their components and isomorphisms between the fundamental groups of their corresponding components. Applying Lemma 4.4.4, we can obtain a splitting $M = N_1 \cup N_2$ with the following properties:

(1) $N_1 \subset \operatorname{int} M'_1$,
(2) $N_0 = N_1 \cap N_2$ is a strong deformation retract of $M'_1 \cap N_2 = N_2 \setminus \operatorname{int} M'_2$,
(3) there exists a homotopy equivalence $\alpha : |K| \to N_0$ such that $\alpha \simeq f$ in M'_1.

It follows from (2) that N_1 is a strong deformation retract of M'_1, that is, there is a strong deformation retraction $r : M'_1 \to N_1$. By (3), α has a homotopy inverse $\beta : N_0 \to |K|$. See the following diagram:

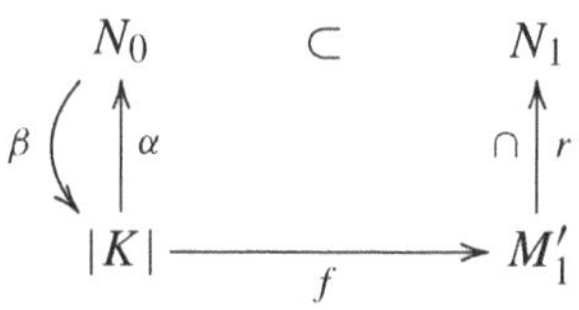

Observe that $rf\beta \simeq r\alpha\beta \simeq r|N_0 = \mathrm{id}$ in N_1, that is, $rf\beta$ is homotopic to the inclusion $N_0 \subset N_1$. Since r and β are homotopy equivalences and f induces a bijection between the components and isomorphisms between the fundamental groups of these corresponding components, the inclusion $N_0 \subset N_1$ induces a bijection between the components and isomorphisms between the fundamental groups of these corresponding components.

Using the same procedure as above, we can obtain a finite simplicial complex L with a homotopy domination $g : |L| \to N_2$ that induces a bijection between their components and isomorphisms between the fundamental groups of their corresponding components, and a splitting $M = M_1 \cup M_2$ with the following properties:

(4) $M_2 \subset \operatorname{int} N_2$, i.e., $\operatorname{int} M_1 \supset N_1$,
(5) $M_0 = M_1 \cap M_2$ is a strong deformation retract of $N_2 \cap M_1 = M_1 \setminus \operatorname{int} N_1$,
(6) there exists a homotopy equivalence $\beta : |L| \to M_0$ such that $\beta \simeq g$ in N_2.

Then, by the same argument as above, the inclusion $M_0 \subset M_2$ induces a bijection between the components and isomorphisms between the fundamental groups of these corresponding components.

We show that the inclusion $M_0 \subset M_1$ induces a bijection between the components, that is, if C be a component of M_1, then $C \cap M_0$ is a component of M_0. Since the inclusion $M_0 \subset M_1 \setminus \operatorname{int} N_1$ is a homotopy equivalence by (5) and $C \cap M_0 = (C \setminus \operatorname{int} N_1) \cap M_0$, it suffices to show that $C \setminus \operatorname{int} N_1$ is a component of $M_1 \setminus \operatorname{int} N_1$. If $C \subset M_1 \setminus N_1$, then $C = C \setminus \operatorname{int} N_1$ is a component of $M_1 \setminus \operatorname{int} N_1$. When $C \cap N_1 \neq \emptyset$, each component of $C \cap N_1$ is a component of N_1, which meets N_0 because the inclusion $N_0 \subset N_1$ induces a surjection between the components. Therefore,

$$C \setminus \operatorname{int} N_1 = C \cap (M_1 \setminus \operatorname{int} N_1) \supset C \cap N_0 \neq \emptyset.$$

Assume that $C \setminus \operatorname{int} N_1$ is disconnected. Since C is a path-component of M_1 because M_1 is locally path-connected, we have a path $\alpha : \mathbf{I} \to C$ such that $\alpha(0)$ and $\alpha(1)$ belong to different path-components of $C \setminus \operatorname{int} N_1$. Then, we can assume that $\alpha(\mathbf{I}) \subset C \cap N_1$. Indeed, let C_0 be a path-component of $C \setminus \operatorname{int} N_1$ with $\alpha(0) \in C_0$ and replace $\mathbf{I}$ with $[t_0, t_1]$, where

$$t_0 = \sup\left\{t \in \mathbf{I} \,\middle|\, \alpha(t) \in C_0\right\} \text{ and } t_1 = \inf\left\{t > t_0 \,\middle|\, \alpha(t) \in C \setminus \operatorname{int} N_1\right\}.$$

Since the inclusion $N_0 \subset N_1$ induces an injection between the components (= the path-components), we have a path $\beta : \mathbf{I} \to N_0$ such that $\beta(0) = \alpha(0)$ and $\beta(1) = \alpha(1)$, where $\beta(\mathbf{I}) \subset C$ because C is a path-component of M_1. So, $\alpha(0)$ and $\alpha(1)$ are connected by a path in $C \cap N_0 \subset C \setminus \operatorname{int} N_1$, hence $\alpha(1) \in C_0$ which is a contradiction. Consequently, $C \setminus \operatorname{int} N_1$ is connected. Then, it follows that $C \setminus \operatorname{int} N_1$ is a component of $M_1 \setminus \operatorname{int} N_1$.

Now, we show that M_0, M_1, and M_2 are connected. Since the inclusions $M_0 \subset M_i$, $i = 1, 2$, induce bijections between the components, it suffices to show that M_0 is path-connected. Let $x, y \in M_0$. Since $M = h(X \times \boldsymbol{Q} \times \mathbb{R})$, there is a path $\gamma : \mathbf{I} \to M$ such that $\gamma(0) = x$ and $\gamma(1) = y$. Because M_0 is bi-collared in M, we have a neighborhood U of M_0 in M and a strong deformation retraction $r : U \to M_0$ (i.e., r is a retraction such that $r \simeq \mathrm{id}$ rel. M_0 in U). Then, we have $0 = t_0 < t_1 < \cdots < t_n = 1$ such that if $\gamma([t_{i-1}, t_i]) \not\subset U$, then $\gamma([t_{i-1}, t_i]) \subset M_1$ or $\gamma([t_{i-1}, t_i]) \subset M_2$. When $\gamma([t_{i-1}, t_i]) \not\subset U$, since the inclusions $M_0 \subset M_1$ and $M_0 \subset M_2$ induce injections between the components, then there exists a path $\gamma_i : [t_{i-1}, t_i] \to M_0$ with $\gamma_i(t_{i-1}) = \gamma(t_{i-1})$ and $\gamma_i(t_i) = \gamma(t_i)$. We can define a path $\gamma' : \mathbf{I} \to M_0$ as follows:

$$\gamma'|[t_{i-1}, t_i] = \begin{cases} r\gamma|[t_{i-1}, t_i] & \text{if } \gamma([t_{i-1}, t_i]) \subset U, \\ \gamma_i & \text{otherwise.} \end{cases}$$

Then, $\gamma'(0) = \gamma(0) = x$ and $\gamma'(1) = \gamma(1) = y$. Thus, M_0 is path-connected, so connected.

It remains to prove that the inclusion $M_0 \subset M_1$ induces an isomorphism between the fundamental groups. Note that the inclusion $M_0 \subset M_1$ is a composition of two inclusions $M_0 \subset M_1 \setminus \operatorname{int} N_1 \subset M_1$, where the first inclusion is homotopy equivalence by (5). Thus, it suffices to show that the inclusion $M_1 \setminus \operatorname{int} N_1 \subset M_1$ induces an isomorphism between the fundamental groups. Observe that

$$N_1 \cap (M_1 \setminus \operatorname{int} N_1) = N_1 \cap N_2 = N_0 \text{ and } N_1 \cup (M_1 \setminus \operatorname{int} N_1) = M_1.$$

Recall that the inclusion $N_0 \subset N_1$ induces a bijection between the components and isomorphisms between the fundamental groups of these corresponding components. By connecting all components of N_0 by a tree in $M_1 \setminus \operatorname{int} N_1$, we can reduce N_0 and N_1 to connected, where the above equalities are kept. Then, their fundamental groups are the free products of the fundamental groups of their components and the new inclusion induces an isomorphism between their fundamental groups. Thus, it can be assumed that N_0 and N_1 are connected. Applying the Seifert–van Kampen Theorem 4.2.6, we can obtain a unique homomorphism $\phi : \pi_1(M_1) \to \pi_1(M_1 \setminus \operatorname{int} N_1)$ such that $\phi j_2 = \mathrm{id}$ and $\phi j_1 = i_2 i_1^{-1}$, which implies that j_2 is a monomorphism. See the following commutative diagram:

$$
\begin{array}{ccccc}
\pi_1(N_0) & \xrightarrow[\cong]{i_1} & \pi_1(N_1) & \xrightarrow{i_2 i_1^{-1}} & \\
\downarrow{\scriptstyle i_2} & & \downarrow{\scriptstyle j_1} & & \\
\pi_1(M_1 \setminus \operatorname{int} N_1) & \xrightarrow{j_2} & \pi_1(M_1) & \overset{\phi}{\dashrightarrow} & \pi_1(M_1 \setminus \operatorname{int} N_1) \\
 & & \underset{\mathrm{id}}{=} & &
\end{array}
$$

where i_1, i_2, j_1, and j_2 are homomorphisms induced by the inclusions. Moreover, $\pi_1(M_1)$ is generated by $j_1(\pi_1(N_1)) \cup j_2(\pi_1(M_1 \setminus \operatorname{int} N_1))$. Since i_1 is an isomorphism, it follows that

$$j_1(\pi_1(N_1)) = j_1 i_1(\pi_1(N_0)) = j_2 i_2(\pi_1(N_0)) \subset j_2(\pi_1(M_1 \setminus \operatorname{int} N_1)).$$

Hence, $\pi_1(M_1) = j_2(\pi_1(M_1 \setminus \operatorname{int} N_1))$, which means that j_2 is an epimorphism. Thus, j_2 is an isomorphism. □

Now, using Theorem 4.1.6, we can prove the Splitting Theorem 4.4.1.

Proof (Splitting Theorem 4.4.1) By Proposition 4.4.5, we have a splitting $M = M'_1 \cup M'_2$ such that $M'_0 = M'_1 \cap M'_2$, M'_1, and M'_2 are connected and the inclusions $M'_0 \subset M'_i$, $i = 1, 2$, induce isomorphisms between their fundamental groups. Then, we apply the Seifert–van Kampen Theorem 4.2.6 to show that the inclusions $M'_i \subset M$, $i = 1, 2$, induce isomorphisms between their fundamental groups. Indeed, using the same argument as in the final paragraph of the proof of Proposition 4.4.5, since i_1 (resp. i_2) is an isomorphism, it follows that j_1 (resp. j_2) is also an isomorphism. See the following commutative diagram:

$$
\begin{array}{ccc}
\pi_1(M'_0) & \xrightarrow[\cong]{i_1} & \pi_1(M'_1) \\
{\scriptstyle i_2}\downarrow{\scriptstyle \cong} & & \downarrow{\scriptstyle j_1} \\
\pi_1(M'_2) & \xrightarrow[j_2]{} & \pi_1(M)
\end{array}
$$

Since $\pi_1(M'_1) \cong \pi_1(M) \cong \pi_1(X)$ $(M \approx X \times \boldsymbol{Q} \times \mathbb{R})$, it follows that $\pi_1(M'_1)$ is also free or free Abelian. Due to Lemma 4.4.3, M'_1 is homotopy dominated by a finite simplicial complex. We can apply Theorem 4.1.6 to obtain a finite complex K and a homotopy equivalence $f : |K| \to M'_1$. Applying Lemma 4.4.4, we can obtain a splitting $M = N_1 \cup N_2$ with the following properties:

(1) $N_1 \subset \operatorname{int} M_1'$,
(2) $N_0 = N_1 \cap N_2$ is a strong deformation retract of $M_1' \cap N_2 = N_2 \setminus \operatorname{int} M_2'$,
(3) there exists a homotopy equivalence $\alpha : |K| \to N_0$ such that $\alpha \simeq f$ in M_1'.

By the same the same argument as in the proof of Proposition 4.4.5, we can see that the inclusion $N_0 \subset N_1$ is homotopic to $rf\beta$, where $\beta : N_0 \to |K|$ is a homotopy inverse of α and $r : M_1' \to N_1$ is a strong deformation retraction. Since f is a homotopy equivalence, the inclusion $N_0 \subset N_1$ is a homotopy equivalence, which means that N_0 is a strong deformation retract of N_1 (Proposition 1.13.15). This implies that N_2 is a strong deformation retract of M, that is, the inclusion $N_2 \subset M$ is a homotopy equivalence. Therefore, $N_2 \simeq X$, so N_2 has the homotopy type of a finite simplicial complex.

Again using the above procedure, we can obtain a new splitting, $M = M_1 \cup M_2$, such that

(4) $M_2 \subset \operatorname{int} N_2$, i.e., $\operatorname{int} M_1 \supset N_1$,
(5) $M_0 = M_1 \cap M_2 \subset N_2 \cap M_1 = M_1 \setminus \operatorname{int} N_1$ is a homotopy equivalence,
(6) $M_0 \subset M_2$ is a homotopy equivalence, equivalently M_0 is a strong deformation retract of M_2.

As observed above, N_0 is a strong deformation retract of N_1, from which it follows that $M_1 \setminus \operatorname{int} N_1$ is a strong deformation retract of M_1. Thus, $M_1 \setminus \operatorname{int} N_1 \subset M_1$ is a homotopy equivalence. Combining this with (5), $M_0 \subset M_1$ is a homotopy equivalence, which means that M_0 is a strong deformation retract of M_1 (Proposition 1.13.15). It follows from this with (6) that M_0 is a strong deformation retract of M, that is, the inclusion $M_0 \subset M$ is a homotopy equivalence. □

4.5 An Immersion of a Punctured n-Torus into $\mathbb{R}^n$

A map $\alpha : X \to Y$ is called an **immersion** if α is locally an open embedding, that is, each $x \in X$ has an open neighborhood U in X such that $\alpha|U$ is an open embedding. The ***n*-torus (*n*-dimensional torus)**[17] $\mathbf{T}^n$ is defined as the n-hold product of the unit sphere $\mathbf{S}^1$, that is,

$$\mathbf{T}^n = (\mathbf{S}^1)^n = \underbrace{\mathbf{S}^1 \times \cdots \times \mathbf{S}^1}_{n \text{ many}}.$$

Let $e : \mathbb{R} \to \mathbf{S}^1$ be the covering projection defined by $e(t) = (\cos \pi t, \sin \pi t)$ for every $t \in \mathbb{R}$. When every $(x, y) \in \mathbb{R}^2$ is identified with $x + iy \in \mathbb{C}$ (i.e., we regard $\mathbb{R}^2 = \mathbb{C}$), we can describe as follows:

$$e(t) = \exp(\pi i t) = \cos \pi t + i \sin \pi t \quad \text{for every } t \in \mathbb{R}.$$

[17]The plural is **tori**.

Note that $e([-1, 1]) = \mathbf{S}^1$ and $e|(-1, 1)$ is injective. The covering projection $e^n : \mathbb{R}^n \to \mathbf{T}^n$ is defined by $e^n(x) = (e(x(1)), \ldots, e(x(n)))$, where $e^n([-1, 1]^n) = \mathbf{T}^n$ and $e^n|(-1, 1)^n$ is injective. We will use the **punctured torus** $\mathbf{T}^n_0 = \mathbf{T}^n \setminus \{p_0\}$, where $p_0 = e^n(1, \ldots, 1) \notin e^n([-1/4, 1/4]^n)$. Immersions of $\mathbf{T}^n_0$ play a key role in our proof of the Handle Straightening Theorem in the next section. In this section, we will prove the following:

Theorem 4.5.1 *There exists an immersion* $\alpha : \mathbf{T}^n_0 \to \mathbb{R}^n$ *such that*

$$\alpha e^n|[-1/4, 1/4]^n = \mathrm{id}.$$

For the two-dimensional punctured torus $\mathbf{T}^2_0$, such an immersion can be constructed as indicated in Fig. 4.7. But, it is hard to extend such a construction to the n-dimensional punctured tori $\mathbf{T}^n_0$, $n > 2$. To prove Theorem 4.5.1, we start at the following lemma, which can be easily proved:

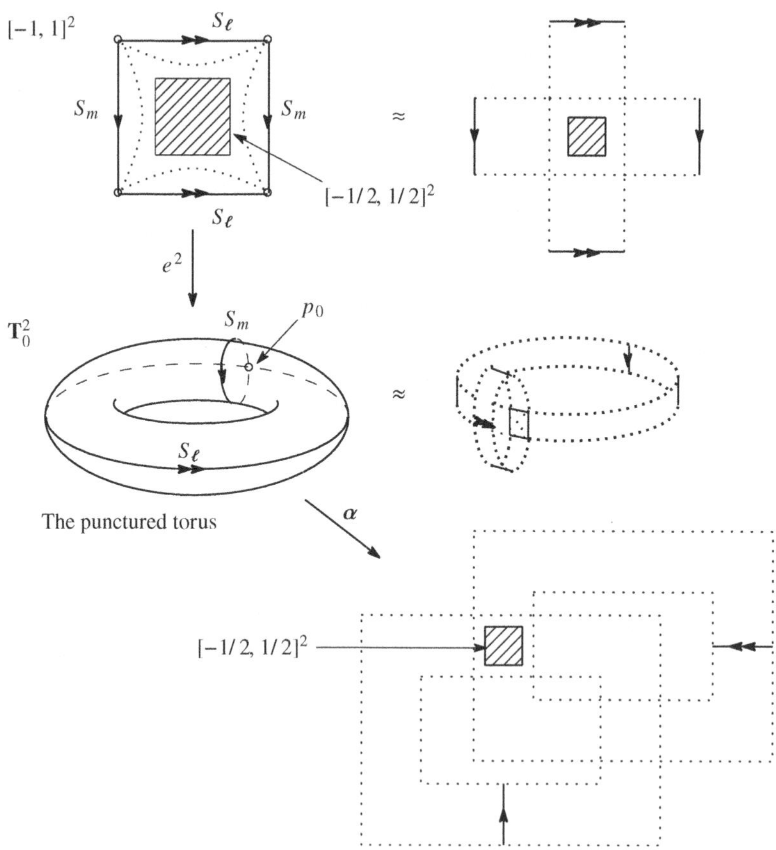

Fig. 4.7 An immersion $\alpha : \mathbf{T}^2_0 \to \mathbb{R}^2$

Lemma 4.5.2 *Let $f : \mathbf{T}^n \times [-1, 1] \to \mathbb{R}^{n+1}$ be an embedding such that $0 < \mathrm{pr}_{n+1} f(x, t) < c$ for every $(x, t) \in \mathbf{T}^n \times [-1, 1]$. Then, the map $f^* : \mathbf{T}^{n+1} \times [-1, 1] \to \mathbb{R}^{n+2}$ defined in the following formula is an embedding such that $\mathrm{pr}_{n+2} f^*(x, t) > 0$ for every $(x, t) \in \mathbf{T}^{n+1} \times [-1, 1]$:*

$$\begin{aligned} f^*(x, e(s), t) = (&\mathrm{pr}_1 f(x, t), \ldots, \mathrm{pr}_n f(x, t), \\ &\mathrm{pr}_{n+1} f(x, t) \cos \pi s, \mathrm{pr}_{n+1} f(x, t) \sin \pi s + c), \end{aligned}$$

where $(x, e(s), t) \in \mathbf{T}^n \times \mathbf{S}^1 \times [-1, 1] = \mathbf{T}^{n+1} \times [-1, 1]$ and $s \in [-1, 1]$.

Proof Since $\mathbf{T}^{n+1} \times [-1, 1]$ is compact, it suffices to show that f^* is a continuous injection. The continuity of f^* is obvious from the definition. To see that f^* is injective, assume that $f^*(x, e(s), t) = f^*(x', e(s'), t')$, where $(x, e(s), t), (x', e(s'), t') \in \mathbf{T}^n \times \mathbf{S}^1 \times [-1, 1]$ and $s, s' \in [-1, 1]$. Then, $\mathrm{pr}_i f(x, t) = \mathrm{pr}_i f(x', t')$ for each $i = 1, \ldots, n$ and

$$\begin{aligned} \mathrm{pr}_{n+1} f(x, t) &= \sqrt{(\mathrm{pr}_{n+1} f(x, t) \cos \pi s)^2 + (\mathrm{pr}_{n+1} f(x, t) \sin \pi s)^2} \\ &= \sqrt{(\mathrm{pr}_{n+1} f(x', t') \cos \pi s')^2 + (\mathrm{pr}_{n+1} f(x', t') \sin \pi s')^2} \\ &= \mathrm{pr}_{n+1} f(x', t'), \end{aligned}$$

hence $f(x, t) = f(x', t')$. Because f is injective, we have $(x, t) = (x', t')$. Since $\mathrm{pr}_{n+1} f(x, t) = \mathrm{pr}_{n+1} f(x', t')$, it follows that

$$e(s) = (\cos \pi s, \sin \pi s) = (\cos \pi s', \sin \pi s') = e(s').$$

Therefore, $(x, e(s), t) = (x', e(s'), t')$. Thus, f^* is an injection. Moreover,

$$\begin{aligned} \mathrm{pr}_{n+2} f^*(x, e(s), t) &= \mathrm{pr}_{n+1} f(x, t) \sin \pi s + c \\ &\geqslant c - \mathrm{pr}_{n+1} f(x, t) > 0. \end{aligned}$$

Hence, $\mathrm{pr}_{n+2} f^*(x, t) > 0$ for every $(x, t) \in \mathbf{T}^{n+1} \times [-1, 1]$. □

We will define immersions $h_n : \mathbb{R}^n \times [-1, 1] \to \mathbb{R}^{n+1}$, $n \in \mathbb{N}$, which induce embeddings $\tilde{h}_n : \mathbf{T}^n \times [-1, 1] \to \mathbb{R}^{n+1}$ such that $h_n = \tilde{h}_n(e^n \times \mathrm{id})$. First, we inductively define maps $f_n, g_n : \mathbb{R}^n \times [-1, 1] \to \mathbb{R}$, $n \in \mathbb{N}$, as follows:

$$\begin{aligned} f_1(x_1, t) = (2 + t) \cos \pi x_1, \ g_1(x_1, t) &= (2 + t) \sin \pi x_1 + 2^2, \\ f_n(x_1, \ldots, x_n, t) &= g_{n-1}(x_1, \ldots, x_{n-1}, t) \cos \pi x_n, \text{ and} \\ g_n(x_1, \ldots, x_n, t) &= g_{n-1}(x_1, \ldots, x_{n-1}, t) \sin \pi x_n + 2^{n+1}. \end{aligned}$$

Let $h_n : \mathbb{R}^n \times [-1, 1] \to \mathbb{R}^{n+1}$, $n \in \mathbb{N}$, be maps defined as follows:

$$h_n(x_1, \ldots, x_n, t) = \big(f_1(x_1, t), f_2(x_1, x_2, t), \\ \ldots, f_n(x_1, \ldots, x_n, t), g_n(x_1, \ldots, x_n, t)\big).$$

In other words, h_n can be described as follows:

$$\begin{aligned} h_1(x_1, t) &= ((2+t)\cos \pi x_1, (2+t)\sin \pi x_2 + 4); \\ h_2(x_1, x_2, t) &= ((2+t)\cos \pi x_1, ((2+t)\sin \pi x_1 + 4)\cos \pi x_2, \\ &\qquad ((2+t)\sin \pi x_1 + 4)\sin \pi x_2 + 8); \\ h_3(x_1, x_2, x_3, t) &= ((2+t)\cos \pi x_1, ((2+t)\sin \pi x_1 + 4)\cos \pi x_2, \\ &\qquad (((2+t)\sin \pi x_1 + 4)\sin \pi x_2 + 8)\cos \pi x_3, \\ &\qquad (((2+t)\sin \pi x_1 + 4)\sin \pi x_2 + 8)\sin \pi x_3 + 16); \\ &\cdots. \end{aligned}$$

Then, as is easily observed, h_n induces the map $\tilde{h}_n : \mathbf{T}^n \times [-1, 1] \to \mathbb{R}^{n+1}$ such that $h_n = \tilde{h}_n(e^n \times \mathrm{id})$, that is,

$$h_n(x_1, \ldots, x_n, t) = \tilde{h}_n(e^n(x_1, \ldots, x_n), t) = \tilde{h}_n(e(x_1), \ldots, e(x_n), t).$$

By using Lemma 4.5.2, we can easily show that $\tilde{h}_n$ is an embedding, which is called the **standard embedding** of $\mathbf{T}^n \times [-1, 1]$ into $\mathbb{R}^{n+1}$ (Fig. 4.8). Hence, h_n is an immersion.

We have also the following:

$$\begin{aligned} g_1(x_1, t) &= (2+t)\sin \pi x_1 + 2^2; \\ g_2(x_1, x_2, t) &= (2+t)\sin \pi x_1 \sin \pi x_2 + 2^2 \sin \pi x_2 + 2^3; \\ g_3(x_1, x_2, x_3, t) &= (2+t)\sin \pi x_1 \sin \pi x_2 \sin \pi x_3 \\ &\qquad + 2^2 \sin \pi x_2 \sin \pi x_3 + 2^3 \sin \pi x_3 + 2^4; \\ &\cdots. \end{aligned}$$

We are ready to start proving Theorem 4.5.1.

Proof (Theorem 4.5.1) Recall $p_0 = e^n(1, \ldots, 1)$ $(= e^n(\pm 1, \ldots, \pm 1))$ and let

$$S = \big\{x \in \mathbb{R}^n \bigm| x(i) = 0 \text{ for some } i = 1, \ldots, n\big\} \text{ and}$$
$$S_\varepsilon = \big\{x \in \mathbb{R}^n \bigm| |x(i)| < \varepsilon \text{ for some } i = 1, \ldots, n\big\},\ 0 < \varepsilon < 1.$$

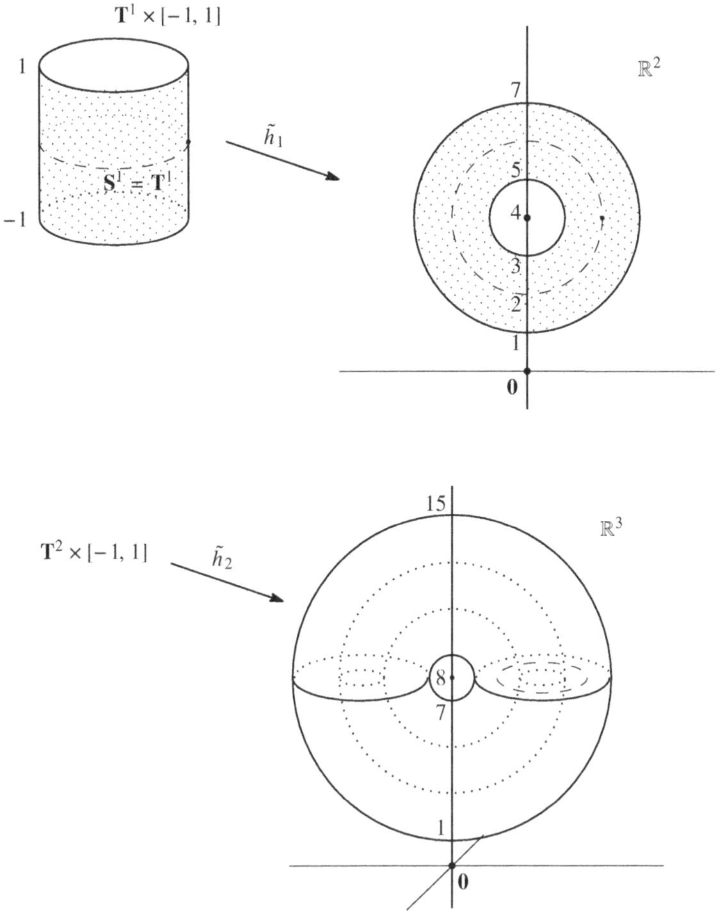

Fig. 4.8 The standard embedding $\tilde{h}_n : \mathbf{T}^n \times [-1, 1] \to \mathbb{R}^{n+1}$

Then, $\mathbf{T}^n_0 = \mathbf{T}^n \setminus \{p_0\} \approx e^n(S_\varepsilon)$. Indeed,

$$\begin{aligned}\mathbf{T}^n \setminus e^n(S_\varepsilon) &= \left\{ e^n(x) \in \mathbf{T}^n \;\middle|\; |x(i)| \geqslant \varepsilon \text{ for each } i = 1, \ldots, n \right\} \\ &= e^n\big([1-\varepsilon, 1+\varepsilon]^n\big).\end{aligned}$$

Let $\varphi : \mathbb{R}^n \to \mathbb{R}$ be the map defined by

$$\begin{aligned}\varphi(x) &= 2^{-n} \sin \pi x_1 \cdots \sin \pi x_n + 2^{-n+1} \sin \pi x_2 \cdots \sin \pi x_n \\ &\qquad\qquad + \cdots + 2^{-2} \sin \pi x_{n-1} \sin \pi x_n + 2^{-1} \sin \pi x_n \\ &= 2^{-(n+1)} g_n(x, 0) - 1,\end{aligned}$$

where $x = (x_1, \dots, x_n) \in \mathbb{R}^n$. Using maps $f_1, \dots, f_n$ defined above and φ, we define $f_\varphi : \mathbb{R}^n \to \mathbb{R}^n$ as follows:

$$f_\varphi(x) = \big(f_1(x_1, \varphi(x)),\ f_2(x_1, x_2, \varphi(x)), \dots, f_n(x_1, \dots, x_n, \varphi(x))\big)$$
$$\text{for each } x = (x_1, \dots, x_n) \in \mathbb{R}^n.$$

In other words, $f_\varphi(x) = ph_n(x)$, where $p : \mathbb{R}^{n+1} = \mathbb{R}^n \times \mathbb{R} \to \mathbb{R}^n$ is the projection and $h_n : \mathbb{R}^n \times [-1, 1] \to \mathbb{R}^{n+1}$ is the immersion defined above.

Now, using elementary properties of determinations, we can compute the Jacobian determinant of h as follows:

$$\begin{aligned}
\det J_{f_\varphi} &= \det\left(\frac{\partial f_i}{\partial x_j} + \frac{\partial f_i}{\partial t}\frac{\partial \varphi}{\partial x_j}\right) \\
&= \det\left(\begin{array}{c|c} \dfrac{\partial f_i}{\partial x_j} + \dfrac{\partial f_i}{\partial t}\dfrac{\partial \varphi}{\partial x_j} & \mathbf{0} \\ \hline \dfrac{\partial \varphi}{\partial x_j} & 1 \end{array}\right) = \det\left(\begin{array}{c|c} \dfrac{\partial f_i}{\partial x_j} & -\dfrac{\partial f_i}{\partial t} \\ \hline \dfrac{\partial \varphi}{\partial x_j} & 1 \end{array}\right) \\
&= \det\left(\begin{array}{c|c} \dfrac{\partial f_i}{\partial x_j} & -\dfrac{\partial f_i}{\partial t} \\ \hline \dfrac{\partial \varphi}{\partial x_j} & 0 \end{array}\right) + \det\left(\begin{array}{c|c} \dfrac{\partial f_i}{\partial x_j} & -\dfrac{\partial f_i}{\partial t} \\ \hline \mathbf{0} & 1 \end{array}\right).
\end{aligned}$$

Observe that $\partial f_i/\partial x_j = 0$ if $i < j$, and $\partial f_i/\partial x_i$ has a factor of $\sin \pi x_i$. In the last part of the above equality, on S, the upper left-hand corner of the second matrix is triangular with at least one zero on the diagonal, hence the determinant of this matrix is zero. Since $\varphi(x) = 2^{-(n+1)} g_n(x, 0)$, it follows that $\partial\varphi/\partial x_j = 2^{-(n+1)}\partial g_n/\partial x_j$ at $x \in S$ and $t = 0$. Moreover, $\partial g_n/\partial t = \sin \pi x_1 \cdots \sin \pi x_n = 0$ on S. Hence, at $x \in S$ and $t = 0$,

$$\det J_{f_\varphi} = \det\left(\begin{array}{c|c} \dfrac{\partial f_i}{\partial x_j} & -\dfrac{\partial f_i}{\partial t} \\ \hline \dfrac{\partial \varphi}{\partial x_j} & 0 \end{array}\right) = -2^{-(n+1)} \det\left(\begin{array}{c|c} \dfrac{\partial f_i}{\partial x_j} & \dfrac{\partial f_i}{\partial t} \\ \hline \dfrac{\partial g_n}{\partial x_j} & \dfrac{\partial g_n}{\partial t} \end{array}\right) = -\frac{\det J_{h_n}}{2^{n+1}}.$$

Since h_n is an immersion, it follows that $\det J_{f_\varphi} = -2^{-(n+1)} \det J_{h_n} \neq 0$ at $x \in S$ and $t = 0$. Then, each point $x \in S$ has a neighborhood U_x in $\mathbb{R}^n$ such that $f_\varphi|U_x$ is an embedding. Since S is compact, we can choose small $\varepsilon > 0$ so that $f_\varphi|S_\varepsilon$ is an immersion. Recall that $\mathbf{T}_0 \approx S_\varepsilon$. Hence, we have an immersion $\alpha' : \mathbf{T}_0 \to \mathbb{R}^n$.

Finally, we modify this immersion α' to an immersion $\alpha : \mathbf{T}_0^n \to \mathbb{R}^n$ so that $\alpha e^n|[-1/4, 1/4]^n = \mathrm{id}$. Choose a sufficiently small $\delta > 0$ so that $\alpha' e^n|[-\delta, \delta]^n : [-\delta, \delta]^n \to \mathbb{R}^n$ is an embedding. We have a homeomorphism $\beta : [-1, 1]^n \to [-1, 1]^n$ such that $\beta|\partial[-1, 1]^n = \mathrm{id}$ and $\beta([-1/4, 1/4]^n) = [-\delta/2, \delta/2]^n$. This β induces a homeomorphism $\tilde{\beta} : \mathbf{T}^n \to \mathbf{T}^n$ such that $\tilde{\beta} e^n|[-1, 1]^n = e^n\beta$, where

$\tilde{\beta}(\mathbf{T}_0^n) = \mathbf{T}_0^n$ because

$$\tilde{\beta}(p_0) = \tilde{\beta}e^n(1, \ldots, 1) = e^n\beta(1, \ldots, 1) = e^n(1, \ldots, 1) = p_0.$$

Then, $\alpha'\tilde{\beta}|\mathbf{T}_0^n : \mathbf{T}_0^n \to \mathbb{R}^n$ is an immersion and the following is an embedding:

$$\alpha'\tilde{\beta}e^n|[-1/4, 1/4]^n = \alpha'e^n\beta|[-1/4, 1/4]^n : [-1/4, 1/4]^n \to \mathbb{R}^n.$$

Since $\alpha'e^n(\partial[-\delta/2, \delta/2]^n)$ is bi-collared in $\mathbb{R}^n$, the generalized Schoenflies Theorem 2.7.7 provides a homeomorphism $\psi : \mathbb{R}^n \to \mathbb{R}^n$ such that $\psi([-1/4, 1/4]^n) = \alpha'e^n([-\delta/2, \delta/2]^n)$. Since $\psi^{-1}\alpha'\tilde{\beta}e^n(\partial[-1/4, 1/4]^n) = \partial[-1/4, 1/4]^n$, the homeomorphism

$$\psi^{-1}\alpha'\tilde{\beta}e^n|[-1/4, 1/4]^n : [-1/4, 1/4]^n \to [-1/4, 1/4]^n$$

can be extended to a homeomorphism $\varphi : \mathbb{R}^n \to \mathbb{R}^n$ by

$$\varphi(x) = 4\|x\|_\infty\psi^{-1}\alpha'\tilde{\beta}e^n\big((4\|x\|_\infty)^{-1}x\big) \text{ for } x \in \mathbb{R}^n \setminus [-1/4, 1/4]^n,$$

where $\|x\|_\infty = \max\{|x(1)|, \cdots, |x(n)|\}$. Then, $\varphi^{-1}\psi^{-1}\alpha'\tilde{\beta}|\mathbf{T}_0^n : \mathbf{T}_0^n \to \mathbb{R}^n$ is the desired immersion. □

4.6 The Handle Straightening Theorem

A compact subpolyhedron Y of a polyhedron X is said to be **straight** in X provided that $\operatorname{bd} Y$ is PL collared in both Y and $X \setminus \operatorname{int} Y$, that is, there exists a PL bi-collar $k : \operatorname{bd} Y \times [-1, 1] \to X$ such that $k(\operatorname{bd} Y \times [-1, 0]) \subset Y$ and $k(\operatorname{bd} Y \times [0, 1]) \subset X \setminus \operatorname{int} Y$.

We use the same notational convention as Sect. 4.4, that is, for each $n \in \mathbb{N}$, we write $\boldsymbol{Q} = [-1, 1]^n \times \boldsymbol{Q}_n$ and $p_n : X \times \boldsymbol{Q} \to X \times [-1, 1]^n$ denotes the projection defined as follows:

$$p_n(x, y) = (x, y(1), \ldots, y(n)) \text{ for each } (x, y) \in X \times \boldsymbol{Q}.$$

Lemma 4.6.1 *Let X be a polyhedron. Given a compactum A in $X \times \boldsymbol{Q}$ and an open set U in $X \times \boldsymbol{Q}$ with $A \subset U$, there exist $n \in \mathbb{N}$ and a straight compact subpolyhedron Y of $X \times [-1, 1]^n$ such that $A \subset \operatorname{int} p_n^{-1}(Y) \subset p_n^{-1}(Y) \subset U$.*

Proof For each $x \in A$, we can take $n(x) \in \mathbb{N}$ and an open neighborhood V_x of $p_{n(x)}(x)$ in $X \times [-1, 1]^{n(x)}$ so that $p_{n(x)}^{-1}(V_x) \subset U$. Since A is compact, there are $x_1, \ldots, x_k \in A$ such that $A \subset \bigcup_{i=1}^k p_{n(x_i)}^{-1}(V_{x_i})$. Let $n = \max\{n(x_1), \ldots, n(x_k)\}$.

Then we have the following open set in $X \times [-1, 1]^n$:

$$V = \bigcup_{i=1}^{k} V_{x_i} \times [-1, 1]^{n-n(x_i)},$$

where $A \subset p_n^{-1}(V) \subset U$. Take a triangulation K of $X \times [-1, 1]^n$ so that

$$K \prec \{V, (X \times [-1, 1]^n) \setminus p_n(A)\}.$$

Let L be the subcomplex of K with $|L| = |K[p_n(A)]|$. Then, we have $|L| \subset V$. Choose $0 < \varepsilon < 1$ so that $|N_\varepsilon(L, K)| \subset V$, where $N_\varepsilon(L, K)$ is the ε-neighborhood of L in K.[18] Then, $Y = |N_\varepsilon(L, K)|$ is the desired one. In fact, since Y is a regular neighborhood of $|L|$ in $|K| = X \times [-1, 1]^n$ by Lemma 1.9.4, it follows from Theorem 1.9.6 that $\operatorname{bd} Y$ is PL collared in both Y and $(X \times [-1, 1]^n) \setminus \operatorname{int} Y$. □

The following Handle Straightening Theorem is the main theorem of this section, which we will apply to prove the topological invariance of Whitehead torsion mentioned in Sect. 4.1.

Theorem 4.6.2 (Handle Straightening) *Let X be a polyhedron and $n \in \mathbb{N}$. Given an open embedding $h : \mathbb{R}^n \times \boldsymbol{Q} \to X \times \boldsymbol{Q}$, there exists a homeomorphism $g : X \times \boldsymbol{Q} \to X \times \boldsymbol{Q}$ and a straight subpolyhedron Y of $X \times [-1, 1]^m$ for some $m \in \mathbb{N}$ such that*

$$g|(X \times \boldsymbol{Q}) \setminus h([-1/4, 1/4]^n \times \boldsymbol{Q}) = \mathrm{id} \;\; \textit{and}$$

$$gh([-1/8, 1/8]^n \times \boldsymbol{Q}) = p_m^{-1}(Y) = Y \times \boldsymbol{Q}_m.$$

We first treat only the case $n = 1$ that is simpler than the case $n \geqslant 2$.

Proof (The Case n = 1 of Theorem 4.6.2) Let us consider the restriction of the open embedding $h : \mathbb{R} \times \boldsymbol{Q} \to X \times \boldsymbol{Q}$ to the following open subspace:

$$\begin{aligned}\big((-1/4, 1/4) \setminus [-1/8, 1/8]\big) \times \boldsymbol{Q} &= \big((-1/4, -1/8) \times \boldsymbol{Q}\big) \cup \big((1/8, 1/4) \times \boldsymbol{Q}\big)\\ &\approx \big(\{-1/4\} \times \boldsymbol{Q} \times \mathbb{R}\big) \cup \big(\{1/4\} \times \boldsymbol{Q} \times \mathbb{R}\big).\end{aligned}$$

Applying the Splitting Theorem 4.4.1 to each component, we can obtain splittings

$$h((-1/4, -1/8) \times \boldsymbol{Q}) = M_{-1} \cup M_{-2} \;\text{ and }\; h((1/8, 1/4) \times \boldsymbol{Q}) = M_1 \cup M_2$$

such that $M_{\pm 1} \cap M_{\pm 2}$ are strong deformation retracts of $M_{\pm 1} \cup M_{\pm 2}$ (so they are contractible) and $M_{\pm 1} \cap M_{\pm 2} = A_\pm \times \boldsymbol{Q}_m$ for some compact contractible PL bi-

[18] For the ε-neighborhood $N_\varepsilon(L, K)$, see p. 47.

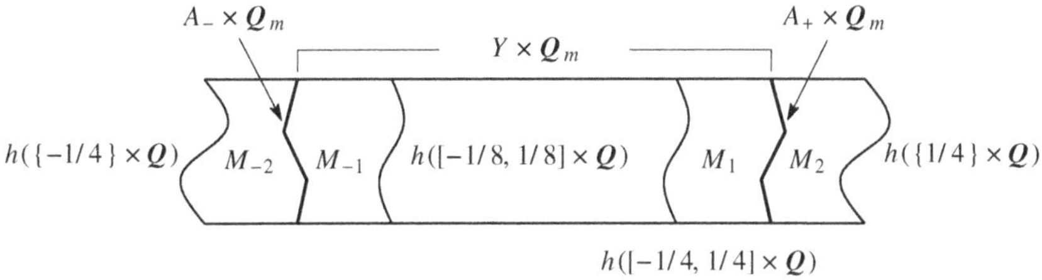

Fig. 4.9 The polyhedron $Y \subset X \times [-1, 1]^m$ when $n = 1$

collared subpolyhedra $A_\pm \subset X \times [-1, 1]^m$, where we can assume that

$$\operatorname{cl} M_{\pm 1} \supset h(\{\pm 1/8\} \times Q) \quad \Big(\text{equiv. } \operatorname{cl} M_{\pm 2} \supset h(\{\pm 1/4\} \times Q)\Big).$$

See Fig. 4.9. Then, it follows that

$$(A_- \cup A_+) \times Q_m = \operatorname{bd}(h([-1/8, 1/8] \times Q) \cup M_{-1} \cup M_1).$$

Thus, we have a straight subpolyhedron

$$Y = p_m(h([-1/8, 1/8] \times Q) \cup M_{-1} \cup M_1) \subset X \times [-1, 1]^m,$$

where $\operatorname{bd} Y = A_- \cup A_+$ and

$$h([-1/8, 1/8] \times Q) \cup M_{-1} \cup M_1 = Y \times Q_m \subset h([-1/4, 1/4] \times Q).$$

To construct a homeomorphism $g : X \times Q \to X \times Q$, observe that

$$h([-1/4, 1/4] \times Q) \setminus (\operatorname{int} Y \times Q_m) = M_{-2} \cup M_2 \quad \text{and}$$
$$\operatorname{bd} M_{\pm 2} = (A_\pm \times Q_m) \cup h(\{\pm 1/4\} \times Q).$$

Refer to Fig. 4.9. Since $M_{\pm 2}$ are compact contractible Q-manifolds, we have $M_{\pm 2} \approx Q$ by Theorem 3.7.9. Using the Z-Set Unknotting Theorem 2.11.6, we can easily construct homeomorphisms $g_- : h([-1/4, -1/8] \times Q) \to M_{-2}$ and $g_+ : h([1/8, 1/4] \times Q) \to M_2$ so that

$$g_\pm | h(\{\pm 1/4\} \times Q) = \mathrm{id} \quad \text{and} \quad g_\pm(h(\{\pm 1/8\} \times Q) = A_\pm \times Q_m.$$

Since $Y \times Q_m \approx Q$, we use the Z-Set Unknotting Theorem 2.11.6 again to construct $g_0 : h([-1/8, 1/8] \times Q) \to Y \times Q_m$ so that

$$g_0 | h(\{\pm 1/8\} \times Q) = g_\pm | h(\{\pm 1/8\} \times Q).$$

Then, pasting g_0, $g_\pm$, and id, we can obtain the desired homeomorphism $g : X \times \boldsymbol{Q} \to X \times \boldsymbol{Q}$ such that

$$g|(X \times \boldsymbol{Q}) \setminus h([-1/4, 1/4] \times \boldsymbol{Q}) = \mathrm{id} \quad \text{and}$$
$$gh([-1/8, 1/8] \times \boldsymbol{Q}) = p_m^{-1}(Y) = Y \times \boldsymbol{Q}_m.$$

The proof of the case $n = 1$ is completed. □

For the case $n \geqslant 2$, we will use the immersion $\alpha : \mathbf{T}_0^n \to \mathbb{R}^n$ constructed in the previous section and Theorem 4.1.7. To apply Theorem 4.1.7, we need the following:

Lemma 4.6.3 *Let X be a polyhedron and Y a subpolyhedron of $X \times [-1, 1]^k$, and let $\alpha : M \to X \times \boldsymbol{Q}$ be an immersion of a $\boldsymbol{Q}$-manifold M such that $\alpha|\alpha^{-1}(Y \times \boldsymbol{Q}_k) : \alpha^{-1}(Y \times \boldsymbol{Q}_k) \to Y \times \boldsymbol{Q}_k$ is a homeomorphism. Then, for each compactum C in M, there exist a compact polyhedron P, a PL immersion $f : P \to X \times [-1, 1]^m$, $m \geqslant k$, and an embedding $g : P \times \boldsymbol{Q}_m \to M$ such that $g(P \times \boldsymbol{Q}_m)$ is a neighborhood of C in M, $Y \times [-1, 1]^k$ is a subpolyhedron of P, $f|Y \times [-1, 1]^k = \mathrm{id}$, and $\alpha g = f \times \mathrm{id}$, that is, the following diagram commutes:*

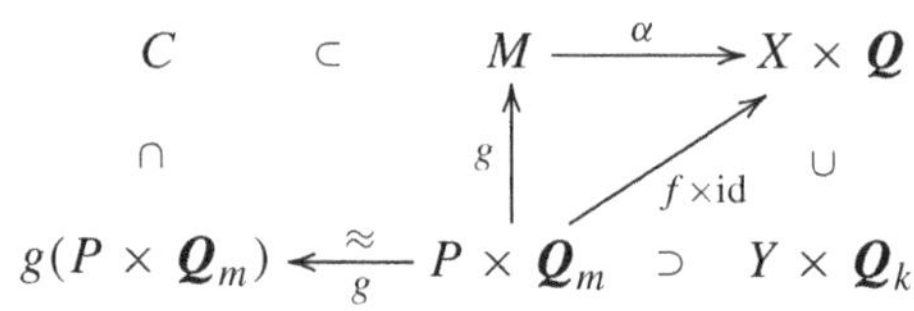

Proof Since α is an immersion, M has an open cover $\mathcal{U}$ such that $\alpha|U$ is an open embedding for each $U \in \mathcal{U}$. Since C is compact, we can easily find finite open sets $V_1, \ldots, V_n$ in M and $m \geqslant k$ such that $C \subset \bigcup_{i=1}^n V_i$, each $\mathrm{cl}\, V_i$ is compact and contained in some $U_i \in \mathcal{U}$, and $\alpha(V_i) = \mathrm{int}\, A_i \times \boldsymbol{Q}_m$ for some compact straight subpolyhedron A_i of $X \times [-1, 1]^m$, where each $\alpha|\,\mathrm{cl}\, V_i : \mathrm{cl}\, V_i \to A_i \times \boldsymbol{Q}_m$ is a homeomorphism. If $\mathrm{cl}\, V_i \cap \mathrm{cl}\, V_j \neq \emptyset$, then

$$A_i \cap A_j = p_m(\alpha(\mathrm{cl}\, V_i) \cap \alpha(\mathrm{cl}\, V_j)) \supset p_m\alpha(\mathrm{cl}\, V_i \cap \mathrm{cl}\, V_j) \neq \emptyset,$$

where it should be remarked that $A_i \cap A_j \neq \emptyset$ does not imply $\mathrm{cl}\, V_i \cap \mathrm{cl}\, V_j \neq \emptyset$. However, the following holds:

$$(Y \times [-1, 1]^{m-k}) \cap A_i \neq \emptyset \;\Leftrightarrow\; \alpha^{-1}(Y \times \boldsymbol{Q}_k) \cap \mathrm{cl}\, V_i \neq \emptyset.$$

Indeed, the left-hand side is equivalent to the following:

$$(Y \times \boldsymbol{Q}_k) \cap (A_i \times \boldsymbol{Q}_m) \neq \emptyset,$$

which is equivalent to the right-hand side because $\alpha^{-1}(A_i \times \boldsymbol{Q}_m) = \operatorname{cl} V_i$. For each $i = 1, \ldots, n$, let

$$J(i) = \{j < i \mid \operatorname{cl} V_i \cap \operatorname{cl} V_j \neq \emptyset\} \text{ and}$$

$$B_i = A_i \cap \Big((Y \times [-1, 1]^{m-k}) \cup \bigcup_{j \in J(i)} A_j\Big).$$

We can inductively define polyhedra $P_0 \subset P_1 \subset \cdots \subset P_n$ as follows:

$$P_0 = Y \times [-1, 1]^{m-k} \text{ and } P_i = P_{i-1} \cup_{B_i} A_i,\text{[19]}$$

where $P_{i-1} \cup_{B_i} A_i = P_{i-1} \oplus A_i$ if $B_i = \emptyset$. Then, $P = P_n$ is the desired polyhedron. The required PL immersion $f : P \to X \times [-1, 1]^m$ can be defined by $f|P_0 = \mathrm{id}$ (i.e., $f|Y \times [-1, 1]^{m-k} = \mathrm{id}$) and $f|A_i = \mathrm{id}$ for each $i = 1, \ldots, n$. We define an embedding $g : P \times \boldsymbol{Q}_m \to M$ as follows:

$$g|P_0 \times \boldsymbol{Q}_m = (\alpha|\alpha^{-1}(Y \times \boldsymbol{Q}_k))^{-1},$$

$$g|A_i \times \boldsymbol{Q} = (\alpha|\operatorname{cl} V_i)^{-1} : A_i \times \boldsymbol{Q}_m \to \operatorname{cl} V_i \text{ for each } i = 1, \ldots, n.$$

Then, $\alpha g = f \times \mathrm{id}$. Moreover, $g(P \times \boldsymbol{Q}_m)$ is a neighborhood of C in M because $\bigcup_{i=1}^n V_i \subset g(P \times \boldsymbol{Q}_m)$. □

Let us start to prove the case $n \geqslant 2$.

Proof (The Case n $\geqslant$ 2 of Theorem 4.6.2) We write

$$B_1 = [-1/8, 1/8]^n,\ B_2 = [-1/4, 1/4]^n \subset \mathbb{R}^n,$$

where $(B_i, \operatorname{bd} B_i) \approx (\mathbf{B}^n, \mathbf{S}^{n-1})$ and $B_1 \subset \operatorname{int} B_2$. Consider the restriction of the open embedding $h : \mathbb{R}^n \times \boldsymbol{Q} \to X \times \boldsymbol{Q}$ to the following subspace:

$$(\operatorname{int} B_2 \setminus B_1) \times \boldsymbol{Q} \approx \operatorname{bd} B_1 \times \boldsymbol{Q} \times \mathbb{R},$$

which is an open embedding into $X \times \boldsymbol{Q}$. Since $\pi_1(\operatorname{bd} B_1) = \mathbb{Z}$ or 0 depending on $n = 2$ or $n > 2$, we can apply the Splitting Theorem 4.4.1 to obtain a splitting

$$h((\operatorname{int} B_2 \setminus B_1) \times \boldsymbol{Q}) = M_1 \cup M_2$$

such that the inclusion $M_1 \cap M_2 \subset M_1 \cup M_2$ is a homotopy equivalence and $M_1 \cap M_2 = A \times \boldsymbol{Q}_m$ for some compact PL bi-collared subpolyhedron $A \subset X \times [-1, 1]^m$,

[19]Given two spaces X and Y containing a common subspace $A \neq \emptyset$, $X \cup_A Y$ denotes the adjunction space $X \cup_{\mathrm{id}_A} Y$, where $\mathrm{id}_A : A \to X$.

where we may assume that $\mathrm{cl}\, M_i \supset h((\mathrm{bd}\, B_i) \times \boldsymbol{Q})$, that is,

$$\mathrm{cl}\, M_i = M_i \cup h((\mathrm{bd}\, B_i) \times \boldsymbol{Q}), \ i = 1, 2.$$

Then, M_i is homotopy dense in $\mathrm{cl}\, M_i$ because $h((\mathrm{bd}\, B_i) \times \boldsymbol{Q})$ is collared in $h((B_2 \setminus \mathrm{int}\, B_1)$. Thus, the following inclusion is also a homotopy equivalence:

$$A \times \boldsymbol{Q}_m = \mathrm{cl}\, M_1 \cap \mathrm{cl}\, M_2 \subset \mathrm{cl}\, M_1 \cup \mathrm{cl}\, M_2,$$

which implies that $A \times \boldsymbol{Q}_m$ is a strong deformation retract of $\mathrm{cl}\, M_1 \cup \mathrm{cl}\, M_2$. Hence, $A \times \boldsymbol{Q}_m$ is a strong deformation retract of each of $\mathrm{cl}\, M_1$ and $\mathrm{cl}\, M_2$.

Let $r : \mathrm{cl}\, M_1 \to A \times \boldsymbol{Q}_m$ be a strong deformation retraction. Note that the following inclusion is a homotopy equivalence

$$h(\mathrm{bd}\, B_1 \times \boldsymbol{Q}) \subset h((B_2 \setminus \mathrm{int}\, B_1) \times \boldsymbol{Q}) = \mathrm{cl}\, M_1 \cup \mathrm{cl}\, M_2,$$

where $\mathrm{cl}\, M_1$ is a strong deformation retract of $\mathrm{cl}\, M_1 \cup \mathrm{cl}\, M_2$ because $A \times \boldsymbol{Q}_m = \mathrm{cl}\, M_1 \cap \mathrm{cl}\, M_2$ is a strong deformation retract of $\mathrm{cl}\, M_2$. Hence, the inclusion $h(\mathrm{bd}\, B_1 \times \boldsymbol{Q}) \subset \mathrm{cl}\, M_1$ is a homotopy equivalence. Thus, the following restriction is a homotopy equivalence:

$$r|h(\mathrm{bd}\, B_1 \times \boldsymbol{Q}) : h(\mathrm{bd}\, B_1 \times \boldsymbol{Q}) \to A \times \boldsymbol{Q}_m.$$

Observe that A is the boundary of the following straight subpolyhedron:

$$Y = p_m(h(B_1 \times \boldsymbol{Q}) \cup M_1) \subset X \times [-1, 1]^m.$$

Then, it follows that

$$h(B_1 \times \boldsymbol{Q}) \cup M_1 = Y \times \boldsymbol{Q}_m, \ h(B_2 \times \boldsymbol{Q}) = (Y \times \boldsymbol{Q}_m) \cup \mathrm{cl}\, M_2 \ \text{ and}$$
$$\mathrm{cl}\, M_1 = (Y \times \boldsymbol{Q}_m) \setminus h(\mathrm{int}\, B_1 \times \boldsymbol{Q}).$$

Since $\mathrm{bd}\, Y \times \boldsymbol{Q}_m = A \times \boldsymbol{Q}_m$ is a strong deformation retract of $\mathrm{cl}\, M_2$, it follows that $Y \times \boldsymbol{Q}_m$ is a strong deformation retract of $h(B_2 \times \boldsymbol{Q})$, which is contractible. As a compact contractible $\boldsymbol{Q}$-manifold is homeomorphic to the Hilbert cube (Theorem 3.7.9), we have $Y \times \boldsymbol{Q}_m \approx \boldsymbol{Q}$.

We have to construct a homeomorphism g of $X \times \boldsymbol{Q}$ onto itself so that

$$g|(X \times \boldsymbol{Q}) \setminus h(B_2 \times \boldsymbol{Q}) = \mathrm{id} \ \text{ and } \ gh(B_1 \times \boldsymbol{Q}) = p_m^{-1}(Y) = Y \times \boldsymbol{Q}_m.$$

As before, let $\mathbf{T}^n_0 = \mathbf{T}^n \setminus \{p_0\}$, where

$$p_0 = e^n(1, \dots, 1) \in \mathbf{T}^n \setminus e^n(B_2),$$

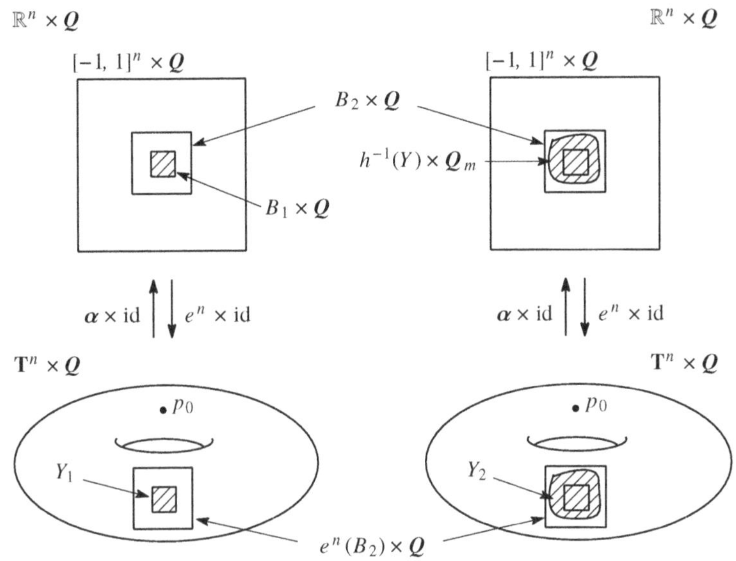

Fig. 4.10 Y_1 and Y_2

and let $\alpha : \mathbf{T}^n_0 \to \mathbb{R}^n$ be the immersion such that $\alpha e^n | B_2 = \mathrm{id}$, which had been obtained in Theorem 4.5.1. Let

$$Y_1 = e^n(B_1) \times \boldsymbol{Q} \text{ and } Y_2 = (e^n \times \mathrm{id})h^{-1}(Y \times \boldsymbol{Q}_m).$$

These are both subsets of $\mathbf{T}^n_0 \times \boldsymbol{Q}$, which are homeomorphic to the Hilbert cube $\boldsymbol{Q}$. Then, $Y_1 \subset \operatorname{int} Y_2 \subset Y_2 \subset e^n(B_2) \times \boldsymbol{Q}$ (cf. Fig. 4.10). Since $e^n\alpha|e^n(B_2) = \mathrm{id}$, we can define a strong deformation retraction r_0 so that the following diagram commutes:

$$\begin{array}{ccc} Y_2 \setminus \operatorname{int} Y_1 & \xrightarrow{r_0} & \operatorname{bd} Y_2 \\ {\scriptstyle h(\alpha \times \mathrm{id})}\downarrow & & \uparrow{\scriptstyle (e^n \times \mathrm{id})h^{-1}} \\ \operatorname{cl} M_1 & \xrightarrow[r]{} & \operatorname{bd} Y \times \boldsymbol{Q}_m \end{array}$$

Then, the restriction $r_0 | \operatorname{bd} Y_1$ is a homotopy equivalence because

$$h(\alpha \times \mathrm{id})(\operatorname{bd} Y_1) = h(\operatorname{bd} B_1 \times \boldsymbol{Q})$$

and $r | h(\operatorname{bd} B_1 \times \boldsymbol{Q})$ is a homotopy equivalence.

Since $\pi_1(\operatorname{bd} Y_1) \cong \pi_1(\mathbf{S}^{n-1}) \cong \mathbb{Z}$ or 0, we can apply Theorem 4.1.7 to obtain a homeomorphism $f_0' : \operatorname{bd} Y_1 \to \operatorname{bd} Y_2$ such that $f_0' \simeq r_0|\operatorname{bd} Y_1$. Recall that $Y_1 \approx Y_2 \approx \boldsymbol{Q}$. Since $\operatorname{bd} Y_1$ and $\operatorname{bd} Y_2$ are Z-sets in Y_1 and Y_2, respectively, f_0' extends to a homeomorphism $f_0 : Y_1 \to Y_2$ (Corollary 2.10.7). Then, $f_0(\operatorname{bd} Y_1) = \operatorname{bd} Y_2$ and $f_0|\operatorname{bd} Y_1 \simeq r_0|\operatorname{bd} Y_1 \simeq \operatorname{id}$ in $Y_2 \setminus \operatorname{int} Y_1$.

In the following diagram, we have just defined the bottom homeomorphism f_0. We will construct upper homeomorphisms step-by-step so that each ladder of the diagram commutes with the bottom. Finally, f_4 will be constructed so that $f_4|\operatorname{bd} B_2^n \times \boldsymbol{Q} = \operatorname{id}$. Then, hf_4h^{-1} can extend to the desired homeomorphism g.

$$
\begin{array}{ccc}
X \times \boldsymbol{Q} & \xrightarrow{\ g\ } & X \times \boldsymbol{Q} \\
{\scriptstyle h}\uparrow & & \uparrow{\scriptstyle h} \\
B_2 \times \boldsymbol{Q} & \xrightarrow{\ f_4\ } & B_2 \times \boldsymbol{Q} \\
\cup & & \cup \\
\operatorname{int} B_2 \times \boldsymbol{Q} & \xrightarrow{\ f_3\ } & \operatorname{int} B_2 \times \boldsymbol{Q} \\
{\scriptstyle \xi \times \mathrm{id}}\downarrow & & \downarrow{\scriptstyle \xi \times \mathrm{id}} \\
\mathbb{R}^n \times \boldsymbol{Q} & \xrightarrow{\ f_2\ } & \mathbb{R}^n \times \boldsymbol{Q} \\
{\scriptstyle e^n \times \mathrm{id}}\downarrow & & \downarrow{\scriptstyle e^n \times \mathrm{id}} \\
\mathbf{T}^n \times \boldsymbol{Q} & \xrightarrow{\ f_1\ } & \mathbf{T}^n \times \boldsymbol{Q} \\
\cup & & \cup \\
Y_1 & \xrightarrow{\ f_0\ } & Y_2
\end{array}
$$

Step 1 We will extend f_0 to a homeomorphism $f_1 : \mathbf{T}^n \times \boldsymbol{Q} \to \mathbf{T}^n \times \boldsymbol{Q}$ so that $f_1 \simeq \operatorname{id}$. Here, we want to use Theorem 4.1.7, but it can be applied to $\boldsymbol{Q}$-manifolds being the products of compact polyhedra with $\boldsymbol{Q}$ whose fundamental groups are free or free Abelian. Consider the following $\boldsymbol{Q}$-manifolds (cf. Fig. 4.10):

$$\big(\mathbf{T}^n \times \boldsymbol{Q}\big) \setminus \operatorname{int} Y_1 = \big(\mathbf{T}^n \setminus e^n(\operatorname{int} B_1)\big) \times \boldsymbol{Q} \text{ and}$$
$$\big(\mathbf{T}^n \times \boldsymbol{Q}\big) \setminus \operatorname{int} Y_2 = \big(\mathbf{T}^n \times \boldsymbol{Q}\big) \setminus \operatorname{int}(e^n \times \operatorname{id})h^{-1}(Y \times \boldsymbol{Q}_m).$$

Here, $\mathbf{T}^n \setminus e^n(\operatorname{int} B_1)$ can be regarded as a compact polyhedron and the first one is the product of this polyhedron and $\boldsymbol{Q}$. However, for now, it is not known whether the second one can be regarded as the product of a polyhedron and $\boldsymbol{Q}$ or not.

We take the following neighborhoods of $p_0 = e^n(1, \ldots, 1) \in \mathbf{T}^n$:

$$D_1 = e^n([7/8, 9/8]^n),\ D_2 = e^n([3/4, 5/4]^n) \subset \mathbf{T}^n \setminus e^n(B_2),$$

where $(D_i, \operatorname{bd} D_i) \approx (\mathbf{B}^n, \mathbf{S}^{n-1})$ and $D_1 \subset \operatorname{int} D_2$. Then, we have a polyhedron $P = \mathbf{T}^n \setminus (e^n(\operatorname{int} B_1) \cup \operatorname{int} D_1)$ such that

$$P \times \boldsymbol{Q} = \big(\mathbf{T}^n \times \boldsymbol{Q}\big) \setminus \big(\operatorname{int} Y_1 \cup (\operatorname{int} D_1 \times \boldsymbol{Q})\big),$$

where it should be remarked that $P \simeq \mathbf{T}^n \setminus \{p_0, \mathbf{0}\}$.

$$\pi_1(P) \cong \begin{cases} \mathbb{Z} * \mathbb{Z} * \mathbb{Z} & \text{if } n = 2, \\ \mathbb{Z} \oplus \cdots \oplus \mathbb{Z} \quad (n \text{ copies}) & \text{if } n > 2. \end{cases}$$

Consider the following $\boldsymbol{Q}$-manifold:

$$W = \big(\mathbf{T}^n \times \boldsymbol{Q}\big) \setminus \big(\operatorname{int} Y_2 \cup (\{p_0\} \times \boldsymbol{Q})\big).$$

Then, $\operatorname{bd} Y_2 \subset W$, $W \setminus (\operatorname{int} D_1 \times \boldsymbol{Q})$ is compact and the following composition is an immersion:

$$W \subset \mathbf{T}^n_0 \times \boldsymbol{Q} \xrightarrow{\alpha \times \mathrm{id}} \mathbb{R}^n \times \boldsymbol{Q} \xrightarrow{\ h\ } X \times \boldsymbol{Q},$$

where $h(\alpha \times \mathrm{id})(\operatorname{bd} Y_2) = A \times \boldsymbol{Q}_m$ and $A = \operatorname{bd} Y$ is a compact PL bi-collared subpolyhedron of $X \times [-1, 1]^m$. Applying Lemma 4.6.3, we can obtain a compact polyhedron Z, a PL immersion $\alpha' : Z \to X \times [-1, 1]^{m'}$, $m' \geqslant m$, and an embedding $\varphi : Z \times \boldsymbol{Q}_{m'} \to W$ such that $\varphi(Z \times \boldsymbol{Q}_{m'})$ is a neighborhood of $W \setminus (\operatorname{int} D_1 \times \boldsymbol{Q})$, $A \times [-1, 1]^{m'-m}$ is a subpolyhedron of Z, $\alpha' | A \times \boldsymbol{Q}_m = \mathrm{id}$, and

$$h(\alpha \times \mathrm{id})\varphi = \alpha' \times \mathrm{id} : Z \times \boldsymbol{Q}_m \to X \times [-1, 1]^m \times \boldsymbol{Q}_m.$$

Then, it follows that

$$\varphi(A \times \boldsymbol{Q}_m) = (e^n \times \mathrm{id})h^{-1}(\operatorname{bd} Y \times \boldsymbol{Q}_m) = \operatorname{bd} Y_2,$$
$$(\operatorname{int} D_2 \setminus D_1) \times \boldsymbol{Q} \subset \varphi(Z \times \boldsymbol{Q}_m),$$

and $\psi = \varphi^{-1} | (\operatorname{int} D_2 \setminus D_1) \times \boldsymbol{Q}$ is an open embedding into $Z \times \boldsymbol{Q}_m$, where

$$(\operatorname{int} D_2 \setminus D_1) \times \boldsymbol{Q} \approx \operatorname{bd} D_1 \times \boldsymbol{Q} \times \mathbb{R} \approx \mathbf{S}^{n-1} \times \boldsymbol{Q} \times \mathbb{R}.$$

Since $\pi_1(\mathbf{S}^{n-1}) \cong \mathbb{Z}$ or 0, we can apply the Splitting Theorem 4.4.1 to obtain a splitting

$$\psi((\operatorname{int} D_2 \setminus D_1) \times \boldsymbol{Q}) = A_1 \cup A_2$$

such that the inclusion $A_1 \cap A_2 \subset A_1 \cup A_2$ is a homotopy equivalence and $A_1 \cap A_2 = B \times \boldsymbol{Q}_k$ for some PL bi-collared subpolyhedron B of $Z \times [-1, 1]^k$ for some $k \geqslant m'$,

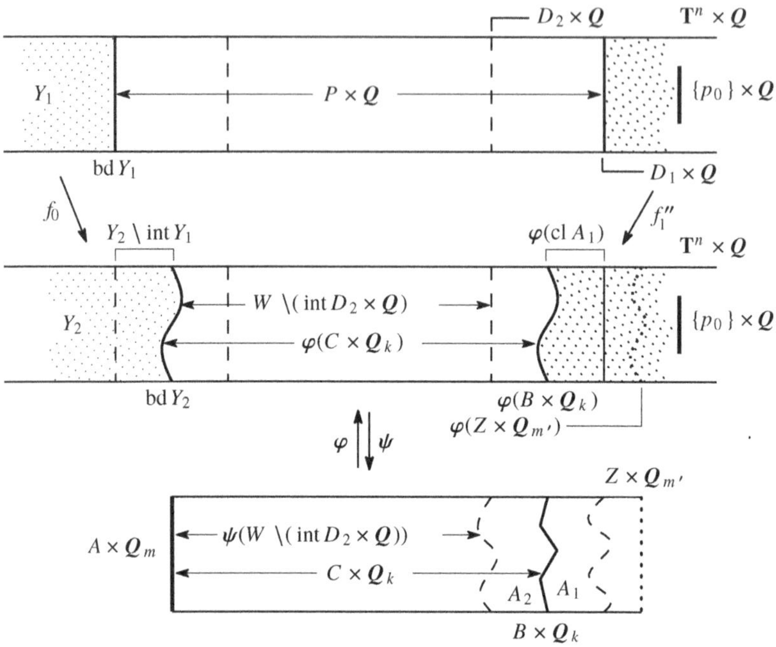

Fig. 4.11 Y_1 and Y_2

where we may assume $\mathrm{cl}\, A_1 \supset \psi(\mathrm{bd}\, D_1 \times \boldsymbol{Q})$, that is,

$$\mathrm{cl}\, A_i = A_i \cup \psi(\mathrm{bd}\, D_i \times \boldsymbol{Q}), \ i = 1, 2.$$

Then, $B \times \boldsymbol{Q}_k$ is a strong deformation retract of each of $\mathrm{cl}\, A_1$ and $\mathrm{cl}\, A_2$. See Fig. 4.11.

Let $r' : \mathrm{cl}\, A_1 \to B \times \boldsymbol{Q}_k$ be a strong deformation retraction. Then, by the same argument as r, we can see that the following restriction is a homotopy equivalence:

$$r'|\psi(\mathrm{bd}\, D_1 \times \boldsymbol{Q}) : \psi(\mathrm{bd}\, D_1 \times \boldsymbol{Q}) \to B \times \boldsymbol{Q}_k.$$

Since $\varphi\psi|\varphi(\operatorname{int} D_2 \setminus D_1) \times \boldsymbol{Q} = \mathrm{id}$, we can define a strong deformation retraction r_0 so that the following diagram commutes:

$$\begin{array}{ccc} \varphi(\mathrm{cl}\, A_1) & \xrightarrow{r_1} & \varphi(B \times \boldsymbol{Q}_k) \\ \psi \big\downarrow & & \big\uparrow \varphi \\ \mathrm{cl}\, A_1 & \xrightarrow[r']{} & B \times \boldsymbol{Q}_k \end{array}$$

Then, the restriction $r_1|\operatorname{bd} D_1 \times \boldsymbol{Q}$ is a homotopy equivalence. In the same way as f_0', we can apply Theorem 4.1.7 to obtain a homeomorphism $f_1' : \operatorname{bd} D_1 \times \boldsymbol{Q} \to \varphi(B \times \boldsymbol{Q}_k)$ such that $f_1' \simeq r_1|\operatorname{bd} D_1 \times \boldsymbol{Q}$.

Since $B \times \boldsymbol{Q}_k$ is a strong deformation retract of $\operatorname{cl} A_2$, it follows that $\operatorname{cl} A_1 \cup (D_1 \times \boldsymbol{Q})$ is a strong deformation retract of $D_2 \times \boldsymbol{Q}$, so it is contractible. Hence, we have $\operatorname{cl} A_1 \cup (D_1 \times \boldsymbol{Q}) \approx \boldsymbol{Q}$ by Theorem 3.7.9. Observe that $B \times \boldsymbol{Q}_k$ is a Z-set in $\operatorname{cl} A_1 \cup (D_1 \times \boldsymbol{Q})$. Since $D_1 \times \boldsymbol{Q} \approx \boldsymbol{Q}$ and $\operatorname{bd} D_1 \times \boldsymbol{Q}$ is a Z-set in $D_1 \times \boldsymbol{Q}$, f_1' extends to a homeomorphism $f_1'' : D_1 \times \boldsymbol{Q} \to \operatorname{cl} A_1 \cup (D_1 \times \boldsymbol{Q})$ by Corollary 2.10.7. Again see Fig. 4.11.

Observe that B is the boundary of the following subpolyhedron:

$$C = p_k\big(\psi(W \setminus (\operatorname{int} D_2 \times \boldsymbol{Q})) \cup A_2\big) \subset Z \times [-1,1]^k,$$

where we have

$$\begin{aligned}\varphi(C \times \boldsymbol{Q}_k) &= \big(W \setminus (\operatorname{int} D_2 \times \boldsymbol{Q})\big) \cup \varphi(\operatorname{cl} A_2)\\ &= W \setminus \big((\operatorname{int} D_1 \times \boldsymbol{Q}) \cup \varphi(\operatorname{int} A_1)\big)\\ &= (\mathbf{T}^n \times \boldsymbol{Q}) \setminus \big(\operatorname{int} Y_2 \cup (\operatorname{int} D_1 \times \boldsymbol{Q}) \cup \varphi(\operatorname{int} A_1)\big).\end{aligned}$$

Then, it follows that

$$\begin{aligned}&(Y_2 \setminus \operatorname{int} Y_1) \cup \varphi(C \times \boldsymbol{Q}_k) \cup \varphi(\operatorname{cl} A_1)\\ &\qquad\qquad = (\mathbf{T}^n \times \boldsymbol{Q}) \setminus (\operatorname{int} Y_1 \cup (\operatorname{int} D_1 \times \boldsymbol{Q}))\\ &\qquad\qquad = P \times \boldsymbol{Q}.\end{aligned}$$

Moreover, we have

$$\begin{aligned}&(Y_2 \setminus \operatorname{int} Y_1) \cap \varphi(C \times \boldsymbol{Q}_k) = \operatorname{bd} Y_2 \quad \text{and}\\ &\varphi(C \times \boldsymbol{Q}_k) \cap \varphi(\operatorname{cl} A_1) = \varphi(B \times \boldsymbol{Q}_k),\end{aligned}$$

where $\operatorname{bd} Y_2$ and $\varphi(B \times \boldsymbol{Q}_k)$ are Z-sets in $\varphi(C \times \boldsymbol{Q}_k)$. Already we obtained two strong deformation retractions

$$r_0 : Y_2 \setminus \operatorname{int} Y_1 \to \operatorname{bd} Y_2 \quad \text{and} \quad r_1 : \varphi(\operatorname{cl} A_1) \to \varphi(B \times \boldsymbol{Q}_k),$$

which extend to a strong deformation retraction $r_2 : P \times \boldsymbol{Q} \to \varphi(C \times \boldsymbol{Q}_k)$. Then, we can apply Theorem 4.1.7 to see that r_2 is homotopic to a homeomorphism f_1''. Already, we obtained two homeomorphisms

$$f_0' : \operatorname{bd} Y_1 \to \operatorname{bd} Y_2 \quad \text{and} \quad f_1' : \operatorname{bd} D_1 \times \boldsymbol{Q} \to \varphi(B \times \boldsymbol{Q}_k)$$

such that $f_0' \simeq r_0|\operatorname{bd} Y_1$ and $f_1' \simeq r_1|\operatorname{bd} D_1 \times \boldsymbol{Q}$, where $\operatorname{bd} Y_1$ and $\operatorname{bd} D_1 \times \boldsymbol{Q}$ are Z-sets in $P \times \boldsymbol{Q}$. Since $f_0' \simeq f_1''|\operatorname{bd} Y_1$ and $f_1' \simeq f_1''|\operatorname{bd} D_1 \times \boldsymbol{Q}$, we can apply the Z-set Unknotting Theorem 2.11.6 to obtain a homeomorphism

$$f_1^* : P \times \boldsymbol{Q} \to \varphi(C \times \boldsymbol{Q}_k)$$

such that $f_1^*|\operatorname{bd} Y_1 = f_0'$, $f_1^*|\operatorname{bd} D_1 \times \boldsymbol{Q} = f_1'$, and $f_1^* \simeq f_1'' \simeq r_2$. Finally, we can extend f_1^* to a homeomorphism $f_1 : \mathbf{T}^n \to \mathbf{T}^n$ by $f_1|Y_1 = f_0$ and $f_1|D_1 \times \boldsymbol{Q} = f_1'$.[20]

Step 2 (*Construction of* f_2) Note that $\pi_1(\mathbb{R}^n \times \boldsymbol{Q}, (\mathbf{0}, \mathbf{0})) = 0$. By the Lifting Criterion Theorem 4.2.5, $f_1(e^n \times \mathrm{id})$ can be lifted to a map f_2 with $f_2(\mathbf{0}, \mathbf{0}) = (\mathbf{0}, \mathbf{0})$. Namely, f_2 is a unique map such that the following diagram commutes:

$$\begin{array}{ccc} \mathbb{R}^n \times \boldsymbol{Q} & \xrightarrow{f_2} & \mathbb{R}^n \times \boldsymbol{Q} \\ {\scriptstyle e^n \times \mathrm{id}}\downarrow & & \downarrow{\scriptstyle e^n \times \mathrm{id}} \\ \mathbf{T}^n \times \boldsymbol{Q} & \xrightarrow[f_1]{} & \mathbf{T}^n \times \boldsymbol{Q} \end{array}$$

Then, f_2 is a homeomorphism. Indeed, $f_1^{-1}(e^n \times \mathrm{id})$ can be lifted to a map f_2' with $f_2'(\mathbf{0}, \mathbf{0}) = (\mathbf{0}, \mathbf{0})$. Since $f_2 f_2'$ and $f_2' f_2$ are liftings of id, it follows from uniqueness of liftings that $f_2 f_2' = \mathrm{id}$ and $f_2' f_2 = \mathrm{id}$.

By uniqueness of liftings, we have

$$f_2|B_1 \times \boldsymbol{Q} = (\alpha \times \mathrm{id}) f_1(e^n \times \mathrm{id})|B_1 \times \boldsymbol{Q}$$

because both are liftings of $f_1(e^n \times \mathrm{id})|B_1 \times \boldsymbol{Q}$ and

$$f_2(\mathbf{0}, \mathbf{0}) = (\mathbf{0}, \mathbf{0}) = (\alpha \times \mathrm{id}) f_1(e^n \times \mathrm{id})(\mathbf{0}, \mathbf{0}).$$

Then, observe

$$\begin{aligned} f_2(B_1 \times \boldsymbol{Q}) &= (\alpha \times \mathrm{id}) f_1(e^n(B_1) \times \boldsymbol{Q}) \\ &= (\alpha \times \mathrm{id}) f_1(Y_1) = (\alpha \times \mathrm{id})(Y_2) \\ &= (\alpha \times \mathrm{id})(e^n \times \mathrm{id}) h^{-1}(Y \times \boldsymbol{Q}_m) \\ &= h^{-1}(Y \times \boldsymbol{Q}_m). \end{aligned}$$

[20] In Chapman's Lecture Notes [(2)], it is asserted that $f_1 \simeq \mathrm{id}$ and its proof is left to the reader as an exercise. However, the author doesn't know its proof.

Note that $2\mathbb{Z}^n = \{v \in \mathbb{R}^n \mid e^n(v) = e^n(\mathbf{0})\}$ $(= (e^n)^{-1}(e^n(\mathbf{0})))$ and

$$\mathbb{R}^n = \bigcup_{v \in 2\mathbb{Z}^n} \prod_{i=1}^{n} [v(i) - 1, v(i) + 1].$$

Let $x \in \prod_{i=1}^{n} [v(i) - 1, v(i) + 1]$, $v \in 2\mathbb{Z}^n$, and $y \in \boldsymbol{Q}$. Then, $x - v \in [-1, 1]^n$. Since $v(i) \in 2\mathbb{Z}$, it follows that

$$\begin{aligned} e(x(i) - v(i)) &= (\cos \pi (x(i) - v(i)), \sin \pi (x(i) - v(i))) \\ &= (\cos \pi x(i), \sin \pi x(i)) = e(x(i)), \end{aligned}$$

hence $e^n(x - v) = e^n(x)$. Replacing x by $\mathrm{pr}_{\mathbb{R}^n} f_2(x, y)$, we also have

$$e^n(\mathrm{pr}_{\mathbb{R}^n} f_2(x, y) - v) = e^n \mathrm{pr}_{\mathbb{R}^n} f_2(x, y).$$

Since f_2 is a lifting of $f_1(e^n \times \mathrm{id})$, it follows that

$$\begin{aligned} &(e^n \times \mathrm{id}) f_2(x - v, y) = f_1(e^n \times \mathrm{id})(x - v, y) = f_1(e^n \times \mathrm{id})(x, y) \quad \text{and} \\ &(e^n \times \mathrm{id})(\mathrm{pr}_{\mathbb{R}^n} f_2(x, y) - v, \mathrm{pr}_{\boldsymbol{Q}} f_2(x, y)) \\ &\qquad = (e^n \times \mathrm{id}) f_2(x, y) = f_1(e^n \times \mathrm{id})(x, y). \end{aligned}$$

By uniqueness of liftings, we have

$$f_2(x - v, y) = f_2(x, y) = (\mathrm{pr}_{\mathbb{R}^n} f_2(x, y) - v, \mathrm{pr}_{\boldsymbol{Q}} f_2(x, y)).$$

Then, it follows that

$$\begin{aligned} \|\mathrm{pr}_{\mathbb{R}^n} f_2(x, y) - x\|_\infty &= \|(\mathrm{pr}_{\mathbb{R}^n} f_2(x, y) - v) - (x - v)\|_\infty \\ &= \|\mathrm{pr}_{\mathbb{R}^n} f_2(x - v, y) - (x - v)\|_\infty. \end{aligned}$$

Consequently, we have

$$\sup_{(x,y) \in \mathbb{R}^n \times \boldsymbol{Q}} \|\mathrm{pr}_{\mathbb{R}^n} f_2(x, y) - x\|_\infty = \sup_{(x,y) \in [-1,1]^n \times \boldsymbol{Q}} \|\mathrm{pr}_{\mathbb{R}^n} f_2(x, y) - x\|_\infty.$$

Because of the compactness of $[-1, 1]^n \times \boldsymbol{Q}$, we have

$$c_0 = \sup_{(x,y) \in [-1,1]^n \times \boldsymbol{Q}} \|\mathrm{pr}_{\mathbb{R}^n} f_2(x, y) - x\|_\infty < \infty.$$

Step 3 (*Construction of* f_3) Choose $1/8 < t_0 < 1/4$ so that

$$f_2(B_1 \times \boldsymbol{Q}) = h^{-1}(Y \times \boldsymbol{Q}_m) \subset [-t_0, t_0]^n \times \boldsymbol{Q}.$$

We define a homeomorphism $\gamma : [0, 1/4) \to \mathbb{R}_+$ as follows:

$$\gamma(t) = \begin{cases} t & \text{if } t \leqslant t_0, \\ \dfrac{t(1-4t_0)}{1-4t} & \text{if } t \geqslant t_0. \end{cases}$$

Let $\xi : \operatorname{int} B_2 \to \mathbb{R}^n$ be a homeomorphism defined by $\xi(x) = \gamma(\|x\|_\infty)\|x\|_\infty^{-1}x$, where $\|x\|_\infty = \max\{|x(1), \dots, |x(n)|\}$. Note that $\|\xi(x)\|_\infty = \gamma(\|x\|_\infty)$ for each $x \in \mathbb{R}^n$. The homeomorphism f_3 is defined by the following commutative diagram:

$$\begin{array}{ccc} \operatorname{int} B_2 \times \boldsymbol{Q} & \xrightarrow{f_3} & \operatorname{int} B_2 \times \boldsymbol{Q} \\ {\scriptstyle \xi \times \mathrm{id}}\downarrow & & \downarrow{\scriptstyle \xi \times \mathrm{id}} \\ \mathbb{R}^n \times \boldsymbol{Q} & \xrightarrow[f_2]{} & \mathbb{R}^n \times \boldsymbol{Q} \end{array}$$

Since $\xi|[-t_0, t_0]^n = \mathrm{id}$, we have

$$f_3|B_1 \times \boldsymbol{Q} = f_2|B_1 \times \boldsymbol{Q} = (\alpha \times \mathrm{id}) f_1(e^n \times \mathrm{id})|B_1 \times \boldsymbol{Q}.$$

Observe that ξ^{-1} is defined by $\xi^{-1}(y) = \gamma^{-1}(\|y\|_\infty)\|y\|_\infty^{-1}y$, where

$$\gamma^{-1}(s) = \begin{cases} s & \text{if } s \leqslant t_0, \\ \dfrac{s}{1+4(s-t_0)} & \text{if } s \geqslant t_0. \end{cases}$$

For each $s \geqslant t_0$ and $s' \geqslant 1$, since $1/8 < t_0 < 1/4$, it follows that

$$\begin{aligned} \left|\gamma^{-1}(s) - \gamma^{-1}(s')\right| &= \left|\frac{s}{1+4(s-t_0)} - \frac{s'}{1+4(s'-t_0)}\right| \\ &= \frac{|s-s'|(1-4t_0)}{(1+4(s-t_0))(1+4(s'-t_0))} \\ &< \frac{|s-s'|}{32ss'} \leqslant \frac{|s-s'|}{32s}. \end{aligned}$$

For each $x, x' \in \mathbb{R}^n$ with $\|x\|_\infty \geqslant 1 + c_0$, and $\|x - x'\|_\infty \leqslant c_0$, since

$$\big|\|x\|_\infty - \|x'\|_\infty\big| \leqslant \|x - x'\|_\infty \leqslant c_0,$$

it follows that $\|x'\|_\infty \geqslant \|x\|_\infty - c_0 \geqslant 1$ and

$$\begin{aligned}\left\|\xi^{-1}(x) - \xi^{-1}(x')\right\|_\infty &= \left\|\gamma^{-1}(\|x\|_\infty)\|x\|_\infty^{-1}x - \gamma^{-1}(\|x'\|_\infty)\|x'\|_\infty^{-1}x'\right\|_\infty \\ &\leqslant \gamma^{-1}(\|x\|_\infty)\|x\|_\infty^{-1}\|x - x'\|_\infty \\ &\quad + \gamma^{-1}(\|x\|_\infty)\left|\|x\|_\infty^{-1} - \|x'\|_\infty^{-1}\right|\|x'\|_\infty \\ &\quad + \left|\gamma^{-1}(\|x\|_\infty) - \gamma^{-1}(\|x'\|_\infty)\right|\|x'\|_\infty^{-1}\|x'\|_\infty \\ &< \frac{c_0}{4\|x\|_\infty} + \frac{\left|\|x'\|_\infty - \|x\|_\infty\right|}{4\|x\|_\infty} + \frac{\left|\|x'\|_\infty - \|x\|_\infty\right|}{32\|x\|_\infty} \\ &\leqslant \frac{17c_0}{32\|x\|_\infty} < \frac{c_0}{\|x\|_\infty}.\end{aligned}$$

Thus, for each $(x, y) \in \operatorname{int} B_2 \times \boldsymbol{Q}$ with $\|x\|_\infty \geqslant \gamma^{-1}(1 + c_0)$ (i.e., $\|\xi(x)\|_\infty = \gamma(\|x\|_\infty) \geqslant 1 + c_0$), we have

$$\begin{aligned}\|\mathrm{pr}_{\operatorname{int} B_2} f_3(x, y) - x\|_\infty &= \|\xi^{-1}\mathrm{pr}_{\mathbb{R}^n} f_2(\xi(x), y) - \xi^{-1}\xi(x)\|_\infty \\ &< \frac{c_0}{\|\xi(x)\|_\infty} = \frac{c_0}{\gamma(\|x\|_\infty)}.\end{aligned}$$

Step 4 (*Construction of f_4*) Consider the product $(B_2 \times \boldsymbol{Q})_{\mathrm{bd}\, B_2}$ reduced over $\mathrm{bd}\, B_2$.[21] We can extend f_3 to a homeomorphism $\bar{f}_3 : (B_2 \times \boldsymbol{Q})_{\mathrm{bd}\, B_2} \to (B_2 \times \boldsymbol{Q})_{\mathrm{bd}\, B_2}$ defined by $\bar{f}_3 | \mathrm{bd}\, B_2 = \mathrm{id}$.

$$\begin{array}{ccc} (B_2 \times \boldsymbol{Q})_{\mathrm{bd}\, B_2} & \xrightarrow{\bar{f}_3} & (B_2 \times \boldsymbol{Q})_{\mathrm{bd}\, B_2} \\ \cup & & \cup \\ \operatorname{int} B_2 \times \boldsymbol{Q} & \xrightarrow[f_3]{} & \operatorname{int} B_2 \times \boldsymbol{Q} \end{array}$$

To check the continuity of $\bar{f}_3$ at $x \in \mathrm{bd}\, B_2$, for each $\varepsilon > 0$, let

$$\delta = \min\left\{1/4 - \gamma^{-1}(\max\{1 + c_0, 2c_0/\varepsilon\}), \varepsilon/2\right\} > 0.$$

For each and $(x', y') \in \operatorname{int} B_2 \times \boldsymbol{Q}$ with $\|x - x'\|_\infty < \delta$,

$$\|x'\|_\infty \geqslant \|x\|_\infty - \|x - x'\|_\infty > 1/4 - \delta \geqslant \gamma^{-1}(\max\{1 + c_0, 2c_0/\varepsilon\}),$$

[21]For the reduced product, refer to p. 108. In this case, $(B_2 \times \boldsymbol{Q})_{\mathrm{bd}\, B_2} \approx \mathrm{bd}\, B_2 \cup_{\mathrm{pr}_{\mathrm{bd}\, B_2}} B_2 \times \boldsymbol{Q}$, where $\mathrm{pr}_{\mathrm{bd}\, B_2} : \mathrm{bd}\, B_2 \times \boldsymbol{Q} \to \mathrm{bd}\, B_2$ is the projection.

that is, $\gamma(\|x'\|_\infty) \geqslant \max\{1 + c_0, 2c_0/\varepsilon\}$, so $c_0/\gamma(\|x'\|_\infty) \leqslant \varepsilon/2$. Therefore,

$$\begin{aligned}\|\mathrm{pr}_{\mathrm{int}\,B_2} f_3(x', y') - x\|_\infty &\leqslant \|\mathrm{pr}_{\mathrm{int}\,B_2} f_3(x', y') - x'\|_\infty + \|x' - x\|_\infty \\ &< \frac{c_0}{\gamma(\|x'\|_\infty)} + \delta \leqslant \varepsilon/2 + \varepsilon/2 = \varepsilon.\end{aligned}$$

Thus, $\bar{f}_3$ is continuous at each point of bd B_2, Since $\bar{f}_3$ is a continuous bijection, it is a homeomorphism.

The following N is a compact $\boldsymbol{Q}$-manifold:

$$N = \big((B_2 \setminus (-t_0, t_0)^n) \times \boldsymbol{Q}\big)_{\mathrm{bd}\,B_2} = (B_2 \times \boldsymbol{Q})_{\mathrm{bd}\,B_2} \setminus (-t_0, t_0)^n) \times \boldsymbol{Q}.$$

Indeed, since $B_2 \setminus (-t_0, t_0)^n \approx \mathrm{bd}\,B_2 \times \mathbf{I}$, we have

$$N \approx ((\mathrm{bd}\,B_2 \times \mathbf{I}) \times \boldsymbol{Q})_{\mathrm{bd}\,B_2 \times \{0\}} \approx \mathrm{bd}\,B_2 \times (\mathbf{I} \times \boldsymbol{Q})_{\{0\}} = \mathrm{bd}\,B_2 \times C(\boldsymbol{Q}),$$

where the cone $C(\boldsymbol{Q})$ is homeomorphic to $\boldsymbol{Q}$ (Theorem 2.7.5). Hence, N is a $\boldsymbol{Q}$-manifold by the Edwards Factor Theorem 3.8.1.

This can also be seen as follows: Observe that N is homeomorphic to the mapping cylinder $M(\mathrm{pr}_{\mathrm{bd}\,B_2})$ of the projection $\mathrm{pr}_{\mathrm{bd}\,B_2} : \mathrm{bd}\,B_2 \times \boldsymbol{Q} \to \mathrm{bd}\,B_2$, where the projection $\mathrm{pr}_{\mathrm{bd}\,B_2}$ is cell-like. From the Mapping Cylinder Theorem 3.7.3, it follows that N is a $\boldsymbol{Q}$-manifold.

Let $q : (B_2 \setminus (-t_0, t_0)^n) \times \boldsymbol{Q} \to N$ be a map defined by

$$q|(\mathrm{int}\,B_2 \setminus (-t_0, t_0)^n) \times \boldsymbol{Q} = \mathrm{id} \quad \text{and} \quad q|\,\mathrm{bd}\,B_2 \times \boldsymbol{Q} = \mathrm{pr}_{\mathrm{bd}\,B_2}.$$

Because q is cell-like, it is a near-homeomorphism by Theorem 3.7.8. Combining this with the Z-set Unknotting Theorem 2.11.6, we can obtain a homeomorphism $\eta' : (B_2 \setminus (-t_0, t_0)^n) \times \boldsymbol{Q} \to N$ such that $\eta'(x, \mathbf{0}) = x$ for each $x \in \mathrm{bd}\,B_2$,

$$\eta'|\partial[-t_0, t_0]^n \times \boldsymbol{Q} = \mathrm{id} \quad \text{and} \quad \eta' \simeq q \;\; \text{rel.}\; \partial[-t_0, t_0]^n \times \boldsymbol{Q}$$

(cf. Theorem 1.13.10). Then, η' can be extended to a homeomorphism $\eta : B_2 \times \boldsymbol{Q} \to (B_2 \times \boldsymbol{Q})_{\mathrm{bd}\,B_2}$ by $\eta|[-t_0, t_0]^n \times \boldsymbol{Q} = \mathrm{id}$. We define a homeomorphism f_4' by the following commutative diagram:

$$\begin{array}{ccc} B_2 \times \boldsymbol{Q} & \xrightarrow{\;f_4'\;} & B_2 \times \boldsymbol{Q} \\ \eta \downarrow & & \downarrow \eta \\ (B_2 \times \boldsymbol{Q})_{\mathrm{bd}\,B_2} & \xrightarrow[\;\bar{f}_3\;]{} & (B_2 \times \boldsymbol{Q})_{\mathrm{bd}\,B_2} \end{array}$$

Because $\eta|[-t_0, t_0]^n \times \boldsymbol{Q} = \mathrm{id}$, we have

$$f_4'|B_1 \times \boldsymbol{Q} = \bar{f}_3|B_1 \times \boldsymbol{Q} = f_3|B_1 \times \boldsymbol{Q} = (\alpha \times \mathrm{id}) f_1(e^n \times \mathrm{id})|B_1 \times \boldsymbol{Q}.$$

Since $\eta|\operatorname{bd} B_2 \times \boldsymbol{Q} = \eta'|\operatorname{bd} B_2 \times \boldsymbol{Q} \simeq q|\operatorname{bd} B_2 \times \boldsymbol{Q}$ in N, it follows that

$$\begin{aligned} f_4'|\operatorname{bd} B_2 \times \boldsymbol{Q} &= \eta^{-1}\bar{f}_3\eta|\operatorname{bd} B_2 \times \boldsymbol{Q} \\ &\simeq \eta^{-1}\bar{f}_3 q|\operatorname{bd} B_2 \times \boldsymbol{Q} = \eta^{-1} q|\operatorname{bd} B_2 \times \boldsymbol{Q} \\ &\simeq \mathrm{id} \quad \text{in } (B_2 \setminus (-t_0, t_0)^n) \times \boldsymbol{Q}. \end{aligned}$$

Thus, we can apply the Z-set Unknotting Theorem 2.11.6 to obtain a homeomorphism $f_4 : B_2 \times \boldsymbol{Q} \to B_2 \times \boldsymbol{Q}$ such that

$$f_4|\operatorname{bd} B_2 \times \boldsymbol{Q} = \mathrm{id} \ \text{ and } \ f_4|[-t_0, t_0]^n \times \boldsymbol{Q} = f_4'|[-t_0, t_0]^n \times \boldsymbol{Q}.$$

The above second condition implies that

$$f_4|B_1 \times \boldsymbol{Q} = (\alpha \times \mathrm{id}) f_1(e^n \times \mathrm{id})|B_1 \times \boldsymbol{Q}.$$

This completes the proof. □

The following is the $\boldsymbol{Q}$-factored version of the Alexander Trick (Theorem 3.6.7), which will be used with the Handle Straightening Theorem 4.6.2 to characterize simple homotopy equivalences:

Theorem 4.6.4 ($\boldsymbol{Q}$-FACTORED ALEXANDER TRICK) *Let $f : \mathbf{B}^n \times \boldsymbol{Q} \to \mathbf{B}^n \times \boldsymbol{Q}$ be a homeomorphism such that $f|\mathbf{S}^{n-1} \times \boldsymbol{Q} = \mathrm{id}$. Then, there is an isotopy $h : \mathbf{B}^n \times \boldsymbol{Q} \times \mathbf{I} \to \mathbf{B}^n \times \boldsymbol{Q}$ with $h_0 = \mathrm{id}$ and $h_1 = f$.*

Proof As defined in Note (p. 118), we have a coordinate-switching pseudo-isotopy

$$\theta : \mathbf{B}^n \times \boldsymbol{Q} \times \boldsymbol{Q} \times \mathbf{I} \to \mathbf{B}^n \times \boldsymbol{Q}$$

such that for every $(x, y, z) \in \mathbf{B}^n \times \boldsymbol{Q} \times \boldsymbol{Q}$,

$$\begin{aligned} \theta_1(x, y, z) &= (x, z(1), y(1), z(2), y(2), z(3), y(3), \dots), \\ \theta_0(x, y, z) &= (x, y) = (x, y(1), y(2), y(3), y(4), \dots), \\ \theta_t(x, y, z) &= (x, y(1), \cdots, y(k), *, *, \cdots) \ \text{ if } t \leqslant 2^{-k}. \end{aligned}$$

By the same way, we can define a pseudo-isotopy

$$\theta' : \mathbf{B}^n \times \boldsymbol{Q} \times \boldsymbol{Q} \times \mathbf{I} \to \mathbf{B}^n \times \boldsymbol{Q}$$

such that for every $(x, y, z) \in \mathbf{B}^n \times \boldsymbol{Q} \times \boldsymbol{Q}$,

$$\theta'_1(x, y, z) = (x, z(1), y(1), z(2), y(2), z(3), y(3), \ldots),$$
$$\theta'_0(x, y, z) = (x, z) = (x, z(1), z(2), z(3), z(4), \ldots),$$
$$\theta'_t(x, y, z) = (x, z(1), \cdots, z(k), *, *, \cdots) \text{ if } t \leqslant 2^{-k}.$$

Then, we define a homotopy

$$\vartheta : \mathbf{B}^n \times \boldsymbol{Q} \times \boldsymbol{Q} \times \mathbf{I} \to \mathbf{B}^n \times \boldsymbol{Q}$$

as follows: for $(x, y, z, t) \in \mathbf{B}^n \times \boldsymbol{Q} \times \boldsymbol{Q} \times \mathbf{I}$,

$$\vartheta_t(x, y, z) = \begin{cases} \theta_{2-2t}(tx, y, z) & \text{if } t \geqslant 2^{-1}, \\ \theta'_{2t}(tx, y, z), & \text{if } t \leqslant 2^{-1}. \end{cases}$$

If $0 < t < 1$, then $\vartheta_t : \mathbf{B}^n \times \boldsymbol{Q} \times \boldsymbol{Q} \to t\mathbf{B}^n \times \boldsymbol{Q}$ is a homeomorphism and $\vartheta_t(\mathbf{S}^{n-1} \times \boldsymbol{Q} \times \boldsymbol{Q}) = t\mathbf{S}^{n-1} \times \boldsymbol{Q}$. As is easily observed, ϑ induces the following $(0, 1)$-preserving homeomorphism:

$$\hat{\vartheta} : \mathbf{B}^n \times \boldsymbol{Q} \times \boldsymbol{Q} \times (0, 1) \to \bigcup_{0<t<1} t\mathbf{B}^n \times \boldsymbol{Q} \times \{t\},$$
$$(x, y, z, t) \mapsto \bar{\vartheta}(x, y, z, t) = (\vartheta_t(x, y, z), t),$$

which has the following inverse:

$$\hat{\vartheta}^{-1} : \bigcup_{0<t<1} t\mathbf{B}^n \times \boldsymbol{Q} \times \{t\} \to \mathbf{B}^n \times \boldsymbol{Q} \times \boldsymbol{Q} \times (0, 1)$$
$$(x, y, t) \mapsto \hat{\vartheta}^{-1}(x, y, t) = (\vartheta_t^{-1}(x, y), t).$$

Observe that for $(x, y, z) \in \mathbf{B}^n \times \boldsymbol{Q} \times \boldsymbol{Q}$,

$$\vartheta_1(x, y, z) = \theta_0(x, y, z) = (x, y) = (x, y(1), y(2), y(3), y(4), \ldots),$$
$$\vartheta_0(x, y, z) = \theta'_0(x, y, z) = (\mathbf{0}, z) = (\mathbf{0}, z(1), z(2), z(3), z(4), \ldots),$$
$$\vartheta_t(x, y, z) = (tx, y(1), \cdots, y(k), *, *, \cdots) \text{ if } t \geqslant 1 - 2^{-(k+1)},$$
$$\vartheta_t(x, y, z) = (tx, z(1), \cdots, z(k), *, *, \cdots) \text{ if } t \leqslant 2^{-(k+1)}.$$

The desired isotopy h can be defined by $h_0 = f$, $h_1 = \mathrm{id}$, and for $0 < t < 1$ and $(x, y) \in \mathbf{B}^n \times \boldsymbol{Q}$,

$$h_t(x, y) = \begin{cases} \vartheta_t(f \times \mathrm{id}_{\boldsymbol{Q}})\vartheta_t^{-1}(x, y) & \text{if } x \in t\mathbf{B}^n, \\ (x, y) & \text{otherwise,} \end{cases}$$

where the above first line means that the following diagram commutes:

$$\begin{array}{ccc} t\mathbf{B}^n \times \boldsymbol{Q} & \xrightarrow{\ h_t\ } & t\mathbf{B}^n \times \boldsymbol{Q} \\ \uparrow{\scriptstyle \vartheta_t} & & \uparrow{\scriptstyle \vartheta_t} \\ \mathbf{B}^n \times \boldsymbol{Q} \times \boldsymbol{Q} & \xrightarrow[f \times \mathrm{id}_{\boldsymbol{Q}}]{} & \mathbf{B}^n \times \boldsymbol{Q} \times \boldsymbol{Q} \end{array}$$

Since $f|\mathbf{S}^{n-1} \times \boldsymbol{Q} = \mathrm{id}$, we have

$$\vartheta_t(f \times \mathrm{id}_{\boldsymbol{Q}})\vartheta_t^{-1}|t\mathbf{S}^{n-1} \times \boldsymbol{Q} = \mathrm{id} \quad \text{for every } 0 < t < 1.$$

Hence, it follows that $h|\mathbf{B}^n \times \boldsymbol{Q} \times (0, 1)$ is continuous.

To verify the continuity of h at points of $\mathbf{B}^n \times \boldsymbol{Q} \times \{0, 1\}$, let $\varepsilon > 0$ and $d \in \mathrm{Metr}(\mathbf{B}^n \times \boldsymbol{Q})$. Since $\vartheta(f \times \mathrm{id}_{\boldsymbol{Q}} \times \mathrm{id}_{\mathbf{I}})$ is uniformly continuous, we can choose $k \in \mathbb{N}$ so that

$$\begin{aligned} & t, t' \in \mathbf{I},\ x, x' \in \mathbf{B}^n,\ y, y', z \in \boldsymbol{Q},\ |t - t'| < 2^{-k}, \\ & \qquad \|x - x'\| < 2^{-k+1}, \text{ and } y(1) = y'(1), \dots, y(k) = y'(k) \\ & \qquad \Rightarrow d(\vartheta_t(f(x, y), z), \vartheta_t(f(x', y'), z)),\ d((x, y), (x', y')) < \varepsilon. \end{aligned} \tag{*}$$

If $1 - 2^{-(k+1)} \leqslant t < 1$, $(x, y) \in t\mathbf{B}^n \times \boldsymbol{Q}$, and $\vartheta_t^{-1}(x, y) = (x', y', z)$, then

$$\begin{gathered} x' = t^{-1}x,\ y'(1) = y(1), \dots, y'(k) = y(k), \\ \|x' - x\| = (t^{-1} - 1)\|x\| \leqslant 1 - t \leqslant 2^{-(k+1)} < 2^{-k}, \\ h_t(x, y) = \vartheta_t(f \times \mathrm{id}_{\boldsymbol{Q}})\vartheta_t^{-1}(x, y) = \vartheta_t(f(t^{-1}x, y'), z), \text{ and} \\ f(x, y) = \vartheta_1(f(x, y), z), \end{gathered}$$

hence $d(h_t(x, y), f(x, y)) < \varepsilon$ by (*). If $0 < t \leqslant 2^{-(k+1)}$, $(x, y) \in t\mathbf{B}^n \times \boldsymbol{Q}$, $\vartheta_t^{-1}(x, y) = (x', y', z)$, and $f(x', y') = (x'', y'')$, then

$$\begin{gathered} y(1) = z(1), \dots, y(k) = z(k), \\ \|tx'' - x\| \leqslant t\|x''\| + \|x\| \leqslant 2t < 2^{-k+1}, \text{ and} \\ h_t(x, y) = \vartheta_t(x'', y'', z) = (tx'', z(1), \cdots, z(k), *, *, \cdots), \end{gathered}$$

hence $d(h_t(x, y), (x, y)) < \varepsilon$. □

Now, we can prove the following characterization of simple homotopy equivalences, which implies the TOPOLOGICAL INVARIANCE OF WHITEHEAD TORSION (Theorem 4.1.3):

Theorem 4.6.5 *Let X and Y be compact polyhedra. A map $f : X \to Y$ is a simple homotopy equivalence if and only if $f \times \mathrm{id} : X \times \boldsymbol{Q} \to Y \times \boldsymbol{Q}$ is homotopic to a homeomorphism. In particular, if f is a homeomorphism, then f is a simple homotopy equivalence.*

Proof The "only if" part is Theorem 4.1.2. Hence, it suffices to prove the "if" part.

(*The "if" Part*) We use induction on $\dim X$. When $\dim X = 0$, since $Y \times \boldsymbol{Q}$ is the topological sum of finite copies of $\boldsymbol{Q}$, Y is a finite union of contractible components and f gives a bijection between the vertices of X and these components of Y. Applying Theorem 4.1.5 to each component of Y, we can conclude that f is a simple homotopy equivalence.

Assuming the theorem is true in all cases in which $\dim X < n$, we consider the n-dimensional case. An n-dimensional compact polyhedron X is the union of $|K^{(n-1)}|$ and finite n-simplexes, where K is a triangulation of X. The n-dimensional case can be proved by induction on number of n-simplexes. Thus, assuming the theorem is valid for a compact polyhedron X', we prove the "if" part for $X = X' \cup \sigma$, where σ is an n-simplex with $\sigma \cap X' = \partial\sigma$. Note that the theorem is true when $X' = |K^{(n-1)}|$ by the inductive assumption. Let $f : X \to Y$ be a map such that $f \times \mathrm{id} : X \times \boldsymbol{Q} \to Y \times \boldsymbol{Q}$ is homotopic to a homeomorphism h. Observe that f is homotopic to the map $h_0 : X \to Y$ induced by h, that is, the following diagram commutes:

$$\begin{array}{ccc} X & \xrightarrow{\;h_0\;} & Y \\ {\scriptstyle \times 0}\downarrow & & \uparrow{\scriptstyle \mathrm{pr}_Y} \\ X \times \boldsymbol{Q} & \xrightarrow[\;h\;]{} & Y \times \boldsymbol{Q} \end{array}$$

Let $\varphi : \mathbb{R}^n \to \operatorname{rint}\sigma$ be a homeomorphism such that $\varphi([-1/4, 1/4]^n)$ is a regular neighborhood of the barycenter $\hat{\sigma}$ (so $\sigma \setminus \varphi((-1/4, 1/4)^n)$ is a regular neighborhood of the boundary $\partial\sigma$). Applying the Handle Straightening Theorem 4.6.2 to the open embedding $\varphi \times \mathrm{id} : \mathbb{R}^n \times \boldsymbol{Q} \to X \times \boldsymbol{Q}$, we can find a homeomorphism $g : Y \times \boldsymbol{Q} \to Y \times \boldsymbol{Q}$, $m \in \mathbb{N}$, and a straight subpolyhedron Y_1 of $Y \times [-1, 1]^m$ such that

$$g|(Y \times \boldsymbol{Q}) \setminus h(\varphi([-1/2, 1/2]^n) \times \boldsymbol{Q}) = \mathrm{id} \;\text{ and}$$

$$gh(\varphi([-1/4, 1/4]^n) \times \boldsymbol{Q}) = p_m^{-1}(Y_1) = Y_1 \times \boldsymbol{Q}_m,$$

which implies that

$$g(h(\varphi([-1/2, 1/2]^n) \times \boldsymbol{Q})) = h(\varphi([-1/2, 1/2]^n) \times \boldsymbol{Q}) \text{ and}$$
$$g|h(\varphi(\partial[-1/2, 1/2]^n) \times \boldsymbol{Q}) = \mathrm{id}.$$

Then, applying the $\boldsymbol{Q}$-factored Alexander Trick 4.6.4, we can see that $g \simeq \mathrm{id}$.

We can write $Y \times [-1, 1]^m = Y_1 \cup Y_2$, where

$$Y_2 = (Y \times [-1, 1]^m) \setminus \operatorname{int} Y_1.$$

Similarly, we can write $X = X_1 \cup X_2$, where

$$X_1 = \varphi([-1/4, 1/4]^n) \text{ and } X_2 = X \setminus \varphi((-1/4, 1/4)^n).$$

Let $X_0 = X_1 \cap X_2 = \varphi(\partial[-1/4, 1/4]^n)$ and $Y_0 = Y_1 \cap Y_2 = \operatorname{bd} Y_1 = \operatorname{bd} Y_2$. Then, $gh(X_i \times \boldsymbol{Q}) = Y_i \times \boldsymbol{Q}_m$ for each $i = 0, 1, 2$. Moreover, there are homotopy equivalences $\alpha : X \to Y \times [-1, 1]^m$ and $\alpha_i : X_i \to Y_i$, $i = 0, 1, 2$, such that the following diagrams commute:

$$\begin{array}{ccc} X & \xrightarrow{\alpha} & Y \times [-1,1]^m \\ {\scriptstyle \times 0}\downarrow & & \uparrow{\scriptstyle p_m} \\ X \times \boldsymbol{Q} & \xrightarrow[gh]{} & Y \times \boldsymbol{Q} \end{array} \qquad \begin{array}{ccc} X_i & \xrightarrow{\alpha_i} & Y_i \\ {\scriptstyle \times 0}\downarrow & & \uparrow{\scriptstyle \mathrm{pr}_{Y_i}} \\ X_i \times \boldsymbol{Q} & \xrightarrow[gh]{} & Y_i \times \boldsymbol{Q}_m \end{array}$$

In fact, each α_i is a restriction of α.

Since Y_1 is contractible, it follows from Theorem 4.1.5 that $\alpha_1 : X_1 \to Y_1$ is a simple homotopy equivalence. When $n = 1$, Y_0 is the union of two contractible components. Applying Theorem 4.1.5 to each component of Y_0, we can see that $\alpha_0 : X_0 \to Y_0$ is a simple homotopy equivalence. When $n \geqslant 2$, $\pi_1(Y_0) \cong \mathbb{Z}$ or 0 because $Y_0 \simeq \mathbf{S}^{n-1}$. Again by Theorem 4.1.5, $\alpha_0 : X_0 \to Y_0$ is a simple homotopy equivalence.

Since $\sigma \setminus \varphi((-1/4, 1/4)^n)$ is a regular neighborhood of the boundary $\partial\sigma$, it is easy to see that $\sigma \setminus \varphi((-1/4, 1/4)^n)$ collapses to $\partial\sigma$. Hence, X_2 collapses to X', which implies that the inclusion $X' \subset X_2$ is a simple homotopy equivalence. From the "only if" part, it follows that the inclusion $X' \times \boldsymbol{Q} \subset X_2 \times \boldsymbol{Q}$ is homotopic to a homeomorphism. Observe that $\alpha_2 \times \mathrm{id} : X_2 \times \boldsymbol{Q} \to Y_2 \times \boldsymbol{Q}$ is homotopic to the homeomorphism $gh|X_2 \times \boldsymbol{Q} : X_2 \times \boldsymbol{Q} \to Y_2 \times \boldsymbol{Q}$. Therefore, the restriction $\alpha_2 \times \mathrm{id}|X' \times \boldsymbol{Q} : X' \times \boldsymbol{Q} \to Y_2 \times \boldsymbol{Q}$ is homotopic to a homeomorphism. By the assumption that the theorem is true for X', the restriction $\alpha_2|X'$ (i.e., the composition $X' \subset X_2 \xrightarrow{\alpha_2} Y_2$) is a simple homotopy equivalence. Hence, it follows from Proposition 4.1.1 that $\alpha_2 : X_2 \to Y_2$ is also a simple homotopy equivalence.

Now, applying Theorem 4.1.4, we conclude that $\alpha : X \to Y \times [-1, 1]^m$ is a simple homotopy equivalence. Since the projection $\mathrm{pr}_Y : Y \times [-1, 1]^m \to Y$ is clearly a simple homotopy equivalence, it follows from Proposition 4.1.1 that $\mathrm{pr}_Y\alpha : X \to Y$ is a simple homotopy equivalence. Observe the following commutative diagram:

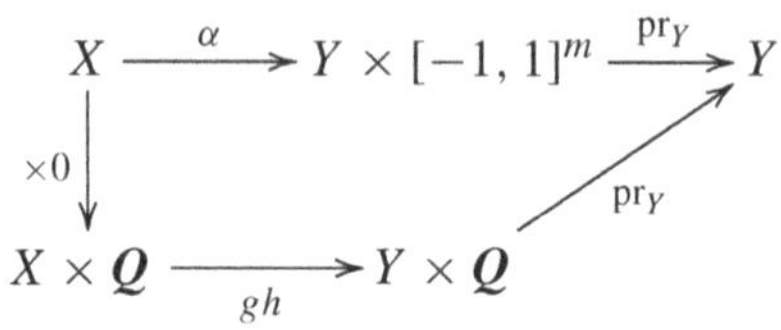

Since $gh \simeq h$, it follows that $\mathrm{pr}_Y\alpha \simeq h_0 \simeq f$. Therefore, f is a simple homotopy equivalence. □

4.7 The Triangulation Theorem for $\boldsymbol{Q}$-Manifolds

In this section, we prove the following Triangulation Theorem for $\boldsymbol{Q}$-manifolds:

Theorem 4.7.1 (TRIANGULATION) *For every $\boldsymbol{Q}$-manifold M, there exists a locally finite simplicial complex K such that $M \approx |K| \times \boldsymbol{Q}$.*

We use the same notational convention as the previous two sections: $\boldsymbol{Q} = [-1, 1]^n \times \boldsymbol{Q}_n$ and $p_n : X \times \boldsymbol{Q} \to X \times [-1, 1]^n$ is the projection,. Then, for $A \subset X \times [-1, 1]^n$, $p_n^{-1}(A) = A \times \boldsymbol{Q}^n$.

Lemma 4.7.2 *Let $M_0 \subset M$ be compact $\boldsymbol{Q}$-manifolds such that* $\mathrm{bd}_M M_0$ *is a Z-set in M_0 and*

$$\big(M \setminus \mathrm{int}_M M_0, \mathrm{bd}_M M_0\big) \approx \big(\mathbf{B}^n \times \boldsymbol{Q}, \mathbf{S}^{n-1} \times \boldsymbol{Q}\big).$$

If $M \approx X \times \boldsymbol{Q}$ for some compact polyhedron X, then there exists $m \in \mathbb{N}$ and a subpolyhedron X_0 of $X \times [-1, 1]^m$ such that $M_0 \approx X_0 \times \boldsymbol{Q}$.

Proof First, we will construct an open embedding $h : \mathbb{R}^n \times \boldsymbol{Q} \to M$ such that $h((-1, 1)^n \times \boldsymbol{Q}) = M \setminus M_0$, that is, $M_0 = M \setminus h((-1, 1)^n \times \boldsymbol{Q})$. By the condition, we have a homeomorphism

$$\psi : \big([-1, 1]^n \times \boldsymbol{Q}, \partial[-1, 1]^n \times \boldsymbol{Q}\big) \to (M \setminus \mathrm{int}\, M_0, \mathrm{bd}\, M_0).$$

Since $\mathbb{R}^n \setminus (-1, 1)^n \approx \partial[-1, 1]^n \times [0, 1)$, we have a homeomorphism

$$\theta : (\mathbb{R}^n \setminus (-1, 1)^n) \times \boldsymbol{Q} \to \partial[-1, 1]^n \times \boldsymbol{Q} \times [0, 1).$$

Since $\mathrm{bd}_M M_0$ is a Z-submanifold of M_0, it has a collar in M_0 by the Collaring Theorem 2.11.1. Using this collar and the homeomorphism ψ, we can easily define an open embedding $k : \partial[-1, 1]^n \times \boldsymbol{Q} \times [0, 1) \to M_0$ such that

$$k(x, y, 0) = \psi(x, y) \quad \text{for each} \quad (x, y) \in \partial[-1, 1]^n \times \boldsymbol{Q}.$$

The desired open embedding $h : \mathbb{R}^n \times \boldsymbol{Q} \to M$ can be defined as follows: $h|[-1, 1]^n \times \boldsymbol{Q} = \psi$ and $h|(\mathbb{R}^n \setminus (-1, 1)^n) \times \boldsymbol{Q} = k\theta$. See the following diagram:

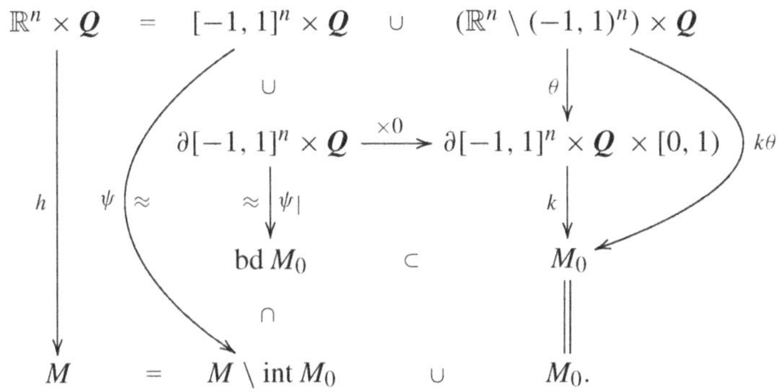

Then, it follows that $M_0 = M \setminus h((-1, 1)^n \times \boldsymbol{Q})$.

Now, let $f : M \to X \times \boldsymbol{Q}$ be a homeomorphism. Then, $fh : \mathbb{R}^n \times \boldsymbol{Q} \to X \times \boldsymbol{Q}$ is an open embedding and

$$f(M_0) = (X \times \boldsymbol{Q}) \setminus fh((-1, 1)^n \times \boldsymbol{Q}).$$

Applying the Handle Straightening Theorem 4.6.2, we can obtain a homeomorphism $g : X \times \boldsymbol{Q} \to X \times \boldsymbol{Q}$ and a straight subpolyhedron $Y \subset X \times [-1, 1]^m$ for some $m \in \mathbb{N}$ such that $gfh([-1, 1]^m \times \boldsymbol{Q}) = Y \times \boldsymbol{Q}_m$. Then, $X_0 = (X \times [-1, 1]^m) \setminus \operatorname{int} Y$ is a subpolyhedron of $X \times [-1, 1]^m$. Moreover,

$$\begin{aligned} M_0 \approx gf(M_0) &= g(X \times \boldsymbol{Q}) \setminus gfh((-1, 1)^m \times \boldsymbol{Q}) \\ &= (X \times \boldsymbol{Q}) \setminus (\operatorname{int} Y \times \boldsymbol{Q}_m) = \big((X \times [-1, 1]^m) \setminus \operatorname{int} Y\big) \times \boldsymbol{Q}_m \\ &= X_0 \times \boldsymbol{Q}_m \approx X_0 \times \boldsymbol{Q}. \end{aligned}$$

Thus, we have the result. □

Now, combining Corollary 4.3.9 with Lemma 4.7.2, we can prove the Finite Triangulation Theorem for compact $\boldsymbol{Q}$-manifolds:

Theorem 4.7.3 (FINITE TRIANGULATION) *For every compact $\boldsymbol{Q}$-manifold M, there exists a finite simplicial complex K such that $M \approx |K| \times \boldsymbol{Q}$.*

Proof By virtue of Corollary 4.3.9 we have a tower of compact $\boldsymbol{Q}$-manifolds $M = M_0 \subset \cdots \subset M_k$ and $0 < n_1 \leqslant n_2 \leqslant \cdots \leqslant n_k$ such that $M_k \approx \boldsymbol{Q}$, each M_{i-1} is a

Z-submanifold of M_i, and

$$\big(M_i \setminus \operatorname{int} M_{i-1}, \operatorname{bd}_{M_i} M_{i-1}\big) \approx \big(\mathbf{B}^{n_i} \times \boldsymbol{Q}, \mathbf{S}^{n_i-1} \times \boldsymbol{Q}\big).$$

Then, using Lemma 4.7.2 iteratively, we can easily obtain the result. □

For every compact ANR X, $X \times \boldsymbol{Q}$ is a compact $\boldsymbol{Q}$-manifold by the Edwards Factor Theorem 3.8.1. Then, applying the above Finite Triangulation Theorem, we have a finite simplicial complex K such that $X \times \boldsymbol{Q} \approx |K| \times \boldsymbol{Q}$. Thus, the Borsuk conjecture is true, that is:

Corollary 4.7.4 *Every compact ANR has the homotopy type of a finite simplicial complex.* □

To prove the Triangulation Theorem 4.7.1, we need the following:

Lemma 4.7.5 *Let $M_0 \subset M$ be compact $\boldsymbol{Q}$-manifolds such that M_0 is a Z-set in M (i.e., M_0 is a Z-submanifold of M). Given a compact polyhedron X_0 with a homeomorphism $h_0 : X_0 \times \boldsymbol{Q} \to M_0$, there exists a compact polyhedron X with a homeomorphism $h : X \times \boldsymbol{Q} \to M$ such that X_0 is a subpolyhedron of X and $h|X_0 \times \boldsymbol{Q} = h_0$.*

Proof By the Finite Triangulation Theorem 4.7.3, there exists a compact polyhedron X_1 with a homeomorphism $h_1 : X_1 \times \boldsymbol{Q} \to M$. Let $i_j : X_j \to X_j \times \boldsymbol{Q}$, $j = 0, 1$, be the injection defined by $i_j(x) = (x, \mathbf{0})$ for each $x \in X_j$. We have simplicial complexes K_0, K_1 with $|K_0| = X_0$, $|K_1| = X_1$, and a simplicial approximation $\varphi : K_0 \to K_1$ of the following composition of maps:

$$X_0 \xrightarrow{i_0} X_0 \times \boldsymbol{Q} \xrightarrow{h_0} M_0 \subset M \xrightarrow{h_1^{-1}} X_1 \times \boldsymbol{Q} \xrightarrow{\mathrm{pr}_1} X_1.$$

Let $X = |Z_\varphi|$ be the polyhedron of the simplicial mapping cylinder Z_φ of φ. Since the simplicial collapsing map $\bar{c}_\varphi : X \to X_1$ is a cell-like map, we can apply Theorem 3.7.8 to obtain a homeomorphism $g : X \times \boldsymbol{Q} \to X_1 \times \boldsymbol{Q}$ such that $g \simeq \bar{c}_\varphi \times \mathrm{id}$. Observe that

$$\varphi \times \mathrm{id}_{\boldsymbol{Q}} \simeq (\varphi \times \mathrm{id}_{\boldsymbol{Q}}) i_0 \mathrm{pr}_1 = i_1 \varphi \mathrm{pr}_1 \simeq i_1 \mathrm{pr}_1 h_1^{-1} h_0 i_0 \mathrm{pr}_1 \simeq h_1^{-1} h_0.$$

Then, it follows that

$$h_1 g | X_0 \times \boldsymbol{Q} \simeq h_1(\bar{c}_\varphi \times \mathrm{id}) | X_0 \times \boldsymbol{Q} = h_1(\varphi \times \mathrm{id}) \simeq h_0.$$

Since $X_0 = |K_0|$ is a Z-set in $X = |Z_\varphi|$, $h_1 g(X_0 \times \boldsymbol{Q})$ is a Z-set in M, which means that $h_1 g|X_0 \times \boldsymbol{Q}$ is a Z-embedding into M. See the following diagram:

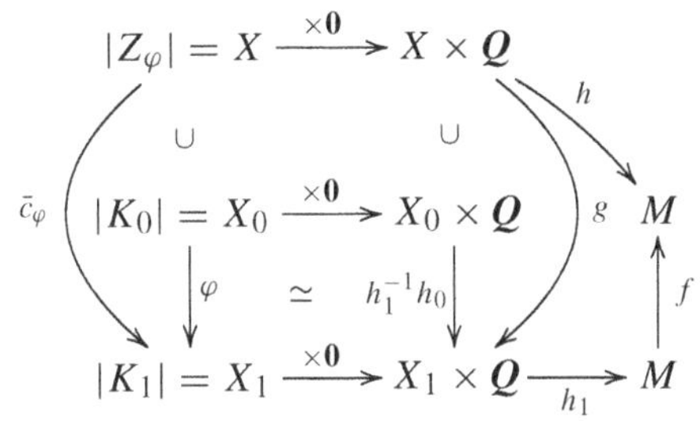

Using the Z-Set Unknotting Theorem 2.11.6, we can obtain a homeomorphism $f : M \to M$ such that $fh_1g|X_0 \times \boldsymbol{Q} = h_0$. Then, $h = fh_1g : X \times \boldsymbol{Q} \to M$ is a homeomorphism such that $h|X_0 \times \boldsymbol{Q} = h_0$. □

Lemma 4.7.6 *Let A_1 and A_2 be compacta in a $\boldsymbol{Q}$-manifold M. Suppose that there exist compact polyhedra X_1 and X_2 with embeddings $g_i : X_i \times \boldsymbol{Q} \to M$, $i = 1, 2$, such that $A_i \subset \operatorname{int} g_i(X_i \times \boldsymbol{Q})$. Then, there exists a polyhedron Y with an embedding $h : Y \times \boldsymbol{Q} \to M$ such that $A_1 \cup A_2 \subset \operatorname{int} h(Y \times \boldsymbol{Q})$.*

Proof Since $g_1^{-1}(A_1 \cap A_2)$ is a compactum in $X_1 \times \boldsymbol{Q}$, we can apply Lemma 4.6.1 to obtain $n_0 \in \mathbb{N}$ and a straight subpolyhedron Y_0 of $X_1 \times [-1, 1]^{n_0}$ such that

$$g_1^{-1}(A_1 \cap A_2) \subset Y_0 \times \boldsymbol{Q}_{n_0} \subset (X_1 \times \boldsymbol{Q}) \cap g_1^{-1}(\operatorname{int} g_2(X_2 \times \boldsymbol{Q})).$$

Regarding $X_1 \times [-1, 1]^{n_0}$ as X_1, we can assume that Y_0 is a straight subpolyhedron of X_1. Then,

$$A_1 \cap A_2 \subset g_1(Y_0 \times \boldsymbol{Q}) \subset g_1(X_1 \times \boldsymbol{Q}) \cap \operatorname{int} g_2(X_2 \times \boldsymbol{Q}).$$

Since $g_2(X_2 \times \boldsymbol{Q}) \setminus \operatorname{int} g_1(Y_0 \times \boldsymbol{Q})$ and $\operatorname{bd} g_1(Y_0 \times \boldsymbol{Q})$ are compact $\boldsymbol{Q}$-manifolds and $\operatorname{bd} g_1(Y_0 \times \boldsymbol{Q})$ is collared in $g_2(X_2 \times \boldsymbol{Q}) \setminus \operatorname{int} g_1(Y_0 \times \boldsymbol{Q})$, we can apply Lemma 4.7.5 to replace X_2 with a new polyhedron such that Y_0 is a subpolyhedron of X_2 and $g_2|Y_0 \times \boldsymbol{Q} = g_1|Y_0 \times \boldsymbol{Q}$, where $X_1 \cap X_2 = Y_0$. There exist disjoint open sets U_i in M, $i = 1, 2$, such that

$$A_i \setminus \operatorname{int} g_1(Y_0 \times \boldsymbol{Q}) \subset U_i \subset g_i(X_i \times \boldsymbol{Q})$$

$$\left(\text{i.e., } g_1^{-1}(A_i) \setminus (\operatorname{int} Y_0 \times \boldsymbol{Q}) \subset g_1^{-1}(U_i) \subset X_i \times \boldsymbol{Q}\right).$$

Again we apply Lemma 4.6.1 to obtain $n_i \in \mathbb{N}$ and a straight subpolyhedron Y_i of $X_i \times [-1, 1]^{n_i}$, $i = 1, 2$, such that

$$A_i \setminus \operatorname{int} g_1(Y_0 \times \boldsymbol{Q}) \subset \operatorname{int} g_i(Y_i \times \boldsymbol{Q}_{n_i}) \subset g_i(Y_i \times \boldsymbol{Q}_{n_i}) \subset U_i,$$

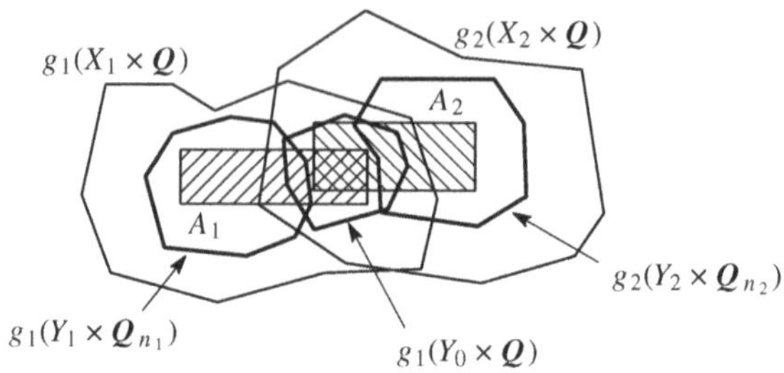

Fig. 4.12 Construction of Y and $h : Y \times Q \to M$

where $(Y_0 \times [-1, 1]^{n_i}) \cup Y_i$ is a subpolyhedron of $X_i \times [-1, 1]^{n_i}$ because it is a union of two subpolyhedra of $X_i \times [-1, 1]^{n_i}$. Let $n = \max\{n_1, n_2\}$. Then, we have the following polyhedron:

$$Y = \left(Y_0 \times [-1, 1]^n\right) \cup \left(Y_1 \times [-1, 1]^{n-n_1}\right) \cup \left(Y_2 \times [-1, 1]^{n-n_2}\right),$$

where $Y_i \times [-1, 1]^{n-n_i} = Y_i$ if $n - n_i = 0$ (cf. Fig. 4.12). Since $g_2|Y_0 \times Q = g_1|Y_0 \times Q$, we can define an embedding $h : Y \times Q_n \to M$ as follows:

$$h|Y_0 \times Q = g_1|Y_0 \times Q \text{ and}$$
$$h|Y_i \times Q_{n_i} = g_i|Y_i \times Q_{n_i}, i = 1, 2.$$

By replacing Q_n with Q, the desired embedding $h : Y \times Q \to M$ can be obtained. □

Lemma 4.7.7 *Let A be a compactum in a Q-manifold M. For each open neighborhood U of A in M, there exists a compact Q-manifold neighborhood N in M such that $N \subset U$ and* bd N *is a Q-manifold being bi-collared in M, that is, it is collared in both N and $M \setminus$* int N.

Proof For each point $x \in A$, we have an open embedding $g_x : \mathbb{R}_+ \times Q \to M$ (Proposition 2.1.5), where we can take g_x so that $x \in g_x([0, 1/2))$. Since A is compact, there are $x_1, \ldots, x_n \in A$ such that $A \subset \bigcup_{i=1}^n g_{x_i}([0, 1/2) \times Q)$ For each $i = 1, \ldots, n$, let $A_i = A \cap g_{x_i}([0, 1/2] \times Q)$ and $g_i = g_{x_i}|\mathbf{I} \times Q$. Then, $A = \bigcup_{i=1}^n A_i$ and $A_i \subset g_i([0, 1) \times Q) = \operatorname{int} g_i(\mathbf{I} \times Q)$. By induction, we can apply Lemma 4.7.6 to obtain a polyhedron X and an embedding $h : X \times Q \to M$ such that $A \subset \operatorname{int} h(X \times Q)$. Moreover, we apply Lemma 4.6.1 to obtain $m \in \mathbb{N}$ and a straight subpolyhedron Y of $X \times [-1, 1]^m$ such that $h^{-1}(A) \subset Y \times Q_m \subset h^{-1}(U)$. Thus, we have a compact Q-manifold $N = h(Y \times Q_m)$. Then, $\operatorname{bd} N = h(\operatorname{bd} Y \times Q_m)$ is a Q-manifold being bi-collared in M. □

Now, we can prove the Triangulation Theorem 4.7.1:

Proof (Triangulation Theorem 4.7.1) We may assume that M is connected and non-compact. Then, we can write $M = \bigcup_{i \in \mathbb{N}} X_i$, where $X_i \subset \operatorname{int} X_{i+1}$. Using

Lemma 4.7.7, we can obtain compact $\boldsymbol{Q}$-manifolds N_i, $i \in \mathbb{N}$, such that $\operatorname{bd} N_i$ is a $\boldsymbol{Q}$-manifold being bi-collared in M and $X_i \cup N_{i-1} \cup \operatorname{int} N_i$, where $N_0 = \emptyset$. Then, each $M_i = N_i \setminus \operatorname{int} N_{i-1}$ is also a $\boldsymbol{Q}$-manifold and $\operatorname{bd} M_i = \operatorname{bd} N_i \cup \operatorname{bd} N_{i-1}$ is a Z-submanifold of M_i. Due to Finite Triangulation Theorem 4.7.3, we have polyhedra Y_i and homeomorphisms $g_i : Y_i \times \boldsymbol{Q} \to \operatorname{bd} N_i$, $i \in \mathbb{N}$. Applying Lemma 4.7.5, we have polyhedra X_i and homeomorphisms $h_i : X_i \times \boldsymbol{Q} \to M_i$, $i \in \mathbb{N}$, such that each $Y_i \cup Y_{i-1}$ is a subpolyhedron of X_i, $h_i|Y_i = g_i$, and $h_i|Y_{i-1} = g_{i-1}$, where $Y_0 = \emptyset$. Then, $X = \bigcup_{i \in \mathbb{N}} X_i$ is a polyhedron and a homeomorphism $h : X \times \boldsymbol{Q} \to M$ can be defined by $h|X_i = h_i$. This completes the proof. □

4.8 Further Results on $\boldsymbol{Q}$-Manifolds

In this section, we will introduce some further results on $\boldsymbol{Q}$-manifolds without proofs. Fathi and Visetti [61] strengthened the Handle Straightening Theorem 4.6.2 as follows:

Theorem 4.8.1 (Strong Handle Straightening) *Let $\mathcal{E}$ be the space of all open embeddings of $\mathbf{B}^k \times 6(\mathbf{B}^n \setminus \mathbf{S}^{n-1}) \times \boldsymbol{Q}$ into $\mathbf{B}^k \times \mathbb{R}^n \times \boldsymbol{Q}$ with the following property:*

(∗) $h|\mathbf{S}^{k-1} \times 6(\mathbf{B}^n \setminus \mathbf{S}^{n-1}) \times \boldsymbol{Q} = \mathrm{id}$,

where $\mathcal{E}$ admits the compact-open topology. Then, there exists a neighborhood $\mathcal{U}$ of the inclusion

$$i : \mathbf{B}^k \times 6(\mathbf{B}^n \setminus \mathbf{S}^{n-1}) \times \boldsymbol{Q} \subset \mathbf{B}^k \times \mathbb{R}^n \times \boldsymbol{Q} \quad (\in \mathcal{E})$$

and a deformation $\varphi : \mathcal{U} \to \mathcal{E}$ such that $\varphi_1(i) = i$, $\varphi_1(h)|\mathbf{B}^k \times \mathbf{B}^n \times \boldsymbol{Q} = \mathrm{id}$, and

$$\varphi_1(h)|\mathbf{B}^k \times (\mathbb{R}^n \setminus 5(\mathbf{B}^n \setminus \mathbf{S}^{n-1})) \times \boldsymbol{Q} = h|\mathbf{B}^k \times (\mathbb{R}^n \setminus 5(\mathbf{B}^n \setminus \mathbf{S}^{n-1})) \times \boldsymbol{Q}.$$

Using the Handle Straightening Theorems 4.6.2 and 4.8.1 above, they applied the techniques developed by Edwards, Kirby, and Siebenmann in the famous two papers [60] and [133] to prove the following theorem:

Theorem 4.8.2 (Fathi–Visetti Deformation) *Let M be a $\boldsymbol{Q}$-manifold, C be a closed set in M, and U be an open set with $C \subset U$. Then, every open cover $\mathcal{U}$ of M has an open refinement $\mathcal{V}$ such that, if an open embedding $g : U \to M$ is $\mathcal{V}$-close to the inclusion, then $g|C$ extends to a homeomorphism $\tilde{g} : M \to M$ such that $\tilde{g}|M \setminus U = \mathrm{id}$ and $\tilde{g}|U$ is $\mathcal{U}$-homotopic to g.*

Applying these results, Ferry [64] proved the following Q-manifold version of Theorem 2.6.3, which is also a generalization of Theorem 3.7.8 (cf. Theorem 1.15.9):

Theorem 4.8.3 (Homeomorphism Approximation) *Let M and N be Q-manifolds. Then, each open cover $\mathcal{U}$ of N has an open refinement $\mathcal{V}$ such that any proper $\mathcal{V}$-homotopy equivalence $f : M \to N$ is $\mathcal{U}$-close to a homeomorphism.*

Using this theorem, he proved in [64] that the homeomorphism group $\mathrm{Homeo}(M)$ is an ANR for every compact Q-manifold M (this result was also proved in [141] by Toruńczyk independently). Combining this result with Geoghegan's result in [68] (or Keesling's result in [91]) and Toruńczyk Factor Theorem 2.2.14, we can obtain the following Q-manifold version of the Homeomorphism Group Problem (cf. p. 266).

Theorem 4.8.4 (Homeomorphism Group) *For every compact Q-manifold M, the homeomorphism group* $\mathrm{Homeo}(M)$ *is an ℓ^2-manifold.*

Other than above, there are many interesting results on Q-manifolds, which are not discussed in this chapter nor any other part. As one of them, Chapman and Siebenmann studied in [47] the case when a non-compact Q-manifold has a Q-manifold compactification with a Z-set reminder.

Notes for Chapter 4

This chapter (except the first two sections) was written based on Chapman's Lecture Notes [(2)] on Q-manifolds. However, we give every possible detail.

In Sect. 4.1, we gave a brief explanation of the simple homotopy type. As a combinatorial version of the homotopy type, the simple homotopy type was introduced by J.H.C. Whitehead in [154], where the term the "*nucleus*" was used instead of the "simple homotopy type." Since simplicial complexes are much too hard to deal with in this context, Whitehead introduced CW complexes in his paper [155] and then succeeded in developing his theory in [157]. But, in order not to add extra sections, we avoided introducing CW complexes. Simple Homotopy Theory is a part of PL Topology (Combinatorial Topology), which is also useful. For this theory, Cohen's book [(20)] is an excellent textbook. But this book avoids proving Theorem 4.1.5 because of difficulty. So it is necessary to refer to the paper of H. Bass, A. Heller, and R. Swan [25].

We also need Theorem 4.1.6, which is a result on Wall's finiteness obstruction and is related to Algebraic K-theory. Wall's finiteness obstruction theory was founded by C.T.C. Wall in [147]. For this theory, we refer the reader to surveys by G. Mislin [111] or Ferry and Ranicki [66]. In [65], S. Ferry gave a geometric approach to Wall's finiteness obstruction.

Concerning contents of Sect. 4.2, any textbook on Algebraic Topology can be referred to, e.g., [(22)], [(23)], etc. The fundamental groups and covering spaces are contained even in Munkres' textbook [(7)]. Theorem 4.3.1 is standard in Algebraic Topology,

The results of Sect. 4.4 were obtained by T.A. Chapman in [44], where the Splitting Theorem 4.4.1 appears explicitly but the procedures given in Theorems 4.4.4 and 4.4.5 appear implicitly.

The proof of Theorem 4.5.1 is due to S. Ferry [63]. The Handle Straightening Theorem 4.6.2 was established in [46]. Lemma 4.6.3 appears in [44]. The torus idea is the torus wrapping-unwrapping idea of Kirby and Siebenmann in [94], which was used to triangulate finite-dimensional manifolds. The Alexander Trick appeared in the paper [1]. There are various versions other than the Q-factored version 4.6.4.

Chapter 5
Manifolds Modeled on Homotopy Dense Subspaces of Hilbert Spaces

Consecutively in this chapter, we set the following convention:

- *let* Γ *be an infinite set (or an infinite discrete space) with* $\operatorname{card}\Gamma = \tau$.

Recall that $\ell_f^2(\Gamma)$ is a dense linear subspace of Hilbert space $\ell^2(\Gamma)$ defined as the linear span of the orthonormal basis $\mathbf{e}_\gamma$, $\gamma \in \Gamma$, that is,

$$\ell_f^2(\Gamma) = \left\{x \in \ell^2(\Gamma) \;\middle|\; x(\gamma) = 0 \text{ except for finitely many } \gamma \in \Gamma \right\},$$

where we write $\ell_f^2(\mathbb{N}) = \ell_f^2$. Note that $\prod_{n\in\mathbb{N}}[-2^{-n}, 2^{-n}] \subset \ell^2$ is a natural copy of the Hilbert cube $\boldsymbol{Q} = [-1, 1]^{\mathbb{N}}$. Let ℓ_Q^2 be a dense linear subspace of ℓ^2 defined as the linear span of $\prod_{n\in\mathbb{N}}[-2^{-n}, 2^{-n}]$, that is,

$$\ell_Q^2 = \bigcup_{m\in\mathbb{N}} m \prod_{n\in\mathbb{N}} [-2^{-n}, 2^{-n}] = \left\{x \in \ell^2 \;\middle|\; \sup_{n\in\mathbb{N}} 2^n |x(n)| < \infty\right\}.$$

It will be proved that ℓ_Q^2 is homeomorphic to $B(\boldsymbol{Q})$, the **pseudo-boundary** of $\boldsymbol{Q}$, which is the complement of $I(\boldsymbol{Q}) = (-1, 1)^{\mathbb{N}}$, the **pseudo-interior** of $\boldsymbol{Q}$. Moreover, let $\boldsymbol{Q}_f = [-1, 1]_f^{\mathbb{N}}$.

Given a pair (E, F) of a space and its subspace, an $(\boldsymbol{E}, \boldsymbol{F})$**-manifold pair** (or a **manifold pair modeled on** (E, F)) is a pair (M, N) of a paracompact space and its subspace having an open cover $\mathcal{U} \in \operatorname{cov}(M)$ with open embeddings $\varphi_U : U \to E$, $U \in \mathcal{U}$, such that $\varphi_U(U \cap N) = \varphi_U(U) \cap F$. Then, M is an E-manifold and N is an F-manifold.

In this chapter, introducing f.d.cap sets and cap sets for ℓ^2 or $\boldsymbol{Q}$ due to R.D. Anderson, we characterize the pairs (ℓ^2, ℓ_f^2), (ℓ^2, ℓ_Q^2), $(\boldsymbol{Q}, \boldsymbol{Q}_f)$, and $(\boldsymbol{Q}, B(\boldsymbol{Q}))$, and manifold pairs modeled on such pairs. Proving the topological uniqueness of f.d.cap sets and cap sets, we show that $(\ell_f^2)_f^{\mathbb{N}} \approx \ell_f^2 \approx \boldsymbol{Q}_f$ and $(\ell_Q^2)_f^{\mathbb{N}} \approx \ell_Q^2 \approx B(\boldsymbol{Q})$. So, all results in Chap. 2 are valid for ℓ_f^2- and ℓ_Q^2-manifolds.

K. Sakai, *Topology of Infinite-Dimensional Manifolds*, Springer Monographs in Mathematics, https://doi.org/10.1007/978-981-15-7575-4_5

As non-separable generalization, we introduce absorption bases due to J.E. West and characterize the following pairs and manifold pairs modeled on them:

$$\big(\ell^2(\Gamma), \ell^2_f(\Gamma)\big), \ \big(\ell^2(\Gamma) \times \ell^2, \ell^2_f(\Gamma) \times \ell^2_Q\big),$$
$$\big(\ell^2(\Gamma) \times \ell^2, \ell^2_f(\Gamma) \times \ell^2\big), \text{ and } \big(\ell^2(\Gamma) \times \ell^2, \ell^2(\Gamma) \times \ell^2_f\big).$$

The topological uniqueness of absorption bases is proved and it is shown that all results in Chap. 2 are valid for manifolds modeled on

$$\ell^2_f(\Gamma), \ \ell^2_f(\Gamma) \times \ell^2_Q, \ \ell^2_f(\Gamma) \times \ell^2, \text{ and } \ell^2(\Gamma) \times \ell^2_f.$$

Moreover, introducing absorbing sets in $\ell^2(\Gamma)$ due to Bestvina and Mogilski,[1] which are non-F_σ version of cap sets or absorption bases, we show the existence of universal spaces for absolute Borel classes and give their topological characterizations. Manifolds modeled on such universal spaces are also discussed.

5.1 Cap Sets and F.D.Cap Sets in ℓ^2 and Q

A subset M of a metrizable space X is called a **cap set**[2] for X if there is a tower $M_1 \subset M_2 \subset \cdots$ of compact Z-sets in X with $M = \bigcup_{n\in\mathbb{N}} M_n$ and the following property:

(cap) for each $\mathcal{U} \in \mathrm{cov}(X)$, each $m \in \mathbb{N}$, and each compact set $C \subset X$, there exist $n \in \mathbb{N}$ and an embedding $h : C \to M_n$ such that $h|C \cap M_m = \mathrm{id}$ and h is $\mathcal{U}$-close to id.

In the above, if all M_n, $n \in \mathbb{N}$, and C are finite-dimensional, M is called an **f.d.cap set**.[3] The property (cap) is called the **compact absorption property** (abbrev. **cap**) for X. The property (cap) restricting C to finite-dimensional is written (f.d.cap) and called the **finite-dimensional compact absorption property** (abbrev. **f.d.cap**) for X. We use the convention that "(f.d.)cap" represents two alternatives in an obvious way, one being cap and the other being f.d.cap. It is easy to see the following:

Fact (1) *Every (f.d.)cap set for X is dense in X.*[4]

(2) *If $f : X \to Y$ is a homeomorphism and M is an (f.d.)cap set for X, then $f(M)$ is an (f.d.)cap set for Y.*

(3) *Any non-separable metrizable space has no (f.d.)cap sets.*

[1] Bestvina and Mogilski treated only the separable case, which will be extended to the non-separable case.

[2] In the literature, it is also called a cap-set.

[3] In the literature, it is also called an f-d cap-set.

[4] In fact, every (f.d.)cap set for X is homotopy dense in X (cf. Proposition 5.3.5).

In the case where X is an ℓ^2-manifold, let us remember that every compact set in X is a Z-set (Proposition 3.2.10). Hence, every compact set in $\boldsymbol{s}$ ($\approx I(\boldsymbol{Q})$) is a Z-set in $\boldsymbol{s}$ ($\approx I(\boldsymbol{Q})$). Moreover, every compact set in $I(\boldsymbol{Q})$ is also a Z-set in $\boldsymbol{Q}$ (Theorem 2.10.6). In this section, we give various (f.d.)cap sets in ℓ^2, $\boldsymbol{s}$, and $\boldsymbol{Q}$, and show that they are homeomorphic to the linear subspace ℓ^2_Q (or ℓ^2_f) of the Hilbert space ℓ^2 by proving the topological uniqueness of them for ℓ^2-manifolds and $\boldsymbol{Q}$-manifolds. Recall that ℓ^2_f is the linear subspace of ℓ^2 with $\ell^2_f = \boldsymbol{s}_f$ ($= \mathbb{R}^{\mathbb{N}}_f$) as sets (as linear spaces). It is also shown that $(\ell^2_f)^{\mathbb{N}}_f \approx \ell^2_f$ and $(\ell^2_Q)^{\mathbb{N}}_f \approx \ell^2_Q$, which means that all results in Chap. 2 are valid for ℓ^2_f-manifolds and ℓ^2_Q-manifolds.

Since $\boldsymbol{Q}$ is a convex set in $\boldsymbol{s}$, the radial interior can be defined as follows:

$$\begin{aligned}\operatorname{rint} \boldsymbol{Q} &= \{x \in \boldsymbol{Q} \mid \forall y \in \boldsymbol{Q}, \exists \delta > 0 \text{ such that } (1+\delta)x - \delta y \in \boldsymbol{Q}\} \\ &= \{x \in \boldsymbol{Q} \mid \forall y \in \boldsymbol{Q}, \exists z \in \boldsymbol{Q} \text{ such that } x \in \operatorname{rint}\langle y, z\rangle\}.\end{aligned}$$

It should be noted that $I(\boldsymbol{Q})_f \subset \operatorname{rint} \boldsymbol{Q} \subset I(\boldsymbol{Q})$, where $I(\boldsymbol{Q})_f = (-1, 1)^{\mathbb{N}}_f$.

Lemma 5.1.1 *The radial interior* rint $\boldsymbol{Q}$ *of* $\boldsymbol{Q}$ *can be represented as follows:*

$$\operatorname{rint} \boldsymbol{Q} = \Big\{x \in \boldsymbol{Q} \;\Big|\; \sup_{n\in\mathbb{N}} |x(n)| < 1\Big\} = \bigcup_{n\in\mathbb{N}} \Big[-\frac{n}{n+1}, \frac{n}{n+1}\Big]^{\mathbb{N}}.$$

Proof The second equality is trivial. If $x \in \operatorname{rint} \boldsymbol{Q}$, then $(1+\delta)x \in \boldsymbol{Q}$ for some $\delta > 0$. Since $(1+\delta)|x(n)| \leqslant 1$ for every $n \in \mathbb{N}$, it follows that $\sup_{n\in\mathbb{N}} |x(n)| \leqslant (1+\delta)^{-1} < 1$. Conversely, let $x \in \boldsymbol{Q}$ with $\alpha = \sup_{n\in\mathbb{N}} |x(n)| < 1$. Then, $\delta = (1-\alpha)/2 > 0$. For each $y \in \boldsymbol{Q}$, since $\sup_{n\in\mathbb{N}} |x(n) - y(n)| < 2$, it follows that

$$\begin{aligned}\sup_{n\in\mathbb{N}} |(1+\delta)x(n) - \delta y(n)| &\leqslant \sup_{n\in\mathbb{N}} |x(n)| + \delta \sup_{n\in\mathbb{N}} |x(n) - y(n)| \\ &< \alpha + 2\delta = 1.\end{aligned}$$

Hence, $(1+\delta)x - \delta y \in \boldsymbol{Q}$, which means that $x \in \operatorname{rint} \boldsymbol{Q}$. □

Let $\boldsymbol{s}_Q$ be the linear subspace of $\boldsymbol{s}$ that is the linear span of $\boldsymbol{Q}$. Then,

$$\boldsymbol{s}_Q = \bigcup_{n\in\mathbb{N}} n\boldsymbol{Q} = \bigcup_{n\in\mathbb{N}} [-n, n]^{\mathbb{N}} = \Big\{x \in \boldsymbol{s} \;\Big|\; \sup_{n\in\mathbb{N}} |x(n)| < \infty\Big\},$$

where $\boldsymbol{s}_Q = \ell^\infty$ as sets. Moreover, let $\boldsymbol{s}_f = \mathbb{R}^{\mathbb{N}}_f$ (the space of finite sequences). As is easily observed, we have

$$(\boldsymbol{s}, \boldsymbol{s}_Q, \boldsymbol{s}_f) \approx (I(\boldsymbol{Q}), \operatorname{rint} \boldsymbol{Q}, I(\boldsymbol{Q})_f).$$

The linear subspaces $\ell^1_f \subset \ell^1_Q$ of ℓ^1 is defined just as $\ell^2_f \subset \ell^2_Q$, where $\ell^1_f = \ell^2_f$ $(= \mathbf{s}_f)$, $\ell^1_Q = \ell^2_Q$ as sets (as linear spaces). Then, we have $\ell^1_f \subset \ell^1_Q \subset \ell^1$. For any $p > 2$, since $\prod_{n\in\mathbb{N}}[-2^{-n}, 2^{-n}] \subset \ell^p$, we can similarly obtain $\ell^p_f \subset \ell^p_Q \subset \ell^p$, where $\ell^p_Q = \ell^1_Q$ and $\ell^p_f = \ell^1_f$ as sets (as linear spaces).

Note Taking $a_i > 0$, $i \in \mathbb{N}$, so that $\sum_{i\in\mathbb{N}} a_i^p < \infty$, we have the subspace $\prod_{i\in\mathbb{N}}[-a_i, a_i] \subset \ell^p$, which is a copy of the Hilbert cube $\boldsymbol{Q}$. As the linear span of $\prod_{i\in\mathbb{N}}[-a_i, a_i]$, we can define ℓ^p_Q, that is,

$$\ell^p_Q = \left\{x \in \ell^p \;\middle|\; \sup_{i\in\mathbb{N}} a_i^{-1}|x(i)| < \infty\right\},$$

which is different from the above ℓ^1_Q as sets.

Lemma 5.1.2 *Let $X = \ell^p$, $p \in \mathbb{N}$, and C be a compact set in X. For each $\varepsilon > 0$, there exists $n \in \mathbb{N}$ such that $\sum_{i>n} |x(i)|^p < \varepsilon^p$ for every $x \in C$.*

Proof Since C is compact, we can take $x_1, \dots, x_k \in C$ such that $C \subset \bigcup_{i=1}^k \mathrm{B}(x_i, \varepsilon/2)$. For each $j = 1, \dots, k$, there is $n_j \in \mathbb{N}$ such that $\sum_{i>n} |x_j(i)|^p < (\varepsilon/2)^p$. Let $n_0 = \max\{n_1, \dots, n_k\}$. Then, each $x \in C$ is contained in some $\mathrm{B}(x_j, \varepsilon/2)$, from which it follows that

$$\begin{aligned}\Big(\sum_{i>n} |x_j(i)|^p\Big)^{1/p} &\leqslant \Big(\sum_{i>n} |x(i) - x_j(i)|^p\Big)^{1/p} + \Big(\sum_{i>n} |x(i)|^p\Big)^{1/p} \\ &< \|x - x_j\| + \varepsilon/2 < \varepsilon.\end{aligned}$$

Therefore, $\sum_{i>n} |x(i)|^p < \varepsilon^p$. □

Recall that $I(\boldsymbol{Q}) = (-1, 1)^{\mathbb{N}}$ and $B(\boldsymbol{Q}) = \boldsymbol{Q} \setminus I(\boldsymbol{Q})$ are the **pseudo-interior** and the **pseudo-boundary** of $\boldsymbol{Q}$, respectively. Moreover, $\boldsymbol{Q}_f = [-1, 1]^{\mathbb{N}}_f$ and $I(\boldsymbol{Q})_f = (-1, 1)^{\mathbb{N}}_f$. Now, we show the following:

Proposition 5.1.3

(1) rint $\boldsymbol{Q}$ *is a cap set for* $\boldsymbol{Q}$.
(2) $B(\boldsymbol{Q})$ *is a cap set for* $\boldsymbol{Q}$.
(3) $\boldsymbol{Q}_f$ *is an f.d.cap set for* $\boldsymbol{Q}$.
(4) $\boldsymbol{Q}_f \times \boldsymbol{Q}$ *is a cap set for* $\boldsymbol{Q}^2 = \boldsymbol{Q} \times \boldsymbol{Q}$.
(5) $\mathbf{s}_Q$ *is a cap set for* $\mathbf{s}$, *so* rint $\boldsymbol{Q}$ *is a cap set for* $I(\boldsymbol{Q})$.
(6) $\mathbf{s}_f$ *is an f.d.cap set for* $\mathbf{s}$, *so* $I(\boldsymbol{Q})_f$ *is an f.d.cap set for* $I(\boldsymbol{Q})$.
(7) $\mathbf{s}_f \times \boldsymbol{Q}$ *is a cap set for* $\mathbf{s} \times \boldsymbol{Q}$.
(8) ℓ^1_Q *is a cap set for* ℓ^1.
(9) ℓ^1_f *is an f.d.cap set for* ℓ^1.

Proof We use the following metric $d \in \mathrm{Metr}(\boldsymbol{s})$:

$$d(x, y) = \sup_{n\in\mathbb{N}} \min\left\{n^{-1}, |x(n) - y(n)|\right\}.$$

Here, we regard that $\boldsymbol{Q}$ inherits this metric. In the following, we use the following norm for $\mathbb{R}^n$, $n \in \mathbb{N}$:

$$\|x\| = \max\left\{|x(1)|, \ldots, |x(n)|\right\}.$$

Then, it should be remarked that $[-1, 1]^n$ and $\partial[-1, 1]^n$ are the unit closed ball and the unit sphere of $\mathbb{R}^n$ with respect to this norm.

(1): Due to Lemma 5.1.1, rint $\boldsymbol{Q}$ is the union of the following tower, whose terms are compact Z-sets in $\boldsymbol{Q}$ by Theorem 2.10.6:

$$\left(\left[-\frac{n}{n+1}, \frac{n}{n+1}\right]^{\mathbb{N}}\right)_{n\in\mathbb{N}}.$$

So, we may show that this tower has the cap for $\boldsymbol{Q}$. For each $\varepsilon > 0$ and $m \in \mathbb{N}$, choose $n \in \mathbb{N}$ so that $n > m$ and $1/n < \varepsilon$. We define an embedding $h : \boldsymbol{Q} \to [-n/(n+1), n/(n+1)]^{\mathbb{N}}$ as follows:

$$h(x)(i) = \begin{cases} x(i) & \text{if } |x(i)| \leqslant \dfrac{n-1}{n}, \\[2ex] \dfrac{n-1}{n} + \dfrac{1}{n+1}\left(x(n) - \dfrac{n-1}{n}\right) & \text{if } x(i) \geqslant \dfrac{n-1}{n}, \\[2ex] -\dfrac{n-1}{n} + \dfrac{1}{n+1}\left(x(n) + \dfrac{n-1}{n}\right) & \text{if } x(i) \leqslant -\dfrac{n-1}{n}. \end{cases}$$

Then, $h|[-m/(m+1), m/(m+1)]^{\mathbb{N}} = \mathrm{id}$. For each $x \in \boldsymbol{Q}$ and $i \in \mathbb{N}$, $|h(x)(i) - x(i)| < 1/n < \varepsilon$, so $d(h, \mathrm{id}) < \varepsilon$. Hence, rint $\boldsymbol{Q}$ is a cap set for $\boldsymbol{Q}$.

(2): For each $n \in \mathbb{N}$, let

$$M_n = \bigcup_{i=1}^{n} \mathrm{pr}_i^{-1}(\{\pm 1\}) = \partial[-1, 1]^n \times \boldsymbol{Q}.$$

Then, $M_1 \subset M_2 \subset \cdots$ is a tower of compact Z-sets in $\boldsymbol{Q}$ with $B(\boldsymbol{Q}) = \bigcup_{n\in\mathbb{N}} M_n$. We show that this has the cap for $\boldsymbol{Q}$. Let C be a compact set in $\boldsymbol{Q}$. For each $m \in \mathbb{N}$ and $\varepsilon > 0$, choose $n \in \mathbb{N}$ so that $n > m$ and $n^{-1} < \varepsilon$. As in Fig. 5.1, we define a retraction

$$\begin{aligned} r : [-1, 1]^n = [-1, 1]^{n-1} \times [-1, 1] &\to \partial[-1, 1]^{n-1} \times [-1, 1] \cup [-1, 1]^{n-1} \times \{1\} \\ &= \partial[-1, 1]^n \setminus \mathrm{rint}[-1, 1]^{n-1} \times \{-1\} \end{aligned}$$

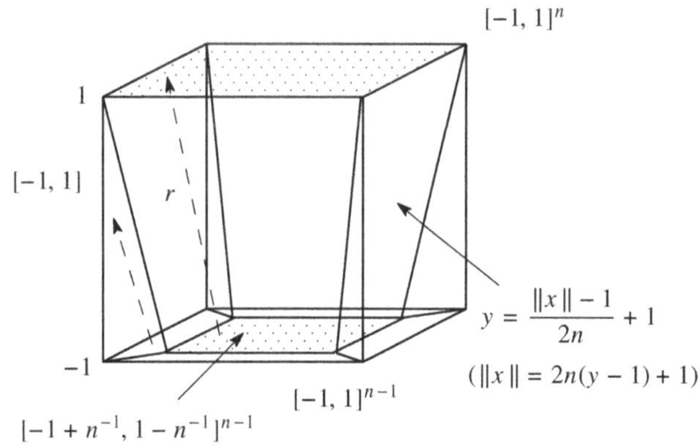

Fig. 5.1 The retraction r

such that $\|\mathrm{pr}_1 r(x, y) - x\| < n^{-1}$ $(< \varepsilon)$ for each $(x, y) \in [-1, 1]^{n-1} \times [-1, 1]$, where $\mathrm{pr}_1 : [-1, 1]^{n-1} \times [-1, 1] \to [-1, 1]^{n-1}$ is the projection. Indeed, such r can be defined as follows:

$$r(x, y) = \begin{cases} \left(\dfrac{2nx}{y - 1 + 2n}, 1\right) & \text{if } \|x\| \leqslant \dfrac{y-1}{2n} + 1, \\ \left(\dfrac{x}{\|x\|}, y + 2n(1 - \|x\|)\right) & \text{if } \|x\| \geqslant \dfrac{y-1}{2n} + 1, \end{cases}$$

Let $\theta : \boldsymbol{Q}^2 \times \mathbf{I} \to \boldsymbol{Q}$ be the coordinate-switching pseudo-isotopy defined in Note on p. 118 (where X is now considered as a singleton or $\boldsymbol{Q}$). Then, $\theta_0 = \mathrm{pr}_1$ and $\theta_t : \boldsymbol{Q}^2 \to \boldsymbol{Q}$ is a homeomorphism for each $t > 0$. Let $p_n : \boldsymbol{Q} = [-1, 1]^n \times \boldsymbol{Q} \to [-1, 1]^n$ and $q_n : \boldsymbol{Q} = [-1, 1]^n \times \boldsymbol{Q} \to \boldsymbol{Q}$ be the projections, that is,

$$p_n(x) = (x(1), \dots, x(n)) \text{ and } q_n(x) = (x(n + i))_{i \in \mathbb{N}},$$

and define a map $\alpha : C \to \mathbf{I}$ by $\alpha(x) = 1 - \|p_{n-1}(x)\|$. Then,

$$\alpha^{-1}(0) = C \cap (\partial[-1, 1]^{n-1} \times \boldsymbol{Q}) = C \cap M_{n-1}.$$

By the universality of $\boldsymbol{Q}$, we have an embedding $g : C \to \boldsymbol{Q}$. Then, we define a map $h : C \to M_n = \partial[-1, 1]^n \times \boldsymbol{Q}$ as follows:

$$h(x) = (rp_n(x), \theta_{\alpha(x)}(q_n(x), g(x))).$$

For each $x \in C \cap M_{n-1}$, since $\alpha(x) = 0$,

$$h(x) = (rp_n(x), \theta_0(q_n(x), g(x))) = (p_n(x), q_n(x)) = x.$$

Therefore, $h|C \cap M_{n-1} = \mathrm{id}$. Because $M_m \subset M_{n-1}$, we have $h|C \cap M_m = \mathrm{id}$. Moreover, for every $x \in C$,

$$\|p_{n-1}h(x) - p_{n-1}(x)\| = \|rp_n(x) - p_{n-1}(x)\| \leqslant n^{-1},$$

which means that $d(h, \mathrm{id}) \leqslant n^{-1} < \varepsilon$.

To verify that h is an embedding, let $x \neq y \in C$. If $\alpha(x) = \alpha(y) = 0$, then $h(x) = x \neq y = h(y)$ as seen above. When $\alpha(x) = \alpha(y) = t > 0$, since θ_t is a homeomorphism and g is an embedding, it follows that

$$q_n h(x) = \theta_t(q_n(x), g(x)) \neq \theta_t(q_n(y), g(y)) = q_n h(y),$$

which implies $h(x) \neq h(y)$. If $\alpha(x) \neq \alpha(y)$, then

$$\|p_{n-1}h(x)\| = \|p_{n-1}(x)\| = 1 - \alpha(x) \neq 1 - \alpha(y) = \|p_{n-1}(y)\| = \|p_{n-1}h(y)\|,$$

hence $h(x) \neq h(y)$.

(3): Each $[-1, 1]^n$ is identified with $\{x \in \boldsymbol{Q} \mid x(i) = 0 \text{ if } i > n\}$, which is a compact Z-set in $\boldsymbol{Q}$ by Theorem 2.10.6. We may show that the tower $([-1, 1]^n)_{n\in\mathbb{N}}$ has the f.d.cap for $\boldsymbol{Q}$. Let C be a finite-dimensional compact set in $\boldsymbol{Q}$ and $k = 2\dim C + 1$. For each $\varepsilon > 0$ and $m \in \mathbb{N}$, choose $n \in \mathbb{N}$ so that $n > m$ and $n^{-1} < \varepsilon$. By the Embedding Theorem 1.12.14, there exists an embedding $g : C \to \mathbf{I}^k$. We define an embedding $h : C \to [-1, 1]^{n-1} \times \mathbf{I} \times \mathbf{I}^k \subset [-1, 1]^{n+k-1}$ as follows:

$$h(x) = (x(1), \ldots, x(n-1), \alpha(x), \alpha(x)g(x)),$$

where $\alpha : C \to \mathbf{I}$ is the map defined by $\alpha(x) = d(x, C \cap [-1, 1]^{n-1})$. For each $x \in C \cap [-1, 1]^{n-1}$, since $\alpha(x) = 0$, we have

$$h(x) = (x(1), \ldots, x(n-1), 0, 0, \ldots) = x,$$

hence $h|C \cap [-1, 1]^{n-1} = \mathrm{id}$, so $h|C \cap [-1, 1]^m = \mathrm{id}$. Since $p_{n-1}h(x) = p_{n-1}(x)$ for every $x \in C$, it follows that $d(h, \mathrm{id}) \leqslant n^{-1} < \varepsilon$. To see that h is an embedding, let $x \neq y \in C$. If $\alpha(x) = \alpha(y) = 0$, then $h(x) = x \neq y = h(y)$. If $\alpha(x) = \alpha(y) = t > 0$, then $tg(x) \neq tg(y)$, which implies that $h(x) \neq h(y)$. If $\alpha(x) \neq \alpha(y)$, then $\mathrm{pr}_n h(x) = \alpha(x) \neq \alpha(y) = \mathrm{pr}_n h(y)$, hence $h(x) \neq h(y)$.

(4): As in the proof of (3), we regard $[-1, 1]^n \subset \boldsymbol{Q}$. We show that the tower $([-1, 1]^n \times \boldsymbol{Q})_{n\in\mathbb{N}}$ has the cap for $\boldsymbol{Q}^2$. We use the metric $\rho \in \mathrm{Metr}(\boldsymbol{Q}^2)$ defined as follows:

$$\rho((x, y), (x', y')) = \max\left\{d(x, x'), d(y, y')\right\}.$$

Let C be a compact set in $\boldsymbol{Q}^2$. For each $\varepsilon > 0$ and $m \in \mathbb{N}$, choose $n \in \mathbb{N}$ so that $n > m$ and $n^{-1} < \varepsilon$. As in the proof of (2), take the coordinate-switching pseudo-isotopy $\theta : \boldsymbol{Q}^2 \times \mathbf{I} \to \boldsymbol{Q}$ defined in Note on p. 118. By the universality of $\boldsymbol{Q}$, we have an embedding $g : C \to \boldsymbol{Q}$. We define an embedding $h : C \to [-1, 1]^{n-1} \times \mathbf{I} \times \boldsymbol{Q} \subset \boldsymbol{Q}^2$ as follows:

$$h(x, y) = (x(1), \dots, x(n-1), \alpha(x, y), \theta_{\alpha(x,y)}(y, g(x, y))),$$

where $\alpha : C \to \mathbf{I}$ is the map defined by

$$\alpha(x, y) = \min\left\{2^{-n}, \rho((x, y), C \cap ([-1, 1]^{n-1} \times \boldsymbol{Q}))\right\}.$$

Since $\alpha(x, y) \leqslant 2^{-n}$, it follows from (♯) on p. 119 that $p_n\theta_{\alpha(x,y)}(y, g(x, y)) = p_n(y)$. For each $(x, y) \in C \cap ([-1, 1]^{n-1} \times \boldsymbol{Q})$, since $\alpha(x, y) = 0$, we have

$$\begin{aligned} h(x, y) &= (x(1), \dots, x(n-1), 0, \theta_0(y, g(x, y))) \\ &= (x(1), \dots, x(n-1), 0, 0, \dots, y) = (x, y), \end{aligned}$$

hence $h|C \cap ([-1, 1]^{n-1} \times \boldsymbol{Q}) = \mathrm{id}$, so $h|C \cap ([-1, 1]^m \times \boldsymbol{Q}) = \mathrm{id}$. For every $(x, y) \in C$, since $p_{n-1}\mathrm{pr}_1 h(x, y) = p_{n-1}(x)$ and $p_n\theta_{\alpha(x,y)}(y, g(x, y)) = p_n(y)$,

$$\rho(h(x, y), (x, y)) \leqslant \max\left\{n^{-1}, (n+1)^{-1})\right\} \leqslant n^{-1} < \varepsilon.$$

To see that h is an embedding, let $(x, y) \neq (x', y') \in C$. If $\alpha(x, y) = \alpha(x', y') = 0$, then $h(x, y) = (x, y) \neq (x', y') = h(x', y')$. When $\alpha(x, y) = \alpha(x', y') = t > 0$, since θ_t is a homeomorphism and g is an embedding, it follows that

$$\mathrm{pr}_2 h(x, y) = \theta_t(y, g(x, y)) \neq \theta_t(y', g(x', y')) = \mathrm{pr}_2 h(x', y'),$$

which implies that $h(x, y) \neq h(x', y')$. If $\alpha(x, y) \neq \alpha(x', y')$, then

$$\mathrm{pr}_n h(x, y) = \alpha(x, y) \neq \alpha(x', y') = \mathrm{pr}_n h(x', y'),$$

so we have $h(x, y) \neq h(x', y')$.

(5): Recall that every compact set in $\boldsymbol{s}$ is a Z-set. It suffices to show that the tower $([-n, n]^{\mathbb{N}})_{n\in\mathbb{N}}$ has the cap for $\boldsymbol{s}$, where $[-1, 1]^n$ is identified with $\{x \in \boldsymbol{s} \mid x(i) = 0 \text{ if } i > n\}$. Let C be a compact set in $\boldsymbol{s}$, $\varepsilon > 0$, and $m \in \mathbb{N}$. Choose $n_0 \in \mathbb{N}$ so that $n_0 > m$ and $n_0^{-1} < \varepsilon$. Furthermore, choose $n_1 \in \mathbb{N}$ so that $n_1 \geqslant n_0$ and $\mathrm{pr}_i(C) \subset [-n_1, n_1]$ for each $i < n_0$, where $\mathrm{pr}_i : \boldsymbol{s} = \mathbb{R}^{\mathbb{N}} \to \mathbb{R}$ is the projection onto the i-th factor. We define an embedding $h : C \to [-n_1-1, n_1+1]^{\mathbb{N}}$ as follows:

$$h(x) = \begin{cases} x(i) & \text{if } i < n_0 \text{ or } |x(i)| \leqslant n_1, \\ n_1 + \dfrac{x(i) - n_1}{x(i)} & \text{if } i \geqslant n_0 \text{ and } x(i) \geqslant n_1, \\ -n_1 - \dfrac{x(i) + n_1}{x(i)} & \text{if } i \geqslant n_0 \text{ and } x(i) \leqslant -n_1. \end{cases}$$

Then, $h|C \cap [-n_0, n_0]^{\mathbb{N}} = \mathrm{id}$, so $h|C \cap [-m, m]^{\mathbb{N}} = \mathrm{id}$. Since $h(x)(i) = x(i)$ for every $x \in C$ and $i < n_0$, it follows that $d(h, \mathrm{id}) \leqslant n_0^{-1} < \varepsilon$. Hence, $\boldsymbol{s}_Q$ is a cap set for $\boldsymbol{s}$.

(6): As with (5), it suffices to show that the tower $([-n, n]^n)_{n\in\mathbb{N}}$ has the f.d.cap for $\boldsymbol{s}$. Let C be a finite-dimensional compact set in $\boldsymbol{s}$ and $k = 2\dim C + 1$. For each $\varepsilon > 0$ and $m \in \mathbb{N}$, take the same $n_0, n_1 \in \mathbb{N}$ as chosen in (5) and an embedding $g : C \to \mathbf{I}^k$ as in (3) (Theorem 1.12.14). Let $n = \max\{n_1, n_0 + k\}$. We define an embedding $h : C \to [-n_1, n_1]^{n_0} \times \mathbf{I} \times \mathbf{I}^k \subset [-n, n]^n$ as follows:

$$h(x) = (x(1), \ldots, x(n_0), \alpha(x), \alpha(x)g(x)),$$

where $\alpha : C \to \mathbf{I}$ is a map defined by $\alpha(x) = d(x, C \cap [-n_0, n_0]^{n_0})$. Then, $h|C \cap [-n_0, n_0]^{n_0} = \mathrm{id}$, so $h|C \cap [-m, m]^m = \mathrm{id}$. Since $h(x)(i) = x(i)$ for every $x \in C$ and $i < n_0$, it follows that $d(h, \mathrm{id}) \leqslant n_0^{-1} < \varepsilon$. Hence, $\boldsymbol{s}_f$ is an f.d.cap set for $\boldsymbol{s}$.

(7): We may show that the tower $([-n, n]^n \times \boldsymbol{Q})_{n\in\mathbb{N}}$ has the cap for $\boldsymbol{s} \times \boldsymbol{Q}$. The proof is the same as (4), where $\boldsymbol{Q} \times \boldsymbol{Q}$ and $[-1, 1]^{n-1}$ should be replaced with $\boldsymbol{s} \times \boldsymbol{Q}$ and $[-(n-1), n-1]^{n-1}$, respectively. Details are left to the reader.

(8): It is enough to prove that the tower $(n \prod_{i\in\mathbb{N}}[-2^{-i}, 2^{-i}])_{n\in\mathbb{N}}$ has the cap for ℓ^1. Let C be a compact set in ℓ^1, $0 < \varepsilon < 1$, and $m \in \mathbb{N}$. By Lemma 5.1.2, we can choose $n_0 \in \mathbb{N}$ so that $n_0 > m$ and $\sum_{i>n_0} |x(i)| < \varepsilon$ for every $x \in C$. Moreover, choose $n_1 \in \mathbb{N}$ so that $n_1 \geqslant n_0$ and $\mathrm{pr}_i(C) \subset n_1[-2^{-i}, 2^{-i}]$ for each $i \leqslant n_0$, where $\mathrm{pr}_i : \ell^1 \to \mathbb{R}$ is the projection onto the i-th factor. Note that if $x \in C$ and $i > n_0$, then $|x(i)| < \varepsilon < 1$ and $2^{-i}m < 2^{-n_0}n_0 < 1$. We define an embedding $h : C \to n_1 \prod_{i\in\mathbb{N}}[-2^{-i}, 2^{-i}]$ as follows:

$$h(x)(i) = \begin{cases} x(i) & \text{if } i \leqslant n_0 \text{ or } |x(i)| \leqslant 2^{-i}m, \\ 2^{-i}m + \dfrac{2^{-i}(n_1 - m)(x(i) - 2^{-i}m)}{1 - 2^{-i}m} & \\ & \text{if } i > n_0 \text{ and } 2^{-i}m \leqslant x(i) < 1, \\ -2^{-i}m + \dfrac{2^{-i}(n_1 - m)(x(i) + 2^{-i}m)}{1 - 2^{-i}m} & \\ & \text{if } i > n_0 \text{ and } -1 < x(i) \leqslant -2^{-i}m. \end{cases}$$

Then, $h|C \cap m \prod_{i\in\mathbb{N}}[-2^{-i}, 2^{-i}] = \mathrm{id}$. For each $x \in C$,

$$\begin{aligned} \|h(x) - x\| &= \sum_{i>n_0} |h(x)(i) - x(i)| = \sum_{i>n_0} \big(|x(i)| - |h(x)(i)|\big) \\ &\leqslant \sum_{i>n_0} |x(i)| < \varepsilon, \end{aligned}$$

hence $d(h, \mathrm{id}) < \varepsilon$. Thus, ℓ^1_Q is a cap set for ℓ^1.

(9): We show that the tower $([-n, n]^n)_{n\in\mathbb{N}}$ has the f.d.cap for ℓ^1. Let C be a finite-dimensional compact set in ℓ^1 and $k = 2\dim C + 1$. For each $0 < \varepsilon < 1$ and $m \in \mathbb{N}$, choose $n_0 \in \mathbb{N}$ so that $n_0 > m$ and $\sum_{i>n_0} |x(i)| < \varepsilon/2$ for every $x \in C$ (the same as in (8)). Moreover, choose $n_1 \in \mathbb{N}$ so that $n_1 \geqslant n_0 + k + 1$ and $\mathrm{pr}_i(C) \subset [-n_1, n_1]$ for each $i = 1, \ldots, n_0$. Let $\alpha : C \to \mathbf{I}$ be a map defined by

$$\alpha(x) = \min\left\{1, d(x, C \cap [-m, m]^m)\right\},$$

and let $\delta = \varepsilon/2(k+1)$. By the Embedding Theorem 1.12.14, we have an embedding $g : C \to [0, \delta]^k$. We define an embedding $h : C \to [-n_1, n_1]^{n_0} \times [0, \delta]^{k+1} \subset [-n_1, n_1]^{n_1}$ as follows:

$$h(x) = (x(1), \ldots, x(n_0), \alpha(x)\delta, g(x)).$$

Then, $h|C \cap [-m, m]^m = \mathrm{id}$. For each $x \in C$,

$$\begin{aligned} \|h(x) - x\| = \sum_{i>n_0} |h(x)(i) - x(i)| &\leqslant \sum_{i>n_0} |h(x)(i)| + \sum_{i>n_0} |x(i)| \\ &\leqslant (k + 1)\delta + \varepsilon/2 < \varepsilon, \end{aligned}$$

hence $d(h, \mathrm{id}) < \varepsilon$. Thus, ℓ^1_f is an f.d.cap set for ℓ^1. □

Remark 5.1 (1): Recall that $B(\boldsymbol{Q}) = \bigcup_{n\in\mathbb{N}} \mathrm{pr}_n^{-1}(\{\pm 1\})$, where $\mathrm{pr}_n : \boldsymbol{Q} = [-1, 1]^{\mathbb{N}} \to [-1, 1]$ is the projection onto the n-th factor. The proof of (2) can be adjusted to see that the subset $\bigcup_{n\in\mathbb{N}} \mathrm{pr}_n^{-1}(1) \subset \boldsymbol{Q}$ is also a cap set for $\boldsymbol{Q}$. Details are left as an exercise.

(2): We can show that $I(\boldsymbol{Q})_f = (-1, 1)^{\mathbb{N}}_f$ is an f.d.cap set for $\boldsymbol{Q}$, that is, the following tower has the f.d.cap for $\boldsymbol{Q}$:

$$\left(\left[-\frac{n}{n+1}, \frac{n}{n+1}\right]^n\right)_{n\in\mathbb{N}}.$$

Indeed, combining methods of (1) and (3), for each $m \in \mathbb{N}$ and $\varepsilon > 0$, we can choose $m > n$ and construct an embedding $h : C \to [-n/(n+1), n/(n+1)]^n$ such that $h|C \cap [-m/(m+1), m/(m+1)]^m = \mathrm{id}$ and $d(h, \mathrm{id}) < \varepsilon$. Filling in details is a good exercise for the reader.

(3): It can be shown that $\boldsymbol{Q}^{\mathbb{N}}_f$ is a cap set for $\boldsymbol{Q}^{\mathbb{N}}$, that is, the tower $(\boldsymbol{Q}^n)_{n\in\mathbb{N}}$ has the cap for $\boldsymbol{Q}^{\mathbb{N}}$, where $\boldsymbol{Q}^n$ is identified with $\{x \in \boldsymbol{Q}^{\mathbb{N}} \mid x(i) = \mathbf{0} \text{ if } i > n\} \subset \boldsymbol{Q}^{\mathbb{N}}$. The proof is left to the reader.

(4): Similar to (8) and (9), for every $p \in \mathbb{N}$, ℓ^p_Q is a cap set and ℓ^p_f is an f.d.cap set for ℓ^p, where the copy $\prod_{n\in\mathbb{N}}[-2^{-n}, 2^{-n}] \subset \ell^p$ of the Hilbert cube $\boldsymbol{Q}$ can be replaced with another suitable copy. The proof is a good exercise for the reader.

As we saw above, there are various (f.d.)cap sets. However, (f.d.)cap sets for an ℓ^2-manifold and a $\boldsymbol{Q}$-manifold are topologically unique, that is:

Theorem 5.1.4 *Let X be an ℓ^2-manifold or a $\boldsymbol{Q}$-manifold, and let M and N be (f.d.)cap sets for X. For each open cover $\mathcal{U}$ of X, there exists a homeomorphism $h : X \to X$ such that $h(M) = N$ and h is $\mathcal{U}$-close to* id.

Proof Let $M = \bigcup_{i\in\mathbb{N}} M_i$ and $N = \bigcup_{j\in\mathbb{N}} N_j$, where $M_1 \subset M_2 \subset \cdots$ and $N_1 \subset N_2 \subset \cdots$ are towers of (finite-dimensional) compact Z-sets in X and they have the (f.d.)cap. For each open cover $\mathcal{U} \in \operatorname{cov}(X)$, we have a complete metric $d \in \operatorname{Metr}(X)$ such that $\{\overline{\mathrm{B}}_d(x, 1) \mid x \in X\} \prec \mathcal{U}$ (Proposition 1.3.22 (1)).

We will inductively construct increasing sequences $1 = i(1) < i(2) < \cdots$, $j(1) < j(2) < \cdots \in \mathbb{N}$, and homeomorphisms $h_n : X \to X$, $n \in \mathbb{N}$, so that the following conditions are satisfied:

(1) $d(h_n, h_{n-1}) < 2^{-n}$, $d(h_n^{-1}, h_{n-1}^{-1}) < 2^{-n}$,
(2) $h_n|M_{i(n-1)} = h_{n-1}|M_{i(n-1)}$, and
(3) $N_{j(n-1)} \subset h_n(M_{i(n)}) \subset N_{j(n)}$, so $h_n^{-1}(N_{j(n-1)}) \subset M_{i(n)} \subset h_n^{-1}(N_{j(n)})$,

where $j(0) = 0$, $N_0 = \emptyset$, and $h_0 = \mathrm{id}$. Then, (1) implies that $(h_n)_{n\in\mathbb{N}}$ and $(h_n^{-1})_{n\in\mathbb{N}}$ uniformly converge to maps $h, h' : X \to X$ such that $d(h, \mathrm{id}) < 1$ and $d(h', \mathrm{id}) < 1$, hence h and h' are $\mathcal{U}$-close to id. Since $h_{n+1}^{-1}|h_{n+1}(M_{i(n)}) = h_n^{-1}|h_n(M_{i(n)})$ by (2), it follows from (3) that $h_{n+1}^{-1}|N_{j(n-1)} = h_n^{-1}|N_{j(n-1)}$. Hence, for each $n \in \mathbb{N}$, $h|M_{i(n)} = h_n|M_{i(n)}$ and $h'|N_{j(n-1)} = h_n^{-1}|N_{j(n-1)}$, which implies that

$$h'h|M_{i(n)} = (h'|N_{j(n)})(h_n|M_{i(n)}) = (h_{n+1}^{-1}|N_{j(n)})(h_{n+1}|M_{i(n)}) = \mathrm{id} \quad \text{and}$$

$$hh'|N_{j(n-1)} = (h|M_{i(n)})(h_n^{-1}|N_{j(n-1)}) = (h_n|M_{i(n)})(h_n^{-1}|N_{j(n-1)}) = \mathrm{id}.$$

Moreover, it follows from (3) that

$$h(M) = \bigcup_{n\in\mathbb{N}} h_n(M_{i(n)}) = \bigcup_{n\in\mathbb{N}} N_{j(n)} = N \quad \text{and}$$

$$h'(N) = \bigcup_{n\in\mathbb{N}} h'(N_{j(n)}) = \bigcup_{n\in\mathbb{N}} M_{i(n)} = M.$$

Thus, $h'h|M = \mathrm{id}$ and $hh'|N = \mathrm{id}$, which means that $h'h = \mathrm{id}$ and $hh' = \mathrm{id}$ because M and N are dense in X. Therefore, $h : X \to X$ is a homeomorphism with $h^{-1} = h'$ that is the desired one.

Now, assume that $h_0, \ldots, h_{n-1} : X \to X$, $1 = i(1) < \cdots < i(n-1)$, and $0 < j(1) < \cdots < j(n-1)$ have been obtained so as to satisfy the conditions (1), (2), and (3). As is easily observed, $(h_{n-1}(M_i))_{i\in\mathbb{N}}$ is a tower of (finite-dimensional) compact Z-sets in X with the (f.d.)cap. Take $\mathcal{V} \in \operatorname{cov}(X)$ with $\operatorname{mesh} \operatorname{st} \mathcal{V} < 2^{n-1}$ and $\operatorname{mesh} h_{n-1}^{-1}(\operatorname{st} \mathcal{V}) < 2^{-n-1}$. Since $N_{j(n-1)}$ is a (finite-dimensional) compact Z-set in X and $N_{j(n-1)} \cap h_{n-1}(M_{i(n-1)}) = h_{n-1}(M_{i(n-1)})$, there exist $i(n) > i(n-1)$ and an embedding $g : N_{j(n-1)} \to h_{n-1}(M_{i(n)})$ such that $g|h_{n-1}(M_{i(n-1)}) = \mathrm{id}$ and

g is $\mathcal{V}$-homotopic to id. Then, $g(N_{j(n-1)})$ is a Z-set in X. By the Z-Set Unknotting Theorem 2.5.12 or 2.11.6, g can be extended to a homeomorphism $\tilde{g} : X \to X$ that is st$\mathcal{V}$-isotopic to id. Then,

$$N_{j(n-1)} \subset \tilde{g}^{-1}(h_{n-1}(M_{i(n)})), \ \tilde{g}^{-1}|h_{n-1}(M_{i(n-1)}) = \text{id},$$
$$d(\tilde{g}^{-1}, \text{id}) < 2^{-n-1}, \text{ and } d(h_{n-1}^{-1}\tilde{g}, h_{n-1}^{-1}) < 2^{-n-1}.$$

Note that $\tilde{g}^{-1}(h_{n-1}(M_{i(n)}))$ is a (finite-dimensional) compact Z-set in X. Take $\mathcal{W} \in \text{cov}(X)$ with mesh st$\mathcal{W} < 2^{n-1}$ and mesh $h_{n-1}^{-1}\tilde{g}(\text{st}\,\mathcal{W}) < 2^{-n-1}$. Now, by the (f.d.)cap of $(N_n)_{n\in\mathbb{N}}$, we have $j(n) > j(i-1)$ and an embedding $f : \tilde{g}^{-1}(h_{n-1}(M_{i(n)})) \to N_{j(n)}$ such that $f|N_{j(n-1)} = \text{id}$ and f is $\mathcal{W}$-homotopic to id. Again using the Z-Set Unknotting Theorem 2.5.12 or 2.11.6, we can extend f to a homeomorphism $\tilde{f} : X \to X$ that is st$\mathcal{W}$-isotopic to id. Then,

$$N_{j(n-1)} = \tilde{f}(N_{j(n-1)}) \subset \tilde{f}\tilde{g}^{-1}(h_{n-1}(M_{i(n)})) \subset N_{j(n)},$$
$$\tilde{f}\tilde{g}^{-1}|h_{n-1}(M_{i(n-1)}) = \tilde{f}|h_{n-1}(M_{i(n-1)}) = \text{id},$$
$$d(\tilde{f}, \text{id}) < 2^{-n-1}, \text{ and } d(h_{n-1}^{-1}\tilde{g}\tilde{f}^{-1}, h_{n-1}^{-1}\tilde{g}) < 2^{-n-1}.$$

Moreover, it follows that

$$d(\tilde{f}\tilde{g}^{-1}, \text{id}) \leqslant d(\tilde{f}, \text{id}) + d(\tilde{g}^{-1}, \text{id}) < 2^{-n} \text{ and}$$
$$d(h_{n-1}^{-1}\tilde{g}\tilde{f}^{-1}, h_{n-1}^{-1}) \leqslant d(h_{n-1}^{-1}\tilde{g}\tilde{f}^{-1}, h_{n-1}^{-1}\tilde{g}) + d(h_{n-1}^{-1}\tilde{g}, h_{n-1}^{-1}) < 2^{-n}.$$

Thus, the homeomorphism $h_n = \tilde{f}\tilde{g}^{-1}h_{n-1} : X \to X$ satisfies the conditions (1), (2), and (3). By induction, the proof is completed. □

The following is a direct consequence of the above theorem:

Corollary 5.1.5 *Let X and Y be together ℓ^2- or $\boldsymbol{Q}$-manifolds such that $X \approx Y$, and let M and N be (f.d.)cap sets for X and Y, respectively. Then, $(X, M) \approx (Y, N)$, that is, there exists a homeomorphism $h : X \to Y$ with $h(M) = N$, where h can be taken arbitrarily close to a given homeomorphism $f : X \to Y$.* □

Since $\boldsymbol{s} \approx \ell^p$, $p \in \mathbb{N}$, by the Mazur Theorem 1.2.3 and Kadec–Anderson Theorem 3.5.6, the following is obtained from Proposition 5.1.3 with Remark 5.1(4) and Corollary 5.1.5:

Corollary 5.1.6 *The pairs (ℓ^p, ℓ^p_Q), $p \in \mathbb{N}$, are homeomorphic to $(\boldsymbol{s}, \boldsymbol{s}_Q)$ and the pairs (ℓ^p, ℓ^p_f), $p \in \mathbb{N}$, are homeomorphic to $(\boldsymbol{s}, \boldsymbol{s}_f)$.* □

By Proposition 5.1.3 and Remark 5.1, we have the following homeomorphisms among pairs:

Corollary 5.1.7 (1) $(\boldsymbol{Q}, \operatorname{rint} \boldsymbol{Q}) \approx (\boldsymbol{Q}, B(\boldsymbol{Q})) \approx (\boldsymbol{Q}^2, \boldsymbol{Q}_f \times \boldsymbol{Q}) \approx (\boldsymbol{Q}^{\mathbb{N}}, \boldsymbol{Q}_f^{\mathbb{N}})$; (2) $(\boldsymbol{Q}, \boldsymbol{Q}_f) \approx (\boldsymbol{Q}, I(\boldsymbol{Q})_f)$; (3) $(s \times \boldsymbol{Q}, s_f \times \boldsymbol{Q}) \approx (s, s_{\boldsymbol{Q}}) \approx (\ell^p \times \boldsymbol{Q}, \ell^p_f \times \boldsymbol{Q}) \approx (\ell^p, \ell^p_{\boldsymbol{Q}})$, $p \in \mathbb{N}$. □

The above homeomorphisms give the following homeomorphisms:

Corollary 5.1.8 (1) $\operatorname{rint} \boldsymbol{Q} \approx B(\boldsymbol{Q}) \approx \boldsymbol{Q}_f \times \boldsymbol{Q} \approx \boldsymbol{Q}_f^{\mathbb{N}} \approx s_f \times \boldsymbol{Q} \approx s_{\boldsymbol{Q}} \approx \ell^p_f \times \boldsymbol{Q} \approx \ell^p_{\boldsymbol{Q}}$, $p \in \mathbb{N}$; (2) $\boldsymbol{Q}_f \approx s_f \approx \ell^p_f$, $p \in \mathbb{N}$. □

In Corollary 5.1.7 (1), the first homeomorphism gives $(\boldsymbol{Q}, \partial \boldsymbol{Q}) \approx (\boldsymbol{Q}, I(\boldsymbol{Q}))$.[5] Hence, we have the following:

Corollary 5.1.9 $\partial \boldsymbol{Q} \approx I(\boldsymbol{Q}) \approx s \approx \ell^p$, $p \in \mathbb{N}$. □

It is easy to see that $(\boldsymbol{Q}_f^{\mathbb{N}})_f^{\mathbb{N}} \approx \boldsymbol{Q}_f^{\mathbb{N}}$ and $(\mathbb{R}_f^{\mathbb{N}})_f^{\mathbb{N}} \approx \mathbb{R}_f^{\mathbb{N}}$, where $\boldsymbol{Q}_f^{\mathbb{N}} \approx \ell^2_Q$ and $\mathbb{R}_f^{\mathbb{N}} \approx \ell^2_f$. Thus, ℓ^2_Q and ℓ^2_f are normed linear spaces such that

$$(\ell^2_Q)_f^{\mathbb{N}} \approx \ell^2_Q \quad \text{and} \quad (\ell^2_f)_f^{\mathbb{N}} \approx \ell^2_f,$$

which means the following:

Theorem 5.1.10 *All results on E-manifolds in Chap. 2 are valid for ℓ^2_Q- and ℓ^2_f-manifolds.* □

Since every ℓ^2_f-manifold has the homotopy type of a countable locally finite simplicial complex by Theorem 1.13.22, the following Triangulation Theorem for ℓ^2_f-manifolds is obtained as a combination of the Classification Theorem 2.6.1 and Proposition 2.2.17:

Theorem 5.1.11 (Triangulation) *Every ℓ^2_f-manifold is homeomorphic to $|K| \times \ell^2_f$ for some countable locally finite simplicial complex K.* □

Concerning the Triangulation Theorem for ℓ^2_Q-manifolds, we discuss the proof in the next section.

In Theorem 3.9.2, it was proved that every infinite-dimensional compact convex set C in a metrizable topological linear space is homeomorphic to the Hilbert cube $\boldsymbol{Q}$ if C is an AR. Concerning the radial interior $\operatorname{rint} C$, we have the following:

Theorem 5.1.12 *Let C be an infinite-dimensional compact convex set C in a metrizable topological linear space E. If C is an AR and* $\operatorname{rint} C \neq \emptyset$, *then* $(C, \operatorname{rint} C) \approx (\boldsymbol{Q}, \operatorname{rint} \boldsymbol{Q})$.

Proof As mentioned above, $C \approx \boldsymbol{Q}$ by Theorem 3.9.2. By virtue of Corollary 5.1.5, it is enough to show that $\operatorname{rint} C$ is a cap set for C. By a translation, we may assume that $\mathbf{0} \in \operatorname{rint} C$ without loss of generality. Due to Theorem 1.4.15, E has an admissible F-norm $\|\cdot\|$.

[5] $\partial \boldsymbol{Q} = \boldsymbol{Q} \setminus \operatorname{rint} \boldsymbol{Q}$ is the radial boundary (cf. p. 22).

First, we will show that aC is a Z-set in C for each $0 < a < 1$. For each map $f : \mathbf{I}^n \to C$ and $\varepsilon > 0$, we just have to construct a map $f' : \mathbf{I}^n \to \partial C$ ε-close to f (cf. Theorem 2.8.6). Since C is convex, we can define maps $\vartheta_k : C^{k+1} \times \Delta^k \to C$, $k = 1, \dots, n$, as follows:

$$\vartheta_k(x_1 \dots, x_{k+1}; z) = \sum_{i=1}^{k+1} z(i)x_i \quad \text{for each } (x_1 \dots, x_{k+1}; z) \in C^{k+1} \times \Delta^k,$$

where Δ^k is the standard k-simplex. Note that $\vartheta_k(x, \dots, x; z) = x$ for each $x \in C$ and $z \in \Delta^k$. Each $x \in C$ has an open neighborhood U_x in C such that $\operatorname{diam} \vartheta(U_x^{k+1} \times \Delta^k) < \varepsilon/4$ for every $k = 1, \dots, n$. Let $\delta > 0$ be a Lebesgue number for the open cover $\{U_x \mid x \in C\}$ of C, where it should be noted that $\delta < \varepsilon/4$ because $\operatorname{diam} U_x < \varepsilon/4$ for each $x \in C$. Take a triangulation K of $\mathbf{I}^n$ (i.e., $\mathbf{I}^n = |K|$) so that $\operatorname{diam} f(\sigma) < \delta$ for each $\sigma \in K$. Then, $f|K^{(0)}$ is uniquely extends to a map $g : \mathbf{I}^n = |K| \to C$ that is affine on each simplex of K, that is,

$$g\left(\textstyle\sum_{i=1}^{k+1} z(i)v_i\right) = \textstyle\sum_{i=1}^{k+1} z(i) f(v_i) = \vartheta(f(v_1), \dots, f(v_{k+1}); z)$$
$$\text{for each } \langle v_1, \dots, v_{k+1}\rangle \in K \text{ and } z \in \Delta^k.$$

For each $\sigma = \langle v_1, \dots, v_{k+1}\rangle \in K$, since $\operatorname{diam} g(\sigma^{(0)}) = \operatorname{diam} f(\sigma^{(0)}) < \delta$, it follows that $g(\sigma^{(0)})$ is contained in some U_x, whence

$$\begin{aligned}\operatorname{diam} g(\sigma) &= \operatorname{diam} \vartheta\left(\{(f(v_1), \dots, f(v_{k+1}))\} \times \Delta^k\right)\\ &\leqslant \operatorname{diam} \vartheta(U_x^{k+1} \times \Delta^k) < \varepsilon/4.\end{aligned}$$

Hence, we can easily see that g is $\varepsilon/2$-close to f.

Each $x \in C$ has an open neighborhood U_x in E with an open circled neighborhood V_x of $\mathbf{0}$ in E such that $\operatorname{diam}(U_x + V_x) \cap C < \varepsilon/2$. Since C is compact, there are finite $x_1, \dots, x_m \in C$ such that $C \subset \bigcup_{i=1}^m U_{x_i}$. Then, $V_0 = \bigcap_{i=1}^m V_{x_i}$ is an open circled neighborhood of $\mathbf{0}$ in E and

$$\operatorname{diam}(x + V_0) \cap C < \varepsilon/2 \quad \text{for each } x \in C.$$

Since $C - C$ is compact and $\operatorname{fl} C$ is an infinite-dimensional linear subspace of E, $V_0 \cap \operatorname{fl} C \not\subset C - C$, that is, there exists $v \in V_0 \cap \operatorname{fl} C \setminus (C - C)$. Then, $x + v \notin C$ for each $x \in C$. Indeed, if $x + v = y \in C$, then $v = y - x \in C - C$. Let $F = \operatorname{fl}(g(K^{(0)}) \cup \{\mathbf{0}, v\})$. Note that F is a finite-dimensional linear subspace of E and $F = \operatorname{fl}(F \cap C)$ because

$$F = \operatorname{fl}\left(g(K^{(0)}) \cup \{\mathbf{0}, v\}\right) \subset \operatorname{fl}(F \cap C) \subset F.$$

Then, applying Propositions 1.4.3 and 1.4.4, we have

$$\mathbf{0} \in F \cap \operatorname{rint} C = \operatorname{rint}(F \cap C) = \operatorname{core}_F(F \cap C).$$

Since the Minkowski functional $p = p_{F\cap C} : F \to \mathbb{R}_+$ for $F \cap C$ is continuous (Proposition 1.4.6), we can define a map $\lambda : \mathbf{I}^n \times \mathbf{I} \to \mathbb{R}$ as follows:

$$\lambda(x,t) = p(g\varphi(x) + tv) \text{ for each } (x,t) \in \mathbf{I}^n \times \mathbf{I}.$$

Then, $\lambda(x,0) = p(g\varphi(x)) < 1$ and $\lambda(x,1) = p(g\varphi(x) + v) > 1$ because

$$g(\mathbf{I}^n) \subset F \cap \operatorname{rint} C = \operatorname{rint}(F \cap C) \text{ and}$$
$$g(x) + v \notin C = \operatorname{rcl} C \text{ for each } x \in \mathbf{I}^n.$$

By the intermediate value theorem, we have $0 < \alpha(x) < 1$ such that $\lambda(g(x), \alpha(x)) = 1$, that is, $g\varphi(x) + \alpha(x)v \in \partial(F \cap C)$. Due to Proposition 1.4.2,

$$g\varphi(x) + sv \in \partial(F \cap C),\ 0 < s < 1$$
$$\Rightarrow \forall t \in [0,s),\ g\varphi(x) + tv \in \operatorname{rint}(F \cap C) = p^{-1}([0,1)).$$

Hence, $\alpha(x)$ is uniquely determined for each $x \in \mathbf{I}^n$. Then, we can define $f' : \mathbf{I}^n \to \partial(F \cap C) \subset \partial C$ by $f'(x) = g(x) + \alpha(x)v$. Since V_0 is circled and $v \in V_0$,

$$g(x),\ f'(x) = g(x) + \alpha(x)v \in (g(x) + V_0) \cap C,$$

which means that f' is $\varepsilon/2$-close to g because $\operatorname{diam}(g(x) + V_0) \cap C < \varepsilon/2$. Thus, f is ε-close to f'.

For the continuity of f', we just have to show that $\alpha : \mathbf{I}^n \to \mathbb{R}$ is continuous. Suppose $\alpha(x) < s$. Then, $g(x) + sv \notin F \cap C$. Since C is closed, if $x' \in \mathbf{I}^n$ is sufficiently close to x, then $g(x')+sv \notin F\cap C$, which implies that $\alpha(x') < s$. Hence, α is upper semi-continuous. Now, suppose $\alpha(x) > s$. Then, $g(x)+sv \in \operatorname{rint}(F\cap C)$. Note that $g(\mathbf{I}^n) + \mathbb{R}v \subset F$. As observed above,

$$g(x) + sv \in \operatorname{rint}(F \cap C) = \operatorname{int}_F(F \cap C).$$

So, if $x' \in \mathbf{I}^n$ is sufficiently close to x then $g(x')+sv \in \operatorname{int}_F(F\cap C)$, which implies that $\alpha(x') > s$. Hence, α is lower semi-continuous. Consequently, α is continuous.

Now, we have Z-sets $(1 - 2^{-n})C$, $n \in \mathbb{N}$, in C, which are homeomorphic to $C \approx \boldsymbol{Q}$. Then, $\bigcup_{n\in\mathbb{N}}(1 - 2^{-n})C = \operatorname{rint} C$. Indeed, for each $x \in \operatorname{rint} C$, we can find $t > 1$ so that $tx = y \in C$. Choose $n \in \mathbb{N}$ so that $2^{-n} < 1 - 1/t$, i.e., $1/t < 1 - 2^{-n}$. Then, we have

$$x = (1/t)y \in \langle \mathbf{0}, (1 - 2^{-n})y \rangle \subset (1 - 2^{-n})C.$$

Note that $(1-2^{-m})C$ is a Z-set in $(1-2^{-n})C$ for any $m > n$ because

$$(1-2^{-m})C = (1-2^{-n})\frac{1-2^{-m}}{1-2^{-n}}C.$$

To see that the tower $((1-2^{-n})C)_{n\in\mathbb{N}}$ has the property (cap), let $\varepsilon > 0$ and $m \in \mathbb{N}$. We may construct an embedding $h : C \to (1-2^{-n})C$ for some $n > m$ so that $h|(1-2^{-m})C = f|(1-2^{-m})C = \mathrm{id}$ and h is ε-close to id. Since C is compact, we can choose $n \in \mathbb{N}$ so that $n > m$ and

$$\|x-(1-t)x\| = \|tx\| < \varepsilon/2 \text{ for every } x \in C \text{ and } 0 \leqslant t \leqslant 2^{-n}.$$

Let $\varphi : C \to (1-2^{-n})C$ be a map defined by $\varphi(x) = (1-2^{-n})x$ for each $x \in C$. Then, $\varphi|(1-2^{-m})C \simeq_{\varepsilon/2} \mathrm{id}$ in $(1-2^{-n})C$ by the straight line homotopy

$$(x,t) \mapsto (1-t)\varphi(x) + tx = (1-2^{-n})(1-t)x + tx = (1-2^{-n}+2^{-n}t)x.$$

Applying the Homotopy Extension Theorem 1.13.11, we can obtain a map $f : C \to (1-2^{-n})C$ such that $f|(1-2^{-m})C = \mathrm{id}$ and $f \simeq_{\varepsilon/2} \varphi$. By the strong $\mathfrak{M}_0$-universality of $\boldsymbol{Q}$ (Theorem 2.10.10), f is $\varepsilon/3$-close to a Z-embedding $h : C \to (1-2^{-n})C$ such that $h|(1-2^{-m})C = f|(1-2^{-m})C = \mathrm{id}$. Then, h is ε-close to id. This completes the proof. □

5.2 Manifold Pairs Modeled on the Pair of ℓ^2 (or $\boldsymbol{Q}$) and its (F.D.)Cap Set

At the end of the previous section, we saw that all results in Chap. 2 are valid for ℓ^2_Q- and ℓ^2_f-manifolds. In this section, using (f.d.)cap sets, we characterize manifold pairs modeled on the pairs (ℓ^2, ℓ^2_f), (ℓ^2, ℓ^2_Q), $(\boldsymbol{Q}, \boldsymbol{Q}_f)$, and $(\boldsymbol{Q}, \operatorname{rint} \boldsymbol{Q})$, and then prove the Open Embedding Theorem, the Classification Theorem, and the Triangulation Theorem for these manifold pairs. The Triangulation Theorem for ℓ^2_Q-manifolds will be obtained as a corollary of the one for (ℓ^2, ℓ^2_Q)-manifold pairs.

First, we show that every ℓ^2-manifold and $\boldsymbol{Q}$-manifold has (f.d.)cap sets, and that every ℓ^2_Q-manifold (resp. ℓ^2_f-manifold) M can be embedded in an ℓ^2-manifold and a $\boldsymbol{Q}$-manifold as a cap set (resp. an f.d.cap set). To this end, we need the following:

Lemma 5.2.1 *Let M be an (f.d.)cap set for a metrizable space X. For every open set U in X, $M \cap U$ is an (f.d.)cap set for U.*

Proof Let $M = \bigcup_{n\in\mathbb{N}} M_n$, where $M_1 \subset M_2 \subset \cdots$ is a tower of (finite-dimensional) compact Z-sets in X that has the (f.d.)cap for X. We can write $U = \bigcup_{n\in\mathbb{N}} F_n$, where each F_n is closed in X and $F_n \subset \operatorname{int} F_{n+1}$. Then, $M \cap U =$

$\bigcup_{n\in\mathbb{N}}(M_n \cap F_n)$ and each $M_n \cap F_n$ is a (finite-dimensional) compact Z-set in U. We show that the tower $(M_n \cap F_n)_{n\in\mathbb{N}}$ has the (f.d.)cap for U.

Let C be a (finite-dimensional) compact set in U. Because of compactness, $C \subset \operatorname{int} F_k$ for some $k \in \mathbb{N}$. For each $\mathcal{U} \in \operatorname{cov}(U)$, let

$$\mathcal{V} = \mathcal{U} \cup \{X \setminus F_k\} \in \operatorname{cov}(X).$$

For each $m \in \mathbb{N}$, there exists $n \in \mathbb{N}$ and an embedding $h : C \to M_n$ such that $n > m$, $h|C \cap M_m = \mathrm{id}$, and h is $\mathcal{V}$-close to id. Then, $h(C) \subset U$. Otherwise, there is some $x \in C$ such that $h(x) \notin U$, hence $x, h(x) \in X \setminus F_k$, which contradicts $C \subset \operatorname{int} F_k$. Because of compactness, $h(C) \subset \operatorname{int} F_{n_0}$ for some $n_0 \geqslant \max\{n, k\}$. Thus, we have embedding $h : C \to M_{n_0} \cap F_{n_0}$ that is $\mathcal{U}$-close to id. □

Proposition 5.2.2 *Every ℓ^2-manifold and Q-manifold has (f.d.)cap sets.*

Proof By the Open Embedding Theorem 2.5.10, every ℓ^2-manifold M is homeomorphic to an open set U in ℓ^2, that is, there is a homeomorphism $f : M \to U$. Due to Lemma 5.2.1, $\ell^2_Q \cap U$ (or $\ell^2_f \cap U$) is an (f.d.)cap set for U. Then, $f^{-1}(\ell^2_Q \cap U)$ (or $f^{-1}(\ell^2_f \cap U)$) is an (f.d.)cap set for M.

For a Q-manifold M, $M \times [0, 1)$ is homeomorphic to an open set U in Q by the Open Embedding Theorem 2.11.4. As above, $M \times [0, 1)$ has an (f.d.)cap set $N = \bigcup_{n\in\mathbb{N}} N_n$, where $N_1 \subset N_2 \subset \cdots$ is a tower of (finite-dimensional) compact Z-sets in $M \times [0, 1)$ that has the (f.d.)cap for $M \times [0, 1)$. Then, each N_n is a Z-set in $M \times \mathbf{I}$ and the tower $(N_n)_{n\in\mathbb{N}}$ has the (f.d.)cap for $M \times \mathbf{I}$. Indeed, for each (finite-dimensional) compact Z-set C in $M \times \mathbf{I}$, $m \in \mathbb{N}$, and $\mathcal{U} \in \operatorname{cov}(M \times \mathbf{I})$, choose $\mathcal{V} \in \operatorname{cov}(M \times \mathbf{I})$ and $k \in \mathbb{N}$ so that $\operatorname{st}\mathcal{V} \prec \mathcal{U}$, $N_m \subset M \times [0, 1 - 2^{-k}]$, and

$$\big\{\{x\} \times [1 - 2^{-k}, 1] \bigm| (\{x\} \times [1 - 2^{-k}, 1]) \cap C \neq \emptyset\big\} \prec \mathcal{V}.$$

Let $f : \mathbf{I} \to [0, 1 - 2^{-k-1}]$ be a PL homeomorphism defined by $f|[0, 1 - 2^{-k}] = \mathrm{id}$ and

$$f(t) = \frac{t + 1 - 2^{-k}}{2} \quad \text{for } 1 - 2^{-k} \leqslant t \leqslant 1.$$

Then, $(\mathrm{id}_M \times f)|C$ is $\mathcal{V}$-close to id, $(\mathrm{id}_M \times f)|C \cap N_m = \mathrm{id}$, and $(\mathrm{id}_M \times f)(C)$ is a (finite-dimensional) compact Z-set in $M \times [0, 1)$. Hence, there exist $n \in \mathbb{N}$ and an embedding $g : (\mathrm{id}_M \times f)(C) \to N_n$ such that $g|(\mathrm{id}_M \times f)(C) \cap N_m = \mathrm{id}$ and g is $\mathcal{V}$-close to id. Note that $C \cap N_m \subset (\mathrm{id}_M \times f)(C) \cap N_m$. Thus, we have an embedding $g(\mathrm{id}_M \times f)|C : C \to N_n$ that is $\mathcal{U}$-close to id and

$$g(\mathrm{id}_M \times f)|C \cap N_m = (g|(\mathrm{id}_M \times f)(C) \cap N_m)(\mathrm{id}_M \times f)|C \cap N_m = \mathrm{id}.$$

Therefore, $M \times \mathbf{I}$ has an (f.d.)cap set N. Since $M \approx M \times \mathbf{I}$, M has an (f.d.)cap set. □

Theorem 5.2.3 *Every ℓ^2_Q-manifold (resp. ℓ^2_f-manifold) M can be embedded in an ℓ^2-manifold and a $\boldsymbol{Q}$-manifold as a cap set (resp. an f.d.cap set).*

Proof First, recall that $\ell^2_Q \approx \operatorname{rint} \boldsymbol{Q}$ and $\boldsymbol{Q}_f \approx \ell^2_f$ (Corollary 5.1.8), and also that $\ell^2 \approx \boldsymbol{s}$ (the Kadec–Anderson Theorem 3.5.6). Now, as mentioned at the end of previous section, the Open Embedding Theorem 2.5.10 is valid for ℓ^2_Q-manifolds (resp. ℓ^2_f-manifolds). Hence, M is homeomorphic to an open set U in $\operatorname{rint} \boldsymbol{Q}$ (resp. in $\boldsymbol{s}_f$). Take an open set V in $\boldsymbol{Q}$ such that $U = V \cap \operatorname{rint} \boldsymbol{Q}$ (resp. $U = V \cap \boldsymbol{s}_f$). Then, V is a $\boldsymbol{Q}$-manifold and $V \cap \boldsymbol{s}$ is an ℓ^2-manifold. Due to Lemma 5.2.1, U is a cap set (resp. an f.d.cap set) in V and $V \cap \boldsymbol{s}$. □

We show that a subset of an ℓ^2- or $\boldsymbol{Q}$-manifold is an (f.d.)cap set if it is an (f.d.)cap set locally, that is:

Proposition 5.2.4 *Let X be an ℓ^2- or $\boldsymbol{Q}$-manifold and $M \subset X$. If every $x \in X$ has an open neighborhood U in X such that $M \cap U$ is an (f.d.)cap set for U, then M is an (f.d.)cap set for X.*

Proof We apply Michael's Theorem 1.3.20 on local properties. Let $\mathcal{P}$ be the property of open sets in X defined as follows:

$$U \text{ has property } \mathcal{P} \underset{\text{def}}{\Leftrightarrow} M \cap U \text{ is an (f.d.)cap set for } U.$$

It suffices to show that property $\mathcal{P}$ is G-hereditary, where (G-1) is Lemma 5.2.1. Because of separability of X, any discrete collection of open sets is countable. Then, (G-3) can be easily proved.

To see (G-2), let U and V be open sets in X such that $M \cap U$ and $M \cap V$ are (f.d.)cap sets for U and V, respectively. On the other hand, $U \cup V$ has an (f.d.)cap set N by Proposition 5.2.2. We will construct a homeomorphism $h : U \cup V \to U \cup V$ such that $h(N) = M \cap (U \cup V)$, which implies that $M \cap (U \cup V)$ is an (f.d.)cap set for $U \cup V$. Take an open cover $\mathcal{U} \in \operatorname{cov}(U)$ that is fitting in $U \cup V$ (Lemma 2.1.16). Since $N \cap U$ is an (f.d.)cap set for V by Lemma 5.2.1, we can apply Theorem 5.1.4 to obtain a homeomorphism $f : U \to U$ such that $f(N \cap U) = M \cap U$ and f is $\mathcal{U}$-close to id. Since $\mathcal{U}$ is fitting in $U \cup V$, f can be extended to a homeomorphism $\tilde{f} : U \cup V \to U \cup V$ by $\tilde{f}|V \setminus U = \mathrm{id}$. Then, we have

$$\tilde{f}(N) = (M \cap U) \cup (N \cap V \setminus U),$$

which is an (f.d.)cap set for $U \cup V$. In the same way as above, we can obtain a homeomorphism $\tilde{g} : U \cup V \to U \cup V$ such that $\tilde{g}|U \setminus V = \mathrm{id}$.

$$\begin{aligned}\tilde{g}\tilde{f}(N) &= (M \cap V) \cup (\tilde{f}(N) \cap U \setminus V)\\ &= (M \cap V) \cup (M \cap U \setminus V) = M \cap (U \cup V).\end{aligned}$$

Then, $h = \tilde{g}\tilde{f}$ is the desired one. □

Now we can prove the following characterization:

Theorem 5.2.5 *Let M be an ℓ^2-manifold (resp. a $\boldsymbol{Q}$-manifold) and $N \subset M$.*

(1) *(M, N) is an (ℓ^2, ℓ^2_f)-manifold pair (resp. a $(\boldsymbol{Q}, \boldsymbol{Q}_f)$-manifold pair) if and only if N is an f.d.cap set for M.*
(2) *(M, N) is an (ℓ^2, ℓ^2_Q)-manifold pair (resp. a $(\boldsymbol{Q}, \operatorname{rint} \boldsymbol{Q})$-manifold pair) if and only if N is a cap set for M.*

Proof Because of similarity, we give only the proof of (1).

First, assume that (M, N) is an (ℓ^2, ℓ^2_f)-manifold pair (resp. a $\boldsymbol{Q}$-manifold). Then, every $x \in M$ has an open neighborhood U in M such that $N \cap U$ is an f.d.cap set for U. It follows from Proposition 5.2.4 that N is an f.d.cap set for M.

Conversely, assume that N is an f.d.cap set for M. Then, every $x \in M$ has an open neighborhood U in M that is homeomorphic to an open set V in ℓ^2 (resp. $\boldsymbol{Q}$). Then, $\ell^2_f \cap V$ (resp. $\boldsymbol{Q}_f \cap V$) is an f.d.cap set for V by Lemma 5.2.1. By virtue of Theorem 5.1.4, $(U, N \cap U) \approx (V, \ell^2_f \cap V)$ (resp. $(U, N \cap U) \approx (V, \boldsymbol{Q}_f \cap V)$), which means that (M, N) is an (ℓ^2, ℓ^2_f)-manifold pair (resp. a $(\boldsymbol{Q}, \boldsymbol{Q}_f)$-manifold pair). □

Using the Open Embedding Theorem 2.5.10 and the Topological Uniqueness of (f.d.)cap sets for ℓ^2-manifolds (Theorem 5.1.4), we can easily obtain the following:

Theorem 5.2.6 (Open Embedding) *For each (ℓ^2, ℓ^2_f)-manifold pair (resp. (ℓ^2, ℓ^2_Q)-manifold pair) (M, N), there exists an open embedding $g : M \to \ell^2$ such that $g(N) = g(M) \cap \ell^2_f$ (resp. $g(N) = g(M) \cap \ell^2_Q$).* □

Combining the Classification Theorem 2.6.1 and Corollary 5.1.5, we also have the following:

Theorem 5.2.7 (Classification) *Let (M, N) and (X, Y) be (ℓ^2, ℓ^2_f)-manifold pairs (or (ℓ^2, ℓ^2_Q)-manifold pairs). Then, $(M, N) \approx (X, Y)$ if and only if $M \simeq X$ or $N \simeq Y$.* □

The following is a combination of the Classification Theorem 3.8.8 and Corollary 5.1.5:

Theorem 5.2.8 (Classification) *Let (M, N) and (X, Y) be $(\boldsymbol{Q}, \boldsymbol{Q}_f)$-manifold pairs (or $(\boldsymbol{Q}, \operatorname{rint} \boldsymbol{Q})$-manifold pairs). Then, $(M, N) \approx (X, Y)$ if and only if M and X are CE equivalent.* □

For the products, we have the following:

Lemma 5.2.9 *Let M be a cap set for a metrizable space X. If Y is a separable locally compact ANR, then $M \times Y$ is a cap set for $X \times Y$.*

Proof Let $M = \bigcup_{n \in \mathbb{N}} M_n$, where $M_1 \subset M_2 \subset \cdots$ is a tower of compact Z-sets in X that has the cap for X. We can write $Y = \bigcup_{n \in \mathbb{N}} Y_n$, where each Y_n is a compact set in Y and $Y_n \subset \operatorname{int} Y_{n+1}$. Then, $M \times Y = \bigcup_{n \in \mathbb{N}} (M_n \cap Y_n)$ and each $M_n \times Y_n$ is a compact Z-set in $X \times Y$. We show that the tower $(M_n \cap Y_n)_{n \in \mathbb{N}}$ has the cap for

$X \times Y$. Here, we use the metric d for $X \times Y$ defined as follows:

$$d((x, y), (x', y')) = \max\{d_X(x, x'), d_Y(y, y')\},$$

where $d_X \in \mathrm{Metr}(X)$ and $d_Y \in \mathrm{Metr}(Y)$.

Let C be a compact set in $X \times Y$, $m \in \mathbb{N}$, and $\varepsilon > 0$. Then, we have $n_1 \in \mathbb{N}$ and an embedding $f : \mathrm{pr}_1(C) \to M_{n_1}$ such that $n_1 > m$, $f|\mathrm{pr}_1(C) \cap M_m = \mathrm{id}$ and $d_X(f, \mathrm{id}) < \varepsilon$. Choose $n \in \mathbb{N}$ so that $n \geqslant n_1$ and $\mathrm{pr}_2(C) \subset Y_n$. Since $C \subset \mathrm{pr}_1(C) \times \mathrm{pr}_2(C)$, we have an embedding $f \times \mathrm{id}|C : C \to M_n \times Y_n$, where $f \times \mathrm{id}|C \cap (M_m \times Y_m) = \mathrm{id}$ and $d(f \times \mathrm{id}, \mathrm{id}) < \varepsilon$. □

The above proof cannot be applied to the f.d.cap set case because the finite-dimensionality of C does not imply that of $\mathrm{pr}_1(C)$ even if Y is finite-dimensional.

Lemma 5.2.10 *Let M be an f.d.cap set for an ℓ^2- or $\boldsymbol{Q}$-manifold X. If Y is a separable locally finite-dimensional locally compact ANR, then $M \times Y$ is an f.d.cap set for $X \times Y$.*

Proof As in the proof of Lemma 5.2.9, let $M = \bigcup_{n\in\mathbb{N}} M_n$, where $M_1 \subset M_2 \subset \cdots$ is a tower of finite-dimensional compact Z-sets in X that has the f.d.cap for X. We can write $Y = \bigcup_{n\in\mathbb{N}} Y_n$, where each Y_n is a finite-dimensional compact set in Y and $Y_n \subset \mathrm{int}\, Y_{n+1}$. Then, $M \times Y = \bigcup_{n\in\mathbb{N}}(M_n \cap Y_n)$ and each $M_n \times Y_n$ is a finite-dimensional compact Z-set in $X \times Y$ (the Product Theorem 1.12.12). We show that the tower $(M_n \cap Y_n)_{n\in\mathbb{N}}$ has the f.d.cap for $X \times Y$. We use the same metric as in the proof of Lemma 5.2.9.

Let C be a finite-dimensional compact set in $X \times Y$, $m \in \mathbb{N}$, and $\varepsilon > 0$. By Proposition 2.1.17, we have an open cover $\mathcal{U}$ of $X \setminus M_m$ such that $\mathrm{mesh}\, \mathcal{U} < \varepsilon$ and $\mathcal{U}$ is fitting in X. Since $X \setminus M_m$ is an ℓ^2- or $\boldsymbol{Q}$-manifold, there exists an closed embedding $f : C \setminus (M_m \times Y) \to X \setminus M_m$ that is $\mathcal{U}$-close to pr_1 (the Strong Universality Theorem 2.3.15 or 2.10.11). Since $\mathcal{U}$ is fitting in X, f can be extended to a map $\tilde{f} : C \to X$ by $\tilde{f}|C \cap (M_m \times Y) = \mathrm{pr}_1|C \cap (M_m \times Y)$. Then, $d_X(\tilde{f}, \mathrm{pr}_1) < \varepsilon$ and

$$\tilde{f}(C) = f(C \setminus (M_m \times Y)) \cup (\mathrm{pr}_1(C) \cap X_m).$$

From the Addition Theorem 1.12.11, it follows that $\tilde{f}(C)$ is a finite-dimensional. By the f.d.cap of $(M_n)_{n\in\mathbb{N}}$ for X, there exist $n_1 \in \mathbb{N}$ and an embedding $g : \tilde{f}(C) \to M_{n_1}$ such that $n_1 > m$ and $g|\tilde{f}(C) \setminus M_m = \mathrm{id}$. Choose $n \in \mathbb{N}$ so that $n \geqslant n_1$ and $C \subset X \times Y_n$. We can define an embedding $h : C \to M_n \times Y_n$ by

$$h(x) = (g\tilde{f}(x), \mathrm{pr}_2(x)) \quad \text{for each } x \in C.$$

Obviously, $d(h, \mathrm{id}) < \varepsilon$. For each $x \in C \cap (M_m \times Y_m)$, it follows that

$$h(x) = (g\tilde{f}(x), \mathrm{pr}_2(x)) = (\tilde{f}(x), \mathrm{pr}_2(x)) = (\mathrm{pr}_1(x), \mathrm{pr}_2(x)) = x,$$

hence $h|C \cap (M_m \times Y_m) = \mathrm{id}$. □

The following is obtained from Lemmas 5.2.9 and 5.2.10 above:

Proposition 5.2.11 *Let K be a countable locally finite simplicial complex and let M be an (f.d.)cap set for ℓ^2 or $\boldsymbol{Q}$. Then, $|K| \times M$ is an (f.d.)cap set for $|K| \times \ell^2$ or $|K| \times \boldsymbol{Q}$.* □

In the above, $|K| \times \ell^2$ and $|K| \times \boldsymbol{Q}$ are respectively an ℓ^2-manifold and a $\boldsymbol{Q}$-manifold. Combining Proposition 5.2.11, Theorem 5.1.4, and the Triangulation Theorems 2.6.6 and 4.7.1, we have the following Triangulation Theorem:

Theorem 5.2.12 (TRIANGULATION) *Let (E, F) be one of (ℓ^2, ℓ^2_f), (ℓ^2, ℓ^2_Q), $(\boldsymbol{Q}, \boldsymbol{Q}_f)$, or $(\boldsymbol{Q}, \operatorname{rint} \boldsymbol{Q})$. For each (E, F)-manifold pair (M, N), there exists a countable locally finite simplicial complex K such that*

$$(M, N) \approx (|K| \times E, |K| \times F).$$ □

The following Triangulation Theorem for ℓ^2_Q-manifolds can also be obtained by combining Theorems 5.2.12 and 5.2.3.

Theorem 5.2.13 (TRIANGULATION) *Each ℓ^2_Q-manifold is homeomorphic to $|K| \times \ell^2_Q$ for some countable locally finite simplicial complex K.* □

The Triangulation Theorem for ℓ^2_f-manifolds (Theorem 5.1.11) can also be proved in the same way. We can generalize Proposition 5.1.3(4), (7), and Corollary 5.1.7(1), (3) as follows:

Theorem 5.2.14 *If M is an f.d.cap set for an ℓ^2- or $\boldsymbol{Q}$-manifold X, then $M \times \boldsymbol{Q}$ is a cap set for $X \times \boldsymbol{Q}$. In other words, if (X, M) is an (ℓ^2, ℓ^2_f)- or $(\boldsymbol{Q}, \boldsymbol{Q}_f)$-manifold pair, then $(X \times \boldsymbol{Q}, M \times \boldsymbol{Q})$ is an (ℓ^2, ℓ^2_Q)- or $(\boldsymbol{Q}, \operatorname{rint} \boldsymbol{Q})$-manifold pair.*

Proof Because of the similarity, we only give the proof for the ℓ^2-manifold case. Each point $x \in X$ has an open neighborhood U in X that is homeomorphic to an open set in ℓ^2. Due to Lemma 5.2.1, $M \cap U$ is an f.d.cap set for U. Because $\ell^2 \approx \boldsymbol{s}$, we have an open embedding $f : U \to \boldsymbol{s}$. Then, $\boldsymbol{s}_f \cap f(U)$ is an f.d.cap set for $f(U)$ by Lemma 5.2.1. By virtue of Corollary 5.1.5, we have $(U, M \cap U) \approx (f(U), \boldsymbol{s}_f \cap f(U))$, hence

$$\left(U \times \boldsymbol{Q}, (M \cap U) \times \boldsymbol{Q}\right) \approx \left(f(U) \times \boldsymbol{Q}, (\boldsymbol{s}_f \cap f(U)) \times \boldsymbol{Q}\right).$$

Since $\boldsymbol{s}_f \times \boldsymbol{Q}$ is a cap set for $\boldsymbol{s} \times \boldsymbol{Q}$ by Proposition 5.1.3(7), the following is a cap set for $f(U) \times \boldsymbol{Q}$ by Lemma 5.2.1:

$$(\boldsymbol{s}_f \times \boldsymbol{Q}) \cap (f(U) \times \boldsymbol{Q}) = (\boldsymbol{s}_f \cap f(U)) \times \boldsymbol{Q}.$$

Therefore, the following is a cap set for $U \times \boldsymbol{Q}$:

$$(M \times \boldsymbol{Q}) \cap (U \times \boldsymbol{Q}) = (M \cap U) \times \boldsymbol{Q}.$$

We can apply Proposition 5.2.4 to obtain the result. □

Note In the above theorem, the ℓ^2-manifold case can also be obtained by combining the Open Embedding Theorem 5.2.6, Lemma 5.2.1, Corollary 5.1.5, and Proposition 5.1.3(7).

Combining Theorems 5.2.14 and 5.2.3, we have the following:

Corollary 5.2.15 *For every ℓ^2_f-manifold M, $M \times \boldsymbol{Q}$ is an ℓ^2_Q-manifold.* □

In the rest of this section, we show the following:

Proposition 5.2.16 *Let M be an (f.d.)cap set for an ℓ^2- or $\boldsymbol{Q}$-manifold X.*

(1) *For any Z-set A, $M \setminus A$ is an (f.d.)cap set for X.*
(2) *If L is a countable union of (finite-dimensional) compact Z-sets and $M \subset L$, then L is an (f.d.)cap set for X.*

Proof Let $M = \bigcup_{n\in\mathbb{N}} M_n$, where $M_1 \subset M_2 \subset \cdots$ is a tower of (finite-dimensional) compact Z-sets in X that has the (f.d.)cap for X. Take $d \in \operatorname{Metr}(X)$.

(1): For each $n \in \mathbb{N}$, let $L_n = \{x \in M_n \mid d(x, A) \geqslant n^{-1}\}$. Then, $M \setminus A = \bigcup_{n\in\mathbb{N}} L_n$ and each L_n is a (finite-dimensional) compact Z-set in X as a closed subset of M_n. Let C be a (finite-dimensional) compact Z-set in X, $m \in \mathbb{N}$, and $\varepsilon > 0$. Since A is a Z-set in X, $X \setminus A$ is homotopy dense in X (Theorem 2.8.6), that is, there is a homotopy $h : X \times \mathbf{I} \to X$ such that $h_0 = \mathrm{id}$ and $h(X \times (0, 1]) \subset X \setminus A$. Since C is compact, we can find $\delta > 0$ such that $\operatorname{diam} h(\{x\} \times [0, \delta]) < \varepsilon/2$ for every $x \in C$. We define a map $\lambda : C \to [0, \delta]$ as follows:

$$\lambda(x) = \min\{\delta, d(x, L_m)\} \text{ for each } x \in C.$$

Then, we have a map $f' : C \to X\setminus A$ defined by $f'(x) = h(x, \lambda(x))$ for each $x \in C$, where $f' \simeq_{\varepsilon/2} \mathrm{id}$ and $f'|C \cap L_m = \mathrm{id}$. By the Strong Universality Theorem 2.9.6 or 2.10.10, we can easily obtain a Z-embedding $f : C \to X \setminus A$ such that $f|C \cap L_m = \mathrm{id}$ and $d(f, f') < \varepsilon/2$. Using the (f.d.)cap of $(M_n)_{n\in\mathbb{N}}$, we can obtain $n' > m$ and an embedding $g : f(C) \to M_{n'}$ such that $g|f(C) \cap M_m = \mathrm{id}$ and

$$d(g, \mathrm{id}) < \min\{\varepsilon/2, \operatorname{dist}(f(C), A)\}.$$

Then, it follows that

$$d(gf, \mathrm{id}) \leqslant d(g, \mathrm{id}) + d(f, \mathrm{id}) < \varepsilon/2 + \varepsilon/2 = \varepsilon.$$

Since $C \cap L_m \subset f(C) \cap M_m$, we have $gf|C \cap L_m = g|C \cap L_m = \mathrm{id}$. For each $x \in C$,

$$d(gf(x), f(x)) \leqslant d(g, \mathrm{id}) < \mathrm{dist}(f(C), A) \leqslant d(f(x), A)$$
$$\leqslant d(f(x), gf(x)) + d(gf(x), A),$$

hence $d(gf(x), A) > 0$. Since C is compact, we can find $n \geqslant n'$ such that $d(fg(x), A) > n^{-1}$ for every $x \in C$. Then, $fg(C) \subset L_n$. Thus, we have an embedding $gf : C \to L_n$ such that $gf|C \cap L_m = \mathrm{id}$ and $d(gf, \mathrm{id}) < \varepsilon$.

(2): First, we will show that if C is a (finite-dimensional) compact Z-set in X, then $M \cup C$ is an (f.d.)cap set for X. From the (f.d.)cap of $(M_n)_{n\in\mathbb{N}}$, there exists an embedding $h : C \to M_n$ for some $n \in \mathbb{N}$ such that $h \simeq \mathrm{id}$. Since M_n is a Z-set in X, h is a Z-embedding. By the Z-Set Unknotting Theorem 2.9.7 or 2.11.6, we can extend h to a homeomorphism $\tilde{h} : X \to X$. It follows from (2) that $M \setminus \tilde{h}(C)$ and $\tilde{h}(M) \setminus \tilde{h}(C)$ are (f.d.)cap sets for X, which are (f.d.)cap sets for $X \setminus \tilde{h}(C)$ by Lemma 5.2.1. Taking an open cover of $X \setminus \tilde{h}(C)$ fitting in X (Lemma 2.1.16), we can apply the Topological Uniqueness of (f.d.)cap sets (Theorem 5.1.4) to obtain a homeomorphism $g : X \to X$ such that $g(\tilde{h}(M) \setminus \tilde{h}(C)) = M \setminus \tilde{h}(C)$ and $g|\tilde{h}(C) = \mathrm{id}$. Thus, we have a homeomorphism $g\tilde{h} : X \to X$ such that

$$g\tilde{h}(M \cup C) = g(\tilde{h}(M) \setminus \tilde{h}(C)) \cup g\tilde{h}(C))$$
$$= (M \setminus \tilde{h}(C)) \cup \tilde{h}(C) = M \cup \tilde{h}(C) = M,$$

where the last equality comes from $\tilde{h}(C) \subset M$. Consequently, $M \cup C$ is an (f.d.)cap set for X.

Using the above special case, we show the general case. Let $L = \bigcup_{n\in\mathbb{N}} L_n$, where $N_1 \subset N_2 \subset \cdots$ are (finite-dimensional) compact Z-sets in X. We will construct homeomorphisms $h_n : X \to X$, $n \in \mathbb{N}$, so as to satisfy the following conditions:

(1) $h_n(M) = M \cup L_n$,
(2) $d(h_n, h_{n-1}) < 2^{-n}$ and $d(h_n^{-1}, h_{n-1}^{-1}) < 2^{-n}$,
(3) $h_n|M_{n-1} \cup h_{n-1}^{-1}(L_{n-1}) = h_{n-1}|M_{n-1} \cup h_{n-1}^{-1}(L_{n-1})$,

where $d \in \mathrm{Metr}(X)$ is complete and $M_0 = L_0 = \emptyset$. Then, a homeomorphism $h : X \to X$ can be obtained as the uniform limit of $(h_n)_{n\in\mathbb{N}}$, whose inverse h^{-1} is the uniform limit of $(h_n^{-1})_{n\in\mathbb{N}}$. Observe that

$$h(M) = h\big(\bigcup_{n\in\mathbb{N}} M_n\big) = \bigcup_{n\in\mathbb{N}} h_n(M_n \cup h_n^{-1}(L_n))$$
$$= \bigcup_{n\in\mathbb{N}} h_n(M_n) \cup \bigcup_{n\in\mathbb{N}} L_n = L.$$

Hence, L is an (f.d.)cap set for X.

Now, assuming that h_{n-1} has been obtained, we define h_n as follows: For the sake of simplicity, let

$$A = M_{n-1} \cup h_{n-1}^{-1}(L_{n-1}) \text{ and } B = h_{n-1}(M_{n-1}) \cup L_{n-1} = h_{n-1}(A).$$

From Lemma 5.2.1 and Proposition 5.2.16, $M \setminus A$ and $(M \cup L_n) \setminus B$ are (f.d.)cap sets for $X \setminus A$ and $X \setminus B$, respectively. Let $\mathcal{U}$ be an open cover of $X \setminus B$ fitting in X that refines the covers $\{\mathrm{B}_d(x, 2^{-n-1}) \mid x \in X \setminus A\}$ and $\{h_{n-1}(\mathrm{B}_d(x, 2^{-n-1})) \mid x \in X \setminus B\}$ (cf. Proposition 2.1.17), which means mesh $\mathcal{U} < 2^{-n}$ and mesh $h_{n-1}^{-1}(\mathcal{U}) < 2^{-n}$. By Corollary 5.1.5, we can obtain a homeomorphism $h' : X \setminus A \to X \setminus B$ such that $h'(M \setminus A) = (M \cup L_n) \setminus B$ and h' is $\mathcal{U}$-close to h_{n-1}. Since $\mathcal{U}$ is fitting in X, we can extend h' to a homeomorphism $h_n : X \to X$ so as to satisfy (3), that is, $h_n|A = h_{n-1}|A$. Then, we show (1) as follows:

$$h_n(M) = h'(M \setminus A) \cup h_{n-1}(A) = ((M \cup L_n) \setminus B) \cup B = M \cup L_n.$$

Moreover, $d(h_n, h_{n-1}) < 2^{-n}$ and

$$d(h_n^{-1}, h_{n-1}^{-1}) = d(\mathrm{id}, h_{n-1}^{-1}h_n) = d(h_{n-1}^{-1}h_{n-1}, h_{n-1}^{-1}h_n) < 2^{-n}.$$

Thus, (2) is also satisfied. □

5.3 Absorption Property and Absorption Bases

In the previous two sections, introducing (f.d.)cap sets for Hilbert space ℓ^2 or ℓ^2-manifolds, we characterized the pairs (ℓ^2, ℓ^2_f) and (ℓ^2, ℓ^2_Q), and manifold pairs modeled on these pairs. We want to generalize results to the pairs $(\ell^2(\Gamma), \ell^2_f(\Gamma))$ and $(\ell^2(\Gamma) \times \ell^2, \ell^2_f(\Gamma) \times \ell^2_Q)$, and manifold pairs modeled on these pairs, where it should be noted that $(\ell^2 \times \ell^2, \ell^2_f \times \ell^2_Q) \approx (\ell^2, \ell^2_Q)$.

An (f.d.)cap set is defined as a set with a tower of members of $\mathfrak{M}_0$ (or $\mathfrak{M}_0^{\mathrm{fd}}$) that has the (finite-dimensional) compact absorption property (cap) (or (f.d.cap)). Although (cap) and (f.d.cap) are properties of a tower $M_1 \subset M_2 \subset \cdots$ in X with $M = \bigcup_{n \in \mathbb{N}} M_n$, we can translate them into properties of a subset $M \subset X$ itself. In this section, we introduce the absorption property for a general class $\mathfrak{C}$ other than $\mathfrak{M}_0$ (or $\mathfrak{M}_0^{\mathrm{fd}}$). As mentioned in Fact (3) on p. 354, it should be noted that any non-separable metrizable space has no (f.d.)cap sets.

Hereafter, in this chapter:

- *let $\mathfrak{C}$ be a closed hereditary, additive topological class of metrizable spaces,*

where $\mathfrak{C}$ is said to be:

- **topological** if $\mathfrak{C}$ contains any space homeomorphic to a member of $\mathfrak{C}$;
- **closed hereditary** if $\mathfrak{C}$ contains any closed subspace of a member of $\mathfrak{C}$;
- **additive** if $\mathfrak{C}$ contains any space that is the union of two closed subspaces belonging to $\mathfrak{C}$.

It should be remarked that $\mathfrak{C}$ contains singletons because it is closed hereditary. Hence, $\mathfrak{C}$ also contains every finite discrete space because it is additive.

For a class $\mathfrak{C}$ ($\subset \mathfrak{M}$), we will use the following notation:

- $\mathfrak{C}(\tau)$ — the subclass of $\mathfrak{C}$ consisting of spaces of weight $\leqslant \tau$;
- $\mathfrak{C}(\aleph_0)$ — the subclass of $\mathfrak{C}$ consisting of separable spaces;
- $\mathfrak{C}_\sigma$ — the class consisting of all metrizable spaces that can be expressed as countable unions of *closed* subspaces belonging to $\mathfrak{C}$;
- $\mathfrak{C}^{(n)}$ — the subclass of $\mathfrak{C}$ consisting of spaces of dimension $\leqslant n$;
- $\mathfrak{C}^{\mathrm{fd}}$ — the subclass of $\mathfrak{C}$ consisting of finite-dimensional spaces, i.e., $\mathfrak{C}^{\mathrm{fd}} = \bigcup_{n\in\omega} \mathfrak{C}^{(n)}$;
- $\mathfrak{C}^{\mathrm{scd}}$ — the subclass of $\mathfrak{C}$ consisting of strongly countable-dimensional spaces;
- $\oplus\mathfrak{C}$ — the class consisting of topological sums (spaces expressed as discrete unions) of members of $\mathfrak{C}$;
- $\tau\mathfrak{C}$ — the class of topological sums (spaces expressed as discrete unions) of at most τ many members of $\mathfrak{C}$.[6]

These classes are closed hereditary and topological. They are also additive, except for the last two classes, $\oplus\mathfrak{C}$ and $\tau\mathfrak{C}$. In fact, $\aleph_0\mathfrak{M}_0$ is not additive. For example,

$$X = \bigcup_{n\in\omega}[2n, 2n+1],\ Y = \bigcup_{n\in\mathbb{N}}[2n-1, 2n] \in \aleph_0\mathfrak{M}_0,$$

but $X \cup Y = \mathbb{R}_+ \notin \aleph_0\mathfrak{M}_0$. Evidently, $\mathfrak{C}_\sigma$, $\tau\mathfrak{C}$, $(\tau\mathfrak{C})_\sigma \subset \mathfrak{M}(\tau)$ if $\mathfrak{C} \subset \mathfrak{M}(\tau)$. The class $(\mathfrak{M}_0)_\sigma$ is the class consisting of all σ-compact metrizable spaces,

$$\mathfrak{M}_0(\aleph_0) = \mathfrak{M}_0 \text{ and } (\mathfrak{M}_0)_\sigma(\aleph_0) = (\mathfrak{M}_0)_\sigma.$$

It should be noted that $(\mathfrak{C}_\sigma)^{\mathrm{fd}} \neq (\mathfrak{C}^{\mathrm{fd}})_\sigma = (\mathfrak{C}_\sigma)^{\mathrm{scd}}$. The last equality can be shown as follows: Evidently, $(\mathfrak{C}^{\mathrm{fd}})_\sigma \subset (\mathfrak{C}_\sigma)^{\mathrm{scd}}$. Every $X \in (\mathfrak{C}_\sigma)^{\mathrm{scd}}$ can be written as $X = \bigcup_{i\in\mathbb{N}} X_i = \bigcup_{n\in\mathbb{N}} A_n$, where each X_i and A_n is closed in X, $X_i \in \mathfrak{C}$, and $\dim A_n < \infty$. Then, $X = \bigcup_{i,n\in\mathbb{N}}(X_i \cap A_n)$, where each $X_i \cap A_n$ is closed in X and $\dim X_i \cap A_n \leqslant \dim A_n < \infty$ by the Subset Theorem 1.12.7. Moreover, $X_i \cap A_n \in \mathfrak{C}$ because $\mathfrak{C}$ is closed hereditary. Thus, $X_i \cap A_n \in \mathfrak{C}^{\mathrm{fd}}$ for each $i, n \in \mathbb{N}$, which implies that $X \in (\mathfrak{C}^{\mathrm{fd}})_\sigma$. As is easily observed, $(\mathfrak{C}_\sigma)^{\mathrm{scd}} = (\mathfrak{C}^{\mathrm{scd}})_\sigma$, which allows us to write this without parentheses, $\mathfrak{C}_\sigma^{\mathrm{scd}}$.

[6]In general, $(\oplus\mathfrak{C})(\tau) \subset \tau\mathfrak{C}$. Note that $(\oplus\mathfrak{C})(\tau) = \tau\mathfrak{C}$ if and only if $\mathfrak{C} = \mathfrak{C}(\tau)$.

For a metrizable space X, we define

$$\mathfrak{C}_Z(X) = \{ C \in \mathfrak{C} \mid C \text{ is a } Z\text{-set in } X \}.$$

It is said that $M \subset X$ is **$\mathfrak{C}$-absorptive** in X if it satisfies the following condition called the **$\mathfrak{C}$-absorption property**:

(ab$_{\mathfrak{C}}$) If $C \in \mathfrak{C}_Z(X)$, $D \subset M \cap C$ is closed in X and $\mathcal{U} \in \operatorname{cov}(X)$, then there is a Z-embedding $f : C \to X$ such that f is $\mathcal{U}$-close to id, $f|D = \text{id}$, and $f(C) \subset M$.[7]

The following can be easily shown:

Fact (1) *If every singleton in X is a Z-set, then every $\mathfrak{C}$-absorptive set in X is dense in X.*[8]

(2) *If $M \subset X$ is $\mathfrak{C}$-absorptive and $f : X \to Y$ is a homeomorphism, then $f(M)$ is $\mathfrak{C}$-absorptive in Y.*

Note that the $\mathfrak{M}_0$- (or $\mathfrak{M}_0^{\text{fd}}$-)absorption property is a property for subsets but the (finite-dimensional) compact absorption property (cap) (or (f.d.cap)) is a property for towers of subsets. However, we have the following:

Proposition 5.3.1 *Let X be an ℓ^2- or Q-manifold. Then, every (f.d.)cap set M in X is $\mathfrak{M}_0$- (or $\mathfrak{M}_0^{\text{fd}}$-)absorptive in X.*

Proof Let $M = \bigcup_{n\in\mathbb{N}} M_n$, where $M_1 \subset M_2 \subset \cdots$ is a tower of (finite-dimensional) compact Z-sets in X that has the (f.d.)cap for X. Let $C \in \mathfrak{M}_0$ (or $\in \mathfrak{M}_0^{\text{fd}}$) be a Z-set in X and $D \subset M \cap C$ a closed subset. For each $\mathcal{U} \in \operatorname{cov}(X)$, choose $d \in \operatorname{Metr}(X)$ so that $\{\mathrm{B}_d(x, 1) \mid x \in X\} \prec \mathcal{U}$ (Proposition 1.3.22(1)). We can take a tower $C_1 \subset C_2 \subset \cdots$ of closed sets in C such that $C \setminus D = \bigcup_{n\in\mathbb{N}} C_n$. Then, we can inductively obtain $1 = n(1) < n(2) < \cdots \in \mathbb{N}$ and Z-embeddings $h_i : D \cup C_i \to X$, $i \in \mathbb{N}$, such that

$$h_i(C_i) \subset M_{n(i)},\ h_i|D = \text{id},\ h_i|C_{i-1} = h_{i-1}|C_{i-1}, \text{ and } d(h_i, h_{i-1}) < 2^{-i},$$

where $h_0 = \text{id}$. Indeed, assume that h_{i-1} has been obtained. Since $h_{i-1}(C_i)$ misses $D = h_{i-1}(D)$, we can choose

$$0 < \delta < \min\{2^{-(i+1)}, \operatorname{dist}(h_{i-1}(C_i), D)\}.$$

[7] In [150], the $\mathfrak{C}$-absorption property was defined by replacing a Z-embedding $f : C \to X$ in (ab$_{\mathfrak{C}}$) with a homeomorphism $f : X \to X$. However, when X is an $\ell^2(\Gamma)$-manifold, it will be shown in Lemma 5.3.2 that these are coincident.

[8] It will be shown in Proposition 5.3.5 that if $\mathfrak{M}_0^{\text{fd}} \subset \mathfrak{C}$, then every $\mathfrak{C}$-absorptive set in X is homotopy dense in X.

Applying the (f.d.)cap for X, we have $n(i) > n(i-1)$ and an embedding $f : h_{i-1}(C_i) \to M_{n(i)}$ such that $f|h_{i-1}(C_i) \cap M_{n(i-1)} = \mathrm{id}$ and $f \simeq_\delta \mathrm{id}$. Because $h_{i-1}(C_{i-1}) \subset M_{n(i-1)}$, we have $f|h_{i-1}(C_{i-1}) = \mathrm{id}$. Since $d(f, \mathrm{id}) < \delta$, it follows that $D \cap f(h_{i-1}(C_i)) = \emptyset$. Hence, we can define a Z-embedding $g : B \cup C_i \to X$ by $g|B = \mathrm{id}$ and $g|C_i = fh_{i-1}|C_i$. Since $g \simeq_\delta h_{i-1}|D \cup C_i$, applying the Homotopy Extension Theorem 1.13.11, we can extend g to a map $\tilde{g} : h_{i-1}(C) \to X$ such that $\tilde{g} \simeq_\delta h_{i-1}$. By the Strong Universality of ℓ^2- or $\boldsymbol{Q}$-manifolds (Theorem 2.9.6 or 2.10.10), we can obtain a Z-embedding $h_i : C \to X$ such that $h_i|D \cup C_i = \tilde{g}|D \cup C_i = g$ and $d(h_i, h_{i-1}) < \delta \leqslant 2^{-i}$. Then, $h_i(C_i) = g(C_i) = fh_{i-1}(C_i) \subset M_{n(i)}$, $h_i|D = g|D = \mathrm{id}$, and $h_i|C_{i-1} = g|C_{i-1} = fh_{i-1}|C_{i-1} = h_{i-1}|C_{i-1}$.

Because $C = D \cup \bigcup_{i \in \mathbb{N}} C_i$ and $h_j|C_i = h_i|C_i$ for $j > i$, we can define a map $h : C \to X$ by $h|D = \mathrm{id}$ and $h|C_i = h_i$ for $i \in \mathbb{N}$. The continuity follows from the fact that h is the uniform limit of h_i. In fact, $d(h, h_i) < 2^{-i}$ for each $i \in \mathbb{N}$ because if $x \in D \cup C_i$ then $h(x) = h_i(x)$, and if $x \in C_j$, $j > i$, then $h(x) = h_j(x)$ and

$$\begin{aligned} d(h_j(x), h_i(x)) &\leqslant d(h_j, h_{j-1}) + \cdots + d(h_{i+1}, h_i) \\ &< 2^{-j} + \cdots + 2^{-(i+1)} < 2^{-i}. \end{aligned}$$

In particular, $d(h, \mathrm{id}) = d(h, h_0) < 1$, which implies that h is $\mathcal{U}$-close to id. For each $x \neq y \in A$, we can find $i \in \mathbb{N}$ such that $x, y \in D \cup C_i$. Then, $h(x) = h_i(x) \neq h_i(y) = h(y)$. Hence, h is injective, which means that h is an embedding because C is compact. Moreover,

$$h(C \setminus D) = \bigcup_{i \in \mathbb{N}} h(C_i) \subset \bigcup_{i \in \mathbb{N}} M_{n(i)} = M.$$

Thus, we have the result. □

Remark 5.2 In the above proof, when M is an f.d.cap set for X, it is enough to assume that $\dim C \setminus D < \infty$ instead of $\dim C < \infty$.

It is said that $M \subset X$ is **strongly $\mathfrak{C}$-absorptive** in X if it satisfies the following stronger condition than (ab$_{\mathfrak{C}}$), which is called the **strong $\mathfrak{C}$-absorption property**:

(sab$_{\mathfrak{C}}$) If $C \in \mathfrak{C}_Z(X)$, $D \subset M \cap C$ is closed in X, and U is an open neighborhood of C in X with $\mathcal{U} \in \mathrm{cov}(U)$, then there is a homeomorphism $f : X \to X$ such that f is $\mathcal{U}$-isotopic to id (hence $f|X \setminus U = \mathrm{id}$), $f|D = \mathrm{id}$, and $f(C) \subset M$.

Lemma 5.3.2 *Every $\mathfrak{C}$-absorptive set M in an $\ell^2(\Gamma)$-manifold X is strongly $\mathfrak{C}$-absorptive in X.*

Proof Let C, D, U, and $\mathcal{U}$ be as in (sab$_{\mathfrak{C}}$). Take $\mathcal{V} \in \mathrm{cov}(U)$ so that $\mathrm{st}\,\mathcal{V} \prec \mathcal{U}$ and $\mathrm{st}\,\mathcal{V}$ is fitting in X (cf. Proposition 2.1.17). Moreover, take an open neighborhood V of C in X so that $\mathrm{cl}\, V \subset U$, and $\mathcal{W} \in \mathrm{cov}(X)$ so that

$$\mathcal{W} \underset{h}{\prec} \mathcal{V}[C] \cup \{X \setminus \mathrm{cl}\, V\}.$$

Since M is $\mathfrak{C}$-absorptive, we have a Z-embedding $f : C \to X$ such that f is $\mathcal{W}$-close to id, $f|D = \text{id}$, and $f(C) \subset M$. Then, $f \simeq_{\mathcal{V}} \text{id}$. We can apply the Z-Set Unknotting Theorem 2.9.7 to extend f to a homeomorphism $\tilde{f} : U \to U$ such that $\tilde{f}$ is $\operatorname{st}\mathcal{V}$-isotopic to id. Since $\operatorname{st}\mathcal{V}$ is fitting in X, $\tilde{f}$ can be extended over X. Thus, we have the desired homeomorphism of X onto itself. □

Remark 5.3 Lemma 5.3.2 is valid even when X is a $\boldsymbol{Q}$-manifold. Indeed, in the above proof, $\mathcal{V}$ can be taken so that $\mathcal{V}$ is locally finite in U and each member of $\mathcal{V}$ has the compact closure in X. Then, the homotopy $f \simeq_{\mathcal{V}} \text{id}$ is proper by Proposition 1.3.9, that is, $f \simeq^{p}_{\mathcal{V}} \text{id}$. Hence, we can apply the Z-Set Unknotting Theorem 2.11.6 instead of 2.9.7.

The following lemma is useful and will be used frequently:

Lemma 5.3.3 *Let $X = (X, d)$ be a complete metric space. Given a sequence $(f_n)_{n\in\mathbb{N}}$ of homeomorphisms of X onto itself and a sequence $(\mathcal{U}_n)_{n\in\mathbb{N}}$ of open covers of X with the following conditions:*

(i) *f_n is $\mathcal{U}_n$-close to* id_X,
(ii) $\operatorname{mesh}\mathcal{U}_n < 2^{-n}$ *and* $\operatorname{mesh} f_1^{-1} \cdots f_n^{-1}(\mathcal{U}_{n+1}) < 2^{-n}$,

the sequence $(f_n \cdots f_1)_{n\in\mathbb{N}}$ converges to a homeomorphism $f : X \to X$ with $d(f, \text{id}_X) < 1$.

Proof Recall that $\mathrm{C}(X, X)$ is complete with respect to the sup-metric. Since $(f_n \cdots f_1)_{n\in\mathbb{N}}$ is uniformly Cauchy, it converges to a map $f \in \mathrm{C}(X, X)$, where

$$d(f, \text{id}_X) \leqslant \sum_{n=1}^{\infty} d(f_n, \text{id}_X) < \sum_{n=1}^{\infty} 2^{-n} = 1.$$

For each $x \in X$, we have $U \in \mathcal{U}_{n+1}$ such that $f_{n+1}^{-1}(x),\ x = f_{n+1} f_{n+1}^{-1}(x) \in U$, and then

$$d(f_1^{-1} \cdots f_n^{-1}(x), f_1^{-1} \cdots f_n^{-1} f_{n+1}^{-1}(x)) \leqslant \operatorname{diam} f_1^{-1} \cdots f_n^{-1}(U) < 2^{-n}.$$

Thus, $(f_1^{-1} \cdots f_n^{-1})_{n\in\mathbb{N}}$ is also uniformly Cauchy, hence it converges to a map $g \in \mathrm{C}(X, X)$.

It remains to show that $fg = gf = \text{id}_X$, which means that f is a homeomorphism with $f^{-1} = g$. For each $x \in X$ and $\varepsilon > 0$, by the continuity of f at $g(x)$, there is $\delta > 0$ so that $d(g(x), x') < \delta$ implies $d(fg(x), f(x')) < \varepsilon/2$. Choose $m \in \mathbb{N}$ so that $n \geqslant m$ implies

$$d(f, f_n \cdots f_1) < \varepsilon/2 \text{ and } d(g, f_1^{-1} \cdots f_n^{-1}) < \delta.$$

Then, it follows that

$$\begin{aligned} d(fg(x), x) &\leqslant d(fg(x), f(f_1^{-1} \cdots f_n^{-1}(x))) \\ &\quad + d(f(f_1^{-1} \cdots f_n^{-1}(x)), f_n \ldots f_1(f_1^{-1} \cdots f_n^{-1}(x))) \\ &< \varepsilon/2 + \varepsilon/2 = \varepsilon. \end{aligned}$$

This means that $fg(x) = x$. Thus, we have $fg = \mathrm{id}_X$. In the same way as above, we can see that $gf = \mathrm{id}_X$. □

Lemma 5.3.4 *Let M be a $\mathfrak{C}$-absorptive set in an $\ell^2(\Gamma)$-manifold X. For each Z_σ-set $C \in \mathfrak{C}_\sigma$ in X, each open set W in X and $\mathcal{U} \in \mathrm{cov}(W)$, there exists a homeomorphism $f : X \to X$ such that f is $\mathcal{U}$-close to* id *(so $f|X \setminus W = \mathrm{id}$) and $f(C \cap W) \subset M$.*

Proof By Proposition 2.1.17, we may assume that $\mathcal{U}$ is fitting in X. Since W is completely metrizable (1.3.16), W has an admissible complete metric such that $\{\overline{\mathrm{B}}_d(x, 1) \mid x \in W\} \prec \mathcal{V}$ (1.3.22(1)). Then, it suffices to construct a homeomorphism $f : W \to W$ such that $f(C \cap W) \subset M \cap W$ and $d(f, \mathrm{id}) < 1$.

Since $C \in \mathfrak{C}_\sigma$ is a Z_σ-set in X, $\mathfrak{C}$ is closed hereditary and additive, and W is F_σ in X, we can write $C \cap W = \bigcup_{n \in \mathbb{N}} C_n$, where $C_1 \subset C_2 \subset \cdots$ and $C_n \in \mathfrak{C}_Z(X)$ for every $n \in \mathbb{N}$. Because the class $\mathfrak{C}$ is topological, using Lemma 5.3.2, we can inductively define homeomorphisms $f_n : X \to X$ and $\mathcal{U}_n \in \mathrm{cov}(W)$, $n \in \mathbb{N}$, so as to satisfy the following conditions:

(1) f_n is $\mathcal{U}_n$-close to id (which implies that $f_n|X \setminus U = \mathrm{id}$),
(2) $f_n|f_{n-1} \cdots f_1(C_{n-1}) = \mathrm{id}$, $f_n(f_{n-1} \cdots f_1(C_n)) \subset M$,
(3) $\mathrm{mesh}\, \mathcal{U}_n < 2^{-n}$ and $\mathrm{mesh}\, f_1^{-1} \cdots f_n^{-1}(\mathcal{U}_{n+1}) < 2^{-n}$,

where $C_0 = \emptyset$ and $f_0 = \mathrm{id}$, i.e., the second condition for $n = 1$ means $f_1(C_1) \subset M$. Then, by Lemma 5.3.3, the sequence $(f_n \cdots f_1|U)_{n \in \mathbb{N}}$ converges to a homeomorphism $f : W \to W$ with $d(f, \mathrm{id}) < 1$. For each $m > n$, since $f_m \cdots f_{n+1}|f_n \cdots f_1(C_n) = \mathrm{id}$, it follows that $f|C_n = f_n \cdots f_1|C_n$. Hence, $f(C_n) = f_n \cdots f_1(C_n) \cap f(W) \subset M \cap W$. Since $C \cap W = \bigcup_{n \in \mathbb{N}} C_n$, we have $f(C \cap W) \subset M \cap W$. □

Remark 5.4 Lemma 5.3.4 is valid even when X is a $\boldsymbol{Q}$-manifold because so is Lemma 5.3.2 (cf. Remark 5.3).

Proposition 5.3.5 *Let X be an $\ell^2(\Gamma)$-manifold. If $\mathfrak{M}_0^{\mathrm{fd}} \subset \mathfrak{C}$, then every $\mathfrak{C}$-absorptive set M in X is homotopy dense in X.*

Proof To show that M is homotopy dense in X, we verify condition (b) in Corollary 1.15.5. Let U be an open set in X and let $\alpha : \mathbf{S}^{n-1} \to U \cap M$ be a map that extends to a map $\beta : \mathbf{B}^n \to U$. Since $\alpha(\mathbf{S}^{n-1})$ is a Z-set in U, the complement $U \setminus \alpha(\mathbf{S}^{n-1})$ is homotopy dense in U. Hence, we can assume that $\beta(\mathbf{B}^n \setminus \mathbf{S}^{n-1}) \subset U \setminus \alpha(\mathbf{S}^{n-1})$. Take an open cover $\mathcal{V}$ of $U \setminus \alpha(\mathbf{S}^{n-1})$ so that $\mathrm{st}\, \mathcal{V}$ is fitting in U (cf. Proposition 2.1.17). By the strong $\mathfrak{M}_1$-universality of $\ell^2(\Gamma)$-

manifolds, we have a (strong) Z-embedding $\gamma : \mathbf{B}^n \setminus \mathbf{S}^{n-1} \to U \setminus \alpha(\mathbf{S}^{n-1})$ that is $\mathcal{V}$-close to $\beta|\mathbf{B}^n \setminus \mathbf{S}^{n-1}$. Using Lemma 5.3.4, we can obtain a homeomorphism $f : U \setminus \alpha(\mathbf{S}^{n-1}) \to U \setminus \alpha(\mathbf{S}^{n-1})$ such that $f\gamma(\mathbf{B}^n \setminus \mathbf{S}^{n-1}) \subset U \cap M$ and f is $\mathcal{V}$-close to id. Since $f\gamma : \mathbf{B}^n \setminus \mathbf{S}^{n-1} \to U \cap M$ is st$\mathcal{V}$-close to β, $f\gamma$ can be extended to a map $\tilde{\alpha} : \mathbf{B}^n \to U \cap M$ defined by $\tilde{\alpha}|\mathbf{S}^{n-1} = \beta|\mathbf{S}^{n-1} = \alpha$, which is an extension of α. Thus, it follows from Corollary 1.15.5 that M is homotopy dense in X. □

Remark 5.5 In the above proof, the strong $\mathfrak{M}_1$-universality can be replaced with the proper $\mathfrak{C}$-universality. Hence, Proposition 5.3.5 is valid even when X is a $\boldsymbol{Q}$-manifold because so is Lemma 5.3.4 (cf. Remark 5.4).

Remark 5.6 Since every n-dimensional compactum can be embedded into $\mathbf{I}^{2n+1}$ (Embedding Theorem 1.12.14), we have the following:

$$\mathfrak{M}_0^{\mathrm{fd}} \subset \mathfrak{C} \;\Leftrightarrow\; \mathfrak{C} \text{ contains } n\text{-cubes } \mathbf{I}^n,\ n \in \mathbb{N}.$$

Corollary 5.3.6 *Suppose that $\mathfrak{M}_0^{\mathrm{fd}} \subset \mathfrak{C}$. Every $\mathfrak{C}$-absorptive set M in an $\ell^2(\Gamma)$-manifold X is an ANR, every Z-set in M is a strong Z-set in M and every compact set in M is a strong Z-set in M.*

Proof By Proposition 5.3.5 above, M is homotopy dense in X. Since X is an ANR, it follows from Theorem 1.13.26 that M is an ANR. Since every Z-set in X is a strong Z-set in X (Theorem 2.9.3), it follows from Proposition 2.9.4 that every Z-set in M is a strong Z-set in M.

Recall that every compact set in an $\ell^2(\Gamma)$-manifold is a strong Z-set (Proposition 3.2.10 or Corollary 3.3.14 with Toruńczyk Characterization Theorem 3.4.1). Hence, we can apply the argument in the proof of Proposition 2.9.4 to prove that every compact set in M is a strong Z-set. The details are left to the reader. □

Recall that every Z-set in each $\boldsymbol{Q}$-manifold is a strong Z-set because of the local compactness of $\boldsymbol{Q}$-manifolds (Proposition 2.8.1). As seen in Remark 5.5, Proposition 5.3.5 is valid for a $\boldsymbol{Q}$-manifold. Moreover, applying Theorem 1.13.26 as in the above proof, we can prove the following:

Proposition 5.3.7 *Suppose that $\mathfrak{M}_0^{\mathrm{fd}} \subset \mathfrak{C}$. Every $\mathfrak{C}$-absorptive set M in a $\boldsymbol{Q}$-manifold X is homotopy dense in X and it is an ANR.* □

Now, we can prove the following characterization of $\mathfrak{C}$-absorptive sets in $\ell^2(\Gamma)$-manifolds:

Theorem 5.3.8 *Let X be an $\ell^2(\Gamma)$-manifold. For $M \subset X$, the following are equivalent:*

(a) *M is $\mathfrak{C}$-absorptive in X.*
(b) *$M \cap W$ is $\mathfrak{C}$-absorptive in every open set $W \subset X$.*

(c) *For each open set U in X, $\mathcal{V} \in \operatorname{cov}(U)$, and each Z_σ-set C in U with $C \in \mathfrak{C}_\sigma$, there exists a homeomorphism $f : U \to U$ such that f is $\mathcal{V}$-close to* id *and $f(C) \subset M \cap U$.*

(d) *For each $C \in \mathfrak{C}_Z(X)$, each open set U in X and $\mathcal{V} \in \operatorname{cov}(U)$, there exists a homeomorphism $f : U \to U$ such that f is $\mathcal{V}$-close to* id *and $f(C \cap U) \subset M \cap U$.*

Proof The implication (b) $\Rightarrow$ (a) is obvious. The implications (a) $\Rightarrow$ (d) and (b) $\Rightarrow$ (c) are direct consequences of Lemma 5.3.4 above. Since every Z_σ-set in an open set $U \subset X$ is a Z_σ-set in X, the implication (a) $\Rightarrow$ (c) also follows from Lemma 5.3.4. Moreover, in condition (d), $C \cap U$ is a Z_σ-set in U and $C \cap U \in \mathfrak{C}_\sigma$ because U is F_σ in X and $\mathfrak{C}$ is closed hereditary. So, (c) $\Rightarrow$ (d) is trivial. Thus, we only have to show the implications (d) $\Rightarrow$ (a) and (c) $\Rightarrow$ (b).

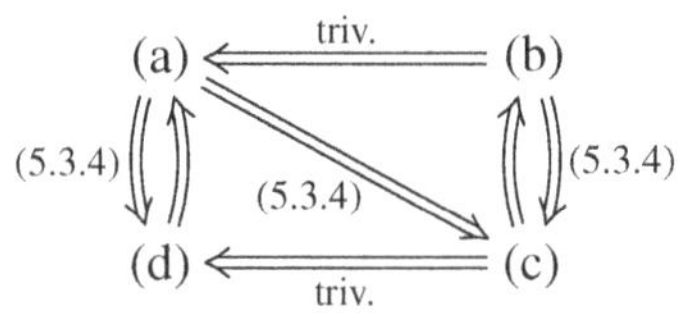

(d) $\Rightarrow$ (a) (or (c) $\Rightarrow$ (b)): Let $C \in \mathfrak{C}_Z(X)$ (or $\in \mathfrak{C}_Z(W)$) with $D \subset M \cap C$ a closed set in X (where $C \setminus D \in \mathfrak{C}_\sigma$ is F_σ in X). For each open neighborhood U of C in X (or in W) with $\mathcal{V} \in \operatorname{cov}(U)$, take an open cover $\mathcal{W}$ of the open set $U \setminus D$ such that $\mathcal{W} \prec \mathcal{V}|(U \setminus D)$ and $\mathcal{W}$ is fitting in X (Proposition 2.1.17). By (d) (or (c)), we have a homeomorphism $f : U \setminus D \to U \setminus D$ such that f is $\mathcal{W}$-close to id and $f(C \cap (U \setminus D)) \subset M$. Then, f extends to a homeomorphism $\tilde{f} : X \to X$ (or $\tilde{f} : W \to W$) by $\tilde{f}|X \setminus (U \setminus D) = \mathrm{id}$ (or $\tilde{f}|W \setminus (U \setminus D) = \mathrm{id}$), whence $\tilde{f}$ is $\mathcal{V}$-close to id, $f|D = \mathrm{id}$ and $f(C) \subset M$. □

Remark 5.7 Theorem 5.3.8 is also valid for a $\boldsymbol{Q}$-manifold X because Lemma 5.3.4 is valid for a $\boldsymbol{Q}$-manifold X (Remark 5.4).

Corollary 5.3.9 *Every $\mathfrak{C}$-absorptive set M in an $\ell^2(\Gamma)$-manifold X is $\tau\mathfrak{C}$-absorptive in X.*

Proof Let $C = \bigoplus_{\lambda \in \Lambda} C_\lambda \in (\tau\mathfrak{C})_Z(X)$, where card $\Lambda \leqslant \tau$ and $C_\lambda \in \mathfrak{C}$ for each $\lambda \in \Lambda$. For a closed subset $D \subset C \cap M$ and each $\lambda \in \Lambda$, let $D_\lambda = D \cap C_\lambda$. Then, $C_\lambda \in \mathfrak{C}_Z(X)$ and D_λ is a closed subset of $C_\lambda \cap M$. We have open sets W_λ, $\lambda \in \Lambda$, in X such that $(W_\lambda)_{\lambda \in \Lambda}$ is discrete in X. By Theorem 5.3.8, $M \cap W_\lambda$ is $\mathfrak{C}$-absorptive in W_λ. For every $\mathcal{U} \in \operatorname{cov}(X)$, there are Z-embeddings $f_\lambda : C_\lambda \to X$, $\lambda \in \Lambda$, such that f_λ is $\mathcal{U}$-close to id, $f_\lambda|D_\lambda = \mathrm{id}$, and $f_\lambda(C_\lambda) \subset M \cap W_\lambda$. We can define a Z-embedding $f : C \to X$ by $f|C_\lambda = f_\lambda$ for each $\lambda \in \Lambda$. Then, f is $\mathcal{U}$-close to id, $f|D = \mathrm{id}$, and $f_\lambda(C_\lambda) \subset M$, Thus, M is $\tau\mathfrak{C}$-absorptive in X. □

To treat non-separable spaces, we need the following:

Proposition 5.3.10 *Let M be a $\mathfrak{C}$-absorptive set in an $\ell^2(\Gamma)$-manifold X. Suppose that $\{U_\lambda \mid \lambda \in \Lambda\}$ is a pairwise disjoint open collection in X, L is a Z_σ-set in $\bigcup_{\lambda\in\Lambda} U_\lambda$, and $L \cap U_\lambda \in \mathfrak{C}_\sigma$ for each $\lambda \in \Lambda$. Then, for every open set W in X and $\mathcal{V} \in \operatorname{cov}(W)$, there is a homeomorphism $f : X \to X$ such that f is $\mathcal{V}$-close to* id *(so $f|X \setminus W = \mathrm{id}$) and $f(L \cap W \cap U_\lambda) \subset M \cap W \cap U_\lambda$ for each $\lambda \in \Lambda$.*

Proof Due to Proposition 2.1.17, there exist $\mathcal{V}_\lambda \in \operatorname{cov}(W \cap U_\lambda)$, $\lambda \in \Lambda$, such that $\mathcal{V}_\lambda \prec \mathcal{V}$ and $\mathcal{V}_\lambda$ is fitting in X. Then,

$$\bigcup_{\lambda\in\Lambda} \mathcal{V}_\lambda \in \operatorname{cov}\left(W \cap \bigcup\nolimits_{\lambda\in\Lambda} U_\lambda\right),$$

which is fitting in X. Since L is a Z_σ-set in $\bigcup_{\lambda\in\Lambda} U_\lambda$, each $L \cap U_\lambda$ is a Z_σ-set in U_λ (Corollary 2.8.7(2)). Applying Lemma 5.3.4, we can obtain homeomorphisms $f_\lambda : W \cap U_\lambda \to W \cap U_\lambda$, $\lambda \in \Lambda$, such that f_λ is $\mathcal{V}_\lambda$-close to id and $f_\lambda(L \cap W \cap U_\lambda) \subset M \cap W \cap U_\lambda$. Since $\bigcup_{\lambda\in\Lambda} \mathcal{V}_\lambda$ is fitting in X, the desired homeomorphism $f : X \to X$ can be defined by

$$f|W \cap U_\lambda = f_\lambda,\ \lambda \in \Lambda, \text{ and } f|X \setminus \left(W \cap \bigcup\nolimits_{\lambda\in\Lambda} U_\lambda\right) = \mathrm{id}. \qquad \square$$

A **$\mathfrak{C}$-absorption base** M for X is defined as a $\mathfrak{C}$-absorptive set M in X that is a σ-discrete union of members of $\mathfrak{C}_Z(X)$, that is, $M = \bigcup_{n\in\mathbb{N}} M_n$ and each M_n is a discrete union of members of $\mathfrak{C}_Z(X)$. Then, each M_n is a Z-set in X as a discrete union of Z-sets (Corollary 2.8.7(1)), which means that $M_n \in \oplus\mathfrak{C}_Z(X)$, so $M_n \in \tau\mathfrak{C}_Z(X)$ if $w(X) \leqslant \tau$. Thus, every M is a Z_σ-set in X. Evidently, if $M \subset X$ is a $\mathfrak{C}$-absorption base for X and $f : X \to Y$ is a homeomorphism, then $f(M)$ is a $\mathfrak{C}$-absorption base for Y. It is easy to show that if each M_n is a discrete union of at most countably many members of $\mathfrak{C}_Z(X)$, then M is a countable union of members of $\mathfrak{C}_Z(X)$. In particular, when X is separable, then M is a countable union of members of $\mathfrak{C}_Z(X)$. Thus, we have the following:

Proposition 5.3.11 *Let X be a separable metrizable space and $M \subset X$. Then, M is a $\mathfrak{C}$-absorption base for X if and only if M is $\mathfrak{C}$-absorptive in X and written as a countable union of members of $\mathfrak{C}_Z(X)$.* $\square$

By the following proposition, a $\mathfrak{C}$-absorption base can be defined as a $\mathfrak{C}$-absorptive Z_σ-set that belongs to $(\tau\mathfrak{C})_\sigma$:

Proposition 5.3.12 *Let M be a $\mathfrak{C}$-absorptive set in a (separable) metrizable space X. Then, M is a $\mathfrak{C}$-absorption base for X if and only if M is a Z_σ-set in X and $M \in (\tau\mathfrak{C})_\sigma$ (or $M \in \mathfrak{C}_\sigma$).*

Proof The "only if" part is obvious. We show the "if" part. Since M is a Z_σ-set in X and $M \in (\tau\mathfrak{C})_\sigma$, we can write $M = \bigcup_{i\in\mathbb{N}} K_i = \bigcup_{n\in\mathbb{N}} M_n$, where each K_i is a Z-set in X and $M_n \in \tau\mathfrak{C}$ for each $n \in \mathbb{N}$. Then, each $K_i \cap M_n$ is a Z-set in X because it is closed in K_i. For each $n \in \mathbb{N}$, $M_n = \bigoplus_{\alpha \in A(n)} C_\alpha$, where $C_\alpha \in \mathfrak{C}$, $\alpha \in A(n)$, and $\operatorname{card} A(n) \leqslant \tau$. Then, $\{C_\alpha \mid \alpha \in A(n)\}$ is discrete in X because it is discrete in M_n and M_n is closed in X. Since each C_α is closed in M, $K_i \cap C_\alpha$ is closed in K_i, so a Z-set in X. Since $\mathfrak{C}$ is closed hereditary, it follows that $K_i \cap C_\alpha \in \mathfrak{C}_Z(X)$. Thus, M is a σ-discrete union of members of $\mathfrak{C}_Z(X)$. □

Combining the above with Corollary 5.3.9, we have:

Corollary 5.3.13 *Every $\mathfrak{C}$-absorption base M for an $\ell^2(\Gamma)$-manifold X is a $\tau\mathfrak{C}$-absorption base for X.* □

When an $\ell^2(\Gamma)$-manifold X has a $\mathfrak{C}$-absorption base, it is topologically unique, that is:

Theorem 5.3.14 (Uniqueness of Absorption Bases) *Let M and N be $\mathfrak{C}$-absorption bases for an $\ell^2(\Gamma)$-manifold X. For each open cover $\mathcal{U}$ of X, there exists a homeomorphism $f : X \to X$ such that $f(M) = N$ and f is $\mathcal{U}$-close to* id.

Proof First, let

$$M = \bigcup_{n \in \mathbb{N}} M_n, \; M_n = \bigcup_{\alpha \in A(n)} C_\alpha, \; N = \bigcup_{n \in \mathbb{N}} N_n, \text{ and } N_n = \bigcup_{\beta \in B(n)} D_\beta,$$

where $\{C_\alpha \mid \alpha \in A(n)\}, \{D_\beta \mid \beta \in B(n)\} \subset \mathfrak{C}_Z(X)$, $n \in \mathbb{N}$, are discrete in X. For each $n \in \mathbb{N}$, let $M_n^* = \bigcup_{i \leqslant n} M_i$ and $N_n^* = \bigcup_{i \leqslant n} N_i$. As discrete unions of Z-sets, $M_n = \bigcup_{\alpha \in A(n)} C_\alpha$ and $N_n = \bigcup_{\beta \in B(n)} D_\beta$ are Z-sets in X (Corollary 2.8.7(1)). Hence, M_n^* and N_n^* are also Z-sets in X.

We use a complete metric $d \in \operatorname{Metr}(X)$ such that $\{\overline{\mathrm{B}}_d(x, 1) \mid x \in U\} \prec \mathcal{U}$ (1.3.22(1)). By induction, we shall construct homeomorphisms $f_n, g_n : X \to X$ and $\mathcal{U}_n \in \operatorname{cov}(X)$, $n \in \mathbb{N}$, so as to satisfy the following conditions:

(1) $f_n(g_{n-1}^{-1} f_{n-1} \cdots g_0^{-1} f_0(M_n^*)) \subset N$,
(2) $g_n(N_n^*) \subset f_n(g_{n-1}^{-1} f_{n-1} \cdots g_0^{-1} f_0(M))$
(i.e., $N_n^* \subset g_n^{-1} f_n(g_{n-1}^{-1} f_{n-1} \cdots g_0^{-1} f_0(M))$),
(3) $f_n | g_{n-1}^{-1} f_{n-1} \cdots g_0^{-1} f_0(M_{n-1}^*) \cup N_{n-1}^* = \mathrm{id}$,
(4) $g_n | f_n g_{n-1}^{-1} f_{n-1} \cdots g_0^{-1} f_0(M_n^*) \cup N_{n-1}^* = \mathrm{id}$,
(5) $g_n^{-1} f_n$ is $\operatorname{st} \mathcal{U}_n$-close to id,
(6) $\operatorname{mesh} \operatorname{st} U_n < 2^{-n}$ and $\operatorname{mesh} f_1^{-1} g_1 \cdots f_{n-1}^{-1} g_{n-1}(\operatorname{st} \mathcal{U}_n) < 2^{-n}$,

where $f_0 = g_0 = \mathrm{id}$ and $M_0^* = N_0^* = \emptyset$. When $n = 1$, the condition (3) is nothing, whereas the conditions (1), (2), (4) and (6) mean $f_1(M_1) \subset N$, $g_1(N_1) \subset f_1(M)$, $g_1 | f_1(M_1) = \mathrm{id}$, and $\operatorname{mesh} \operatorname{st} \mathcal{U}_1 < 1/2$, respectively.

Suppose that $f_0, g_0, \ldots, f_{n-1}, g_{n-1}$ have been obtained. Then, we can take $\mathcal{U}_n \in \operatorname{cov}(X)$ so as to satisfy the condition (6). Since $\{C_\alpha \mid \alpha \in A(n)\}$ and $\{D_\beta \mid \beta \in B(n)\}$ are discrete in X, we have pairwise disjoint open collections $\{U_\alpha \mid \alpha \in A(n)\}$

and $\{V_\beta \mid \beta \in B(n)\}$ in X such that $C_\alpha \subset U_\alpha$ and $D_\beta \subset V_\beta$. For each $\alpha \in A(n)$, let

$$U_\alpha^* = U_\alpha \setminus \big(g_{n-1}^{-1} f_{n-1} \cdots g_0^{-1} f_0(M_{n-1}^*) \cup N_{n-1}^*\big).$$

Since N is a $\mathfrak{C}$-absorptive in X, we can apply Proposition 5.3.10 to obtain a homeomorphism $f_n : X \to X$ such that

$$f_n(g_{n-1}^{-1} f_{n-1} \cdots g_0^{-1} f_0(C_\alpha) \cap U_\alpha^*)) \subset N \cap U_\alpha^* \quad \text{for each } \alpha \in A(n),$$

$f_n|X \setminus \bigcup_{\alpha \in A(n)} U_\alpha^* = \mathrm{id}$, and f_n is $\mathcal{U}_n$-close to id. Then, f_n satisfies (1) and (3). Note that $f_n g_{n-1}^{-1} f_{n-1} \cdots g_0^{-1} f_0(M)$ is $\mathfrak{C}$-absorptive. For each $\beta \in B(n)$, let

$$V_\beta^* = V_\beta \setminus \big(f_n g_{n-1}^{-1} f_{n-1} \cdots g_0^{-1} f_0(M_n^*) \cup N_{n-1}^*\big).$$

Again, applying Proposition 5.3.10, we can obtain a homeomorphism $g_n : X \to X$ such that

$$g_n(D_\beta \cap V_\beta^*) \subset f_n g_{n-1}^{-1} f_{n-1} \cdots g_0^{-1} f_0(M) \cap V_\beta^* \quad \text{for each } \beta \in B(n),$$

$g_n|X \setminus \bigcup_{\beta \in B(n)} V_\beta^* = \mathrm{id}$, and g_n is $\mathcal{U}_n$-close to id. Then, g_n satisfies (2) and (4). Since f_n and g_n is $\mathcal{U}_n$-close to id, we have the condition (5).

Now, due to Lemma 5.3.3, the sequence $(g_n^{-1} f_n g_{n-1}^{-1} f_{n-1} \cdots g_0^{-1} f_0)_{n \in \mathbb{N}}$ converges to a homeomorphism $h : X \to X$ with $d(h, \mathrm{id}) < 1$. Since $\{\overline{\mathrm{B}}_d(x, 1) \mid x \in U\} \prec \mathcal{U}$, h is $\mathcal{U}$-close to id. It remains to show that $h(M) = N$. By (1), (4) and (3), we have

$$h(M_n^*) = f_n g_{n-1}^{-1} f_{n-1} \cdots g_0^{-1} f_0(M_n^*) \subset N \quad \text{for each } n \in \omega.$$

Hence, $h(M) = \bigcup_{n \in \omega} h(M_n^*) \subset N$. Moreover, by (2), (3) and (4), we have

$$\begin{aligned} N_n^* &\subset g_n^{-1} f_n g_{n-1}^{-1} f_{n-1} \cdots g_0^{-1} f_0(M) \\ &= g_m^{-1} f_m \cdots g_0^{-1} f_0(M) \quad \text{for each } m > n \in \mathbb{N}. \end{aligned}$$

Then, it follows that $N = \bigcup_{n \in \mathbb{N}} N_n^* \subset h(M)$. Consequently, we have $h(M) = N$. □

Remark 5.8 When X is a $\boldsymbol{Q}$-manifold, Theorem 5.3.14 can be proved by the same proof, that is, an $\mathfrak{M}_0$- (or $\mathfrak{M}_0^{\mathrm{fd}}$-)absorption base for a $\boldsymbol{Q}$-manifold X is topologically unique.

Using the above theorem, we can show that an $\mathfrak{M}_0$- (or $\mathfrak{M}_0^{\mathrm{fd}}$-)absorption base for an ℓ^2-manifold X is none other than an (f.d.)cap set for X, that is:

Proposition 5.3.15 *Let X be an ℓ^2- or $\boldsymbol{Q}$-manifold. Then, $M \subset X$ is an $\mathfrak{M}_0$- (or $\mathfrak{M}_0^{\mathrm{fd}}$-)absorption base for X if and only if M is an (f.d.)cap set for X.*

Proof First, we show the "if" part. As seen in Proposition 5.3.1, an (f.d.)cap set M for X is $\mathfrak{M}_0$- (or $\mathfrak{M}_0^{\mathrm{fd}}$-)absorptive in X. Since M is a countable union of (finite-dimensional) compact Z-sets in X, M is an $\mathfrak{M}_0$- (or $\mathfrak{M}_0^{\mathrm{fd}}$-)absorption base for X by Proposition 5.3.11.

The "only if" part follows from Theorem 5.3.14 with Remark 5.8. Indeed, X has an (f.d.)cap set N (Proposition 5.2.2). Due to the "if" part proved above, N is an $\mathfrak{M}_0$- (or $\mathfrak{M}_0^{\mathrm{fd}}$-)absorption base for X. If M is an $\mathfrak{M}_0$- (or $\mathfrak{M}_0^{\mathrm{fd}}$-)absorption base for X, then there is a homeomorphism $h : X \to X$ such that $f(N) = M$. Therefore, M is also an (f.d.)cap set for X. □

5.4 Absorption Bases and Strong Universality

In Sect. 3.1, we introduced two kinds of universality for a class $\mathfrak{C}$ of spaces. Recall that X is said to be **universal for** $\mathfrak{C}$ or **$\mathfrak{C}$-universal** if any map $f : Y \to X$ from $Y \in \mathfrak{C}$ can be approximated by closed embeddings. And, it is said that X is **strongly universal for** $\mathfrak{C}$ or **strongly $\mathfrak{C}$-universal** when the following condition is satisfied:

($\mathrm{SU}_{\mathfrak{C}}$) For each $C \in \mathfrak{C}$ and each closed set $D \subset C$, if $f : C \to X$ is a map such that $f|D$ is a strong Z-embedding, then, for each open cover $\mathcal{U}$ of X, there is a strong Z-embedding $h : C \to X$ such that $h|D = f|D$ and h is $\mathcal{U}$-close to f.

Here, we introduce one more universality. It is said that X is **fine $\mathfrak{C}$-universal** or **fine universal for** $\mathfrak{C}$ if any map $f : Y \to X$ from $Y \in \mathfrak{C}$ can be approximated by strong Z-embeddings. Then, the following implications hold:

$$\boxed{\text{strong } \mathfrak{C}\text{-universality}} \Longrightarrow \boxed{\text{fine } \mathfrak{C}\text{-universality}} \Longrightarrow \boxed{\mathfrak{C}\text{-universality}}$$

Recall the following results in Sects. 3.1 and 3.2:

- *For an ANR $X \in \mathfrak{M}_1(\tau)$, the $\mathfrak{M}_1(\tau)$-universality and the strong $\mathfrak{M}_1(\tau)$-universality are equivalent* (Theorem 3.1.19).
- *For an ANR $X \in \mathfrak{L}(\tau)$, the $\mathfrak{M}_0$-universality, the strong $\mathfrak{M}_0$-universality, the proper $\mathfrak{L}(\tau)$-universality, and the strong proper $\mathfrak{L}(\tau)$-universality are equivalent* (Theorem 3.2.9).

In this section, we show several properties of the strong $\mathfrak{C}$-universality available to $\mathfrak{C}$-absorption bases. In fact, the strong $\mathfrak{C}$-universality is related to the $\mathfrak{C}$-absorption property as in the following proposition:

Proposition 5.4.1 *Let X be an $\ell^2(\Gamma)$-manifold and $M \subset X$.*

(1) *If M is strongly $\mathfrak{C}(\tau)$-universal and homotopy dense in X, then M is $\mathfrak{C}$-absorptive in X.*
(2) *If $\mathfrak{M}_0^{\mathrm{fd}} \subset \mathfrak{C} \subset \mathfrak{M}_1(\tau)$ and M is $\mathfrak{C}$-absorptive in X, then M is strongly $\mathfrak{C}$-universal.*

Proof (1): Let $C \in \mathfrak{C}_Z(X)$ with $D \subset M \cap C$ a closed subset. For each $\mathcal{U} \in \operatorname{cov}(X)$, let $\mathcal{V} \in \operatorname{cov}(X)$ be a star-refinement of $\mathcal{U}$. Since M is homotopy dense in X, we have a map $f : C \to M$ such that $f|D = \mathrm{id}$ and f is $\mathcal{V}$-close to id. Then, $f(D) = D$ is a strong Z-set in M (cf. Corollary 5.3.6). Because $C \in \mathfrak{C}(\tau)$ and M is strongly $\mathfrak{C}(\tau)$-universal, f is $\mathcal{V}$-close to a strong Z-embedding $h : C \to M$ with $h|D = f|D = \mathrm{id}$. Then, h is $\mathcal{U}$-close to id. Hence, M is $\mathfrak{C}$-absorptive in X.

(2): Let $C \in \mathfrak{C}$ with $D \subset C$ a closed subset, and let $f : C \to M$ be a map such that $f|D$ is a strong Z-embedding in M.[9] Note that M is an ANR (Corollary 5.3.6). By Proposition 2.8.12, we may assume that $f(C \setminus D) \cap f(D) = \emptyset$. Since M is homotopy dense in X (Proposition 5.3.5) and $f(D)$ is a Z-set in M, it follows that $\operatorname{cl} f(D)$ is a Z-set in X. Then, $\operatorname{cl} f(D) \setminus f(D)$ is F_σ in X, i.e., $f(D)$ is G_δ in X because $f(D) \approx D \in \mathfrak{M}_1(\tau)$. Hence, $\operatorname{cl} f(D) \setminus f(D)$ is a Z_σ-set in X. By the Negligibility Theorem 2.9.9, we have $Y = X \setminus (\operatorname{cl} f(D) \setminus f(D)) \approx X$, where $f(D)$ is a Z-set in Y.[10]

For each open cover $\mathcal{U} \in \operatorname{cov}(M)$, take $\mathcal{V} \in \operatorname{cov}(M)$ such that $\operatorname{st} \mathcal{V} \prec \mathcal{U}$. For each $V \in \mathcal{V}$, let $\widetilde{V}$ be an open set in Y such that $\widetilde{V} \cap M = V$. Then, $Y_0 = \bigcup_{V \in \mathcal{V}} \widetilde{V}$ is an open set in Y, which is an $\ell^2(\Gamma)$-manifold. By the strong $\mathfrak{M}_1(\tau)$-universality of Y_0 (Theorem 3.4.2), we have a Z-embedding $g : C \to Y_0$ such that $g|D = f|D$ and g is $\widetilde{\mathcal{V}}$-close to f, where $\widetilde{\mathcal{V}} = \{\widetilde{V} \mid V \in \mathcal{V}\} \in \operatorname{cov}(Y_0)$.

Observe that $Y_0 \setminus g(D) \subset Y \setminus g(D) = X \setminus \operatorname{cl} f(D)$. Since Y_0 is open in Y, $Y_0 \setminus g(D)$ is open in X. Then, $g(C \setminus D)$ is a Z_σ-set in $Y_0 \setminus g(D)$ and $g(C \setminus D) \in \mathfrak{C}_\sigma$. By Proposition 2.1.17, we can take $\mathcal{W} \in \operatorname{cov}(Y_0 \setminus g(D))$ so that $\mathcal{W} \prec \widetilde{\mathcal{V}}$ and $\mathcal{W}$ is fitting in Y_0. Applying condition (c) of Theorem 5.3.8, we have a homeomorphism $h : Y_0 \setminus g(D) \to Y_0 \setminus g(D)$ such that h is $\mathcal{W}$-close to id and $h(g(C \setminus D)) \subset M \cap (Y_0 \setminus g(D))$. Then, h can be extended to a homeomorphism $\tilde{h} : Y_0 \to Y_0$ by $\tilde{h}|g(D) = \mathrm{id}$. Thus, we have a closed embedding $\tilde{h}g : C \to M \cap Y_0$ that is $\operatorname{st}\mathcal{V}$-close to $f|C \setminus D$, where $(\operatorname{st} \widetilde{\mathcal{V}})|M \cap Y_0 = \operatorname{st} \mathcal{V}$ because $M \cap Y_0$ is dense in Y_0. Since $\operatorname{st} \mathcal{V} \prec \mathcal{U}$, $\tilde{h}g$ is $\mathcal{U}$-close to f. Hence, M is strongly $\mathfrak{C}$-universal. □

Remark 5.9 In the above proof of (2), if $f(D)$ is closed in X, then the proof can be performed without the Negligibility Theorem 2.9.9, where $Y = X$.

As a combination of Propositions 5.4.1, 5.3.5 and 5.3.12, we have the following characterization of the $\mathfrak{C}$-absorption property:

Corollary 5.4.2 *Let X be an $\ell^2(\Gamma)$-manifold, $M \subset X$, and $\mathfrak{M}_0^{\mathrm{fd}} \subset \mathfrak{C} \subset \mathfrak{M}_1(\tau)$.*

[9] Note that $f(D)$ is not a (strong) Z-set in X unless it is closed in X.

[10] It is unknown whether M is $\mathfrak{C}$-absorptive in Y, or not.

(1) *M is $\mathfrak{C}$-absorptive in X if and only if M is strongly $\mathfrak{C}$-universal and homotopy dense in X.*
(2) *M is a $\mathfrak{C}$-absorption base for X if and only if M is a strongly $\mathfrak{C}$-universal and homotopy dense Z_σ-set in X and $M \in (\tau\mathfrak{C})_\sigma$.* □

The same proof as Proposition 5.4.1(1) is available even when X is a $\boldsymbol{Q}$-manifold. If $\mathfrak{M}_0^{\mathrm{fd}} \subset \mathfrak{C} \subset \mathfrak{M}_0$, then $f(D)$ is closed in X in the proof of Proposition 5.4.1(2). Then, as mentioned in Remark 5.9, we have the proof without the Negligibility Theorem 2.9.9, which is valid even when X is a $\boldsymbol{Q}$-manifold. Thus, we have the following $\boldsymbol{Q}$-manifold versions of Proposition 5.4.1 and Corollary 5.4.2:

Proposition 5.4.3 *Let X be a $\boldsymbol{Q}$-manifold and $M \subset X$.*

(1) *If M is strongly $\mathfrak{C}$-universal and homotopy dense in X, then M is $\mathfrak{C}$-absorptive in X.*
(2) *If $\mathfrak{M}_0^{\mathrm{fd}} \subset \mathfrak{C} \subset \mathfrak{M}_0$ and M is $\mathfrak{C}$-absorptive in X, then M is strongly $\mathfrak{C}$-universal.* □

Corollary 5.4.4 *Let X be a $\boldsymbol{Q}$-manifold, $M \subset X$, and $\mathfrak{M}_0^{\mathrm{fd}} \subset \mathfrak{C} \subset \mathfrak{M}_0$.*

(1) *M is $\mathfrak{C}$-absorptive in X if and only if M is strongly $\mathfrak{C}$-universal and homotopy dense in X.*
(2) *M is a $\mathfrak{C}$-absorption base for X if and only if M is a strongly $\mathfrak{C}$-universal and homotopy dense Z_σ-set in X and $M \in \mathfrak{C}_\sigma$.* □

Combining Corollaries 5.4.2, 5.4.4 with Proposition 5.3.15, we have:

Proposition 5.4.5 *Let X be an ℓ^2- or $\boldsymbol{Q}$-manifold and $M \subset X$. Then, M is an (f.d.)cap set for X if and only if M is a (strongly countable-dimensional) σ-compact,*[11] *strongly $\mathfrak{M}_0$- (or $\mathfrak{M}_0^{\mathrm{fd}}$-)universal homotopy dense Z_σ-set in X.* □

It is said that $\mathfrak{C}$ is **I-stable** if $C \times \mathbf{I} \in \mathfrak{C}$ for every $C \in \mathfrak{C}$, which implies that $C \times \mathbf{I}^n$ for every $C \in \mathfrak{C}$ and $n \in \mathbb{N}$. If $\mathfrak{C}$ is $\mathbf{I}$-stable, then $\mathfrak{C}$ contains all cubes $\mathbf{I}^n$, $n \in \mathbb{N}$, which implies that $\mathfrak{M}_0^{\mathrm{fd}} \subset \mathfrak{C}$. A class $\mathfrak{C}$ said to be **multiplicative** if $C_1 \times C_2 \in \mathfrak{C}$ for every $C_1, C_2 \in \mathfrak{C}$. Then, we have the following:

Fact *When $\mathfrak{C}$ is multiplicative,*

$$\mathfrak{M}_0^{\mathrm{fd}} \subset \mathfrak{C} \ \Leftrightarrow\ \mathfrak{C} \textit{ is } \mathbf{I}\textit{-stable}.$$

We regard Γ as a discrete space with $\operatorname{card} \Gamma = \tau$. The following is obvious:

Proposition 5.4.6 *If $\mathbf{I}^n \times \Gamma \in \mathfrak{C}$ for each $n \in \mathbb{N}$, then every $\mathfrak{C}$-universal space has the τ-discrete cells property. In particular, every $\tau\mathfrak{C}$-universal space has the τ-discrete cells property if $\mathfrak{C}$ is $\mathbf{I}$-stable.*

[11] A strongly countable-dimensional σ-compact metrizable space can be written as a countable union of finite-dimensional compact subsets.

We prove the following:

Proposition 5.4.7 *Suppose that $\mathfrak{C}$ is* **I***-stable. Let X and Y be ANRs such that every Z-set in $X \times Y$ is a strong Z-set. If X is strongly $\mathfrak{C}$-universal, then $X \times Y$ is strongly $\mathfrak{C}$-universal, too.*

Proof Let $f : C \to X \times Y$ be a map of $C \in C$ such that $f|D$ is a Z-embedding for a closed set $D \subset C$. Then, $f(D)$ is a strong Z-set in $X \times Y$. By Proposition 2.8.12, we may assume that $f(C \setminus D) \cap f(D) = \emptyset$ and f is closed over $f(D)$. Given $d \in \mathrm{Metr}(X)$ and $d' \in \mathrm{Metr}(Y)$, we define $\rho \in \mathrm{Metr}(X \times Y)$ as follows:

$$\rho((x, y), (x', y')) = d(x, x') + d'(y, y').$$

We inductively apply the strong $\mathfrak{C}$-universality of X to obtain strong Z-embeddings $g_n : C \to X$, $n \in \mathbb{N}$, such that

$$g_n(C) \cap g_m(C) = \emptyset \quad \text{for } n \neq m \text{ and } \quad g_n \simeq_{1/4(n+1)} \mathrm{pr}_X f,$$

where $\mathrm{pr}_X : X \times Y \to X$ is the projection. Then, we have $g_n \simeq_{1/2(n+1)} g_{n+1}$. Since $C \times [n, n+1] \in \mathfrak{C}$, we can apply the strong $\mathfrak{C}$-universality of X to obtain a map $g : C \times [1, \infty) \to X$ such that:

(1) $g|C \times [n, n+1]$ is a strong Z-embedding,
(2) $g(x, n) = g_n(x)$ for $x \in C$,
(3) $\mathrm{diam}\, g(\{x\} \times [n, n+1]) < 3/4(n+1)$ for $x \in C$,
(4) $g(C \times [n-1, n]) \cap g(C \times [n, n+1]) = g_n(C)$,
(5) $g(C \times [n, n+1]) \cap g(C \times [m, m+1]) = \emptyset$ if $|n - m| \geqslant 2$.

Now, we can define a homotopy $h : C \times \mathbf{I} \to X \times Y$ as follows:

$$h_0 = f \quad \text{and} \quad h_t(x) = (g(x, t^{-1}), \mathrm{pr}_Y f(x)) \quad \text{for } t > 0,$$

where $\mathrm{pr}_Y : X \times Y \to Y$ is the projection. Then, $\rho(f(x), h(x, t)) \leqslant t$ for every $(x, t) \in C \times \mathbf{I}$. Indeed, if $1/(n+1) < t \leqslant 1/n$, then

$$\begin{aligned} \rho(f(x), h(x, t)) &\leqslant d(f(x), g_n(x)) + \mathrm{diam}\, g(\{x\} \times [n, n+1]) \\ &< 1/4(n+1) + 3/4(n+1) = 1/(n+1) < t. \end{aligned}$$

For each map $\alpha : X \times Y \to (0, 1]$, let $\gamma : X \times Y \to \mathbf{I}$ be the map defined as follows:

$$\gamma(z) = \tfrac{1}{2} \min\{\alpha(z),\ \rho(z, f(D))\} \quad \text{for each } z \in X \times Y.$$

We define a map $f' : C \to X \times Y$ by

$$f'(x) = (\mathrm{pr}_X h(x, \gamma(f(x))),\ \mathrm{pr}_Y f(x)) \quad \text{for each } x \in C.$$

Then, $\rho(f'(x), f(x)) \leqslant \gamma(f(x)) < \alpha(f(x))$ for each $x \in C$. Since $\gamma^{-1}(0) = f(D)$, we have $f'|D = f|D$. For each $x \notin D$, $\mathrm{pr}_X f'(x) = g(x, \gamma(f(x))^{-1})$, where $\mathrm{pr}_X : X \times Y \to X$ is the projection. Hence, $\mathrm{pr}_X f'|C \setminus D$ is injective. Since $f'(C \setminus D) \cap f'(D) = \emptyset$, it follows that f' is injective.

To see that f' is closed, let $x_n \in C$, $n \in \mathbb{N}$, such that $f'(x_n) \to z \in X \times Y$ $(n \to \infty)$. When $z \in f(D)$, we have $\rho(f(x_n), z) \leqslant 2\rho(f'(x_n), z)$. Indeed,

$$\begin{aligned}\rho(f(x_n), z) &\leqslant \rho(f(x_n), f'(x_n)) + \rho(f'(x_n), z)\\ &\leqslant \gamma(f(x_n)) + \rho(f'(x_n), z)\\ &\leqslant \tfrac{1}{2}\rho(f(x_n), f(D)) + \rho(f'(x_n), z)\\ &\leqslant \tfrac{1}{2}\rho(f(x_n), z) + \rho(f'(x_n), z).\end{aligned}$$

Hence, $f(x_n) \to z$, which means $x_n \to f^{-1}(z)$ because f is closed over $f(D)$. When $z \notin f(D)$, if $\liminf_{n\in\mathbb{N}} \gamma(f(x_n)) = 0$, replacing $(x_n)_{n\in\mathbb{N}}$ by a subsequence, we can assume that $\lim_{n\to\infty} \gamma(f(x_n)) = 0$, which implies that $f(x_n) \to z$. Then, $\gamma(f(x_n)) \to \gamma(z) > 0$, which is a contradiction. Hence, $\liminf_{n\in\mathbb{N}} \gamma(f(x_n)) > 0$, then $1/(k+1) \leqslant \liminf_{n\in\mathbb{N}} \gamma(f(x_n)) < 1/k$ for some $k \in \mathbb{N}$. Replacing $(x_n)_{n\in\mathbb{N}}$ by a subsequence, we assume that $1/(k+1) \leqslant \gamma(f(x_n)) < 1/k$ for every $n \in \mathbb{N}$. Then,

$$h(x_n, \gamma(f(x_n))) = g(x_n, \gamma(f(x_n))^{-1}) \to \mathrm{pr}_Y(z)\ (n \to \infty).$$

Since $g|C \times [k, k+1]$ is a closed embedding, $(x_n)_{n\in\mathbb{N}}$ is convergent. Thus, $f' : C \to X \times Y$ is a closed embedding.

To see that $f'(C)$ is a Z-set in $X \times Y$, let $k : \boldsymbol{Q} \to X \times Y$ be a map and $\varepsilon > 0$. Since $f'(D) = f(D)$ is a Z-set in $X \times Y$, we have a map $k' : \boldsymbol{Q} \to X \times Y$ such that $k'(\boldsymbol{Q}) \cap f'(D) = \emptyset$ and $\rho(k', k) < \varepsilon/2$. Then, there is some $n_0 \in \mathbb{N}$ such that

$$x \in C,\ f'(x) \in k'(\boldsymbol{Q}) \Rightarrow \gamma(f(x)) > 1/n_0.$$

Otherwise, for each $n \in \mathbb{N}$, there is $x_n \in C$ such that $f'(x_n) \in k'(\boldsymbol{Q})$ and $\gamma(f(x_n)) \leqslant 1/n$. Then, $(x_n)_{n\in\mathbb{N}}$ has a subsequence $(x_{n_i})_{i\in\mathbb{N}}$ such that $f'(x_{n_i})$ converges to some $z_0 \in k'(\boldsymbol{Q})$. Since $\gamma(z_0) = \lim_{i\to\infty} \gamma(f(x_{n_i})) = 0$ and $\alpha(z_0) > 0$, it follows that $\rho(z_0, f(D)) = 0$, which means that $z_0 \in f(D) = f'(D)$. This contradicts the fact that $k'(\boldsymbol{Q}) \cap f'(D) = \emptyset$.

Now, let $D' = (\gamma f)^{-1}([0, 1/n_0])$. Then, $D \subset D'$ and D' is closed in C. In the above, we actually proved that $f'(D') \cap k'(\boldsymbol{Q}) = \emptyset$, where $\mathrm{dist}(k'(\boldsymbol{Q}), f'(D')) > 0$ because $k'(\boldsymbol{Q})$ is compact. Let

$$\delta = \tfrac{1}{2}\min\left\{\varepsilon,\ \mathrm{dist}(k'(\boldsymbol{Q}), f'(D'))\right\} > 0.$$

Since $g(C \times [1, n_0])$ is a strong Z-set in X, we have a map $k'' : \boldsymbol{Q} \to X$ such that $d(k'', \mathrm{pr}_X k') < \delta$ and $k''(\boldsymbol{Q}) \cap g(C \times [1, n_0]) = \emptyset$. Thus, we have the map

$k^* : \boldsymbol{Q} \to X \times Y$ defined by $k^*(x) = (k''(x), \mathrm{pr}_Y k'(x))$. Then,

$$\rho(k^*, k) \leqslant \rho(k^*, k') + \rho(k', k) = d(k'', \mathrm{pr}_X k') + \rho(k', k) < \delta + \varepsilon/2 \leqslant \varepsilon.$$

For each $x \in C$, if $\gamma(f(x)) > 1/n_0$, then

$$\mathrm{pr}_X f'(x) = g(x, \gamma(f(x))^{-1}) \in g(C \times [1, n_0]) \subset X \setminus k''(\boldsymbol{Q}),$$

which implies that $f'(x) \notin k^*(\boldsymbol{Q})$. If $\gamma(f(x)) \leqslant 1/n_0$, then $x \in D'$, so it follows that

$$\begin{aligned} \rho(f'(x), k^*(\boldsymbol{Q})) &\geqslant \rho(f'(x), k'(\boldsymbol{Q})) - \rho(k^*, k') \\ &= \rho(f'(x), k'(\boldsymbol{Q})) - d(k'', \mathrm{pr}_X k') \\ &\geqslant \mathrm{dist}(f'(D'), k'(\boldsymbol{Q})) - \delta \\ &> \mathrm{dist}(k'(\boldsymbol{Q}), f'(D'))/2 > 0, \end{aligned}$$

which means that $f'(x) \notin k^*(\boldsymbol{Q})$. Consequently, $k^*(\boldsymbol{Q}) \cap f'(C) = \emptyset$. Hence, f' is a Z-embedding. □

Concerning the product of an absorption base for an $\ell^2(\Gamma)$-manifold and a completely metrizable ANR, we have the following:

Proposition 5.4.8 *Suppose that $\mathfrak{M}_0^{\mathrm{fd}} \subset \mathfrak{C} \subset \mathfrak{M}_1(\tau)$ and $\mathfrak{C}$ is multiplicative. Let $M \in (\tau\mathfrak{C})_\sigma$ be a homotopy dense Z_σ-set in an $\ell^2(\Gamma)$-manifold X, and let $Y \in \mathfrak{C}$ be an ANR with $w(Y) \leqslant \tau$. If one of M and Y is strongly $\mathfrak{C}$-universal, then $M \times Y$ is a $\mathfrak{C}$-absorption base for $X \times Y$.*

Proof By using the Toruńczyk Factor Theorem 2.2.14, it is easy to see that $X \times Y$ is an $\ell^2(\Gamma)$-manifold. Then, every Z-set in $X \times Y$ is a strong Z-set. Since $\mathfrak{C}$ is multiplicative, $\mathfrak{M}_0^{\mathrm{fd}} \subset \mathfrak{C}$ means that $\mathfrak{C}$ is $\mathbf{I}$-stable, and also we have $M \times Y \in (\tau\mathfrak{C})_\sigma$. Since M is a homotopy dense Z_σ-set in X, it follows that $M \times Y$ is a homotopy dense Z_σ-set in $X \times Y$. Because M or Y is strongly $\mathfrak{C}$-universal, $M \times Y$ is strongly $\mathfrak{C}$-universal by Proposition 5.4.7. Applying Corollary 5.4.2(2), we have the result. □

For a non-degenerate metrizable space X, let $\mathcal{F}_X$ be the class of spaces that are homeomorphic to closed sets in X (equivalently they can be embedded into X as closed sets). The following should be remarked:

- If $\mathfrak{C} \subset \mathcal{F}_X$ and X is strongly $\mathcal{F}_X$-universal, then X is strongly $\mathfrak{C}$-universal.

Lemma 5.4.9 *For a non-degenerate metrizable space X, the following hold:*

(1) *$\mathcal{F}_X$ is topological and closed hereditary.*
(2) *If $X \approx X \times Y$ for some non-degenerate path-connected space Y, then $\mathcal{F}_X$ is $\mathbf{I}$-stable.*
(3) *If X is an AR and $X \approx X \times X$, then $\mathcal{F}_X$ is additive and multiplicative.*

(4) *If* $w(X) = \tau$ *and* $X \approx X \times Y$ *for some non-compact space* Y, *then* $\Gamma \in \mathcal{F}_X$.

Proof Item (1) follows from the definition and (2) follows from the fact that Y contains an arc (i.e., a copy of $\mathbf{I}$) by Proposition 1.1.1.

(3): Since $X \approx X \times X$, it easily follows that $\mathcal{F}_X$ is multiplicative. To see the additivity, let $C = A \cup B$, where $A, B \in \mathcal{F}_X$ are closed in C. Since X is an AR, we have maps $f, g : C \to X$ such that $f|A$ and $g|B$ are closed embeddings. Moreover, there exists an embedding $\varphi : \mathbf{I} \to X$. Let $k : C \to \mathbf{I}$ be a map with $k^{-1}(0) = A$. Then, we can define a closed embedding $h : C \to X \times X \times X$ by $h(x) = (f(x), g(x), \varphi(k(x)))$. Since $X \approx X \times X \times X$, it follows that $\mathcal{F}_X$ is additive.

(4): We may assume that $X = (X, d)$ is a metric space. Note that X is non-compact. For each $n \in \mathbb{N}$, we can use Zorn's Lemma to find a maximal 2^{-n}-discrete set X_n in X. Since $X_* = \bigcup_{n\in\mathbb{N}} X_n$ is dense in X, we have $\sup_{n\in\mathbb{N}} \operatorname{card} X_n = w(X)$ because

$$w(X) = \operatorname{dens} X \leqslant \operatorname{card} X_* = \sup_{n\in\mathbb{N}} \operatorname{card} X_n \leqslant c(X) = w(X).$$

On the other hand, since Y is a non-compact metric space, it has a countable discrete set $\{y_n \mid n \in \mathbb{N}\}$, where $y_n \neq y_m$ if $n \neq m$. Then, $Z = \bigcup_{n\in\mathbb{N}} X_n \times \{y_n\}$ is discrete in $X \times Y$ and $\operatorname{card} Z = \sup_{n\in\mathbb{N}} \operatorname{card} X_n = w(X)$. Since $\Gamma \approx Z$ and $X \times Y \approx X$, it follows that $\Gamma \in \mathcal{F}_X$. □

Given a base point $* \in X$, we define

$$X^{\mathbb{N}}_f = \left\{x \in X^{\mathbb{N}} \;\middle|\; x(n) = * \text{ except for finitely many } n \in \mathbb{N} \right\},$$

which has the base point $* = (*, *, \dots) \in X^{\mathbb{N}}_f$. Then, we have $(X^{\mathbb{N}}_f)^{\mathbb{N}}_f$, which is homeomorphic to $X^{\mathbb{N}}_f$. If X is an AR, $X^{\mathbb{N}}_f$ is also an AR. Indeed, by the Arens–Eells Embedding Theorem 1.13.2, X can be regarded as a retract of a normed linear space E, where it can be assumed that $* = 0$ by translation. Then, $X^{\mathbb{N}}_f$ is a retract of $E^{\mathbb{N}}_f$. Since $E^{\mathbb{N}}_f$ is an AE (so an AR) by the Dugundji Extension Theorem 1.13.1, it follows that $X^{\mathbb{N}}_f$ is an AR. Hence, the following is obtained from Lemma 5.4.9 above:

Proposition 5.4.10 *Let* X *be a non-degenerate AR with* $w(X) = \tau$ *(and* $* \in X$ *a base point). Then, the classes* $\mathcal{F}_{X^{\mathbb{N}}}$ *and* $\mathcal{F}_{X^{\mathbb{N}}_f}$ *are topological, closed hereditary, additive, multiplicative, and contain* $\mathbf{I}^n \times \Gamma$ *for every* $n \in \mathbb{N}$.

Proof The last assertion follows from the fact that $\mathbf{I}$ and Γ belong to the classes $\mathcal{F}_{X^{\mathbb{N}}}$ and $\mathcal{F}_{X^{\mathbb{N}}_f}$. □

Concerning the strong universality of the spaces $X^{\mathbb{N}}$ and $X^{\mathbb{N}}_f$, we have the following:

Proposition 5.4.11 *Let* X *be a non-degenerate AR (with* $* \in X$ *a base point). Then,* $X^{\mathbb{N}}$ *is strongly* $\mathcal{F}_{X^{\mathbb{N}}}$*-universal and* $X^{\mathbb{N}}_f$ *is strongly* $\mathcal{F}_{X^{\mathbb{N}}_f}$*-universal.*

Proof Because of similarity, we give a proof for the space $X_f^{\mathbb{N}}$. We may assume that X is non-degenerate. Since $X_f^{\mathbb{N}} \approx (X_f^{\mathbb{N}})_f^{\mathbb{N}}$, it suffices to show that $(X_f^{\mathbb{N}})_f^{\mathbb{N}}$ is strongly $\mathcal{F}_{X_f^{\mathbb{N}}}$-universal. For each $k \in \mathbb{N}$, let $\mathrm{pr}_k : (X_f^{\mathbb{N}})_f^{\mathbb{N}} \to X_f^{\mathbb{N}}$ be the projection onto the k-th factor. We use a metric $d \in \mathrm{Metr}((X_f^{\mathbb{N}})_f^{\mathbb{N}})$ such that $d(x, x') \leqslant 2^{-k}$ if $x(i) = x'(i)$ for every $i < k$. Let $f : C \to (X_f^{\mathbb{N}})_f^{\mathbb{N}}$ be a map of $C \in \mathcal{F}_{X_f^{\mathbb{N}}}$ such that $f|D$ is a Z-embedding for a closed set $D \subset C$. By Theorem 2.8.6, we may assume that $f(C \setminus D) \cap f(D) = \emptyset$. Observe that

$$X_f^{\mathbb{N}} \approx \{*'\} \times X_f^{\mathbb{N}} \times \{(*, *, \dots)\} \subset X \times X_f^{\mathbb{N}} \times X_f^{\mathbb{N}} \approx X_f^{\mathbb{N}},$$

hence $X_f^{\mathbb{N}}$ can be embedded into itself as a Z-set that misses $*$. Then, there is a Z-embedding $g : C \to X_f^{\mathbb{N}}$ with $* \notin g(C)$. Since $X_f^{\mathbb{N}}$ is an AR, it has an equi-connecting map $\lambda : X_f^{\mathbb{N}} \times X_f^{\mathbb{N}} \times \mathbf{I} \to X_f^{\mathbb{N}}$, that is, $\lambda(x, y, 0) = x$, $\lambda(x, y, 1) = y$ and $\lambda(x, x, t) = x$ for every $t \in \mathbf{I}$. We define a homotopy $h : C \times \mathbf{I} \to (X_f^{\mathbb{N}})_f^{\mathbb{N}}$ as follows: $h_0 = f$, and for $2^{-k} \leqslant t \leqslant 2^{-k+1}$, $k \in \mathbb{N}$,

$$\begin{aligned} h(x, t) = (&f_1(x), \dots, f_{k-1}(x), \lambda(f_k(x), g(x), 2^k t - 1), \\ &g(x), g(x), \lambda(g(x), *, 2^k t - 1), *, *, \dots), \end{aligned}$$

where $f_k = \mathrm{pr}_k f : C \to X_f^{\mathbb{N}}$. Then, $d(f(x), h(x, t)) \leqslant t$ for every $(x, t) \in C \times \mathbf{I}$. Indeed, if $2^{-k} < t \leqslant 2^{-k+1}$ then $d(f(x), h(x, t)) \leqslant 2^{-k} < t$. Observe

$$\mathrm{pr}_{k+1} h|C \times [2^{-k-1}, 2^{-k+1}] = g\,\mathrm{pr}_C,$$

where $\mathrm{pr}_C : C \times \mathbf{I} \to C$ is the projection. Moreover,

$$\begin{aligned} &h(C \times [2^{-k}, 1]) \cap h(C \times [2^{-k-2}, 2^{-k-1}]) \\ &\quad \subset (X_f^{\mathbb{N}})^{k+3} \times \{*\} \times \{*\} \times \{(*, *, \dots)\} \\ &\qquad \cap (X_f^{\mathbb{N}})^{k+3} \times g(C) \times X_f^{\mathbb{N}} \times \{(*, *, \dots)\} = \emptyset. \end{aligned}$$

For each map $\alpha : (X_f^{\mathbb{N}})_f^{\mathbb{N}} \to (0, 1]$, let $\gamma : (X_f^{\mathbb{N}})_f^{\mathbb{N}} \to \mathbf{I}$ be the map defined by

$$\gamma(y) = \tfrac{1}{2} \min\{\alpha(y),\ d(y, f(D))\} \ \text{ for each } y \in (X_f^{\mathbb{N}})_f^{\mathbb{N}}.$$

We define a map $f' : C \to (X_f^{\mathbb{N}})_f^{\mathbb{N}}$ by

$$f'(x) = h(x, \gamma(f(x))) \ \text{ for each } \in C.$$

Then, $f'|D = f|D$ and $d(f'(x), f(x)) \leqslant \gamma(f(x)) < \alpha(f(x))$ for every $x \in C$. Since $f'(C \setminus D) \cap f'(D) = \emptyset$ and $f'|C \setminus D$ is injective, it follows that f' is injective.

To see that f' is closed, let $x_n \in C$, $n \in \mathbb{N}$, such that $f'(x_n) \to y \in (X_f^{\mathbb{N}})_f^{\mathbb{N}}$ $(n \to \infty)$. When $y \in f(D)$, we have

$$\begin{aligned} d(f(x_n), f'(x_n)) &\leqslant \gamma(f(x_n)) \leqslant \tfrac{1}{2} d(f(x_n), f(D)) \\ &\leqslant d(f(x_n), f(D)) - d(f(x_n), f'(x_n)) \\ &\leqslant d(f(x_n'), f(D)) \to 0, \end{aligned}$$

hence $f(x_n) \to y$. Since f is closed over $f(D)$, we have $x_n \to f^{-1}(y)$. When $y \notin f(D)$, if $\liminf_{n\in\mathbb{N}} \gamma(f(x_n)) = 0$, replacing $(x_n)_{n\in\mathbb{N}}$ by a subsequence, we assume that $\lim_{n\to\infty} \gamma(f(x_n)) = 0$, which implies that $f(x_n) \to y$. Then, $\gamma(f(x_n)) \to \gamma(y) > 0$, which is a contradiction. Hence, $\liminf_{n\in\mathbb{N}} \gamma(f(x_n)) > 0$. Choose $k \in \mathbb{N}$ so that $2^{-k} \leqslant \liminf_{n\in\mathbb{N}} \gamma(f(x_n)) < 2^{-k+1}$. Replacing $(x_n)_{n\in\mathbb{N}}$ by a subsequence, we assume that $2^{-k} \leqslant \gamma(f(x_n)) < 2^{-k+1}$ for every $n \in \mathbb{N}$. Then, $g(x_n) \to y(k+1)$ $(n \to \infty)$. Since g is a closed embedding, $(x_n)_{n\in\mathbb{N}}$ is convergent. Thus, $f' : C \to (X_f^{\mathbb{N}})_f^{\mathbb{N}}$ is a closed embedding.

Since $g(X)$ is a Z-set in $X_f^{\mathbb{N}}$ and $* \in X_f^{\mathbb{N}} \setminus g(X)$, there are maps of $(X_f^{\mathbb{N}})_f^{\mathbb{N}}$ into $(X_f^{\mathbb{N}} \setminus g(X))_f^{\mathbb{N}}$ arbitrarily close to id. Therefore, $f'(X)$ is a Z-set in $(X_f^{\mathbb{N}})_f^{\mathbb{N}}$, that is, f' is a Z-embedding. □

As a corollary of the above proposition, we have the following:

Corollary 5.4.12 *Let X be a non-degenerate AR (with $* \in X$ a base point). Then, $X^{\mathbb{N}}$ (resp. $X_f^{\mathbb{N}}$) is strongly $\mathfrak{C}$-universal for any class $\mathfrak{C} \subset \mathcal{F}_{X^{\mathbb{N}}}$ (resp. $\mathfrak{C} \subset \mathcal{F}_{X_f^{\mathbb{N}}}$).* □

Lemma 5.4.13 *Assume that $\mathfrak{C}$ is multiplicative.*

(1) *If $X, Y \in (\tau\mathfrak{C})_\sigma$, then $X \times Y \in (\tau\mathfrak{C})_\sigma$.*
(2) *If $X = (X, *) \in (\tau\mathfrak{C})_\sigma$, then $X_f^{\mathbb{N}} \in (\tau\mathfrak{C})_\sigma$.*

Proof (1): We can write $X = \bigcup_{n\in\mathbb{N}} X_n$ and $Y = \bigcup_{n\in\mathbb{N}} Y_n$, where each X_n and Y_n is closed in X and Y respectively, $X_n = \bigoplus_{\alpha\in A(n)} C_\alpha$, $Y_n = \bigoplus_{\beta\in B(n)} D_\beta$, $C_\alpha, D_\beta \in \mathfrak{C}$, and $\operatorname{card} A(n), \operatorname{card} B(n) \leqslant \tau$. Then,

$$X \times Y = \bigcup_{(n,m)\in\mathbb{N}^2} X_n \times Y_m \text{ and}$$

$$X_n \times Y_m = \bigoplus_{(\alpha,\beta)\in A(n)\times B(m)} C_\alpha \times D_\beta \text{ for each } (n, m) \in \mathbb{N}^2,$$

where $X_n \times Y_m$ is closed in $X \times Y$, $C_\alpha \times D_\beta \in \mathfrak{C}$, and $\operatorname{card} A(n) \times B(m) \leqslant \tau$. Hence, $X \times Y \in (\tau\mathfrak{C})_\sigma$.

(2): Let $X = \bigcup_{n\in\mathbb{N}} X_n$, where each X_n is closed in X, $X_n = \bigoplus_{\alpha\in A(n)} C_\alpha$, $C_\alpha \in \mathfrak{C}$, and $\operatorname{card} A(n) \leqslant \tau$. Identifying $X^k = \{x \in X^{\mathbb{N}} \mid x(i) = * \text{ for } i > k\}$, we have $X_f^{\mathbb{N}} = \bigcup_{k\in\mathbb{N}} X^k$. Then, as is easily observed, we have

$$X_f^{\mathbb{N}} = \bigcup_{k\in\mathbb{N}} \bigcup_{\nu\in\mathbb{N}^k} \prod_{i=1}^{k} X_{\nu(i)} \text{ and}$$

$$\prod_{i=1}^{k} X_{\nu(i)} = \bigoplus_{\alpha\in\prod_{i=1}^{k} A(\nu(i))} \prod_{i=1}^{k} C_{\alpha(i)} \text{ for each } \nu \in \mathbb{N}^k, k \in \mathbb{N},$$

where $\prod_{i=1}^{k} X_{\nu(i)}$ is closed in $X_f^{\mathbb{N}}$, $\prod_{i=1}^{k} C_{\alpha(i)} \in \mathfrak{C}$, and $\operatorname{card} \prod_{i=1}^{k} A(\nu(i)) \leqslant \tau$. Hence, $X_f^{\mathbb{N}} \in (\tau\mathfrak{C})_\sigma$. □

If M and N are homotopy dense in X and Y respectively, then $M \times N$ is homotopy dense in $X \times Y$. If A is a Z-set (resp. a Z_σ-set) in an ANR X and B is a closed set (resp. an F_σ set) in Y, then $A \times B$ is a Z-set (resp. a Z_σ-set) in $X \times Y$. By combining Corollary 5.4.2(2), Proposition 5.4.7, and Lemma 5.4.13(1), we have the following generalization of Proposition 5.4.8:

Proposition 5.4.14 *Suppose that $\mathfrak{M}_0^{\mathrm{fd}} \subset \mathfrak{C} \subset \mathfrak{M}_1(\tau)$ and $\mathfrak{C}$ is multiplicative. Let X be an $\ell^2(\Gamma)$-manifold and $Y \in \mathfrak{M}_1(\tau)$ be an ANR. Let M and N be homotopy dense F_σ sets in X and Y respectively and $M, N \in (\tau\mathfrak{C})_\sigma$. If M (or N) is a $\mathfrak{C}$-absorption base for X (or Y),*[12] *then $M \times N$ is a $\mathfrak{C}$-absorption base for $X \times Y$.* □

Concerning simplicial complexes, we have the following:

Proposition 5.4.15 *For an arbitrary simplicial complex K with $\operatorname{card} K \leqslant \tau$, $|K|_{\mathrm{m}} \in (\tau\mathfrak{M}_0^{\mathrm{fd}})_\sigma$. Actually, $|K|_{\mathrm{m}} = \bigcup_{i\in\mathbb{N}} D_i$, where each D_i is a discrete union of n_i-cells.*

Proof For each n-simplex $\sigma \in K$ and $k \in \mathbb{N}$, let

$$\sigma_k = \left\{x \in \sigma \,\middle|\, \rho_K(x, \partial\sigma) \geqslant (n+k)^{-1}\right\} \subset \operatorname{rint}\sigma.$$

Then, $\sigma_k \in \mathfrak{M}_0^{\mathrm{fd}}$ and $\operatorname{rint}\sigma = \bigcup_{k\in\mathbb{N}} \sigma_k$. For each $k \in \mathbb{N}$, $\{\sigma_k \mid \sigma \in K(n)\}$ is discrete in $|K|_{\mathrm{m}}$, where $K(n) = K^{(n)} \setminus K^{(n-1)}$ is the set of all n-simplexes of K. Hence, $M_{n,k} = \bigcup_{\sigma\in K(n)} \sigma_k \in \tau\mathfrak{M}_0^{(n)} = (\tau\mathfrak{M}_0)^{(n)}$ is closed in $|K|_{\mathrm{m}}$. Thus, we have $|K|_{\mathrm{m}} = \bigcup_{n,k\in\mathbb{N}} M_{n,k} \in ((\tau\mathfrak{M}_0)^{\mathrm{fd}})_\sigma$. □

Remark 5.10 Evidently, $(\tau\mathfrak{M}_0)^{\mathrm{fd}} \subset \tau\mathfrak{M}_0^{\mathrm{fd}}$. But the equality does not hold. In fact, $\bigoplus_{n\in\mathbb{N}} \mathbf{I}^n \in \aleph_0\mathfrak{M}_0^{\mathrm{fd}} \setminus (\aleph_0\mathfrak{M}_0)^{\mathrm{fd}}$. Nevertheless, $(\tau\mathfrak{M}_0^{\mathrm{fd}})_\sigma = ((\tau\mathfrak{M}_0)^{\mathrm{fd}})_\sigma = (\tau\mathfrak{M}_0)_\sigma^{\mathrm{scd}}$.[13] Indeed, $((\tau\mathfrak{M}_0)^{\mathrm{fd}})_\sigma \subset (\tau\mathfrak{M}_0^{\mathrm{fd}})_\sigma$ is trivial. Each $X \in (\tau\mathfrak{M}_0^{\mathrm{fd}})_\sigma$ can

[12] Or, M is strongly $\mathfrak{C}$-universal and N is a Z_σ-set in Y, or otherwise N is strongly $\mathfrak{C}$-universal and M is a Z_σ-set in X.

[13] For the notation $\mathfrak{C}_\sigma^{\mathrm{scd}}$, see p. 377.

be written as $X = \bigcup_{i \in \mathbb{N}} X_i$, where $X_i \in \tau\mathfrak{M}_0^{\mathrm{fd}}$. Each X_i can be written as $X_i = \bigoplus_{\lambda \in \Lambda_i} X_{i,\lambda}$, where $\dim X_{i,\lambda} < \infty$ and $\operatorname{card} \Lambda_i \leqslant \tau$. Let $\Lambda_{i,n} = \{\lambda \in \Lambda_i \mid \dim X_{i,\lambda} = n\}$. Then, we have $X_{i,n} = \bigoplus_{\lambda \in \Lambda_{i,n}} X_{i,\lambda} \in (\tau\mathfrak{M}_0)^{(n)}$ by the Locally Finite Sum Theorem 1.12.9. Hence, $X = \bigcup_{i,n \in \mathbb{N}} X_{i,n} \in ((\tau\mathfrak{M}_0)^{\mathrm{fd}})_\sigma$.

For a locally finite-dimensional simplicial complex K with $\operatorname{card} K \leqslant \tau$, $|K|_{\mathrm{m}}$ is a completely metrizable ANR (cf. Theorems 1.7.2, 1.13.4) and $|K|_{\mathrm{m}} \in (\tau\mathfrak{M}_0^{\mathrm{fd}})_\sigma$ as above. Thus, we have the following corollary:

Corollary 5.4.16 *Suppose that $\mathfrak{M}_0^{\mathrm{fd}} \subset \mathfrak{C} \subset \mathfrak{M}_1(\tau)$ and $\mathfrak{C}$ is multiplicative. Let M be a $\mathfrak{C}$-absorption base for an $\ell^2(\Gamma)$-manifold X. Then, for each locally finite-dimensional simplicial complex K with* card $K \leqslant \tau$, *$|K|_{\mathrm{m}} \times X$ is an $\ell^2(\Gamma)$-manifold and $|K|_{\mathrm{m}} \times M$ is a $\mathfrak{C}$-absorption base for $|K|_{\mathrm{m}} \times X$.* □

Remark 5.11 In the above, it suffices to assume that $\mathfrak{C}$ is **I**-stable instead of assumption that $\mathfrak{C}$ is multiplicative. Indeed, $|K|_{\mathrm{m}} = \bigcup_{i \in \mathbb{N}} D_i$, where each D_i is a discrete union of n_i-cells. If $\mathfrak{C}$ is **I**-stable, then $B \times C \in \mathfrak{C}$ for each cell B and $C \in \mathfrak{C}$, which implies $|K|_{\mathrm{m}} \times M \in (\tau\mathfrak{C})_\sigma$.

Moreover, we have the following:

Proposition 5.4.17 *Let X be a completely metrizable AR with $* \in X$ a base point and $w(X) = \tau$. Suppose that $\mathfrak{C}$ is multiplicative. If M is a homotopy dense F_σ set in X, $* \in M \in (\tau\mathfrak{C})_\sigma$, and $\mathfrak{C} \subset \mathcal{F}_{M_f^{\mathbb{N}}}$, then $M_f^{\mathbb{N}}$ is a $\mathfrak{C}$-absorption base for $X^{\mathbb{N}}$ $(\approx \ell^2(\Gamma))$.*

Proof Note that $X^{\mathbb{N}} \approx \ell^2(\Gamma)$ by Theorem 3.5.15. Since M is homotopy dense in X, $M = (M, *)$ is also a pointed non-degenerate AR and $M_f^{\mathbb{N}}$ is homotopy dense in $X^{\mathbb{N}}$ (Proposition 1.13.27). Identifying $X^n = X^n \times \{(*, *, \cdots)\} \subset X^{\mathbb{N}}$, X^n is a Z-set in $X^{\mathbb{N}}$ (e.g., it is easy to check condition (b) of Theorem 2.8.6). Hence, $X_f^{\mathbb{N}}$ is a Z_σ-set in $X^{\mathbb{N}}$. Since $M_f^{\mathbb{N}}$ is F_σ in $X_f^{\mathbb{N}}$, it easily follows that $M_f^{\mathbb{N}}$ is a Z_σ-set in $X^{\mathbb{N}}$. By Lemma 5.4.13(2), $M_f^{\mathbb{N}} \in (\tau\mathfrak{C})_\sigma$. Moreover, due to Corollary 5.4.12, the strong $\mathfrak{C}$-universality of $M_f^{\mathbb{N}}$ comes from $\mathfrak{C} \subset \mathcal{F}_{M_f^{\mathbb{N}}}$. Hence, $M_f^{\mathbb{N}}$ is $\mathfrak{C}$-absorptive by Proposition 5.4.1(1). Thus, it follows from Corollary 5.4.2(2) that $M_f^{\mathbb{N}}$ is a $\mathfrak{C}$-absorption base for $X^{\mathbb{N}}$. □

We will show that the strong $\mathfrak{C}$-universality is open hereditary, that is,

Proposition 5.4.18 *Let X be an ANR. If X is strongly $\mathfrak{C}$-universal, then every open set in X is strongly $\mathfrak{C}$-universal.*

Proof Let U be an open set in X and $f : C \to U$ be a map of $C \in \mathfrak{C}$ such that $f|D$ is a strong Z-embedding for some closed set D in C. Take $d \in \operatorname{Metr}(X)$ and define $U_n = \{x \in U \mid d(x, X \setminus U) \geqslant 2^{-n}\}$ for each $n \in \mathbb{N}$. Then, each U_n is closed in U, $U_n \subset \operatorname{int} U_{n+1}$ and $U = \bigcup_{n\in\mathbb{N}} U_n$. For each $\mathcal{V} \in \operatorname{cov}(U)$, inductively choose $\mathcal{V}_n \in \operatorname{cov}(X)$, $n \in \mathbb{N}$, so that

$$\operatorname{mesh} \mathcal{V}_n < 2^{-n}, \quad \operatorname{st} \mathcal{V}_n \prec \mathcal{V}_{n-1} \quad \text{and} \quad \operatorname{st}(U_{n+1}, \mathcal{V}_n) \subset \operatorname{int} U_{n+2},$$

where $\mathcal{V}_0 = \mathcal{V} \cup \{X \setminus U_1\}$. Let $A_n = f^{-1}(U_n)$ and $B_n = f^{-1}(U \setminus \operatorname{int} U_{n+1})$. Then, each A_n and B_n are disjoint closed sets in C,

$$A_n \subset \operatorname{int} A_{n+1}, \quad \operatorname{int} A_{n+1} \cup B_n = C \quad \text{and} \quad C = \bigcup_{n\in\mathbb{N}} A_n.$$

We shall inductively construct maps $f_n : C \to U$, $n \in \mathbb{N}$, satisfying the following conditions:

(1) $f_n|A_{n-1} \cup B_n \cup D = f_{n-1}|A_{n-1} \cup B_n \cup D$,
(2) $f_n|A_n \cup D : A_n \cup D \to U$ is a Z-embedding,
(3) f_n is $\mathcal{V}_n$-close to f_{n-1}, and
(4) $f_n(A_{n+2}) \subset U_{n+2}$,

where $A_0 = \emptyset$, $B_0 = C$ and $f_0 = f$.

Assume that f_{n-1} has been obtained. Since X is a strongly $\mathfrak{C}$-universal ANR, we have a $\mathcal{V}_n$-homotopy $h : C \times \mathbf{I} \to X$ such that $h_0 = f_{n-1}$, $h_1 : C \to X$ is a Z-embedding,

$$h_1|A_{n-1} \cup (D \setminus \operatorname{int} B_n) = f_{n-1}|A_{n-1} \cup (D \setminus \operatorname{int} B_n) \quad \text{and}$$
$$h_t|D \setminus \operatorname{int} B_n = f|D \setminus \operatorname{int} B_n \quad \text{for every } t \in \mathbf{I}.$$

Taking an Urysohn map $\lambda : C \to \mathbf{I}$ with $\lambda(B_n) = 0$ and $\lambda(A_n) = 1$, we define $f_n : C \to X$ by $f_n(x) = h(x, \lambda(x))$. Then, f_n satisfies the conditions (1)–(3) by definition. To see (4), let $x \in A_{n+2}$. If $x \in A_{n+1}$ then $f_{n-1}(x) \in U_{n+1}$, hence $f_n(x) \in \operatorname{st}(U_{n+1}, \mathcal{V}_n) \subset U_{n+1}$. If $x \in A_{n+2} \setminus A_{n+1} \subset B_n$ then $f_n(x) = f(x) \in U_{n+2}$. Thus, f_n satisfies (4), and then $f_n(C) \subset U$.

Since $C = \bigcup_{n\in\mathbb{N}} \operatorname{int} A_n$, we can define a map $g : C \to U$ by $g|A_n = f_n|A_n$. Then, $g|D = f|D$ by (1). It is easy to see that g is $\mathcal{V}$-close to f. By (1) and (2), g is injective. For each closed set F in C, observe that

$$g(F) = \bigcup_{n\in\mathbb{N}} g(F \cap (A_n \setminus \operatorname{int} A_{n-1})) = \bigcup_{n\in\mathbb{N}} f_n(F \cap (A_n \setminus \operatorname{int} A_{n-1})),$$

which is a locally finite union of closed sets in U, hence $g(F)$ is closed in U. Thus, g is a closed embedding. Since $g(C) = \bigcup_{n\in\mathbb{N}} f_n(A_n \setminus \operatorname{int} A_{n-1})$ is a locally finite union of Z-sets in U, it follows that $g(C)$ is a Z-set in U. Hence, g is a Z-embedding. □

Using the above proposition, we can prove the following:

Theorem 5.4.19 *If M is a $\mathfrak{C}$-absorption base for an $\ell^2(\Gamma)$-manifold X, then $M \cap U$ is a $\mathfrak{C}$-absorption base for every open set U in X.*

Proof Since M is homotopy dense in X, M is an ANR (Theorem 1.13.26) and $M \cap U$ is homotopy dense in U (Corollary 1.15.6). Since M is extensively $\mathfrak{C}$-universal, so is $M \cap U$ by Proposition 5.4.18. Hence, $M \cap U$ is $\mathfrak{C}$-absorptive by Proposition 5.4.1(1). We can write $M = \bigcup_{n \in \mathbb{N}} M_n$, where each M_n is a discrete union of members of $\mathfrak{C}_Z(X)$, Because U is F_σ in X, we can write $U = \bigcup_{n \in \mathbb{N}} A_n$, where each A_n is closed in X. Since $\mathfrak{C}$ is closed hereditary, it follows that $M_n \cap A_m$ is a discrete union of members of $\mathfrak{C}_Z(U)$, and then $M \cap U = \bigcup_{n,m \in \mathbb{N}} (M_n \cap A_m)$. Thus, $M \cap U$ is a $\mathfrak{C}$-absorption base for U. □

Combining the above theorem with the Open Embedding Theorem 2.5.10, we have the following:

Corollary 5.4.20 *Suppose that $\ell^2(\Gamma)$ has a $\mathfrak{C}$-absorption base. Then, every $\ell^2(\Gamma)$-manifold has a $\mathfrak{C}$-absorption base.* □

At the end of this section, we shall show the following:

Proposition 5.4.21 *Suppose that $\mathfrak{M}_0^{\mathrm{fd}} \subset \mathfrak{C} \subset \mathfrak{M}_1(\tau)$. If M is a $\mathfrak{C}$-absorption base for an $\ell^2(\Gamma)$-manifold X, then M is strongly $\tau\mathfrak{C}$-universal.*

Proof First, note that every Z-set in M is a strong Z-set (Corollary 5.3.6). Let $C = \bigoplus_{\lambda \in \Lambda} C_\lambda \in \tau\mathfrak{C}$, where $C_\lambda \in \mathfrak{C}$ for each $\lambda \in \Lambda$ and card $\Lambda \leqslant \tau$. Let $f : C \to M$ be a map such that $f|D$ is a Z-embedding, where D is a closed set in C. For each $\mathcal{U} \in \mathrm{cov}(M)$, let $\mathcal{V} \in \mathrm{cov}(M)$ be a star-refinement of $\mathcal{U}$. Then, X has an open collection $\widetilde{\mathcal{V}}$ such that $\mathcal{V} = \widetilde{\mathcal{V}}|M$. Due to Proposition 5.4.19, M is a $\mathfrak{C}$-absorption base for $\bigcup \widetilde{\mathcal{V}}$. Hence, replacing X with the open set $\bigcup \widetilde{\mathcal{V}}$, we can regard $\widetilde{\mathcal{V}} \in \mathrm{cov}(X)$.

Since $\mathfrak{C} \subset \mathfrak{M}_1(\tau)$, it follows that $\tau\mathfrak{C} \subset \mathfrak{M}_1(\tau)$. By the strongly $\mathfrak{M}_1(\tau)$-universality of X, we have a strong Z-embedding $g : C \to X$ such that $g|D = f|D$ and g is $\widetilde{\mathcal{V}}$-close to f. Then, we have a discrete open collection $\{W_\lambda \mid \lambda \in \Lambda\}$ in X such that $h(C_\lambda) \subset W_\lambda$ for each $\lambda \in \Lambda$. Due to Theorem 5.3.8, every $M \cap W_\lambda$ is $\mathfrak{C}$-absorptive in W_λ. Hence, we have Z-embeddings $h_\lambda : C_\lambda \to M \cap W_\lambda$, $\lambda \in \Lambda$, such that $h_\lambda|D_\lambda = \mathrm{id}$ and h_λ is $\widetilde{\mathcal{V}}$-close to id, where $D_\lambda = D \cap C_\lambda$ for each $\lambda \in \Lambda$. We define a Z-embedding $h : C \to \bigcup_{\lambda \in \Lambda} W_\lambda$ by $h|C_\lambda = h_\lambda$, $\lambda \in \Lambda$. Then, $hg : C \to M$ is a Z-embedding such that $hg|D = f|D$ and hg is $\mathcal{U}$-close to f, where $\mathrm{st}\,\widetilde{\mathcal{V}}|M = \mathrm{st}\,\mathcal{V} \prec \mathcal{U}$. Consequently, M is strongly $\tau\mathfrak{C}$-universal. □

5.5 Four Types of Absorption Bases for $\ell^2(\Gamma)$

In this section, we discuss four types of model spaces being absorption bases for Hilbert space $\ell^2(\Gamma)$, that is, an $\mathfrak{M}_0$-absorption base, an $\mathfrak{M}_0^{\mathrm{fd}}$-absorption base, an $\mathfrak{M}_1(\aleph_0)$-absorption base, and an $\mathfrak{M}_1(\tau)$-absorption base for $\ell^2(\Gamma)$.

Theorem 5.5.1 *The subspace $\ell^2_f(\Gamma) \subset \ell^2(\Gamma)$ is an $\mathfrak{M}_0^{\mathrm{fd}}$-absorption base for $\ell^2(\Gamma)$, that is, an $\mathfrak{M}_0^{\mathrm{fd}}$-absorptive set which is a σ-discrete union of members of $\mathfrak{M}_0^{\mathrm{fd}}(\ell^2(\Gamma))$.*

Proof By virtue of Theorem 1.2.3, $(\ell^2(\Gamma), \ell^2_f(\Gamma))$ can be replaced with $(\ell^1(\Gamma), \ell^1_f(\Gamma))$. This replacement makes calculation of distance easy.

(*The $\mathfrak{M}_0^{\mathrm{fd}}$-Absorption Property*) Let $C \in \mathfrak{M}_0^{\mathrm{fd}}(\ell^1(\Gamma))$ and U be an open set in $\ell^1(\Gamma)$ with $\mathcal{V} \in \mathrm{cov}(U)$. We may assume that $\mathcal{V}$ is fitting in $\ell^1(\Gamma)$. It suffices to construct an injective map $h : C \cap U \to U \cap \ell^1_f(\Gamma)$ that is $\mathcal{V}$-close to id. Indeed, since $\mathcal{V}$ is fitting in $\ell^1(\Gamma)$, h extends a map $\bar{h} : C \to U$ by $\bar{h}|C \setminus U = \mathrm{id}$, that is a closed embedding because $\bar{h}$ is injective and C is compact. Since a compact set in $\ell^1(\Gamma)$ is a Z-set in $\ell^1(\Gamma)$ by Proposition 3.2.10 or Corollary 3.3.14, C and $\bar{h}(C)$ are Z-sets in $\ell^1(\Gamma)$, hence $C \cap U$ and $h(C \cap U) = \bar{h}(C) \cap U$ are Z-sets in U.

Now, we shall construct h. First, we write $C \setminus U = \bigcup_{i \in \mathbb{N}} C_i$ where each C_i is compact and $C_i \subset \mathrm{int}_C\, C_{i+1}$. We can choose $\delta_1 > \delta_2 > \cdots > 0$ so that

$$\left\{\mathrm{B}(x, \delta_i) \;\middle|\; x \in C_i \setminus \mathrm{int}\, C_{i-1}\right\} \prec \mathcal{V} \quad \text{for every } i \in \mathbb{N},$$

where $C_0 = \emptyset$. Indeed, for each $x \in C_i \setminus \mathrm{int}\, C_{i-1}$, choose $\delta_x > 0$ so that $\mathrm{B}(x, \delta_x)$ is contained in some $V_x \in \mathcal{V}$. Since $C_i \setminus \mathrm{int}\, C_{i-1}$ is compact, $C_i \setminus \mathrm{int}\, C_{i-1} \subset \bigcup_{j=1}^{m} \mathrm{B}(x_j, \delta_{x_j}/2)$ for some $x_1, \ldots, x_m \in C_i \setminus \mathrm{int}\, C_{i-1}$. Then, the desired $\delta_i > 0$ can be chosen as follows:

$$0 < \delta_i < \min\left\{\delta_{i-1},\ \delta_{x_1}/2, \ldots, \delta_{x_m}/2\right\}.$$

For each $x \in C_i \setminus \mathrm{int}\, C_{i-1}$, choose $F_x \in \Gamma$ so that $\sum_{\gamma \in \Gamma \setminus F_x} |x(\gamma)| < \delta_i/4$. Since $C_i \setminus \mathrm{int}\, C_{i-1}$ is compact, we have a finite set $A_i \subset C_i \setminus \mathrm{int}\, C_{i-1}$ such that $C_i \setminus \mathrm{int}\, C_{i-1} \subset \bigcup_{a \in A_i} \mathrm{B}(a, \delta_i/4)$. Let $F_i = \bigcup_{a \in A_i} F_a$. Then, the following holds:

$$\sum_{\gamma \in \Gamma \setminus F_i} |x(\gamma)| < \delta_i/2 \quad \text{for every } x \in C_i \setminus \mathrm{int}\, C_{i-1}.$$

Indeed, x is contained in $\mathrm{B}(a, \delta_i/4)$ for some $a \in A_i$, whence it follows that

$$\begin{aligned}
\sum_{\gamma \in \Gamma \setminus F_i} |x(\gamma)| &\leqslant \sum_{\gamma \in \Gamma \setminus F_i} |x(\gamma) - a(\gamma)| + \sum_{\gamma \in \Gamma \setminus F_i} |a(\gamma)| \\
&\leqslant \|x - a\| + \sum_{\gamma \in \Gamma \setminus F_a} |a(\gamma)| \\
&< \delta_i/4 + \delta_i/4 = \delta_i/2.
\end{aligned}$$

Let $\Gamma_0 = F_1$, $\Lambda_0 = \emptyset$ and take $\Lambda_i, \Gamma_{i+1} \in \mathrm{Fin}(\Gamma)$, $i \in \mathbb{N}$, as follows:

$$\Gamma_i = \Gamma_{i-1} \cup \Lambda_{i-1} \cup F_{i+1},\ \ \Lambda_i \cap \Gamma_i = \emptyset \ \text{ and } \operatorname{card} \Lambda_i = 2 \dim C + 2.$$

For each $i \in \Lambda$, choose any $\lambda_i \in \Lambda_i$. By Embedding Theorem 1.12.14, we have embeddings $h_i : C \to \mathbf{I}^{\Lambda_i}$ such that $h_i(x)(\lambda_i) = 1$ for every $x \in C$. For each $i \in \mathbb{N}$, we have a map $\varphi_i : C \to \mathbf{I}$ with $\varphi_i^{-1}(1) = C \setminus \operatorname{int}_C C_i$ and $\varphi_i^{-1}(0) = C_{i-1}$. Let $\varepsilon_i = \delta_{i+1}/6(2\dim C + 2)$. We define $h : C \setminus U \to \ell_f^1(\Gamma)$ as follows:

$$h(x)(\gamma) = \begin{cases} \varphi_i(x)x(\gamma) & \text{if } \gamma \in \Gamma_i \setminus \big(\Gamma_{i-1} \cup \Lambda_{i-1}\big), \\ \varepsilon_i(1 - \varphi_{i+1}(x))\varphi_{i-1}(x)h_i(x)(\gamma) & \\ \qquad\qquad + \varphi_{i+1}(x)x(\gamma) & \text{if } \gamma \in \Lambda_i, \\ 0 & \text{if } \gamma \in \Gamma \setminus \bigcup_{i\in\mathbb{N}} \Gamma_i. \end{cases}$$

For each $x \in C_i \setminus \operatorname{int}_C C_{i-1}$, since $\varphi_j(x) = 0$ for $j > i$ and $\varphi_j(x) = 1$ for $j < i$, we have the following:

$$h(x)(\gamma) = \begin{cases} \varepsilon_{i+1}\varphi_i(x)h_{i+1}(x)(\gamma) & \text{if } \gamma \in \Lambda_{i+1}, \\ 0 & \text{if } \gamma \in \Gamma_{i+1} \setminus \big(\Gamma_i \cup \Lambda_i\big), \\ \varepsilon_i h_i(x)(\gamma) & \text{if } \gamma \in \Lambda_i, \\ \varphi_i(x)x(\gamma) & \text{if } \gamma \in \Gamma_i \setminus \big(\Gamma_{i-1} \cup \Lambda_{i-1}\big), \\ \varepsilon_{i-1}(1 - \varphi_i(x))h_{i-1}(x)(\gamma) + \varphi_i(x)x(\gamma) & \text{if } \gamma \in \Lambda_{i-1}, \\ x(\gamma) & \text{if } \gamma \in \Gamma_{i-1}, \\ 0 & \text{if } \gamma \in \Gamma \setminus (\Gamma_{i+1} \cup \Lambda_{i+1}). \end{cases}$$

Since $F_i \subset \Gamma_{i-1}$, it follows that

$$\begin{aligned} \|x - h(x)\| &\leqslant \sum_{\gamma\in\Gamma\setminus\Gamma_{i-1}} |x(\gamma) - h(x)(\gamma)| \\ &\leqslant \sum_{\gamma\in\Gamma\setminus(\Gamma_{i-1}\cup\Lambda_i\cup\Lambda_{i-1})} |x(\gamma)| + \sum_{\gamma\in\Lambda_i\cup\Lambda_{i-1}} (1 - \varphi_i(x))|x(\gamma)| \\ &\qquad + \sum_{\gamma\in\Lambda_{i+1}} \varepsilon_{i+1}\varphi_i(x)|h_{i+1}(x)(\gamma)| + \sum_{\gamma\in\Lambda_i} \varepsilon_i|h_i(x)(\gamma)| \\ &\qquad\qquad + \sum_{\gamma\in\Lambda_{i-1}} \varepsilon_{i-1}(1 - \varphi_i(x)|h_{i-1}(x)(\gamma)| \\ &\leqslant \sum_{\Gamma\setminus\Gamma_{i-1}} |x(\gamma)| + \varepsilon_{i+1}\operatorname{card}\Lambda_{i+1} + \varepsilon_i\operatorname{card}\Lambda_i + \varepsilon_{i-1}\operatorname{card}\Lambda_{i-1} \\ &\leqslant \delta_i/2 + 3(2\dim C + 2)\varepsilon_{i-1} \leqslant \delta_i/2 + \delta_i/2 = \delta_i \end{aligned}$$

Therefore, h is $\mathcal{V}$-close to id.

To see that h is injective, let $x \neq y \in C \setminus U$, where $x \in C_i \setminus \operatorname{int}_C C_{i-1}$ and $y \in C_j \setminus \operatorname{int}_C C_{j-1}$. We may assume that $i \leqslant j$ without loss of generality. When $i + 1 < j$, we have $h(x)(\lambda_j) = 0 < \varepsilon_j = h(y)(\lambda_j)$, so $h(x) \neq h(y)$. When $i = j$, since h_i is an embedding,

$$h(x)(\gamma) = \varepsilon_i h_i(x)(\gamma) \neq \varepsilon_i h_i(y)(\gamma) = h(y)(\gamma) \text{ for some } \gamma \in \Lambda_i,$$

hence $h(x) \neq h(y)$. When $j = i + 1$, if $x \in C_j \setminus \operatorname{int} C_{j-1}$ then $h(x) \neq h(y)$ as in the case $i = j$. If $x \in \operatorname{int} C_i$ then $\varphi_i(x) < 1$, hence

$$h(x)(\lambda_{i+1}) = \varepsilon_{i+1}\varphi_i(x) < \varepsilon_{i+1} = h(y)(\lambda_{i+1}),$$

which implies that $h(x) \neq h(y)$.

(A Cell Complex Structure of $\ell^1_f(\Gamma)$) For each $v \in \mathbb{Z}^\Gamma_f$ and $F \in \operatorname{Fin}(\Gamma)$, we define

$$\begin{aligned} Q_{v,F} &= v + \sum_{\gamma \in F} \mathbf{I}\mathbf{e}_\gamma \\ &= \big\{ x \in \ell^1_f(\Gamma) \bigm| x(\gamma) = v(\gamma) \text{ if } \gamma \in \Gamma \setminus F, \\ &\qquad\qquad v(\gamma) \leqslant x(\gamma) \leqslant v(\gamma) + 1 \text{ if } \gamma \in F \big\} \subset \ell^1_f(\Gamma), \end{aligned}$$

which is an n-cell (an n-cube), where $n = \operatorname{card} F$. Any face of $Q_{v,F}$ can be written as $Q_{v',F''}$, where

$$v' = v + \sum_{\gamma' \in F'} \mathbf{e}_{\gamma'}, \ F' \subset F, \ F'' \subset F \setminus F'.$$

Moreover, when $Q_{v,F} \cap Q_{v',F'} \neq \emptyset$, we have the following:

(1) if $\gamma \notin F \cup F'$, then $v(\gamma) = v'(\gamma)$;
(2) if $\gamma \in F \setminus F'$, then $v'(\gamma) = v(\gamma)$ or $v'(\gamma) = v(\gamma) + 1$, where

$$[v(\gamma), v(\gamma) + 1] \cap \{v'(\gamma)\} = \big\{\max\{v(\gamma), v'(\gamma)\}\big\};$$

(3) if $\gamma \in F' \setminus F$, then $v(\gamma) = v'(\gamma)$ or $v'(\gamma) + 1$, where

$$[v'(\gamma), v'(\gamma) + 1] \cap \{v(\gamma)\} = \big\{\max\{v(\gamma), v'(\gamma)\}\big\};$$

(4) if $\gamma \in F \cap F'$ and $v(\gamma) \neq v'(\gamma)$, then $|v(\gamma) - v'(\gamma)| = 1$, where

$$[v(\gamma), v(\gamma) + 1] \cap [v'(\gamma), v'(\gamma) + 1] = \big\{\max\{v(\gamma), v'(\gamma)\}\big\}.$$

Then, it follows that $Q_{v,F} \cap Q_{v',F'} = Q_{v'',F''}$, where

$$v''(\gamma) = \max\{v(\gamma), v'(\gamma)\} \text{ for } \gamma \in \Gamma \text{ and}$$
$$F'' = \{\gamma \in F \cap F' \mid v(\gamma) = v'(\gamma)\}.$$

Thus, we have the following cell complex:

$$L = \mathbb{Z}_f^\Gamma \cup \{Q_{v,F} \mid v \in \mathbb{Z}_f^\Gamma,\ F \in \mathrm{Fin}(\Gamma)\},$$

where $L^{(0)} = \mathbb{Z}_f^\Gamma$. Then, $|L| = \ell_f^1(\Gamma)$ as sets. Indeed, for each $x \in \ell_f^1(\Gamma)$, $F = \{\gamma \in \Gamma \mid x(\gamma) \neq 0\}$ is finite. Let $v \in \mathbb{Z}_f^\Gamma$ be defined as follows:

$$v(\gamma) = \begin{cases} \lfloor x(\gamma) \rfloor & \text{if } \gamma \in F, \\ 0 & \text{otherwise,} \end{cases}$$

where $\lfloor r \rfloor$ is the largest integer not greater than r. Then, $x \in v + \sum_{\gamma \in F} \mathbf{I}\mathbf{e}_\gamma = Q_{v,F}$.

($\ell_f^1(\Gamma)$ *is a* σ*-Discrete Union of Members of* $\mathfrak{M}_0^{\mathrm{fd}}(\ell^1(\Gamma))$) For each $v \in \mathbb{Z}_f^\Gamma$, $F \in \mathrm{Fin}(\Gamma)$, and $k \in \mathbb{N}$, let

$$Q_{v,F,k} = v + \sum_{\gamma \in F} [2^{-k}, 1 - 2^{-k}]\mathbf{e}_\gamma.$$

Then, rint $Q_{v,F} = \bigcup_{k \in \mathbb{N}} Q_{v,F,k}$. We prove that $\{Q_{v,F,k} \mid v \in \mathbb{Z}_f^\Gamma, F \in \mathrm{Fin}(\Gamma)\}$ is discrete in $\ell_f^1(\Gamma)$. Let $x \in Q_{v,F,k}$ and $x' \in Q_{v',F',k}$, where $(v, F) \neq (v', F')$.

When $F \neq F'$, there exists $\gamma \in (F \setminus F') \cup (F' \setminus F)$. Then,

$$v(\gamma) + 2^{-k} \leqslant x(\gamma) \leqslant v(\gamma) + 1 - 2^{-k} \text{ and } x'(\gamma) = v'(\gamma) \in \mathbb{Z}, \text{ or}$$
$$v'(\gamma) + 2^{-k} \leqslant x'(\gamma) \leqslant v'(\gamma) + 1 - 2^{-k} \text{ and } x(\gamma) = v(\gamma) \in \mathbb{Z}.$$

In either case, we have $\|x - x'\| \geqslant |x(\gamma) - x'(\gamma)| \geqslant 2^{-k}$.

When $F = F'$, it follows that $v \neq v' \in \mathbb{Z}_f^\Gamma$, hence $|v(\gamma) - v'(\gamma)| \geqslant 1$ for some $\gamma \in \Gamma$. When $\gamma \notin F = F'$, since $x(\gamma) = v(\gamma)$ and $x'(\gamma) = v'(\gamma)$, it follows that

$$|x(\gamma) - x'(\gamma)| = |v(\gamma) - v'(\gamma)| \geqslant 1.$$

When $\gamma \in F = F'$ and $v(\gamma) < v'(\gamma)$, since $x(\gamma) \leqslant v(\gamma) + 1 - 2^{-k}$ and $v'(\gamma) + 2^{-k} \leqslant x'(\gamma)$, it follows that

$$x'(\gamma) - x(\gamma) \geqslant v'(\gamma) + 2^{-k} - (v(\gamma) + 1 - 2^{-k}) \geqslant 2^{-k+1}.$$

When $\gamma \in F = F'$ and $v'(\gamma) < v(\gamma)$, since $x'(\gamma) \leqslant v'(\gamma) + 1 - 2^{-k}$ and $v(\gamma) + 2^{-k} \leqslant x(\gamma)$, it follows that

$$x(\gamma) - x'(\gamma) \geqslant v(\gamma) + 2^{-k} - (v'(\gamma) + 1 - 2^{-k}) \geqslant 2^{-k+1}.$$

In either case, we have $\|x - x'\| \geqslant |x(\gamma) - x'(\gamma)| \geqslant 2^{-k+1}$.

Consequently, if $Q_{v,F,k} \neq Q_{v',F',k}$, then

$$\mathrm{dist}(Q_{v,F,k}, Q_{v',F',k}) = \inf\{\|x - x'\| \mid x \in Q_{v,F,k}, x' \in Q_{v',F',k}\} \geqslant 2^{-k},$$

which means that $\{Q_{v,F,k} \mid v \in \mathbb{Z}_f^\Gamma, F \in \mathrm{Fin}(\Gamma)\}$ is discrete in $\ell_f^1(\Gamma)$. Observe

$$\ell_f^1(\Gamma) = |L| = \bigcup_{\substack{v \in \mathbb{Z}_f^\Gamma \\ F \in \mathrm{Fin}(\Gamma)}} \mathrm{rint}\, Q_{v,F} = \bigcup_{k \in \mathbb{N}} \bigcup_{\substack{v \in \mathbb{Z}_f^\Gamma \\ F \in \mathrm{Fin}(\Gamma)}} Q_{v,F,k},$$

where every $Q_{v,F,k}$ is a Z-set in $\ell^1(\Gamma)$ because it is compact. Therefore, $\ell_f^1(\Gamma)$ is a σ-discrete union of members of $\mathfrak{M}_0^{\mathrm{fd}}(\ell^1(\Gamma)$. □

Now, we can characterize $\ell_f^2(\Gamma) \subset \ell^2(\Gamma)$ as an $\mathfrak{M}_0^{\mathrm{fd}}$-absorption base, that is, the following theorem holds:

Theorem 5.5.2 *For $F \subset \ell^2(\Gamma)$, the following are equivalent:*

(a) $(\ell^2(\Gamma), F) \approx (\ell^2(\Gamma), \ell_f^2(\Gamma))$;
(b) *F is an $\mathfrak{M}_0^{\mathrm{fd}}$-absorption base for $\ell^2(\Gamma)$;*
(c) *For each open cover $\mathcal{U}$ of $\ell^2(\Gamma)$, there is a homeomorphism $f : \ell^2(\Gamma) \to \ell^2(\Gamma)$ such that $f(F) = \ell_f^2(\Gamma)$ and f is $\mathcal{U}$-close to* id.

Proof The implication (c) ⇒ (a) is obvious. Since $\ell_f^2(\Gamma)$ is an $\mathfrak{M}_0^{\mathrm{fd}}$-absorption base for $\ell^2(\Gamma)$ by Theorem 5.5.1, we have (a) ⇒ (b). The implication (b) ⇒ (c) follows from Theorem 5.3.14. □

Remark 5.12 Due to Theorems 1.2.3 and 5.5.2 above is valid for any other $(\ell^p(\Gamma), \ell_f^p(\Gamma))$, $1 \leqslant p < \infty$.

We can show the following $\mathfrak{M}_0$-version of Theorem 5.5.2:

Theorem 5.5.3 *For $F \subset \ell^2(\Gamma)$, the following are equivalent:*

(a) $(\ell^2(\Gamma), F) \approx (\ell^2(\Gamma) \times \boldsymbol{Q}, \ell_f^2(\Gamma) \times \boldsymbol{Q})$;
(b) *F is an $\mathfrak{M}_0$-absorption base for $\ell^2(\Gamma)$;*
(c) *For each open cover $\mathcal{U}$ of $\ell^2(\Gamma)$, there is a homeomorphism $f : \ell^2(\Gamma) \times \boldsymbol{Q} \to \ell^2(\Gamma)$ such that $f(\ell_f^2(\Gamma) \times \boldsymbol{Q}) = F$ and f is $\mathcal{U}$-close to* pr_1.

Proof Since $\boldsymbol{Q}$ is strongly $\mathfrak{M}_0$-universal, we can apply Proposition 5.4.8 and Corollary 5.4.2 to see that $\ell_f^2(\Gamma) \times \boldsymbol{Q}$ is an $\mathfrak{M}_0$-absorption base for $\ell^2(\Gamma) \times \boldsymbol{Q}$.

By the Toruńczyk Factor Theorem 2.2.12 and Corollary 2.3.10, the projection $\mathrm{pr}_1 : \ell^2(\Gamma) \times \boldsymbol{Q} \to \ell^2(\Gamma)$ is a near-homeomorphism. Since there is a homeomorphism $h : \ell^2(\Gamma) \times \boldsymbol{Q} \to \ell^2(\Gamma)$ arbitrarily close to pr_1, the implication (a) $\Rightarrow$ (b) is trivial and (b) $\Rightarrow$ (c) can be obtained by using Theorem 5.3.14. By Theorem 5.3.14, we have the implication (c) $\Rightarrow$ (a). □

Note that the pair $(\ell^2(\Gamma) \times \boldsymbol{Q}, \ell^2_f(\Gamma) \times \boldsymbol{Q})$ is not a pair of linear spaces but it is homeomorphic to a pair of normed linear spaces, that is:

Theorem 5.5.4 *The pair* $(\ell^2(\Gamma) \times \boldsymbol{Q}, \ell^2_f(\Gamma) \times \boldsymbol{Q})$ *is homeomorphic to the pair* $(\ell^2(\Gamma) \times \ell^2, \ell^2_f(\Gamma) \times \ell^2_Q)$.

Proof Because $\operatorname{card}\Gamma = \operatorname{card}\Gamma + \aleph_0$, we have

$$\big(\ell^2(\Gamma), \ell^2_f(\Gamma)\big) \approx \big(\ell^2(\Gamma) \times \ell^2, \ell^2_f(\Gamma) \times \ell^2_f\big).$$

Moreover, $(\ell^2 \times \boldsymbol{Q}, \ell^2_f \times \boldsymbol{Q}) \approx (\ell^2, \ell^2_Q)$ by Corollary 5.1.7(3). Hence,

$$\begin{aligned}\big(\ell^2(\Gamma) \times \boldsymbol{Q}, \ell^2_f(\Gamma) \times \boldsymbol{Q}\big) &\approx \big(\ell^2(\Gamma) \times \ell^2 \times \boldsymbol{Q}, \ell^2_f(\Gamma) \times \ell^2_f \times \boldsymbol{Q}\big) \\ &\approx \big(\ell^2(\Gamma) \times \ell^2, \ell^2_f(\Gamma) \times \ell^2_Q\big).\end{aligned}$$ □

Note Applying Proposition 5.4.14, we can show that $\ell^2_f(\Gamma) \times \ell^2_Q$ is an $\mathfrak{M}_0$-absorption base for $\ell^2(\Gamma) \times \ell^2$. The above theorem is also a corollary of Theorem 5.5.3.

Next, we will consider the following pairs:

$$\big(\ell^2(\Gamma) \times \ell^2, \ell^2_f(\Gamma) \times \ell^2\big) \text{ and } \big(\ell^2(\Gamma) \times \ell^2, \ell^2(\Gamma) \times \ell^2_f\big),$$

where $\ell^2(\Gamma) \times \ell^2 \approx \ell^2(\Gamma)$.

First, replacing $\boldsymbol{Q}$ and $\mathfrak{M}_0$ with ℓ^2 and $\mathfrak{M}_1(\aleph_0)$ in Theorem 5.5.3 and its proof, we can obtain the following theorem (the details are left to the reader):

Theorem 5.5.5 *For* $F \subset \ell^2(\Gamma)$, *the following are equivalent:*

(a) $(\ell^2(\Gamma), F) \approx (\ell^2(\Gamma) \times \ell^2, \ell^2_f(\Gamma) \times \ell^2)$;
(b) *F is an $\mathfrak{M}_1(\aleph_0)$-absorption base for $\ell^2(\Gamma)$;*
(c) *For each open cover $\mathcal{U}$ of $\ell^2(\Gamma)$, there is a homeomorphism $f : \ell^2(\Gamma) \times \ell^2 \to \ell^2(\Gamma)$ such that $f(\ell^2_f(\Gamma) \times \ell^2) = F$ and f is $\mathcal{U}$-close to* pr_1.

The following is also obtained:

Theorem 5.5.6 *The following are homeomorphic to* $(\ell^2(\Gamma) \times \ell^2, \ell^2_f(\Gamma) \times \ell^2)$:

$$\big(\ell^2(\Gamma) \times \ell^2 \times \ell^2, \ell^2_f(\Gamma) \times \ell^2_Q \times \ell^2\big) \text{ and } \big(\ell^2(\Gamma) \times \boldsymbol{Q} \times \ell^2, \ell^2_f(\Gamma) \times \boldsymbol{Q} \times \ell^2\big).$$

Proof By Theorem 5.5.4 and the fact $\boldsymbol{Q} \times \ell^2 \approx \ell^2$, we have the following:

$$\begin{aligned}\left(\ell^2(\Gamma) \times \ell^2 \times \ell^2, \ell^2_f(\Gamma) \times \ell^2_Q \times \ell^2\right) &\approx \left(\ell^2(\Gamma) \times \boldsymbol{Q} \times \ell^2, \ell^2_f(\Gamma) \times \boldsymbol{Q} \times \ell^2\right)\\ &\approx \left(\ell^2(\Gamma) \times \ell^2, \ell^2_f(\Gamma) \times \ell^2\right). \qquad \square\end{aligned}$$

For the second pair, since $\ell^2(\Gamma)$ is strongly $\mathfrak{M}_1(\tau)$-universal, we can apply Proposition 5.4.8 and Corollary 5.4.2(2) to see that $\ell^2(\Gamma) \times \ell^2_f$ is an $\mathfrak{M}_1(\tau)$-absorption base for $\ell^2(\Gamma) \times \ell^2$. Because the projection $\mathrm{pr}_1 : \ell^2(\Gamma) \times \ell^2 \to \ell^2(\Gamma)$ is a near-homeomorphism, the following can be proved in the same way as Theorem 5.5.3 (the details are left to the reader):

Theorem 5.5.7 *For $F \subset \ell^2(\Gamma)$, the following are equivalent:*

(a) $(\ell^2(\Gamma), F) \approx (\ell^2(\Gamma) \times \ell^2, \ell^2(\Gamma) \times \ell^2_f)$;
(b) *F is an $\mathfrak{M}_1(\tau)$-absorption base for $\ell^2(\Gamma)$;*
(c) *For each open cover $\mathcal{U}$ of $\ell^2(\Gamma)$, there is a homeomorphism $f : \ell^2(\Gamma) \times \ell^2 \to \ell^2(\Gamma)$ such that $f(\ell^2(\Gamma) \times \ell^2_f) = F$ and f is $\mathcal{U}$-close to* pr_1.

Concerning the pairs $(\ell^2(\Gamma) \times \ell^2, \ell^2(\Gamma) \times \ell^2_Q)$ and $(\ell^2(\Gamma)^{\mathbb{N}}, \ell^2(\Gamma)^{\mathbb{N}}_f)$, we have the following:

Theorem 5.5.8 *The pairs $(\ell^2(\Gamma) \times \ell^2, \ell^2(\Gamma) \times \ell^2_Q)$ and $(\ell^2(\Gamma)^{\mathbb{N}}, \ell^2(\Gamma)^{\mathbb{N}}_f)$ are homeomorphic to $(\ell^2(\Gamma) \times \ell^2, \ell^2(\Gamma) \times \ell^2_f)$.*

Proof First, the following is easy:

$$\begin{aligned}\left(\ell^2(\Gamma) \times \ell^2, \ell^2(\Gamma) \times \ell^2_Q\right) &\approx \left(\ell^2(\Gamma) \times \ell^2 \times \boldsymbol{Q}, \ell^2(\Gamma) \times \ell^2_f \times \boldsymbol{Q}\right)\\ &\approx \left(\ell^2(\Gamma) \times \ell^2, \ell^2(\Gamma) \times \ell^2_f\right).\end{aligned}$$

For the pair $(\ell^2(\Gamma)^{\mathbb{N}}, \ell^2(\Gamma)^{\mathbb{N}}_f)$, note that $\ell^2(\Gamma)^{\mathbb{N}}_f = \bigcup_{n \in \mathbb{N}} \ell^2(\Gamma)^n$, where each $\ell^2(\Gamma)^n$ is a completely metrizable Z-set in $\ell^2(\Gamma)^{\mathbb{N}}$ with $w(\ell^2(\Gamma)^n) = \tau$, i.e., a member of $(\mathfrak{M}_1(\tau))(\ell^2(\Gamma)^{\mathbb{N}})$. By Proposition 1.13.27, $\ell^2(\Gamma)^{\mathbb{N}}_f$ is homotopy dense in $\ell^2(\Gamma)^{\mathbb{N}}$. Since $\mathfrak{M}_1(\tau) = \mathcal{F}_{\ell^2(\Gamma)} \subset \mathcal{F}_{\ell^2(\Gamma)^{\mathbb{N}}_f}$, it follows from Corollary 5.4.12 that $\ell^2(\Gamma)^{\mathbb{N}}_f$ is strongly $\mathfrak{M}_1(\tau)$-universal. Hence, $\ell^2(\Gamma)^{\mathbb{N}}_f$ is an $\mathfrak{M}_1(\tau)$-absorption base for $\ell^2(\Gamma)^{\mathbb{N}}$ (Corollary 5.4.2(2)). Since $\ell^2(\Gamma)^{\mathbb{N}} \approx \ell^2(\Gamma)$ (Corollary 2.2.13), we have

$$\left(\ell^2(\Gamma)^{\mathbb{N}}, \ell^2(\Gamma)^{\mathbb{N}}_f\right) \approx \left(\ell^2(\Gamma) \times \ell^2,\ \ell^2(\Gamma) \times \ell^2_f\right). \qquad \square$$

Remark 5.13 For $\Gamma' \subset \Gamma$ with $\operatorname{card} \Gamma' = \tau' > \aleph_0$, we can consider the spaces $\ell^2_f(\Gamma) \times \ell^2(\Gamma')$ and $\ell^2(\Gamma) \times \ell^2_f(\Gamma')$. The former is an $\mathfrak{M}_1(\tau')$-absorption base for $\ell^2(\Gamma) \times \ell^2(\Gamma')$ but the latter is an $\mathfrak{M}_1(\tau)$-absorption base for $\ell^2(\Gamma) \times \ell^2(\Gamma')$, which is the same as $\ell^2_f(\Gamma) \times \ell^2$ and $\ell^2(\Gamma) \times \ell^2_f$. Then, for the former pair, the same result as Theorem 5.5.5 holds. For the latter pair, we have

$$\big(\ell^2(\Gamma) \times \ell^2(\Gamma'),\ \ell^2(\Gamma) \times \ell^2_f(\Gamma')\big) \approx \big(\ell^2(\Gamma) \times \ell^2, \ell^2(\Gamma) \times \ell^2_f\big).$$

Combining Proposition 5.4.17 with the above results, we obtain the following theorem:

Theorem 5.5.9

(1) *The space* $\ell^2_f(\Gamma)^{\mathbb{N}}_f$ *is an* $\mathfrak{M}_0^{\mathrm{fd}}$*-absorption base for* $\ell^2(\Gamma)^{\mathbb{N}}$*, hence*

$$\big(\ell^2(\Gamma)^{\mathbb{N}}, \ell^2_f(\Gamma)^{\mathbb{N}}_f\big) \approx \big(\ell^2(\Gamma), \ell^2_f(\Gamma)\big).$$

(2) *The space* $(\ell^2_f(\Gamma) \times \ell^2_Q)^{\mathbb{N}}_f$ *is an* $\mathfrak{M}_0$*-absorption base for* $(\ell^2(\Gamma) \times \ell^2)^{\mathbb{N}}$*, hence*

$$\big((\ell^2(\Gamma) \times \ell^2)^{\mathbb{N}}, (\ell^2_f(\Gamma) \times \ell^2_Q)^{\mathbb{N}}_f\big) \approx \big(\ell^2(\Gamma) \times \ell^2, \ell^2_f(\Gamma) \times \ell^2_Q\big).$$

(3) *The space* $(\ell^2_f(\Gamma) \times \ell^2)^{\mathbb{N}}_f$ *is an* $\mathfrak{M}_1(\aleph_0)$*-absorption base for* $(\ell^2(\Gamma) \times \ell^2)^{\mathbb{N}}$*, hence*

$$\big((\ell^2(\Gamma) \times \ell^2)^{\mathbb{N}}, (\ell^2_f(\Gamma) \times \ell^2)^{\mathbb{N}}_f\big) \approx \big(\ell^2(\Gamma) \times \ell^2, \ell^2_f(\Gamma) \times \ell^2\big).$$

(4) *The space* $(\ell^2(\Gamma) \times \ell^2_f)^{\mathbb{N}}_f$ *is an* $\mathfrak{M}_1(\tau)$*-absorption base for* $(\ell^2(\Gamma) \times \ell^2)^{\mathbb{N}}$*, hence*

$$\big((\ell^2(\Gamma) \times \ell^2)^{\mathbb{N}}, (\ell^2(\Gamma) \times \ell^2_f)^{\mathbb{N}}_f\big) \approx \big(\ell^2(\Gamma) \times \ell^2, \ell^2(\Gamma) \times \ell^2_f\big).$$ [14] □

By the above theorem, we have

$$\ell^2_f(\Gamma)^{\mathbb{N}}_f \approx \ell^2_f(\Gamma),\ (\ell^2_f(\Gamma) \times \ell^2_Q)^{\mathbb{N}}_f \approx \ell^2_f(\Gamma) \times \ell^2_Q,$$
$$(\ell^2_f(\Gamma) \times \ell^2)^{\mathbb{N}}_f \approx \ell^2_f(\Gamma) \times \ell^2, \text{ and } (\ell^2(\Gamma) \times \ell^2_f)^{\mathbb{N}}_f \approx \ell^2(\Gamma) \times \ell^2_f.$$

Thus, we can conclude as follows:

Theorem 5.5.10 *All results on E-manifolds in Chap. 2 are valid for* $\ell^2_f(\Gamma)$*-,* $(\ell^2_f(\Gamma) \times \ell^2_Q)$*-,* $(\ell^2_f(\Gamma) \times \ell^2)$*-, and* $(\ell^2(\Gamma) \times \ell^2_f)$*-manifolds.* □

In particular, the Open Embedding Theorem 2.5.10 and the Classification Theorem 2.6.1 are valid for these manifolds.

---

[14] This is equivalent to $((\ell^2(\Gamma)^{\mathbb{N}})^{\mathbb{N}}, (\ell^2(\Gamma)^{\mathbb{N}}_f)^{\mathbb{N}}_f) \approx (\ell^2(\Gamma)^{\mathbb{N}}, \ell^2(\Gamma)^{\mathbb{N}}_f)$, which is easily seen. In fact, $\big((X^{\mathbb{N}})^{\mathbb{N}}, (A^{\mathbb{N}}_f)^{\mathbb{N}}_f\big) \approx \big(X^{\mathbb{N}\times\mathbb{N}}, A_f^{\mathbb{N}\times\mathbb{N}}\big) \approx \big(X^{\mathbb{N}}, A^{\mathbb{N}}_f\big)$ for an arbitrary pair (X, A) of spaces.

Remark 5.14 In the separable case, $\boldsymbol{s} \times \boldsymbol{s}_f$, $\boldsymbol{s} \times \boldsymbol{s}_Q$, and $\boldsymbol{s}_f^{\mathbb{N}}$ are $\mathfrak{M}_1(\aleph_0)$-absorption bases for $\boldsymbol{s} \times \boldsymbol{s}$, $\boldsymbol{s} \times \boldsymbol{s}$, and $\boldsymbol{s}^{\mathbb{N}}$, respectively. Hence, we have

$$\begin{aligned}(\ell^2 \times \ell^2, \ell^2 \times \ell_f^2) &\approx (\ell^2 \times \ell^2, \ell^2 \times \ell_Q^2) \approx ((\ell^2)^{\mathbb{N}}, (\ell^2)_f^{\mathbb{N}}) \\ &\approx (\boldsymbol{s} \times \boldsymbol{s}, \boldsymbol{s} \times \boldsymbol{s}_f) \approx (\boldsymbol{s} \times \boldsymbol{s}, \boldsymbol{s} \times \boldsymbol{s}_Q) \approx (\boldsymbol{s}^{\mathbb{N}}, \boldsymbol{s}_f^{\mathbb{N}}).\end{aligned}$$

Since every closed set in $\boldsymbol{Q}$ is compact, $\boldsymbol{Q}$ has no $\mathfrak{M}_1(\aleph_0)$-absorption bases homeomorphic to $\boldsymbol{s}_f^{\mathbb{N}}$. Although $I(\boldsymbol{Q}) \times \boldsymbol{Q}_f \approx \boldsymbol{s} \times \boldsymbol{s}_f$ ($\approx \boldsymbol{s}_f^{\mathbb{N}}$) and $I(\boldsymbol{Q}) \times \boldsymbol{Q}_f$ is homotopy dense in $\boldsymbol{Q} \times \boldsymbol{Q}$ ($\approx \boldsymbol{Q}$), $I(\boldsymbol{Q}) \times \boldsymbol{Q}_f$ is not a Z_σ-set in $\boldsymbol{Q} \times \boldsymbol{Q}$ because it is not σ-compact.

Remark 5.15 Although $\ell_f^2(\Gamma)$, $\ell_f^2(\Gamma) \times \ell_Q^2$, and $\ell_f^2(\Gamma) \times \ell^2$ are not homeomorphic to each other, their countable products are homeomorphic to each other. Indeed, since $\ell_f^2(\Gamma) \approx \ell_f^2(\Gamma) \times \mathbb{R}$, we have the following:

$$\begin{aligned}\ell_f^2(\Gamma)^{\mathbb{N}} &\approx \ell_f^2(\Gamma)^{\mathbb{N}} \times \mathbb{R}^{\mathbb{N}} = \ell_f^2(\Gamma)^{\mathbb{N}} \times \boldsymbol{s} \\ &\approx \ell_f^2(\Gamma)^{\mathbb{N}} \times \boldsymbol{s}^{\mathbb{N}} \approx (\ell_f^2(\Gamma) \times \boldsymbol{s})^{\mathbb{N}} \approx (\ell_f^2(\Gamma) \times \ell^2)^{\mathbb{N}}.\end{aligned}$$

Moreover, $\ell^2 \approx \ell^2 \times \boldsymbol{Q}$. Then, it follows that

$$\begin{aligned}(\ell_f^2(\Gamma) \times \ell^2)^{\mathbb{N}} &\approx (\ell_f^2(\Gamma) \times \ell^2 \times \boldsymbol{Q})^{\mathbb{N}} \approx (\ell_f^2(\Gamma) \times \ell^2)^{\mathbb{N}} \times \boldsymbol{Q}^{\mathbb{N}} \\ &\approx \ell_f^2(\Gamma)^{\mathbb{N}} \times \boldsymbol{Q}^{\mathbb{N}} \approx (\ell_f^2(\Gamma) \times \boldsymbol{Q})^{\mathbb{N}} \approx (\ell_f^2(\Gamma) \times \ell_Q^2)^{\mathbb{N}}.\end{aligned}$$

Additionally, the countable power $(\ell^2(\Gamma) \times \ell_f^2)^{\mathbb{N}}$ is also homeomorphic to $\ell_f^2(\Gamma)^{\mathbb{N}}$. More generally, let F be a $\mathfrak{C}$-absorption base for $\ell^2(\Gamma)$, where $\mathfrak{M}_0^{\mathrm{fd}} \subset \mathfrak{C} \subset \mathfrak{M}_1(\tau)$. Then, we can prove $F^{\mathbb{N}} \approx \ell_f^2(\Gamma)^{\mathbb{N}}$. This will be done in Sect. 5.10 (see Theorem 5.12.10).

5.6 $\mathfrak{C}$-Universality and $\mathfrak{C}_\sigma$-Universality

In the previous section, we have considered four types of absorption bases for Hilbert space $\ell^2(\Gamma)$ and manifolds modeled on those spaces. To consider a more general situation, the strong $\mathfrak{C}_\sigma$-universality is important. In this section, we discuss when the strong $\mathfrak{C}$-universality implies the strong $\mathfrak{C}_\sigma$-universality. Evidently, $\mathfrak{C}_\sigma$ is closed hereditary, additive, and topological because $\mathfrak{C}$ is.

A class $\mathfrak{C}$ is said to be **open hereditary** if $\mathfrak{C}$ contains any open subspace of a member of $\mathfrak{C}$.

Proposition 5.6.1 *The class $\mathfrak{C}_\sigma$ is open hereditary.*

Proof Let $X \in \mathfrak{C}_\sigma$, that is, $X = \bigcup_{n\in\mathbb{N}} X_n$, where $X_n \in \mathfrak{C}$, $n \in \mathbb{N}$. Each open subspace U of X is F_σ in X, that is, there are closed sets F_i, $i \in \mathbb{N}$, in X such that $U = \bigcup_{i\in\mathbb{N}} F_i$. Since $\mathfrak{C}$ is closed hereditary, $X_n \cap F_i \in \mathfrak{C}$ for each $n, i \in \mathbb{N}$. Then, $U = \bigcup_{n\in\mathbb{N}} \bigcup_{i\in\mathbb{N}} (X_n \cap F_i) \in \mathfrak{C}_\sigma$. □

By virtue of the above proposition, the following two propositions can be applied to the class $\mathfrak{C}_\sigma$ even if $\mathfrak{C}$ is not open hereditary.

Proposition 5.6.2 *Suppose that $\mathfrak{C}$ is open hereditary and let X be an ANR. Then, X is strongly $\mathfrak{C}$-universal if and only if every open subspace of X is fine $\mathfrak{C}$-universal.*

Proof The "only if" part follows from Proposition 5.4.18. To see the "if" part, let $f : C \to X$ be a map of $C \in \mathfrak{C}$ such that $f|D$ is a strong Z-embedding for a closed set D in C. By Proposition 2.8.12, f is arbitrarily close to a map $g : C \to X$ such that $g|D = f|D$, $g(C \setminus D) \cap f(D) = \emptyset$ and g is closed over $g(D) = f(D)$. Then, $C \setminus D \in \mathfrak{C}$ because $\mathfrak{C}$ is open hereditary. By the fine $\mathfrak{C}$-universality of $X \setminus g(D)$, $g|C \setminus D$ is arbitrarily close to a strong Z-embedding $h' : C \setminus D \to X \setminus g(D)$. If h' is sufficiently close to $g|C \setminus D$, then the extension $h : C \to X$ of h' defined by $h|D = g|D$ is continuous and closed over $g(D)$. Thus, we have a strong Z-embedding h arbitrarily close to f such that $h|D = f|D$. This means that X is strongly $\mathfrak{C}$-universal. □

Remark 5.16 As already seen in Proposition 5.4.18, the converse implication in the above proposition also holds even if $\mathfrak{C}$ is not open hereditary.

Applying Michael's Theorem 1.3.20 on local property, we prove the following proposition:

Proposition 5.6.3 *Let X be an ANR such that every Z-set is a strong Z-set. If each point of X has a strongly $\mathfrak{C}$-universal neighborhood, then X is strongly $\mathfrak{C}$-universal.*

Proof To apply Michael's Theorem 1.3.20, it is enough to prove the following:

(G-1) For open sets $U \supset V$ in X, if U is strongly $\mathfrak{C}$-universal, then V is strongly $\mathfrak{C}$-universal;
(G-2) For two open sets U_1 and U_2 in X, if each U_i is strongly $\mathfrak{C}$-universal, then $U_1 \cup U_2$ is strongly $\mathfrak{C}$-universal;
(G-3) For a discrete collection $\{U_\lambda\}_{\lambda\in\Lambda}$ of open sets in X, if each U_λ is strongly $\mathfrak{C}$-universal, then $\bigcup_{\lambda\in\Lambda} U_\lambda$ is strongly $\mathfrak{C}$-universal.

Since (G-1) is a direct consequence of Proposition 5.4.18 and (G-3) is trivial, we only have to show (G-2).

We write $U = U_1 \cup U_2$. Let $f : A \to U$ be a map from $A \in \mathfrak{C}$ such that $f|B$ is a Z-embedding for a closed set B in A. Choose open sets W_i in U, $i = 0, 1, 2$ so that

$$U_1 \setminus U_2 \subset W_0 \subset \mathrm{cl}_U W_0 \subset W_1 \subset \mathrm{cl}_U W_1 \subset W_2 \subset \mathrm{cl}_U W_2 \subset U_1.$$

For each open cover $\mathcal{U}$ of U, let $\mathcal{V}$ be an open star-refinement of $\mathcal{U}$ such that

$$\mathrm{st}(\mathrm{cl}_U W_1, \mathcal{V}) \subset W_2, \ \mathrm{st}(U \setminus W_0, \mathcal{V}) \subset U_2 \ \text{ and}$$
$$\mathrm{st}(U \setminus W_1, \mathrm{st}^2 \mathcal{V}) \subset U \setminus \mathrm{cl}_U W_0.$$

See Fig. 5.2. Since $\mathfrak{C}$ is closed hereditary, we have

$$C = (B \cap f^{-1}(\mathrm{cl}_U W_2)) \cup f^{-1}(\mathrm{cl}_U W_1) \in \mathfrak{C},$$

which is closed in A. Note that $f|B \cap f^{-1}(\mathrm{cl}_U W_2)$ is a strong Z-embedding into U_1. Since U_1 is strongly $\mathfrak{C}$-universal, we have a strong Z-embedding $g : C \to U_1$ such that

$$g|B \cap f^{-1}(\mathrm{cl}_U W_2) = f|B \cap f^{-1}(\mathrm{cl}_U W_2) \ \text{ and}$$
$$g \simeq_{\mathcal{V}} f|C \ \text{ rel. } B \cap f^{-1}(\mathrm{cl}_U W_2).$$

Since $\mathrm{st}(\mathrm{cl}_U W_1, \mathcal{V}) \subset W_2$, it follows that $g(f^{-1}(\mathrm{cl}_U W_1)) \subset W_2$, so

$$g(f^{-1}(\mathrm{cl}_U W_1)) \cap f(f^{-1}(U \setminus W_2)) = \emptyset.$$

Using the Homotopy Extension Theorem 1.13.11, we can extend g to a map $\tilde{g} : A \to U$ such that

$$\tilde{g}|B \cup f^{-1}(U \setminus W_2) = f|B \cup f^{-1}(U \setminus W_2) \ \text{ and}$$
$$\tilde{g} \simeq_{\mathcal{V}} f \ \text{ rel. } B \cup f^{-1}(U \setminus W_2).$$

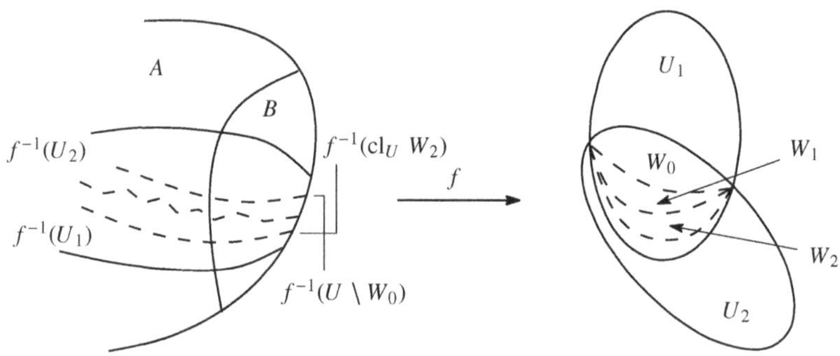

Fig. 5.2 Approximating a map by using Z-embeddings

Then, $\tilde{g}|B \cup C$ is a strong Z-embedding because $\tilde{g}|C = g|C$ and $\tilde{g}|B = f|B$ are strong Z-embeddings and

$$\begin{aligned}\tilde{g}(C) \cap \tilde{g}(B) &= \big(g(B \cap f^{-1}(\mathrm{cl}_U W_2)) \cup g(f^{-1}(\mathrm{cl}_U W_1))\big) \cap f(B)\\ &= \big(f(B \cap f^{-1}(\mathrm{cl}_U W_2)) \cap f(B)\big) \cup \big(g(f^{-1}(\mathrm{cl}_U W_1)) \cap f(B)\big)\\ &= f(B \cap f^{-1}(\mathrm{cl}_U W_2)) \cup \big(g(f^{-1}(\mathrm{cl}_U W_1)) \cap f(B \cap f^{-1}(W_2))\big)\\ &= f(B \cap f^{-1}(\mathrm{cl}_U W_2)) \cup \big(g(f^{-1}(\mathrm{cl}_U W_1)) \cap g(B \cap f^{-1}(W_2))\big)\\ &= g(B \cap f^{-1}(\mathrm{cl}_U W_2)) \cup g(B \cap f^{-1}(\mathrm{cl}_U W_1))\big)\\ &= g(B \cap f^{-1}(\mathrm{cl}_U W_2)) = \tilde{g}(C \cap B)\end{aligned}$$

Since $\mathfrak{C}$ is closed hereditary, we have $f^{-1}(U \setminus W_0) \in \mathfrak{C}$. Observe that

$$\tilde{g}(f^{-1}(U \setminus W_0)) \subset \mathrm{st}(U \setminus W_0, \mathcal{V}) \subset U_2.$$

So, the restriction $\tilde{g}|f^{-1}(U \setminus W_0)$ is a map into U_2. Let

$$D = (B \cup C) \cap f^{-1}(U \setminus W_0) = (B \setminus f^{-1}(W_0)) \cup f^{-1}(\mathrm{cl}_U W_1 \setminus W_0).$$

Then, $\tilde{g}|D$ is a strong Z-embedding into U_2. Since U_2 is strongly $\mathfrak{C}$-universal, we have a strong Z-embedding $h : f^{-1}(U \setminus W_0) \to U_2$ such that

$$h|D = \tilde{g}|D \ \text{ and } \ h \simeq_{\mathcal{V}} \tilde{g}|f^{-1}(U \setminus W_0) \ \text{ rel. } D.$$

Since $h \simeq_{\mathrm{st}\mathcal{V}} f|f^{-1}(U \setminus W_0)$, we have $h(f^{-1}(U \setminus W_1)) \subset \mathrm{st}(U \setminus W_1, \mathrm{st}\,\mathcal{V})$. Moreover, $g(f^{-1}(\mathrm{cl}_U W_0)) \subset \mathrm{st}(\mathrm{cl}_U W_0, \mathcal{V})$. Observe that

$$\mathrm{st}(U \setminus W_1, \mathrm{st}^2 \mathcal{V}) \subset U \setminus \mathrm{cl}_U W_0 \ \Rightarrow\ \mathrm{st}(U \setminus W_1, \mathrm{st}\,\mathcal{V}) \cap \mathrm{st}(\mathrm{cl}_U W_0, \mathcal{V}) = \emptyset.$$

Thus, it follows that

$$h(f^{-1}(U \setminus W_1)) \cap g(f^{-1}(\mathrm{cl}\, W_0)) = \emptyset.$$

Since $h|f^{-1}(\mathrm{cl}_U W_1 \setminus W_0) = g|f^{-1}(\mathrm{cl}_U W_1 \setminus W_0)$, we can extend h to a map $\tilde{h} : A \to U$ by $\tilde{h}|f^{-1}(\mathrm{cl}_U W_0) = g|f^{-1}(\mathrm{cl}_U W_0)$. Then, $\tilde{h}$ is a strong Z-embedding because $\tilde{h}|f^{-1}(U \setminus W_1) = h|f^{-1}(U \setminus W_1)$ and $\tilde{h}|f^{-1}(\mathrm{cl}_U W_1) = g|f^{-1}(\mathrm{cl}_U W_1)$ are strong Z-embeddings and

$$\begin{aligned}\tilde{h}(f^{-1}(U \setminus W_1)) \cap \tilde{h}(f^{-1}(\mathrm{cl}_U W_1)) &= h(f^{-1}(U \setminus W_1)) \cap g(f^{-1}(\mathrm{cl}_U W_1))\\ &= h(f^{-1}(U \setminus W_1)) \cap g(f^{-1}(\mathrm{cl}_U W_1 \setminus W_0))\\ &= h(f^{-1}(U \setminus W_1)) \cap h(f^{-1}(\mathrm{cl}_U W_1 \setminus W_0))\end{aligned}$$

$$= h(f^{-1}(U \setminus W_1) \cap f^{-1}(\mathrm{cl}_U W_1 \setminus W_0))$$
$$= \tilde{h}(f^{-1}(U \setminus W_1) \cap f^{-1}(\mathrm{cl}_U W_1).$$

Since $\mathrm{st}\,\mathcal{V} \prec \mathcal{U}$, $\tilde{h}$ is $\mathrm{st}\,\mathcal{V}$-close to f. Moreover, $\tilde{h}|B = f|B$ because

$$\tilde{h}|B \cap f^{-1}(U \setminus W_0) = \tilde{g}|B \cap f^{-1}(U \setminus W_0) = f|B \cap f^{-1}(U \setminus W_0) \text{ and}$$
$$\tilde{h}|B \cap f^{-1}(\mathrm{cl}_U W_0) = g|B \cap f^{-1}(\mathrm{cl}_U W_0) = f|B \cap f^{-1}(\mathrm{cl}_U W_0).$$

This completes the proof. □

A metrizable space X is called a **Z_σ-space** (resp. a **strong Z_σ-space**) is such that X itself is a Z_σ-set (resp. a strong Z_σ-set) in X. Since every closed set in a strong Z_σ-space X is a strong Z_σ-set in X, we have the following from Proposition 3.2.11:

Lemma 5.6.4 *Let X be an ANR with $w(X) = \tau$. If X is a strong Z_σ-space with the τ-discrete cells property, then every Z-set in X is a strong Z-set.* □

For a strong Z_σ-space being an ANR, a combination of the strong $\mathfrak{C}$-universality and the τ-discrete cells property implies the strong $\mathfrak{C}_\infty$-universality, that is:

Proposition 5.6.5 *Let X be a strong Z_σ-space that is an ANR with the τ-discrete cells property, where $w(X) = \tau$. If X is strongly $\mathfrak{C}$-universal, then X is strongly $\mathfrak{C}_\sigma$-universal.*

Proof Since $\mathfrak{C}_\sigma$ is open hereditary, Proposition 5.6.2 can be applied to $\mathfrak{C}_\sigma$. Hence, it suffices to show that each non-empty open set U in X satisfies condition ($\mathrm{au}_{\mathfrak{C}_\sigma}$) of Proposition 5.6.2. Observe that U is an ANR with $w(U) = \tau$,[15] and a strongly $\mathfrak{C}$-universal strong Z_σ-space by Propositions 5.4.18 and 2.8.14. Moreover, U has the τ-discrete cells property by Proposition 3.2.3. Thus, U satisfies the same condition as X. Consequently, it is enough to show that X is fine $\mathfrak{C}_\sigma$-universal. Since X has the τ-discrete cells property, every Z-set in X is a strong Z-set by Lemma 5.6.4. Hence, "strong Z-embedding" can be replaced by "Z-embedding."

Let $f : C \to X$ be a map of $C \in \mathfrak{C}_\sigma$. For each $\mathcal{U} \in \mathrm{cov}(X)$, by 1.3.22(1), we can choose $d \in \mathrm{Metr}(X)$ so that $\{\overline{\mathrm{B}}_d(x, 1) \mid x \in X\} \prec \mathcal{U}$. Then, it suffices to construct a Z-embedding $f_* : C \to X$ such that $d(f_*, f) < 1$. We write $X = \bigcup_{i \in \mathbb{N}} X_i$, where $\emptyset = X_1 \subset X_2 \subset \cdots$ are strong Z-sets in X.

(*Particular Case*) If C is an open set in some member of $\mathfrak{C}$, we can write $C = \bigcup_{i \in \mathbb{N}} C_i$, where each C_i is closed in C, $C_i \in \mathfrak{C}$ and $C_i \subset \mathrm{int}\, C_{i+1}$. We shall inductively construct maps $f_i : C \to X$, $\alpha_i : X \to (0, 2^{-i}]$, $i \in \mathbb{N}$, such that

(1) $f_i|C_i$ is a Z-embedding,

[15] Indeed, $w(U) \leqslant w(X) \leqslant \tau$ and, using the τ-discrete cells property of X, we can easily prove $w(U) \geqslant \tau$.

(2) $f_i|C_{i-1} = f_{i-1}|C_{i-1}$,
(3) $\operatorname{cl} f_i(C \setminus \operatorname{int} C_{i+1}) \cap X_{i+1} = \emptyset$ and $\operatorname{cl} f_i(C \setminus \operatorname{int} C_i) \cap X_i = \emptyset$,
(4) $f_i \in N_{\alpha_i}(f_{i-1})$,
(5) $\alpha_i(x) \leqslant \frac{1}{2}d(x, X_i)$ for each $x \in \operatorname{cl} f_{i-1}(C \setminus \operatorname{int} C_i)$ and
$\alpha_i(x) \leqslant \frac{1}{2}d(x, X_{i-1})$ for each $x \in \operatorname{cl} f_{i-1}(C \setminus \operatorname{int} C_{i-1})$,

where $f_0 = f$ and $C_0 = \emptyset$.

Assume that $f_0,\ldots,f_{i-1}$ have been obtained. By (3), we can apply the Tietze Extension Theorem 1.3.1 to obtain maps $\beta, \beta' : X \to (0, \infty)$ such that

$$\beta(x) = d(x, X_i) \quad \text{for } x \in \operatorname{cl} f_{i-1}(C \setminus \operatorname{int} C_i),$$
$$\beta'(x) = d(x, X_{i-1}) \quad \text{for } x \in \operatorname{cl} f_{i-1}(C \setminus \operatorname{int} C_{i-1}).$$

We define a map $\alpha_i : X \to (0, 2^{-i}]$ by

$$\alpha_i(x) = \min\left\{2^{-i}, \tfrac{1}{2}\beta(x), \tfrac{1}{2}\beta'(x)\right\} \quad \text{for each } x \in X.$$

Then, α_i satisfies (5). By Proposition 3.1.2, we have $\mathcal{V} \in \operatorname{cov}(X)$ such that $(\operatorname{st}\mathcal{V})(f_{i-1}) \subset N_{\alpha_i}(f_{i-1})$. By the fine $\mathfrak{C}$-universality of X, we have a Z-embedding $v_i : C_i \to X$ such that $v_i|C_{i-1} = f_{i-1}|C_{i-1}$, and $v_i \simeq_{\mathcal{V}} f_{i-1}|C_i$. By the Homotopy Extension Theorem 1.13.11, v_i extends to a map $\tilde{v}_i : C \to X$, that is $\mathcal{V}$-homotopic to f_{i-1}. Since X_{i+1} is a strong Z-set in X, we have a $\mathcal{V}$-homotopy $h : X \times \mathbf{I} \to X$ such that $\operatorname{cl}_X h_1(X) \cap X_{i+1} = \emptyset$. We define $f_i : C \to X$ by

$$f_i(x) = h_{\lambda(x)}(\tilde{v}_i(x)) \quad \text{for each } x \in C,$$

where $\lambda : X \to \mathbf{I}$ is an Urysohn map with $\lambda(C_i) = 0$ and $\lambda(C \setminus C_{i+1}) = 1$. Then, f_i satisfies the conditions (1)–(4).

By (2) and (4), we have a map $f_* : C \to X$ such that $f_*|C_i = f_i|C_i$ for each $i \in \mathbb{N}$ and $d(f_*, f) \leqslant 1$, hence f_* is $\mathcal{U}$-close to f. By (3) and (5), we can see

$$\operatorname{cl} f_*(C \setminus \operatorname{int} C_i) \cap X_i = \emptyset \quad \text{for each } i \in \mathbb{N}.$$

Thus, f_* is an embedding and

$$f_*(C) = \bigcup_{i\in\mathbb{N}} f_*(C_i \setminus \operatorname{int} C_{i-1}) = \bigcup_{i\in\mathbb{N}} f_i(C_i \setminus \operatorname{int} C_{i-1})$$

is a locally finite union of Z-sets in X, hence it is a Z-set in X. Therefore, f_* is a Z-embedding.

(*General Case*) We now consider the general case $C \in \mathfrak{C}_\sigma$, that is, $C = \bigcup_{i\in\mathbb{N}} C_i$, where $C_1 \subset C_2 \subset \cdots$ are closed in C and $C_i \in \mathfrak{C}$. In this case, it cannot be assumed that $C_i \subset \operatorname{int} C_{i+1}$. For each $k \in \mathbb{N}$, since $\mathrm{C}(\mathbf{I}^k, X)$ with the sup-metric has the same weight as X, there is a map $g_k : \mathbf{I}^k \times \Gamma \to X$ such that $\{g_{k,\gamma} \mid \gamma \in \Gamma\}$ is dense in $\mathrm{C}(\mathbf{I}^k, X)$, where $g_{k,\gamma} : \mathbf{I}^k \to X$ is defined by $g_{k,\gamma}(x) = g_k(x, \gamma)$. Given an open cover $\mathcal{U}$ of X, let $\mathcal{U}_0$ be an open star-refinement of $\mathcal{U}$. By induction, we shall construct maps $f_i : C \to X$, $g_k^i : \mathbf{I}^k \times \Gamma \to X$ ($k \leqslant i$), and open covers $\mathcal{U}_i$ of $X \setminus (f_i(C_i) \cup X_i)$, $i \in \mathbb{N}$, such that:

(6) $f_i|C_i$ is a Z-embedding,
(7) $f_i|C_{i-1} = f_{i-1}|C_{i-1}$,
(8) $f_i(C \setminus C_i) \cap f_i(C_i) = \emptyset$,
(9) f_i is closed over $f_i(C_i) \cup X_i$,
(10) $f_i|C \setminus C_{i-1}$ is $\mathcal{U}_{i-1}$-close to $f_{i-1}|C \setminus C_{i-1}$,
(11) $\operatorname{cl} f_i(C \setminus C_{i-1}) \cap \big(X_i \setminus (f_{i-1}(C_{i-1}) \cup X_{i-1})\big) = \emptyset$,
(12) $\operatorname{st} \mathcal{U}_i \prec \mathcal{U}_{i-1}$,
(13) $\operatorname{diam} U < \min\{2^{-(i+1)}, \frac{1}{2}d(U, f_i(C_i) \cup X_i)\}$ for each $U \in \mathcal{U}_i$,
(14) $g_k^i(\mathbf{I}^k \times \Gamma)$ is a Z-set in X,
(15) $f_i(C) \cap \bigcup_{k \leqslant j \leqslant i} g_k^j(\mathbf{I}^k \times \Gamma) = \emptyset$,
(16) $\{g_{k,\gamma}^i \mid \gamma \in \Gamma\}$ is 2^{-i}-dense in $\mathrm{C}(\mathbf{I}^k, X)$, that is, each $g \in \mathrm{C}(\mathbf{I}^k, X)$ is 2^{-i}-close to some $g_{k,\gamma}^i$,

where $f_0 = f$ and $C_0 = X_0 = \emptyset$.

Assume that f_{i-1}, g_k^{i-1} ($k \leqslant i-1$) and $\mathcal{U}_{i-1}$ have been obtained. By (6), $f_{i-1}(C_{i-1})$ is a Z-set in X, so a strong Z-set in X. Using the τ-discrete cells property, we have a map $g_k^i : \mathbf{I}^k \times \Gamma \to X$ ($k \leqslant i$) such that

$$g_k^i(\mathbf{I}^k \times \Gamma) \cap f_{i-1}(C_{i-1}) = \emptyset,$$

and each g_k^i is $2^{-(i+1)}$-close to g_k, hence it satisfies (16). Since X is a Z_σ-space, every closed set is a Z_σ-set in X. Because of compactness, $g_k^i(\mathbf{I}^k \times \{\gamma\})$ is a Z-set in X for each $\gamma \in \Gamma$ (Proposition 2.8.8). Their discrete union $g_k^i(\mathbf{I}^k \times \Gamma)$ is also a Z-set in X (Corollary 2.8.7). Thus, g_k^i satisfies (14) too.

Now, we write

$$W = X \setminus (f_{i-1}(C_{i-1}) \cup X_{i-1}).$$

Then, $\mathcal{U}_{i-1}$ is an open cover of W. Let $\mathcal{V}$ be an open star-refinement of $\mathcal{U}_{i-1}$. The open set W in X is a strong Z_σ-space and it has τ-discrete cells property (Proposition 3.2.3), hence each Z-set in W is a strong Z-set by Proposition 3.2.11. Note that $X_i \cap W$ is a strong Z-set in W by Proposition 2.8.14 and W is strongly $\mathfrak{C}$-universal by Proposition 5.4.18. We apply the particular case to the open set $C_i \setminus C_{i-1}$ in $C_i \in \mathfrak{C}$, and use the Homotopy Extension Theorem 1.13.11 to construct a map $h : C \setminus C_{i-1} \to W$ such that:

(17) $h|C_i \setminus C_{i-1}$ is a Z-embedding,
(18) h is $\mathcal{V}$-close to $f_{i-1}|C \setminus C_{i-1}$,
(19) $\operatorname{cl} h(C \setminus C_{i-1}) \cap W \cap \big(X_i \cup \bigcup_{k \leqslant j \leqslant i} g_k^j(\mathbf{I}^k \times \Gamma)\big) = \emptyset$.

Since $h(C_i \setminus C_{i-1}) \cup (X_i \cap W)$ is a strong Z-set in W, we apply Proposition 2.8.12 to obtain a map $\tilde{h} : C \setminus C_{i-1} \to W$ such that:

(20) $\tilde{h}$ is $\mathcal{V}$-close to h, hence it is $\mathcal{U}_{i-1}$-close to $f_{i-1}|C \setminus C_{i-1}$ by (18),
(21) $\operatorname{cl} \tilde{h}(C \setminus C_{i-1}) \cap W \cap \big(X_i \cup \bigcup_{k \leqslant j \leqslant i} g_k^j(\mathbf{I}^k \times \Gamma)\big) = \emptyset$,
(22) $\tilde{h}|C_i \setminus C_{i-1} = h|C_i \setminus C_{i-1}$,
(23) $\tilde{h}(C \setminus C_i) \cap \tilde{h}(C_i \setminus C_{i-1}) = \emptyset$,
(24) $\tilde{h}$ is closed over $\tilde{h}(C_i \setminus C_{i-1}) \cup (X_i \cap W)$.

For each $z \in C_{i-1}$ and $\varepsilon > 0$, since f_{i-1} is continuous, we have a neighborhood V of z in C such that $y \in V$ implies $d(f_{i-1}(y), f_{i-1}(z)) < \varepsilon/2$. For each $y \in V \setminus C_{i-1}$, choose $U \in \mathcal{U}_{i-1}$ so that $\tilde{h}(y), f_{i-1}(y) \in U$, whence we have $d(\tilde{h}(y), f_{i-1}(y)) < \frac{1}{2} d(f_{i-1}(y), f_{i-1}(z))$ by (13) for $i-1$. Then, we have

$$\begin{aligned} d(\tilde{h}(y), f_{i-1}(z)) &\leqslant d(\tilde{h}(y), f_{i-1}(y)) + d(f_{i-1}(y), f_{i-1}(z)) \\ &< \tfrac{3}{2} d(f_{i-1}(y), f_{i-1}(z)) < \varepsilon. \end{aligned}$$

Therefore, as an extension of $\tilde{h}$, we can obtain the map $f_i : C \to X$ satisfying (7), that clearly satsfies (8), (10), (11), and (15) (cf. (23), (20), (21)).

Since $f_i|C_{i-1} = f_{i-1}|C_{i-1}$ and $f_i|C_i \setminus C_{i-1} = h|C_i \setminus C_{i-1}$ are injective and

$$f_i(C_i \setminus C_{i-1}) \cap f_{i-1}(C_{i-1}) = \tilde{h}(C_i \setminus C_{i-1}) \cap f_{i-1}(C_{i-1}) = \emptyset,$$

it follows that $f_i|C_i$ is injective. If f_i satisfies (9), that is, f_i is closed over $f_i(C_i) \cup X_i$, then $f_i|C_i$ is an embedding.

Suppose that f_i is not closed over $f_i(C_i) \cup X_i$. Then, there exist $a \in f_i(C_i) \cup X_i$, a neighborhood U of $f_i^{-1}(a)$ in C (we allow $U = f_i^{-1}(a) = \emptyset$) and a sequence $(z_n)_{n \in \mathbb{N}}$ in $C \setminus U$ with $\lim_{n \to \infty} f_i(z_n) = a$. Since $f_i|C_{i-1} = f_{i-1}|C_{i-1}$ is a closed embedding into X by (6) for $i-1$, we have $z_n \in C \setminus C_{i-1}$ for sufficiently large $n \in \mathbb{N}$. Since $f_i|C \setminus C_{i-1}$ is closed over $f_i(C_i \setminus C_{i-1}) \cup (X_i \cap W)$ by (24), it follows that $a \notin f_i(C_i \setminus C_{i-1}) \cup (X_i \cap W)$. Recall $a \in f_i(C_i) \cup X_i$. Then, we have

$$a \in f_i(C_{i-1}) \cup (X_i \setminus W) = f_{i-1}(C_{i-1}) \cup X_{i-1}.$$

For sufficiently large $n \in \mathbb{N}$, we can choose $U_n \in \mathcal{U}_{i-1}$ so that $f_i(z_n), f_{i-1}(z_n) \in U_n$ by (10), whence

$$d(f_{i-1}(z_n), a) \leqslant d(f_i(z_n), f_{i-1}(z_n)) + d(f_i(z_n), a) < \tfrac{3}{2} d(f_i(z_n), a).$$

Then, $\lim_{n \to \infty} f_{i-1}(z_n) = a$, which implies that $f_{i-1}^{-1}(a) \neq \emptyset$ by (9) for $i-1$. Since $f_{i-1}^{-1}(a) \subset C_{i-1}$ by (8) for $i-1$, it follows from (7) that $f_{i-1}^{-1}(a) \subset f_i^{-1}(a) \subset U$. Again by (9) for $i-1$, we have a neighborhood V of a in X such that $f_{i-1}^{-1}(V) \subset U$. For sufficiently large $n \in \mathbb{N}$, $f_{i-1}(z_n) \in V$, hence $z_n \in f_{i-1}^{-1}(V) \subset U$. This is a contradiction. Therefore, f_i satisfies (9).

To see (6), it suffices to show that $f_i(C_i)$ is a Z-set in X. Observe that

$$X \setminus (f_i(C_i) \cup X_{i-1}) = W \setminus h(C_i \setminus C_{i-1}),$$

which is open in W. Then, $f_i(C_i) \cup X_{i-1}$ is closed in X, hence $f_i(C_i) \cup X_i$ is also closed in X. Since $f_i(C_i \setminus C_{i-1}) = h(C_i \setminus C_{i-1})$ is a Z-set in W and $X \setminus W = (f_{i-1}(C_{i-1}) \cup X_i)$ is a Z-set in X, it follows that $f_i(C_i) \cup X_i$ is a Z-set in X. By (8) and (9), we can see that $f_i(C_i)$ is closed in $f_i(C_i) \cup X_i$. Therefore, $f_i(C_i)$ is a Z-set in X.

Finally, by choosing an open cover $\mathcal{U}_i$ of $X \setminus (f_i(C_i) \cup X_i)$ so as to satisfy (12) and (13), we can obtain f_i, g^i_k $(k \leqslant i)$ and $\mathcal{U}_i$ that satisfy all conditions (6)–(16).

By (7), we can define $f_* : C \to X$ defined by $f_*|C_i = f_i|C_i$. It follows from (10) and (13) that f_* is 2^{-i}-close to f_i. Thus, f_* is the uniform limit of $(f_i)_{i\in\mathbb{N}}$, so f_* is continuous. By (6), f_* is injective. Then, to see that f_* is a Z-embedding, it remains to be shown that f_* is a closed map and $f_*(C)$ is a Z-set in X.

Now, assume that f_* is not closed. Then, we have a sequence $(z_n)_{n\in\mathbb{N}}$ in C such that $(z_n)_{n\in\mathbb{N}}$ has no convergent subsequences but $(f_*(z_n))_{n\in\mathbb{N}}$ converges to some $a \in X$. Let $a \in X_m \setminus X_{m-1}$. Then, $z_n \in C \setminus C_m$ for sufficiently large $n \in \mathbb{N}$. Otherwise, C_m contains a subsequence of $(z_n)_{n\in\mathbb{N}}$, that is convergent because $f_*|C_m = f_m|C_m$ is a closed embedding. From (7), (10) and (12), it follows that $f_*|C \setminus C_m$ is st $\mathcal{U}_m$-close to $f_m|C \setminus C_m$. By (13), we have $x_n, y_n \in X$ for sufficiently large $n \in \mathbb{N}$ such that

$$d(f_*(z_n), x_n) < \tfrac{1}{2}d(f_*(z_n), a),$$
$$d(x_n, y_n) < \tfrac{1}{2}d(x_n, a), \text{ and}$$
$$d(y_n, f_m(z_n)) < \tfrac{1}{2}d(y_n, a).$$

Then, $(f_m(z_n))_{n\in\mathbb{N}}$ also converges to a, hence

$$a \in \operatorname{cl} f_m(C \setminus C_m) \subset \operatorname{cl} f_m(C \setminus C_{m-1}),$$

which implies that $a \in f_{m-1}(C_{m-1})$ by (11). By (6), (7) and (8), there is unique $c \in C_{m-1}$ such that $f_m(c) = f_{m-1}(c) = a$, where $f_m^{-1}(a) = \{c\}$ by (8). Since $(z_n)_{n\in\mathbb{N}}$ does not converge to c and f_m is closed over $f_m(C_m)$ by (9), we have a neighborhood V of a in X such that infinitely many z_n are not contained in $f_m^{-1}(V)$, that is, infinitely many $f_m(z_n)$ are not contained in V. This is a contradiction. Therefore, f_* is a closed map.

To see that $f_*(C)$ is a Z-set in X, let $g : \mathbf{I}^k \to X$ be a map and $\varepsilon > 0$. Choose $j \geqslant k$ so that $2^{-j} < \varepsilon$. Then, g is ε-close to some $g^j_{k,\gamma}$ by (16), whence $f_i(C_i) \cap g^j_{k,\gamma}(\mathbf{I}^k) = \emptyset$ for every $i \geqslant j$ by (15). Since $f_*(C) = \bigcup_{i \geqslant j} f_i(C_i)$, it follows that $f_*(C) \cap g^j_{k,\gamma}(\mathbf{I}^k) = \emptyset$. Hence, $f_*(C)$ is a Z-set in X. □

As a corollary, we have the following:

Corollary 5.6.6 *Suppose that* $\mathbf{I}^n \times \Gamma \in \mathfrak{C}$ *for every* $n \in \mathbb{N}$. *Let* X *be an ANR that is a strong* Z_σ*-space. Then, the following are equivalent:*

(a) *X is strongly $\mathfrak{C}$-universal;*
(b) *X is strongly $\mathfrak{C}_\sigma$-universal;*
(c) *Every open set in X is strongly $\mathfrak{C}_\sigma$-universal;*
(d) *Every open set in X is fine $\mathfrak{C}_\sigma$-universal.*

Proof By Propositions 5.6.5 and 5.4.18, we have (a) $\Rightarrow$ (b) $\Rightarrow$ (c). The implications (b) $\Rightarrow$ (a) and (c) $\Rightarrow$ (d) are trivial. Since $\mathfrak{C}_\sigma$ is open hereditary, the implication (d) $\Rightarrow$ (b) follows from Proposition 5.6.2. □

In Proposition 5.6.5, it is assumed that $\mathbf{I}^n \times \Gamma \in \mathfrak{C}$ for every $n \in \mathbb{N}$. However, this assumption is not necessary when $\mathfrak{C}$ is a class of separable metrizable spaces and X is separable. The following are the separable versions of Proposition 5.6.5 and Corollary 5.6.6:

Proposition 5.6.7 *Let* X *be a separable ANR and* $\mathfrak{C} \subset \mathfrak{M}(\aleph_0)$. *If* X *is a strongly* $\mathfrak{C}$*-universal strong* Z_σ*-space, then* X *is strongly* $\mathfrak{C}_\sigma$*-universal.*

Corollary 5.6.8 *Let* X *be a separable ANR that is a strong* Z_σ*-space and* $\mathfrak{C} \subset \mathfrak{M}(\aleph_0)$. *Then, the following are equivalent:*

(a) *X is strongly $\mathfrak{C}$-universal;*
(b) *X is strongly $\mathfrak{C}_\sigma$-universal;*
(c) *Every open set in X is strongly $\mathfrak{C}_\sigma$-universal;*
(d) *Every open set in X is fine $\mathfrak{C}_\sigma$-universal.*

Indeed, in the separable case, Lemma 5.6.4 holds without assuming the discrete cells property, that is, the following is obtained from Proposition 3.2.12:

Lemma 5.6.9 *Let* X *be a separable ANR. If* X *is a strong* Z_σ*-space, then every* Z*-set in* X *is a strong* Z*-set.* □

Under the assumption that $\mathfrak{C}$ is $\mathbf{I}$-stable and $\Gamma \in \mathfrak{C}$, the $\mathfrak{C}$-universality of X implies the τ-discrete cells property of X (Proposition 5.4.6). However, when X is separable, the discrete cells property is not necessary because Lemma 5.6.9 above can be used instead of Lemma 5.6.4. Thus, for a separable ANR X, the first half (before the general case) of the proof of Proposition 5.6.5 is valid without assuming that $\mathfrak{C}$ is $\mathbf{I}$-stable and $\Gamma \in \mathfrak{C}$.

In the second half (the general case) of the proof, using the condition that $\mathbf{I}^k \times \Gamma \in \mathfrak{C}$, we obtained a dense set $\{g_{k,\gamma} \mid \gamma \in \Gamma\}$ in $\mathrm{C}(\mathbf{I}^k, X)$. However, when X is separable, the space $\mathrm{C}(\mathbf{I}^k, X)$ with the sup-metric is also separable, that is, it has a countable dense set $\{g_{k,n} \mid n \in \mathbb{N}\}$. In the proof, using $g_{k,n}$ instead of $g_{k,\gamma}$, we can inductively construct maps $g^i_{k,n} : \mathbf{I}^k \to X$, $k, n \leqslant i$, so as to satisfy the following conditions:

(14′) $g^i_{k,n}(\mathbf{I}^k)$ is a Z-set (hence $\bigcup_{k,n \leqslant j \leqslant i} g^j_{k,n}(\mathbf{I}^k)$ is a Z-set in X),
(15′) $f_i(C) \cap \bigcup_{k,n \leqslant j \leqslant i} g^j_{k,n}(\mathbf{I}^k) = \emptyset$,

(16′) $d(g^i_{k,n}, g_{k,n}) < 2^{-(i+1)}$.

In the inductive step, since $f_{i-1}(C_{i-1})$ is a Z-set in X, we can find maps $g^i_{k,n} : \mathbf{I}^k \to X \setminus f_{i-1}(C_{i-1})$, $k, n \leqslant i$, so as to satisfy (16′). Then, $g^i_{k,n}(\mathbf{I}^k)$ is a compact (strong) Z_σ-set in X because X is a strong Z_σ-space. Hence, $g^i_{k,n}(\mathbf{I}^k)$ is a Z-set in X by Proposition 2.8.8, so (14′) is also satisfied. Moreover, the discrete cells property of W $(= X\setminus(f_{i-1}(C_{i-1}) \cup X_{i-1}))$ is not necessary for the same reason as X mentioned above. Thus, for a separable ANR X, we can construct maps $f_i : C \to X$, $g^i_{k,n} : \mathbf{I}^k \to X$, $k, n \leqslant i$, and open covers $\mathcal{U}_i$ of $X \setminus (f_i(C_i) \cup X_i)$, $i \in \mathbb{N}$, so as to satisfy conditions (6)–(13) and (14′)–(16′) without assuming that $\mathfrak{C}$ is $\mathbf{I}$-stable and $\Gamma \in \mathfrak{C}$.

By the same definition as in the proof of Proposition 5.6.5, we define $f_* : C \to X$. By the same proof, we can show that f_* is an closed embedding and $d(f_*, f) < 1$. To see that $f_*(C)$ is a Z-set in X, let $g : \mathbf{I}^k \to X$ be a map and $\varepsilon > 0$. Since $\{g_{k,n} \mid n \in \mathbb{N}\}$ is dense in $\mathrm{C}(\mathbf{I}^k, X)$, g is $\varepsilon/2$-close to some $g_{k,n}$. Choose $j \geqslant k, n$ so that $2^{-j} < \varepsilon$. Since $d(g^j_{k,n}, g_{k,n}) < 2^{-(j+1)} < \varepsilon/2$, we have $d(g^j_{k,n}, g) < \varepsilon$. Moreover, $f_i(C_i) \cap g^j_{k,n}(\mathbf{I}^k) = \emptyset$ for every $i \geqslant j$ by (15′). Since $f_*(C) = \bigcup_{i \geqslant j} f_i(C_i)$, it follows that $f_*(C) \cap g^j_{k,n}(\mathbf{I}^k) = \emptyset$. Hence, $f_*(C)$ is a Z-set in X. Thus, Proposition 5.6.7 can be proved.

5.7 Manifolds Modeled on Absorption Bases for $\ell^2(\Gamma)$

In Sect. 5.5, we considered four types of absorption bases for Hilbert space $\ell^2(\Gamma)$ and manifolds modeled on those spaces. It was seen that all results in Chap. 2 are valid for such manifolds. In this section, given a $\mathfrak{C}$-absorption base F for $\ell^2(\Gamma)$, we discuss F-manifolds and $(\ell^2(\Gamma), F)$-manifold pairs, where $\mathfrak{C}$ is a closed hereditary, additive topological class of metrizable spaces with weight $\leqslant \tau$.

First, applying Michael's Theorem 1.3.20, we show the following theorem:

Theorem 5.7.1 *Let X be an $\ell^2(\Gamma)$-manifold and $M \subset X$. If X has an open cover $\mathcal{U}$ such that $M \cap U$ is a $\mathfrak{C}$-absorption base for every $U \in \mathcal{U}$, then M is a $\mathfrak{C}$-absorption base for X.*

Proof To apply Michael's Theorem 1.3.20, it is enough to show the following:

(G-1) If U is an open set in X and $M \cap U$ is a $\mathfrak{C}$-absorption base for U, then $M \cap V$ is a $\mathfrak{C}$-absorption base for any open set V in U;

(G-2) If U and V are open sets in X, $M \cap U$ and $M \cap V$ are $\mathfrak{C}$-absorption bases for U and V respectively, then $M \cap (U \cup V)$ is a $\mathfrak{C}$-absorption base for $U \cup V$;

(G-3) If $\{U_\lambda \mid \lambda \in \Lambda\}$ is a discrete collection of open sets in X and $M \cap U_\lambda$ is a $\mathfrak{C}$-absorption base for every $\lambda \in \Lambda$, then $M \cap \bigcup_{\lambda \in \Lambda} U_\lambda$ is a $\mathfrak{C}$-absorption base for $\bigcup_{\lambda \in \Lambda} U_\lambda$.

Since (G-1) is a direct consequence of Theorem 5.4.19 and (G-3) is trivial, we only have to prove (G-2). Due to Corollary 5.4.20, X has a $\mathfrak{C}$-absorption base N. Since

$N \cap (U \cup V)$ is a $\mathfrak{C}$-absorption base for $U \cap V$ by Theorem 5.4.19, it suffices to construct a homeomorphism $h : U \cup V \to U \cup V$ such that $h(N \cap (U \cup V)) = M \cap (U \cup V)$.

Take $\mathcal{U} \in \operatorname{cov}(U)$ that is fitting in X. Since $N \cap U$ and $M \cap U$ are $\mathfrak{C}$-absorption bases for U and U is an $\ell^2(\Gamma)$-manifold, we can apply Theorem 5.3.14 to obtain a homeomorphism $f : U \to U$ such that $f(N \cap U) = M \cap U$ and f is $\mathcal{U}$-close to id. Since $\mathcal{U}$ is fitting in X, f can be extended to a homeomorphism $\tilde{f} : X \to X$ with $\tilde{f}|X \setminus U = \text{id}$. Then, $\tilde{f}(N)$ is a $\mathfrak{C}$-absorption base for X, hence $\tilde{f}(N) \cap V$ is a $\mathfrak{C}$-absorption base for V by Theorem 5.4.19. Now, take $\mathcal{V} \in \operatorname{cov}(V)$ that is fitting in X. As above, we have a homeomorphism $g : V \to V$ such that $g(\tilde{f}(N) \cap V) = M \cap V$ and g is $\mathcal{V}$-close to id, so g can be extended to a homeomorphism $\tilde{g} : X \to X$ with $\tilde{g}|X \setminus V = \text{id}$. Then, we have the desired homeomorphism $h = \tilde{g}\tilde{f}|U \cup V : U \cup V \to U \cup V$. Indeed,

$$\begin{aligned} h(N \cap (U \cup V)) &= \tilde{g}\tilde{f}((N \cap U \setminus V) \cup (N \cap V)) \\ &= \tilde{g}((M \cap U \setminus V) \cup \tilde{g}(\tilde{f}(N) \cap V)) \\ &= (M \cap U \setminus V) \cup (M \cap V) \\ &= M \cap (U \cup V). \end{aligned}$$ □

Using the above theorem, we can characterize manifold pairs as follows:

Theorem 5.7.2 (Characterization) *Suppose that $\mathfrak{M}_0^{\mathrm{fd}} \subset \mathfrak{C} \subset \mathfrak{M}_1(\tau)$. Let F be a $\mathfrak{C}$-absorption base for $\ell^2(\Gamma)$. For an $\ell^2(\Gamma)$-manifold M and $N \subset M$, the following are equivalent:*

(a) *the pair (M, N) is an $(\ell^2(\Gamma), F)$-manifold pair;*
(b) *N is a $\mathfrak{C}$-absorption base for M;*
(c) *$N \in (\tau\mathfrak{C})_\sigma$ and N is a strongly $\mathfrak{C}$-universal and homotopy dense Z_σ-set in M.*

Proof Since the equivalence (b) ⇔ (c) is Corollary 5.4.2, we prove the equivalence (a) ⇔ (b).

(a) ⇒ (b): Assume that (M, N) is an $(\ell^2(\Gamma), F)$-manifold pair, that is, there exists an open cover $\mathcal{U} \in \operatorname{cov}(M)$ with open embeddings $\varphi_U : U \to \ell^2(\Gamma)$, $U \in \mathcal{U}$, such that $\varphi_U(U \cap N) = \varphi_U(U) \cap F$. Then, each $U \cap N$ is a $\mathfrak{C}$-absorption base for U because $\varphi_U(U \cap N)$ is a $\mathfrak{C}$-absorption base for $\varphi_U(U)$ by Theorem 5.4.19. Hence, N is a $\mathfrak{C}$-absorption base for M by Theorem 5.7.1.

(b) ⇒ (a): Assume that N is a $\mathfrak{C}$-absorption base for M. Since M is an $\ell^2(\Gamma)$-manifold, we have an open cover $\mathcal{U} \in \operatorname{cov}(M)$ with open embeddings $\varphi_U : U \to \ell^2(\Gamma)$, $U \in \mathcal{U}$. Then, $U \cap N$ is a $\mathfrak{C}$-absorption base for U by Theorem 5.4.19. Hence, $\varphi_U(U \cap N)$ is a $\mathfrak{C}$-absorption base for $\varphi_U(U)$. On the other hand, $\varphi_U(U) \cap F$ is a $\mathfrak{C}$-absorption base for $\varphi_U(U)$ by Theorem 5.4.19. Using the topological uniqueness of absorption bases for $\ell^2(\Gamma)$-manifolds (Theorem 5.3.14), we have a homeomorphism $h_U : \varphi_U(U) \to \varphi_U(U)$ such that $h_U(\varphi_U(U \cap N)) = \varphi_U(U) \cap F$. Then, $h_U\varphi_U :$

$U \to \ell^2(\Gamma)$ is an open embedding such that $h_U\varphi_U(U \cap N)) = h_U\varphi_U(U) \cap F$. Therefore, (M, N) is an $(\ell^2(\Gamma), F)$-manifold pair. □

Applying Michael's Theorem 1.3.20, we can also prove the following:

Theorem 5.7.3 *Suppose that $\mathfrak{M}_0^{\rm fd} \subset \mathfrak{C} \subset \mathfrak{M}_1(\tau)$. Let F be a $\mathfrak{C}$-absorption base for $\ell^2(\Gamma)$. Then, every F-manifold X can be embedded into an $\ell^2(\Gamma)$-manifold M as a $\mathfrak{C}$-absorption base (i.e., (M, X) is an $(\ell^2(\Gamma), F)$-manifold pair).*

Proof To apply Michael's Theorem 1.3.20, we prove the following:

(G-1) For open sets $U \supset V$ in X, if U can be embedded into an $\ell^2(\Gamma)$-manifold as a $\mathfrak{C}$-absorption base, then V can be embedded into an $\ell^2(\Gamma)$-manifold as a $\mathfrak{C}$-absorption base;

(G-2) For two open sets U_1 and U_2 in X, if each U_i can be embedded into an $\ell^2(\Gamma)$-manifold as a $\mathfrak{C}$-absorption base, then $U_1 \cup U_2$ can be embedded into an $\ell^2(\Gamma)$-manifold as a $\mathfrak{C}$-absorption base;

(G-3) For a discrete collection $\{U_\lambda\}_{\lambda\in\Lambda}$ of open sets in X, if each U_λ can be embedded into an $\ell^2(\Gamma)$-manifold as a $\mathfrak{C}$-absorption base, then $\bigcup_{\lambda\in\Lambda} U_\lambda$ can be embedded into an $\ell^2(\Gamma)$-manifold as a $\mathfrak{C}$-absorption base.

Since (G-1) easily follows from Proposition 5.4.18 and (G-3) is trivial because $\operatorname{card}\Lambda \leqslant w(X) = \tau$, it suffices to show (G-2).

Let M_i be an $\ell^2(\Gamma)$-manifold such that U_i is a $\mathfrak{C}$-sbsorption base for M_i. Then, M_i has an open set V_i such that $V_i \cap U_i = U_1 \cap U_2$. Applying the Lavrentieff Homeomorphism Extension Theorem 1.3.19, we have G_δ sets G_i in V_i, $i = 1, 2$, and a homeomorphism $h_0 : G_1 \to G_2$ such that $U_1 \cap U_2 \subset G_i$ and $h_0|U_1 \cap U_2 = \mathrm{id}$. Observe that $V_i \setminus G_i$ is F_σ in M_i. Refer to Fig. 5.3. Since U_i is homotopy dense in M_i and $V_i \setminus G_i \subset M_i \setminus U_i$, it follows that $V_i \setminus G_i$ is Z_σ-set in M_i. Hence, $M_i' = M_i \setminus (V_i \setminus G_i) \approx M_i$ by the Negligibility Theorem 3.4.8. Then, G_i is open in M_i' because $V_i \cap M_i' = V_i \setminus (V_i \setminus G_i) = G_i$. Identifying G_1 and G_2 by the homeomorphism h_0, we have an $\ell^2(\Gamma)$-manifold $M = M_1' \cup_{G_1 = G_2} M_2'$, where the topology of M is generated by the open sets of M_1' and M_2'. Observe that U_i is a homotopy dense Z_σ-set in M_i'. Hence, we apply Corollary 5.4.2(2) to see that U_i is a $\mathfrak{C}$-absorption base for M_i' for $i = 1, 2$. Therefore, $U_1 \cup U_2$ is a $\mathfrak{C}$-absorption base for M by Theorem 5.7.1.

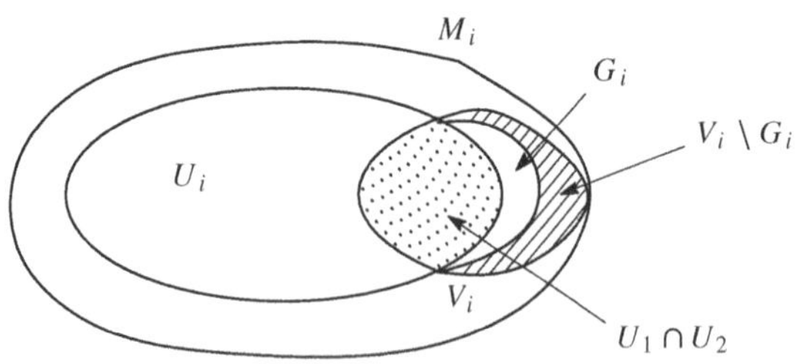

Fig. 5.3 Constructing an $\ell^2(\Gamma)$-manifold containing $U_1 \cup U_2$

Each point $x \in M$ has an open neighborhood U that is homeomorphic to an open set in F. So, we have an open set V in $\ell^2(\Gamma)$ such that $V \cap F \approx U$, where $V \cap F$ is a $\mathfrak{C}$-absorption base for V by Theorem 5.4.19. Applying Michael's Theorem 1.3.20, we have the result. □

According to Theorem 5.7.3 above, Corollary 5.3.6 can be rewritten as follows:

Corollary 5.7.4 *Suppose that $\mathfrak{M}_0^{\mathrm{fd}} \subset \mathfrak{C} \subset \mathfrak{M}_1(\tau)$. Let F be a $\mathfrak{C}$-absorption base for $\ell^2(\Gamma)$. Every F-manifold M is an ANR such that $M \in (\tau\mathfrak{C})_\sigma$, every Z-set in M is a strong Z-set in M, and every compact set in M is a strong Z-set in M.* □

Combining Theorem 5.7.2 with results on $\ell^2(\Gamma)$-manifolds, we can obtain the following results:

Theorem 5.7.5 *Suppose that $\mathfrak{M}_0^{\mathrm{fd}} \subset \mathfrak{C} \subset \mathfrak{M}_1(\tau)$. Let F be a $\mathfrak{C}$-absorption base for $\ell^2(\Gamma)$.*

(1) (OPEN EMBEDDING) *For each $(\ell^2(\Gamma), F)$-manifold pair (M, N), there exists an open embedding $g : M \to \ell^2(\Gamma)$ such that $g(N) = g(M) \cap F$.*

Sketch of Proof. Combine the Open Embedding Theorem 2.5.10 and Theorem 5.3.14.

(2) (CLASSIFICATION) *For $(\ell^2(\Gamma), F)$-manifold pairs (M, N) and (X, Y), $(M, N) \approx (X, Y)$ if and only if $M \simeq X$ or $N \simeq Y$.*

Sketch of Proof. Combine the Classification Theorem 2.6.1 and Theorem 5.3.14.

(3) (TRIANGULATION) *For each $(\ell^2(\Gamma), F)$-manifold pair (M, N), there exists a locally finite-dimensional simplicial complex K such that*

$$(M, N) \approx (|K|_{\mathrm{m}} \times \ell^2(\Gamma), |K|_{\mathrm{m}} \times F).$$

Sketch of Proof. Combine the above (2) with the Triangulation Theorem 2.6.6 and Corollary 5.4.16.

(4) *Let (M, N) and (X, Y) be $(\ell^2(\Gamma), F)$-manifold pairs. For each fine homotopy equivalence $f : M \to X$ and $\mathcal{U} \in \mathrm{cov}(X)$, there is a homeomorphism $h : (M, N) \to (X, Y)$ such that h is $\mathcal{U}$-close to f.*

Sketch of Proof. Combine the Homeomorphism Approximation Theorem 2.6.3 and Theorem 5.3.14.

By virtue of Theorem 5.7.3, we have also the following:

Corollary 5.7.6 *Suppose that $\mathfrak{M}_0^{\mathrm{fd}} \subset \mathfrak{C} \subset \mathfrak{M}_1(\tau)$. Let F be a $\mathfrak{C}$-absorption base for $\ell^2(\Gamma)$.*

(1) (OPEN EMBEDDING) *Every F-manifold can be embedded into F as an open set.*
(2) (CLASSIFICATION) *For F-manifolds M and N, $M \approx N$ if and only if $M \simeq N$.*
(3) (TRIANGULATION) *For each F-manifold M, there exists a locally finite-dimensional simplicial complex K such that $M \approx |K|_{\mathrm{m}} \times F$.*

(4) *For F-manifolds M and N, every fine homotopy equivalence $f : M \to N$ is a near-homeomorphism.*

We can also prove the following ANR Factor Theorem:

Theorem 5.7.7 (ANR Factor) *Suppose that $\mathfrak{M}_0^{\mathrm{fd}} \subset \mathfrak{C} \subset \mathfrak{M}_1(\tau)$ and $\mathfrak{C}$ is multiplicative. Let F be a $\mathfrak{C}$-absorption base for $\ell^2(\Gamma)$. Then, each ANR $X \in (\tau\mathfrak{C})_\sigma$ embedded into an ANR $\widetilde{X} \in \mathfrak{M}_1(\tau)$ such that $(\widetilde{X} \times \ell^2(\Gamma), X \times F)$ is an $(\ell^2(\Gamma), F)$-manifold pair. In particular, $X \times F$ is an F-manifold for any ANR $X \in (\tau\mathfrak{C})_\sigma$.*

Proof Since $X \in (\tau\mathfrak{C})_\sigma$, we can write $X = \bigcup_{i\in\mathbb{N}} X_i$, where each X_i is closed in X and $X_i \in \tau\mathfrak{C}$. Because $\mathfrak{C} \subset \mathfrak{M}_i(\tau)$, each X_i is completely metrizable. Using Hausdorff's Metric Extension Theorem 1.13.8, we have a metric $d_i \in \mathrm{Metr}(X)$ such that X_i is d_i-complete and $d_i \leqslant 1$. We define a metric $d \in \mathrm{Metr}(X)$ as follows:

$$d(x, y) = \sum_{i\in\mathbb{N}} 2^{-i} d_i(x, y).$$

Then, every X_i is d-complete. By virtue of Theorem 1.13.37, there is a metric $\rho \in \mathrm{Metr}(X)$ such that $\rho \geqslant d$ and (X, ρ) is a uniform ANR. Due to Theorem 1.13.35, the metric completion $\widetilde{X} = (\widetilde{X}, \tilde{d})$ is an ANR and X is homotopy dense in $\widetilde{X}$. Then, $\widetilde{X} \times \ell^2(\Gamma)$ is an $\ell^2(\Gamma)$-manifold by the Toruńczyk Factor Theorem 2.2.14 ($\ell^2(\Gamma) \approx \ell^1(\Gamma)$). Observe that each X_i is ρ-complete, which means X_i is closed in $\widetilde{X}$. Hence, X is F_σ in $\widetilde{X}$. From Proposition 5.4.14, it follows that $X \times F$ is a $\mathfrak{C}$-absorption base for $\widetilde{X} \times \ell^2(\Gamma)$, which implies that $(\widetilde{X} \times \ell^2(\Gamma), X \times F)$ is an $(\ell^2(\Gamma), F)$-manifold pair by Theorem 5.7.2. □

The following Stability Theorem is a direct consequence of the above:

Theorem 5.7.8 (Stability) *Suppose that $\mathfrak{M}_0^{\mathrm{fd}} \subset \mathfrak{C} \subset \mathfrak{M}_1(\tau)$ and $\mathfrak{C}$ is multiplicative. Let F be a $\mathfrak{C}$-absorption base for $\ell^2(\Gamma)$. For each F-manifold M, the projection $\mathrm{pr} : M \times F \to M$ is a near-homeomorphism. Consequently, every F-manifold M is F-stable, that is, $M \times F \approx M$.* □

We have the following characterization:

Theorem 5.7.9 (Characterization) *Suppose that $\mathfrak{M}_0^{\mathrm{fd}} \subset \mathfrak{C} \subset \mathfrak{M}_1(\tau)$ and $\mathfrak{C}$ is multiplicative. Let F be a $\mathfrak{C}$-absorption base for $\ell^2(\Gamma)$. An ANR X is an F-manifold if and only if $X \in (\tau\mathfrak{C})_\sigma$ and $X \times F \approx X$.* □

Proof The "if" part comes from the Factor Theorem 5.7.7. The "only if" part is a combination of Corollary 5.7.4 and the Stability Theorem 5.7.8. □

To characterize manifolds modeled on absorption bases for $\ell^2(\Gamma)$, we generalize the discrete n-cells property. The following property of X is called the **τ-discrete approximation property** for a class $\mathfrak{C}$ (abbrev. τ-DAP($\mathfrak{C}$)):

(τ-DAP) Let $f : A = \bigoplus_{\gamma\in\Gamma} A_\lambda \to X$ be a map, where $A_\gamma \in \mathfrak{C}$ for each $\gamma \in \Gamma$. Then, for each open cover $\mathcal{U}$ of X, there is a map $g : A \to X$ such that g is $\mathcal{U}$-close to f and $(g(A_\gamma))_{\gamma\in\Gamma}$ is discrete in X.

The τ-discrete n-cells property is the τ-DAP for the class of spaces homeomorphic to $\mathbf{I}^n$. For each $n \in \omega$, let $\mathfrak{M}_0^{(n)}$ be the class of compacta with dim $\leqslant n$. Then, $\mathfrak{M}_0^{\mathrm{fd}} = \bigcup_{n\in\omega} \mathfrak{M}_0^{(n)}$. Evidently, the τ-DAP($\mathfrak{M}_0^{(n)}$) implies the τ-discrete n-cells property.

In the above (τ-DAP), the term "discrete" can be replaced with "locally finite" by the following lemma, which can be proved in the same way as Proposition 3.2.1:

Lemma 5.7.10 *Let $\mathfrak{C} \subset \mathfrak{M}_0$. A space X has the τ-DAP($\mathfrak{C}$) if X has the following weaker property:*

- *Let $f : A = \bigoplus_{\gamma\in\Gamma} A_\gamma \to X$ be a map, where $A_\gamma \in \mathfrak{C}$ for each $\gamma \in \Gamma$. Then, for each open cover $\mathcal{U}$ of X, there is a map $g : A \to X$ such that g is $\mathcal{U}$-close to f and $(g(A_\gamma))_{\gamma\in\Gamma}$ is locally finite in X.*

Proof Give a well-order $\leqslant$ on Γ so that

$$\operatorname{card}\{\gamma' \in \Gamma \mid \gamma' < \gamma\} < \operatorname{card}\Gamma = \tau \quad \text{for every } \gamma \in \Gamma.$$

Let $f : A = \bigoplus_{\gamma\in\Gamma} A_\gamma \to X$ be a map, where $A_\gamma \in \mathfrak{C}$ for each $\gamma \in \Gamma$. Then, consider the map $f\mathrm{pr}_A : A \times \Gamma \to X$, where $\mathrm{pr}_A : A \times \Gamma \to A$ is the projection. Since $A \times \Gamma = \bigoplus_{(\gamma,\gamma')\in\Gamma^2} A_\gamma \times \{\gamma'\}$ and $\operatorname{card}\Gamma^2 = \operatorname{card}\Gamma = \tau$, there is a map $\bar{g} : A \times \Gamma \to X$ such that $\bar{g}$ is $\mathcal{U}$-close to $f\mathrm{pr}_A$ and $(\bar{g}(A_\gamma \times \{\gamma'\}))_{(\gamma,\gamma')\in\Gamma^2}$ is locally finite in X. Note that each $\bar{g}(A_\gamma \times \{\gamma'\})$ meets only finite many other $\bar{g}(A_\delta \times \{\delta'\})$. By transfinite induction, we can choose $\delta(\gamma) \in \Gamma$ for each $\gamma \in \Gamma$ so that

$$\bar{g}(A_\gamma \times \{\delta(\gamma)\}) \cap \bar{g}(A_{\gamma'} \times \{\delta(\gamma')\}) = \emptyset \quad \text{for } \gamma' < \gamma.$$

Then, $(\bar{g}(A_\gamma \times \{\delta(\gamma)\}))_{\gamma\in\Gamma}$ is discrete in X because it is locally finite and mutually disjoint. Hence, the desired map $g : A = \bigoplus_{\gamma\in\Gamma} A_\gamma \to X$ can be defined by $g(z) = g(z, \delta(\gamma))$ for each $z \in A_\gamma$, $\gamma \in \Gamma$. □

In the following, we show three more lemmas.

Lemma 5.7.11 *Let X be an ANR with the τ-discrete $(2n + 1)$-cells property. Suppose that W is an open set in X that is contractible in X. Then, W has the τ-DAP($\mathfrak{M}_0^{(n)}$).*

Proof[16] Let $f : A = \bigoplus_{\gamma\in\Gamma} A_\gamma \to W$ be a map, where $A_\gamma \in \mathfrak{M}_0^{(n)}$ for each $\gamma \in \Gamma$. By the Embedding Theorem 1.12.14, we can assume that $A \subset \mathbf{I}^{2n+1} \times \Lambda$ and $A_\lambda \subset \mathbf{I}^{2n+1} \times \{\lambda\}$. Since W is contractible in X, $f : A \to X$ is null-homotopic. Using the Homotopy Extension Theorem 1.13.11, we can extend f to a map $\tilde{f}$:

[16] Cf. Proof of Lemma 3.4.4.

$\mathbf{I}^{2n+1} \times \Lambda \to X$. Then, we have an open neighborhood V of A in $\mathbf{I}^{2n+1} \times \Lambda$ such that $\operatorname{cl} V \subset \tilde{f}^{-1}(W)$. We can write $W = \bigcup_{i \in \mathbb{N}} W_i$, where each W_i is open in X and $\operatorname{cl} W_i \subset W_{i+1}$. For each $\mathcal{U} \in \operatorname{cov}(W)$, choose $\mathcal{U}' \in \operatorname{cov}(W)$ so that

$$\operatorname{st} \mathcal{U}' \prec \mathcal{U} \text{ and } \mathcal{U}' \prec \{W_i \setminus \operatorname{cl} W_{i-2} \mid i \in \mathbb{N}\},$$

where $W_0 = W_{-1} = \emptyset$. For each $i \in \mathbb{N}$, let

$$R_i = f^{-1}(\operatorname{cl} W_i \setminus W_{i-1}) \text{ and } \mathcal{U}_i = \mathcal{U}'|W_{2i} \cup \{X \setminus \operatorname{cl} W_{2i-1}\} \in \operatorname{cov}(X).$$

Using the τ-discrete $(2n + 1)$-cells property of X, we can obtain a map $g_i : \mathbf{I}^{2n+1} \times \Lambda \to X$ such that $g_i \simeq_{\mathcal{U}_i} \tilde{f}$ and $(g_i(\mathbf{I}^{2n+1} \times \{\gamma\}))_{\gamma \in \Gamma}$ is discrete in X. Then, $g_i|R_{2i-1} \simeq_{\mathcal{U}'} f|R_{2i-1}$ for every $i \in \mathbb{N}$. Using the Homotopy Extension Theorem 1.13.11, we can obtain a map $g : A \to W$ such that $g \simeq_{\mathcal{U}'} f$ and $g|R_{2i-1} = g_i|R_{2i-1}$ for each $i \in \mathbb{N}$. It is easy to see that $(g(A_\gamma \cap R_{2i-1}))_{\gamma \in \Gamma}$ is discrete in $W_{2i} \setminus \operatorname{cl} W_{2i-3}$. Hence, $(g(A_\gamma \cap R_{2i-1}))_{(\gamma,i) \in \Gamma \times \mathbb{N}}$ is locally finite in W.

Next, we can choose $\mathcal{U}'' \in \operatorname{cov}(W)$ so that $\mathcal{U}'' \prec \mathcal{U}'$ and if a map $h : A \to W$ is $\mathcal{U}''$-close to g, then $(h(A_\gamma \cap R_{2i-1}))_{(\gamma,i) \in \Gamma \times \mathbb{N}}$ is locally finite in W. By the same construction as g, we can obtain a map $h : A \to W$ such that $h \simeq_{\mathcal{U}''} g$ and $(h(A_\gamma \cap R_{2i}))_{(\gamma,i) \in \Gamma \times \mathbb{N}}$ is locally finite in W. Then, by our choice of $\mathcal{U}''$, $(h(A_\gamma \cap R_{2i-1}))_{(\gamma,i) \in \Gamma \times \mathbb{N}}$ is locally finite in W. Therefore, $(h(A_\gamma \cap R_i))_{(\gamma,i) \in \Gamma \times \mathbb{N}}$ is locally finite in W, which implies that $(h(A_\gamma))_{\gamma \in \Gamma}$ is locally finite in W. Since $h \simeq_{\mathcal{U}''} g \simeq_{\mathcal{U}'} f$, it follows that $h \simeq_{\mathcal{U}} f$. Thus, W has the property of Lemma 5.7.10, which implies τ-DAP($\mathfrak{M}_0^{(n)}$). □

Lemma 5.7.12 *If ANR X has τ-DAP($\mathfrak{M}_0^{(n)}$), then X has the following stronger property:*

- *Let $f : A = \bigoplus_{\gamma \in \Gamma} A_\gamma \to X$ be a map and $B = \bigoplus_{\gamma \in \Gamma} B_\gamma$ a closed subset of A such that $(f(B_\gamma))_{\gamma \in \Gamma}$ is discrete in X and each $f(B_\gamma)$ is strong Z-set in X, where $A_\gamma \in \mathfrak{M}_0^{(n)}$ and $B_\gamma = A_\gamma \cap B$ for each $\gamma \in \Gamma$. Then, for each open cover $\mathcal{U}$ of X, there is a map $g : A \to X$ such that $g|B = f|B$, $g \simeq_{\mathcal{U}} f$ rel. B, and $(g(A_\gamma))_{\gamma \in \Gamma}$ is discrete in X.*

Proof Let $\mathcal{U}_1 \in \operatorname{cov}(X)$ be a star-refinement of $\mathcal{U}$. Since $(f(B_\gamma))_{\gamma \in \Gamma}$ is discrete in X, we can choose open sets U_γ, $\gamma \in \Gamma$, so that $(U_\gamma)_{\gamma \in \Gamma}$ is discrete in X and $f(B_\gamma) \subset U_\gamma$. Choose $\mathcal{U}_2 \in \operatorname{cov}(X)$ so that

$$\mathcal{U}_2 \prec \mathcal{U}_1 \wedge \{U_\gamma,\ X \setminus f(B) \mid \gamma \in \Gamma\}.$$

Due to Corollary 2.8.16, as a discrete union of strong Z-sets, $f(B) = \bigcup_{\gamma \in \Gamma} f(B_\gamma)$ is a strong Z-set in X. So, there is a $\mathcal{U}_2$-homotopy $h : X \times \mathbf{I} \to X$ such that $h_0 = \operatorname{id}$ and $\operatorname{cl} h_1(X) \cap f(B) = \emptyset$. Then, choose an open set G in X such that $f(B) \subset G \subset \operatorname{cl} G \subset \bigcup_{\gamma \in \Gamma} U_\gamma$ and $h_1(X) \cap G = \emptyset$. For each $\gamma \in \Gamma$, let $G_\gamma = U_\gamma \cap G$.

Since $h(f(B_\gamma) \times \mathbf{I}) \subset U_\gamma$, we can easily find an open neighborhood V_γ of B_γ in A_γ such that $h(f(V_\gamma) \times \mathbf{I}) \subset U_\gamma$. Let $\alpha : A \to \mathbf{I}$ be an Urysohn map with $\alpha(B) = 0$ and $\alpha(A \setminus V) = 1$, where $V = \bigcup_{\gamma \in \Gamma} V_\gamma$. We define a map $f' : A \to X$ by $f'(x) = h(f(x), \alpha(x))$ for each $x \in A$. Then, $f'|B = h_0 f|B = f|B$ and $f' \simeq_{\mathcal{U}_2} h_0 f = f$ rel. B. Since $f'(A \setminus V) \subset h_1(X)$, $G_\gamma \subset U_\gamma$, and $U_\gamma \cap \bigcup_{\gamma' \neq \gamma} U_{\gamma'} = \emptyset$, it follows that

$$
\begin{aligned}
f'(A \setminus V_\gamma) = f'(A \setminus V) \cup \bigcup_{\gamma' \neq \gamma} f'(V_{\gamma'}) &\subset h_1(X) \cup \bigcup_{\gamma' \neq \gamma} U_{\gamma'} \\
&\subset (X \setminus G) \cup \bigcup_{\gamma' \neq \gamma} G_{\gamma'} \cup \bigcup_{\gamma' \neq \gamma} U_{\gamma'} \\
&= (X \setminus G_\gamma) \cup \bigcup_{\gamma' \neq \gamma} U_{\gamma'} = X \setminus G_\gamma.
\end{aligned}
$$

Therefore, $f'(A \setminus V_\gamma) \cap G_\gamma = \emptyset$ for each $\gamma \in \Gamma$.

For each $\gamma \in \Gamma$, choose an open neighborhood H_γ of $f(B_\gamma)$ in X so that $\operatorname{cl} H_\gamma \subset G_\gamma$ and let $H = \bigcup_{\gamma \in \Gamma} H_\gamma$. Then, $\operatorname{cl} H = \bigcup_{\gamma \in \Gamma} \operatorname{cl} H_\gamma$ because $(H_\gamma)_{\gamma \in \Gamma}$ is discrete in X. Now, choose $\mathcal{U}_3 \in \operatorname{cov}(X)$ so that

$$
\mathcal{U}_3 \prec \mathcal{U}_1 \wedge \{H_\gamma,\ G_\gamma \setminus f(B_\gamma),\ X \setminus \operatorname{cl} H \mid \gamma \in \Gamma\}.
$$

Using the τ-DAP($\mathfrak{M}^{(n)}$) of X, we can obtain a $\mathcal{U}_3$-homotopy $h' : A \times \mathbf{I} \to X$ such that $h'_0 = f'$ and $(h'_1(A_\gamma))_{\gamma \in \Gamma}$ is discrete in X. Since $h'_0|B = f'|B = f|B$, it follows that $h'(B_\gamma \times \mathbf{I}) \subset H_\gamma$ for each $\gamma \in \Gamma$. Each B_γ has an open neighborhood W_γ in A_γ such that $h'(W_\gamma \times \mathbf{I}) \subset H_\gamma$. Let $\alpha' : A \to \mathbf{I}$ be an Urysohn map with $\alpha'(B) = 0$ and $\alpha'(A \setminus W) = 1$, where $W = \bigcup_{\gamma \in \Gamma} W_\gamma$. Now, the desire map $g : A \to X$ can be defined by $g(x) = h'(x, \alpha'(x))$ for each $x \in A$. Indeed, $g|B = h'_0|B = f'|B = f|B$ and $g \simeq_{\mathcal{U}_3} h'_0 = f' \simeq_{\mathcal{U}_2} f$ rel. B. Since $\mathcal{U}_2, \mathcal{U}_3 \prec \mathcal{U}_1$ and $\operatorname{st} \mathcal{U}_1 \prec \mathcal{U}$, it follows that $g \simeq_{\mathcal{U}} f$ rel. B. Thus, it remains to be shown that $(g(A_\gamma))_{\gamma \in \Gamma}$ is discrete in X.

To see the discreteness of $(g(A_\gamma))_{\gamma \in \Gamma}$, note that $g(A_\gamma \cap W_\gamma) \subset H_\gamma$ for each $\gamma \in \Gamma$, and that $(g(A_\gamma \setminus W_\gamma))_{\gamma \in \Gamma}$ is discrete in X because $g(A_\gamma \setminus W_\gamma) \subset h'_1(A_\gamma)$ for each $\gamma \in \Gamma$. When $x \in \bigcup_{\gamma \in \Gamma} g(A_\gamma \setminus W_\gamma)$ and $x \in \operatorname{cl} H = \bigcup_{\gamma \in \Gamma} \operatorname{cl} H_\gamma$, we have $\gamma_0, \gamma_1 \in \Gamma$ such that $x \in g(A_{\gamma_0} \setminus W_{\gamma_0})$ and $x \in \operatorname{cl} H_{\gamma_1}$, where $x = g(a)$ for some $a \in A_{\gamma_0} \setminus W_{\gamma_0}$. If $\gamma_0 \neq \gamma_1$, then $a \in A_{\gamma_0} \subset A \setminus V_{\gamma_1}$. Since $g(a) = x \in \operatorname{cl} H_{\gamma_1} \subset G_{\gamma_1}$ and $g \simeq_{\mathcal{U}_3} f'$, it follows that $f'(a) \in G_{\gamma_1}$ but $f'(a) \in f'(A \setminus V_{\gamma_1}) \subset X \setminus G_{\gamma_1}$, which is a contradiction. Therefore, $\gamma_0 = \gamma_1$. Let

$$
U_x = X \setminus \left(\bigcup_{\gamma \neq \gamma_0} g(A_\gamma \setminus W_\gamma) \cup \bigcup_{\gamma \neq \gamma_0} \operatorname{cl} H_\gamma \right).
$$

Then, U_x is an open neighborhood in X that meets at most one of $g(A_\gamma)$.

When $x \in X \setminus \bigcup_{\gamma\in\Gamma} g(A_\gamma \setminus W_\gamma)$, since $(\mathrm{cl}\, H_\gamma)_{\gamma\in\Gamma}$ is discrete in X, x has a neighborhood U_x in X such that U_x does not meet any $g(A_\gamma \setminus W_\gamma)$ and U_x meets at most one of $\mathrm{cl}\, H_\gamma$. Then, U_x meets at most one of $g(A_\gamma)$. When $x \in X \setminus \mathrm{cl}\, H$, x has a neighborhood U_x in X such that U_x does not meet any $\mathrm{cl}\, H_\gamma$ and U_x meets at most one of $g(A_\gamma \setminus W_\gamma)$. Then, U_x meets at most one of $g(A_\gamma)$. □

Observe that $(\tau\mathfrak{M}_0)^{\mathrm{fd}} = \bigcup_{n\in\omega} \tau\mathfrak{M}_0^{(n)}$.

Lemma 5.7.13 *Let X be a strongly $\mathfrak{M}_0^{\mathrm{fd}}$-universal ANR. If X has the τ-DAP($\mathfrak{M}_0^{(n)}$) for every $n \in \omega$, then X is strongly $(\tau\mathfrak{M}_0)^{\mathrm{fd}}$-universal.*

Proof Let $f : A = \bigoplus_{\gamma\in\Gamma} A_\gamma \to X$ be a map and $B = \bigoplus_{\gamma\in\Gamma} B_\gamma$ a closed subset A such that $f|B$ is a strong Z-embedding into X, where $A_\gamma \in \mathfrak{M}_0^{(n)}$ and $B_\gamma = A_\gamma \cap B$ for each $\gamma \in \Gamma$. For each $\mathcal{U} \in \mathrm{cov}(X)$, let $\mathcal{V} \in \mathrm{cov}(X)$ be a star-refinement of $\mathcal{U}$. Since $(f(B_\gamma))_{\gamma\in\Gamma}$ is discrete in X, we can apply Lemma 5.7.12 to obtain a map $f' : A \to X$ such that $f'|B = f|B$, $f' \simeq_{\mathcal{V}} f$ rel. B, and $(f'(A_\gamma))_{\gamma\in\Gamma}$ is discrete in X. Then, X has open sets U_γ, $\gamma \in \Gamma$, such that $f'(A_\gamma) \subset U_\gamma$ and $(U_\gamma)_{\gamma\in\Gamma}$ is discrete in X. Note that $f'(A) = \bigcup_{\gamma\in\Gamma} f'(A_\gamma)$ is closed in X. We can choose $\mathcal{W} \in \mathrm{cov}(X)$ so that

$$\mathcal{W} \prec \mathcal{V} \wedge \{U_\gamma,\ X \setminus f'(A)\}.$$

Now, we can use the strong $\mathfrak{M}_0^{\mathrm{fd}}$-universality of X to obtain strong Z-embeddings $g_\gamma : A_\gamma \to X$ such that $g_\gamma|B_\gamma = f'|B_\gamma = f|B_\gamma$ and $g_\gamma \simeq_{\mathcal{W}} f'|A_\gamma$ rel. B_γ. Then, $g_\gamma(A_\gamma) \subset U_\gamma$ for every $\gamma \in \Gamma$. Since $(U_\gamma)_{\gamma\in\Gamma}$ is discrete in X, the map $g = \bigoplus_{\gamma\in\Gamma} g_\gamma : A = \bigoplus_{\gamma\in\Gamma} A_\gamma \to X$ is a strong Z-embedding and $g|B = f'|B = f|B$. Then, $g \simeq_{\mathcal{W}} f' \simeq_{\mathcal{V}} f$, which implies $g \simeq_{\mathcal{U}} f$. Therefore, X is strongly $\tau\mathfrak{M}_0^{(n)}$-universal for every $n \in \omega$, which means that X is strongly $(\tau\mathfrak{M}_0)^{\mathrm{fd}}$-universal. □

Recall that $(\mathfrak{C}^{\mathrm{fd}})_\sigma = \mathfrak{C}_\sigma^{\mathrm{scd}}$ $\left(= (\mathfrak{C}_\sigma)^{\mathrm{scd}} = (\mathfrak{C}^{\mathrm{scd}})_\sigma\right)$ (cf. p. 377), where $\mathfrak{C}^{\mathrm{scd}}$ (resp. $\mathfrak{C}^{\mathrm{fd}}$) is the subclass of $\mathfrak{C}$ consisting of strongly countable-dimensional (resp. finite-dimensional) spaces. In Remark 5.10 (p. 396), we observed that

$$(\tau\mathfrak{M}_0^{\mathrm{fd}})_\sigma = ((\tau\mathfrak{M}_0)^{\mathrm{fd}})_\sigma = (\tau\mathfrak{M}_0)_\sigma^{\mathrm{scd}}.$$

Now, we have the following theorem:

Theorem 5.7.14 *Let X be a strongly $\mathfrak{M}_0^{\mathrm{fd}}$-universal ANR with $w(X) = \tau$ and a strong Z_σ-space. If X has the τ-discrete cells property, then X is strongly $(\tau\mathfrak{M}_0)_\sigma^{\mathrm{scd}}$-universal.*

Proof Note that every Z-set in X is a strong Z-set by Lemma 5.6.4. By virtue of Proposition 5.6.3, it suffices so show that each point $x \in X$ has a strongly $(\tau\mathfrak{M}_0)_\sigma^{\mathrm{scd}}$-universal neighborhood. Since X is an ANR, each $x \in X$ has an open neighborhood U that is contractible in X, where $w(U) = \tau$.[17] Then, we can apply Lemma 5.7.11 to see that U has the τ-DAP($\mathfrak{M}_0^{(n)}$) for every $n \in \omega$. Since U is strongly $\mathfrak{M}_0^{\mathrm{fd}}$-universal by Proposition 5.4.18, it follows from Lemma 5.7.13 that U is strongly $(\tau\mathfrak{M}_0)^{\mathrm{fd}}$-universal, where $(\tau\mathfrak{M}_0)^{\mathrm{fd}}$ is $\mathbf{I}$-stable and $\Gamma \in (\tau\mathfrak{M}_0)^{\mathrm{fd}}$. Since $((\tau\mathfrak{M}_0)^{\mathrm{fd}})_\sigma = (\tau\mathfrak{M}_0)_\sigma^{\mathrm{scd}}$ as observed above, it follows from Proposition 5.6.5 that U is strongly $(\tau\mathfrak{M}_0)_\sigma^{\mathrm{scd}}$-universal. □

As seen in Proposition 5.4.15 with Remark 5.10, $|K|_{\mathrm{m}} \in (\tau\mathfrak{M}_0^{\mathrm{fd}})_\sigma = (\tau\mathfrak{M}_0)_\sigma^{\mathrm{scd}}$ for an arbitrary simplicial complex K with $\operatorname{card} K \leqslant \tau$. Hence, if X is strongly $(\tau\mathfrak{M}_0)_\sigma^{\mathrm{scd}}$-universal, then every map $f : |K|_{\mathrm{m}} \to X$ can be approximated by strong Z-embeddings, which implies the discrete (or locally finite) τ-polyhedral sequence property of X, that is, for any sequence $(K_i)_{i\in\mathbb{N}}$ of finite-dimensional simplicial complexes with $\operatorname{card} K_i \leqslant \tau$, each map $f : \bigoplus_{i\in\mathbb{N}} |K_i|_{\mathrm{m}} \to X$ can be approximated by maps $g : \bigoplus_{i\in\mathbb{N}} |K_i|_{\mathrm{m}} \to X$ such that $(g(|K_i|))_{i\in\mathbb{N}}$ is discrete (or locally finite) in X.[18] Thus, we have the following corollary:

Corollary 5.7.15 *Let X be a strongly $\mathfrak{M}_0^{\mathrm{fd}}$-universal ANR with $w(X) = \tau$ and a strong Z_σ-space. If X has the τ-discrete cells property, then X has the discrete (or locally finite) τ-polyhedral sequence property.* □

Recall that the following is the discrete (or locally finite) cell-tower property,[19] which is a little stronger than the countable discrete cells property:

- Each map $f : \bigoplus_{n\in\mathbb{N}} \mathbf{I}^n \to X$ can be approximate by maps $g : \bigoplus_{n\in\mathbb{N}} \mathbf{I}^n \to X$ such that $(g(\mathbf{I}^n))_{n\in\mathbb{N}}$ is discrete (or locally finite).

The following is the separable version of Corollary 5.7.15:

Proposition 5.7.16 *Let X be a separable ANR that is a strong Z_σ-space. If X is strongly $\mathfrak{M}_0^{\mathrm{fd}}$-universal, then X has the discrete cell-tower property.*

Proof Due to Proposition 5.6.7, X is strongly $(\mathfrak{M}_0^{\mathrm{fd}})_\sigma$-universal. Since $\bigoplus_{n\in\mathbb{N}} \mathbf{I}^n \in (\mathfrak{M}_0^{\mathrm{fd}})_\sigma$, every map $f : \bigoplus_{n\in\mathbb{N}} \mathbf{I}^n \to X$ can be approximate by embeddings, which implies the discrete cell-tower property of X. □

Now, we can obtain the following characterization of manifolds modeled on an absorption base for $\ell^2(\Gamma)$:

Theorem 5.7.17 *Suppose that $\mathfrak{M}_0^{\mathrm{fd}} \subset \mathfrak{C} \subset \mathfrak{M}_1(\tau)$. Let F be a $\mathfrak{C}$-absorption base for $\ell^2(\Gamma)$. For a space X, the following are equivalent:*

(a) *X is an F-manifold;*
(b) *X can be embedded into an $\ell^2(\Gamma)$-manifold $\widetilde{X}$ such that $(\widetilde{X}, X)$ is an $(\ell^2(\Gamma), F)$-manifold pair;*

[17] See Footnote 15 (p. 412).

[18] For the discrete (or locally finite) τ-polyhedral sequence property, refer to Sect. 3.3.

[19] The discrete cell-tower property is known as the discrete approximation property.

(c) *X can be embedded into an $\ell^2(\Gamma)$-manifold as a $\mathfrak{C}$-absorption base;*
(d) *$X \in (\tau\mathfrak{C})_\sigma$ and X is a strongly $\mathfrak{C}$-universal ANR that is a strong Z_σ-space with the τ-discrete cells property.*

Proof The first three conditions are equivalent. Indeed, the equivalence (b) $\Leftrightarrow$ (c) follows from Theorem 5.7.2, the implication (a) $\Rightarrow$ (c) is Theorem 5.7.3, and (b) $\Rightarrow$ (a) is obvious. We will show the equivalent (c) $\Leftrightarrow$ (d).

(c) $\Rightarrow$ (d): Assume that X is a $\mathfrak{C}$-absorption base for an $\ell^2(\Gamma)$-manifold M. By Corollary 5.3.6, X is an ANR and every Z-set in X is a strong Z-set. Moreover, $X \in (\tau\mathfrak{C})_\sigma(\tau)$ and X is a strongly $\mathfrak{C}$-universal homotopy dense Z_σ-set in M by Theorem 5.7.2. Since X is homotopy dense in M, every Z-set in M contained in X is a Z-set in X, so a strong Z-set in X. Hence, X is a strong Z_σ-set in itself, that is, it is a strong Z_σ-space.

We will show that X has the τ-discrete cells property. For each $n \in \mathbb{N}$, let $f : \mathbf{I}^n \times \Gamma \to X$ be a map. For each $\mathcal{U} \in \operatorname{cov}(X)$, let $\mathcal{V}$ be an open collection in M such that $\operatorname{st}\mathcal{V}|X \prec \mathcal{U}$. Then, $N = \bigcup\mathcal{V}$ is an $\ell^2(\Gamma)$-manifold as an open set in M. Since N has the τ-discrete cells property, we have a map $g : \mathbf{I}^n \times \Gamma \to N$ such that g is $\mathcal{V}$-close to f and $\{g(\mathbf{I}^n \times \{\gamma\}) \mid \gamma \in \Gamma\}$ is discrete in N, that is, there is open sets $W_\gamma, \gamma \in \Gamma$, such that $g(\mathbf{I}^n \times \{\gamma\}) \subset W_\gamma$ for each $\gamma \in \Gamma$ and $Y = \bigcup_{\gamma\in\Gamma} g(\mathbf{I}^n \times \{\gamma\})$ is closed in N. Choose an open set W in M so that

$$Y \subset W \subset \operatorname{cl} W \subset \bigcup_{\gamma\in\Gamma} W_\gamma.$$

and then, choose $\mathcal{V}' \in \operatorname{cov}(N)$ so that

$$\mathcal{V}' \prec \mathcal{V} \wedge \Big(\big\{W_\gamma \cap W \bigm| \gamma \in \Gamma\big\} \cup \big\{X \setminus Y\big\}\Big).$$

Since X is homotopy dense in N, we have a map $h : N \to X$ that is $\mathcal{V}'$-close to id. Then, we have a map $hg : \mathbf{I}^n \times \Gamma \to X$ that is $\mathcal{U}$-close to id. Since $hg(\mathbf{I}^n \times \{\gamma\}) \subset W_\gamma \cap W$ for each $\gamma \in \Gamma$, it follows that $\{hg(\mathbf{I}^n \times \{\gamma\}) \mid \gamma \in \Gamma\}$ is discrete in X. Thus, X has the τ-discrete cells property.

(d) $\Rightarrow$ (c): Because $X \in (\tau\mathfrak{C})_\sigma$, X has closed sets $X_i \in \tau\mathfrak{C}$, $i \in \mathbb{N}$, with $X = \bigcup_{i\in\mathbb{N}} X_i$. Since $\tau\mathfrak{C} \subset \mathfrak{M}_1(\tau)$, each X_i is completely metrizable. Applying Hausdorff's Metric Extension Theorem 1.13.8, we can obtain metrics $d_i \in \operatorname{Metr}(X)$, $i \in \mathbb{N}$, such that each X_i is d_i-complete and $d_i \leqslant 1$. Then, we can define $d \in \operatorname{Metr}(X)$ by $d(x, y) = \sum_{i\in\mathbb{N}} 2^{-i} d_i(x, y)$ for each $x, y \in X$. Now, applying Theorem 1.13.37, we can obtain $\rho \in \operatorname{Metr}(X)$ such that $\rho \geqslant d$ such that (X, ρ) is a uniform ANR. Then, it follows from Theorem 1.13.35 that the metric completion $\widetilde{X}$ of (X, ρ) is an ANR and X is homotopy dense in $\widetilde{X}$. Since X_i is ρ-complete because $2^{-i} d_i \leqslant \rho$, each X_i is closed in $\widetilde{X}$.

We show that $\widetilde{X}$ is an $\ell^2(\Gamma)$-manifold. First, note that $\widetilde{X}$ is completely metrizable ANR. Since X is homotopy dense in $\widetilde{X}$ and it has the τ-discrete cells property, $\widetilde{X}$ has also the τ-discrete cells property and $w(\widetilde{X}) = w(X) \leqslant \tau$. Because $\mathfrak{M}_0^{\mathrm{fd}} \subset \mathfrak{C}$, the strong $\mathfrak{C}$-universality implies the strong $\mathfrak{M}_0^{\mathrm{fd}}$-universality. Hence, we can apply Corollary 5.7.15 to see that X has the discrete τ-polyhedral sequence property of X, which implies the discrete τ-polyhedral sequence property of $\widetilde{X}$ because X is homotopy dense in $\widetilde{X}$. By virtue of the Toruńczyk Characterization (Theorem 3.4.2), $\widetilde{X}$ is an $\ell^2(\Gamma)$-manifold.

Recall that each X_i in the above is closed in $\widetilde{X}$. Since X is a strong Z_σ-space, there are strong Z-sets A_n, $n \in \mathbb{N}$, in X with $X = \bigcup_{n\in\mathbb{N}} A_n$. Since X is homotopy dense in $\widetilde{X}$ and A_n is a Z-set in X, it follows that each $A_n \cap X_i$ is a Z-set in $\widetilde{X}$. Hence, X is a homotopy dense Z_σ-set in $\widetilde{X}$. Since $X \in (\tau\mathfrak{C})_\sigma$ and X is strongly $\mathfrak{C}$-universal, X is a $\mathfrak{C}$-absorption base for $\widetilde{X}$ by Corollary 5.4.2(2). The proof is complete. □

Remark 5.17 In Theorem 5.7.17, when $\mathbf{I}^n \times \Gamma \in \mathfrak{C}$ for every $n \in \mathbb{N}$, the τ-discrete cells property can be removed from condition (d) because the strong $\mathfrak{C}$-universality implies the τ-discrete cells property.

The following is the separable version of Theorem 5.7.17 above:

Theorem 5.7.18 *Suppose that $\mathfrak{M}_0^{\mathrm{fd}} \subset \mathfrak{C} \subset \mathfrak{M}_1(\aleph_0)$. Let F be a $\mathfrak{C}$-absorption base for ℓ^2. For a space X, the following are equivalent:*

(a) *X is an F-manifold;*
(b) *X can be embedded into an ℓ^2-manifold $\widetilde{X}$ such that $(\widetilde{X}, X)$ is an (ℓ^2, F)-manifold pair;*
(c) *X can be embedded into an ℓ^2-manifold as a $\mathfrak{C}$-absorption base;*
(d) *$X \in \mathfrak{C}_\sigma$ and X is a strongly $\mathfrak{C}$-universal ANR that is a strong Z_σ-space.*

Note that the discrete cells property is removed from the above (d), where it should also be noted that $(\aleph_0\mathfrak{C})_\sigma = \mathfrak{C}_\sigma$. For the proof, we may only adjust the second paragraph of the above proof of Theorem 5.7.17 (d) ⇒ (c). To show that $\widetilde{X}$ is an ℓ^2-manifold, it suffices to prove that $\widetilde{X}$ has the discrete cell-tower property. This is easy. Indeed, X is homotopy dense in $\widetilde{X}$ as seen in the first paragraph and X has the discrete cell-tower property by Proposition 5.7.16 because X is strongly $\mathfrak{M}_0^{\mathrm{fd}}$-universal.

5.8 Homotopy Dense Embedding into an $\ell^2(\Gamma)$-Manifold

In the previous section, we considered manifolds modeled on a $\mathfrak{C}$-absorption base F for $\ell^2(\Gamma)$. To characterize such manifolds, the homotopy dense embeddability into an $\ell^2(\Gamma)$-manifold is critical. In this section, we characterize spaces which can be embedded into an $\ell^2(\Gamma)$-manifold as a homotopy dense set, that is, we prove the following theorem:

Theorem 5.8.1 (T. Banakh) *A space X can be embedded into an $\ell^2(\Gamma)$-manifold as a homotopy dense set if and only if X is an ANR with $w(X) = \tau$ and X has both the τ-locally finite cells property and the locally finite displacing property.*[20]

To prove the above theorem, we need the following lemma:

Lemma 5.8.2 *Let $X = (X, d)$ be a metric ANR and $A = \bigoplus_{\lambda\in\Lambda} A_\lambda$ a topological sum of spaces. Suppose that each map $f : A \times \mathbf{I} \to X$ can be approximated by maps $g : A \times \mathbf{I} \to X$ such that $(g(A_\lambda \times \mathbf{I}))_{\lambda\in\Lambda}$ is locally finite in X. Then, for each map $f : A \to X$, there exists a map $\bar{f} : A \times \mathbf{I} \to X$ such that:*

(i) *$d(\bar{f}(x,t), f(x)) < t$ for each $(x,t) \in A \times \mathbf{I}$;*
(ii) *$(\bar{f}(A_\lambda \times [t,1]))_{\lambda\in\Lambda}$ is locally finite in X for each $t > 0$.*

Proof For each map $f : A \to X$, let $f_0 = f\mathrm{pr}_A : A \times \mathbf{I} \to X$. Since X is an ANR, we can apply the assumption to obtain a map $g : A \times \mathbf{I} \to X$ such that $f_0 \simeq_{2^{-2}} g$ and $(g(A_\lambda \times \mathbf{I}))_{\lambda\in\Lambda}$ is locally finite in X. Then, we can easily construct a map $f_1 : A\times\mathbf{I} \to X$ such that $f_1|A\times[0,1/2] = f_0|A\times[0,1/2]$, $f_1|A\times\{1\} = g|A\times\{1\}$, and $d(f_1(x,t), f(x)) = d(f_1(x,t), f_0(x,t)) < 2^{-2}$ for every $(x,t) \in A \times \mathbf{I}$.

By induction, we can construct maps $f_n : A \times \mathbf{I} \to X$, $n \in \mathbb{N}$, so that

$$(f_n(A_\lambda \times [2^{-n+1}, 1]))_{\lambda\in\Lambda} \text{ is locally finite in } X,$$

$$f_n|A \times ([0, 2^{-n}] \cup [2^{-n+2}, 1]) = f_{n-1}|A \times ([0, 2^{-n}] \cup [2^{-n+2}, 1]), \text{ and}$$

$$d(f_n(x,t), f_{n-1}(x,t)) < 2^{-n-1} \text{ for every } (x,t) \in A \times \mathbf{I},$$

where $[2^{-n+2}, 1] = \emptyset$ if $n = 1$. Indeed, assume that f_n has been constructed. Since $(f_n(A_\lambda \times [2^{-n+1}, 1]))_{\lambda\in\Lambda}$ is locally finite in X, we have an open cover $\mathcal{U}$ such that each $U \in \mathcal{U}$ meets only finitely many $f_n(A_\lambda \times [2^{-n+1}, 1])$, $\lambda \in \Lambda$. Take $\mathcal{V} \in \mathrm{cov}(X)$ so that $\mathrm{st}\,\mathcal{V} \prec \mathcal{U}$ and $\mathrm{mesh}\,\mathcal{V} < 2^{-n-2}$. Then, each $V \in \mathcal{V}$ meets only finitely many $\mathrm{st}(f_n(A_\lambda \times [2^{-n+1}, 1]), \mathcal{V})$, $\lambda \in \Lambda$, which implies that $(\mathrm{st}(f_n(A_\lambda \times [2^{-n+1}, 1]), \mathcal{V}))_{\lambda\in\Lambda}$ is also locally finite in X. For each $\lambda \in \Lambda$, choose an open neighborhood W_λ of $A_\lambda \times [2^{-n+1}, 1]$ in $A_\lambda \times \mathbf{I}$ so that

$$f_n(W_\lambda) \subset \mathrm{st}(f_n(A_\lambda \times [2^{-n+1}, 1]), \mathcal{V}).$$

Since X is an ANR, we can apply the assumption to obtain a map $g : A \times \mathbf{I} \to X$ such that $f_n \simeq_{\mathcal{V}} g$ and $(g(A_\lambda \times \mathbf{I}))_{\lambda\in\Lambda}$ is locally finite in X. Then, we can easily construct a map $f_{n+1} : A \times \mathbf{I} \to X$ such that f_{n+1} is $\mathcal{V}$-close to f_n,

$$f_{n+1}|A \times ([0, 2^{-n-1}] \cup [2^{-n+1}, 1]) = f_n|A \times ([0, 2^{-n-1}] \cup [2^{-n+1}, 1]), \text{ and}$$

$$f_{n+1}|(A_\lambda \times [2^{-n}, 1]) \setminus W_\lambda = g|(A_\lambda \times [2^{-n}, 1]) \setminus W_\lambda \text{ for each } \lambda \in \Lambda,$$

[20]This is also called the countable locally finite approximation property—see the footnote on p. 238.

Since mesh $\mathcal{V} < 2^{-n-2}$, it follows that

$$d(f_{n+1}(x,t), f_n(x,t)) < 2^{-n-2} \text{ for every } (x,t) \in A \times \mathbf{I}.$$

Moreover, for each $\lambda \in \Lambda$,

$$f_{n+1}(A_\lambda \times [2^{-n}, 1]) \subset g(A_\lambda \times \mathbf{I}) \cup \mathrm{st}(f_n(A_\lambda \times [2^{-n+1}, 1], \mathcal{V}),$$

which implies that $(f_{n+1}(A_\lambda \times [2^{-n}, 1]))_{\lambda\in\Lambda}$ is locally finite in X. Thus, f_{n+1} satisfies the desired properties.

Finally, let $\bar{f} = \lim_{n\to\infty} f_n : A \times \mathbf{I} \to X$ be the uniform limit of $(f_n)_{n\in\mathbb{N}}$. For each $t > 0$, take $n \in \mathbb{N}$ so that $2^{-n} < t \leqslant 2^{-n+1}$. Since $\bar{f}|A \times [2^{-n}, 1] = f_{n+1}|A \times [2^{-n}, 1]$ and $(f_{n+1}(A_\lambda \times [2^{-n}, 1]))_{\lambda\in\Lambda}$ is locally finite in X. Hence, $(\bar{f}(A_\lambda \times [t, 1])_{\lambda\in\Lambda}$ is also locally finite in X. Moreover, observe

$$f_{n-1}|A \times [0, 2^{-n+1}] = \cdots = f_0|A \times [0, 2^{-n+1}] = f\mathrm{pr}_A|A \times [0, 2^{-n+1}].$$

Then, it follows that

$$\begin{aligned} d(\bar{f}(x,t), f(x)) &= d(f_{n+1}(x,t), f_n(x,t)) + d(f_n(x,t), f_{n-1}(x,t)) \\ &< 2^{-n-2} + 2^{-n-1} < 2^{-n} < t. \end{aligned}$$

Thus, $\bar{f}$ satisfies (1) and (2). □

Proof (Theorem 5.8.1) (*The "only if" Part*) Assume that X is a homotopy dense set in an $\ell^2(\Gamma)$-manifold M. Then, $w(X) = w(M) = \tau$ and X is an ANR because every homotopy dense set in an ANR is also an ANR (Theorem 1.13.26). We will show that X has the τ-locally finite cells property (resp. the locally finite displacing property). Let $f : \mathbf{I}^n \times \Gamma \to X$ be a map. For each $\mathcal{U} \in \mathrm{cov}(X)$, there is an open collection $\widetilde{\mathcal{U}}$ in M such that $\widetilde{\mathcal{U}}|X = \mathcal{U}$. Then, $N = \bigcup \widetilde{\mathcal{U}}$ is an $\ell^2(\Gamma)$-manifold as an open set in M. For a star-refinement $\mathcal{V} \in \mathrm{cov}(N)$ of $\widetilde{\mathcal{U}}$, f is $\mathcal{V}$-close to a map $g : \mathbf{I}^n \times \Gamma \to X \to N$ (resp. $\mathrm{pr}_N : N \times \mathbb{N} \to N$ is $\mathcal{V}$-close to a map $g : N \times \mathbb{N} \to N$) such that $(g_\gamma(\mathbf{I}^n))_{\gamma\in\Gamma}$ (resp. $(g(N \times \{n\}))_{n\in\mathbb{N}}$) is locally finite in N. Then, $\mathcal{V}$ has an open refinement $\mathcal{W} \in \mathrm{cov}(N)$ such that the star $\mathrm{st}(W, \mathcal{W})$ of each $W \in \mathcal{W}$ meets only finitely many $g_\gamma(\mathbf{I}^n)$, $\gamma \in \Gamma$ (resp. $g(N \times \{n\})$, $n \in \mathbb{N}$), which means that each $W \in \mathcal{W}$ meets only finitely many $\mathrm{st}(g_\gamma(\mathbf{I}^n), \mathcal{W})$, $\gamma \in \Gamma$ (resp. $\mathrm{st}(g(N \times \{n\}), \mathcal{W})$, $n \in \mathbb{N}$). Since X is homotopy dense in M, there is a map $h : M \to X$ that is $\mathcal{W}$-close to id. Thus, we have a map $hg : \mathbf{I}^n \times \Gamma \to X$ (resp. $hg|X \times \mathbb{N} : X \times \mathbb{N} \to X$) that is $\mathcal{V}$-close to f (resp. $\mathrm{pr}_X = \mathrm{pr}_N|X \times \mathbb{N}$). Since each $W \in \mathcal{W}$ meets only finitely many $hg_\gamma(\mathbf{I}^n)$, $\gamma \in \Gamma$ (resp. $hg(X \times \{n\})$, $n \in \mathbb{N}$) $(hg_\gamma(\mathbf{I}^n))_{\gamma\in\Gamma}$ (resp. $(hg(X \times \{n\}))_{n\in\mathbb{N}}$) is locally finite in X.

(*The "if" Part*) Assume that X is an ANR with $w(X) = \tau$ and X has both the τ-locally finite cells property and the locally finite τ-skeletal tower property. By combination Theorems 1.3.10, 1.3.17 (2), and Proposition 1.13.36 (or

Theorems 1.13.37 and 1.13.35), X can be embedded into a completely metrizable ANR $\widetilde{X}$ as a homotopy dense set, where $w(\widetilde{X}) = w(X) = \tau$. Let $d \in \mathrm{Metr}(\widetilde{X})$.

For each $n \in \mathbb{N}$, let $\mathcal{F}_n$ be a dense subset of $\mathrm{C}(\mathbf{I}^n, X)$ with

$$\operatorname{card}\mathcal{F}_n = \operatorname{dens}\mathrm{C}(\mathbf{I}^n, X) = w(\mathrm{C}(\mathbf{I}^n, X) = w(X) = \operatorname{card}\Gamma$$

(cf. Proposition 1.1.2 (5)). Taking a bijection $\eta_n : \Gamma \to \Gamma \times \mathcal{F}_n$, we define a map $f_n : \mathbf{I}^n \times \Gamma \to X$ by

$$f_n(x, \gamma) = \mathrm{pr}_{\mathcal{F}_n}\eta_n(\gamma)(x) \ \text{ for each } (x, \gamma) \in \mathbf{I}^n \times \Gamma.$$

Since X has the τ-locally finite cells property, we can apply Lemma 5.8.2 to obtain a map $\bar{f}_n : \mathbf{I}^n \times \Gamma \times \mathbf{I} \to X$ such that:

(1) $d(\bar{f}_n(x, \gamma, t), f_n(x, \gamma)) < t$ for each $(x, \gamma, t) \in \mathbf{I}^n \times \Gamma \times \mathbf{I}$,
(2) for every $t > 0$, $(\bar{f}_n(\mathbf{I}^n \times \{\gamma\} \times [t, 1]))_{\gamma\in\Gamma}$ is locally finite in X.

Then, for each $m \in \mathbb{N}$, we can find an open neighborhood U^n_m of X in $\widetilde{X}$ so that

(3) $(\bar{f}_n(\mathbf{I}^n \times \{\gamma\} \times [1/m, 1]))_{\gamma\in\Gamma}$ is locally finite in U^m_n.

Since X has the locally finite displacing property, we can again apply Lemma 5.8.2 to obtain a map $p : X \times \mathbb{N} \times \mathbf{I} \to X$ such that:

(4) $d(p(x, n, t), x) < t$ for each $(x, n, t) \in X \times \mathbb{N} \times \mathbf{I}$,
(5) for every $t > 0$, $(p(X \times \{n\} \times [t, 1]))_{n\in\mathbb{N}}$ is locally finite in X.

Again, for each $m \in \mathbb{N}$, we can find an open neighborhood V_m of X in $\widetilde{X}$ so that

(6) $(p(X \times \{n\} \times [1/m, 1]))_{n\in\mathbb{N}}$ is locally finite in V_m.

We will show that $M = \bigcap_{n,m\in\mathbb{N}}(U^n_m \cap V_m)$ is the required $\ell^2(\Gamma)$-manifold. First, M is completely metrizable as a G_δ set in $\widetilde{X}$ (Theorem 1.3.17(2)). Since X is homotopy dense in $\widetilde{X}$, X is also homotopy dense in M. From Theorem 1.13.26 it follows that M is an ANR.

To prove the τ-locally finite cells property of M, let $g : \mathbf{I}^n \times \Gamma \to M$ be a map. For each $\gamma \in \Gamma$, let $g_\gamma : \mathbf{I}^n \to M$ be the map defined by $g_\gamma(x) = g(x, \gamma)$ for $x \in \mathbf{I}^n$. For each $\mathcal{U} \in \mathrm{cov}(M)$, let $\mathcal{V} \in \mathrm{cov}(M)$ be a star-refinement of $\mathcal{U}$. Since X is homotopy dense in M, we have a map $h : M \to X$ that is $\mathcal{V}$-close to id_M, which implies that hg is $\mathcal{V}$-close to g. Because $\mathcal{F}_n$ is dense in $\mathrm{C}(\mathbf{I}^n, X)$, each hg_γ is $\mathcal{V}$-close to a map $k_\gamma \in \mathcal{F}_n$. Then, hg is $\mathcal{V}$-close to the map $k : \mathbf{I}^n \times \Gamma \to X$ defined by $k(x, \gamma) = k_\gamma(x)$ for each $(x, \gamma) \in \mathbf{I}^n \times \Gamma$. Moreover, we define an injection $\chi : \Gamma \to \Gamma \times \mathcal{F}_n$ by $\chi(\gamma) = (\gamma, k_\gamma)$ for each $\gamma \in \Gamma$. Then, $\xi = \eta_n^{-1}\chi : \Gamma \to \Gamma$ is an injection and $f_n(\mathrm{id}_{\mathbf{I}^n} \times \xi) = k$. Indeed, for each $(x, \gamma) \in \mathbf{I}^n \times \Gamma$,

$$\begin{aligned} f_n(\mathrm{id}_{\mathbf{I}^n} \times \xi)(x, \gamma) &= f_n(x, \xi(\gamma)) = f_n(x, \eta_n^{-1}(\gamma, k_\gamma)) \\ &= \mathrm{pr}_{\mathcal{F}_n}\eta_n(\eta_n^{-1}(\gamma, k_\gamma))(x) = k_\gamma(x) = k(x, \gamma). \end{aligned}$$

By virtue of Lemma 3.1.3, we can take a Lipschitz map $\alpha : M \to (0, 1)$ with the Lipschitz constant $\leqslant 1$ so that $N_\alpha(k) \subset \mathcal{V}(k)$. We define a map $k^* : \mathbf{I}^n \times \Gamma \to M$ as follows:

$$k^*(x, \gamma) = \bar{f}_n(x, \xi(\gamma), \alpha k(x, \gamma)) \text{ for each } (x, \gamma) \in \mathbf{I}^n \times \Gamma.$$

Then, it follows from (1) that for each $(x, \gamma) \in \mathbf{I}^n \times \Gamma$,

$$d(k^*(x, \gamma), k(x, \gamma)) = d(\bar{f}_n(x, \xi(\gamma), \alpha k(x, \gamma)), f_n(x, \xi(\gamma))) < \alpha k(x, \gamma).$$

Therefore, $k^* \in N_\alpha(k) \subset \mathcal{V}(k)$, so k^* is $\mathcal{V}$-close to k. Since $\operatorname{st}\mathcal{V} \prec \mathcal{U}$, it follows that k^* is $\mathcal{U}$-close to g. Moreover, it should be noticed that

$$\begin{aligned} \alpha k^*(x, \gamma) - \alpha k(x, \gamma) &\leqslant \left|\alpha k^*(x, \gamma) - \alpha k(x, \gamma)\right| \\ &\leqslant d(k^*(x, \gamma), k(x, \gamma)) < \alpha k(x, \gamma). \end{aligned}$$

Thus, we have $\alpha k^*(x, \gamma) < 2\alpha k(x, \gamma)$.

We will show that $(k^*(\mathbf{I}^n \times \{\gamma\}))_{\gamma\in\Gamma}$ is locally finite in M. For each $x \in M$, $U_x = \{y \in M \mid \alpha(y) > \alpha(x)/2\}$ is a neighborhood of x in M. For $(y, \gamma) \in \mathbf{I}^n \times \Gamma$, if $k^*(y, \gamma) \in U_x$, then $\alpha k^*(y, \gamma) > \alpha(x)/2$, which implies that $\alpha k(y, \gamma) > \alpha(x)/4$ because $\alpha k^*(x, \gamma) < 2\alpha k(x, \gamma)$. Hence,

$$U_x \cap k^*(\mathbf{I}^n \times \{\gamma\}) \subset U_x \cap \bar{f}_n(\mathbf{I}^n \times \{\xi(\gamma)\} \times [\alpha(x)/4, 1]).$$

Choose $m \in \mathbb{N}$ so that $1/m \leqslant \alpha(x)/4$. By (3), x has a neighborhood V_x in M which meets only finitely many $\bar{f}_n(\mathbf{I}^n \times \{\gamma\} \times [1/m, 1]$, $\gamma \in \Gamma$. Since $\xi : \Gamma \to \Gamma$ is injective, $U_x \cap V_x$ meets only finitely many $k^*(\mathbf{I}^n \times \{\gamma\})$, $\gamma \in \Gamma$. Therefore, $(k^*(\mathbf{I}^n \times \{\gamma\}))_{\gamma\in\Gamma}$ is locally finite in M.

We finally show the locally finite displacing property of M. For each $\mathcal{U} \in \operatorname{cov}(M)$, let $\alpha : M \to (0, 1)$ be a map such that $N_\alpha(\mathrm{pr}_M) \subset \mathcal{U}(\mathrm{pr}_M)$, where $\mathrm{pr}_M : M \times \mathbb{N} \to M$ is the projection. We can take a map $h : M \to X$ such that $d(h(x), x) < \alpha(x)/2$ because X is homotopy dense in M. We define $g : M \times \mathbb{N} \to M$ as follows:

$$g(x, n) = \bar{p}(h(x), n, \alpha(x)/2) \text{ for each } (x, n) \in M \times \mathbb{N}.$$

From (4), it follows that for each $(x, n) \in M \times \mathbb{N}$,

$$\begin{aligned} d(g(x, n), \mathrm{pr}_M(x, n)) &\leqslant d(p(h(x), n, \alpha(x)/2, h(x)) + d(h(x), x) \\ &< \alpha(x)/2 + \alpha(x)/2 = \alpha(x), \end{aligned}$$

which means that $g \in N_\alpha(\mathrm{pr}_M) \subset \mathcal{U}(\mathrm{pr}_M)$, that is, g is $\mathcal{U}$-close to pr_M. For each $x \in M$, $U_x = \{y \in M \mid \alpha(y) > \alpha(x)/2\}$ is a neighborhood of x in M. For $y \in M$, if $g(y,n) \in U_x$, then $\alpha g(y,n) > \alpha(x)/2$. Hence,

$$U_x \cap g(M \times \{n\}) \subset U_x \cap \bar{p}(h(M) \times \{n\} \times [\alpha(x)/2, 1]).$$

Choose $m \in \mathbb{N}$ so that $1/m \leqslant \alpha(x)/2$. By (6), x has a neighborhood V_x in M which meets only finitely many $\bar{p}(h(M) \times \{n\} \times [1/m, 1])$, $n \in \mathbb{N}$. Hence, $U_x \cap V_x$ meets only finitely many $g(M \times \{n\})$, $n \in \mathbb{N}$. Thus, we have shown the locally finite displacing property of M. □

Since the locally finite cell-tower property is equivalent to the countable locally finite Hilbert cubes property, the following can be proved in the same way as the above. The details of the proof are left to the reader.

Theorem 5.8.3 *A space X can be embedded into an ℓ^2-manifold as a homotopy dense set if and only if X is a separable ANR and X has the locally finite cell-tower property.*

In an $\ell^2(\Gamma)$-manifold M, every Z-set is a strong Z-set (Theorem 2.9.3). Due to Proposition 2.9.4, every homotopy dense subset X of M has the same property. Thus, we have the following:

Corollary 5.8.4 *Let X be an ANR. In the following cases, every Z-set in X is a strong Z-set:*

(1) $w(X) = \tau > \aleph_0$, *and X has both the τ-locally finite cells property and the locally finite displacing property;*
(2) *X is separable (i.e, $w(X) = \aleph_0$) and X has the locally finite cell-tower property.* □

5.9 Four Types of Infinite-Dimensional Manifolds

In this section, we characterize four types of manifolds considered in Sect. 5.5, that is, manifolds modeled on the following spaces:

$$\ell^2_f(\Gamma),\ \ell^2_f(\Gamma) \times \ell^2_Q,\ \ell^2_f(\Gamma) \times \ell^2, \text{ or } \ell^2(\Gamma) \times \ell^2_f.$$

These model spaces can be embedded into $\ell^2(\Gamma)$ as absorption bases for the classes $\mathfrak{M}_0^{\mathrm{fd}}$, $\mathfrak{M}_0$, $\mathfrak{M}_1(\aleph_0)$, or $\mathfrak{M}_1(\tau)$. Due to Theorem 5.7.17, these manifolds can be characterized as strongly $\mathfrak{C}$-universal ANRs with the τ-discrete cells property which are strong Z_σ-spaces in the class $(\tau\mathfrak{C})_\sigma$, where $\mathfrak{C}$ is one of $\mathfrak{M}_0^{\mathrm{fd}}$, $\mathfrak{M}_0$, $\mathfrak{M}_1(\aleph_0)$, or $\mathfrak{M}_1(\tau)$. Thus, we have to characterize spaces belonging to the classes $(\tau\mathfrak{M}_0^{\mathrm{fd}})_\sigma$, $(\tau\mathfrak{M}_0)_\sigma$, $(\tau\mathfrak{M}_1(\aleph_0))_\sigma$, and $\mathfrak{M}_1(\tau)_\sigma$ $(= (\mathfrak{M}_1)_\sigma(\tau))$.

It is said that X is **σ-completely metrizable** if $X \in (\mathfrak{M}_1)_\sigma$, that is, it is a countable union of completely metrizable *closed* subspaces, where $(\mathfrak{M}_1)_\sigma(\tau) = \mathfrak{M}_1(\tau)_\sigma$. A space X is said to be **locally separable** if each point $x \in X$ has a separable neighborhood in X. Evidently, every space in $\tau(\mathfrak{M}_1(\aleph_0))$ is locally separable. It is said that X is **σ-locally separable** if it is a countable union of locally separable metrizable closed subspaces.

Proposition 5.9.1 *Every locally separable metrizable space X is a topological sum (i.e., a discrete union) of separable metrizable spaces.*

Proof Since X is paracompact and locally separable, X has a locally finite open cover $\mathcal{U}$ such that each $U \in \mathcal{U}$ is separable. Then, $U \in \mathcal{U}$ has a countable dense subset D. Since each member of $\mathcal{U}[U]$ contains some $x \in D$ and each $x \in D$ is contained in at most finitely many members of $\mathcal{U}[U]$, it follows that $\mathcal{U}[U]$ is countable. We define an equivalence relation $\sim$ on X as follows:

$$x \sim y \underset{\text{def}}{\Longleftrightarrow} \exists U_0, U_1, \ldots U_n \in \mathcal{U} \text{ such that } x \in U_0,\ y \in U_n,\ U_{i-1} \cap U_i \neq \emptyset.$$

This equivalence relation decompose X into disjoint equivalence classes X_λ, $\lambda \in \Lambda$. Since each X_λ is open in X, $\{X_\lambda \mid \lambda \in \Lambda\}$ is a discrete collection of closed sets in X. Thus, we have $X = \bigoplus_{\lambda \in \Lambda} X_\lambda$. It is easy to see that each X_λ is a union of countably many members of $\mathcal{U}$. Hence, each X_λ is separable. □

As a corollary, we have the following:

Corollary 5.9.2 *In order that $X \in \tau(\mathfrak{M}_1(\aleph_0))_\sigma$, it is necessary and sufficient that X is a σ-locally separable and σ-completely metrizable space with $w(X) \leqslant \tau$.* □

Note that locally compact metrizable spaces are locally separable. Concerning locally compact spaces, we have the following:

Proposition 5.9.3 *Every locally compact metrizable space X is a countable union of closed sets X_n, $n \in \mathbb{N}$, such that each X_n is a discrete union of compacta, that is, $X \in (\oplus\mathfrak{M}_0)_\sigma$, where $(\oplus\mathfrak{M}_0)_\sigma(\tau) = (\tau\mathfrak{M}_0)_\sigma$. Moreover, if X is strongly countable-dimensional, then the condition* $\dim X_n \leqslant n$ *can be required, hence $X \in (\oplus\mathfrak{M}_0^{\mathrm{fd}})_\sigma$, where $(\oplus\mathfrak{M}_0^{\mathrm{fd}})_\sigma(\tau) = (\tau\mathfrak{M}_0^{\mathrm{fd}})_\sigma = (\tau\mathfrak{M}_0)_\sigma^{\mathrm{scd}}$.*

Proof By the same reasoning as Proposition 5.9.1, X has a locally finite open cover $\mathcal{U}$ such that $\operatorname{cl} U$ is compact for each $U \in \mathcal{U}$. As is easily observed, $\mathcal{U}$ is star-finite, that is, $\mathcal{U}[U]$ is finite for each $U \in \mathcal{U}$ (cf. Theorem 4.9.10 in [GAGT]). Then, using this cover $\mathcal{U}$, we can define an equivalence relation $\sim$ on X as in the proof of Proposition 5.9.1, which decomposes X into disjoint equivalence classes X_λ, $\lambda \in \Lambda$. Thus, we have $X = \bigoplus_{\lambda \in \Lambda} X_\lambda$, where each X_λ is a union of countably many members of $\mathcal{U}$. Then, we can write $X_\lambda = \bigcup_{n \in \mathbb{N}} \operatorname{cl} U_{\lambda,n}$, where $U_{\lambda,n} \in \mathcal{U}$. For each $n \in \mathbb{N}$, $\{\operatorname{cl} U_{\lambda,n} \mid \lambda \in \Lambda\}$ is a discrete collection of compacta in X. Then, we have $X_n = \bigcup_{\lambda \in \Lambda} \operatorname{cl} U_{\lambda,n} \in \oplus\mathfrak{M}_0$, $n \in \mathbb{N}$, and $X = \bigcup_{n \in \mathbb{N}} X_n \in (\oplus\mathfrak{M}_0)_\sigma$.

If X is strongly countable-dimensional, then there is a tower $A_1 \subset A_2 \subset \cdots$ of closed sets in X such that $X = \bigcup_{n\in\mathbb{N}} A_n$ and $\dim A_n \leqslant n$ (cf. the Countable Sum Theorem 1.12.8 or the Locally Finite Sum Theorem 1.12.9). By replacing each X_n by $A_n \cap \bigcup_{i\leqslant n} X_i$, the condition $\dim X_n \leqslant n$ can be satisfied. □

Proposition 5.9.3 can be stated as follows:

$$\mathfrak{L}(\tau) \subset (\tau\mathfrak{M}_0)_\sigma \text{ and } \mathfrak{L}(\tau)^{\mathrm{scd}} \subset (\tau\mathfrak{M}_0)_\sigma^{\mathrm{scd}},$$

where the equalities do not hold. In fact, $(\mathfrak{M}_0)_\sigma \not\subset \mathfrak{L}$ because $\mathbb{Q}$, the space of rationals, is σ-compact but nowhere locally compact.

A metrizable space said to be **σ-locally compact** if it is a countable union of locally compact metrizable *closed* subspaces. Thus, $\mathfrak{L}_\sigma$ ($\mathfrak{L}_\sigma(\tau)$) is the class of all σ-locally compact metrizable spaces (with weight $\leqslant \tau$). The following easily follows from Proposition 5.9.3:

$$\mathfrak{L}(\tau)_\sigma \subset (\tau\mathfrak{M}_0)_\sigma \text{ and } \mathfrak{L}(\tau)_\sigma^{\mathrm{scd}} \subset (\tau\mathfrak{M}_0)_\sigma^{\mathrm{scd}}.$$

Since every member of $\oplus\mathfrak{M}_0$ (or $\tau\mathfrak{M}_0$) is locally compact, the converse inclusion also holds, so we have

$$(\tau\mathfrak{M}_0)_\sigma = \mathfrak{L}(\tau)_\sigma \text{ and } (\tau\mathfrak{M}_0^{\mathrm{fd}})_\sigma = \mathfrak{L}(\tau)_\sigma^{\mathrm{scd}}.$$

Corollary 5.9.4 *The class $(\oplus\mathfrak{M}_0)_\sigma$ (or $(\tau\mathfrak{M}_0)_\sigma$) is equal to the class of all σ-locally compact metrizable spaces (with weight $\leqslant \tau$) and $(\oplus\mathfrak{M}_0^{\mathrm{fd}})_\sigma$ (or $(\tau\mathfrak{M}_0^{\mathrm{fd}})_\sigma$) is the class of all strongly countable-dimensional and σ-locally compact metrizable spaces (with weight $\leqslant \tau$).* □

The above results are summarized as follows:

- $(\tau\mathfrak{M}_0^{\mathrm{fd}})_\sigma = \mathfrak{L}(\tau)_\sigma^{\mathrm{scd}}$—the class of all strongly countable-dimensional σ-locally compact metrizable spaces with weight $\leqslant \tau$;
- $(\tau\mathfrak{M}_0)_\sigma = \mathfrak{L}(\tau)_\sigma$—the class of all σ-locally compact metrizable spaces with weight $\leqslant \tau$;
- $(\tau\mathfrak{M}_1(\aleph_0))_\sigma$—the class of all σ-locally separable and σ-completely metrizable spaces with weight $\leqslant \tau$;
- $\mathfrak{M}_1(\tau)_\sigma$—the class of all σ-completely metrizable spaces with weight $\leqslant \tau$.

Combining the above characterizations of four classes with Theorem 5.7.17, we can characterize manifolds modeled on the following spaces:

$$\ell_f^2(\Gamma),\ \ell_f^2(\Gamma) \times \ell_Q^2,\ \ell_f^2(\Gamma) \times \ell^2, \text{ or } \ell^2(\Gamma) \times \ell_f^2.$$

Theorem 5.9.5 *For a space X, the following are equivalent:*

(a) *X is an $\ell_f^2(\Gamma)$-manifold;*

(b) *X can be embedded into an $\ell^2(\Gamma)$-manifold $\widetilde{X}$ such that the pair $(\widetilde{X}, X)$ is an $(\ell^2(\Gamma), \ell^2_f(\Gamma))$-manifold pair;*
(c) *X can be embedded into an $\ell^2(\Gamma)$-manifold as an $\mathfrak{M}_0^{\mathrm{fd}}$-absorption base;*
(d) *X is a strongly $\mathfrak{M}_0^{\mathrm{fd}}$-universal, strongly countable-dimensional, and σ-locally compact ANR that is a strong Z_σ-space with the τ-discrete cells property.*

Theorem 5.9.6 *For a space X, the following are equivalent:*

(a) *X is an $(\ell^2_f(\Gamma) \times \ell^2_Q)$-manifold;*
(b) *X can be embedded into an $\ell^2(\Gamma)$-manifold $\widetilde{X}$ such that the pair $(\widetilde{X}, X)$ is an $(\ell^2(\Gamma) \times \ell^2, \ell^2_f(\Gamma) \times \ell^2_Q)$-manifold pair;*
(c) *X can be embedded into an $\ell^2(\Gamma)$-manifold as an $\mathfrak{M}_0$-absorption base;*
(d) *X is a strongly $\mathfrak{M}_0$-universal σ-locally compact ANR that is a strong Z_σ-space with the τ-discrete cells property.*

Remark 5.18 In condition (d) of Theorem 5.9.6 (5.9.5), X is a strong Z_σ-space if and only if every (finite-dimensional) compact set in X is a strong Z-set. Note that every compact set in X is a Z-set in X when X is an ANR with the τ-discrete cells property (Proposition 3.2.10).

Theorem 5.9.7 *For a space X, the following are equivalent:*

(a) *X is an $(\ell^2_f(\Gamma) \times \ell^2)$-manifold;*
(b) *X can be embedded into an $\ell^2(\Gamma)$-manifold $\widetilde{X}$ such that the pair $(\widetilde{X}, X)$ is an $(\ell^2(\Gamma) \times \ell^2, \ell^2_f(\Gamma) \times \ell^2)$-manifold pair;*
(c) *X can be embedded into an $\ell^2(\Gamma)$-manifold as an $\mathfrak{M}_1(\aleph_0)$-absorption base;*
(d) *X is a strongly $\mathfrak{M}_1(\aleph_0)$-universal, σ-locally separable, and σ-completely metrizable ANR that is a strong Z_σ-space with the τ-discrete cells property.*

Remark 5.19 In the above (d), when $\tau > \aleph_0$, X is a strong Z_σ-space if and only if every separable completely metrizable closed set in X is a strong Z-set. Note that every closed set in X with weight $< \tau$ is a Z-set in X when X is an ANR with the τ-discrete cells property (Proposition 3.2.10).

Theorem 5.9.8 *For a space X, the following are equivalent:*

(a) *X is an $(\ell^2(\Gamma) \times \ell^2_f)$-manifold;*
(b) *X can be embedded into an $\ell^2(\Gamma)$-manifold $\widetilde{X}$ such that the pair $(\widetilde{X}, X)$ is an $(\ell^2(\Gamma) \times \ell^2, \ell^2(\Gamma) \times \ell^2_f)$-manifold pair;*
(c) *X can be embedded into an $\ell^2(\Gamma)$-manifold as an $\mathfrak{M}_1(\tau)$-absorption base;*
(d) *X is a strongly $\mathfrak{M}_1(\tau)$-universal, σ-completely metrizable ANR that is a strong Z_σ-space.*

Remark 5.20 It should be noted that the τ-discrete cells property is removed from (d) of Theorem 5.9.8 different from the other three theorems. In fact, since $\mathbf{I}^n \times \Gamma \in \mathfrak{M}_1(\tau)$ for every $n \in \mathbb{N}$, the strong $\mathfrak{M}_1(\tau)$-universality implies the τ-discrete cells property (cf. Remark 5.17).

Remark 5.21 As mentioned in Remark 5.13 (p. 406), given an uncountable cardinal $\tau' < \tau$, we can consider an $\mathfrak{M}_1(\tau')$-absorption base for $\ell^2(\Gamma)$. Let $\Gamma' \subset \Gamma$ with $\operatorname{card}\Gamma' = \tau'$. Then, $\ell^2_f(\Gamma) \times \ell^2(\Gamma')$ is an $\mathfrak{M}_1(\tau')$-absorption base for $\ell^2(\Gamma) \times \ell^2(\Gamma')$ ($\approx \ell^2(\Gamma)$). We can obtain the same characterization of $\ell^2_f(\Gamma) \times \ell^2(\Gamma')$-manifolds as Theorem 5.9.7.

In the separable case (i.e., $\tau = \aleph_0$), we have the following:

- $(\aleph_0\mathfrak{M}_0^{\mathrm{fd}})_\sigma = (\mathfrak{M}_0)_\sigma^{\mathrm{scd}}$—the class of all strongly countable-dimensional σ-compact metrizable spaces;
- $(\aleph_0\mathfrak{M}_0)_\sigma = (\mathfrak{M}_0)_\sigma$—the class of all σ-compact metrizable spaces;
- $(\aleph_0\mathfrak{M}_1(\aleph_0))_\sigma = \mathfrak{M}_1(\aleph_0)_\sigma$—all separable σ-completely metrizable spaces.

Then, we can use Theorem 5.7.18 instead of Theorem 5.7.17. So, the discrete cells property can be removed from condition (d) of each theorem. For the separable version of Theorem 5.9.6, recall that $(\ell^2 \times \ell^2, \ell^2_f \times \ell^2_Q) \approx (\ell^2, \ell^2_Q)$, hence $\ell^2_f \times \ell^2_Q \approx \ell^2_Q$. Moreover, Theorems 5.9.7 and 5.9.8 have the same separable version. The following are separable versions of the above theorems:

Theorem 5.9.9 *For a space X, the following are equivalent:*

(a) *X is an ℓ^2_f-manifold;*
(b) *X can be embedded into an ℓ^2-manifold $\widetilde{X}$ such that the pair $(\widetilde{X}, X)$ is an (ℓ^2, ℓ^2_f)-manifold pair;*
(c) *X can be embedded into an ℓ^2-manifold as an f.d.cap set;*
(d) *X is a strongly $\mathfrak{M}_0^{\mathrm{fd}}$-universal, strongly countable-dimensional, and σ-compact ANR that is a strong Z_σ-space.*

Theorem 5.9.10 *For a space X, the following are equivalent:*

(a) *X is an ℓ^2_Q-manifold;*
(b) *X can be embedded into an ℓ^2-manifold $\widetilde{X}$ such that the pair $(\widetilde{X}, X)$ is an (ℓ^2, ℓ^2_Q)-manifold pair;*
(c) *X can be embedded into an ℓ^2-manifold as a cap set;*
(d) *X is a strongly $\mathfrak{M}_0$-universal σ-compact ANR that is a strong Z_σ-space.*

In condition (d) of Theorem 5.9.10 (5.9.9), "that is a strong Z_σ-space" can be replaced by "such that every (finite-dimensional) compact set in X is a strong Z-set." Indeed, a (strongly countable-dimensional) σ-compact metrizable space X is a strong Z_σ-space if and only if every (finite-dimensional) compact set in X is a strong Z-set. Thus, we have the following MOGILSKI CHARACTERIZATIONS of ℓ^2_f- and ℓ^2_Q-manifolds:

Theorem 5.9.11 (MOGILSKI) *A space X is an ℓ^2_Q- (or ℓ^2_f-)manifold if and only if X is a strongly $\mathfrak{M}_0$-(or $\mathfrak{M}_0^{\mathrm{fd}}$-)universal (strongly countable-dimensional) σ-compact ANR and every (finite-dimensional) compact set in X is a strong Z-set.* □

Theorem 5.9.12 *For a space X, the following are equivalent:*

(a) *X is an $(\ell^2_f \times \ell^2)$-manifold;*
(b) *X can be embedded into an ℓ^2-manifold $\widetilde{X}$ such that the pair $(\widetilde{X}, X)$ is an $(\ell^2 \times \ell^2, \ell^2_f \times \ell^2)$-manifold pair;*
(c) *X can be embedded into an ℓ^2-manifold as an $\mathfrak{M}_1(\aleph_0)$-absorption base;*
(d) *X is a strongly $\mathfrak{M}_1(\aleph_0)$-universal, separable σ-completely metrizable ANR that is a strong Z_σ-space.*

Concerning a local-compactification of an ℓ^2_Q- (or ℓ^2_f-)manifold, we have the following:

Theorem 5.9.13 *Let $\widetilde{X}$ is a local-compactification of an ℓ^2_Q- (or ℓ^2_f-)manifold X. If X is homotopy dense in $\widetilde{X}$, then $\widetilde{X}$ is a $\boldsymbol{Q}$-manifold and X is an (f.d.)cap set for $\widetilde{X}$.*

Proof Since X is homotopy dense in $\widetilde{X}$ and X is an ANR, it follows that $\widetilde{X}$ is an ANR (Theorem 1.13.26). Using the strong $\mathfrak{M}_0$- (or $\mathfrak{M}_0^{\mathrm{fd}}$-)universality of X, we can easily prove that $\widetilde{X}$ has the disjoint cells property. Hence, $\widetilde{X}$ is a $\boldsymbol{Q}$-manifold by the Toruńczyk Characterization of $\boldsymbol{Q}$-manifolds (Theorem 3.8.6). Due to Proposition 5.4.5, X is an (f.d.)cap set for $\widetilde{X}$. □

Remark 5.22 A completion $\widetilde{X}$ of an ℓ^2_Q- (or ℓ^2_f-)manifold X is not necessary to be an ℓ^2-manifold even if X is homotopy dense in $\widetilde{X}$. For example, consider the example defined on p. 162 (see Fig. 2.18), that is,

$$X = \mathbf{I} \times \{0\} \cup \{n^{-1} \mid n \in \mathbb{N}\} \times \mathbf{I} \subset \mathbb{R}^2 \quad \text{and} \quad A = \{(0,0)\} \subset X.$$

Then, X is a separable completely metrizable AR, in which A is a Z-set but not a strong Z-set. As stated in Remark 2.18, $(X \times \ell^2)_A$ is not an ℓ^2-manifold. Since $\dim X = 1$ and X is σ-compact, it follows that

$$(X \times \ell^2_f)_A \setminus A = (X \setminus A) \times \ell^2_f \approx \ell^2_f,$$

which is homotopy dense in $(X \times \ell^2)_A$ because ℓ^2_f is a homotopy dense in ℓ_2 and A is a Z-set in $(X \times \ell^2)_A$. Thus, $(X \times \ell^2)_A$ is a completion of $(X \times \ell^2_f)_A \setminus A$ such that $(X \times \ell^2_f)_A \setminus A$ is homotopy dense in $(X \times \ell^2)_A$ but $(X \times \ell^2)_A$ is not an ℓ^2-manifold. Similarly, $(X \times \ell^2_Q)_A \setminus A \approx \ell^2_Q$ which is homotopy dense in $(X \times \ell^2)_A$.

In the remaining part of this section, we introduce an abosorptively universal tower characterizing ℓ^2_Q- (or ℓ^2_f-)manifolds. For a subclass $\mathfrak{C}$ of $\mathfrak{M}_0$, a tower $X_1 \subset X_2 \subset \cdots$ of subsets of a space X is said to be **absorptively universal** for $\mathfrak{C}$ or **absorptively $\mathfrak{C}$-universal** in X if it has the following property:[21]

[21] In [51], the word "strongly" is used instead of "absorptively."

(au$_{\mathfrak{C}}$) For each $C \in \mathfrak{C}$ and each closed set $D \subset C$, if $f : C \to X$ is a map such that $f|D$ is an embedding into some X_m, then, for each open cover $\mathcal{U}$ of X, there is an embedding $h : C \to X_n$ for some $n > m$ such that $h|D = f|D$ and h is $\mathcal{U}$-close to f.

Evidently, the property (au$_{\mathfrak{M}_0}$) (resp. (au$_{\mathfrak{M}_0^{\mathrm{fd}}}$)) implies (cap) (resp. (f.d.cap)). When X is strongly $\mathfrak{M}_0$- (resp. $\mathfrak{M}_0^{\mathrm{fd}}$-)universal and $X_1 \subset X_2 \subset \cdots$ is a tower of strong Z-sets, the property (cap) (resp. (f.d.cap)) implies (au$_{\mathfrak{M}_0}$) (resp. (au$_{\mathfrak{M}_0^{\mathrm{fd}}}$)). For a subclass $\mathfrak{C} \subset \mathfrak{M}_0$, the existence of an absorptively $\mathfrak{C}$-universal tower of strong Z-sets implies the strong $\mathfrak{C}$-universality, that is:

Proposition 5.9.14 *Suppose that $\mathfrak{C} \subset \mathfrak{M}_0$ is closed hereditary and topological. An ANR X is strongly $\mathfrak{C}$-universal if X contains an absorptively $\mathfrak{C}$-universal tower $(X_n)_{n\in\mathbb{N}}$ of strong Z-sets.*

Proof Let $f : C \to X$ be a map of $C \in \mathfrak{C}$ with a closed set D such that $f|D$ is a strong Z-embedding. We write $C \setminus D = \bigcup_{i\in\mathbb{N}} C_i$, where $C_1 \subset C_2 \subset \cdots$ are closed sets in C. Then, $C_i \in \mathfrak{C}$ for every $i \in \mathbb{N}$ because $\mathfrak{C}$ is closed hereditary. For each $\mathcal{U} \in \operatorname{cov}(X)$, let $d \in \operatorname{Metr}(X)$ such that $\{\overline{\mathrm{B}}_d(x, 1) \mid x \in X\} \prec \mathcal{U}$ (Proposition 1.3.22 (1)). We will inductively construct maps $g_i : C \to X$, $i \in \mathbb{N}$, with $n(1) < n(2) < \cdots$ so as to satisfy the following:

(1) $g_i(C_i) \subset X_{n(i)}$,
(2) $g_i|C_i \cup D$ is an embedding,
(3) $g_i|C_{i-1} \cup D = g_{i-1}|C_{i-1} \cup D$,
(4) $d(g_i, g_{i-1}) < 2^{-i}$,

where $g_0 = f$ and $C_0 = \emptyset$. In the above (2), $g_i|C_i \cup D$ is a strong Z-embedding. Indeed, $g_i(C_i)$ is a strong Z-set in X because $g_i(C_i)$ is a closed subset of a strong Z-set $X_{n(i)}$ in X. Hence, $g_i(C_i \cup D) = g_i(C_i) \cup f(D)$ is a strong Z-set in X.

Suppose g_{i-1} has been constructed. We can apply Proposition 2.8.12 to obtain a map $g' : C \to X$ such that:

(5) g' is $2^{-(i+1)}$-close to g_{i-1},
(6) $g'|C_{i-1} \cup D = g_{i-1}|C_{i-1} \cup D$,
(7) $g'(C \setminus (C_{i-1} \cup D)) \subset X \setminus g_{i-1}(D)$,
(8) g' is closed over $g_{i-1}(D)$.

Since $g'|C_{i-1} = g_{i-1}|C_{i-1}$ by (6), it follows from (8) and (7) that

$$\begin{aligned} g_{i-1}(D) \cap \operatorname{cl} g'(C_i) &= g_{i-1}(D) \cap g'(C_i) \\ &= (g_{i-1}(D) \cap g'(C_i \setminus C_{i-1})) \cup (g_{i-1}(D) \cap g_{i-1}(C_{i-1})) \\ &= g_{i-1}(D \cap C_{i-1}) = \emptyset. \end{aligned}$$

Hence, we can take $\mathcal{V} \in \operatorname{cov}(X)$ so that $\operatorname{mesh} \mathcal{V} < 2^{-(i+1)}$ and

$$\operatorname{st}(g'(C_i), \mathcal{V}) \cap g_{i-1}(D) = \emptyset.$$

Using ($\mathrm{au}_{\mathfrak{C}}$) for $(X_n)_{n\in\mathbb{N}}$, we can obtain an embedding $g'' : C_i \to X_{n(i)}$ for some $n(i) > n(i-1)$ such that $g''|C_{i-1} = g'|C_{i-1}$ and $g'' \simeq_{\mathcal{V}} g'|C_i$. Since $g''(C_i) \subset \mathrm{st}(g'(C_i), \mathcal{V})$, it follows that $g''(C_i) \cap g_{i-1}(D) = \emptyset$. Hence, g'' extends to an embedding $g^* : C_i \cup D \to X$ defined by $g^*|D = g_{i-1}|D$, where $g^* \simeq_{\mathcal{V}} g'|C_i \cup D$. Using the Homotopy Extension Theorem 1.13.11, we can obtain a map $g_i : C \to X$ such that $g_i|C_i \cup D = g^*$ is an embedding, and $g_i \simeq_{\mathcal{V}} g'$. Then, $g_i(C_i) = g''(C_i) \subset X_{n(i)}$ and

$$d(g_i, g_{i-1}) \leqslant d(g_i, g') + d(g', g_{i-1}) < 2^{-i}.$$

Observe that $g_i|C_{i-1} = g''|C_{i-1} = g'|C_{i-1} = g_{i-1}|C_{i-1}$ and $g_i|D = g^*|D = g_{i-1}|D$. Thus, we have $g_i|C_{i-1} \cup D = g_{i-1}|C_{i-1} \cup D$.

Now, we have obtained maps $g_i : C \to X$, $i \in \mathbb{N}$, with $n(1) < n(2) < \cdots$ satisfying (1)–(4). Then, the required embedding $g : C \to X$ can be obtained as the uniform limit $g = \lim_{i\to\infty} g_i$. Indeed, $g|D = f|D$ because $g_i|D = f|D$ for every $i \in \mathbb{N}$. Moreover,

$$d(g, f) = d(g, g_0) \leqslant \sum_{i\in\mathbb{N}} d(g_i, g_{i-1}) < \sum_{i\in\mathbb{N}} 2^{-i} = 1,$$

which implies that g is $\mathcal{U}$-close to f. Since $C = \bigcup_{i\in\mathbb{N}}(C_i \cup D)$ and each $g|C_i \cup D = g_i|C_i \cup D$ is an embedding, it follows that g is injective, which means that g is an embedding because C is compact. Because each $g_i(C_i \cup D)$ is a strong Z-set in X, $g(C) = \bigcup_{i\in\mathbb{N}} g_i(C_i \cup D)$ is a strong Z_σ-set in X. Due to Corollary 2.8.12, every compact strong Z_σ-set is a strong Z-set. Thus, $g(C)$ is a strong Z-embedding. □

Using the above proposition, we can characterize ℓ^2_Q- (or ℓ^2_f-)manifolds as follows:

Theorem 5.9.15 *A space X is an ℓ^2_Q- (or ℓ^2_f-)manifold if and only if X is a (strongly countable-dimensional) σ-compact ANR that is a strong Z_σ-space and contains an absorptively $\mathfrak{M}_0$- (or $\mathfrak{M}_0^{\mathrm{fd}}$-)universal tower.*

Proof (*The "if" Part*) Since X contains an absorptively $\mathfrak{M}_0$- (or $\mathfrak{M}_0^{\mathrm{fd}}$-)universal tower, X is strongly $\mathfrak{M}_0$- (or $\mathfrak{M}_0^{\mathrm{fd}}$-)universal by Proposition 5.9.14. Hence, X is an ℓ^2_Q- (or ℓ^2_f-)manifold by Theorem 5.9.10 (or 5.9.9).

(*The "only if" Part*) Due to Theorem 5.9.10 (or 5.9.9), an ℓ^2_Q- (or ℓ^2_f-) manifold X can be embedded into an ℓ^2-manifold $\widetilde{X}$ as an (f.d.)cap set, where X is homotopy dense in $\widetilde{X}$ (Propositions 5.3.1 and 5.3.5). Hence, there is a tower $(X_n)_{n\in\mathbb{N}}$ of (finite-dimensional) compact Z-sets in $\widetilde{X}$ such that $X = \bigcup_{n\in\mathbb{N}} X_n$ and $(X_n)_{n\in\mathbb{N}}$ has the property (cap) (or (f.d.cap)). Since every Z-set in an ℓ^2-manifold is a strong Z-set, each X_n is a strong Z-set in $\widetilde{X}$, which implies that each X_n is a strong Z-set in X (Proposition 2.9.4). Since $\widetilde{X}$ is strongly $\mathfrak{M}_0$-universal, the property (cap) (or (f.d.cap)) implies ($\mathrm{au}_{\mathfrak{M}_0}$) (resp. ($\mathrm{au}_{\mathfrak{M}_0^{\mathrm{fd}}}$)) in $\widetilde{X}$, which implies that $(X_n)_{n\in\mathbb{N}}$ has the property ($\mathrm{au}_{\mathfrak{M}_0}$) (resp. ($\mathrm{au}_{\mathfrak{M}_0^{\mathrm{fd}}}$)) in X. Hence, $(X_n)_{n\in\mathbb{N}}$ is an absorptively $\mathfrak{M}_0$- (or $\mathfrak{M}_0^{\mathrm{fd}}$-)universal tower in X. □

To apply the above theorem to convex sets, we show the following:

Lemma 5.9.16 *Let C be a separable convex set in a metrizable topological linear space E.*

(1) *If C is infinite-dimensional, then C has an absorptively $\mathfrak{M}_0^{\mathrm{fd}}$-universal tower.*
(2) *If C contains a convex subset affinely homeomorphic to $\boldsymbol{Q}$, then C has an absorptively $\mathfrak{M}_0$-universal tower.*

Proof Due to Theorem 1.4.15, E has an admissible F-norm $\|\cdot\|$. Since C is separable, it has a countable dense subset $\{a_i \mid i \in \mathbb{N}\}$, where we may assume that $a_1 = \mathbf{0}$.

(1): For each $n \in \mathbb{N}$, let $C_n = \langle a_1, \ldots, a_n \rangle$. Since C is infinite-dimensional and $\{a_i \mid i \in \mathbb{N}\}$ is dense in C, there is some $m > n$ such that $a_m \in C \setminus \mathrm{fl}\, C_n$, and then $\dim C_m > \dim C_n$. Thus, $\sup_{n\in\mathbb{N}} \dim C_n = \infty$.

To see the absorptive $\mathfrak{M}_0^{\mathrm{fd}}$-universality of $(C_n)_{n\in\mathbb{N}}$, let $f : X \to C$ be a map of $X \in \mathfrak{M}_0^{\mathrm{fd}}$ such that $f|A : A \to C_m$ is an embedding for some $m \in \mathbb{N}$ and a closed set $A \subset X$. For each $\varepsilon > 0$, we will find $n' \geqslant m$ and construct a map $f' : X \to C_{n'}$ so that f' is $\varepsilon/2$-close to f and $f'|A = f|A$. Since C_m is an AR, $f|A$ extends to a map $f'' : X \to C_m$. Then, A has an open neighborhood U such that

$$\|f(x) - f''(x)\| < \varepsilon/8 \ \text{ for every } x \in U.$$

Let $\alpha : X \to \mathbf{I}$ be a map such that $\alpha^{-1}(0) = A$ and $\alpha^{-1}(1) = X \setminus U$. By Theorem 1.13.21 with Remark 1.13, we have a finite simplicial complex K and maps $\varphi : X \to |K|$ and $\psi : |K| \to C$ such that $f \simeq_{\varepsilon/8} \psi\varphi$. Let $k = \dim K$. Then, subdividing K if necessary, we can assume that $\operatorname{diam} \psi(\sigma) < \varepsilon/8(k+1)$ for every $\sigma \in K$. For each $v \in K^{(0)}$, choose $i(v) \in \mathbb{N}$ so that $\|\psi(v) - a_{i(v)}\| < \varepsilon/8(k+1)$, and let $n' = \max\{n,\ i(v) \mid v \in K^{(0)}\}$. We can define $\psi' : |K| \to C_{n'}$ as follows:

$$\psi'(y) = \sum_{v\in K^{(0)}} \beta_v^K(y) a_{i(v)}.$$

Then, for each $y \in |K|$,

$$\begin{aligned}
\|\psi(y) - \psi'(y)\| &\leqslant \sum_{v\in c_K(y)^{(0)}} \|\beta_v^K(y)\psi(y) - \beta_v^K(y) a_{i(v)}\| \\
&\leqslant \sum_{v\in c_K(y)^{(0)}} \|\psi(y) - a_{i(v)}\| \\
&\leqslant \sum_{v\in c_K(y)^{(0)}} \|\psi(y) - \psi(v)\| + \sum_{v\in c_K(y)^{(0)}} \|\psi(v) - a_{i(v)}\| \\
&< (k+1) \operatorname{diam} \psi(c_K(y)) + \varepsilon/8 < \varepsilon/4,
\end{aligned}$$

where $c_K(y) \in K$ is the carrier of y in K. Thus, we have a map $\psi'\varphi : X \to C_{n'}$ that is $3\varepsilon/8$-close to f. Indeed, for each $x \in X$,

$$\begin{aligned}\|f(x) - \psi'\varphi(x)\| &\leqslant \|f(x) - \psi\varphi(x)\| + \|\psi\varphi(x) - \psi'\varphi(x)\| \\ &< \varepsilon/8 + \varepsilon/4 = 3\varepsilon/8.\end{aligned}$$

The desired map $f' : X \to C_{n'}$ can be defined as follows:

$$f'(x) = (1 - \alpha(x))f''(x) + \alpha(x)\psi'\varphi(x) \quad \text{for each } x \in X.$$

Indeed, $f'|A = f''|A = f|A$ and $f'|X \setminus U = \psi'\varphi|X \setminus U$, so $f'|X \setminus U$ is $\varepsilon/2$-close to $f|X \setminus U$. For each $x \in U$,

$$\begin{aligned}\|f'(x) - f(x)\| &= \|(1 - \alpha(x))f''(x) + \alpha(x)\psi'\varphi(x) - f(x)\| \\ &\leqslant \|(1 - \alpha(x))(f''(x) - f(x))\| + \|\alpha(x)(\psi'\varphi(x) - f(x))\| \\ &\leqslant \|f''(x) - f(x)\| + \|\psi'\varphi(x) - f(x)\| \\ &< \varepsilon/8 + 3\varepsilon/8 = \varepsilon/2.\end{aligned}$$

Now, by Theorem 1.12.14, we have an embedding $h : X \to \mathbf{I}^{2\dim X+1}$. Taking a map $\lambda : X \to \mathbf{I}$ with $\alpha^{-1}(0) = A$, we define a map $g : X \to C_{n'} \times \mathbf{I}^{2\dim X+2}$ as follows:

$$g(x) = (f'(x), \lambda(x), \lambda(x)h(x)) \quad \text{for each } x \in X.$$

Then, it is easy to see that $g(x) = (f'(x), 0, \mathbf{0})$ for any $x \in A$ and g is injective, which means that g is an embedding. Choose $n > n'$ so that

$$\dim C_n \geqslant \dim C_{n'} + 2\dim X + 2.$$

Since the boundary of each cell is collared in the cell, there exists an embedding $e : C_{n'} \times \mathbf{I}^{2\dim X+2} \to C_n$ such that

$$e(x, \mathbf{0}) = x \ \text{ and } \ \operatorname{diam} e(\{x\} \times \mathbf{I}^{2\dim X+2}) < \varepsilon/2 \ \text{ for each } x \in C_m.$$

Then, $eg : X \to C_n$ is an embedding, $eg|A = f'|A = f|A$, and eg is $\varepsilon/2$-close to f', so ε-close to f.

(2): By the assumption, we have an affine embedding $h : \boldsymbol{Q} \to C$, where we can assume that $h(\mathbf{0}) = \mathbf{0} \in C$. Let

$$C_0 = \{h(x) \mid x \in \boldsymbol{Q} \text{ and } x(2i) = 0 \text{ for every } i \in \mathbb{N}\}.$$

Then, $\{h(\mathbf{e}_{2n}) \mid n \in \mathbb{N}\}$ is a linearly independent subset of $E \setminus \operatorname{fl} C_0$, where $\operatorname{fl} C_0$ is a linear subspace of E because $\mathbf{0} \in C_0$. Hence, we can inductively choose v_n, $n \in \mathbb{N}$, so that

$$v_n \in E \setminus \operatorname{fl}(C_0 \cup \{a_1, \dots, a_n, v_1, \dots, v_{n-1}\}).$$

For each $n \in \mathbb{N}$, let

$$D_n = \langle C_0 \cup \{a_1, \dots, a_n, v_1, \dots, v_{n-1}\}\rangle \text{ and}$$
$$C_n = \langle C_0 \cup \{a_1, \dots, a_n, v_1, \dots, v_n\}\rangle.$$

Then, $v_n \in E \setminus \operatorname{fl} D_{n-1}$ and C_n is the cone over D_n with v_n the vertex. Since $C_{n-1} \subset D_n$ and D_n is collared in C_n, it follows that C_{n-1} is a Z-set in C_n. We will show that every C_n is affinely homeomorphic to a convex set in a normed linear space, which implies that C_n is an AR. Then, applying Keller's Theorem 3.9.2, we will conclude that C_n is homeomorphic to $\boldsymbol{Q}$.

Assume that C_{n-1} is affinely homeomorphic to a convex set in a normed linear space L. Since C_n is a cone over D_n, it suffices to show that D_n is affinely homeomorphic to a convex set in a normed linear space. Let $\varphi : C_{n-1} \to L$ be an affinely embedding of C_{n-1} into L. Then, φ extends to a linear isomorphism $\tilde{\varphi} : \operatorname{fl} C_{n-1} \to L$, where $\tilde{\varphi}$ is not a homeomorphism in general. When $a_n \in \operatorname{fl} C_{n-1}$, we have $D_n \subset \operatorname{fl} C_{n-1}$. There exist maps $q : C_{n-1} \times \mathbf{I} \to D_n$ and $p : C_{n-1} \times \mathbf{I} \to L$ defined as follows:

$$q(x, t) = (1 - t)a_n + tx \quad \text{and} \quad p(x, t) = (1 - t)\tilde{\varphi}(a_n) + t\varphi(x),$$

where q is a quotient map because C_{n-1} is compact. Since $\tilde{\varphi}q = p$, $\tilde{\varphi}|D_n : D_n \to \tilde{\varphi}(D_n)$ is a continuous bijection, which is a homeomorphism because D_n is compact. Thus, D_n is affinely homeomorphic to a convex set in L. When $a_n \notin \operatorname{fl} C_{n-1}$, D_n is the cone over C_{n-1} with a_n the vertex. Hence, every $x \in D_n \setminus \{a_n\}$ is uniquely written as $x = (1 - t)a_n + ty$, $y \in C_{n-1}$, $0 < t \leqslant 1$. We define an affine map $\varphi^* : D_n \to L \times \mathbb{R}$ by

$$\varphi^*((1 - t)a_n + ty) = (t\varphi(y), 1 - t) \text{ for each } y \in C_{n-1} \text{ and } t \in \mathbf{I}.$$

Then, φ^* is a continuous injection, which is an embedding because D_n is compact. Hence, D_n is affinely homeomorphic to a convex set in $L \times \mathbb{R}$. Consequently, C_n is affinely homeomorphic to a convex set in a normed linear space.

We have just shown that each C_n is a Z-set in C_{n+1} and homeomorphic to $\boldsymbol{Q}$. Now, using this fact, we will show that the tower $(C_n)_{n\in\mathbb{N}}$ is absorptively $\mathfrak{M}_0$-universal. Let $f : X \to C$ be a map of $X \in \mathfrak{M}_0$ such that $f|A : A \to C_m$ is an embedding for some $m \in \mathbb{N}$ and a closed set $A \subset X$. By the same argument as (1), for each $\varepsilon > 0$, we can find $n > m$ and construct a map $f' : X \to C_n$ so that f' is $\varepsilon/2$-close to f and $f'|A = f|A$. Since C_m is a Z-set in C_n, it follows that $f'|A = f|A : A \to C_n \approx \boldsymbol{Q}$ is a Z-embedding. Applying the Strong Universality

Theorem 2.3.16, we can obtain an embedding $f'' : X \to C_n$ such that f'' is $\varepsilon/2$-close to f' and $f''|A = f'|A$, so f'' is ε-close to f and $f''|A = f|A$. □

Lemma 5.9.17 *Let C be a separable convex set in a metrizable topological linear space E. If C is an AR and some point of C has no totally bounded neighborhood in C, then every compact set in C is a strong Z-set in C.*

Proof Since C has the discrete cell-tower property by Lemma 3.5.8, it follows from Corollary 5.8.4 that every compact set in C is a strong Z-set in C. □

A convex set C in a metric linear space $E = (E, d)$ is said to be **locally complete at** $x \in C$ if x has a complete neighborhood in C.

Proposition 5.9.18 *Let C be an infinite-dimensional convex set in a metric linear space $E = (E, d)$. If C is locally complete at some point, then C contains a convex set A such that A is affinely homeomorphic to the Hilbert cube $\boldsymbol{Q}$.*

Proof We may assume that d is induced from an F-norm $\|\cdot\|$ and C is locally complete at $\mathbf{0} \in C$, that is, there exists $\varepsilon > 0$ such that every Cauchy sequence in $C \cap \overline{\mathrm{B}}(\mathbf{0}, \varepsilon)$ converges in C. Since C is infinite-dimensional, C has a linearly independent sequence $(x_i)_{i\in\mathbb{N}}$ such that $\|x_i\| < \varepsilon$ for each $i \in \mathbb{N}$. We can inductively choose $0 < a_i < 1$, $i \in \mathbb{N}$, so that $\|a_i x_i\| < 2^{-(i-k+1)}\delta_k$, $k = 0, \ldots, i-1$, where $\delta_0 = 2\varepsilon$ and for $k > 0$,

$$\delta_k = \inf\left\{ \left\| \sum_{i=1}^k s_i x_i - \sum_{i=1}^k t_i x_i \right\| \;\middle|\; (s_i)_{i=1}^k, (t_i)_{i=1}^k \in \prod_{i=1}^k [0, a_i], \right.$$
$$\left. \exists i = 1, \ldots, k \text{ such that } |s_i - t_i| \geqslant \min\{1/k, a_i\} \right\}.$$

Indeed, we have an embedding $\varphi_k : \prod_{i=1}^k [0, a_i] \to C$ defined by $\varphi((t_i)_{i=1}^k) = \sum_{i=1}^k t_i x_i$. Then, it follows that $\delta_k > 0$. Hence, we can proceed the inductive choice of a_i.

For each $k \in \mathbb{N}$, let $\mathrm{pr}_k : \prod_{i\in\mathbb{N}} [0, a_i] \to \prod_{i=1}^k [0, a_i]$ be the projection. For each $(t_i)_{i=1}^k \in \prod_{i=1}^k [0, a_i]$,

$$\|\varphi((t_i)_{i=1}^k)\| = \left\| \sum_{i=1}^k t_i x_i \right\| \leqslant \sum_{i=1}^k \|t_i x_i\| \leqslant \sum_{i=1}^k \|a_i x_i\| < \sum_{i=1}^k 2^{-(i+1)}\delta_0 < \varepsilon.$$

Then, we have maps $\varphi_k \mathrm{pr}_k : \prod_{i\in\mathbb{N}} [0, a_i] \to C \cap \overline{\mathrm{B}}(\mathbf{0}, \varepsilon)$, $k \in \mathbb{N}$. For each $(t_i)_{i\in\mathbb{N}} \in \prod_{i\in\mathbb{N}} [0, a_i]$,

$$\begin{aligned} \left\| \varphi_{k+1}\mathrm{pr}_{k+1}((t_i)_{i\in\mathbb{N}}) - \varphi_k \mathrm{pr}_k((t_i)_{i\in\mathbb{N}}) \right\| &= \|t_{k+1}x_{k+1}\| \\ &\leqslant \|a_{k+1}x_{k+1}\| < 2^{-(k+2)}\delta_0 = 2^{-(k+1)}\varepsilon, \end{aligned}$$

which means that $(\varphi_k \mathrm{pr}_k)_{k\in\mathbb{N}}$ is uniformly Cauchy. Hence, $(\varphi_k \mathrm{pr}_k)_{k\in\mathbb{N}}$ uniformly converges to the map $\varphi : \prod_{i\in\mathbb{N}}[0, a_i] \to C \cap \overline{\mathrm{B}}(\mathbf{0}, \varepsilon)$. Then,

$$\varphi((t_i)_{i\in\mathbb{N}}) = \sum_{i\in\mathbb{N}} t_i x_i \quad \text{for each } (t_i)_{i\in\mathbb{N}} \in \prod_{i\in\mathbb{N}}[0, a_i].$$

Hence, φ is an affine map. Moreover, for each $k \in \mathbb{N}$ and $(t_i)_{i\in\mathbb{N}} \in \prod_{i\in\mathbb{N}}[0, a_i]$,

$$\begin{aligned}\left\|\varphi((t_i)_{i\in\mathbb{N}}) - \varphi_k \mathrm{pr}_k((t_i)_{i\in\mathbb{N}})\right\| &= \left\|\sum_{i>k} t_i x_i\right\| \leqslant \sum_{i>k} \|t_i x_i\| \\ &\leqslant \sum_{i>k} \|a_i x_i\| < \sum_{i>k} 2^{-(i-k+1)}\delta_k = \delta_k/2.\end{aligned}$$

To see that φ is injective, let $(t_i)_{i\in\mathbb{N}} \neq (s_i)_{i\in\mathbb{N}} \in \prod_{i\in\mathbb{N}}[0, a_i]$. Then, $t_i \neq s_i$ for some $i \in \mathbb{N}$. Choose $k \geqslant i$ so that $1/k \leqslant |s_i - t_i|$ $(\leqslant a_i)$. It follows that

$$\begin{aligned}\left\|\varphi((s_i)_{i\in\mathbb{N}}) - \varphi((t_i)_{i\in\mathbb{N}})\right\| &\geqslant \left\|\varphi_k \mathrm{pr}_k((s_i)_{i\in\mathbb{N}}) - \varphi_k \mathrm{pr}_k((t_i)_{i\in\mathbb{N}})\right\| \\ &\quad - \left\|\varphi((s_i)_{i\in\mathbb{N}}) - \varphi_k \mathrm{pr}_k((s_i)_{i\in\mathbb{N}})\right\| \\ &\quad - \left\|\varphi_k \mathrm{pr}_k((t_i)_{i\in\mathbb{N}}) - \varphi((t_i)_{i\in\mathbb{N}})\right\| \\ &> \left\|\sum_{i=1}^{k} s_i x_i - \sum_{i=1}^{k} t_i x_i\right\| - \delta_k/2 - \delta_k/2 \\ &\geqslant \delta_k - \delta_k = 0,\end{aligned}$$

which means that $\varphi((s_i)_{i\in\mathbb{N}}) \neq \varphi((t_i)_{i\in\mathbb{N}})$. Thus, φ is injective, so it is an affine embedding. Therefore, the image of φ is a convex set in E, which is affinely homeomorphic to $\prod_{i\in\mathbb{N}}[0, a_i]$ and so to $\boldsymbol{Q}$. □

Now, we can prove the following:

Theorem 5.9.19 *Let C be a σ-compact convex set in a metrizable topological linear space E such that some point of C has no totally bounded neighborhood in C.*

(1) *$C \approx \ell^2_f$ if C is strongly countable-dimensional.*
(2) *$C \approx \ell^2_Q$ if C is an AR and contains an infinite-dimensional compact convex set.*

Proof In Lemma 5.9.17, we have already shown that every compact set in C is a strong Z-set in C. Hence, C is a strong Z_σ-space.

(1): As a convex set in a topological linear space, C is contractible and locally contractible. Since C is countable-dimensional, it follows that C is an AR (Haver's Theorem 1.13.38 and Proposition 1.13.3(2)). By virtue of Lemma 5.9.16(1), C

contains an absorptively $\mathfrak{M}_0^{\mathrm{fd}}$-universal tower. Thus, applying Theorem 5.9.15, we have (1).

(2): As we saw in Proposition 5.9.18, C contains a convex subset affinely homeomorphic to the Hilbert cube $\boldsymbol{Q}$. By Lemma 5.9.16(2), C has an absorptively $\mathfrak{M}_0$-universal tower. Again, applying Theorem 5.9.15, we have (2). □

Recall that a topological linear space E is finite-dimensional if and only if $\mathbf{0} \in E$ has a totally bounded neighborhood in E (Theorem 1.4.12). As a special case of Theorem 5.9.19 above, we have the following:

Theorem 5.9.20 *Let E be a metrizable topological linear space.*

(1) *$E \approx \ell_f^2$ if E is $\aleph_0$-dimensional, that is, E has a countable Hamel basis.*
(2) *$E \approx \ell_Q^2$ if E is a σ-compact AR and contains an infinite-dimensional compact convex set.* □

5.10 Absorbing Sets

In this chapter, as before,

- *let $\mathfrak{C}$ be a closed hereditary, additive topological class of metrizable spaces.*

A strongly $\mathfrak{C}$-universal homotopy dense set X in M is called an **absorbing set** in M for $\mathfrak{C}$ or simply a **$\mathfrak{C}$-absorbing set** if X is a Z_σ-space and $X \in \mathfrak{C}_\sigma$, equivalently X is a countable union of members of $\mathfrak{C}_Z(X)$, where $\mathfrak{C}_Z(X)$ is the collection of Z-sets in X that is a member of $\mathfrak{C}$.

Indeed, let $X = \bigcup_{i\in\mathbb{N}} A_i = \bigcup_{j\in\mathbb{N}} C_j$, where each A_i is a Z-set in X, each C_j is closed in X and $C_j \in \mathfrak{C}$. Then, $X = \bigcup_{i,j\in\mathbb{N}} A_i \cap C_j$. Since $\mathfrak{C}$ is closed hereditary and any closed subset of a Z-set in X is a Z-set in X, it follows that $A_i \cap C_j \in \mathfrak{C}_Z(X)$ for every $i, j \in \mathbb{N}$.

The following is a direct consequence of Proposition 5.4.1(1):

Lemma 5.10.1 *Every $\mathfrak{C}$-absorbing set in an $\ell^2(\Gamma)$-manifold M is $\mathfrak{C}$-absorptive in M.* □

When $\mathfrak{M}_0^{\mathrm{fd}} \subset \mathfrak{C} \subset \mathfrak{M}_1(\tau)$, we can characterize $\mathfrak{C}$-absorbing sets in an $\ell^2(\Gamma)$-manifold as follows:

Proposition 5.10.2 *Let M be an $\ell^2(\Gamma)$-manifold with $X \subset M$. When $\mathfrak{M}_0^{\mathrm{fd}} \subset \mathfrak{C} \subset \mathfrak{M}_1(\tau)$, X is a $\mathfrak{C}$-absorbing set in M if and only if X is a $\mathfrak{C}$-absorptive Z_σ-space and $X \in \mathfrak{C}_\sigma$.*

Proof The "only if" part follows from Lemma 5.10.1. By Propositions 5.3.5 and 5.4.1(2), a $\mathfrak{C}$-absorptive X in M is a strongly $\mathfrak{C}$-universal homotopy dense set in M, which means the "if" part. □

The following differences should be noted:

- A $\mathfrak{C}$-absorption base for M is a Z_σ-set *in M* but a $\mathfrak{C}$-absorbing set in M is a Z_σ-set *in itself*.
- Every $\mathfrak{C}$-absorption base for an $\ell^2(\Gamma)$-manifold M is *σ-completely metrizable* (i.e., it belongs to $\mathfrak{M}_1(\tau)_\sigma$), but a $\mathfrak{C}$-absorbing set in M is *not* necessary σ-completely metrizable.

We have to keep in mind the following:

- It is not required for a $\mathfrak{C}$-absorbing set to be F_σ in M.

In fact, we have the following:

Proposition 5.10.3 *Let M be an $\ell^2(\Gamma)$-manifold.*

(1) *A $\mathfrak{C}$-absorbing (or $\tau\mathfrak{C}$-absorbing) set X in M is a $\mathfrak{C}$-absorption base for M if and only if X is F_σ in M.*
(2) *When $\mathfrak{M}_0^{\mathrm{fd}} \subset \mathfrak{C} \subset \mathfrak{M}_1(\tau)$, every $\mathfrak{C}$-absorption base X for M is a $\tau\mathfrak{C}$-absorbing F_σ set in M, which is a $\mathfrak{C}$-absorbing set in M if $(\tau\mathfrak{C})_\sigma = \mathfrak{C}_\sigma$.*

Proof (1): The "only if" part is trivial. To see the "if" part, note that X is $\mathfrak{C}$-absorptive in M (Lemma 5.10.1). Since X is F_σ in M, there are closed sets A_n, $n \in \mathbb{N}$, in M such that $X = \bigcup_{n\in\mathbb{N}} A_n$. As a $\mathfrak{C}$-absorbing set, we can write $X = \bigcup_{n\in\mathbb{N}} X_n$, where $X_n \in \mathfrak{C}_Z(X)$ (or $X_n \in \tau\mathfrak{C}_Z(X)$) for each $n \in \mathbb{N}$. Then, every $A_i \cap X_n$ is closed in M. Since X is homotopy dense in X and each X_n is a Z-set in X, it is easy to see that every $A_i \cap X_n$ is a Z-set in M. Because $\mathfrak{C}$ is closed hereditary, $A_i \cap X_n \in \mathfrak{C}_Z(M)$ (or $A_i \cap X_n \in \tau\mathfrak{C}_Z(M)$) for each $i, n \in \mathbb{N}$. Thus, X is a $\mathfrak{C}$-absorption base for M.

(2): Combining Corollaries 5.3.9 and 5.4.2(2), we can see that X is a strongly $\tau\mathfrak{C}$-universal homotopy dense Z_σ-set in M and $X \in (\tau\mathfrak{C})_\sigma$. Since X is homotopy dense in M, if $A \subset X$ is a Z-set in M, then A is a Z-set in X. Hence, X is a Z_σ-space. Thus, X is a $\tau\mathfrak{C}$-absorbing set in M. □

In particular, an $\mathfrak{M}_0$- (or $\mathfrak{M}_0^{\mathrm{fd}}$-)absorption base for an $\ell^2(\Gamma)$-manifold M is an $\tau\mathfrak{M}_0$- (or $\tau\mathfrak{M}_0^{\mathrm{fd}}$-)absorbing set in M, but an $\mathfrak{M}_0$- (or $\mathfrak{M}_0^{\mathrm{fd}}$-)absorption base for an ℓ^2-manifold M (i.e., an (f.d.)cap set in M) is an $\mathfrak{M}_0$- (or $\mathfrak{M}_0^{\mathrm{fd}}$-) absorbing set in M. Moreover, an $\mathfrak{M}_1(\tau)$- (or $\mathfrak{M}_1(\aleph_0)$-)absorption base for an $\ell^2(\Gamma)$-manifold M is an $\mathfrak{M}_1(\tau)$- (or $\tau(\mathfrak{M}_1(\aleph_0))$-)absorbing set in M. An $\mathfrak{M}_1(\aleph_0)$-absorption base for an ℓ^2-manifold M is an $\mathfrak{M}_1(\aleph_0)$-absorbing set in M.

Due to Theorem 3.4.9, if N is a homotopy dense G_δ set in an $\ell^2(\Gamma)$-manifold M, then the inclusion $N \hookrightarrow M$ is a near-homeomorphism. Hence, the following lemma can be easily obtained, which will be frequently used:

Lemma 5.10.4 *Let X be a $\mathfrak{C}$-absorbing set in an $\ell^2(\Gamma)$-manifold M. If $\widetilde{X}$ is a G_δ set in M with $X \subset \widetilde{X}$, then $\widetilde{X}$ is an $\ell^2(\Gamma)$-manifold, X is a $\mathfrak{C}$-absorbing set in $\widetilde{X}$, and the inclusion $\widetilde{X} \hookrightarrow M$ is a near-homeomorphism.* □

We have the following uniqueness of absorbing sets in an $\ell^2(\Gamma)$-manifold, which is weaker than the uniqueness of absorption bases for an $\ell^2(\Gamma)$-manifold (Theorem 5.3.14):

Theorem 5.10.5 (Uniqueness of Absorbing Sets) *Let X and Y be two $\mathfrak{C}$-absorbing sets in an $\ell^2(\Gamma)$-manifold M. Then, for each open cover $\mathcal{U}$ of M, there is a homeomorphism $f : X \to Y$ that is $\mathcal{U}$-close to the inclusion $X \subset M$.*

Proof We can write $X = \bigcup_{n\in\mathbb{N}} X_n$, $Y = \bigcup_{n\in\mathbb{N}} Y_n$, where $X_n \in \mathfrak{C}_Z(X)$ and $Y_n \in \mathfrak{C}_Z(Y)$ for each $n \in \mathbb{N}$. Take open covers of M as follows:

$$\mathcal{U} \overset{*}{\succ} \mathcal{U}_1 \overset{*}{\succ} \mathcal{U}_2 \overset{*}{\succ} \cdots \quad \text{and} \quad \operatorname{mesh} \mathcal{U}_n < 2^{-n}.$$

We shall inductively construct homeomorphisms $f_n : G_n \to H_n$, $g_n : H'_n \to G'_n$ between G_δ-sets in M such that $X \subset G_n \cap G'_n$, $Y \subset H_n \cap H'_n$ and the following are satisfied:

(1) $f_n|X$ is $\mathcal{U}_n$-close to $f_{n-1}|X$ and $g_n|Y$ is $\mathcal{U}_n$-close to $g_{n-1}|Y$;
(2) $f_n|X_{n-1} = f_{n-1}|X_{n-1}$ and $g_n|Y_{n-1} = g_{n-1}|Y_{n-1}$;
(3) $f_n(X_n)$ is a Z-set in Y and $g_n(Y_n)$ is a Z-set in X;
(4) $g_n f_n|X_n = \mathrm{id}$ and $f_n g_n|Y_n = \mathrm{id}$.

Then, we have the maps $f = \lim_{n\to\infty} f_n|X : X \to Y$ and $g = \lim_{n\to\infty} g_n|Y : Y \to X$, where $fg = \mathrm{id}_Y$ and $gf = \mathrm{id}_X$, that is, f is a homeomorphism with $f^{-1} = g$.

Let $G_0 = G'_0 = H_0 = H'_0 = M$ and $f_0 = g_0 = \mathrm{id}_M$. Assume that $f_i : G_i \to H_i$, $g_i : H'_i \to G'_i$ have been constructed for $i < n$ so as to satisfy (1)–(4). By using the strong $\mathfrak{C}$-universality of Y, we have a Z-embedding $h : X_n \cup g_{n-1}(Y_{n-1}) \to Y$ such that

$$h|X_{n-1} \cup g_{n-1}(Y_{n-1}) = f_{n-1}|X_{n-1} \cup g_{n-1}(Y_{n-1}) \quad \text{and}$$
$$h \simeq_{\mathcal{U}_{n+5}} f_{n-1}|X_n \cup g_{n-1}(Y_{n-1}).$$

By Lavrentieff's Homeomorphism Extension Theorem 1.3.19, h extends to a homeomorphism $\tilde{h} : A \to B$ between G_δ-sets A and B of G_{n-1} and H_{n-1}, respectively, such that

$$X_{n-1} \cup g_{n-1}(Y_{n-1}) \subset A \subset \mathrm{cl}_{G_{n-1}} \big(X_{n-1} \cup g_{n-1}(Y_{n-1})\big) \quad \text{and}$$
$$h(X_{n-1} \cup g_{n-1}(Y_{n-1})) \subset B \subset \mathrm{cl}_{H_{n-1}} h(X_{n-1} \cup g_{n-1}(Y_{n-1})).$$

By using Lavrentieff's G_δ Extension Theorem 1.3.18, we may assume that

$$\tilde{h} \simeq_{\mathcal{U}_{n+5}} f_{n-1}|A \ \text{ in } H_{n-1}.$$

We can define G_δ-sets G_n and H_n in M as follows:

$$G_n = G_{n-1} \setminus \big(\mathrm{cl}_{G_{n-1}}(X_{n-1} \cup g_{n-1}(Y_{n-1})) \setminus A\big),$$
$$H_n = H_{n-1} \setminus \big(\mathrm{cl}_{H_{n-1}} h(X_{n-1} \cup g_{n-1}(Y_{n-1})) \setminus B\big).$$

Then, by Lemma 5.10.4, G_n, G_{n-1}, H_n and H_{n-1} are homeomorphic to M and also the inclusions $G_n \hookrightarrow G_{n-1}$ and $H_n \hookrightarrow H_{n-1}$ are near-homeomorphisms. So, we have a homeomorphism $\alpha : G_{n-1} \to G_n$ so close to $\mathrm{id}_{G_{n-1}}$ that $f_{n-1}\alpha^{-1} \simeq_{\mathcal{U}_{n+6}} f_{n-1}|G_n$ in H_{n-1}. We have also a homeomorphism $\beta : H_{n-1} \to H_n$ such that $\beta \simeq_{\mathcal{U}_{n+6}} \mathrm{id}_{H_{n-1}}$ in H_{n-1}. Observe that A and B are Z-sets in G_n and H_n, respectively, hence $\beta f_{n-1}\alpha^{-1}|A$ and $\tilde{h}$ are Z-embedding into H_n. Since $\beta f_{n-1}\alpha^{-1}|A \simeq_{\mathcal{U}_{n+5}} f_{n-1}|A$ in H_{n-1} and H_n is homotopy dense in H_{n-1}, it follows that $\beta f_{n-1}\alpha^{-1}|A \simeq_{\mathcal{U}_{n+5}} f_{n-1}|A$ in H_n. Similarly, $\tilde{h} \simeq_{\mathcal{U}_{n+5}} f_{n-1}|A$ in H_n. Hence, $\beta f_{n-1}\alpha^{-1}|A \simeq_{\mathcal{U}_{n+4}} \tilde{h}$ in H_n. By the Z-Set Unknotting Theorem 2.9.7, we have a homeomorphism $\gamma : H_n \to H_n$ such that $\gamma\beta f_{n-1}\alpha^{-1}|A = \tilde{h}$ and $\gamma \simeq_{\mathcal{U}_{n+5}} \mathrm{id}$. The homeomorphism $f_n = \gamma\beta f_{n-1}\alpha^{-1} : G_n \to H_n$ satisfies (1)–(3).

The homeomorphism $g_n : H'_n \to G'_n$ can be similarly constructed so as to satify (1)–(4). Thus, the proof is completed. □

Remark 5.23 In the above, the homeomorphism $f : X \to Y$ can be extended to a homeomorphism $\tilde{f} : \tilde{X} \to \tilde{Y}$ between G_δ sets in M with $X \subset \tilde{X}$ and $Y \subset \tilde{Y}$. However, $\tilde{X} \neq \tilde{Y}$ in general. Thus, Theorem 5.10.5 above is weaker than Theorem 5.3.14.

Corollary 5.10.6 *Let X and Y be $\mathfrak{C}$-absorbing sets in $\ell^2(\Gamma)$-manifolds M and N, respectively. If X and Y have the same homotopy type (equiv. M and Y have the same homotopy type), then they are homeomorphic, i.e.,*

$$X \simeq Y \ (or\ \ M \simeq N) \ \Rightarrow \ X \approx Y.$$

Proof By the Classification Theorem 2.6.1, we have $M \approx N$, that is, there is a homeomorphism $f : M \to N$. Then, $f(X)$ is a $\mathfrak{C}$-absorbing set in N. By Theorem 5.10.5 above, we have $f(X) \approx Y$. Therefore, $X \approx Y$. □

Theorem 5.10.7 *Let X and Y be $\mathfrak{C}$-absorbing sets in $\ell^2(\Gamma)$-manifolds M and N, respectively. Then, every fine homotopy equivalence $f : X \to Y$ is a near-homeomorphism.*

Proof For each $\mathcal{U} \in \mathrm{cov}(Y)$, let $\mathcal{V} \in \mathrm{cov}(Y)$ be a star-refinement of $\mathcal{U}$. We have an open set $\tilde{Y}$ in N and $\tilde{\mathcal{V}} \in \mathrm{cov}(\tilde{Y})$ such that $Y \subset \tilde{Y}$ and $\tilde{\mathcal{V}}|Y = \mathcal{V}$. By Lavrentieff's G_δ Extension Theorem 1.3.18, f extends to a map $\tilde{f} : \tilde{X} \to \tilde{Y}$ over a G_δ set $\tilde{X}$ in M. Then, $\tilde{X}$ and $\tilde{Y}$ are $\ell^2(\Gamma)$-manifolds by Lemma 5.10.4. Moreover, $\tilde{f}$ is a fine homotopy equivalence. Indeed, by Theorem 1.15.1, we may show that $\tilde{f}$ is a hereditary weak homotopy equivalence. For each open set U in $\tilde{Y}$, since X and Y are homotopy dense in $\tilde{X}$ and $\tilde{Y}$ respectively, the inclusions $f^{-1}(U \cap Y) = \tilde{f}^{-1}(U) \cap X \hookrightarrow \tilde{f}^{-1}(U)$ and $U \cap Y \hookrightarrow Y$ are (weak) homotopy equivalences. Since $f|f^{-1}(U \cap Y) : f^{-1}(U \cap Y) \to U \cap Y$ is a (weak) homotopy equivalence, so is $\tilde{f}|\tilde{f}^{-1}(U) : \tilde{f}^{-1}(U) \to U$:

$$\begin{array}{ccccc} f^{-1}(U \cap Y) & = & \tilde{f}^{-1}(U) \cap X & \subset & \tilde{f}^{-1}(U) \\ & \searrow^{f|} & & & \downarrow \tilde{f}| \\ & & U \cap Y & \subset & U \end{array}$$

By the Homeomorphism Approximation Theorem 2.6.3, $\tilde{f} : \widetilde{X} \to \widetilde{Y}$ is a near-homeomorphism. Hence, we have a homeomorphism $g : \widetilde{X} \to \widetilde{Y}$ that is $\widetilde{\mathcal{V}}$-close to $\tilde{f}$. Since $g(X)$ and Y are $\mathfrak{C}$-absorbing sets in $\widetilde{Y}$, we have a homeomorphism $h : g(X) \to Y$ that is $\widetilde{\mathcal{V}}$-close to id by Theorem 5.10.5. Then, $hg : X \to Y$ is a homeomorphism that is $\mathcal{U}$-close to f. □

Concerning strong Z_σ-spaces, we have the following:

Lemma 5.10.8 *Let X be an ANR that is a strong Z_σ-space. Every open set in X is a strong Z_σ-space.*

Proof Let $X = \bigcup_{n\in\mathbb{N}} A_n$, where each A_n is a strong Z-set in X. Due to Proposition 2.8.14, each $A_n \cap U$ is a strong Z-set in U. So, $U = \bigcup_{n\in\mathbb{N}}(A_n \cap U)$ is a strong Z_σ-space. □

Concerning open subsets of a $\mathfrak{C}$-absorbing set, the following holds:

Proposition 5.10.9 *Let X be a $\mathfrak{C}$-absorbing set in an $\ell^2(\Gamma)$-manifold M. For each open set U in M, $X \cap U$ is a $\mathfrak{C}$-absorbing set in U.*

Proof From Propositions 5.4.18 and 1.13.6(3), $X \cap U$ is also a strongly $\mathfrak{C}$-universal ANR. Due to Corollary 1.15.6, $X \cap U$ is homotopy dense in U. Since X is a strong Z_σ-space, $X \cap U$ is a strong Z_σ-space by Lemma 5.10.8. Because $\mathfrak{C}_\sigma$ is open hereditary, it follows that $X \cap U \in \mathfrak{C}_\sigma$. Thus, $X \cap U$ is a $\mathfrak{C}$-absorbing set in U. □

Combining the above proposition with the Open Embedding Theorem 2.5.10, we have the following:

Corollary 5.10.10 *If $\ell^2(\Gamma)$ has a $\mathfrak{C}$-absorbing set Ω, then every $\ell^2(\Gamma)$-manifold has a $\mathfrak{C}$-absorbing set.* □

We can consider manifolds modeled on a $\mathfrak{C}$-absorbing set in $\ell^2(\Gamma)$.

Corollary 5.10.11 *Let Ω be a $\mathfrak{C}$-absorbing set in $\ell^2(\Gamma)$. If X is a $\mathfrak{C}$-absorbing set in some $\ell^2(\Gamma)$-manifold M, then X is an Ω-manifold.*

Proof Each point $x \in X$ has an open neighborhood U in M that is homeomorphic to an open ball in $\ell^2(\Gamma)$, which means $U \approx \ell^2(\Gamma)$. Then, $X \cap U$ is a $\mathfrak{C}$-absorbing set in U. By Corollary 5.10.6, we have $X \cap U \approx \Omega$. Therefore, X is an Ω-manifold. □

We prove that an ANR is a strong Z_σ-space if it is a local strong Z_σ-space, that is:

Proposition 5.10.12 *Let X be an ANR. If each point of X has an open neighborhood being a strong Z_σ-space, then X is a strong Z_σ-space.*

Proof Applying Michael's Theorem 1.3.20, we can show the proposition, where it suffices to prove the following:

(G-1) For open sets $U \supset V$ in X, if U is a strong Z_σ-space, then V is a strong Z_σ-space;
(G-2) For two open sets U and V in X, if both U and V are strong Z_σ-spaces, then $U \cup V$ is a strong Z_σ-space;
(G-3) For a discrete collection $\{U_\lambda\}_{\lambda \in \Lambda}$ of open sets in X, if each U_λ is a strong Z_σ-space, then $\bigcup_{\lambda \in \Lambda} U_\lambda$ is a strong Z_σ-space.

Since (G-1) is a direct consequence of Lemma 5.10.8, we have to show (G-2) and (G-3), but we only prove (G-2) because of similarity. For (G-3), the proof is left to the reader.

Let $U = \bigcup_{n \in \mathbb{N}} A_n$ and $V = \bigcup_{n \in \mathbb{N}} B_n$, where each A_n and B_n is a strong Z-set in U and V, respectively. Since U and V are F_σ in X, we can assume that each A_n and B_n is closed in X. Then, each A_n and B_n is a strong Z-set in X by Proposition 2.8.9(2). Hence, $A_n \cup B_n$ is a strong Z-set in X by Proposition 2.8.9(1). Due to Proposition 2.8.14, $A_n \cup B_n$ is a strong Z-set in $U \cup V$. Thus, $U \cup V = \bigcup_{n \in \mathbb{N}} (A_n \cup B_n)$ is a strong Z_σ-space. □

For the class $(\tau\mathfrak{C})_\sigma$, we show the following:

Proposition 5.10.13 *Let X be a metrizable space with $w(X) \leqslant \tau$. If each point of X has an open neighborhood $U \in (\tau\mathfrak{C})_\sigma$, then $X \in (\tau\mathfrak{C})_\sigma$.*

Proof Since the result can be obtained by using Michael's Theorem 1.3.20, it suffices to prove the following:

(G-1) For open sets $U \supset V$ in X, if $U \in (\tau\mathfrak{C})_\sigma$, then $V \in (\tau\mathfrak{C})_\sigma$;
(G-2) For two open sets U and V in X, if $U, V \in (\tau\mathfrak{C})_\sigma$, then $U \cup V \in (\tau\mathfrak{C})_\sigma$;
(G-3) For a discrete collection $(U_\lambda)_{\lambda \in \Lambda}$ of open sets in X, if $U_\lambda \in (\tau\mathfrak{C})_\sigma$ for each $\lambda \in \Lambda$, then $\bigcup_{\lambda \in \Lambda} U_\lambda \in (\tau\mathfrak{C})_\sigma$.

Since $(\tau\mathfrak{C})_\sigma$ is open hereditary (Proposition 5.6.1), (G-1) holds.

For (G-2), let $U = \bigcup_{n \in \mathbb{N}} A_n$ and $V = \bigcup_{n \in \mathbb{N}} B_n$, where $A_n, B_n \in \tau\mathfrak{C}$ and each A_n and B_n is closed in U and V, respectively. Since U and V are F_σ in X, we can assume that each A_n and B_n is closed in X. Then, $A_n \cup B_n$ is closed in X, hence in $U \cup V$. Since $\tau\mathfrak{C}$ is additive, $A_n \cup B_n \in \tau\mathfrak{C}$. Hence, $U \cup V = \bigcup_{n \in \mathbb{N}} (A_n \cup B_n) \in (\tau\mathfrak{C})_\sigma$.

For (G-3), since $(U_\lambda)_{\lambda \in \Lambda}$ is discrete in X, it follows that $\operatorname{card} \Lambda \leqslant w(X) \leqslant \tau$. We can write $U_\lambda = \bigcup_{n \in \mathbb{N}} C_{\lambda,n}$, where $C_{\lambda,n} \in \tau\mathfrak{C}$ and each $C_{\lambda,n}$ is closed in U_λ. Then, $C_n = \bigcup_{\lambda \in \Lambda} C_{\lambda,n} \in \tau\mathfrak{C}$ and C_n is closed in $\bigcup_{\lambda \in \Lambda} U_\lambda$ because $C_n \cap U_\lambda = C_{\lambda,n}$. Therefore, $\bigcup_{\lambda \in \Lambda} U_\lambda = \bigcup_{n \in \mathbb{N}} C_n \in (\tau\mathfrak{C})_\sigma$. □

Using the above two propositions, we can prove the following:.

Proposition 5.10.14 *Suppose that $\tau\mathfrak{C} = \mathfrak{C}$. Let Ω be a $\mathfrak{C}$-absorbing set in $\ell^2(\Gamma)$. Then, every Ω-manifold X is a strongly $\mathfrak{C}$-universal ANR that is a strong Z_σ-space, $X \in \mathfrak{C}_\sigma$, and every Z-set in X is a strong Z-set.*

Proof Each point of X has an open neighborhood U that is homeomorphic to an open set in Ω, that is, $U \approx W \cap \Omega$ for some open set W in $\ell^2(\Gamma)$. Then, $W \cap \Omega$ is a $\mathfrak{C}$-absorbing set in W by Proposition 5.10.9. Hence, U is a $\mathfrak{C}$-absorbing set in some $\ell^2(\Gamma)$-manifold $\widetilde{U}$, which implies that U is a strongly $\mathfrak{C}$-universal ANR that is a strong Z_σ-space and $U \in \mathfrak{C}_\sigma$. Hence, X is an ANR by the Hanner Theorem (Proposition 1.13.6(4)). Form Proposition 5.10.12, it follows that X is a strong Z_σ-space.

We show that every Z-set A in X is a strong Z-set. As we saw above, each $a \in A$ has an open neighborhood U in X that is a $\mathfrak{C}$-absorbing set in some $\ell^2(\Gamma)$-manifold $\widetilde{U}$. Then, we have a closed set A_U in $\widetilde{U}$ such that $A_U \cap U = A \cap U$. Since U is homotopy dense in $\widetilde{U}$ and $A \cap U$ is a Z-set in U (Corollary 2.8.7(3)), it follows that A_U is a Z-set in an $\ell^2(\Gamma)$-manifold $\widetilde{U}$, so a strong Z-set in $\widetilde{U}$. Hence, $A \cap U = A_U \cap U$ is a strong Z-set in U. Applying Proposition 2.8.15, we can conclude that A is a strong Z-set in X.

We can now apply Proposition 5.6.3 to prove that X is strongly $\mathfrak{C}$-universal. Moreover, because $\tau\mathfrak{C} = \mathfrak{C}$, we have $X \in \mathfrak{C}_\sigma$ by Proposition 5.10.13. The proof is complete. □

Now, we can obtain the following characterization:

Theorem 5.10.15 (Characterization) *Suppose that $\tau\mathfrak{C} = \mathfrak{C}$ and $\mathfrak{C}$ is* **I**-*stable.*[22] *Let Ω be a $\mathfrak{C}$-absorbing set in $\ell^2(\Gamma)$. For a space X, the following are equivalent:*

(a) *X is an Ω-manifold;*
(b) *X can be embedded into an $\ell^2(\Gamma)$-manifold as a $\mathfrak{C}$-absorbing set;*
(c) *$X \in \mathfrak{C}_\sigma$ is a strongly $\mathfrak{C}$-universal ANR with $w(X) = \tau$ that is a strong Z_σ-space.*

Proof (a) ⇒ (c): There exists $\mathcal{U} \in \operatorname{cov}(X)$ such that each $U \in \mathcal{U}$ is homeomorphic to an open set in Ω. Since Ω is a strong Z_σ-space, each U is also a strong Z_σ-space by Lemma 5.10.8. By virtue of Proposition 5.10.12, X is a strong Z_σ-space. Since Ω is homotopy dense in $\ell^2(\Gamma)$, it is an AR, hence each $U \in \mathcal{U}$ is an ANR (Proposition 1.13.6(3)). By Hanner Theorem (Proposition 1.13.6(4)), X is also an ANR. Since $\mathfrak{C}_\sigma$ is open hereditary and $\Omega \in \mathfrak{C}_\sigma$, it follows that $U \in \mathfrak{C}_\sigma$. Applying Proposition 5.10.13, we have $X \in (\tau\mathfrak{C})_\sigma = \mathfrak{C}_\sigma$. Because of the definition of manifolds, $w(X) = w(\Omega) = \tau$.

For each $U \in \mathcal{U}$, take an open set $\widetilde{U}$ in $\ell^2(\Gamma)$ such that $U \approx \widetilde{U} \cap \Omega$. Since $\widetilde{U} \cap \Omega$ is a $\mathfrak{C}$-absorbing set in $\widetilde{U}$ (Proposition 5.10.9), $\widetilde{U} \cap \Omega$ is a strongly $\mathfrak{C}$-universal homotopy dense set in $\widetilde{U}$. Hence, each $U \in \mathcal{U}$ is strongly $\mathfrak{C}$-universal. Due to Proposition 5.10.14, each Z-set in X is a strong Z-set. Hence, we can apply Proposition 5.6.3 to see that X is strongly $\mathfrak{C}$-universal.

---

[22]Note that $\tau\mathfrak{C} = \mathfrak{C}$ implies $\Gamma \in \mathfrak{C}$. Since $\mathfrak{C}$ is **I**-stable, $\mathbf{I}^n \times \Gamma \in \mathfrak{C}$ for every $n \in \mathbb{N}$. Then, the strong $\mathfrak{C}$-universality implies the τ-discrete cells property.

(c) ⇒ (b): Due to the definition of a $\mathfrak{C}$-absorbing set, it suffices to show that X can be embedded into an $\ell^2(\Gamma)$-manifold as a homotopy dense set. This can be done as in the proof of Theorem 5.7.17 (d) ⇒ (c).

For readers' convenience, we repeat the proof. First, take $d \in \mathrm{Metr}(X)$ such that (X, d) is a uniform ANR (Theorem 1.13.37). Then, the metric completion $\widetilde{X}$ of (X, d) is an ANR and X is homotopy dense in $\widetilde{X}$ by Theorem 1.13.35. Because X has the τ-discrete cells property, $\widetilde{X}$ has the τ-discrete cells property. Since $\mathbf{I}^n \times \Gamma \in \mathfrak{C}$ for every $n \in \mathbb{N}$, it follows that $\mathfrak{M}_0^{\mathrm{fd}} \subset \mathfrak{C}$. So, the strong $\mathfrak{C}$-universality of X implies the strong $\mathfrak{M}_0^{\mathrm{fd}}$-universality. By Corollary 5.7.15 X has the discrete τ-polyhedral sequence property of X, which implies the discrete τ-polyhedral sequence property of $\widetilde{X}$ because X is homotopy dense in $\widetilde{X}$. By virtue of the Toruńczyk Characterization (Theorem 3.4.2), $\widetilde{X}$ is an $\ell^2(\Gamma)$-manifold.

(b) ⇒ (a): Let M be an $\ell^2(\Gamma)$-manifold in which X is a $\mathfrak{C}$-absorbing set. By the Open Embedding Theorem 2.5.10, M can be regarded as an open set in $\ell^2(\Gamma)$. Then, $\Omega \cap M$ is a $\mathfrak{C}$-absorbing set in M (Proposition 5.10.9). By the Uniqueness of Absorbing Sets (Theorem 5.10.5), X is homeomorphic to the open subset $\Omega \cap M$ of Ω. Therefore, X is an Ω-manifold. □

The following is a separable version of Theorem 5.10.15 above:

Theorem 5.10.16 (CHARACTERIZATION) *Let Ω be a $\mathfrak{C}$-absorbing set in ℓ^2.*[23] *For a space X, the following are equivalent:*

(a) *X is an Ω-manifold;*
(b) *X can be embedded into an ℓ^2-manifold as a $\mathfrak{C}$-absorbing set;*
(c) *$X \in \mathfrak{C}_\sigma$ is a strongly $\mathfrak{C}$-universal ANR that is a strong Z_σ-space.*[24]

For manifolds modeled on a $\mathfrak{C}$-absorbing set in $\ell^2(\Gamma)$, the Classification and the Open Embedding Theorems hold:

Theorem 5.10.17 *Suppose that $\tau\mathfrak{C} = \mathfrak{C}$. Let Ω be a $\mathfrak{C}$-absorbing set in $\ell^2(\Gamma)$.*

(1) (CLASSIFICATION) *For two Ω-manifolds X and Y, $X \approx Y$ if and only if $X \simeq Y$.*

Sketch of Proof. This is just combination of Theorem 5.10.15 with Corollary 5.10.6.

(2) (OPEN EMBEDDING) *Every Ω-manifold can be embedded into Ω as an open set.*

Sketch of Proof. Embed each Ω-manifold M into an $\ell^2(\Gamma)$-manifold $\widetilde{M}$ as a $\mathfrak{C}$-absorbing set (Theorem 5.10.15). Then, $\widetilde{M}$ is homeomorphic to an open set U in $\ell^2(\Gamma)$ (the Open Embedding Theorem 2.5.10), where $\Omega \cap U$ is a $\mathfrak{C}$-absorbing set in U (Proposition 5.10.9) Since the image of M is also a $\mathfrak{C}$-absorbing set in U, we can apply Theorem 5.10.5 to obtain the result.

(3) *For two Ω-manifolds X and Y, every fine homotopy equivalence $f : X \to Y$ is a near-homeomorphism.*

Sketch of Proof. This is just combination of Theorem 5.10.15 with Theorem 5.10.7.

---

[23]This implies $\mathfrak{C} \subset \mathfrak{M}(\aleph_0)$.

[24]Since $\mathfrak{C}_\sigma \subset \mathfrak{M}(\aleph_0)$, it follows that X is separable.

To prove the Triangulation Theorem, we need the following:

Theorem 5.10.18 *Suppose that $\tau\mathfrak{C} = \mathfrak{C}$ and $\mathfrak{C}$ is* **I**-*stable. Let Ω be a $\mathfrak{C}$-absorbing set in $\ell^2(\Gamma)$. For an arbitrary simplicial complex K with* $\operatorname{card} K^{(0)} \leqslant \tau$, $|K|_m \times \Omega$ *is an Ω-manifold.*

Proof First, note that $|K|_m \times \Omega$ is an ANR and

$$w(|K|_m \times \Omega) = w(\Omega)w(|K|_m) \leqslant \tau\aleph_0 \operatorname{card} K^{(0)} \leqslant \tau^2 = \tau.$$

Since Ω is strongly $\mathfrak{C}$-universal, $|K|_m \times \Omega$ is strongly $\mathfrak{C}$-universal (Proposition 5.4.7). By Theorem 5.4.15, we can write $|K|_m = \bigcup_{i\in\mathbb{N}} D_i$, where each D_i is a discrete union of at most τ many n_i-cells. Let $\Omega = \bigcup_{j\in\mathbb{N}} C_j$, where $C_j \in \mathfrak{C}_Z(\Omega)$ for each $j \in \mathbb{N}$. Since $\mathfrak{C}$ is **I**-stable, $D_i \times C_j \in \tau\mathfrak{C} = \mathfrak{C}$ for each $i, j \in \mathbb{N}$. Thus, $|K|_m \times \Omega = \bigcup_{i,j\in\mathbb{N}} D_i \times C_j \in \mathfrak{C}_\sigma$. It is easy to see that each $|K|_m \times C_j$ is a strong Z-set in $|K|_m \times \Omega$. Hence, $|K|_m \times \Omega$ is a strong Z_σ-space. Therefore, $|K|_m \times \Omega$ is an Ω-manifold by Theorem 5.10.15. □

Remark 5.24 In the above theorem, it is not assumed that K is locally finite-dimensional. Then, $|K|_m$ is not necessarily completely metrizable (cf. Theorem 1.7.2), that is, $|K|_m \times \ell^2(\Gamma)$ is not necessarily an $\ell^2(\Gamma)$-manifold.[25]

Now, we prove the Triangulation Theorem:

Theorem 5.10.19 (TRIANGULATION) *Suppose that $\tau\mathfrak{C} = \mathfrak{C}$ and $\mathfrak{C}$ is* **I**-*stable. Let Ω be a $\mathfrak{C}$-absorbing set in $\ell^2(\Gamma)$. For each Ω-manifold M, there exists a locally finite-dimensional simplicial complex K such that $M \approx |K|_m \times \Omega$.*

Proof Since M is an ANR (Theorem 5.10.15), there exists a locally finite-dimensional simplicial complex K such that

$$\operatorname{card} K^{(0)} \leqslant w(M) = w(\Omega) = \tau \text{ and } M \simeq |K|_m$$

(Theorem 1.13.22). Due to Theorem 5.10.18, $|K|_m \times \Omega$ is an Ω-manifold. Then, applying the Classification Theorem 5.10.17, we have the result. □

When $\mathfrak{C}$ is multiplicative, Theorem 5.10.18 can be generalized as follows:

Theorem 5.10.20 (ANR FACTOR) *Suppose that $\tau\mathfrak{C} = \mathfrak{C}$,* **I** $\in \mathfrak{C}$, *and $\mathfrak{C}$ is multiplicative.*[26] *Let Ω be a $\mathfrak{C}$-absorbing set in $\ell^2(\Gamma)$. For each ANR $X \in \mathfrak{C}_\sigma$, $X \times \Omega$ is an Ω-manifold. Moreover, if X is homotopy dense in an ANR $\widetilde{X} \in \mathfrak{M}_1(\tau)$, then $X \times \Omega$ is a $\mathfrak{C}$-absorbing set in an $\ell^2(\Gamma)$-manifold $\widetilde{X} \times \ell^2(\Gamma)$.*

Proof First note that $X \times \Omega$ is an ANR. Since Ω is strongly $\mathfrak{C}$-universal, $X \times \Omega$ is strongly $\mathfrak{C}$-universal (Proposition 5.4.7). Since $\mathfrak{C}$ is multiplicative and $X, \Omega \in \mathfrak{C}_\sigma$,

[25]It is proved in [120] that the completion of $(|K|, \rho_K)$ is an ANR in which $|K|$ is homotopy dense, where ρ_K is the metric defined in Sect. 1.7.

[26]Hence, $\mathfrak{C}$ is **I**-stable.

it is easy to see $X \times \Omega \in \mathfrak{C}_\sigma$. Since Ω is a strong Z_σ-space, it follows that $X \times \Omega$ is also a strong Z_σ-space. Thus, $X \times \Omega$ is an Ω-manifold by Theorem 5.10.15. The additional statement can be obtained in combination with the Toruńczyk Factor Theorem 2.2.14. □

We have the following corollary (cf. Theorem 5.10.17(1)):

Corollary 5.10.21 *Suppose that $\tau\mathfrak{C} = \mathfrak{C}$, $\mathbf{I} \in \mathfrak{C}$, and $\mathfrak{C}$ is multiplicative. Let Ω be a $\mathfrak{C}$-absorbing set in $\ell^2(\Gamma)$. For each AR $X \in \mathfrak{C}_\sigma$, $X \times \Omega \approx \Omega$. Moreover, if X is homotopy dense in an AR $\widetilde{X} \in \mathfrak{M}_1(\tau)$, then $X \times \Omega$ is a $\mathfrak{C}$-absorbing set in $\widetilde{X} \times \ell^2(\Gamma) \approx \ell^2(\Gamma)$.* □

Combining the ANR Factor Theorems 5.10.20 and 5.10.17(3), we have the following:

Theorem 5.10.22 (Stability) *Suppose that $\tau\mathfrak{C} = \mathfrak{C}$, $\mathbf{I} \in \mathfrak{C}$, and $\mathfrak{C}$ is multiplicative. Let Ω be a $\mathfrak{C}$-absorbing set in $\ell^2(\Gamma)$. For each Ω-manifold M, the projection $\mathrm{pr}_M : M \times \Omega \to M$ is a near-homeomorphism.* □

We prove the Z-Set Unknotting Theorem:

Theorem 5.10.23 (Z-Set Unknotting) *Suppose that $\mathbf{I}^n \times \Gamma \in \mathfrak{C}$ for every $n \in \mathbb{N}$. Let X be a $\mathfrak{C}$-absorbing set in an $\ell^2(\Gamma)$-manifold M and $\mathcal{U}$, $\mathcal{V}$ open covers of X. If $f : A \to B$ is a homeomorphism between Z-sets in X and f is $\mathcal{U}$-homotopic to the inclusion $A \subset X$, then f extends to a homeomorphism $h : X \to X$ that is $\mathrm{st}(\mathcal{U}, \mathcal{V})$-close to* id.

Proof First, assume that $A \cap B = \emptyset$. Then, removing $\mathrm{cl}_M A \cap \mathrm{cl}_M B$ from M, we may assume that $\mathrm{cl}_M A \cap \mathrm{cl}_M B = \emptyset$. Let $\mathcal{W} \in \mathrm{cov}(X)$ be a star-refinement of $\mathcal{V}$. Take open collections $\widetilde{\mathcal{U}}$ and $\widetilde{\mathcal{W}}$ in M such that $\widetilde{\mathcal{U}}|X = \mathcal{U}$ and $\widetilde{\mathcal{W}}|X = \mathcal{W}$. Let

$$M' = \left(\bigcup \widetilde{\mathcal{U}} \cap \bigcup \widetilde{\mathcal{W}}\right) \setminus (\mathrm{cl}_M A \cap \mathrm{cl}_M B).$$

Then, $X \subset M'$ and M' is G_δ in M. By Lavrentieff's Theorem 1.3.19, f extends to a homeomorphism $\tilde{f} : \widetilde{A} \to \widetilde{B}$ between G_δ-sets in M' such that $\tilde{f}$ is $\widetilde{\mathcal{U}}$-homotopic to id in M', where $A \subset \widetilde{A} \subset \mathrm{cl}_{M'} A$ and $B \subset \widetilde{B} \subset \mathrm{cl}_{M'} B$. Let

$$\widetilde{X} = M' \setminus \left((\mathrm{cl}_{M'} A \setminus \widetilde{A}) \cup (\mathrm{cl}_{M'} B \setminus \widetilde{B})\right).$$

Then, $X \subset \widetilde{X}$ and $\widetilde{X}$ is a homotopy dense G_δ-set in M, hence $\widetilde{X}$ is an $\ell^2(\Gamma)$-manifold. By pushing the homotopy from $\tilde{f}$ to id into $\widetilde{X}$, $\tilde{f}$ is $\mathrm{st}(\widetilde{\mathcal{U}}, \widetilde{\mathcal{W}})$-homotopic to id in $\widetilde{X}$, Moreover, $\widetilde{A} = \mathrm{cl}_{\widetilde{X}} A$ and $\widetilde{B} = \mathrm{cl}_{\widetilde{X}} B$, which are Z-sets in $\widetilde{X}$. Applying the Z-Set Unknotting Theorem 2.9.7, we can obtain a homeomorphism $g : \widetilde{X} \to \widetilde{X}$ such that $g|\widetilde{A} = \tilde{f}$ and g is $\mathrm{st}(\mathrm{st}(\widetilde{\mathcal{U}}, \widetilde{\mathcal{W}}), \widetilde{\mathcal{W}})$-close to id. Then, $g(X)$ is a $\mathfrak{C}$-absorbing set in $\widetilde{X}$. By Propositions 5.4.18 and 5.6.2, $X \setminus \widetilde{B}$ and $g(X) \setminus \widetilde{B}$ are $\mathfrak{C}$-absorbing sets in $\widetilde{X} \setminus \widetilde{B}$, where $X \setminus \widetilde{B} = X \setminus B$ and $g(X) \setminus \widetilde{B} = g(X) \setminus B$. Let $\mathcal{W}'$ be an open cover of $\widetilde{X} \setminus \widetilde{B}$ such that $\mathcal{W}' \prec \mathcal{W}$ and $\mathrm{diam}\, W < \mathrm{dist}(W, \widetilde{X} \setminus \widetilde{B})$ for each $W \in \mathcal{W}'$, where $\widetilde{X}$ is given some $d \in \mathrm{Metr}(\widetilde{X})$. By Theorem 5.10.5, we have a homeomorphism $h : g(X) \setminus B \to X \setminus B$ that is $\mathcal{W}'$-close to id. We define $\tilde{h} : g(X) \to X$ as follows:

$$\tilde{h}(x) = \begin{cases} h(x) & \text{if } x \in g(X) \setminus B, \\ x & \text{if } x \in B. \end{cases}$$

Then, $\tilde{h}$ is a homeomorphism that is $\widetilde{\mathcal{W}}$-close to id. Thus, we have a homeomorphism $\tilde{h}g|X : X \to X$ that is $\mathrm{st}(\widetilde{\mathcal{U}}, \mathrm{st}\,\widetilde{\mathcal{W}})$-close to id, hence it is $\mathrm{st}(\mathcal{U}, \mathcal{V})$-close to id.

Now, applying the above case, we show the general case. Choose $\mathcal{W}, \mathcal{W}' \in \mathrm{cov}(X)$ so that

$$\mathcal{V} \succ \mathrm{st}^2\,\mathcal{W} \text{ and } \mathcal{W} \underset{h}{\succ} \mathcal{W}'.$$

Since $\mathbf{I}^n \times \Gamma \in \mathfrak{C}$ for every $n \in \mathbb{N}$ and X is a strong Z_σ-space, the strong $\mathfrak{C}$-universality of X implies the strong $\mathfrak{C}_\sigma$-universality (Proposition 5.6.5). Since $A \cup B$ is a Z-set in X and $A \in \mathfrak{C}_\sigma$ as a closed set in $X \in \mathfrak{C}_\sigma$, we have a Z-embedding $g : A \to X \setminus (A \cup B)$ that is $\mathcal{W}'$-close to id, so g and g^{-1} are $\mathcal{W}$-homotopic to id. Applying the above, we have homeomorphisms $h_1 : X \to X$ such that $h_1|A = g$ and h_1 is $\mathrm{st}\,\mathcal{W}$-close to id. Observe that $fg^{-1} : g(A) \to B$ is $\mathrm{st}(\mathcal{U}, \mathcal{W})$-homotopic to id. Again applying the above we have homeomorphisms $h_2 : X \to X$ such that $h_2|g(A) = fg^{-1}$ and h_2 is $\mathrm{st}(\mathrm{st}(\mathcal{U}, \mathcal{W}), \mathcal{W})$-close to id. Then, $h_2 h_1 : X \to X$ is the desired homeomorphism. □

The following two lemmas are trivial.

Lemma 5.10.24 *If X is homotopy dense in $\ell^2(\Gamma)$, then $X^{\mathbb{N}}$ and $X^{\mathbb{N}}_f$ are homotopy dense in $\ell^2(\Gamma)^{\mathbb{N}}$.* □

In the above, $X^{\mathbb{N}}_f$ is homotopy dense in $X^{\mathbb{N}}$ (cf. Proposition 1.13.27).

Lemma 5.10.25 *Let X be a strong Z_σ-space. Then, $X \times Y$ is a strong Z_σ-space for any Y. In particular, $X^{\mathbb{N}} = X \times X^{\mathbb{N}\setminus\{1\}}$ is a strong Z_σ-space. If $X = (X, *)$ is pointed, then $X^{\mathbb{N}}_f = X \times X^{\mathbb{N}\setminus\{1\}}_f$ is a strong Z_σ-space.* □

Combining Proposition 5.4.11 with the above two lemmas, we can obtain the following:

Proposition 5.10.26 *Let X be a homotopy dense set in $\ell^2(\Gamma)$ which is a strong Z_σ-space (and $\mathbf{0} \in X$). Then, $X^{\mathbb{N}}$ (resp. $X^{\mathbb{N}}_f$) is an absorbing set in $\ell^2(\Gamma)^{\mathbb{N}}$ for the class $\mathcal{F}_{X^{\mathbb{N}}}$ (resp. $\mathcal{F}_{X^{\mathbb{N}}_f}$).* □

In the above, note that $X^{\mathbb{N}}$ and $X^{\mathbb{N}}_f$ are Z_σ-sets in itself. Every Z-set in $X^{\mathbb{N}}$ and $X^{\mathbb{N}}_f$ belongs to the classes $\mathcal{F}_{X^{\mathbb{N}}}$ and $\mathcal{F}_{X^{\mathbb{N}}_f}$, respectively.

5.11 Absolute Borel Classes

For a metrizable space X, let $\mathfrak{A}_0(X)$ and $\mathfrak{M}_0(X)$ be the collections consisting of all open sets and of all closed sets in X, respectively. For each $n \in \mathbb{N}$, we inductively define $\mathfrak{A}_n(X)$ and $\mathfrak{M}_n(X)$ as the collections consisting of all countable unions of sets in $\mathfrak{M}_{n-1}(X)$ and of all countable intersections of sets in $\mathfrak{A}_{n-1}(X)$, respectively.

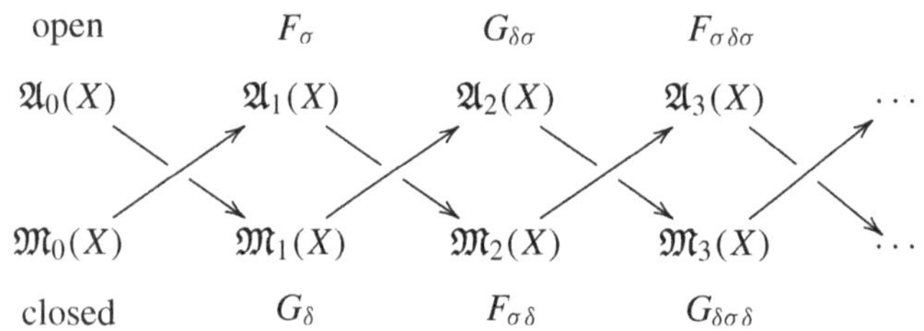

Then, $\mathfrak{M}_1(X)$ and $\mathfrak{A}_1(X)$ are the collections of all G_δ sets and of all F_σ sets in X, respectively. Sets contained in $\mathfrak{A}_2(X)$, $\mathfrak{M}_2(X)$, $\ldots$ are said to be $G_{\delta\sigma}$, $F_{\sigma\delta}$, $\ldots$ in X, respectively.

We can determine the notation of a set of $\mathfrak{M}_n(X)$ or $\mathfrak{A}_n(X)$ as follows: First write the alternate sequence of letters δ and σ with n length and the last letter is δ or σ according to $\mathfrak{M}_n(X)$ or $\mathfrak{A}_n(X)$. It is decided by the first letter (δ or σ) whether it is $G_{\delta\sigma\cdots}$ or $F_{\sigma\delta\cdots}$.

Since X is a metrizable space, open sets and closed sets in X are F_σ and G_δ respectively, that is, $\mathfrak{A}_0(X) \subset \mathfrak{A}_1(X)$ and $\mathfrak{M}_0(X) \subset \mathfrak{M}_1(X)$. By induction, we have $\mathfrak{A}_{n-1}(X) \subset \mathfrak{A}_n(X)$ and $\mathfrak{M}_{n-1}(X) \subset \mathfrak{M}_n(X)$ for every $n \in \mathbb{N}$. Hence, it follows that

$$\bigcup_{i=1}^{n-1} \mathfrak{A}_i(X) \cup \mathfrak{M}_i(X) \subset \mathfrak{A}_n(X) \cap \mathfrak{M}_n(X).$$

For infinite ordinals α, by transfinite induction, we define $\mathfrak{A}_\alpha(X)$ (resp. $\mathfrak{M}_\alpha(X)$) as the collection consisting of all countable unions (resp. countable intersections) of sets in $\bigcup_{\beta<\alpha} \mathfrak{M}_\beta(X)$ (resp. $\bigcup_{\beta<\alpha} \mathfrak{A}_\beta(X)$), where $\bigcup_{\beta<\alpha} \mathfrak{M}_\beta(X)$ (resp. $\bigcup_{\beta<\alpha} \mathfrak{A}_\beta(X)$) can be replaced with $\bigcup_{\beta<\alpha} \mathfrak{A}_\beta(X) \cup \mathfrak{M}_\beta(X)$. When $\alpha = \beta + 1$, $\mathfrak{A}_\alpha(X)$ and $\mathfrak{M}_\alpha(X)$ can be defined as the collections consisting of all countable unions of sets in $\mathfrak{M}_\beta(X)$ and of all countable intersections of sets in $\mathfrak{A}_\beta(X)$, respectively. It should be noted that

$$A \in \mathfrak{A}_\alpha(X) \;\Leftrightarrow\; X \setminus A \in \mathfrak{M}_\alpha(X).$$

The following collection is the σ-field on X generated by open sets and closed sets in X, which is called the **Borel field** on X and whose member is called a **Borel set** in X:

$$\bigcup_{\operatorname{card}\alpha<\aleph_1} \mathfrak{A}_\alpha(X) \cup \mathfrak{M}_\alpha(X).$$

It should be noted that countable unions and finite intersections of sets in $\mathfrak{A}_\alpha(X)$ belong to $\mathfrak{A}_\alpha(X)$ and countable intersections and finite unions of sets in $\mathfrak{M}_\alpha(X)$ belong to $\mathfrak{M}_\alpha(X)$. For each map $f : X \to Y$, $f^{-1}(\mathfrak{A}_\alpha(Y)) \subset \mathfrak{A}_\alpha(X)$ and $f^{-1}(\mathfrak{M}_\alpha(Y)) \subset \mathfrak{M}_\alpha(X)$. In particular, for $X \subset Y$, $\mathfrak{A}_\alpha(Y)|X \subset \mathfrak{A}_\alpha(X)$ and $\mathfrak{M}_\alpha(Y)|X \subset \mathfrak{M}_\alpha(X)$. For $\alpha > 0$, if X is open or closed in Y then $\mathfrak{A}_\alpha(X) \subset \mathfrak{A}_\alpha(Y)$ and $\mathfrak{M}_\alpha(X) \subset \mathfrak{M}_\alpha(Y)$. Moreover, if $A \in \mathfrak{A}_\alpha(X)$ (resp. $A \in \mathfrak{M}_\alpha(X)$) then both open sets and closed sets in A belong to $\mathfrak{A}_\alpha(X)$ (resp. $\mathfrak{M}_\alpha(X)$). So, it follows that Borel sets in a Borel set in X are also Borel sets in X.

For an infinite cardinal τ and each ordinal α, let $\mathfrak{A}_\alpha(\tau)$ (resp. $\mathfrak{M}_\alpha(\tau)$) denote the class of all metrizable spaces X with $w(X) \leqslant \tau$ such that $X \in \mathfrak{A}_\alpha(Y)$ (resp. $X \in \mathfrak{M}_\alpha(Y)$) whenever X is embedded in a metrizable space Y. When $\tau = \aleph_0$, we write $\mathfrak{A}_\alpha(\aleph_0) = \mathfrak{A}_\alpha$ and $\mathfrak{M}_\alpha(\aleph_0) = \mathfrak{M}_\alpha$. Note that the classes $\mathfrak{A}_\alpha(\tau)$ and $\mathfrak{M}_\alpha(\tau)$, $\alpha > 0$, are closed hereditary and open hereditary, that is, if $X \in \mathfrak{A}_\alpha(\tau)$ (resp. $X \in \mathfrak{M}_\alpha(\tau)$), then both open sets and closed sets in X belong to $\mathfrak{A}_\alpha(\tau)$ (resp. $\mathfrak{M}_\alpha(\tau)$). Similar to absolutely G_δ, a metrizable space X is said to be **absolutely F_σ** if $X \in \mathfrak{A}_1(w(X))$, **absolutely $F_{\sigma\delta}$** if $X \in \mathfrak{M}_2(w(X))$, **absolutely $G_{\delta\sigma}$** if $X \in \mathfrak{A}_2(w(X))$, ...[27] A metrizable space X is called an **absolutely Borel space** if X is a Borel set in Y whenever X is embedded in a metrizable space Y.[28]

It is natural to call a metrizable space X **absolutely open** (resp. **absolutely closed**) if $X \in \mathfrak{A}_0(w(X))$ (resp. $X \in \mathfrak{M}_0(w(X))$). The only empty set is absolutely open (i.e., $\mathfrak{A}_0 = \mathfrak{A}_0(\tau) = \{\emptyset\}$ for every $\tau \geqslant \aleph_0$).[29] The absolutely closed spaces are the compacta, namely $\mathfrak{M}_0 = \mathfrak{M}_0(\tau)$ is the class of all compacta for every $\tau \geqslant \aleph_0$).

Every compactum is clearly absolutely closed. Any non-compact metrizable space X contains a countably infinite discrete set D. By Hausdorff's Metric Extension Theorem 1.13.8, X has an admissible metric d such that D is isometric to $\{n^{-1} \mid n \in \mathbb{N}\}$ with the usual metric. Let $\tilde{X} = (\tilde{X}, \tilde{d})$ be the metric completion of $X = (X, d)$. Then, D has an accumulation point x in $\tilde{X}$, where $x \notin X$. Hence, X is not closed in $\tilde{X}$.

Lemma 5.11.1 *Let X be a G_δ subspace of a metrizable space Y. Then, $\mathfrak{M}_\alpha(X) \subset \mathfrak{M}_\alpha(Y)$ for $\alpha > 0$ and $\mathfrak{A}_\alpha(X) \subset \mathfrak{A}_\alpha(Y)$ for $\alpha > 1$.*

Proof It is easy to see $\mathfrak{M}_1(X) \subset \mathfrak{M}_1(Y)$. We shall show $\mathfrak{M}_2(X) \subset \mathfrak{M}_2(Y)$. Then, the result could be obtained by transfinite induction. To see $\mathfrak{M}_2(X) \subset \mathfrak{M}_2(Y)$, it suffices to prove that each F_σ-set in X is $F_{\sigma\delta}$ in Y. Let $A = \bigcup_{n\in\mathbb{N}} A_n \subset X$, where

[27] In many articles, they are also called absolute G_δ, absolute F_σ, absolute $F_{\sigma\delta}$, etc.

[28] This is also an absolute Borel space in many articles.

[29] If $X \neq \emptyset$, then $X \approx X \times \{0\}$ is not open in $X \times \mathbf{I}$.

each A_n is closed in X. We can write $X = \bigcap_{i\in\mathbb{N}} U_i$, where each U_i is open in Y. Then, it follows that

$$A = X \cap \bigcup_{n\in\mathbb{N}} \mathrm{cl}_Y A_n = \bigcap_{i\in\mathbb{N}} \left(U_i \cap \bigcup_{n\in\mathbb{N}} \mathrm{cl}_Y A_n \right) = \bigcap_{i\in\mathbb{N}} \bigcup_{n\in\mathbb{N}} U_i \cap \mathrm{cl}_Y A_n.$$

Since each U_i is F_σ in Y, each $U_i \cap \mathrm{cl}_Y A_n$ is also F_σ in Y, whence A is $F_{\sigma\delta}$ in Y. □

Theorem 5.11.2 *Let X be a metrizable space with $w(X) \leqslant \tau$ and $\alpha > 0$ (resp. $\alpha > 1$). Then, $X \in \mathfrak{M}_\alpha(\tau)$ (resp. $X \in \mathfrak{A}_\alpha(\tau)$) if and only if $X \in \mathfrak{M}_\alpha(Y)$ (resp. $X \in \mathfrak{A}_\alpha(Y)$) for some completely metrizable space Y containing X as a subspace.*

Proof Since the "only if" part is trivial, we shall show the "if" part. Suppose that X is a subspace of a metrizable space Z. We show that $X \in \mathfrak{M}_\alpha(Z)$ (resp. $X \in \mathfrak{A}_\alpha(Z)$). By replacing Z with the metric completion of Z with respect to some $d \in \mathrm{Metr}(Z)$, we may assume that Z is completely metrizable. By Lavrentieff's Theorem 1.3.19, we have a homeomorphism $h : G \to H$ between G_δ-sets $G \subset Y$ and $H \subset Z$ such that $h|X = \mathrm{id}$. Since $X \in \mathfrak{M}_\alpha(Y)$ (resp. $X \in \mathfrak{A}_\alpha(Y)$), it follows that $X \in \mathfrak{M}_\alpha(G)$ (resp. $X \in \mathfrak{A}_\alpha(G)$), hence $X \in \mathfrak{M}_\alpha(H)$ (resp. $X \in \mathfrak{A}_\alpha(H)$). By Lemma 5.11.1 above, we have $X \in \mathfrak{M}_\alpha(Z)$ (resp. $X \in \mathfrak{A}_\alpha(Z)$). □

Remark 5.25 Although both Lemma 5.11.1 and Theorem 5.11.2 is valid for $\mathfrak{M}_1$, they are not valid for $\mathfrak{A}_1$. In fact, $\nu^0 = \mathbb{R} \setminus \mathbb{Q}$ is G_δ in $\mathbb{R}$ and F_σ in itself but it is not F_σ in $\mathbb{R}$ because $\mathbb{R}$ is a Baire space (the Baire Category Theorem 1.3.15).

Notice that every separable metrizable space can be embedded into a compactum (e.g., the Hilbert cube). So, we have the following:

Corollary 5.11.3 *For a separable metrizable space X and $\alpha > 0$, $X \in \mathfrak{M}_\alpha$ (resp. $X \in \mathfrak{A}_\alpha$) if and only if $X \in \mathfrak{M}_\alpha(Y)$ (resp. $X \in \mathfrak{A}_\alpha(Y)$) for some compactum Y containing X as a subspace.* □

Concerning absolutely Borel spaces, we have the following:

Corollary 5.11.4 *Let X be a metrizable space with $w(X) \leqslant \tau$. For $\alpha > 1$ (resp. $\alpha > 2$), $X \in \mathfrak{A}_\alpha(\tau)$ (resp. $X \in \mathfrak{M}_\alpha(\tau)$) if and only if $X = \bigcup_{i\in\mathbb{N}} X_i$ (resp. $X = \bigcap_{i\in\mathbb{N}} X_i$), where $X_i \in \bigcup_{\beta<\alpha} \mathfrak{M}_\beta(\tau)$ (resp. $X_i \in \bigcup_{\beta<\alpha} \mathfrak{A}_\beta(\tau)$) for each $i \in \mathbb{N}$.*

Proof We can embed X into a completely metrizable space Y (e.g., the metric completion with respect some $d \in \mathrm{Metr}(X)$).

We show the "only if" part. Since $X \in \mathfrak{A}_\alpha(Y)$ (resp. $X \in \mathfrak{M}_\alpha(Y)$), we can write $X = \bigcup_{i\in\mathbb{N}} X_i$ (resp. $X = \bigcap_{i\in\mathbb{N}} X_i$), where $X_i \in \bigcup_{\beta<\alpha} \mathfrak{M}_\beta(Y)$ (resp. $X_i \in \bigcup_{\beta<\alpha} \mathfrak{A}_\beta(Y)$) for each $i \in \mathbb{N}$. By Theorem 5.11.2, $X_i \in \bigcup_{\beta<\alpha} \mathfrak{M}_\beta(\tau)$ (resp. $X_i \in \bigcup_{\beta<\alpha} \mathfrak{A}_\beta(\tau)$) for each $i \in \mathbb{N}$.

To see the "if" part, let $X = \bigcup_{i\in\mathbb{N}} X_i$ (resp. $X = \bigcap_{i\in\mathbb{N}} X_i$), where $X_i \in \bigcup_{\beta<\alpha} \mathfrak{M}_\beta(\tau)$ (resp. $X_i \in \bigcup_{\beta<\alpha} \mathfrak{A}_\beta(\tau)$) for each $i \in \mathbb{N}$. Then, $X_i \in \bigcup_{\beta<\alpha} \mathfrak{M}_\beta(Y)$

(resp. $X_i \in \bigcup_{\beta<\alpha} \mathfrak{A}_\beta(Y)$) for each $i \in \mathbb{N}$. Hence, $X \in \mathfrak{A}_\alpha(Y)$ (resp. $X \in \mathfrak{M}_\alpha(Y)$), which implies $X \in \mathfrak{A}_\alpha(\tau)$ (resp. $X \in \mathfrak{M}_\alpha(\tau)$) by Theorem 5.11.2. □

Due to Theorem 1.3.16, a metrizable space is absolutely G_δ if and only if it is completely metrizable. Hence, the following is obtained from Corollary 5.11.4:

Corollary 5.11.5 *Every σ-completely metrizable space is absolutely $G_{\delta\sigma}$, that is, $\mathfrak{M}_1(\tau)_\sigma \subset \mathfrak{A}_2(\tau)$. Actually, a metrizable space is absolutely $G_{\delta\sigma}$ if and only if it is a countable union of completely metrizable subsets.* □

In the above, "absolutely $G_{\delta\sigma}$" does not mean "σ-completely metrizable." Recall that a metrizable space X is σ-completely metrizable if it is a countable union of completely metrizable *closed* subsets (cf. p. 435).

The following sequence of spaces and maps is called an **inverse sequence**:

$$X_1 \xleftarrow{f_1} X_2 \xleftarrow{f_2} X_3 \xleftarrow{f_3} \cdots,$$

where each f_i is called a **bonding map**. The **inverse limit** of the above sequence is defined as the following subspace of the product space $\prod_{i\in\mathbb{N}} X_i$:

$$\varprojlim(X_i, f_i) = \left\{(x_i)_{i\in\mathbb{N}} \in \textstyle\prod_{i\in\mathbb{N}} X_i \;\middle|\; \forall i \in \mathbb{N},\ x_i = f_i(x_{i+1})\right\},$$

which is simply written as $\varprojlim X_i$. The inverse sequence $X_1 \supset X_2 \supset \cdots$ is called a **nested sequence** in X, where every space X_i is a subspace of X_1 and inclusions are bonding maps. For such a sequence, $\varprojlim X_i \approx \bigcap_{i\in\mathbb{N}} X_i$. For inverse sequences and the inverse limits, refer to [GAGT, Sect. 4.10].

For any inverse sequence $X_1 \xleftarrow{f_1} X_2 \xleftarrow{f_2} \cdots$, the nested sequence $X_1^* \supset X_2^* \supset \cdots$ in $\prod_{i\in\mathbb{N}} X_i$ is defined as follows:

$$X_n^* = \left\{(x_i)_{i\in\mathbb{N}} \in \textstyle\prod_{i\in\mathbb{N}} X_i \;\middle|\; \forall i < n,\ x_i = f_i(x_{i+1})\right\},$$

where $X_1^* = \prod_{i\in\mathbb{N}} X_i$. Then, we have $\varprojlim X_i = \bigcap_{i\in\mathbb{N}} X_i^*$. The following is the counterpart of Corollary 5.11.4.

Corollary 5.11.6 *Let X be a metrizable space with $w(X) \leqslant \tau$. For $\alpha > 2$, $X \in \mathfrak{M}_\alpha(\tau)$ if and only if $X \approx \varprojlim X_i$, where $X_i \in \bigcup_{\beta<\alpha} \mathfrak{A}_\beta(\tau)$ for each $i \in \mathbb{N}$.*

Proof To see the "only if" part, embed X into a completely metrizable space Y. Since $X \in \mathfrak{M}_\alpha(Y)$, we can write $X = \bigcap_{i\in\mathbb{N}} X_i$, where $X_i \in \bigcup_{\beta<\alpha} \mathfrak{A}_\beta(Y)$ for each $i \in \mathbb{N}$. For each $i \in \mathbb{N}$, $X_i^* = \bigcap_{j=1}^{i} X_j \in \bigcup_{1<\beta<\alpha} \mathfrak{A}_\beta(Y)$, whence $X_i^* \in \bigcup_{1<\beta<\alpha} \mathfrak{A}_\beta(\tau)$ by Theorem 5.11.2. Then, $X = \bigcap_{i\in\mathbb{N}} X_i^*$ is the inverse limit of the nested sequence $X_1^* \supset X_2^* \supset \cdots$.

To see the "if" part, let $X = \varprojlim X_i$, where $X_i \in \bigcup_{\beta<\alpha} \mathfrak{A}_\beta(\tau)$ for each $i \in \mathbb{N}$. Embed X_i into a completely metrizable space Y_i. Then, $Y = \prod_{i\in\mathbb{N}} Y_i$ is completely metrizable. Since $X_i \in \bigcup_{\beta<\alpha} \mathfrak{A}_\beta(Y_i)$, we have $\mathrm{pr}_i^{-1}(X_i) \in \bigcup_{\beta<\alpha} \mathfrak{A}_\beta(Y)$, where $\mathrm{pr}_i : Y \to Y_i$ is the projection. Then, it follows that

$$\prod_{i\in\mathbb{N}} X_i = \bigcap_{i\in\mathbb{N}} \mathrm{pr}_i^{-1}(X_i) \in \mathfrak{M}_\alpha(Y).$$

Since $X = \varprojlim X_i$ is a closed set in $\prod_{i\in\mathbb{N}} X_i$, we have $X \in \mathfrak{M}_\alpha(Y)$, which implies $X \in \mathfrak{M}_\alpha(\tau)$ by Theorem 5.11.2. □

Remark 5.26 In the above, the "if" part is also valid for $\alpha = 2$, but the "only if" part does not hold (cf. Corollary 5.12.6).

In the rest of this section, we shall characterize the class $\mathfrak{A}_1(\tau)$, that is, the class of absolutely F_σ spaces with weight $\leqslant \tau$. A subset A of X is said to be **locally closed** in X if each $x \in A$ has a neighborhood U in X such that $A \cap U$ is closed in U, equivalently $A = W \cap \mathrm{cl}_X A$ for some open set in W in X. A locally closed set can also be defined as the intersection of an open set and a closed set, or the difference between two closed sets. A metrizable space X is said to be **absolutely locally closed** if X is locally closed in an arbitrary metrizable space that contains X as a subspace. This concept characterizes the local compactness, that is:

Theorem 5.11.7 *A metrizable space X is absolutely locally closed if and only if it is locally compact.*

Proof If X is a locally compact subspace in a metrizable space Y, then each $x \in X$ has a neighborhood U in Y such that $X \cap U$ is compact, whence $X \cap U$ is closed in U. Thus, X is locally closed in Y.

On the contrary, assume that $X = (X, d)$ is a metric space that is not locally compact, that is, X has a point x_0 such that $\overline{\mathrm{B}}(x_0, \varepsilon)$ is not compact in X for every $\varepsilon > 0$. Then, $\overline{\mathrm{B}}(x_0, 1)$ contains a countably infinite discrete set D_1, where we can assume that $x_0 \notin D_1$. Let $\delta_1 = d(x_0, D_1) > 0$. Then, $\overline{\mathrm{B}}(x_0, \delta_1/2)$ contains a countably infinite discrete set D_2. Inductively, we can obtain countably infinite discrete sets D_n, $n \in \mathbb{N}$, such that $\delta_n = d(x_0, D_n) > 0$ and $D_{n+1} \subset \overline{\mathrm{B}}(x_0, \delta_n/2)$. Since $D = \{x_0\} \cup \bigcup_{n\in\mathbb{N}} D_n$ is closed in X, we can apply Hausdorff's Metric Extension Theorem 1.13.8 to obtain an admissible metric ρ for X such that D is isometric to the following subset of Euclidean plane:

$$\{(0,0)\} \cup \{(n^{-1}, (n+i-1)^{-1}) \mid n, i \in \mathbb{N}\} \subset \mathbb{R}^2,$$

where $(0, 0)$ is corresponding to x_0 and each $\{(n^{-1}, (n+i-1)^{-1}) \mid i \in \mathbb{N}\}$ is corresponding to D_n. Let $\widetilde{X} = (\widetilde{X}, \tilde{\rho})$ be the metric completion of $X = (X, \rho)$. Then, each D_n has an accumulation point x_n in $\widetilde{X}$, where $x_n \notin X$ and $\tilde{\rho}(x_0, x_n) = 1/n$. Observe that $\lim_{n\to\infty} x_n = x_0$. Hence, x_0 has no neighborhoods U in $\widetilde{X}$ such that $X \cap U$ is closed in U. Therefore, X is not locally closed in $\widetilde{X}$. □

Note that every locally closed set is an F_σ set because it is the intersection of an open set and a closed set. Hence, we have the following:

Corollary 5.11.8 *Every absolutely locally closed metrizable space is absolutely F_σ. In other words, every locally compact metrizable space is absolutely F_σ. Thus, $\mathfrak{L}(\tau) \subset \mathfrak{A}_1(\tau)$.* □

We can characterize absolutely F_σ spaces as follows:

Proposition 5.11.9 *A metrizable space X is absolutely F_σ if and only if, for each $d \in \mathrm{Metr}(X)$, X is a countable union of d-complete subsets.*

Proof To see the "if" part, assume that X is a subset of a metric space $Y = (Y, d)$. By the condition, we can write $X = \bigcup_{n\in\mathbb{N}} X_n$, where each X_n is d-complete, so it is closed in Y. Therefore, X is an F_σ-set in Y.

To see the "only if" part, assume that X is absolutely F_σ. For every $d \in \mathrm{Metr}(X)$, let $\widetilde{X}^d$ be the metric completion of (X, d). Since X is an F_σ set in $\widetilde{X}^d$, we can write $X = \bigcup_{n\in\mathbb{N}} X_n$, where each X_n is closed in $\widetilde{X}^d$, that is, X_n is d-complete. □

Proposition 5.11.10 *Every F_σ set A in an absolutely F_σ metrizable space X is absolutely F_σ.*

Proof First, we show the case where A is closed in X. By Hausdorff's Metric Extension Theorem 1.13.8, every admissible metric d for A extends to an admissible metric $\tilde{d}$ for X. By Proposition 5.11.9, we can write $X = \bigcup_{n\in\mathbb{N}} X_n$, where each X_n is $\tilde{d}$-complete. Then, $A = \bigcup_{n\in\mathbb{N}} A \cap X_n$ and each $A \cap X_n$ is $\tilde{d}$-complete. By Proposition 5.11.9 again, A is absolutely F_σ.

Now, we show the case where A is an F_σ set in X. Let $A = \bigcup_{n\in\mathbb{N}} A_n$, where each A_n is closed in X. We have shown that A_n is absolutely F_σ. For every admissible metric d for A, we can write $A_n = \bigcup_{m\in\mathbb{N}} A_{n,m}$, where each $A_{n,m}$ is d-complete. Then, $A = \bigcup_{n,m\in\mathbb{N}} A_{n,m}$. By Proposition 5.11.9 again, A is absolutely F_σ. □

Lemma 5.11.11 *Let X be a metrizable space. If each $x \in X$ has a neighborhood with weight $\leqslant \tau$, then X is a disjoint union of open sets with weight $\leqslant \tau$.*

Proof We may assume $\tau \geqslant \aleph_0$. By the paracompactness of X (Theorem 1.3.2), X has a locally finite open cover $\mathcal{U}$ such that $w(U) \leqslant \tau$ for every $U \in \mathcal{U}$. For each $U \in \mathcal{U}$, we define $U^* = \bigcup_{n\in\mathbb{N}} \mathrm{st}^n(U, \mathcal{U})$, where

$$\mathrm{st}^n(U, \mathcal{U}) = \mathrm{st}(\mathrm{st}^{n-1}(U, \mathcal{U})), \ \mathrm{st}^0(U, \mathcal{U}) = U.$$

Since $\mathcal{U}$ is locally finite, if $w(\mathrm{st}^{n-1}(U, \mathcal{U})) \leqslant \tau$ then $\operatorname{card} \mathcal{U}[\mathrm{st}^{n-1}(U, \mathcal{U})] \leqslant \tau$, hence $w(\mathrm{st}^n(U, \mathcal{U})) \leqslant \tau$. By induction, $w(\mathrm{st}^n(U, \mathcal{U})) \leqslant \tau$ for every $n \in \mathbb{N}$. Hence, we have $w(U^*) \leqslant \tau$. Observe that

$$U_1, U_2 \in \mathcal{U}, \ U_1 \cap U_2 \neq \emptyset \Rightarrow U_1^* = U_2^*,$$

which means that $\{U^* \mid U \in \mathcal{U}\}$ is pairwise disjoint. Thus, $X = \bigcup_{U\in\mathcal{U}} U^*$ is a disjoint union of open sets with weight $\leqslant \tau$. □

Recall that X is σ-locally compact if X is a countable union of locally compact *closed* subsets,[30] where "closed" can be removed by virtue of Corollary 5.11.8, that is:

Lemma 5.11.12 *A metrizable space is σ-locally compact if and only if is a countable union of locally compact subsets.* □

Lemma 5.11.13 *Every locally separable*[31] *absolutely F_σ space X is σ-locally compact.*

Proof By Lemma 5.11.11 above, we can write $X = \bigcup_{\lambda\in\Lambda} X_\lambda$, where each X_λ is a separable open set in X and $X_\lambda \cap X_{\lambda'} = \emptyset$ if $\lambda \neq \lambda'$. Then, each X_λ can be embedded in the Hilbert cube $\mathbf{I}^{\mathbb{N}}$. Since X_λ is an F_σ-set in $\mathbf{I}^{\mathbb{N}}$ by Proposition 5.11.10, we can write $X_\lambda = \bigcup_{n\in\mathbb{N}} F_{\lambda,n}$, where each $F_{\lambda,n}$ is closed in $\mathbf{I}^{\mathbb{N}}$, hence it is compact. For each $n \in \mathbb{N}$, since $\{F_{\lambda,n} \mid \lambda \in \Lambda\}$ is discrete in X, $F_n = \bigcup_{\lambda\in\Lambda} F_{\lambda,n}$ is locally compact and closed in X. Then, $X = \bigcup_{n\in\mathbb{N}} F_n$ is σ-locally compact. □

Lemma 5.11.14 *Let X be a non-separable completely metrizable space. If every non-empty open set in X is non-separable then X contains a closed set that is homeomorphic to $\nu^0 = \mathbb{R} \setminus \mathbb{Q}$, the space of irrationals.*

Proof We may assume that $X = (X, d)$ be a complete metric space. First, we show the following claim:

Claim *For each $\varepsilon > 0$, every non-empty open set U in X contains an infinite open collection $\mathcal{U}$ with $0 < \delta < \varepsilon$ such that* $\operatorname{mesh}\mathcal{U} < \delta$ *and* $\operatorname{dist}(U, U') > \delta$ *for each $U \neq U' \in \mathcal{U}$.*

Indeed, for each $n \in \mathbb{N}$, we apply Zorn's Lemma to obtain a maximal ε/n-discrete set D_n in U. Since each D_n is ε/n-dense in U, $\bigcup_{n\in\mathbb{N}} D_n$ is dense in U, whence $\operatorname{card}\bigcup_{n\in\mathbb{N}} D_n \geqslant \operatorname{dens} U > \aleph_0$. Then, $\operatorname{card} D_n > \aleph_0$ for some $n \in \mathbb{N}$. Let $\delta = \varepsilon/3n > 0$. The open collection $\mathcal{U} = \{\mathrm{B}(x, \delta) \mid x \in D_n\}$ is the desired one.

Using the above claim, we can inductively construct discrete open collections $\{U(a) \mid a \in \mathbb{N}^n\}$, $n \in \mathbb{N}$, such that $U(a, i) \subset U(a)$ and $\operatorname{diam} U(a, i) < \frac{1}{2} \operatorname{diam} U(a)$ for each $(a, i) \in \mathbb{N}^n \times \mathbb{N} = \mathbb{N}^{n+1}$. Since X is complete, we have a map $f : \mathbb{N}^{\mathbb{N}} \to X$ such that

$$\{f(a)\} = \bigcap_{n\in\mathbb{N}} U(a(1), \ldots, a(n)) \quad \text{for each } a \in \mathbb{N}^{\mathbb{N}}.$$

Then, it is easy to see that f is a closed embedding. □

Lemma 5.11.15 *Every completely metrizable absolutely F_σ space X is a countable union of locally separable F_σ-sets.*

---

[30]This definition is given on p. 436.

[31]For the definition, see p. 435.

Proof Let $U_0 = \emptyset$ and $K_0 = X$. When $U_\alpha, K_\alpha \subset X$ have been defined for all ordinals $\alpha < \beta$, we define U_β as the union of all separable open sets in $\bigcap_{\alpha<\beta} K_\alpha$ and $K_\beta = \bigcap_{\alpha<\beta} K_\alpha \setminus U_\beta$. Then, each K_α is closed in X, each U_α is F_σ in X and if $\bigcap_{\alpha<\beta} K_\alpha \neq \emptyset$ then $U_\beta \neq \emptyset$, whence $K_\alpha \supsetneqq K_\beta$ for every $\alpha < \beta$. Indeed, $K = \bigcap_{\alpha<\beta} K_\alpha$ is closed in X, hence it is completely metrizable and absolutely F_σ. If every non-empty open set in K is non-separable, then $\nu^0 = \mathbb{R} \setminus \mathbb{Q}$ can be embedded in K as a closed set by Lemma 5.11.14. By Lemma 5.11.10, ν^0 is absolutely F_σ, hence ν^0 is an F_σ-set in $\mathbb{R}$, which contradicts the Baire Category Theorem 1.3.15. Therefore, $U_\alpha \neq \emptyset$. Thus, $\bigcap_{\alpha<\beta^*} K_\alpha = \emptyset$ for sufficiently large ordinal β^*, which means $X = \bigcup_{\alpha<\beta^*} U_\alpha$.

Let $d \in \mathrm{Metr}(X)$. For each $\alpha < \beta^*$ and $n \in \mathbb{N}$, we define

$$V_{\alpha,n} = \left\{x \in U_\alpha \;\middle|\; d(x, K_\alpha) \geqslant 2^{-n}\right\} \subset U_\alpha.$$

Note that $\bigcup_{n\in\mathbb{N}} V_{\alpha,n} = U_\alpha$. We now define $X_n = \bigcup_{\alpha<\beta^*} V_{\alpha,n}$. Then, X_n is an F_σ-set in X, which is locally separable because each $V_{\alpha,n}$ is locally separable and $\mathrm{dist}(V_{\alpha,n}, V_{\beta,n}) \geqslant 2^{-n}$ for $\alpha < \beta < \beta^*$. Moreover, we have

$$\bigcup_{n\in\mathbb{N}} X_n = \bigcup_{n\in\mathbb{N}} \bigcup_{\alpha<\beta^*} V_{\alpha,n} = \bigcup_{\alpha<\beta^*} \bigcup_{n\in\mathbb{N}} V_{\alpha,n} = \bigcup_{\alpha<\beta^*} U_\alpha = X.$$

Thus, we have result. □

Now, we prove the following characterization of absolutely F_σ spaces:

Theorem 5.11.16 *A metrizable space X is absolutely F_σ if and only if X is σ-locally compact. Consequently, $\mathfrak{A}_1(\tau) = \mathfrak{L}(\tau)_\sigma$.*

Proof The "if" part follows from Corollary 5.11.8. We shall prove the "only if" part. By Proposition 5.11.9, we can write $X = \bigcup_{n\in\mathbb{N}} X_n$, where each X_n is a completely metrizable closed set in X, which is absolutely F_σ by Proposition 5.11.10. By Lemma 5.11.15, we can write $X_n = \bigcup_{i\in\mathbb{N}} X_{n,i}$, where each $X_{n,i}$ is a locally separable F_σ-set in X_n, which is absolutely F_σ by Proposition 5.11.10. By Lemma 5.11.13, we can write $X_{n,i} = \bigcup_{j\in\mathbb{N}} X_{n,i,j}$, where each $X_{n,i,j}$ is locally compact. □

Recall that we proved $\mathfrak{L}(\tau)_\sigma = (\tau\mathfrak{M}_0)_\sigma$ in Corollary 5.9.4. Therefore, we have

$$\mathfrak{A}_1(\tau) = \mathfrak{L}(\tau)_\sigma = (\tau\mathfrak{M}_0)_\sigma.$$

Since every separable locally compact space is σ-compact (or $(\aleph_0\mathfrak{M}_0)_\sigma = (\mathfrak{M}_0)_\sigma$), the following can be obtained as a corollary of Theorem 5.11.16:

Theorem 5.11.17 *A separable metrizable space X is absolutely F_σ if and only if X is σ-compact. Namely, $\mathfrak{A}_1 = \mathfrak{A}_1(\aleph_0) = (\mathfrak{M}_0)_\sigma$.* □

In the next section, we will show that a metrizable space X is absolutely $F_{\sigma\delta}$ if and only if X is homeomorphic to the inverse limit of an inverse sequence of σ-completely metrizable spaces (Corollary 5.12.6).

Summary The results of this section can be summarized as follows:

- absolutely open = $\emptyset$ $\quad\therefore\ \mathfrak{A}_0(\tau) = \mathfrak{A}_0 = \{\emptyset\}$;
- absolutely closed = compact $\quad\therefore\ \mathfrak{M}_0(\tau) = \mathfrak{M}_0$;
- absolutely locally closed = locally compact
  (a locally closed set = an intersection of an open set and a closed set);
- absolutely F_σ = σ-locally compact = σ-discrete compact
  $\therefore\ \mathfrak{A}_1(\tau) = \mathfrak{L}(\tau)_\sigma = (\tau\mathfrak{M}_0)_\sigma$
  (absolutely F_σ + separable = σ-compact $\quad\therefore\ \mathfrak{A}_1 = (\mathfrak{M}_0)_\sigma$);
- absolutely G_δ = completely metrizable;
- absolutely $F_{\sigma\delta}$ = the inverse limit of σ-completely metrizable spaces;
- absolutely $G_{\delta\sigma}$ = a countable union of σ-completely metrizable subspaces.

In the following diagrams, the up-right arrow and the down-right arrow mean the operations of the countable union and the inverse limit, respectively. The difference between the separable case and the non-separable case is only the first step.

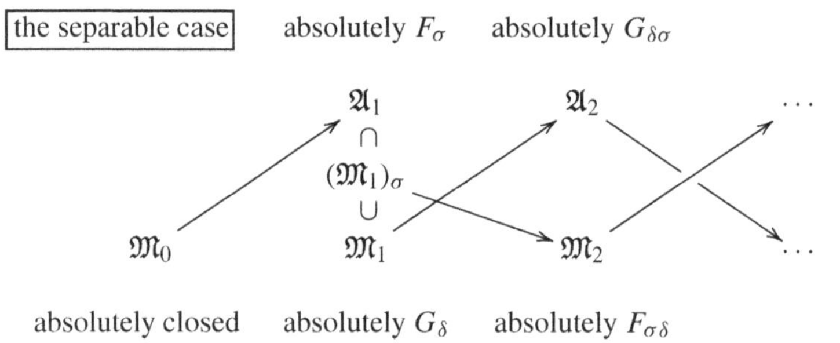

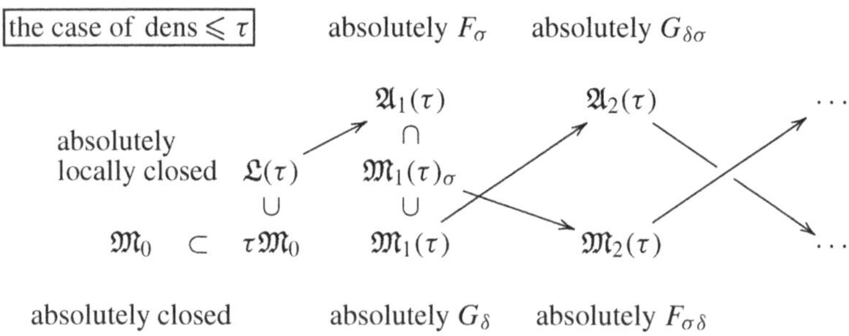

Since $(\tau\mathfrak{M}_0)_\sigma = \mathfrak{L}(\tau)_\sigma = \mathfrak{A}_1(\tau)$, the following comes from Corollary 5.6.6:

Proposition 5.11.18 *For an ANR that is a strong Z_σ-set, the strong $\tau\mathfrak{M}_0$-universality, the strong $\mathfrak{L}(\tau)$-universality, and the strong $\mathfrak{A}_1(\tau)$-universality are equivalent.* □

By this proposition, in an $\ell^2(\Gamma)$-manifold, a $\tau\mathfrak{M}_0$-absorbing set, an $\mathfrak{L}(\tau)$-absorbing set, and an $\mathfrak{A}_1(\tau)$-absorbing set are the same thing. In particular, $\ell^2_f(\Gamma) \times \ell^2_Q$ is not only a $\tau\mathfrak{M}_0$-absorbing set but also an $\mathfrak{A}_1(\tau)$-absorbing set in $\ell^2(\Gamma) \times \ell^2$ ($\approx \ell^2(\Gamma)$), which is an $\mathfrak{L}(\tau)$-absorbing set at the same time. Since $\ell^2_f(\Gamma) \times \ell^2_Q$ itself is a member of $\mathfrak{A}_1(\tau)$, it is a **universal space** for the class $\mathfrak{A}_1(\tau)$ in the sense of a space belonging to $\mathfrak{A}_1(\tau)$ containing all members of $\mathfrak{A}_1(\tau)$ as closed sets, that is, $\mathcal{F}_{\ell^2_f(\Gamma)\times\ell^2_Q} = \mathfrak{A}_1(\tau)$. In the same sense, $\boldsymbol{Q}$ and $\ell^2(\Gamma)$ are universal spaces for the classes $\mathfrak{M}_0$ and $\mathfrak{M}_1(\tau)$, respectively. Indeed, $\mathcal{F}_{\boldsymbol{Q}} = \mathfrak{M}_0$ and $\mathcal{F}_{\ell^2(\Gamma)} = \mathfrak{M}_1(\tau)$.

5.12 Universal Spaces for the Absolute Borel Classes

In Sect. 5.9, we characterized manifolds modeled on the spaces $\ell^2_f(\Gamma)$, $\ell^2_f(\Gamma) \times \ell^2_Q$, $\ell^2_f(\Gamma) \times \ell^2$, and $\ell^2(\Gamma)^{\mathbb{N}}_f$ ($\approx \ell^2(\Gamma) \times \ell^2_f$). These model spaces are absorption bases for the classes $\mathfrak{M}^{\mathrm{fd}}_0$, $\mathfrak{M}_0$, $\mathfrak{M}_1(\aleph_0)$, and $\mathfrak{M}_1(\tau)$, respectively, which are absorbing sets for the classes $\tau\mathfrak{M}^{\mathrm{fd}}_0$, $\tau\mathfrak{M}_0$, $\tau\mathfrak{M}_1(\aleph_0)$, and $\mathfrak{M}_1(\tau)$, respectively. As noted at the end of the previous section, $\ell^2_f(\Gamma) \times \ell^2_Q$ is also an $\mathfrak{A}_1(\tau)$-absorbing set in $\ell^2(\Gamma) \times \ell^2$ ($\approx \ell^2(\Gamma)$) and it is a **universal space** for the class $\mathfrak{A}_1(\tau)$ in the sense of $\mathcal{F}_{\ell^2_f(\Gamma)\times\ell^2_Q} = \mathfrak{A}_1(\tau)$.

In this section, we show that $\ell^2_f(\Gamma)^{\mathbb{N}}$ and $(\ell^2(\Gamma) \times \ell^2_f)^{\mathbb{N}}$ are $\mathfrak{M}_2(\tau)$-absorbing sets in $\ell^2(\Gamma)^{\mathbb{N}}$ and $(\ell^2(\Gamma) \times \ell^2)^{\mathbb{N}}$, respectively. Consequently,

$$(\ell^2(\Gamma) \times \ell^2_f)^{\mathbb{N}} \approx \ell^2_f(\Gamma)^{\mathbb{N}} \approx (\ell^2_f(\Gamma) \times \ell^2_Q)^{\mathbb{N}} \approx (\ell^2_f(\Gamma) \times \ell^2)^{\mathbb{N}},$$ [32]

which are universal spaces for the class $\mathfrak{M}_2(\tau)$. More generally, it will be shown that if $\mathfrak{M}^{\mathrm{fd}}_0 \subset \mathfrak{C} \subset \mathfrak{M}_1(\tau)$ and F is a $\mathfrak{C}$-absorbing set in $\ell^2(\Gamma)$, then $F^{\mathbb{N}} \approx \ell^2_f(\Gamma)^{\mathbb{N}}$. In addition, we inductively construct universal spaces $\Omega_\alpha(\tau)$, $\Lambda_\alpha(\tau)$ in $\ell^2(\Gamma)$ for the absolute Borel classes $\mathfrak{M}_\alpha(\tau)$, $\mathfrak{A}_\alpha(\tau)$, $\alpha > 1$, respectively, where $\Omega_2(\tau) \approx \ell^2_f(\Gamma)^{\mathbb{N}}$ and $\Lambda_2(\tau) \approx \ell^2(\Gamma)^{\mathbb{N}}_f$. As mentioned above, $\Lambda_1(\tau) \approx \ell^2_f(\Gamma) \times \ell^2_Q$ and $\Lambda_1(\tau)^{\mathbb{N}} \approx \Omega_2(\tau)$. Concerning manifolds modeled on these universal spaces, all the results obtained in Sect. 5.10 can be applied.

First of all, we note the following:

Fact *If A is F_σ in X, then $A^{\mathbb{N}}$ is $F_{\sigma\delta}$ in $X^{\mathbb{N}}$.*

This is the simplest case of the following lemma:

[32] For the last two homeomorphisms, see Note on p. 408.

Lemma 5.12.1 *Let* $X = \prod_{i\in\mathbb{N}} X_i$ *be the countable product space of metrizable spaces. If* $A_i \in \mathfrak{A}_{\beta_i}(X_i)$ *and* $\beta_i < \alpha$, *then* $\prod_{i\in\mathbb{N}} A_i \in \mathfrak{M}_\alpha(X)$. *In particular, if* $A \in \mathfrak{A}_\alpha(Y)$, *then* $A^{\mathbb{N}} \in \mathfrak{M}_{\alpha+1}(X^{\mathbb{N}})$.

Proof For each $i \in \mathbb{N}$, let $\mathrm{pr}_i : X \to X_i$ is the projection. Since $\mathrm{pr}_i^{-1}(A_i) \in \mathfrak{A}_{\alpha_i}(X)$, it follows that

$$\prod_{i\in\mathbb{N}} A_i = \bigcap_{i\in\mathbb{N}} \mathrm{pr}_i^{-1}(A_i) \in \mathfrak{M}_\alpha(X). \qquad \square$$

We have the following embedding lemma for F_σ sets in $\ell^2(\Gamma)$:

Lemma 5.12.2 *For each* F_σ *set* X *in* $\ell^2(\Gamma)$, *there exists a closed embedding* $\varphi : \ell^2(\Gamma) \to \ell^2(\Gamma) \times \ell^2$ *such that* $\varphi^{-1}(\ell^2(\Gamma) \times \ell^2_f) = X$.

Proof Since $(\ell^2, \ell^2_f) \approx (\boldsymbol{s}, \boldsymbol{s}_f) = (\mathbb{R}^{\mathbb{N}}, \mathbb{R}^{\mathbb{N}}_f)$ (Corollary 5.1.6), we may construct a closed embedding $\varphi : \ell^2(\Gamma) \to \ell^2(\Gamma) \times \boldsymbol{s}$ so that $\varphi^{-1}(\ell^2(\Gamma) \times \boldsymbol{s}_f) = X$. Let $X = \bigcup_{j\in\mathbb{N}} A_i$, where each A_i is closed in $\ell^2(\Gamma)$. We have maps $\alpha_i : \ell^2(\Gamma) \to \mathbb{R}$, $i \in \mathbb{N}$, such that $\alpha_i^{-1}(0) = A_i$. We define a map $\varphi : \ell^2(\Gamma) \to \ell^2(\Gamma) \times \boldsymbol{s}$ as follows:

$$\varphi(x) = (x, (\alpha_i(x))_{i\in\mathbb{N}}) \quad \text{for each } x \in \ell^2(\Gamma).$$

Then, φ is a closed embedding because $\mathrm{pr}_1\varphi = \mathrm{id}$, where $\mathrm{pr}_1 : \ell^2(\Gamma) \times \boldsymbol{s} \to \ell^2(\Gamma)$ is the projection. If $x \in A_i$, then $\alpha_j(x) = 0$ for every $j \geqslant i$. Hence, we have $\varphi(X) \subset \ell^2(\Gamma) \times \boldsymbol{s}_f$. If $x \notin X$, then $\alpha_i(x) \neq 0$ for every $i \in \mathbb{N}$. Thus, $\varphi(\ell^2(\Gamma) \setminus X) \subset \ell^2(\Gamma) \times (\boldsymbol{s} \setminus \boldsymbol{s}_f)$. Consequently, $\varphi^{-1}(\ell^2(\Gamma) \times \boldsymbol{s}_f) = X$. $\square$

Using the above lemma, we can prove that $\ell^2(\Gamma) \times \ell^2_f$ is a universal space for σ-completely metrizable spaces with weight $\leqslant \tau$, that is:

Proposition 5.12.3 $\mathfrak{M}_1(\tau)_\sigma = \mathcal{F}_{\ell^2(\Gamma)\times\ell^2_f}$.

Proof Since $\ell^2(\Gamma) \times \ell^2_f \in \mathfrak{M}_1(\tau)_\sigma$ and $\mathfrak{M}_1(\tau)_\sigma$ is closed hereditary, it follows that $\mathcal{F}_{\ell^2(\Gamma)\times\ell^2_f} \subset \mathfrak{M}_1(\tau)_\sigma$. To see $\mathfrak{M}_1(\tau)_\sigma \subset \mathcal{F}_{\ell^2(\Gamma)\times\ell^2_f}$, it suffices to show $\mathfrak{M}_1(\tau)_\sigma \setminus \mathfrak{M}_1(\tau) \subset \mathcal{F}_{\ell^2(\Gamma)\times\ell^2_f}$ because $\mathfrak{M}_1(\tau) \subset \mathcal{F}_{\ell^2(\Gamma)} \subset \mathcal{F}_{\ell^2(\Gamma)\times\ell^2_f}$. Note that every infinite-dimensional Banach space with weight τ is homeomorphic to $\ell^2(\Gamma)$ (Theorem 3.5.17). Thus, for each $X \in \mathfrak{M}_1(\tau)_\sigma \setminus \mathfrak{M}_1(\tau)$, we may find an infinite-dimensional Banach space E with $w(E) = \tau$ such that X can be embedded into the product space $E \times \ell^2_f$ as a closed set.

For each $X \in \mathfrak{M}_1(\tau)_\sigma \setminus \mathfrak{M}_1(\tau)$, we have a tower $X_1 \subsetneqq X_2 \subsetneqq X_3 \subsetneqq \cdots$ of completely metrizable closed sets in X such that $X = \bigcup_{i\in\mathbb{N}} X_i$. Using Hausdorff's Metric Extension Theorem 1.13.8, we have $d_i \in \mathrm{Metr}(X)$ such that X_i is d_i-complete (i.e., the restriction $d_i|X_i^2$ is complete) and $\mathrm{diam}_{d_i} X \leqslant 1$. We can define $d \in \mathrm{Metr}(X)$ by

$$d(x, y) = \sum_{i\in\mathbb{N}} 2^{-i} d_i(x, y).$$

Then, every X_i is d-complete. By the Arens–Eells Embedding Theorem 1.13.2, X can be isometrically embedded into a Banach space $E = (E, \|\cdot\|)$ with $\operatorname{dens} E = \aleph_0 \operatorname{dens} X$. Let $f : X \to E$ be an isometric embedding. Then, we can define an embedding $h : X \to E \times \ell^2_f$ as follows:

$$h(x) = \big(f(x), (d(x, X_i))_{i\in\mathbb{N}}\big).$$

To show that $h(X)$ is closed in $E \times \ell^2_f$, let $(y, z) \in \operatorname{cl}_{E\times\ell^2_f} h(X)$. Then, we have $x_n \in X$, $n \in \mathbb{N}$, such that $(h(x_n))_{n\in\mathbb{N}}$ converges to (y, z). Because $z \in \ell^2_f$, we have $i_0 = \min\{i \in \mathbb{N} \mid z(i) = 0\}$. Since $\lim_{n\to\infty} d(x_n, X_{i_0}) = z(i_0) = 0$, we can choose $x'_n \in X_{i_0}$, $n \in \mathbb{N}$, so that $d(x_n, x'_n) \to 0$ $(n \to \infty)$. Because f is an isometry, we have $\|f(x_n) - f(x'_n)\| = d(x_n, x'_n) \to 0$. Since $\lim_{n\to\infty} f(x_n) = y$, it follows that

$$\|f(x'_n) - y\| \leqslant \|f(x'_n) - f(x_n)\| + \|f(x_n) - y\| \to 0 \quad (n \to \infty).$$

Recall that X_{i_0} is d-complete, that is, $f(X_{i_0})$ is $\|\cdot\|$-complete, which means that $f(X_{i_0})$ is closed in E. Therefore, $y \in f(X_{i_0})$, that is, $y = f(x)$ for some $x \in X_{i_0}$. Then, $x'_n \to x$ $(n \to \infty)$. Since $d(x_n, x'_n) \to 0$, it follows that $x_n \to x$ $(n \to \infty)$. Hence, $h(x) = (y, z)$. Thus, $h(X)$ is closed in $E \times \ell^2_f$. □

The following is an embedding lemma for $F_{\sigma\delta}$ sets in $\ell^2(\Gamma)$:

Lemma 5.12.4 *For each $F_{\sigma\delta}$ set X in $\ell^2(\Gamma)$, there exists a closed embedding $\varphi : \ell^2(\Gamma) \to (\ell^2(\Gamma) \times \ell^2)^{\mathbb{N}}$ such that $\varphi^{-1}((\ell^2(\Gamma) \times \ell^2_f)^{\mathbb{N}}) = X$.*

Proof Write $X = \bigcap_{i\in\mathbb{N}} X_i$, where each X_i is F_σ in $\ell^2(\Gamma)$. By Lemma 5.12.2, we have closed embeddings $\varphi_i : \ell^2(\Gamma) \to \ell^2(\Gamma) \times \ell^2$, $i \in \mathbb{N}$, such that $\varphi_i^{-1}(\ell^2(\Gamma) \times \ell^2_f) = X_i$. We define $\varphi : \ell^2(\Gamma) \to (\ell^2(\Gamma) \times \ell^2)^{\mathbb{N}}$ by $\varphi(x) = (\varphi_i(x))_{i\in\mathbb{N}}$ for each $x \in \ell^2(\Gamma)$. Then, φ is a closed embedding and

$$\varphi^{-1}\big((\ell^2(\Gamma) \times \ell^2_f)^{\mathbb{N}}\big) = \bigcap_{i\in\mathbb{N}} \varphi_i^{-1}(\ell^2(\Gamma) \times \ell^2_f) = \bigcap_{i\in\mathbb{N}} X_i = X.$$

□

Now, we can prove that $(\ell^2(\Gamma) \times \ell^2_f)^{\mathbb{N}}$ is a universal space for absolutely $F_{\sigma\delta}$ spaces with weight $\leqslant \tau$, that is,

Proposition 5.12.5 $\mathcal{F}_{(\ell^2(\Gamma)\times\ell^2_f)^{\mathbb{N}}} = \mathfrak{M}_2(\tau)$.

Proof Because ℓ^2_f is F_σ in ℓ^2, $\ell^2(\Gamma) \times \ell^2_f$ is F_σ in $\ell^2(\Gamma) \times \ell^2$. From Lemma 5.12.1, it follows that $(\ell^2(\Gamma) \times \ell^2_f)^{\mathbb{N}}$ is $F_{\sigma\delta}$ in $(\ell^2(\Gamma) \times \ell^2)^{\mathbb{N}}$, which means $(\ell^2(\Gamma) \times \ell^2_f)^{\mathbb{N}} \in \mathfrak{M}_2(\tau)$. Therefore, $\mathcal{F}_{(\ell^2(\Gamma)\times\ell^2_f)^{\mathbb{N}}} \subset \mathfrak{M}_2(\tau)$.

Conversely, let $X \in \mathfrak{M}_2(\tau)$. Then, X can be regarded as an $F_{\sigma\delta}$ set in $\ell^2(\Gamma)$. Applying Lemma 5.12.4, we can embed X into $(\ell^2(\Gamma) \times \ell^2_f)^{\mathbb{N}}$ as a closed set. Hence, $\mathfrak{M}_2(\tau) \subset \mathcal{F}_{(\ell^2(\Gamma)\times\ell^2_f)^{\mathbb{N}}}$. □

As a corollary of the above proposition, we have the following characterization of absolutely $F_{\sigma\delta}$ spaces:

Corollary 5.12.6 *A metrizable space X is absolutely $F_{\sigma\delta}$ (i.e., $X \in \mathfrak{M}_2(\tau)$) if and only if $X \approx \varprojlim X_n$, where each X_n is σ-completely metrizable ($X_n \in \mathfrak{M}_1(\tau)_\sigma$).*

Proof (*The "only if" part*) Let $X \in \mathfrak{M}_2(\tau)$. Since $\mathcal{F}_{(\ell^2(\Gamma)\times\ell^2_f)^{\mathbb{N}}} = \mathfrak{M}_2(\tau)$ by Proposition 5.12.5, we can assume that X is a closed set in $(\ell^2(\Gamma) \times \ell^2_f)^{\mathbb{N}}$. Then, $(\ell^2(\Gamma) \times \ell^2_f)^{\mathbb{N}}$ is regarded as the inverse limit of the following inverse sequence:

$$\ell^2(\Gamma) \times \ell^2_f \longleftarrow (\ell^2(\Gamma) \times \ell^2_f)^2 \longleftarrow (\ell^2(\Gamma) \times \ell^2_f)^3 \longleftarrow \cdots .$$

For each $n \in \mathbb{N}$, let $p_n : (\ell^2(\Gamma) \times \ell^2_f)^{\mathbb{N}} \to (\ell^2(\Gamma) \times \ell^2_f)^n$ be the projection defined by $p_n(x) = (x(1), \ldots, x(n))$. We define $X_n = \operatorname{cl} p_n(X) \subset (\ell^2(\Gamma) \times \ell^2_f)^n$. Since $(\ell^2(\Gamma) \times \ell^2_f)^n \approx \ell^2(\Gamma) \times \ell^2_f$, each X_n is σ-completely metrizable by Proposition 5.12.3. By virtue of [GAGT, Proposition 4.10.8(2)], X is the inverse limit of the upper sequence of the following commutative diagram:

$$\begin{array}{ccccccc}
p_1(X) & \longleftarrow & p_2(X) & \longleftarrow & p_3(X) & \longleftarrow & \cdots \\
\cap & & \cap & & \cap & & \\
X_1 & \longleftarrow & X_2 & \longleftarrow & X_3 & \longleftarrow & \cdots
\end{array}$$

Then, as is easily observed, $X \subset \varprojlim X_n \subset \operatorname{cl} X$. Since X is closed, it follows that $X = \varprojlim X_n$.

(*The "if" part*) Let $X = \varprojlim X_n$, where each X_n is σ-completely metrizable. Due to Proposition 5.12.3, $X_n \in \mathfrak{M}_1(\tau)_\sigma = \mathcal{F}_{\ell^2(\Gamma)\times\ell^2_f}$. Hence, $\prod_{n\in\mathbb{N}} X_n \in \mathcal{F}_{(\ell^2(\Gamma)\times\ell^2_f)^{\mathbb{N}}} = \mathfrak{M}_2(\tau)$ by Proposition 5.12.5. Since $X = \varprojlim X_n$ is a closed subset of $\prod_{n\in\mathbb{N}} X_n$, we have $X \in \mathfrak{M}_2(\tau)$, which means that X is absolutely $F_{\sigma\delta}$. □

The following proposition means that $\ell^2_f(\Gamma)^{\mathbb{N}}$ is also a universal space for absolutely $F_{\sigma\delta}$ spaces with weight $\leqslant \tau$:

Proposition 5.12.7 $\mathcal{F}_{\ell^2_f(\Gamma)^{\mathbb{N}}} = \mathfrak{M}_2(\tau)$.

Proof Because $\ell^2_f(\Gamma)$ is F_σ in $\ell^2(\Gamma)$, $\ell^2_f(\Gamma)^{\mathbb{N}}$ is $F_{\sigma\delta}$ in $\ell^2(\Gamma)^{\mathbb{N}}$ by Lemma 5.12.1. Hence, $\ell^2_f(\Gamma)^{\mathbb{N}} \in \mathfrak{M}_2(\tau)$, which implies that $\mathcal{F}_{\ell^2_f(\Gamma)^{\mathbb{N}}} \subset \mathfrak{M}_2(\tau)$.

To see $\mathfrak{M}_2(\tau) \subset \mathcal{F}_{\ell^2_f(\Gamma)^{\mathbb{N}}}$, since $\mathfrak{M}_2(\tau) = \mathcal{F}_{(\ell^2(\Gamma)\times\ell^2_f)^{\mathbb{N}}}$ (Proposition 5.12.5), it suffices to show that $(\ell^2(\Gamma) \times \ell^2_f)^{\mathbb{N}} \in \mathcal{F}_{\ell^2_f(\Gamma)^{\mathbb{N}}}$. Note that $\ell^2(\Gamma)$ can be embedded in to $J(\Gamma)^{\mathbb{N}}$ (Theorem 1.3.5),[33] where $J(\Gamma)$ is a closed subspace of $\ell^2_f(\Gamma)$ called the

[33]Due to Corollary 3.5.16, $J(\Gamma)^{\mathbb{N}} \approx \ell^2(\Gamma)$.

hedgehog. Since ℓ_f^2 can be regarded as a closed subspace of $\ell_f^2(\Gamma)$, $\ell^2(\Gamma) \times \ell_f^2$ can be embedded into $\ell_f^2(\Gamma)^{\mathbb{N}} \times \ell_f^2(\Gamma) \approx \ell_f^2(\Gamma)^{\mathbb{N}}$ as a closed set. Hence, $\ell^2(\Gamma) \times \ell_f^2$ can be embedded into $(\ell_f^2(\Gamma)^{\mathbb{N}})^{\mathbb{N}} \approx \ell_f^2(\Gamma)^{\mathbb{N}}$ as a closed set, that is, $(\ell^2(\Gamma) \times \ell_f^2)^{\mathbb{N}} \in \mathcal{F}_{\ell_f^2(\Gamma)^{\mathbb{N}}}$. □

Proposition 5.12.8 *Let E and F be homotopy dense sets in $\ell^2(\Gamma)$ (with $\mathbf{0} \in E \cap F$) which are (strong) Z_σ-spaces. If $E^{\mathbb{N}} \in \mathcal{F}_{F^{\mathbb{N}}}$ and $F^{\mathbb{N}} \in \mathcal{F}_{E^{\mathbb{N}}}$ (resp. $E_f^{\mathbb{N}} \in \mathcal{F}_{F_f^{\mathbb{N}}}$ and $F_f^{\mathbb{N}} \in \mathcal{F}_{E_f^{\mathbb{N}}}$), then $E^{\mathbb{N}} \approx F^{\mathbb{N}}$ (resp. $E_f^{\mathbb{N}} \approx F_f^{\mathbb{N}}$), where $E^{\mathbb{N}}$ and $F^{\mathbb{N}}$ (resp. $E_f^{\mathbb{N}}$ and $F_f^{\mathbb{N}}$) are $\mathcal{F}_{E^{\mathbb{N}}}$-absorbing (resp. $\mathcal{F}_{E_f^{\mathbb{N}}}$-absorbing) sets in $\ell^2(\Gamma)^{\mathbb{N}}$ ($\approx \ell^2(\Gamma)$).*

Proof The assumption means that $\mathcal{F}_{E^{\mathbb{N}}} = \mathcal{F}_{F^{\mathbb{N}}}$ (resp. $\mathcal{F}_{E_f^{\mathbb{N}}} = \mathcal{F}_{F_f^{\mathbb{N}}}$), which are topological, closed hereditary, and additive (Proposition 5.4.10). Due to Proposition 5.10.26, $E^{\mathbb{N}}$ and $F^{\mathbb{N}}$ (resp. $E_f^{\mathbb{N}}$ and $F_f^{\mathbb{N}}$) are $\mathcal{F}_{E^{\mathbb{N}}}$-absorbing (resp. $\mathcal{F}_{E_f^{\mathbb{N}}}$-absorbing) sets in $\ell^2(\Gamma)^{\mathbb{N}}$. By the Uniqueness of Absorbing Sets (Theorem 5.10.5), we have $E^{\mathbb{N}} \approx F^{\mathbb{N}}$ (resp. $E_f^{\mathbb{N}} \approx F_f^{\mathbb{N}}$). □

Note that $\ell^2(\Gamma) \times \ell_f^2$ and $\ell_f^2(\Gamma)$ are homotopy dense in $\ell^2(\Gamma) \times \ell^2$ and $\ell^2(\Gamma)$, respectively. Combining Propositions 5.12.5, 5.12.7, and 5.12.8, we can obtain the following:

Theorem 5.12.9 *The spaces $(\ell^2(\Gamma) \times \ell_f^2)^{\mathbb{N}}$ and $\ell_f^2(\Gamma)^{\mathbb{N}}$ are homeomorphic, which are $\mathfrak{M}_2(\tau)$-absorbing sets in $(\ell^2(\Gamma) \times \ell^2)^{\mathbb{N}}$ and $\ell^2(\Gamma)^{\mathbb{N}}$, respectively.* □

Theorem 5.12.9 is generalized as follows:

Theorem 5.12.10 *Let $\mathfrak{M}_0^{\mathrm{fd}} \subset \mathfrak{C} \subset \mathfrak{M}_1(\tau)$. For every $\mathfrak{C}$-absorbing set F in $\ell^2(\Gamma)^{\mathbb{N}}$, $F^{\mathbb{N}}$ is homeomorphic to $\ell_f^2(\Gamma)^{\mathbb{N}}$.*

Proof By virtue of Proposition 5.12.8, it suffices to show $\mathcal{F}_{\ell_f^2(\Gamma)^{\mathbb{N}}} = \mathcal{F}_{F^{\mathbb{N}}}$.

Recall that $\ell^2(\Gamma) \times \ell_f^2$ is strongly $\mathfrak{M}_1(\tau)$-universal (cf. Proposition 5.4.7). Because $\mathbf{I}^n \times \Gamma \in \mathfrak{M}_1(\tau)$ for every $n \in \mathbb{N}$, we can apply Proposition 5.6.5 to show that $\ell^2(\Gamma) \times \ell_f^2$ is strongly $\mathfrak{M}_1(\tau)_\sigma$-universal. Since $F \in \mathfrak{C}_\sigma \subset \mathfrak{M}_1(\tau)_\sigma$, F can be embedded into $\ell^2(\Gamma) \times \ell_f^2$ as a closed set, which implies that $F^{\mathbb{N}} \in \mathcal{F}_{(\ell^2(\Gamma) \times \ell_f^2)^{\mathbb{N}}}$. Hence, $\mathcal{F}_{F^{\mathbb{N}}} \subset \mathcal{F}_{(\ell^2(\Gamma) \times \ell_f^2)^{\mathbb{N}}} = \mathcal{F}_{\ell_f^2(\Gamma)^{\mathbb{N}}}$.

Due to Proposition 5.10.2, F is $\mathfrak{C}$-absorptive in $\ell^2(\Gamma)$. Because $\mathfrak{M}_0^{\mathrm{fd}} \subset \mathfrak{C}$, F is $\mathfrak{M}_0^{\mathrm{fd}}$-absorptive in $\ell^2(\Gamma)$, which implies that F is $\tau\mathfrak{M}_0^{\mathrm{fd}}$-absorptive in $\ell^2(\Gamma)$. Combining Propositions 5.4.1(2) and 5.6.5, we can see that F is strongly $(\tau\mathfrak{M}_0^{\mathrm{fd}})_\sigma$-universal. Therefore, $\ell_f^2(\Gamma)$ can be embedded into F as a closed set, so $\ell_f^2(\Gamma)^{\mathbb{N}}$ can be embedded into $F^{\mathbb{N}}$ as a closed set. Thus, it follows that $\mathcal{F}_{\ell_f^2(\Gamma)^{\mathbb{N}}} \subset \mathcal{F}_{F^{\mathbb{N}}}$. □

Note By the above theorem, we can show that if $\mathfrak{M}_0^{\mathrm{fd}} \subset \mathfrak{C} \subset \mathfrak{M}_1(\tau)$ and F is a $\mathfrak{C}$-absorption base for $\ell^2(\Gamma)$, then $F^{\mathbb{N}} \approx \ell_f^2(\Gamma)^{\mathbb{N}}$. (This was mentioned at the end of Sect. 5.5, p. 408.) Indeed, $\mathfrak{M}_0^{\mathrm{fd}} \subset \tau\mathfrak{C} \subset \mathfrak{M}_1(\tau)$ and every $\mathfrak{C}$-absorption base for $\ell^2(\Gamma)$ is $\tau\mathfrak{C}$-absorbing in $\ell^2(\Gamma)$ by Proposition 5.10.3(2).

Construction of Universal Spaces, $\Lambda_\alpha(\tau)$ and $\Omega_\alpha(\tau)$

Step 1 (*Construction of* $\Lambda_1(\tau)$) Recall that $\ell^2_f(\Gamma) \times \ell^2_Q$ is a universal space for $\mathfrak{A}_1(\tau)$, which is homotopy dense in $\ell^2(\Gamma) \times \ell^2$. Let $\psi_1 : \ell^2(\Gamma) \times \ell^2 \to \ell^2(\Gamma)$ be a homeomorphism. We define

$$\Lambda_1(\tau) = \psi_1(\ell^2_f(\Gamma) \times \ell^2_Q) \subset \ell^2(\Gamma).$$

Then, $\Lambda_1(\tau)$ is a universal space for $\mathfrak{A}_1(\tau)$, which is homotopy dense in $\ell^2(\Gamma)$. As is previously mentioned, $\Lambda_1(\tau)$ is an $\mathfrak{A}_1(\tau)$-absorbing set in $\ell^2(\Gamma)$.

In the following, using transfinite induction, we will construct an $\mathfrak{A}_\alpha(\tau)$-absorbing set and an $\mathfrak{M}_\alpha(\tau)$-absorbing set in $\ell^2(\Gamma)$ for every countable ordinal $\alpha > 1$, which belong to $\mathfrak{A}_\alpha(\tau)$ and $\mathfrak{M}_\alpha(\tau)$, so they are universal spaces for the classes $\mathfrak{M}_\alpha(\tau)$ and $\mathfrak{A}_\alpha(\tau)$.

Step 2 (*Construction of* $\Omega_2(\tau)$) Let $\psi_2 : (\ell^2(\Gamma) \times \ell^2)^{\mathbb{N}} \to \ell^2(\Gamma)$ be a homeomorphism. We define

$$\Omega_2(\tau) = \psi_2((\ell^2(\Gamma) \times \ell^2_f)^{\mathbb{N}}) \subset \ell^2(\Gamma),$$

where it should be noted that

$$\Omega_2(\tau) \approx (\ell^2(\Gamma) \times \ell^2_f)^{\mathbb{N}} \approx (\ell^2(\Gamma) \times \ell^2_Q)^{\mathbb{N}} = \Lambda_1(\tau)^{\mathbb{N}}.$$

From Theorem 5.12.9 and Proposition 5.12.5, it follows that $\Omega_2(\tau)$ is an $\mathfrak{M}_2(\tau)$-absorbing set in $\ell^2(\Gamma)$ and $\Omega_2(\tau) \in \mathfrak{M}_2(\tau)$. Thus, $\Omega_2(\tau)$ is a universal space for $\mathfrak{M}_2(\tau)$. Then, Lemma 5.12.4 can be rewritten as follows:

(M_2) *For each* $X \in \mathfrak{M}_2(\ell^2(\Gamma))$, *there exists a closed embedding* $\varphi_{2,X} : \ell^2(\Gamma) \to \ell^2(\Gamma)$ *such that* $\varphi^{-1}_{2,X}(\Omega_2(\tau)) = X$.

Indeed, the embedding $\varphi_{2,X}$ can be defined as a composition of the embedding $\varphi : \ell^2(\Gamma) \to (\ell^2(\Gamma) \times \ell^2)^{\mathbb{N}}$ obtained by Lemma 5.12.4 and the homeomorphism $\psi_2 : (\ell^2(\Gamma) \times \ell^2)^{\mathbb{N}} \to \ell^2(\Gamma)$ used in the definition of $\Omega_2(\tau)$.

Furthermore, we can assume that $\mathbf{0} \notin \Omega_2(\tau)$ because $\ell^2(\Gamma)$ is homogeneous.

Step 3 (*Construction of* $\Lambda_\alpha(\tau)$, $\alpha > 1$) Assume that an $\mathfrak{M}_\alpha(\tau)$-absorbing set Ω_α in $\ell^2(\Gamma)$ is defined so that $\mathbf{0} \notin \Omega_\alpha \in \mathfrak{M}_\alpha(\tau)$ and:

(M_α) *For each* $X \in \mathfrak{M}_\alpha(\ell^2(\Gamma))$, *there exists a closed embedding* $\varphi_{\alpha,X} : \ell^2(\Gamma) \to \ell^2(\Gamma)$ *such that* $\varphi^{-1}_{\alpha,X}(\Omega_\alpha(\tau)) = X$.

Let $\psi : \ell^2(\Gamma)^{\mathbb{N}} \to \ell^2(\Gamma)$ be a homeomorphism. We define

$$\Lambda_\alpha(\tau) = \psi\big((\ell^2(\Gamma) \setminus \Omega_\alpha(\tau))^{\mathbb{N}}_f\big).$$

Then, $\Lambda_\alpha(\tau)$ is an $\mathcal{F}_{\Lambda_\alpha(\tau)}$-absorbing set in $\ell^2(\Gamma)$ by Proposition 5.10.26. Since $\ell^2(\Gamma) \setminus \Omega_\alpha(\tau) \in \mathfrak{A}_\alpha(\ell^2(\Gamma))$ because $\Omega_\alpha(\tau) \in \mathfrak{M}_\alpha(\ell^2(\Gamma))$, it follows that $\Lambda_\alpha(\tau) \in \mathfrak{A}_\alpha(\ell^2(\Gamma))_\sigma = \mathfrak{A}_\alpha(\ell^2(\Gamma))$. Moreover, the following holds:

(A_α) *For each $X \in \mathfrak{A}_\alpha(\ell^2(\Gamma))$, there exists a closed embedding $\psi_{\alpha,X} : \ell^2(\Gamma) \to \ell^2(\Gamma)$ such that $\psi_{\alpha,X}^{-1}(\Lambda_\alpha(\tau)) = X$.*

Indeed, for each $X \in \mathfrak{A}_\alpha(\ell^2(\Gamma))$, there exists a closed embedding $\varphi_{\alpha,\ell^2(\Gamma)\setminus X} : \ell^2(\Gamma) \to \ell^2(\Gamma)$ such that $\varphi_{\alpha,X}^{-1}(\Omega_\alpha(\tau)) = \ell^2(\Gamma) \setminus X$, which implies that $\varphi_{\alpha,X}^{-1}(\ell^2(\Gamma) \setminus \Omega_\alpha(\tau)) = X$. Identifying $\ell^2(\Gamma) = \ell^2(\Gamma) \times \{\mathbf{0}\} \subset \ell^2(\Gamma)^{\mathbb{N}}$, we have a closed embedding $\psi_{\alpha,X} = \psi\varphi_{\alpha,X} : \ell^2(\Gamma) \to \ell^2(\Gamma)$. Then, we have

$$\psi_{\alpha,X}^{-1}(\Lambda_\alpha(\tau)) = \psi_{\alpha,X}^{-1}\big(\psi(\ell^2(\Gamma) \setminus \Omega_\alpha(\tau))_f^{\mathbb{N}}\big) = \varphi_{\alpha,X}^{-1}\big(\ell^2(\Gamma) \setminus \Omega_\alpha(\tau)\big) = X.$$

Thus, (A_α) holds. Here, observe

$$X \approx \varphi_{\alpha,X}(X) = \varphi_{\alpha,X}\big(\ell^2(\Gamma)) \cap \Lambda_\alpha(\tau)\big),$$

which means that X can be embedded into $\Lambda_\alpha(\tau)$ as a closed set. Hence, $\mathfrak{A}_\alpha(\tau) \subset \mathcal{F}_{\Lambda_\alpha(\tau)}$. Because $\mathfrak{A}_\alpha(\tau)$ is closed hereditary and $\Lambda_\alpha(\tau) \in \mathfrak{A}_\alpha(\tau)$, we have $\mathcal{F}_{\Lambda_\alpha(\tau)} \subset \mathfrak{A}_\alpha(\tau)$. Hence, $\mathcal{F}_{\Lambda_\alpha(\tau)} = \mathfrak{A}_\alpha(\tau)$. Consequently, $\Lambda_\alpha(\tau)$ is an $\mathfrak{A}_\alpha(\tau)$-absorbing set in $\ell^2(\Gamma)$.

Step 4 (*Construction of $\Omega_\alpha(\tau)$, $\alpha > 2$*) Assume that $\mathfrak{A}_\beta(\tau)$-absorbing sets Λ_β, $1 < \beta < \alpha$ in $\ell^2(\Gamma)$ are defined so that $\Lambda_\beta \in \mathfrak{A}_\beta(\tau)$ and the above (A_β) holds. We define

$$\Omega_\alpha(\tau) = \begin{cases} \psi(\Lambda_\beta(\tau)^{\mathbb{N}}) & \text{if } \alpha = \beta + 1, \\ \psi_\alpha\big(\big(\prod_{0<\beta<\alpha} \Lambda_\beta(\tau)\big)^{\mathbb{N}}\big) & \text{if } \alpha \text{ is a limit ordinal,} \end{cases}$$

where $\psi : \ell^2(\Gamma)^{\mathbb{N}} \to \ell^2(\Gamma)$ and $\psi_\alpha : (\ell^2(\Gamma)^{\alpha\setminus 1})^{\mathbb{N}} \to \ell^2(\Gamma)$ are homeomorphisms.[34] Then, $\Omega_\alpha(\tau)$ is an $\mathcal{F}_{\Omega_\alpha(\tau)}$-absorbing set in $\ell^2(\Gamma)$ by Proposition 5.10.26. Since $\Lambda_\beta(\tau) \in \mathfrak{A}_\beta(\tau)$ for $\beta < \alpha$, it follows from Lemma 5.12.1 that $\Omega_\alpha(\tau) \in \mathfrak{M}_\alpha(\tau)$. Because $\mathfrak{M}_\alpha(\tau)$ is closed hereditary, we have $\mathfrak{F}_{\Omega_\alpha(\tau)} \subset \mathfrak{M}_\alpha(\tau)$. To see that $\Omega_\alpha(\tau)$ is an $\mathfrak{M}_\alpha(\tau)$-absorbing set in $\ell^2(\Gamma)$, we have to verify $\mathfrak{M}_\alpha(\tau) \subset \mathcal{F}_{\Omega_\alpha(\tau)}$. To complete induction, we have to prove condition (M_α).

When $\alpha = \beta + 1$, let $X = \bigcap_{i\in\mathbb{N}} X_i \in \mathfrak{M}_\alpha(\ell^2(\Gamma))$, where $X_i \in \mathfrak{A}_\beta(\ell^2(\Gamma))$ for each $i \in \mathbb{N}$. By (A_β), we have closed embeddings $\psi_{\beta,X_i} : \ell^2(\Gamma) \to \ell^2(\Gamma)$, $i \in \mathbb{N}$, such that $\psi_{\beta,X_i}^{-1}(\Lambda_\beta(\tau)) = X_i$. We define $\varphi_{\alpha,X} : \ell^2(\Gamma) \to \ell^2(\Gamma)$ as follows:

$$\varphi_{\alpha,X}(x) = \psi\big(\big(\psi_{\beta,X_i}(x)\big)_{i\in\mathbb{N}}\big)$$

[34]The ordinal α is defined as the set $\alpha = \{\beta \mid \beta < \alpha\}$, e.g., $0 = \emptyset$, $1 = \{0\} = \{\emptyset\}$, $2 = \{0, 1\} = \{\emptyset, \{\emptyset\}\}$, and so on. Hence, $\alpha \setminus 1 = \{\beta \mid 0 < \beta < \alpha\}$. Moreover, $\beta + 1 = \beta \cup \{\beta\}$.

Then, $\varphi_{\alpha,X}$ is a closed embedding and

$$\varphi_{\alpha,X}^{-1}(\Omega_\alpha(\tau)) = \varphi_{\alpha,X}^{-1}\big(\psi(\Lambda_\beta(\tau)^{\mathbb{N}})\big) = \bigcap_{n\in\mathbb{N}} \psi_{\beta,X_i}^{-1}(\Lambda_\beta(\tau)) = \bigcap_{i\in\mathbb{N}} X_i = X.$$

Thus, (M_α) holds. Here, observe

$$X \approx \varphi_{\alpha,X}(X) = \varphi_{\alpha,X}(\ell^2(\Gamma) \cap \Omega_\alpha(\tau)),$$

which means that X can be embedded into $\Omega_\alpha(\tau)$ as a closed set. Hence, $\mathfrak{M}_\alpha(\tau) \subset \mathcal{F}_{\Omega_\alpha(\tau)}$.

When α is a limit ordinal, let $X \in \mathfrak{M}_\alpha(\ell^2(\Gamma))$. Then, we can write $X = \bigcap_{i\in\mathbb{N}} X_i$, where $X_i \in \mathfrak{A}_{\beta_i}(\ell^2(\Gamma))$ and $\beta_i < \alpha$. By (A_{β_i}), we have a closed embedding $\psi_{\beta_i,X_i} : \ell^2(\Gamma) \to \ell^2(\Gamma)$ such that $\psi_{\beta_i,X_i}^{-1}(\Lambda_{\beta_i}(\tau)) = X_i$. Then, we have a closed embedding $\phi_X : \ell^2(\Gamma) \to \ell^2(\Gamma)^{\mathbb{N}}$ defined by

$$\phi_X(x) = (\psi_{\beta_i,X_i}(x))_{i\in\mathbb{N}},$$

where it follows that $\phi_X^{-1}(\prod_{i\in\mathbb{N}} \Lambda_{\beta_i}(\tau)) = \bigcap_{i\in\mathbb{N}} X_i = X$. Take $e_\beta \in \Lambda_\beta(\tau)$, $0 < \beta < \alpha$, and define $h : \ell^2(\Gamma)^{\mathbb{N}} \to (\ell^2(\Gamma)^{\alpha\setminus 1})^{\mathbb{N}}$ as follows:

$$\mathrm{pr}_\beta \mathrm{pr}_i h((x_i)_{i\in\mathbb{N}}) = \begin{cases} x_i & \text{if } \beta = \beta_i, \\ e_\beta & \text{otherwise.} \end{cases}$$

Let $p : (\ell^2(\Gamma)^{\alpha\setminus 1})^{\mathbb{N}} \to \ell^2(\Gamma)^{\mathbb{N}}$ be a map defined by $\mathrm{pr}_i p = \mathrm{pr}_{\beta_i} \mathrm{pr}_i$ for each $i \in \mathbb{N}$, that is, $p((x_i)_{i\in\mathbb{N}}) = (x_i(\beta_i))_{i\in\mathbb{N}}$. Then, h is a closed embedding and $ph = \mathrm{id}$ because $\mathrm{pr}_i ph = \mathrm{pr}_{\beta_i} \mathrm{pr}_i h = \mathrm{pr}_i$ for each $i \in \mathbb{N}$. Now, we have a closed embedding $\varphi_{\alpha,X} = \psi h \phi_X : \ell^2(\Gamma) \to \ell^2(\Gamma)$. Then, it follows that

$$\begin{aligned} \varphi_{\alpha,X}^{-1}(\Omega_\alpha(\tau)) &= \varphi_{\alpha,X}^{-1}\Big(\psi\Big(\big(\textstyle\prod_{0<\beta<\alpha} \Lambda_\beta(\tau)\big)^{\mathbb{N}}\Big)\Big) \\ &= \phi_X^{-1}\Big(h^{-1}\Big(\big(\textstyle\prod_{0<\beta<\alpha} \Lambda_\beta(\tau)\big)^{\mathbb{N}}\Big)\Big) \\ &= \phi_X^{-1}\big(\textstyle\prod_{i\in\mathbb{N}} \Lambda_{\beta_i}(\tau)\big) = X. \end{aligned}$$

Thus, (M_α) holds. Here, observe

$$X \approx \varphi_{\alpha,X}(X) = \varphi_{\alpha,X}(\ell^2(\Gamma) \cap \Omega_\alpha(\tau)),$$

which means that X can be embedded into $\Omega_\alpha(\tau)$ as a closed set. Hence, $\mathfrak{M}_\alpha(\tau) \subset \mathcal{F}_{\Omega_\alpha(\tau)}$.

The following is now obtained:

Theorem 5.12.11 *For every countable ordinal $\alpha > 1$, there exist an $\mathfrak{M}_\alpha(\tau)$-absorbing set $\Omega_\alpha(\tau) \in \mathfrak{M}_\alpha(\tau)$ and an $\mathfrak{A}_\alpha(\tau)$-absorbing set $\Lambda_\alpha(\tau) \in \mathfrak{A}_\alpha(\tau)$ in $\ell^2(\Gamma)$.* □

Now, we have the following:

Theorem 5.12.12 *For every countable ordinal $\alpha > 1$,*

$$\Omega_\alpha(\tau) \approx \ell^2(\Gamma) \times \Omega_\alpha \quad \textit{and} \quad \Lambda_\alpha(\tau) \approx \ell^2(\Gamma) \times \Lambda_\alpha.$$

Proof Observe the following homeomorphy:

$$\begin{aligned}\big(\ell^2(\Gamma), \Omega_2(\tau)\big) &\approx \big((\ell^2(\Gamma) \times \ell^2)^{\mathbb{N}}, (\ell^2(\Gamma) \times \ell^2_f)^{\mathbb{N}}\big)\\ &\approx \big(\ell^2(\Gamma)^{\mathbb{N}} \times (\ell^2)^{\mathbb{N}}, \ell^2(\Gamma)^{\mathbb{N}} \times (\ell^2_f)^{\mathbb{N}}\big)\\ &\approx \big(\ell^2(\Gamma) \times \ell^2, \ell^2(\Gamma) \times \Omega_2\big),\end{aligned}$$

which implies that $\Omega_2(\tau) \approx \ell^2(\Gamma) \times \Omega_2$. For a countable ordinal $\alpha > 1$, assume that $\Omega_\alpha(\tau) \approx \ell^2(\Gamma) \times \Omega_\alpha$. Then, $\ell^2(\Gamma) \times \Omega_\alpha$ is homotopy dense in $\ell^2(\Gamma) \times \ell^2 \approx \ell^2(\Gamma)$. Since $\ell^2(\Gamma)^{\mathbb{N}}_f \approx \ell^2(\Gamma) \times \ell^2_f$ by Theorem 5.5.8 and $\ell^2_f \times \Lambda_\alpha \approx \Lambda_\alpha$ by Corollary 5.10.21, it follows that

$$\begin{aligned}\Lambda_\alpha(\tau) &\approx \big((\ell^2(\Gamma) \times \ell^2) \setminus (\ell^2(\Gamma) \times \Omega_\alpha)\big)^{\mathbb{N}}_f \approx \ell^2(\Gamma)^{\mathbb{N}}_f \times (\ell^2 \setminus \Omega_\alpha)^{\mathbb{N}}_f\\ &\approx \ell^2(\Gamma) \times \ell^2_f \times \Lambda_\alpha \approx \ell^2(\Gamma) \times \Lambda_\alpha.\end{aligned}$$

Now, assume that $\Lambda_\beta(\tau) \approx \ell^2(\Gamma) \times \Lambda_\beta$ for $1 < \beta < \alpha$, where $\ell^2(\Gamma) \times \Lambda_\beta$ is homotopy dense in $\ell^2(\Gamma) \times \ell^2 \approx \ell^2(\Gamma)$. If $\alpha = \beta + 1$, then

$$\Omega_\alpha(\tau) \approx \Lambda_\beta(\tau)^{\mathbb{N}} \approx \big(\ell^2(\Gamma) \times \Lambda_\beta\big)^{\mathbb{N}} \approx \ell^2(\Gamma)^{\mathbb{N}} \times \Lambda_\beta^{\mathbb{N}} \approx \ell^2(\Gamma) \times \Omega_\alpha.$$

If α is a limit ordinal, then

$$\begin{aligned}\Omega_\alpha(\tau) &\approx \Big(\prod_{1<\beta<\alpha} \Lambda_\beta(\tau)\Big)^{\mathbb{N}} \approx \Big(\prod_{1<\beta<\alpha} \ell^2(\Gamma) \times \Lambda_\beta\Big)^{\mathbb{N}}\\ &\approx \ell^2(\Gamma)^{\mathbb{N}} \times \Big(\prod_{1<\beta<\alpha} \Lambda_\beta\Big)^{\mathbb{N}} \approx \ell^2(\Gamma) \times \Omega_\alpha.\end{aligned}$$

Thus, we have the result by transfinite induction. □

The results on Ω-manifolds obtained in Sect. 5.10 can be applied to $\Omega_\alpha(\tau)$-manifolds and $\mathfrak{A}_\alpha(\tau)$-manifolds. In particular, Theorem 5.10.15 can be rewritten for $\Omega_\alpha(\tau)$- (or $\mathfrak{A}_\alpha(\tau)$-)manifolds as follows:

Theorem 5.12.13 (Characterization) *Let $\alpha > 1$ be a countable ordinal. A space X is an $\Omega_\alpha(\tau)$- (or $\Lambda_\alpha(\tau)$-)manifold if and only if $X \in \mathfrak{M}_\alpha(\tau)$ (or $\in \mathfrak{A}_\alpha(\tau)$) is a strongly $\mathfrak{M}_\alpha(\tau)$- (or $\mathfrak{A}_\alpha(\tau)$)-universal ANR that is a strong Z_σ-space.* □

Combining the above with Theorem 5.10.17(1), we have the following characterization of $\Omega_\alpha(\tau)$ (and $\Lambda_\alpha(\tau)$):

Theorem 5.12.14 (Characterization) *Let $\alpha > 1$ be a countable ordinal. A space X is homeomorphic to $\Omega_\alpha(\tau)$ (or $\Lambda_\alpha(\tau)$) if and only if $X \in \mathfrak{M}_\alpha(\tau)$ (or $\in \mathfrak{A}_\alpha(\tau)$) is a strongly $\mathfrak{M}_\alpha(\tau)$- (or $\mathfrak{A}_\alpha(\tau)$)-universal AR that is a strong Z_σ-space.* □

Remark 5.27 In Theorems 5.12.13 and 5.12.14, the condition $X \in \mathfrak{M}_\alpha(\tau)$ (or $\in \mathfrak{A}_\alpha(\tau)$) can be replaced by the condition $X \in \mathfrak{M}_\alpha(\tau)_\sigma$ (or $\in \mathfrak{A}_\alpha(\tau)_\sigma$) according to Theorem 5.10.15. However, $\Omega_\alpha(\tau) \in \mathfrak{M}_\alpha(\tau)$ (or $\Lambda_\alpha(\tau) \in \mathfrak{A}_\alpha(\tau)$) and every $\Omega_\alpha(\tau)$- (or $\Lambda_\alpha(\tau)$-)manifold can be embbeded into $\Omega_\alpha(\tau)$ (or $\Lambda_\alpha(\tau)$) as a closed set.

Notes for Chapter 5

This chapter is based on results of three papers, [41] by T.A. Chapman, [150] by J.E. West, and [29] by M. Bestvina and J. Mogilski. The paper [150] is a non-separable generalization of [41]. We presented here a non-separable version of [29], which was obtained in papers [97], [108], and [129]. For other results concerning absorbing sets that we did not discuss here, one can refer to the book [[5]] of T. Banakh, T. Radul and M. Zarichnyi. For Borel sets discussed in Sect. 5.11, Kecheris' book [[16]] is an excellent reference but it treats only the separable case. For non-separable Borel sets, refer to the article of A.H. Stone [135] (cf. [136]).

The main results in Sects. 5.1 and 5.2 were obtained by T.A. Chapman [41] and H. Toruńczyk [137]. An (f.d.)cap set was introduced by R.D. Anderson [10] (cf. [8]) and the resembling concept of skeletoids was introduced by Bessaga and Pełczyński [28]. The former concept is used in [41] and the latter is in [137].

In Sects. 5.3 and 5.4, we discussed $\mathfrak{C}$-absorption bases, which was introduced by J.E. West [150] as a $\mathfrak{C}$-absorptive set with a $\mathfrak{C}$-complex structure. Our definition is a little different from his original definition in [150]. In fact, we does not use $\mathfrak{C}$-complex structures. As K. Koshino pointed out in [97, Proposition 8.3], a metrizable space X has a $\mathfrak{C}$-complex structure if and only if X is a countable union of closed subsets which are discrete unions of members of $\mathfrak{C}$. Moreover, in conformity to the definition of (f.d.)cap sets, we adopt Z-embeddings instead of ambient homeomorphisms in the definition of the $\mathfrak{C}$-absorption property. Lemma 5.3.3 is due to Anderson and Bing [12]. The Uniqueness of Absorption Bases (Theorem 5.3.14) was established by J.E. West [150]. The main results of Sect. 5.4 are non-separable versions of results in [29], which were obtained in [129].

Theorems 5.5.1 and 5.5.2 were established by West [150], and were generalized to the four types of absorption bases for $\ell^2(\Gamma)$ in [129], [108], [97].

Proposition 5.6.7 and Corollary 5.6.8 were established by Bestvina and Mogilski [29], and were generalized in [129] to the non-separable versions, Proposition 5.6.5 and Corollary 5.6.6.

The case $F = \ell_2(\Gamma)$ of Theorem 5.7.2 was obtained by West [150]. The τ-discrete approximation property for a class $\mathfrak{C}$ was introduced by K. Koshino [97] and Lemmas 5.7.10–5.7.13 were proved by him, but his formulations are a little different from ours. Lemma 5.7.12 is a special case of Proposition 4.4 in [97].

Theorem 5.8.1 was established in [18] as a non-separable version of Theorem 5.8.3 [[5]].

Theorems 5.9.5 and 5.9.6 were obtained in [97] by improving results of [129]. Their origin, Theorem 5.9.11, was established by Mogilski [112]. Propositions 5.9.14, 5.9.18, and Theorem 5.9.19 were established by Curtis, Dobrowolski, and Mogilski [51]. Theorem 5.9.15 was not formulated but implicitly stated in the paper [51].

The results in Sect. 5.10 are due to Bestvina and Mogilski [29]. In [29], the separability is assumed but it is observed in [129] that these are valid without the separability.

Theorem 5.11.7, Propositions 5.11.9, 5.11.10, Lemmas 5.11.11, 5.11.13, 5.11.15, and Theorem 5.11.16 were shown by A.H. Stone [136].

The results of Sect. 5.12 are due to [108]. Lemma 5.12.2 is implicitly proved in [161]. Theorem 5.12.12 was pointed out by T. Banakh.

As mentioned in the first paragraph, for further results, one can refer to the book [(5)].

Chapter 6
Manifolds Modeled on Direct Limits and Combinatorial ∞-Manifolds

Let $\varinjlim \mathbb{R}^n = \mathbb{R}^\infty$ and $\varinjlim \boldsymbol{Q}^n = \boldsymbol{Q}^\infty$ denote the direct limits of the following towers:

$$\mathbb{R} \subset \mathbb{R}^2 \subset \mathbb{R}^3 \subset \cdots ; \quad \boldsymbol{Q} \subset \boldsymbol{Q}^2 \subset \boldsymbol{Q}^3 \subset \cdots .$$

As naturally expected, $\mathbb{R}^\infty$ is a locally convex topological linear space (Theorem 6.1.4). The space $\boldsymbol{Q}^\infty$ is also homeomorphic to a locally convex topological linear space. In fact, it is known as the Heisey Theorem (Theorem 6.2.3) that $\boldsymbol{Q}^\infty$ is homeomorphic to the dual space of any separable infinite-dimensional Banach space with the bounded weak-star topology. In this chapter, we give characterizations of $\mathbb{R}^\infty$-manifolds and $\boldsymbol{Q}^\infty$-manifolds. Fundamental results on these manifolds are also demonstrated. An $\mathbb{R}^\infty$-manifold is triangulable, that is, it is homeomorphic to the polyhedron of some simplicial complex. In the second half of this chapter, we study triangulations of $\mathbb{R}^\infty$-manifolds. Generalizing combinatorial manifolds to the infinite-dimensional case, we can introduce combinatorial ∞-manifolds. For these manifolds, the Hauptvermutung can be proved, though that does not hold for combinatorial n-manifolds if $3 < n < \infty$. It is also proved that every triangulation of an $\mathbb{R}^\infty$-manifold is a combinatorial ∞-manifold. We also consider combinatorial ∞-manifolds with the metric topology, and show their relation with manifolds modeled on ℓ^2_f and ℓ^2.

6.1 Preliminaries for Direct Limits

In this section, we review basic results on the direct limits which will be used in this chapter. For their proofs, refer to [GAGT, Sect. 2.8].

K. Sakai, *Topology of Infinite-Dimensional Manifolds*, Springer Monographs in Mathematics, https://doi.org/10.1007/978-981-15-7575-4_6

Let $X_1 \subset X_2 \subset \cdots$ be a tower of spaces, where each X_n is a subspace of X_{n+1}. The **direct limit** $\varinjlim X_n$ of this tower is the space $\bigcup_{n\in\mathbb{N}} X_n$ endowed with the weak topology with respect to the tower $(X_n)_{n\in\mathbb{N}}$, that is,

$$U \subset \varinjlim X_n \text{ is open in } \varinjlim X_n \;\Leftrightarrow\; \forall n \in \mathbb{N},\, U \cap X_n \text{ is open in } X_n$$

$$\Bigl(\text{equiv. } A \subset \varinjlim X_n \text{ is closed in } \varinjlim X_n \;\Leftrightarrow\; \forall n \in \mathbb{N},\, A \cap X_n \text{ is closed in } X_n\Bigr).$$

In other words, the topology of $\varinjlim X_n$ is the finest topology such that every X_n is a subspace of $\varinjlim X_n$, which is equal to the finest topology such that every inclusion $X_n \subset \varinjlim X_n$ is continuous. For an arbitrary space Y,

$$f : \varinjlim X_n \to Y \text{ is continuous } \Leftrightarrow \forall n \in \mathbb{N},\, f|X_n \text{ is continuous.}$$

It should also be noted that the direct limit $\varinjlim X_n$ is T_1 but, in general, non-Hausdorff for a tower of Hausdorff spaces $X_1 \subset X_2 \subset \cdots$ even if each X_n is closed in X_{n+1}. Such an example can be found in [GAGT, 2.10.3].

As is easily observed, $\varinjlim X_{n(i)} = \varinjlim X_n$ for any increasing sequence $n(1) < n(2) < \cdots$ in $\mathbb{N}$. It is also easy to prove the following proposition:

Proposition 6.1.1 *Let $X_1 \subset X_2 \subset \cdots$ and $Y_1 \subset Y_2 \subset \cdots$ be towers of spaces. Suppose that there exist $n(1) < n(2) < \cdots$, $m(1) < m(2) < \cdots \in \mathbb{N}$ and maps $f_i : X_{n(i)} \to Y_{m(i)}$ and $g_i : Y_{m(i)} \to X_{n(i+1)}$ such that $g_i f_i = \mathrm{id}_{X_{n(i)}}$ and $f_{i+1} g_i = \mathrm{id}_{Y_{m(i)}}$, that is, the following diagram is commutative:*

$$\begin{array}{ccccccc}
X_{n(1)} & \subset & X_{n(2)} & \subset & X_{n(3)} & \subset & \cdots \\
{\scriptstyle f_1}\downarrow & \nearrow^{g_1} & {\scriptstyle f_2}\downarrow & \nearrow^{g_2} & {\scriptstyle f_3}\downarrow & & \\
Y_{m(1)} & \subset & Y_{m(2)} & \subset & Y_{m(3)} & \subset & \cdots
\end{array}$$

Then, f_i, $i \in \mathbb{N}$, induce the homeomorphism $f_\infty : \varinjlim X_n \to \varinjlim Y_n$, whose inverse f_∞^{-1} is induced by g_i, $i \in \mathbb{N}$. □

The following can be easily proved:

Proposition 6.1.2 *Let $X = \varinjlim X_n$ be the direct limit of a tower $X_1 \subset X_2 \subset X_3 \subset \cdots$ of spaces.*

(1) *Each closed (or open) subspace Y of X is the direct limit of the tower $X_1 \cap Y \subset X_2 \cap Y \subset X_3 \cap Y \subset \cdots$.* [2.8.2]
(2) *For a tower $U_1 \subset U_2 \subset U_3 \subset \cdots$, if each U_n is an open subspace of X_n, then $U = \varinjlim U_n$ is an open subspace of X.*
(3) *Each compact subset of X is contained in some X_n.* [2.8.6(1)]

The above (1) and (3) are quoted from [GAGT]. It should be remarked that (2) comes from (1) if $U_n = U \cap X_n$ for each $n \in \mathbb{N}$, but $U_n \neq U \cap X_n$ in general.

To see (2), it suffices to show that $V \subset U$ is open in U if and only if it is open in X. If V is open in X then each $V \cap U_n = V \cap X_n \cap U_n$ is open in U_n because $V \cap X_n$ is open in X_n and U_n is an open subspace of X_n. Hence, V is open in U. Conversely, assume that V is open in U. Then, for each $n \in \mathbb{N}$, $V \cap X_n = \bigcup_{i \geqslant n}(V \cap U_i \cap X_n)$ is open in X_n because each $V \cap U_i$ is open in an open subspace U_i of X_i and X_n is a subspace of X_i. Then, V is open in X.

For the product of the direct limits, we have the following results (Propositions 2.8.3 and 2.8.4 in [GAGT]):

Proposition 6.1.3 *Let $X_1 \subset X_2 \subset X_3 \subset \cdots$ and $Y_1 \subset Y_2 \subset Y_3 \subset \cdots$ be towers of spaces.*

(1) *For any locally compact space Y,*

$$\big(\varinjlim X_n\big) \times Y = \varinjlim (X_n \times Y) \quad \text{as spaces.} \qquad \boxed{2.8.3}$$

(2) *If each X_n and Y_n are locally compact, then*

$$\varinjlim X_n \times \varinjlim Y_n = \varinjlim (X_n \times Y_n) \quad \text{as spaces.} \qquad \boxed{2.8.4}$$

The space $\mathbb{R}^\infty$ has the naturally defined addition $\mathfrak{a} : \mathbb{R}^\infty \times \mathbb{R}^\infty \to \mathbb{R}^\infty$ and scalar multiplication $\mathfrak{m} : \mathbb{R}^\infty \times \mathbb{R} \to \mathbb{R}^\infty$, that is, the following diagrams commute for each $n \in \mathbb{N}$:

$$\begin{array}{ccc} \mathbb{R}^n \times \mathbb{R}^n & \xrightarrow{\ \mathfrak{a}_n\ } & \mathbb{R}^n \\ \cap\big\downarrow & & \big\downarrow\cap \\ \mathbb{R}^\infty \times \mathbb{R}^\infty & \xrightarrow[\ \mathfrak{a}\ ]{} & \mathbb{R}^\infty \end{array} \qquad \begin{array}{ccc} \mathbb{R}^n \times \mathbb{R} & \xrightarrow{\ \mathfrak{m}_n\ } & \mathbb{R}^n \\ \cap\big\downarrow & & \big\downarrow\cap \\ \mathbb{R}^\infty \times \mathbb{R} & \xrightarrow[\ \mathfrak{m}\ ]{} & \mathbb{R}^\infty \end{array}$$

where $\mathfrak{a}_n$ and $\mathfrak{m}_n$ are the usual addition and scalar multiplication. Applying Proposition 6.1.3, we have

$$\mathbb{R}^\infty \times \mathbb{R}^\infty = \varinjlim (\mathbb{R}^n \times \mathbb{R}^n) \text{ and } \mathbb{R} \times \mathbb{R}^\infty = \varinjlim (\mathbb{R} \times \mathbb{R}^n).$$

Hence, $\mathfrak{a}$ and $\mathfrak{m}$ are continuous. Thus, $\mathbb{R}^\infty$ is a topological linear space.

Theorem 6.1.4 *The space $\mathbb{R}^\infty$ is a locally convex topological linear space with the naturally defined addition and scalar multiplication.*

Proof It has been proved that $\mathbb{R}^\infty$ is a topological linear space. To see that $\mathbb{R}^\infty$ is locally convex, let U be a neighborhood of $\mathbf{0} \in \mathbb{R}^\infty = \bigcup_{n \in \mathbb{N}} \mathbb{R}^n$. We can inductively choose $\varepsilon_n > 0$, $n \in \mathbb{N}$, so that $\prod_{i=1}^n [-\varepsilon_1, \varepsilon_1] \subset U \cap \mathbb{R}^n$ for each $n \in \mathbb{N}$. (The proof is left to readers as an exercise.) Then, $\varinjlim \prod_{i=1}^n (-\varepsilon_1, \varepsilon_1) = \bigcup_{n \in \mathbb{N}} \prod_{i=1}^n (-\varepsilon_1, \varepsilon_1)$ is a convex open neighborhood of $\mathbf{0}$ that is contained in U. Hence, $\mathbb{R}^\infty$ is locally convex. □

Proposition 6.1.5 *Every separable locally compact metrizable space X is the direct limit of a tower of compacta.*

Proof Since X has a countable open cover $\{U_i \mid i \in \mathbb{N}\}$ such that every $\mathrm{cl}\, U_i$ is compact, we define $X_n = \bigcup_{i=1}^n \mathrm{cl}\, U_i$ for each $n \in \mathbb{N}$. Then, it is easy to see that $X = \varinjlim X_n$, where $X_1 \subset X_2 \subset \cdots$ are compacta. □

Since $\mathbb{R}^\infty$ is a locally convex topological linear space (Theorem 6.1.4), it is an AE by the Dugundji Theorem 1.13.1. Note that $\boldsymbol{Q}^\infty$ is a convex set in the linear space $\boldsymbol{s}^\infty = \varinjlim \boldsymbol{s}^n$ $(= \varinjlim(\mathbb{R}^{\mathbb{N}})^n)$ and it inherits the addition and scalar multiplication from $\boldsymbol{s}^\infty$. By Proposition 6.1.3, the following operation is continuous:

$$\boldsymbol{Q}^\infty \times \boldsymbol{Q}^\infty \times \mathbf{I} \ni (x, y, t) \mapsto (1-t)x + ty \in \boldsymbol{Q}^\infty,$$

where it should be remarked that the scalar multiplication is continuous on $\boldsymbol{s}^\infty$ by Proposition 6.1.3(1) but the addition is not continuous on $\boldsymbol{s}^\infty$ (see Remark 6.2 below). It is easy to see that $\boldsymbol{Q}^\infty$ is locally convex. So, by the same proof as the Dugundji Theorem 1.13.1 (cf. [GAGT, Proof of Theorem 6.1.1]), we can be prove that $\boldsymbol{Q}^\infty$ is an AE.[1] Thus, we have the following:

Proposition 6.1.6 *The spaces $\mathbb{R}^\infty$ and $\boldsymbol{Q}^\infty$ are AEs.* □

The above proposition can also be obtained by the following general theorem [GAGT, Theorem 6.1.8]:

Theorem 6.1.7 *For a closed tower $X_1 \subset X_2 \subset \cdots$ of ANEs (resp. AEs), the direct limit $\varinjlim X_n$ is an ANE (resp. an AE).* [2.1.8]

Remark 6.1 The AE case is not contained in the statement of Theorem 6.1.8 in [GAGT]. In the proof, each N_n can be taken X_n if each Y_n is an AE. However, $X \neq \bigcup_{n\in\mathbb{N}} X_n$ in general. Accordingly, we redefine

$$X_n = \{x \in X \mid d(x, A_n) \leqslant d(x, A \setminus A_n\} \cup \{x \in X \mid d(x, A) \geqslant 1/n\}.$$

Then, each X_n is closed in X, $X_n \cap A = A_n$, and $X = \bigcup_{n\in\mathbb{N}} X_n = \bigcup_{n\in\mathbb{N}} \mathrm{int}\, X_n$. Thus, we can modify the proof of Theorem 6.1.8 in [GAGT] for the AE case.

Due to Proposition 1.13.5(4), a paracompact space is an ANE if it is locally an ANE (i.e., each point has an ANE neighborhood). Hence, we have the following:

Proposition 6.1.8 *Every $\mathbb{R}^\infty$- and $\boldsymbol{Q}^\infty$-manifold are ANEs.* □

Let $\mathbf{I}^\infty = \varinjlim \mathbf{I}^n$ and $\Delta^\infty = \varinjlim \Delta^n$ denote the direct limits of the towers

$$\mathbf{I} \subset \mathbf{I}^2 \subset \mathbf{I}^3 \subset \cdots; \ \Delta^1 \subset \Delta^2 \subset \Delta^3 \subset \cdots.$$

[1] In the next section, it will be proved that $\boldsymbol{Q}^\infty$ is homeomorphic to a locally convex topological linear space.

Let $B = \{\mathbf{e}_i \mid i \in \mathbb{N}\}$ be the canonical orthonormal basis of ℓ^2, that is, $\mathbf{e}_n$ is the unit vector defined by $\mathbf{e}_n(n) = 1$, $\mathbf{e}_n(i) = 0$ for $i \neq n$. Then, Δ^∞ is the polyhedron of the full simplicial complex $\Delta(B)$ (cf. [GAGT, p. 143]). Since each Δ^n is called the standard n-simplex, we call Δ^∞ the **standard ∞-simplex**. Later, it will be shown that the spaces $\mathbf{I}^\infty$ and Δ^∞ are homeomorphic to $\mathbb{R}^\infty$ (Corollary 6.4.3).

Remark 6.2 The direct limit $s^\infty = \varinjlim s^n$ $(= \varinjlim(\mathbb{R}^{\mathbb{N}})^n)$ is not a topological linear space. In fact, the addition is not continuous. To see this fact, let

$$D = \big\{(k\mathbf{e}_n, \mathbf{0}, \ldots, \mathbf{0}, \underset{\substack{\uparrow \\ \text{the } n\text{-th coordinate}}}{k^{-1}\mathbf{e}_1}, \mathbf{0}, \mathbf{0}, \ldots) \bigm| k, n \in \mathbb{N}\big\} \subset s^\infty = s_f^{\mathbb{N}},$$

where $\mathbf{e}_i \in s = \mathbb{R}^{\mathbb{N}}$ be the unit vector defined by $\mathbf{e}_i(i) = 1$ and $\mathbf{e}_i(j) = 0$ for $j \neq i$. For each $n \in \mathbb{N}$,

$$D \cap s^n = \bigcup_{i=1}^{n} \big\{(k\mathbf{e}_i, \mathbf{0}, \ldots, \mathbf{0}, k^{-1}\mathbf{e}_1) \in s^i \bigm| k \in \mathbb{N}\big\},$$

which is discrete in s^n. Hence, D is closed in $s^\infty = \varinjlim s^n$. Thus, we obtain an open neighborhood $U = s^\infty \setminus D$ of $\mathbf{0}$ in s^∞. Then, $V + V \not\subset U$ for any neighborhood V of $\mathbf{0}$ in s^∞, that is, the addition is not continuous. Indeed, we have $n \in \mathbb{N}$ and an open neighborhood W of $\mathbf{0}$ in $\mathbb{R}^n$ such that $p_n^{-1}(W) \subset V \cap s$, where $s = s^1 \subset s^\infty = \varinjlim s^n$ and $p_n : s = \mathbb{R}^{\mathbb{N}} \to \mathbb{R}^n$ be the projection defined by $p_n(x) = (x(1), \ldots, x(n))$. Note that $\mathbb{R}_+\mathbf{e}_{n+1} \subset p_n^{-1}(W) \subset V \cap s$. Choose $k \in \mathbb{N}$ so that

$$(\mathbf{0}, \ldots, \mathbf{0}, k^{-1}\mathbf{e}_1) \in V \cap s^{n+1}.$$

Then, it follows that

$$x = (k\mathbf{e}_{n+1}, \mathbf{0}, \mathbf{0}, \ldots),\ y = (\mathbf{0}, \ldots, \mathbf{0}, \underset{\substack{\uparrow \\ \text{the } (n+1)\text{-th coordinate}}}{k^{-1}\mathbf{e}_1}, \mathbf{0}, \mathbf{0}, \ldots) \in V,$$

but $x + y \in D$, that is, $x + y \notin U$.

6.2 The Bounded Weak-Star Topology

In this section, we prove the Heisey Theorem asserting that the direct limit $Q^\infty = \varinjlim Q^n$ is homeomorphic to a locally convex topological linear space, which is the space E^* of continuous linear functionals on a separable normed linear space $E = (E, \|\cdot\|)$ admitting the bounded weak-star topology.

For each normed linear space $E = (E, \|\cdot\|)$, let E^* be the dual (or conjugate) space of E, which is the normed linear space of continuous linear functionals on E with the norm

$$\|f\| = \sup_{x\in E\setminus\{\mathbf{0}\}} \frac{|f(x)|}{\|x\|} = \sup_{x\in \mathbf{S}_E} |f(x)| = \sup_{x\in \mathbf{B}_E} |f(x)|,$$ [2]

where $\mathbf{B}_E$ and $\mathbf{S}_E$ are the unit closed ball and the unit sphere of E, respectively. Then, no matter whether E is complete or not, the normed linear space E^* is complete, so it is always a Banach space.

This can be shown as follows: For any Cauchy sequence $(f_n)_{n\in\mathbb{N}}$ in the normed linear space E^*, $(f_n|\mathbf{S}_E)_{n\in\mathbb{N}}$ is uniformly Cauchy in $\mathrm{C}(\mathbf{S}_E, \mathbb{R})$, hence it converges to a map $f \in \mathrm{C}(\mathbf{S}_E, \mathbb{R})$ because $\mathrm{C}(\mathbf{S}_E, \mathbb{R})$ is complete with respect to the sup-metric. Then, f extends to a linear functional $\tilde{f} : E \to \mathbb{R}$ as follows:

$$\tilde{f}(x) = \begin{cases} \|x\| f(\|x\|^{-1}x) & \text{if } x \neq 0, \\ 0 & \text{if } x = 0. \end{cases}$$

Indeed, for each $x, y \in E \setminus \{\mathbf{0}\}$,

$$\begin{aligned} \tilde{f}(x+y) &= \|x+y\| f(\|x+y\|^{-1}(x+y)) \\ &= \|x+y\| \lim_{n\to\infty} f_n(\|x+y\|^{-1}(x+y)) \\ &= \lim_{n\to\infty} f_n(x+y) = \lim_{n\to\infty} f_n(x) + \lim_{n\to\infty} f_n(y) \\ &= \|x\| \lim_{n\to\infty} f_n(\|x\|^{-1}x) + \|y\| \lim_{n\to\infty} f_n(\|y\|^{-1}y) \\ &= \|x\| f(\|x\|^{-1}x) + \|y\| f(\|y\|^{-1}y) \\ &= \tilde{f}(x) + \tilde{f}(y). \end{aligned}$$

Moreover, for each $x \in E \setminus \{0\}$ and $t \in \mathbb{R} \setminus \{0\}$,

$$\tilde{f}(tx) = \|tx\| f(\|tx\|^{-1}tx) = t\|x\| f(\|x\|^{-1}x) = t\tilde{f}(x).$$

By the definition, we have $\lim_{n\to\infty} \|f_n - \tilde{f}\| = 0$, that is, $(f_n)_{n\in\mathbb{N}}$ converges to $\tilde{f}$ in the normed linear space E^*.

We can regard $E^* \subset \mathbb{R}^E$, E^* has the topology inherited from the product topology of $\mathbb{R}^E$, which is different from the norm topology (the topology induced the norm) and called the **weak-star topology**. Let $E^*(w^*)$ denote the space E^* with the weak-star topology. Then, $E^*(w^*)$ is a locally convex topological linear space as a linear subspace of the topological linear space $\mathbb{R}^E$ with the product topology. The weak-star topology is the coarsest topology such that the projection $\mathrm{pr}_x : E^* \ni$

[2]Recall that a linear functional $f : E \to \mathbb{R}$ is continuous if and only if $f(V)$ is bounded for some neighborhood V of $\mathbf{0} \in E$ [GAGT, Proposition 3.4.4]. Since V contains $r\mathbf{B}_E$ for some $r > 0$, $f(\mathbf{B}_E) = r^{-1}f(r\mathbf{B}_E) \subset r^{-1}f(V)$ is also bounded.

$f \to f(x) \in \mathbb{R}$ (i.e., the evaluation) is continuous for every $x \in E$, which is also called the **point convergence topology**. Note that $\mathrm{id}_{E^*} : (E^*, \|\cdot\|) \to E^*(w^*)$ is continuous.

Theorem 6.2.1 *Let $E = (E, \|\cdot\|)$ be a separable infinite-dimensional normed linear space and $\mathbf{B}_{E^*}$ the unit closed ball of the dual E^*. Then, $\mathbf{B}_{E^*}$ with the weak-star topology is homeomorphic to the Hilbert cube.*

Proof Note that $\mathbf{B}_{E^*}$ is an infinite-dimensional convex set in a locally convex topological linear space $E^*(w^*)$. Then, by Keller's Theorem 3.9.2, it suffices to show that $\mathbf{B}_{E^*}$ is compact and metrizable.

(*Compactness*) Since $|f(x)| \leqslant \|x\|$ for every $f \in \mathbf{B}_{E^*}$ and $x \in E$, the space $\mathbf{B}_{E^*}$ with the weak-star topology is a subspace of the product space $\prod_{x\in E}[-\|x\|, \|x\|]$. To prove the compactness of $\mathbf{B}_{E^*}$ it suffices to show that $\mathbf{B}_{E^*}$ is closed in the compact space $\prod_{x\in E}[-\|x\|, \|x\|]$. For a linear functional $f : E \to \mathbb{R}$, if $|f(x)| \leqslant \|x\|$ for each $x \in E$, then f is continuous because $f(\mathbf{B}_E) \subset [-1, 1]$. Then, it follows that

$$\mathbf{B}_{E^*} = \prod_{x\in E}[-\|x\|, \|x\|] \cap \{f \in \mathbb{R}^E \mid f \text{ is linear}\}.$$

Let $\alpha \in \prod_{x\in E}[-\|x\|, \|x\|] \setminus \mathbf{B}_{E^*}$ $(\subset \mathbb{R}^E)$. Then, $\alpha : E \to \mathbb{R}$ is non-linear, which means that $\alpha(x + y) \neq \alpha(x) + \alpha(y)$ for some $x, y \in E$ or $\alpha(tz) \neq t\alpha(z)$ for some $z \in E$ and $t \in \mathbb{R} \setminus \{0\}$. In the former case, choose open sets U, V_x, V_y of $\mathbb{R}$ so that $\alpha(x) \in V_x$, $\alpha(y) \in V_y$, $\alpha(x + y) \in U$ and $U \cap (V_x + V_y) = \emptyset$. Then, $\mathrm{pr}_{x+y}^{-1}(U) \cap \mathrm{pr}_x^{-1}(V_x) \cap \mathrm{pr}_y^{-1}(V_y)$ is a neighborhood of α in $\mathbb{R}^E$ whose member is non-linear. In the latter case, choose open sets U_z, V_z of $\mathbb{R}$ so that $\alpha(tz) \in U_z$, $\alpha(z) \in V_z$ and $U_z \cap tV_z = \emptyset$. Then, $\mathrm{pr}_{tz}^{-1}(U_z) \cap \mathrm{pr}_x^{-1}(V_z)$ is a neighborhood of α in $\mathbb{R}^E$ whose member is non-linear. Thus, α has a neighborhood in $\prod_{x\in E}[-\|x\|, \|x\|]$ missing $\mathbf{B}_{E^*}$, which means that $\mathbf{B}_{E^*}$ is closed in $\prod_{x\in E}[-\|x\|, \|x\|]$.

(*Metrizability*) Since E is separable, it has a countable dense set D. Then, we define the map $\varphi : \mathbb{R}^E \to \mathbb{R}^D$ by $\varphi(\alpha) = \alpha|D$ for each $\alpha \in E^*$, where the both $\mathbb{R}^E$ and $\mathbb{R}^D$ have the product topologies. Because $\mathbf{B}_{E^*}$ is compact, the restriction $\varphi|\mathbf{B}_{E^*}$ is a closed map. Moreover, if $\alpha \neq \alpha' \in \mathbf{B}_{E^*}$, then $\varphi(\alpha) = \alpha|D \neq \alpha'|D = \varphi(\alpha')$ because α and α' are continuous and D is dense in E. Hence, the restriction $\varphi|\mathbf{B}_{E^*}$ is an embedding. Since $\mathbb{R}^D \approx \mathbb{R}^{\mathbb{N}}$ is metrizable, it follows that $\mathbf{B}_{E^*}$ is metrizable. □

Remark 6.3 As seen in the above proof, $\mathbf{B}_{E^*}$ with the weak-star topology is compact for any normed linear space E, but for its metrizability, it is necessary to assume that E is separable.

We can introduce another topology of E^* called the **bounded weak-star topology**, which is defined as follows: $A \subset E^*$ is closed in the bounded weak-star topology if $A \cap B$ is weak-star closed in each bounded set $B \subset E^*$. Let $E^*(bw^*)$ denote the space E^* with the bounded weak-star topology. Note that

$E^* = \bigcup_{n\in\mathbb{N}} n\mathbf{B}_{E^*}$. Since each $n\mathbf{B}_{E^*}$ is bounded and any bounded set in E^* is contained in some $n\mathbf{B}_{E^*}$, the topology of $E^*(bw^*)$ is equal to the weak topology with respect to the tower $\mathbf{B}_{E^*} \subset 2\mathbf{B}_{E^*} \subset 3\mathbf{B}_{E^*} \subset \cdots$, which means that $E^*(bw^*) = \varinjlim n\mathbf{B}_{E^*}$, where each $n\mathbf{B}_{E^*}$ admits the weak-star topology.

Proposition 6.2.2 *For every normed linear space E, the space $E^*(bw^*)$ is also a locally convex topological linear space.*

Proof It suffices to show that the addition $\alpha : E^*(bw^*) \times E^*(bw^*) \to E^*(bw^*)$ and the multiplication $\mu : \mathbb{R} \times E^*(bw^*) \to E^*(bw^*)$ are continuous. With respect to the weak-star topology, the following restrictions are continuous:

$$\alpha|n\mathbf{B}_{E^*} \times n\mathbf{B}_{E^*} : n\mathbf{B}_{E^*} \times n\mathbf{B}_{E^*} \to 2n\mathbf{B}_{E^*} \text{ and}$$
$$\mu|[-n,n] \times n\mathbf{B}_{E^*} : [-n,n] \times n\mathbf{B}_{E^*} \to n^2\mathbf{B}_{E^*}.$$

The following can be obtained by Proposition 6.1.3:

$$E^*(bw^*) \times E^*(bw^*) = \varinjlim(n\mathbf{B}_{E^*} \times n\mathbf{B}_{E^*}) \text{ and}$$
$$\mathbb{R} \times E^*(bw^*) = \varinjlim([-n,n] \times n\mathbf{B}_{E^*}).$$

Hence, the addition α and the multiplication μ are continuous with the bounded weak-star topologies.

To see the local convexity of $E^*(bw^*)$, let U be a neighborhood of $\mathbf{0}$ in $E^*(bw^*)$. We will inductively define convex open neighborhoods V_n of $\mathbf{0}$ in $n\mathbf{B}_{E^*}$, $n \in \mathbb{N}$, so that $\mathrm{cl}_{w^*} V_n \subset U \cap n\mathbf{B}_{E^*}$ and $V_n \subset V_{n+1}$, where $\mathrm{cl}_{w^*} = \mathrm{cl}_{E^*(w^*)}$.

First, choose a convex open neighborhood V_1 of $\mathbf{0}$ in $\mathbf{B}_{E^*}$ so that $\mathrm{cl}_{w^*} V_1 \subset U \cap \mathbf{B}_{E^*}$. Suppose that a convex open neighborhood V_n of $\mathbf{0}$ in $n\mathbf{B}_{E^*}$ has been chosen. Then, $\mathrm{cl}_{w^*} V_n \subset U \cap (n+1)\mathbf{B}_{E^*}$ and $\mathrm{cl}_{w^*} V_n$ is compact because $n\mathbf{B}_{E^*}$ is compact. For each $x \in \mathrm{cl}_{w^*} V_n$, there exists a neighborhood V_x of $\mathbf{0}$ in $E^*(w^*)$ such that

$$(x + V_x) \cap (n+1)\mathbf{B}_{E^*} \subset U \cap (n+1)\mathbf{B}_{E^*}.$$

Since $E^*(w^*)$ is a linear topological space, there is an open neighborhood W_x of $\mathbf{0}$ in $E^*(w^*)$ such that $W_x + W_x \subset V_x$. Since $\mathrm{cl}\, V_n$ is compact, we have $x_1, \ldots, x_p \in \mathrm{cl}\, V_n$ such that $\mathrm{cl}_{w^*} V_n \subset \bigcup_{i=1}^p (x_i + W_{x_i})$. Choose a convex open neighborhood W_0 of $\mathbf{0}$ in $E^*(w^*)$ so that $\mathrm{cl}_{w^*} W_0 \subset \bigcap_{i=1}^p W_{x_i}$. We define

$$V_{n+1} = (V_n + W_0) \cap (n+1)\mathbf{B}_{E^*}.$$

Then, V_{n+1} is a convex open neighborhood of $\mathbf{0}$ in $(n+1)\mathbf{B}_{E^*}$ and $V_n \subset V_{n+1}$.

To see that $\mathrm{cl}_{w^*} V_{n+1} \subset U \cap (n+1)\mathbf{B}_{E^*}$. we show that

$$\mathrm{cl}_{w^*} V_{n+1} = (\mathrm{cl}_{w^*} V_n + \mathrm{cl}_{w^*} W_0) \cap (n+1)\mathbf{B}_{E^*}.$$

Since the addition is continuous, the right hand term is contained in the left hand term. To see the converse inclusion, it suffices to show that $\mathrm{cl}_{w^*} V_n + \mathrm{cl}_{w^*} W_0$ is closed in $E^*(w^*)$. Let $z \in E^* \setminus (\mathrm{cl}_{w^*} V_n + \mathrm{cl}_{w^*} W_0)$, For each $y \in \mathrm{cl}_{w^*} V_n$, $z - y \in E^* \setminus \mathrm{cl}_{w^*} W_0$, so $V_y - U_y \subset E^* \setminus \mathrm{cl}_{w^*} W_0$ for some open neighborhoods U_y and V_y of y and z in $E^*(w^*)$, respectively Since $\mathrm{cl}_{w^*} V_n$ is compact, we have $y_1, \ldots, y_q \in \mathrm{cl}_{w^*} V_n$ such that $\mathrm{cl}_{w^*} V_n \subset \bigcup_{i=1}^q U_{y_i}$. Then, $V_z = \bigcap_{i=1}^q V_{y_i}$ is an open neighborhood of z in (E^*, w^*). Each $y \in \mathrm{cl}_{w^*} V_n$ is contained in some U_{y_i}, where $V_z - y \subset V_{y_i} - U_{y_i} \subset E^* \setminus \mathrm{cl}_{w^*} W_0$. It follows that $V_z - \mathrm{cl}_{w^*} V_n \subset E^* \setminus \mathrm{cl}_{w^*} W_0$, which means that $V_z \subset E^* \setminus (\mathrm{cl}_{w^*} V_n + \mathrm{cl}_{w^*} W_0)$. Hence, $E^* \setminus (\mathrm{cl}_{w^*} V_n + \mathrm{cl}_{w^*} W_0)$ is open in $E^*(w^*)$, that is, $\mathrm{cl}_{w^*} V_n + \mathrm{cl}_{w^*} W_0$ is closed in $E^*(w^*)$.

The inclusion $\mathrm{cl}_{w^*} V_{n+1} \subset U \cap (n+1)\mathbf{B}_{E^*}$ can be obtained as follows:

$$\begin{aligned}
\mathrm{cl}_{w^*} V_{n+1} &\subset \mathrm{cl}_{w^*}(V_n + W_0) \cap (n+1)\mathbf{B}_{E^*} \\
&\subset (\mathrm{cl}_{w^*} V_n + \mathrm{cl}_{w^*} W_0) \cap (n+1)\mathbf{B}_{E^*} \\
&\subset \bigcup_{i=1}^{p} (x_i + W_{x_i} + \mathrm{cl}_{w^*} W_0) \cap (n+1)\mathbf{B}_{E^*} \\
&\subset \bigcup_{i=1}^{p} (x_i + W_{x_i} + W_{x_i}) \cap (n+1)\mathbf{B}_{E^*} \\
&\subset \bigcup_{i=1}^{p} (x_i + V_{x_i}) \cap (n+1)\mathbf{B}_{E^*} \subset U \cap (n+1)\mathbf{B}_{E^*}.
\end{aligned}$$

Now, we define $V = \bigcup_{n \in \mathbb{N}} V_n \subset E^*$. Clearly, V is convex and $V \subset U$. For each $n \in \mathbb{N}$, $V \cap n\mathbf{B}_{E^*} = \bigcup_{m \geqslant n} V_m \cap n\mathbf{B}_{E^*}$ is open in $n\mathbf{B}_{E^*}$ because V_m is open $m\mathbf{B}_{E^*}$ for each $m \geqslant n$. Hence, V is open in $E^*(bw^*)$. Thus, U contains a convex open neighborhood of $\mathbf{0}$ in $E^*(bw^*)$. □

Now, we prove the following Heisey Theorem:

Theorem 6.2.3 (Heisey) *For every separable infinite-dimensional Banach space E, the space $E^*(bw^*)$ is homeomorphic to $\boldsymbol{Q}^\infty$.*

Proof In the above, we saw that $E^*(bw^*) = \varinjlim n\mathbf{B}_{E^*}$, where $n\mathbf{B}_{E^*} \approx \mathbf{B}_{E^*} \approx \boldsymbol{Q}$ by Theorem 6.2.1 and each $n\mathbf{B}_{E^*}$ is a Z-set in $(n+1)\mathbf{B}_{E^*}$ because $n\mathbf{B}_{E^*} \subset \mathrm{rint}(n+1)\mathbf{B}_{E^*}$ (Theorems 2.10.6 and 5.1.12). Applying the Z-set Unknotting Theorem 2.11.6 repeatedly, we can obtain homeomorphisms $h_n : n\mathbf{B}_{E^*} \to \boldsymbol{Q}^n$, $n \in \mathbb{N}$, such that $h_{n+1}|n\mathbf{B}_{E^*} = h_n$, that is, we can obtain the following commutative diagram of homeomorphisms:

$$\begin{array}{ccccccc}
\mathbf{B}_{E^*} & \subset & 2\mathbf{B}_{E^*} & \subset & 3\mathbf{B}_{E^*} & \subset & \cdots \\
\downarrow{\scriptstyle h_1} & & \downarrow{\scriptstyle f_2} & & \downarrow{\scriptstyle f_3} & & \\
\boldsymbol{Q} & \subset & \boldsymbol{Q}^2 & \subset & \boldsymbol{Q}^3 & \subset & \cdots
\end{array}$$

where $\boldsymbol{Q}^n$ is identified with $\boldsymbol{Q}^n \times \{0\} \subset \boldsymbol{Q}^{n+1}$. Then, homeomorphisms h_n, $n \in \mathbb{N}$, induce the homeomorphism $h_\infty : E^*(bw^*) \to \boldsymbol{Q}^\infty$. □

Thus, the space $\boldsymbol{Q}^\infty$ is homeomorphic to a locally convex topological linear space.

6.3 Embedding Neighborhood Extension Properties

In this section, we show that $\mathbb{R}^\infty$-manifolds and $\boldsymbol{Q}^\infty$-manifolds have the following embedding neighborhood extension properties:

($\mathcal{E}_f$) Each embedding from a closed subset A of any finite-dimensional compactum B into X can be extended to an embedding from some neighborhood of A in B into X.

($\mathcal{E}$) Each embedding from a closed subset A of any compactum B into X can be extended to an embedding from some neighborhood of A in B into X.

It will be proved that for every ANE X, $X \times \mathbb{R}^\infty$ and $X \times \boldsymbol{Q}^\infty$ have these properties, respectively. As particular cases, $\mathbb{R}^\infty$ and $\boldsymbol{Q}^\infty$ have these properties, respectively. The manifold cases can be obtained by the following lemma:

Lemma 6.3.1 *As properties of open sets in a space X, properties ($\mathcal{E}_f$) and ($\mathcal{E}$) are G-hereditary.*

Proof Because of similarity, we prove only that property ($\mathcal{E}_f$) is G-hereditary, that is, property ($\mathcal{E}_f$) satisfies (G-1), (G-2) and (G-3). Obviously, property ($\mathcal{E}_f$) satisfies (G-1), that is, for open sets $V \subset U$ in X, if U has property ($\mathcal{E}_f$), V has too. It is easy to see that property ($\mathcal{E}_f$) satisfies (G-3), that is, for a discrete collection $\{U_\lambda \mid \lambda \in \Lambda\}$ of open sets in X, if each U_λ has property ($\mathcal{E}_f$), then $\bigcup_{\lambda \in \Lambda} U_\lambda$ has too.

To show (G-2), assume that open sets U and V in X have property ($\mathcal{E}_f$). Let $f : A \to U \cup V$ be an embedding from a closed subset A of a finite-dimensional compactum B. Then, $A \setminus f^{-1}(U)$ has open neighborhoods W_1, W_2 in B such that

$$\operatorname{cl} W_1 \subset W_2 \quad \text{and} \quad \operatorname{cl} W_2 \cap (A \setminus f^{-1}(V)) = \emptyset.$$

Since $f(A \setminus W_1) \subset U$ and U has property ($\mathcal{E}_f$), we have a closed neighborhood N_1 of the closed set $A \setminus W_1$ in the finite-dimensional compactum $B \setminus W_1$, and an embedding $f_1 : N_1 \to U$ such that $f_1|A \setminus W_1 = f|A \setminus W_1$. Observe that

$$f_1(A \setminus W_2) \cap f(A \cap \operatorname{cl} W_1) = \emptyset \quad \text{and} \quad A \setminus W_2 \subset N_1.$$

We can find a closed neighborhood N_2 of $A \setminus W_2$ in B such that $N_2 \subset N_1 \setminus \operatorname{cl} W_1$ and $f_1(N_2) \cap f(A \cap \operatorname{cl} W_1) = \emptyset$. Then, we have an embedding $f_2 : A \cup N_2 \to U \cup V$ defined by $f_2|A = f$ and $f_2|N_2 = f_1|N_2$ (Fig. 6.1).

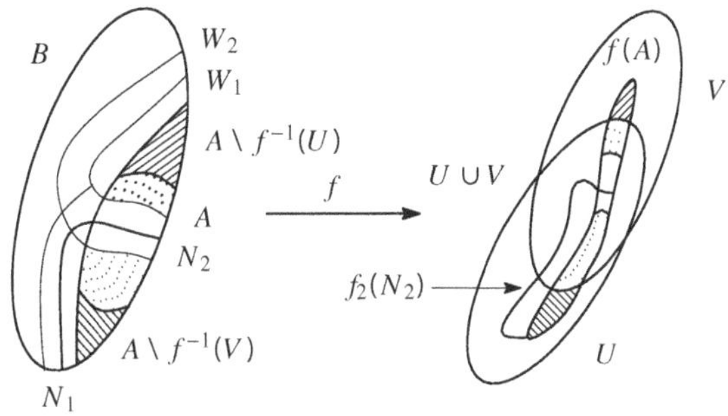

Fig. 6.1 Construction of f_2

In the above arguments, W_2 can be chosen so small as to be contained in a given neighborhood of $A \setminus f^{-1}(U)$ in B. Taking account of this observation, we can apply the above arguments to f_2 to obtain an open neighborhood W_3 of $(A \cup N_2) \setminus f_2^{-1}(V)$ in B, a closed neighborhood N_3 of $(A \cup N_2) \setminus W_3$ in B and an embedding $f_3 : A \cup N_2 \cup N_3 \to U \cup V$ such that $f_3|A \cup N_2 = f_2$ such that $W_3 \cap A \subset \operatorname{int} N_2$, which implies that

$$A \setminus \operatorname{int} N_2 \subset A \setminus W_3 \subset (A \cup N_2) \setminus W_3 \subset \operatorname{int} N_3.$$

Then, $N_2 \cup N_3$ is a neighborhood of A in B. Thus, f extends to an embedding of a neighborhood. Therefore, $U \cup V$ has property $(\mathcal{E}_f)$. □

The following is the key lemma of this section:

Lemma 6.3.2 *Let X be an AE (resp. an ANE) and let $f : A \to X$ be an embedding of a closed subset of a finite-dimensional compactum B with $2 \dim B + 2 \leqslant n$. Then, f extends to an embedding $g : B \to X \times \mathbf{I}^n$ (resp. $g : N \to X \times \mathbf{I}^n$ of some neighborhood N of A in B), where X is identified with $X \times \{0\} \subset X \times \mathbf{I}^n$. When f extends to a map $\tilde{f} : B \to X$, for each open cover $\mathcal{U}$ of $X \times \mathbf{I}^n$, there exists an embedding $g : B \to X \times \mathbf{I}^n$ such that $g \simeq_{\mathcal{U}} \tilde{f}$ rel. A.*

Proof Since X is an AE (resp. an ANE), f extends to a map $\tilde{f} : B \to X$ (resp. $\tilde{f} : N \to X$ of some closed neighborhood N of A in B). Hence, it suffices to prove the second half.

Let $\mathcal{U} \in \operatorname{cov}(X \times \mathbf{I}^n)$. Since $\tilde{f}(B)$ is compact, we have $0 < r \leqslant 1$ such that for each $y \in \tilde{f}(B)$, $\{y\} \times [0, r]^n$ is contained in some member of $\mathcal{U}$. Since $2 \dim B + 1 \leqslant n - 1$, there is an embedding $h : B \to \mathbf{I}^{n-1}$ by the Embedding Theorem 1.12.14. Take a map $k : B \to [0, r]$ such that $A = k^{-1}(0)$ and define a map $g : B \to X \times \mathbf{I}^n$ $(= X \times \mathbf{I} \times \mathbf{I}^{n-1})$ as follows:

$$g(x) = (\tilde{f}(x), k(x), k(x)h(x)) \text{ for each } x \in B.$$

Then, $g|A = f$. It is easy to see that g is injective, so an embedding. Since $g(x) \in \{\tilde{f}(x)\} \times [0, r]^n$ for each $x \in B$, it follows that $g \simeq_{\mathcal{U}} \tilde{f}$ rel. A. □

Remark 6.4 In the above, it suffices to assume that X is an AE(m) (resp. an ANE(m)) and $\dim B \leqslant m$.

Proposition 6.3.3 *Let $X_1 \subset X_2 \subset X_3 \subset \cdots$ be a closed tower of spaces. Suppose that there are embeddings $k_i : X_i \times \mathbf{I} \to X_{i+1}$, $i \in \mathbb{N}$, such that $k_i(x, 0) = x$ for each $x \in X_i$, and each X_i is LC^{i-1} (equiv. an ANE(i)). Then, every embedding $f : A \to X_m$ of a closed subset of a compactum B with $\dim B \leqslant m$ extends to an embedding $g : N \to X_{m+n}$ from some neighborhood N of A in B, where $n = 2\dim B + 2$. If X_m is $(m-1)$-connected, then f is extends to an embedding $g : B \to X_{m+n}$.*

Proof Note that X_m is an ANE(m) [GAGT, Theorem 6.12.3], and if X_m is $(m-1)$-connected then X_m is AE(m) [GAGT, Theorem 6.12.5]. There exists an embedding $\varphi : X_m \times \mathbf{I}^n \to X_{m+n}$ such that $\varphi(x, \mathbf{0}) = x$ for each $x \in X_m$. In fact, φ can be defined as follows:

$$\varphi(x, t_1, \ldots, t_n) = k_{m+n-1}(\cdots(k_{m+1}(k_m(x, t_1), t_2), \ldots, t_n).$$

Applying Lemma 6.3.2, we can obtain an embedding $g' : N \to X_m \times \mathbf{I}^n$ from some neighborhood N of A in B. Then, $g = \varphi g' : N \to X_{m+n}$ is the desired embedding. □

Remark 6.5 By Proposition 6.3.3, $X \times \mathbf{I}^n$ in Lemma 6.3.2 can be replaced with $X \times \mathbb{R}^n$, $X \times \Delta^n$, $X \times \mathbf{B}^n$ or $X \times \mathbf{S}^n$.

Since every compactum can be embedded into $\boldsymbol{Q}$ (Theorem 1.3.5) and $\mathbf{I} \times \boldsymbol{Q} \approx \boldsymbol{Q}$, replacing $\mathbf{I}^{n-1}$ with $\boldsymbol{Q}$, the above proof can be performed without dimensional restriction. Thus, we have the following infinite-dimensional version of Lemma 6.3.2:

Lemma 6.3.4 *Let X be an AE (resp. an ANE) and let $f : A \to X$ be an embedding of a closed subset of a compactum B. Then, f extends to an embedding $g : B \to X \times \boldsymbol{Q}$ (resp. $g : N \to X \times \boldsymbol{Q}$ of some neighborhood N of A in B), where X is identified with $X \times \{0\} \subset X \times \boldsymbol{Q}$. When f extends to a map $\tilde{f} : B \to X$, for each open cover $\mathcal{U}$ of $X \times \boldsymbol{Q}$, there exists an embedding $g : B \to X \times \boldsymbol{Q}$ such that $g \simeq_{\mathcal{U}} \tilde{f}$ rel. A.* □

By the same proof of [GAGT, Theorem 5.8.10] (Theorem 1.12.15(2) on 65), the following infinite-dimensional version can be proved:

Theorem 6.3.5 *For every locally compact separable metrizable space X, $\mathrm{Emb}(X, \boldsymbol{Q} \times (0, 1])$ is dense in the space $\mathrm{C}^P(X, \boldsymbol{Q} \times (0, 1])$ with the limitation topology, that is, for each open cover $\mathcal{U}$ of $\boldsymbol{Q} \times (0, 1]$, every proper map $f : X \to \boldsymbol{Q} \times (0, 1]$ is $\mathcal{U}$-close to a closed embedding $h : X \to \boldsymbol{Q} \times (0, 1]$.*

Indeed, the General Position Lemma 1.8.10 is valid for $\boldsymbol{Q} \times (0, 1]$, that is, for every countable open collection $\{U_i \mid i \in \mathbb{N}\}$ in $\boldsymbol{Q} \times (0, 1]$, there is a countable affinely independent set $\{x_i \mid i \in \mathbb{N}\} \subset \boldsymbol{Q} \times (0, 1]$ ($\subset \mathbb{R}^{\mathbb{N}} \times \mathbb{R}$) such that $x_i \in U_i$ for each $i \in \mathbb{N}$. Hence, ignoring the order of open covers and dimension of the nerve, we can perform the proof of [GAGT, Theorem 5.8.10] for $\boldsymbol{Q} \times (0, 1]$.

Applying Theorem 6.3.5, we can prove the following analog of Lemma 6.3.4:

Lemma 6.3.6 *Every embedding $f : A \to \boldsymbol{Q}$ of a closed subset of a compactum B extends to an embedding $g : B \to \boldsymbol{Q} \times \mathbf{I}$, where $\boldsymbol{Q}$ is identified with $\boldsymbol{Q} \times \{0\} \subset \boldsymbol{Q} \times \mathbf{I}$.*

Proof Since $\boldsymbol{Q}$ is an AE, f extends to a map $\tilde{f} : B \to X$. Take a map $k : B \to \mathbf{I}$ with $k^{-1}(0) = A$ and define a map $f' : B \to \boldsymbol{Q} \times \mathbf{I}$ as follows:

$$f'(x) = (\tilde{f}(x), k(x)) \quad \text{for each} \quad x \in B.$$

Then, the restriction $f'|B \setminus A : B \setminus A \to \boldsymbol{Q} \times (0, 1]$ is a proper map. By Lemma 2.1.16, we have an open cover $\mathcal{U}$ of $\boldsymbol{Q} \times (0, 1]$ which is fitting in $\boldsymbol{Q} \times \mathbf{I}$. We apply Theorem 6.3.5 to obtain a closed embedding $g' : B \setminus A \to \boldsymbol{Q} \times (0, 1]$ $\mathcal{U}$-close to $f'|B \setminus A$. Since $\mathcal{U}$ is fitting in $\boldsymbol{Q} \times \mathbf{I}$, g' is extended to a map $g : B \to \boldsymbol{Q} \times \mathbf{I}$ by $g|A = f'|A = f$. Then, $g|B \setminus A = g'$ and $g|A = f$ are injective and

$$g(B \setminus A) \cap g(A) \subset (\boldsymbol{Q} \times (0, 1]) \cap (\boldsymbol{Q} \times \{0\}) = \emptyset,$$

hence g is injective, so g is an embedding because B is compact. □

Remark 6.6 The analog of Lemma 6.3.6 can be similarly proved, that is, we can apply Theorem 1.12.15 to extend an embedding $f : A \to \mathbf{I}^m$ to an embedding $g : B \to \mathbf{I}^m \times \mathbf{I} = \mathbf{I}^{m+1}$, where $m = 2 \dim B + 1$.

Due to Proposition 6.1.2(3), applying Lemma 6.3.2 (Remark 6.5) or Lemma 6.3.4 to the case where X is replaced by $X \times \mathbb{R}^m$ or $X \times \boldsymbol{Q}^m$, we can easily prove the following:

Proposition 6.3.7 *Let X be an ANE. Then, the product spaces $X \times \mathbb{R}^\infty$ and $X \times \boldsymbol{Q}^\infty$ have properties ($\mathcal{E}_f$) and ($\mathcal{E}$), respectively. In particular, the spaces $\mathbb{R}^\infty$ and $\boldsymbol{Q}^\infty$ have properties ($\mathcal{E}_f$) and ($\mathcal{E}$), respectively, where every embedding from a closed subset A of any finite-dimensional compactum B into $\mathbb{R}^\infty$ can be extended to an embedding of B into $\mathbb{R}^\infty$ and every embedding from a closed subset A of any compactum B into $\boldsymbol{Q}^\infty$ can be extended to an embedding of B into $\boldsymbol{Q}^\infty$.* □

By virtue of Proposition 6.3.3, the following can be similarly proved:

Proposition 6.3.8 *Suppose that X is one of $\mathbf{I}^\infty$, Δ^∞, $\mathbf{B}^\infty$, and $\mathbf{S}^\infty$. Then, each embedding from a closed subset A of any finite-dimensional compactum B into X can be extended to an embedding from B.* □

The following can be obtained from Lemma 6.3.6:

Proposition 6.3.9 *Each embedding from a closed subset A of any compactum B into $\boldsymbol{Q} \times \mathbf{I}^\infty$ can be extended to an embedding from B.* □

Since $\mathbb{R}^\infty$-manifolds and $\boldsymbol{Q}^\infty$-manifolds are paracompact and they have properties $(\mathcal{E}_f)$ and $(\mathcal{E})$ locally by Proposition 6.3.7 above, the following proposition follows from Lemma 6.3.1 and Michael's Theorem 1.3.20:

Proposition 6.3.10 *Every $\mathbb{R}^\infty$-manifold has property $(\mathcal{E}_f)$ and every $\boldsymbol{Q}^\infty$-manifold has property $(\mathcal{E})$.* □

Here, we introduce the following embedding approximation properties:

$(\mathcal{A}_f)$ Let A be a closed set in a finite-dimensional compactum B and $f : B \to X$ be a map such that $f|A$ is an embedding. Then, for each open cover $\mathcal{U}$ of X, there exists an embedding $g : B \to X$ such that $g|A = f|A$ and $g \simeq_{\mathcal{U}} f$.

$(\mathcal{A})$ Let A be a closed set in a compactum B and $f : B \to X$ be a map such that $f|A$ is an embedding. Then, for each open cover $\mathcal{U}$ of X, there exists an embedding $g : B \to X$ such that $g|A = f|A$ and $g \simeq_{\mathcal{U}} f$.

As in Proposition 6.3.7, we have also the following:

Proposition 6.3.11 *The spaces $\mathbb{R}^\infty$ and $\boldsymbol{Q}^\infty$ have also properties $(\mathcal{A}_f)$ and $(\mathcal{A})$, respectively.* □

Remark 6.7 The manifold version of the above also holds just like Proposition 6.3.10. That would follow from Michael's Theorem 1.3.20 if it could be proved that properties $(\mathcal{A}_f)$ and $(\mathcal{A})$ are G-hereditary. However, unlike Lemma 6.3.1, it is difficult to prove (G-2). Since open sets in $\mathbb{R}^\infty$ and $\boldsymbol{Q}^\infty$ have these properties, that will be obtained as a corollary of the Open Embedding Theorem for these manifolds.

6.4 Characterization of $\mathbb{R}^\infty$- and $\boldsymbol{Q}^\infty$-Manifolds

Recall that we assume the **separability** for $\mathbb{R}^\infty$-manifolds and $\boldsymbol{Q}^\infty$-manifolds (p. 81), equivalently they have only at most countable many components. In this section, it is proved that $\mathbb{R}^\infty$-manifolds are characterized as the direct limits of towers of finite-dimensional compacta having property $(\mathcal{E}_f)$ and $\boldsymbol{Q}^\infty$-manifolds are characterized as the direct limits of towers of compacta having property $(\mathcal{E})$. It is also proved that $\mathbb{R}^\infty$-manifolds and $\boldsymbol{Q}^\infty$-manifolds can be embedded in $\mathbb{R}^\infty$ and $\boldsymbol{Q}^\infty$ as open sets, respectively. First of all, we will show the following:

Theorem 6.4.1 *The direct limit X of a tower of finite-dimensional compacta can be embedded into $\mathbb{R}^\infty$ as an open subset if X has property $(\mathcal{E}_f)$.*

Proof We write $X = \varinjlim X_n$, where $X_1 \subset X_2 \subset X_3 \subset \cdots$ is a tower of finite-dimensional compacta. Let $n(1) = 1$ and $m(1) = 2 \dim X_{n(1)} + 1$. Then, there is an embedding $f_1 : X_{n(1)} \to \mathbb{R}^{m(1)}$. Using property $(\mathcal{E}_f)$ of X, we have an open neighborhood U_1 of $f_1(X_{n(1)})$ in $\mathbb{R}^{m(1)}$ and an embedding $g_1 : \operatorname{cl} U_1 \to X$ such that $\operatorname{cl} U_1$ is compact and $g_1|f_1(X_{n(1)}) = f_1^{-1}$, that is, $g_1 f_1 = \mathrm{id}$. By compactness, we can find $n(2) > n(1)$ such that $g_1(\mathrm{cl}\, U_1) \subset X_{n(2)}$. Thus, we have an embedding $g_1 : \mathrm{cl}\, U_1 \to X_{n(2)}$ such that $g_1 f_1$ is the inclusion of $X_{n(1)}$ into $X_{n(2)}$. Let

$$m(2) = m(1) + 2 \dim X_{n(2)} + 2.$$

Applying Lemma 6.3.2, we can extend the embedding $g_1^{-1} : g_1(\mathrm{cl}\, U_1) \to \mathbb{R}^{m(2)}$ to an embedding $f_2 : X_{n(2)} \to \mathbb{R}^{m(2)}$.

As above, using property $(\mathcal{E}_f)$ of X, we have an open neighborhood U_2 of $f_2(X_{n(2)})$ in $\mathbb{R}^{m(2)}$ and an embedding $g_2 : \mathrm{cl}\, U_2 \to X_{n(3)}$ such that $\mathrm{cl}\, U_2$ is compact, $n(3) > n(2)$ and $g_2 f_2$ is the inclusion of $X_{n(2)}$ into $X_{n(3)}$. Let

$$m(3) = m(2) + 2 \dim X_{n(3)} + 2.$$

Applying Lemma 6.3.2 again, we can extend the embedding $g_2^{-1} : g_2(\mathrm{cl}\, U_2) \to \mathbb{R}^{m(3)}$ to an embedding $f_3 : X_{n(3)} \to \mathbb{R}^{m(3)}$.

Iterating these constructions, we have increasing sequences $n(1) < n(2) < \cdots$, $m(1) < m(2) < \cdots$, open sets U_i in $\mathbb{R}^{m(i)}$, $i \in \mathbb{N}$, and the following commutative diagram of embeddings:

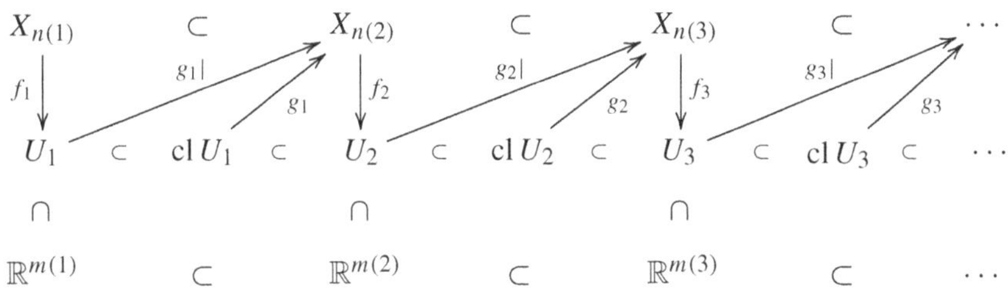

Due to Proposition 6.1.1, f_i, $i \in \mathbb{N}$, induce the homeomorphism $f_\infty : \varinjlim X_{n(i)} \to \varinjlim U_i$, where $\varinjlim X_{n(i)} = \varinjlim X_n = X$ and $\varinjlim U_i$ is an open subspace of $\varinjlim \mathbb{R}^{m(i)} = \varinjlim \mathbb{R}^n = \mathbb{R}^\infty$ (Proposition 6.1.2(2)). Consequently, $f_\infty : X \to \mathbb{R}^\infty$ is an open embedding. □

Remark 6.8 Replacing $\mathbb{R}$ with $\mathbf{I}$ in the above, it follows that if the direct limit X of finite-dimensional compacta has property $(\mathcal{E}_f)$, then X can be embedded into $\mathbf{I}^\infty = \varinjlim \mathbf{I}^n$.

The following theorem can be similarly proved:

Theorem 6.4.2 *The direct limit X of a tower of finite-dimensional compacta is homeomorphic to $\mathbb{R}^\infty$ if each embedding from a closed subset A of any finite-dimensional compactum B into X can be extended to an embedding from B.*

Proof Now, using the property stronger than $(\mathcal{E}_f)$, we can take $U_i = \mathbb{R}^{m(i)}$ in the proof of Theorem 6.4.1, that is, we have increasing sequences $n(1) < n(2) < \cdots$, $m(1) < m(2) < \cdots$, and the following commutative diagram of embeddings:

$$
\begin{array}{ccccccc}
X_{n(1)} & \subset & X_{n(2)} & \subset & X_{n(3)} & \subset & \cdots \\
{\scriptstyle f_1}\downarrow & \overset{g_1}{\nearrow} & {\scriptstyle f_2}\downarrow & \overset{g_2}{\nearrow} & {\scriptstyle f_3}\downarrow & \overset{g_3}{\nearrow} & \\
\mathbb{R}^{m(1)} & \subset & \mathbb{R}^{m(2)} & \subset & \mathbb{R}^{m(3)} & \subset & \cdots
\end{array}
$$

Hence, we have $X \approx \mathbb{R}^\infty$ by Proposition 6.1.1. □

By Proposition 6.3.8, we have the following:

Corollary 6.4.3 *The spaces* $\mathbf{I}^\infty$, Δ^∞, $\mathbf{B}^\infty$, *and* $\mathbf{S}^\infty$ *are homeomorphic to* $\mathbb{R}^\infty$. □

Using Lemma 6.3.4 instead of Lemma 6.3.2, we can prove the following theorem in the same way as above:

Theorem 6.4.4 *The direct limit X of a tower of compacta can be embedded into $\boldsymbol{Q}^\infty$ as an open subset if X has property* $(\mathcal{E})$. □

Similarly, we can obtain the following:

Theorem 6.4.5 *The direct limit X of a tower of compacta is homeomorphic to $\boldsymbol{Q}^\infty$ if each embedding from a closed subset A of any compactum B into X can be extended to an embedding from B.* □

Since $\mathbf{I}^\infty \approx \mathbb{R}^\infty$, the following is a combination of Theorem 6.4.5 and Proposition 6.3.9:

Corollary 6.4.6 *The product spaces $\boldsymbol{Q} \times \mathbf{I}^\infty$ and $\boldsymbol{Q} \times \mathbb{R}^\infty$ are homeomorphic to $\boldsymbol{Q}^\infty$.* □

Now, we show that $\mathbb{R}^\infty$-manifolds and $\boldsymbol{Q}^\infty$-manifolds can be written as the direct limits like the model $\mathbb{R}^\infty$ and $\boldsymbol{Q}^\infty$, that is:

Proposition 6.4.7 *Every $\mathbb{R}^\infty$-manifold is the direct limit of a tower of finite-dimensional compacta and every $\boldsymbol{Q}^\infty$-manifold is the direct limit of a tower of compacta.*

Proof Because of similarity, we only show the proposition for an $\mathbb{R}^\infty$-manifold M. We consider the property of closed sets in M being the direct limit of a tower of finite-dimensional compacta. Since $\mathbb{R}^\infty$ has this property, M has this property locally by Proposition 6.1.2(1). We will show that this property is F-hereditary. Then, the result follows from Michael's Theorem 1.3.21. Proposition 6.1.2(1) means that this satisfies (F-1).

To see (F-2), let A and B be closed sets in M which are the direct limits of towers of finite-dimensional compacta, $A_1 \subset A_2 \subset A_3 \subset \cdots$ and $B_1 \subset B_2 \subset B_3 \subset \cdots$. Then, each $A_n \cup B_n$ is a compact set in M and it is metrizable (Proposition 1.3.6(2)) and finite-dimensional (Theorem 1.12.8 or 1.12.9). Thus, we have the following tower of finite-dimensional compacta:

$$A_1 \cup B_1 \subset A_2 \cup B_2 \subset A_3 \cup B_3 \subset \cdots,$$

and $A \cup B = \bigcup_{n\in\mathbb{N}}(A_n \cup B_n)$. For $C \subset A \cup B$, if C is closed in $A \cup B$, then

$$C \cap (A_n \cup B_n) = (C \cap A_n) \cup (C \cap B_n)$$

is closed in $A_n \cup B_n$ for each $n \in \mathbb{N}$. Conversely, suppose that $C \cap (A_n \cup B_n)$ is closed in $A_n \cup B_n$ for each $n \in \mathbb{N}$. Since each $(C \cap A) \cap A_n = C \cap A_n$ is closed in A_n, $C \cap A$ is closed in A, hence it is closed in M. Similarly, $C \cap B$ is closed in M. Hence, $C = (C \cap A) \cup (C \cap B)$ is closed in M, so it is closed in $A \cup B$. Therefore, $A \cup B = \varinjlim(A_n \cup B_n)$.

For (F-3), recall that we assume that M is separable. Then, a discrete collection of subsets of M has at most countable members. Let $\{A_i \mid i \in \mathbb{N}\}$ be discrete collection closed sets in M such that $A_i = \varinjlim A_{i,n}$, where $A_{i,1} \subset A_{i,2} \subset A_{i,3} \subset \cdots$ is a tower of finite-dimensional compacta. For each $n \in \mathbb{N}$, let $B_n = \bigcup_{i=1}^n A_{i,n+1-i}$, that is,

$$B_1 = A_{1,1},\ B_2 = A_{1,2} \cup A_{2,1},\ B_3 = A_{1,3} \cup A_{2,2} \cup A_{3,1},\ \cdots.$$

Then, each B_n is compact metrizable (Proposition 1.3.6(2)) and finite-dimensional (Theorem 1.12.8 or 1.12.9). Namely, $B_1 \subset B_2 \subset B_3 \subset \cdots$ is a tower of finite-dimensional compacta. It is easy to see that $\bigcup_{n\in\mathbb{N}} A_n = \varinjlim B_i$. □

Summing up Propositions 6.3.10 and 6.4.7 and Theorems 6.4.1 and 6.4.4, we can obtain the following characterizations of $\mathbb{R}^\infty$-manifolds and $\boldsymbol{Q}^\infty$-manifolds and the Open Embedding Theorem for these manifolds:

Theorem 6.4.8 (Characterization) *A space X is an $\mathbb{R}^\infty$-manifold if and only if X is the direct limit of a tower of finite-dimensional compacta and has property $(\mathcal{E}_f)$. A space X is an $\boldsymbol{Q}^\infty$-manifold if and only if X is the direct limit of a tower of compacta and has property $(\mathcal{E})$.* □

Theorem 6.4.9 (Open Embedding) *Every $\mathbb{R}^\infty$-manifold can be embedded into $\mathbb{R}^\infty$ as an open set and every $\boldsymbol{Q}^\infty$-manifold can be embedded into $\boldsymbol{Q}^\infty$ as an open set.* □

As mentioned at the end of the previous section, the Open Embedding Theorem 6.4.9 yields the following manifold version of Proposition 6.3.11, which will be used in the next section:

Theorem 6.4.10 *Every $\mathbb{R}^\infty$-manifold has property $(\mathcal{A}_f)$ and every $\boldsymbol{Q}^\infty$-manifold has property $(\mathcal{A})$.* □

We also apply the Open Embedding Theorem 6.4.9 to prove the following theorem:

Theorem 6.4.11 *Let M be an $\mathbb{R}^\infty$-manifold or a $\boldsymbol{Q}^\infty$-manifold. Then, M has the homotopy type of a countable (locally finite) simplicial complex.*

Proof Due to the Whitehead–Milnor Theorem 1.10.4, it suffices to show that M is homotopy dominated by a simplicial complex. By the Open Embedding

Theorem 6.4.9, it can be assumed that M is an open set in $\mathbb{R}^\infty$ or in $\boldsymbol{Q}^\infty$. Then, M has an open cover $\mathcal{U}$ consisting of convex open sets in $\mathbb{R}^\infty$ or in $\boldsymbol{Q}^\infty$.[3] Because of the paracompactness of M, $\mathcal{U}$ has an open star-refinement $\mathcal{V}$ (Theorem 1.3.11). Let $N(\mathcal{V})$ be the nerve of $\mathcal{V}$. Again because of the paracompactness of M, we have a canonical map $\varphi : M \to |N(\mathcal{V})|$ (Theorem 1.11.4). Let $\psi_0 : \mathcal{V} \to M$ be a choice function, that is, $\psi_0(V) \in V$ for each $V \in \mathcal{V}$. For each simplex $\sigma \in N(\mathcal{V})$, let $\sigma^{(0)} = \{V_0, \ldots, V_n\} \subset \mathcal{V} = N(\mathcal{V})^{(0)}$. Then, $\bigcap_{i=0}^n V_i \neq \emptyset$, which implies that $\psi_0(\sigma^{(0)}) \subset \bigcup_{i=0}^n V_i \subset \mathrm{st}(V_0, \mathcal{V})$. Since $\mathrm{st}\,\mathcal{V} \prec \mathcal{U}$, $\psi_0(\sigma^{(0)})$ is contained in some $U_\sigma \in \mathcal{U}$ (i.e., ψ_0 is a partial $\mathcal{U}$-realization of $N(\mathcal{V})$), and then, the convex hull $\langle\psi_0(\sigma^{(0)})\rangle$ is also contained in U_σ because of convexity of U_σ. Therefore, ψ_0 has an extension $\psi : |N(\mathcal{V})| \to M$ which is affine on each simplex of $N(\mathcal{V})$. Note that $\{\psi(\sigma) \mid \sigma \in N(\mathcal{V})\} \prec \mathcal{U}$ (i.e., ψ is a full $\mathcal{U}$-realization of $N(\mathcal{V})$).

To see that $\psi\varphi \simeq \mathrm{id}$, let $x \in M$. Since φ is a canonical map, every vertex V of the carrier $c_{N(\mathcal{V})}(\varphi(x))$ is a member of $\mathcal{V}[x]$. Due to the definition of ψ, $\psi(c_{N(\mathcal{V})}(\varphi(x)))$ is contained in some $U_x \in \mathcal{U}$, where $\mathrm{st}(V_x, \mathcal{V}) \subset U_x$ for some $V_x \in c_{N(\mathcal{V})}(\varphi(x))^{(0)}$. Since $x \in V_x$, it follows that $x, \psi(\varphi(x)) \in U_x$. Therefore, $\psi\varphi$ is $\mathcal{U}$-close to id, so $\psi\varphi \simeq_{\mathcal{U}} \mathrm{id}$ by the straight line homotopy because every member of $\mathcal{U}$ is convex. □

Applying the Characterization Theorem 6.4.8, we can prove the following ANE Factor Theorem:

Theorem 6.4.12 (ANE Factor) *Let X be an ANE. If X is the direct limit of a tower of finite-dimensional compacta then $X \times \mathbb{R}^\infty$ is an $\mathbb{R}^\infty$-manifold; and if X is the direct limit of a tower of compacta then $X \times \boldsymbol{Q}^\infty$ is an $\boldsymbol{Q}^\infty$-manifold.*

Proof Because of similarity, we prove only the $\mathbb{R}^\infty$-manifold case. By Proposition 6.3.7, $X \times \mathbb{R}^\infty$ has property $(\mathcal{E}_f)$. Let $X = \varinjlim X_n$, where $X_1 \subset X_2 \subset \cdots$ is the tower of finite-dimensional compacta. Since $\mathbb{R}^\infty = \varinjlim [-n, n]^n$ and $\dim(X_n \times [-n, n]^n) \leqslant n \dim X_n < \infty$ by the Product Theorem of Dimension (Theorem 1.12.12), $X \times \mathbb{R}^\infty = \varinjlim (X_n \times [-n, n]^n)$ is the limit of finite-dimensional compacta by Proposition 6.1.3. Then, the result follows from the Characterization Theorem 6.4.8. □

For a countable simplicial complex K, $|K|$ is an ANE (Theorem 1.13.4) and $|K| = \varinjlim |K_n|$, where $K_1 \subset K_2 \subset \cdots$ is a tower of finite subcomplexes of K. Hence, the following is a corollary of the above:

Corollary 6.4.13 *For a countable simplicial complex K, $|K| \times \mathbb{R}^\infty$ is an $\mathbb{R}^\infty$-manifold and $|K| \times \boldsymbol{Q}^\infty$ is a $\boldsymbol{Q}^\infty$-manifold.* □

---

[3] Recall that $\mathbb{R}^\infty$ is a topological linear space (Theorem 6.1.4) and $\boldsymbol{Q}^\infty$ is a convex set with the following continuous operation:

$$\boldsymbol{Q}^\infty \times \boldsymbol{Q}^\infty \times \mathbf{I} \ni (x, y, t) \mapsto (1 - t)x + ty \in \boldsymbol{Q}^\infty.$$

Or, as we saw in Sect. 6.2, $\boldsymbol{Q}^\infty$ can be regarded as a locally convex topological linear space.

6.5 Classification of $\mathbb{R}^\infty$- and Q^∞-Manifolds

In this section, we prove the following Classification Theorem for $\mathbb{R}^\infty$-manifolds and Q^∞-manifolds:

Theorem 6.5.1 (CLASSIFICATION) *Any two $\mathbb{R}^\infty$-manifolds or any two Q^∞-manifolds are homeomorphic to each other if and only if they have the same homotopy type (equivalently, the same weak homotopy type).*

At the same time, the following is also proved:

Theorem 6.5.2 (HOMEOMORPHISM APPROXIMATION) *Let M and N be both $\mathbb{R}^\infty$-manifolds or Q^∞-manifolds. Then, every fine homotopy equivalence $f : M \to N$ is a near-homeomorphism.*

The Classification Theorem 6.5.1 can be proved by the same method as the proof of the Homeomorphism Approximation Theorem 6.5.2. In practice, under the assumption that f is a homotopy equivalence, ignoring the covering control in the following proof, we get the proof of Theorem 6.5.1.

Proof (Theorem 6.5.2) Because of similarity, we prove only the $\mathbb{R}^\infty$-manifold case.

Write $M = \varinjlim X_n$ and $N = \varinjlim Y_n$, where $X_1 \subset X_2 \subset X_3 \subset \cdots$ and $Y_1 \subset Y_2 \subset Y_3 \subset \cdots$ are towers of finite-dimensional compacta. For each open cover $\mathcal{U}$ of N, we will construct a homeomorphism which is $\mathcal{U}$-close to f. Because of the paracompactness of N, we have the following sequence of open covers of N (Theorem 1.3.11):

$$\mathcal{U} \overset{*}{\succ} \mathcal{U}_1 \overset{*}{\succ} \mathcal{U}_2 \overset{*}{\succ} \mathcal{U}_3 \overset{*}{\succ} \cdots .$$

Then, we inductively define open covers $\mathcal{V}_1, \mathcal{V}_2, \mathcal{V}_3, \ldots$ of N as follows:

$$\mathcal{V}_1 = \mathcal{U}_1 \text{ and } \mathcal{V}_{n+1} = \operatorname{st}^2(\mathcal{V}_n, \mathcal{U}_{n+1}), \ n \in \mathbb{N}.$$

Note that $\mathcal{V}_n \prec \mathcal{U}$ for every $n \in \mathbb{N}$ because

$$\begin{aligned}\operatorname{st}(\mathcal{V}_n, \operatorname{st}\mathcal{U}_{n+1}) \prec \operatorname{st}(\mathcal{V}_n, \mathcal{U}_n) &= \operatorname{st}(\operatorname{st}^2(\mathcal{V}_{n-1}, \mathcal{U}_n), \mathcal{U}_n) = \operatorname{st}(\mathcal{V}_{n-1}, \operatorname{st}\mathcal{U}_n)\\ &\prec \cdots \prec \operatorname{st}(\mathcal{V}_1, \operatorname{st}\mathcal{U}_2) \prec \operatorname{st}(\mathcal{U}_1, \mathcal{U}_1) = \operatorname{st}\mathcal{U}_1 \prec \mathcal{U}.\end{aligned}$$

Let $n(1) = 1$. Since N has property $(\mathcal{A}_f)$ (Theorem 6.4.10), we can take an embedding $f_1 : X_{n(1)} \to N$ such that $f_1 \simeq_{\mathcal{V}_1} f|X_{n(1)}$. By Proposition 6.1.2(3), we can find $m(1) \in \mathbb{N}$ such that $f_1(X_{n(1)}) \subset Y_{m(1)}$.

Since f is a fine homotopy equivalence, f has a $\mathcal{U}_3$-homotopy inverse $g : N \to M$. Then, $g(\mathcal{V}_1) \prec f^{-1}(\operatorname{st}(\mathcal{V}_1, \mathcal{U}_3))$ because fg is $\mathcal{U}_3$-close to id_N. Hence, it follows that

$$gf_1 \simeq_{f^{-1}(\operatorname{st}(\mathcal{V}_1, \mathcal{U}_3))} gf|X_{n(1)} \simeq_{f^{-1}(\mathcal{U}_3)} \mathrm{id},$$

so $g|f_1(X_{n(1)}) \simeq_{f^{-1}(\mathrm{st}^2(\mathcal{V}_1,\mathcal{U}_3))} f_1^{-1}$. Applying the Homotopy Extension Theorem 1.13.11, we can extend f_1^{-1} to a map $g' : Y_{m(1)} \to M$ such that $g' \simeq_{f^{-1}(\mathrm{st}^2(\mathcal{V}_1,\mathcal{U}_3))} g|Y_{m(1)}$. Since M has property $(\mathcal{A}_f)$ (Theorem 6.4.10), we can take an embedding $g_1 : Y_{m(1)} \to M$ such that $g_1|f_1(X_{n(1)}) = g'|f_1(X_{n(1)}) = f_1^{-1}|f_1(X_{n(1)})$ and $g_1 \simeq_{f^{-1}(\mathcal{U}_3)} g'$. Then,

$$fg_1 \simeq_{\mathcal{U}_3} fg' \simeq_{\mathrm{st}^2(\mathcal{V}_1,\mathcal{U}_3)} fg|Y_{m(1)} \simeq_{\mathcal{U}_3} \mathrm{id} \text{ and}$$
$$\mathrm{st}^3(\mathcal{V}_1,\mathcal{U}_3) = \mathrm{st}(\mathcal{V}_1, \mathrm{st}\,\mathcal{U}_3) \prec \mathrm{st}(\mathcal{V}_1,\mathcal{U}_2),$$

which implies that $g_1^{-1} \simeq_{\mathrm{st}(\mathcal{V}_1,\mathcal{U}_2)} f|g_1(Y_{m(1)})$.

Choose $n(2) > n(1)$ so that $g_2(Y_{m(1)}) \subset X_{n(2)}$ (Proposition 6.1.2(3)). Applying the Homotopy Extension Theorem 1.13.11, we can extend g_1^{-1} to a map $f' : X_{n(2)} \to N$ such that $f' \simeq_{\mathrm{st}(\mathcal{V}_1,\mathcal{U}_2)} f|X_{n(2)}$, and using property $(\mathcal{A}_f)$ of N, we obtain an embedding $f_2 : X_{n(2)} \to N$ such that $f_2|g_1(Y_{m(1)}) = g_1^{-1}$ and $f_2 \simeq_{\mathcal{U}_2} f'$. Since $\mathrm{st}^2(\mathcal{V}_1,\mathcal{U}_2) = \mathcal{V}_2$, it follows that $f_2 \simeq_{\mathcal{V}_2} f|X_{n(2)}$. Then, we have $m(2) > m(1)$ such that $f_2(X_{n(2)}) \subset Y_{m(2)}$ (Proposition 6.1.2(3)).

By the same construction as g_1, but now, using a $\mathcal{U}_4$-homotopy inverse of f, we can take an embedding $g_2 : Y_{m(2)} \to M$ such that $g_2|f_2(X_{n(2)}) = f_2^{-1}|f_2(X_{n(2)})$ and $g_2^{-1} \simeq_{\mathrm{st}(\mathcal{V}_2,\mathcal{U}_3)} f|g_2(Y_{m(2)})$.

Thus, inductively we can obtain the following commutative diagram of embeddings:

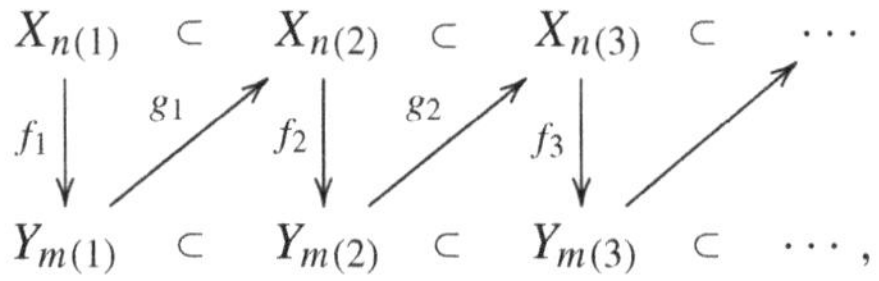

where $f_i \simeq_{\mathcal{V}_i} f|X_{n(i)}$ and $g_i^{-1} \simeq_{\mathrm{st}(\mathcal{V}_i,\mathcal{U}_{i+1})} f|g_i(Y_{m(i)})$ for each $i \in \mathbb{N}$. Then, f_i, $i \in \mathbb{N}$, induce a homeomorphism $f_\infty : M \to N$ (Proposition 6.1.1). Since $f_\infty|X_{n(i)} = f_i$ for each $i \in \mathbb{N}$, it follows that f_∞ is $\mathcal{U}$-close to f. □

The following is a direct consequence of the Classification Theorem 6.5.1:

Corollary 6.5.3 *Every contractible $\mathbb{R}^\infty$-manifold is homeomorphic to $\mathbb{R}^\infty$ and every contractible $\boldsymbol{Q}^\infty$-manifold is homeomorphic to $\boldsymbol{Q}^\infty$.* □

Then, the following is obtained from Proposition 6.1.5, the ANE Factor Theorem 6.4.12, and the above corollary:

Corollary 6.5.4 *For every finite-dimensional locally compact AR X, $X \times \mathbb{R}^\infty \approx \mathbb{R}^\infty$; and for every locally compact AR X, $X \times \boldsymbol{Q}^\infty \approx \boldsymbol{Q}^\infty$.*[4] □

[4]The separability of X comes from the connectedness.

Remark 6.9 In the $\mathbb{R}^\infty$-case of Corollary 6.5.4, the finite-dimensionality cannot be replaced by the strong countable-dimensionality. In fact, the following is a strongly countable-dimensional compact AR:

$$X = (\mathbf{I} \times \{\mathbf{0}\}) \cup \bigcup_{n\in\mathbb{N}} \{1/n\} \times [0, 1/n]^n \subset \mathbf{I}^{\mathbb{N}}.$$

It follows from Proposition 6.1.2(3) that X cannot be the direct limit of a tower of finite-dimensional compacta. Hence, X never be homeomorphic to $\mathbb{R}^\infty$.

Combining Theorem 6.4.11, Corollary 6.4.13, and the Classification Theorem 6.5.1, we have the following:

Theorem 6.5.5 (TRIANGULATION) *Every $\mathbb{R}^\infty$-manifold is homeomorphic to $|K| \times \mathbb{R}^\infty$ for some countable locally finite simplicial complex K, and every $\boldsymbol{Q}^\infty$-manifold is homeomorphic to $|K| \times \boldsymbol{Q}^\infty$ for some countable locally finite simplicial complex K.* □

Applying Theorem 6.5.5, we show the following:

Proposition 6.5.6 *Let M be an $\mathbb{R}^\infty$-manifold or a $\boldsymbol{Q}^\infty$-manifold. Every compact set A in M is a Z-set in M, that is, for each $\mathcal{U} \in \mathrm{cov}(M)$, there is a map $f : M \to M \setminus A$ that is $\mathcal{U}$-close to* id.

Proof We show only the $\mathbb{R}^\infty$-manifold case. By Theorem 6.5.5 and Corollary 6.4.3, we may assume that $M = |K| \times \mathbf{I}^\infty$ for some locally finite simplicial complex. Then, it follows from Proposition 6.1.3 that $M = |K| \times \varinjlim \mathbf{I}^n = \varinjlim(|K| \times \mathbf{I}^n)$. Since A is compact, $A \subset |K| \times \mathbf{I}^n$ for some $n \in \mathbb{N}$. Let $p_{n+1} : M = |K| \times \mathbf{I}^\infty \to |K| \times \mathbf{I}^{n+1}$ be the projection. For each $\mathcal{U} \in \mathrm{cov}(M)$, sliding $(n + 1)$-th coordinate, we can construct a map

$$f : M \to p_{n+1}^{-1}(|K| \times \mathbf{I}^n \times (0, 1]) \subset M \setminus A$$

that is $\mathcal{U}$-close to id. Hence, A is a Z-set in M.

The $\boldsymbol{Q}^\infty$-manifold case can be proved by replacing $|K|$ with $|K| \times \boldsymbol{Q}$ in the above (cf. Corollary 6.4.6). □

Now, we can see the following relation between $\mathbb{R}^\infty$-manifolds and $\boldsymbol{Q}^\infty$-manifolds:

Theorem 6.5.7 *For every $\mathbb{R}^\infty$-manifold M, $M \times \boldsymbol{Q}$ is $\boldsymbol{Q}^\infty$-manifold. Conversely, for every $\boldsymbol{Q}^\infty$-manifold N, there exists an $\mathbb{R}^\infty$-manifold M such that $N \approx M \times \boldsymbol{Q}$.*

Proof By the Triangulation Theorem 6.5.5, every $\mathbb{R}^\infty$-manifold M is homeomorphic to $|K| \times \mathbb{R}^\infty$ for some countable locally finite simplicial complex K. Since $\boldsymbol{Q} \times \mathbb{R}^\infty \approx \boldsymbol{Q}^\infty$ by Corollary 6.4.6, $M \times \boldsymbol{Q} \approx |K| \times \boldsymbol{Q}^\infty$, which is a $\boldsymbol{Q}^\infty$-manifold by Corollary 6.4.13.

By the Triangulation Theorem 6.5.5, every $\boldsymbol{Q}^\infty$-manifold N is homeomorphic to $|K| \times \boldsymbol{Q}^\infty$ for some countable locally finite simplicial complex K. Then, $M = |K| \times \mathbb{R}^\infty$ is $\mathbb{R}^\infty$-manifold by Corollary 6.4.13. Since $\boldsymbol{Q} \times \mathbb{R}^\infty \approx \boldsymbol{Q}^\infty$ by Corollary 6.4.6, we have $M \times \boldsymbol{Q} \approx |K| \times \boldsymbol{Q}^\infty \approx N$. □

The following is a combination of Proposition 6.1.8, the ANE Factor Theorem 6.4.12, and the Homeomorphism Approximation Theorem 6.5.2:

Theorem 6.5.8 (Stability) *For every $\mathbb{R}^\infty$-manifold M, the projection $\mathrm{pr}_M : M \times \mathbb{R}^\infty \to M$ is a near-homeomorphism. For every $\boldsymbol{Q}^\infty$-manifold the projection $\mathrm{pr}_M : M \times \boldsymbol{Q}^\infty \to M$ is a near-homeomorphism.* □

Using Corollary 6.5.4, we can easily prove the following:

Theorem 6.5.9 *For any $\mathbb{R}^\infty$-manifold M and any finite-dimensional locally compact AR X, the projection $\mathrm{pr}_M : M \times X \to M$ is a near-homeomorphism. For any $\boldsymbol{Q}^\infty$-manifold and any locally compact AR X the projection $\mathrm{pr}_M : M \times X \to M$ is a near-homeomorphism.* □

6.6 Simplicial Approximations of PL Embeddings

It is said that a space X is **triangulable** if X is a topological polyhedron, that is, $X \approx |K|$ of some simplicial complex K, where K is called a **triangulation** of X and it is said that K **triangulates** X or X is **triangulated** by K (cf. p. 30). This is the usual sense of a triangulation of a space. Note that triangulable spaces are strongly countable-dimensional because $|K| = \bigcup_{n\in\omega} |K^{(n)}|$ for every simplicial complex K and each $|K^{(n)}|$ is an n-dimensional closed set in $|K|$. Refer to Sect. 1.12.

When the model space is strongly infinite-dimensional like Hilbert space or the Hilbert cube, infinite-dimensional manifolds are never triangulable because it is not countable-dimensional (Theorem 1.12.13). So, the Triangulation Theorems for infinite-dimensional manifolds are formulated as a proposition that an E-manifold M is homeomorphic to $|K| \times E$ for some simplicial complex K (cf. Theorems 2.6.6, 3.8.9, 4.7.1, 5.2.12, 5.10.19, and 6.5.5). However, for $\mathbb{R}^\infty$-manifolds, the model space $\mathbb{R}^\infty$ itself is triangulable. In fact, $\mathbb{R}^\infty$ is triangulated by the countable infinite full complex because $\mathbb{R}^\infty \approx \Delta^\infty$ by Corollary 6.4.3. Since each $\mathbb{R}^\infty$-manifold M is homeomorphic to $|K| \times \mathbb{R}^\infty$ for some countable locally finite simplicial complex K (Theorem 6.5.5), M can be triangulated by the product simplicial complex of K and the countable infinite full complex. Thus, we have the following:

Fact *Every $\mathbb{R}^\infty$-manifold is triangulable in the usual sense.*

By the way, the real line $\mathbb{R}$ has the natural triangulation, which is the simplicial complex with $\mathbb{Z}$ the vertices. Each $\mathbb{R}^n$ is triangulated by the iterated product of this simplicial complex. As their union, we have a natural triangulation of $\mathbb{R}^\infty$. Then, it is appropriate to ask whether this natural triangulation is combinatorially equivalent to the countable infinite full complex or not. If they are combinatorially equivalent, then the same question arises for triangulations of an $\mathbb{R}^\infty$-manifold, that is, the Hauptvermutung for triangulations of $\mathbb{R}^\infty$-manifolds comes out (cf. p. 44), which will be proved in Sect. 6.9 (Theorem 6.9.1).

In the rest of this chapter, we will discuss such a problem and related topics. To handle triangulations of $\mathbb{R}^\infty$-manifolds, we need some preliminaries on PL embeddings and simplicial subdivisions. Let K and L be cell (or simplicial) complexes. Recall that a map $f : |K| \to |L|$ is **PL (piecewise linear)** if K has a subdivision K' such that $f|C$ is affine for each $C \in K'$. Every *proper* PL map $f : |K| \to |L|$ is simplicial with respect to some simplicial subdivisions of K and L (Theorem 1.8.3).[5] Of course, when K is finite (so $|K|$ is compact), every PL map $f : |K| \to |L|$ is simplicial with respect to some simplicial subdivisions of K and L.

The following Simplicial Approximation Theorem for PL Embeddings is our tool to study triangulations of $\mathbb{R}^\infty$-manifolds, which will be proved in the next two sections:

Theorem 6.6.1 (Simplicial Approximation of a PL Embedding) *Let K and L be finite simplicial complexes and J a subcomplex of K. Suppose that $f : |K| \to |L|$ is a PL embedding such that $f||J|$ is simplicial with respect to J and L. Then, there exist simplicial subdivisions $K^* \lhd K$ and $L^* \lhd L$ with an embedding $f^* : |K| \to |L|$ such that $J \subset K^*$, $f(J) = \{f(\sigma) \mid \sigma \in J\} \subset L^*$, f^* is simplicial with respect to K^* and L^*, f^* is contiguous to f with respect to L, and*

$$f^*\big||K \setminus K[J]| \cup |J| = f\big||K \setminus K[J]| \cup |J|,$$

which implies that $f^* \simeq_L f$ rel. $|K \setminus K[J]| \cup |J|$ by the straight line homotopy.[6]

Let K be a finite simplicial complex. For $\Psi \subset K$ (Ψ need not be a subcomplex of K) and $A \subset |K|$, we keep using the following notations:

$$|\Psi| = \bigcup \Psi = \bigcup_{\sigma \in \Psi} \sigma \quad \text{and} \quad \Psi[A] = \big\{\sigma \in \Psi \bigm| \sigma \cap A \neq \emptyset\big\}.$$

Recall that $N(A, K)$ and $C(A, K)$ are the simplicial neighborhood and the simplicial complement of A in K, respectively (p. 31), which are the subcomplexes of K defined as follows:

$$N(A, K) = \big\{\tau \in K \bigm| \exists \sigma \in K[A] \text{ such that } \tau \leqslant \sigma\big\} = \bigcup_{\sigma \in K[A]} F(\sigma) \quad \text{and}$$

---

[5]As mentioned just before Theorem 1.8.3, there exists a PL map $f : |K| \to |L|$ that is not simplicial with respect to any simplicial subdivisions of K and L.

[6]Here, $J \subset K^*$ and $f(J) \subset L^*$ imply that J and $f(J)$ are subcomplexes K^* and L^*, respectively. For "contiguous" and "the straight line homotopy", see p. 32.

$$C(A, K) = K \setminus K[A] = \{\sigma \in K \mid \sigma \cap A = \emptyset\}.$$

Let K' be a subdivision of K satisfying the following condition:

(†) $\sigma_1, \sigma_2 \in K,\ \tau \in K',\ \tau \cap \operatorname{rint} \sigma_i \neq \emptyset,\ i = 1, 2 \Rightarrow \sigma_1 \leqslant \sigma_2$ or $\sigma_2 \leqslant \sigma_1$.

This condition is equivalent to the following:

(‡) $\sigma_1, \sigma_2 \in K,\ \sigma_1 \not\leqslant \sigma_2,\ \sigma_2 \not\leqslant \sigma_1$

$$\Rightarrow \forall \tau \in K',\ \tau \cap \operatorname{rint} \sigma_1 = \emptyset \text{ or } \tau \cap \operatorname{rint} \sigma_2 = \emptyset$$

$$\left(\nexists \tau \in K' \text{ such that } \tau \cap \operatorname{rint} \sigma_i \neq \emptyset, i = 1, 2 \right).$$

Then, any subdivision of K' also satisfies the condition (†) for K. Every derived subdivision of K satisfies this condition (†).[7] When $A = |L|$ for some subcomplex L of K and K' is a derived subdivision relative to $L \cup C(L, K)$, $N(L, K')$ is a derived neighborhood of L in K (cf. Sect. 1.9). Here, we also define

$$\partial N(A, K') = N(A, K') \cap C(A, K').^8$$

For each $\sigma \in K$, we consider the following subpolyhedra of $|K|$:

$$S_{K'}(\sigma) = \left|K'[\operatorname{rint} \sigma]\right| = \left|N(\operatorname{rint} \sigma, K')\right| = \bigcup_{v \in K'^{(0)} \cap \operatorname{rint} \sigma} \left|\operatorname{St}(v, K')\right|,^9$$

$$T_{K'}(\sigma) = \left|K'[\sigma] \setminus K'[\partial \sigma]\right| = S_{K'}(\sigma) \setminus \bigcup_{\sigma' < \sigma} O_{K'}(\sigma'), \text{ and}$$

$$R_{K'}(\sigma) = \left|K'[\operatorname{rint} \sigma] \cap K'[\partial \sigma]\right| = S_{K'}(\sigma) \cap \bigcup_{\sigma' < \sigma} S_{K'}(\sigma'),$$

where obviously $S_{K'}(\sigma) = T_{K'}(\sigma) \cup R_{K'}(\sigma)$ (cf. Fig. 6.2). Note that

$$T_{K'}(v) = S_{K'}(v) = \left|\operatorname{St}(v, K')\right| \text{ for each } v \in K^{(0)} \text{ and}$$

$$T_{K'}(\sigma) \subsetneqq S_{K'}(\sigma) \subsetneqq \left|N(\sigma, K')\right| \text{ for each } \sigma \in K \setminus K^{(0)}.$$

[7]For the definition of a derived subdivision of K, refer to Sect. 1.6.

[8]We adopt the same notation as the boundary of a combinatorial n-manifold. When K is a combinatorial n-manifold (i.e., $|K|$ is a PL n-manifold) and K' is a derived subdivision relative to $L \cup C(L, K)$, $N(L, K')$ is also a combinatorial n-manifold with the boundary (Theorem A.3.1 combined with Theorem A.1.6). As is easily observed, the boundary of the combinatorial n-manifold $N(L, K')$ coincides with the set defined above.

[9]For $\tau \in K'$ and $\sigma \in K$, $\tau \cap \operatorname{rint} \sigma \neq \emptyset \Leftrightarrow \tau^{(0)} \cap \operatorname{rint} \sigma \neq \emptyset$ because $K' \lhd K$.

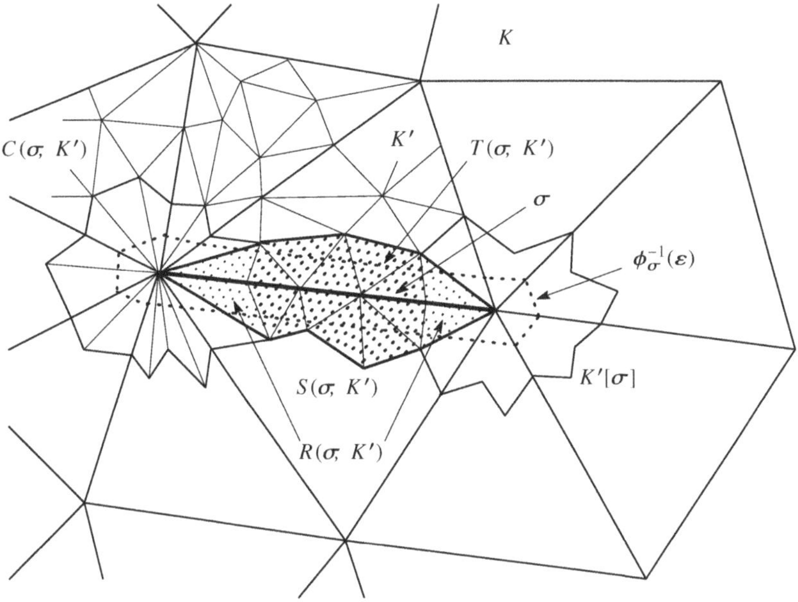

Fig. 6.2 $S(\sigma, K')$, $T(\sigma, K')$, $R(\sigma, K')$ and $\phi_\sigma^{-1}(\varepsilon)$

We also define

$$O_{K'}(\sigma) = \bigcup_{v \in K'^{(0)} \cap \operatorname{rint} \sigma} O_{K'}(v) \quad \text{and}$$

$$\partial S_{K'}(\sigma) = S_{K'}(\sigma) \setminus O_{K'}(\sigma) = S_{K'}(\sigma) \cap \left|C(\operatorname{rint} \sigma, K')\right|.$$

Then, $\operatorname{cl} O_{K'}(\sigma) = S_{K'}(\sigma)$ but $O_{K'}(\sigma) \neq \operatorname{int} S_{K'}(\sigma)$ in general. If σ is principal in K (i.e., it is not a face of any other simplex of K), then $S_{K'}(\sigma) = \sigma$, $O_{K'}(\sigma) = \operatorname{rint} \sigma$, and $\partial S_{K'}(\sigma) = \partial \sigma$. As is easily observed, $S_{K'}(\sigma)$, $T_{K'}(\sigma)$, $R_{K'}(\sigma)$, and $\partial S_{K'}(\sigma)$ are subpolyhedra of $|K|$ triangulated by the following subcomplexes of K', respectively:

$$S(\sigma, K') = N(\operatorname{rint} \sigma, K') = \bigcup_{\tau \in K'[\operatorname{rint} \sigma]} F(\tau) = \bigcup_{v \in K'^{(0)} \cap \operatorname{rint} \sigma} \operatorname{St}(v, K'),$$

$$T(\sigma, K') = \bigcup_{\sigma \in K'[\sigma] \setminus K'[\partial \sigma]} F(\tau) = S(\sigma, K') \cap C(\partial \sigma, K'),$$

$$R(\sigma, K') = \bigcup_{\sigma \in K'[\operatorname{rint} \sigma] \cap K'[\partial \sigma]} F(\tau) = S(\sigma, K') \cap \bigcup_{\sigma' < \sigma} S(\sigma', K'),$$

$$\partial S(\sigma, K') = \partial N(\operatorname{rint} \sigma, K').$$

We apply these notations to cell complexes and cellular subdivisions, too.

Now, let I denote the natural triangulation of the unit interval $\mathbf{I} = [0, 1]$, that is, $I = \{0, 1, \mathbf{I}\}$, For each $\sigma \in K$, let $\phi_\sigma : K \to I$ be the simplicial map uniquely determined by the condition $\phi_\sigma^{-1}(0) = \sigma$ (and $\phi_\sigma^{-1}(1) = |C(\sigma, K)|$). In fact, $\phi_\sigma : |K| \to \mathbf{I}$ can be defined by

$$\phi_\sigma(x) = 1 - \sum_{v \in \sigma^{(0)}} \beta_v^K(x) \ \text{ for each } x \in |K|,$$

where $\beta_v^K(x)$ is the barycentric coordinate of x at $v \in K^{(0)}$ with respect to K. When $0 < \varepsilon_\sigma < \varepsilon_{\sigma'} < 1$ are given for $\sigma' < \sigma$, we define maps $\psi_{\sigma',\sigma}, \psi_{\sigma,\sigma'} : |K| \to \mathbb{R}$ as follows:

$$\psi_{\sigma',\sigma}(x) = \frac{\phi_{\sigma'}(x)}{\varepsilon_{\sigma'}} - \frac{\phi_\sigma(x)}{\varepsilon_\sigma} \ \text{ for each } x \in |K|;$$

$$\psi_{\sigma,\sigma'}(x) = \frac{\phi_\sigma(x)}{\varepsilon_\sigma} - \frac{\phi_{\sigma'}(x)}{\varepsilon_{\sigma'}} \ \text{ for each } x \in |K|.$$

Then, $\psi_{\sigma,\sigma'} = -\psi_{\sigma',\sigma}$ and these maps are affine on each simplex of K.

A simplicial subdivision $K' \lhd K$ is said to be **flattened along** a subcomplex J of K, if K' satisfies the condition (†) and there exist $\varepsilon_\sigma \in (0, 1/2), \sigma \in J$, such that $\varepsilon_\sigma < \varepsilon_{\sigma'}/3$ for any $\sigma' < \sigma$, and the following equalities hold for each $\sigma \in J$:

$$K'^{(0)} \cap S_{K'}(\sigma) = K'^{(0)} \cap \left(\phi_\sigma^{-1}(0) \cup \phi_\sigma^{-1}(\varepsilon_\sigma)\right), \tag{6.1}$$

$$S_{K'}(\sigma) = \phi_\sigma^{-1}([0, \varepsilon_\sigma]) \cap \bigcap_{\sigma' < \sigma} \psi_{\sigma',\sigma}^{-1}(\mathbb{R}_+), \tag{6.2}$$

where $\bigcap_{\sigma' < \sigma} \psi_{\sigma',\sigma}^{-1}(\mathbb{R}_+) = |K|$ if $\dim \sigma = 0$. Then, as is easily observed, we have

$$O_{K'}(\sigma) = \phi_\sigma^{-1}([0, \varepsilon_\sigma)) \cap \bigcap_{\sigma' < \sigma} \psi_{\sigma',\sigma}^{-1}((0, \infty)). \tag{6.3}$$

Evidently, (6.1) means that $K'^{(0)} \cap \phi_\sigma^{-1}((0, \varepsilon_\sigma)) = \emptyset$.

Lemma 6.6.2 *In the above definition, (6.2) comes from the following two equalities:*

$$T_{K'}(\sigma) = \phi_\sigma^{-1}([0, \varepsilon_\sigma]) \cap \bigcap_{\sigma' < \sigma} \phi_{\sigma'}^{-1}([\varepsilon_{\sigma'}, 1]), \tag{6.4}$$

$$R_{K'}(\sigma) = \bigcap_{\sigma' < \sigma} \psi_{\sigma',\sigma}^{-1}(\mathbb{R}_+) \cap \bigcup_{\sigma' < \sigma} \phi_{\sigma'}^{-1}([0, \varepsilon_{\sigma'}]), \tag{6.5}$$

where $\bigcap_{\sigma' < \sigma} \phi_{\sigma'}^{-1}([\varepsilon_{\sigma'}, 1]) = |K|$ *if* $\dim \sigma = 0$.

Proof It suffices to show that the right hand of (6.2) is equal to the sum of the right hand of (6.4) and (6.5), that is, the following holds:

$$\begin{aligned}\phi_\sigma^{-1}([0,\varepsilon_\sigma]) \cap \bigcap_{\sigma'<\sigma} \psi_{\sigma',\sigma}^{-1}(\mathbb{R}_+) &= \left(\phi_\sigma^{-1}([0,\varepsilon_\sigma]) \cap \bigcap_{\sigma'<\sigma} \phi_{\sigma'}^{-1}([\varepsilon_{\sigma'},1])\right) \\ &\quad \cup \left(\bigcap_{\sigma'<\sigma} \psi_{\sigma',\sigma}^{-1}(\mathbb{R}_+) \cap \bigcup_{\sigma'<\sigma} \phi_{\sigma'}^{-1}([0,\varepsilon_{\sigma'}])\right). \qquad (6.6)\end{aligned}$$

First, observe that for $\sigma' < \sigma$,

$$\phi_\sigma(x) \leqslant \varepsilon_\sigma,\ \phi_{\sigma'}(x) \geqslant \varepsilon_{\sigma'} \Rightarrow \psi_{\sigma',\sigma}(x) = \frac{\phi_{\sigma'}(x)}{\varepsilon_{\sigma'}} - \frac{\phi_\sigma(x)}{\varepsilon_\sigma} \geqslant 0,$$

which implies that

$$\phi_\sigma^{-1}([0,\varepsilon_\sigma]) \cap \bigcap_{\sigma'<\sigma} \phi_{\sigma'}^{-1}([\varepsilon_{\sigma'},1]) \subset \bigcap_{\sigma'<\sigma} \psi_{\sigma',\sigma}^{-1}(\mathbb{R}_+).$$

Since $\bigcap_{\sigma'<\sigma} \phi_{\sigma'}^{-1}([\varepsilon_{\sigma'},1]) = |K| \setminus \bigcup_{\sigma'<\sigma} \phi_{\sigma'}^{-1}([0,\varepsilon_{\sigma'}))$, it follows that

$$\left(\phi_\sigma^{-1}([0,\varepsilon_\sigma]) \cap \bigcap_{\sigma'<\sigma} \phi_{\sigma'}^{-1}([\varepsilon_{\sigma'},1])\right) \cup \bigcup_{\sigma'<\sigma} \phi_{\sigma'}^{-1}([0,\varepsilon_{\sigma'}]) = \bigcup_{\sigma'\leqslant\sigma} \phi_{\sigma'}^{-1}([0,\varepsilon_{\sigma'}]).$$

Therefore, the right hand side of (6.6) is equal to the following:

$$\bigcap_{\sigma'<\sigma} \psi_{\sigma',\sigma}^{-1}(\mathbb{R}_+) \cap \bigcup_{\sigma'\leqslant\sigma} \phi_{\sigma'}^{-1}([0,\varepsilon_{\sigma'}]).$$

The left hand side of (6.6) is obviously contained in the above set. If x is a point of the above set, then

$$\forall \sigma' < \sigma,\ \psi_{\sigma',\sigma}(x) = \frac{\phi_{\sigma'}(x)}{\varepsilon_{\sigma'}} - \frac{\phi_\sigma(x)}{\varepsilon_\sigma} \geqslant 0 \ \text{ and}$$
$$\exists \sigma' \leqslant \sigma \ \text{ such that } \ \phi_{\sigma'}(x) \leqslant \varepsilon_{\sigma'},$$

which implies that $\phi_\sigma(x) \leqslant \varepsilon_\sigma$. Hence, the above set is contained in the left hand side of (6.6). Thus, we have the equality (6.6). □

Under the hypothesis of the Simplicial Approximation Theorem 6.6.1, the embedding $f : |K| \to |L|$ is simplicial with respect to simplicial subdivisions

$K' \lhd K$ and $L' \lhd L$. Replacing K' and L' with those barycentric subdivisions $\operatorname{Sd} K'$ and $\operatorname{Sd} L'$ if necessary, we can assume that both K' and L' satisfy condition (†) for K and L, respectively.[10] The restriction $f||J| : J \to L$ is simplicial. Then, $f(J)$ is a subcomplex of L and $f||J| : J \to f(J)$ is a simplicial isomorphism. The Simplicial Approximation Theorem 6.6.1 can be obtained as a combination of the following two propositions:

Proposition 6.6.3 *Let K and L be finite simplicial complexes and J a subcomplex of K. Suppose that $f : |K| \to |L|$ is an embedding such that $f||J|$ is simplicial with respect to J and L and f is simplicial with respect to simplicial subdivisions $K' \lhd K$ and $L' \lhd L$ being flattened along subcomplexes $J \subset K$ and $f(J) \subset L$, respectively. Then, there exist simplicial subdivisions $K^* \lhd K$ and $L^* \lhd L$ with an embedding $f^* : |K| \to |L|$ such that J and $f(J)$ are subcomplexes K^* and L^* respectively, f^* is simplicial with respect to K^* and L^*, f^* is contiguous to f with respect to L, and*

$$f^*\big||K \setminus K'[J]| \cup |J| = f\big||K \setminus K'[J]| \cup |J|.$$

Proposition 6.6.4 *Let K and L be finite simplicial complexes with J a subcomplex of K, $K' \lhd K$, $L' \lhd L$, and let $f : K' \to L'$ be a simplicial embedding such that $f||J|$ is simplicial with respect to J and L. Then, there exist simplicial subdivisions $K'' \lhd K'$ and $L'' \lhd L'$ being flattened along J and $f(J)$, respectively, and a simplicial embedding $f'' : K'' \to L''$ such that $f''(\tau) = f(\tau)$ for each $\tau \in L'$ and*

$$f''\big||K \setminus K'[J]| \cup |J| = f\big||K \setminus K'[J]| \cup |J|.$$

These two propositions will be proved in Sects. 6.7 and 6.8, respectively.

6.7 Retriangulating the Simplicial Neighborhood

In this section, we will prove Proposition 6.6.3. Throughout this section, let K and L be finite simplicial complexes with J a subcomplex of K, and let $f : |K| \to |L|$ be an embedding such that $f||J|$ is simplicial with respect to J and L and f is simplicial with respect to simplicial subdivisions $K' \lhd K$ and $L' \lhd L$ which are flattened along $J \subset K$ and $f(J) \subset L$, and satisfy condition (†) for K and L, respectively. Here, the simplicial embedding $f : K' \to L'$ might have an undesirable shape as in Fig. 6.3. To correct such an undesirable shape, we retriangulate the simplicial neighborhood $N(|J|, K')$ of $|J|$ in K' and similarly

[10]The barycentric subdivisions $\operatorname{Sd} K'$ and $\operatorname{Sd} L'$ satisfy condition (†). For each simplex $\sigma \in K'$, $f(\hat{\sigma})$ is the barycenter of $f(\sigma)$ because f is *injective*. Hence, $f : \operatorname{Sd} K' \to \operatorname{Sd} L'$ is simplicial.

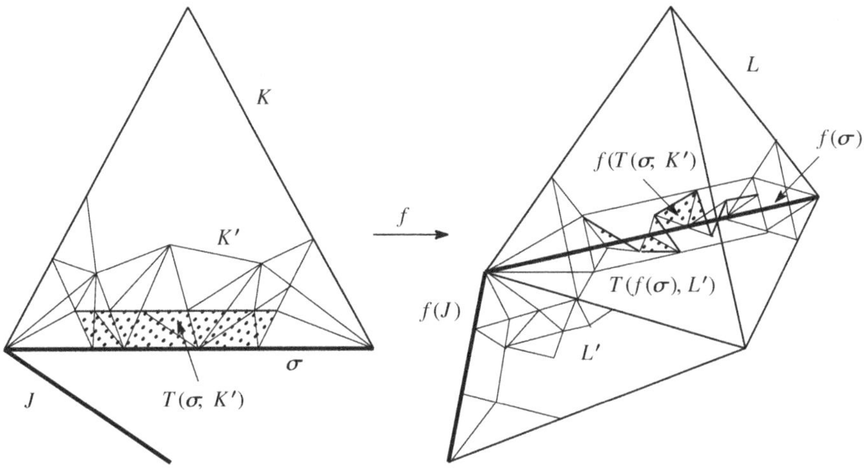

Fig. 6.3 An undesirable shape of $f : K' \to L'$

$N(|f(J)|, L')$ of $|f(J)|$ in L' so as to extend $f||C(|J|, K')| \cup |J|$ to a desirable f^*.

For each $\sigma \in J$, we define

$$D_{K'}(\sigma) = \left| K'[\operatorname{rint}\sigma] \setminus \bigcup_{\sigma < \sigma' \in J} K'[\operatorname{rint}\sigma'] \right|,$$

which is a polyhedron triangulated by the following subcomplex of K':

$$D(\sigma, K') = \bigcup_{\tau \in K'[\operatorname{rint}\sigma] \setminus \bigcup_{\sigma < \sigma' \in J} K'[\operatorname{rint}\sigma']} F(\tau).$$

Then, it follows that

$$N(|J|, K') = \bigcup_{\sigma \in J} D(\sigma, K') \text{ and } \left|N(|J|, K')\right| = \left|K'[|J|]\right| = \bigcup_{\sigma \in J} D_{K'}(\sigma).$$

Additionally, we define

$$B_{K'}(\sigma) = D_{K'}(\sigma) \cap |\partial N(|J|, K')| = D_{K'}(\sigma) \cap |C(|J|, K')|,$$

which is a polyhedron triangulated by the following subcomplex of K':

$$B(\sigma, K') = D(\sigma, K') \cap \partial N(|J|, K') = D(\sigma, K') \cap C(|J|, K').$$

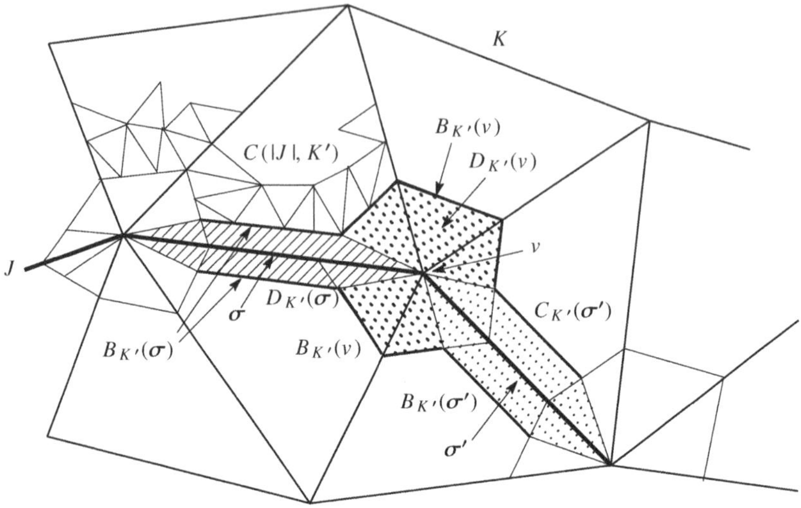

Fig. 6.4 $D_{K'}(\sigma)$ and $B_{K'}(\sigma)$

Then, it follows that

$$\partial N(|J|, K') = \bigcup_{\sigma \in J} B(\sigma, K') \text{ and } \left|\partial N(|J|, K')\right| = \bigcup_{\sigma \in J} B_{K'}(\sigma).$$

Observe Fig. 6.4.

Lemma 6.7.1 *For each $\sigma \in J$,*

$$D_{K'}(\sigma) = S_{K'}(\sigma) \setminus \bigcup_{\sigma < \sigma' \in J} O_{K'}(\sigma') \tag{6.7}$$

$$= \phi_\sigma^{-1}([0, \varepsilon_\sigma]) \cap \bigcap_{\sigma \gtrless \sigma' \in J} \psi_{\sigma',\sigma}^{-1}(\mathbb{R}_+); \tag{6.8}$$

$$B_{K'}(\sigma) = \phi_\sigma^{-1}(\varepsilon_\sigma) \cap \bigcap_{\sigma \gtrless \sigma' \in J} \psi_{\sigma',\sigma}^{-1}(\mathbb{R}_+). \tag{6.9}$$

Proof To see (6.7), let

$$\tau \in K'[\operatorname{rint} \sigma] \setminus \bigcup_{\sigma < \sigma' \in J} K'[\operatorname{rint} \sigma'].$$

Then, we have $\tau \subset |K'[\operatorname{rint}\sigma]| = S_{K'}(\sigma)$ and $\tau \cap \bigcup_{\sigma < \sigma' \in J} \operatorname{rint}\sigma' = \emptyset$. If $\tau \cap O_{K'}(v) \neq \emptyset$ for some $v \in K'^{(0)} \cap \bigcup_{\sigma < \sigma' \in J} \operatorname{rint}\sigma'$, then $\tau \cap \operatorname{rint}\tau' \neq \emptyset$ for some

$\tau' \in \mathrm{St}(v, K')$, which means that $v \in \tau' \leqslant \tau$. This is a contradiction. Thus, it follows that $\tau \subset S_{K'}(\sigma) \setminus \bigcup_{\sigma<\sigma'\in J} O_{K'}(\sigma')$. Consequently,

$$D_{K'}(\sigma) \subset S_{K'}(\sigma) \setminus \bigcup_{\sigma<\sigma'\in J} O_{K'}(\sigma').$$

To see the converse inclusion, let $x \in S_{K'}(\sigma) \setminus \bigcup_{\sigma<\sigma'\in J} O_{K'}(\sigma')$. Since $S_{K'}(\sigma) = |K'[\operatorname{rint}\sigma]|$, x is contained in some $\tau \in K'[\operatorname{rint}\sigma]$. Let $\tau \in K'[\operatorname{rint}\sigma]$ be the smallest such simplex. Assume that $\tau \in \bigcup_{\sigma<\sigma'\in J} K'[\operatorname{rint}\sigma']$, that is, $\tau \in K'[\operatorname{rint}\sigma']$ for some $\sigma' \in J$ with $\sigma < \sigma'$. Then, τ has a vertex $v' \in \operatorname{rint}\sigma'$. Since $x \notin O_{K'}(\sigma')$, it follows that $\beta^{K'}_{v'}(x) = 0$, which means that x is contained in the opposite face $\tau' < \tau$ of v'. This contradicts the minimality of τ. Therefore, $\tau \in K'[\operatorname{rint}\sigma] \setminus \bigcup_{\sigma<\sigma'\in J} K'[\operatorname{rint}\sigma']$, which implies that $x \in D_{K'}(\sigma)$. Thus, (6.7) holds.

To show (6.8), note that

$$|K| \setminus O_{K'}(\sigma') \supset \psi^{-1}_{\sigma,\sigma'}((-\infty, 0]) = \psi^{-1}_{\sigma',\sigma}(\mathbb{R}_+) \quad \text{for each } \sigma < \sigma' \in J,$$

because $O_{K'}(\sigma') \subset \psi^{-1}_{\sigma,\sigma'}((0,\infty))$ by (6.3). From the first equality and (6.2), it follows that

$$\begin{aligned} D_{K'}(\sigma) &= S_{K'}(\sigma) \cap \bigcap_{\sigma<\sigma'\in J} \big(|K| \setminus O_{K'}(\sigma')\big) \\ &\supset \phi^{-1}_{\sigma}([0, \varepsilon_\sigma]) \cap \bigcap_{\sigma'<\sigma} \psi^{-1}_{\sigma',\sigma}(\mathbb{R}_+) \cap \bigcap_{\sigma<\sigma'\in J} \psi^{-1}_{\sigma',\sigma}(\mathbb{R}_+) \\ &= \phi^{-1}_{\sigma}([0, \varepsilon_\sigma]) \cap \bigcap_{\sigma\gtrless\sigma'\in J} \psi^{-1}_{\sigma',\sigma}(\mathbb{R}_+). \end{aligned}$$

For the converse inclusion, note that

$$D_{K'}(\sigma) \subset S_{K'}(\sigma) = \phi^{-1}_{\sigma}([0, \varepsilon_\sigma]) \cap \bigcap_{\sigma'<\sigma} \psi^{-1}_{\sigma',\sigma}(\mathbb{R}_+).$$

Hence, it suffices to show that $\psi_{\sigma',\sigma}(x) \geqslant 0$ for each $x \in D_{K'}(\sigma)$ and $\sigma < \sigma' \in J$. On the contrary, assume that $\psi_{\sigma',\sigma}(x_0) < 0$ for some $x_0 \in D_{K'}(\sigma)$ and $\sigma < \sigma' \in J$. Let $\sigma' \in J$ be the smallest such simplex, that is, $\psi_{\sigma',\sigma}(x_0) < 0$ but $\psi_{\sigma'',\sigma}(x_0) \geqslant 0$ if $\sigma < \sigma'' < \sigma'$. Then, $\psi_{\sigma,\sigma'}(x_0) = -\psi_{\sigma',\sigma}(x_0) > 0$ and

$$\psi_{\sigma'',\sigma'}(x_0) = \psi_{\sigma'',\sigma}(x_0) + \psi_{\sigma,\sigma'}(x_0) > 0 \quad \text{if } \sigma < \sigma'' < \sigma'.$$

For each $\sigma'' < \sigma$, since $D_{K'}(\sigma) \subset S_{K'}(\sigma) \subset \psi^{-1}_{\sigma'',\sigma}(\mathbb{R}_+)$, it follows that $\psi_{\sigma'',\sigma}(x_0) \geqslant 0$, which implies that

$$\psi_{\sigma'',\sigma'}(x_0) = \psi_{\sigma'',\sigma}(x_0) + \psi_{\sigma,\sigma'}(x_0) > 0.$$

Moreover, since $x_0 \in \phi_\sigma^{-1}([0, \varepsilon_\sigma])$ and $\psi_{\sigma',\sigma}(x_0) < 0$, we have

$$\frac{\phi_{\sigma'}(x_0)}{\varepsilon_{\sigma'}} < \frac{\phi_\sigma(x_0)}{\varepsilon_\sigma} \leqslant 1.$$

Hence, $\phi_{\sigma'}(x_0) < \varepsilon_{\sigma'}$. Thus, it follows from (6.3) that

$$x_0 \in \phi_{\sigma'}^{-1}([0, \varepsilon_{\sigma'})) \cap \bigcap_{\sigma'' < \sigma'} \psi_{\sigma'',\sigma'}^{-1}((0, \infty)) = O_{K'}(\sigma'),$$

which contradicts (6.7). Consequently, we have the equality (6.8).

Finally, we show (6.9). Let $x \in D_{K'}(\sigma)$. By (6.8), it suffices to show the following equivalence:

$$\phi_\sigma(x) = \varepsilon_\sigma \ \Leftrightarrow\ x \in |C(|J|, K)|.$$

To see ($\Leftarrow$), let $x \in |C(|J|, K)|$. Since $x \notin \operatorname{rint} \tau$, it follows that $\tau \cap \sigma < \tau$ and x is contained in the opposite face $\tau' < \tau$ of $\tau \cap \sigma$. Hence, by (6.1), $\phi_\sigma(x) = \varepsilon_\sigma$, that is, $x \in \phi_\sigma^{-1}(0)$.

To show ($\Rightarrow$), let $x \in \phi_\sigma^{-1}(\varepsilon_\sigma)$. By (6.1), the carrier $c_{K'}(x)$ is a face of τ and $c_{K'}(x)^{(0)} \subset \phi_\sigma^{-1}(\varepsilon_\sigma)$. Then, $c_{K'}(x) \cap \sigma = \emptyset$. If $c_{K'}(x) \cap \operatorname{rint} \sigma' \neq \emptyset$ for some $\sigma' \in J$, then $\tau \cap \operatorname{rint} \sigma' \neq \emptyset$, which contradicts (‡). Hence, $c_{K'}(x) \cap \operatorname{rint} \sigma' = \emptyset$ for any $\sigma' \in J$, which means that $c_{K'}(x) \cap |J| = \emptyset$. Thus, we have $x \in |C(|J|, K')|$. The proof is complete. □

Lemma 6.7.2 *For each $\sigma \in J$, $D_{K'}(\sigma) \cap |J| = \sigma$.*

Proof Since $\sigma \subset D_{K'}(\sigma) \cap |J|$, it suffices to show that $D_{K'}(\sigma) \cap |J| \subset \sigma$. Each $x \in D_{K'}(\sigma) \cap |J|$ is contained in some

$$\tau \in K'[\operatorname{rint} \sigma] \setminus \bigcup_{\sigma < \sigma' \in J} K'[\operatorname{rint} \sigma'].$$

Then, $\sigma \not< c_J(x)$ because $\tau \cap c_J(x) \neq \emptyset$. Additionally, $\tau \cap \operatorname{rint} c_J(x) \neq \emptyset$. So, it follows from (†) that $c_J(x) \leqslant \sigma$, which implies that $x \in \sigma$. □

Lemma 6.7.3 *For each $\sigma, \sigma' \in J$, the following hold:*

(1) $\sigma \not\leqslant \sigma',\ \sigma' \not\leqslant \sigma \ \Rightarrow\ D_{K'}(\sigma) \cap D_{K'}(\sigma') = \sigma \cap \sigma'$;
(2) $\sigma < \sigma' \ \Rightarrow\ D_{K'}(\sigma) \cap D_{K'}(\sigma') = D_{K'}(\sigma) \cap \psi_{\sigma',\sigma}^{-1}(0) = D_{K'}(\sigma') \cap \psi_{\sigma,\sigma'}^{-1}(0)$.
(3) $\sigma < \sigma' \ \Rightarrow\ D_{K'}(\sigma) \cap \psi_{\sigma',\sigma}^{-1}(0) \cap |J| = D_{K'}(\sigma') \cap \psi_{\sigma,\sigma'}^{-1}(0) \cap |J| = \sigma$.

Proof (1): When $\sigma \cap \sigma' = \emptyset$, we may show that $D_{K'}(\sigma) \cap D_{K'}(\sigma') = \emptyset$. If $x \in D_{K'}(\sigma) \cap D_{K'}(\sigma')$, then $\phi_\sigma(x) < \varepsilon_\sigma < 1/2$ and $\phi_{\sigma'}(x) < \varepsilon_{\sigma'} < 1/2$ by (6.8), which implies

$$\begin{aligned}\sum_{v \in K^{(0)}} \beta_v^K(x) &\geqslant \sum_{v \in \sigma^{(0)}} \beta_v^K(x) + \sum_{v \in \sigma'^{(0)}} \beta_v^K(x) \\ &= (1 - \phi_\sigma(x)) + (1 - \phi_{\sigma'}(x)) \\ &> (1 - 1/2) + (1 - 1/2) = 1.\end{aligned}$$

This is a contradiction. Therefore, $D_{K'}(\sigma) \cap D_{K'}(\sigma') = \emptyset$.

When $\sigma \cap \sigma' \neq \emptyset$, the inclusion $\sigma'' = \sigma \cap \sigma' \subset D_{K'}(\sigma) \cap D_{K'}(\sigma')$ is evident. To see the converse inclusion, let $x \in D_{K'}(\sigma) \cap D_{K'}(\sigma')$. Since $\sigma'' < \sigma,\ \sigma'$, it follows from (6.8) that $\psi_{\sigma'',\sigma}(x),\ \psi_{\sigma'',\sigma'}(x) \geqslant 0$, which means that

$$\frac{\phi_{\sigma''}(x)}{\varepsilon_{\sigma''}} \geqslant \frac{\phi_\sigma(x)}{\varepsilon_\sigma},\ \frac{\phi_{\sigma'}(x)}{\varepsilon_{\sigma'}}.$$

Recall that $\varepsilon_\sigma < \varepsilon_{\sigma''}/3$. Hence, we have

$$\phi_\sigma(x) \leqslant \frac{\varepsilon_\sigma}{\varepsilon_{\sigma''}} \phi_{\sigma''}(x) < \phi_{\sigma''}(x)/3.$$

Similarly, $\phi_{\sigma'}(x) < \phi_{\sigma''}(x)/3$. So, it follows that

$$\begin{aligned}1 = \sum_{v \in K^{(0)}} \beta_v^K(x) &\geqslant \sum_{v \in \sigma^{(0)}} \beta_v^K(x) + \sum_{v \in \sigma'^{(0)}} \beta_v^K(x) - \sum_{v \in \sigma''^{(0)}} \beta_v^K(x) \\ &= (1 - \phi_\sigma(x)) + (1 - \phi_{\sigma'}(x)) - (1 - \phi_{\sigma''}(x)) \\ &> 1 - \phi_{\sigma''}(x)/3 - \phi_{\sigma''}(x)/3 + \phi_{\sigma''}(x) = 1 + \phi_{\sigma''}(x)/3,\end{aligned}$$

which implies that $\phi_{\sigma''}(x) = 0$, that is, $x \in \sigma''$.

(2): For each $x \in D_{K'}(\sigma) \cap D_{K'}(\sigma')$, $\psi_{\sigma',\sigma}(x) \geqslant 0$ and $\psi_{\sigma,\sigma'}(x) \geqslant 0$ by (6.8). Because $\psi_{\sigma,\sigma'}(x) = -\psi_{\sigma',\sigma}(x)$, we have $\psi_{\sigma',\sigma}(x) = \psi_{\sigma,\sigma'}(x) = 0$. Therefore,

$$D_{K'}(\sigma) \cap D_{K'}(\sigma') \subset D_{K'}(\sigma) \cap \psi_{\sigma',\sigma}^{-1}(0),\ D_{K'}(\sigma') \cap \psi_{\sigma,\sigma'}^{-1}(0).$$

To see the converse inclusion, it suffices to show that

$$D_{K'}(\sigma) \cap \psi_{\sigma',\sigma}^{-1}(0) = D_{K'}(\sigma') \cap \psi_{\sigma,\sigma'}^{-1}(0).$$

Due to (6.8), the above equality can be rewritten as follows:

$$\psi_{\sigma',\sigma}^{-1}(0) \cap \phi_\sigma^{-1}([0,\varepsilon_\sigma]) \cap \bigcap_{\sigma \gtrless \sigma'' \in J\setminus\{\sigma'\}} \psi_{\sigma'',\sigma}^{-1}(\mathbb{R}_+)$$
$$= \psi_{\sigma,\sigma'}^{-1}(0) \cap \phi_{\sigma}^{-1}([0,\varepsilon_{\sigma'}]) \cap \bigcap_{\sigma' \gtrless \sigma'' \in J\setminus\{\sigma\}} \psi_{\sigma'',\sigma'}^{-1}(\mathbb{R}_+). \qquad (*)$$

Observe the following equivalence for $x \in |K|$,

$$x \in \psi_{\sigma',\sigma}^{-1}(0) \left(= \psi_{\sigma,\sigma'}^{-1}(0)\right) \Leftrightarrow \frac{\phi_{\sigma'}(x)}{\varepsilon_{\sigma'}} = \frac{\phi_\sigma(x)}{\varepsilon_\sigma},$$

Hence, we have the following:

$$\psi_{\sigma',\sigma}^{-1}(0) \cap \phi_\sigma^{-1}([0,\varepsilon_\sigma]) = \psi_{\sigma,\sigma'}^{-1}(0) \cap \phi_{\sigma'}^{-1}([0,\varepsilon_{\sigma'}]).$$

In each case of $\sigma'' < \sigma$ and $\sigma' < \sigma'' \in J$, for each $x \in \psi_{\sigma',\sigma}^{-1}(0)$,

$$\psi_{\sigma'',\sigma}(x) = \frac{\phi_{\sigma''}(x)}{\varepsilon_{\sigma''}} - \frac{\phi_\sigma(x)}{\varepsilon_\sigma} = \frac{\phi_{\sigma''}(x)}{\varepsilon_{\sigma''}} - \frac{\phi_{\sigma'}(x)}{\varepsilon_{\sigma'}} = \psi_{\sigma'',\sigma'}(x),$$

which implies that

$$\psi_{\sigma',\sigma}^{-1}(0) \cap \psi_{\sigma'',\sigma}^{-1}(\mathbb{R}_+) = \psi_{\sigma,\sigma'}^{-1}(0) \cap \psi_{\sigma'',\sigma'}^{-1}(\mathbb{R}_+).$$

Thus, $(*)$ follows from the following two inclusions:

$$\psi_{\sigma',\sigma}^{-1}(0) \cap \bigcap_{\sigma''<\sigma} \psi_{\sigma'',\sigma}^{-1}(\mathbb{R}_+) \subset \bigcap_{\sigma''<\sigma'} \psi_{\sigma'',\sigma'}^{-1}(\mathbb{R}_+); \qquad (*\mathrm{i})$$

$$\psi_{\sigma,\sigma'}^{-1}(0) \cap \bigcap_{\sigma'<\sigma''\in J} \psi_{\sigma'',\sigma'}^{-1}(\mathbb{R}_+) \subset \bigcap_{\sigma<\sigma''\in J} \psi_{\sigma'',\sigma}^{-1}(\mathbb{R}_+). \qquad (*\mathrm{ii})$$

To verify $(*\mathrm{i})$, take a point x in the left hand set of $(*\mathrm{i})$. For each $\sigma'' < \sigma'$, $\psi_{\sigma\cap\sigma'',\sigma}(x) \geqslant 0$. Observe that

$$1 - \phi_{\sigma''}(x) = \sum_{v\in\sigma''^{(0)}} \beta_v^K(x) \leqslant 1 - \sum_{v\in\sigma^{(0)}} \beta_v^K(x) + \sum_{v\in\sigma^{(0)}\cap\sigma''^{(0)}} \beta_v^K(x)$$
$$= \phi_\sigma(x) + 1 - \phi_{\sigma\cap\sigma''}(x).$$

Hence, $\phi_{\sigma''}(x) \geqslant \phi_{\sigma\cap\sigma''}(x) - \phi_{\sigma}(x)$. As we saw above, $x \in \psi^{-1}_{\sigma',\sigma}(0)$ means that $\phi_{\sigma'}(x)/\varepsilon_{\sigma'} = \phi_{\sigma}(x)/\varepsilon_{\sigma}$. Since $\varepsilon_{\sigma}, \varepsilon_{\sigma''} < \varepsilon_{\sigma\cap\sigma''}/3$, we have $\varepsilon_{\sigma} + \varepsilon_{\sigma''} < \varepsilon_{\sigma\cap\sigma''}$. Then, it follows that

$$\begin{aligned}
\psi_{\sigma'',\sigma'}(x) &= \frac{\phi_{\sigma''}(x)}{\varepsilon_{\sigma''}} - \frac{\phi_{\sigma'}(x)}{\varepsilon_{\sigma'}} \\
&\geqslant \frac{\phi_{\sigma'}(x) + \phi_{\sigma\cap\sigma''}(x) - \phi_{\sigma}(x)}{\varepsilon_{\sigma''}} - \frac{\phi_{\sigma}(x)}{\varepsilon_{\sigma}} \\
&\geqslant \frac{\phi_{\sigma\cap\sigma''}(x)}{\varepsilon_{\sigma''}} - \frac{(\varepsilon_{\sigma} + \varepsilon_{\sigma''})\phi_{\sigma}(x)}{\varepsilon_{\sigma''}\varepsilon_{\sigma}} \\
&\geqslant \frac{\phi_{\sigma\cap\sigma''}(x)}{\varepsilon_{\sigma''}} - \frac{\varepsilon_{\sigma\cap\sigma''}\phi_{\sigma}(x)}{\varepsilon_{\sigma''}\varepsilon_{\sigma}} \\
&\geqslant \frac{\varepsilon_{\sigma\cap\sigma''}}{\varepsilon_{\sigma''}}\left(\frac{\phi_{\sigma\cap\sigma''}(x)}{\varepsilon_{\sigma\cap\sigma''}} - \frac{\phi_{\sigma}(x)}{\varepsilon_{\sigma}}\right) \\
&= \frac{\varepsilon_{\sigma\cap\sigma''}}{\varepsilon_{\sigma''}}\psi_{\sigma\cap\sigma'',\sigma}(x) \geqslant 0.
\end{aligned}$$

Therefore, $x \in \bigcap_{\sigma''<\sigma'} \psi^{-1}_{\sigma'',\sigma}(\mathbb{R}_+)$.

To show (∗ii), take a point x in the left hand set of (∗ii). For each $\sigma < \sigma'' \in J$, $\psi_{\sigma\cap\sigma'',\sigma}(x) \geqslant 0$. Observe that

$$\begin{aligned}
1 - \phi_{\sigma''}(x) = \sum_{v\in\sigma''^{(0)}} \beta_v^K(x) &\leqslant 1 - \sum_{v\in\sigma'^{(0)}} \beta_v^K(x) + \sum_{v\in\sigma'^{(0)}\cap\sigma''^{(0)}} \beta_v^K(x) \\
&= \phi_{\sigma'}(x) + 1 - \phi_{\sigma'\cap\sigma''}(x).
\end{aligned}$$

Hence, $\phi_{\sigma''}(x) \geqslant \phi_{\sigma'\cap\sigma''}(x) - \phi_{\sigma'}(x)$. As in (∗i), $\phi_{\sigma'}(x)/\varepsilon_{\sigma'} = \phi_{\sigma}(x)/\varepsilon_{\sigma}$. Since $\varepsilon_{\sigma'}, \varepsilon_{\sigma''} < \varepsilon_{\sigma'\cap\sigma''}/3$, we have $\varepsilon_{\sigma'} + \varepsilon_{\sigma''} < \varepsilon_{\sigma'\cap\sigma''}$. Then, it follows that

$$\begin{aligned}
\psi_{\sigma'',\sigma'}(x) &= \frac{\phi_{\sigma''}(x)}{\varepsilon_{\sigma''}} - \frac{\phi_{\sigma'}(x)}{\varepsilon_{\sigma'}} \\
&\geqslant \frac{\phi_{\sigma'\cap\sigma''}(x) - \phi_{\sigma'}(x)}{\varepsilon_{\sigma''}} - \frac{\phi_{\sigma'}(x)}{\varepsilon_{\sigma'}} \\
&= \frac{\phi_{\sigma'\cap\sigma''}(x)}{\varepsilon_{\sigma''}} - \frac{(\varepsilon_{\sigma'} + \varepsilon_{\sigma''})\phi_{\sigma'}(x)}{\varepsilon_{\sigma''}\varepsilon_{\sigma'}} \\
&\geqslant \frac{\phi_{\sigma'\cap\sigma''}(x)}{\varepsilon_{\sigma''}} - \frac{\varepsilon_{\sigma'\cap\sigma''}\phi_{\sigma}(x)}{\varepsilon_{\sigma''}\varepsilon_{\sigma}} \\
&= \frac{\varepsilon_{\sigma'\cap\sigma''}}{\varepsilon_{\sigma''}}\left(\frac{\phi_{\sigma'\cap\sigma''}(x)}{\varepsilon_{\sigma'\cap\sigma''}} - \frac{\phi_{\sigma}(x)}{\varepsilon_{\sigma}}\right) \\
&= \frac{\varepsilon_{\sigma'\cap\sigma''}}{\varepsilon_{\sigma''}}\psi_{\sigma\cap\sigma'',\sigma}(x) \geqslant 0.
\end{aligned}$$

Therefore, $x \in \bigcap_{\sigma''<\sigma'} \psi_{\sigma'',\sigma}^{-1}(\mathbb{R}_+)$.

(3): Because $\phi_\sigma^{-1}(0) = \sigma \subset \sigma' = \phi_{\sigma'}^{-1}(0)$, we have $\sigma \subset \psi_{\sigma',\sigma}^{-1}(0) = \psi_{\sigma,\sigma'}^{-1}(0)$. Hence, the following is obtained from Lemma 6.7.2 and (2) above:

$$\sigma = D_{K'}(\sigma) \cap \psi_{\sigma',\sigma}^{-1}(0) \cap |J| = D_{K'}(\sigma') \cap \psi_{\sigma,\sigma'}^{-1}(0) \cap |J|.$$

The proof of the lemma is completed. □

Although each $D_{K'}(\sigma)$ is not a (convex linear) cell, the boundary $\partial D_{K'}(\sigma)$ can be defined as follows:

$$\begin{aligned}\partial D_{K'}(\sigma) &= B_{K'}(\sigma) \cup \bigcap_{\sigma \gneqq \sigma' \in J} \big(D_{K'}(\sigma') \cap D_{K'}(\sigma)\big) \\ &= \big(D_{K'}(\sigma) \cap \phi_\sigma^{-1}(\varepsilon_\sigma)\big) \cup \bigcap_{\sigma \gneqq \sigma' \in J} \big(D_{K'}(\sigma) \cap \psi_{\sigma',\sigma}^{-1}(0)\big),\end{aligned}$$

where it should be remarked that $D_{K'}(\sigma)$ can be represented as (6.8).

Remark 6.10 For each $\sigma \in J$ and $\varrho \in K \setminus J$ with $\sigma < \varrho$, the following is a (convex linear) cell:

$$D_{K'}^{\varrho}(\sigma) = \varrho \cap D_{K'}(\sigma) = \varrho \cap \phi_\sigma^{-1}([0, \varepsilon_\sigma]) \cap \bigcap_{\sigma \gneqq \sigma' \in J \cap F(\varrho)} \psi_{\sigma',\sigma}^{-1}(\mathbb{R}_+),$$

because $\phi_\sigma|\varrho$ and $\psi_{\sigma',\sigma}|\varrho$ are affine functionals for $\sigma \gneqq \sigma' \in J \cap F(\varrho)$. Then, the boundary of $D_{K'}^{\varrho}(\sigma)$ can be written as follows:

$$\partial D_{K'}^{\varrho}(\sigma) = \big(\partial D_{K'}(\sigma) \cap \varrho\big) \cup \sigma.$$

By virtue of Lemma 6.7.3, for each $\sigma, \sigma' \in J$ and $\varrho, \varrho' \in K \setminus J$ with $\sigma < \varrho$ and $\sigma' < \varrho'$, the following holds:

$$D_{K'}^{\varrho}(\sigma) \cap D_{K'}^{\varrho'}(\sigma') \neq \emptyset \Rightarrow D_{K'}^{\varrho}(\sigma) \cap D_{K'}^{\varrho'}(\sigma') < D_{K'}^{\varrho}(\sigma),\ D_{K'}^{\varrho'}(\sigma').$$

Hence, $\bigcup_{\sigma\in J} \bigcup_{\sigma<\varrho\in K\setminus J} F(D_{K'}^{\varrho}(\sigma))$ is a cell complex, where $F(D_{K'}^{\varrho}(\sigma))$ is the cell complex consisting of all faces of $D_{K'}^{\varrho}(\sigma)$. Although we will not use this complex, it will help us understand the following arguments—See Fig. 6.5.

Now, we prove Proposition 6.6.3.

Proof (Proposition 6.6.3) The desired simplicial subdivision K^* will be obtained by replacing $N(|J|, K')$ with a new triangulation $M(J, K')$ of $|N(|J|, K')|$. We will construct a triangulation $M(J, K')$ of $|N(|J|, K')|$ so that

$$J \cup \partial N(|J|, K') \subset M(J, K') \text{ and } M(J, K')^{(0)} = J^{(0)} \cup \partial N(|J|, K')^{(0)}.$$

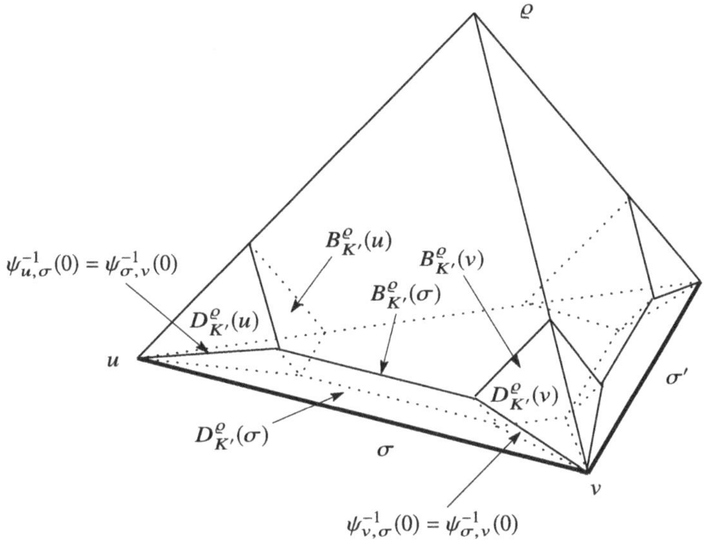

Fig. 6.5 $D^{\varrho}_{K'}(\sigma)$ and $B^{\varrho}_{K'}(\sigma)$

First, we introduce some notation. Let

$$B(J^{(n)}, K') = \bigcup_{\sigma \in J^{(n)}} B(\sigma, K'), \; n = 0, 1, \ldots, \dim J,$$

which are subcomplexes of $\partial N(|J|, K')$ and

$$\partial N(|J|, K') = \bigcup_{n=0}^{\dim J} B(J^{(n)}, K').$$

Give a total order $\leqslant$ on $J^{(0)}$ and for each $\sigma \in J \setminus J^{(0)}$, let $v_\sigma = \max \sigma^{(0)}$ and $\sigma^* = \langle \sigma^{(0)} \setminus \{v_\sigma\} \rangle < \sigma$, i.e., σ^* is the opposite face of the vertex v_σ. Then, we can write $\sigma = v_\sigma \sigma^*$. Additionally, we consider the following two subpolyhedra of $D_{K'}(\sigma)$:

$$P_{K'}(\sigma) = B_{K'}(\sigma) \cup \bigcup_{\sigma' \leqslant \sigma^*} \big(D_{K'}(\sigma') \cap D_{K'}(\sigma)\big);$$

$$Q_{K'}(\sigma) = \bigcup_{\sigma \gneqq \sigma' \in J[v_\sigma]} \big(D_{K'}(\sigma') \cap D_{K'}(\sigma)\big).$$

Then, we have $\partial D_{K'}(\sigma) = P_{K'}(\sigma) \cup Q_{K'}(\sigma)$.

For each vertex $v \in J$, $D(v, K')$ is a simplicial complex, which is the cone over $B(v, K')$ with the vertex v, that is,

$$D(v, K') = \{v\} \cup B(v, K') \cup \{v\tau \mid \tau \in B(v, K')\}.$$

Thus, we have a triangulation $M(v, K') = D(v, K')$ of $D_{K'}(v)$ such that

$$v \in M(v, K'), \ B(v, K') \subset M(v, K'), \text{ and}$$
$$M(v, K')^{(0)} = \{v\} \cup B(v, K')^{(0)}.$$

Since $D_{K'}(v) \cap D_{K'}(v') = \emptyset$ if $v \neq v' \in J^{(0)}$, $M(J^{(0)}, K') = \bigcup_{v \in J^{(0)}} M(v, K')$ is a triangulation $M(J^{(0)}, K')$ of $D_{K'}(J^{(0)})$.

Suppose that a triangulation $M(J^{(n-1)}, K')$ of $D_{K'}(J^{(n-1)})$ have been defined as $\bigcup_{\sigma \in J^{(n-1)}} M(\sigma, K')$, where each $M(\sigma, K')$ is a triangulation of $D_{K'}(\sigma)$ satisfying the following conditions:

(1) $\sigma \in M(\sigma, K')$, $B(\sigma, K') \subset M(\sigma, K')$, $M(\sigma, K')^{(0)} = \sigma^{(0)} \cup B(\sigma, K')^{(0)}$,
(2) $M(\sigma, K')$ is the cone with the vertex v_σ over the triangulation of $P_{K'}(\sigma)$ induced by $M(J^{(n-1)}, K')$.

Then, for each n-simplex $\sigma \in J$ (i.e., $\sigma \in J^{(n)} \setminus J^{(n-1)}$), we define $M(\sigma, K')$ by the above (2), which evidently satisfies (1). Observe that if $\sigma' \in J^{(n-1)}[v_\sigma]$ and $D_{K'}(\sigma') \cap D_{K'}(\sigma) \neq \emptyset$, then $M(\sigma', K')$ and $M(\sigma, K')$ induce the same triangulation of $D_{K'}(\sigma') \cap D_{K'}(\sigma)$ because $M(\sigma', K')$ satisfies (2). Hence, $M(\sigma, K')$ induces the same triangulation of $Q_{K'}(\sigma)$ as the one induced by $M(J^{(n-1)}, K')$. Since $D_{K'}(\sigma) \cap D_{K'}(\sigma') = \sigma \cap \sigma'$ for every $\sigma, \sigma' \in J^{(n)} \setminus J^{(n-1)}$, we can define a triangulation $M(J^{(n)}, K')$ of $D_{K'}(J^{(n)})$ as $M(J^{(n)}, K') = \bigcup_{\sigma \in J^{(n)}} M(\sigma, K')$.

By induction on n, we can obtain a triangulation $M(J, K') = M(J^{(\dim J)})$ of $|N(|J|, K')| = D_{K'}(J^{(\dim J)})$ such that

$$J \cup \partial N(|J|, K') \subset M(J, K') \ \text{ and } \ M(J, K')^{(0)} = J^{(0)} \cup \partial N(|J|, K')^{(0)}.$$

Then, the simplicial subdivision $K^* \lhd K$ can be defined as follows:

$$K^* = C(|J|, K') \cup M(J, K').$$

Using the total order $\leqslant$ on $f(J)^{(0)} = f(J^{(0)})$ so that $f|J^{(0)} : J^{(0)} \to f(J)^{(0)}$ is an order isomorphism, we can similarly define a triangulation $M(f(J), L')$ of $|N(|f(J)|, L')|$ such that

$$f(J) \cup \partial N(|f(J)|, L') \subset M(f(J), L') \ \text{ and}$$
$$M(f(J), L')^{(0)} = f(J)^{(0)} \cup \partial N(|f(J)|, L')^{(0)}.$$

Then, the simplicial subdivision $L^* \lhd L$ can be defined as follows:

$$L^* = C(|f(J)|, L') \cup M^*(f(J), L') \lhd L.$$

Note that $f|J^{(0)}$ is an order isomorphism and the restriction $f||\partial N(|J|, K')| : \partial N(|J|, K') \to \partial N(|f(J)|, L')$ is a simplicial embedding. Let $f' : M(J, K') \to M(f(J), L')$ be a simplicial map induced by

$$f|M(J, K')^{(0)} : M(J, K')^{(0)} \to M(f(J), L')^{(0)}.$$

Then, $f'||\partial N(|J|, K')| = f||\partial N(|J|, K')|$ and $f'||J| = f||J|$ because $f'|J^{(0)} = f|J^{(0)}$. Since $f'|M(J, K')^{(0)} = f|M(J, K')^{(0)}$ is injective, f' is a simplicial embedding. Note

$$|C(|J|, K')| \cap |N(|J|, K')| = |\partial N(|J|, K')|.$$

So, we can define a simplicial embedding $f^* : K^* \to L^*$ by

$$f^*||C(|J|, K')| = f||C(|J|, K')| \text{ and } f^*||N(|J|, K')| = f'||N(|J|, K')|.$$

Then, $f^*||J| = f'||J| = f||J|$. By the definition, f^* is L-close to f, that is, f^* is contiguous to $f(x)$ with respect to L. □

6.8 Flattening Simplicial Subdivisions

In this section, we will prove Proposition 6.6.4. Throughout this section, let K and L be finite simplicial complexes with J a subcomplex of K, and let $f : |K| \to |L|$ be an embedding such that $f||J|$ is simplicial with respect to J and L and f is simplicial with respect to simplicial subdivisions $K' \lhd K$ and $L' \lhd L$, which satisfy condition (†) for K and L, respectively.

We will inductively make subdivisions $K' \rhd K_0 \rhd K_1 \rhd \cdots \rhd K_n$, $n = \dim J$, so that each K_i is flattened along $J^{(i)}$ and apply the same process to L' to make subdivisions $L' \rhd L_0 \rhd L_1 \rhd \cdots \rhd L_n$ together with simplicial maps $f_i : K_i \to L_i$, $i = 0, \ldots, n$. For such inductive construction, we need the following lemma:

Lemma 6.8.1 *Let $K' \lhd K$ be flattened along J. For each $\sigma \in J$,*

$$\left|K'[\partial\sigma]\right| = \bigcup_{\sigma' < \sigma} \phi_{\sigma'}^{-1}([0, \varepsilon_{\sigma'}]) \text{ and} \tag{6.10}$$

$$\left|K' \setminus K'[\partial\sigma]\right| = \bigcap_{\sigma' < \sigma} \phi_{\sigma'}^{-1}([\varepsilon_{\sigma'}, 1]). \tag{6.11}$$

Proof From (6.2), it follows that

$$\begin{aligned}|K'[\partial\sigma]| &= \bigcup_{\sigma'<\sigma} |K'[\operatorname{rint}\sigma']| = \bigcup_{\sigma'<\sigma} |S(\sigma', K')| \\ &= \bigcup_{\sigma'<\sigma} \left(\phi_{\sigma'}^{-1}([0, \varepsilon_{\sigma'}]) \cap \bigcap_{\sigma''<\sigma'} \psi_{\sigma'',\sigma'}^{-1}(\mathbb{R}_+)\right) \subset \bigcup_{\sigma'<\sigma} \phi_{\sigma'}^{-1}([0, \varepsilon_{\sigma'}]).\end{aligned}$$

For each $x \in \bigcup_{\sigma'<\sigma} \phi_{\sigma'}^{-1}([0, \varepsilon_{\sigma'}])$, let σ' be a minimal proper face of σ such that $x \in \phi_{\sigma'}^{-1}([0, \varepsilon_{\sigma'}])$. Then, we have

$$\psi_{\sigma'',\sigma'}(x) = \frac{\phi_{\sigma''}(x)}{\varepsilon_{\sigma''}} - \frac{\phi_{\sigma'}(x)}{\varepsilon_{\sigma'}} > 0 \text{ for any } \sigma'' < \sigma'$$

because $\phi_{\sigma''}(x) > \varepsilon_{\sigma''}$ and $\phi_{\sigma'}(x) \leqslant \varepsilon_{\sigma'}$. It follows from (6.2) that

$$x \in \phi_{\sigma'}^{-1}([0, \varepsilon_{\sigma'}]) \cap \bigcap_{\sigma''<\sigma'} \psi_{\sigma'',\sigma'}^{-1}(\mathbb{R}_+) = |S(\sigma', K')| \subset |K'[\partial\sigma]|,$$

where $\bigcap_{\sigma''<\sigma'} \psi_{\sigma'',\sigma'}^{-1}(\mathbb{R}_+) = |K|$ if $\dim \sigma' = 0$. Therefore, we have $|K'[\partial\sigma]| = \bigcup_{\sigma'<\sigma} \phi_{\sigma'}^{-1}([0, \varepsilon_{\sigma'}])$, which is (6.10).

To see (6.11), observe $|K'| = |K' \setminus K'[\partial\sigma]| \cup |K'[\partial\sigma]|$. For each $\tau \in K' \setminus K'[\partial\sigma]$, $\operatorname{rint}\tau$ misses any $\tau' \in K'[\partial\sigma]$, which means that $\operatorname{rint}\tau \subset |K'| \setminus |K'[\partial\sigma]|$ and $\tau \subset \operatorname{cl}(|K'| \setminus |K'[\partial\sigma]|)$. Hence, it follows that

$$\begin{aligned}|K' \setminus K'[\partial\sigma]| &= \operatorname{cl}\left(|K'| \setminus |K'[\partial\sigma]|\right) = \operatorname{cl}\left(|K'| \setminus \bigcup_{\sigma'<\sigma} \phi_{\sigma'}^{-1}([0, \varepsilon_{\sigma'}])\right) \\ &= \operatorname{cl} \bigcap_{\sigma'<\sigma} \phi_{\sigma'}^{-1}((\varepsilon_{\sigma'}, 1]) = \bigcap_{\sigma'<\sigma} \phi_{\sigma'}^{-1}([\varepsilon_{\sigma'}, 1]).\end{aligned}$$

The proof is complete. □

The following lemma is easily seen:

Lemma 6.8.2 *Let K' be a simplicial subdivision of K with the condition (†) and $\sigma \in K$. Then, the following hold:*

(1) *For each $\tau \in K'[\operatorname{rint}\sigma]$ with $\tau \not\subset \sigma$, $\tau \cap \sigma$ is a proper face of τ denoted by τ_σ, which has the opposite face $\tau_\sigma^* < \tau$. Then, $\tau_\sigma^* = \tau \cap \partial S_{K'}(\sigma) \in C(\sigma, K')$ and τ can be written as the join $\tau = \tau_\sigma \tau_\sigma^*$. If π is a hyperplane (i.e., a maximal proper flat) in the flat $\operatorname{fl}\tau$ such that $\pi \cap \tau \neq \emptyset$ and $\pi \cap (\tau_\sigma \cup \tau_\sigma^*) = \emptyset$, then π cuts τ into two cells which have a common face $\tau \cap \pi$.*
(2) *For $\tau \in K'[\operatorname{rint}\sigma] \cap K'[\partial\sigma]$ with $\tau \not\subset \sigma$, $\tau \cap \partial\sigma$ is a proper face of τ denoted by $\tau_{\partial\sigma}$, which has the opposite face $\tau_{\partial\sigma}^* < \tau$ and $\tau_{\partial\sigma}^* = \tau \cap T_{K'}(\sigma) \in K'[\sigma] \setminus$*

$K'[\partial\sigma]$. In fact, $\tau_{\partial\sigma} = \tau_{\sigma_0}$, where σ_0 is the largest proper face σ such that $\tau \cap \operatorname{rint}\sigma_0 \neq \emptyset$ (i.e., $\tau \in K'[\operatorname{rint}\sigma_0]$).[11] *If π is a hyperplane in the flat $\operatorname{fl}\tau^*_{\partial\sigma}$ such that $\pi \cap \tau^*_{\partial\sigma} \neq \emptyset$ and $\pi \cap ((\tau^*_{\partial\sigma})_\sigma \cup (\tau^*_{\partial\sigma})^*_\sigma) = \emptyset$, then $\tilde{\pi} = \operatorname{fl}(\tau_{\partial\sigma} \cup \pi)$ is a hyperplane in the flat $\operatorname{fl}\tau$ cutting τ into two cells which have a common face $\tau \cap \tilde{\pi} = \langle \tau_{\partial\sigma} \cup (\tau^*_{\partial\sigma} \cap \pi)\rangle$.*

In the above (2), each point of the join $\tau_{\partial\sigma}\tau^*_{\partial\sigma}$ can be written as

$$(1-t)x + ty, \ x \in \tau_{\partial\sigma}, \ y \in \tau^*_{\partial\sigma}, \ t \in \mathbf{I},$$

which is unique if $0 < t < 1$. Given a cell $C \subset \tau_T$, we have a cell $\langle \tau_{\partial\sigma} \cup C\rangle \subset \tau$, which can be written as $\tau_{\partial\sigma}C$ because

$$\langle \tau_{\partial\sigma} \cup C\rangle = \big\{(1-t)x + ty \ \big| \ x \in \tau_{\partial\sigma}, \ y \in C, \ t \in \mathbf{I}\big\}.$$

When $\sigma \in K$ is not principal, we define

$$\delta(\sigma, K') = \min\big\{\phi_\sigma(v) \ \big| \ v \in C(\sigma, K')^{(0)}\big\} > 0.$$

Because of the condition (†), $\delta(\sigma, K') < 1$. Let $0 < \varepsilon < \delta(\sigma, K')$. For each simplex $\tau \in K'[\sigma] \setminus K'[\partial\sigma]$ ($\tau \subset T_{K'}(\sigma)$), $\tau \cap \phi_\sigma^{-1}(\varepsilon)$ is the intersection of τ and a hyperplane in $\operatorname{fl}\tau$, which is a cell cutting τ into two cells $\tau \cap \phi_\sigma^{-1}([0, \varepsilon])$ and $\tau \cap \phi_\sigma^{-1}([\varepsilon, 1])$ as observed in Lemma 6.8.2(1). Note that $\tau_\sigma = \tau \cap \phi_\sigma^{-1}(0)$, $\tau^*_\sigma < \tau \cap \phi_\sigma^{-1}([\varepsilon, 1])$, and $\tau \cap \phi_\sigma^{-1}(1) = \emptyset$. Then, we have the following cellular subdivision of $F(\tau)$:

$$F_\varepsilon(\tau) = F(\tau \cap \phi_\sigma^{-1}([0, \varepsilon])) \cup F(\tau \cap \phi_\sigma^{-1}([\varepsilon, 1])),$$

where $\tau_\sigma, \tau^*_\sigma \in F_\varepsilon(\tau)$.

Due to Lemma 6.8.2(2), each $\tau \in K'[\operatorname{rint}\sigma] \cap K'[\partial\sigma]$ ($\tau \subset R_{K'}(\sigma)$) can be written as the join $\tau = \tau_{\partial\sigma}\tau^*_{\partial\sigma}$. Since the cell $\tau^*_{\partial\sigma} \cap \phi_\sigma^{-1}(\varepsilon)$ is the intersection of $\tau^*_{\partial\sigma}$ and a hyperplane in the flat $\operatorname{fl}\tau^*_{\partial\sigma}$, we have a cell

$$\tau_{\partial\sigma}(\tau^*_{\partial\sigma} \cap \phi_\sigma^{-1}(\varepsilon)) = \langle \tau_{\partial\sigma} \cup (\tau^*_{\partial\sigma} \cap \phi_\sigma^{-1}(\varepsilon))\rangle,$$

which is the intersection of τ and a hyperplane in the flat $\operatorname{fl}\tau$ cutting τ into two cells $\tau_{\partial\sigma}(\tau^*_{\partial\sigma} \cap \phi_\sigma^{-1}([0, \varepsilon])$ and $\tau_{\partial\sigma}(\tau^*_{\partial\sigma} \cap \phi_\sigma^{-1}([\varepsilon, 1])$. Then, we have the following cellular subdivision of $F(\tau)$:

$$F_\varepsilon(\tau) = F(\tau_{\partial\sigma}(\tau^*_{\partial\sigma} \cap \phi_\sigma^{-1}([0, \varepsilon]))) \cup F(\tau_{\partial\sigma}(\tau^*_{\partial\sigma} \cap \phi_\sigma^{-1}([\varepsilon, 1]))),$$

where $\tau_\sigma = \tau_{\partial\sigma}(\tau^*_{\partial\sigma})_\sigma$, $\tau \cap \partial S_{K'}(\sigma) = \tau_{\partial\sigma}(\tau^*_{\partial\sigma})^*_\sigma \in F_\varepsilon(\tau)$. Refer to Fig. 6.6.

---

[11] The condition (†) means that $\leqslant$ is a total order on $\{\sigma' < \sigma \mid \tau \cap \operatorname{rint}\sigma'\}$.

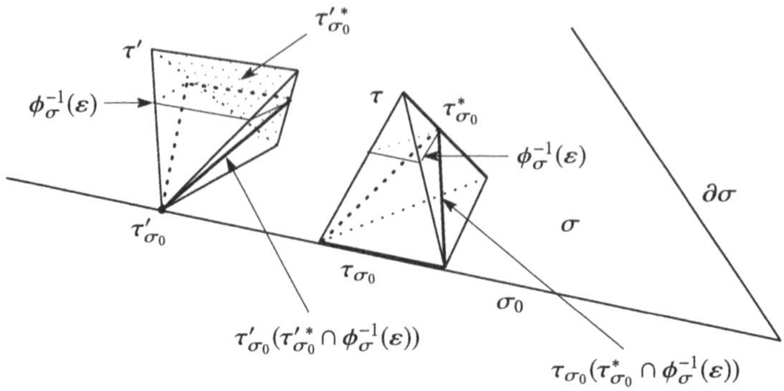

Fig. 6.6 $\tau_{\sigma_0}(\tau^*_{\sigma_0} \cap \phi_\sigma^{-1}([0, \varepsilon])$

As is easily observed, the following are cellular subdivisions of $K'_{T(\sigma)}$, $K'_{R(\sigma)}$, and $K'_{S(\sigma)}$:

$$T_\varepsilon(\sigma, K') = \bigcup_{\tau \in K'[\sigma] \setminus K'[\partial\sigma]} F_\varepsilon(\tau) = \big\{\tau_\sigma,\ \tau^*_\sigma,\ \tau \cap \phi_\sigma^{-1}(\varepsilon),\ \tau \cap \phi_\sigma^{-1}([0, \varepsilon]),$$
$$\tau \cap \phi_\sigma^{-1}([\varepsilon, 1]) \ \big|\ \tau \in K'[\sigma] \setminus K'[\partial\sigma]\big\},$$
$$R_\varepsilon(\sigma, K') = \bigcup_{\tau \in K'[\operatorname{rint}\sigma] \cap K'[\partial\sigma]} F_\varepsilon(\tau) = \big\{\tau_{\partial\sigma},\ \tau',\ \tau_{\partial\sigma}\tau' \ \big|\ \tau' \in F_\varepsilon(\tau^*_{\partial\sigma}),$$
$$\tau \in K'[\operatorname{rint}\sigma] \cap K'[\partial\sigma]\big\},$$
$$S_\varepsilon(\sigma, K') = T_\varepsilon(\sigma, K') \cup R_\varepsilon(\sigma, K'),$$

where $\partial S(\sigma, K')$ is a subcomplex of $S_\varepsilon(\sigma, K')$ and

$$S_\varepsilon(\sigma, K')^{(0)} = S(\sigma, K')^{(0)} \cup \big(\big|T(\sigma, K')^{(1)}\big| \cap \phi_\sigma^{-1}(\varepsilon)\big).$$

Giving a total order on $S_\varepsilon(\sigma, K')^{(0)}$, we can apply Theorem 1.6.1 to obtain a simplicial subdivision $S'_\varepsilon(\sigma, K') \lhd S_\varepsilon(\sigma, K')$ with $S'_\varepsilon(\sigma, K')^{(0)} = S_\varepsilon(\sigma, K')^{(0)}$. It should be noted that $\partial S(\sigma, K')$ is a subcomplex of $S'_\varepsilon(\sigma, K')$.

Lemma 6.8.3 *Let K' be a simplicial subdivision of K with the condition (†) and J a subcomplex of K. Suppose that K' is flattened along $J^{(n-1)}$, that is, there are $\varepsilon_\sigma \in (0, 1/2)$, $\sigma \in J^{(n-1)}$, such that $\varepsilon_{\sigma'} > \varepsilon_\sigma$ if $\sigma' < \sigma$ and (6.1)–(6.5) hold. For each n-simplexes of J, let*

$$0 < \varepsilon_\sigma < \min\big\{\delta(\sigma, K'),\ \varepsilon_{\sigma'} \ \big|\ \sigma' < \sigma\big\}.$$

If K'' is a simplicial subdivision of K' obtained by replacing $S(\sigma, K')$ with $S'_{\varepsilon_\sigma}(\sigma, K')$ for every n-simplex $\sigma \in J$, that is,

$$K'' = \Big(K' \setminus \bigcup_{\sigma \in J(n)} S(\sigma, K')\Big) \cup \bigcup_{\sigma \in J(n)} S'_{\varepsilon_\sigma}(\sigma, K'),$$

then K'' is flattened along $J^{(n)}$, where $J(n) = J^{(n)} \setminus J^{(n-1)}$ is the set of all n-simplexes of J.

Proof First of all, K'' satisfies (6.1) because

$$\begin{aligned} S(\sigma, K'')^{(0)} &= S'_{\varepsilon_\sigma}(\sigma, K')^{(0)} = S_{\varepsilon_\sigma}(\sigma, K')^{(0)} \\ &= S(\sigma, K')^{(0)} \cup \big(\big|T(\sigma, K')^{(1)}\big| \cap \phi_\sigma^{-1}(\varepsilon)\big). \end{aligned}$$

By virtue of Lemma 6.6.2, it is not necessary to show the equality (6.2), which comes from (6.4) and (6.5).

To prove (6.4) for $\sigma \in J(n)$ and K'', we show

$$\big|K'[\sigma] \setminus K'[\partial\sigma]\big| = \big|K'[\sigma]\big| \cap \big|K' \setminus K'[\partial\sigma]\big|.$$

The left hand set is obviously contained in the right hand. To see the opposite inclusion, let $x \in |K'[\sigma]| \cap |K' \setminus K'[\partial\sigma]|$. Then, we take $\tau \in K'[\sigma]$ and $\tau' \in K' \setminus K'[\partial\sigma]$ such that $x \in \operatorname{rint}\tau$ and $x \in \operatorname{rint}\tau'$. Since $\operatorname{rint}\tau \cap \operatorname{rint}\tau' \neq \emptyset$, it follows that $\tau = \tau' \in K'[\sigma] \setminus K'[\partial\sigma]$. Hence, $x \in |K'[\sigma] \setminus K'[\partial\sigma]|$.

Note that $\phi_\sigma^{-1}([0, \varepsilon_\sigma]) \subset |K'[\sigma]|$ because $\varepsilon_\sigma < \delta(\sigma, K')$. Now, from the definition of $T_{\varepsilon_\sigma}(\sigma, K')$ and the equality (6.11), it follows that

$$\begin{aligned} T_{K''}(\sigma) &= \bigcup_{\tau \in K'[\sigma] \setminus K'[\partial\sigma]} (\tau \cap \phi_\sigma^{-1}([0, \varepsilon_\sigma])) = \big|K'[\sigma] \setminus K'[\partial\sigma]\big| \cap \phi_\sigma^{-1}([0, \varepsilon_\sigma]) \\ &= \phi_\sigma^{-1}([0, \varepsilon_\sigma]) \cap \big|K' \setminus K'[\partial\sigma]\big| = \phi_\sigma^{-1}([0, \varepsilon_\sigma]) \cap \bigcap_{\sigma' < \sigma} \phi_{\sigma'}^{-1}([\varepsilon_{\sigma'}, 1]). \end{aligned}$$

To prove (6.5) for $\sigma \in J(n)$ and K'', note that

$$K''[\operatorname{rint}\sigma] \cap K''[\partial\sigma] \prec K'[\operatorname{rint}\sigma] \cap K'[\partial\sigma].$$

So, the following can be obtained from (6.10):

$$R_{K''}(\sigma) \subset R_{K'}(\sigma) \subset \big|K'[\partial\sigma]\big| = \bigcup_{\sigma' < \sigma} \phi_{\sigma'}^{-1}([0, \varepsilon_{\sigma'}]). \tag{6.12}$$

For each $\tau \in K'[\operatorname{rint}\sigma] \cap K'[\partial\sigma]$, we can write

$$\tau = \tau_{\partial\sigma}(\tau^*_{\partial\sigma} \cap \phi_\sigma^{-1}([0, \varepsilon_\sigma])) \cup \tau_{\partial\sigma}(\tau^*_{\partial\sigma} \cap \phi_\sigma^{-1}([\varepsilon_\sigma, 1])),$$

where

$$\tau_{\partial\sigma}(\tau^*_{\partial\sigma} \cap \phi_\sigma^{-1}([0, \varepsilon_\sigma]) \in K''[\operatorname{rint}\sigma] \cap K''[\partial\sigma] \text{ but}$$
$$\tau_{\partial\sigma}(\tau^*_{\partial\sigma} \cap \phi_\sigma^{-1}([\varepsilon_\sigma, 1]) \notin K''[\operatorname{rint}\sigma].$$

Hence, it follows that

$$R_{K''}(\sigma) = \bigcup_{\tau \in K'[\operatorname{rint}\sigma] \cap K'[\partial\sigma]} \tau_{\partial\sigma}(\tau^*_{\partial\sigma} \cap \phi_\sigma^{-1}([0, \varepsilon_\sigma])). \tag{6.13}$$

For each $\tau \in K'[\operatorname{rint}\sigma] \cap K'[\partial\sigma]$, let σ_0 be a maximal proper face of σ such that $\tau \cap \operatorname{rint}\sigma_0 \neq \emptyset$. Then, by (6.1), we have

$$\tau_{\partial\sigma} = \tau_{\sigma_0} = \tau \cap \phi_{\sigma_0}^{-1}(0) \text{ and } \tau^*_{\partial\sigma} = \tau^*_{\sigma_0} = \tau \cap \phi_{\sigma_0}^{-1}(\varepsilon_{\sigma_0}).$$

Moreover, $\tau_{\sigma_0} \subset \psi_{\sigma_0,\sigma}^{-1}(0)$ and $\tau^*_{\sigma_0} \cap \phi_\sigma^{-1}([0, \varepsilon_n]) \subset \psi_{\sigma_0,\sigma}^{-1}(\mathbb{R}_+)$ because

$$\psi_{\sigma_0,\sigma}(y) = \frac{\phi_{\sigma_0}(y)}{\varepsilon_{\sigma_0}} - \frac{\phi_\sigma(y)}{\varepsilon_\sigma} = 0 \text{ for each } y \in \tau_{\sigma_0} \text{ and}$$
$$\psi_{\sigma_0,\sigma}(z) = \frac{\phi_{\sigma_0}(z)}{\varepsilon_{\sigma_0}} - \frac{\phi_\sigma(z)}{\varepsilon_\sigma} \geqslant 0 \text{ for each } z \in \tau^*_{\sigma_0} \cap \phi_\sigma^{-1}([0, \varepsilon_\sigma]).$$

Hence, it follows that

$$\tau_{\partial\sigma}(\tau^*_{\partial\sigma} \cap \phi_\sigma^{-1}([0, \varepsilon_\sigma])) = \langle \tau_{\sigma_0} \cup (\tau^*_{\sigma_0} \cap \phi_\sigma^{-1}([0, \varepsilon_\sigma])) \rangle \subset \tau \cap \psi_{\sigma_0,\sigma}^{-1}(\mathbb{R}_+).$$

Similarly, we can obtain

$$\tau_{\partial\sigma}(\tau^*_{\partial\sigma} \cap \phi_\sigma^{-1}([\varepsilon_\sigma, 1])) \subset \tau \cap \psi_{\sigma_0,\sigma}^{-1}(-\mathbb{R}_+),$$

where $-\mathbb{R}_+ = (-\infty, 0]$. Consequently, we have

$$\tau_{\partial\sigma}(\tau^*_{\partial\sigma} \cap \phi_\sigma^{-1}([0, \varepsilon_\sigma])) = \tau \cap \psi_{\sigma_0,\sigma}^{-1}(\mathbb{R}_+). \tag{6.14}$$

Now, we show that

$$\tau \cap \psi_{\sigma_0,\sigma}^{-1}(\mathbb{R}_+) \subset \psi_{\sigma',\sigma}^{-1}(\mathbb{R}_+) \text{ for any } \sigma' < \sigma,$$

Let $x \in \tau \cap \psi_{\sigma_0,\sigma}^{-1}(\mathbb{R}_+)$. When $\sigma' < \sigma_0$, by (6.2) for σ_0 and K', we have

$$x \in \tau \subset \left|K'[\operatorname{rint}\sigma_0]\right| = S_{K'}(\sigma_0) \subset \psi_{\sigma',\sigma_0}^{-1}(\mathbb{R}_+).$$

Then, it follows that

$$\begin{aligned}\psi_{\sigma',\sigma}(x) &= \frac{\phi_{\sigma'}(x)}{\varepsilon_{\sigma'}} - \frac{\phi_\sigma(x)}{\varepsilon_\sigma} \\ &= \left(\frac{\phi_{\sigma'}(x)}{\varepsilon_{\sigma'}} - \frac{\phi_{\sigma_0}(x)}{\varepsilon_{\sigma_0}}\right) + \left(\frac{\phi_{\sigma_0}(x)}{\varepsilon_{\sigma_0}} - \frac{\phi_\sigma(x)}{\varepsilon_n}\right) \\ &= \psi_{\sigma',\sigma_0}(x) + \psi_{\sigma_0,\sigma}(x) \geqslant 0.\end{aligned}$$

To see that $\psi_{\sigma',\sigma}(x) \geqslant 0$ for any $\sigma' < \sigma$, assume the contrary, that is, $\psi_{\sigma',\sigma}(x) < 0$ for some $\sigma' < \sigma$. Let $\sigma_1 < \sigma$ be such a minimal proper face. Then, as we saw above, $\sigma_1 \not\leqslant \sigma_0$, which implies that $\tau \in K' \setminus K'[\operatorname{rint}\sigma_1]$. The following can be obtained by (6.2):

$$\tau \subset \left|K' \setminus K'[\operatorname{rint}\sigma_1]\right| = \phi_{\sigma_1}^{-1}([\varepsilon_{\sigma_1}, 1]) \cup \bigcup_{\sigma' < \sigma_1} \psi_{\sigma',\sigma_1}^{-1}(-\mathbb{R}_+),$$

where $\bigcup_{\sigma' < \sigma_1} \psi_{\sigma',\sigma_1}^{-1}(-\mathbb{R}_+) = \emptyset$ if $\dim \sigma_1 = 0$. If $x \in \phi_{\sigma_1}^{-1}([\varepsilon_{\sigma_1}, 1])$, that is, $\phi_{\sigma_1}(x) \geqslant \varepsilon_{\sigma_1}$, we have

$$\psi_{\sigma_1,\sigma}(x) = \frac{\phi_{\sigma_1}(x)}{\varepsilon_{\sigma_1}} - \frac{\phi_\sigma(x)}{\varepsilon_\sigma} \geqslant 0,$$

which is a contradiction. Hence, $x \in \bigcup_{\sigma' < \sigma_1} \psi_{\sigma',\sigma_1}^{-1}(-\mathbb{R}_+)$, that is, $\psi_{\sigma',\sigma_1}(x) \leqslant 0$ for some $\sigma' < \sigma_1$. By the minimality of σ_1, we have $\psi_{\sigma',\sigma}(x) \geqslant 0$. Then, it follows that

$$\psi_{\sigma_1,\sigma}(x) = \psi_{\sigma_1,\sigma'}(x) + \psi_{\sigma',\sigma}(x) = -\psi_{\sigma',\sigma_1}(x) + \psi_{\sigma',\sigma}(x) \geqslant 0.$$

This is a contradiction because $\psi_{\sigma_1,\sigma}(x) < 0$.

We have just obtained $\tau \cap \psi_{\sigma_0,\sigma}^{-1}(\mathbb{R}_+) \subset \bigcap_{\sigma' < \sigma} \psi_{\sigma',\sigma}^{-1}(\mathbb{R}_+)$. From (6.14), it follows that

$$\tau_{\partial\sigma}(\tau_{\partial\sigma}^* \cap \phi_\sigma^{-1}([0, \varepsilon_\sigma])) = \tau \cap \psi_{\sigma_0,\sigma}^{-1}(\mathbb{R}_+) = \tau \cap \bigcap_{\sigma' < \sigma} \psi_{\sigma',\sigma}^{-1}(\mathbb{R}_+).$$

Combining this with (6.12) and (6.13), we have the following inclusion:

$$R_{K''}(\sigma) \subset \bigcap_{\sigma' < \sigma} \psi_{\sigma',\sigma}^{-1}(\mathbb{R}_+) \cap \bigcup_{\sigma' < \sigma} \phi_{\sigma'}^{-1}([0, \varepsilon_{\sigma'}]).$$

To see the converse inclusion, let

$$x \in \bigcap_{\sigma'<\sigma} \psi_{\sigma',\sigma}^{-1}((0,\infty)) \cap \bigcup_{\sigma'<\sigma} \phi_{\sigma'}^{-1}((0,\varepsilon_{\sigma'})).$$

Since $\bigcup_{\sigma'<\sigma} \phi_{\sigma'}^{-1}([0,\varepsilon_{\sigma'}]) = |K'[\partial\sigma]|$ (6.10), $x \in \operatorname{rint}\tau$ for some $\tau \in K'[\partial\sigma]$, where $\tau^{(0)} \setminus \partial\sigma \neq \emptyset$. Let $\sigma_0 < \sigma$ be a maximal proper face such that $\tau \cap \operatorname{rint}\sigma_0 \neq \emptyset$, which means that $\tau \cap \partial\sigma = \tau \cap \sigma_0$. We show that $\tau \cap \operatorname{rint}\sigma \neq \emptyset$. On the contrary, assume $\tau \cap \operatorname{rint}\sigma = \emptyset$. Then, $\tau \cap \sigma = \tau \cap \sigma_0$, which implies that $\tau^{(0)} \cap \sigma = \tau^{(0)} \cap \sigma_0$. For each $v \in \tau^{(0)} \cap \sigma$,

$$\psi_{\sigma_0,\sigma}(v) = \frac{\phi_{\sigma_0}(v)}{\varepsilon_{\sigma_0}} - \frac{\phi_\sigma(v)}{\varepsilon_\sigma} = 0.$$

For each $v \in \tau^{(0)} \setminus \sigma$, $\phi_{\sigma_0}(v) = \varepsilon_{\sigma_0}$ by (6.1) and $\phi_\sigma(v) \geqslant \delta(\sigma, K') > \varepsilon_\sigma$. Hence,

$$\psi_{\sigma_0,\sigma}(v) = \frac{\phi_{\sigma_0}(v)}{\varepsilon_{\sigma_0}} - \frac{\phi_\sigma(v)}{\varepsilon_\sigma} < 0.$$

Since $x \in \operatorname{rint}\tau$, we have $\psi_{\sigma_0,\sigma}(x) < 0$, which contradicts $x \in \bigcap_{\sigma'<\sigma} \psi_{\sigma',\sigma}^{-1}((0,\infty))$. Consequently, $\tau \in K'[\operatorname{rint}\sigma] \cap K'[\partial\sigma]$. Thus, it follows that

$$x \in \tau \cap \psi_{\sigma_0,\sigma}^{-1}(\mathbb{R}_+) \subset R_{K''}(\sigma).$$

Now, we have shown that

$$\bigcap_{\sigma'<\sigma} \psi_{\sigma',\sigma}^{-1}((0,\infty)) \cap \bigcup_{\sigma'<\sigma} \phi_{\sigma'}^{-1}((0,\varepsilon_{\sigma'})) \subset R_{K''}(\sigma).$$

Since $|R(\sigma, K'')|$ is closed in $|K|$, it follows that

$$\bigcap_{\sigma'<\sigma} \psi_{\sigma',\sigma}^{-1}(\mathbb{R}_+) \cap \bigcup_{\sigma'<\sigma} \phi_{\sigma'}^{-1}([0,\varepsilon_{\sigma'}]) \subset R_{K''}(\sigma).$$

Concerning (6.4) and (6.5) for $\sigma' \in J^{(n-1)}$ and K'', note that $T_{K''}(\sigma') = T_{K'}(\sigma')$ and $R_{K''}(\sigma') = R_{K'}(\sigma')$ if $\sigma' \in J^{(n-1)}$ and $\sigma' \not< \sigma$ for any $\sigma \in J(n)$. In this case, (6.4) and (6.5) are contained in our assumption. When $\sigma' < \sigma$ for some $\sigma \in J(n)$, to verify (6.4) and (6.5) for σ' and K'', it suffices to prove that $T_{K''}(\sigma') = T_{K'}(\sigma')$ and $R_{K''}(\sigma') = R_{K'}(\sigma')$.

To see $T_{K''}(\sigma') = T_{K'}(\sigma')$, let $\tau \in K'[\sigma'] \setminus K'[\partial\sigma']$ If $\tau \notin K'[\operatorname{rint}\sigma]$, then $\tau \in K''[\sigma'] \setminus K''[\partial\sigma']$. When $\tau \in K'[\operatorname{rint}\sigma]$, since $\tau \in K'[\operatorname{rint}\sigma] \cap K'[\partial\sigma]$, it follows that

$$\tau = \tau_{\partial\sigma}\tau_{\partial\sigma}^* = \tau_{\partial\sigma}(\tau_{\partial\sigma}^* \cap \phi_\sigma^{-1}([0,\varepsilon_\sigma])) \cup \tau_{\partial\sigma}(\tau_{\partial\sigma}^* \cap \phi_\sigma^{-1}([\varepsilon_\sigma, 1])),$$

where $\tau^*_{\partial\sigma} \notin K'[\sigma']$ because $\tau^*_{\partial\sigma} \in K'[\operatorname{rint}\sigma] \setminus K'[\partial\sigma]$. Since $\tau \in K'[\sigma'] \setminus K'[\partial\sigma']$, we have $\tau_{\partial\sigma} \in K''[\sigma'] \setminus K''[\partial\sigma']$. Hence,

$$\tau_{\partial\sigma}(\tau^*_{\partial\sigma} \cap \phi_\sigma^{-1}([0, \varepsilon_\sigma])),\ \tau_{\partial\sigma}(\tau^*_{\partial\sigma} \cap \phi_\sigma^{-1}([\varepsilon_\sigma, 1]) \in K''[\sigma'] \setminus K''[\partial\sigma'].$$

Consequently, we have

$$T_{K''}(\sigma') = \left|K''[\sigma'] \setminus K''[\partial\sigma']\right| = \left|K'[\sigma'] \setminus K'[\partial\sigma']\right| = T_{K'}(\sigma').$$

To see $R_{K''}(\sigma') = R_{K'}(\sigma')$, let $\tau \in K'[\operatorname{rint}\sigma'] \cap K'[\partial\sigma']$. If $\tau \notin K'[\operatorname{rint}\sigma]$, then $\tau \in K''[\operatorname{rint}\sigma'] \cap K''[\partial\sigma']$. When $\tau \in K'[\operatorname{rint}\sigma]$, as above,

$$\tau = \tau_{\partial\sigma}\tau^*_{\partial\sigma} = \tau_{\partial\sigma}(\tau^*_{\partial\sigma} \cap \phi_\sigma^{-1}([0, \varepsilon_\sigma])) \cup \tau_{\partial\sigma}(\tau^*_{\partial\sigma} \cap \phi_\sigma^{-1}([\varepsilon_\sigma, 1])),$$

where $\tau^*_{\partial\sigma} \notin K'[\sigma']$. Since $\tau \in K'[\operatorname{rint}\sigma'] \cap K'[\partial\sigma']$, we have $\tau_{\partial\sigma} \in K''[\operatorname{rint}\sigma'] \cap K''[\partial\sigma']$. Hence,

$$\tau_{\partial\sigma}(\tau^*_{\partial\sigma} \cap \phi_\sigma^{-1}([0, \varepsilon_\sigma])),\ \tau_{\partial\sigma}(\tau^*_{\partial\sigma} \cap \phi_\sigma^{-1}([\varepsilon_\sigma, 1]) \in K''[\operatorname{rint}\sigma'] \cap K''[\partial\sigma'].$$

Consequently, we have

$$R_{K''}(\sigma') = \left|K''[\operatorname{rint}\sigma'] \cap K''[\partial\sigma']\right| = \left|K'[\operatorname{rint}\sigma'] \cap K'[\partial\sigma']\right| = R_{K'}(\sigma').$$

The proof is completed. □

Let $\sigma \in J$ be fixed. Then, $f(\sigma) \in f(J)$ need not be principal in L even if σ is principal in K. Since $f : K' \to L'$ is simplicial and injective, the following inclusions between subcomplexes follow from the definitions:

$$f(S(\sigma, K')) \subset S(f(\sigma), L'),\ f(\partial S(\sigma, K')) \subset \partial S(f(\sigma), L'),$$
$$f(T(\sigma, K')) \subset T(f(\sigma), L'),\ \text{and}\ f(R(\sigma, K')) \subset R(f(\sigma), L').$$

For each $\tau \in S(\sigma, K')$, $f|\tau : \tau \to f(\tau)$ is an affine homeomorphism. Moreover, if $\tau \in T(\sigma, K')$, then $f(\tau_\sigma) = f(\tau)_{f(\sigma)}$ and $f(\tau^*_\sigma) = f(\tau)^*_{f(\sigma)}$. If $\tau \in R(\sigma, K')$, then $f(\tau_{\partial\sigma}) = f(\tau)_{\partial f(\sigma)}$ and $f(\tau^*_{\partial\sigma}) = f(\tau)^*_{\partial f(\sigma)}$.

Let $0 < \varepsilon < \min\{\delta(\sigma, K'), \delta(f(\sigma), L')\}$. Then, we define an injective function $f_{\sigma,\varepsilon} : S_\varepsilon(\sigma, K') \to S_\varepsilon(f(\sigma), L')$ as follows:

On $T_\varepsilon(\sigma, K')$ (cf. Fig. 6.7), for each $\tau \in K'[\sigma] \setminus K'[\partial\sigma]$,

$$f_{\sigma,\varepsilon}(\tau_\sigma) = f(\tau)_{f(\sigma)},\ f_{\sigma,\varepsilon}(\tau^*_\sigma) = f(\tau)^*_{f(\sigma)},$$
$$f_{\sigma,\varepsilon}(\tau \cap \phi_\sigma^{-1}(\varepsilon)) = f(\tau) \cap \phi_{f(\sigma)}^{-1}(\varepsilon),$$
$$f_{\sigma,\varepsilon}(\tau \cap \phi_\sigma^{-1}([0, \varepsilon])) = f(\tau) \cap \phi_{f(\sigma)}^{-1}([0, \varepsilon]),$$
$$f_{\sigma,\varepsilon}(\tau \cap \phi_\sigma^{-1}([\varepsilon, 1])) = f(\tau) \cap \phi_{f(\sigma)}^{-1}([\varepsilon, 1]).$$

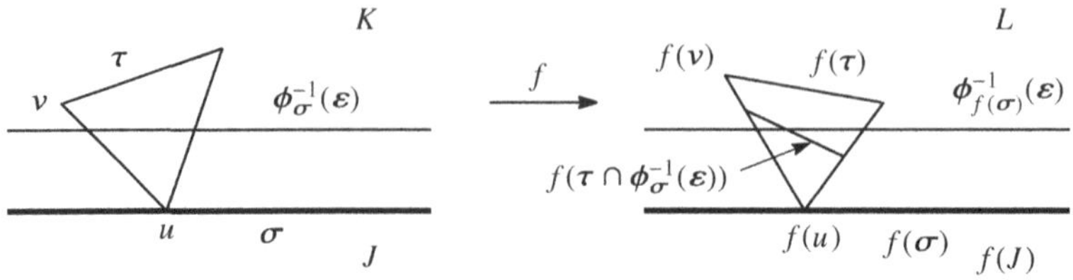

Fig. 6.7 $f'_{\sigma,\varepsilon}$ on $T_\varepsilon(\sigma, K')$

On $R_\varepsilon(\sigma, K')$, for each $\tau \in K'[\operatorname{rint}\sigma] \cap K'[\partial\sigma]$ and $\tau' \in F_\varepsilon(\tau^*_{\partial\sigma})$,

$$f_{\sigma,\varepsilon}(\tau_{\partial\sigma}) = f(\tau)_{\partial f(\sigma)}, \; f_{\sigma,\varepsilon}(\tau_{\partial\sigma}\tau') = f(\tau)_{\partial f(\sigma)} f'_{\sigma,\varepsilon}(\tau'),$$

where $f_{\sigma,\varepsilon}$ is already defined on $F_\varepsilon(\tau^*_{\partial\sigma}) \subset T_\varepsilon(\sigma, K')$.

It should be noticed that $\partial S(\sigma, K')$ and $\partial S(f(\sigma), L')$ are subcomplexes of $S_\varepsilon(\sigma, K')$ and $S_\varepsilon(f(\sigma), L')$, respectively. Then, $f_{\sigma,\varepsilon}(\tau) = f(\tau)$ if $\tau \in \partial S(\sigma, K')$ or $\tau \in K'$ with $\tau \subset \sigma$. Moreover, $f_{\sigma,\varepsilon}(S(\sigma, K')^{(0)}) \subset S(f(\sigma), L')^{(0)}$ and

$$f_{\sigma,\varepsilon}\big(\big|T(\sigma, K')^{(1)}\big| \cap \phi_\sigma^{-1}(\varepsilon)\big) \subset \big|T(f(\sigma), L')^{(1)}\big| \cap \phi_{f(\sigma)^{-1}}(\varepsilon).$$

We can give total orders on $S_\varepsilon(\sigma, K')^{(0)}$ and $S_\varepsilon(f(\sigma), L')^{(0)}$ so that $f_{\sigma,\varepsilon}|S(\sigma, K')^{(0)}$ is order preserving. Using these orders, we can apply Theorem 1.6.1 to obtain simplicial subdivisions

$$S'_\varepsilon(\sigma, K') \lhd S_\varepsilon(\sigma, K') \text{ and } S'_\varepsilon(f(\sigma), L') \lhd S_\varepsilon(f(\sigma), L'),$$

which contain $\partial S(\sigma, K')$ and $\partial S(f(\sigma), L')$ as subcomplexes, respectively. By Proposition 1.8.2, we can obtain a simplicial map $f'_{\sigma,\varepsilon} : S'_\varepsilon(\sigma, K') \to S'_\varepsilon(f(\sigma), L')$ as an extension of $f_{\sigma,\varepsilon}|S(\sigma, K')^{(0)}$, where

$$f'_{\sigma,\varepsilon}||\partial S(\sigma, K')| \cup \sigma = f||\partial S(\sigma, K')| \cup \sigma$$

because $f_{\sigma,\varepsilon}|\partial S(\sigma, K')^{(0)} = f|\partial S(\sigma, K')^{(0)}$ and $f_{\sigma,\varepsilon}|K'^{(0)} \cap \sigma = f|K'^{(0)} \cap \sigma$. Moreover,

$$f'_{\sigma,\varepsilon}(\tau) = f(\tau) \text{ for each } \tau \in S_\varepsilon(\sigma, K').$$

Since $f'_{\sigma,\varepsilon}|S(\sigma, K')^{(0)} = f_{\sigma,\varepsilon}|S(\sigma, K')^{(0)}$ is an injection, $f'_{\sigma,\varepsilon}$ is a simplicial embedding. Additionally, $f'_{\sigma,\varepsilon}$ can be extended to a simplicial embedding

$$f' : \big(K' \setminus S(\sigma, K')\big) \cup S_\varepsilon(\sigma, K') \to \big(L' \setminus S(\sigma, L')\big) \cup S_\varepsilon(\sigma, L')$$

by $f'|L' \setminus S(\sigma, L') = f|L' \setminus S(\sigma, L')$ because $f'_{\sigma,\varepsilon}||\partial S(\sigma, K')| = f||\partial S(\sigma, K')|$. Then, it should be noted that

$$f'(\tau) = f(\tau) \quad \text{for each } \tau \in S_\varepsilon(\sigma, K').$$

We now prove Proposition 6.6.4.

Proof (Proposition 6.6.4) As mentioned before, we can assume that both K' and L' satisfy condition (†) for K and L, respectively. Let

$$0 < \varepsilon_0 < \min\left\{\delta(v, K'),\ \delta(f(v), L') \;\middle|\; v \in J^{(0)}\right\}.$$

For each $v \in J^{(0)}$, we have cellular subdivisions $S_{\varepsilon_0}(v, K')$ and $S_{\varepsilon_0}(f(v), L')$ of $S(v, K')$ and $S(f(v), L')$, which contain $\partial S(v, K')$ and $\partial S(f(v), L')$ as subcomplexes, respectively. We also have an injection $f_{v,\varepsilon_0} : S_{\varepsilon_0}(v, K') \to S_{\varepsilon_0}(f(v), L')$ induced by the simplicial embedding f. As mentioned just before the lemma, giving total orders on $S_{\varepsilon_0}(v, K')^{(0)}$ and $S_{\varepsilon_0}(f(v), L')^{(0)}$ so that $f_{v,\varepsilon_0}|S(v, K')^{(0)}$ is order preserving, we can define simplicial subdivisions

$$S'_{\varepsilon_0}(v, K') \lhd S_{\varepsilon_0}(v, K') \quad \text{and} \quad S'_{\varepsilon_0}(f(v), L') \lhd S_{\varepsilon_0}(f(v), L'),$$

and a simplicial embedding $f'_{v,\varepsilon_0} : S'_{\varepsilon_0}(v, K') \to S'_{\varepsilon_0}(f(v), L')$ such that

$$f'_{v,\varepsilon_0}||\partial S(v, K')| \cup \{v\} = f||\partial S(v, K')| \cup \{v\} \quad \text{and}$$
$$f'_{v,\varepsilon_0}(\tau) = f(\tau) \quad \text{for each } \tau \in S(v, K'),$$

where $\partial S(v, K') \subset S'_{\varepsilon_0}(v, K')$ and $\partial S(f(v), L') \subset S'_{\varepsilon_0}(f(v), L')$.

We can define $K_0 \lhd K'$ and $L_0 \lhd L'$ as follows:

$$K_0 = \left(K' \setminus \bigcup_{v \in J^{(0)}} S(v, K')\right) \cup \bigcup_{v \in J^{(0)}} S'_{\varepsilon_0}(v, K') \quad \text{and}$$

$$L_0 = \left(L' \setminus \bigcup_{v \in J^{(0)}} S(f(v), L')\right) \cup \bigcup_{v \in J^{(0)}} S'_{\varepsilon_0}(f(v), L'),$$

that is, they are obtained by replacing each $S(v, K')$ and $S(f(v), L')$ with $S'_{\varepsilon_0}(v, K')$ and $S'_{\varepsilon_0}(f(v), L')$, respectively. Then, they are flattened along $J^{(0)}$ and $f(J)^{(0)}$, respectively. Indeed, for each $v \in J^{(0)}$,

$$S(v, K_0)^{(0)} = T(v, K_0)^{(0)} = \left|T(v, K')^{(1)}\right| \cap \phi_v^{-1}(\varepsilon_0),$$
$$\left|S(v, K_0)\right| = \left|T(v, K_0)\right| = \phi_v^{-1}([0, \varepsilon_0]),$$

$$S(f(v), L_0)^{(0)} = T(f(v), L_0)^{(0)} = \left|T(f(v), L')^{(1)}\right| \cap \phi_{f(v)}^{-1}(\varepsilon_0), \text{ and}$$
$$\left|S(f(v), L_0)\right| = \left|T(f(v), L_0)\right| = \phi_{f(v)}^{-1}([0, \varepsilon_0]).$$

Hence, K_0 and L_0 satisfy (6.1), (6.2), and (6.4). For (6.5), there is nothing to be verified because $R(v, K_0) = R(f(v), L_0) = \emptyset$ for any $v \in J^{(0)}$. Thus, $K_0 \lhd K'$ and $L_0 \lhd L'$ are flattened along $J^{(0)}$ and $f(J)^{(0)}$, respectively. Additionally, we can define a simplicial embedding $f_0 : K_0 \to L_0$ as follows:

$$f_0|S'_{\varepsilon_0}(v, K') = f'_{v,\varepsilon_0} \quad \text{for each } v \in J^{(0)} \text{ and}$$
$$f_0\big|\left|K \setminus K'[J^{(0)}]\right| = f\big|\left|K \setminus K'[J^{(0)}]\right|.$$

Then, it follows that $f_0||J| = f||J|$ and $f_0(\tau) = f(\tau)$ for each $\tau \in K'[J^{(0)}]$.

Suppose that $\varepsilon_0 > \varepsilon_1 > \cdots > \varepsilon_{n-1} > 0$, $K_{n-1} \lhd K'$, $L_{n-1} \lhd L'$, and a simplicial embedding $f_{n-1} : K_{n-1} \to L_{n-1}$ have been defined so that (6.1)–(6.5) hold for $(\varepsilon_i)_{i=0}^{n-1}$, $\sigma \in J^{(n-1)}$ and K_{n-1}, and for $(\varepsilon_i)_{i=0}^{n-1}$, $f(\sigma) \in f(J)^{(n-1)}$ and L_{n-1} (i.e., K_{n-1} and L_{n-1} are flattened along $J^{(n-1)}$ and $f(J)^{(n-1)}$, respectively),

$$f_{n-1}\big|\left|K \setminus K'[J^{(n-1)}]\right| \cup |J| = f\big|\left|K \setminus K'[J^{(n-1)}]\right| \cup |J|, \text{ and}$$
$$f_{n-1}(\tau) = f(\tau) \quad \text{for each } \tau \in K'[J^{(n-1)}].$$

Then, we take

$$0 < \varepsilon_n < \min\left\{\varepsilon_{n-1},\ \delta(\sigma, K_{n-1}),\ \delta(f(\sigma), L_{n-1}) \;\middle|\; \sigma \in J(n)\right\},$$

where we write $J(n) = J^{(n)} \setminus J^{(n-1)}$. For each $\sigma \in J(n)$, we have cellular subdivisions $S_{\varepsilon_n}(\sigma, K_{n-1})$ and $S_{\varepsilon_n}(f(\sigma), L_{n-1})$ of $S(\sigma, K_{n-1})$ and $S(f(\sigma), L_{n-1})$, which contain $\partial S(\sigma, K_{n-1})$ and $\partial S(f(\sigma), L_{n-1})$ as subcomplexes, respectively. Moreover, we have an injection $f_{\sigma,\varepsilon_n} : S_{\varepsilon_n}(\sigma, K_{n-1}) \to S_{\varepsilon_n}(f(\sigma), L_{n-1})$ induced by the simplicial embedding f_{n-1}. As mentioned just before the lemma, giving total orders on $S_{\varepsilon_n}(\sigma, K_{n-1})^{(0)}$ and $S_{\varepsilon_n}(f(\sigma), L_{n-1})^{(0)}$ so that $f_{\sigma,\varepsilon_n}|S(v, K_{n-1})^{(0)}$ is order preserving, we can define simplicial subdivisions

$$S'_{\varepsilon_n}(\sigma, K_{n-1}) \lhd S_{\varepsilon_n}(\sigma, K_{n-1}) \text{ and } S'_{\varepsilon_n}(f(\sigma), L_{n-1}) \lhd S_{\varepsilon_n}(f(\sigma), L_{n-1}),$$

and a simplicial embedding $f'_{\sigma,\varepsilon_n} : S'_{\varepsilon_n}(\sigma, K_{n-1}) \to S'_{\varepsilon_n}(f(\sigma), L_{n-1})$ such that

$$f'_{\sigma,\varepsilon_n}||\partial S(\sigma, K_{n-1})| \cup \sigma = f||\partial S(\sigma, K_{n-1})| \cup \sigma \quad \text{and}$$
$$f'_{\sigma,\varepsilon_n}(\tau) = f(\tau) \quad \text{for each } \tau \in S(\sigma, K_{n-1}),$$

where $\partial S(\sigma, K_{n-1}) \subset S'_{\varepsilon_n}(\sigma, K_{n-1})$ and $\partial S(f(\sigma), L_{n-1}) \subset S'_{\varepsilon_n}(f(\sigma), L_{n-1})$.

We can define $K_n \lhd K_{n-1}$ and $L_n \lhd L_{n-1}$ as follows:

$$K_n = \Big(K_{n-1} \setminus \bigcup_{\sigma \in J^{(n)}} S(\sigma, K_{n-1})\Big) \cup \bigcup_{\sigma \in J^{(n)}} S'_{\varepsilon_n}(\sigma, K_{n-1}) \text{ and}$$

$$L_n = \Big(L_{n-1} \setminus \bigcup_{\sigma \in J^{(n)}} S(f(\sigma), L_{n-1})\Big) \cup \bigcup_{\sigma \in J^{(n)}} S'_{\varepsilon_n}(f(\sigma), L_{n-1}),$$

that is, they are obtained by replacing each $S(\sigma, K_{n-1})$ and $S(f(\sigma), L_{n-1})$ with $S'_{\varepsilon_n}(\sigma, K_{n-1})$ and $S'_{\varepsilon_n}(f(\sigma), L_{n-1})$, respectively. Then, by Lemma 6.8.3, (6.1)–(6.5) hold for $(\varepsilon_i)_{i=0}^n$, $\sigma \in J^{(n)}$ and K_n, and for $(\varepsilon_i)_{i=0}^n$, $f(\sigma) \in f(J)^{(n)}$ and L_n (i.e., K_n and L_n are flattened along $J^{(n)}$ and $f(J)^{(n)}$, respectively). Additionally, we can define a simplicial embedding $f_n : K_n \to L_n$ as follows:

$$f_n|S'_{\varepsilon_n}(\sigma, K_{n-1}) = f'_{\sigma,\varepsilon_n} \quad \text{for each } \sigma \in J(n) \text{ and}$$

$$f_n\big||K \setminus K'[J^{(n)}]\big| = f\big||K \setminus K'[J^{(n)}]\big|.$$

As is easily observed, $f_n||J| = f_{n-1}||J| = f||J|$ and

$$f_n(\tau) = f_{n-1}(\tau) = f(\tau) \quad \text{for each } \tau \in K'[J^{(n)}].$$

By induction, we can obtain $\varepsilon_0 > \varepsilon_1 > \cdots > \varepsilon_{\dim J} > 0$, $K'' = K_{\dim J} \lhd K'$, $L'' = L_{\dim J} \lhd L'$, and a simplicial embedding $f'' = f_{\dim J} : K^\flat \to L^\flat$ such that (6.1)–(6.5) hold for $(\varepsilon_i)_{i=0}^{\dim J}$, $\sigma \in J$ and K'', and for $(\varepsilon_i)_{i=0}^{n-1}$, $f(\sigma) \in f(J)$ and L'', that is, K'' and L'' are flattened along J and $f(J)$, respectively,

$$f''\big||K \setminus K'[J]\big| \cup |J| = f\big||K \setminus K'[J]\big| \cup |J|, \text{ and}$$

$$f''(\tau) = f(\tau) \quad \text{for each } \tau \in K'[J^{(n)}].$$

The proof is completed. □

6.9 Combinatorial ∞-Manifolds and PL ∞-Manifolds

Recall that simplicial complexes K and L are **combinatorially equivalent** (written as $K \cong L$) if they admit simplicially isomorphic subdivisions, equivalently $|K|$ and $|L|$ are PL homeomorphic. An ***n*-dimensional combinatorial manifold** (or a **combinatorial *n*-manifold**) is a simplicial complex K such that the star of every vertex is combinatorially equivalent to the standard n-simplex Δ^n (precisely, the natural triangulation of Δ^n, i.e., the simplicial complex $F(\Delta^n)$ consisting of all faces of Δ^n). The polyhedron $|K|$ of a combinatorial n-manifold K is an n-manifold because $\Delta^n \approx \mathbf{I}^n$. However, there exists an n-manifold that is not triangulated by any combinatorial n-manifold. In fact, it is known that a non-triangulable n-manifold exists. In Geometric Topology, there had been a very big longstanding conjecture called the **Hauptvermutung**,[12] which claims that any two triangulations of a triangulable space have a common subdivision, equivalently any two homeomorphic simplicial complexes are combinatorially equivalent. This conjecture is false even for combinatorial n-manifolds—Refer to [36, Section 9].

The **standard ∞-simplex** Δ^∞ is defined as the convex hall of the unit vectors $\mathbf{e}_i \in \mathbb{R}^{\mathbb{N}}$, $i \in \mathbb{N}$, defined by $\mathbf{e}_i(i) = 1$ and $\mathbf{e}_i(j) = 0$ for $j \neq i$. Then, $F(\Delta^\infty)$ is the countable full simplicial complex with vertices $\mathbf{e}_i$, $i \in \mathbb{N}$. A simplicial complex K is called an **infinite-dimensional combinatorial manifold** (or a **combinatorial ∞-manifold**) if the star of every vertex is combinatorially equivalent to $F(\Delta^\infty)$, i.e., $\mathrm{St}(v, K) \cong F(\Delta^\infty)$ for every $v \in K^{(0)}$. We define an **infinite-dimensional PL manifold** (or a **PL ∞-manifold**) as a polyhedron M such that each point $x \in M$ has a polyhedral neighborhood N (i.e., N is a subpolyhedron of M with $x \in \operatorname{int} N$) that is PL homeomorphic to Δ^∞. Evidently, the polyhedron $|K|$ of a combinatorial ∞-manifold K is a PL ∞-manifold. Since $\Delta^\infty \approx \mathbb{R}^\infty$ by Corollary 6.4.3, every PL ∞-manifold is an $\mathbb{R}^\infty$-manifold. The converse implications are also true. In Sect. 6.11, we will actually prove that every triangulation of each $\mathbb{R}^\infty$-manifold is a combinatorial ∞-manifold (Theorem 6.11.8). Consequently, for a simplicial complex K, the following are equivalent (cf. Theorem 6.11.1):

- K is a combinatorial ∞-manifold;
- $|K|$ is a PL ∞-manifold;
- $|K|$ is an $\mathbb{R}^\infty$-manifold.

Then, it follows that every simplicial subdivision of a combinatorial ∞-manifold is a combinatorial ∞-manifold (Corollary 6.11.2). It should be noted that this fact is non-trivial.

For combinatorial ∞-manifolds, the Hauptvermutung holds, that is:

Theorem 6.9.1 (Hauptvermutung) *Any two homeomorphic combinatorial ∞-manifolds are combinatorially equivalent.*

In this section, we will prove the following strong version:

Theorem 6.9.2 (Strong Hauptvermutung) *Let K and L be triangulations of PL ∞-manifolds. If $|K|$ and $|L|$ have the same homotopy type, then K and L are combinatorially equivalent. In other words, the following implication holds:*

$$|K| \simeq |L| \;\Rightarrow\; K \cong L$$

[12] German for *main conjecture* (cf. p. 44).

First, we show the following:

Lemma 6.9.3 *Let K be a triangulation of a PL ∞-manifold M and $x \in M$. For each neighborhood U of x in M, there exists a simplicial subdivision $K' \lhd K$ and $x \in K'^{(0)}$ such that $|\operatorname{St}(x, K')| \subset U$ and $\operatorname{St}(x, K')$ is combinatorially equivalent to $F(\Delta^\infty)$.*

Proof Let N be a polyhedral neighborhood of x in M that is PL homeomorphic Δ^∞. By virtue of Propositions 1.6.6, 1.6.8, and Theorem 1.8.7, we have simplicial subdivisions $K' \lhd K$ and $L \lhd F(\Delta^\infty)$, a subcomplex $K'_0 \subset K'$ with $|K_0| = N$, and a simplicial isomorphism $\varphi : K'_0 \to L$. Let $v_0 = \varphi(x)$. Changing the coordinates if necessary, we may assume that only the first $(n+1)$-coordinates of v_0 are positive (and further coordinates are all zero), that is, Δ^n is the carrier of v_0 in $F(\Delta^\infty)$. Let $S = F(\Delta^n)_{v_0}$ be the starring of $F(\Delta^n)$ at v_0.[13] Choose $s \in (0, 1)$ so that

$$D_0 = (\beta^S_{v_0})^{-1}([s, 1]) \subset \varphi(U \cap N) \cap \operatorname{int} |L[v_0]| \cap \Delta^n,$$

where D_0 is an n-simplex. We define a cellular subdivision S_0 of $F(\Delta^n)$ as follows:

$$S_0 = \left\{ v_0\sigma \cap (\beta^S_{v_0})^{-1}(\tau) \;\middle|\; \sigma < \Delta^n, \tau \in I_s \right\},$$

where $I_s = \{0, s, 1, [0, s], [s, 1]\}$. Then, $\operatorname{St}(v_0, S_0)$ is a simplicial complex with $|\operatorname{St}(v_0, S_0)| = D_0$. For each $i \in \mathbb{N}$, inductively choose $s_i \in (0, 1)$ so that

$$D_i = ((1 - s_i)v_0 + s_i\mathbf{e}_{i+1})D_{i-1} \subset \varphi(U \cap N) \cap \operatorname{int} |L[v_0]| \cap \Delta^{n+i+1},$$

where D_i is the join of $v_i = (1 - s_i)v_0 + s_i\mathbf{e}_{i+1}$ and D_{i-1}. Then, each D_i is an $(n+i)$-simplex and $D_\infty = \bigcup_{i\in\mathbb{N}} D_i$ is affinely homeomorphic to Δ^∞. We inductively define cellular subdivisions S_i of $F(\Delta^{n+i})$, $i \in \mathbb{N}$, as follows:

$$S_i = S_{i-1} \cup (F(\partial\Delta^{n+i}) \setminus F(\Delta^{n+i-1})) \\ \cup \left\{ v_i\sigma \;\middle|\; \sigma \in S_{i-1} \cup (F(\partial\Delta^{n+i}) \setminus F(\Delta^{n+i-1})) \right\}.$$

Then, each $\operatorname{St}(v_0, S_i)$ is a simplicial complex with $|\operatorname{St}(v_0, S_i)| = D_i$. Moreover, $S_\infty = \bigcup_{i\in\omega} S_i$ is a subdivision of $F(\Delta^\infty)$ and $|\operatorname{St}(v_0, S_\infty)| = D_\infty$. Due to Theorem 1.6.4, L and S_∞ have the following common subdivision:

$$L' = \left\{ \sigma \cap C \;\middle|\; \sigma \in L,\ C \in S_\infty,\ \sigma \cap C \neq \emptyset \right\}.$$

Since $|\operatorname{St}(v_0, S_i)| = D_i \subset |L[v_0]| \cap \varphi(U \cap N)$ for each $i \in \omega$, it follows that $|\operatorname{St}(v_0, L')| = |\operatorname{St}(v_0, S_\infty)| \subset \varphi(U \cap N)$. Observe that $\operatorname{St}(v_0, L')$ is a simplicial subdivision of $F(D_\infty)$, where $F(D_\infty)$ is simplicially isomorphic to $F(\Delta^\infty)$. Hence,

[13] Cf. Sect. 1.5.

$\mathrm{St}(v_0, L')$ is combinatorially equivalent to $F(\Delta^\infty)$. We apply Theorem 1.6.1 to obtain a simplicial subdivision $L'' \lhd L'$ with the same vertices, which is the desired one because $\mathrm{St}(v_0, L'') = \mathrm{St}(v_0, L')$. □

To prove Theorem 6.9.2, we need the following lemma:

Lemma 6.9.4 *Let M be a PL ∞-manifold and let P be a compact polyhedron with a subpolyhedron Q. If $f : P \to M$ is a map such that $f|Q$ is a PL embedding, then f is homotopic to a PL embedding $g : P \to M$ relative to Q.*

Proof Since P is compact, using Lemma 6.9.3, we can obtain finite subpolyhedra $N_1, \ldots, N_n$ of M such that $f(P) \subset \bigcup_{i=1}^n \operatorname{int} N_i$ and each N_i is PL homeomorphic to Δ^∞. For each $i = 1, \ldots, n$, let $\varphi_i : N_i \to \Delta^\infty$ be a PL homeomorphism. Let K be a triangulation of P with a subcomplex J such that $|J| = Q$ and, for each $\sigma \in K$, $|K[\sigma]|$ is contained in some $f^{-1}(\operatorname{int} N_i)$, where $|K[\sigma]| = \bigcup_{v \in \sigma^{(0)}} |\mathrm{St}(v, K)|$. For each $i = 1, \ldots, n$, let

$$K_i = \left\{\sigma \in K \;\middle|\; |K[\sigma]| \subset f^{-1}(\operatorname{int} N_i)\right\}.$$

Then, $K = \bigcup_{i=1}^n K_i$, K_i is a subcomplex of K, and

$$|K_i| \subset |K[|K_i|]| = \bigcup_{v \in K_i^{(0)}} |\mathrm{St}(v, K)| \subset f^{-1}(\operatorname{int} N_i).$$

Note that $C(|K_i|, K) = K \setminus K[|K_i|]$ is a subcomplex of K with $|C(|K_i|, K)| \cap |K_i| = \emptyset$.

First, we will construct a map $g_1 : |K| \to M$ such that $g_1||J \cup K_1|$ is a PL embedding and

$$g_1 \simeq f \ \text{ rel. } |J| \cup |C(|K_1|, K)|,$$

where non-degenerate homotopy tracks are contained in $N_1 \subset U_1$. Let J_1' be a simplicial subdivision of the subcomplex $J_1 = J \cap K_1$ of K_1 such that $\varphi_1 f|\sigma$ is affine for each $\sigma \in J_1'$. We can extend J_1' to a simplicial subdivision K_1' of K_1 (cf. Proposition 1.6.8). Because $\Delta^\infty = \bigcup_{i \in \mathbb{N}} \Delta^i$, the compact set $\varphi_1(f(|J|) \cap N_1$ is contained in some Δ^k. Then, $\varphi_1 f||J_1|$ can be extended to a map $g_1' : |K_1| \to \Delta^\infty$ such that $g_1'|K_1'^{(0)} \setminus J_1'^{(0)}$ is an injection into $\{\mathbf{e}_i \mid i > k+1\}$ $(= F(\Delta^\infty)^{(0)} \setminus \Delta^k)$ and $g_1'|\sigma$ is affine for each $\sigma \in K_1'$. Since $g_1'|K_1'^{(0)}$ is injective and $g_1'(\operatorname{rint}\sigma) \cap g_1'(\sigma') = \emptyset$ for each $\sigma \neq \sigma' \in K_1'$, it follows that g_1' is a PL embedding. Then, $\varphi_1^{-1} g_1' : |K_1| \to N_1$ is also a PL embedding, $\varphi_1^{-1} g_1'||J_1'| = f||J_1'|$, and

$$\varphi_1^{-1} g_1'(|K_1'| \setminus |J_1'|) \cap f(|J|) = \emptyset.$$

Because $\varphi_1^{-1} g_1' \simeq f||K_1'|$ rel. $|J_1|$ in N_1 $(\subset M)$, we can apply the Homotopy Extension Theorem 1.5.7 to extend $\varphi_1^{-1} g_1'$ to the desired map $g_1 : |K| \to M$.

By using the same construction inductively (i.e., in the above, replacing J, K_1, and f with $J \cup K_1 \cup \cdots \cup K_{i-1}$, K_i, and g_{i-1}, respectively), we can obtain maps $g_i : |K| \to M, i = 1, \ldots, n$, such that

$$g_i||J \cup K_1 \cup \cdots \cup K_i| \text{ is a PL embedding and}$$
$$g_i \simeq g_{i-1} \text{ rel. } |J \cup K_1 \cup \cdots \cup K_{i-1}| \cup |C(|K_i|, K)|,$$

where $g_0 = f$ and non-degenerate homotopy tracks of $g_i \simeq g_{i-1}$ are contained in $N_i \subset U_i$. Then, g_n is the desired one. □

Remark 6.11 In the above proof, given an open cover $\mathcal{U} \in \operatorname{cov}(M)$, we can take each N_i so as to be contained in some $U_i \in \mathcal{U}$. Then, it follows that $f \simeq_{\mathcal{U}} g$ rel. Q.

Combining Lemma 6.9.4 with Theorem 6.6.1, we have the following proposition:

Proposition 6.9.5 *Let K be a finite simplicial complex with a subcomplex J and let L be a triangulation of a PL ∞-manifold. If $f : |K| \to |L|$ is a map such that $f||J|$ is an embedding and simplicial with respect to J and L, then f is homotopic to an embedding $g : |K| \to |L|$ relative to $|J|$, which is simplicial with respect to simplicial subdivisions $K' \lhd K$ and $L' \lhd L$ with $J \subset K'$ and $f(J) \subset L'$.* □

Remark 6.12 Due to Remark 6.11, in Proposition 6.9.5 above, given an open cover $\mathcal{U} \in \operatorname{cov}(|L|)$, g can be taken so that $f \simeq_{\mathcal{U}} g$ rel. $|J|$.

Now, we can prove the Strong Hauptvermutung for PL ∞-manifolds (Theorem 6.9.2), which states that for any triangulations of PL ∞-manifolds K and L,

$$|K| \simeq |L| \Rightarrow K \cong L.$$

Proof (Theorem 6.9.2) Let $f : |K| \to |L|$ be a homotopy equivalence with a homotopy inverse $g : |L| \to |K|$. Write $K = \bigcup_{n \in \mathbb{N}} K_n$ and $L = \bigcup_{n \in \mathbb{N}} L_n$, where $K_1 \subset K_2 \subset \cdots$ and $L_1 \subset L_2 \subset \cdots$ are finite subcomplexes of K and L, respectively. Put $n_1 = 1$. Applying Proposition 6.9.5, we have simplicial subdivisions $K^*_{n_1} \lhd K_{n_1}$, $L' \lhd L$, and a simplicial embedding $f_1 : K^*_{n_1} \to L'$ such that $f_1 \simeq f||K_{n_1}|$, where $f_1(K^*_{n_1})$ is a subcomplex of L'. By Proposition 1.6.8, we have a simplicial subdivision $K' \lhd K$ such that $K^*_{n_1}$ is a subcomplex of K'.

Choose $n_2 > n_1$ so that $f_1(|K_{n_1}|) \subset |L_{n_2}|$. Let L'_{n_2} be the subdivision of L_{n_2} induced by $L' \lhd L$. Note that $f_1^{-1} : f_1(|K_{n_1}|) \to |K|$ is homotopic to $g|f_1(|K_{n_1}|)$. Indeed, $f_1^{-1} \simeq (gf||K_{n_1}|)f_1^{-1} \simeq gf_1f_1^{-1} = g|f_1(|K_{n_1}|)$. By the Homotopy Extension Theorem 1.5.7, f_1^{-1} can be extended to a map $g'_1 : |L'| \to |K'|$ such that $g'_1 \simeq g$. Then, $g'_1|f_1(|K_{n_1}|) = f_1^{-1} : f_1(K^*_{n_1}) \to K'$ is a simplicial embedding. Again applying Proposition 6.9.5, we have simplicial subdivisions $L^*_{n_2} \lhd L'_{n_2}$, $K'' \lhd K'$, and a simplicial embedding $g_1 : L^*_{n_2} \to K''$ such that $f_1(K^*_{n_1})$ is a subcomplex of $L^*_{n_2}$, $g_1|f_1(|K_{n_1}|) = f_1^{-1}$ and $g_1 \simeq g'_1||L_{n_2}| \simeq g||L_{n_2}|$. Then, $g_1(L^*_{n_2})$ is a subcomplex of K'', hence $K^*_{n_1}$ is so. By Proposition 1.6.8, we have a simplicial subdivision $L'' \lhd L'$ such that $L^*_{n_2}$ is a subcomplex of L''.

Choose $n_3 > n_2$ so that $g_1(|L_{n_2}|) \subset |K_{n_3}|$. Similarly applying Proposition 6.9.5 as in the above, we have simplicial subdivisions $K^*_{n_3} \lhd K''_{n_3}$ and $L''' \lhd L''$, and simplicial embedding $f_2 : K^*_{n_3} \to L'''$ such that $g_1(L^*_{n_2})$ is a subcomplex of $K^*_{n_3}$, $f_2|g_1(|L_{n_2}|) = g_1^{-1}$, and $f_2 \simeq f||K_{n_3}|$.

Iterating this process, we can obtain $n_1 < n_2 < n_3 < \cdots \in \mathbb{N}$ and the following commutative diagram of simplicial embeddings of simplicial subdivisions $K^*_{n_{2i-1}} \lhd K_{n_{2i-1}}$ and $L^*_{n_{2i}} \lhd L_{n_{2i}}$:

$$
\begin{array}{ccccccccc}
K^*_{n_1} & \subset & K^*_{n_3} & \subset & K^*_{n_5} & \subset & \cdots \\
{\scriptstyle f_1}\downarrow & \overset{g_1}{\nearrow} & {\scriptstyle f_2}\downarrow & \overset{g_2}{\nearrow} & {\scriptstyle f_3}\downarrow & \nearrow & \\
L^*_{n_2} & \subset & L^*_{n_4} & \subset & L^*_{n_6} & \subset & \cdots .
\end{array}
$$

Then, $K^* = \bigcup_{i\in\mathbb{N}} K^*_{n_{2i-1}}$ and $L^* = \bigcup_{i\in\mathbb{N}} L^*_{n_{2i}}$ are simplicial subdivisions of K and L, respectively. We can define simplicial maps $f_* : K^* \to L^*$ and $g_* : L^* \to K^*$ by $f_*|K_{n_{2i-1}} = f_i$ and $g_*|L_{n_{2i}} = g_i$ for each $i \in \mathbb{N}$. Then, f_* is a simplicial isomorphism and g_* is the inverse of f_*. Hence, K^* and L^* are simplicially isomorphic. Consequently, K and L are combinatorially equivalent. □

In the above proof, if f is a homeomorphism, then the simplicial isomorphism $f_* : K^* \to L^*$ can be taken so as to be arbitrarily close to f, that is, for each $\mathcal{U} \in \operatorname{cov}(|L|)$, there exist simplicial subdivisions $K^* \lhd K$, $L^* \lhd L$, and a simplicial isomorphism $f_* : K^* \to L^*$ that is $\mathcal{U}$-close to f. To show this, note that $g = f^{-1}$ in the above proof. Take a sequence of open covers $\mathcal{V}_n \in \operatorname{cov}(|L|)$, $n \in \mathbb{N}$ so that $\operatorname{st}\mathcal{V}_n \prec \mathcal{V}_{n-1}$, where $\mathcal{V}_0 = \mathcal{U}$. Inductively define $\mathcal{W}_{n+1} = \operatorname{st}(\mathcal{W}_n, \mathcal{V}_{n+1})$, $n \in \mathbb{N}$, where $\mathcal{W}_1 = \mathcal{V}_1$. Then, $\mathcal{W}_n \prec \mathcal{U}$ for every $n \in \mathbb{N}$. Indeed, $\operatorname{st}(\mathcal{W}_1, \mathcal{V}_1) = \operatorname{st}\mathcal{V}_1 \prec \mathcal{U}$, and $\operatorname{st}(\mathcal{W}_n, \mathcal{V}_n) \prec \mathcal{U}$ implies

$$
\begin{aligned}
\operatorname{st}(\mathcal{W}_{n+1}, \mathcal{V}_{n+1}) &= \operatorname{st}(\operatorname{st}(\mathcal{W}_n, \mathcal{V}_{n+1}), \mathcal{V}_{n+1}) \\
&\prec \operatorname{st}(\mathcal{W}_n, \operatorname{st}\mathcal{V}_{n+1}) \prec \operatorname{st}(\mathcal{W}_n, \mathcal{V}_n) \prec \mathcal{U}.
\end{aligned}
$$

Due to Remark 6.12, f_1 can be taken so that $f_1 \simeq_{\mathcal{V}_1} f||K_{n_1}|$. Since $f_1^{-1} = f^{-1}ff_1^{-1} \simeq_{f^{-1}(\mathcal{W}_1)} f^{-1}f_1f_1^{-1} = f^{-1}|f_1(|K_{n_1}|) = g|f_1(|K_{n_1}|)$, f_1^{-1} can be extended to a map g_1' such that $g_1' \simeq_{f^{-1}(\mathcal{W}_1)} g = f^{-1}$. Then, g_1 can be taken so that $g_1 \simeq_{f^{-1}(\mathcal{V}_2)} g_1'||L_{n_2}|$, which implies that $g_1 \simeq_{f^{-1}(\mathcal{W}_2)} f^{-1}|L_{n_2}|$. Since $g_1^{-1} = ff^{-1}g_1^{-1} \simeq_{\mathcal{W}_2} fg_1g_1^{-1} = f|g_1(|L_{n_2})|$, g_1^{-1} can be extended to a map f_2' such that $f_2' \simeq_{\mathcal{W}_2} f$. Then, f_2 can be taken so that $f_2 \simeq_{\mathcal{V}_3} f_2'||K_{n_3}|$, which implies that $f_2 \simeq_{\mathcal{W}_3} f||K_{n_3}|$. Thus, each f_i can be inductively constructed so that $f_i \simeq_{\mathcal{W}_{2i-1}} f||K_{n_{2i-1}}|$. Since $f_*||K_{n_{2i-1}}| = f_i$ and $\mathcal{W}_{2i-1} \prec \mathcal{U}$, it follows that f_* is $\mathcal{U}$-close to f. Recall that every PL ∞-manifold is an $\mathbb{R}^\infty$-manifold, and that every fine homotopy equivalence between $\mathbb{R}^\infty$-manifolds is a near-homeomorphism (Theorem 6.5.2). So, we conclude as follows:

Theorem 6.9.6 *Every fine homotopy equivalence between PL ∞-manifolds is arbitrarily close to PL homeomorphisms.* □

The following is a direct consequence of Theorem 6.9.2:

Corollary 6.9.7 *Every contractible PL ∞-manifold is PL homeomorphic to Δ^∞. In other words, any triangulation of a contractible PL ∞-manifold is combinatorially equivalent to $F(\Delta^\infty)$.* □

6.10 Characterization of PL ∞-Manifolds

In this section, we prove the following characterization of PL ∞-manifolds:

Theorem 6.10.1 (CHARACTERIZATION) *For a separable polyhedron X, the following are equivalent:*

(a) *X is a PL ∞-manifold;*
(b) *Each PL embedding $f : Q \to X$ of a subpolyhedron of a compact polyhedron P into X extends to a PL-embedding $\tilde{f} : N \to X$ of a polyhedral neighborhood N of Q in P;*[14]
(c) *For each compact subpolyhedron $Y \subset X$, there exists a PL embedding $h : Y \times \mathbf{I} \to X$ such that $h(y, 0) = y$ for every $y \in Y$.*

For any countable simplicial complex K, the polyhedron $|K| \times \Delta^\infty$ satisfies condition (c) above because each compact subset is contained in some $|K| \times \Delta^n$. Hence, $|K \times_s F(\Delta^\infty)| = |K| \times \Delta^\infty$ is a PL ∞-manifold. Thus, the following can be obtained by Theorem 6.9.2:

Theorem 6.10.2 (STABLE HAUPTVERMUTUNG) *Let K and L be countable simplicial complexes. If $|K|$ and $|L|$ have the same homotopy type, then $K \times_s F(\Delta^\infty)$ and $L \times_s F(\Delta^\infty)$ are combinatorially equivalent. In other words, the following implication holds:*

$$|K| \simeq |L| \Rightarrow K \times_s F(\Delta^\infty) \cong L \times_s F(\Delta^\infty). \quad \square$$

Let $I = \{0, 1, \mathbf{I}\}$ and $R = \{n, [n, n+1] \mid n \in \mathbb{Z}\}$ be ordered simplicial complexes with the ordinal order. Using the product simplicial complexes of ordered simplicial complexes, we inductively define I^n and R^n, $n \in \mathbb{N}$, as follows:

$$I^{n+1} = I^n \times_s I \text{ and } R^{n+1} = R^n \times_s R,$$

[14] A subpolyhedron of P is closed in P.

where $I^1 = I$ and $R^1 = R$ (cf. Sect. 1.5). We define

$$I^\infty = \bigcup_{n\in\mathbb{N}} I^n \text{ and } R^\infty = \bigcup_{n\in\mathbb{N}} R^n,$$

where I^n and R^n are identified with $I^n \times_s \{0\} \subset I^{n+1}$ and $R^n \times_s \{0\} \subset R^{n+1}$, respectively. For any countable simplicial complex K, the following polyhedra also satisfy condition (c) above:

$$|K \times_s I^\infty| = |K| \times |I^\infty| \text{ and } |K \times_s R^\infty| = |K| \times \mathbb{R}^\infty$$

because their compact subset is contained in some $|K| \times |I^n|$ and $|K| \times \mathbb{R}^n$, respectively. Hence, these polyhedra are also PL ∞-manifolds. Thus, Theorem 6.10.2 above holds if $F(\Delta^\infty)$ is replaced with I^∞ or R^∞. Furthermore, it is also true even if only $K \times_s F(\Delta^\infty)$ is replaced with $K \times_s I^\infty$ or $K \times_s R^\infty$. As a special case that $K = L$ is a singleton, we have the following:

Corollary 6.10.3 *The simplicial complexes I^∞ and R^∞ are combinatorially equivalent to $F(\Delta^\infty)$.* □

In advance of proving Theorem 6.10.1, we prove the following characterization of polyhedra being PL homeomorphic to Δ^∞:

Theorem 6.10.4 (PL CHARACTERIZATION OF Δ^∞) *For a separable polyhedron X, the following are equivalent:*

(a) *X is PL homeomorphic to Δ^∞;*
(b) *Each PL embedding $f : Q \to X$ of a subpolyhedron of a compact polyhedron P extends to a PL-embedding $\tilde{f} : P \to X$;*
(c) *X is contractible and, for each compact subpolyhedron $Y \subset X$, there exists a PL embedding $h : Y \times \mathbf{I} \to X$ such that $h(y, 0) = y$ for every $y \in Y$.*

Proof Since Δ^∞ is an AE, Δ^∞ satisfies condition (b) by Lemma 6.9.4. Because the latter half of (c) is a special case of (b), it follows that Δ^∞ also satisfies condition (c). Hence, we have the implications (a) $\Rightarrow$ (b) and (a) $\Rightarrow$ (c). Thus, it is sufficient to show the implications (c) $\Rightarrow$ (b) $\Rightarrow$ (a).

(c) $\Rightarrow$ (b): Let $f : Q \to X$ be a PL embedding of a closed subpolyhedron of a compact polyhedron P. Because X is an AE (Proposition 1.13.3 (1)), we can apply the PL Approximation Theorem 1.8.9 to obtain a PL map $f' : P \to X$ that is an extension of f. Let $\dim P = n$. We have a simplicial triangulation L of P with a subcomplex $L_0 \subset L$ such that $|L_0| = Q$ and $hf|\sigma$ is affine for each $\sigma \in L_0$. By the General Position Lemma 1.8.10 there exist a countable set $A \subset \mathbf{I}^{2n+1}$ such that any $2n + 2$ many points of $A \cup \{\mathbf{0}\}$ are affinely independent. Triangulate P by a simplicial complex such that its subcomplex triangulates Q and f' is affine on each simplex, and let $P^{(0)}$ denote the 0-skeleton of this triangulation. There exists a PL-map $g : P \to \mathbf{I} \times \mathbf{I}^{2n+1} = \mathbf{I}^{2n+2}$ such that g is affine on each simplex of this triangulation, $g|P^{(0)} \setminus Q$ is an injection into $\{1\} \times A$, and $g(Q) = \{0\} \times \{\mathbf{0}\}$.

Observe that $g|P \setminus Q$ is injective. By the assumption, there exists a PL embedding $h : f(P) \times \mathbf{I}^{2n+2} \to X$ such that $h(f(x), \mathbf{0}) = f(x)$ for every $x \in P$. We can define a PL map $\tilde{f} : P \to X$ as follows:

$$\tilde{f}(x) = h(f(x), g(x)) \text{ for each } x \in P.$$

Then, $\tilde{f}|Q = f|Q$ and $\tilde{f}$ is an embedding because it is a continuous injection defined on a compactum. Thus, f extends to a PL embedding $\tilde{f} : P \to X$.

(b) ⇒ (a): We adopt the same method as Theorem 6.9.2 but the Homotopy Extension Theorem 1.5.7 is now not necessary. Note that every contractible polyhedron is an AE (Theorem 1.13.4 and Proposition 1.13.3(1)). Let L be a simplicial complex triangulating X (i.e., $|L| = X$). By virtue of Theorem 6.6.1, it follows from (b) that L satisfies the following condition:

(♯) Let K be a finite simplicial complex with a subcomplex K_0 and let $f : K_0 \to L'$ be a simplicial embedding of K_0 into a simplicial subdivision $L' \lhd L$. Then, there exist simplicial subdivisions $K' \lhd K$, $L'' \lhd L'$, and a simplicial embedding $\tilde{f} : K' \to L''$ such that K_0 is a subcomplex of K' and $\tilde{f}||K_0| = f$ (so $f(K_0)$ and $\tilde{f}(K')$ are subcomplexes of L'').

Note that $F(\Delta^\infty)$ also satisfies (♯) and $F(\Delta^\infty) = \bigcup_{n\in\mathbb{N}} F(\Delta^n)$. Take a tower $L_1 \subset L_2 \subset \cdots$ of finite subcomplexes of L such that $L = \bigcup_{n\in\mathbb{N}} L_n$.

Put $n_1 = 1$. Since $F(\Delta^\infty)$ satisfies (♯), there are simplicial subdivisions $L^*_{n_1} \lhd L_{n_1}$, $F(\Delta^\infty)' \lhd F(\Delta^\infty)$, and a simplicial embedding $f_1 : L^*_{n_1} \to F(\Delta^\infty)'$.[15] It should be noted that $f_1(L^*_{n_1})$ is a subcomplex of $F(\Delta^\infty)'$. Due to Proposition 1.6.8, $L^*_{n_1}$ is a subcomplex of some simplicial subdivision $L' \lhd L$.

Choose $n_2 > n_1$ so that $f_1(|L^*_{n_1}|) \subset \Delta^{n_2}$. Let $F(\Delta^{n_2})'$ be the subdivision of $F(\Delta^{n_2})$ induced by $F(\Delta^\infty)'$. Then, $f_1(L^*_{n_1})$ is a subcomplex of $F(\Delta^{n_2})'$. Since L satisfies (♯), we have simplicial subdivisions $F(\Delta^{n_2})^* \lhd F(\Delta^{n_2})'$, $L'' \lhd L'$, and a simplicial embedding $g_1 : F(\Delta^{n_2})^* \to L''$ such that $f_1(L^*_{n_1})$ is a subcomplex of $F(\Delta^{n_2})^*$ and $g_1|f_1(|L_{n_1}|) = f_1^{-1}$, where $g_1(F(\Delta^{n_2})^*)$ and $L^*_{n_1}$ are subcomplexes of L''. Due to Proposition 1.6.8, $F(\Delta^{n_2})^*$ is a subcomplex of some simplicial subdivision $F(\Delta^\infty)'' \lhd F(\Delta^\infty)'$.

Choose $n_3 > n_2$ so that $g_1(\Delta^{n_2}) \subset |L_{n_3}|$. Let L''_{n_3} be the subdivision of L_{n_3} induced by L''. Again using (♯) for $F(\Delta^\infty)$, we can obtain simplicial subdivisions $L^*_{n_3} \lhd L''_{n_3}$, $F(\Delta^\infty)''' \lhd F(\Delta^\infty)''$, and a simplicial embedding $f_2 : K^*_{n_3} \to F(\Delta^\infty)'''$ such that $g_1(F(\Delta^{n_2})^*)$ is a subcomplex of $K^*_{n_3}$ and $f_2|g_1(\Delta^{n_2}) = g_1^{-1}$.

[15]For this first step, we do not need (♯). This can be easily shown by using the General Position Lemma 1.8.10 and Theorem 1.8.3.

Iterating this process, we can construct the following commutative diagram of simplicial embeddings of simplicial subdivisions $K^*_{n_{2i-1}} \lhd K_{n_{2i-1}}$ and $F(\Delta^{n_{2i}})^* \lhd F(\Delta^{n_{2i}})$:

$$\begin{array}{ccccccc}
K^*_{n_1} & \subset & K^*_{n_3} & \subset & K^*_{n_5} & \subset & \cdots \\
{\scriptstyle f_1}\downarrow & \overset{g_1}{\nearrow} & {\scriptstyle f_2}\downarrow & \overset{g_2}{\nearrow} & {\scriptstyle f_3}\downarrow & \nearrow & \\
F(\Delta^{n_2})^* & \subset & F(\Delta^{n_4})^* & \subset & F(\Delta^{n_6})^* & \subset & \cdots.
\end{array}$$

Then, we can obtain simplicial subdivisions

$$K^* = \bigcup_{i\in\mathbb{N}} K^*_{n_{2i-1}} \lhd K \quad \text{and} \quad F(\Delta^\infty)^* = \bigcup_{i\in\mathbb{N}} F(\Delta^{n_{2i}})^* \lhd F(\Delta^\infty).$$

Those are simplicially isomorphic. Consequently, K is combinatorially equivalent to $F(\Delta^\infty)$, that is, $X = |K|$ is PL homeomorphic to Δ^∞. □

We have the following corollary:

Corollary 6.10.5 *Let $K_1 \subset K_2 \subset \cdots$ be a tower of combinatorial n_i-manifolds K_i, $i \in \mathbb{N}$, where $n_1 < n_2 < \cdots$. Then, $K = \bigcup_{i\in\mathbb{N}} K_i$ is a combinatorial ∞-manifold.*

Proof We may show that the star $\mathrm{St}(v, K)$ of each vertex $v \in K^{(0)}$ satisfies condition (b) of Theorem 6.10.4. Let $f : Q \to |\,\mathrm{St}(v, K)|$ be a PL embedding of a subpolyhedron Q of a compact polyhedron P. Note that $|\,\mathrm{St}(v, K)| = \varinjlim |\,\mathrm{St}(v, K_n)|$, where $v \in K_n^{(0)}$ for sufficiently large $n \in \mathbb{N}$. Choose $i \in \mathbb{N}$ so that $f(Q) \subset |\,\mathrm{St}(v, K_i)|$ and $n_i \geqslant 2\dim P + 1$. Since K_i is a combinatorial n_i-manifold, $|\,\mathrm{St}(v, K_i)|$ is a PL n_i-ball, that is, there is a PL homeomorphism $h : |\,\mathrm{St}(v, K_i)| \to \mathbf{I}^{n_i}$. We have a simplicial triangulation L of P with a subcomplex $L_0 \subset L$ such that $|L_0| = Q$ and $hf|\sigma$ is affine for each $\sigma \in L_0$. Using the General Position Lemma 1.8.10, we can find a countable infinite set $A \subset \mathbf{I}^{n_i} \setminus hf(L_0^{(0)})$ so that any $2\dim P + 2$ many points of $A \cup hf(L_0^{(0)})$ are affinely independent. Then, we can define a PL embedding $g : P = |L| \to \mathbf{I}^{n_i}$ as follows: $g|Q = f$ and $g|L^{(0)} \setminus L_0 : L^{(0)} \setminus L_0 \to A$ is an injection, and $g|\sigma$ is affine for each $\sigma \in L \setminus L_0$. Observe that

$$\sigma \neq \sigma' \in K \;\Rightarrow\; g(\operatorname{rint}\sigma) \cap g(\operatorname{rint}\sigma) = \emptyset,$$

which implies that g is injective. The result follows from Theorem 6.10.4. □

Let P be a polyhedron and $A \subset P$. A neighborhood N of A in P is called a **polyhedral neighborhood** if N is a subpolyhedron of P, that is, there is a triangulation K of P and a subcomplex $L \subset K$ such that $N = |L|$. In particular, $|K[A]|$ $(= \bigcup\{\sigma \in K \mid \sigma \cap A \neq \emptyset\})$ is a polyhedral neighborhood of A. Indeed, $|K[A]| = |N(A, K)|$, where $N(A, K)$ is the simplicial neighborhood of

A in K.[16] Observe that $N(\{v\}, K) = \mathrm{St}(v, K)$ for $v \in K^{(0)}$, and $N(|L|, K) = \bigcup_{v\in L^{(0)}} \mathrm{St}(v, K)$ for a subcomplex L of K. The following is proved in the appendix as Theorem A.3.1 (p. 593):

Proposition 6.10.6 *Let L be a full subcomplex of a combinatorial n-manifold K. Then, every derived neighborhood $N(L, K')$ of L in K is a combinatorial n-manifold, where K' is a derived subdivision of K relative to $L \cup C(L, K)$.*

The following lemma can be proved in the same way as [GAGT, Lemma 4.7.2]. For readers' convenience, we give a proof.

Lemma 6.10.7 *Let K be a finite simplicial complex and L a subcomplex of K. For any open neighborhood U of $|L|$ in $|K|$, there is a simplicial subdivision K' of K such that L is a subcomplex of K' and $|N(L, K')| \subset U$.*

Proof Replacing K by $\mathrm{Sd}_L K$, we may assume that L is full in K. Recall the metric ρ_K on $|K|$ is defined as follows:

$$\rho_K(x, y) = \left\|\beta^K(x) - \beta^K(y)\right\|_1 = \sum_{v\in K^{(0)}} \left|\beta_v^K(x) - \beta_v^K(y)\right|,$$

where $x = \sum_{v\in K^{(0)}} \beta_v^K(x)v$, $y = \sum_{v\in K^{(0)}} \beta_v^K(y)$, and

$$\beta^K(x) = (\beta_v^K(x))_{v\in K^{(0)}}, \beta^K(y) = (\beta_v^K(y))_{v\in K^{(0)}} \in \ell^1(K^{(0)}).$$

Since $|L|$ and $|K| \setminus U$ are disjoint closed sets in the compact metric space $|K| = (|K|, \rho_K)$, we have

$$\delta = \mathrm{dist}_{\rho_K}(|L|, |K| \setminus U) > 0.$$

For each simplex $\sigma \in K[|L|] \setminus L$ choose $v(\sigma) \in \mathrm{rint}\,\sigma$ so that $\mathrm{dist}_{\rho_K}(v(\sigma), \sigma \cap |L|) < \delta$. For each simplex $\sigma \in C(|L|, K)$, let $v(\sigma)$ be the barycenter $\hat{\sigma}$ of σ. Using these points, we define the derived subdivision K' of K relative to L (cf. Proposition 1.6.8). Then, L is a subcomplex of K'.

To see that $|N(L, K')| \subset U$, let $x \in |N(L, K')|$. Then, we have $\sigma_0 < \sigma_1 < \cdots < \sigma_n \in K$ such that

$$\sigma_0 = \langle v_1, \ldots, v_m\rangle \in L,\ \sigma_1, \ldots, \sigma_n \in K \setminus L,\ \text{and}$$
$$x \in \langle v_1, \ldots, v_m, v(\sigma_1), \ldots, v(\sigma_n)\rangle,$$

[16] See p. 31.

where x can be written as follows:

$$x = \sum_{i=1}^{m} z(i)v_i + \sum_{j=1}^{n} z(j+m)v(\sigma_j) \quad \text{for some } z \in \Delta^{m+n-1}.$$

For each $j = 1, \dots, n$, because $\sigma_j \cap |L| \neq \emptyset$, we can choose $u_j \in \sigma_j \cap |L|$ so that $\rho_K(v(\sigma_j), u_j) < \delta$. Since L is full in K, $\sigma_j \cap |L| = \langle \sigma_j^{(0)} \cap |L| \rangle \in L$ for each $j = 1, \dots, n$. Note that $\sigma_1 \cap |L| < \cdots < \sigma_n \cap |L|$. Hence, we have

$$y = \sum_{i=1}^{m} z(i)v_i + \sum_{j=1}^{n} z(j+m)u_j \in \sigma_n \cap |L| \in L.$$

Since $\beta^K|\sigma_n$ is affine, it follows that

$$\begin{aligned}
\rho_K(x, y) &= \left\| \beta^K(x) - \beta^K(y) \right\|_1 \\
&= \left\| \sum_{j=1}^{n} z(j+m)\beta^K(v(\sigma_j)) - \sum_{j=1}^{n} z(j+m)\beta^K(u_j) \right\|_1 \\
&= \sum_{j=1}^{n} z(j+m) \left\| \beta^K(v(\sigma_j)) - \beta^K(u_j) \right\|_1 \\
&= \sum_{j=1}^{n} z(j+m)\rho_K(v(\sigma_j), u_j) < \delta.
\end{aligned}$$

Thus, $\rho_K(x, |L|) < \delta$, which means that $x \in U$. □

Remark 6.13 Lemma 6.10.7 is valid for a locally finite simplicial complex K and a finite subcomplex L of K. Indeed, the proof can be performed if K is replaced with a finite subcomplex K_0 of K such that $|L| \subset \operatorname{int}|K_0|$. The simplicial neighborhood $N(L, K)$ of L in K can be taken as such K_0.

Now, we are ready to prove Theorem 6.10.1.

Proof (Theorem 6.10.1) The implication (b) ⇒ (c) is trivial. Since X is an ANE and compact polyhedra are metrizable, the implication (a) ⇒ (b) follows from Lemma 6.9.4. We have to prove the implications (c) ⇒ (b) ⇒ (a).

(c) ⇒ (b): Let $f : Q \to X$ be PL embedding of a closed subpolyhedron Q of a compact polyhedron P into X and let $n = \dim P$. Since X is an ANE and P is metrizable, we can apply the PL Approximation Theorem 1.8.9 to extend f to a PL map $f' : N \to X$ of a polyhedral neighborhood N of Q in P. In the same way as in the proof of (c) ⇒ (b) in Theorem 6.10.4, using the General Position Lemma 1.8.10, we can construct a PL embedding $\tilde{f} : N \to X$ extending of f.

(b) $\Rightarrow$ (a): Let L be a triangulation of X (i.e., $|L| = X$). In condition (b), let K be a triangulation of P with a subcomplex K_0 such that $Q = |K_0|$ and f is simplicial with respect to K_0 and a simplicial subdivision $L' \lhd L$. When f is extended to a PL embedding $f' : N \to X$ of a polyhedral neighborhood N of Q in P, we apply Lemma 6.10.7 with Remark 6.13 to obtain a simplicial subdivision $K' \lhd K$ such that K_0 is a subcomplex of K' and $|N(Q, K')| \subset N$. Then, by virtue of Theorem 6.6.1, it follows from (b) that L satisfies the following condition:

($\natural$) Let K be a finite simplicial complex with a subcomplex K_0 and let $f : K_0 \to L'$ be a simplicial embedding of K_0 into a simplicial subdivision $L' \lhd L$. Then, there exist simplicial subdivisions $K' \lhd K$ and $L'' \lhd L'$ such that K_0 is a subcomplex of K' and f extends to a simplicial embedding $\tilde{f} : N(K_0, K') \to L''$ (so $f(K_0)$ and $\tilde{f}(N(K_0, K'))$ are subcomplexes of L'').

Take a tower $L_1 \subset L_2 \subset \cdots$ of finite subcomplexes of L such that $L = \bigcup_{n\in\mathbb{N}} L_n$. As mentioned after Theorem 6.10.4, the complex R^∞ is combinatorially equivalent to $F(\Delta^\infty)$. Hence, R^∞ satisfies condition ($\sharp$) in the proof of Theorem 6.10.4, which is ($\natural$) with $N(K_0, K')$ replaced by K'.

Put $n_1 = 1$. Since R^∞ satisfies ($\sharp$), there exist simplicial subdivisions $L^*_{n_1} \lhd L_{n_1}$, $(R^\infty)' \lhd R^\infty$, and a simplicial embedding $f_1 : L^*_{n_1} \to (R^\infty)'$. Note that $f_1(L^*_{n_1})$ is a subcomplex of $(R^\infty)'$. Due to Proposition 1.6.8, $L^*_{n_1}$ is a subcomplex of some simplicial subdivision $L' \lhd L$.

Choose $n_2 > n_1$ so that $f_1(|L^*_{n_1}|) \subset \mathbb{R}^{n_2}$. Let $(R^{n_2})'$ be the subdivision of R^{n_2} induced by $(R^\infty)'$. Since L satisfies ($\natural$), we have simplicial subdivisions $(R^{n_2})'' \lhd (R^{n_2})'$ and $L'' \lhd L'$ such that $f_1^{-1} : f_1(L^*_{n_1}) \to L^*_{n_1}$ can be extended to a simplicial embedding $g_1 : N(f_1(|L^*_{n_1}|), (R^{n_2})'') \to L''$. Then, by Proposition 6.10.6, $N_1 = N(f_1(|L^*_{n_1}|), (R^{n_2})'')$ is a combinatorial n_2-manifold which is a subcomplex of $(R^{n_2})''$. Due to Proposition 1.6.8, $(R^{n_2})''$ is a subcomplex of some simplicial subdivision $(R^\infty)'' \lhd (R^\infty)'$.

Choose $n_3 > n_2$ so that $g_1(|N_1|) \subset |L_{n_3}|$. Let L''_{n_3} be the subdivision of L_{n_3} induced by L''. Again using ($\sharp$) for R^∞, we have simplicial subdivisions $L^*_{n_3} \lhd L''_{n_3}$ and $(R^\infty)''' \lhd (R^\infty)''$ such that $g_1^{-1} : g_1(N_1) \to N_1$ can be extended to a simplicial embedding $f_2 : L^*_{n_3} \to (R^\infty)'''$.

Iterating this process, we can obtain the following commutative diagram of simplicial embeddings of simplicial subdivisions $L^*_{n_{2i-1}} \lhd L_{n_{2i-1}}$ and combinatorial n_{2i}-manifolds N_i, $i \in \mathbb{N}$:

$$\begin{array}{ccccccc}
L^*_{n_1} & \subset & L^*_{n_3} & \subset & L^*_{n_5} & \subset & \cdots \\
{\scriptstyle f_1}\downarrow & \nearrow^{g_1} & {\scriptstyle f_2}\downarrow & \nearrow^{g_2} & {\scriptstyle f_3}\downarrow & \nearrow & \\
N_1 & \subset & N_2 & \subset & N_3 & \subset & \cdots .
\end{array}$$

Then, $L^* = \bigcup_{i\in\mathbb{N}} L^*_{n_{2i-1}} \lhd L$ and $N = \bigcup_{i\in\mathbb{N}} N_i$ are combinatorial ∞-manifolds by Corollary 6.10.5, which are simplicially isomorphic. Thus, L has a simplicial subdivision that is a combinatorial ∞-manifold, hence $X = |L|$ is a PL ∞-manifold.[17] □

6.11 Characterization of Combinatorial ∞-Manifolds

In this section, we prove the following characterization of combinatorial manifolds:

Theorem 6.11.1 (CHARACTERIZATION) *For a countable simplicial complex K, the following equivalent:*

(a) *K is a combinatorial ∞-manifold;*
(b) *$|K|$ is a PL ∞-manifold;*
(c) *$|K|$ is an $\mathbb{R}^\infty$-manifold;*
(d) *Every simplex of K is a Z-set in $|K|$;*
(e) *Every simplex of K is a Z-set in $|K|_{\mathrm{m}}$.*

As mentioned in the introductory part of Sect. 6.9, the implications (a) $\Rightarrow$ (b) $\Rightarrow$ (c) are trivial. Due to Proposition 6.5.6, every compact set in an $\mathbb{R}^\infty$-manifold is a Z-set, we have the implication (c) $\Rightarrow$ (d). The implication (d) $\Rightarrow$ (e) follows from the fact that $\mathrm{id} : |K| \to |K|_{\mathrm{m}}$ is a fine homotopy equivalence (Theorem 1.7.6). We will prove the implications (e) $\Rightarrow$ (b) and (c) $\Rightarrow$ (a), which correspond to Theorems 6.11.6 and 6.11.8, respectively.

From Theorem 6.11.1 above, the following can be easily obtained:

Corollary 6.11.2 (1) *Every simplicial subdivision of a combinatorial ∞-manifold is also a combinatorial ∞-manifold.*
(2) *A simplicial complex is a combinatorial ∞-manifold if it has a combinatorial ∞-manifold subdivision.* □

Recall that a simplicial complex K is countable if and only if $|K|$ is separable (Proposition 1.5.5(2)). As stated as above, every compact set in an $\mathbb{R}^\infty$-manifold is a Z-set. Hence, the equivalence (c) $\Leftrightarrow$ (d) in Theorem 6.11.1 can be restated as follows:

Corollary 6.11.3 *A separable space X is an $\mathbb{R}^\infty$-manifold if and only if X is triangulable and every compact set in X is a Z-set in X.* □

Lemma 6.11.4 *Let K be a simplicial complex. If each simplex $\sigma \in K$ is a Z-set in $|K|_{\mathrm{m}}$, then every skeleton $|K^{(n)}|$ is a Z-set in $|K|_{\mathrm{m}}$.*

Proof For each $\mathcal{U} \in \mathrm{cov}(|K|_{\mathrm{m}})$, take $\mathcal{V} \in \mathrm{cov}(|K|_{\mathrm{m}})$ so that $\mathrm{st}^n \mathcal{V} \prec \mathcal{U}$ For each $i = 0, 1, \ldots, n$, $|K^{(i)}| \setminus |K^{(i-1)}|$ is closed in $|K|_{\mathrm{m}} \setminus |K^{(i-1)}|$, where $K^{(-1)} = \emptyset$.

[17] It has not been proved that a simplicial complex is a combinatorial ∞-manifold if it has a simplicial subdivision that is a combinatorial ∞-manifold.

Each $x \in |K^{(i)}| \setminus |K^{(i-1)}|$ is contained in $\operatorname{rint}\sigma$ for some $\sigma \in K^{(i)} \setminus K^{(i-1)}$. Then, $O_{\operatorname{Sd} K}(\hat{\sigma})$ is a neighborhood of x in $|K|_{\mathrm{m}} \setminus |K^{(i-1)}|$ and

$$(|K^{(i)}| \setminus |K^{(i-1)}|) \cap O_{\operatorname{Sd} K}(\hat{\sigma}) = \operatorname{rint}\sigma,$$

which is a Z-set in $O_{\operatorname{Sd} K}(\hat{\sigma})$ by Corollary 2.8.7(ii). It follows from Corollary 2.8.7(iii) that $|K^{(i)}| \setminus |K^{(i-1)}|$ is a Z-set in $|K|_{\mathrm{m}} \setminus |K^{(i-1)}|$. Hence, there is a map $f_i : |K|_{\mathrm{m}} \setminus |K^{(i-1)}| \to |K|_{\mathrm{m}} \setminus |K^{(i)}|$ that is $\mathcal{V}$-close to id. Thus, we have a map $f_n \cdots f_1 f_0 : |K|_{\mathrm{m}} \to |K|_{\mathrm{m}} \setminus |K^{(n)}|$ that is $\mathcal{U}$-close to id. □

To prove the implication (e) ⇒ (b) in Theorem 6.11.1, we need the following two lemmas:

Lemma 6.11.5 *Let K be a finite simplicial complex with* $\dim K \geqslant n$. *Then,* $|K| \setminus |K^{(n-1)}|$ *contains an infinite point set A such that for each $\sigma \in K \setminus K^{(n-1)}$, $A \cap \operatorname{rint}\sigma$ is infinite and any $n+1$ many points in $\sigma^{(0)} \cup (\sigma \cap A)$ are affinely independent.*

Proof First note that for each $\sigma \in K \setminus K^{(n-1)}$, any $n+1$ many points in $\sigma^{(0)}$ are affinely independent. Write $K \setminus K^{(n-1)} = \{\sigma_1, \ldots, \sigma_m\}$. We can inductively choose $x_k \in |K| \setminus |K^{(n-1)}|$, $k \in \mathbb{N}$, so that:

(i) $x_{jm+i} \in \operatorname{rint}\sigma$ for each $i = 1, \ldots, m$ and $j \in \omega$,
(ii) any $n+1$ many points in $\sigma^{(0)} \cup (\sigma \cap \{x_k \mid k \in \mathbb{N}\})$ are affinely independent.

Indeed, assume x_k has been chosen for $k < jm+i$ ($i = 1, \ldots, m$). We can choose $x_{jm+i} \in \operatorname{rint}\sigma_i$ not lying on any hyperplane spanned by n points in $\sigma_{i'}^{(0)} \cup (\sigma_{i'} \cap \{x_k \mid k < jm+i\})$, $\sigma_{i'} \geqslant \sigma_i$. Then, $A = \{x_k \mid k \in \mathbb{N}\}$ is the desired one. □

The following corresponds to the implication (e) ⇒ (b) in Theorem 6.11.1:

Theorem 6.11.6 *Let K be a countable simplicial complex. If each simplex of K is a Z-set in $|K|_{\mathrm{m}}$ (or $|K|$), then $|K|$ is a PL ∞-manifold.*

Proof We may prove that $|K|$ satisfies condition (c) in Theorem 6.10.1. For each compact subpolyhedron Y of $|K|$, there is a finite subcomplex L of K such that $Y \subset |L|$. Then, we will construct a PL embedding $h : |L| \times \mathbf{I} \to |K|$ such that $h(x, 0) = x$ for every $x \in |L|$.

Put $\dim L = n$. Since $|K^{(2n+3)}|$ is a Z-set in $|K|_{\mathrm{m}}$ by Lemma 6.11.4, we can take a map $f : |L| \to |K|_{\mathrm{m}} \setminus |K^{(2n+3)}|$ so close to id that

(1) $f(|\operatorname{St}(\hat{\sigma}, \operatorname{Sd}^2 L)|) \subset O_{\operatorname{Sd} K}(\hat{\sigma})$ for each $\sigma \in L$.

Since $\{O_{\operatorname{Sd} K}(\hat{\sigma}) \mid \sigma \in K \setminus K^{(2n+3)}\}$ is an open cover of $|K| \setminus |K^{(2n+3)}|$, we have a subdivision $L' \lhd \operatorname{Sd}^2 L$ such that

$$\big\{|\operatorname{St}(v, L)| \bigm| v \in L'^{(0)}\big\} \prec \big\{f^{-1}(O_{\operatorname{Sd} K}(\hat{\sigma}) \bigm| \sigma \in K \setminus K^{(2n+3)}\big\}.$$

Then, for each $v \in L'^{(0)}$, choose $\sigma_v \in K \setminus K^{(2n+3)}$ so that

(2) $f(|\operatorname{St}(v, L')| \subset O_{\operatorname{Sd} K}(\hat{\sigma}_v)$.

Let K_0 be a finite subcomplex of K such that $L \cup \{\sigma_v \mid v \in L'^{(0)}\} \subset K_0$. Applying Lemma 6.11.5 to K_0, we can define an injection $g : L'^{(0)} \to |K_0|$ so as to satisfy the following conditions:

(3) $g(v) \in \operatorname{rint} \sigma_v$ for each $v \in L'^{(0)}$, and
(4) for each $\sigma \in K_0 \setminus K_0^{(2n+3)}$, any $2n+4$ many points in $\sigma^{(0)} \cup (\sigma \cap g(L'^{(0)}))$ are affinely independent.

Let $\tilde{L}$ be a triangulation of $|L| \times \mathbf{I}$ defined as follows:

$$\tilde{L} = \big\{\langle(\hat{\sigma}_0, 0), \ldots, (\hat{\sigma}_p, 0), (\hat{\sigma}_{p+1}, 1), \ldots, (\hat{\sigma}_q, 1)\rangle \mid \\ \sigma_0 < \cdots < \sigma_p \leqslant \sigma_{p+1} < \cdots < \sigma_q \in L\big\}.$$

Then, $\operatorname{Sd} L \times \{0, 1\}$ is as the subcomplex of $\tilde{L}$. Additionally, let $\tilde{L}'$ be a subdivision of $\tilde{L}$ defined as follows:

$$\begin{aligned}\tilde{L}' = \operatorname{Sd} L \times \{0\} \cup L' \times \{1\} \\ \cup \big\{\langle(\hat{\sigma}_0, 0), \ldots, (\hat{\sigma}_p, 0), (v_0, 1), \ldots, (v_k, 1)\rangle \mid \\ \sigma_0 < \cdots < \sigma_p \leqslant \sigma_{p+1} < \cdots < \sigma_q \in L, \\ \tau = \langle v_0, \ldots, v_k\rangle \in L',\ \tau \subset \langle\hat{\sigma}_{p+1}, \ldots, \hat{\sigma}_q\rangle\big\},\end{aligned}$$

where $\tilde{L}'^{(0)} = \{(\hat{\sigma}, 0) \mid \sigma \in L\} \cup \{(v, 1) \mid v \in L'^{(0)}\}$. We define an injection $h_0 : \tilde{L}'^{(0)} \to |K_0|$ as follows:

$$h_0(\hat{\sigma}, 0) = \hat{\sigma} \text{ for } \sigma \in L \text{ and } h_0(v, 1) = g(v) \text{ for } v \in L'^{(0)}.$$

Let $\sigma_0 < \cdots < \sigma_p \leqslant \sigma_{p+1} < \cdots < \sigma_q \in L$ and $\tau = \langle v_0, \ldots, v_k\rangle \in L'$ such that $\tau \subset \langle\hat{\sigma}_{p+1}, \ldots, \hat{\sigma}_q\rangle$. Since $\tau \subset \bigcap_{i=0}^k |\operatorname{St}(v_i, L')|$, it follows from (2) that $f(\tau) \subset \bigcap_{i=0}^k O_{\operatorname{Sd} K}(\hat{\sigma}_{v_i})$. Hence, $\hat{\sigma}_{v_0}, \ldots, \hat{\sigma}_{v_k}$ span a simplex of $\operatorname{Sd} K$. Then, we can assume that $\sigma_{v_0} \leqslant \cdots \leqslant \sigma_{v_k}$. On the other hand, $\tau \subset |\operatorname{St}(\hat{\sigma}_r, \operatorname{Sd}^2 L)|$ for some $r = p+1, \ldots, q$ because $L' \lhd \operatorname{Sd}^2 L$ and $\tau \subset \langle\hat{\sigma}_{p+1}, \ldots, \hat{\sigma}_q\rangle$. Then, $f(\tau) \subset O_{\operatorname{Sd} K}(\hat{\sigma}_r)$ by (1). Thus, it follows that $O_{\operatorname{Sd} K}(\hat{\sigma}_{v_0}) \cap O_{\operatorname{Sd} K}(\hat{\sigma}_r) \neq \emptyset$, which means that $\hat{\sigma}_{v_0}$ and $\hat{\sigma}_r$ span a simplex of $\operatorname{Sd} K$, that is, $\sigma_{v_0} \leqslant \sigma_r$ or $\sigma_r \leqslant \sigma_{v_0}$. Since $\dim \sigma_r \leqslant \dim L = n$ and $\dim \sigma_{v_0} > 2n+3$, we have $\sigma_r < \sigma_{v_0}$. Consequently, $\sigma_p \leqslant \sigma_r < \sigma_{v_0} \leqslant \cdots \leqslant \sigma_{v_k}$. It follows that $\hat{\sigma}_0, \ldots, \hat{\sigma}_p \in \sigma_p \subset \sigma_{v_k}$ and $g(v_0), \ldots, g(v_k) \in \sigma_{v_k}$ by (3). Hence, h_0 can be extended to a PL map $h : |\tilde{L}'| = |L| \times \mathbf{I} \to |K_0| \subset |K|$ that is affine on each simplex of $\tilde{L}'$. By the definition, $h(x, 0) = x$ for every $x \in |L|$.

In the above, note that $\sigma_p^{(0)} \subset \sigma_{v_k}^{(0)}$ and $\sigma_{v_k} \in K_0 \setminus K_0^{(2n+3)}$. Since $\tau \subset \langle \hat{\sigma}_{p+1}, \dots, \hat{\sigma}_q \rangle$ and $q \leqslant \dim \sigma_q \leqslant \dim L = n$, it follows that

$$\begin{aligned}(p+1)+(k+1) = p + \dim \tau + 2 &\leqslant p + \dim \langle \hat{\sigma}_{p+1}, \dots, \hat{\sigma}_q \rangle + 2 \\ &\leqslant p + (q-p) + 2 \leqslant q + 2 \leqslant n + 2 (< 2n+4).\end{aligned}$$

By (4), $\sigma_p^{(0)} \cup \{g(v_0), \dots, g(v_k)\}$ is an affinely independent set in σ_{v_k}. Then, $\hat{\sigma}_0, \dots, \hat{\sigma}_p$ are affinely independent and $\mathrm{fl}\{\hat{\sigma}_0, \dots, \hat{\sigma}_p\} \subset \mathrm{fl}\, \sigma_p^{(0)}$. Since $g(v_0) \notin \mathrm{fl}\, \sigma_p^{(0)}$ and

$$g(v_i) \notin \mathrm{fl}(\sigma_p^{(0)} \cup \{g(v_0), \dots, g(v_{i-1})\}) \ \text{ for } i = 1, \dots, k,$$

it follows that $g(v_0) \notin \mathrm{fl}\{\hat{\sigma}_0, \dots, \hat{\sigma}_p\}$ and

$$g(v_i) \notin \mathrm{fl}\{\hat{\sigma}_0, \dots, \hat{\sigma}_p, g(v_0), \dots, g(v_{i-1})\} \ \text{ for } i = 1, \dots, k.$$

This means that $\{\hat{\sigma}_0, \dots, \hat{\sigma}_p, g(v_0), \dots, g(v_k)\}$ is affinely independent. Thus, for each simplex $\sigma \in \tilde{L}'$, $h(\sigma)$ is contained in a simplex of $K \setminus K^{(2n+3)}$ and $h(\sigma^{(0)})$ is affinely independent, hence $h|\sigma$ is an embedding.

If h is not an embedding, then there exist two simplexes $\sigma, \tau \in \tilde{L}'$ such that $h(\mathrm{rint}\, \sigma) \cap h(\mathrm{rint}\, \tau) \neq \emptyset$. In this case, $h(\sigma) \cup h(\tau)$ is contained in a simplex of $K \setminus K^{(2n+3)}$. Using (4) in the same way as above, we can prove that $h(\sigma^{(0)}) \cup h(\tau^{(0)})$ is affinely independent. This is a contradiction. Therefore, h is an embedding. □

Since any triangulation of a contractible PL ∞-manifold is combinatorially equivalent to $F(\Delta^\infty)$ by Corollary 6.9.7, the following is a direct consequence of Theorem 6.11.6:

Corollary 6.11.7 *A countable simplicial complex K is combinatorially equivalent to $F(\Delta^\infty)$ if $|K|_m$ is contractible and each simplex of K is a Z-set in $|K|_m$ (or $|K|$ is contractible and each simplex of K is a Z-set in $|K|$).*

Using the above corollary, we prove the following theorem, which contains the implication (c) $\Rightarrow$ (a) in Theorem 6.11.1:

Theorem 6.11.8 *Every $\mathbb{R}^\infty$-manifold is triangulable and any triangulation of an $\mathbb{R}^\infty$-manifold is a combinatorial ∞-manifold.*

To prove this theorem, we need the following lemma:

Lemma 6.11.9 *Let K be a simplicial complex. For every vertex $v \in K^{(0)}$, $|\mathrm{Lk}(v, K)|$ is a strong Z-set in $|\mathrm{St}(v, K)|$.*

Proof Let $h : |\mathrm{St}(v, K)| \times \mathbf{I} \to |\mathrm{St}(v, K)|$ be the straight line homotopy defined as follows:

$$h(x, t) = (1-t)x + tv \ \text{ for each } (x, t) \in |\mathrm{St}(v, K)| \times \mathbf{I}.$$

Then, $h_0 = \mathrm{id}$ and $h_t(|\operatorname{St}(v, K)|) \subset O_K(v)$ if $t > 0$. For each closed set A in $|\operatorname{St}(v, K)| \times \mathbf{I}$, $h(A)$ is closed in $|\operatorname{St}(v, K)|$. Indeed, for each $\sigma \in \operatorname{St}(v, K)$, if $\sigma \in K[v]$, then $h(A) \cap \sigma = h(A \cap (\sigma \times \mathbf{I}))$ is closed in σ; if $\sigma \in \operatorname{Lk}(v, K)$, then $h(A) \cap \sigma = \operatorname{pr}_1(A \cap (\sigma \times \{0\}))$ is closed in σ. For each open cover $\mathcal{U} \in \operatorname{cov}(|\operatorname{St}(v, K)|)$, $h^{-1}(\mathcal{U})$ is an open cover of $|\operatorname{St}(v, K)| \times \mathbf{I}$. By Lemma 2.5.2, we have a map $\alpha : |\operatorname{St}(v, K)| \to (0, 1] \subset \mathbf{I}$ such that

$$\big\{\{x\} \times [0, \alpha(x)] \bigm| x \in |\operatorname{St}(v, K)|\big\} \prec h^{-1}(\mathcal{U}).$$

Then, we define a map $f : |\operatorname{St}(v, K)| \to |\operatorname{St}(v, K)|$ as follows:

$$f(x) = h(x, \alpha(x)) \quad \text{for each } x \in |\operatorname{St}(v, K)|,$$

which is $\mathcal{U}$-close to id. Since $\alpha(x) > 0$ for every $x \in |\operatorname{St}(v, K)|$, $f(|\operatorname{St}(v, K)|) \subset O_K(v) = |\operatorname{St}(v, K)| \setminus |\operatorname{Lk}(v, K)|$. Let $\operatorname{Gr}(\alpha)$ be the graph of α, that is,

$$\operatorname{Gr}(\alpha) = \big\{(x, \alpha(x)) \bigm| x \in |\operatorname{St}(v, K)|\big\} \subset |\operatorname{St}(v, K)| \times \mathbf{I}.$$

Since $\operatorname{Gr}(\alpha)$ is closed in $|\operatorname{St}(v, K)| \times \mathbf{I}$, $f(|\operatorname{St}(v, K)|) = h(\operatorname{Gr}(\alpha))$ is closed in $|\operatorname{St}(v, K)|$. Hence,

$$|\operatorname{Lk}(v, K)| \cap \operatorname{cl} f(|\operatorname{St}(v, K)|) = |\operatorname{Lk}(v, K)| \cap f(|\operatorname{St}(v, K)|) = \emptyset.$$

This means that $|\operatorname{Lk}(v, K)|$ is a strong Z-set in $|\operatorname{St}(v, K)|$. □

Remark 6.14 The above proof shows that the open star $O_K(v)$ is homotopy dense in $|\operatorname{St}(v, K)|$. Moreover, since $h_1(|\operatorname{St}(v, K)|) = \{v\}$ in the above proof, $|\operatorname{St}(v, K)|$ is contractible.

Let K be a simplicial complex. For a simplex $\sigma \in K$, let $K_{\hat{\sigma}}$ be the starring of K at the barycenter $\hat{\sigma}$ of σ (cf. Sect. 1.5). Then, $|\operatorname{St}(\sigma, K)| = |\operatorname{St}(\hat{\sigma}, K_{\hat{\sigma}})|$ and $|\operatorname{Lk}(\sigma, K)| \subset |\operatorname{Lk}(\hat{\sigma}, K_{\hat{\sigma}})|$.

Note that if $\dim \sigma > 0$ then $|\operatorname{Lk}(\sigma, K)| \neq |\operatorname{Lk}(\hat{\sigma}, K_{\hat{\sigma}})|$ because $\partial\sigma \subset |\operatorname{Lk}(\hat{\sigma}, K_{\hat{\sigma}})|$ but $\partial\sigma \cap |\operatorname{Lk}(\sigma, K)| = \emptyset$.

By this observation, Lemma 6.11.9 can be generalized as follows:

Proposition 6.11.10 *Let K be a simplicial complex. For any simplex $\sigma \in K$, $|\operatorname{Lk}(\sigma, K)|$ is a strong Z-set in $|\operatorname{St}(\sigma, K)|$ and $|\operatorname{St}(\sigma, K)|$ is contractible.* □

Now, we prove Theorem 6.11.8.

Proof (Theorem 6.11.8) Since the first statement was already seen in the second paragraph of Sect. 6.6, p. 500. So, it suffices to see the second statement.

Let K be a simplicial complex such that $|K|$ is an $\mathbb{R}^\infty$-manifold. By virtue of Corollary 6.11.7, to prove that K is a combinatorial ∞-manifold, it suffices to show that for every $v \in K^{(0)}$, every $\sigma \in \operatorname{St}(v, K)$ is a Z-set in $|\operatorname{St}(v, K)|$. Since every compact set in an $\mathbb{R}^\infty$-manifold is a Z-set (Proposition 6.5.6), every $\sigma \in K$ is a Z-set in $|K|$.

Let $v \in K^{(0)}$ and $\sigma \in \mathrm{St}(v, K)$. For each open cover $\mathcal{U} \in \mathrm{cov}(|\mathrm{St}(v, K)|)$, let $\mathcal{V} \in \mathrm{cov}(|\mathrm{St}(v, K)|)$ be a star-refinement of $\mathcal{U}$. Due to Lemma 6.11.9, $|\mathrm{Lk}(v, K)|$ is a strong Z-set in $f(|\mathrm{St}(v, K)|$, that is, there is a map $f : |\mathrm{St}(v, K)| \to |\mathrm{St}(v, K)|$ such that $|\mathrm{Lk}(v, K)| \cap \mathrm{cl}\, f(|\mathrm{St}(v, K)|) = \emptyset$ and f is $\mathcal{V}$-close to id. Then, $\mathrm{cl}\, f(|\mathrm{St}(v, K)|) \subset O_K(v)$, $\mathrm{cl}\, f(|\mathrm{St}(v, K)|)$ is closed in $|K|$ and $O_K(v)$ is open in $|K|$. Let

$$\mathcal{W} = \mathcal{V}|O_K(v) \cup \big\{|K| \setminus \mathrm{cl}\, f(|\mathrm{St}(v, K)|)\big\} \in \mathrm{cov}(|K|).$$

Since σ is a Z-set in $|K|$, there is a map $g : |K| \to |K| \setminus \sigma$ that is $\mathcal{W}$-close to id. Then, $g(\mathrm{cl}\, f(|\mathrm{St}(v, K)|)) \subset O_K(v)$ and $g|\mathrm{cl}\, f(|\mathrm{St}(v, K)|)$ is $\mathcal{V}$-close to id. Thus, we have a map $fg : |\mathrm{St}(v, K)| \to O_K(v) \setminus \sigma$ that is $\mathcal{U}$-close to id because $\mathrm{st}\,\mathcal{V} \prec \mathcal{U}$. Therefore, σ is a Z-set in $|\mathrm{St}(v, K)|$. □

As mentioned before and after Theorem 6.10.2, for any countable simplicial complex K, the polyhedra $|K \times_s I^\infty|$, $|K \times_s R^\infty|$, and $|K \times_s F(\Delta^\infty)|$ are PL ∞-manifolds. Hence, we have the following:

Corollary 6.11.11 *For every countable simplicial complex K, the product simplicial complexes $K \times_s F(\Delta^\infty)$, $K \times_s I^\infty$, and $K \times_s R^\infty$ are combinatorial ∞-manifolds, which are combinatorially equivalent.* □

In the Characterization Theorem 6.11.1, the following condition can be added to the list of equivalent conditions:

- $|K| \times \boldsymbol{Q}$ is a $\boldsymbol{Q}^\infty$-manifold.

In other words, the hollowing holds:

Theorem 6.11.12 *Let K be a countable simplicial complex. Then, K is a combinatorial ∞-manifold if and only if $|K| \times \boldsymbol{Q}$ is a $\boldsymbol{Q}^\infty$-manifold.*

Proof If K is a combinatorial ∞-manifold, then $|K|$ is a $\mathbb{R}^\infty$-manifold, hence $|K| \times \boldsymbol{Q}$ is a $\boldsymbol{Q}^\infty$-manifold by Theorem 6.5.7. Conversely, if $|K| \times \boldsymbol{Q}$ is a $\boldsymbol{Q}^\infty$-manifold, then every compact subset of $|K| \times \boldsymbol{Q}$ is a Z-set by Theorem 6.5.6. For each simplex $\sigma \in K$, $\sigma \times \boldsymbol{Q}$ is a Z-set in $|K| \times \boldsymbol{Q}$. For each $\mathcal{U} \in \mathrm{cov}(|K|)$, let

$$\mathcal{U}^* = \{U \times \boldsymbol{Q} \mid U \in \mathcal{U}\} \in \mathrm{cov}(|K| \times \boldsymbol{Q}).$$

We have a map $f : |K| \times \boldsymbol{Q} \to |K| \times \boldsymbol{Q} \setminus \sigma \times \boldsymbol{Q}$ that is $\mathcal{U}^*$-close to id. Let $i_{\mathbf{0}} : |K| \to |K| \times \boldsymbol{Q}$ be the natural injection defined by $i_{\mathbf{0}}(x) = (x, \mathbf{0})$, and $\mathrm{pr}_1 : |K| \times \boldsymbol{Q} \to |K|$ be the projection onto $|K|$. Then, $\mathrm{pr}_1 f i_{\mathbf{0}} : |K| \to |K| \setminus \sigma$ that is $\mathcal{U}$-close to id. Hence, σ is a Z-set in $|K|$. By the Characterization Theorem 6.11.1, K is a combinatorial ∞-manifold. □

6.12 Simplicial Complexes with Contractible Links

In the previous section, we established the Characterization Theorem for combinatorial ∞-manifolds (Theorem 6.11.1). In particular, a combinatorial ∞-manifold has been characterized as a countable simplicial complex K such that every $\sigma \in K$ is a Z-set in $|K|$. In this section, we prove the following theorem:

Theorem 6.12.1 *For a simplicial complex K, every simplex of K is a Z-set in $|K|_{\mathrm{m}}$ if and only if the link* $\mathrm{Lk}(\sigma, K)$ *at every simplex $\sigma \in K$ is non-empty and contractible.*

Combining this theorem with the Characterization Theorem 6.11.1, a combinatorial ∞-manifold can be characterized as follows:

Theorem 6.12.2 (Characterization) *A countable simplicial complex K is a combinatorial ∞-manifold if and only if the link* $\mathrm{Lk}(\sigma, K)$ *of each simplex $\sigma \in K$ is non-empty and contractible.* □

Given two joinable simplicial complexes K and L in the same linear space (p. 29), the **join** $K * L$ is defined as follows:

$$K * L = K \cup L \cup \{\sigma\tau \mid \sigma \in K, \tau \in L\}.$$

The following is evident:

Fact *For each simplex $\sigma \in K$, the subcomplexes $F(\sigma)$ and* $\mathrm{Lk}(\sigma, K)$ *are joinable, and* $\mathrm{St}(\sigma, K) = F(\sigma) * \mathrm{Lk}(\sigma, K)$, *where* $F(\sigma) = \{\tau \mid \tau \leqslant \sigma\}$.

We define the **join** of spaces X and Y as the following space:

$$X * Y = X \cup Y \cup (X \times Y \times (0, 1)),$$

which has the topology generated by open sets in $X \times Y \times \mathbf{I}$ and sets

$$U \cup U \times Y \times (0, t), \ \ V \cup X \times V \times (t, 1),$$

where U is open in X, V is open in Y, and $0 < t < 1$. When X and Y are metrizable, the join $X * Y$ is also metrizable. This can be easily shown by using the Bing Metrization Theorem 1.3.3.

Proposition 6.12.3 *For any two joinable simplicial complexes K and L, $|K|_{\mathrm{m}} * |L|_{\mathrm{m}} \approx |K * L|_{\mathrm{m}}$ by a homeomorphism $f : |K|_{\mathrm{m}} * |L|_{\mathrm{m}} \to |K * L|_{\mathrm{m}}$ such that* $f||K| \cup |L| = \mathrm{id}$.

Proof For each $(x, y, t) \in |K| \times |L| \times (0, 1)$, $c_K(x) \in K$ and $c_L(y) \in L$ are joinable, and $(1 - t)x + ty \in c_K(x)c_L(y) \in K * L$. So, we can define $f : |K|_{\mathrm{m}} * |L|_{\mathrm{m}} \to |K * L|_{\mathrm{m}}$ by $f||K| \cup |L| = \mathrm{id}$ and

$$f(x, y, t) = (1 - t)x + ty \ \text{ for } (x, y, t) \in |K| \times |L| \times (0, 1).$$

Each $z \in |K * L| \setminus (|K| \cup |L|)$ can be uniquely written as follows:

$$z = \sum_{v \in (K*L)^{(0)}} \beta_v^{K*L}(z) v$$

and $(K * L)^{(0)} = K^{(0)} \cup L^{(0)}$. Let $\alpha : |K * L|_{\mathrm{m}} \setminus (|K| \cup |L|) \to (0, 1)$ be a map defined by $\alpha(z) = \sum_{v \in L^{(0)}} \beta_v^{K*L}(z)$. Then, $1 - \alpha(z) = \sum_{v \in K^{(0)}} \beta_v^{K*L}(z)$. We define $g : |K * L|_{\mathrm{m}} \to |K|_{\mathrm{m}} * |L|_{\mathrm{m}}$ defined by $g||K| \cup |L| = \mathrm{id}$ and

$$g(z) = \left(\sum_{v \in K^{(0)}} \frac{\beta_v^{K*L}(z)}{1 - \alpha(z)} v,\ \sum_{v \in L^{(0)}} \frac{\beta_v^{K*L}(z)}{\alpha(z)} v,\ \alpha(z) \right) \text{ for } z \in |K * L| \setminus (|K| \cup |L|).$$

For each $z \in |K * L| \setminus (|K| \cup |L|)$,

$$\begin{aligned} fg(z) &= (1 - \alpha(z)) \sum_{v \in K^{(0)}} \frac{\beta_v^{K*L}(z)}{1 - \alpha(z)} v + \alpha(z) \sum_{v \in L^{(0)}} \frac{\beta_v^{K*L}(z)}{\alpha(z)} v \\ &= \sum_{v \in (K*L)^{(0)}} \beta_v^{K*L}(z) v = z. \end{aligned}$$

Since $f||K| \cup |L| = g||K| \cup |L| = \mathrm{id}$, we have $fg = \mathrm{id}$. For each $(x, y, t) \in |K| \times |L| \times (0, 1)$, let $z = f(x, y, t) = (1 - t)x + ty$. Then,

$$z = (1 - t) \sum_{v \in K^{(0)}} \beta_v^K(x) v + t \sum_{v \in L^{(0)}} \beta_v^L(y) v,$$

which means that $\beta_v^{K*L}(z) = (1 - t)\beta_v^K(x)$ if $v \in K^{(0)}$, and $\beta_v^{K*L}(z) = t\beta_v^L(y)$ if $v \in L^{(0)}$. Since $\alpha(z) = \sum_{v \in L^{(0)}} t\beta_v^L(y) = t$, it follows that

$$\sum_{v \in K^{(0)}} \frac{\beta_v^{K*L}(z)}{1 - \alpha(z)} v = \sum_{v \in K^{(0)}} \beta_v^K(x) v = x \text{ and } \sum_{v \in L^{(0)}} \frac{\beta_v^{K*L}(z)}{\alpha(z)} v = \sum_{v \in L^{(0)}} \beta_v^L(y) v = y,$$

hence $gf(x, y, t) = g(z) = (x, y, t)$. Since $f||K| \cup |L| = g||K| \cup |L| = \mathrm{id}$, we have $gf = \mathrm{id}$. Consequently, f is a bijection and $f^{-1} = g$. The continuity of f and g remains to be proved.

To see the continuity of $f||K|_{\mathrm{m}} \times |L|_{\mathrm{m}} \times (0, 1)$, let

$$(x, y, t),\ (x', y', t') \in |K| \times |L| \times (0, 1),\ f(x, y, t) = z \text{ and } f(x', y', t') = z'.$$

As we saw above,

$$\begin{aligned} v \in K^{(0)} &\Rightarrow \beta_v^{K*L}(z) = (1 - t)\beta_v^K(x),\ \beta_v^{K*L}(z') = (1 - t')\beta_v^K(x'); \\ v \in L^{(0)} &\Rightarrow \beta_v^{K*L}(z) = t\beta_v^K(y),\ \beta_v^{K*L}(z') = t'\beta_v^K(y'). \end{aligned}$$

Hence, we have

$$\sum_{v\in K^{(0)}} \left| \frac{\beta_v^{K*L}(z)}{1-t} - \frac{\beta_v^{K*L}(z')}{1-t'} \right| = \sum_{v\in K^{(0)}} \left|\beta_v^K(x) - \beta_v^K(x')\right| \text{ and}$$

$$\sum_{v\in L^{(0)}} \left| \frac{\beta_v^{K*L}(z)}{t} - \frac{\beta_v^{K*L}(z')}{t'} \right| = \sum_{v\in L^{(0)}} \left|\beta_v^L(y) - \beta_v^L(y')\right|.$$

Then, it follows that

$$\begin{aligned}
\rho_{K*L}(z,z') &= \sum_{v\in (K*L)^{(0)}} \left|\beta_v^{K*L}(z) - \beta_v^{K*L}(z')\right| \\
&= \sum_{v\in K^{(0)}} \left|\beta_v^{K*L}(z) - \beta_v^{K*L}(z')\right| + \sum_{v\in L^{(0)}} \left|\beta_v^{K*L}(z) - \beta_v^{K*L}(z')\right| \\
&\leqslant (1-t) \sum_{v\in K^{(0)}} \left| \frac{\beta_v^{K*L}(z)}{1-t} - \frac{\beta_v^{K*L}(z')}{1-t'} \right| \\
&\qquad\qquad + \left|(1-t)-(1-t')\right| \sum_{v\in K^{(0)}} \frac{\beta_v^{K*L}(z')}{1-t'} \\
&\quad + t \sum_{v\in L^{(0)}} \left| \frac{\beta_v^{K*L}(z)}{t} - \frac{\beta_v^{K*L}(z')}{t'} \right| + |t-t'| \sum_{v\in L^{(0)}} \frac{\beta_v^{K*L}(z')}{t'} \\
&= (1-t) \sum_{v\in K^{(0)}} \left|\beta_v^K(x) - \beta_v^K(x')\right| + |t-t'| \sum_{v\in K^{(0)}} \beta_v^K(x') \\
&\qquad + t \sum_{v\in L^{(0)}} \left|\beta_v^L(y) - \beta_v^L(y')\right| + |t-t'| \sum_{v\in L^{(0)}} \beta_v^L(y') \\
&= (1-t)\rho_K(x,x') + t\rho_L(y,y') + 2|t-t'|.
\end{aligned}$$

Since $|K|_{\mathrm{m}} \times |L|_{\mathrm{m}} \times (0,1)$ is an open subspace of $|K|_{\mathrm{m}} * |L|_{\mathrm{m}}$, f is continuous at each $(x,y,t) \in |K| \times |L| \times (0,1)$.

To verify the continuity of f at $x \in |K|$, let $\varepsilon > 0$. The following is an open neighborhood $f(x) = x$ in $|K|_{\mathrm{m}} * |L|_{\mathrm{m}}$:

$$U = \mathrm{B}_{\rho_K}(x,\varepsilon/3) \cup \mathrm{B}_{\rho_K}(x,\varepsilon/3) \times |L| \times (0,\varepsilon/3).$$

Then, we may show that $f(U) \subset \mathrm{B}_{\rho_{K*L}}(x,\varepsilon)$. First, it follows that

$$f(\mathrm{B}_{\rho_K}(x,\varepsilon/3)) = \mathrm{B}_{\rho_K}(x,\varepsilon/3) \subset \mathrm{B}_{\rho_{K*L}}(x,\varepsilon/3).$$

For each $(x', y, t) \in \mathrm{B}_{\rho_K}(x, \varepsilon/3) \times |L| \times (0, \varepsilon/3)$,

$$\begin{aligned}\rho_K(f(x), f(x', y, t)) &= \rho_K(x, (1-t)x' + ty)\\ &= \sum_{v\in K^{(0)}} |\beta_v^K(x) - (1-t)\beta_v^K(x')| + \sum_{u\in L^{(0)}} t\beta_u^L(y)\\ &\leqslant \sum_{v\in K^{(0)}} |\beta_v^K(x) - \beta_v^K(x')| + \sum_{v\in K^{(0)}} t\beta_v^K(x') + \sum_{u\in L^{(0)}} t\beta_u^L(y)\\ &= \rho_K(x, x') + 2t = \varepsilon/3 + 2\varepsilon/3 = \varepsilon.\end{aligned}$$

Hence, $f(U) \subset \mathrm{B}_{\rho_{K*L}}(x, \varepsilon)$. Similarly, the continuity at $y \in |L|$ can be verified. Thus, the continuity of f has been proved.

From the definition, it follows that $g||K * L|_{\mathrm{m}} \setminus (|K| \cup |L|)$ is continuous. Since $|K * L|_{\mathrm{m}} \setminus (|K| \cup |L|)$ is an open subspace of $|K * L|_{\mathrm{m}}$, g is continuous at each $z \in |K * L| \setminus (|K| \cup |L|)$.

To verify the continuity of g at $x \in |K|$, let U be an open neighborhood of $g(x) = x$ in $|K|_{\mathrm{m}} * |L|_{\mathrm{m}}$ with $U \cap |L| = \emptyset$. We can take $0 < \varepsilon < 1$ such that

$$\mathrm{B}_{\rho_K}(x, \varepsilon) \cup \mathrm{B}_{\rho_K}(x, \varepsilon) \times Y \times (0, \varepsilon) \subset U.$$

Then, we may show that $g(\mathrm{B}_{\rho_{K*L}}(x, \varepsilon/3)) \subset U$. First, it follows that

$$g(\mathrm{B}_{\rho_{K*L}}(x, \varepsilon/3) \cap |K|) = g(\mathrm{B}_{\rho_K}(x, \varepsilon/3)) = \mathrm{B}_{\rho_K}(x, \varepsilon/3) \subset U.$$

For each $z \in \mathrm{B}_{\rho_{K*L}}(x, \varepsilon/3) \setminus |K|$,[18]

$$g(z) = \left(\sum_{v\in K^{(0)}} \frac{\beta_v^{K*L}(z)}{1-\alpha(z)} v,\ \sum_{v\in L^{(0)}} \frac{\beta_v^{K*L}(z)}{\alpha(z)} v,\ \alpha(z) \right),$$

where $\alpha(z) = \sum_{v\in L^{(0)}} \beta_v^{K*L}(z)$. Observe that

$$\begin{aligned}\rho_{K*L}(x, z) &= \sum_{v\in (K*L)^{(0)}} \left|\beta_v^{K*L}(x) - \beta_v^{K*L}(z)\right|\\ &= \sum_{v\in K^{(0)}} \left|\beta_v^K(x) - \beta_v^{K*L}(z)\right| + \sum_{v\in L^{(0)}} \beta_v^{K*L}(z).\end{aligned}$$

Since $\rho_{K*L}(x, z) < \varepsilon/3$, it follows that

$$\alpha(z) = \sum_{v\in L^{(0)}} \beta_v^{K*L}(z) < \varepsilon/3 < \varepsilon \ \text{ and } \ \sum_{v\in K^{(0)}} |\beta_v^K(x) - \beta_v^{K*L}(z)| < \varepsilon/3.$$

[18]Note that $\mathrm{B}_{\rho_{K*L}}(x, \varepsilon/3) \cap |L| = \emptyset$ because $\mathrm{dist}_{\rho_{K*L}}(|K|, |L|) = 2$.

Here, we define

$$x' = \sum_{v\in K^{(0)}} \frac{\beta_v^{K*L}(z)}{1-\alpha(z)} v.$$

Then, it follows that

$$\begin{aligned}\rho_K(x,x') &= \sum_{v\in K^{(0)}} \left|\beta_v^K(x) - \beta_v^K(x')\right| \\ &= \sum_{v\in K^{(0)}} \left|\beta_v^K(x) - \frac{\beta_v^K(x)}{1-\alpha(z)}\right| + \sum_{v\in K^{(0)}} \left|\frac{\beta_v^K(x)}{1-\alpha(z)} - \frac{\beta_v^{K*L}(z)}{1-\alpha(z)}\right| \\ &= \sum_{v\in K^{(0)}} \frac{\alpha(z)}{1-\alpha(z)}\beta_v^K(x) + \frac{1}{1-\alpha(z)} \sum_{v\in K^{(0)}} \left|\beta_v^K(x) - \beta_v^{K*L}(z)\right| \\ &< \frac{\alpha(z)}{1-\alpha(z)} + \frac{\varepsilon/3}{1-\alpha(z)} < \frac{2\varepsilon/3}{1-1/3} = \varepsilon,\end{aligned}$$

which implies that $g(z) \in \mathrm{B}_{\rho_K}(x,\varepsilon) \times Y \times (0,\varepsilon) \subset U$. Consequently, we have $g(\mathrm{B}_{\rho_{K*L}}(x,\varepsilon/3)) \subset U$. Similarly, the continuity of g at $y \in |L|$ can be verified. Thus, the continuity of g has been proved. The proof is complete. □

As a corollary of Proposition 6.12.3, we have the following:

Corollary 6.12.4 *Let K be a simplicial complex. For each simplex $\sigma \in K$, $\sigma * |\operatorname{Lk}(\sigma,K)|_{\mathrm{m}} \approx |\operatorname{St}(\sigma,K)|_{\mathrm{m}}$ by a homeomorphism $f : \sigma * |\operatorname{Lk}(\sigma,K)|_{\mathrm{m}} \to |\operatorname{St}(\sigma,K)|_{\mathrm{m}}$ such that $f|\sigma \cup |\operatorname{Lk}(\sigma,K)| = \mathrm{id}$.*

We show some propositions of the join $X * Y$.

Proposition 6.12.5 *If X is compact and Y is contractible, then X is a strong Z-set in $X * Y$.*

Proof Let $\mathcal{U}$ be an open cover of $X * Y$. From the compactness of X, we can find $0 < r < 1$ such that

$$\big\{\{x\} \cup \{x\} \times Y \times (0,r] \bigm| x \in X\big\} \prec \mathcal{U}.$$

Let $h : Y \times \mathbf{I} \to Y$ be a homotopy such that $h_0(Y) = \{y_0\}$ for some $y_0 \in Y$ and $h_1 = \mathrm{id}$. We can define a map $f : X * Y \to X * Y$ as follows:

$$f(x) = (x, y_0, r) \text{ for every } x \in X,\ f|Y \cup X \times Y \times [r,1) = \mathrm{id}, \text{ and}$$
$$f(x,y,t) = (x, h(y,t/r)) \text{ for every } (x,y,t) \in X \times Y \times (0,r].$$

Then, f is $\mathcal{U}$-close to id and $X \cap \operatorname{cl} f(X * Y) = \emptyset$. □

Proposition 6.12.6 *Let A be a (strong) Z-set in a compact ANR X. Then, for any space Y, A is a (strong) Z-set in $X * Y$.*

Proof Let $\mathcal{U}$ be an open cover of $X * Y$. From the compactness of X, we can find an open cover $\mathcal{V} \in \operatorname{cov}(X)$ and $0 < r < 1$ such that

$$\{V \cup V \times Y \times (0, r] \mid V \in \mathcal{V}\} \prec \mathcal{U}.$$

Since A is a (strong) Z-set in an ANR X, we have a $\mathcal{V}$-homotopy $h : X \times \mathbf{I} \to X$ such that $A \cap h_0(X) = \emptyset$ ($A \cap \operatorname{cl} h_0(X) = \emptyset$) and $h_1 = \operatorname{id}$. We can define a map $f : X * Y \to X * Y$ as follows: $f|X = h_0$, $f|Y = \operatorname{id}$, and

$$f(x, y, t) = \begin{cases} (h_0(x), y, t) & \text{if } t \leqslant r/2, \\ (h(x, 2t/r - 1), y, t) & \text{if } r/2 \leqslant t \leqslant r, \\ (x, y, t) & \text{if } t \geqslant r. \end{cases}$$

Then, f is $\mathcal{U}$-close to id and $A \cap f(X * Y) = \emptyset$ ($A \cap \operatorname{cl} f(X * Y) = \emptyset$). □

Proposition 6.12.7 *For any spaces X and Y, X is a strong deformation retract of $X * Y \setminus Y$.*

Proof We define a homotopy $h : (X * Y \setminus Y) \times I \to (X * Y \setminus Y)$ as follows: $h_t|X = \operatorname{id}$ for every $t \in \mathbf{I}$, and for each $(x, y, s) \in X \times Y \times (0, 1)$,

$$h_t(x, y, s) = \begin{cases} (x, y, ts) & \text{if } t > 0. \\ x & \text{if } t = 0. \end{cases}$$

Then, $h_0 : X * Y \setminus Y \to X$ is retraction, $h_1 = \operatorname{id}$, and $h_t|X = \operatorname{id}$ for every $t \in \mathbf{I}$, which means that X is a strong deformation retract of $X * Y \setminus Y$. □

Now, we prove Theorem 6.12.1.

Proof (Theorem 6.12.1) By virtue of Corollary 6.12.4, $|\operatorname{St}(\sigma, K)|_{\mathrm{m}}$ can be identified with $\sigma * |\operatorname{Lk}(\sigma, K)|_{\mathrm{m}}$.

(*The "only if" Part*) Let $\sigma \in K$. Since σ is a Z-set in $|K|_{\mathrm{m}}$, σ is not principal in K, that is, a face of some other simplex of K. Hence, $\operatorname{Lk}(\sigma, K) \neq \emptyset$. We can write $\operatorname{rint} \sigma = \bigcup_{i \in \mathbb{N}} C_i$, where each C_i is compact. Since σ is a Z-set in $|K|_{\mathrm{m}}$, each C_i is also a Z-set in $|K|_{\mathrm{m}}$. Note that each C_i is contained in the interior $\operatorname{int}_{|K|_{\mathrm{m}}} |\operatorname{St}(\sigma, K)|$. Then, combining Corollary 2.8.7 (ii) and (iii), we can see that each C_i is a Z-set in $|\operatorname{St}(\sigma, K)|_{\mathrm{m}}$. It follows from Proposition 6.12.6 that $\partial\sigma$ is a Z-set in $|\operatorname{St}(\sigma, K)|_{\mathrm{m}}$. Thus, σ is a Z_σ-set in $\operatorname{St}(\sigma, K)|_{\mathrm{m}}$. Since a compact Z_σ-set is a Z-set by Proposition 2.8.8, σ is a Z-set in $|\operatorname{St}(\sigma, K)|_{\mathrm{m}}$. Hence, $|\operatorname{St}(\sigma, K)|_{\mathrm{m}} \setminus \sigma \simeq |\operatorname{St}(\sigma, K)|_{\mathrm{m}}$. Moreover, by Proposition 6.12.7, we have

$$|\operatorname{St}(\sigma, K)|_{\mathrm{m}} \setminus \sigma \approx \sigma * |\operatorname{Lk}(\sigma, K)|_{\mathrm{m}} \setminus \sigma \simeq |\operatorname{Lk}(\sigma, K)|_{\mathrm{m}}.$$

Therefore, $|\operatorname{St}(\sigma, K)|_{\mathrm{m}} \simeq |\operatorname{Lk}(\sigma, K)|_{\mathrm{m}}$. Since $|\operatorname{St}(\sigma, K)|_{\mathrm{m}}$ is contractible (cf. Corollary 6.12.4), it follows that $|\operatorname{Lk}(\sigma, K)|_{\mathrm{m}}$ is also contractible.

(*The "if" Part*) For each $v \in K^{(0)}$, $\{v\}$ is a Z-set in $|\operatorname{St}(v, K)|_{\mathrm{m}}$ and $|\operatorname{St}(v, K)|_{\mathrm{m}}$ is a neighborhood of v in $|K|_{\mathrm{m}}$, hence $\{v\}$ is a Z-set in $|K|_{\mathrm{m}}$ by Corollary 2.8.7 (iii). Now, assume that every $(n-1)$-simplex of K is a Z-set in $|K|_{\mathrm{m}}$, and let σ be an n-simplex of K. Then, $\partial\sigma$ is a Z-set in $|K|_{\mathrm{m}}$ because it is a finite union of Z-sets (Corollary 2.8.7 (i)). Since $|\operatorname{Lk}(\sigma, K)|_{\mathrm{m}}$ is contractible, it follows from Proposition 6.12.5 that σ is a Z-set in $|\operatorname{St}(\sigma, K)|_{\mathrm{m}}$. As above, write $\operatorname{rint}\sigma = \bigcup_{i\in\mathbb{N}} C_i$, where each C_i is compact. Then, each C_i is a Z-set in $|\operatorname{St}(\sigma, K)|_{\mathrm{m}}$. Since $|\operatorname{St}(\sigma, K)|_{\mathrm{m}}$ is a neighborhood of C_i in $|K|_{\mathrm{m}}$, it follows from Corollary 2.8.7 (iii) that C_i is a Z-set in $|K|_{\mathrm{m}}$. Therefore, σ is a Z_σ-set in $|K|_{\mathrm{m}}$ Since a compact Z_σ-set is a Z-set by Proposition 2.8.8, σ is a Z-set in $|K|_{\mathrm{m}}$. By induction, every simplex of K is a Z-set in $|K|_{\mathrm{m}}$. □

6.13 Metric Combinatorial ∞-Manifolds

In Sect. 6.11, it is proved that a countable simplicial complex K is a combinatorial ∞-manifold if and only if the polyhedron $|K|$ is an $\mathbb{R}^\infty$-manifold, where $|K|$ has the Whitehead (or weak) topology. Here, we consider the polyhedron $|K|_{\mathrm{m}}$ with the metric topology (cf. Sect. 1.7). The topology of $|K|_{\mathrm{m}}$ is induced by the following metric ρ_K:

$$\rho_K(x, y) = \|\beta^K(x) - \beta^K(y)\| = \sum_{v\in K^{(0)}} \left|\beta_v^K(x) - \beta_v^K(y)\right|,$$

where $\|\cdot\|$ is the norm of the Banach space $\ell^1(K^{(0)})$ and $\beta^K : |K| \to \ell^1(K^{(0)})$ is the canonical representation of K (defined by $\beta^K(x) = (\beta_v^K(x))_{v\in K^{(0)}}$). To Theorem 6.11.1 we can add one more condition that $|K|_{\mathrm{m}}$ is an ℓ_f^2-manifold, that is:

Theorem 6.13.1 *A countable simplicial complex K is a combinatorial ∞-manifold if and only if $|K|_{\mathrm{m}}$ is an ℓ_f^2-manifold.*

To prove the above theorem, we first prove the following:

Theorem 6.13.2 *Let M be an $\mathbb{R}^\infty$-manifold and let Y be a metrizable space. If there exists a bijective fine homotopy equivalence $\varphi : M \to Y$ and every compact set in Y is a strong Z-set in Y, then Y is an ℓ_f^2-manifold.*

Proof To prove that Y is an ℓ_f^2-manifold, we apply Theorem 5.9.9. First, note that M is an ANE and Y is $\mathcal{U}$-homotopy dominated by M for every $\mathcal{U} \in \operatorname{cov}(X)$. Hence, it follows from Hanner's Characterization of ANR (Theorem 1.13.20) that Y is an ANR. Due to Proposition 6.4.7, there is a tower $M_1 \subset M_2 \subset \cdots$ of finite-dimensional compacta with $M = \varinjlim M_n$. Note that each $\varphi|M_n : M_n \to \varphi(M_n)$ is a homeomorphism. Thus, $Y = \bigcup_{n\in\mathbb{N}} \varphi(M_n)$ is σ-compact and strongly countable-dimensional. Since every compact set in Y is a strong Z-set, it follows that Y is a strong Z_σ-space. It remains to show that Y is strongly $\mathfrak{M}_0^{\mathrm{fd}}$-universal.

To see the strong $\mathfrak{M}_0^{\mathrm{fd}}$-universality of Y, let $f : A \to Y$ be a map from a finite-dimensional compactum A such that $f|B$ is an embedding, where $B \subset A$ is a closed subset of A. We show that f is approximated by embeddings $g : A \to Y$ with $g|B = f|B$. Then, by Proposition 2.8.12, it can be assumed that $f(A \setminus B) \subset Y \setminus f(B)$. For each $\mathcal{U} \in \mathrm{cov}(Y)$, let $\mathcal{V} \in \mathrm{cov}(Y)$ be a star-refinement of $\mathcal{U}$. We can take $d \in \mathrm{Metr}(Y)$ so that $\{\overline{\mathrm{B}}_d(x, 1) \mid x \in Y\} \prec \mathcal{V}$ (Proposition 1.3.22(1)).

Due to Theorem 6.4.10, M has the embedding approximation property $(\mathcal{A}_f)$. We write $A \setminus B = \bigcup_{n\in\mathbb{N}} A_n$, where each A_n is closed in A and $A_n \subset \operatorname{int} A_{n+1}$. We inductively construct maps $f_n : A \to Y$, $n \in \mathbb{N}$, with an increasing sequence $m(1) < m(2) < m(3) < \cdots \in \mathbb{N}$ such that:

(1) $d(f_n, f_{n-1}) < 2^{-n}$,
(2) $f_n|A_{n-1} = f_{n-1}|A_{n-1}$, $f_n|B = f|B$,
(3) $f_n|A_n \cup B$ is an embedding,
(4) $f_n(A \setminus B) \subset Y \setminus f(B)$,
(5) $f_n(A_n) \subset \varphi(M_{m(n)})$,

where $f_0 = f$ and $A_0 = \emptyset$.

Suppose that f_{n-1} has been obtained. Since $f_{n-1}(A_n) \cap f(B) = \emptyset$, we can take $\mathcal{W}_n \in \mathrm{cov}(Y)$ so that

$$\operatorname{st}\mathcal{W}_n \prec \{Y \setminus f(A_n), Y \setminus f(B)\} \text{ and } \operatorname{mesh}\operatorname{st}^2\mathcal{W}_n < 2^{-n-1}.$$

Since $\varphi : M \to Y$ is a fine homotopy equivalence, there is a map $\psi_n' : Y \to M$ such that $\varphi\psi_n' \simeq_{\mathcal{W}_n} \mathrm{id}$ and $\psi_n'\varphi \simeq_{\varphi^{-1}(\mathcal{W}_n)} \mathrm{id}$. Note that $\varphi|M_{m(n-1)} : M_{m(n-1)} \to \varphi(M_{m(n-1)})$ is a homeomorphism and

$$\psi_n'|\varphi(M_{m(n-1)}) \simeq_{\varphi^{-1}(\mathcal{W}_n)} (\varphi|M_{m(n-1)})^{-1}.$$

By the Homotopy Extension Theorem 1.13.11, we have a map $\psi_n : Y \to M$ such that

$$\psi_n|\varphi(M_n) = (\varphi|M_{m(n-1)})^{-1} \text{ and } \psi_n \simeq_{\varphi^{-1}(\mathcal{W}_n)} \psi_n'.$$

Applying $(\mathcal{A}_f)$ to $\psi_n f_{n-1} : A \to M$, we can obtain an embedding $g_n : A \to M$ such that

$$g_n|A_{n-1} = \psi_n f_{n-1}|A_{n-1} \text{ and } g_n \simeq_{\varphi^{-1}(\mathcal{W}_n)} \psi_n f_{n-1}.$$

Then, $\varphi g_n \simeq_{\operatorname{st}\mathcal{W}_n} f_{n-1}$ and $\varphi g_n|A_{n-1} = f_{n-1}|A_{n-1}$. Indeed,

$$\varphi g_n \simeq_{\mathcal{W}_n} \varphi\psi_n f_{n-1} \simeq_{\mathcal{W}_n} \varphi\psi_n' f_{n-1} \simeq_{\mathcal{W}_n} \varphi(\varphi|M_{m(n-1)})^{-1} f_{n-1} = f_{n-1} \text{ and}$$

$$\varphi g_n|A_{n-1} = \varphi\psi_n f_{n-1}|A_{n-1} = \varphi(\varphi|M_{m(n-1)})^{-1} f_{n-1}|A_{n-1} = f_{n-1}|A_{n-1}.$$

Note that $\varphi g_n|A_n$ is an embedding and

$$\varphi g_n(A_n) \subset \mathrm{st}(f_{n-1}(A_n), \mathrm{st}\,\mathcal{W}_n) \subset Y \setminus f(B).$$

Since $g_n(A_n)$ is compact, it is contained in $M_{m(n)}$ for some $m(n) > m(n-1)$. Then, $\varphi g_n(A_n) \subset M_{m(n)}$. Using the Homotopy Extension Theorem 1.13.11, we can construct a map $f_n' : A \to Y$ such that $f_n'|A_n = \varphi g_n|A_n$, $f_n'|B = f|B$, and $f_n' \simeq_{\mathrm{st}\,\mathcal{W}_n} f_{n-1}$. Then, $f_n'|A_n \cup B$ is an embedding, which is a strong Z-embedding because every compact set in Y is a strong Z-set. So, using Proposition 2.8.12, we can obtain a map $f_n : A \to Y$ such that $f_n \simeq_{\mathcal{W}_n} f_n'$, $f_n|A_n \cup B = f_n'|A_n \cup B$, and $f_n(A \setminus (A_n \cup B) \subset Y \setminus f_n(A_n \cup B)$. Then, f_n satisfies (3) and (4). Since $f_n \simeq_{\mathrm{st}^2\,\mathcal{W}_n} f_{n-1}$, we have $d(f_n, f_{n-1}) < 2^{-n-1}$, which means (1). By the definition, f_n satisfies (2) and (5).

Now, we define $g : A \to Y$ as follows:

$$g(x) = \begin{cases} f_n(x) & \text{if } x \in A_n, n \in \mathbb{N}, \\ f(x) & \text{if } x \in B. \end{cases}$$

Observe that $d(f_n, g) < 2^{-n}$ for each $n \in \mathbb{N}$, which means that $(f_n)_{n\in\mathbb{N}}$ uniformly converges to g. Hence, g is continuous. From (3) and the definition, it easily follows that g is injective. Consequently, g is an embedding. By the definition, $g|B = f|B$. Since $d(f, g) < 2^0 = 1$, g is $\mathcal{U}$-close to f. □

Now, using Theorem 6.13.2, we prove Theorem 6.13.1.

Proof (Theorem 6.13.1) To see the "if" part, suppose that $|K|_{\mathrm{m}}$ is an ℓ_f^2-manifold. Since every compact set in $|K|_{\mathrm{m}}$ is a strong Z-set (Theorem 5.9.5 with Remark 5.18), every simplex of K is a strong Z-set in $|K|_{\mathrm{m}}$, which means that K is a combinatorial ∞-manifold (Theorem 6.11.1).

To see the "only if" part, suppose that K is a combinatorial ∞-manifold. Then, every simplex of K is a strong Z-set in $|K|_{\mathrm{m}}$ (Theorem 6.11.1), which implies that $|K|_{\mathrm{m}}$ is a strong Z_σ-space. Since a compact Z_σ-set in an ANR is a Z-set (Proposition 2.8.8) and a Z-set in a separable ANR is a strong Z-set if it is a strong Z-set, every compact set in $|K|_{\mathrm{m}}$ is a strong Z-set in $|K|_{\mathrm{m}}$. Recall that the identity map $\mathrm{id} : |K| \to |K|_{\mathrm{m}}$ is a fine homotopy equivalence (Theorem 1.7.6). Thus, the "only if" part is a direct consequence of Theorem 6.13.2. □

In exactly the same way as Theorem 6.13.2, applying Theorem 5.9.10 instead of Theorem 5.9.9,[19] we can prove the following $\boldsymbol{Q}^\infty$-manifold version of Theorem 6.13.2:

[19] In the proof of Theorem 6.13.2, replace $\mathfrak{M}_0^{\mathrm{fd}}$ and $(\mathcal{A}_f)$ with $\mathfrak{M}_0$ and $(\mathcal{A})$ respectively and remove the term "finite-dimensional."

Theorem 6.13.3 *Let M be a $\boldsymbol{Q}^\infty$-manifold and let Y be a metrizable space. If there exists a bijective fine homotopy equivalence $f : M \to Y$ and every compact set in Y is a strong Z-set in Y, then Y is an ℓ^2_Q-manifold.* □

In the same way as for Theorem 6.13.1, we can prove the following (the details are left to the reader):

Theorem 6.13.4 *A countable simplicial complex K is a combinatorial ∞-manifold if and only if $|K|_{\mathrm{m}} \times \boldsymbol{Q}$ is an ℓ^2_Q-manifold.*

For a simplicial complex K, the completion of $|K| = (|K|, \rho_K)$ is called the **$\boldsymbol{\ell^1}$-completion** of $|K|$ and denoted by $\overline{|K|}^{\ell^1}$, which can be identified with the closure of $\beta^K(|K|)$ in $\ell^1(K^{(0)})$, that is,

$$\overline{|K|}^{\ell^1} = \mathrm{cl}_{\ell^1(K^{(0)})}\, \beta^K(|K|).$$

Theorem 6.13.5 *A countable simplicial complex K is a combinatorial ∞-manifold if and only if the ℓ^1-completion $\overline{|K|}^{\ell^1}$ of $|K|$ is an ℓ^2-manifold. In this case, $|K|_{\mathrm{m}}$ is an f.d.cap set for $\overline{|K|}^{\ell^1}$, that is, $(\overline{|K|}^{\ell^1}, |K|_{\mathrm{m}})$ is an (ℓ^2, ℓ^2_f)-manifold pair.*

To prove Theorem 6.13.5, we need some preliminaries. For each $x \in \ell^1(K^{(0)})$ $(\subset \mathbb{R}^\Gamma)$, let $V_x = \{v \in K^{(0)} \mid x(v) \neq 0\}$. Then, V_x is countable. For any V, let $\Delta(V)$ be the full simplicial complex with $\Delta(V)^{(0)} = V$.

Lemma 6.13.6 *Let K be a simplicial complex. For $x \in \ell^1(K^{(0)})$, $x \in \overline{|K|}^{\ell^1}$ if and only if $x(v) \geqslant 0$ for all $v \in K^{(0)}$, $\|x\| = \sum_{v \in V_x} x(v) = 1$, and $\Delta(V_x) \subset K$.*

Proof (*The "only if" Part*) Let $x \in \overline{|K|}^{\ell^1}$. Then, for each $v \in K^{(0)}$, $x(v) = \mathrm{pr}_v(x) \geqslant 0$, because the topology of $\ell^1(K^{(0)})$ is finer than the one inherited from the product topology of $\mathbb{R}^{K^{(0)}}$. Moreover, $\|x\| = 1$ by the continuity of the norm $\|\cdot\|$. For each $\sigma \in \Delta(V_x)$, choose $\varepsilon > 0$ so that $\varepsilon < x(v)$ for every $v \in \sigma^{(0)}$. There is some $y \in |K|$ with $\|x - y\| < \varepsilon$. Then, for each $v \in \sigma^{(0)}$,

$$y(v) \geqslant x(v) - |x(v) - y(v)| > x(v) - \varepsilon > 0.$$

Hence, $\sigma \leqslant c_K(y)$, which implies $\sigma \in K$. Thus, $\Delta(V_x) \subset K$.

(*The "if" Part*) If V_x is finite, then $\langle V_x \rangle \in \Delta(V_x) \subset K$, which implies that $x \in \langle V_x \rangle \subset |K| \subset \overline{|K|}^{\ell^1}$. When V_x is infinite, for each $\varepsilon > 0$, we can take $v_1, \ldots, v_n \in V_x$ such that

$$1 - \sum_{i=1}^{n} x(v_i) = \|x\| - \sum_{i=1}^{n} x(v_i) < \varepsilon/2.$$

Then, $\sigma = \langle x_, \ldots, x_n \rangle \in \Delta(V_x) \subset K$. We define $y \in \sigma \subset |K|$ as follows:

$$y(v) = \begin{cases} x(v_1) + 1 - \sum_{i=1}^{n} x(v_i) & \text{if } v = v_1, \\ x(v_i) & \text{if } v = v_2, \ldots, v_n, \\ 0 & \text{otherwise.} \end{cases}$$

Observe that

$$\begin{aligned} \|x - y\| &= \sum_{v \in K^{(0)}} |x(v) - y(v)| = \sum_{i=1}^{n} |x(v_i) - y(v_i)| + \left(1 - \sum_{i=1}^{n} x(v_i)\right) \\ &= 2\left(1 - \sum_{i=1}^{n} x(v_i)\right) < \varepsilon. \end{aligned}$$

Therefore, $x \in \overline{|K|}^{\ell^1}$. □

As we saw in the above lemma, $\overline{|K|}^{\ell^1}$ is a subspace of the unit sphere $\mathbf{S}_{\ell^1 K^{(0)}}$ of $\ell^1(K^{(0)})$. But we also remark that $\overline{|K|}^{\ell^1} \subset \mathbf{I}^{K^{(0)}}$. Due to Proposition 1.2.2 (Remark 1.2), the unit sphere $\mathbf{S}_{\ell^1(K^{(0)})}$ is also a subspace of the product space $\mathbb{R}^{K^{(0)}}$. Hence, $\overline{|K|}^{\ell^1}$ can also be regarded as a subspace of the product space $\mathbf{I}^{K^{(0)}}$. Then, $\overline{|K|}^{\ell^1} \subset \mathrm{cl}_{\mathbf{I}^{K^{(0)}}} |K|$.

Let K be a countable infinite simplicial complex with $K^{(0)} = \{v_i \mid i \in \mathbb{N}\}$, where $v_i \neq v_j$ if $i \neq j$. Identifying each vertex $v_i \in K^{(0)}$ with the unit vector $\mathbf{e}_i \in \mathbb{R}^{\mathbb{N}}$, we regard $|K| \subset \mathbf{I}^{\mathbb{N}}$ $(\subset \mathbb{R}^{\mathbb{N}})$, that is, each $x = \sum_{i \in \mathbb{N}} x(v_i) v_i \in |K|$ is regarded as the point $x \in \mathbf{I}^{\mathbb{N}}$, where $x(i) = x(v_i)$, $i \in \mathbb{N}$. The topology $|K|_{\mathrm{m}}$ is clearly the subspace topology of $|K|$ inherited from $\mathbf{I}^{\mathbb{N}}$. Thus, $|K|_{\mathrm{m}}$ can be regarded as a subspace of $\mathbf{I}^{\mathbb{N}}$. Then, $\mathrm{cl}_{\mathbf{I}^{\mathbb{N}}} |K|$ is a compactification of $|K|_{\mathrm{m}}$. Let $Q_0 = \mathbf{I}^{\mathbb{N}} \setminus \{\mathbf{0}\}$. The following is a local-compactification of $|K|_{\mathrm{m}}$:

$$\overline{|K|}^{Q_0} = \mathrm{cl}_{Q_0} |K| = \mathrm{cl}_{\mathbf{I}^{\mathbb{N}}} |K| \setminus \{\mathbf{0}\},$$

where $\mathbf{0} \in \mathrm{cl}_{\mathbf{I}^{\mathbb{N}}} |K^{(0)}| \subset \mathrm{cl}_{\mathbf{I}^{\mathbb{N}}} |K|$. Besides, note that $\overline{|K|}^{\ell^1} \subset \overline{|K|}^{Q_0}$. For each $x \in \boldsymbol{Q}$, let $N_x = \{i \in \mathbb{N} \mid x(i) \neq 0\}$. We use the metric d_p for $\mathbf{I}^{\mathbb{N}}$ defined as follows:

$$d_p(x, y) = \sup_{i \in \mathbb{N}} \min\left\{|x(i) - y(i)|,\ 1/i\right\}.$$

Lemma 6.13.7 *Let K be a countable simplicial complex without principal simplexes*[20] *and let $K^{(0)} = \{v_i \mid i \in \mathbb{N}\}$ ($v_i \neq v_j$ if $i \neq j$). For $x \in \mathbf{I}^{\mathbb{N}}$, $x \in \mathrm{cl}_{\mathbf{I}^{\mathbb{N}}} |K|$ if and only if $\sum_{i \in \mathbb{N}} x(i) \leqslant 1$, and $\langle v_{i_1}, \ldots, v_{i_n} \rangle \in K$ if $i_1, \ldots, i_n \in N_x$.*

Proof The "only if" part is trivial, which is valid for any countable infinite simplicial complex K. To see the "if" part, suppose that $x \in \mathbf{I}^{\mathbb{N}}$, $\sum_{i \in \mathbb{N}} x(i) \leqslant 1$, and $\langle v_{i_1}, \ldots, v_{i_n} \rangle \in K$ if $i_1, \ldots, i_n \in V_x$. For each $n \in \mathbb{N}$, we write $\{i \in N_x \mid i < n\} = \{i_1, \ldots, i_m\}$, where $i_1 < \cdots < i_m$. Then, $\langle v_{i_1}, \ldots, v_{i_m} \rangle \in K$. Since K has no principal simplexes, we have $i_{m+1} > i_m$ such that $\langle v_{i_1}, \ldots, v_{i_m}, v_{i_{m+1}} \rangle \in K$. Note that $\sum_{j=1}^{m} x(i_j) \leqslant 1$. We can define $y \in |K|$ as follows:

$$y(i) = \begin{cases} x(i_j) & i = i_j,\ j = 1, \ldots, m, \\ 1 - \sum_{j=1}^{m} x(i_j) & i = i_{m+1}, \\ 0 & \text{otherwise.} \end{cases}$$

Then, $d_p(x, y) < 1/(i_n + 1) < 1/n$. Hence, $x \in \mathrm{cl}_{\mathbf{I}^{\mathbb{N}}} |K|$. □

Combining the above two lemmas, we have the following:

Lemma 6.13.8 *For each countable simplicial complex K without principal simplexes,*

$$\overline{|K|}^{Q_0} = (0, 1] \cdot \overline{|K|}^{\ell^1} = \left\{ tx \;\middle|\; x \in \overline{|K|}^{\ell^1},\ 0 < t \leqslant 1 \right\}. \qquad \square$$

Recall that X is **locally equi-connected** (LEC) if there exists a neighborhood U of the diagonal Δ_X in X^2 and a map $\lambda : U \times \mathbf{I} \to X$ called an **equi-connecting map** such that

$$\lambda(x, y, 0) = x,\ \lambda(x, y, 1) = y \ \text{ for all } (x, y) \in U,$$
$$\lambda(x, x, t) = x \ \text{ for all } x \in X \text{ and } t \in \mathbf{I},$$

where X is **equi-connected** (EC) if $U = X^2$. Then, $A \subset X$ is **λ-convex** if $A^2 \subset U$ and $\lambda(A^2 \times \mathbf{I}) \subset A$. Due to Corollary 1.14.4, a metrizable LEC (resp. EC) space X is an ANR (resp. AR) if X has an equi-connecting map λ such that each point of X has arbitrarily small λ-convex neighborhoods.

Theorem 6.13.9 *Let K be a countable simplicial complex without principal simplexes. Then, $\mathrm{cl}_{\mathbf{I}^{\mathbb{N}}} |K|$ is an AR; hence $\overline{|K|}^{Q_0}$ is an ANR.*

Proof Let $\mu : (\mathbf{I}^{\mathbb{N}})^2 \to \mathbf{I}^{\mathbb{N}}$ be a map defined by

$$\mu(x, y)(i) = \min\left\{x(i),\ y(i)\right\},\ i \in \mathbb{N},$$

[20] A **principal simplex** of K means a simplex of K that is not a face of any other simplex of K.

and define a map $\lambda : (\mathbf{I}^{\mathbb{N}})^2 \times \mathbf{I} \to \mathbf{I}^{\mathbb{N}}$ as follows:

$$\lambda(x, y, t) = \begin{cases} (1-2t)x + 2t\mu(x, y) & \text{if } 0 \leqslant t \leqslant 1/2, \\ (2t-1)y + (2-2t)\mu(x, y) & \text{if } 1/2 \leqslant t \leqslant 1. \end{cases}$$

Then, $\lambda(x, y, 0) = x$, $\lambda(x, y, 1) = y$, $\lambda(x, x, t) = x$ for any $x, y \in \mathbf{I}^{\mathbb{N}}$ and $t \in \mathbf{I}$, that is, λ is an equi-connecting map for $\mathbf{I}^{\mathbb{N}}$. For $x, y \in \mathbf{I}^{\mathbb{N}}$ with $\sum_{i\in\mathbb{N}} x(i)$, $\sum_{i\in\mathbb{N}} y(i) < \infty$,

$$N_{\mu(x,y)} = N_x \cap N_y \ \text{ and } \ \sum_{i\in\mathbb{N}} \mu(x, y)(i) \leqslant \sum_{i\in\mathbb{N}} x(i), \sum_{i\in\mathbb{N}} y(i).$$

Then, it follows that

$$N_{\lambda(x,y,t)} \subset N_x \text{ if } 0 \leqslant t \leqslant 1/2; \ N_{\lambda(x,y,t)} \subset N_y \text{ if } 1/2 \leqslant t \leqslant 1;$$
$$\sum_{i\in\mathbb{N}} \mu(x, y)(i) \leqslant \sum_{i\in\mathbb{N}} x(i), \sum_{i\in\mathbb{N}} y(i).$$

Hence, $\lambda((\mathrm{cl}_{\mathbf{I}^{\mathbb{N}}} |K|)^2 \times \mathbf{I}) \subset \mathrm{cl}_{\mathbf{I}^{\mathbb{N}}} |K|$ by Lemma 6.13.7, Thus, the restriction $\lambda|(\mathrm{cl}_{\mathbf{I}^{\mathbb{N}}} |K|)^2 \times \mathbf{I}$ is an equi-connecting map for $\mathrm{cl}_{\mathbf{I}^{\mathbb{N}}} |K|$.

Let $x, y, z \in \mathrm{cl}_{\mathbf{I}^{\mathbb{N}}} |K|$ with $d_p(x, z), d_p(y, z) < 1/n$ and let $t \in \mathbf{I}$. Then, for each $i = 1, \ldots, n$,

$$|x(i) - z(i)|, \ |y(i) - z(i)| < 1/n \ (\leqslant 1/i),$$

which implies that $|\mu(x, y)(i) - z(i)| < 1/n$. Hence,

$$|\lambda(x, y, t)(i) - z(i)| < 1/n \ \text{ for each } i = 1, \ldots, n.$$

Thus, we have

$$d_p(\lambda(x, y, t), z) = \sup_{i\in\mathbb{N}} \min \left\{ |\lambda(x, y, t)(i) - z(i)|, \ 1/i \right\} < 1/n,$$

which means that the $1/n$-neighborhood $\mathrm{B}_{d_p}(z, 1/n)$ of z in $\mathrm{cl}_{\mathbf{I}^{\mathbb{N}}} |K|$ is λ-convex. By Corollary 1.14.4, $\mathrm{cl}_{\mathbf{I}^{\mathbb{N}}} |K|$ is an AR, and its open subset $\overline{|K|}^{Q_0} = \mathrm{cl}_{\mathbf{I}^{\mathbb{N}}} |K| \setminus \{\mathbf{0}\}$ is an ANR. □

Theorem 6.13.10 *For each combinatorial ∞-manifold K, $\overline{|K|}^{Q_0}$ is an $\mathbb{R}_+$-stable Q-manifold, for which $|K|_{\mathrm{m}}$ is an f.d.cap set and $\overline{|K|}^{Q_0} \setminus \overline{|K|}^{\ell^1}$ is a cap set.*

Proof Since $|K|_{\mathrm{m}}$ is an ℓ^2_f-manifold (Theorem 6.13.1), if $|K|_{\mathrm{m}}$ is homotopy dense in $\overline{|K|}^{Q_0}$, then the result follows from Theorem 5.9.13. Since $\overline{|K|}^{Q_0}$ is an ANR

(Theorem 6.13.9), we can apply Proposition 1.15.8 to show that $|K|_{\mathrm{m}}$ is homotopy dense in $\overline{|K|}^{Q_0}$.

Let C be a compact set in $\overline{|K|}^{Q_0}$ and $\varepsilon > 0$. Since C is closed in $\mathbf{I}^{\mathbb{N}}$ and $\mathbf{0} \notin C$, we can choose $n \in \mathbb{N}$ so that $1/n < \min\{\varepsilon/3, d_p(\mathbf{0}, C)\}$. For each $x \in C$,

$$d_p(\mathbf{0}, x) = \sup_{i \in \mathbb{N}} \min\big\{|x(i)|,\ 1/i\big\} > 1/n,$$

which means that $x(i) > 1/n$ for some $i = 1, \ldots, n$. Let L be the maximal subcomplex of K with $L^{(0)} = \{v_1, \ldots, v_n\}$, where $K^{(0)} = \{v_i \mid i \in \mathbb{N}\}$ and the maximality of L means that L is full in K. Let

$$|L|^* = \big\{tx \bigm| x \in |L|,\ t \in [1/n, 1]\big\} \subset \overline{|K|}^{Q_0}$$

and define a map $f : C \to |L|^*$ as follows:

$$f(x)(i) = \begin{cases} x(i) & \text{if } i \leqslant n, \\ 0 & \text{if } i > n. \end{cases}$$

Then, $d_p(f, \mathrm{id}) < 1/n < \varepsilon/3$.

Since $|L|$ is a Z-set in the ℓ^2_f-manifold $|K|_{\mathrm{m}}$ because of compactness, we have a map $g : |K|_{\mathrm{m}} \to |K|_{\mathrm{m}} \setminus |L|$ with $d_p(g, \mathrm{id}) < \varepsilon/3$. We can define a map $g^* : |L|^* \to \overline{|K|}^{Q_0}$ by $g(tx) = tg(x)$ because $|L| \times [1/n, 1] \approx |L|^*$ by the following correspondence:

$$|L| \times [1/n, 1] \ni (x, t) \mapsto tx \in |L|^*.$$

Then, $d_p(g^*, \mathrm{id}) \leqslant d_p(g, \mathrm{id}) < \varepsilon/3$. Observe that

$$g^*(|L|^*) \cap \big\{tx \bigm| x \in |L|,\ t \in [0, 1]\big\} = \emptyset,$$

which means that for each $y \in g^*(|L|^*)$, there is some $i(y) > n$ such that $\mathrm{pr}_{i(y)}(y) = y(i(y)) > 0$. Because of the compactness of $g^*(|L|^*)$, we have $y_1, \ldots, y_\ell \in g^*(|L|^*)$ such that $g^*(|L|^*) \subset \bigcup_{j=1}^{\ell} \mathrm{pr}_{i(y_j)}((0, \infty))$. Let $m = \max\{i(y_1), \ldots, i(y_\ell)\}$. Then, each $y \in g^*(|L|^*)$ has a positive coordinate $y(i) > 0$ for some $i = n+1, \ldots, m$. We can define a map $k : g^*(|L|^*) \to |K|_{\mathrm{m}}$ as follows:

$$k(y)(i) = \begin{cases} y(i) & \text{if } i \leqslant n, \\ \dfrac{\big(1 - \sum_{j=1}^{n} y(j)\big) y(i)}{\sum_{j=n+1}^{m} y(j)} & \text{if } n < i \leqslant m, \\ 0 & \text{if } i > m. \end{cases}$$

Then, $d_p(k, \mathrm{id}) < 1/n < \varepsilon/3$. Thus, we have a map $kg^* f : C \to |K|_{\mathrm{m}}$, where

$$d_p(kg^* f, \mathrm{id}) \leqslant d_p(k, \mathrm{id}) + d_p(g^*, \mathrm{id}) + d_p(f, \mathrm{id}) < \varepsilon.$$

From Theorem 5.9.13, it follows that $\overline{|K|}^{\boldsymbol{Q}_0}$ is a $\boldsymbol{Q}$-manifold and $|K|_{\mathrm{m}}$ is an f.d.cap set for $\overline{|K|}^{\boldsymbol{Q}_0}$.

Next, we show that $\overline{|K|}^{\boldsymbol{Q}_0}$ is $\mathbb{R}_+$-stable, where we use Theorem 2.11.13. It can be assume that K is connected. Since $|K|_{\mathrm{m}}$ is connected, so is $\overline{|K|}^{\boldsymbol{Q}_0}$. It suffices to show that $\overline{|K|}^{\boldsymbol{Q}_0}$ is properly contractible to infinity, that is, for each compact set C in $\overline{|K|}^{\boldsymbol{Q}_0}$, there is a proper homotopy $h : \overline{|K|}^{\boldsymbol{Q}_0} \times \mathbf{I} \to \overline{|K|}^{\boldsymbol{Q}_0}$ with $h_0 = \mathrm{id}$ and $h_1(\overline{|K|}^{\boldsymbol{Q}_0}) \subset \overline{|K|}^{\boldsymbol{Q}_0} \setminus C$. For each compact set C in $\overline{|K|}^{\boldsymbol{Q}_0}$, choose $n \in \mathbb{N}$ so that $1/n < d_p(\mathbf{0}, C)$. Then,

$$\overline{\mathrm{B}}_{d_p}(\mathbf{0}, 1/n) = \left\{x \in \mathbf{I}^{\mathbb{N}} \mid x(i) \leqslant 1/n \ \text{ if } i \leqslant n\right\} \subset \mathbf{I}^{\mathbb{N}} \setminus C.$$

We define a homotopy $h : \mathbf{I}^{\mathbb{N}} \times \mathbf{I} \to \mathbf{I}^{\mathbb{N}}$ as follows:

$$h(x,t) = ((1-t) + t/n)x \ \text{ for each } (x,t) \in \mathbf{I}^{\mathbb{N}} \times \mathbf{I},$$

where $h_0 = \mathrm{id}$ and $h_1(\mathbf{I}^{\mathbb{N}}) = [0, 1/n]^{\mathbb{N}} \subset \overline{\mathrm{B}}_{d_p}(\mathbf{0}, 1/n) \subset \mathbf{I}^{\mathbb{N}} \setminus C$. For each $x \in \overline{|K|}^{\boldsymbol{Q}_0}$ and $t \in \mathbf{I}$,

$$\sum_{i \in \mathbb{N}} h_t(x)(i) = ((1-t) + t/n) \sum_{i \in \mathbb{N}} x(i) \leqslant \sum_{i \in \mathbb{N}} x(i) \leqslant 1.$$

Since $(1-t) + t/n > 0$ for each $t \in \mathbf{I}$, we have

$$x(i) \neq 0 \Leftrightarrow h_t(x)(i) = ((1-t) + t/n)x(i) \neq 0,$$

which implies $V_x = V_{h_t(x)}$. By Lemma 6.13.7, we have $h(\overline{|K|}^{\boldsymbol{Q}_0} \times \mathbf{I}) \subset \overline{|K|}^{\boldsymbol{Q}_0}$. By restricting h, the desired proper homotopy can be obtained.

Finally, we prove that $\overline{|K|}^{\boldsymbol{Q}_0} \setminus \overline{|K|}^{\ell^1}$ is a cap set for $\overline{|K|}^{\boldsymbol{Q}_0}$. From Lemma 6.13.8, it follows that

$$\overline{|K|}^{\boldsymbol{Q}_0} \setminus \overline{|K|}^{\ell^1} = \bigcup_{n \in \mathbb{N}} \left(1 - \frac{1}{n}\right) \cdot \overline{|K|}^{\boldsymbol{Q}_0}.$$

Each $(1 - 1/n) \cdot \overline{|K|}^{\boldsymbol{Q}_0}$ is a $\boldsymbol{Q}$-manifold because it is homeomorphic to $\overline{|K|}^{\boldsymbol{Q}_0}$, where $(1 - 1/n) \cdot \overline{|K|}^{\boldsymbol{Q}_0}$ is a Z-set in $(1 - 1/(n+1)) \cdot \overline{|K|}^{\boldsymbol{Q}_0}$. More generally, for each $0 < t < 1$, $t \cdot \overline{|K|}^{\boldsymbol{Q}_0}$ is a Z-set in $\overline{|K|}^{\boldsymbol{Q}_0}$. This can be shown by using Theorem 2.8.6 as follows: For each map $f : \boldsymbol{Q} \to \overline{|K|}^{\boldsymbol{Q}_0}$ and $\varepsilon > 0$, letting $C = f(\boldsymbol{Q})$ in the above, we can obtain a map $h : f(\boldsymbol{Q}) \to |K|_{\mathrm{m}}$ with $d_p(h, \mathrm{id}) < \varepsilon$. Then, $hf : \boldsymbol{Q} \to |K|_{\mathrm{m}} \subset \overline{|K|}^{\boldsymbol{Q}_0} \setminus t \cdot \overline{|K|}^{\boldsymbol{Q}_0}$ and $d_p(hf, f) < \varepsilon$.

It remains to show that the tower $\left((1-1/n)\cdot\overline{|K|}^{Q_0}\right)_{n\in\mathbb{N}}$ has the cap. Let C be a compact set in $\overline{|K|}^{Q_0}$ and $m \in \mathbb{N}$. For each $\varepsilon > 0$, choose $n > m$ so that $1/n < \varepsilon/6$. As we saw above, we can obtain a map $f : C \to |K|_{\mathrm{m}}$ such that $f \simeq_{\varepsilon/6} \mathrm{id}$. Then, we have a map

$$f' : C \to (1-1/n)\cdot|K|_{\mathrm{m}} \subset (1-1/n)\cdot\overline{|K|}^{Q_0}$$

defined by $f'(x) = (1-1/n)f(x)$ for each $x \in C$, where $f' \simeq_{\varepsilon/6} f$. Since $(1-1/n)\cdot\overline{|K|}^{Q_0}$ is a $\boldsymbol{Q}$-manifold, f' is $\varepsilon/6$-homotopic to a Z-embedding $g : C \to (1-1/n)\cdot\overline{|K|}^{Q_0}$ (Strong Universality Theorem 2.10.10). Then, $g \simeq_{\varepsilon/2} \mathrm{id}$, $C\cap(1-1/m)\cdot\overline{|K|}^{Q_0}$ is a Z-set in $(1-1/n)\cdot\overline{|K|}^{Q_0}$, and $g|C\cap(1-1/m)\cdot\overline{|K|}^{Q_0}$ is a Z-embedding into $(1-1/n)\cdot\overline{|K|}^{Q_0}$ Applying the Z-Set Unknotting Theorem 2.11.6, we have a homeomorphism $h : (1-1/n)\cdot\overline{|K|}^{Q_0} \to (1-1/n)\cdot\overline{|K|}^{Q_0}$ such that $h \simeq_{\varepsilon/2} \mathrm{id}$ and $h|C\cap(1-1/m)\cdot\overline{|K|}^{Q_0} = g|C\cap(1-1/m)\cdot\overline{|K|}^{Q_0}$. Then, $h^{-1}g : C \to (1-1/n)\cdot\overline{|K|}^{Q_0}$ is a Z-embedding such that $h^{-1}g$ is ε-close to id and $h^{-1}g|C\cap(1-1/m)\cdot\overline{|K|}^{Q_0} = \mathrm{id}$. □

Below, it will be proved in Theorem 6.13.13 that $|K|_{\mathrm{m}}$ is homotopy dense in $\overline{|K|}^{\ell^1}$ for an arbitrary simplicial complex K. Combining this with Theorem 6.13.10, we can easily obtain the following:

Corollary 6.13.11 *For each combinatorial ∞-manifold K, $\overline{|K|}^{\ell^1}$ is an ℓ^2-manifold, for which $|K|_{\mathrm{m}}$ is an f.d.cap set.* □

Corollary 6.13.12 *For each contractible combinatorial ∞-manifold K,*

$$(\mathrm{cl}_{\mathbf{I}^{\mathbb{N}}}|K|, |K|_{\mathrm{m}}) \approx (\boldsymbol{Q}, \boldsymbol{Q}_f),$$

$$(\mathrm{cl}_{\mathbf{I}^{\mathbb{N}}}|K|, \mathrm{cl}_{\mathbf{I}^{\mathbb{N}}}|K| \setminus \overline{|K|}^{\ell^1}) \approx (\boldsymbol{Q}, B(\boldsymbol{Q})), \quad \textit{and}$$

$$(\mathrm{cl}_{\mathbf{I}^{\mathbb{N}}}|K|, \overline{|K|}^{\ell^1}) \approx (\boldsymbol{Q}, I(\boldsymbol{Q})).$$

Proof Combining Theorems 6.13.10 and 2.11.11, we have $\overline{|K|}^{Q_0} \approx \boldsymbol{Q} \times \mathbb{R}_+ \approx \boldsymbol{Q} \setminus \{\mathbf{0}\}$. Since $\mathrm{cl}_{\mathbf{I}^{\mathbb{N}}}|K|$ is the one-point compactification of $\overline{|K|}^{Q_0}$, it follows that $\mathrm{cl}_{\mathbf{I}^{\mathbb{N}}}|K| \approx \boldsymbol{Q}$. Since the one-point set in $\boldsymbol{Q}$ is a Z-set in $\boldsymbol{Q}$, $\overline{|K|}^{Q_0}$ is homotopy dense in $\mathrm{cl}_{\mathbf{I}^{\mathbb{N}}}|K|$. Hence, $|K|_{\mathrm{m}}$ is homotopy dense in $\mathrm{cl}_{\mathbf{I}^{\mathbb{N}}}|K|$. Since $|K|_{\mathrm{m}} \approx \ell^2_f$, it follows that $|K|_{\mathrm{m}}$ is an f.d.cap set for $\mathrm{cl}_{\mathbf{I}^{\mathbb{N}}}|K|$, that is, $(\mathrm{cl}_{\mathbf{I}^{\mathbb{N}}}|K|, |K|_{\mathrm{m}}) \approx (\boldsymbol{Q}, \boldsymbol{Q}_f)$. As in Lemma 6.13.8, we have $\mathrm{cl}_{\mathbf{I}^{\mathbb{N}}}|K| = \mathbf{I}\cdot\overline{|K|}^{\ell^1}$. By the same proof as the last part of Theorem 6.13.10, it can be proved that $\mathrm{cl}_{\mathbf{I}^{\mathbb{N}}}|K| \setminus \overline{|K|}^{\ell^1}$ is a cap set for $\mathrm{cl}_{\mathbf{I}^{\mathbb{N}}}|K|$, that is,

$$(\mathrm{cl}_{\mathbf{I}^{\mathbb{N}}}|K|, \mathrm{cl}_{\mathbf{I}^{\mathbb{N}}}|K| \setminus \overline{|K|}^{\ell^1}) \approx (\boldsymbol{Q}, B(\boldsymbol{Q})) \text{ and}$$

$$(\mathrm{cl}_{\mathbf{I}^{\mathbb{N}}}|K|, \overline{|K|}^{\ell^1}) \approx (\boldsymbol{Q}, I(\boldsymbol{Q})).$$

□

Here, we give a proof of Theorem 6.13.5:

Proof (Theorem 6.13.5) The "only if" part is Theorem 6.13.10 (cf. Corollary 6.13.11), To see the "if" part, assume that $\overline{|K|}^{\ell^1}$ is an ℓ^2-manifold. Since every compact set in an ℓ^2-manifold is a Z-set (Proposition 3.2.10), each simplex $\sigma \in K$ is a Z-set in $\overline{|K|}^{\ell^1}$. As mentioned above, it will be proved that $|K|_{\mathrm{m}}$ is homotopy dense in $\overline{|K|}^{\ell^1}$ (Theorem 6.13.13). Then, it follows that each simplex $\sigma \in K$ is a Z-set in $|K|_{\mathrm{m}}$. Hence, we can conclude that K is a combinatorial ∞-manifold by the Characterization Theorem 6.11.1 for combinatorial ∞-manifolds. □

Theorem 6.13.13 *For every simplicial complex K, the ℓ^1-completion $\overline{|K|}^{\ell^1}$ of $|K|$ is an ANR and $|K|_{\mathrm{m}}$ is homotopy dense in $\overline{|K|}^{\ell^1}$.*

Proof To prove that $\overline{|K|}^{\ell^1}$ is an ANR, it suffices to construct an equi-connecting map λ such that each point of $\overline{|K|}^{\ell^1}$ has arbitrarily small λ-convex neighborhoods. We first define $\mu : \ell^1(K^{(0)})^2 \to \ell^1(K^{(0)})$ by

$$\mu(x, y)(v) = \min\big\{|x(v)|,\ |y(v)|\big\},\ v \in K^{(0)}.$$

Then, μ is continuous. Indeed,

$$\begin{aligned}\big|\min\big\{|x(v)|,\ |y(v)|\big\} &- \min\big\{|x'(v)|,\ |y'(v)|\big\}\big| \\ &\leqslant \max\big\{\big||x(v)| - |x'(v)|\big|,\ \big||y(v)| - |y'(v)|\big|\big\} \\ &\leqslant \max\big\{\big|x(v) - x'(v)\big|,\ \big|y(v) - y'(v)\big|\big\} \\ &\leqslant \big|x(v) - x'(v)\big| + \big|y(v) - y'(v)\big|,\end{aligned}$$

which implies that

$$\|\mu(x, y) - \mu(x', y')\| \leqslant \|x - x'\| + \|y - y'\|.$$

From the definition, it follows that $\mu(x, y) = 0$ if and only if $x(v) = 0$ or $y(v) = 0$ for every $v \in K^{(0)}$, which implies that $\|x - y\| = \|x\| + \|y\|$. Hence, $\|x - y\| < \|x\| + \|y\|$ implies $\mu(x, y) \neq 0$. Observe that $V_{\mu(x,y)} = V_x \cap V_y$ for each $(x, y) \in \ell^1(K^{(0)})^2$.

We define an open neighborhood U of the diagonal $\Delta\overline{|K|}^{\ell^1}$ in $\ell^1(K^{(0)})^2$ as follows:

$$U = \Big\{(x, y) \in \big(\overline{|K|}^{\ell^1}\big)^2 \ \big|\ \|x - y\| < 2\Big\}.$$

By the above observation, $\mu(x, y) \neq 0$ for each $(x, y) \in U$. It is easy to see that

$$x,\ \frac{\mu(x, y)}{\|\mu(x, y)\|} \in \overline{|\Delta(V_x)|}^{\ell^1} \ \text{ and }\ y,\ \frac{\mu(x, y)}{\|\mu(x, y)\|} \in \overline{|\Delta(V_y)|}^{\ell^1},$$

where $\overline{|\Delta(V_x)|}^{\ell^1}$ and $\overline{|\Delta(V_y)|}^{\ell^1}$ are convex sets in $\ell^1(K^{(0)})$ and they are contained in $\overline{|K|}^{\ell^1}$. Hence, we can define an equi-connecting map $\lambda : U \times \mathbf{I} \to \overline{|K|}^{\ell^1}$ as follows:

$$\lambda(x, y, t) = \begin{cases} (1-2t)x + \dfrac{2t\mu(x, y)}{\|\mu(x, y)\|} & \text{if } 0 \leqslant t \leqslant 1/2, \\ (2t-1)y + \dfrac{(2-2t)\mu(x, y)}{\|\mu(x, y)\|} & \text{if } 1/2 \leqslant t \leqslant 1. \end{cases}$$

Now, we show that each point $z \in \overline{|K|}^{\ell^1}$ has arbitrarily small λ-convex neighborhoods. For each $\varepsilon > 0$, choose $v_1, \ldots, v_n \in V_z$ so that

$$1 - \sum_{i=1}^{n} z(v_i) = \|z\| - \sum_{i=1}^{n} z(v_i) < \varepsilon/2.$$

For each $i = 1, \ldots, n$, choose $0 < \alpha_i < z(v_i)$ so that $1 - \sum_{i=1}^{n} \alpha_i < \varepsilon/2$ and define

$$W = \left\{x \in \overline{|K|}^{\ell^1} \;\middle|\; x(v_i) > \alpha_i \text{ for all } i = 1, \ldots, n\right\}.$$

Then, W is an open neighborhood of z in $\overline{|K|}^{\ell^1}$ with $\operatorname{diam} W < \varepsilon$. Indeed, for each $x, y \in W$.

$$\begin{aligned} \|x - y\| &\leqslant \sum_{i=1}^{n} |x(v_i) - y(v_i)| + \left(1 - \sum_{i=1}^{n} x(v_i)\right) + \left(1 - \sum_{i=1}^{n} y(v_i)\right) \\ &\leqslant \sum_{i=1}^{n} \big(x(v_i) - \alpha_i\big) + \sum_{i=1}^{n} \big(y(v_i) - \alpha_i\big) + 2 - \sum_{i=1}^{n} x(v_i) - \sum_{i=1}^{n} y(v_i) \\ &\leqslant 2 - 2\sum_{i=1}^{n} \alpha_i < \varepsilon. \end{aligned}$$

To see the λ-convexity of W, let $x, y \in W$ and $t \in \mathbf{I}$. Then, $\mu(x, y)(v_i) = \min\{x(v_i), y(v_i)\}$ and $\|\mu(x, y)\| \leqslant 1$. When $t \leqslant 1/2$, for each $i = 1, \ldots, n$,

$$\begin{aligned} \lambda(x, y, t)(v_i) &= (1-2t)x(v_i) + \frac{2t \min\{x(v_i), y(v_i)\}}{\|\mu(x, y)\|} \\ &\geqslant \min\{x(v_i), y(v_i)\} > \alpha_i, \end{aligned}$$

which means $\lambda(x, y, t) \in W$. Similarly, when $t \geqslant 1/2$, we have $\lambda(x, y, t) \in W$. Therefore, W is λ-convex. Thus, it follows from Theorem 1.14.4 that $\overline{|K|}^{\ell^1}$ is an ANR.

Next, applying Corollary 1.15.5, we show that $|K|_{\mathrm{m}}$ is homotopy dense in $\overline{|K|}^{\ell^1}$. For each $z \in \overline{|K|}^{\ell^1}$ and $\varepsilon > 0$, choose $v_1, \ldots, v_n \in V_y$ ($\subset K^{(0)}$) so that

$$1 - \sum_{i=1}^{n} z(v_i) = \|y\| - \sum_{i=1}^{n} z(v_i) < \varepsilon/2.$$

Then, $\sigma_0 = \langle v_1, \ldots, v_n \rangle \in \Delta(V_z) \subset K$ by Lemma 6.13.6. Take $\delta > 0$ so that

$$\delta < \varepsilon,\ z(v_i)/2,\ i = 1, \ldots, n.$$

Let $\alpha : \mathbf{S}^{k-1} \to |K|_{\mathrm{m}} \cap \mathrm{B}(z, \delta)$ be a map. Then, $\alpha(\mathbf{S}^{k-1}) \subset |\operatorname{St}(\sigma_0, K)|$. Otherwise, $\sigma_0 \not\leqslant c_K(x)$ for some $x \in \alpha(\mathbf{S}^{k-1})$, that is, $x(v_i) = 0$ for some $i = 1, \ldots, n$. Then, $\|x - z\| \geqslant |x(v_i) - z(v_i)| = z(v_i) > \delta$, which is a contradiction. We define $y \in \operatorname{rint} \sigma_0$ as follows:

$$y(v) = \begin{cases} z(v_i) & \text{if } v = v_i \text{ for some } i < n, \\ z(v_n) + \left(1 - \sum_{i=1}^{n} z(v_i)\right) & \text{if } v = v_n, \\ 0 & \text{otherwise.} \end{cases}$$

Then, it follows that

$$\|y - z\| = |y(v_n) - z(v_n)| + 1 - \sum_{i=1}^{n} z(v_i) = 2\left(1 - \sum_{i=1}^{n} z(v_i)\right) < \varepsilon.$$

We can now define a homotopy $h : \mathbf{S}^{k-1} \times \mathbf{I} \to |K|_{\mathrm{m}}$ as follows:

$$h(x, t) = (1 - t)\alpha(x) + ty \in c_K(x) \subset |\operatorname{St}(\sigma_0, K)| \subset |K|,$$

where $h_0 = \alpha$ and $h_1(\mathbf{S}^{k-1}) = \{y\}$. Observe that

$$\|h(x, t) - z\| \leqslant (1 - t)\|\alpha(x) - z\| + t\|y - z\| \leqslant (1 - t)\delta + t\varepsilon < \varepsilon,$$

which means that $h(\mathbf{S}^{k-1} \times \mathbf{I}) \subset |K|_{\mathrm{m}} \cap \mathrm{B}(z, \varepsilon)$. Hence, α is null-homotopic in $|K|_{\mathrm{m}} \cap \mathrm{B}(z, \varepsilon)$. Thus, we can apply Corollary 1.15.5 to conclude that $|K|_{\mathrm{m}}$ is homotopy dense in $\overline{|K|}^{\ell^1}$. □

Summing up Theorems 6.11.1, 6.11.12, 6.12.2, 6.13.1, 6.13.4, and 6.13.5, the following are equivalent for a countable simplicial complex K:

(a) K is a combinatorial ∞-manifold;
(b) $|K|$ is a PL ∞-manifold;
(c) $|K|$ is an $\mathbb{R}^\infty$-manifold;

(c') $|K| \times \boldsymbol{Q}$ is a $\boldsymbol{Q}^\infty$-manifold;
(d) Every simplex of K is a Z-set in $|K|$;
(e) Every simplex of K is a Z-set in $|K|_{\mathrm{m}}$;
(f) $\mathrm{Lk}(\sigma, K)$ is non-empty and contractible for each simplex $\sigma \in K$;
(g) $|K|_{\mathrm{m}}$ is an ℓ^2_f-manifold;
(g') $|K|_{\mathrm{m}} \times \boldsymbol{Q}$ is an $\ell^2_{\boldsymbol{Q}}$-manifold;
(h) $\overline{|K|}^{\ell^1}$ is an ℓ^2-manifold.

It is unknown whether the converse of Theorem 6.13.10 is true or not, that is:

? Open Problem

For a countable simplicial complex K, if $\overline{|K|}^{\boldsymbol{Q}_0}$ is an $\mathbb{R}_+$-stable $\boldsymbol{Q}$-manifold, then is K a combinatorial ∞-manifold?

Remark 6.15 Let K be a simplicial complex satisfying the following condition:

(Lk-τ) $\operatorname{card} \mathrm{Lk}(\sigma, K)^{(0)} = \tau$ and $|\mathrm{Lk}(\sigma, K)| \simeq 0$ for every $\sigma \in K$.

Due to Theorem 6.12.1, the second condition of the above (Lk-τ) is equivalent to the following:

- Every $\sigma \in K$ is a Z-set in $|K|$.

? Open Problems

For an uncountable simplicial complex K, the following questions occur:

1. Does the condition (Lk-τ) imply that $\mathrm{St}(v, K) \cong F(\Gamma)$ for each $v \in K^{(0)}$?
2. Is $|K|_{\mathrm{m}}$ an $\ell^2_f(\Gamma)$-manifold?
3. Is $\overline{|K|}^{\ell^1}$ an $\ell^2(\Gamma)$-manifold?
4. Is $|K|_{\mathrm{m}}$ an $\mathfrak{M}^{\mathrm{fd}}_0$-absorption base for $\overline{|K|}^{\ell^1}$?
5. Let K and L be simplicial complexes satisfying (Lk-τ). Does $|K| \approx |L|$ imply $K \cong L$?

6.14 Bi-topological Infinite-Dimensional Manifolds

A **bi-topological space** is a space $X = (X, \mathfrak{T}_1, \mathfrak{T}_2)$ given two topologies $\mathfrak{T}_1$ and $\mathfrak{T}_2$, or a couple (X_1, X_2) of spaces $X_i = (X, \mathfrak{T}_i)$, $i = 1, 2$. For example, the polyhedron $|K|$ of a simplicial complex K has two natural topologies, namely the weak (or Whitehead) topology $\mathfrak{T}_{\mathrm{w}}$ and the metric topology $\mathfrak{T}_{\mathrm{m}}$. Then, $(|K|_{\mathrm{w}}, |K|_{\mathrm{m}})$

(or $(|K|, \mathfrak{T}_{\mathrm{w}}, \mathfrak{T}_{\mathrm{m}})$) is a bi-topological space. The underlying sets of $\mathbb{R}^\infty = \varinjlim \mathbb{R}^n$ and $\boldsymbol{Q}^\infty = \varinjlim \boldsymbol{Q}^n$ are the same as the subspaces $\mathbb{R}^{\mathbb{N}}_f$ $(= \boldsymbol{s}_f)$ and $\boldsymbol{Q}^{\mathbb{N}}_f$ of the countable product spaces $\mathbb{R}^{\mathbb{N}}$ $(= \boldsymbol{s})$ and $\boldsymbol{Q}^{\mathbb{N}}$, respectively. Thus, $(\mathbb{R}^\infty, \mathbb{R}^{\mathbb{N}}_f)$ and $(\boldsymbol{Q}^\infty, \boldsymbol{Q}^{\mathbb{N}}_f)$ are also bi-topological spaces, where

$$\mathbb{R}^{\mathbb{N}}_f \approx \ell^2_f \quad \text{and} \quad \boldsymbol{Q}^{\mathbb{N}}_f \approx \ell^2_Q.$$

Let $X = (X, \mathfrak{T}_1, \mathfrak{T}_2)$ and $Y = (Y, \mathfrak{T}'_1, \mathfrak{T}'_2)$ be bi-topological spaces. A function $f : X \to Y$ is called a **bi-topological map**, a **bi-topological embedding**, or a **bi-topological homeomorphism** if both $f : (X, \mathfrak{T}_1) \to (Y, \mathfrak{T}'_1)$ and $f : (X, \mathfrak{T}_2) \to (Y, \mathfrak{T}'_2)$ are continuous, embeddings, or homeomorphisms, respectively. A **bi-topological open** (resp. **closed**) **embedding** means a bi-topological embedding $f : X \to Y$ such that $h(X)$ is open (resp. closed) with respect to both topologies $\mathfrak{T}'_1$ and $\mathfrak{T}'_2$.

A bi-topological space $(M, \mathfrak{T}_{\mathrm{w}}, \mathfrak{T}_{\mathrm{m}})$ is called a **bi-topological manifold modeled on** $(\mathbb{R}^\infty, \mathbb{R}^{\mathbb{N}}_f)$ (resp. $(\boldsymbol{Q}^\infty, \boldsymbol{Q}^{\mathbb{N}}_f)$) or simply an **$(\mathbb{R}^\infty, \mathbb{R}^{\mathbb{N}}_f)$-manifold** (resp. a **$(\boldsymbol{Q}^\infty, \boldsymbol{Q}^{\mathbb{N}}_f)$-manifold**) if the topology $\mathfrak{T}_{\mathrm{m}}$ is metrizable, $\mathfrak{T}_{\mathrm{m}} \subset \mathfrak{T}_{\mathrm{w}}$ and each point of M has a $\mathfrak{T}_{\mathrm{m}}$-open neighborhood U with a bi-topological open embedding

$$\varphi : (U, \mathfrak{T}_{\mathrm{w}}, \mathfrak{T}_{\mathrm{m}}) \to (\mathbb{R}^\infty, \mathbb{R}^{\mathbb{N}}_f) \quad \big(\text{resp. } \varphi : (U, \mathfrak{T}_{\mathrm{w}}, \mathfrak{T}_{\mathrm{m}}) \to (\boldsymbol{Q}^\infty, \boldsymbol{Q}^{\mathbb{N}}_f)\big).$$

Then, $(M, \mathfrak{T}_{\mathrm{w}})$ is an $\mathbb{R}^\infty$-manifold (resp. $\boldsymbol{Q}^\infty$-manifold) and $(M, \mathfrak{T}_{\mathrm{n}})$ is an $\mathbb{R}^{\mathbb{N}}_f$-manifold (resp. $\boldsymbol{Q}^{\mathbb{N}}_f$-manifold).

In this section, we show fundamental properties of bi-topological manifolds modeled on $(\mathbb{R}^\infty, \mathbb{R}^{\mathbb{N}}_f)$ and $(\boldsymbol{Q}^\infty, \boldsymbol{Q}^{\mathbb{N}}_f)$. The following theorem is their source:

Theorem 6.14.1 *Let $M = (M, \mathfrak{T}_{\mathrm{w}}, \mathfrak{T}_{\mathrm{m}})$ and $N = (N, \mathfrak{T}'_{\mathrm{w}}, \mathfrak{T}'_{\mathrm{m}})$ be bi-topological spaces satisfying the following conditions:*

(i) *$(M, \mathfrak{T}_{\mathrm{w}})$ and $(N, \mathfrak{T}'_{\mathrm{w}})$ are $\mathbb{R}^\infty$-manifolds (resp. $\boldsymbol{Q}^\infty$-manifolds),*
(ii) *$(M, \mathfrak{T}_{\mathrm{m}})$ and $(N, \mathfrak{T}'_{\mathrm{m}})$ are $\mathbb{R}^{\mathbb{N}}_f$-manifolds (resp. $\boldsymbol{Q}^{\mathbb{N}}_f$-manifolds),*
(iii) *$\mathrm{id}_M : (M, \mathfrak{T}_{\mathrm{w}}) \to (M, \mathfrak{T}_{\mathrm{m}})$ and $\mathrm{id}_N : (N, \mathfrak{T}'_{\mathrm{w}}) \to (N, \mathfrak{T}'_{\mathrm{m}})$ are fine homotopy equivalences (so $\mathfrak{T}_{\mathrm{m}} \subset \mathfrak{T}_{\mathrm{w}}$).*

Suppose that $(M, \mathfrak{T}_{\mathrm{m}}) \approx (N, \mathfrak{T}'_{\mathrm{m}})$. Then, for each $\mathfrak{T}'_{\mathrm{m}}$-open cover $\mathcal{U}$ of N, each homeomorphism $f : (M, \mathfrak{T}_{\mathrm{m}}) \to (N, \mathfrak{T}'_{\mathrm{m}})$ is $\mathcal{U}$-close to a bi-topological homeomorphism $h : M \to N$.

Remark 6.16 The following example shows that the condition (iii) in Theorem 6.14.1 is essential. Let $(M, \mathfrak{T}_{\mathrm{m}}) = (N, \mathfrak{T}'_{\mathrm{m}}) = \mathbf{I}^2 \times \mathbb{R}^{\mathbb{N}}_f$, $(M, \mathfrak{T}_{\mathrm{w}}) = \mathbf{I}^2 \times \mathbb{R}^\infty$, and $(N, \mathfrak{T}'_{\mathrm{w}}) = \varinjlim X_n \times \mathbb{R}^\infty = \varinjlim (X_n \times \mathbb{R}^n)$, where $X_1 \subset X_2 \subset \cdots$ are defined as follows:

$$X_n = \{0\} \times \mathbf{I} \cup \mathbf{I} \times \partial \mathbf{I} \cup [1/n, 1] \times \mathbf{I} \subset \mathbf{I}^2, \; n \in \mathbb{N}.$$

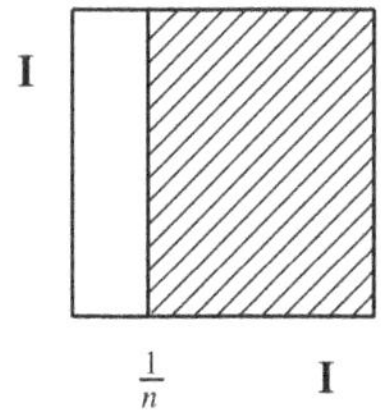

Fig. 6.8 $X_n = \{0\} \times \mathbf{I} \cup \mathbf{I} \times \partial\mathbf{I} \cup [1/n, 1] \times \mathbf{I}$

Then, $(N, \mathfrak{T}'_{\mathrm{w}}) \approx \mathbf{S}^1 \times \mathbb{R}^\infty$ which is not homotopic to $(M, \mathfrak{T}_{\mathrm{w}})$. See Fig. 6.8.

Before proving Theorem 6.14.1, we will derive fundamental properties of $(\mathbb{R}^\infty, \mathbb{R}^{\mathbb{N}}_f)$- and $(\boldsymbol{Q}^\infty, \boldsymbol{Q}^{\mathbb{N}}_f)$-manifolds from Theorem 6.14.1. To this end, the following proposition is necessary:

Proposition 6.14.2 *For each $(\mathbb{R}^\infty, \mathbb{R}^{\mathbb{N}}_f)$- or $(\boldsymbol{Q}^\infty, \boldsymbol{Q}^{\mathbb{N}}_f)$-manifold $(M, \mathfrak{T}_{\mathrm{w}}, \mathfrak{T}_{\mathrm{m}})$, the identity* $\mathrm{id} : (M, \mathfrak{T}_{\mathrm{w}}) \to (M, \mathfrak{T}_{\mathrm{m}})$ *is a fine homotopy equivalence.*

Proof Recall that an $\mathbb{R}^\infty$-manifold is an ANE (Proposition 6.1.8). Due to Theorem 1.15.1 with Remark 1.15, it suffices to show that $\mathrm{id} : (M, \mathfrak{T}_{\mathrm{w}}) \to (M, \mathfrak{T}_{\mathrm{m}})$ is a local $*$-connection. Any $\mathfrak{T}_{\mathrm{m}}$-neighborhood of each $x \in M$ contains a $\mathfrak{T}_{\mathrm{m}}$-open neighborhood U of x with a bi-topological open embedding

$$\varphi : (U, \mathfrak{T}_{\mathrm{w}}, \mathfrak{T}_{\mathrm{m}}) \to (\mathbb{R}^\infty, \mathbb{R}^{\mathbb{N}}_f) \quad \big(\text{or } \varphi : (U, \mathfrak{T}_{\mathrm{w}}, \mathfrak{T}_{\mathrm{m}}) \to (\boldsymbol{Q}^\infty, \boldsymbol{Q}^{\mathbb{N}}_f)\big)$$

such that $\varphi(U)$ is convex, where the following operation is continuous with respect to the topology inherited from $\mathbb{R}^\infty$ (resp. $\boldsymbol{Q}^\infty$):

$$\varphi(U)^2 \times \mathbf{I} \ni (y, z, t) \mapsto (1-t)y + tz \in \varphi(U).$$

Then, it follows that $\varphi(U)$ is contractible as a subspace of $\mathbb{R}^\infty$ (resp. $\boldsymbol{Q}^\infty$). Hence, U is contractible with respect to $\mathfrak{T}_{\mathrm{w}}$, which means that id is a local $*$-connection.

□

For $(\mathbb{R}^\infty, \mathbb{R}^{\mathbb{N}}_f)$- and $(\boldsymbol{Q}^\infty, \boldsymbol{Q}^{\mathbb{N}}_f)$-manifolds, the Characterization, Open Embedding, Classification, and Triangulation Theorems are easily obtained:

Theorem 6.14.3 (Characterization) *A bi-topological space $(M, \mathfrak{T}_{\mathrm{w}}, \mathfrak{T}_{\mathrm{m}})$ is an $(\mathbb{R}^\infty, \mathbb{R}^{\mathbb{N}}_f)$-manifold (resp. $(\boldsymbol{Q}^\infty, \boldsymbol{Q}^{\mathbb{N}}_f)$-manifold) if and only if $(M, \mathfrak{T}_{\mathrm{w}})$ is an $\mathbb{R}^\infty$-manifold (resp. $\boldsymbol{Q}^\infty$-manifold), $(M, \mathfrak{T}_{\mathrm{m}})$ is an $\mathbb{R}^{\mathbb{N}}_f$-manifold (resp. $\boldsymbol{Q}^{\mathbb{N}}_f$-manifold), and* $\mathrm{id}_M : (M, \mathfrak{T}_{\mathrm{w}}) \to (M, \mathfrak{T}_{\mathrm{m}})$ *is a fine homotopy equivalence.*

Proof The "only if" part follows from Proposition 6.14.2. We show the "if" part. Since $(M, \mathfrak{T}_{\mathrm{m}})$ is an $\mathbb{R}^{\mathbb{N}}_f$-manifold (resp. a $\boldsymbol{Q}^{\mathbb{N}}_f$-manifold), there exists an open embedding $g : (M, \mathfrak{T}_{\mathrm{m}}) \to \mathbb{R}^{\mathbb{N}}_f$ (resp. $g : (M, \mathfrak{T}_{\mathrm{m}}) \to \boldsymbol{Q}^{\mathbb{N}}_f$) by the Open Embedding Theorem 2.5.10 with Theorem 5.5.10 (or combination of Theorems 5.2.3, 5.2.5, 5.2.6). Then, $g(M)$ is not only open in $\mathbb{R}^{\mathbb{N}}_f$ (resp. $\boldsymbol{Q}^{\mathbb{N}}_f$) but also open in $\mathbb{R}^\infty$ (resp. $\boldsymbol{Q}^\infty$), where we write the former by N_{m} and the latter by N_{w}. Thus, we have an $(\mathbb{R}^\infty, \mathbb{R}^{\mathbb{N}}_f)$-manifold (resp. $(\boldsymbol{Q}^\infty, \boldsymbol{Q}^{\mathbb{N}}_f)$-manifold) $(N_{\mathrm{w}}, N_{\mathrm{m}})$. Note that $\mathrm{id} : N_{\mathrm{w}} \to N_{\mathrm{m}}$ is a fine homotopy equivalence. By Theorem 6.14.1, we have a bi-topological homeomorphism $h : (M, \mathfrak{T}_{\mathrm{w}}, \mathfrak{T}_{\mathrm{m}}) \to (N_{\mathrm{w}}, N_{\mathrm{m}})$. Hence, $(M, \mathfrak{T}_{\mathrm{w}}, \mathfrak{T}_{\mathrm{m}})$ is an $(\mathbb{R}^\infty, \mathbb{R}^{\mathbb{N}}_f)$-manifold (resp. $(\boldsymbol{Q}^\infty, \boldsymbol{Q}^{\mathbb{N}}_f)$-manifold). □

In the above proof, h is a bi-topological open embedding into $(\mathbb{R}^\infty, \mathbb{R}^{\mathbb{N}}_f)$ (resp. $(\boldsymbol{Q}^\infty, \boldsymbol{Q}^{\mathbb{N}}_f)$). Thus, we have also the following Open Embedding Theorem:

Theorem 6.14.4 (Open Embedding) *For each $(\mathbb{R}^\infty, \mathbb{R}^{\mathbb{N}}_f)$-manifold (resp. $(\boldsymbol{Q}^\infty, \boldsymbol{Q}^{\mathbb{N}}_f)$-manifold) $(M, \mathfrak{T}_{\mathrm{w}}, \mathfrak{T}_{\mathrm{m}})$, there exists a bi-topological open embedding*

$$h : (M, \mathfrak{T}_{\mathrm{w}}, \mathfrak{T}_{\mathrm{m}}) \to (\mathbb{R}^\infty, \mathbb{R}^{\mathbb{N}}_f) \ \big(\textit{resp. } (\boldsymbol{Q}^\infty, \boldsymbol{Q}^{\mathbb{N}}_f)\big).$$

Namely, every $(\mathbb{R}^\infty, \mathbb{R}^{\mathbb{N}}_f)$-manifold (resp. $(\boldsymbol{Q}^\infty, \boldsymbol{Q}^{\mathbb{N}}_f)$-manifold) can be bi-topologically embedded into $(\mathbb{R}^\infty, \mathbb{R}^{\mathbb{N}}_f)$ (resp. $(\boldsymbol{Q}^\infty, \boldsymbol{Q}^{\mathbb{N}}_f)$) as an open set. □

By combining the Classification Theorem 2.6.1 for $\mathbb{R}^{\mathbb{N}}_f$- and $\boldsymbol{Q}^{\mathbb{N}}_f$-manifolds (cf. Theorem 5.1.10) and Theorem 6.14.1, the following Classification Theorem can be obtained:

Theorem 6.14.5 (Classification) *Suppose that both $(M, \mathfrak{T}_{\mathrm{w}}, \mathfrak{T}_{\mathrm{m}})$ and $(N, \mathfrak{T}'_{\mathrm{w}}, \mathfrak{T}'_{\mathrm{m}})$ are $(\mathbb{R}^\infty, \mathbb{R}^{\mathbb{N}}_f)$- or $(\boldsymbol{Q}^\infty, \boldsymbol{Q}^{\mathbb{N}}_f)$-manifolds. If $(M, \mathfrak{T}_{\mathrm{m}}) \simeq (N, \mathfrak{T}'_{\mathrm{m}})$, then M and N are bi-topologically homeomorphic, which means that there exists a bi-topological homeomorphism $h : M \to N$.* □

In the above, $(M, \mathfrak{T}_{\mathrm{m}}) \simeq (N, \mathfrak{T}'_{\mathrm{m}})$ can be replaced with $(M, \mathfrak{T}_{\mathrm{w}}) \simeq (N, \mathfrak{T}'_{\mathrm{w}})$.

Theorem 6.14.6 *Any $\mathbb{R}^\infty$- or $\mathbb{R}^{\mathbb{N}}_f$-manifold has the unique $(\mathbb{R}^\infty, \mathbb{R}^{\mathbb{N}}_f)$-manifold structure and any $\boldsymbol{Q}^\infty$- or $\boldsymbol{Q}^{\mathbb{N}}_f$-manifold has the unique $(\boldsymbol{Q}^\infty, \boldsymbol{Q}^{\mathbb{N}}_f)$-manifold structure.*

Proof By the Classification Theorem 6.14.5, it suffices to show that any $\mathbb{R}^\infty$- or $\mathbb{R}^{\mathbb{N}}_f$-manifold has an $(\mathbb{R}^\infty, \mathbb{R}^{\mathbb{N}}_f)$-manifold structure and any $\boldsymbol{Q}^\infty$- or $\boldsymbol{Q}^{\mathbb{N}}_f$-manifold has a $(\boldsymbol{Q}^\infty, \boldsymbol{Q}^{\mathbb{N}}_f)$-manifold structure. As is seen by easy observation, the proof of Theorem 6.14.3 shows that every $\mathbb{R}^{\mathbb{N}}_f$-manifold (resp. every $\boldsymbol{Q}^{\mathbb{N}}_f$-manifold) has an $(\mathbb{R}^\infty, \mathbb{R}^{\mathbb{N}}_f)$-manifold structure (resp. a $(\boldsymbol{Q}^\infty, \boldsymbol{Q}^{\mathbb{N}}_f)$-manifold structure).

By Theorem 6.11.8, any $\mathbb{R}^\infty$-manifold M is homeomorphic to $|K|$ for some combinatorial ∞-manifold K. Then, $|K|_{\mathrm{m}}$ is an $\mathbb{R}^{\mathbb{N}}_f$-manifold (Theorem 6.13.1) and $\mathrm{id} : |K| \to |K|_{\mathrm{m}}$ is a fine homotopy equivalence (Theorem 1.7.6). From the Characterization Theorem 6.14.3, it follows that $(|K|, |K|_{\mathrm{m}})$ is an $(\mathbb{R}^\infty, \mathbb{R}_f^{\mathbb{N}})$-manifold. Thus, M has an $(\mathbb{R}^\infty, \mathbb{R}_f^{\mathbb{N}})$-manifold structure.

For any Q^∞-manifold M, there is some combinatorial ∞-manifold K such that $M \approx |K| \times Q$ (Theorem 6.11.12). Then, $|K|_{\mathrm{m}} \times Q$ is an $Q_f^{\mathbb{N}}$-manifold (Theorem 6.13.4). Moreover, $\mathrm{id} : |K| \times Q \to |K|_{\mathrm{m}} \times Q$ is a fine homotopy equivalence (Theorem 1.7.6 and Corollary 1.15.4 with Remark 1.16). Hence, by the Characterization Theorem 6.14.3, $(|K| \times Q, |K|_{\mathrm{m}} \times Q)$ is a $(Q^\infty, Q_f^{\mathbb{N}})$-manifold. Thus, M has a $(Q^\infty, Q_f^{\mathbb{N}})$-manifold structure. □

The above proof of the $\mathbb{R}^\infty$- or Q^∞-manifold case can be easily modified to a proof of the following Triangulation Theorem:

Theorem 6.14.7 (Triangulation) *For each $(\mathbb{R}^\infty, \mathbb{R}_f^{\mathbb{N}})$-manifold (resp. $(Q^\infty, Q_f^{\mathbb{N}})$-manifold) $(M, \mathfrak{T}_{\mathrm{w}}, \mathfrak{T}_{\mathrm{m}})$, there exists a simplicial complex K with a bi-topological homeomorphism*

$$h : (M, \mathfrak{T}_{\mathrm{w}}, \mathfrak{T}_{\mathrm{m}}) \to (|K|, |K|_{\mathrm{m}}) \big(\textit{resp. } (|K| \times Q, |K|_{\mathrm{m}} \times Q)\big),$$

where K must be a combinatorial ∞-manifold. □

The following are also easily obtained:

Theorem 6.14.8 (Stability) *For each $(\mathbb{R}^\infty, \mathbb{R}_f^{\mathbb{N}})$-manifold (resp. $(Q^\infty, Q_f^{\mathbb{N}})$-manifold) $(M, \mathfrak{T}_{\mathrm{w}}, \mathfrak{T}_{\mathrm{m}})$, there exists a bi-topological homeomorphism*

$$h : (M_{\mathrm{w}} \times \mathbb{R}^\infty, M_{\mathrm{m}} \times \mathbb{R}_f^{\mathbb{N}}) \to (M, \mathfrak{T}_{\mathrm{w}}, \mathfrak{T}_{\mathrm{m}})$$

$$\big(\textit{resp. } h : (M_{\mathrm{w}} \times Q^\infty, M_{\mathrm{m}} \times Q_f^{\mathbb{N}}) \to (M, \mathfrak{T}_{\mathrm{w}}, \mathfrak{T}_{\mathrm{m}})\big).$$

Given a $\mathfrak{T}_{\mathrm{m}}$-open cover $\mathcal{U}$ of M, h can be taken $\mathcal{U}$-close to the projection.

Sketch of Proof. By Corollary 1.15.4 with Remark 1.16, $\mathrm{id} : (M, \mathfrak{T}_{\mathrm{w}}) \times \mathbb{R}^\infty \to (M, \mathfrak{T}_{\mathrm{m}}) \times \mathbb{R}_f^{\mathbb{N}}$ (resp. $\mathrm{id} : (M, \mathfrak{T}_{\mathrm{w}}) \times Q^\infty \to (M, \mathfrak{T}_{\mathrm{m}}) \times Q_f^{\mathbb{N}}$) is a fine homotopy equivalence. Then, the result can be obtained by combining the Stability Theorem 2.3.7 for ℓ_f^2-manifolds (resp. ℓ_Q^2-manifolds) (cf. Theorem 5.1.10) and Theorem 6.14.1.

Theorem 6.14.9 (Negligibility) *Let $(M, \mathfrak{T}_{\mathrm{w}}, \mathfrak{T}_{\mathrm{m}})$ be an $(\mathbb{R}^\infty, \mathbb{R}_f^{\mathbb{N}})$- or $(Q^\infty, Q_f^{\mathbb{N}})$-manifold. If A is a Z-set in $(M, \mathfrak{T}_{\mathrm{m}})$, then there exists a bi-topological homeomorphism $h : M \setminus A \to M$. Given a $\mathfrak{T}_{\mathrm{m}}$-open cover $\mathcal{U}$ of M, h can be taken $\mathcal{U}$-close to the identity.*

Sketch of Proof. Let $\mathfrak{T}'_{\mathrm{w}} = \mathfrak{T}_{\mathrm{w}}|M \setminus A$ and $\mathfrak{T}'_{\mathrm{m}} = \mathfrak{T}_{\mathrm{m}}|M \setminus A$. Then, $(M \setminus A, \mathfrak{T}'_{\mathrm{w}}, \mathfrak{T}'_{\mathrm{m}})$ is also an $(\mathbb{R}^\infty, \mathbb{R}_f^{\mathbb{N}})$-manifold (resp. a $(Q^\infty, Q_f^{\mathbb{N}})$-manifold). Combining the Negligibility Theorem 2.9.8 for ℓ_f^2-manifolds (resp. ℓ_Q^2-manifolds) (cf. Theorem 5.1.10) and Theorem 6.14.1, we can obtain the result.

Theorem 6.14.10 ($\mathfrak{T}_{\mathrm{w}}$-Compact Set Unknotting) *Let $(M, \mathfrak{T}_{\mathrm{w}}, \mathfrak{T}_{\mathrm{m}})$ be an $(\mathbb{R}^\infty, \mathbb{R}_f^{\mathbb{N}})$- or $(Q^\infty, Q_f^{\mathbb{N}})$-manifold and let $f : A \to B$ be a homeomorphism* *between $\mathfrak{T}_w$-compact sets in M. If f is homotopic to* id *in $(M, \mathfrak{T}_m)$, then f extends to a bi-topological homeomorphism $\tilde{f} : M \to M$. Given a $\mathfrak{T}_m$-open cover $\mathcal{U}$ of M, if $f \simeq_{\mathcal{U}}$* id *in $(M, \mathfrak{T}_m)$, then $\tilde{f}$ can be taken $\mathcal{U}$-close to* id.

Sketch of Proof. Note that A and B are Z-sets in $(M, \mathfrak{T}_w)$. Combining Z-Set Unknotting Theorem 2.9.7 for ℓ_f^2-manifolds (resp. ℓ_Q^2-manifolds) (cf. Theorem 5.1.10) and Theorem 6.14.1, we can obtain the result.

The following is a key lemma to prove Theorem 6.14.1:

Lemma 6.14.11 *Let $\varphi : X \to Y$ is a bijective fine homotopy equivalence from an $\mathbb{R}^\infty$-manifold (resp. a $\boldsymbol{Q}^\infty$-manifold) X onto an $\mathbb{R}_f^{\mathbb{N}}$-manifold (resp. a $\boldsymbol{Q}_f^{\mathbb{N}}$-manifold) Y. Given finite-dimensional compacta $A \subset X$ and $B \subset Y$ (there is no dimensional assumption in the $(\boldsymbol{Q}^\infty, \boldsymbol{Q}_f^{\mathbb{N}})$-case) so that $\varphi(A) \subset B$, for each $\mathcal{U} \in \operatorname{cov}(Y)$, there exists a $\mathcal{U}$-homotopy inverse $h : Y \to X$ of φ such that $h\varphi|A = \mathrm{id}$ (i.e., $h|\varphi(A) = \varphi^{-1}|\varphi(A)$) and $h|B$ is an embedding.*

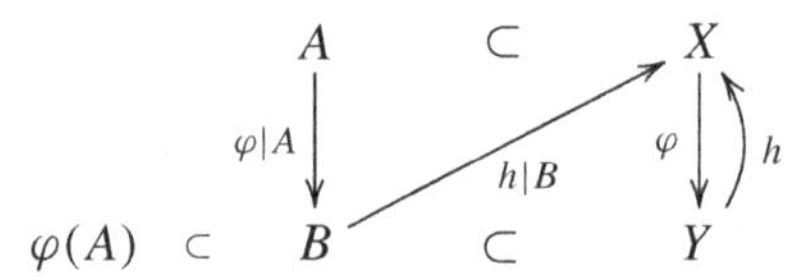

Proof Choose $\mathcal{V} \in \operatorname{cov}(Y)$ so that $\operatorname{st}\mathcal{V} \prec \mathcal{U}$ and let $f : Y \to X$ be a $\mathcal{V}$-homotopy inverse of φ. Then, $\varphi^{-1}|\varphi(A)$ is continuous and $\varphi^{-1}|\varphi(A) \simeq_{\varphi^{-1}(\mathcal{V})} f|\varphi(A)$ because $f\varphi \simeq_{\varphi^{-1}(\mathcal{V})} \mathrm{id}_Y$. Since X is an ANE (Theorem 6.1.8), we can apply the Homotopy Extension Theorem 1.13.11 to obtain a map $g : Y \to X$ such that $g|\varphi(A) = \varphi^{-1}|\varphi(A)$ and $g \simeq_{\varphi^{-1}(\mathcal{V})} f$. Since X has property $(\mathcal{A}_f)$ (resp. $(\mathcal{A})$) (Theorem 6.4.10), there exists an embedding $j : B \to X$ such that $j|\varphi(A) = g|\varphi(A) = \varphi^{-1}|\varphi(A)$ and $j \simeq_{\varphi^{-1}(\mathcal{V})} g|B$. Again using the Homotopy Extension Theorem 1.13.11, we can extend j to a map $h : Y \to X$ such that $h \simeq_{\varphi^{-1}(\mathcal{V})} g$. Then, $h|\varphi(A) = j|\varphi(A) = \varphi^{-1}|\varphi(A)$ and $h|B = j$ is an embedding. Moreover,

$$\varphi h \simeq_{\mathcal{V}} \varphi g \simeq_{\mathcal{V}} \varphi f \simeq_{\mathcal{V}} \mathrm{id}_X \quad \text{and}$$

$$h\varphi \simeq_{\varphi^{-1}(\mathcal{V})} g\varphi \simeq_{\varphi^{-1}(\mathcal{V})} f\varphi \simeq_{\varphi^{-1}(\mathcal{V})} \mathrm{id}_X,$$

which implies that h is a $\mathcal{U}$-homotopy inverse of φ. □

Proof (Theorem 6.14.1) For simplicity, let

$$X = (M, \mathfrak{T}_m),\ X_\infty = (M, \mathfrak{T}_w),\ Y = (N, \mathfrak{T}'_m),\ Y_\infty = (N, \mathfrak{T}'_w),$$

$$\varphi = \mathrm{id}_M : X_\infty \to X, \text{ and } \quad \psi = \mathrm{id}_N : Y_\infty \to Y,$$

where X_∞ and Y_∞ are $\mathbb{R}^\infty$-manifolds (resp. $\boldsymbol{Q}^\infty$-manifolds), X and Y are $\mathbb{R}_f^{\mathbb{N}}$-manifolds (resp. $\boldsymbol{Q}_f^{\mathbb{N}}$-manifolds), and, φ and ψ are fine homotopy equivalences. Then, a homeomorphism $f : X \to Y$ and $\mathcal{U} \in \operatorname{cov}(Y)$ are given. Due to

Proposition 1.3.22(1), there is $\rho \in \mathrm{Metr}(Y)$ such that $\{\overline{\mathrm{B}}_\rho(y, 1) \mid y \in Y\} \prec \mathcal{U}$. We will construct a homeomorphism $f_\infty : X \to Y$ that is also a homeomorphism from X_∞ onto Y_∞ and $\rho(f_\infty, f) < 1$.

We can write $X_\infty = \varinjlim X_n$ and $Y_\infty = \varinjlim Y_n$, where $X_1 \subset X_2 \subset \cdots$ and $Y_1 \subset Y_2 \subset \cdots$ of finite-dimensional compacta (resp. compacta) in X and Y, respectively. Then, we can assume that X_1 is a singleton. In the following, let $d \in \mathrm{Metr}(X)$. Recall that any compact set in an $\mathbb{R}^{\mathbb{N}}_f$-manifold (resp. a $\boldsymbol{Q}^{\mathbb{N}}_f$-manifold) is a Z-set.

Put $m(1) = 1$ and $f_1 = f$. Since $X_{m(1)}$ is a singleton, $f_1(X_{m(1)}) \subset Y_{n(1)}$ for some $n(1) \in \mathbb{N}$. Choose $\mathcal{U}_1 \in \mathrm{cov}(Y)$ so that

$$\operatorname{mesh} \operatorname{st} \mathcal{U}_1 < 2^{-2} \quad \text{and} \quad \operatorname{mesh} f_1^{-1}(\operatorname{st} \mathcal{U}_1) < 2^{-2}.$$

Since $f_1\varphi$ is a bijective fine homotopy equivalence, we can apply Lemma 6.14.11 to obtain a $\mathcal{U}_1$-homotopy inverse j_1 of $f_1\varphi$ such that $j_1 f_1|X_{m(1)} = j_1 f_1 \varphi|X_{m(1)} = \mathrm{id}$ and $j_1|Y_{n(1)}$ is an embedding. Then, $j_1(Y_{n(1)}) \subset X_{m(2)}$ for some $m(2) > m(1)$ (Proposition 6.1.2 (3)).

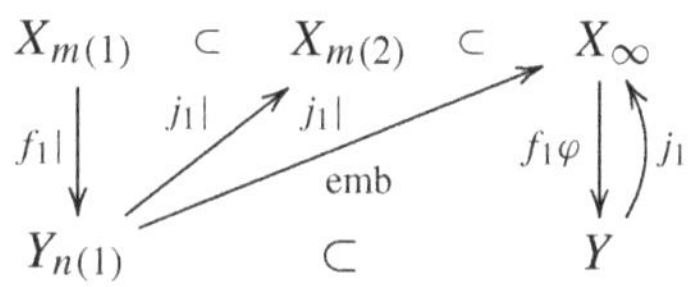

Since $\varphi(f_1\varphi)^{-1}(\mathcal{U}_1) = f_1^{-1}(\mathcal{U}_1)$, it follows that

$$j_1 f_1|f_1^{-1}(Y_{n(1)}) = \varphi j_1 f_1 \varphi|f_1^{-1}(Y_{n(1)}) \simeq_{f_1^{-1}(\mathcal{U}_1)} \varphi|f_1^{-1}(Y_{n(1)}) = \mathrm{id},$$

where $f_1^{-1}(Y_{n(1)})$ is a Z-set in X and $j_1 f_1|f_1^{-1}(Y_{n(1)})$ is a Z-embedding into X. We can apply the Z-Set Unknotting Theorem 2.9.7 for ℓ^2_f-manifolds (resp. ℓ^2_Q-manifolds) (cf. Theorem 5.1.10) to obtain a homeomorphism $h_1 : X \to X$ such that $h_1|f_1^{-1}(Y_{n(1)}) = j_1 f_1|f_1^{-1}(Y_{n(1)})$ and $h_1 \simeq_{f_1^{-1}(\operatorname{st}\mathcal{U}_1)} \mathrm{id}$. Thus, we have a homeomorphism $g_1 = h_1 f_1^{-1} : Y \to X$ such that

$$g_1|Y_{n(1)} = h_1 f_1^{-1}|Y_{n(1)} = j_1|Y_{n(1)} \quad \text{and} \quad g_1 \simeq_{f_1^{-1}(\operatorname{st}\mathcal{U}_1)} f_1^{-1}.$$

Then, $g_1(Y_{n(1)}) = j_1(Y_{n(1)}) \subset X_{m(2)}$. Because $f_1(X_{m(1)}) \subset Y_{n(1)}$, we have $g_1 f_1|X_{m(1)} = j_1 f_1|X_{m(1)} = \mathrm{id}$.

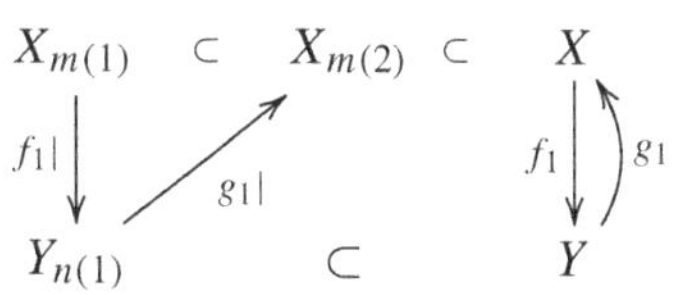

Moreover, it follows that

$$d(g_1, f_1^{-1}) \leqslant \operatorname{mesh} f_1^{-1}(\operatorname{st}\mathcal{U}_1) < 2^{-2} \text{ and}$$
$$\rho(f_1, g_1^{-1}) = \rho(f_1 g_1, \mathrm{id}) \leqslant \operatorname{mesh}\operatorname{st}\mathcal{U}_1 < 2^{-2}.$$

Next, choose $\mathcal{V}_1 \in \operatorname{cov}(X)$ so that

$$\operatorname{mesh}\operatorname{st}\mathcal{V}_1 < 2^{-2} \text{ and } \operatorname{mesh} g_1^{-1}(\operatorname{st}\mathcal{V}_1) < 2^{-2}.$$

In the same way as the previous paragraph, we can apply Lemma 6.14.11 to obtain a $\mathcal{V}_1$-homotopy inverse k_1 of $g_1\psi$ such that $k_1 g_1|Y_{n(1)} = k_1 g_1 \psi|Y_{n(1)} = \mathrm{id}$ and $k_1|X_{m(2)}$ is an embedding. Then, $k_1(X_{m(2)}) \subset Y_{n(2)}$ for some $n(2) > n(1)$.

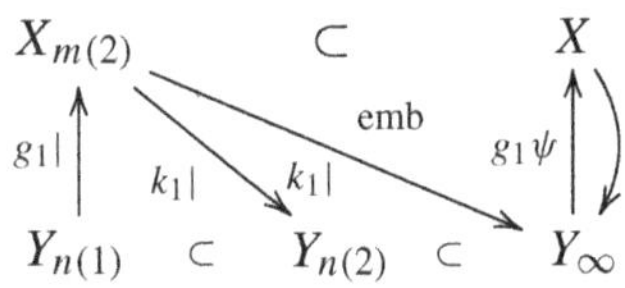

By the same argument as the previous paragraph, we can apply the Z-Set Unknotting Theorem 2.9.7 to obtain a homeomorphism $h'_2 : Y \to Y$ such that $h'_2|g_1^{-1}(X_{m(2)}) = \psi k_1 g_1|g_1^{-1}(X_{m(2)})$ and $h'_2 \simeq_{g_1^{-1}(\operatorname{st}\mathcal{V}_1)} \mathrm{id}$. Then, $f_2 = h'_1 g_1^{-1} : X \to Y$ is a homeomorphism and

$$f_2|X_{m(2)} = h'_1 g_1^{-1}|X_{m(2)} = k_1|X_{m(2)} \text{ and } f_2 \simeq_{g_1^{-1}(\operatorname{st}\mathcal{V}_1)} g_1^{-1}.$$

Then, $f_2(X_{m(2)}) = k_1(X_{m(2)}) \subset Y_{n(2)}$ and $f_2 g_1|Y_{n(1)} = k_1 g_1|Y_{n(1)} = \mathrm{id}$.

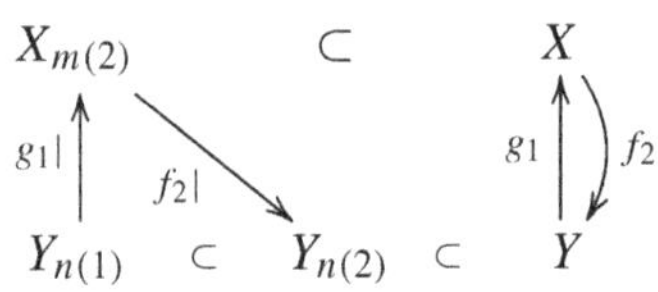

Moreover, it follows that

$$\rho(f_2, g_1^{-1}) \leqslant \operatorname{mesh} g_1^{-1}(\operatorname{st}\mathcal{V}_1) < 2^{-2} \text{ and}$$
$$d(g_1, f_2^{-1}) = d(g_1 f_2, \mathrm{id}) \leqslant \operatorname{mesh}\operatorname{st}\mathcal{V}_1 < 2^{-2}.$$

Therefore, we have

$$\rho(f_1, f_2) \leqslant \rho(f_1, g_1^{-1}) + \rho(f_2, g_1^{-1}) < 2^{-2} + 2^{-2} = 2^{-1}.$$

Now, in the construction of g_1, replacing $f_1, m(1), n(1)$ with $f_2, m(2), n(2)$, and choosing $\mathcal{U}_2 \in \operatorname{cov}(Y)$ instead of $\mathcal{U}_1$ so that

$$\operatorname{mesh} \operatorname{st} \mathcal{U}_2 < 2^{-3} \quad \text{and} \quad \operatorname{mesh} f_2^{-1}(\operatorname{st} \mathcal{U}_2) < 2^{-3},$$

we can obtain a homeomorphism $g_2 : Y \to X$ and $m(3) > m(2)$ such that

$$g_2 f_2 | X_{m(2)} = \mathrm{id}, \; g_2(Y_{n(2)}) \subset X_{m(3)},$$
$$d(g_2, f_2^{-1}) < 2^{-3}, \text{ and } \rho(f_2, g_2^{-1}) < 2^{-3}.$$

Then, it follows that we have

$$d(g_1, g_2) \leqslant d(g_1, f_2^{-1}) + d(g_2, f_2^{-1}) < 2^{-2} + 2^{-3} < 2^{-1}.$$

By repeating these constructions, we can obtain homeomorphisms $f_i : X \to Y$, $g_i : Y \to X$, $i \in \mathbb{N}$, and increasing sequences $m(1) < m(2) < \cdots$, $n(1) < n(2) < \cdots$ such that

$$f_i(X_{m(i)}) \subset Y_{n(i)}, \; g_i(Y_{n(i)}) \subset X_{m(i+1)},$$
$$g_i f_i | X_{m(i)} = \mathrm{id}, \; f_{i+1} g_i | Y_{n(i)} = \mathrm{id},$$
$$d(g_i, f_i^{-1}) < 2^{-(i+1)}, \; \rho(f_i, g_i^{-1}) < 2^{-(i+1)},$$
$$\rho(f_{i+1}, g_i^{-1}) < 2^{-(i+1)}, \text{ and } d(g_i, f_{i+1}^{-1}) < 2^{-(i+1)}.$$

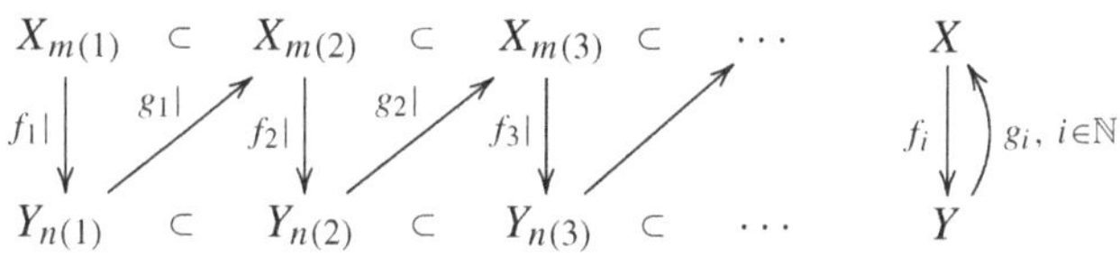

Observe that

$$f_{i+1}|X_{m(i)} = f_{i+1} g_i f_i | X_{m(i)} = (f_{i+1} g_i | Y_{n(i)}) f_i | X_{m(i)} = f_i | X_{m(i)} \text{ and}$$
$$g_{i+1}|Y_{n(i)} = g_{i+1} f_{i+1} g_i | Y_{n(i)} = (g_{i+1} f_{i+1} | X_{m(i+1)}) g_i | Y_{n(i)} = g_i | Y_{n(i)}.$$

Hence, we can define $f_\infty : X \to Y$ and $g_\infty : Y \to X$ by $f_\infty | X_{m(i)} = f_i | X_{m(i)}$ and $g_\infty | Y_{n(i)} = g_i | Y_{n(i)}$ for each $i \in \mathbb{N}$. Due to Proposition 6.1.1, $f_\infty : X_\infty = \varinjlim X_i \to Y_\infty = \varinjlim Y_i$ is a homeomorphism and $f_\infty^{-1} = g_\infty$. Moreover,

$$\rho(f_i, f_{i+1}) \leqslant \rho(f_i, g_i^{-1}) + \rho(f_{i+1}, g_i^{-1}) < 2^{-(i+1)} + 2^{-(i+1)} = 2^{-i} \text{ and}$$
$$d(g_i, g_{i+1}) \leqslant d(g_i, f_{i+1}^{-1}) + d(g_{i+1}, f_{i+1}^{-1}) < 2^{-(i+1)} + 2^{-(i+2)} < 2^{-i}.$$

Then, it follows that

$$\rho(f_n, f_\infty) \leqslant \sum_{i \geqslant n} \rho(f_i, f_{i+1}) < \sum_{i \geqslant n} 2^{-i} = 2^{-n+1} \quad \text{and}$$

$$d(g_n, g_\infty) \leqslant \sum_{i \geqslant n} d(g_i, g_{i+1}) \sum_{i \geqslant n} 2^{-i} = 2^{-n+1},$$

which means that f_n and g_n, $n \in \mathbb{N}$, uniformly converge to f_∞ and g_∞, respectively. Therefore, $f_\infty : X \to Y$ and $g : Y \to X$ are also continuous, that is, $f_\infty : X \to Y$ is a homeomorphism and $f_\infty^{-1} = g_\infty$. Additionally, $\rho(f, f_\infty) = \rho(f_1, f_\infty) < 1$. The proof is completed. □

In the rest of this section, we will give some variations of Theorem 6.14.3. As shown in the example of Remark 6.16, it is essential in Theorem 6.14.3 that $\mathrm{id}_M : (M, \mathfrak{T}_\mathrm{w}) \to (M, \mathfrak{T}_\mathrm{m})$ is a fine homotopy equivalence. To see this condition, we introduce some notion. Let $(X_n)_{n\in\mathbb{N}}$ be a tower of subset of a metric space $X = (X, d)$. It is said that $(X_n)_{n\in\mathbb{N}}$ has the **mapping absorption property for compacta** in X if for any compactum $A \subset X$, $\varepsilon > 0$, and $n \in \mathbb{N}$, there is a map $f : A \to X_m$ for some $m \geqslant n$ such that $f|A \cap X_n = \mathrm{id}$ and $d(f, \mathrm{id}) < \varepsilon$. Then, it should be noted that $\bigcup_{n\in\mathbb{N}} X_n$ is dense in X. We have the following:

Lemma 6.14.12 *Let $(X_n)_{n\in\mathbb{N}}$ be a tower of compacta of an ANR X and $\varphi = \mathrm{id}_X : \varinjlim X_n \to X$ be the map induced by the identity. If $(X_n)_{n\in\mathbb{N}}$ has the mapping absorption property for compacta in X, then $\varphi : \varinjlim X_n \to X$ is a fine homotopy equivalence.*

Proof For each $\mathcal{U} \in \mathrm{cov}(X)$, Proposition 1.3.22 provides us with $d \in \mathrm{Metr}(X)$ such that

$$\left\{ \overline{\mathrm{B}}_d(x, 1) \mid x \in X \right\} \prec \mathcal{U}.$$

Due to Hanner's characterization of ANRs (Theorem 1.13.20), X is 2^{-2}-homotopy dominated by a simplicial complex K, that is, there are maps $f : X \to |K|$ and $g : |K| \to X$ such that $gf \simeq_{2^{-2}} \mathrm{id}$. We can inductively construct maps $g_i : |K| \to X$, $i \in \omega$, such that $g_i||K^{(i-1)}| = g_{i-1}||K^{(i-1)}|$, $g_i \simeq_{2^{-(i+3)}} g_{i-1}$ rel. $|K^{(i-1)}|$, and g_i maps each i-simplex $\sigma \in K$ into some $X_{n(\sigma)}$. Indeed, let $g_{-1} = g$ and $K^{(-1)} = \emptyset$, and assume that g_{i-1} is constructed. Then, for each i-simplex $\sigma \in K$, $g_{i-1}(\partial\sigma) \subset X_{n(\partial\sigma)}$, where $n(\partial\sigma) = \max_{\sigma' < \sigma} n(\sigma')$. Using the mapping absorption property for compacta, we can easily obtain a map $g_\sigma : \sigma \to X_{n(\sigma)}$ such that $g_\sigma|\partial\sigma = g_{i-1}|\partial\sigma$, $g_\sigma \simeq_{2^{-(i+3)}} g_{i-1}|\sigma$ rel. $\partial\sigma$, and $g_\sigma(\sigma)$ is contained in some $X_{n(\sigma)}$. We define a map $g'_i : |K^{(i)}| \to X$ by $g'_i||K^{(i-1)}| = g_{i-1}||K^{(i-1)}|$, and $g'_i|\sigma = g_\sigma$ for each i-simplex $\sigma \in K$. Observe that $g'_i \simeq_{2^{-(i+3)}} g_{i-1}||K^{(i)}|$ rel. $|K^{(i-1)}|$. Applying the Homotopy Extension Theorem 1.5.7 (or 1.13.11), we can extend g'_i to a map $g_i : |K| \to X$ such that $g_i \simeq_{2^{-(i+3)}} g_{i-1}$ rel. $|K^{(i-1)}|$.

Now, we define a map $g_\infty : |K| \to \bigcup_{n\in\mathbb{N}} X_n \subset X$ by $g_\infty||K^{(i)}| = g_i||K^{(i)}|$ for each $i \in \omega$. It easily follows that $g_\infty \simeq_{2^{-2}} g$ in X. Since g_∞ maps each i-simplex $\sigma \in K$ into some $X_{n(\sigma)}$, $g_\infty : |K| \to \varinjlim X_n$ is also a map. Thus, we have a map $\psi = g_\infty f : X \to \varinjlim X_n$. Then, each $\psi(X_n)$ is contained in some X_m (Proposition 6.1.2 (3)). Moreover, observe that

$$\varphi\psi = \varphi g_\infty f = g_\infty f \simeq_{2^{-2}} gf \simeq_{2^{-2}} \mathrm{id}_X.$$

Hence, $\varphi\psi \simeq_{\mathcal{U}} \mathrm{id}$. Let $h : X \times \mathbf{I} \to X$ be a 2^{-1}-homotopy with $h_0 = \mathrm{id}_X$ and $h_1 = \varphi\psi$. Note that each $h(X_n \times \{0, 1\})$ is contained in some X_m, so h_1 can be regarded as a map $h_1 = \psi\varphi : \varinjlim X_n \to \varinjlim X_n$. Again using the mapping absorption property for compacta and the Homotopy Extension Theorem 1.13.11 (cf. the construction of g_i above), we can inductively construct maps $h'_n : X \times \mathbf{I} \to X$, $n \in \mathbb{N}$, such that

$$h'_n|X_{n-1} \times \mathbf{I} = h'_{n-1}|X_{n-1} \times \mathbf{I},\ h'_n|X \times \{0, 1\} = h|X \times \{0, 1\},$$
$$h'_n \simeq_{2^{-(n+1)}} h'_{n-1} \ \text{ rel. } (X_{n-1} \times \mathbf{I}) \cup (X \times \{0, 1\}),$$

and $h'_n(X_n \times \mathbf{I})$ is contained in some X_m, where $h'_0 = h$ and $X_0 = \emptyset$. Since $(\varinjlim X_n) \times \mathbf{I} = \varinjlim(X_n \times \mathbf{I})$ (Proposition 6.1.3 (1)), we can define a homotopy $h^* : (\varinjlim X_n) \times \mathbf{I} \to \varinjlim X_n$ by $h^*|X_n \times \mathbf{I} = h'_n|X_n \times \mathbf{I}$ for each $n \in \mathbb{N}$. Then, $h^*_0 = h_0 = \mathrm{id}$ and $h^*_1 = h_1 = \psi\varphi$. Observe that

$$d(\varphi h^*(x, t), h(x, t)) \leqslant \sum_{n=1}^{\infty} d(h'_n(x, t), h'_{n-1}(x, t)) \leqslant \sum_{n=1}^{\infty} 2^{-(n+1)} = 2^{-1}.$$

It follows that φh^* is a 1-homotopy, which implies that h^* is a $\varphi^{-1}(\mathcal{U})$-homotopy. Hence, $\psi\varphi \simeq_{\varphi^{-1}(\mathcal{U})} \mathrm{id}$. Consequently, ψ is a $\mathcal{U}$-homotopy inverse of φ. □

Theorem 6.14.13 *Let $M = (M, \mathfrak{T}_{\mathrm{w}}, \mathfrak{T}_{\mathrm{m}})$ be a bi-topological space such that $\mathfrak{T}_{\mathrm{m}}$ is metrizable and $\mathfrak{T}_{\mathrm{m}} \subset \mathfrak{T}_{\mathrm{w}}$. The following conditions are equivalent:*

(a) *M is a $(\boldsymbol{Q}^\infty, \boldsymbol{Q}^{\mathbb{N}}_f)$- (or $(\mathbb{R}^\infty, \mathbb{R}^{\mathbb{N}}_f)$-)manifold;*
(b) *$(M, \mathfrak{T}_{\mathrm{w}})$ is a $\boldsymbol{Q}^\infty$- (or $\mathbb{R}^\infty$-)manifold, $\mathrm{id}_M : (M, \mathfrak{T}_{\mathrm{w}}) \to (M, \mathfrak{T}_{\mathrm{m}})$ is a fine homotopy equivalence, and every compact set in $(M, \mathfrak{T}_{\mathrm{m}})$ is a strong Z-set;*
(c) *$(M, \mathfrak{T}_{\mathrm{w}})$ is a $\boldsymbol{Q}^\infty$- (or $\mathbb{R}^\infty$-)manifold, $(M, \mathfrak{T}_{\mathrm{m}})$ is an ANR, every compact set in $(M, \mathfrak{T}_{\mathrm{m}})$ is a strong Z-set, and*

   ($*_{\mathrm{c}}$) *given a $\mathfrak{T}_{\mathrm{m}}$-compactum C in M and a $\mathfrak{T}_{\mathrm{w}}$-compactum D so that $D \subset C$, for each $\mathfrak{T}_{\mathrm{m}}$-open cover $\mathcal{U}$ of M, there is a map $f : C \to (M, \mathfrak{T}_{\mathrm{w}})$ such that $f|D = \mathrm{id}$ and f is $\mathcal{U}$-close to* id.

Proof Since the mapping absorption property for compacta clearly implies the condition ($*_{\mathrm{c}}$), the implication (c) ⇒ (b) follows from Lemma 6.14.12 above. We will show the implications (a) ⇒ (b) ⇒ (c), from which the equivalence (b) ⇔ (c) will be obtained. Finally, we will show that the combination (b) and (c) implies (a), that is, (b) + (c) ⇒ (a). Thus, the three conditions (a), (b), and (c) are equivalent.

(a) ⇒ (b): Obviously, $(M, \mathfrak{T}_w)$ is a $\boldsymbol{Q}^\infty$- (or $\mathbb{R}^\infty$-)manifold. It was shown in Proposition 6.14.2 that $\mathrm{id}_M : (M, \mathfrak{T}_w) \to (M, \mathfrak{T}_m)$ is a fine homotopy equivalence. Recall that $\boldsymbol{Q}_f^{\mathbb{N}}$ (or $\mathbb{R}_f^{\mathbb{N}}$) is homeomorphic to an (f.d.)cap set ℓ_Q^2 (resp. ℓ_f^2) for ℓ_2 (Corollary 5.1.8, Proposition 5.1.3 and Remark 5.1). From Proposition 5.3.1 and Corollary 5.3.6, it follows that every compact set in $(M, \mathfrak{T}_m)$ is a strong Z-set.

(b) ⇒ (c): As a $\boldsymbol{Q}^\infty$- (or $\mathbb{R}^\infty$-)manifold, $(M, \mathfrak{T}_w)$ is an ANE (Proposition 6.1.8). Since $\mathrm{id}_M : (M, \mathfrak{T}_w) \to (M, \mathfrak{T}_m)$ is a fine homotopy equivalence, $(M, \mathfrak{T}_m)$ is ε-homotopy dominated by $(M, \mathfrak{T}_w)$ for every $\varepsilon > 0$ with respect to some admissible metric of $(M, \mathfrak{T}_m)$, which means that $(M, \mathfrak{T}_m)$ is an ANR by Hanner's Characterization of ANRs 1.13.20.

To verify the condition $(*_c)$, for each $\mathfrak{T}_m$-open cover $\mathcal{U}$ of M, let $\mathcal{V}$ be a $\mathfrak{T}_m$-open star-refinement of $\mathcal{U}$. Then, $\varphi = \mathrm{id}_M : (M, \mathfrak{T}_w) \to (M, \mathfrak{T}_m)$ has a $\mathcal{V}$-homotopy inverse $\psi : (M, \mathfrak{T}_m) \to (M, \mathfrak{T}_w)$. Let $D \subset C$ be as in $(*_c)$. Since $\psi|D = \psi\varphi|D \simeq_{\mathcal{V}} \mathrm{id}$, we can apply the Homotopy Extension Theorem 1.13.11 to obtain a map $f : C \to (M, \mathfrak{T}_w)$ such that $f|D = \mathrm{id}$ and $f \simeq_{\mathcal{V}} \psi|D$. Since ψ is $\mathcal{V}$-close to id and $\mathrm{st}\,\mathcal{V} \prec \mathcal{U}$, it follows that f is $\mathcal{U}$-close id.

(b) + (c) ⇒ (a): Due to the Characterization Theorem 6.14.3, it suffices to show that $(M, \mathfrak{T}_m)$ is an $\boldsymbol{Q}_f^{\mathbb{N}}$- (or $\mathbb{R}_f^{\mathbb{N}}$-)manifold, where $\boldsymbol{Q}_f^{\mathbb{N}} \approx \ell_Q^2$ and $\mathbb{R}_f^{\mathbb{N}} \approx \ell_f^2$. To this end, Theorem 5.9.15 can be applied. Due to Proposition 6.4.7, we can write $(M, \mathfrak{T}_w) = \varinjlim X_n$, where $(X_n)_{n\in\mathbb{N}}$ is a tower of (finite-dimensional) compacta. Because $\mathfrak{T}_m \subset \mathfrak{T}_w$, each X_n is compact in $(M, \mathfrak{T}_m)$, hence it is a strong Z-set in $(M, \mathfrak{T}_m)$ by (b). We will show that $(X_n)_{n\in\mathbb{N}}$ has the absorptively $\mathfrak{M}_0$-universal property $(\mathrm{au}_{\mathfrak{M}_0})$ (resp. the absorptively $\mathfrak{M}_0^{\mathrm{fd}}$-universal property $(\mathrm{au}_{\mathfrak{M}_0^{\mathrm{fd}}})$) in X (cf. p. 440). Then, applying Theorem 5.9.15, we can conclude that $(M, \mathfrak{T}_m)$ is a $\boldsymbol{Q}_f^{\mathbb{N}}$- (or $\mathbb{R}_f^{\mathbb{N}}$-)manifold.

Let $C \in \mathfrak{M}_0$ with a closed subset $D \subset C$, and let $f : C \to (M, \mathfrak{T}_m)$ be a map such that $f|D : D \to X_m$ is an embedding into some $m \in \mathbb{N}$. For each open cover $\mathcal{U} \in \mathrm{cov}(M, \mathfrak{T}_m)$, let $\mathcal{V} \in \mathrm{cov}(M, \mathfrak{T}_m)$ be a star-refinement of $\mathcal{U}$. Since $\varphi = \mathrm{id}_M : (M, \mathfrak{T}_w) \to (M, \mathfrak{T}_m)$ is a fine homotopy equivalence, it has a $\mathcal{V}$-homotopy inverse $\psi : (M, \mathfrak{T}_m) \to (M, \mathfrak{T}_w)$, that is, $\varphi\psi \simeq_{\mathcal{V}} \mathrm{id}_M$ in $(M, \mathfrak{T}_m)$ and $\psi\varphi \simeq_{\mathcal{V}} \mathrm{id}_M$ in $(M, \mathfrak{T}_w)$, where $\mathcal{V} \in \mathrm{cov}(M, \mathfrak{T}_w)$ because $\mathfrak{T}_m \subset \mathfrak{T}_w$. Then, $f|D : D \to X_m \subset (M, \mathfrak{T}_w)$ is an embedding and

$$f|D \simeq_{\mathcal{V}} \psi\varphi f|D = \psi f|D \ \text{ in } (M, \mathfrak{T}_w).$$

Since $(M, \mathfrak{T}_w)$ is an ANE (Proposition 6.1.8), we have a map $f' : C \to (M, \mathfrak{T}_w)$ such that $f'|D = f|D$ and $f' \simeq_{\mathcal{V}} \psi f$ in $(M, \mathfrak{T}_w)$. Since $(M, \mathfrak{T}_w)$ has property $(\mathcal{A})$ (or $(\mathcal{A}_f)$) (Theorem 6.4.10), there is an embedding $g : C \to (M, \mathfrak{T}_w)$ such that $g|D = f'|D = f|D$ and $g \simeq_{\mathcal{V}} f'$. Since $g(C)$ is a compact subset of $(M, \mathfrak{T}_w) =$

$\varinjlim X_n$, it follows that $g(C) \subset X_n$ for some $n > m$. Thus, we have an embedding $\varphi g : C \to X_n \subset (M, \mathfrak{T}_{\mathrm{m}})$ such that $\varphi g|D = \varphi f|D = f|D$ and

$$\varphi g \simeq_{\mathcal{V}} \varphi f' \simeq_{\mathcal{V}} \varphi\psi f \simeq_{\mathcal{V}} f.$$

Since $\operatorname{st}\mathcal{V} \prec \mathcal{U}$, we have $\varphi g \simeq_{\mathcal{U}} f$. □

Notes for Chapter 6

For results in Sect. 6.1, Sect. 2.8 in [GAGT] can be referred to as mentioned there. The Heisey Theorem 6.2.3 was established in [71]. The results in Sect. 6.3 were obtained in [117].

The Characterization Theorem 6.4.8 for $\mathbb{R}^\infty$- and $\boldsymbol{Q}^\infty$-manifolds was established by K. Sakai [117]. The Open Embedding Theorem 6.4.9, the Stability Theorem 6.5.8, the Classification Theorem 6.5.1, and the Triangulation Theorem 6.5.5 were all established by R.E. Heisey in his papers [72–74, 76]. However, they are easily derived from the Characterization Theorem 6.4.8.

The Homeomorphism Approximation Theorem 6.5.2 was obtained by Vo Thanh Liem as a stronger form, the so-called α-approximation theorem in [98, 99]. The proof presented here was given by K. Sakai [117].

The Simplicial Approximation Theorem 6.6.1 for PL embeddings was established in [121] but there were some gaps in the proof. In the present book, we fixed them and gave details in each step of the proof. We dedicated three sections (Sects. 6.6–6.8) to proving Theorem 6.6.1. In [121], a combinatorial ∞-manifold was introduced and the Hauptvermutung for combinatorial ∞-manifolds (Theorem 6.9.1) was proved as the strong version (Theorem 6.9.2).

The Characterization Theorem 6.10.1 was given as a characterization of combinatorial ∞-manifolds in [121], where it was implicitly used without proof that a simplicial complex is a combinatorial ∞-manifold if it is combinatorially equivalent to a combinatorial ∞-manifold. So, here is introduced a PL ∞-manifold. The Characterization Theorem 6.11.1 jointed with Theorem 6.12.2 was established in [125]. In [125], a characterization for combinatorial ∞-manifolds by using Homology was also given. Propositions 6.12.5 are 6.12.6 are proved in [125]. In [87] and in Appendix of [125], Z-sets in polyhedra are discussed.

Theorems 6.13.1, 6.13.2, and 6.13.4 were established in [122], but our proof of Theorem 6.13.2 is different from the one given there. Theorems 6.13.5, 6.13.9, 6.13.10, and Corollaries 6.13.11, 6.13.12 were demonstrated in two papers [123] and [124]. Theorem 6.13.13 was obtained in [120].

The results of Sect. 6.14 were obtained in [127] and [23]. Theorem 6.14.1 was proved in [127] without condition (iii) but Lemma 2 in [127] was disproved in [23], whose counter-example is the example of Remark 6.16 given in [23]. Lemma 6.14.11 is a reformulation of Lemma 3 in [127] and Key Lemma in [23]. Lemma 6.14.12 is Lemma 1 in [127]. Theorem 6.14.13 is one of characterizations obtained in [23]. For other characterizations, refer to the paper [23].

LF-Spaces

The inductive limit of Fréchet spaces in the category of topological linear spaces is called an **LF-space**. The space $\mathbb{R}^\infty$ is an LF-space. However, in general, the inductive limits in the category of topological linear spaces is not equal to the inductive limits in the category of topological spaces. We did not mention manifolds modeled on LF-spaces in this chapter; P. Mankiewicz [101] gave the topological classification for LF-spaces, that is, every LF-space F is homeomorphic to one of the spaces, $\mathbb{R}^\infty$, $\ell^2(\Gamma)$, or $\sum_{i\in\mathbb{N}} \ell^2(\Gamma_i)$, where $\operatorname{card}\Gamma = \operatorname{dens} F$ and $\aleph_0 < \operatorname{card}\Gamma_1 < \operatorname{card}\Gamma_2 < \cdots$ with $\sup_{i\in\mathbb{N}} \operatorname{card}\Gamma_i = \operatorname{dens} F$. Moreover, Banakh and Repovš [22] gave a topological characterization of LF-spaces (or their open subsets). For open sets in an LF-space $\ell^2(\Gamma) \times \mathbb{R}^\infty$ ($\operatorname{card}\Gamma \geqslant \aleph_0$), the Classification and the Triangulation Theorems were established in [109]. For open sets in an LF-space $\sum_{i\in\mathbb{N}} \ell^2(\Gamma_i)$ ($\aleph_0 < \operatorname{card}\Gamma_1 < \operatorname{card}\Gamma_2 < \cdots$), the Triangulation Theorem was established in [110].

Piecewise Linear $\mathbb{R}^\infty$-Manifolds

In [75], Heisey introduced a piecewise linear $\mathbb{R}^\infty$-manifold which is different from the PL ∞-manifold defined in Sect. 6.9. A map $f : U \to V$ between open sets in $\mathbb{R}^\infty$ is called an **$\mathbb{R}^\infty$-piecewise linear map** if for every compact polyhedron $C \subset U$ and for each $n \in \mathbb{N}$ such that $f(C) \subset V \cap \mathbb{R}^n$, the restriction $f|C : C \to V \cap \mathbb{R}^n$ is PL in the usual sense. Using $\mathbb{R}^\infty$-piecewise linear maps, he introduced a **piecewise linear $\mathbb{R}^\infty$-structure** as in differential topology. A **piecewise linear $\mathbb{R}^\infty$-manifold** is a paracompact space together with a piecewise linear $\mathbb{R}^\infty$-structure. An **$\mathbb{R}^\infty$-piecewise linear map** and an **$\mathbb{R}^\infty$-piecewise isomorphism** between piecewise linear $\mathbb{R}^\infty$-manifolds are defined like a differential map and a diffeomorphism between differential manifolds. He proved that every piecewise linear $\mathbb{R}^\infty$-manifold can be embedded into $\mathbb{R}^\infty$ as a closed piecewise linear $\mathbb{R}^\infty$-manifold. In [119], it was proved that two piecewise linear $\mathbb{R}^\infty$-manifolds are $\mathbb{R}^\infty$-piecewise isomorphic if they have the same homotopy type. Combining this result with the Open embedding Theorem 6.4.9, we can conclude that each $\mathbb{R}^\infty$-manifold has a unique piecewise linear $\mathbb{R}^\infty$-structure.

Products of Absorbing Sets and $\mathbb{R}^\infty$

Let Ω be an absorbing set in the Hilbert cube $\boldsymbol{Q}$. Although we did not mention in the text, $\Omega \times \mathbb{R}^\infty$ is also an interesting object of study (cf. [19]).

Appendix A
PL n-Manifolds and Combinatorial n-Manifolds

A separable polyhedron M is called a **PL n-manifold** if each $x \in M$ has a neighborhood that is PL homeomorphic to $\mathbf{I}^n$. Then, M is an n-dimensional polyhedron, that is, $M = |K|$ for some n-dimensional cell (or simplicial) complex K. So, M is also called an **n-dimensional PL manifold**. Obviously, every PL n-manifold M is an n-manifold. PL n-manifolds are the main subjects for PL Topology. A countable simplicial complex K is called a **combinatorial n-manifold** if $\mathrm{St}(v, K) \cong F(\Delta^n)$ for every $v \in K^{(0)}$. Then, K is an n-dimensional complex. So, K is also called an **n-dimensional combinatorial manifold**. In this Appendix, we give some fundamental results on PL n-manifolds and combinatorial n-manifolds. Among them, we show that a simplicial complex is a combinatorial n-manifold if and only if it is a triangulation of a PL n-manifold. We also prove two results, the first is Proposition 6.10.6 used in the proof of Theorem 6.10.1, and the second is the PL manifold version of Theorem 1.8.11. For the contents of this Appendix, the reader can refer to the book [(17)].

A.1 Characterizations of Combinatorial n-Manifolds

In this section, we show that a simplicial complex K is a combinatorial n-manifold if and only if $|K|$ is a PL n-manifold (Theorem A.1.6), and also give characterizations of combinatorial n-manifolds (Theorem A.1.10).

First of all, we show the following:

Proposition A.1.1 *For every $n \in \omega$, $F(\Delta^n)$ and $F(\partial\Delta^{n+1})$ are combinatorial n-manifolds.*

Proof It is obvious that $F(\Delta^n)$ is a combinatorial n-manifold. For each vertex $v \in F(\partial\Delta^{n+1})^{(0)}$, let $\sigma < \Delta^{n+1}$ be the opposite face to v, that is, $\Delta^{n+1} = v\sigma$. Then, it

K. Sakai, *Topology of Infinite-Dimensional Manifolds*, Springer Monographs in Mathematics, https://doi.org/10.1007/978-981-15-7575-4

follows that

$$\begin{aligned}\mathrm{St}(v, F(\partial\Delta^{n+1})) &= F(\partial\Delta^{n+1}) \setminus \{\sigma\} = v * F(\partial\sigma)\\ &\equiv \hat{\sigma} * F(\partial\sigma) \cong F(\sigma) \equiv F(\Delta^n),\end{aligned}$$

where $\hat{\sigma}$ is the barycenter of σ. Therefore, $F(\partial\Delta^{n+1})$ is a combinatorial n-manifold. □

Next, we will prove that every two n-cells are PL homeomorphic to each other. To this end, we have to describe each cell as a suitable cell complex. Let C be an n-cell in $\mathbb{R}^n$ with $\mathbf{0} \in \operatorname{int} C$. For each $D < C$, we have a cone $\mathbb{R}_+ D = \{tx \mid t \geqslant 0,\ x \in D\} \subset E$ with $\mathbf{0}$ the vertex. Then, $\mathbb{R}_+ D \cap C$ is a cell and

$$F(\partial(\mathbb{R}_+ D \cap C)) = F(D) \cup \{\mathbf{0}\} \cup \{\mathbb{R}_+ D' \cap C \mid D' < D\}.$$

It is easy to see that the following is a cell complex subdividing $F(C)$:

$$K_C = \{\mathbb{R}_+ D \cap C \mid D < C\} \cup F(\partial C) \cup \{\mathbf{0}\}.$$

Note that $K_C = \mathrm{St}(\mathbf{0}, K_C)$ and $F(\partial C) = \mathrm{Lk}(\mathbf{0}, K_C)$.

Proposition A.1.2 *Every two n-cells C and D are PL homeomorphic to each other, that is,*

$$(F(C), F(\partial C)) \cong (F(D), F(\partial D)).$$

Proof Although C and D are contained in some linear spaces, $\mathrm{fl}\, C$ and $\mathrm{fl}\, D$ are affinely homeomorphic to $\mathbb{R}^n$. Hence, it can be assumed that $C, D \subset \mathbb{R}^n$. Moreover, using translation, we may assume that $\mathbf{0} \in \operatorname{int} C \cap \operatorname{int} D$. Since C and D are compact, we can choose $r > 0$ so that $C \subset rD$ and $D \subset rC$. Then, $rK_C = \{rC' \mid C' \in K_C\}$ and $rK_D = \{rD' \mid D' \in K_D\}$ are cell complexes congruent to K_C and K_D, respectively. Due to Proposition 1.6.3, the following are cell complexes:

$$\begin{aligned}S_C &= \{C' \cap rD' \mid C' \in K_C,\ D' \in K_D,\ C' \cap D' \neq \emptyset\}\\ &= \{\mathbb{R}_+ C' \cap \mathbb{R}_+ D' \cap C \mid C' < C,\ D' < D\} \cup L_C \cup \{\mathbf{0}\};\\ L_C &= \{C' \cap \mathbb{R}_+ D' \mid C' < C,\ D' < D,\ C' \cap \mathbb{R}_+ D' \neq \emptyset\};\\ S_D &= \{rC' \cap D' \mid C' \in K_C,\ D' \in K_D,\ C' \cap D' \neq \emptyset\}\\ &= \{\mathbb{R}_+ C' \cap \mathbb{R}_+ D' \cap D \mid C' < C,\ D' < D\} \cup L_D \cup \{\mathbf{0}\};\\ L_D &= \{\mathbb{R}_+ C' \cap D' \mid C' < C,\ D' < D,\ \mathbb{R}_+ C' \cap D' \neq \emptyset\},\end{aligned}$$

which subdivide $F(C)$, $F(\partial C)$, $F(D)$, and $F(\partial D)$. We can define a bijection $\eta : S_C \to S_D$ as follows:

$$\eta(\mathbb{R}_+C' \cap \mathbb{R}_+D' \cap C) = \mathbb{R}_+C' \cap \mathbb{R}_+D' \cap D \ \text{ for } C' < C,\, D' < D,$$
$$\eta(\mathbb{R}_+D' \cap C') = \mathbb{R}_+C' \cap D' \ \text{ for } C' < C,\, D' < D \text{ with } C' \cap \mathbb{R}_+D' \neq \emptyset,$$

where $\mathbb{R}_+C' \cap D' = \mathbb{R}_+C' \cap \mathbb{R}_+D' \cap D'$. Evidently, $\eta(L_C) = L_D$. The following can be easily verified:

$$C', C'' \in S_C,\ C'' < C' \ \Leftrightarrow\ \eta(C'') < \eta(C').$$

Hence, we have $(S_C, L_C) \equiv (S_D, L_D)$, which implies

$$(F(C), F(\partial C)) \cong (F(D), F(\partial D)).$$ □

Using the same technique as the above, we can prove the following:

Lemma A.1.3 *Let K be a locally finite simplicial complex and $K' \lhd K$. For each $v \in K^{(0)}$,*

$$\big(\operatorname{St}(v, K'), \operatorname{Lk}(v, K')\big) \cong \big(\operatorname{St}(v, K), \operatorname{Lk}(v, K)\big).$$

Proof Since $\operatorname{St}(v, K)$ is finite, we may assume that $|\operatorname{St}(v, K)| \subset \mathbb{R}^n$ and $v = \mathbf{0}$. Then, note the following equality:

$$\mathbb{R}_+|\operatorname{St}(v, K)| = \mathbb{R}_+|\operatorname{Lk}(v, K)| = \mathbb{R}_+|\operatorname{St}(v, K')| = \mathbb{R}_+|\operatorname{Lk}(v, K')|.$$

Since $|\operatorname{St}(v, K)|$ is compact (so bounded in $\mathbb{R}^n$) and $v = \mathbf{0}$ is an interior point of $|\operatorname{St}(v, K')|$ in $\mathbb{R}_+|\operatorname{St}(v, K)|$, we can choose $r > 0$ so that

$$|\operatorname{St}(v, K)| \subset r|\operatorname{St}(v, K')| = |r\operatorname{St}(v, K')|,$$

where $r\operatorname{St}(v, K') = \{r\sigma \mid \sigma \in \operatorname{St}(v, K')\}$ is a simplicial complex and $r\operatorname{St}(v, K') \equiv \operatorname{St}(v, K')$. Then, for each $\sigma \in \operatorname{Lk}(v, K')$ and $\tau \in \operatorname{Lk}(v, K)$ with $v\sigma \cap v\tau \neq \emptyset$, we have the following cells:

$$v\sigma \cap v\tau = \mathbb{R}_+\sigma \cap \mathbb{R}_+\tau \cap |\operatorname{St}(v, K')| > \sigma \cap v\tau = \mathbb{R}_+\sigma \cap \mathbb{R}_+\tau \cap |\operatorname{Lk}(v, K')|;$$
$$v(r\sigma) \cap v\tau = \mathbb{R}_+\sigma \cap \mathbb{R}_+\tau \cap |\operatorname{St}(v, K)| > v(r\sigma) \cap \tau = \mathbb{R}_+\sigma \cap \mathbb{R}_+\tau \cap |\operatorname{Lk}(v, K)|.$$

Applying Proposition 1.6.3, we have the following cell complexes:

$$\begin{aligned} S_0 &= \big\{\sigma \cap \tau \bigm| \sigma \in \operatorname{St}(v, K'),\ \tau \in \operatorname{St}(v, K),\ \sigma \cap \tau \neq \emptyset\big\} \\ &= \big\{\mathbb{R}_+\sigma \cap \mathbb{R}_+\tau \cap |\operatorname{St}(v, K')| \bigm| \sigma \in \operatorname{Lk}(v, K'),\ \tau \in \operatorname{Lk}(v, K)\big\} \cup L_0 \cup \{\mathbf{0}\}; \end{aligned}$$

$$L_0 = \{\sigma \cap \mathbb{R}_+\tau \mid \sigma \in \mathrm{Lk}(v, K'),\ \tau \in \mathrm{Lk}(v, K),\ \sigma \cap \mathbb{R}_+\tau \neq \emptyset\};$$
$$S_1 = \{r\sigma \cap \tau \mid \sigma \in \mathrm{St}(v, K'),\ \tau \in \mathrm{St}(v, K),\ r\sigma \cap \tau \neq \emptyset\}$$
$$= \{\mathbb{R}_+\sigma \cap \mathbb{R}_+\tau \cap |\mathrm{St}(v, K)| \mid \sigma \in \mathrm{Lk}(v, K'),\ \tau \in \mathrm{Lk}(v, K)\} \cup L_1 \cup \{\mathbf{0}\};$$
$$L_1 = \{\mathbb{R}_+\sigma \cap \tau \mid \sigma \in \mathrm{Lk}(v, K'),\ \tau \in \mathrm{Lk}(v, K),\ \sigma \cap \mathbb{R}_+\tau \neq \emptyset\},$$

which are subdivisions of $\mathrm{St}(v, K')$, $\mathrm{Lk}(v, K')$, $\mathrm{St}(v, K)$, and $\mathrm{Lk}(v, K)$, respectively. Then, in the same way that we showed $(S_C, L_C) \equiv (S_D, L_D)$ in the proof of Proposition A.1.2, we can show $(S_0, L_0) \equiv (S_1, L_1)$. Thus, we have

$$(\mathrm{St}(v, K'), \mathrm{Lk}(v, K')) \cong (\mathrm{St}(v, K), \mathrm{Lk}(v, K)). \qquad \square$$

Lemma A.1.3 can be generalized as follows:

Proposition A.1.4 *Let K and L be locally finite simplicial complexes. For $v \in K^{(0)}$ and $u \in L^{(0)}$, if $(K, v) \cong (L, u)$, then*

$$(\mathrm{St}(v, K), \mathrm{Lk}(v, K)) \cong (\mathrm{St}(u, L)\, \mathrm{Lk}(u, L)).$$

Proof There exist subdivisions K' and L' of K and L respectively such that $(K', v) \equiv (L', u)$. Then,

$$(\mathrm{St}(v, K'), \mathrm{Lk}(v, K')) \equiv (\mathrm{St}(u, L'), \mathrm{Lk}(u, L')).$$

Applying Lemma A.1.3, we have

$$\begin{aligned}(\mathrm{St}(v, K), \mathrm{Lk}(v, K)) &\cong (\mathrm{St}(v, K'), \mathrm{Lk}(v, K')) \\ &\equiv (\mathrm{St}(u, L'), \mathrm{Lk}(u, L')) \\ &\cong (\mathrm{St}(u, L), \mathrm{Lk}(u, L)).\end{aligned} \qquad \square$$

By combining the following proposition with Theorem 1.8.7, a simplicial complex is a combinatorial n-manifold if it is PL homeomorphic to $\mathbf{I}^n$.

Proposition A.1.5 *Let K be a simplicial complex with a PL homeomorphism $h : |K| \to \Delta^n$. For each vertex $v \in K^{(0)}$,*

$$(\mathrm{St}(v, K), \mathrm{Lk}(v, K)) \cong \begin{cases}(F(\Delta^n), F(\partial\Delta^n)) & \text{if } h(v) \in \operatorname{rint}\Delta^n, \\ (F(\Delta^n), F(\Delta^{n-1})) & \text{if } h(v) \in \partial\Delta^n,\end{cases}$$

where $\Delta^{n-1} = \langle \mathbf{e}_1, \ldots, \mathbf{e}_n\rangle \subset \Delta^n = \langle \mathbf{e}_1, \ldots, \mathbf{e}_n, \mathbf{e}_{n+1}\rangle$.

Proof When $h(v) \in \operatorname{rint} \Delta^n$, let $L = F(\Delta^n)_{h(v)}$, the starring of $F(\Delta^n)$ at $h(v)$, that is,

$$L = \{h(v)\sigma \mid \sigma \in F(\partial\Delta^n)\} \cup F(\partial\Delta^n) \cup \{h(v)\}.$$

Then, $|L| = \Delta^n$, $h(v) \in L^{(0)}$, $L = \mathrm{St}(h(v), L)$, and $F(\partial\Delta^n) = \mathrm{Lk}(h(v), L)$. Since $(K, v) \cong (L, h(v))$ by Theorem 1.8.7 with Remark 1.5, we can apply Proposition A.1.4 to obtain

$$\begin{aligned}\big(\mathrm{St}(v, K), \mathrm{Lk}(v, K)\big) &\cong \big(\mathrm{St}(h(v), L), \mathrm{Lk}(h(v), L)\big)\\ &\cong \big(F(\Delta^n), F(\partial\Delta^n)\big).\end{aligned}$$

When $h(v) \in \partial\Delta^n$, it is easy to see that $(\Delta^n, h(v))$ is PL homeomorphic to $(\Delta^n, \mathbf{e}_{n+1})$. Then, it follows that $(K, v) \cong (F(\Delta^n), \mathbf{e}_{n+1})$. So, we can apply Proposition A.1.4 to obtain

$$\begin{aligned}\big(\mathrm{St}(v, K), \mathrm{Lk}(v, K)\big) &\cong \big(\mathrm{St}(\mathbf{e}_{n+1}, F(\Delta^n)), \mathrm{Lk}(\mathbf{e}_{n+1}, F(\Delta^n))\big)\\ &= \big(F(\Delta^n), F(\Delta^{n-1})\big).\end{aligned}$$ □

Now, we can prove the following characterization:

Theorem A.1.6 *A simplicial complex K is a combinatorial n-manifold if and only if $|K|$ is a PL n-manifold.*

Proof (*The "if" part*) Each $x \in |K|$ is contained in the open star $O_K(v)$ of some vertex $v \in K^{(0)}$. Then, $|\mathrm{St}(v, K)|$ is neighborhood of x in $|K|$. Since $\mathrm{St}(v, K) \cong F(\Delta^n)$, $|\mathrm{St}(v, K)|$ is PL homeomorphic to $\mathbf{I}^n$ by Theorem 1.8.7 and Proposition A.1.2. Therefore, $|K|$ is a PL n-manifold.

(*The "if" part*) Each $x \in |K|$ has a neighborhood which is PL homeomophic to $\mathbf{I}^n$. Applying Proposition 1.6.6, we have a simplicial subdivision $K' \lhd K$ and a subcomplex L of K' such that $|L|$ is a neighborhood of x in $|K|$ being PL homeomorphic to $\mathbf{I}^n$. By the starring of K at x, we can take K' so that $x \in K'^{(0)}$, so $x \in L^{(0)}$. Because $x \in \operatorname{int}_{|K'|}|L|$, it follows that $\mathrm{St}(x, K') = \mathrm{St}(x, L)$. Due to Proposition A.1.2, $|L|$ is PL homeomorphic to Δ^n. Combining Lemma A.1.3 and Proposition A.1.5, we have $\mathrm{St}(x, K) \cong \mathrm{St}(x, K') = \mathrm{St}(x, L) \cong F(\Delta^n)$. □

Evidently, a polyhedron is a PL n-manifold if it is PL homeomorphic to a PL n-manifold. Hence, we have the following:

Corollary A.1.7 *A simplicial complex is a combinatorial n-manifold if it is combinatorially equivalent to a combinatorial n-manifold.* □

Let K and L be joinable simplicial complexes. Then, every two simplicial subdivisions $K' \lhd K$ and $L' \lhd L$ are joinable and $K' * L' \lhd K * L$. Hence, the following is trivial:

Lemma A.1.8 *For joinable simplicial complexes K_i and L_i, $i = 1, 2$,*

$$K_1 \cong K_2,\ L_1 \cong L_2 \ \Rightarrow\ K_1 * L_1 \cong K_2 * L_2,$$

where any two PL homeomorphisms $f : |K_1| \to |K_2|$ and $g : |L_1| \to |L_2|$ extend to a PL homeomorphism $h : |K_1 * L_1| \to |K_2 * L_2|$. □

For joinable simplexes, we have the following:

Proposition A.1.9 *Let σ and τ be joinable simplexes with $\sigma = n$ and $\tau = m$. Then, the following fold:*[1]

(i) $\big(F(\sigma) * F(\tau), F(\sigma) * F(\partial\tau) \cup F(\partial\sigma) * F(\tau)\big)$
$$= \big(F(\sigma\tau), F(\partial(\sigma\tau)\big) \equiv \big(F(\Delta^{n+m+1}), F(\partial\Delta^{n+m+1})\big);$$

(ii) $\big(F(\sigma) * F(\partial\tau), F(\partial\sigma) * F(\partial\tau)\big) \cong \big(F(\Delta^{n+m}), F(\partial\Delta^{n+m})\big)$.

Proof Since the join $\sigma\tau$ is defined as the simplex $\langle\sigma^{(0)} \cup \tau^{(0)}\rangle$, (i) is trivial.

To see (ii), for each $i = 1, \ldots, n+m$, let $v_i = \mathbf{e}_i \in \mathbb{R}^{n+m}$, where $\mathbf{e}_i$ is the unit vector defined by $\mathbf{e}_i(i) = 1$ and $\mathbf{e}_i(j) = 0$ for $j \neq i$. Moreover, let

$$v_0 = -(v_1 + \cdots + v_n) \ \text{ and } \ v_{n+m+1} = -(v_{n+1} + \cdots + v_{n+m}) \in \mathbb{R}^{n+m}.$$

Then, the following can be easily seen:

(1) $v_0, v_1, \ldots, v_{n+m}$ are affinely independent.
(2) For each $k = n+1, \ldots, n+m$, $v_0, v_1, \ldots, v_{k-1}, v_{k+1}, \ldots, v_{n+m+1}$ are also affinely independent.

Indeed, to see (1), assume that $\sum_{i=0}^{n+m} t_i v_i = \mathbf{0}$ and $\sum_{i=0}^{n+m} t_i = 0$. Then,

$$\sum_{i=1}^{n}(t_i - t_0)\mathbf{e}_i + \sum_{i=n+1}^{n+m} t_i\mathbf{e}_i = \mathbf{0},$$

which implies that $t_1 = \cdots = t_n = t_0$ and $t_{n+1} = \cdots = t_{n+m} = 0$. It follows that $(n+1)t_0 = \sum_{i=0}^{n} t_i = \sum_{i=0}^{n+m} t_i = 0$. Hence, $t_0 = t_1 = \cdots = t_n = 0$.

To see (2), assume that $\sum_{i\neq k} t_i v_i = \mathbf{0}$ and $\sum_{i\neq k} t_i = 0$. Then,

$$\sum_{i=1}^{n}(t_i - t_0)\mathbf{e}_i + \sum_{i=n+1}^{k-1}\big(t_i - t_{n+m+1}\big)\mathbf{e}_i - t_{n+m+1}\mathbf{e}_k + \sum_{i=k+1}^{n+m}\big(t_i - t_{n+m+1}\big)\mathbf{e}_i = \mathbf{0},$$

which implies that $t_1 = \cdots = t_n = t_0$, $t_{n+1} = \cdots = t_{k-1} = t_{n+m+1} = 0$, and $t_{k+1} = \cdots = t_{n+m} = t_{n+m+1} = 0$. It follows that $(n+1)t_0 = \sum_{j=0}^{n} t_j = \sum_{j\neq k} t_j = 0$. Hence, $t_0 = t_1 = \cdots = t_n = 0$.

[1] Recall $\sigma\emptyset = \emptyset\sigma = \sigma$, where $\emptyset\emptyset = \emptyset$.

From (1) and (2), the following simplicial complexes are joinable:

$$K = F(\langle v_0, v_1, \ldots, v_n\rangle) \ \text{ and } \ L = F(\partial\langle v_{n+1}, \ldots, v_{n+m+1}\rangle).$$

Then, $K * L \equiv F(\sigma) * F(\partial\tau)$ because $K \equiv F(\sigma)$ and $L \equiv F(\partial\tau)$. We will show that $|K * L|$ is an $(n+m)$-cell, that is,

$$|K * L| = \langle v_0, v_1, \ldots, v_{n+m+1}\rangle \subset \mathbb{R}^{n+m}.$$

The left hand side is obviously contained in the right hand side. To see the converse, let

$$x = \sum_{i=0}^{n+m+1} s_i v_i \in \langle v_0, v_1, \ldots, v_{n+m+1}\rangle,$$

where $s_0, s_1, \ldots, s_{n+m+1} \geqslant 0$ and $\sum_{j=0}^{n+m+1} s_j = 1$. Choose $k = n+1, \cdots, n+m+1$ so that

$$s_k = \min\{s_{n+1}, \ldots, s_{n+m+1}\}.$$

For each $i = 0, 1, \ldots, n+m+1$, define $t_i \geqslant 0$ as follows:

$$t_i = \begin{cases} s_i + ms_k/(n+1) & i \leqslant n, \\ s_i - s_k & i > n, \end{cases}$$

where $t_k = 0$. Since $\sum_{i=0}^{n} v_i = \mathbf{0}$ and $\sum_{i=n+1}^{n+m+1} v_i = \mathbf{0}$, it follows that

$$\begin{aligned} \sum_{i \neq k} t_i v_i &= \sum_{i=0}^{n} s_i v_i + \frac{ms_k}{n+1} \sum_{i=0}^{n} v_i + \sum_{i=n+1}^{n+m+1} s_i v_i - s_k \sum_{i=n+1}^{n+m+1} v_i \\ &= \sum_{i=0}^{n+m+1} s_i v_i = x. \end{aligned}$$

Moreover, we have

$$\sum_{i \neq k} t_i = \sum_{i=0}^{n+m+1} s_i + (n+1)\frac{ms_k}{n+1} - ms_k = 1.$$

Therefore,

$$\begin{aligned} x &\in \langle v_0, \dots, v_{k-1}, v_{k+1}, \dots, v_{n+m+1}\rangle \\ &= \langle v_0, \dots, v_n\rangle\langle v_{n+1}, \dots, v_{k-1}, v_{k+1}, \dots, v_{n+m+1}\rangle \subset |K * L|. \end{aligned}$$

Now, we saw that $|K * L|$ is an $(n + m)$-cell. Hence, $K * L \cong F(\Delta^{n+m})$ by Proposition A.1.2. Consequently, $F(\sigma) * F(\partial\tau) \cong F(\Delta^{n+m})$.

As is easily observed,

$$(K * L)^{(n+m-1)} = K^{(n-1)} * L \cup K * L^{(m-2)}.$$

The $(n + m)$-simplexes of $K * L$ are the following:

$$\langle v_0, \dots, v_{k-1}, v_{k+1}, \dots, v_{n+m+1}\rangle,\ k = n + 1, \dots, n + m + 1.$$

Every $(n+m-1)$-simplex of $K^{(n-1)} * L$ is a face of only one of the above $(n+m)$-simplexes and every $(n+m-1)$-simplexes of $K * L^{(m-2)}$ is a common face of two of the above $(n + m)$-simplexes. Therefore, $|K^{(n-1)} * L|$ is the boundary of the cell $|K * L|$. Then, it follows from Proposition A.1.2 that

$$\begin{aligned} \big(K * L, K^{(n-1)} * L\big) &\cong \big(F(|K * L|), F(\partial|K * L|)\big) \\ &\cong \big(F(\Delta^{n+m}, F(\partial\Delta^{n+m})\big). \end{aligned}$$

By the way, observe

$$K^{(n-1)} = F(\partial\langle v_0, v_1, \dots, v_n\rangle) \equiv F(\partial\Delta^n) \equiv F(\partial\sigma).$$

Thus, we have (ii). □

Now, we have the following characterization of combinatorial manifolds:

Theorem A.1.10 *For a simplicial complex K, the following are equivalent:*

(a) *K is a combinatorial n-manifold;*
(b) $(\mathrm{St}(v, K), \mathrm{Lk}(v, K)) \cong (F(\Delta^n, F(\partial\Delta^n))$ *or* $(F(\Delta^n, F(\Delta^{n-1}))$ *for each vertex $v \in K^{(0)}$;*
(c) $\mathrm{Lk}(v, K) \cong F(\partial\Delta^n)$ *or* $F(\Delta^{n-1})$ *for each vertex $v \in K^{(0)}$;*
(d) $\mathrm{Lk}(\sigma, K) \cong F(\partial\Delta^{n-k})$ *or* $F(\Delta^{n-k-1})$ *for each k-simplex $\sigma \in K$;*
(e) $\mathrm{St}(\sigma, K) \cong F(\Delta^n)$ *for each simplex $\sigma \in K$.*

Proof The implications (e) ⇒ (a) and (b) ⇒ (c) are obvious ((d) ⇒ (c) and (b) ⇒ (a) are obvious too). Since $\mathrm{St}(\sigma, K) = \mathrm{Lk}(\sigma, K) * F(\sigma)$ for each i-simplex $\sigma \in K$, the implications (c) ⇒ (a) and (d) ⇒ (e) come from Proposition A.1.9 (cf. Lemma A.1.8). Then, it suffices to show the implications (a) ⇒ (b) and (c) ⇒ (d).

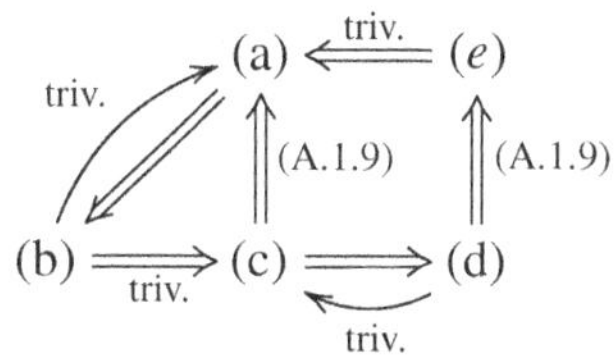

(a) $\Rightarrow$ (b): For each $v \in K^{(0)}$, let $S = \mathrm{St}(v, K)$. Then, $v \in S^{(0)}$ and

$$(\mathrm{St}(v, S), \mathrm{Lk}(v, S)) = (\mathrm{St}(v, K), \mathrm{Lk}(v, K)).$$

Since $S \cong F(\Delta^n)$ by (a), there exists a PL homeomorphism $h : |S| \to \Delta^n$. Applying Proposition A.1.5, we can obtain

$$\big(\mathrm{St}(v, S), \mathrm{Lk}(v, S)\big) \cong \begin{cases} \big(F(\Delta^n), F(\partial\Delta^n)\big) & \text{if } h(v) \in \operatorname{rint} \Delta^n, \\ \big(F(\Delta^n), F(\Delta^{n-1})\big) & \text{if } h(v) \in \partial\Delta^n. \end{cases}$$

(c) $\Rightarrow$ (d): By induction on n, we show this implication on n. The case $n = 0$ is obvious. Assuming the implication (c) $\Rightarrow$ (d) for $n - 1$, we show the case n. The assumption means the theorem holds for $n - 1$. For $k > 0$, each k-simplex σ can be written as $\sigma = v\tau$, where $v \in \sigma^{(0)}$ and τ is the opposite face of σ to v, i.e., $\tau^{(0)} = \sigma^{(0)} \setminus \{v\}$. Then, as is easily observed, $\mathrm{Lk}(\sigma, K) = \mathrm{Lk}(\tau, \mathrm{Lk}(v, K))$. Due to Proposition A.1.1 and Corollary A.1.7, (c) implies that $\mathrm{Lk}(v, K)$ is a combinatorial $(n - 1)$-manifold. Since τ is an $(k - 1)$-simplex, it follows from the inductive assumption that

$$\mathrm{Lk}(\sigma, K) = \mathrm{Lk}(\tau, \mathrm{Lk}(v, K)) \cong F(\partial\Delta^{n-k}) \text{ or } F(\Delta^{n-k-1}).$$

Hence, (c) $\Rightarrow$ (d) is valid for n. Thus, we have the implication (c) $\Rightarrow$ (d) by induction. □

As a corollary, we have the following:

Corollary A.1.11 *Let K be a combinatorial n-manifold.*

(1) *Every $\sigma \in K$ is a face of an n-simplex of K. Equivalently, every k-simplex $\sigma \in K$ is joinable to some $(n - k - 1)$-simplex of K in K.*
(2) *Each $(n - 1)$-simplex $\sigma \in K$ is a common face of distinct two n-simplexes of K or a face of exactly one n-simplex of K. Equivalently, each $(n - 1)$-simplex $\sigma \in K$ is joinable to one or two vertices in K.* □

A.2 The Boundary of a Combinatorial n-Manifold

Every PL n-manifold M is an n-manifold. The interior $\operatorname{Int} M$ and the boundary ∂M of M can be defined similarly to those of an n-manifold. Actually, $x \in M$ is called an **interior point** of M if it has a neighborhood N with a PL homeomorphism $h : N \to \mathbf{I}^n$ such that $h(x) \in (0, 1)^n$. The set $\operatorname{Int} M$ of all interior points of M is called the **interior** of M, which is open in M. We call $x \in M$ a **boundary point** of M if it is not an interior point, that is, it has a neighborhood N with a PL homeomorphic $h : N \to \mathbf{I}^n$ such that $h(x) \in \partial\mathbf{I}^n$. We define the **boundary** ∂M of M as the set of boundary points. A PL n-manifold M with $\partial M = \emptyset$ is called a **PL n-manifold without boundary**. A PL n-manifold without boundary is called a **closed PL n-manifold** if it is compact.

For a combinatorial n-manifold K, we define

$$\partial K = \{\sigma \in K \mid \operatorname{Lk}(\sigma, K) \cong F(\Delta^{n-\dim\sigma-1})\},$$

which is called the **boundary** of K. For every n-simplex $\sigma \in K$, $\operatorname{Lk}(\sigma, K) = \emptyset$ because σ is principal. Hence, $\partial K \subset K^{(n-1)}$. In fact, $\dim \partial K = n - 1$ if $\partial K \neq \emptyset$.

Indeed, assume $\partial K \neq \emptyset$. Then, we have $v \in (\partial K)^{(0)}$. Since $\operatorname{Lk}(v, K) \cong F(\Delta^{n-1})$, it is easily seen that $\operatorname{Lk}(v, K)$ contains an $(n-2)$-simplex σ which is a face of only one $(n-1)$-simplex of $\operatorname{Lk}(v, K)$. Consider the $(n-1)$-simplex $v\sigma \in K$. Then, it follows that

$$\operatorname{Lk}(v\sigma, K) = \operatorname{Lk}(\sigma, \operatorname{Lk}(v, K)) \cong F(\Delta^0) = F(\Delta^{n-(n-1)-1}),$$

which means that $v\sigma \in \partial K$. Therefore, $\dim \partial K = n - 1$.

Due to Corollary A.1.11(2), the following fact is trivial:

Fact *For a combinatorial n-manifold K, an $(n-1)$-simplex $\sigma \in K$ belongs to the boundary ∂K if and only if σ is a face of only one n-simplex of K, equivalently σ is joinable to only one vertex in K.*

We will show that $\partial|K| = |\partial K|$ for every combinatorial n-manifold. For the simplest case, the following holds:

Lemma A.2.1 *For every $n \in \mathbb{N}$, $\partial F(\Delta^n) = F(\partial\Delta^n)$ and $\partial F(\partial\Delta^n) = \emptyset$. Consequently, $|\partial F(\Delta^n)| = \partial\Delta^n$.*

Proof Obviously, $\operatorname{Lk}(\Delta^n, F(\Delta^n)) = \emptyset$, which means that $\Delta^n \notin \partial F(\Delta^n)$. For each $\sigma \in F(\partial\Delta^n)$, let τ be the opposite face of Δ^n to σ, that is, $\Delta^n = \sigma\tau$. Then, $\operatorname{Lk}(\sigma, F(\Delta^n)) = F(\tau)$, which means $\sigma \in \partial F(\Delta^n)$. Therefore, $\partial F(\Delta^n) = F(\partial\Delta^n)$. Additionally, $\operatorname{Lk}(\sigma, F(\partial\Delta^n)) = F(\partial\tau)$, which means $\sigma \notin \partial F(\partial\Delta^n)$. Consequently, $\partial F(\partial\Delta^n) = \emptyset$. □

A combinatorial n-manifold K with $\partial K = \emptyset$ is called a **combinatorial n-manifold without boundary**. Concerning the boundary of a combinatorial n-manifold, we have the following:

Proposition A.2.2 *For a combinatorial n-manifold K with $\partial K \neq \emptyset$, ∂K is a combinatorial $(n-1)$-manifold without boundary.*

Proof First, we have to show that ∂K is a subcomplex of K, that is, if $\sigma \in \partial K$ and $\tau < \sigma$, then $\tau \in \partial K$. For $\tau < \sigma \in K$, it suffices to prove that $\tau \notin \partial K$ implies $\sigma \notin \partial K$, that is, $\mathrm{Lk}(\tau, K) \cong F(\partial\Delta^{n-\dim\tau})$ implies $\mathrm{Lk}(\sigma, K) \cong F(\partial\Delta^{n-\dim\sigma})$ (cf. Theorem A.1.10 (c)). We may show the case when $\dim\sigma = \dim\tau + 1$, that is, $\sigma = v\tau$, $v \in K^{(0)}$. Then, by induction, the general case can be obtained. When $\sigma = v\tau$, we have $\mathrm{Lk}(\sigma, K) = \mathrm{Lk}(v, \mathrm{Lk}(\tau, K))$, where $\mathrm{Lk}(\tau, K) \cong F(\partial\Delta^{n-\dim\tau})$ by the assumption. Due to Proposition A.1.1 and Lemma A.2.1, $F(\partial\Delta^{n-\dim\tau})$ is a combinatorial $(n-\dim\tau-1)$-manifold without boundary. Hence,

$$\mathrm{Lk}(\sigma, K) = \mathrm{Lk}(v, \mathrm{Lk}(\tau, K)) \cong F(\partial\Delta^{n-\dim\tau-1}) = F(\partial\Delta^{n-\dim\sigma}).$$

Next, we will show that ∂K is a combinatorial $(n-1)$-manifold without boundary. For each k-simplex $\sigma \in \partial K$, obviously $\mathrm{Lk}(\sigma, \partial K) \subset \mathrm{Lk}(\sigma, K)$. Due to Lemma A.2.1, $\mathrm{Lk}(\sigma, K) \cong F(\Delta^{n-k-1})$ is a combinatorial $(n-k-1)$-manifold with the following boundary:

$$\partial\,\mathrm{Lk}(\sigma, K) \cong \partial F(\Delta^{n-k-1}) = F(\partial\Delta^{n-k-1}).$$

For an $(n-k-2)$-simplex $\tau \in \mathrm{Lk}(\sigma, K)$, $\tau \in \partial\,\mathrm{Lk}(\sigma, K)$ if and only if τ is joinable to only one vertex $v \in \mathrm{Lk}(\sigma, K)^{(0)}$ in $\mathrm{Lk}(\sigma, K)$. Then, $\tau\sigma \in K$ is an $(n-1)$-simplex which is joinable to v in K but not joinable to any other vertex of K in K, which means that $\tau\sigma \in \partial K$. Thus, for an $(n-k-2)$-simplex $\tau \in \mathrm{Lk}(\sigma, K)$, $\tau \in \partial\,\mathrm{Lk}(\sigma, K)$ if and only if $\tau \in \mathrm{Lk}(\sigma, \partial K)$. For $i < n-k-2$, every i-simplex of $\mathrm{Lk}(\sigma, K)$ is a face of an $(n-k-2)$-simplex of $\mathrm{Lk}(\sigma, K)$ because $\mathrm{Lk}(\sigma, K)$ is a combinatorial $(n-k-1)$-manifold. Therefore,

$$\mathrm{Lk}(\sigma, \partial K) = \partial\,\mathrm{Lk}(\sigma, K) \cong F(\partial\Delta^{n-k-1}).$$

Consequently, ∂K is a combinatorial $(n-1)$-manifold without boundary. □

For a combinatorial n-manifold K, $|K|$ is a PL n-manifold (Theorem A.1.6). Concerning boundaries ∂K and $\partial|K|$, we have the following:

Theorem A.2.3 *For every combinatorial n-manifold K, $|\partial K| = \partial|K|$.*

Proof For $x \in |K|$, we may show that $x \in |\partial K|$ if and only if $x \in \partial|K|$.

($\subset$): For each $x \in |\partial K|$, $c_K(x) = c_{\partial K}(x) \in \partial K$, which means

$$\mathrm{Lk}(c_K(x), K) \cong F(\Delta^{n-k-1}),\ k = \dim c_K(x).$$

Then, by Lemma A.1.8, it follows that

$$\begin{aligned}\big(\operatorname{St}(c_K(x), K), F(c_K(x))\big) &= \big(F(c_K(x)) * \operatorname{Lk}(c_K(x), K), F(c_K(x))\big)\\ &\cong \big(F(\Delta^k) * F(\Delta^{n-k-1}), F(\Delta^k)\big).\end{aligned}$$

Moreover, due to Proposition A.1.9(i),

$$\begin{aligned}\big(F(\Delta^k) * F(\Delta^{n-k-1}), F(\partial\Delta^k) * F(\Delta^{n-k-1)} \cup F(\Delta^k) * F(\partial\Delta^{n-k-1})\big)\\ \cong \big(F(\Delta^n), F(\partial\Delta^n)\big).\end{aligned}$$

Hence, there is a PL homeomorphism $h : |\operatorname{St}(c_K(x), K)| \to \Delta^n$ such that $h(x) \in h(c_K(x)) \subset \partial\Delta^n$. Since $|\operatorname{St}(c_K(x), K)|$ is a neighborhood of x in $|K|$, we have $x \in \partial|K|$.

($\supset$): For each $x \in |K| \setminus |\partial K|$, $c_K(x) \notin \partial K$, that is,

$$\operatorname{Lk}(c_K(x), K) \cong F(\partial\Delta^{n-k}),\ k = \dim c_K(x).$$

By Lemma A.1.8, we have a PL homeomorphism

$$f : \big|\operatorname{St}(c_K(x), K)\big| = \big|F(c_K(x)) * \operatorname{Lk}(c_K(x), K)\big| \to \big|F(\Delta^k) * F(\partial\Delta^{n-k})\big|$$

such that $f(c_K(x)) = \Delta^k$ and $f(\partial c_K(x)) = \partial\Delta^k$. Since $f(x) \in f(\operatorname{rint} c_K(x)) = \operatorname{rint}\Delta^k$, it follows that $f(x) \notin |F(\partial\Delta^k) * F(\partial\Delta^{n-k})|$. By the way, due to Proposition A.1.9(ii),

$$\begin{aligned}\big(F(\Delta^k) * F(\partial\Delta^{n-k}), F(\partial\Delta^k) * F(\partial\Delta^{n-k})\big) &\cong \big(F(\Delta^n), F(\partial\Delta^n)\big)\\ &\cong \big(F(\mathbf{I}^n), F(\partial\mathbf{I}^n)\big).\end{aligned}$$

Thus, we have a PL homeomorphism $h : |\operatorname{St}(c_K(x), K)| \to \mathbf{I}^n$ such that $h(x) \in h(\operatorname{rint} c_K(x)) = \operatorname{rint}\mathbf{I}^n = (0, 1)^n$. Since $|\operatorname{St}(c_K(x), K)|$ is a neighborhood of x in $|K|$, we have $x \in \operatorname{Int}|K| = |K| \setminus \partial|K|$. □

A.3 Regular Neighborhoods in PL n-Manifolds

In this section, we will prove Proposition 6.10.6.

Let K be a simplicial complex and L a subcomplex of K. First, we recall the ε-neighborhood of L in K. In Sect. 1.9, we defined a simplicial map $\varphi_L : K \to I = \{0, 1, \mathbf{I}\}$ such that $\varphi_L(L^{(0)}) = 0$ and $\varphi_L(K^{(0)} \setminus L^{(0)}) = 1$. Due to Proposition 1.9.1, L is full in K if and only if $|L| = \varphi_L^{-1}(0)$. Then, for $0 < \varepsilon < 1$, we define the following cell complexes:

$$N_\varepsilon(L,K) = \{\sigma \cap \varphi_L^{-1}(\tau) \mid \sigma \in K[|L|],\ \tau \leqslant [0,\varepsilon]\},$$
$$C_\varepsilon(L,K) = \{\sigma \cap \varphi_L^{-1}(\tau) \mid \sigma \in K \setminus L,\ \tau \leqslant [\varepsilon,1]\},$$
$$K_L^\varepsilon = N_\varepsilon(L,K) \cup C_\varepsilon(L,K),$$

where $N_\varepsilon(L,K)$ is the ε-neighborhood of L in K, $|N_\varepsilon(L,K)| = \varphi_L^{-1}([0,\varepsilon])$, $|C_\varepsilon(L,K)| = \varphi_L^{-1}([\varepsilon,1])$, and $\mathrm{bd}\,|N_\varepsilon(L,K)| = \varphi_L^{-1}(\varepsilon)$. Applying Proposition 1.6.8, we can obtain a derived subdivision K' relative to $L \cup C(L,K)$ such that $N(L,K') \lhd N_\varepsilon(L,K)$ (cf. p. 47).

Theorem A.3.1 *Let M be a PL n-manifold and P a subpolyhedron of M. Then, a regular neighborhood N of P in M is a PL n-manifold with boundary. If $P \subset \mathrm{Int}\, M = M \setminus \partial M$, then $\partial N = \mathrm{bd}_M N$.*

Proof Let K be a triangulation of M (i.e., $M = |K|$) such that $P = |L|$ for some full subpolyhedron L of K and $N = N(L,K')$ for some derived subdivision K' of K relative to $L \cup C(L,K)$. Then, K is a combinatorial n-manifold by Theorem A.1.6. By virtue of the uniqueness of regular neighborhoods (Theorem 1.9.5), we may assume that $N(L,K') \rhd N_\varepsilon(L,K)$, $0 < \varepsilon < 1$.

As is easily observed,

$$N = |N_\varepsilon(L,K)| = \bigcup_{v \in L^{(0)}} O_K(v) \cap |N_\varepsilon(L,K)|,$$

where $O_K(v)$ is the open star at v. For each $x \in N$, choose $v \in L^{(0)}$ so that $x \in O_K(v)$. Then, $|\,\mathrm{St}(v,K)| \cap N$ is a neighborhood of x in N and

$$\big|\,\mathrm{St}(v,K)\big| \cap N = \big|\,\mathrm{St}(v,K)\big| \cap \big|N_\varepsilon(L,K)\big| = \big|\,\mathrm{St}(v,K_L^\varepsilon)\big|.$$

It should be remarked that K_L^ε is not a simplicial subdivision of K nor $|\,\mathrm{St}(v,K')| \neq |\,\mathrm{St}(v,K_L^\varepsilon)|$. However, due to Theorem 1.6.1, giving an order on $(K_L^\varepsilon)^{(0)}$ so that $C^{(0)}$ has the maximum v_C for each $C \in K_L^\varepsilon$, we can define a simplicial subdivision of K_L^ε with the same set of vertices as K_L^ε. Giving such an order on $(K_L^\varepsilon)^{(0)}$ satisfying the additional condition that v is the maximum of $C^{(0)}$ for every $C \in K_L^\varepsilon[v]$ ($\subset \mathrm{St}(v,K_L^\varepsilon)$), we apply Theorem 1.6.1 to define a simplicial subdivision $K^v \lhd K_L^\varepsilon$. Then, it follows that $|\,\mathrm{St}(v,K^v)| = |\,\mathrm{St}(v,K_L^\varepsilon)|$. Since $\mathrm{St}(v,K^v) \cong \mathrm{St}(v,K) \cong F(\Delta^n)$ by Lemma A.1.3, $|\,\mathrm{St}(v,K_L^\varepsilon)|$ is PL homeomorphic to $\mathbf{I}^n$. Therefore, $N = |N(L,K')|$ is a PL n-manifold.

In the above, assume that $P \subset \mathrm{Int}\, M$. For each $v \in L^{(0)}$ ($\subset P$), it follows from Theorem A.1.10 that

$$O_K(v) = |\,\mathrm{St}(v,K)| \setminus |\,\mathrm{Lk}(v,K)| \approx \Delta^n \setminus \partial\Delta^n = \mathrm{rint}\,\Delta^n.$$

Then, we have $O_K(v) \subset \operatorname{Int} M$. Hence, $N \subset \bigcup_{v \in L^{(0)}} O_K(v) \subset \operatorname{Int} M$. Each $x \in N$ has a polyhedral neighborhood B in M with a PL homeomorphism $h : B \to \mathbf{I}^n$ such that $h(x) \in (0, 1)^n$.

If $x \in \operatorname{int}_M N$, then the above B can be taken so small that $B \subset \operatorname{int}_M N$, which implies $x \in \operatorname{Int} N$. Conversely, if $x \in \operatorname{Int} N$, then there is a polyhedral neighborhood D in N with a PL homeomorphism $g : D \to \mathbf{I}^n$ such that $g(x) \in (0, 1)^n$, where D can be taken so small that $D \subset B$. By virtue of the Invariance of Domain (Corollary 1.12.5), $fg^{-1}((0, 1)^n)$ is open in $(0, 1)^n$, which implies that D is open in B. Therefore, $x \in \operatorname{int}_M N$. Thus, we have $\operatorname{int}_M N = \operatorname{Int} N$, which means

$$\partial N = N \setminus \operatorname{Int} N \subset N \setminus \operatorname{int}_M N = \operatorname{bd}_M N. \qquad \square$$

The setting at the beginning of the above proof shows that Theorem A.3.1 is a restatement of Proposition 6.10.6. Thus, we have Proposition 6.10.6.

A.4 PL Embedding Approximation Theorem

In this section, we give the PL manifold version of Theorem 1.8.11, in which $Q \times \mathbf{I}^n$ is replaced with a PL n-manifold. Namely, we prove the following PL EMBEDDING APPROXIMATION THEOREM:

Theorem A.4.1 (PL EMBEDDING APPROXIMATION FOR PL MANIFOLDS) *Let P be a compact polyhedron with a subpolyhedron $P_0 \subset P$ and let M be a PL n-manifold. Suppose that $n \geqslant 2 \dim P + 1$ and $f : P \to M$ is a map such that $f|P_0$ is a PL embedding. Then, for each open cover $\mathcal{U}$ of M, there exists a PL embedding $h : P \to M$ such that $h|P_0 = f|P_0$ and $h \simeq_{\mathcal{U}} f$ rel. P_0.*

Proof Take subpolyhedra $C_1, \ldots, C_k \subset M$ so that $f(P) \subset \bigcup_{i=1}^k \operatorname{int} C_i$ and each C_i is PL homeomorphic to $\mathbf{I}^n$. Let $\varphi_i : \mathbf{I}^n \to C_i$, $i = 1, \ldots, k$, be PL embeddings. There is a sequence of subpolyhedra $P_0 \subset P_1 \subset \cdots \subset P_k = P$ such that $f(\operatorname{cl}(P_i \setminus P_{i-1})) \subset \operatorname{int} C_i$ for each $i = 1, \ldots, k$. Due to Proposition 1.3.22 (1), M has an admissible metric d such that $\{\overline{\mathrm{B}}_d(x, 1) \mid x \in X\} \prec \mathcal{U}$.

By induction, we will construct maps $f_i : P \to M$, $i = 1, \ldots, k$, such that $f_i|P_i$ is a PL embedding, $f_i|P_{i-1} = f_{i-1}|P_{i-1}$, $f_i \simeq_{1/k} f_{i-1}$ rel. P_{i-1}, and $f_i(\operatorname{cl}(P_j \setminus P_{j-1})) \subset \operatorname{int} C_j$ for each $j > i$, where $f_0 = f$. Suppose f_{i-1} have been constructed. Choose $\varepsilon > 0$ so that $\varepsilon < 1/k$ and

$$\mathrm{N}(f_{i-1}(\operatorname{cl}(P_j \setminus P_{j-1})), \varepsilon) \subset \operatorname{int} C_j \quad \text{for each } j > i.$$

By the uniform continuity of φ_i, we have $\delta > 0$ such that

$$x, y \in \mathbf{I}^n, \ \|x - y\| < \delta \ \Rightarrow \ d(\varphi_i(x), \varphi_i(y)) < \varepsilon.$$

Since $f_{i-1}(P_{i-1})$ is a polyhedron and $f_{i-1}(P_{i-1}) \setminus \operatorname{int} C_i$ is its subpolyhedron, $\operatorname{cl}(P_i \setminus f_{i-1}^{-1}(C_i)) = P_{i-1} \setminus \operatorname{int} f_{i-1}^{-1}(C_i)$ is a subpolyhedron of P. It follows that $P_i \cap f_{i-1}^{-1}(C_i)$ is also a subpolyhedron of P and $P_{i-1} \cap f_{i-1}^{-1}(C_i)$ is its subpolyhedron. Consider the following map:

$$\varphi_i^{-1} f | P_i \cap f_{i-1}^{-1}(C_i) : P_i \cap f_{i-1}^{-1}(C_i) \to \mathbf{I}^n,$$

where $\varphi_i^{-1} f_{i-1} | P_{i-1} \cap f_{i-1}^{-1}(C_i)$ is a PL embedding. Applying the PL Embedding Approximation Theorem 1.8.11 (the case Q is a singleton), we can obtain a PL embedding $g : P_i \cap f_{i-1}^{-1}(C_i) \to \mathbf{I}^n$ such that

$$g | P_{i-1} \cap f_{i-1}^{-1}(C_i) = \varphi_i^{-1} f_{i-1} | P_{i-1} \cap f_{i-1}^{-1}(C_i) \text{ and}$$
$$g \simeq_\delta \varphi_i^{-1} f_{i-1} \text{ rel. } P_{i-1} \cap f_{i-1}^{-1}(C_i).$$

Then, a PL embedding $f' : P_i \to M$ can be defined by $f' | P_{i-1} = f_{i-1} | P_{i-1}$ and $f' | P_i \cap f_{i-1}^{-1}(C_i) = \varphi_i g$. Observe that $f' \simeq_\varepsilon f_{i-1} | P_i$ rel. P_{i-1}. Applying the Homotopy Extension Theorem 1.5.7, we can extend f' to a map $f_i : P \to M$ such that $f_i \simeq_\varepsilon f_{i-1}$ rel. P_{i-1}. Then, $f_i | P_i = f'$ is a PL embedding, $f_i | P_{i-1} = f_{i-1} | P_{i-1}$, $f_i \simeq_{1/k} f_{i-1}$ rel. P_{i-1}, and $f_i(\operatorname{cl}(P_j \setminus P_{j-1})) \subset \operatorname{int} C_j$ for each $j > i$. Refer to Fig. A.1.

The map $h = f_k : P = P_k \to M$ is the desired PL embedding. Indeed, $h | P_0 = f_k | P_0 = f_0 | P_0 = f | P_0$ and

$$h = f_k \simeq_{1/k} f_{k-1} \simeq_{1/k} \cdots \simeq_{1/k} f_0 = f \text{ rel. } P_0,$$

which implies that $h \simeq_{\mathcal{U}} f$ rel. P_0. □

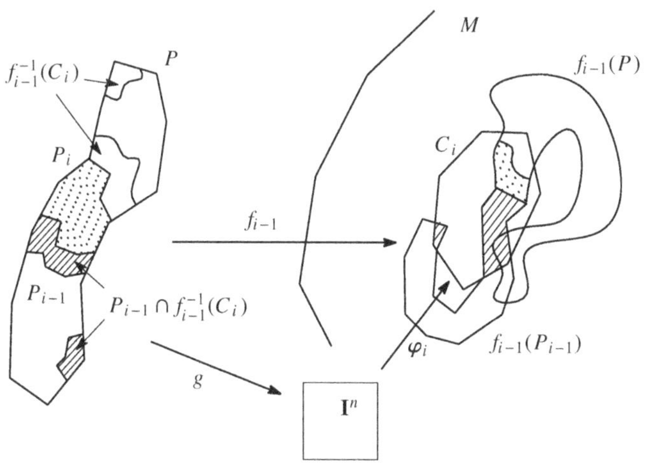

Fig. A.1 $P_i \setminus \operatorname{int} f_{i-1}^{-1}(C_i)$ and $f_{i-1}(P_{i-1}) \setminus \operatorname{int} C_i$

Epilogue

In 1964, answering Klee's question, R.D. Anderson showed in [3] that the product of a triod, T, and the Hilbert cube $\boldsymbol{Q}$ is homeomorphic to $\boldsymbol{Q}$. This is the starting point of Infinite-Dimensional Topology.[1] And in 1966, answering the longstanding Fréchet's question, he proved in [4] that Hilbert space ℓ^2 is homeomorphic to $\mathbb{R}^{\mathbb{N}}$. Since then, Infinite-Dimensional Topology has been developed rapidly and remarkable results have been obtained in a short period (cf. Anderson's essay [9]). The first book [(1)] on Infinite-Dimensional Topology was prepared by C. Bessaga and A. Pełczynski in 1975. From that book, one could learn basic properties of $\boldsymbol{Q}$ and fundamental results on manifolds modeled on infinite-dimensional metric linear spaces. For instance, Keller's Theorem, the topological homogeneity of $\boldsymbol{Q}$, the Classification and Open Embedding Theorems for Hilbert manifolds are covered in that book. In the same year, T.A. Chapman gave us his lecture notes [(2)] on Hilbert cube manifolds, by which we could learn fundamental results on $\boldsymbol{Q}$-manifolds containing the Classification and the Triangulation Theorems for $\boldsymbol{Q}$-manifolds.

In 1980 and 1981, two celebrated papers [143, 145] by H. Toruńczyk were published, in which Hilbert cube manifolds and Hilbert manifolds are characterized topologically. With these results, the theories of Hilbert cube manifolds and Hilbert manifolds were almost completed. However, Toruńczyk's characterizations can be expected to apply to various spaces appearing in many branches of Mathematics. On the other hand, manifolds modeled on incomplete metric linear spaces have also been studied and given characterizations. Thus, in order to learn those infinite-dimensional manifolds, some self-contained textbook has been long-expected.[2]

In 1989, J. van Mill prepared the excellent self-contained textbook [(3)], which saves us a lot of time in learning Toruńczyk's characterization of compact $\boldsymbol{Q}$-manifolds. However, it is not sufficient to cover the non-compact case. In his

[1] Cf. Footnote 4 in Preface (p. vii).

[2] For the history of Infinite-Dimensional Topology, refer to the article of T. Koetsier and J. van Mill [96, Sect. 4].

K. Sakai, *Topology of Infinite-Dimensional Manifolds*, Springer Monographs in Mathematics, https://doi.org/10.1007/978-981-15-7575-4

paper [148], J.J. Walsh provided an alternative proof of the characterization of $\boldsymbol{Q}$-manifolds but that paper is not self-contained. The paper [30] coauthored by P. Bowers, M. Bestvina, J. Mogilski, and J.J. Walsh provided an alternative proof of the characterization of ℓ_2-manifolds but is not so easy for a graduate student to read. Concerning Toruńczyk's characterization of non-separable Hilbert manifolds, we have to read his paper together with references. Since van Mill's book was published, over thirty years have passed but there are no books containing these results.

In 1996, two books [(4)] and [(5)] were published. The second chapter of Chigogidze's book [(4)] is devoted to infinite-dimensional manifolds but it is a survey article and not suitable for graduate students to study them. The book [(5)] by Banakh, Radul, and Zarichnyi is a good reference for absorbing sets in infinite-dimensional manifolds but the reader is required to have background knowledge of infinite-dimensional manifolds, that is, fundamental results containing characterization of ℓ_2- and $\boldsymbol{Q}$-manifolds. Thus, for graduate students, it is hard to study infinite-dimensional manifolds.

This book was designed for graduate students to study fundamental results on infinite-dimensional manifolds and related topics. The origin of this project goes back to the time when van Mill's book published. In every year, for a graduate course in the University of Tsukuba, I had given lectures on various topics, which were (1) basic properties of paracompact spaces, simplicial complexes, and dimensions, (2) basic properties of ANRs, Michael's Selection Theorem, and Toruńczyk's Factor Theorem, and (3) fundamental results on infinite-dimensional manifolds and the characterization of Hilbert manifolds. In 1991, combining these notes, I had written up my personal lecture notes entitled *Lectures on ANR's and Infinite-Dimensional Manifolds* consisting of three parts:

Part I: Paracompact Spaces, Simplicial Complexes, and Dimensions,
Part II: Absolute Neighborhood Retracts,
Part III: Infinite-Dimensional Manifolds,

where $\boldsymbol{Q}$-manifolds were not contained in Part III because students can refer to Chapman's lecture notes [(2)] and van Mill's book [(3)].

Since then, adding new sections and rearranging contents, Parts I and II were finally divided into five chapters:

Chap. 1: Metrization and Paracompact Spaces,
Chap. 2: Topology of Linear Spaces and Convex Sets,
Chap. 3: Simplicial Complexes and Polyhedra,
Chap. 4: Dimensions of Spaces,
Chap. 5: Retracts and Extensors.

Each of these chapters had been used as not only a text for a class of the graduate course but also a text of seminars of graduate students or advanced college seniors. In 2007, these chapters were almost completed and the title was changed to *Lectures on Topology toward Infinite-Dimensional Spaces.*

At that time, the lecture notes had been growing into a thick book because the first five chapters covered not only almost all material to learn about infinite-dimensional manifolds but also basic materials that will be necessary to study related topics. Besides, there were many materials that should be contained in the part on infinite-dimensional manifolds. With the joint work with Yaguchi [129] and Mine's work [108], Bestvina–Mogilski's results in [29] on absorbing sets have been generalized to the non-separable case, which should also be added to the book. Since van Mill's book is not enough for the characterization of non-compact $\boldsymbol{Q}$-manifolds, that should be included. Thus, it seemed better to divide the book into two volumes, the first one treating subjects corresponding to the above five chapters and the second volume treating infinite-dimensional manifolds directly. Actually, the contents of the first volume looked to be useful for students to study any other subject in Topology.

By the way, in 1988, A.N. Dranishnikov [56] solved the Alexandroff Problem affirmatively, that is, he showed the existence of an infinite-dimensional compactum with finite cohomological dimension. That means that the CE Problem was answered affirmatively, that is, there exists a cell-like map of a finite-dimensional compactum onto an infinite-dimensional compactum. In 1994, using that result, R. Cauty [38] constructed a linear metric space which is not an AR. To make the book more publishable, the chapter on cell-like maps was added as the last chapter, in which these distinguished results were contained. But, unfortunately this part is not self-contained. Required definitions and basic results on homotopy groups were added in the chapter on simplicial complexes.

Since cohomological dimension was contained in the last chapter, it became possible to give a proof of $\dim X \times \mathbf{I} = \dim X + 1$ for every metrizable space X. Thus, the first volume became more useful for students in other branches in Topology and publishable independently from the second volume. Finally, the title of this first volume was changed to *Geometric Aspects of General Topology* and was published by Springer in 2013. That book is now a textbook on Dimension Theory and ANR Theory, and emphasizes the relation between them. That is also a good textbook on non-locally finite simplicial complexes, and a useful reference.

After publication of the first book, the remaining part, the second volume, needed to be reformed as a book independent from the first, and was entitled *Topology of Infinite-Dimensional Manifolds.* At that time, writing the chapter on absorbing sets was in progress. Concerning the Triangulation Theorem for $\boldsymbol{Q}$-manifolds, I was considering that it is not necessary to add it because Chapman's lecture notes can be referred to. I had not yet started to write the chapter on $\mathbb{R}^\infty$- and $\boldsymbol{Q}^\infty$-manifolds. Moreover, preliminaries and background results needed to be added. Lots of work was required. Because of personal reasons, I could not continue writing for a couple of years.

In the autumn of 2015, I resumed writing. In the first half of the last chapter, basic properties of $\mathbb{R}^\infty$- and $\boldsymbol{Q}^\infty$-manifolds, was written in 2016. In 2017, changing my mind, I added Chapman's proof of the Triangulation Theorem for $\boldsymbol{Q}$-manifolds based on his lecture notes together with necessary preliminaries. Except for some algebraic preliminary result, each proof was written in detail as much as possible. Then, the proof of Borsuk's conjecture and the topological invariance of Whitehead torsion was included. By reforming the chapter on absorbing sets, West's absorption bases were added in 2018. Finally, infinite-dimensional combinatorial manifolds were added to the last chapter. Then, I was able to fix gaps and to give details of the original proof of the Simplicial Approximation Theorem for PL Embeddings.

While my colleagues and graduate students were expecting earlier publication of this book, it has taken a long time to get this book ready. Over 30 years have passed since I wrote up the first version of the lecture notes. Time flies!

Tsukuba, Japan
May 2020

Katsuro Sakai

Bibliography

Books and Texts

Infinite-Dimensional Topology

(1) C. Bessaga and A. Pełczyński, *Selected Topics in Infinite-Dimensional Topology,* MM **58** (Polish Scientific, Warsaw, 1975)
(2) T.A. Chapman, *Lectures on Hilbert Cube Manifolds,* CBMS Regional Conf. Ser. in Math. **28**, (Amer. Math. Soc., Providence, 1975)
(3) J. van Mill, *Infinite-Dimensional Topology, Prerequisites and Introduction,* North-Holland Math. Library **43** (Elsevier Sci. Publ. B.V., Amsterdam, 1989)
(4) A. Chigogidze, *Inverse Spectra,* North-Holland Math. Library **53** (Elsevier Sci. Publ. B.V., Amsterdam, 1996)
(5) T. Banakh, T. Radul and M. Zarichnyi, *Absorbing Sets in Infinite-Dimensional Manifolds,* Math. Studies, Monog. Ser. **1** (VNTL Publishers, Lviv, 1996)
(6) J. van Mill, *The Infinite-Dimensional Topology of Function Spaces,* North-Holland Math. Library **64** (Elsevier Sci. Publ. B.V., Amsterdam, 2002)

General Topology (Containing Dimension and ANR Theories)

Topological Spaces

(7) J.R. Munkres, *Topology,* 2nd edn. (Prentice Hall, Upper Saddle River, 2000)
(8) J. Dugundji, *Topology* (Allyn and Bacon, Boston, 1966)
(9) R. Engelking, *General Topology, Revised and Complete Edition,* SSPM 6 (Heldermann Verlag, Berlin, 1989)

Dimension Theory

(10) W. Hurewicz, H. Wallman, *Dimension Theory* (Princeton University Press, Princeton, 1941)
(11) K. Nagami, *Dimension Theory* (Academic Press, New York, 1970)

K. Sakai, *Topology of Infinite-Dimensional Manifolds*, Springer Monographs in Mathematics, https://doi.org/10.1007/978-981-15-7575-4

(12) R. Engelking, *Theory of Dimensions, Finite and Infinite* SSPM 10 (Heldermann Verlag, Berlin, 1995)

ANR Theory

(13) K. Borsuk, *Theory of Retracts,* MM 44 (Polish Scientific, Warsaw, 1967)
(14) S.-T. Hu, *Theory of Retracts* (Wayne State University Press, Detroit, 1965)

Dimension and ANR Theories

(15) K. Sakai, *Geometric Aspects of General Topology*, SMM (Springer, Tokyo, 2013)

Descriptive Set Theory

(16) A.S. Kechris, *Classical Descriptive Set Theory*, GTM 156 (Springer, New York, 1995)

Piecewise-Linear Topology (Combinatorial Topology)

(17) C.P. Rourke, B.J. Sanderson, *Introduction to Piecewise-Linear Topology*, Springer Study Edition (Springer, Berlin, 1972, 1982)
(18) J.F.P. Hudson, *Piecewise-Linear Topology* (W.A. Bebjamin, New York, 1969)
(19) E.C. Zeeman, *Seminar on Combinatorial Topology*. Mimeographed Notes (Inst. Hautes Études Sci., Paris, 1963)
(20) M.M. Cohen, *A Course in Simple-Homotopy Theory*, GTM 10 (Springer, New York, 1973)

Topological Linear Spaces

(21) G. Köthe, *Topological Vector Spaces I*, English edition, GTM 159 (Springer, New York, 1969)

Algebraic Topology

(22) A. Hatcher, *Algebraic Topology* (Cambridge University Press, Cambridge, 2002)
(23) W.S. Massey, *A Basic Course in Algebraic Topology*, GTM 127 (Springer, New York, 1991)
(24) G.W. Whitehead, *Homotopy Theory* (MIT Press, Cambridge, 1966)

References

1. J.W. Alexander, On the deformation of an n-cell. Proc. Natl. Acad. Sci. USA **9**, 406–407 (1923)
2. J.W. Alexander, An example of a simply connected surface bounding a region which is not simply connected. Proc. Natl. Acad. Sci. USA **10**, 8–10 (1924)
3. R.D. Anderson, The Hilbert cube as a product of dendrons. Not. Am. Math. Soc. **11**, 572 (1964)
4. R.D. Anderson, Hilbert space is homeomorphic to the countable infinite product of lines. Bull. Am. Math. Soc. **72**, 515–519 (1966)
5. R.D. Anderson, Topological properties of the Hilbert cubes and the infinite product of open intervals. Trans. Am. Math. Soc. **126**, 200–216 (1967)
6. R.D. Anderson, On topological infinite deficiency. Michigan Math. J. **14**, 365–383 (1967)
7. R.D. Anderson, Strongly negligible sets of Fréchet manifolds. Bull. Am. Math. Soc. **75**, 64–67 (1969)
8. R.D. Anderson, A characterization of apparent boundaries of the Hilbert cube. Not. Am. Math. Soc. **16**, 429, Abstract #697-G17 (1969)
9. R.D. Anderson, The early development of infinite dimensional topology, in *Recent Progress in General Topology, II,* ed. by M. Hušek, J. van Mill (Elsevier Scientific, 2002), pp. 547–548
10. R.D. Anderson, On sigma-compact subsets of infinite-dimensional spaces. Unpublished manuscript
11. R.D. Anderson, Spaces of homeomorphisms of finite graphs. Unpublished manuscript
12. R.D. Anderson, R.H. Bing, A complete elementary proof that Hilbert space is homeomorphic to the countable infinite product of lines. Bull. Am. Math. Soc. **74**, 771–792 (1968)
13. R.D. Anderson, T.A. Chapman, Extending homeomorphisms to Hilbert cube manifolds. Pac. J. Math. **38**, 281–293 (1971)
14. R.D. Anderson, J.D. McCharen, On extending homeomorphisms to Fréchet manifolds. Proc. Am. Math. Soc. **25**, 283–289 (1970)
15. R.D. Anderson, R.M. Schori, Factors of infinite-dimensional manifolds. Trans. Am. Math. Soc. **142**, 315–330 (1969)
16. R.D. Anderson, D.W. Curtis, J. van Mill, A fake topological Hilbert space. Trans. Am. Math. Soc. **272**, 311–321 (1982)
17. R.D. Anderson, D.W. Henderson, J.E. West, Negligible subsets of infinite-dimensional manifolds. Compos. Math. **21**, 143–150 (1969)
18. T. Banakh, Characterization of spaces admitting a homotopy dense embedding into Hilbert manifolds. Topol. Appl. **86**, 123–131 (1998)

K. Sakai, *Topology of Infinite-Dimensional Manifolds*, Springer Monographs in Mathematics, https://doi.org/10.1007/978-981-15-7575-4

19. T. Banakh, On hyperspaces and homeomorphism groups homeomorphic to products of absorbing sets and $\mathbb{R}^\infty$. Tsukuba J. Math. **23**, 1–10 (1999)
20. I. Banakh, T. Banakh, Constructing non-compact operators into c_0. Stud. Math. **201**, 65–67 (2010)
21. T. Banakh, R. Cauty, Topological classification of closed convex sets in Fréchet spaces. Stud. Math. **205**, 1–11 (2011)
22. T. Banakh, D. Repovš, A topological characterization of LF-spaces. Topol. Appl. **159**, 1475–1488 (2012)
23. T. Banakh, K. Sakai, Characterizations of $(\mathbb{R}^\infty, \sigma)$- or (Q^∞, Σ)-manifolds and their applications. Topol. Appl. **106**, 115–134 (2000)
24. T. Banakh, I. Zarichnyy, Topological groups and convex sets homeomorphic to non-separable Hilbert spaces. Cent. Eur. J. Math. **6**, 77–86 (2008)
25. H. Bass, A. Heller, R. Swan, The Whitehead group of a polynomial extension. Hautes Etudes Sci. **22**, 61–79 (1964)
26. C. Bessaga, On topological classification of complete linear metric space. Fund. Math. **55**, 251–288 (1965)
27. C. Bessaga, A. Pełczyński, Some remarks on homeomorphisms of F-spaces. Bull. Acad. Polon. Sci., Sér. Sci. Math. Astr. Phys. **10**, 265–270 (1962)
28. C. Bessaga, A. Pełczyński, Estrimated extension theorem, homogeneous collections and skeletons, and their applications to topological classifications of linear metric spaces and convex sets. Fund. Math. **69**, 153–190 (1970)
29. M. Bestvina, J. Mogilski, Characterizing certain incomplete infinite-dimensional absolute retracts. Michigan Math. J. **33**, 291–313 (1986)
30. M. Bestvina, P. Bowers, J. Mogilski, J. Walsh, Characterization of Hilbert space manifolds revisited. Topol. Appl. **24**, 53–69 (1986)
31. R.H. Bing, A homeomorphism between the 3-sphere and the sum of two solid horned spheres. Ann. Math. **56**, 354–362 (1952)
32. C.D. Borges, Negligibility in F-spaces. Math. Japon. **32**, 521–530 (1987)
33. K. Borsuk, Sur l'élimination de phenomènes paradoxaux en topologie générale. Proc. Intern. Cong. Math. vol. I, Amsterdam, 1954, pp. 197–208
34. M. Brown, A proof of the generalized Schoenfies theorem. Bull. Am. Math. Soc. **66**, 74–76 (1960)
35. M. Brown, Locally flat imbeddings of topological manifolds. Ann. Math. (2) **75**, 331–342 (1962)
36. J.L. Bryant, Piecewise linear topology, in *Handbook of Geometric Topology*, ed. by R.J. Daverman, R.B. Sher (Elsevier, Amsterdam, 2002), pp. 219–259
37. J.W. Cannon, Shrinking Cell-Like Decompositions of Manifolds, Codimension Three. Ann. Math. (2) **110**, 83–112 (1979)
38. R. Cauty, Un espace métrique linéaire qui n'est pas un rétracte absolu. Fund. Math. **146**, 85–99 (1994)
39. A.V. Černavskiĭ, Local contractibility of the group of homeomorphisms of a manifold (Russian). Mat. Sbornik **79**(121), 307–356 (1969); English trans. in Math. USSR Sb. **8**, 287–333 (1969)
40. T.A. Chapman, Deficiency in infinite-dimensional manifolds. Gen. Topol. Appl. **1**, 263–272 (1971)
41. T.A. Chapman, Dense sigma-compact subsets of infinite-dimensional manifolds. Trans. Am. Math. Soc. **154**, 399–426 (1971)
42. T.A. Chapman, On some applications of infinite-dimensional manifolds to the theory of shape. Fund. Math. **76**, 181–193 (1972)
43. T.A. Chapman, On the structure of Hilbert cube manifolds. Compos. Math. **24**, 329–353 (1972)
44. T.A. Chapman, Surgery and handle straightening in Hilbert cube manifolds. Pac. J. Math. **45**, 59–79 (1973)

45. T.A. Chapman, Topological invariance of Whitehead torsion. Am. J. Math. **96**, 488–497 (1974)
46. T.A. Chapman, Concordance of Hilbert cube manifolds. Trans. Am. Math. Soc. **219**, 253–268 (1976)
47. T.A. Chapman, L.C. Siebenmann, Finding a boundary for a Hilbert cube manifold. Acta Math. **137**, 171–208 (1976)
48. R. Connelly, A new proof of Brown's collaring theorem. Proc. Am. Math. Soc. **27**, 180–182 (1971)
49. R.H. Crowell, Invertible isotopies. Proc. Am. Math. Soc. **14**, 658–664 (1963)
50. D.W. Curtis, Some theorems and examples on locally equiconnectedness and its applications. Fund. Math. **72**, 101–113 (1971)
51. D. Curtis, T. Dobrowolski, J. Mogilski, Some applications of the topological characterizations of the sigma-compact spaces ℓ^2_f and Σ. Trans. Am. Math. Soc. **284**, 837–846 (1984)
52. D.W. Curtis, R.M. Schori, Hyperspaces of Peano continua are Hilbert cubes. Fund. Math. **101**, 19–38 (1978)
53. W.H. Cutler, Negligible subsets of infinite-dimensional Fréchet manifolds. Proc. Am. Math. Soc. **23**, 668–675 (1969)
54. T. Dobrowolski, H. Toruńczyk, On metric linear spaces homeomorphic to ℓ_2 and compact convex sets homeomorphic to Q. Bull. Acad. Polon. Sci. Sér. Sci. Math. **27**, 883–887 (1979)
55. T. Dobrowolski , H. Toruńczyk, Separable complete ANR's admitting a group structure are Hilbert manifolds. Topol. Appl. **12**, 229–235 (1981)
56. A.N. Dranishnikov, On a problem of P.S. Aleksandrov (in Russian). Mat. Sb. (N.S.) **135**(177)(4), 551–557, 560 (1988): English trans. in Math. USSR-Sb. **63**, 539–545 (1989)
57. J. Dugundji, Locally equiconnected spaces and absolute neighborhood retracts. Fund. Math. **57**, 187–193 (1965)
58. E. Dyer, S. Eilenberg, An adjunction theorem for locally equiconnected spaces. Pac. J. Math. **41**, 669–685 (1972)
59. R.D. Edwards, Characterizing infinite-dimensional manifolds topologically. Séminaire Bourbaki 540. Lecture Notes in Mathematics, vol. 842 (Springer, Berlin, 1979), pp. 278–302
60. R.D. Edwards, R.C. Kirby, Deformations of spaces of imbeddings. Ann. Math. (2) **93**, 63–88 (1971)
61. A. Fathi, Y.M. Visetti, Deformation of open embeddings of Q-manifolds. Trans. Am. Math. Soc. **224**, 427–436 (1976)
62. A. Fathi, A. Marin, Y.M. Visetti, New proofs of Chapman's CE mapping theorem and West's mapping cylinder theorem. Proc. Am. Math. Soc. **67**, 327–334 (1977)
63. S. Ferry, An immersion of $T^n - D^n$ into R^n. Enseignement Math. **20**, 177–178 (1974)
64. S. Ferry, The homeomorphism group of a compact Hilbert cube manifold is an ANR. Ann. Math. **106**, 101–119 (1977)
65. S. Ferry, A simple-homotopy approach to the finiteness obstruction, in *Shape Theory and Geometric Topology*, ed. by S. Mardešić, J. Segal. Lectures Notes in Mathematics, vol. 870 (Springer, Berlin, 1981), pp. 73–81
66. S. Ferry, A. Ranicki, A survey of Wall's finiteness obstruction, in *Surveys on Surgery Theory*, vol. 2, ed. by S. Cappell, A. Ranicki, J. Rosenberg. American Mathematics Studies vol. 149 (Princeton University Press, Princeton, 2001), pp. 63–80.
67. R.H. Fox, On fiber spaces, II. Bull. Am. Math. Soc. **49**, 733–735 (1943)
68. R. Geoghegan, On spaces of homeomorphisms, embeddings and functions. Topology **11**, 159–177 (1972)
69. R. Geoghegan, D.W. Henderson, Stable function spaces. Am. J. Math. **95**, 461–470 (1973)
70. W.E. Haver, A near-selection theorem. Gen. Topol. Appl. **9**, 117–124 (1978)
71. R.E. Heisey, Contracting spaces of maps on the countable direct limit of a space. Trans. Am. Math. Soc. **193**, 389–411 (1974)
72. R.E. Heisey, Stability, open embeddings, and classification of bounded weak-* manifolds. Gen. Topol. Appl. **6**, 327–338 (1976)

73. R.E. Heisey, Manifolds modelled on the direct limit of Hilbert cubes, in *Geometric Topology*, ed. by J.C. Cantrell (Academic Press, New York, 1979), pp. 609–619.
74. R.E. Heisey, Stability, classification, open embeddings, and triangulation of $\mathbb{R}^\infty$-manifolds, in *Proceedings of the International Conference on Geometric Topology*, ed. by K. Borsuk, A. Kirkor (Polish Scientific, Warsaw, 1980), pp. 193–196
75. R.E. Heisey, Embedding piecewise linear $\mathbb{R}^\infty$-manifold into $\mathbb{R}^\infty$. Topol. Proc. **6**, 317–328 (1981)
76. R.E. Heisey, Manifolds modelled on the direct limit of lines. Pac. J. Math. **102**, 47–54 (1982)
77. D.W. Henderson, Open subsets of Hilbert space. Compos. Math. **21**, 312–318 (1969)
78. D.W. Henderson, Infinite-dimensional manifolds are open subsets of Hilbert space. Bull. Am. Math. Soc. **75**, 759–762 (1969)
79. D.W. Henderson, Infinite-dimensional manifolds are open subsets of Hilbert space. Topology **9**, 25–33 (1970)
80. D.W. Henderson, Stable classification of infinite-dimensional manifolds by homotopy type. Invent. Math. **12**, 45–56 (1971)
81. D.W. Henderson, Corrections and extensions of two papers about infinite-dimensional manifolds. Gen. Topol. Appl. **1**, 321–327 (1971)
82. D.W. Henderson, A simplicial complex whose product with any ANR is a simplicial complex. Gen. Topol. Appl. **3**, 81–83 (1973)
83. D.W. Henderson, Z-sets in ANR's. Trans. Am. Math. Soc. **213**, 205–216 (1975)
84. D.W. Henderson, R.M. Schori, Topological classification of infinite-dimensional manifolds by homotopy type. Bull. Am. Math. Soc. **76**, 121–124 (1970)
85. D.W. Henderson, J.E. West, Triangulated infinite-dimensional manifolds. Bull. Am. Math. Soc. **76**, 655–660 (1970)
86. C.J. Himmelberg, Some theorems on equiconnected and locally equiconnected spaces. Trans. Am. Math. Soc. **115**, 43–53 (1965)
87. H. Hosokawa, Z-sets in polyhedra. Bull. Tokyo Gakugei Univ. Ser. IV **27**, 23–34 (1975)
88. W. Jakobsche, The space of homeomorphisms of a 2-dimensional polyhedron is an ℓ_2-manifold. Bull. Acad. Polon. Sci. Sér. Sci. Math. **28**, 71–75 (1980)
89. M.I. Kadec, On topological equivalence of separable Banach spaces (in Russian). Dokl. Akad. Nauk SSSR **167**, 23–25 (1966); English trans. in Soviet Math. Dokl. **7**, 319–322 (1966)
90. M.I. Kadec, Proof of the topological equivalence of all separable infinite-dimensional Banach spaces (in Russian). Funkt. Anal. Ego Pril. **1**, 61–70 (1967); English trans. in Funct. Anal. Appl. **1**, 53–62 (1967)
91. J. Keesling, Using flows to construct Hilbert space factors of function spaces. Trans. Am. Math. Sco. **161**, 1–24 (1971)
92. O.H. Keller, Die Homoiomorphie der kompakten konvexen Mengen im Hilbertschen Raum. Math. Ann. **105**, 748–758 (1931)
93. J.L. Kelley, Hyperspaces of a continuum. Trans. Am. Math. Sco. **52**, 22–36 (1942)
94. R.C. Kirby, L.C. Siebenmann, On the triangulation of manifolds and the Hauptvermutung. Bull. Am. Math. Soc. **75**, 742–749 (1969)
95. V. Klee, Some topological properties of convex sets. Trans. Am. Math. Soc. **78**, 30–45 (1955)
96. T. Koetsier, J. van Mill, By their fruits ye shall know them: some remarks on interaction of general topology with other areas of mathematics, in *History of Topology*, ed. by I.M. James (Elsevier, Amsterdam, 1999), pp. 199–239
97. K. Koshino, Characterizing non-separable sigma-locally compact infinite-dimensional manifolds and its applications. J. Math. Soc. Jpn. **66**, 1155–1189 (2014)
98. V.T. Liem, An α-approximation theorem Q^∞-manifolds. Topol. Appl. **12**, 289–304 (1981)
99. V.T. Liem, An α-approximation theorem $\mathbb{R}^\infty$-manifolds. Rocky Mountain J. Math. **17**, 393–419 (1987). https://doi.org/10.1216/RMJ-1987-17-2-393
100. R. Luke, W.K. Mason, The space of homeomorphisms on a compact two-manifold is an absolute neighborhood retract. Trans. Am. Math. Soc. **164**, 275–285 (1972)
101. P. Mankiewicz, On topological, Lipschitz, and uniform classification of LF-spaces. Stud. Math. **52**, 109–142 (1974)

102. S. Mardešić, Absolute neighborhood retracts and shape theory, in *History of Topology*, ed. by I.M. James (Elsevier, Amsterdam, 1999), pp. 241–269
103. A. Marin, Y.M. Visetti, A general proof of Bing's shrinking criterion. Proc. Am. Math. Soc. **53**, 501–507 (1975)
104. S. Mazur, Une remarque sur l'homéomorphie des champs fonctionnels. Stud. Math. **1**, 83–85 (1929)
105. B. Mazur, On embeddings of spheres. Bull. Am. Math. Soc. **65**, 59–65 (1959)
106. W.D. McIntosh, Adjunction of locally equiconnected spaces. Fund. Math. **63**, 175–192 (1968)
107. J.W. Milnor, Two complexes which are homeomorphic but combinatorially distinct. Ann. Math. **74**, 575–590 (1961)
108. K. Mine, Universal spaces of non-separable absolute Borel classes. Tsukuba J. Math. **30**, 137–148 (2006)
109. K. Mine, K. Sakai, Open subsets of LF-spaces. Bull. Polish Acad. Sci. Math. **56**, 25–37 (2008)
110. K. Mine, K. Sakai, Simplicial complexes and open subsets of non-separable LF-spaces. Can. J. Math. **63**, 436–459 (2011)
111. G. Mislin, Wall's finiteness obstruction, in *Handbook of Algebraic Topology*, ed. by I.M. James (Elsevier, Amsterdam, 1995), pp. 1259–1291
112. J. Mogilski, Characterizing the topology of infinite-dimensional σ-compact manifolds. Proc. Am. Math. Soc. **92**, 111–118 (1984)
113. A.A. Ranicki, On the Hauptvermutung, in *The Hauptvermutung Book*. A Collection of Papers of the Topology of Manifolds, ed. by A.A. Ranicki (Kluwer, Dordrecht, 1996), pp. 3–31
114. K. Sakai, Embeddings of infinite-dimensional manifold pairs and remarks on stability and deficiency. J. Math. Soc. Jpn. **29**, 261–280 (1977)
115. K. Sakai, Stability for infinite-dimensional fiber bundles. Tsukuba J. Math. **5**, 85–99 (1981)
116. K. Sakai, Homeomorphisms of infinite-dimensional fiber bundles. Tsukuba J. Math. **6**, 21–33 (1982)
117. K. Sakai, On $\mathbb{R}^\infty$-manifolds and Q^∞-manifolds. Topol. Appl. **18**, 69–79 (1984)
118. K. Sakai, A characterization of local equi-connectedness. Pac. J. Math. **111**, 231–241 (1984)
119. K. Sakai, Each $\mathbb{R}^\infty$-manifold has a unique piecewise linear $\mathbb{R}^\infty$-structure. Proc. Am. Math. Soc. **90**, 616–618 (1984)
120. K. Sakai, Completions of metric simplicial complexes by using ℓ_p-norms. Topol. Proc. **11**, 177–208 (1986)
121. K. Sakai, Combinatorial infinite-dimensional manifolds and $\mathbb{R}^\infty$-manifolds. Topol. Appl. **26**, 43–64 (1987)
122. K. Sakai, On topologies of triangulated infinite-dimensional manifolds. J. Math. Soc. Jpn. **39**, 287–300 (1987)
123. K. Sakai, The ℓ_1-completion of a metric combinatorial ∞-manifold. Proc. Am. Math. Soc. **100**, 574–578 (1987)
124. K. Sakai, The Q-manifold local-compactification of a metric combinatorial ∞-manifold. Proc. Am. Math. Soc. **100**, 775–780 (1987)
125. K. Sakai, Simplicial complexes triangulating infinite-dimensional manifolds. Topol. Appl. **29**, 167–183 (1988)
126. K. Sakai, The space of cross sections of a bundle. Proc. Am. Math. Soc. **103**, 956–960 (1988)
127. K. Sakai, Connecting direct limit topologies with metrics on infinite-dimensional manifolds. Compos. Math. **81**, 237–245 (1992); *A corringendum,* Compos. Math. **102**, 357 (1996)
128. K. Sakai, R.Y. Wong, Conjugating homeomorphisms to uniform homeomorphisms. Trans. Am. Math. Soc. **311**, 337–356 (1989)
129. K. Sakai, M. Yaguchi, Characterizing manifolds modeled on certain dense subspaces of non-separable Hilbert spaces. Tsukuba J. Math. **27**, 143–159 (2003)
130. A. Schoenflies, Beiträge zur Theorie der Punkymengen. III. Math. Ann. **62**, 286–328 (1906)
131. R.M. Schori, Topological stability for infinite-dimensional manifolds. Compos. Math. **23**, 87–100 (1971)
132. R.M. Schori, J.E. West, The hyperspace of closed interval is a Hilbert cube. Trans. Am. Math. Soc. **213**, 217–235 (1975)

133. L.C. Siebenmann, Deformation of homeomorphisms on stratified sets. Commun. Math. Helv. **47**, 123–163 (1972)
134. J. Stallings, Whitehead torsion of free products. Ann. Math. **82**, 354–363 (1965)
135. A.H. Stone, Non-separable Borel sets. Rozprawy Mat. (= Dissertationes Math.) **28**, 40pp. (1962)
136. A.H. Stone, Absolute F_σ spaces. Proc. Am. Math. Soc. **13**, 495–499 (1962)
137. H. Toruńczyk, Skeletonized sets in complete metric spaces and homeomorphisms of Hilbert cube. Bull. Acad. Polon. Sci. Sér. Sci. Math. **18**, 119–126 (1970)
138. H. Toruńczyk, Compact absolute retracts as factors of the Hilbert space. Fund. Math. **83**, 75–84 (1973)
139. H. Toruńczyk, Absolute retracts as factors of normed linear spaces. Fund. Math. **86**, 53–67 (1974)
140. H. Toruńczyk, On Cartesian factors and the topological classification of linear metric spaces. Fund. Math. **88**, 71–86 (1975)
141. H. Toruńczyk, Homeomorphism groups of compact Hilbert cube manifolds which are manifolds. Bull. Acad. Polon. Sci. Sér. Sci. Math. Astr. Phys. **25**, 401–408 (1977)
142. H. Toruńczyk, Concerning locally homotopy negligible sets and characterization of ℓ_2-manifolds. Fund. Math. **101**, 93–110 (1978)
143. H. Toruńczyk, On CE-images of the Hilbert cube and characterization of Q-manifolds. Fund. Math. **106**, 31–40 (1980)
144. H. Toruńczyk, Characterization of infinite-dimensional manifolds, in *Proceedings of the International Conference on Geometric Topology*, ed. by K. Borsuk, A. Kirkor (Polish Scientific, Warsaw, 1980), 431–437.
145. H. Toruńczyk, Characterizing Hilbert space topology. Fund. Math. **111**, 247–262 (1981)
146. H. Toruńczyk, A correction of two papers concerning Hilbert manifolds. Fund. Math. **125**, 89–93 (1985)
147. C.T.C. Wall, Finiteness conditions for CW complexes. Ann. Math. **81**, 55–69 (1965)
148. J.J. Walsh, Characterization of Hilbert cube manifolds: an alternative proof, in *Geometric and Algebraic Topology*. Banach Center Publications, vol. 18 (Polish Scientific, Warsaw, 1986), pp. 153–160
149. J.E. West, Infinite products which are Hilbert cubes. Trans. Am. Math. Soc. **150**, 1–25 (1970)
150. J.E. West, The ambient homeomorphy of an incomplete subspace of infinite-dimensional Hilbert spaces. Pac. J. Math. **34**, 257–267 (1970)
151. J.E. West, Mapping cylinder of Hilbert cube factors. Gen. Topol. Appl. **1**, 111–125 (1971)
152. J.E. West, Compact ANR's have finite type. Bull. Am. Math. Soc. **81**, 163–165 (1975)
153. J.E. West, Mapping Hilbert cube manifolds to ANR's: a solution of a conjecture of Borsuk. Ann. Math. **106**, 1–18 (1977)
154. J.H.C. Whitehead, Simplicial spaces, nuclei and m-groups. Proc. Lond. Math. Soc. **45**, 243–327 (1939)
155. J.H.C. Whitehead, Combinatorial homotopy, I. Bull. Am. Math. Soc. **55**, 213–245 (1949)
156. J.H.C. Whitehead, A certain exact sequence. Ann. Math. **52**, 51–110 (1950)
157. J.H.C. Whitehead, Simple homotopy types. Am. J. Math. **50**, 1–57 (1950)
158. M. Wojdysławsky, Sur la contractilité des hyperespaces de continus localement connexes Fund. Math. **30**, 247–252 (1938)
159. M. Wojdysławsky, Rétractes absolus et hyperespaces des continus. Fund. Math. **32**, 184–192 (1939)
160. R.Y.T. Wong, Non-compact Hilbert cube manifolds. Unpublished manuscript
161. T. Yagasaki, Infinite-dimensional manifold triples of homeomorphism groups. Topol. Appl. **76**, 261–281 (1997)
162. A. Yamashita, Non-separable Hilbert manifolds of continuous mappings. Topol. Appl. **222**, 177–199 (2017)

Index

K. Sakai, *Topology of Infinite-Dimensional Manifolds*, Springer Monographs in Mathematics, https://doi.org/10.1007/978-981-15-7575-4

The manufacturer's authorised representative in the EU is Springer Nature Customer Service Centre GmbH, Europaplatz 3, 69115 Heidelberg, Germany. If you have any concerns regarding our products, please contact ProductSafety@springernature.com

Printed and bound by CPI Group (UK) Ltd, Croydon, CR0 4YY
01/10/2026
02226249-0010